श्री BalaJi

PROBLEMS IN PHYSICAL CHEMISTRY

for JEE

Main & Advanced

Narendra Avasthi
Director,
Vibrant Academy India (P) Ltd.
KOTA

19th EDITION

SHRI BALAJI PUBLICATIONS
(EDUCATIONAL PUBLISHERS & DISTRIBUTORS)

- Published by:

SHRI BALAJI PUBLICATIONS

(EDUCATIONAL PUBLISHERS & DISTRIBUTORS)

6, Gulshan Vihar,
Opp. Mahalaxmi Enclave,
Jansath Road, Muzaffarnagar (U.P.)
website : www.shribalajibooks.com
email : sbjpub@gmail.com

- ISBN : 978-93-6802-928-1

- First edition : 2008
- Nineteenth edition : 2026

- **Price : ₹ 748.00**

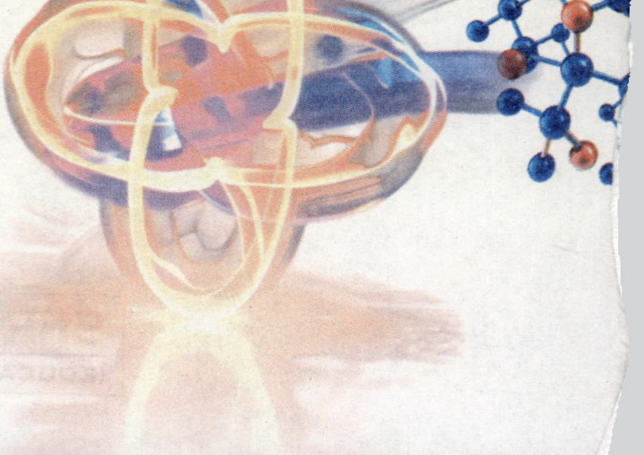

Dedicated to
my
Beloved Parents
&
Elder Brother
for
their blessings and support

Preface

It is a matter of great pleasure for me to present the Nineteenth edition of **"Problems in Physical Chemistry"** for JEE aspirants. This book brings out the experience gained during many years of teaching to the JEE aspirants. The objective of this book is to provide proper guidance and relevant material, which is really needed for the preparation of JEE.

In the book, very useful brief theory of each chapter provided and each chapter consists of three levels of problems to cover the wide subject of chemistry in a nut shell. The level of problems given in this book is essentially required for JEE aspirants.

LEVEL-1: Problems based on basic concepts and are useful just to begin the topic.

LEVEL-2: Challenging problems based on twists and wide applications of facts.

LEVEL-3: Problems based on Comprehensions, Problems with One or More than one Correct Option, Matching Type Problems, Assertion – Reason Type Problems to make the students familiar with current JEE Pattern.

LEVEL-4: Subjective Problems (Integer Type Problems) and Numerical Problems are given.

The problems are completely supported by answers. In the last, hints and solution have also been provided wherever necessary, to save precious time of students.

I hope that this effort will cater to the needs of JEE aspirants and as a matter of facts they will really enjoy the subjects with the problems given. I would feel rewarded if you achieve your goal with the help of this book.

All attempts have been made to make it free from errors. In the last constructive criticism and valuable suggestion from the readers are most welcome to make this effort more useful.

Narendra Avasthi
avasthisir@gmail.com

Acknowledgement

I wish to acknowledge my indebtedness to Mr. Nitin Jain, Mr. N.K. Sethia, Mr. M.S. Chouhan, Mr. V.K. Jaiswal, Mr. Vikas Gupta and Mr. Pankaj Joshi for their enthusiastic support.

I would like to thank Mr. Gyan Gautam sir, Mr. Pushkar Rai, Mr. Samvit Bissa, Mr. Akhilesh Kanther, Dr. Gopal Chaturvedi, Mr. M.K. Chouhan, Dr. Gautam Bothra, Mr. Navneet Bhargawa, Mr. Ajay Bairagi, Mr. Rohit Chouhan, Mr. Kuldeep Goyal and all faculty members for their comments and valuable suggestions towards the improvement of this book.

Finally, this part of book will remain incomplete without co-operation of co-author Mrs. Priti Avasthi, whose time was spent during this job. I admire for their patience, understanding and support.

I also pay my sincere thanks to all the esteemed members of **M/s Shri Balaji Publications** in bringing out this book in such a nice form.

There are undoubtedly many other who are learning their indelible mark on this book. Thanks to every one for their assistance.

Narendra Avasthi

Contents

GROW GREEN

SAVE NATURE

CHAPTER 1

Mole Concept

Laws of Chemical Combination

Chemical reactions take place according to certain laws. These laws are called the Laws of Chemical Combination. “These are no longer useful in chemical calculations now but gives an idea of earlier methods of analysing and relating compounds by mass.”

- **Law of Conservation of Mass [Lavoisier (1774)]**
  During any physical or chemical change, the sum of masses of all substances present in reactions vessel remain conserved.
- **Law of Constant Composition or Definite Proportions [Proust (1799)]**
  In a given chemical compound, the elements are always combined in the same proportions by mass.
- **Law of Multiple Proportions [Dalton (1803)]**
  Whenever two elements form more than one compound, the different masses of one element that combine with the same mass of the other element are in the ratio of small whole numbers.
- **Law of Reciprocal Proportions [Richter (1792)]**
  When two elements combine separately with a fixed mass of a third elements then the ratio of their masses in which they do so is either same or some whole number multiple of the ratio in which they combine with each other.
- **Gay-Lussac’s Law of combining volumes**
  According to Gay-Lussac’s law of combining volume, when gases react together, they always do so in volumes which bear a simple ratio to one another and to the volumes of the products, if these are also gases, provided all measurements of volumes are done under similar conditions of temperature and pressure.
- **Avogadro’s Law**
  The volume of a gas (at fixed pressure and temperature) is proportional to the number of moles (or molecules of gas present). Mathematically we can write

$$V \propto n$$

Or Equal volumes of all gases under similar conditions of temperature and pressure contain equal number of molecules.

Dalton's Atomic Theory

- Matter consists of tiny particles called atoms.
- Atoms are indestructible. In chemical reactions, the atoms rearrange but they do not themselves break apart.
- In any sample of a pure element, all the atoms are identical in mass and other properties.
- The atoms of different elements differ in mass and other properties.
- When atoms of different elements combine to form compounds, new and more complex particles form. However, in a given compound the constituent atoms are always present in the same fixed numerical ratio.

Modern Atomic Theory

- **Atom is no longer considered to be indivisible :** It has been found that an atom has a complex structure and is composed of sub-atomic particles such as electrons, protons and neutrons.
- **Atoms of the same element may not be similar in all respects :**
  Ex. Isotopes ($^{23}_{11}Na$, $^{24}_{11}Na$)
- **Atoms of different elements may be similar in one or more respects :**
  Ex. Isobars. ($^{40}_{20}Ca$, $^{40}_{18}Ar$)
- **Atom is the smallest unit which takes part in chemical reactions :** Although atom is composed of sub-atomic particles, yet it is the smallest particle which takes part in chemical reactions.
- **The ratio in which atoms unite may be fixed and integral but may not be simple :** For example, in sugar molecule ($C_{12}H_{22}O_{11}$), the ratio of C, H and O atoms is 12 : 22 : 11 which is not simple.
- **Atoms of one element can be changed into atoms of other element :**
  Ex. Artificial Nuclear Reactions
- **The mass of atom can be changed into energy :** According to Einstein's equation $E = mc^2$ (E = Energy, m = mass, c = the velocity of light, *i.e.*, 3×10^{10} cm sec^{-1}), mass and energy are inter-convertible.

Mole Concept

(a) Definition of one mole : One mole is a collection of that many entities as there are number of atoms exactly in 12 gm of C-12 isotope.

(b) $1u = 1\,\text{amu} = \left(\frac{1}{12}\right)$ of mass of 1 atom of $C^{12} = \frac{1\text{g}}{N_A} = 1.66 \times 10^{-24}$ g

(c) For Elements :

- 1 g atom = 1 mole of atoms = N_A atoms.
- g atomic mass (GAM) = mass of N_A atoms in g.
- Mole of atoms = $\dfrac{\text{Mass }(g)}{\text{GAM or molar mass}}$

(d) For molecules :

- 1 g molecule = 1 mole of molecule = N_A molecule.
- g molecular mass (GMM) = mass of N_A molecule in g.
- Mole of molecule = $\dfrac{\text{Mass }(g)}{\text{GMM or molar mass}}$

(e) For ionic compounds :

- 1 g formula unit = 1 mole of formula unit = N_A formula unit.
- g formula mass (GFM) = mass of N_A formula unit in g.

- Mole of formula unit $= \dfrac{\text{Mass }(g)}{\text{GFM or molar mass}}$

(f) 1 mole of a substance →
- Contains 6.022×10^{23} particles
- Weighs as much as molecular weight/atomic ionic/weight in grams.
- If it is a gas, one mole occupies a volume of 22.4 L at 1 atm & 273 K

(g) **Average or mean atomic mass :** Average atomic mass of element

$$A_{\text{Avg.}} = \frac{A_1x_1 + A_2x_2 + \ldots\ldots}{x_1 + x_2 + \ldots\ldots}$$

Here A_1, A_2 are isotopic mass of element and x_1, x_2 are natural abundance of isotopes.

(h) **Average or mean molar mass :** The average molar mass of the different substance present in the container $M_{\text{Avg.}} = \dfrac{M_1n_1 + M_2n_2 + \ldots\ldots}{n_1 + n_2 + \ldots\ldots}$

Here M_1, M_2 are molar mass of substances and n_1, n_2 are mole of substances present in the container.

Empirical Formula, Molecular Formula

(a) **Empirical formula :** Formula depicting constituent atom in their simplest ratio.

Molecular formula : Formula depicting actual number of atoms in one molecule of the compound.

(b) **Relation between molecular formula and empirical formula :**

$$n = \frac{\text{Molecular mass}}{\text{Empirical Formula mass}}$$

(c) **Densities :**

- Density $= \dfrac{\text{Mass}}{\text{Volume}}$, Unit: g/cc
- Relative density $= \dfrac{\text{Density of any substance}}{\text{Density of reference substance}}$
- Specific gravity $= \dfrac{\text{Density of any substance}}{\text{Density of water at } 4°\text{C}}$
- Vapour density : Ratio of density of vapour to the density of hydrogen at similar pressure and temperature.

$$\text{Vapour density} = \frac{\text{Density of vapour at some temperature and pressure}}{\text{Density of } H_2 \text{ gas at same temperature and pressure}}$$

$$\text{Vapour density} = \frac{\text{Molecular mass}}{2}$$

Stoichiometry

Stoichiometry pronounced ("stoy – key – om – e – tree") is the calculations of the quantities of reactants and products involved in a chemical reaction. Following methods can be used for solving problems.

(a) Mole Method (Balancing is required)

For Ex. : $2KClO_3 \xrightarrow{\Delta} 2KCl + 3O_2$

$$\frac{\text{Mole of } KClO_3}{2} = \frac{\text{Mole of KCl}}{2} = \frac{\text{Mole of } O_2}{3}$$

(b) Principle of Atom Conservation (P.O.A.C.) method (Balancing is not required)

For Ex. : $KClO_3 \xrightarrow{\Delta} KCl + O_2$

POAC for K : $1 \times$ mole of $KClO_3 = 1 \times$ mole of KCl

POAC for Cl : $1 \times$ mole of $KClO_3 = 1 \times$ mole of KCl

POAC for O : $3 \times$ mole of $KClO_3 = 2 \times$ mole of O_2

Concept of Limiting Reagent

(a) Limiting Reagent : It is very important concept in chemical calculation. It refers to reactant which is present in minimum stoichiometry quantity for a chemical reaction. It is reactant consumed fully in a chemical reaction. So all calculations related to various products or in sequence of reactions are made on the basis of limiting reagent.

(b) Calculation of Limiting Reagent : Divide given moles of each reactant by their stoichiometric coefficient, the one with least ratio is limiting reagent.

Percentage Yield

$$\text{The percentage yield of product} = \frac{\text{Actual yield}}{\text{theoretical maximum yield}} \times 100$$

Concentration Terms

(a) For solutions (homogeneous mixture) :

- If the mixture is not homogeneous, then none of them is applicable.

(i) $\%$ by mass $\left(\frac{w}{W}\right) = \frac{\text{Wt. of solute}}{\text{Wt. of solution}} \times 100$

[$X\%$ by mass means 100 gm solution contains X gm solute ; $\therefore$ $(100 - X)$ gm solvent]

(ii) $\% \left(\frac{w}{V}\right) = \frac{\text{Wt. of solute}}{\text{Volume of solution}} \times 100$

[$X\% \left(\frac{w}{V}\right)$ means 100 mL solution contains X gm solute]

(iii) $\% \left(\frac{V}{V}\right) = \frac{\text{Volume of solute}}{\text{Volume of solution}} \times 100$

- For gases % by volume is same as mole %

(iv) Mole % $= \frac{\text{Moles of solute}}{\text{Total moles}} \times 100$

(v) Mole fraction $(X) = \frac{\text{Moles of solute}}{\text{Total moles}}$

(vi) Molarity $(M) = \frac{\text{Moles of solute}}{\text{Volume of solution (in litre)}}$

(vii) Molality $(m) = \frac{\text{Moles of solute}}{\text{Mass of solvent (in kg)}}$

(viii) Parts per million (ppm) $= \frac{\text{Mass of solute}}{\text{Mass of solvent}} \times 10^6 \cong \frac{\text{Mass of solute}}{\text{Mass of solution}} \times 10^6$

(ix) Formality $(F) = \frac{\text{No. of formula unit}}{\text{Volume of solution (in litre)}}$

(b) (i) On adding solvent in a solution (dilution) : Number of mole of solute remains constant

$$M_f V_f = M_i V_i$$

(ii) Mixing of two solutions of same solute

$$M_f V_f = M_1 V_1 + M_2 V_2 + \ldots$$

(c) Volume strength of H_2O_2:

H_2O_2 (aq) solution labelled as 'x V' 'volume H_2O_2 (for *e.g.,* '20 V' H_2O_2), it means x volume of O_2 (in litre) at 1 atm and 273K that can be obtained from 1 litre of such a sample when it decomposes according to

$$H_2O_2 \longrightarrow H_2O + \frac{1}{2}O_2$$

$$\text{Molarity of } H_2O_2 = \frac{\text{Volume strength of } H_2O_2}{11.2}$$

Eudiometry

(For reactions involving gaseous reactants and products)

- Eudiometry or gas analysis involves the calculations based on gaseous reactions or the reactions in which at least two components are gaseous, in which the amounts of gases are represented by their volumes, measured at the same pressure and temperature.
- Gay-Lussac's law of volume combination holds good.
- Problem may be solved directly is terms of volume, in place of mole.

The volume of gases produced is often given by mentioning certain solvent which absorb contain gases.

Solvent	**gas(es) absorb**
KOH (*aq.*)	CO_2, SO_2, Cl_2
Ammon Cu_2Cl_2	CO
Turpentine oil	O_3
Alkaline pyrogallol	O_2
Water	NH_3, HCl
Anhydrous $CuSO_4/CaCl_2$	H_2O

Level-1

1. The relative atomic mass of sodium is:
(a) 23 g (b) 23 a.m.u.
(c) 23 kg (d) 23

2. The molecular mass of oxygen is:
(a) 16 g (b) 32 g
(c) 16 a.m.u. (d) 32 a.m.u.

3. The number of O atoms in 98 amu of H_2SO_4 is :
(a) 1 mole (b) 4 moles
(c) 4 (d) 1

4. 4.6×10^{22} atoms of an element weight 13.8 g. The atomic mass of the element is :
(a) 120 a.m.u. (b) 35.5 a.m.u.
(c) 108 a.m.u. (d) 180 a.m.u.

5. Number of electrons in 168 amu of ${}^{56}_{26}Fe^{2+}$ ions will be :
(a) 75 N_A (b) 78
(c) 72 (d) 72 N_A

6. The number of neutrons present 124 amu of P_4 is (Given : ${}^{31}_{15}P$) :
(a) 16 (b) 16 N_A
(c) 64 N_A (d) 64

7. Calculate number of neutrons present in 12×10^{25} atoms of oxygen (${}_8O^{17}$) :
(Given: $N_A = 6 \times 10^{23}$)
(a) 1800 (b) 1600
(c) 1800 N_A (d) 3200 N_A

8. If mass of one atom is 3.32×10^{-23} g, then calculate number of nucleons (neutrons and protons) present in 2 atoms of the element :
(a) 40 (b) 20
(c) 10 (d) 40 N_A

9. What is the number of moles of O-atom in 126 amu of HNO_3 ?
(a) 2 (b) $\frac{2}{N_A}$
(c) 6 (d) $\frac{6}{N_A}$

10. A sample of sodium has a mass of 46 g. What is the mass of the same number of calcium atoms as sodium atoms present in given sample ?
(a) 46 g (b) 20 g
(c) 40 g (d) 80 g

11. What is the charge of 96 amu of S^{2-} ?
(a) 2 C (b) 3.2×10^{-19} C
(c) 9.6×10^{-19} C (d) 6 C

12. Calculate number of electrons present in 9.5 g of PO_4^{3-} :
(a) 6 (b) 5 N_A
(c) 0.1 N_A (d) 4.7 N_A

13. An atom of an element is 3.5 times heavier as compared to one atom of ^{12}C isotope. The molar mass of element (g/mol) is :

(a) 3.5 (b) 12

(c) 42 (d) None of these

14. Number of electrons in 28 g of $_{26}Fe^{2+}$ ions will be :

(a) 12 (b) 0.5

(c) 12 N_A (d) 0.5 N_A

15. The total no. of neutrons present in 54 mL H_2O (*l*) are:

(a) 3 N_A (b) 30 N_A

(c) 24 N_A (d) none of these

16. Total no. of electrons present in 48 g Mg^{2+} are:

(a) 24 N_A (b) 2 N_A

(c) 20 N_A (d) none of these

17. The number of neutron in 5 g of D_2O (D is $_1^2H$) are:

(a) 0.25 N_A (b) $2 \cdot 5\, N_A$

(c) $1 \cdot 1\, N_A$ (d) none of these

18. 3.011×10^{22} atoms of an element weight 1.15 gm. The atomic mass of the element is:

(a) 10 amu (b) 2.3 amu

(c) 35.5 amu (d) 23 amu

19. 10 mole of CO_2 contains :

(a) 220 electrons (b) 440 mole protons

(c) 220 mole neutrons (d) $220 \times 6.02 \times 10^{23}$ nucleons

20. Mass of oxygen present in 750 amu sample of $CuSO_4.5H_2O$ is : (Given Cu = 64)

(a) 64 amu (b) 144 amu

(c) 80 amu (d) None of these

21. How many moles of $Ca_3(PO_4)_2$ will contain 0.20 mol of oxygen atoms?

(a) 0.25 (b) 0.025

(c) 40 (d) None of these

22. How many moles of proton weight one mg ? (Weight of one proton = 1.66×10^{-24} gm)

(a) 10^{-3} (b) 6.022×10^{-3}

(c) 6.022×10^{3} (d) 1.66×10^{-3}

23. How many hydrogen atoms are present in 0.1 mole of mohr's salt ($FeSO_4.(NH_4)_2SO_4.6H_2O$)?

(a) $10 \times N_A$ (b) $14 \times N_A$

(c) $2 \times N_A$ (d) None of these

24. The number of electrons present in 3.2 g-ion of O^{2-} is :

(a) 32 N_A (b) 3.2 N_A

(c) N_A (d) 2.0 N_A

25. Cisplatin, an anticancer drug, has the molecular formula $Pt(NH_3)_2Cl_2$. What is the mass (in gram) of one molecule ? (Atomic weights : Pt = 195, H = 1.0, N = 14, Cl = 35.5)

(a) 4.98×10^{-21} (b) 4.98×10^{-22}

(c) 6.55×10^{-21} (d) 3.85×10^{-22}

26. Aspirin has the formula $C_9H_8O_4$. How many atoms of oxygen are there in a tablet weighing 360 mg?

(a) 1.204×10^{23} (b) 1.08×10^{22}

(c) 1.204×10^{24} (d) 4.81×10^{21}

27. 20 g of an ideal gas contains only atoms of S and O occupies 5.6 L at 1 atm and 273 K. What is the mol. wt. of gas ?

(a) 64 (b) 80

(c) 96 (d) None of these

28. A sample of ammonium phosphate, $(NH_4)_3PO_4$, contains 6 moles of hydrogen atoms. The number of moles of oxygen atoms in the sample is:

(a) 1 (b) 2

(c) 4 (d) 6

29. Total number of moles of oxygen atoms in 3 litre $O_3(g)$ at 27°C and 8.21 atm are:

(a) 3 (b) 1

(c) 1 (d) none of these

30. One atom of an element x weigh 6.643×10^{-23} g. Number of moles of atom in 20 kg is:

(a) 4 (b) 40

(c) 100 (d) 500

31. Mass of one atom of the element A is 3.9854×10^{-23} g. How many atoms are contained in 1 g of the element A?

(a) 2.509×10^{23} (b) 6.022×10^{23}

(c) 12.044×10^{23} (d) None of these

32. Which of the following contains the largest mass of hydrogen atoms?

(a) 5.0 moles $C_2H_2O_4$ (b) 1.1 moles $C_3H_8O_3$

(c) 1.5 moles $C_6H_8O_6$ (d) 4.0 moles $C_2H_4O_2$

33. Which has minimum number of atoms of oxygen ?

(a) 10 mL H_2O (l) (b) 0.1 mole of V_2O_5 (s)

(c) 12 gm $O_3(g)$ (d) 12.044×10^{22} molecules of CO_2

34. Rearrange the following (I to IV) in the order of increasing masses:

(I) 0.5 mole of O_3 (II) 0.5 g-atom of oxygen

(III) 3.011×10^{23} molecules of O_2 (IV) 5.6 litre of CO_2 at 1 atm and 273 K

(a) II < IV < III < I (b) II < I < IV < III

(c) IV < II < III < I (d) I < II < III < IV

35. The volume of a drop of water is 0.0018 mL then the number of water molecules present in two drop of water at room temperature is:

(a) 12.046×10^{19} (b) 1.084×10^{18}

(c) 4.84×10^{17} (d) 6.023×10^{23}

36. It is known that atom contain protons, neutrons and electrons. If the mass of neutron is assumed to half of its original value whereas that of proton is assumed to be twice of its original value then the atomic mass of $^{14}_{6}C$ will be :

(a) same (b) 14.28% less

(c) 14.28% more (d) 28.56% less

37. Number of molecules present in one litre of water is equal to :

(a) $\frac{6.023}{23.4} \times 10^{23}$ (b) $18 \times 6.023 \times 10^{23}$

(c) $\frac{18}{22.4} \times 10^{23}$ (d) $55.5 \times 6.023 \times 10^{23}$

38. The density of water is 1g/mL. What is the volume occupied by 1 molecule of water ?

(a) 1.44×10^{-23} mL
(b) 1mL
(c) 18 mL
(d) 2.88×10^{-23} mL

39. What is the mass % of oxygen in $Fe_2(SO_4)_3$?

(a) 24
(b) 28
(c) 32
(d) 48

40. Which contains maximum mass percentage of Nitrogen ?

(a) NH_3
(b) NH_2CONH_2
(c) CH_3NH_2
(d) $C_2H_5NH_2$

41. Common salt obtained from sea-water contains 8.775% NaCl by mass. The number of formula units of NaCl present in 25 g of this salt is :

(a) 3.367×10^{23} formula units
(b) 2.258×10^{22} formula units
(c) 3.176×10^{23} formula units
(d) 4.73×10^{25} formula units

42. If equal mass of following substance are taken then which will have maximum number of molecules ?

(a) $C_6H_{12}O_6$
(b) $C_{12}H_{22}O_{11}$
(c) C_2H_6
(d) CO_2

43. Which of the following contains largest mass of oxygen.

(a) 5 mole of $C_2H_2O_4$
(b) 1.1 mole of $C_3H_8O_3$
(c) 1.5 mole of $C_6H_8O_6$
(d) 4 mole of $C_2H_4O_2$

44. Caffiene has a molecular weight of 194. If it contains 28.9% by mass of nitrogen, number of atoms of nitrogen in one molecule of caffeine is :

(a) 4
(b) 6
(c) 2
(d) 3

45. The number of hydrogen atoms present in 25.6 g of sucrose ($C_{12}H_{22}O_{11}$) which has a molar mass of 342.3 g is :

(a) 22×10^{23}
(b) 9.91×10^{23}
(c) 11×10^{23}
(d) 44×10^{23}

46. A 25.0 mm × 40.0 mm piece of gold foil is 0.25 mm thick. The density of gold is 19.32 g/cm^3. How many gold atoms are in the sheet ? (Atomic weight : Au = 197.0)

(a) 7.7×10^{23}
(b) 1.5×10^{23}
(c) 4.3×10^{21}
(d) 1.47×10^{22}

47. If average molecular wt. of air is 29, then assuming N_2 and O_2 gases are there which options are correct regarding composition of air:

(i) 75% by mass of N_2 (ii) 75% by moles N_2 (iii) 72.41% by mass of N_2

(a) only (i) is correct
(b) only (ii) is correct
(c) both (ii) and (iii) are correct
(d) both (i) and (ii) are correct

48. A currency counting machine counts 6.023 million notes per day. A bank has many notes as number of oxygen atoms in 24.8 g of $Na_2S_2O_3.5H_2O$. How many days would be required to count this notes ?

(a) 10^{16}
(b) 15×10^{15}
(c) 8×10^{16}
(d) None of these

49. The ratio of masses of oxygen and nitrogen in a particular gaseous mixture is 1 : 4. The ratio of number of their molecules is:

(a) 1 : 4
(b) 7 : 32
(c) 1 : 8
(d) 3 : 16

50. Equal masses of oxygen, helium and sulphur dioxide are taken in a container in identical condition. Find the ratio of their moles :

(a) 1 : 16 : 2 (b) 2 : 16 : 1

(c) 4 : 16 : 1 (d) None of these

51. An unknown amino acid has 0.032% sulphur. If each molecule has one sulphur atom only then number of molecules of amino acid present in 1 gm of amino acid is :

(a) 6.02×10^{18} (b) 6.02×10^{20}

(c) 3.01×10^{18} (d) 1.204×10^{21}

52. An anabolic steroid contains 5% sulphur by mass. If its molar mass is 1280 g, then atomicity of S in the molecule will be :

(a) 2 (b) 4

(c) 8 (d) 16

53. If number of neutrons in 0.5 mole of ${}_2^4He$, 0.20 mole of ${}_3^7Li$, 0.15 mole of H_2O and 5 g-atom of hydrogen are w, x, y and z respectively then select correct relation :

(a) $w < x < y < z$ (b) $z < w < x < y$

(c) $z < x < w < y$ (d) $x < w < y < z$

54. 1 mole of a compound contains 1 mole N and 2 mole of O. The mass of one molecule of the compound is :

(a) 46 g (b) 78 g

(c) 23 amu (d) 46 amu.

55. If 890 kJ heat is evolved in the combustion of 1 mole of $CH_4(g)$. What is the weight of methane required to produce 356 kJ of heat on combustion ?

(a) 16 g (b) 6.4 g

(c) 12.8 g (d) 30.6 g

56. An element *X*-exists as ${}_{49}X$ and ${}_{53}X$ in nature. If atomic mass of *X* is 50 amu then the mole % of heavier isotope in nature, is :

(a) 25% (b) 33.33%

(c) 66.67% (d) 75%

57. Naturally occurring boron consists of two isotopes whose atomic weights are 10 and 11. The atomic weight of natural boron is 10.8. Then the percentage abundance of each isotope in natural boron may be :

(a) 10, 90 (b) 20, 80

(c) 30, 70 (d) 15, 85

58. Average atomic mass of magnesium is 24.31 a.m.u. This magnesium is composed of 79 mole % of ${}^{24}Mg$ and remaining 21 mole % of ${}^{25}Mg$ and ${}^{26}Mg$. Calculate mole % of ${}^{26}Mg$.

(a) 10 (b) 11

(c) 15 (d) 16

59. Indium (atomic weight = 114.82) has two naturally occurring isotopes, the predominant one form has isotopic weight 114.9041 and abundance of 95.72%. Which of the following isotopic weights is the most likely for the other isotope ?

(a) 112.94 (b) 115.90

(c) 113.90 (d) 114.90

60. Calculate density of a gaseous mixture which consist of 3.01×10^{24} molecules of N_2 and 32 g of O_2 gas at 3 atm pressure and 860 K temperature (Given : $R = 1/12$ atm L/mole.K)

(a) 0.6 g/L (b) 1.2 g/L

(c) 0.3 g/L (d) 12 g/L

61. A gaseous mixture of H_2 and CO_2 gas contains 66 mass % of CO_2. The vapour density of the mixture is:
(a) 6.1 (b) 5.4
(c) 2.7 (d) 10.8

62. The vapour density of a mixture containing NO_2 and N_2O_4 is 27.6. The mole fraction of N_2O_4 in the mixture is:
(a) 0.1 (b) 0.2
(c) 0.5 (d) 0.8

63. A gaseous mixture contains N_2 and NH_3 in mole ratio of 1 : 4 respectively. The vapour density of the gaseous mixture is :
(a) 8 (b) 12
(c) 24 (d) None of these

64. Disilane, Si_2H_x, is analysed and found to 90.32% silicon by mass what is the value of "x" ?
(a) 3 (b) 4
(c) 6 (d) 8

65. A mixture of O_2 and gas "Y" (mol. wt. 80) in the mole ratio $a : b$ has a mean molecular weight 40. What would be mean molecular weight, if the gases are mixed in the ratio $b : a$ under identical conditions ? (gases are non-reacting):
(a) 40 (b) 48
(c) 62 (d) 72

66. The hydrated salt $Na_2CO_3{\cdot}xH_2O$ undergoes 63% loss in mass on heating and becomes anhydrous. The value of x is:
(a) 10 (b) 12
(c) 8 (d) 18

67. A 6.85 g sample of the hydrate $Sr(OH)_2{\cdot}xH_2O$ is dried in an oven to give 3.13 g of anhydrous $Sr(OH)_2$. What is the value of x ? (Atomic weights : Sr = 87.60, O = 16.0, H = 1.0)
(a) 8 (b) 12
(c) 10 (d) 6

68. What percentage of oxygen is present in the compound $CaCO_3{\cdot}3Ca_3(PO_4)_2$?
(a) 23.3% (b) 45.36%
(c) 41.94% (d) 17.08%

69. The formula of an acid is HXO_2. The mass of 0.025 mole of acid is 1.70 g. What is the mass of 1 mole of X ?
(a) 35 g (b) 36 g
(c) 68 g (d) None of these

70. If water sample are taken from sea, rivers or lake, they will be found to contain hydrogen and oxygen in the approximate ratio of 1 : 8. This indicates the law of:
(a) Multiple proportion (b) Definite proportion
(c) Reciprocal proportions (d) None of these

71. Hydrogen and oxygen combine to form H_2O_2 and H_2O containing 5.93% and 11.2% hydrogen respectively. The data illustrates :
(a) law of conservation of mass (b) law of constant proportion
(c) law of reciprocal proportion (d) law of multiple proportion

72. Carbon and oxygen combine to form two oxides, carbon monoxide and carbon dioxide in which the ratio of the weights of carbon and oxygen is respectively 12 : 16 and 12 : 32. These figures illustrate the :
(a) Law of multiple proportions (b) Law of reciprocal proportions
(c) Law of conservation of mass (d) Law of constant proportions

73. A sample of calcium carbonate ($CaCO_3$) has the following percentage composition : Ca = 40%, C = 12%, O = 48%. If the law of constant proportions is true, then the weight of calcium in 4 g of a sample of calcium carbonate obtained from another source will be :

(a) 0.016 g (b) 0.16 g
(c) 1.6 g (d) 16 g

74. The law of multiple proportion is illustrated by the two compounds :

(a) Sodium chloride and sodium bromide (b) Ordinary water and heavy water
(c) Caustic soda and caustic potash (d) Sulphur dioxide and sulphur trioxide

75. All the substances listed below are fertilizers that contribute nitrogen to the soil. Which of these is the richest source of nitrogen on a mass percentage basis ?

(a) Urea, $(NH_2)_2CO$ (b) Ammonium nitrate, NH_4NO_3
(c) Nitric oxide, NO (d) Ammonia, NH_3

76. Calculate empirical formula of a compound which contains 4 g atom of carbon, 5 g of H-atom and 6.02×10^{23} molecules of oxygen gas :

(a) C_4H_5O (b) C_2H_5O
(c) CH_2O (d) $C_4H_5O_2$

77. A given sample of pure compound contains 9.81 gm of Zn, 1.8×10^{23} atoms of chromium and 0.60 mole of oxygen atoms. What is the simplest formula ?

(a) $ZnCr_2O_7$ (b) $ZnCr_2O_4$
(c) $ZnCrO_4$ (d) $ZnCrO_6$

78. What is the empirical formula of vanadium oxide, if 2.74 g of the metal oxide contains 1.53 g of metal ?

(a) V_2O_3 (b) VO
(c) V_2O_5 (d) V_2O_7

79. Determine the empirical formula of Kelvar, used in making bullet proof vests, is 70.6% C, 4.2% H, 11.8% N and 13.4% O :

(a) $C_7H_5NO_2$ (b) $C_7H_5N_2O$
(c) C_7H_9NO (d) C_7H_5NO

80. 120 g of an organic compound contains 40% C by mass, 8 g-atom of H and 64 g of oxygen. The molecular formula of compound is :

(a) C_2H_5OH (b) CH_3OCH_3
(c) CH_3COOH (d) CH_3COCH_3

81. In a compound, C is 60% by mass and masses of H & O are in a ratio of 1 : 4. The E.F. of compound is :

(a) $C_5H_8O_2$ (b) C_5H_4O
(c) $C_{10}H_{16}O_4$ (d) $C_{15}H_{24}O_8$

82. Complete combustion of a compound gives 4.4 g of CO_2 and 0.9 g of H_2O. If vapour density of compound is 39 then molecular formula of compound is :

(a) C_2H_2 (b) C_2H_4
(c) C_6H_6 (d) C_8H_8

83. Complete combustion of 0.86 g of compound X gives 2.64 g of CO_2 and 1.26 g of H_2O. The lowest molecular mass X can have :

(a) 43 g (b) 86 g
(c) 129 g (d) 172 g

84. The sulphate of a metal M contains 9.87% of M. This sulphate is isomorphous with $ZnSO_4 \cdot 7H_2O$. The atomic weight of M is :

(a) 40.3 (b) 36.3
(c) 24.3 (d) 11.3

85. In an organic compound of molar mass 108 gm mol^{-1} C, H and N atoms are present in 9 : 1 : 3.5 by weight. Molecular formula can be :

(a) $C_6H_8N_2$ (b) $C_7H_{10}N$
(c) $C_5H_6N_3$ (d) $C_4H_{18}N_3$

86. On analysis, a certain compound was found to contain 254 g of iodine (at. mass 127) and 80 g oxygen (at. mass 16). What is the formula of the compound ?

(a) IO (b) I_2O
(c) I_5O_3 (d) I_2O_5

87. 44 g of a sample on complete combustion gives 88 gm CO_2 and 36 gm of H_2O. The molecular formula of the compound may be:

(a) C_4H_6 (b) C_2H_6O
(c) C_2H_4O (d) C_3H_6O

88. An element A is tetravalent and another element B is divalent. The formula of the compound formed from these elements will be :

(a) A_2B (b) AB
(c) AB_2 (d) A_2B_3

89. A compound used in making nylon, is 43.8% oxygen. There are four oxygen atoms per molecule. What is the molecular weight of compound?

(a) 36 (b) 116
(c) 292 (d) 146

90. One mole of element X has 0.444 times the mass of one mole of element Y. One atom of element X has 2.96 times the mass of one atom of ^{12}C. What is the atomic weight of Y ?

(a) 80 (b) 15.77
(c) 46.67 (d) 40.0

91. Suppose two elements X and Y combine to form two compounds XY_2 and X_2Y_3 when 0.05 mole of XY_2 weighs 5 g while 3.011×10^{23} molecules of X_2Y_3 weighs 85 g. The atomic masses of X and Y are respectively :

(a) 20, 30 (b) 30, 40
(c) 40, 30 (d) 80, 60

92. A gaseous compound is composed of 85.7% by mass carbon and 14.3% by mass hydrogen. It's density is 2.28 g/litre at 300 K and 1.0 atm pressure. Determine the molecular formula of the compound :

(a) C_2H_2 (b) C_2H_4
(c) C_4H_8 (d) C_4H_{10}

93. Two element X (at. mass = 75) and Y (at. mass = 16) combine to give a compound having 75.8% of X. The formula of the compound is:

(a) XY (b) X_2Y
(c) X_2Y_2 (d) X_2Y_3

94. A sample of phosphorus that weighs 12.4 g exerts a pressure 8 atm in a 0.821 litre closed vesel at 527°C. The molecular formula of the phosphorus vapour is:

(a) P_2 (b) P_4
(c) P_6 (d) P_8

95. 40 milligram diatomic volatile substance (X_2) is converted to vapour that displaced 4.92 mL of air at 1 atm and 300 K. Atomic weight of element X is nearly:

(a) 400 (b) 240
(c) 200 (d) 100

96. Manganese forms non-stoichiometric oxides having the general formula MnO_x. The value of x for the compound that analyzed 64% by mass Mn:

(a) 1.16 (b) 1.83
(c) 2 (d) 1.93

97. Total moles of gases produced upon complete decomposition of 102 g NH_3 as following are :

$$NH_3(g) \longrightarrow N_2(g) + H_2(g)$$

(a) 4 (b) 6
(c) 8 (d) 12

98. Sucrose reacts with oxygen to produce CO_2 and H_2O according to the equation :

$$C_{12}H_{22}O_{11} + 12O_2 \longrightarrow 12CO_2 + 11H_2O$$

The mass of CO_2 produced per gram of sucrose is :

(a) 1.54 g (b) 3.08 g
(c) 5 g (d) 7.5 g

99. Carbon reacts with chlorine to form CCl_4. 48 g of carbon was reacted with 142 g of Cl_2. Sum of moles of CCl_4 produced and the remaining reactant are :

(a) 2 (b) 3
(c) 4 (d) None of these

100. The minimum mass of $O_2(g)$ required for complete combustion of 140 grams of C_2H_4, is :

(a) 96 g (b) 960 g
(c) 48 g (d) 480 g

101. Consider the reaction $4A + B + 3C \longrightarrow P + 2Q$. Starting with 6 mole of A, 2 mole of B and 4 mole of C, number of mole of the product Q after reaction would be :

(a) 4/3 (b) 8/3
(c) 2 (d) none of these

102. 56 g N_2 reacts with 179.2 L H_2 gas at 1 atm pressure and 273 K temperature to form NH_3. What is the mass of NH_3 formed?

(a) 34 g (b) 68 g
(c) 85 g (d) 65 g

103. Which statement is false for the balanced equation given below ?

$$CS_2 + 3O_2 \longrightarrow 2SO_2 + CO_2$$

(a) One mole of CS_2 will produce one mole of CO_2
(b) The reaction of 16 g of oxygen produces 7.33 g of CO_2
(c) The reaction of one mole of O_2 will produce 2/3 mole of SO_2
(d) Six molecules of oxygen requires three molecules of CS_2

104. 2.0 g sample contain mixture of SiO_2 and Fe_2O_3, on very strong heating leave a residue weighing 1.96 g. The reaction responsible for loss of weight is

$$Fe_2O_3(s) \longrightarrow Fe_3O_4(s) + O_2(g), \text{ (unbalance equation)}$$

What is the percentage by mass of SiO_2 in original sample ?

(a) 10% (b) 20%
(c) 40% (d) 60%

105. How many moles of P_4 can be produced by reaction of 0.10 moles $Ca_5(PO_4)_3F$, 0.36 moles SiO_2 and 0.90 moles C according to the following reaction ?

$$4\,Ca_5(PO_4)_3F + 18\,SiO_2 + 30\,C \longrightarrow 3P_4 + 2CaF_2 + 18CaSiO_3 + 30\,CO$$

(a) 0.060 (b) 0.030
(c) 0.045 (d) 0.075

106. Some older emergency oxygen masks containing potassium superoxide, KO_2 which reacts with CO_2 and water in exhaled air to produce oxygen according to the given equation. If a person exhales 0.667 g of CO_2 per minute, how many grams of KO_2 are consumed in 5.0 minutes?

$$4KO_2 + 2H_2O + 4CO_2 \longrightarrow 4KHCO_3 + 3O_2$$

(a) 10.7 (b) 0.0757
(c) 1.07 (d) 5.38

107. Calculate the weight of lime (CaO) obtained by heating 200 kg of 95% pure limestone ($CaCO_3$):

(a) 104.4 kg (b) 105.4 kg
(c) 212.8 kg (d) 106.4 kg

108. For the reaction, $2Fe(NO_3)_3 + 3Na_2CO_3 \longrightarrow Fe_2(CO_3)_3 + 6NaNO_3$
Initially if 2.5 mole of $Fe(NO_3)_3$ and 3.6 mole of Na_2CO_3 is taken. If 6.3 mole of $NaNO_3$ is obtained then % yield of given reaction is:

(a) 50 (b) 84
(c) 87.5 (d) 100

109. The mass of N_2F_4 produced by the reaction of 3.4 g of NH_3 and 76 g of F_2 is 7.8 g. What is the per cent yield?

$$2NH_3 + 5F_2 \longrightarrow N_2F_4 + 6HF$$

(a) 79.0 (b) 71.2
(c) 84.6 (d) None of these

110. The % loss in mass upon strong heating of $NaHCO_3$ is nearly :

(a) 37% (b) 63%
(c) 74% (d) Data insufficient

111. 36 g of C(s) is allowed to react with 6 g of $H_2(g)$ according to following reaction

$$2C(s) + 3H_2(g) \xrightarrow{80\%} C_2H_6(g)$$

What mass of C_2H_6 formed ?

(a) 30 g (b) 24 g
(c) 36 g (d) 45 g

112. The molar ratio of Fe^{2+} to Fe^{3+} in a mixture of $FeSO_4$ and $Fe_2(SO_4)_3$ having equal number of sulphate ions in both ferrous and ferric sulphates is :

(a) 1:2 (b) 3:2
(c) 2:3 (d) None of these

113. If 6 g H_2 is mixed with 14 g $N_2(g)$ as per following reaction $N_2 + 3H_2 \longrightarrow 2NH_3$ if after completion of reaction mass of NH_3 produced is 12.75 g then % yield of reaction is :

(a) 50% (b) 75%
(c) 85% (d) 45%

114. Silver oxide (Ag_2O) decomposes at temperature 300°C yielding metallic silver and oxygen gas. A 1.60 g sample of impure silver oxide yields 0.104 g of oxygen gas. What is the per cent by mass of the silver oxide in the sample ?

(a) 5.9 (b) 47.125

(c) 94.25 (d) 88.2

115. An ideal gaseous mixture of ethane (C_2H_6) and ethene (C_2H_4) occupies 28 litre at 1 atm and 273 K. The mixture reacts completely with 128 g O_2 to produce CO_2 and H_2O. Mole fraction at C_2H_6 in the mixture is:

(a) 0.6 (b) 0.4

(c) 0.5 (d) 0.8

116. One commercial system removes SO_2 emission from smoke at 95°C by the following series of reaction.

$$SO_2(g) + Cl_2(g) \longrightarrow SO_2Cl_2(g)$$
$$SO_2Cl_2(g) + H_2O(l) \longrightarrow H_2SO_4 + 2HCl$$
$$H_2SO_4 + Ca(OH)_2 \longrightarrow CaSO_4 + H_2O$$

Assuming that 80% SO_2 can react with Cl_2 then how much $CaSO_4$ can be obtained from 100 g SO_2 ?

(a) 212.5 g (b) 170 g

(c) 172.5 g (d) None of these

117. 9 moles of "*D*" and 14 moles of *E* are allowed to react in a closed vessel according to given reactions. Calculate number of moles of *B* formed in the end of reaction, if 4 moles of *G* are present in reaction vessel. (Percentage yield of reaction is mentioned in the reaction)

Step-1 $3D + 4E \xrightarrow{80\%} 5C + A$

Step-2 $3C + 5G \xrightarrow{50\%} 6B + F$

(a) 2.4 (b) 30

(c) 4.8 (d) 1

118. Mass of 0.1 dm^3 of liquid ethanoic acid (density = 1.2 g/mL) is :

(a) 30g (b) 60g

(c) 90g (d) 120g

119. In a solution 180 g water and 640 g CH_3OH present. The mole fraction of solute is :

(a) $\frac{2}{3}$ (b) $\frac{1}{3}$

(c) $\frac{4}{5}$ (d) $\frac{1}{5}$

120. Amount of sucrose ($C_{12}H_{22}O_{11}$) required to be dissolved in 81.0 g of water to give 10% by mass of sucrose in solution is :

(a) 8.1 g (b) 9 g

(c) 10 g (d) 18 g

121. What is the amount of water that should be mixed with 16g of methanol to make the mole fraction of methanol as 0.25 in solution ?

(a) 27 g (b) 12 g

(c) 18 g (d) 36 g

122. What volume of HCl solution of density 1.2 g/cm^3 and containing 36.5% by weight HCl, must be allowed to react with zinc (Zn) in order to liberate 4.0 g of hydrogen ?

(a) 333.33 mL (b) 500 mL

(c) 614.66 mL (d) None of these

123. The density of a 56.0% by weight aqueous solution of 1-propanol ($CH_3CH_2CH_2OH$) is 0.8975 g/cm^3. What is the mole fraction of the compound ?

(a) 0.292
(b) 0.227
(c) 0.241
(d) 0.276

124. How many grams of KCl should be dissolved in 60 g H_2O to give a solution of 20% by weight KCl (Given molar mass of KCl = 74.5) :

(a) 15 g
(b) 1.5 g
(c) 11.5 g
(d) 31.5 g

125. The density of liquid enthanol (C_2H_5OH) is 0.80 g/mL. If 1.5 mole of ethanol are needed for a particular experiment. What volume of ethanol is required ?

(a) 74.4 mL
(b) 116.25 mL
(c) 55.2 mL
(d) 86.25 mL

126. What volume of ethanol contains the same number of molecules as are present in 144 mL of water? Given : density of $H_2O(l) = 1 g/mL$, density of $C_2H_5OH(l) = 0.80 g/mL$:

(a) 460 mL
(b) 70.43 mL
(c) 368 mL
(d) None of these

127. The molarity of 5.6% *w/V* KOH solution is :

(a) 0.1 *M*
(b) 10 *M*
(c) 0.01 *M*
(d) 1 *M*

128. 1.12 g of KOH dissolved in water and formed 500 mL solution. Calculate molarity of solution :

(a) 22.4
(b) 4×10^{-4} *M*
(c) 0.4 *M*
(d) 0.04 *M*

129. The molality of 10% *w/w* NaOH solution is :

(a) $\frac{25}{10}$
(b) $\frac{25}{9}$
(c) $\frac{1000}{9}$
(d) $\frac{2.5}{9}$

130. 100 g of a solute (molar mass = 80 g/mol) present in 500 mL solution (density = 1.20 g/mL). Calculate molality of solution.

(a) 0.25
(b) 1.5
(c) 2.0
(d) 2.5

131. The density of 1.48% by weight $Ca(OH)_2$ solution is 1.025 g/mL. Molarity of solution is :

(a) 0.205 *M*
(b) 0.102 *M*
(c) 1.02 *M*
(d) 2.04 *M*

132. The density of 2.0 molal aq. solution of NaOH is 1.20 g/mL. What is the molarity of NaOH solution?

(a) 1.57 *M*
(b) 1.11 *M*
(c) 2.4 *M*
(d) 2.22 *M*

133. The density of a solution containing 9.8% *w/w* of H_3PO_4 is 1.1 gm/mL. The molarity of solution is :

(a) 1.1
(b) 11
(c) 9.8
(d) None of these

134. An aqueous solution of ethanol has density 1.025 g/mL and it is 2 *M*. What is the molality of this solution ?

(a) 1.79
(b) 2.143
(c) 1.951
(d) None of these

135. Molarity and molality of a solution of an liquid (mol. wt. = 50) in aqueous solution is 9 and 10 respectively. What is the density of solution?

(a) 1 g/cc
(b) 0.95 g/cc
(c) 1.05 g/cc
(d) 1.35 g/cc

136. The relation between molarity (M) and molality (m) is given by :
(ρ = density of solution (g/mL), M_1 = gram molecular weight of solute)

(a) $m = \dfrac{1000\,M}{1000\,\rho - M_1}$
(b) $m = \dfrac{1000\,\rho M}{1000\,\rho - MM_1}$
(c) $m = \dfrac{1000\,MM}{1000\,\rho - MM_1}$
(d) $m = \dfrac{1000\,M}{1000\,\rho - MM_1}$

137. Calculate the molality of 1 L solution of 80% H_2SO_4 (w/V), given that the density of the solution is 1.80 g mL^{-1}.

(a) 8.16
(b) 8.6
(c) 1.02
(d) 10.8

138. 100 mL of 10% NaOH (w/V) is added to 100 mL of 10% HCl (w/V). The resultant solution becomes:

(a) alkaline
(b) strongly alkaline
(c) acidic
(d) neutral

139. What is the molarity of pure liquid benzene, C_6H_6 (specific gravity = 0.88) ?

(a) 11.28 M
(b) 12.82 M
(c) 48.88 M
(d) None of these

140. If 100 mL of A (d = 0.9 g/mL) and 100 mL of H_2O (d = 1 g/mL), are mixed. The mass percentage of A in resulting solution is :

(a) 90
(b) 47.36
(c) 50
(d) None of these

141. Which of the following is not a dimensionless quantity ?

(a) vapour density
(b) molality
(c) specific gravity
(d) mole fraction

142. If 400 mL of 0.40 M NaOH(aq) solution is diluted to 0.10 M NaOH(aq) solution then final volume of solution is:

(a) 100 mL
(b) 1200 mL
(c) 160 mL
(d) 1600 mL

143. How many mL of water should be added to 200 mL of 0.60 M HCl(aq) solution to dilute into 0.20 M HCl(aq) solution?

(a) 200 mL
(b) 400 mL
(c) 600 mL
(d) None of these

144. What is the molartiy of SO_4^{2-} ion in aqueous solution that contain 34.2 ppm of $Al_2(SO_4)_3$? (Assume complete dissociation and density of solution 1 g/mL)

(a) 3×10^{-4} M
(b) 2×10^{-4} M
(c) 10^{-4} M
(d) None of these

145. 0.2 mole of HCl and 0.2 mole of barium chloride were dissolved in water to produce a 500 mL solution. The molarity of the Cl^- ions is :

(a) 0.06 M
(b) 0.09 M
(c) 1.2 M
(d) 0.80 M

146. 10 mL of 0.1 M Na_2SO_4 is mixed with 10 mL of 0.2 M NaCl solution. If final solution has density 1 gm/mL. Then molarity of Na^+ is:

(a) 4 (b) 5

(c) 0.2 (d) 2

147. Calculate the mass of anhydrous HCl in 10 mL of concentrated HCl (density = 1.2 g / mL) solution having 37% HCl by weight.

(a) 4.44 g (b) 4.44 mg

(c) 4.44×10^{-3} mg (d) 0.444 μg

148. What volume of 5.0 M H_2SO_4 should be mixed with 10 L of 1.0 M H_2SO_4 to make 20.0 L of 3.0 M H_2SO_4 solution ?

(a) 1.7 L (b) 5 L

(c) 8.3 L (d) 10.0 L

149. If 50 mL of 0.5 M HCl (aq.) solution is mixed with 200 mL of 0.5 M H_2SO_4(aq) solution, then concentration of H^+ ion in resultant solution will be?

(a) 0.5 M (b) 0.1 M

(c) 0.45 M (d) 0.9 M

150. 1 M HCl and 2 M HCl are mixed in volume ratio of 4 : 1. What is the final molarity of HCl solution?

(a) 1.5 (b) 1

(c) 1.2 (d) 1.8

151. Three solutions X, Y, Z of HCl are mixed to produce 100 mL of 0.1 M solution. The molarities of X, Y and Z are 0.07 M, 0.12 M and 0.15 M respectively. What respective volumes of X, Y and Z should be mixed ?

(a) 50 mL, 25 mL, 25 mL (b) 20 mL, 60 mL, 20 mL

(c) 40 mL, 30 mL, 30 mL (d) 55 mL, 20 mL, 25 mL

152. Aqueous solution of NaOH containing 20% by weight of NaOH has a specific gravity of 1.25. What volume of this solution will be required to preparation of 250 mL of a 0.5 M NaOH solution ?

(a) 10 mL (b) 20 mL

(c) 40 mL (d) None of these

153. A solution of H_2SO_4 (density = 1.96 g/mL) is labelled as 80% by weight. The volume of same solution is used to make 1000 mL 0.2 M H_2SO_4 solution is :

(a) 15.68 mL (b) 12.5 mL

(c) 12.75 mL (d) None of these

154. $Al_2(SO_4)_3$ solution of 1 molal concentration is present in 1 litre solution of 2.684 g/cc. How many moles of $BaSO_4$ would be precipitated on adding $BaCl_2$ in excess ?

(a) 2 moles (b) 3 moles

(c) 6 moles (d) 12 moles

155. Fluoxymesterone, $C_{20}H_{29}FO_3$, is an anabolic steroid. A 500 mL solution is prepared by dissolving 10.0 mg of the steroid in water, 1.0 mL portion of this solution is diluted to a final volume of 1.00 L. What is the resulting molarity ?

(a) 1.19×10^{-10} (b) 1.19×10^{-7}

(c) 5.95×10^{-8} (d) 2.38×10^{-11}

156. The lead nitrate, $Pb(NO_3)_2$, in 25 mL of a 0.15 M solution reacts with all of the aluminium sulphate, $Al_2(SO_4)_3$, in 20 mL of solution. What is the molar concentration of the $Al_2(SO_4)_3$?

$$3Pb(NO_3)_2(aq) + Al_2(SO_4)_3(aq) \longrightarrow 3PbSO_4(s) + 2Al(NO_3)_3(aq)$$

(a) $6.25 \times 10^{-2}\ M$ (b) $2.421 \times 10^{-2}\ M$
(c) $0.1875\ M$ (d) None of these

157. Concentrated HNO_3 is 63% HNO_3 by mass and has a density of 1.4 g/mL. How many millilitres of this solution are required to prepare 250 mL of a 1.20 M HNO_3 solution?
(a) 18.0 (b) 21.42
(c) 20.0 (d) 14.21

158. Volume (in mL) of 3 M aq. KOH solution required for complete neutralisation of 600 mL, 1 M aq. H_3PO_3 solution, is:
(Given: $2KOH + H_3PO_3 \longrightarrow K_2HPO_3 + 2H_2O$)
(a) 400 (b) 600
(c) 200 (d) 800

159. How many millilitres of 0.1 M H_2SO_4 must be added to 50 mL of 0.1 M NaOH to give a solution that has a concentration of 0.05 M in H_2SO_4?
(a) 400 mL (b) 200 mL
(c) 100 mL (d) None of these

160. Decreasing order (first having highest and then others following it) of mass of pure NaOH in each of the aqueous solution:
(i) 50 g of 40% (*w/W*) NaOH
(ii) 50 mL of 50% (*w/V*) NaOH [$d_{soln.} = 1.2$ g / mL]
(iii) 50 g of 15 M NaOH [$d_{soln.} = 1$ g / mL]
(a) i, ii, iii (b) iii, ii, i
(c) ii, iii, i (d) ii, i, iii

161. A certain public water supply contains 0.10 ppb (part per billion) of chloroform ($CHCl_3$). How many molecules of $CHCl_3$ would be obtained in 0.478 mL drop of this water?
(assumed $d = 1$ g/mL)
(a) $4 \times 10^{-13} \times N_A$ (b) $10^{-3} \times N_A$
(c) $4 \times 10^{-10} \times N_A$ (d) None of these

162. 500 mL H_2O_2 sample produce 22.4 litre of O_2 at 1 atm and 273 K. The volume strength of H_2O_2 sample is :
(a) 22.4 V (b) 44.8 V
(c) 11.2 V (d) 4 V

163. 500 mL of 11.2 V H_2O_2 solution and 500 mL of 22.4 V H_2O_2 solution are mixed together at 1 atm and 273 K. The molarity of final H_2O_2 solution is :
(a) 1 M (b) 2 M
(c) 1.5 M (d) 16.8 M

164. The % (*w/V*) of 44.8 V H_2O_2 solution at 1 atm & 273 K is :
(a) 6.8 (b) 10.2
(c) 13.6 (d) None of these

165. A bottle of an aqueous H_2O_2 solution is labelled as '28.375 *V*' H_2O_2 at 1 bar and 273 K and the density of the solution (in g/mL) is 1.25. Choose the correct option:
(a) Molality of H_2O_2 solution is 2 (b) Molarity of H_2O_2 solution is 5
(c) Molality of H_2O_2 solution is 2.15 (d) None of these

166. % *w/w* of H_2O_2 solution whose 500 mL gives of 45.4 litre of O_2 at 1 bar and 273 K is :
(Given : $d_{solution} = 1.088$ g/mL)
(a) 25 (b) 26.34
(c) 23 (d) 75

167. Calculate the number of sulphate ions in 100 mL of 0.001 *M* H_2SO_4 solution :
(a) $10^{-1} N_A$ (b) N_A
(c) 10^{-4} (d) $10^{-4} N_A$

168. Given :
$$8\,KI(aq) + 5H_2SO_4(aq) \longrightarrow 4K_2SO_4(aq) + 4I_2(aq) + H_2S(g) + 4H_2O(l)$$
The volume of 0.2 *M* H_2SO_4 required to produce 68 g of H_2S is :
(a) 25 Litre (b) 50 Litre
(c) 75 Litre (d) 100 Litre

169. Silver metal reacts with HNO_3 according to equation :
$$3Ag(s) + 4HNO_3(aq) \longrightarrow 3AgNO_3(aq) + NO(g) + 2H_2O(l)$$
What volumes of 1.75 *M* HNO_3 (*aq*) is required to react with 0.756 g of Ag?
(a) 4 mL (b) 16 mL
(c) $\frac{16}{3}$ mL (d) None of these

170. A 10 L gaseous alkene $[C_nH_{2n}]$ on complete combustion with excess of oxygen gives 50 L CO_2. The molecular formula of alkene is :
(a) C_2H_4 (b) C_5H_{10}
(c) C_5H_{12} (d) C_5H_8

171. What volume of $O_2(g)$ measured at 0°C and 1 atm, is needed to burn completely 10 L of ethane gas (C_2H_6) measured under same conditions?
(a) 20 L (b) 35 L
(c) 70 L (d) None of these

172. 10 L of $O_2(g)$ is reacted with 30 L of CO(*g*) at STP. The volume of each gas present at the end of the reaction are :
(a) $O_2 = 10$ L, $CO_2 = 20$ L (b) $CO = 10$ L, $CO_2 = 20$ L
(c) $CO = 20$ L, $CO_2 = 10$ L (d) $CO = 15$ L, $CO_2 = 15$ L

173. 10 mL of a gaseous hydrocarbon gives on complete combustion 40 mL of $CO_2(g)$ and 50 mL of $H_2O(g)$ at same pressure and temperature. The formula of hydrocarbon is?
(a) C_4H_{10} (b) C_3H_4
(c) C_2H_6 (d) None of these

174. The percentage by volume of C_3H_8 in a gaseous mixture of C_3H_8, CH_4 and CO is 20. When 100 mL of the mixture is burnt in excess of O_2, the volume of CO_2 produced is:
(a) 90 mL (b) 160 mL
(c) 140 mL (d) none of these

175. A gaseous mixture of propane and butane of volume 3 litre on complete combustion produces 11.0 litre CO_2 under standard conditions of temperature and pressure.
The ratio of volume of butane to propane is:
(a) 1 : 2 (b) 2 : 1
(c) 3 : 2 (d) 3 : 1

Level-2

1. A mixture of NH_4NO_3 and $(NH_4)_2HPO_4$ contain 30.40% mass per cent of nitrogen. What is the mass ratio of the two components in the mixture ?

(a) 2 : 1
(b) 1 : 2
(c) 3 : 4
(d) 4 : 1

2. What volume of 75% alcohol by weight ($d = 0.80\ g/cm^3$) must be used to prepare 150 cm^3 of 30% alcohol by mass ($d = 0.90\ g/cm^3$)?

(a) 67.5 mL
(b) 56.25 mL
(c) 44.44 mL
(d) None of these

3. Calculate the number of millilitres of NH_3 (*aq*) solution ($d = 0.986$ g/mL) contain 2.5% by mass NH_3, which will be required to precipitate iron as $Fe(OH)_3$ in a 0.8 g sample that contains 50% Fe_2O_3.

(a) 0.344 mL
(b) 3.44 mL
(c) 17.24 mL
(d) 10.34 mL

4. In the preparation of iron from haematite (Fe_2O_3) by the reduction with carbon

$$Fe_2O_3 + C \longrightarrow Fe + CO_2$$

how much 80% pure iron may be produced from 120 kg of 90% pure Fe_2O_3 ?

(a) 94.5 kg
(b) 60.48 kg
(c) 116.66 kg
(d) 120 kg

5. A mineral consists of an equimolar mixture of the carbonates of two bivalent metals. One metal is present to the extent of 12.5% by mass. 2.8 g of the mineral on heating lost 1.32 g of CO_2. What is the % by mass of the other metal ?

(a) 87.5
(b) 35.71
(c) 65.11
(d) 23.21

6. 6.2 g of a sample containing Na_2CO_3, $NaHCO_3$ and non-volatile inert impurity on gentle heating loses 5% of its mass due to reaction $2NaHCO_3 \longrightarrow Na_2CO_3 + H_2O + CO_2$. Residue is dissolved in water and formed 100 mL solution and its 10 mL portion requires 7.5 mL of 0.2 *M* aqueous solution of $BaCl_2$ for complete precipitation of carbonates.

Determine mass (in gram) of Na_2CO_3 in the original sample.

(a) 1.59
(b) 1.06
(c) 0.53
(d) None of these

7. Nitric acid can be produced from NH_3 in three steps process given below

(I) $4NH_3(g) + 5O_2(g) \longrightarrow 4NO(g) + 6H_2O(g)$

(II) $2NO(g) + O_2(g) \longrightarrow 2NO_2(g)$

(III) $3NO_2(g) + H_2O(l) \longrightarrow 2HNO_3(aq) + NO(g)$

percent yield of 1st, 2nd and 3rd steps are respectively 50%, 60% and 80% respectively then what volume of $NH_3(g)$ at 1 atm and 0°C required to produced 1575 g of HNO_3.

(a) 156.25
(b) 350 L
(c) 3500 L
(d) None of these

8. 1 M NaOH solution was slowly added into 1000 mL of 183.75 g impure H_2SO_4 solution and the following plot was obtained. The percentage purity of H_2SO_4 sample and slope of the curve respectively are:

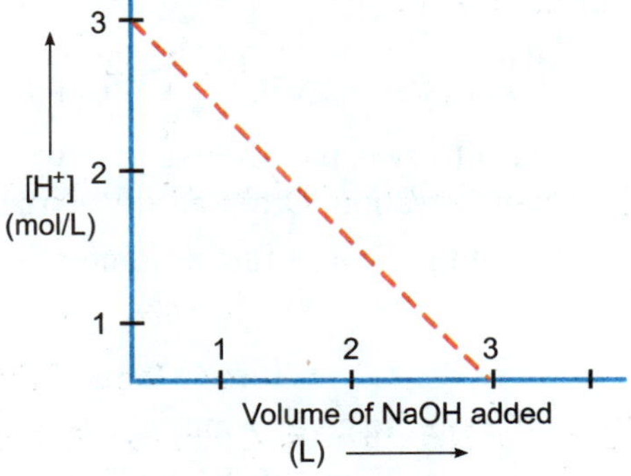

(a) 75%, $-\frac{1}{3}$ (b) 80%, $-\frac{1}{2}$

(c) 80%, -1 (d) None of these

9. MnO_2 on ignition converts into Mn_3O_4. A sample of pyrolusite having 75% MnO_2, 20% inert impurities and rest water is ignited in air to constant mass. What is the percentage of Mn in the ignited sample ?

(a) 24.6% (b) 37%

(c) 55.24% (d) 74.05%

10. A 1.0 g sample of a pure organic compound containing chlorine is fused with Na_2O_2 to convert chlorine to NaCl. The sample is then dissolved in water, and the chloride precipitated with $AgNO_3$, giving 1.96 g of AgCl. If the molecular mass of organic compound is 147, how many chlorine atoms does each molecule contain ?

(a) 1 (b) 2

(c) 3 (d) 4

11. A 0.60 g sample consisting of only CaC_2O_4 and MgC_2O_4 is heated at 500°C, converting the two salts of $CaCO_3$ and $MgCO_3$. The sample then weighs 0.465 g. If the sample had been heated to 900°C, where the products are CaO and MgO. What would the mixtures of oxides have weighed ?

(a) 0.12 g (b) 0.21 g

(c) 0.252 g (d) 0.3 g

12. A metal M forms the sulphate $M_2(SO_4)_3$. A 0.596 gram sample of the sulphate reacts with excess $BaCl_2$ to give 1.220 g $BaSO_4$. What is the atomic mass of M ?

(Atomic mass : S = 32, Ba = 137.3)

(a) 26.9 (b) 69.7

(c) 55.8 (d) 23

13. Urea (H_2NCONH_2) is manufactured by passing $CO_2(g)$ through ammonia solution followed by crystallization. CO_2 for the above reaction is prepared by combustion of hydrocarbons. If combustion of 236 kg of a saturated hydrocarbon (C_nH_{2n+2}) produces as much CO_2 as required for production of 999.6 kg urea then molecular formula of hydrocarbon is:

(a) $C_{10}H_{22}$ (b) $C_{12}H_{26}$

(c) $C_{13}H_{28}$ (d) C_8H_{18}

14. 11.6 g of an organic compound having formula C_nH_{2n+2} is burnt in excess of $O_2(g)$ initially taken in a 22.41 litre steel vessel. Before reaction the gaseous mixture was at 273 K with pressure reading 2 atm. After complete combustion and loss of considerable amount of heat, the mixture of product and excess of O_2 had a temperature of 546 K and 4.6 atm pressure. The formula of organic compound is:

(a) C_2H_6 (b) C_3H_8
(c) C_5H_{12} (d) C_4H_{10}

15. $H_2O_2 + 2KI \xrightarrow{40\%\ \text{yield}} I_2 + 2KOH$

$H_2O_2 + 2KMnO_4 + 3H_2SO_4 \xrightarrow{50\%\ \text{yield}} K_2SO_4 + 2MnSO_4 + 3O_2 + 4H_2O$

150 mL of H_2O_2 sample was divided into two parts. First part was treated with KI and formed KOH required 200 mL of $M/2$ H_2SO_4 for neutralisation. Other part was treated with $KMnO_4$ yielding 6.74 litre of O_2 at 1 atm. and 273 K. Using % yield indicated find volume strength of H_2O_2 sample used.
(a) 5.04 (b) 10.08
(c) 3.36 (d) 33.6

16. SO_2Cl_2 (sulphuryl chloride) reacts with water to given a mixture of H_2SO_4 and HCl. What volume of 0.2 M $Ba(OH)_2$ is needed to completely neutralize 25 mL of 0.2 M SO_2Cl_2 solution:
(a) 25 mL (b) 50 mL
(c) 100 mL (d) 200 mL

17. 5 g sample contain only Na_2CO_3 and Na_2SO_4. This sample is dissolved and the volume made up to 250 mL. 25 mL of this solution neutralizes 20 mL of 0.1 M H_2SO_4. Calculate the % of Na_2SO_4 in the sample.
(a) 42.4 (b) 57.6
(c) 36.2 (d) none of these

18. A silver coin weighing 11.34 g was dissolved in nitric acid. When sodium chloride was added to the solution all the silver (present as $AgNO_3$) was precipitated as silver chloride. The mass of the precipitated silver chloride was 14.35 g. Calculate the percentage of silver in the coin.
(a) 4.8% (b) 95.2%
(c) 90% (d) 80%

19. The conversion of oxygen to ozone occurs to the extent of 15% only. The mass of ozone that can be prepared from 67.2 L of oxygen at 1 atm and 273 K will be :
(a) 14.4 g (b) 96 g
(c) 640 g (d) 64 g

20. During the discharging of lead storage battery, the density of 40% (w/w) $H_2SO_4(aq.)$ solution falls from 1.35 g/mL to 1.225 g/mL solution which is 20% (w/w) $H_2SO_4(aq.)$ solution. Calculate the magnitude of change in molarity of $H_2SO_4(aq.)$ solution in the process:
(a) 2.25 (b) 2.51
(c) 3.01 (d) 5.51

Level-3

Passage-1

Oleum is considered as a solution of SO_3 in H_2SO_4, which is obtained by passing SO_3 in solution of H_2SO_4. When 100 g sample of oleum is diluted with desired mass of H_2O then the total mass of H_2SO_4 obtained after dilution is known as % labelling in oleum.
For example, a oleum bottle labelled as '109% H_2SO_4' means the 109 g total mass of pure H_2SO_4 will be formed when 100 g of oleum is diluted by 9 g of H_2O which combines with all the free SO_3 present in oleum to form H_2SO_4 as $SO_3 + H_2O \longrightarrow H_2SO_4$.

1. What is the % of free SO_3 in an oleum that is labelled as '104.5% H_2SO_4' ?
(a) 10 (b) 20
(c) 40 (d) None of these

2. 9.0 g water is added into oleum sample labelled as "112%" H_2SO_4 then the amount of free SO_3 remaining in the solution is : (STP = 1 atm and 273 K)
(a) 14.93 L at STP (b) 7.46 L at STP
(c) 3.73 L at STP (d) 11.2 L at STP

3. If excess water is added into a bottle sample labelled as "112% H_2SO_4" and is reacted with 5.3 g Na_2CO_3, then find the volume of CO_2 evolved at 1 atm pressure and 300 K temperature after the completion of the reaction :
(a) 2.46 L (b) 24.6 L
(c) 1.23 L (d) 12.3 L

4. 1 g of oleum sample is diluted with water. The solution required 54 mL of 0.4 *N* NaOH for complete neutralization. The % of free SO_3 in the sample is :
(a) 74 (b) 26
(c) 20 (d) None of these

Passage-2

Consider the following series of reactions :

$$Cl_2 + 2NaOH \longrightarrow NaCl + NaClO + H_2O$$
$$3NaClO \longrightarrow 2NaCl + NaClO_3$$
$$4NaClO_3 \longrightarrow 3NaClO_4 + NaCl$$

1. How much Cl_2 is required to prepare 122.5 g of $NaClO_4$ by above sequencial reactions ?
(a) 284 g (b) 213 g
(c) 142 g (d) 71 g

2. How many moles of NaCl will be formed by using 1 mole Cl_2 and other reagents in excess ?
(a) $\frac{1}{12}$ mole (b) 1.67 mole
(c) 1.75 mole (d) 0.75 mole

3. How many moles of $NaClO_3$ obtained after the complection of reaction by taking 1 mole Cl_2 and other reagents in excess ?
(a) $\frac{1}{3}$ mole (b) Zero
(c) $\frac{1}{4}$ mole (d) 1 mole

Passage-3

An alloy of iron and carbon was treated with H_2SO_4 in which only iron reacts as

$$2Fe(s) + 3H_2SO_4(aq) \longrightarrow Fe_2(SO_4)_3(aq) + 3H_2(g)$$

If a sample of alloy weighing 2.5 gm gave 0.03 gm of hydrogen gas.

1. What is the percentage mass of Fe in alloy ?
(a) 22.4 (b) 74.66
(c) 97.2 (d) None of these
2. Average molar mass of alloy sample is :
(a) 14.56 g/mol (b) 20.62 g/mol
(c) 25.38 g/mol (d) None of these

Passage-4

6 gm of an organic compound containing C, H & N only on combustion gives 8.8 gm CO_2 & 7.2 gm water. 24 gm of the same compound on heating liberated NH_3 gas which completely reacts with 25 mL, 98% w/w H_2SO_4 solution (specific gravity 1.6) to form $(NH_4)_2SO_4$.

1. Percentage of Nitrogen in compound is :
(a) 40% (b) 46.67%
(c) 23.33% (d) None of these
2. Empirical formula of the compound is :
(a) CH_4N_2 (b) CH_2N_2
(c) CH_4N (d) C_2H_4N

Passage-5

HNO_3 used as a reagent has specific gravity of 1.42 g/mL and contains 70% (V/V) by strength HNO_3.

1. Molarity of acid is :
(a) 16.78 (b) 15.78
(c) 14.78 (d) 17.78
2. Volume of acid that contains 63 g pure acid is :
(a) 100 mL (b) 40.24 mL
(c) 63.38 mL (d) 70.68 mL

One or More Answer is/are Correct

1. 1 g atom of nitrogen represents :
(a) 6.02×10^{23} N_2 molecules (b) 22.4 litre of N_2 at 1 atm and 273 K
(c) 11.2 litre of N_2 at 1 atm and 273 K (d) 14 g of nitrogen
2. 1 g molecule of V_2O_5 contains :
(a) 5 mole of oxygen atom (b) 2 mole of V atom
(c) 1 mole of oxygen atom (d) 2.5 mole of oxygen atom
3. Select the dimensionless quantity(ies) :
(a) vapour density (b) molality
(c) specific gravity (d) mass fraction

4. Which of the following concentration terms is/are affected by a change in temperature?

(a) Molarity (b) Molality

(c) %(w/V) (d) Specific gravity

5. 1 mole of $Ba(OH)_2$ will exactly neutralize :

(a) 0.5 mole HCl (b) 1 mole of H_2SO_4

(c) 1 mole of H_3PO_3 (d) 2 mole of H_3PO_2

6. The pair of species having different percentage (mass)of carbon is :

(a) CH_3COOH and $C_6H_{12}O_6$ (b) CH_3COOH and C_2H_5OH

(c) $HCOOCH_3$ and HCOOH (d) C_2H_5OH and CH_3OCH_3

7. 30 mL of CH_3OH ($d = 0.8\ g/cm^3$) is mixed with 60 mL of C_2H_5OH ($d = 0.92\ g/cm^3$) at 25°C to form a solution of density 0.88 g/cm^3. Select the correct option :

(a) Molarity and molality of resulting solution are 6.33 and 13.59 respectively

(b) The mole fraction of solute and molality are 0.385 and 13.59 respectively

(c) Molarity and % change in volume are 13.59 and zero respectively

(d) Mole fraction of solvent and molality are 0.615 and 13.59 respectively

8. Which of the following is/are correct for 17 g/L of H_2O_2 solution?

(a) Volume strengths is 5.6 at 273 K and 1 atm

(b) Molarity of solution is 0.5 *M*

(c) 1 mL of this solution gives 2.8 mL O_2 at 273 K and 2 atm

(d) The molarity of solution is 2 *M*

9. Solutions containing 23 g HCOOH is/are :

(a) 46 g of 70% $\left(\frac{w}{V}\right)$ HCOOH ($d_{solution} = 1.40$ g/mL)

(b) 50 g of 10 *M* HCOOH ($d_{solution} = 1$ g/mL)

(c) 50 g of 25% $\left(\frac{w}{w}\right)$ HCOOH

(d) 46 g of 5 *M* HCOOH ($d_{solution} = 1$ g/mL)

10. A sample of H_2O_2 solution labelled as "28 volume" has density of 265 g/L. Mark the correct option(s) representing concentration of same solution in other units :

(a) $M_{H_2O_2} = 2.5$ (b) $\% \frac{w}{V} = 17$

(c) Mole fraction of $H_2O_2 = 0.2$ (d) $m_{H_2O_2} = 13.88$

11. A mixture of 100 mL of CO, CO_2 and O_2 was sparked. When the resulting gaseous mixture was passed through KOH solution, contraction in volume was found to be 80 mL. The composition of initial mixture may be (in the same order) :

(a) 30 mL, 60 mL, 10 mL (b) 30 mL, 50 mL, 20 mL

(c) 50 mL, 30 mL, 20 mL (d) 20 mL, 70 mL, 10 mL

12. The correct statement(s) regarding 1*m* NaOH ($d = 1.144$ g/mL) is:

(a) Molarity of OH^- is 0.9 (b) % (*w*/*V*) of Na^+ is 2.53%

(c) Molarity of NaOH is approximately 1.8 (d) % (*w*/*w*) of OH^- is 2.6%

13. An oxide of nitrogen has 30.43% nitrogen (At .wt. of N = 14) and its one molecule weight is 1.527×10^{-22} g. Which of the following statement regarding the oxide is (are) true?

(a) Its empirical formula is N_2O. (b) Its empirical formula is NO_2.

(c) Its molecular formula is N_2O_4. (d) Its molecular formula is N_4O_2.

14. Some water is added to a mixture of 300 mL of 0.3 M NaOH and 700 mL of 0.2 M NaOH to obtain 0.115 M solution of NaOH as a resulting solution then:

(a) The volume of water added is 1000 mL

(b) Volume of water added is 2000 mL

(c) Total mass of NaOH in the solution is 9.2 g

(d) Total moles of NaOH in the solution is 0.23

15. If 5 mole of A and 9 mole of B reacted as given below

$$2A(g) + 3B(g) \longrightarrow 4D(g) + 2C(g)$$

[Given: Molar mass of A, B, C are 50, 18 and 47 respectively]

Select correct statement(s) from following:

(a) A is limiting reactant.

(b) Vapour density of D is 15.

(c) Maximum 12 mol of D can be formed.

(d) If % yield of reaction is 80% then mol of C will be 4.

Match the Column

Column-I and Column-II contain four entries in each. Entries of Column-I are to be matched with some entries of Column-II. One or more than one entries of Column-I may have the matching with the same entries of Column-II.

1.

Column-I	Column-II
(A) 0.5 mole of $SO_2(g)$	(P) Occupy 11.2 L at 1 atm and 273 K
(B) 1 g of $H_2(g)$	(Q) Weighs 24 g
(C) 0.5 mole of $O_3(g)$	(R) Total no. of atoms $= 1.5 \times N_A$
(D) 1 g molecule of $O_2(g)$	(S) Weighs 32 g

2.

Column-I	Column-II
(A) 44 g CO_2 gas	(P) 1g molecule
(B) 35.2 g of CH_4	(Q) N_A molecule
(C) 48 g of O_3 gas	(R) 22 N_A electrons
(D) 44 g of N_2O gas	(S) 49.28 L at 1 atm and 273 K
	(T) N_A atoms of oxygen

3.

Column-I [Atomic masses (M)]				Column-II (% composition of heavier isotope)
	Isotope-II	Isotope-II	Average	
(A)	$(z-1)$	$(z+3)$	z	(P) 25% by moles
(B)	$(z+1)$	$(z+3)$	$(z+2)$	(Q) 50% by moles
(C)	z	$3z$	$2z$	(R) % by mass dependent on z
(D)	$(z-1)$	$(z+1)$	z	(S) 75% by mass

4.

	Column I		Column II
(A)	10 mL $C_3H_8(g)$ + 110 mL $O_2(g)$	(P)	% increment in volume = 20
(B)	20 mL $C_2H_6O(g)$ + 80 mL $O_2(g)$	(Q)	Remaining volume of O_2 = 20 mL
(C)	10 mL $C_3H_6O_2(g)$ + 50 mL $O_2(g)$	(R)	25% Volume contraction
(D)	40 mL $C_2H_2O_4(g)$ + 60 mL $O_2(g)$	(S)	Required volume of O_2 = 30 mL
		(T)	Produced volume of CO_2 = 30 mL

5.

	Column I		Column II
(A)	3.65% *w/V* of HCl (*aq*) solution	(P)	1.25 *m*
(B)	4% *w/w* of NaOH (*aq*) solution	(Q)	1.04 *m*
(C)	4% *w/V* NaOH (*aq*) solution	(R)	$X_{NaOH} = \frac{3}{163}$
(D)	$\frac{25}{24}$ molal solution of NaOH (*aq*)	(S)	1 *M*
		(T)	1.25 *M*

Assertion-Reason Type Questions

Each question contains STATEMENT-1 (Assertion) and STATEMENT-2 (Reason).
Examine the statements carefully and mark the correct answer according to the instructions given below

(A) If both the statement are TRUE and STATEMENT-2 is the correct explanation of STATEMENT-1

(B) If both the statement are TRUE but STATEMENT-2 is NOT the correct explanation of STATEMENT-1

(C) If STATEMENT-1 is TRUE and STATEMENT-2 is FALSE

(D) If STATEMENT-1 is FALSE and STATEMENT-2 is TRUE

1. **STATEMENT-1 :** Specific gravity is dimensionless.
   STATEMENT-2 : Specific gravity is density of a substance measured w.r.t. density of water at 4°C.
2. **STATEMENT-1 :** Molarity of pure water is 55.55 *M* at 298 K.
   STATEMENT-2 : Molarity is temperature dependent.
3. **STATEMENT-1 :** Gram molecular mass of O_2 is 32.
   STATEMENT-2 : Relative atomic mass of oxygen is 32 a.m.u.

Level-4

Subjective Problems

1. What volume of a liquid (in L) will contain 10 mole ? If molar mass of liquid is 280 and its density is 1.4 g/mL.

2. 16 g of SO_x gas occupies 5.6 L at 1 atm and 273 K. What will be the value of x ?

3. 200 mL of 1 M HCl is mixed with 300 mL of 6 M HCl and the final solution is diluted to 1000 mL. Calculate molar concentration of $[H^+]$ ion.

4. $N_2(g)$ reacts with $H_2(g)$ in either of the following ways depending upon supply of $H_2(g)$:

$$N_2(g) + H_2(g) \longrightarrow N_2H_2(l)$$

$$N_2(g) + 2H_2(g) \longrightarrow N_2H_4(g)$$

If 5 L $N_2(g)$ and 3 L $H_2(g)$ are taken initially (at same temperature and pressure), calculate the contraction in volume after the reaction (in L).

5. One commercial system removes SO_2 emission from smoke at 95° C by the following set of reaction :

$$SO_2(g) + Cl_2(g) \longrightarrow SO_2Cl_2(g)$$

$$SO_2Cl_2(g) + H_2O(l) \longrightarrow H_2SO_4 + HCl$$

$$H_2SO_4 + Ca(OH)_2 \longrightarrow CaSO_4 + H_2O$$

How many grams of $CaSO_4$ may be produced from 3.78 g of SO_2 ?

6. W is the mass of iron (in g) which will be converted into Fe_3O_4 by the action of 18 g of steam on it. What is the value of $W/7$?

$$Fe + H_2O \longrightarrow Fe_3O_4 + H_2$$

7. Calculate the total moles of atoms of each element present in 122.5 g of $KClO_3$.

8. One gram of a metallic chloride was found to contain 0.835 g of chlorine. Its vapour density is 85.5. If its molecular formula is M_xCl_y, then what is value of $(x + y)$?

9. A compound contains three elements C, H & N in mass ratio of 18 : 2 : 7. If molar mass of compound is 108 g/mol then how many moles of nitrogen atoms present in 1 mole of compound ?

10. Hg reacts with Cl_2 to form Hg_2Cl_2 and $HgCl_2$. If 1.5 mole of Hg and 1.0 mole of Cl_2 are added. Calculate the final molar ratio of $HgCl_2$ to Hg_2Cl_2.

11. How many millilitre of 36.5% w/w HCl (aq) solution $(d = 1.20 \text{ g/mL})$ will be required to prepare 1080 mL of decimolar solution $\left(\frac{M}{10}\right)$?

12. Calculate how much mass (in kg) of $(CF_2)_n$ will be produced if 1.52 kg F_2 is consumed in the following reactions ?

$$2CoF_2 + F_2 \longrightarrow 2CoF_3$$

$$(CH_2)_n + 4n\,CoF_3 \longrightarrow (CF_2)_n + 2n\,HF + 4nCoF_2$$

13. A gaseous mixture of propane (C_3H_8) and butane (C_4H_{10}) of volume 3 L on complete combustion produces 10 L of CO_2 at STP.

If $x = \dfrac{\text{Volume of propane}}{\text{Volume of butane}}$ then what is the value of x ?

14. How many mole of NaOH is required to neutralise 198 g of equimolar mixture of H_2SO_4 and H_3PO_4 completely ?

15. $0.1M$ aq. solution of $Al_2(SO_4)_3$ and $0.2M$ solution of $CaCl_2$ are mixed such that ratio of molarity of negative ions to positive ions is 1.6, then calculate ratio of volume in which $Al_2(SO_4)_3$ and $CaCl_2$ solutions are mixed. (Assume no any precipitates is formed)

Numerical Problems

(If the numerical value has more than two decimal places, truncate/round-off the value to TWO decimal places.)

1. Density of a non-reacting gaseous mixture (CH_4 and C_2H_6) at 1 atm and 273 K is 0.75 gm/litre. Calculate mole % of CH_4 in the mixture.

2. 24 g carbon reacts with 48 g O_2 such that no any reactant will be left and simultaneously produced CO and CO_2 gases. Molar mass of final gaseous mixture is :

3. 21 g of solid $MgCO_3$ is dissolved in 200 mL, 7.3% $\left(\frac{w}{V}\right)$ aqueous HCl solution.

Calculate volume of CO_2 gas evolved at 1 atm and 273 K.

$$MgCO_3(s) + 2HCl(aq) \longrightarrow MgCl_2(aq) + H_2O(l) + CO_2(g)$$

4. Alcohol levels in blood can be determined by a redox titration with potassium dichromate according to the balanced equation

$$C_2H_5OH(aq) + 2Cr_2O_7^{2-}(aq) + 16H^+(aq) \longrightarrow 2CO_2(g) + 4Cr^{3+}(aq) + 11H_2O(l)$$

What is the blood alcohol level in mass percent if 10 mL of $0.05\,M\,K_2Cr_2O_7$ is required for titration of a 11.5 g sample of blood?

5. What volume of (in L) 40% $\frac{w}{w}$ aq. NaOH solution ($d = 1.25$ g/mL) is required to prepare $2M$, 5L of NaOH solution?

6. An aqueous KCl solution of density 1.20 g mL^{-1} has a molality of 3.30 mol kg^{-1}.

The molarity of the solution in mol L^{-1} is ________. (Nearest integer)

[Molar mass of KCl = 74.5]

7. If 250 mL of $\left(\frac{M}{2}\right)HNO_3(aq)$ solution is mixed with another 750 mL of 9.45% w/V of $HNO_3(aq)$ solution. Find molarity of resultant solution.

8. A 500 mL fresh solution of $H_2O_2(aq)$ can liberate as much oxygen which is just sufficient for complete combustion of 4 g methane. The volume strength of $H_2O_2(aq)$ solution at STP, is ________.

9. Volume strength of an aqueous solution of $H_2O_2(d = 1.2085$ g/mL) is 2.8375. Calculate molality of H_2O_2 solution.

10. In Carius method for estimation of halogens, 0.2 g of an organic compound gave 0.188 g AgBr. The percentage of bromine in the compound is .... (Nearest integer)

[Atomic mass : Ag = 108, Br = 80]

11. Calculate the mass of SO_2 (in g) produced if 110 g P_4S_3 is mixed with 192 g of O_2 and % yield is 75. Consider the chemical reaction as $P_4S_3 + O_2 \longrightarrow P_4O_{10} + SO_2$ (Unbalanced)

12. Consider following reaction :

$$BaCl_2 + H_2SO_4 \longrightarrow BaSO_4 + 2HCl$$

400 mL of 20.8% $\left(\frac{w}{V}\right)$ $BaCl_2(aq)$ solution and 100 mL of 4.9% $\left(\frac{w}{V}\right)$ $H_2SO_4(aq)$ solution are mixed. If 0.045 mol of $BaSO_4$ obtained then calculate % yield of the reaction.

(At. wt. $Ba = 137, Cl = 35.5, S = 32, H = 1, O = 16$)

13. 50% by volume of C_3H_8 present in a gaseous mixture of C_3H_8, CH_4 and CO. If volume of CO_2 produced is V litre when 600 mL of gaseous mixture treated with excess of O_2. Calculate the value of V.

14. 80 g of a mixture of MgO and CaO when treated with an excess of H_2SO_4, produced $\left(\frac{688}{7}\right)$ g of the mixture of $MgSO_4$ and $CaSO_4$. What is the mass percentage of MgO present in original mixture?

15. NH_3 decompose as following $2NH_3 \longrightarrow N_2 + 3H_2$

Calculate % decrease in vapour density of mixture when 100% reaction takes place.

Answers

Level -1

1. (d)	2. (d)	3. (c)	4. (d)	5. (c)	6. (d)	7. (c)	8. (a)	9. (d)	10. (d)
11. (c)	12. (b)	13. (c)	14. (c)	15. (c)	16. (c)	17. (b)	18. (d)	19. (c)	20. (d)
21. (b)	22. (a)	23. (c)	24. (a)	25. (b)	26. (d)	27. (b)	28. (b)	29. (a)	30. (d)
31. (d)	32. (d)	33. (d)	34. (a)	35. (a)	36. (c)	37. (d)	38. (d)	39. (d)	40. (a)
41. (b)	42. (c)	43. (a)	44. (a)	45. (b)	46. (d)	47. (c)	48. (c)	49. (b)	50. (b)
51. (a)	52. (a)	53. (c)	54. (d)	55. (b)	56. (a)	57. (b)	58. (a)	59. (a)	60. (b)
61. (c)	62. (b)	63. (d)	64. (c)	65. (d)	66. (a)	67. (a)	68. (c)	69. (a)	70. (b)
71. (d)	72. (a)	73. (c)	74. (d)	75. (d)	76. (d)	77. (b)	78. (c)	79. (d)	80. (c)
81. (a)	82. (c)	83. (a)	84. (c)	85. (a)	86. (d)	87. (c)	88. (c)	89. (d)	90. (a)
91. (c)	92. (c)	93. (d)	94. (b)	95. (d)	96. (d)	97. (d)	98. (a)	99. (c)	100. (d)
101. (b)	102. (b)	103. (d)	104. (c)	105. (a)	106. (d)	107. (d)	108. (c)	109. (d)	110. (a)
111. (b)	112. (b)	113. (b)	114. (c)	115. (b)	116. (b)	117. (a)	118. (d)	119. (b)	120. (b)
121. (a)	122. (a)	123. (d)	124. (a)	125. (d)	126. (a)	127. (d)	128. (d)	129. (b)	130. (d)
131. (a)	132. (d)	133. (a)	134. (b)	135. (d)	136. (d)	137. (a)	138. (c)	139. (a)	140. (b)
141. (b)	142. (d)	143. (b)	144. (a)	145. (c)	146. (c)	147. (a)	148. (d)	149. (d)	150. (c)
151. (d)	152. (b)	153. (b)	154. (c)	155. (c)	156. (a)	157. (b)	158. (a)	159. (c)	160. (b)
161. (a)	162. (b)	163. (c)	164. (c)	165. (c)	166. (a)	167. (d)	168. (b)	169. (c)	170. (b)
171. (b)	172. (b)	173. (a)	174. (c)	175. (b)					

Level-2

1. (a)	2. (a)	3. (d)	4. (a)	5. (d)	6. (b)	7. (c)	8. (c)	9. (c)	10. (b)
11. (c)	12. (a)	13. (b)	14. (d)	15. (d)	16. (b)	17. (b)	18. (b)	19. (a)	20. (c)

Level-3

Passage–1	1.	(b)	2.	(c)	3.	(c)	4.	(b)
Passage–2	1.	(a)	2.	(c)	3.	(b)		
Passage–3	1.	(a)	2.	(a)				
Passage–4	1.	(b)	2.	(c)				
Passage–5	1.	(b)	2.	(c)				

One or More Answers is/are correct

1.	(c,d)	2.	(a,b)	3.	(a,c,d)	4.	(a,c,d)	5.	(b,c,d)	6.	(b,c)	7.	(b,d)	8.	(a,b,c)
9.	(a,b)	10.	(a,c,d)	11.	(a,b)	12.	(b)	13.	(b,c)	14.	(a,c,d)	15.	(a,d)		

Match the Column

1. A→ P, R,S; B→ P; C→ P,Q,R; D→ S
2. A→ P,Q,R; B→ R,S ; C→ P,Q; D→ P,Q,R, T
3. A→ P,R; B→ Q,R; C→ Q,S; D→ Q,R
4. A→ R,T; B→ Q ; C→ R, T; D→ D,P
5. A→ S; B→ Q,R ; C→ S; D→ Q,R

Assertion-Reason Type Questions

1.	(A)	2.	(B)	3.	(C)

Level-4

Subjective Problems

1.	2	2.	2	3.	2	4.	6	5.	8	6.	6	7.	5	8.	5	9.	2	10.	1
11.	9	12.	1	13.	2	14.	5	15.	4										

Numerical Problems

1.	94.28	2.	36.00	3.	4.48	4.	0.10	5.	0.80	6.	3.00	7.	1.25	8.	22.70
9.	0.21	10.	40.00	11.	72.00	12.	90.00	13.	1.20	14.	2.50	15.	50.00		

Hints & Solutions

Level 1

1. (d) Relative atomic mass of
$$Na = \frac{\text{Atomic mass of Na (23 a.m.u.)}}{\left(\frac{1}{12}\right)^{th} \text{mass of } {}^{12}_{6}\text{C atom (1 a.m.u.)}}$$

2. (d) Oxygen molecule is O_2 and it's mass of one molecule is 32 a.m.u.

3. (c) No. of molecules of $H_2SO_4 = 1$
No. of oxygen atom = 4

4. (d) Atomic mass $= \frac{13.8}{4.6 \times 10^{22}} \times 6 \times 10^{23}$
$= 180$ amu

5. (c) Each Fe^{2+} contain $26 - 2 = 24\ e^-$
168 amu = $3Fe^{2+}$ ions = 72 electrons

6. (d) 31 amu = 16 neutrons
124 amu $= \frac{16}{31} \times 124 = 64$

7. (c) Number of neutrons $= (17-8) \times 12 \times 10^{25}$
$= 9 \times 2 \times 6 \times 10^{23} \times 100 = 1800\ N_A$

8. (a) Number of nucleons = mass number.
Molar mass of element $= 3.32 \times 10^{-23} \times 6.02 \times 10^{23}$
$\cong 20$ = mass number.
So, number of nucleons in 2 atoms = 40

9. (d) 126 amu HNO_3 = molecules of HNO_3 = 6 atoms of O
Moles of O-atom $= \frac{6}{N_A}$

10. (d) Number of sodium atoms $= \frac{46}{23} = N_A = 2N_A$
Number of calcium atoms = Number of sodium atoms $= 2N_A$
Mass of Ca atoms $= \frac{2N_A}{N_A} \times 40 = 80$ g

11. (c) 96 amu of $S^2 = 3S^{2-}$ ions
Charge on $3S^{2-}$ ions = 3 × charge of 2 electrons $= 3 \times 2 \times 1.6 \times 10^{-19} = 9.6 \times 10^{-19}$ C

12. (b) Moles of $PO_4^{3-} = \frac{9.5}{95} = 0.1$ mole
Number of electrons in 0.1 mole PO_4^{3-}
$= 50 \times 0.1 \times N_A = 5N_A$

13. (c) Molar mass = 3.5 × 12 = 42 g/mol

14. (c) Each Fe^{2+} contain $26 - 2 = 24\ e^-$
So No. of e^- in 28 g $= \frac{28}{56} \times N_A \times 24$

15. (c) No. of moles of $H_2O\ (l) = \frac{54}{18}$
($d = 1.0$ g/mL for H_2O)
H has no neutron
$\therefore$ no. of neutrons in $H_2O = 3 \times 8 \times N_A$

16. (c) Moles of $Mg^{2+} = \frac{48}{24} = 2$ mole
Number of electrons $= 2 \times 10\ N_A = 20\ N_A$

17. (b) Moles of $D_2O = \frac{5}{20} = \frac{1}{4}$
Number of neutrons $= 10 \times \frac{1}{4} \times N_A = 2.5\ N_A$

18. (d) Moles of atoms $= \frac{3.011 \times 10^{22}}{6.02 \times 10^{23}} = \frac{1}{20}$
Atomic mass $= \frac{\text{mass}}{\text{moles}} = \frac{1.15}{1/20} = 23$

19. (c) 1 mole CO_2 contains 22 mole neutrons
$\therefore$ 10 mole CO_2 contains 220 mole neutrons

20. (d) $CuSO_4.5H_2O$ (M.M. = 250)
So 1 molecule ⇒ 9 atom of O
or 250 amu contains ⇒
$9 \times 16 = 144$ amu of O
So, 750 amu ⟶ 3 × 144 = 432 amu of O

21. (b) 8 mole O atoms present in 1 mole compound
$\therefore$ 0.2 mole O atoms present in 0.025 mole compound.

22. (a) $n \times N_A \times 1.66 \times 10^{-24} = 1 \times 10^{-3}$
$n = \frac{10^{21}}{1.66 \times 6.022 \times 10^{23}} = 1 \times 10^{-3}$ mole

23. (c) No. of Hydrogen atoms $= 0.1 \times 20 \times N_A = 2 \times N_A$

24. (a) mole of $O^{2-} = 3.2$.

No. of $O^{2-} = 3.2\ N_A$

No. of $e^- = 3.2\ N_A \times 10 = 32\ N_A$

25. (b) Mass of one molecule $= 300 \times 1.66 \times 10^{-24}$ g

$= 4.98 \times 10^{-22}$ g

26. (d) Moles of $C_9H_8O_4 = \dfrac{360 \times 10^{-3}}{180} = 2 \times 10^{-3}$

Number of O-atoms $= 2 \times 10^{-3} \times N_A \times 4$

$= 8 \times 6.02 \times 10^{20} = 4.81 \times 10^{21}$

27. (b) $M = \dfrac{wRT}{PV} = \dfrac{20 \times 0.0821 \times 273}{5.6 \times 1} = 80$

28. (b) 1 mole $(NH_4)_3PO_4$

= 12 mole H-atoms = 4 mole O-atoms

6 mole H-atoms = 2 mole O-atoms

29. (a) Moles of $O_3 = \dfrac{3 \times 8.21}{0.0821 \times 300} = 1$

Moles of O-atoms = 3

30. (d) Atomic weight of an element

$x = 6.643 \times 10^{-23} \times N_A = 40$

no. of moles of$x = \dfrac{20 \times 1000}{40} = 500$

31. (d) 3.9854×10^{-23} g of element = 1 atom

1 g of element $= \dfrac{1}{3.9854 \times 10^{-23}}$

$= 2.509 \times 10^{22}$

32. (d) 4 mole $C_2H_4O_2$ = 16 mole

H-atoms = 16 g hydrogen

33. (d) 12.044×10^{22} molecules of CO_2 = 0.2 mole

CO_2 = 0.4 mole O-atoms = $0.4\ N_A$ O-atoms

34. (a) (I) 0.5 mole O_3 = 24 g O_3;

(II) 0.5 g atom of oxygen = 8 g

(III) $\dfrac{3.011 \times 10^{23}}{6.022 \times 10^{23}} \times 32 = 16$ g O_2

(IV) $\dfrac{5.6}{22.4} \times 44$ g CO_2 = 11 g CO_2

35. (a) Mass of a drop of water = 0.0018 g

Moles of water $= \dfrac{0.0018}{18} = 10^{-4}$ mole

Number of water molecules in one drop $= 10^{-4}\ N_A$

Number of water molecules in two drops

$= 2 \times 10^{-4} \times 6.023 \times 10^{23} = 12.046 \times 10^{14}$

36. (c) No. of protons in ${}_6C^{14} = 6$;

No. of neutrons in ${}_6C^{14} = 8$;

As per given new atomic mass of

${}_6C^{14} = 12 + 4 = 16$

(As the mass of e^- is negligible as compared to neutron and proton)

% increase in mass $= \dfrac{16 - 14}{14} \times 100 = 14.28$

37. (d) Number of molecules $= \dfrac{1000}{18} \times 6.023 \times 10^{23}$

38. (d) Mass of 1 molecule of water = 18 amu

$= 18 \times 1.66 \times 10^{-24}$ g

Volume of 1 molecule $= \dfrac{\text{mass}}{\text{density}}$

$= 29.88 \times 10^{-24} = 2.988 \times 10^{-23}$ mL

39. (d) Mass % of O $= \dfrac{12 \times 16}{400} \times 100 = 48\%$

40. (a) %N $= \dfrac{14}{17} \times 100\%$ (Maximum)

41. (b) Mass of NaCl $= 25 \times \dfrac{8.775}{100} = 2.193$ g

Number of NaCl units $= \dfrac{2.193}{58.5} \times N_A$

$= 2.25 \times 10^{22}$

42. (c) Greater the molecular mass less will be the number of moles.

43. (a) mass of oxygen = 5 × 4 × 16 = 320 g

(b) mass of oxygen = 1.1 × 3 × 16 = 52.8 g

(c) mass of oxygen = 1.5 × 6 × 16 = 144 g

(d) mass of oxygen = 4 × 2 × 16 = 128 g

44. (a) Let there are x-atoms of N

% of N $= 28.9 = \dfrac{x \times 14}{194} \times 100 \quad \Rightarrow \quad x = 4$

45. (b) Moles of $C_{12}H_{22}O_{11} = \dfrac{25.6}{342.3} = 0.0747$

Number of H-atoms

$= 0.0747 \times 22 \times 6.023 \times 10^{23} = 9.91 \times 10^{23}$

46. (d) Volume of gold foil $= 25 \times 40 \times 0.25\ \text{mm}^3$

$= 250 \times 10^{-3}\ \text{cm}^3$

Mass of gold foil $= 19.32 \times 250 \times 10^{-3}$ g

= 4.83 g

No. of gold atoms $= \dfrac{4.83}{197} \times N_A$

$= 1.47 \times 10^{22}$

47. (c) Let total mass be

100 g

N_2 O_2

x $(100 - x)$

$M_{\text{Avg.}} = \dfrac{\text{Total mass}}{\text{Total moles}}$

$$29 = \frac{100}{\frac{x}{28} + \frac{(100-x)}{32}} \Rightarrow x = 72.41\ \%$$

48. (c) No. of oxygen atoms

$$= \frac{24.8}{248} \times 8 \times 6.023 \times 10^{23} = 0.8 \times N_A$$

$$\text{Day required} = \frac{0.8 \times 6.023 \times 10^{23}}{6.023 \times 10^{6}} = 8 \times 10^{16}$$

49. (b) $\frac{\text{Moles of } O_2}{\text{Moles of } N_2} = \frac{\frac{1}{32}}{\frac{4}{28}} = 7:32$

$$= \frac{\text{Molecules of } O_2}{\text{Molecules of } N_2}$$

50. (b) $n_{O_2} : n_{He} : n_{SO_2}$

$$\frac{x}{2} : \frac{x}{4} : \frac{x}{64}$$

$$64\left(\frac{x}{32} : \frac{x}{4} : \frac{x}{64}\right) \Rightarrow 2:16:1$$

51. (a) $0.032 = \frac{1 \times 32}{\text{GMM}} \times 100$

$\text{GMM} = 10^5$

No. of molecules

$$= \frac{1}{10^5} \times 6.02 \times 10^{23} = 6.02 \times 10^{18}$$

52. (a) Mass of S $= 1280 \times \frac{5}{100} = 64$ g

$$\text{Atomicity} = \frac{64}{32} = 2$$

53. (c) $z = 0$

$x = 0.2 \times N_A \times 2 = 0.8\ N_A$

$w = 0.5 \times N_A \times 2 = N_A$

$y = 0.15 \times 8 \times N_A = 1.2\ N_A$

54. (d) Molar mass of compound

$= 1 \times 14 + 2 \times 16 = 46\ \text{g mol}^{-1}$

$\therefore$ Mass of one molecule of compound = 46 amu.

55. (b) 1 mole $CH_4(g)$ on combution evolved 890 kJ heat

$$\text{Mass of } CH_4(g) \text{ required} = \frac{356}{890} \times 16$$

$$= 6.4 \text{ g}$$

56. (a) $^{49}X = n$ mole

$^{53}X = (100-n)$ mole

$$50 = \frac{(n)(49) + (100-n)(53)}{100}$$

$n = 75$

Heavier isotope $^{53}X = 100 - n = 25\%$

57. (b) Let the percentage of isotope with atomic weight 10 be x.

So, the percentage of isotope with atomic weight 11 will be $100 - x$.

$$\text{Average atomic weight} = \frac{m_1 x_1 + m_2 x_2}{x_1 + x_2}$$

$$= \frac{x \times 10 + (100-x) \times 11}{100} = 10.8 \Rightarrow x = 20$$

$\therefore$% of isotope with atomic weight $10 = 20$

% of isotope with atomic weight $11 = 100 - x = 80$

58. (a) Let mole % of ^{26}Mg be x

$$\therefore \frac{(21-x)\,25 + x\,(26) + 79\,(24)}{100} = 24.31$$

$x = 10\%$

59. (a) Let atomic weight of other isotopoe is M

$$114.82 = \frac{114.9041 \times 95.72 + M \times 4.28}{100}$$

$M = 112.94$

60. (b) Moles of $N_2 = \frac{3.01 \times 10^{24}}{6.02 \times 10^{23}} = 5$

$$\text{Moles of } O_2 = \frac{32}{32} = 1 \text{ mole}$$

$$\text{Average molar mass} = \frac{5 \times 28 + 32 \times 1}{5 + 1}$$

$$= \frac{172}{6} = 28.67$$

$$d_{mix} = \frac{PM_{avg.}}{RT} = \frac{3 \times 28.67}{\frac{1}{12} \times 860} = 1.2 \text{ g/L}$$

61. (c) No. of moles of CO_2 in 100 g mixture

$$= \frac{66}{44} = 1.5$$

No. of moles of H_2 in 100 g mixture

$$= \frac{34}{2} = 17$$

$$M_{average} = \frac{100}{18.5} = 5.40$$

$$\text{V. D.} = \frac{5.4}{2} = 2.7$$

62. (b) Let 1 mole of mixture has x mole N_2O_4

$2 \times 27.6 = x\,(92) + (1-x)\,46;\ x = 0.2$

63. (d) $M_{av} = 2 \times VD_{\text{gas mix}}$;

$$\frac{1 \times 28 + 4 \times 17}{5} = 2 \times \text{VD}\ ;\ \text{VD} = 9.6$$

64. (c) $\frac{2 \times 28}{2 \times 28 + x} \times 100 = 90.32$

$x = 6.02$

65. (d) Let mole fraction of O_2 is x

$40 = 32 \times x + 80\,(1-x)$

or $x = 5/6$

$a : b = x : (1-x) = \frac{5}{6} : \frac{1}{6}$

When ratio is changed

$M_{mixture} = 32 \times \frac{1}{6} + 80 \times \frac{5}{6} = 72$

66. (a) $Na_2CO_3 \cdot xH_2O\ (s) \longrightarrow Na_2CO_3(s) + xH_2O(g)$.

Let 100 g of $Na_2CO_3 \cdot xH_2O$ be present

$\therefore \frac{100\,x}{106 + 18x}$ mole of H_2O formed;

$$63 = \frac{100x}{106 + 18x}(18)$$

$$6678 + 1134x = 1800x$$

$$666x = 6678$$

$$x = 10$$

67. (a) $Sr(OH)_2 \cdot xH_2O \longrightarrow Sr(OH)_2 \quad + \quad xH_2O$

6.85 g 3.13 g 3.72 g

$\%$ of water $= \frac{3.72}{6.85} \times 100 = \frac{18x}{121 + 18x} \times 100$

$\Rightarrow \quad x = 8$

68. (c) % of O $= \frac{16 \times 27}{(100 + 3 \times 310)} \times 100 = 41.94\%$

69. (a) Mass of 1 mole of acid $= \frac{1.70}{0.025} = 68$ g

Mass of $X = 68 -$ mass of H and O

$= 68 - 33 = 35$ g

76. (d) C 4 mole ; H 5 mole ; O 2 mole

77. (b) Moles of Zn $= \frac{9.81}{65} = 0.15$

Moles of Cr $= \frac{1.8 \times 10^{23}}{6.02 \times 10^{23}} = 0.3$

Moles of O $= 0.6$

$Zn_{0.15}Cr_{0.3}O_{0.6}$

$ZnCr_2O_4$

78. (c) Metal oxide = 2.74g;

wt. of vanadium = 1.53 g

% of V $= \frac{1.53}{2.74} \times 100 = 55.83$

Thus, % of O $= 100 - 55.83 = 44.17$

No. of moles of V $= \frac{55.83}{52} = 1.1$

No. of moles of O $= \frac{44.17}{16} = 2.76$

Simplest ratio of V and O = 1:2.5 or 2 :5

Hence, the empirical formula = V_2O_5

79. (d)

	C	H	N	O
Moles	$\frac{70.6}{12}$	$\frac{4.2}{1}$	$\frac{11.8}{14}$	$\frac{13.4}{16}$
Moles	$\frac{5.88}{0.84}$	$\frac{4.2}{0.84}$	$\frac{0.842}{0.84}$	$\frac{0.84}{0.84}$
Simple ratio	7	5	1	1

C_7H_5NO

80. (c) From given info. C : H : O molar ratio $= 4 : 8 : 4$.

81. (a) $\frac{W_H}{W_O} = \frac{1}{4} = \frac{x}{4x}$ so $x + 4x = 40 \Rightarrow x = 8$

So	C	H	O
	60 g	8 g	32 g
	$\frac{60}{12}$	$\frac{8}{1}$	$\frac{32}{16}$

82. (c) $n_C = 1 \times n_{CO_2} = \frac{4.4}{44} = 0.1$

$n_H = 2 \times n_{H_2} = 2 \times \frac{0.9}{18} = 0.1$

EF = CH $\quad n = \frac{39 \times 2}{13} = 6 \quad$ MF = C_6H_6.

83. (a) Moles of C $= n_{CO_2} = \frac{2.64}{44} = 0.06$

$\Rightarrow$ mass of C = 0.72

Moles of H = 2 × moles of H_2O

$= 2 \times \frac{1.26}{18} = 0.14$

mass of H = 0.14

Compound does not contains oxygen.

So EF $\rightarrow C_{0.06}H_{0.14} \Rightarrow C_3H_7$

$\Rightarrow$ Lowest M.M. = 43

84. (c) Formula of sulphate will be —$MSO_4 \cdot 7H_2O$

% of metal

$= \frac{M}{M + 96 + 7 \times 18} \times 100 = 9.87$

$\Rightarrow \quad M = 24.31$

85. (a)

	C	H	N
Moles	$\frac{a}{12}$	$\frac{1}{1}$	$\frac{3.5}{14}$
	$\frac{3}{4}$:	1 :	$\frac{1}{4}$
	3 :	4 :	1

$n = \frac{108}{54} = 2$

$\Rightarrow C_3H_4N \Rightarrow$ M.F. $= (C_3H_4N)_n$

MF $= (C_3H_4N)_2 = C_6H_8N_2$

86. (d)

	I	O
Moles	$\frac{254}{127}$	$\frac{80}{16}$
	2	5

I_2O_5

87. (c) $CO_2 = \frac{88}{44} = 2$ mole of $CO_2 = 2$ mole of C

$H_2O = \frac{36}{18} = 2$ mole of $H_2O = 4$ mole of H

Mass of C + Mass of H + Mass of O = 44

$\Rightarrow 24 + 4 + x = 44; x = 16$

$\therefore$ mole of O = 1 and molecular formula is C_2H_4O.

88. (c) A^{4+} and $2B^{2-}$ will form AB_2 compound. Compound is always electrical neutral.

89. (d) 100 g compound contain 43.8 g oxygen; 1 g oxygen present with $\frac{100}{43.8}$ g compound

16×4 g or 4 mole oxygen present with $\frac{100}{43.8} \times 16 \times 4 \simeq 146$ g compound

$\therefore$ mol. wt. of compound is 146.

90. (a) Molar mass of $x = 0.444 \times$ molar mass of y

Molar mass of $x = 2.96 \times 12$

Molar mass of $y = \frac{2.96 \times 12}{0.444} = 80$

91. (c) Mol. wt. of $XY_2 = \frac{5}{0.05} = 100$

Mol. wt. of $X_2Y_3 = \frac{85}{3.011 \times 10^{23}} \times N_A = 170$

Let molar mass of X and Y are a and b respectively

$\therefore \quad a + 2b = 100$

$2a + 3b = 170;$

$a = 40;$

$b = 30$

92. (c) $d = \frac{PM}{RT} \Rightarrow M = \frac{dRT}{P}$

$= \frac{2.28 \times 0.0821 \times 300}{1} = 56.15$ g/mol

$E.F. = \frac{85.7}{12} : \frac{14.3}{1} = 7.14 : 14.3 = 1 : 2;$

$\therefore$ E.F. is CH_2; M.F. $= (CH_2)_n$

where $n = \frac{56.15}{12 + 2} \simeq 4;$

$\therefore$ M.F. is C_4H_8

93. (d)

X	Y	
$\frac{75.8}{75}$	$\frac{24.2}{16}$	
1	3/2	
2	3	$\Rightarrow X_2Y_3$

94. (b) $M = \frac{WRT}{PV} = \frac{12.4 \times 0.0821 \times 800}{8 \times 0.821} = 124 = P_4$

95. (d) Volume of air displaced = volume of vapour of X_2

$\frac{4.92}{1000} \times 1 = \frac{40 \times 10^{-3}}{M} \times 0.0821 \times 300;$

$M \simeq 200$; Atomic mass of $X = 100$.

96. (d) $\% Mn = \frac{55}{55 + 16x} \times 100 = 64 \Rightarrow x = 1.93$

97. (d)

	$2NH_3(g)$	$\longrightarrow$	$N_2(g)$	+	$3H_2(g)$
	6 mole				
After reaction	0		3		9

Total moles = 12

98. (a) Mass of $CO_2 = \frac{1}{342} \times 12 \times 44 = 1.54$ g

99. (c) $C + 2Cl_2 \longrightarrow CCl_4$

Moles of Carbon, $\frac{48}{12} = 4,$

Moles of Cl_2, $\frac{142}{71} = 2$

Cl_2 is limiting reagent

Moles of Carbon remaining = 3

Moles of CCl_4 formed = 1

$\therefore$ total moles of carbon remaining & CCl_4 formed $= 1 + 3 = 4$

100. (d) $C_2H_4 + 3O_2(g) \longrightarrow 2CO_2(g) + 2H_2O(l)$

$n_{C_2H_4} = \frac{140}{28} = 5$ mole

$n_{O_2} = 5 \times 3 = 15$ mole

Mass of $O_2 = 15 \times 32 = 480$ g

101. (b) $4A + B + 3C \longrightarrow P + 2Q$

$LR = \frac{\text{Moles}}{\text{Stoichiometric Coefficient}} : \frac{6}{4} \; \frac{2}{1} \; \frac{4}{3}$

C is Limiting reagent

$\therefore$ mole of C produced $= \frac{2}{3} \times 4 = \frac{8}{3}$

102. (b)

	N_2	+	$3H_2$	$\longrightarrow$	$2NH_3$
(LR)	(2 mole)		(8 mole 2 mole left)		4 mole formed

$\therefore$ Mass of $NH_3 = 4 \times 17 = 68$ g

103. (d) Six molecules of O_2 required three molecules of CS_2.

104. (c) $3Fe_2O_3(s) \longrightarrow 2Fe_3O_4 + \frac{1}{2}O_2$

480 g Fe_2O_3 provide 16 g O_2. For loss of

$0.04\text{ g } O_2 \longrightarrow 0.04 \times \frac{480}{16} = 1.2\text{ g } Fe_2O_3$

% by mass of $SiO_2 = \frac{0.8}{2.0} \times 100 = 40\%$

105. (a) $4Ca_5(PO_4)_3F + \underset{0.36}{\overset{L.R.}{18SiO_2}} + \underset{0.9}{30C} \longrightarrow 3P_4 + 2CaF_2 + 18CaSiO_3 + 30CO$

18 mole SiO_2 gives 3 mole P_4

0.36 mole SiO_2 will give $= \frac{3}{18} \times 0.36 = 0.06$ mole

106. (d) CO_2 exhaled in 5 minutes $= 0.667 \times 5 = 3.33$ g

moles of $CO_2 = 0.0756$ mole

moles of KO_2 consumed $= 0.0756$ mole

mass of KO_2 consumed $= 5.38$ g

107. (d) Mass of pure $CaCO_3 = 200 \times \frac{95}{100} = 190$ kg

$CaCO_3 \xrightarrow{\Delta} CaO + CO_2$

100 g — 56

1 g — $\frac{56}{100}$

190 kg — $\frac{56}{100} \times 190\text{ kg} = 106.4$ kg

108. (c)

	$2Fe(NO_3)_3$ +	$3Na_2CO_3 \longrightarrow Fe_2(CO_3)_3 + 6NaNO_3$
mole	2.5	3.6
mole/stoichiometric coefficient	1.25	1.2

Limiting reagent is Na_2CO_3 so moles of $NaNO_3$ should be formed $= 3.6 \times 2 = 7.2$

% yield $= \frac{6.3}{7.2} \times 100 = 87.5$

109. (d) $2NH_3 + 5F_2 \longrightarrow N_2F_4 + 6HF$; NH_3 is L.R.

2×17 g NH_3 gives 104 g N_2F_4

3.4 g will gives $= \frac{104}{34} \times 3.4 = 10.4$ g N_2F_4

% yield $= \frac{7.8}{10.4} \times 100 = 75\%$

110. (a) $2NaHCO_3(s) \longrightarrow Na_2CO_3(s) + H_2O(g) + CO_2(g)$

$2(23+1+60)$ ↓ ↓

$= 168$ g $\quad$ 18 g + 44 g = 62 g

% loss $= \frac{62}{168} \times 100 = 36.9 \simeq 37\%$

111. (b) $2C(s) + 3H_2(g) \xrightarrow{80\%} C_2H_6(g)$

3 mole 3 mole

H_2 is L.R. :

Mass $C_2H_6 = \frac{1}{3} \times 3 \times \frac{80}{100} \times 30 = 24$ g

112. (b) In $Fe_2(SO_4)_3$ for 3 mole SO_4^{2-} ion $\longrightarrow$ 2 mole Fe^{3+} present

$\therefore$ In $FeSO_4$, for 3 mole SO_4^{2-} ion $\longrightarrow$ 3 mole Fe^{2+} present

$\therefore$ mole ratio of Fe^{2+} to $Fe^{3+} = 3:2$

113. (b) $N_2 + 3H_2 \longrightarrow 2NH_3$

	N_2	$3H_2$
mass (in g)	14	6
moles	$\frac{14}{28}$	$\frac{6}{2}$
	0.5	3

LR $= N_2(g)$

n_{NH_3} (Theoritical) $= 0.5 \times 2 = 1$

mass of NH_3 (Theoritical) $= 1 \times 17 = 17$ g

% yield $= \frac{12.75}{17} \times 100 = 75\%$

114. (c) Let mass of $Ag_2O = x$ g

$Ag_2O \longrightarrow 2Ag + \frac{1}{2}O_2$

232 g $\quad \frac{1}{2} \times 32 = 16$ g O_2

x g $\quad \frac{16}{232} \times x = 0.104 \Rightarrow x = 1.508$ g

% of $Ag_2O = \frac{1.508}{1.6} \times 100 = 94.25\%$

115. (b) $C_2H_6 + 3.5O_2 \longrightarrow 2CO_2 + 3H_2O$;

$C_2H_4 + 3O_2 \longrightarrow 2CO_2 + 2H_2O$

Let vol. of ethane is x,

$22.4 \times 4 = 3.5x + 3(28 - x)$

$\Rightarrow x = 11.2$ litre

at constant T and P, $V \propto n$;

$\therefore$ Mole fraction of C_2H_6 in mixture

$= \frac{11.2}{28} = 0.4$

116. (b) Mass of $CaSO_4$

$= \frac{80}{64} \times \frac{1}{1} \times \frac{1}{1} \times \frac{1}{1} \times 136 = 170$ g

117. (a) $3D + 4E \xrightarrow{80\%} 5C + A$

9 mole $\quad$ 14 mole $\quad \frac{5}{3} \times 9 \times 0.8 = 12$ mole

$3C + 5G \xrightarrow{50\%} 6B + F$

12 mole $\quad$ 4 mole

Moles of B formed $= \frac{6}{5} \times 4 \times 0.5 = 2.4$

118. (d) Mass of acetic acid $= 0.1 \times 1000 \times 1.2 = 120$

119. (b) $n_{water} = \frac{180}{18} = 10$; $n_{CH_2OH} = \frac{640}{32} = 20$;

$n_{solute} = \frac{10}{30} = \frac{1}{3}$

120. (b) Let mass of sucrose is x

$\frac{x}{x+81} \times 100 = 10$

$x = 9$ g

121. (a) $\frac{n_{H_2O}}{n_{CH_3OH}} = \frac{0.75}{0.25} = 3$

$\Rightarrow w_{H_2O} = \frac{16}{32} \times 3 \times 18 = 27$ g.

122. (a) $Zn + 2HCl \longrightarrow ZnCl_2 + H_2$;

moles of H_2 evolved $= 2$

$\therefore$ moles of HCl required $= 4$

$\therefore \frac{V \times 1.2 \times 0.365}{36.5} = 4$; $\quad V = 333.33$ mL

123. (d) Mass of solution $= 100$ g

Mass of solute $= 56$ g

$\Rightarrow$ moles of solute $= \frac{56}{60} = 0.93$

Mass of solvent $= 44$ g

$\Rightarrow$ moles of solvent $= \frac{44}{18} = 2.44$

$X_{Solute} = \frac{0.93}{0.93 + 2.44} = \frac{0.93}{3.37} = 0.276$

124. (a) Let mass of KCl dissolved is $= x$ g

Total mass of solution $= (60 + x)$ g

$\% \frac{w}{w} = \frac{x}{60 + x} \times 100 = 20$

$x = 15$

125. (d) Wt. of 1.5 mole C_2H_5OH is $1.5 \times 46 = 69$ g

Volume of C_2H_5OH required

$= \frac{69}{0.8} = 86.25$ mL

126. (a) Moles of C_2H_5OH in V mL = Moles of H_2O in 180 mL

$\frac{0.80 \times V}{46} = \frac{1 \times 144}{18}$

127. (d) $0.1 = \frac{M \times 100}{1000} \Rightarrow M = 1\,M$

128. (d) $M_{KOH} = \frac{1.12}{56 \times 0.5} = 0.04\,M$

129. (b) $m = \frac{n}{w} \times 1000 = \frac{0.25}{90} \times 1000 = \frac{25}{9}$

130. (d) Mass of solution $= 500 \times 1.20 = 600$

Mass of solvent $= 600 - 100 = 500$

$M = \frac{100}{80} \times \frac{1}{0.5} = 2.5$

131. (a) 100 g solution contains 1.48 g $Ca(OH)_2$

$\frac{100}{1.025}$ mL solution contains 1.48 g $Ca(OH)_2$

Molarity of solution

$= \frac{10 \times 1.48 \times 1.025}{74} = 0.205\,M$

132. (d) 1000 g solvent present with 2 mol NaOH

$\therefore$ wt of solution $= 1000 + 2 \times 40 = 1080$

$$M = \frac{\text{Moles of NaOH}}{\text{Volume of solution (in mL)}} \times 1000$$

$$= \frac{2}{\left(\frac{1080}{1.2}\right)} \times 1000 = 2.22$$

133. (a) 100 g solution

Contains 9.8 g H_3PO_4

$$n_{H_3PO_4} = \frac{9.8}{98} = 0.1$$

$$V_{\text{solution}} = \frac{100}{1.1}$$

$$M = \frac{0.1}{\left(\frac{100}{1.1}\right)} \times 1000 = 1.1$$

134. (b) 1000 mL solution contain 2 mole of ethanol or 1000 × 1.025 g solution contain 2 mole of ethanol

wt. of solvent $= 1000 \times 1.025 - 2 \times 46$

$$m = \frac{2}{1000 \times 1.025 - 2 \times 46} \times 1000$$

$$m = \frac{2}{933} \times 1000 = 2.143$$

135. (d) $$m = \frac{1000\,M}{1000\,d - M \times M_{\text{Solute}}}$$

$$\Rightarrow d = \frac{1000 \times 9}{10} + 9 \times 50 \times \frac{1}{1000} = 1.35\text{ g/mL}$$

136. (d) Let we have 1 L solution (or 1000 mL)

Wt. of solution (in g) $= 1000 \times \rho$

$$\text{Molarity} = \frac{\text{Moles of solute}}{\text{Volume of solution}} = \frac{\text{Moles of solute}}{1}$$

$$(M) = \frac{\text{Mass of solute}}{\text{Molar mass of solute}}$$

$\therefore$ Mass of solute $= M \times M_{\text{Solute}}$

Mass of solvent (in g) $= 1000\,\rho - M \times M_{\text{Solute}}$

$$\frac{\text{Moles of solute}}{1} = M$$

$$\text{Molality} = \frac{\text{Moles of solute}}{\text{Mass of solvent (in g)}} \times 1000$$

$$= \frac{M \times 1000}{1000\,\rho - M \times M_{\text{Solute}}}$$

or $$\boxed{m = \frac{M \times 1000}{1000\,\rho - M \cdot M_1}}$$

$M_1 =$ Gram molecular wt. of solute.

137. (a) $W_{H_2SO_4} = 800$ g in 1000 mL

Volume of solution $= 1000$ mL

Weight of solution $= 1000.0 \times 1.80$

$= 1800$ g

Weight of water $= 1800 - 800 = 1000$ g

$$\text{Molality} = \frac{800 \times 1000}{98 \times 1000} = 8.16\text{ mol kg}^{-1}$$

138. (c) No. of moles of HCl are more than NaOH, so final solution becomes acidic.

139. (a) Mass of 1000 mL liquid benzene = 880 g

Moles of liquid in 1 L benzene

$$= \frac{880}{78} = 11.28\,M$$

140. (b) $$\%\text{ of A} = \frac{100 \times 0.9}{100 \times 0.9 + 100} \times 100$$

$$= \frac{90}{190} \times 100 = 47.36$$

141. (b) Molality has dimenine mol kg^{-1}

142. (d) $M_1V_1 = M_2V_2$

$0.40 \times 400 = 0.1 \times V_2$

$V_2 = 1600$ mL

143. (b) $M_1V_1 = M_2V_2$; $0.60 \times 200 = 0.20 \times V_2$

$V_2 = 600$ mL

So, water added $= V_2 - V_1 = 600 - 200$

$= 400$ mL

144. (a) $$\frac{\text{mass of } Al_2(SO_4)_3}{\text{mass of water}} \times 10^6 = 34.2$$

1 litre solution contains 1000 g of water

$\Rightarrow$ In 1 litre solution, mass of $Al_2(SO_4)_3$

$$= \frac{34.2 \times 1000}{10^6} = 34.2\text{ mg}$$

$$\text{molarity of } Al_2(SO_4)_3 = \frac{34.2 \times 10^{-3}}{342}\,M$$

$$= 10^{-4}\,M$$

$$Al_2(SO_4)_3(aq) \longrightarrow 2Al^{3+}(aq) + 3SO_4^{2-}(aq)$$

$10^{-4}\,M$ $\quad$ $2 \times 10^{-4}\,M$ $\quad$ $3 \times 10^{-4}\,M$

$[SO_4^{2-}] = 3 \times 10^{-4}\,M$

145. (c) Total $Cl^- = 0.2 + 0.4$ mole (from $BaCl_2$)

$= 0.6$ mole

$$M_{Cl^-} = \frac{0.6}{500} \times 1000 = 1.2$$

146. (c) $$n_{Na^+} = \frac{10 \times 0.2}{1000} + \frac{10 \times 0.2}{1000} = \frac{M \times 20}{1000}$$

$$M = \frac{0.4}{2} = 0.2$$

147. (a) Mass of solution $= 10 \times 1.2 = 12$ g

$$\text{Mass of solute} = 12 \times \frac{37}{100} = 4.44\text{ g}$$

148. (d) Let 'V_1' L of 5.0 $M\,H_2SO_4$ should be mixed.

$\therefore M_1V_1 + M_2V_2 = M_fV_f$

$5 \times V_1 + 1 \times 10 = 3 \times 20$

$\Rightarrow V_1 = 10$ L

149. (d) Mole of HCl $= 50 \times (.5) \times 10^{-3} = 25 \times 10^{-3}$

Mole of $H_2SO_4 = 200(.5) \times 10^{-3}$
$= 100 \times 10^{-3}$

Total mole of

$H^+ = 25 \times 10^{-3} + 2 \times 100 \times 10^{-3}$
$= 225 \times 10^{-3}$

$[H^+] = \frac{225 \times 10^{-3}}{250 \times 10^{-3}} = \frac{9}{10}$

$[H^+] = 0.9\ M$

150. (c) $[HCl] = \frac{1 \times 4V + 2 \times V}{5V} = 1.2\ M$

151. (d) $M_R = \frac{55 \times 0.07 + 20 \times 0.12 + 25 \times 0.15}{55 + 20 + 25} = 0.1$

152. (b) Let volume of Solution is V

$V \times 1.25 \times \frac{20}{100} = 0.5 \times 0.25 \times 40$

$V = 20$ mL

153. (b) $\frac{80 \times 1.96 \times 10}{98} \times V = 1000 \times 0.2$

$V = 12.5$ mL

154. (c) Let no. of moles of $Al_2(SO_4)_3$ in solution is x

$\frac{x}{2684 - x \times 342} \times 1000 = 1$

$x = 2$

$2Al_2(SO_4)_3 \longrightarrow 6$ moles SO_4^{2-}

$\therefore$ 6 moles of $BaSO_4$

155. (c) Initial molarity $(M_1) = \frac{10 \times 10^{-3}}{336 \times 500} \times 1000$
$= 0.0595 \times 10^{-3}$

$M_1V_1 = M_2V_2$

$M_2 = \frac{M_1V_1}{V_2} = \frac{0.0595 \times 10^{-3} \times 1}{1000}$

$= 0.0595 \times 10^{-6} = 5.95 \times 10^{-8}$

156. (a) Molar mass of $Pb(NO_3)_2 = 25 \times 0.15$
$= 3.75$ m. moles

Molar mass of $Al_2(SO_4)_3 = \frac{1}{3} \times 3.75 = M \times 20$

$M = 0.0625 = 6.25 \times 10^{-2}\ M$

157. (b) m-moles of HNO_3 required

$= 250 \times 1.2 = 300$

100 g solution contains 63 g HNO_3 or

$\frac{100}{1.4}$ mL solution contain 1 mole HNO_3

$\therefore$ molarity of HNO_3 solution

$= \frac{1000}{100} \times 1.40 = 14$

$\therefore 14 \times V = 300$ or $V = 21.42$ mL

158. (a) $2KOH + H_3PO_3 \longrightarrow K_2HPO_3 + 2H_2O$

$3 \times V \qquad 1 \times 600$

$\frac{1}{2} \times 3 \times V = 100 = 1 \times 600$

$V = 400$ mL

159. (c) Let vol. is x mL, $x \times 0.1 \times 2 - 50 \times 0.1$

$= (x + 50) \times 0.05 \times 2$

$x = 100$ mL

160. (b) (i) Mass of NaOH $= 50 \times \frac{40}{100} = 20$ g

(ii) $\frac{w}{v}$ NaOH $\rightarrow$ 50 g NaOH in 100 mL

So, In 50 mL mass of NaOH will be = 25 g

(iii) 50 g of 15 M NaOH ($d = 1$ g/mL)

15 M NaOH $\rightarrow$ 15 moles NaOH in 1000 mL solution

Mass of solution $= 1000 \times 1 = 1000$ g

Mass of NaOH $= 15 \times 40 = 600$ g

In 1000 g solution mass of NaOH = 600 g

So, In 50 g solution mass of NaOH

$= \frac{600}{1000} \times 50 = 30$ g

Order (iii) > (ii) > (i)

161. (a) 0.478 mL ≡ 0.478 g of water; 10^9 g water contain 0.10 g $CHCl_3$

$\therefore$ 0.478 g water contain

$\frac{0.1}{10^9} \times 0.478$ g $CHCl_3$

$\therefore n_{CHCl_3} = \frac{0.1}{10^9} \times \frac{0.478}{119.5}$

$\therefore$ No. of molecules $= \frac{0.1}{10^9} \times \frac{0.478}{119.5} \times N_A$

$= 4 \times 10^{-13} \times N_A$

162. (b) $2H_2O_2 \longrightarrow 2H_2O + O_2$

500 mL : 22.4 L

1000 mL : 44.8

163. (c) $\frac{M \times 1000}{1000} = \frac{500 \times 1}{1000} + \frac{500 \times 2}{1000}$; $M = 1.5$

164. (c) $M = \frac{44.8}{11.2} = 4$; $\quad \% \frac{w}{V} = \frac{4 \times 34}{1000} = 100$

165. (c) Molarity $= \frac{28.3 + 5}{11.35} = 2.5\ M$

1000 mL of solution contains 2.5 mole H_2O_2

molality, $m = \frac{2.5}{1250 - 85} \times 1000 = 2.15\ m$

166. (a) $n_{O_2} = \frac{45.4}{22.7} = 2$ mole,

$2H_2O_2 \longrightarrow 2H_2O + O_2$

4 mole $\qquad$ 2 mole

0.5 L solution

$\therefore M_{H_2O_2}$ solution $= \frac{4}{0.5} = 8\,M$

$\therefore \% \, w/w = \frac{4 \times 34}{500 \times 1.088} \times 100 = 25\%$

167. (d) $H_2SO_4(aq) \longrightarrow 2H^+(aq) + SO_4^{2-}(aq)$

$100 \times 10^{-3} \times 0.001$

$= 10^{-4}$ mole

168. (b) $n_{H_2S} = \frac{68}{34} = 2$; $n_{H_2SO_4} = 10 = V \times 0.2$;

$V = 50$ L

169. (c) $n_{HNO_3} = \frac{4}{3} \times n_{Ag}$

$\frac{1.75 \times V}{1000} = \frac{4}{3} \times \frac{0.756}{108}$

$V = \frac{16}{3}$ mL

170. (b) $C_nH_{2n} + \frac{3n}{2}O_2 \rightarrow nCO_2 + nH_2O$

10L (excess) $10\,n = 50$; $n = 5$

$\therefore C_5H_{10}$

171. (b) $C_2H_6(g) + \frac{7}{2}O_2(g) \longrightarrow 2CO_2(g) + 3H_2O(l)$

10 L $3.5 \times 10 = 35$ L

172. (b) $CO + \frac{1}{2}O_2 \longrightarrow CO_2$

30 L 10 L

10 L — 20 L

173. (a) $C_xH_y + \left(x + \frac{y}{4}\right)O_2 \longrightarrow xCO_{2(g)} + \frac{y}{2}H_2O_{(g)}$

10 mL 10 x mole 5 y mole

$10\,x = 40$ $5y = 50$

$x = 4$ ...(1) $y = 10$ ...(2)

$\therefore C_4H_{10}$

174. (c) 100 mL gaseous mixture contain 20 mL C_3H_8

So, volume of CH_4 and CO

$= (100 - 20) = 80$ mL

$C_3H_8 + 5O_2 \longrightarrow 3CO_2 + 4H_2O$

$CH_4 + 2O_2 \longrightarrow CO_2 + 2H_2O$;

$CO + \frac{1}{2}O_2 \longrightarrow CO_2$

80 mL (CH_4 and CO) will produce 80 mL CO_2; C_3H_8 will produce $= 3 \times 20 = 60$ mL

Total CO_2 produce $= 80 + 60 \Rightarrow 140$

175. (b) Suppose, the volume of propane $= V$ L

$\therefore C_3H_8(g) + 5O_2(g) \longrightarrow 3CO_2(g) + 4H_2O(l)$

V $5V$ $3V$

$C_4H_{10}(g) + \frac{13}{2}O_2(g) \longrightarrow 4CO_2(g) + 5H_2O(l)$

$(3 - V)$ $\frac{13}{2}(3 - V)$ $4(3 - V)$

$\because$ Total volume of CO_2 produced $=$ 10 L;

$\because 3V + 4(3 - V) = 11$; $\therefore V = 1$

$\therefore$ Volume of butane $= (3 - 1) = 2$ L

Thus, the ratio of volume of butane to propane $= 2:1$

Level 2

1. (a) Let wt. of NH_4NO_3 and $(NH_4)_2HPO_4$ are x and y gram respectively

$$\frac{\frac{x}{80} \times 2 \times 14 + \frac{y}{132} \times 2 \times 14}{x + y} \times 100 = 30.4$$

$\Rightarrow x : y = 2 : 1$

2. (a) Let V mL of alcohol be required

$\because$ mass of alcohol is same in both solutions

$\therefore \frac{75}{100} \times 0.8 \times V = \frac{30}{100} \times 0.9 \times 150$

$V = 67.5$ mL

3. (d) The precipitation reaction is

$Fe^{3+} + 3NH_3 + 3H_2O \longrightarrow Fe(OH)_3 + 3NH_4^+$

mole of Fe_2O_3 in sample $= \frac{0.80 \times 0.5}{160}$

$= 2.5 \times 10^{-3}$

mole of $Fe^{3+} = 2 \times 2.5 \times 10^{-3} = 5 \times 10^{-3}$

$M_{NH_3} = \frac{0.986 \text{ g/mL} \times 1000 \text{ mL/litre} \times 0.025}{17}$

$\Rightarrow M_{NH_3} = 1.45$

$3 \times$ moles of Fe^{3+} = moles of NH_3

$\Rightarrow 1.45 \times V$ (inL)

$V = \frac{3 \times 5 \times 10^{-3}}{1.45} = 10.34 \times 10^{-3}$ L or 10.34 mL

4. (a) Balanced reaction is

$2F_2O_3 + 3C \longrightarrow 4Fe + 3CO_2$

Number of moles of $Fe_2O_3 = \left(\frac{120 \times 1000}{2 \times 56 + 48}\right) \times \frac{90}{100}$

Mass of 80% pure iron produced

$= \frac{120 \times 1000 \times 0.9}{2 \times 56 + 48} \times \frac{2 \times 56}{0.8}$

$= 94500$ gram or 94.5 kg

5. (d) Let two equimolar carbonates are ACO_3 and BCO_3

mass of metal $A = 2.8 \times 0.125 = 0.35$ g

$ACO_3 + BCO_3 \longrightarrow AO + BO + 2CO_2$

Total two moles CO_3^{2-} provide 2 moles of CO_2 or 60 g CO_3^{2-} provide 44 g of CO_2

Total mass of CO_3^{2-} required to produce 1.32 g of $CO_2 = \frac{60}{44} \times 1.32 = 1.8$ g

mass of metal $(A + B) = 2.8 - 1.8 = 1$ g

mass of $B = 1 - 0.35 = 0.65$ g

$\%$ of $B = \frac{0.65}{2.80} \times 100 = 23.21$

6. (b) Loss in mass due to heating $= 6.2 \times \frac{5}{100}$

$= 0.31$ g

$$2NaHCO_3 \longrightarrow Na_2CO_3 + CO_2 + H_2O$$

2×84 g $\qquad 44 + 18$ g loss

1 mole $NaHCO_3 \longrightarrow$ 31 g loss due to heating

0.31 g loss from 0.01 mole of $NaHCO_3$

Moles of Na_2CO_3 produced $= \frac{0.01}{2} = 0.005$

Total moles of carbonate reacted with $BaCl_2$

$= \left(7.5 \times 0.2 \times \frac{100}{10}\right) \times 10^{-3} = 0.015$

Moles of carbonates in original sample

$= 0.015 - 0.005 = 0.01$

Mass of Na_2CO_3 in original sample

$= 0.01 \times 106 = 1.06$ gram

7. (c) Moles of NO_2 required

$= \left(\frac{1575}{63}\right) \times \frac{3}{2} \times \frac{1}{0.8} = 46.875$

moles of NO required $= \frac{46.875}{0.60}$

moles of NH_3 required $= \frac{46.875}{0.60} \times \frac{1}{0.50}$

$= 156.25$

Volume of NH_3 at STP required

$= 156.25 \times 22.4 = 3500$ L

8. (c) Moles of NaOH consumed when it reacts with $H^+ = 3 \times 1$

moles of H_2SO_4 present in sample $= 1.5$

mass of H_2SO_4 in sample $= 1.5 \times 98 = 147$ g

% purity $= \frac{147}{183.75} \times 100 = 80$

9. (c) Let mass of sample is x g;

mass of MnO_2 is $0.75\,x$ g

moles of $MnO_2 = \frac{0.75x}{87}$

mass of Mn $= \frac{0.75x}{87} \times 55 = 0.474x$

$$3MnO_2 \longrightarrow Mn_3O_4 + O_2$$

mass of Mn_3O_4 produced

$= \frac{0.75x}{87} \times \frac{1}{3} \times 229 = 0.658x$ g

Total mass of residue $= 0.658x + 0.2\,x$

(mass of impurities)

% of Mn in residue $= \frac{0.474x}{0.858x} \times 100$

$= 55.24$

10. (b) Moles of AgCl = moles of chloride

$= \frac{1.96}{143.5} = 0.0136$

Moles of organic compound $= \frac{1}{147}$

$= 6.8 \times 10^{-3}$

Chlorine atoms in each molecule of organic compound $= \frac{0.0136}{6.8 \times 10^{-3}} = 2$

11. (c) Let x g is mass of CaC_2O_4 and $(0.6 - x)$ g mass of MgC_2O_4

$$CaC_2O_4 \xrightarrow{\Delta} CaCO_3 + CO$$

$$MgC_2O_4 \xrightarrow{\Delta} MgCO_3 + CO$$

mass of $CaCO_3$ produced $= \frac{x}{128} \times 100$

mass of $MgCO_3$ produced $= \frac{(0.6 - x)}{112} \times 84$

$\therefore \quad \frac{x}{128} \times 100 + \frac{(0.6 - x)}{112} \times 84 = 0.465$

$x = 0.48$ g

Due to further heating

$$CaCO_3 \xrightarrow{\Delta} CaO + CO_2$$

$- \qquad \frac{x}{128}$

$$MgCO_3 \xrightarrow{\Delta} MgO + CO_2$$

$- \qquad \left(\frac{0.6 - x}{112}\right)$

mass of CaO and MgO produced

$= \frac{0.48}{128} \times 56 + \frac{0.12}{112} \times 40 = 0.252$ g

12. (a) Moles of $BaSO_4 = \frac{1.22}{233.3}$

moles of $M_2(SO_4)_3 = \frac{1.22}{233.3} \times \frac{1}{3}$

$= 1.743 \times 10^{-3}$

wt. of $M_2(SO_4)_3 = 0.596$

$\therefore$ $1.743 \times 10^{-3}\,(2M + 96 \times 3) = 0.596$

$M = 26.9$

13. (b) $2NH_3 + CO_2 \longrightarrow NH_2—CO—NH_2 + H_2O$

moles of urea $= \frac{999.6 \times 10^3}{60} = 16.66 \times 10^3$

$\therefore$ moles of $CO_2 = 16.66 \times 10^3$

$$C_nH_{2n+2} + O_2 \longrightarrow nCO_2 + H_2O$$

$\frac{16.66 \times 10^3}{n} \times [12n + (2n + 2)] = 236 \times 10^3$

$n \cong 12$

$\therefore$ molecular formula of hydrocarbon

$= C_{12}H_{26}$

14. (d) $C_nH_{2n+2} + \left(\frac{3n+1}{2}\right)O_2 \longrightarrow nCO_2 + (n+1)H_2O$

Let initial pressure of C_nH_{2n+2} is P then increase in pressure

$= P\left[(2n+1)-1-\left(\frac{3n+1}{2}\right)\right]$

$= \left(\frac{n-1}{2}\right)P$

546 K and 4.6 atm $\longrightarrow$ 273 K and 2.3 atm;

Increase in pressure = 0.3 atm

$P = \frac{nRT}{V} = \frac{11.6}{M} \times \left(\frac{0.0821 \times 273}{22.41}\right)$

$\left(\frac{n-1}{2}\right) \times \frac{11.6}{(14n+2)} = 0.3$

$\left(\frac{n-1}{14n+2}\right) = \frac{0.6}{11.6} = n = 4$

$\therefore$ Compound is C_4H_{10}

15. (d) Moles of $H_2SO_4 = 0.1$; mole of KOH = 0.2

moles of H_2O_2 used in first reaction

$= \frac{0.2}{2} \times \frac{1}{0.4} = 0.25$

moles of O_2 produce $= \frac{6.74}{22.4} = 0.3$

$\therefore$ moles of H_2O_2 used in second reaction

$= \frac{0.3}{3 \times 0.5} = 0.2$

Total mole of H_2O_2 consumed = 0.45

Molarity of $H_2O_2 = \frac{0.45}{0.15} = 3\ M$

Volume strength = 11.2 × 3 = 33.6 volumes

16. (b) $SO_2Cl_2 + 2H_2O \longrightarrow H_2SO_4 + 2HCl$

m-moles of H_2SO_4 produced = 5; m-moles of HCl produced = 10

m-moles of $Ba(OH)_2$ required

$= 5 \text{ (for } H_2SO_4) + \frac{10}{2} \text{ (for HCl)} = 10$

$\therefore M \times V = 10, 0.2 \times V = 10, V = 50$ mL

17. (b) Only Na_2CO_3 reacts with H_2SO_4;

$Na_2CO_3 + H_2SO_4 \longrightarrow Na_2SO_4 + H_2CO_3$

m-moles of Na_2CO_3 = m-moles of H_2SO_4

$= 20 \times 0.1 = 2$

m-moles of Na_2CO_3 in 250 mL solution

$= \frac{250}{25} \times 2 = 20$

wt. of $Na_2CO_3 = 20 \times 106 \times 10^{-3} = 2.12$ g

% of $Na_2CO_3 = \frac{2.12}{5} \times 100 = 42.4$

$\therefore$ % of $Na_2SO_4 = 100 - 42.4 = 57.6$

18. (b) $Ag + HNO_3 \longrightarrow AgNO_3$

$AgNO_3 + NaCl \longrightarrow NaNO_3 + AgCl$

POAC on Ag

Mole of Ag in coin = mole of Ag in AgCl

$a = \frac{14.35}{143.5} = 0.1$ mole

Mass of Ag in coin = 0.1 × 108 = 10.8 g

$\therefore$ % silver in coin $= \frac{10.8}{11.34} \times 100\% = 95.2\%$

19. (a) Mole of $O_2 = \frac{67.2}{22.4} = 3$ mole

$3O_2 \rightarrow 2O_3$

$\therefore$ Mole of Ozone formed $= \frac{2}{3} \times \frac{15}{100} \times 3$

$= 0.3$ mole

$\therefore$ Mass of Ozone formed = 0.3 × 48 g = 14.4 g

20. (c) Initial $M_1 = \frac{40 \times 1.35 \times 10}{98} = 5.51$

Final $M_2 = \frac{20 \times 1.225 \times 10}{98} = 2.5$

$|M_2 - M_1| = 3.01$

Level 3

Passage-1

1. (b) $H_2O + SO_3 \longrightarrow H_2SO_4$;

18 g water combines with 80 g SO_3

$\therefore$ 4.5 g of H_2O combines with 20 g of SO_3.

$\therefore$ 100 g of oleum contains 20 g of SO_3

or 20% free SO_3.

2. (c) Initial moles of free SO_3 present in oleum

$= \frac{12}{18} = \frac{2}{3}$ moles

= moles of water that can combines with SO_3

moles of free SO_3 combined with water

$= \frac{9}{18} = \frac{1}{2}$ mole

moles of free SO_3 left $= \frac{2}{3} - \frac{1}{2} = \frac{1}{6}$ mole

$\therefore$ volume of free SO_3 at STP $= \frac{1}{6} \times 22.4$

$= 3.73$ L

3. (c) $Na_2CO_3 + H_2SO_4 \longrightarrow Na_2SO_4 + H_2O + CO_2$

moles of CO_2 formed = moles of Na_2CO_3 reacted

$= \frac{5.3}{106} = 0.05$

volume of CO_2 formed at 1 atm pressure and 300 K = 0.05 × 24.63 = 1.23 L

4. (b) eq. of H_2SO_4 + eq. of SO_3 = eq. of NaOH

$\frac{x}{98} \times 2 + \frac{(1-x) \times 2}{80} = 54 \times 0.4 \times 10^{-3}$

$x = 0.74$

% of free $SO_3 = \frac{1 - 0.74}{1} \times 100 = 26\%$

Passage-2

1. (a) $Cl_2 + 2NaOH \longrightarrow NaCl + NaClO + H_2O$

4 mole $\longleftarrow$ 4 mole

$3NaClO \longrightarrow 2NaCl + NaClO_3$

$\frac{3}{1} \times \frac{4}{3} = 4 \text{ mole} \longleftarrow \frac{4}{3} \text{ mole}$

$4NaClO_3 \longrightarrow 3NaClO_4 + NaCl$

$\frac{4}{3} \text{ mole} \longleftarrow 1 \text{ mole}$

Moles of $NaClO_4 = \frac{122.5}{122.5} = 1$ mole

4 mol $Cl_2 = 284$ g Cl_2 will be required.

2. (c) $Cl_2 + 2NaOH \longrightarrow NaCl + NaClO + H_2O$

1 mole 1 mole 1 mole

$3NaClO \longrightarrow 2NaCl + NaClO_3$

1 mole $\frac{2}{3} = 0.67$ mole

$4NaClO_3 \longrightarrow 3NaClO_4 + NaCl$

$\frac{1}{3}$ mole $\frac{1}{4} \times \frac{1}{3} = \frac{1}{12}$ mole $= 0.083$ mole

Total NaCl $= 1 + 0.67 + 0.083 = 1.75$ mole

3. (b) As $NaClO_3$ is intermediate it will be consumed in the complete reaction.

Passage-3

1. (a) % of Fe in alloy $= \frac{\left(\frac{0.03}{2}\right) \times \frac{2}{3} \times 56}{2.5} \times 100\% = 22.4\%$

2. (a) $n_{Fe} = \frac{0.56}{56} = 0.01$

$n_C = \frac{2.5 - 0.56}{12} = 0.162$

Average molecular mass of alloy $= \frac{\text{Total mass}}{\text{Total mole}}$

$= \frac{2.5}{0.01 + 0.162} = 14.56$ g/mol

Passage-4

$$\left.\begin{array}{l} \text{Mol of } CO_2 = \frac{8.8}{44} = 0.2 \text{ mole} \\ \text{Mol of } H_2O = \frac{7.2}{18} = 0.4 \text{ mole} \\ \text{Mol of H} = 2 \times 0.4 = 0.8 \text{ mole} \end{array}\right\} \text{Present in 6 g of compound}$$

N of 24 g compound liberates in the form of NH_3 which reacts with H_2SO_4 as

$2NH_3 + H_2SO_4 \longrightarrow (NH_4)_2SO_4$
0.4 mole

Mass of H_2SO_4 used $= \frac{25 \times 1.6 \times 98}{100}$

Mole of H_2SO_4 used

$= \frac{25 \times 1.6 \times 98}{100 \times 98} = 0.400$ mole

$\therefore$ Mole of NH_3 released $= 0.8$ mole from 24 g compound

$\therefore$ From 6 g compound mol of NH_3 released

$= \frac{0.8}{4} = 0.2$ mole

$\therefore$ Mole of N = Mole of $NH_3 = 0.2$ mole in 6 g

1. (b) $\therefore$ % of N $= \frac{0.2 \times 14}{6} \times 100\% = 46.67\%$

2. (c) Empirical formula

C	H	N
0.2	0.8	.2
1	4	1

Empirical formula is CH_4N

Passage-5

1. (b) Equivalent of HNO_3 in solution $= \frac{70 \times 1.42}{63}$

$M_{HNO_3} = \frac{70 \times 1.42 \times 1000}{63 \times 100} = 15.78$

2. (c) 70×1.42 g pure acid $= 100$ mL

63 g of acid $= \frac{100 \times 63}{70 \times 1.42} = 63.38$ mL

One or More Answers is/are Correct

1. (c, d)

1g-atom of nitrogen = 1 mol *N* atoms

$= \frac{1}{2}$ mol N_2 gas

$\frac{1}{2}$ mol N_2 gas $= 11.2$ L N_2 at STP $= 14$ g N_2

2. (a, b)

1g molecule $V_2O_5 = 1$ mole $V_2O_5 = 2$ mole V atom $= 5$ mole O atom

5. (b, c, d)

Equivalent of $Ba(OH)_2 = 1 \times 2 = 2$

Acids having 2 equivalents will be exactly neutralised by $Ba(OH)_2$.

6. (b, c)

Species having same empirical formula (EF) will have same percentage of C.

	EF
CH_3COOH	CH_2O
$C_6H_{12}O_6$	CH_2O
C_2H_5OH	C_2H_6O
$HCOOCH_3$	CH_2O
HCOOH	CH_2O_2
CH_3OCH_3	C_2H_6O

7. (b, d)

$\because$ both are liquid CH_3OH is solute (less amount)

Mass of $CH_3OH = 30 \times 0.8 = 24$ g,

Mass of $C_2H_5OH = 60 \times 0.92 = 55.2$ g

Mass of solution $= 24 + 55.2 = 79.2$ g

Volume of solution $= \frac{79.2}{0.88} = 90$ mL

$$\text{Molarity} = \frac{n\text{CH}_3\text{OH}}{\text{V(L)}} = \frac{24/32}{90} \times 1000$$

$$= 8.33 \text{ mol L}^{-1}$$

$$\text{Molality} = \frac{n_{\text{Solute}}}{w_{\text{Solvent}}(\text{kg})} = \frac{24/32}{55.2} \times 1000 = 13.59$$

$$\text{Mole fraction of solute} = \frac{\frac{24}{32}}{\frac{24}{32} + \frac{55.2}{46}} = 0.385$$

Mole fraction of solvent $= 1 - 0.385 = 0.615$

8. (a, b, c)

$$17 \text{ g/L } H_2O_2 = \frac{17}{34} \text{ mol L}^{-1} \; H_2O_2$$

$$\text{Molarity of } H_2O_2 = \frac{1}{2}$$

$$M = \frac{V}{11.2} \Rightarrow V = \frac{1}{2} \times 11.2 = 5.6$$

5.6 volume H_2O_2 means 1 mL will give 5.6 mL at 273 K and 1 atm

$$P_1V_1 = P_2V_2$$
$$1 \times 5.6 = 2 \times V_2$$
$$V_2 = 2.8 \text{ mL}$$

9. (a, b)

(a) 46g of 70% $\frac{W}{V}$ HCOOH ($d_{\text{Solution}} = 1.4$ g/mL)

70% $\frac{W}{V}$ HCOOH $\longrightarrow$ 70 g HCOOH in 100 mL solution.

Mass of solution $= 1.4 \times 100 = 140$ g

So, in 140 g solution, mass of HCOOH = 70 g in 46 g, mass of HCOOH $= \frac{70}{140} \times 46 = 23$ g

(b) 10 M HCOOH $\longrightarrow$ 10 mole HCOOH in 1000 mL solution mass of solution = 1000 g

Mass of HCOOH $= 10 \times 46 = 460$ g

So in 50 g solution mass of

$$\text{HCOOH} = \frac{460}{1000} \times 50 = 23 \text{ g}$$

(c) 25% $\frac{W}{W}$ HCOOH $\longrightarrow$ 25 g HCOOH in 100 g solution.

So in 50 g solution, mass of HCOOH = 12.5 g

10. (a,c,d) $V_{\text{strength}} = 56$;

$$\therefore \; M = \frac{28}{11.2} = 2.5$$

$\therefore$ 1 L contain 2.5 moles of H_2O_2

or $2.5 \times 34 = 85$ g H_2O_2

Mass of 1 litre solution = 265 g

($\because d = 265$ g/L)

$\therefore$ $w_{H_2O} = 180$ g or moles of $H_2O = 10$

$$x_{H_2O_2} = \frac{2.5}{2.5 + 10} = 0.2$$

$$\% \frac{w}{\text{v}} = \frac{2.5 \times 34}{1000} \times 100 = 8.5$$

$$m = \frac{2.5}{180} \times 1000 = 13.88$$

11. (a, b)

$$CO + \frac{1}{2}O_2 \longrightarrow CO_2$$

(A) O_2 is L.R.

So, Volume of CO_2 produced = 20 mL

Total volume of $CO_2 = 20 + 60 = 80$ mL

(B) CO is L.R.

Volume of CO_2 produced = 30 mL

Total volume of $CO_2 = 30 + 50 = 80$ mL

(C) O_2 is L.R.

Volume of CO_2 produced = 40 mL

Total volume of $CO_2 = 30 + 40 = 70$ mL

12. (b)

(a) $m \; \frac{M}{d - MM_1/1000} \Rightarrow M\left(1 + \frac{40}{1000}\right) = dm$

$$M = \frac{d}{1.04} = \frac{1.144}{1.04} = 1.1$$

(b) 1000 mL -------- 1.1 moles of Na^+

100 mL ------ 0.11 moles $\equiv 0.11 \times 23 = 2.53$ gm Na^+

(c) Mass of $H_2O = 114.4 - 4.4 = 110$ g

$$n_{H_2O} = \frac{110}{18} = 6.11,$$

$$X_{NaCl} = \frac{0.11}{0.11 + 6.11} = 0.0176$$

(d) 114.4 g solution -------1.87 g OH^-

100 g solution ------ $\frac{1.87}{114.4} \times 100 = 1.635\%$

13. (b, c)

N	:	O
30.43	:	69.57 (mass ratio)
$\frac{30.43}{14}$	:	$\frac{69.57}{16}$ (mol ratio)
2.1736	:	4.348125
1	:	2 $\therefore$ Empirical formula NO_2

Mass of 1 mole molecule $= 1.527 \times 10^{-22} \times 6.022 \times 10^{23}$

$= 91.95594 = 92$ g

$\therefore$ Molecular formula $= N_2O_4$

14. (a, c, d)

$M_1V_1 + M_2V_2 = M_FV_F$

$V_F = \dfrac{0.3 \times 300 + 0.2 \times 700}{0.115} = 2000$ mL

Volume of water added $= 2000 - 1000 = 1000$ mL

Mole of NaOH $= M_1V_1 + M_2V_2 = 0.23$ mole

Mass of NaOH $= 0.23 \times 40 = 9.2$ g

15. (a, d)

(a) $\underset{5\text{mole}}{2A} + \underset{9\text{mole}}{3B} \longrightarrow 2C + 4D$

(b) Molar mass of $D = \dfrac{(2 \times 50 + 3 \times 18 - 2 \times 47)}{4} = 15$

$\therefore$ Vapour density of $D = \dfrac{15}{2} = 7.5$

(c) $2A + 3B \xrightarrow{80\%} 2C + 4D$

$t = 0$, 5 mole, 9 mole, 0, 0

$t = 1$, $5 - 0.8 \times 5$, $9 - \dfrac{3}{2} \times (5 \times 0.8)$, $2 \times (5 \times 0.8)$, $\dfrac{4}{2}(5 \times 0.8)$

$= 5 - 4$, $= 9 - 6 = 3$ mole, $= 8$ mole, $= 8$ mole

(d) Maximum mole of $D = \dfrac{4}{2} \times 5 = 10$ mole

Match the Column

1. (A) 0.5 mole of $SO_2(g) = 32$ g

weight $= 1.5\, N_A$ atoms

$= 11.2$ L at 1 atm and 273 K

(B) 1 g of $H_2(g) = 0.5$ mole

$H_2 = 11.2$ L at 1 atm and 273 K

$= N_A$ atoms

(C) 0.5 moles of $O_3 = 1.5N_A$ atoms $= 11.2$ L at 1 atm and 273 K $= 24$ g weight

(D) 1 g molecules of $O_2 = 1$ mole of $O_2 = 2N_A$ atom $= 22.4$ L at 1 atm and 273 K $= 32$ g weight

2. (A) Moles of $CO_2(g) = \dfrac{44}{44} = 1$

1 mole $CO_2(g) = 1$ g-molecule $CO_2(g)$

(B) Moles of $CH_4(g) = \dfrac{35.2}{16} = 2.2$ moles

$V = 2.2 \times 22.4 = 49.28$ L at 1 atm and 273 K

(C) Moles of $O_3(g) = \dfrac{48}{48} = 1$

(D) Moles of $N_2O(g) = \dfrac{44}{44} = 1$

Number of O atoms $= N_A \times 1 = N_A$

3. Use % by moles $= \dfrac{M_{avg} - M_1}{M_2 - M_1} \times 100$

% by mass = % by moles $\times \dfrac{M_2}{M_{avg}}$

4. (A) $\underset{\substack{10\\0}}{C_3H_8(g)} + \underset{\substack{110\\60}}{5O_2(g)} \longrightarrow \underset{30}{3CO_2(g)} + 4H_2O(l)$

% volume contraction $= \dfrac{30}{120} \times 100 = 25\%$

Volume of CO_2 produced $= 30$ mL

(B) $\underset{\substack{20\\0}}{C_2H_6O(g)} + \underset{\substack{80\\20}}{3O_2(g)} \longrightarrow \underset{20}{2CO_2(g)} + 3H_2O(l)$

Remaining volume of $O_2 = 20$

(C) $\underset{\substack{10\\0}}{C_3H_6O_2(g)} + \underset{\substack{50\\15}}{\frac{7}{2}O_2(g)} \longrightarrow \underset{30}{3CO_2(g)} + 3H_2O(l)\%$

$V_C = \dfrac{15}{60} \times 100 = 25\%$

$V_{CO_2} = 30$ mL

(D) $\underset{\substack{40\\0}}{C_2H_2O_4(g)} + \underset{\substack{60\\40}}{\frac{1}{2}O_2(g)} \longrightarrow \underset{80}{2CO_2(g)} + H_2O(l)$

% increase in vol $= \dfrac{20}{100} \times 100 = 20\%$

5. (A) $M = \dfrac{3.65/36.5}{100} \times 10^3 = 0.1 \times 10 = 1$

(B) $m = \dfrac{4/40}{96} \times 10^3 = 1.04$

$\therefore X_{NaOH} = \dfrac{4/40}{4/40 + \frac{96}{18}} = \dfrac{3}{163}$

(C) $M = \dfrac{4/40}{100} \times 10^3 = 1$

(D) $m_{NaOH} = 1.04$

1000 g solv. $\left(\dfrac{25}{24} \times 40\right)$ g NaOH

$\therefore X_{NaOH} = \dfrac{25/24}{\frac{25}{24} + \frac{500}{g}} = \dfrac{25}{24} \times \left(\dfrac{72}{75 + 4000}\right)$

$= \dfrac{25/24}{56.59} = 0.018 = \dfrac{75}{75 + 4000} = \dfrac{75}{4075} = \dfrac{3}{163}$

Level 4

Subjective Problems

1. Mass of liquid in 10 moles $= 280 \times 10 = 2800$ g

Volume $= \dfrac{2800}{1.4} = 2000$ mL $= 2$L

2. $(M) = \left(\dfrac{16}{5.6}\right)(22.4)$

$M = 64$

$64 = 32 + 16x$

$x = 2$

3. $[H^+] = \dfrac{(200\times1)+(300\times6)}{1000}$

$[H^+] = 2$

4. $N_2(g) + H_2(g) \longrightarrow N_2H_2(l)$

5L	3L	–
2L	–	–

$\Delta V = (8)-(2) = 6$

5. $n_{SO_2} = \dfrac{3.78}{64} = 0.06;$

Let moles of $CaSO_4$ produced $= x$

POAC on S

$SO_2 \longrightarrow CaSO_4$

$(n_S)_{SO_2} = (n_S)_{CaSO_4}$

$0.06\times1 = x\times1 \Rightarrow x = 0.06$

Mass of $CaSO_4 = 0.06\times136 = 8.16$

6. Moles of $H_2O = 1$

$3Fe + 4H_2O \longrightarrow Fe_3O_4 + 4H_2$

Consumed moles of Fe $= \dfrac{3}{4}$

Mass of Fe $= \dfrac{3}{4}\times56 = 42$

7. $n_{KClO_3} = \dfrac{122.5}{122.5} = 1$

Moles of atoms $= 1+1+3 = 5$

8. $M_xCl_y = MCl_y$ ($\because$ valency of Cl $= 1$)

Molar mass of $M_x\ Cl_y = 85.5\times2$

$= 171$ g/mol

$(M)(x)+(35.5)(y) = 171 \quad \ldots(1)$

$\dfrac{\text{Mass of chlorine}}{\text{Mass of metalic chloride}} = \dfrac{0.835}{1}$

$= \dfrac{35.5\,y}{M(x)+(35.5)(y)}$

On solving $y = 4$ and $x = 1$

$x + y = 5$

9. Mass of N present in 1 mole compound $= \dfrac{7}{27}\times108$ g

$\therefore$ Mole of N $= \dfrac{7\times108}{27\times14} = 2$

10. Let a mole & b mole $HgCl_2$ & Hg_2Cl_2 respectively formed.

$Hg + Cl_2 \longrightarrow HgCl_2 + Hg_2Cl_2$

$a + 2b = 1.5 \quad \ldots(1)$ (POAC on Hg)

$2a + 2b = 2 \quad \ldots(2)$ (POAC on Cl)

On solving $a = 0.5, \quad b = 0.5$

$\therefore \quad \dfrac{a}{b} = 1$

11. $V\times\dfrac{36.5\times1.2\times10}{36.5} = 1080\times\dfrac{1}{10}$

$V = 9$ mL

12. $W_{(CF_2)_n} = \left[\dfrac{1.52\times10^3}{38}\times\dfrac{2}{1}\times\dfrac{1}{4n}\right]\times50n$ gm

$= 1\times10^3$ gm or 1 kg.

13. $\underset{x}{C_3H_8(g)} + 5O_2(g) \longrightarrow \underset{3x}{3CO_2(g)} + 4H_2O(l)$

$\underset{3-x}{C_4H_{10}(g)} + \dfrac{13}{2}O_2(g) \longrightarrow \underset{4(3-x)}{4CO_2(g)} + 5H_2O(l)$

$3x + (3-x)4 = 10 \Rightarrow x = 2$

$\therefore \quad \dfrac{V_{C_3H_8}}{V_{C_4H_{10}}} = 2$

14. $\underset{1}{H_2SO_4} + \underset{2}{2NaOH} \longrightarrow Na_2SO_4 + 2H_2O$; 198 g equimolar mixture contains 2 mole H_2SO_4 and 1 mole H_3PO_4

$\underset{1}{H_3PO_4} + \underset{3}{3NaOH} \longrightarrow Na_3PO_4 + 3H_2O$

Total mole of NaOH $= 2 + 3 = 5.0$

15. Let volume of $Al_2(SO_4)_3$ solution $= V_1$ mL

Volume of $CaCl_2$ solution $= V_2$ mL

m-mole of $Al^{3+} = 0.2\,V_1$, $\quad SO_4^{2-} = 0.3\,V_1$,

$Ca^{2+} = 0.2\,V_2$ and $Cl^- = 0.4\,V_2$

$[-] = \dfrac{0.3\,V_1 + 0.4\,V_2}{V_1+V_2}$ and $[+] = \dfrac{0.2\,V_1 + 0.2\,V_2}{V_1+V_2}$

$\dfrac{[-]}{[+]} = 1.6 \Rightarrow \dfrac{0.3\,V_1+0.4\,V_2}{0.2V_1+0.2\,V_2} = 1.6$

$\Rightarrow \dfrac{V_1}{V_2} = 4$

Numerical Problems

1. M_w of mixture $= 22.4\times0.75 = 16.8$

	CH_4	C_2H_6
Mole % :	$x\%$	$(100-x)\%$

$M_w = \dfrac{M_1X_1 + M_2X_2}{X_1+X_2}$

$16.8 = \dfrac{16\times x + 30(100-x)}{100}$

$x = 94.28$

2. Final gaseous mixture contains 1 mol CO and 1 mol CO_2 after reaction

$M_{avg} = \dfrac{1\times28 + 1\times44}{1+1} = 36$ g/mole

3. $n_{MgCO_3} = \dfrac{21}{84}$ mole, $n_{HCl} = \dfrac{200\times\dfrac{7.3}{100}}{36.5} = 0.4$ mole

$MgCO_3(s) + 2HCl(aq) \longrightarrow MgCl_2(aq) + H_2O(l) + CO_2(g)$

$\frac{21}{84} = 0.25 \quad 0.4$

HCl is L.R.

Mole of CO_2 produced $= 0.2$;

Volume $= 0.2 \times 22.4 = 4.48$ L

4. Mass of $C_2H_5OH = \frac{10 \times 0.05 \times 10^{-3}}{2} \times 46$

$= 0.0115$

Mass % of $C_2H_5OH = \frac{0.0115}{11.5} \times 100 = 0.1$

5. Let volume of NaOH solution $= V$ mL

Mass of solution $= 1.25\,V$ g

Mass of NaOH $= 1.25\,V \times \frac{40}{100}$ g

$1.25\,V \times \frac{40}{100} = 2 \times 5 \times 40$

$\Rightarrow \quad V = 800 \text{ mL} = 0.8 \text{ L}$

6. 3.30×74.5 g KCl present with 1000 g water

Mass of solution $= 1245.85$ g ;

Volume of solution $= 1038.2$ mol

Molarity $= \frac{3.3}{1038.2} \times 1000 = 3.17$

7. Molarity of 1st HNO_3 solution $= 0.5\,M$

Molarity of 2nd HNO_3 solution

$= \frac{9.45 \times 10}{63} = 1.5\,M$

$M_1V_1 + M_2V_2 = M_fV_f$

$0.5 \times 250 + 1.5 \times 750 = M_f \times 1000$

Resultant solution have molarity $= 1.25\,M$

8. $CH_4 + 2O_2 \longrightarrow CO_2 + 2H_2O$

4 g

or 0.25 mole

Amount of $O_2 = 0.5$ mole

$H_2O_2(aq) \longrightarrow H_2O\,(l) + \frac{1}{2}O_2(g)$

Mole of $H_2O_2 = 1$ mole

Molarity $= \frac{1}{0.5} = 2$ molar

Volume strength $= 2 \times 11.35 = 22.7$

9. $2.8375 = M \times 11.35$

$M = \frac{1}{4}, \quad V = 1\text{ L}, \quad n = \frac{1}{4},$

Mass of solution $= 1208.5$ g

$m = \frac{0.25}{(1208.5 - 8.5)} \times 1000 = 0.208 \approx 0.21$

10. Mass of Br $= \frac{0.188}{188} \times 80 = 0.08$ g

Mass % of Br $= \frac{0.08}{0.2} \times 100 = 40$

11. $P_4S_3 + 8O_2 \longrightarrow P_4O_{10} + 3SO_2$

Moles of $P_4S_3 = \frac{110}{220} = 0.5$

Moles of $O_2 = \frac{192}{32} = 6$ mole

L.R. is P_4S_3; moles of SO_2 formed $= 0.5 \times 3$

Mass of SO_2 formed $= 0.5 \times 3 \times 64 \times 0.75 = 72$ g

12. Moles of $BaCl_2 = 400 \times \frac{20.8}{100} \times \frac{1}{208} = 0.4$

Moles of $H_2SO_4 = 100 \times \frac{4.9}{100} \times \frac{1}{98} = 0.05$, so H_2SO_4 is limiting reagent

Theoretical moles of H_2SO_4 should be formed $= 0.05$

% yield $= \frac{0.045}{0.05} \times 100 = 90$

13. $\underset{300}{C_3H_8} + \underset{V}{CH_4} + \underset{300-V}{CO} + O_2 \longrightarrow CO_2$

$3 \times 300 + 1 \times V + 1 \times (300 - V) = V_{CO_2}$

$V = 1200$ mL

$V = 1.20$ L

14. $\underset{a \text{ mole}}{MgO} + H_2SO_4 \longrightarrow \underset{a \text{ mole}}{MgSO_4} + H_2O$

$\underset{b \text{ mole}}{CaO} + H_2SO_4 \longrightarrow \underset{b \text{ mole}}{CaSO_4} + H_2O$

$40a + 56b = 40 \quad \text{...(1)}$

$120a + 136b = \frac{688}{7} \quad \text{...(2)}$

By solving eq. (1) and (2)

We get $a = \frac{1}{20}$ mole

$\therefore$ Mass % of MgO $= \frac{1/20 \times 40}{80} \times 100 = 2.50$

15. $(\text{V.D.})_i = \frac{17}{2} = 8.5$

$(\text{V.D.})_f = \frac{17}{2 \times 2} = 4.25$

% change in V.D. $= \frac{8.5 - 4.25}{8.5} \times 100 = 50\,\%$

CHAPTER 2

Atomic Structure

Atomic Number of an Element

Total number of protons present in the nucleus = Total number of electrons present in the atom

Mass number of an element = Number of protons + Number of neutrons.

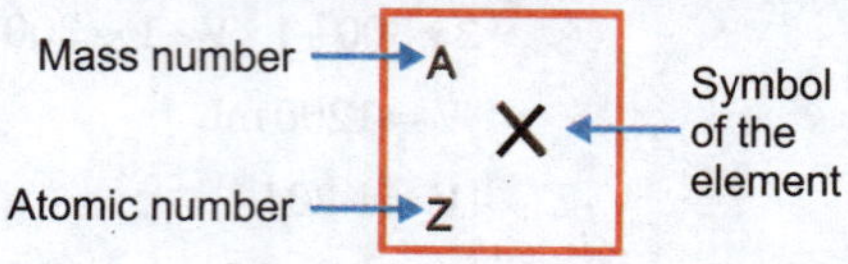

e.g., $^{23}_{11}Na$, $^{35}_{17}Cl$ and so on.

Terms associated with elements

- Isotopes : Atoms having same number of protons but different mass number.
- Isobars : Elements having same mass number.
- Isotones : Elements having same number of neutrons $(A - Z)$.
- Isoelectronic : Species/elements having same number of electrons.
- Isosters : Species having same number of atoms and electrons
- Isodiaphers : Elements having same number of $|N - Z|$ or $|A - 2Z|$
- Paramagnetic : Species having non-zero unpaired electron.
- Diamagnetic : Species having zero unpaired electron.

Rutherford's Model

- Electrons, protons & neutrons are the most important fundamental particles of atoms of all elements (Except hydrogen)
- $^{A}_{Z}X$, Mass number (A) = Atomic number (Z) + number of neutrons (n)
- $R_N = R_0(A)^{1/3}, R_0 = 1.33 \times 10^{-13}$ cm A = mass number, R_N = Radius of nucleus
- $\frac{1}{2} m_\alpha v_\alpha^2 = K \frac{q_1 \times q_2}{r}$; r = distance of closest approach , v_α = Velocity of a α-particle

  m_α = mass of α-particle

  q_1 = charge on α-particle ; q_2 = charge on metal foil.

Size of the Nucleus

The volume of the nucleus is very small and is only a minute fraction of the total volume of the atom. Nucleus has a diameter of the order of 10^{-12} to 10^{-13} cm and the atom has a diameter of the order of 10^{-8} cm.

Thus, diameter (size) of the atom is 1,00,000 times the diameter of the nucleus.

Electromagnetic Spectrum

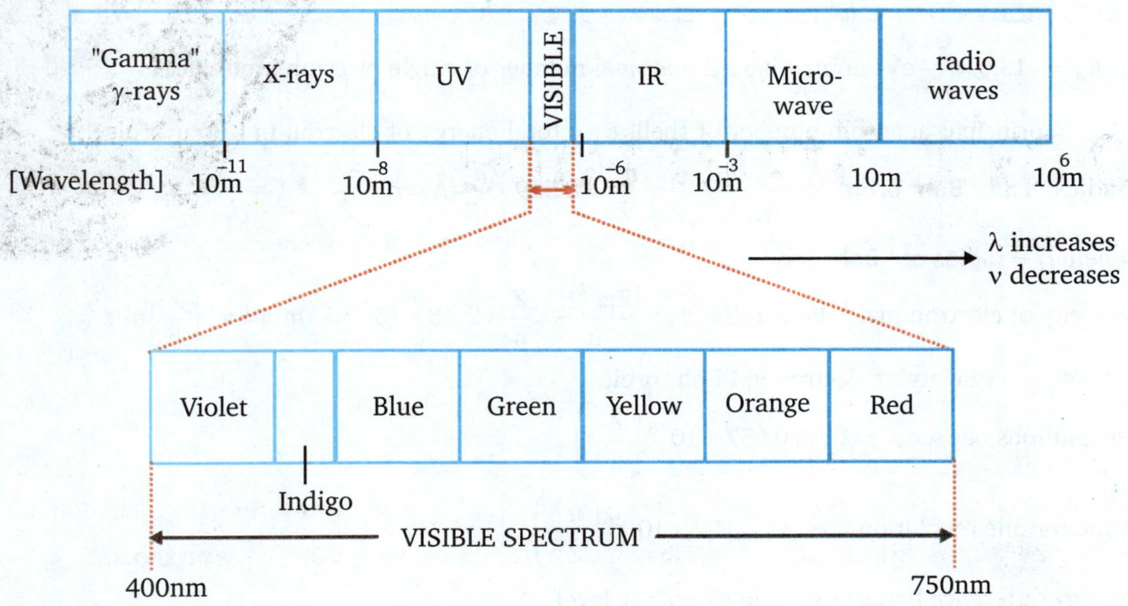

Light

- Photon is considered as massless bundle of energy.
- Energy of light $E = mc^2$, where m = mass of light particle, c = speed of light
- $E_{\text{Photon}} = h\nu = hc/\lambda = hc\bar{\nu} \cong \dfrac{1240\ \text{eV.nm}}{\lambda\,(\text{nm})}$

  where h = Planck constant, λ = wavelength of photon, $\bar{\nu}$ = wave number.
- Quantum efficiency or Quantum Yield $= \dfrac{\text{no. of molecules reacting}}{\text{no. of quanta absorbed}}$

Bohr's Model

- Electrostatic force $= \dfrac{Kq_1q_2}{r^2}$ where $K = \dfrac{1}{4\pi\varepsilon_0} = 9 \times 10^9\ \text{N - m}^2/\text{C}^2$
- Potential energy due to electrostatic force $= \dfrac{Kq_1q_2}{r}$

  q_1 = charge of electron, q_2 = charge of nucleus
- Potential due to a charge (Q) particle at a distance (r) $= \dfrac{KQ}{r}$
- Bohr quantization rule $mvr = n \cdot \dfrac{h}{2\pi} = n.\hbar$
- According to Newtons' second law, in a uniform circular motion resultant of all the forces towards centre must be equal to $\dfrac{mv^2}{r}$.

- $\frac{Kq_1q_2}{r^2} = \frac{mv^2}{r}$,

  where q_1 = charge of electron, q_2 = charge of nucleus, m = mass of electron,
  r = radius of Bohr's orbit
- Total energy of electron in n^{th} Bohr orbit

$$E_n = \frac{E_1}{n^2}Z^2 = -\frac{2\pi^2 me^4K^2}{n^2h^2}Z^2; \qquad E_1 = \frac{-2\pi^2 me^4K^2Z^2}{h^2}$$

$$E_n = -13.6 \times \frac{z^2}{n^2}\text{ eV/atom, where } Z = \text{atomic number of single electron atoms/ion,}$$

n = principal quantum number of shell, E_1 = total energy of electron in I Bohr orbit.

- Radius of nth Bohr orbit, $r_n = \frac{h^2}{4\pi^2e^2mK} \times \frac{n^2}{Z} = 0.529 \times \frac{n^2}{Z}\text{ Å} = r_1\left(\frac{n^2}{Z}\right)\text{Å}$,

  where r_1 = radius of I Bohr orbit.
- Velocity of electron in n^{th} Bohr orbit, $v_n = \frac{2\pi e^2K}{h} \times \frac{Z}{n} = 2.18 \times 10^6 \times \frac{Z}{n}\text{ m/s} = v_1\left(\frac{Z}{n}\right)\text{m/s}$

  where v_1 = velocity of electron in I Bohr orbit.
- Revolutions per sec $= v/2\pi r = 0.657 \times 10^{16}\left(\frac{Z^2}{n^3}\right)$
- Time for one revolution $= 2\pi r/v = 1.52 \times 10^{-16}\left(\frac{n^3}{Z^2}\right)$
- Energy difference between n_1 and n_2 energy level.

$$\Delta E = E_{n_2} - E_{n_1} = 13.6Z^2\left(\frac{1}{n_1^2} - \frac{1}{n_2^2}\right)\text{eV/atom} = \text{IE} \times \left(\frac{1}{n_1^2} - \frac{1}{n_2^2}\right)$$

  where IE = ionization energy of single electron species.

- Ionization energy $= E_\infty - E_{G.S.} = E_{G.S.}$; $E_{G.S.}$ = Energy of electron in ground state
- Total energy of electron in terms of kinetic energy (KE) and potential energy (PE)

$$E_n = KE + PE = -KE = \frac{PE}{2}$$

Spectral Lines

- Rydberg's Equation $\frac{1}{\lambda} = \bar{v} = R_H\left[\frac{1}{n_1^2} - \frac{1}{n_2^2}\right] \times Z^2$; $R_H \cong 109700\text{ cm}^{-1}$ = Rydberg constant
- For First line of a series $n_2 = n_1 + 1$
- Limiting spectral line (series limit) means $n_2 = \infty$
- H_α line means $n_2 = n_1 + 1$; also known as line of longest λ, shortest v, least E
- Similarly H_β line means $n_2 = n_1 + 2$
- When electrons de-excite from higher energy level (n) to ground state in atomic sample, then number of spectral lines observed in the spectrum $= \frac{n(n-1)}{2}$
- When electrons de-excite from higher energy level (n_2) to lower energy level (n_1) in atomic sample, then number of spectral line observed in the spectrum $= \frac{(n_2 - n_1)(n_2 - n_1 + 1)}{2}$

- When electron de-excites from higher energy level (n_2) to lower energy level (n_1) in isolated atom, then number of spectral line observed in the spectrum $= (n_2 - n_1)$
-

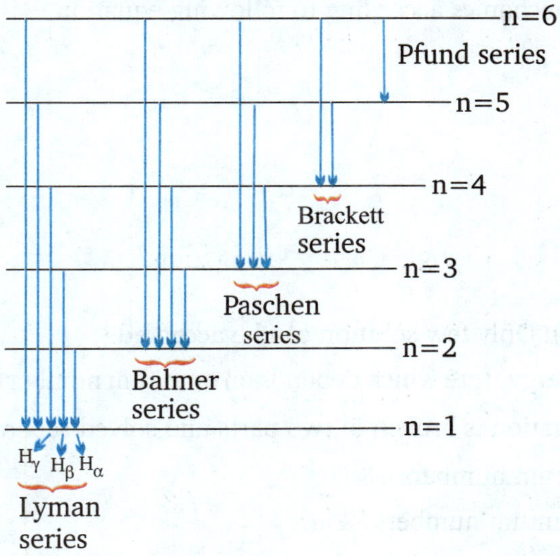

Photoelectric Effect

- When radiation with certain minimum frequency (ν_0 called threshold frequency), strikes the surface of a metal, electrons (called photoelectrons) are ejected from the surface.
- Kinetic energy of photoelectron $= h\nu - w = h\nu - h\nu_0$
  where w = work function
  ν_0 = Threshold frequency
- If $\nu \geq \nu_0$, then photoelectric effect takes place.
- Accelerating potential $= eV = KE = \frac{1}{2}mv^2$

de-Broglie Hypothesis

- All material particles possess wave character as well as particle character.
- $\lambda = h/mv = h/p$
- The circumference of the nth orbit is equal to n times of wavelength of moving electron *i.e.*, $2\pi r_n = n\lambda$
  Number of waves $= n$ = principal quantum number
- Wavelength of electron $(\lambda) \cong \sqrt{\frac{150}{V(\text{volts})}}$ Å
- Wave nature of electron was confirmed by Davisson and Germer experiment.

Heisenberg Uncertainty Principle

- According to this principle, "it is impossible to measure simultaneously the position and momentum of a microscopic particle with absolute accuracy".
  If one of them is measured with greater accuracy, the other becomes less accurate.
- $\Delta x \cdot \Delta p \geq h/4\pi$ or $(\Delta x)(\Delta v) \geq \frac{h}{4\pi m}$ or $(\Delta x)(\Delta\lambda) \geq \frac{\lambda^2}{4\pi}$
  where Δx = Uncertainty in position, Δp = Uncertainty in momentum.
  Δv = Uncertainty in velocity, $\Delta\lambda$ = Uncertainty in wavelength.
  m = mass of microscopic particle, λ = Wavelength of microscopic particle
- Heisenberg replaced the concept of orbit by that of orbital.

Schrodinger Equation

- Schrodinger equation is central equation of wave mechanics according to following equation.

$$\frac{\partial^2\psi}{\partial x^2}+\frac{\partial^2\psi}{\partial y^2}+\frac{\partial^2\psi}{\partial z^2}+\frac{8\pi^2 m}{h^2}(E-V)\psi=0$$

ψ = Wave function = $f(x,y,z)$

E = Total energy of particle

V = Potential energy of particle.

- A Schrodinger equation leads to infinite solutions.
- Most of the solution are not realistic (or acceptable). Only few solution can be accepted.
- Each solution - $\psi(x,y,z)$ correspond to a definite energy state which depends on quantum number n, l & m.

  By proper mathematical manipulation the main equation is broken in two parts and solved separately.

  (i) Radial part contain only 'r', depends on quantum numbers n & l.

  (ii) Angular part contain θ and ϕ, depends on quantum numbers l & m.

  Each value of ψ contain all the information about that particular quantum state.
- **Atomic Orbital :** This is a three dimensional space around the nucleus within which the probability of finding the electron is maximum.
- **Degenerate Orbital :** Orbitals with same value of n and l of same subshell are degenerate orbitals, which have same energy.

  For example, $2p_x, 2p_y, 2p_z$ etc.
- **Radial Probability Density** $= 4\pi r^2 R^2(r)$

  It is the probability of finding electron in the region between r and $r + dr$
- **Radial Node :** It is zero electron density region. $R^2(r)=0$ or $R(r)=0$
- **Nodal Point :** It is a point $(r=0)$ where electron density is zero.
- **Nodal Planes :** It is plane by which two lobes are separated and electron density is zero here.

Quantum Number

There are four of quantum number which are required in order to explain e^- in atom.

- **Principal quantum number (n) :** It determines the size of an orbital. Each value of n represents a shell of orbital. Possible values of n = 1, 2, 3, 4,...........
- **Azimuthal quantum number (l) :** It determines shape of an orbital. Each value of l represents a subshell of an orbital. Possible values of l = 0, 1, 2, ... $(n-1)$
- **Magnetic quantum number (m) :** It decides orientation of orbital in space.

  Possible values of $m = -l, -l+1$, ... 0, 1, 2, l

  Total $(2l+1)$ values are possible form for each value of l.
- **Spin quantum number (s) :** It is intrinsic property of an electron. The electron has two spin states.

  Possible values of $s = +\frac{1}{2}, -\frac{1}{2}$

Shell, sub-shells and orbitals

Shell (n)	Sub-shells (l)	Orbitals (m)
1	0	0
2	0	0
	1	$\pm 1, 0$
3	0	0
	1	$\pm 1, 0$
	2	$\pm 2, \pm 1, 0$

Various Curves

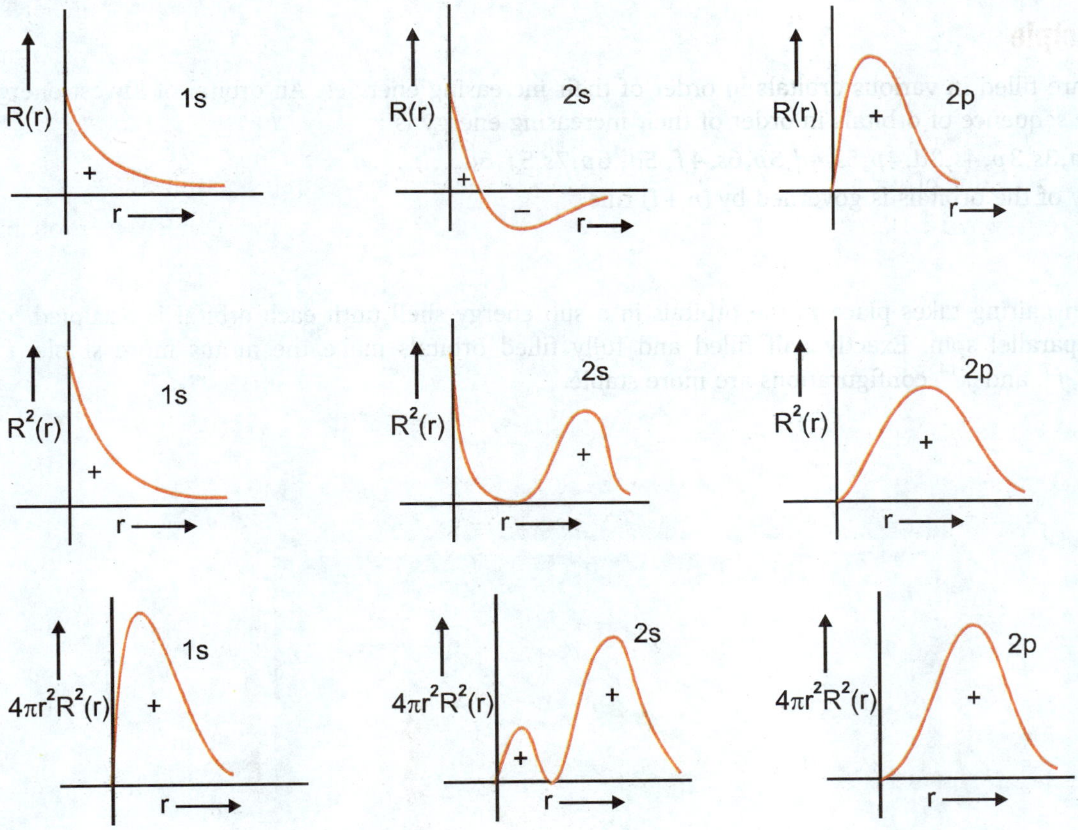

Important Points on Quantum Number

- Orbital angular momentum $= \dfrac{h}{2\pi}\sqrt{l(l+1)}$
- Spin angular momentum $= \dfrac{h}{2\pi}\sqrt{S(S+1)}$
- Spin Magnetic moment $(\mu) = \sqrt{n(n+2)}$ B.M. ; $n =$ number of unpaired electron
- Maximum number of electrons in a shell $= 2n^2$
- Maximum number of electrons in a subshell $= 2(2l+1)$
- Maximum number of electrons in an orbital $= 2$
- Total number of orbitals in a subshell $= 2l + 1$
- Number of subshells in a shell $= n$

- Number of orbitals in a shell $= n^2$
- Radial Nodes $= (n - l - 1)$
- Angular nodes $= l$
- Total nodes $= (n - 1)$
- Azimuthal quantum number / Name of sub-shell

Azimuthal quantum number	0	1	2	3	4
Name of sub-shell	s	p	d	f	g

Pauli's Exclusion Principle

No two electrons in an atom can have the same set of all the four quantum numbers, *i.e.*, an orbital cannot have more than 2 electrons because three quantum numbers (principal, azimuthal and magnetic) at the most may be same but the fourth will be different, *i.e.*, spins will be in opposite directions.

Aufbau Principle

Electrons are filled in various orbitals in order of their increasing energies. An orbital of lowest energy is filled first. The sequence of orbitals in order of their increasing energy is :

$1s, 2s, 2p, 3s, 3p, 4s, 3d, 4p, 5s, 4d, 5p, 6s, 4f, 5d, 6p, 7s, 5f, 6d, ...$

The energy of the orbitals is governed by $(n + l)$ rule.

Hund's Rule

No electron pairing takes place in the orbitals in a sub energy shell until each orbital is occupied by an electron with parallel spin. Exactly half filled and fully filled orbitals make the atoms more stable, *i.e.*, $p^3, p^6, d^5, d^{10}, f^7$ and f^{14} configurations are more stable.

Level-1

1. Among the following statements, that which was not proposed by Dalton was :
   (a) All the atoms of a given element have identical properties including identical mass. Atoms of different elements differ in mass.
   (b) When gases combine or reproduced in a chemical reaction they do so in a simple ratio by volume provided all gases are at the same T & P.
   (c) Chemical reactions involve reorganization of atoms. These are neigher created nor destroyed in a chemical reaction.
   (d) Matter consists of indivisible atoms.
2. Which of the following does not characterise X-rays?
   (a) The radiation can ionise the gas
   (b) It causes fluorescence effect on ZnS
   (c) It is deflected by electric and magnetic fields
   (d) Its wavelength is shorter than ultraviolet rays
3. The increasing order for the values of e/m (charge/mass) is :
   (a) e, p, n, α (b) n, p, e, α
   (c) n, p, α, e (d) n, α, p, e
4. α-particles are represented by:
   (a) Lithium atoms (b) Helium nuclei
   (c) Hydrogen nuclei (d) None of these
5. Rutherford's experiment on scattering of alpha particles showed for the first time that atom has :
   (a) Electrons (b) Protons
   (c) Nucleus (d) Neutrons
6. The volume of nucleus is about :
   (a) 10^{-4} times to that of an atom (b) 10^{-15} times to that of an atom
   (c) 10^{-5} times to that of an atom (d) 10^{-10} times to that of an atom
7. Which of the following statement does not a part of Bohr's model of the hydrogen atom ?
   (a) Angular momentum of the electrons in the orbit is quantized
   (b) The electron in the orbit nearest the nucleus has the lowest energy
   (c) Electrons revolve in different orbits around the nucleus
   (d) The position and velocity of the electrons in the orbit cannot be determined simultaneously
8. In Bohr's stationary orbits:
   (a) Electrons do not move
   (b) Electrons move emitting radiations
   (c) Energy of the electron remains constant
   (d) Angular momentum of the electron is $h/2\pi$
9. Permissible value of orbit angular momentum of electron in a H-like atom will be:
   (a) $1.5\left(\frac{h}{\pi}\right)$ (b) $\frac{h}{4\pi}$
   (c) $\frac{h}{1.5\pi}$ (d) $\frac{h}{3\pi}$
10. On the basis of Bohr's model, the radius of the 3rd orbit is:
    (a) Equal to the radius of first orbit (b) Three times the radius of first orbit
    (c) Five times the radius of first orbit (d) Nine times the radius of first orbit

11. If radius of second stationary orbit (in Bohr's atom) is R. Then radius of third orbit will be :
(a) $R/3$ (b) $9R$
(c) $R/9$ (d) $2.25\,R$

12. Which state of Be^{3+} has the same orbit radius as that of the ground state of hydrogen atom?
(a) 3 (b) 2
(c) 4 (d) 5

13. What atomic number of an element "X" would have to become so that the 4th orbit around X would equal to the 1st Bohr orbit of H atom ?
(a) 3 (b) 4
(c) 16 (d) 25

14. The distance between 4th and 3rd Bohr orbits of He^{+} is :
(a) 2.645×10^{-10} m (b) 1.322×10^{-10} m
(c) 1.851×10^{-10} m (d) None

15. The velocity of the electron (in m/s) in the fifth orbit of Li^{2+} would be :
(a) 2.188×10^{6} (b) $2.188\times10^{6}\left(\frac{3}{5}\right)$
(c) $2.188\times10^{6}\left(\frac{9}{5}\right)$ (d) $2.188\times10^{6}\left(\frac{5}{3}\right)$

16. Select the incorrect graph for velocity of e^- in an orbit *vs.* Z, $\frac{1}{n}$ and n :

(a)
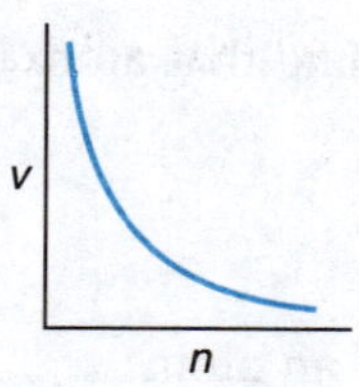

(b)
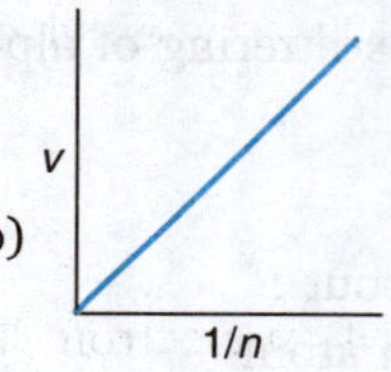

(c)
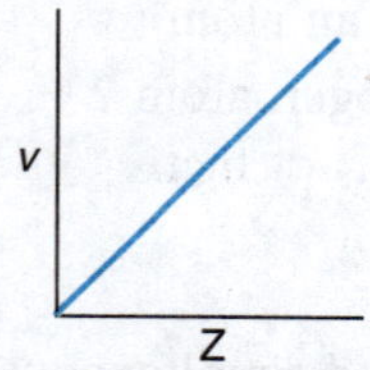

(d)
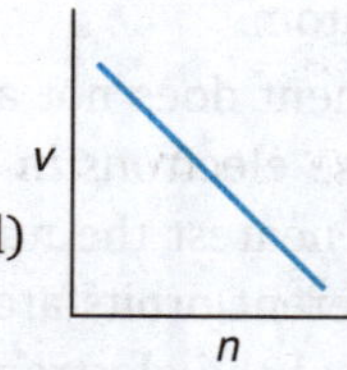

17. For any H like system, the ratio of velocities of I, II & III orbit *i.e.*, $v_1 : v_2 : v_3$ will be:
(a) 1 : 2 : 3 (b) 1 : 1/2 : 1/3
(c) 3 : 2 : 1 (d) 1 : 1 : 1

18. If ε_0 be the permittivity of vacuum and r be the radius of orbit of H -atom in which electron is revolving, then velocity of electron is given by :
(a) $v=\frac{e}{\sqrt{4\pi\varepsilon_0 rm}}$ (b) $v=e\times\sqrt{4\pi\varepsilon_0 rm}$
(c) $v=\frac{4\pi\varepsilon_0 rm}{e}$ (d) $v=\frac{4\pi\varepsilon_0 rm}{e^2}$

19. What is the frequency of revolution of electron present in 2nd Bohr's orbit of H-atom?
(a) $1.016\times10^{16}\ s^{-1}$ (b) $4.065\times10^{16}\ s^{-1}$
(c) $1.626\times10^{15}\ s^{-1}$ (d) $8.2\times10^{14}\ s^{-1}$

20. What is the ratio of time periods (T_1/T_2) in second orbit of hydrogen atom to third orbit of He^+ ion?

(a) 8/27 (b) 32/27

(c) 27/32 (d) None of these

21. If in Bohr's model, for unielectronic atom, time period of revolution is represented as $T_{n,Z}$ where n represents shell no. and Z represents atomic number then the value of $T_{1,2} : T_{2,1}$ will be :

(a) 8 : 1 (b) 1 : 8

(c) 1 : 1 (d) 1 : 32

22. According to Bohr's atomic theory, which of the following is correct ?

(a) Potential energy of electron $\propto \frac{Z^2}{n^2}$

(b) The product of velocity of electron and principle quantum number (n) $\propto Z^2$

(c) Frequency of revolution of electron in an orbit $\propto \frac{Z^2}{n^3}$

(d) Coulombic force of attraction on the electron $\propto \frac{Z^2}{n^2}$

23. The kinetic energy of an electron in an orbit of radius r of Li^{2+} ion is (e = electronic charge) :

(a) $\frac{e^2}{r}$ (b) $\frac{e^2}{3r}$

(c) $\frac{3e^2}{r}$ (d) $\frac{3e^2}{2r}$

24. The potential energy of an electron in the hydrogen atom is –6.8 eV. Indicate in which excited state, the electron is present?

(a) first (b) second

(c) third (d) fourth

25. What is the potential energy of an electron present in N-shell of the Be^{3+} ion?

(a) –3.4 eV (b) –6.8 eV

(c) –13.6 eV (d) –27.2 eV

26. The kinetic and potential energy (in eV) of electron present in third Bohr's orbit of hydrogen atom are respectively :

(a) –1.51, – 3.02 (b) 1.51, – 3.02

(c) –3.02, 1.51 (d) 1.51, – 1.51

27. An electron in an atom jumps in such a way that its kinetic energy changes from x to $\frac{x}{4}$. The change in potential energy will be :

(a) $+\frac{3}{2}x$ (b) $-\frac{3}{8}x$

(c) $+\frac{3}{4}x$ (d) $-\frac{3}{4}x$

28. The correct expression derived for the total energy of an electron in the nth energy level is for H-atom:

(a) $E_n = \frac{2\pi^2 me^4 K^2}{n^2h^2}$ (b) $E_n = -\frac{\pi^2 me^4 K^2}{2n^2h^2}$

(c) $E_n = -\frac{2\pi^2 me^2 K^2}{n^2h^2}$ (d) $E_n = -\frac{2\pi^2 me^4 K^2}{n^2h^2}$

29. Ionization energy for hydrogen atom in joules and eV respectively is:

(a) 218×10^{-20}, 13.6
(b) 218×10^{-20}, 21.8×10^{-13}
(c) 13.6, 21.8×10^{-13}
(d) 13.6, 21.8×10^{-20}

30. If the ionization energy of He^+ is 19.6×10^{-18} J per atom then the energy of Be^{3+} ion in the second stationary state is :

(a) -4.9×10^{-18} J
(b) -44.1×10^{-18} J
(c) -11.025×10^{-18} J
(d) None of these

31. The ionization potential for the electron in the ground state of the hydrogen atom is 13.6 eV atom^{-1}. What would be the ionization potential for the electron in the first excited state of Li^{2+} ?

(a) 3.4 eV
(b) 10.2 eV
(c) 30.6 eV
(d) 6.8 eV

32. What is the separation energy (in eV) for Be^{3+} in the first excited state in eV?

(a) 13.6 eV
(b) 27.2 eV
(c) 40.8 eV
(d) 54.5 eV

33. The ratio of energy of the electron in ground state of H-atom to the electron in third energy state of Be^{3+} is:

(a) 4 : 1
(b) 9 : 4
(c) 1 : 8
(d) 9 : 16

34. If an electron jump from 3rd orbit to 2nd orbit in Li^{2+} then calculate change in energy of electron will be:

(a) 5.4eV
(b) 17eV
(c) 1.8eV
(d) 15eV

35. The energy of the second Bohr orbit in the hydrogen atom is –3.41 eV. The energy of the second Bohr orbit of He^+ ion would be :

(a) –0.85 eV
(b) –13.64 eV
(c) –1.70 eV
(d) –6.82 eV

36. The energy of an electron moving in n^{th} Bohr's orbit of an element is given by $E_n = \frac{-13.6}{n^2} Z^2$ eV/ atom (Z = atomic number). The graph of E vs. Z^2 (keeping "n" constant) will be :

(a)
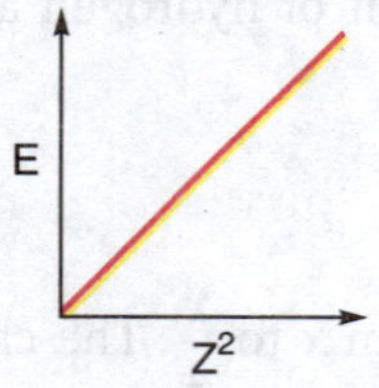

(b)
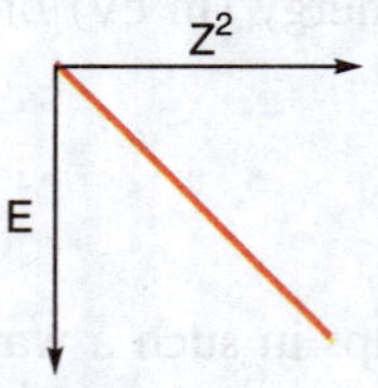

(c)
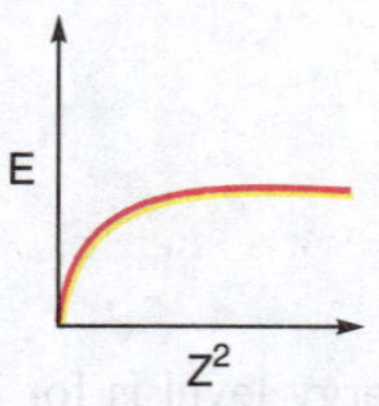

(d)
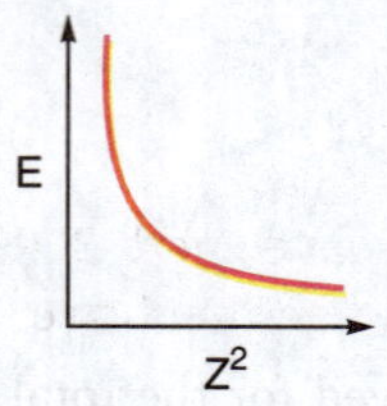

37. Which of the following statement(s) is/are consistent with the Bohr theory of the atom (and no others)?

(1) An electron can remain in a particular orbit as long as it continuously absorbs radiation of a definite frequency.
(2) The lowest energy orbits are those closest to the nucleus.
(3) All electrons can jump from the K shell to the M shell by emitting radiation of a definite frequency.

(a) 1, 2, 3
(b) 2 only
(c) 3 only
(d) 1, 2

38. The mass of an electron is m, charge is e and it is accelerated from rest through a potential difference of V volts. The velocity acquired by electron will be :

(a) $\sqrt{\dfrac{V}{m}}$
(b) $\sqrt{\dfrac{eV}{m}}$
(c) $\sqrt{\dfrac{2eV}{m}}$
(d) zero

39. Which of the options is **not correct** regarding the order of frequency of electromagnetic radiation?

(a) Radio waves < Microwaves < X-rays
(b) Green light < U-V light < γ-radiations
(c) Infrared > Radio waves > γ-radiations
(d) Microwaves < Infrared rays < U-V rays

40. What is the energy content per photon (J) for light of frequency 4.2×10^{14} Hz?

(a) 2.8×10^{-21}
(b) 2.5×10^{-19}
(c) 2.8×10^{-19}
(d) 2.5×10^{-18}

41. Wavelength for high energy EMR transition in H-atom is 91 nm. What energy is needed for this transition?

(a) 1.36 eV
(b) 1240 eV
(c) 13 eV
(d) 13.6 eV

42. Assume that 2×10^{-17} J of light energy is needed by the interior of the human eye to see an object. How many photons of yellow light with $\lambda = 595.2$ nm are needed to generate this minimum energy?

(a) 6
(b) 30
(c) 45
(d) 60

43. Which graph shows how the energy E of a photon of light is related to its wavelengths (λ) ?

(a)
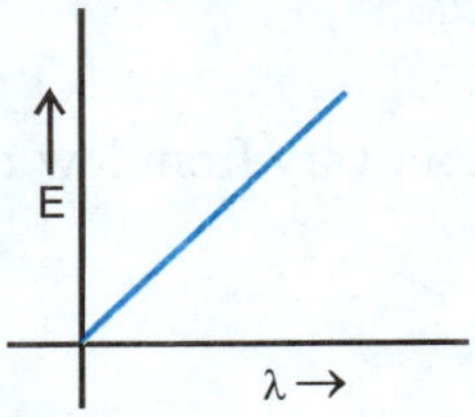

(b)
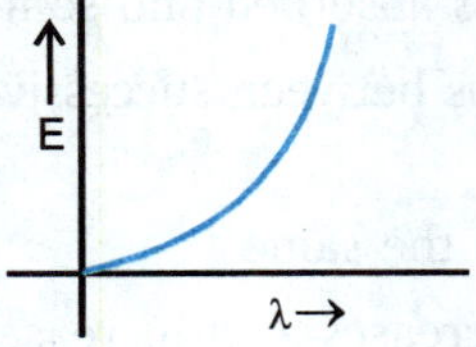

(c)
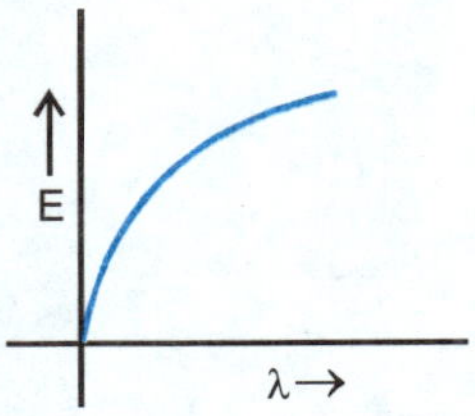

(d)
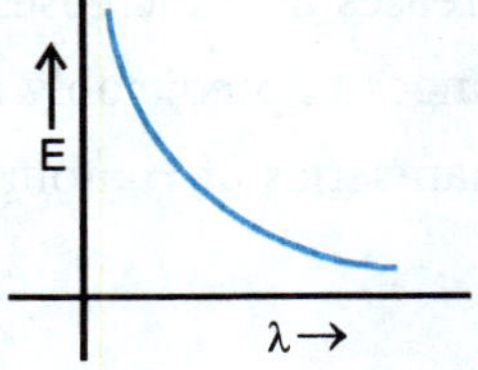

44. The energy of a radiation of wavelength 8000 Å is E_1 and energy of a radiation of wavelength 16000 Å is E_2. What is the relation between these two?

(a) $E_1 = 6E_2$
(b) $E_1 = 2E_2$
(c) $E_1 = 4E_2$
(d) $E_1 = 1/2E_2$

45. Two electromagnetic radiations have wave numbers in the ratio 2:3. Their energies per quanta will be in the ratio:

(a) 3 : 2
(b) 9 : 4
(c) 4 : 9
(d) 2 : 3

46. The number of photons of light having wave number 'x' in 10 J of energy source is :

(a) $10hcx$ (b) $\frac{hc}{10x}$

(c) $\frac{10}{hcx}$ (d) None of these

47. Line spectra is characteristic of :

(a) molecules (b) atoms

(c) radicals (d) none of these

48. The spectrum produced from an element is :

(a) atomic spectrum (b) line spectrum

(c) absorption spectrum (d) any one of the above

49. The H-spectrum show:

(a) Heisenberg's uncertainty principle (b) Diffraction

(c) Polarization (d) Presence of quantized energy level

50. Multiple of fine structure of spectral lines is due to:

(a) Presence of main energy levels (b) Presence of sub-levels

(c) Presence of electronic configuration (d) Is not a characteristics of the atom

51. Splitting of spectral lines under the influence of magnetic field is called:

(a) Zeeman effect (b) Stark effect

(c) Photoelectric effect (d) None of these

52. When an electron jumps from L to K shell :

(a) Energy is absorbed

(b) Energy is released

(c) Energy is neither absorbed nor released

(d) Energy is sometimes absorbed and some times released

53. How do the energy gaps between successive electron energy levels in an atom vary from low to high n values?

(a) All energy gaps are the same

(b) The energy gap decreases as n increases

(c) The energy gap increases as n increases

(d) The energy gap changes unpredictably as n increases

54. Spectrum region of lyman series of H-atom is :

(a) Infrared (b) Visible

(c) Ultra violet (d) None of these

55. What is the energy (kJ/mol) associated with the de-excitation of an electron from $n = 6$ to $n = 2$ in He^+ ion?

(a) 1.36×10^6 (b) 1.36×10^3

(c) 1.16×10^3 (d) 1.78×10^3

56. Which of the following electron transitions in a hydrogen atom will require the largest amount of energy?

(a) from $n = 1$ to $n = 2$ (b) from $n = 2$ to $n = 4$

(c) from $n = 5$ to $n = 1$ (d) from $n = 3$ to $n = 5$

57. What is the shortest wavelength line in the Paschen series of Li^{2+} ion?

(a) $\frac{R}{9}$
(b) $\frac{9}{R}$
(c) $\frac{1}{R}$
(d) $\frac{9R}{4}$

58. What is the maximum wavelength line in the Lyman series of He^+ ion?

(a) $3R$
(b) $\frac{1}{3R}$
(c) $\frac{4}{4R}$
(d) None of these

59. Which electronic transition in a hydrogen atom, starting from the orbit $n=7$, will produce infrared light of wavelength 2170 nm? (Given : $R_H = 1.09677 \times 10^7\ m^{-1}$)

(a) $n=7$ to $n=6$
(b) $n=7$ to $n=5$
(c) $n=7$ to $n=4$
(d) $n=7$ to $n=3$

60. Hydrogen atoms in the ground state are excited by monochromatic radiation of wavelength λ Å. The resulting spectrum consists of maximum 15 different lines. What is the wavelength λ? ($R_H = 109737\ cm^{-1}$)

(a) 937.3 Å
(b) 1025 Å
(c) 1236 Å
(d) None of these

61. Electromagnetic radiation (photon) with highest wavelength results when an electron in the hydrogen atom falls from $n=6$ to :

(a) $n=1$
(b) $n=2$
(c) $n=3$
(d) $n=5$

62. What is the wavelength (in nm) of the spectral line associated with a transition from $n=3$ to $n=2$ for the Li^{2+} ion?

(a) 219
(b) 656
(c) 73.0
(d) 486

63. Find the value of wave number ($\bar{\nu}$) in terms of Rydberg's constant, when transition of electron takes place between two levels of He^+ ion whose sum is 4 and difference is 2.

(a) $\frac{8R}{9}$
(b) $\frac{32R}{9}$
(c) $\frac{3R}{4}$
(d) none of these

64. Which of the following expressions represents the spectrum of Balmer series (If n is the principal quantum number of higher energy level) in Hydrogen atom ?

(a) $\bar{\nu} = \frac{R(n-1)(n+1)}{n^2}$
(b) $\bar{\nu} = \frac{R(n-2)(n+2)}{4n^2}$
(c) $\bar{\nu} = \frac{R(n-2)(n+2)}{n^2}$
(d) $\bar{\nu} = \frac{R(n-1)(n+1)}{4n^2}$

65. Maximum number of radiations of different wavelength are emitted when electrons transfer takes place from fourth energy state to ground state :

(a) 3
(b) 4
(c) 6
(d) 10

66. Electronic transition in He^+ ion takes from n_2 to n_1 shell such that :

$$2n_2 + 3n_1 = 18$$
$$2n_2 - 3n_1 = 6$$

What will be the total number of photons emitted when electrons transit to n_1 shell ?

(a) 21 (b) 15

(c) 20 (d) 10

67. The photoelectric emission from a surface starts only when the light incident upon the surface has certain minimum :

(a) intensity (b) wavelength

(c) frequency (d) velocity

68. In photoelectric effect, the kinetic energy of photoelectrons increases with the

(a) Wavelength of incident light (b) Frequency of incident light

(c) Velocity of incident light (d) Atomic mass of an element

69. According to Einstein's photoelectric equation, the graph between kinetic energy of photoelectrons ejected and the frequency of the incident radiation is :

(a) 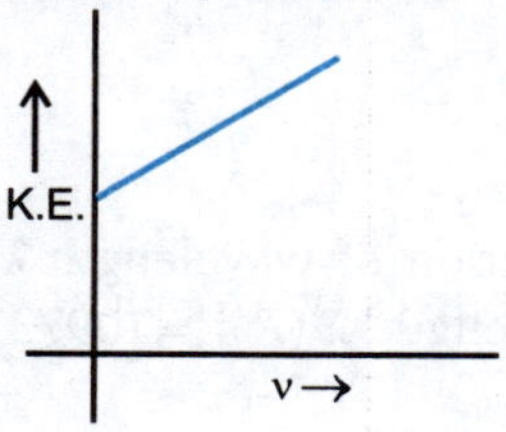

(b) 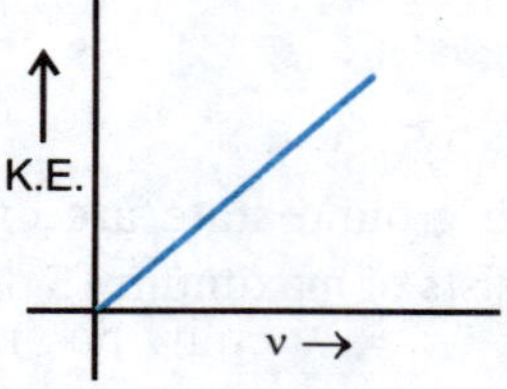

(c) 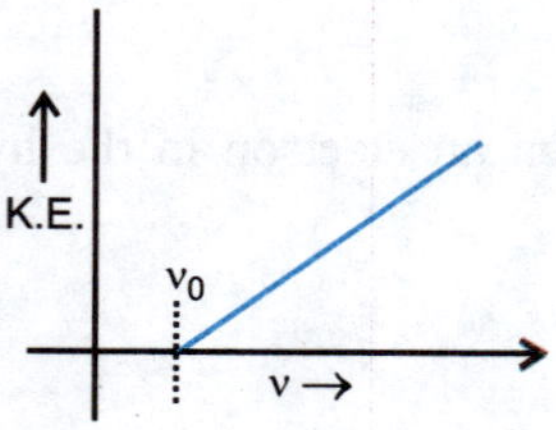

(d) 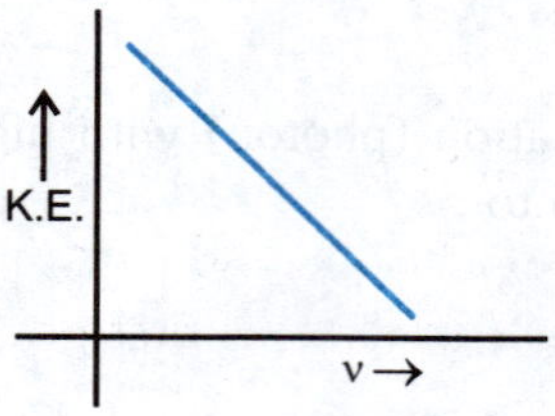

70. If λ_0 and λ be the threshold wavelength and the wavelength of incident light, the velocity of photo-electrons ejected from the metal surface is :

(a) $\sqrt{\frac{2h}{m}(\lambda_0 - \lambda)}$ (b) $\sqrt{\frac{2hc}{m}(\lambda_0 - \lambda)}$

(c) $\sqrt{\frac{2hc}{m}\left(\frac{\lambda_0 - \lambda}{\lambda\lambda_0}\right)}$ (d) $\sqrt{\frac{2h}{m}\left(\frac{1}{\lambda_0} - \frac{1}{\lambda}\right)}$

71. A photon is emitted from a source having energy 10.2 eV. This photon is absorb by a metal of work function 8.7 eV. Calculate kinetic energy of photoelectron (in eV).

(a) 2.5 (b) 1.5

(c) 3.5 (d) 3

72. If the work function of a metal is 6.63×10^{-19} J, the maximum wavelength of the photon required to remove a photoelectron from the metal is .... nm.

[Given : $h = 6.63 \times 10^{-34}$ Js and $c = 3 \times 10^8$ ms^{-1}]

(a) 30 (b) 300

(c) 3000 (d) 1000

73. Select the incorrect statement :

(a) K.E. of photo-electron depends upon the wavelength of incident radiation

(b) Photoelectric current depends on intensity of incident radiation

(c) K.E. of photo-electron depends on frequency of radiation

(d) The value of work function not depends on type of metal

74. A light source of wavelength λ illuminates a metal and ejects photo-electrons with $(K.E.)_{max} = 1\,eV$ Another light source of wavelength $\frac{\lambda}{3}$, ejects photo-electrons from same metal with $(K.E.)_{max} = 4\,eV$

Find the value of work function?

(a) 1 eV (b) 2 eV

(c) 0.5 eV (d) None of these

75. Electromagnetic radiation having $\lambda = 310$ Å is subjected to a metal sheet having work function = 12.8 eV. What will be the velocity of photo-electrons having maximum kinetic energy.

(a) 0, no emission will occur (b) 4.352×10^6 m/s

(c) 3.09×10^6 m/s (d) 8.72×10^6 m/s

76. Slope of V_0 *vs* ν curve is (where V_0 = Stopping potential, ν = subjected frequency)

(a) e (b) $\frac{h}{e}$

(c) ϕ (d) h

77. The ratio of slopes of K_{max} *vs.* ν and V_0 *vs.* ν curves in the photoelectric effect gives (ν = frequency, K_{max} = maximum kinetic energy, V_0 = stopping potential) :

(a) charge of electron

(b) Planck's constant

(c) work function

(d) the ratio of Planck's constant and electronic charge

78. The wave nature of an electron was first given by:

(a) de-Broglie (b) Heisenberg

(c) Moseley (d) Sommerfeld

79. Which is the de-Broglie equation ?

(a) $h = p\lambda$ (b) $h = p\lambda^{-1}$

(c) $h = \lambda p^{-1}$ (d) $h = p + \lambda$

80. The de-Broglie wavelength of a particle with mass 1 g and velocity 100 m/sec is:

(a) 6.63×10^{-33} m (b) 6.63×10^{-34} m

(c) 6.63×10^{-35} m (d) 6.65×10^{-35} m

81. Which of the following has the largest de-Broglie wavelength (all have equal velocity)?

(a) CO_2 molecule (b) NH_3 molecule

(c) Electron (d) Proton

82. The de-Broglie wavelength associated with a particle of mass 10^{-6} kg with a velocity of 10 m s^{-1} is :

(a) 6.63×10^{-22} m (b) 6.63×10^{-29} m

(c) 6.63×10^{-31} m (d) 6.63×10^{-34} m

83. For a diffraction experiment, neutron (mass = 1.67×10^{-27} kg) with a wavelength of 0.150 nm are needed. What velocity must these neutrons have?

(a) 2.64×10^3 m/s (b) 3.0×10^4 m/s

(c) 8×10^3 m/s (d) None of these

84. Which of following graphs correctly represents the variation of particles momentum with de-Broglie wavelength ?

(a) 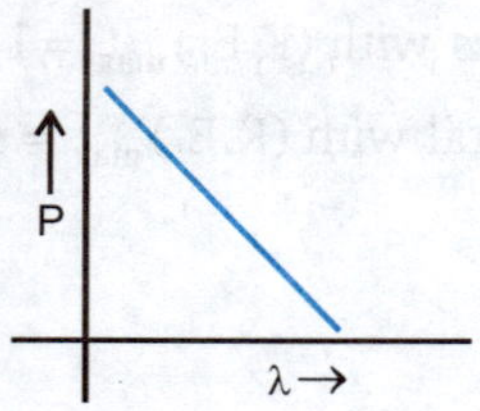

(b) 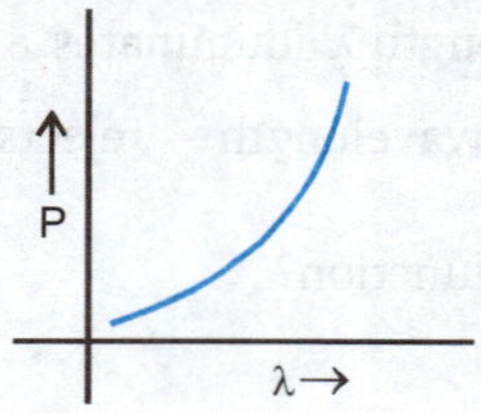

(c) 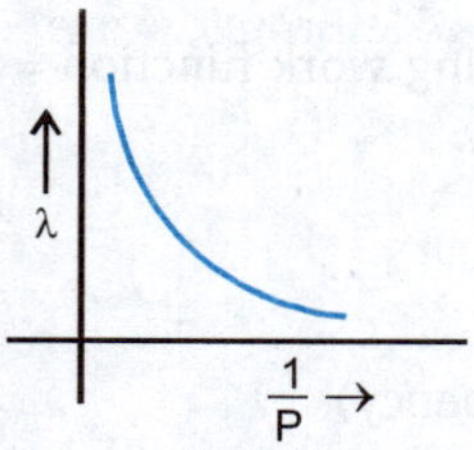

(d) 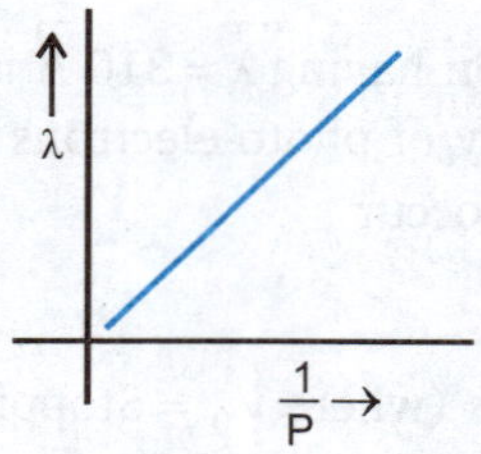

85. For two particles A and B, curves are plotted $\sqrt{V}$ against de-Broglie wavelengths, where V is the potential on the particles. Which of the following relation is correct about the mass of particles?

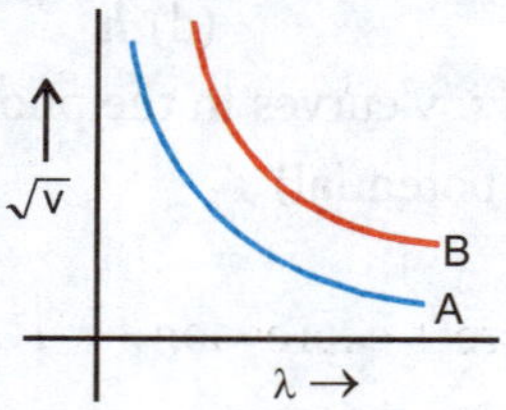

(a) $m_A = m_B$ (b) $m_A > m_B$
(c) $m_A < m_B$ (d) $m_A \le m_B$

86. An excited state of H atom emits a photon of wavelength λ and returns in the ground state. The principal quantum number of excited state is given by :

(a) $\sqrt{\lambda R(\lambda R - 1)}$ (b) $\sqrt{\frac{\lambda R}{(\lambda R - 1)}}$
(c) $\sqrt{\lambda R(\lambda R - 1)}$ (d) $\sqrt{\frac{(\lambda R - 1)}{\lambda R}}$

87. A dye absorbs a photon of wavelength λ and re-emits the same energy into two photons of wavelengths λ_1 and λ_2 respectively. The wavelength λ is related with λ_1 and λ_2 as :

(a) $\lambda = \frac{\lambda_1 + \lambda_2}{\lambda_1 \lambda_2}$ (b) $\lambda = \frac{\lambda_1 \lambda_2}{\lambda_1 + \lambda_2}$
(c) $\lambda = \frac{\lambda_1^2 \lambda_2^2}{\lambda_1 + \lambda_2}$ (d) $\lambda = \frac{\lambda_1 \lambda_2}{(\lambda_1 + \lambda_2)^2}$

88. An electron absorb a photon of wavelength 400 nm and excite to excited state then it emitts 2 photons of wavelength λ_1 and λ_2. If $\lambda_1 = 4\lambda_2$ then the value of λ_1 and λ_2 are:

(a) $\lambda_1 = 2000$ nm, $\lambda_2 = 500$ nm (b) $\lambda_1 = 320$ nm, $\lambda_2 = 80$ nm
(c) $\lambda_1 = 200$ nm, $\lambda_2 = 50$ nm (d) $\lambda_1 = 32$ nm, $\lambda_2 = 8$ nm

89. Be^{3+} and a proton are accelerated by the same potential, their de-Broglie wavelengths have the ratio (assume mass of proton = mass of neutron) :

(a) 1 : 2 (b) 1 : 4
(c) 1 : 1 (d) $1 : 3\sqrt{3}$

90. The de-Broglie wavelength of an electron accelerated by an electric field of V volts is given by :

(a) $\lambda = \frac{1.23}{\sqrt{m}}$ (b) $\lambda = \frac{1.23}{\sqrt{h}} m$

(c) $\lambda = \frac{1.23}{\sqrt{V}}$ nm (d) $\lambda = \frac{1.23}{V}$

91. An electron travels with a velocity of x ms^{-1}. For a proton to have the same de-Broglie wavelength, the velocity will be approximately :

(a) $\frac{1840}{x}$ (b) $\frac{x}{1840}$

(c) $1840\,x$ (d) x

92. The momentum (in kg-m/s) of photon having 6 MeV energy is :

(a) 3.2×10^{-21} (b) 2.0

(c) 1.6×10^{-21} (d) None of these

93. Which of the following relates to photons both as wave motion and as a stream of particles?

(a) Interference (b) $E = mc^2$

(c) Diffraction (d) $E = h\nu$

94. Which of the following is the most correct expression for Heisenberg's uncertainty principle ?

(a) $\Delta x . \Delta p = \frac{h}{4\pi}$ (b) $\Delta x . \Delta p \geq \frac{h}{4\pi}$

(c) $\Delta x . \Delta p \leq \frac{h}{4\pi}$ (d) $\Delta x . \Delta v = \frac{h}{4\pi}$

95. The Heisenberg uncertainty principle can be applied to :

(a) A cricket ball (b) A foot ball

(c) A jet aeroplane (d) An electron

96. If uncertainty in momentum and position for a moving particle of mass m is same than calculate uncertainty in its velocity (h = Planck's constant) :

(a) $\frac{1}{2m}\sqrt{\frac{h}{\pi}}$ (b) $\frac{1}{2\pi}\sqrt{\frac{h}{m}}$

(c) $\frac{1}{m}\sqrt{\frac{h}{2\pi}}$ (d) $\frac{1}{\pi}\sqrt{\frac{h}{2m}}$

97. If an electron is travelling at 200 m/s within 1 m/s uncertainty, what is the theoretical uncertainty in its position in μm (micrometer)?

(a) 14.5 (b) 29

(c) 58 (d) 114

98. An accelerated electron has a speed of 5×10^6 ms^{-1} with an uncertainty of 0.02%. The uncertainty in finding its location while in motion is $y \times 10^{-9}$ m. The value of y is :

[Use mass of electron $= 9.1 \times 10^{-31}$ kg, $h = 6.63 \times 10^{-34}$ Js, $\pi = 3.14$]

(a) 38 (b) 48

(c) 58 (d) 68

99. The mass of a particle is 10^{-10} g and its radius is 2×10^{-4} cm. If its velocity is 10^{-6} cm sec^{-1} with 0.0001% uncertainty in measurement, the uncertainty in its position is :

(a) 5.2×10^{-8} m (b) 5.2×10^{-7} m

(c) 5.2×10^{-6} m (d) 5.2×10^{-9}

100. "The exact path of electron in 2*p*-orbital cannot be determined." The above statement is based upon

(a) Hund's Rule (b) Bohr's Rule

(c) Uncertainty principle (d) Aufbau principle

101. Principal, azimuthal and magnetic quantum numbers are respectively related to:

(a) size, shape and orientation of orbital (b) shape, size and orientation of orbital

(c) size, orientation and shape of orbital (d) none of the above

102. Azimuthal quantum number defines:

(a) e/m ratio of electron (b) spin of electron

(c) angular momentum of electron (d) magnetic momentum of electron

103. Which of the following statements concerning the four quantum numbers is false ?

(a) n gives idea of the size of an orbital

(b) l gives the shape of an orbital

(c) m gives the energy of the electron in the orbital

(d) s gives the direction of spin of the electron in an orbital

104. If the value of azimuthal quantum number is 3, the possible values of magnetic quantum number would be:

(a) 0, 1, 2, 3 (b) $0, -1, -2, -3$

(c) $0, \pm 1, \pm 2, \pm 3$ (d) $\pm 1, \pm 2, \pm 3$

105. Which of the following set of quantum numbers is impossible for an electron?

(a) $n=1, l=0, m_l=0, m_s=+\frac{1}{2}$ (b) $n=9, l=7, m_l=-6, m_s=-\frac{1}{2}$

(c) $n=2, l=1, m_l=0, m_s=+\frac{1}{2}$ (d) $n=3, l=2, m_l=-3, m_s=+\frac{1}{2}$

106. Which of the following sets of quantum numbers is possible?

(a) $n=3;\ l=2;\ m=2$ and $s=+\frac{1}{2}$ (b) $n=3;\ l=4;\ m=0$ and $s=-\frac{1}{2}$

(c) $n=4;\ l=0;\ m=2$ and $s=+\frac{1}{2}$ (d) $n=4;\ l=4;\ m=3$ and $s=+\frac{1}{2}$

107. Which of the following orbitals is not possible ?

(a) $3f$ (b) $4f$

(c) $5f$ (d) $6f$

108. What is the maximum number of electrons which can be accommodated in a shell for which the highest principal quantum number value is 4 ?

(a) 10 (b) 18

(c) 32 (d) 54

109. A subshell $n=5, l=3$ can accommodate :

(a) 10 electrons (b) 14 electrons

(c) 18 electrons (d) None of these

110. For azimuthal quantum number $l=3$, the maximum number of electrons will be:

(a) 2 (b) 6

(c) 0 (d) 14

111. The subshell that arises after f subshell is called g subshell.

What is the total number of orbitals in the shell in which the g subshell first occur?

(a) 9 (b) 16

(c) 25 (d) 36

112. The subshell that arises after f is called the g subshell. How many electrons may occupy the g subshell?
(a) 9 (b) 7
(c) 5 (d) 18

113. How many orbital(s) in an atom can have $n = 3, l = 2$?
(a) 1 (b) 3
(c) 5 (d) 7

114. How many electrons in an atom can have $n = 4, l = 2, m = -2$ and $s = +\frac{1}{2}$?
(a) 1 (b) 2
(c) 5 (d) 10

115. Consider the argon atom. For how many electrons does this atom have $m_l = 1$?
(a) 1 (b) 6
(c) 4 (d) 2

116. An orbital is occupied by an electron with the quantum numbers $n = 4, l = 1$. How many orbitals of this type are found in a multi-electron atom?
(a) $4p, 3$ (b) $4s, 1$
(c) $4d, 5$ (d) $4p, 6$

117. The set of quantum numbers, $n = 2$, $l = 2, m_l = 0$:
(a) describes an electron in a 2s orbital
(b) describes one of the five orbitals of a similar type
(c) describes an electron in a 2p orbital
(d) is not allowed

118. The set of quantum numbers, $n = 3, l = 2, m_l = 0$
(a) describes an electron in a 2s orbital
(b) is not allowed
(c) describes an electron in a 3p orbital
(d) describes one of the five orbitals of same energy

119. What is the maximum number of electrons in an atom that can have the quantum numbers $n = 3$ and $l = 2$?
(a) 2 (b) 5
(c) 6 (d) 10

120. Which of the following statements about an electron with $m_l = +2$ is incorrect?
(a) The electron could be in the third shell (b) The electron is in a non-spherical orbital
(c) The electron may have $m_s = \frac{1}{2}$ (d) The electron is not in a d-orbital

121. Maximum number of electrons in a subshell is given by:
(a) $(2l + 1)$ (b) $2(2l + 1)$
(c) $(2l + 1)^2$ (d) $2(2l + 1)^2$

122. In any subshell, the maximum number of electrons having same value of spin quantum number is :
(a) $\sqrt{l(l + 1)}$ (b) $l + 2$
(c) $2l + 1$ (d) $4l + 2$

123. If n and l are principal and azimuthal quantum numbers respectively, then the expression for calculating the total number of electrons in any energy level is :

(a) $\sum_{l=0}^{l=n} 2(2l+1)$
(b) $\sum_{l=1}^{l=n} 2(2l+1)$
(c) $\sum_{l=0}^{l=n} (2l+1)$
(d) $\sum_{l=0}^{l=n-1} 2(2l+1)$

124. When the azimuthal quantum number has a value of $l = 1$, the shape of the orbital is:
(a) unsymmetrical
(b) spherically symmetrical
(c) dumb-bell
(d) complicated

125. The orbital angular momentum of $3p$ electron is :
(a) $\sqrt{3}\, h$
(b) $\sqrt{6}\, h$
(c) zero
(d) $\sqrt{2}\, \frac{h}{2\pi}$

126. Read the following statements and choose the correct option:
(I) For f-orbital electron, the orbital angular momentum is $\sqrt{3} \cdot \frac{h}{\pi}$.
(II) For s-orbital electron, the orbital angular momentum is zero.
(a) only I is correct
(b) only II is correct
(c) both are correct
(d) both are incorrect

127. Which two orbitals are located along the axis, and not between the axis?
(a) d_{xy}, d_{z^2}
(b) d_{xy}, p_z
(c) d_{yz}, p_x
(d) $p_z, d_{x^2-y^2}$

128. The probability of finding electrons in d_{xy} orbital is :
(a) along X- and Y-axis
(b) along X- and Z-axis
(c) along Y- and Z-axis
(d) at an angle of 45° with X-axis

129. Which of the following statements is correct for an electron having azimuthal quantum number $l = 2$?
(a) The electron may be in the lowest energy shell.
(b) The electron is in a spherical orbital.
(c) The electron must have spin $m_s = +\frac{1}{2}$
(d) The electron may have a magnetic quantum number $= -1$

130. For $4p_y$ orbital: There are
nodal plane = ... and azimuthal quantum number l = ...
(a) 1,0
(b) 0,1
(c) 1,1
(d) 2,1

131. Which of the following statement is correct?
(a) Number of angular nodes $= n - l - 1$
(b) Number of radial nodes $= l$
(c) Total number of nodes $= n - 1$
(d) All of these

132. Which of the following orbitals has two spherical nodes?
(a) $2s$
(b) $4s$
(c) $3d$
(d) $6f$

133. How many radial nodes are present in $4d$-orbital ?

(a) 0 (b) 1

(c) 2 (d) 3

134. How many angular node(s) are present in $3s$-orbital ?

(a) 0 (b) 1

(c) 2 (d) 3

135. The correct order of total number of node of atomic orbitals is:

(a) $4f > 6s > 5d$ (b) $6s > 5d > 4f$

(c) $4f > 5d > 6s$ (d) $5d > 4f > 6s$

136. In a $3d$ subshell, all the five orbitals are degenerate. What does it mean?

(a) All the orbitals have the same orientation.

(b) All the orbitals have the same shape.

(c) All the orbitals have the same energy.

(d) All the orbitals are unoccupied.

137. The energy of an electron of $2p_y$ orbital is:

(a) Greater than $2p_x$ orbital (b) Less than $2p_z$ orbital

(c) Equal to 2s orbital (d) Same as that of $2p_x$ and $2p_z$ orbital

138. Not considering the electron spin, the degeneracy of first excited state ($n = 2$) of H-atom is:

(a) 2 (b) 3

(c) 4 (d) 8

139. Not considering the electron spin, the degeneracy of second excited state of H-atom is 9, while the degeneracy of the first excited state of H^- is:

(a) 1 (b) 2

(c) 3 (d) 4

140. In H-atom energy of electron is determined by :

(a) only n (b) n, l

(c) n, l, m (d) all the four quantum numbers.

141. In a hydrogen atom, which orbital is higher in energy than a $3s$-orbital ?

(a) $2s$ (b) $3p$

(c) $3d$ (d) $4s$

142. The atomic orbitals are progressively filled in order of increasing energy. The principle is called as:

(a) Hund's rule (b) Aufbau principle

(c) Exclusion principle (d) de-Broglie rule

143. The aufbau principle implies that a new electron will enter an orbital for which :

(a) n has a lower value (b) l has a lower value

(c) $(n + l)$ value is maximum (d) $(n + l)$ value is minimum

144. The orbital diagram in which aufbau principle is violated, is :

(a) [↑↓] [↑↑ | ↑ | ↑] (b) [↑] [↑↓ | ↑ | ↑]

(c) [↑↓] [↑ | ↑ | ↑] (d) [↑↓] [↑↓ | ↑↓ | ↑]

145. Pauli's exclusion principle states that:

(a) Nucleus of an atom contains no negative charge

(b) Electrons move in circular orbits around the nucleus

(c) Electrons occupy orbitals of lowest energy

(d) All the four quantum numbers of two electrons in an atom cannot be equal.

146. In a set of degenerate orbitals the electrons distribute themselves to retain similar spins as far as possible. This statement is attributed to:

(a) Pauli's exclusion principle (b) Aufbau principle

(c) Hund's Rule (d) Slater rule

147. Which of the following rules could explain the presence of three unpaired electrons in N-atom?

(a) Hund's rule (b) Aufbau's principle

(c) Heisenberg's uncertainty principle (d) Pauli's exclusion principle

148. If the electronic structure of oxygen atom is written as $1s^2, 2s^2$ [↑↓ | ↑↓ |] (2p) it would violate

(a) Hund's rule (b) Paulis exclusion principle

(c) Both Hund's and Pauli's principles (d) None of these

149. The orbital diagram in which both the Pauli's exclusion principle and Hund's rule are violated, is :

(a) [↑↓] [↓ | ↓ | ↓] (b) [↑↓] [↑↓ | ↑↓ | ↑]

(c) [↑↓] [↑↓ | ↑↓ |] (d) [↑↓] [↑↑ | ↑↓ |]

150. The orbital diagram in which both the Pauli's exclusion principle and Hund's rule are violated is :

(a) [↑↓] [↑↑ | ↑ |] (b) [↑↓] [↑↓ | | ↑↓]

(c) [↑↑] [↑ | ↑ | ↑] (d) [↑↓] [↑↓ | ↑↓ | ↑]

151. It is not possible to explain the Pauli's exclusion principle with the help of this atom.

(a) B (b) Be

(c) C (d) H

152. The number of unpaired valence electrons in an atom of phosphorus is :

(a) 0 (b) 2

(c) 3 (d) 4

153. The possible correct set of quantum numbers for the unpaired electron of Cl atom is :

(a) $2, 0, 0, +\frac{1}{2}$ (b) $2, 1, -1, +\frac{1}{2}$

(c) $3, 1, 1, +\frac{1}{2}$ (d) $3, 0, \pm\frac{1}{2}$

154. The four quantum number for the valence shell electron or last electron of sodium ($Z = 11$) is:

(a) $n = 2, l = 1, m = -1, s = -\frac{1}{2}$ (b) $n = 3, l = 0, m = 0, s = +\frac{1}{2}$

(c) $n = 3, l = 2, m = -2, s = -\frac{1}{2}$ (d) $n = 3, l = 2, m = 2, s = +\frac{1}{2}$

155. Arrange the orbitals of H-atom in the increasing order of their energy:

$3p_x, 2s, 4d_{xy}, 3s, 4p_z, 3p_y, 4s$

(a) $2s < 3s = 3p_x = 3p_y < 4s = 4p_z = 4d_{xy}$ (b) $2s < 3s < 3p_x = 3p_y < 4s = 4p_z = 4d_{xy}$

(c) $2s < 3s < 3p_x = 3p_y < 4s = 4p_z = 4d_{xy}$ (d) $2s < 3s < 3p_x = 3p_y < 4s < 4p_z < 4d_{xy}$

156. The order of filling of electrons in the orbitals of an atom will be:

(a) 3*d*, 4*s*, 4*p*, 4*d*, 5*s* (b) 4*s*, 3*d*, 4*p*, 5*s*, 4*d*

(c) 5*s*, 4*p*, 3*d*, 4*d*, 5*s* (d) 3*d*, 4*p*, 4*s*, 4*d*, 5*s*

157. For which of the following sets of quantum numbers, an electron will have the highest energy?

	n	l	m	s		n	l	m	s
(a)	3	2	1	–1/2	(b)	4	3	–1	+1/2
(c)	4	1	–1	+1/2	(d)	5	0	0	–1/2

158. Which of the following set of quantum numbers belong to highest energy?

(a) $n=4, l=0, m=0, s=+\frac{1}{2}$ (b) $n=2, l=0, m=0, s=+\frac{1}{2}$

(c) $n=3, l=1, m=1, s=+\frac{1}{2}$ (d) $n=3, l=2, m=1, s=+\frac{1}{2}$

159. For which of the following sets of four quantum numbers, an electron will have the highest energy ?

	n	l	m	s
(a)	3	2	1	+1/2
(b)	4	2	1	+1/2
(c)	4	1	0	–1/2
(d)	5	0	0	–1/2

160. The quantum numbers of four electrons ($e1$ to $e4$) are given below :

	n	l	m	s		n	l	m	s
$e1$	3	0	0	+1/2	$e2$	4	0	0	1/2
$e3$	3	2	2	–1/2	$e4$	3	1	–1	1/2

The correct order of decreasing energy of these electrons is :

(a) $e4 > e3 > e2 > e1$ (b) $e2 > e3 > e4 > e1$

(c) $e3 > e2 > e4 > e1$ (d) $e1 > e4 > e2 > e3$

161. Which of the following sets of quantum numbers describes the electron which is removed most easily from a potassium atom in its ground state?

(a) $n=3, l=1, m_l=1, m_s=-\frac{1}{2}$ (b) $n=2, l=1, m_l=0, m_s=-\frac{1}{2}$

(c) $n=4, l=0, m_l=1, m_s=+\frac{1}{2}$ (d) $n=4, l=0, m_l=0, m_s=+\frac{1}{2}$

162. Which of the following elements is represented by the electronic configuration ?

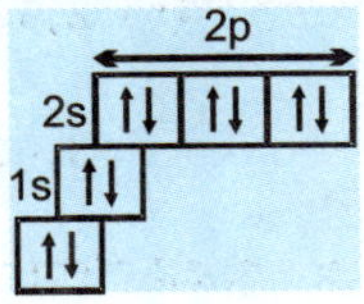

(a) Nitrogen (b) Fluorine

(c) Oxygen (d) Neon

163. The correct ground state electronic configuration of chromium atom is:

(a) $[Ar]3d^5 4s^1$ (b) $[Ar]3d^4 4s^2$

(c) $[Ar]3d^6 4s^0$ (d) $[Ar]4d^5 4s^1$

164. Which of the following electron configurations is correct for iron, (atomic number 26)?

(a) $[Kr]\ 4s^1 3d^6$ (b) $[Kr]\ 4s^1 3d^7$

(c) $[Ar]\ 4s^2 3d^6$ (d) $[Kr]\ 4s^2 3d^6$

165. Which of the following electron configurations is correct for copper, (atomic number 29)?

(a) $[Ar]\,3d^{10}\,4s^1$ (b) $[Kr]\,3d^9\,4s^1$

(c) $[Ar]\,3d^9\,4s^2$ (d) $[Kr]\,3d^{10}\,4s^1$

166. The electronic configurations of $_{24}Cr$ and $_{29}Cu$ are abnormal

(a) Due to extra stability of exactly half filled and exactly fully filled sub shells

(b) Because they belong to *d*-block

(c) Both the above

(d) None of the above

167. Which of the following representation of excited states of atoms is impossible?

(a) $1s^1\,2s^1$ (b) $[Ne]\,3s^2 3p^3 4s^1$

(c) $[Ne]\,3s^2 3p^6 4s^1 3d^6$ (d) $1s^2 2s^2 2p^7 3s^2$

168. Among the following series of transition metal ions, the one where all metal ions have same 3*d* electronic configuration is:

(a) Ti^{2+}, V^{3+}, Cr^{4+}, Mn^{5+} (b) Ti^{3+}, V^{2+}, Cr^{3+}, Mn^{4+}

(c) Ti^{+}, V^{4+}, Cr^{6+}, Mn^{7+} (d) Ti^{4+}, V^{3+}, Cr^{2+}, Mn^{3+}

169. Which of the following has the maximum number of unpaired electrons?

(a) Mn (b) Ti

(c) V (d) Al

170. The ratio of magnetic moments of Fe (III) and Co (II) is :

(a) $\sqrt{5} : \sqrt{7}$ (b) $\sqrt{35} : \sqrt{15}$

(c) 7 : 3 (d) $\sqrt{24} : \sqrt{15}$

171. A compound of vanadium has a magnetic moment (μ) of 1.73 BM. If the vanadium ion in the compound is present as V^{x+}, then, the value of *x* is?

(a) 1 (b) 2

(c) 3 (d) 4

172. Wave mechanical model of the atom depends upon:

(a) de-Broglie concept of dual nature of electron

(b) Heisenberg uncertainty principle

(c) Schrödinger equation

(d) All of these

173. Select the incorrect statement :

(a) ψ is mathematical function of the coordinates of the electron.

(b) ψ does not carry any physical meaning.

(c) ψ is related with all four quantum numbers.

(d) The probability density function (ψ^2) is zero on the plane when electron finding probability is zero.

174. $\psi^2(r,\theta,\phi)$ represents: (for schrödinger wave mechanical model)

(a) Amplitude of electron wave

(b) Probability density of electron

(c) Total probability of finding electron around nucleus

(d) Orbit

175. The correct representation for schrodinger wave function, ψ_{nlm} is:

(a) ψ_{110} (b) ψ_{231}

(c) ψ_{101} (d) ψ_{420}

176. Radial amplitude of electron wave can be represented by:

(a) $R(r)$ (b) $R^2(r)$

(c) $4\pi r^2$ (d) $4\pi r^2 R^2(r)$

177. Which orbital has only positive values of wave function at all distances from the nucleus?

(a) 1*s* (b) 2*s*

(c) 3*p* (d) 3*d*

178. Select correct graph for 2*s* orbital.

(a) 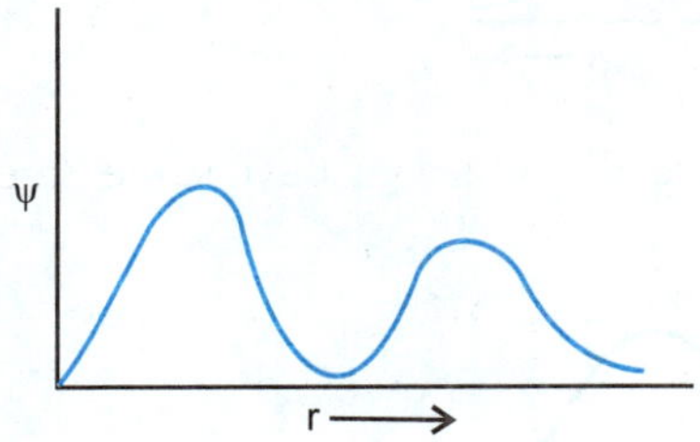

(b) 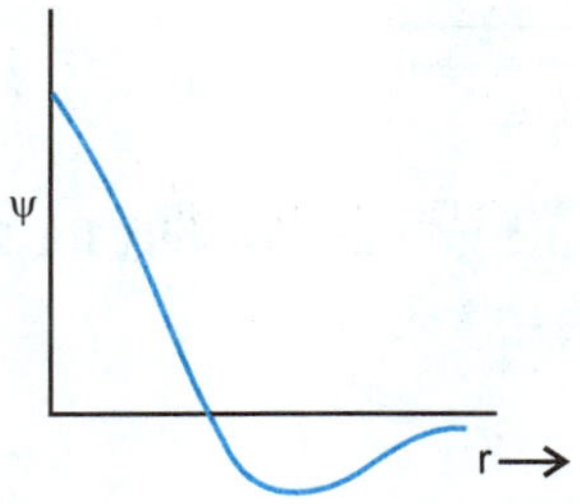

(c) 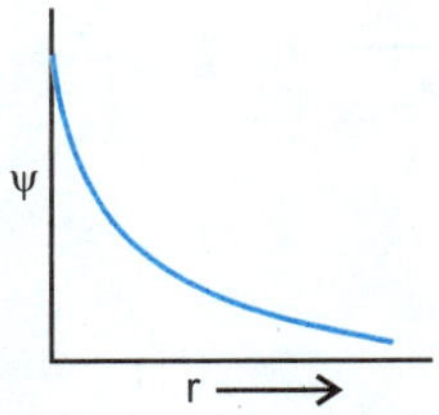

(d) 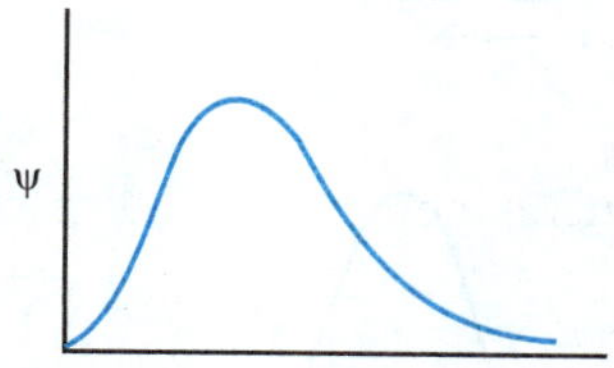

179. Wave function of an orbital is plotted against the distance from nucleus. The graphical representation is of:

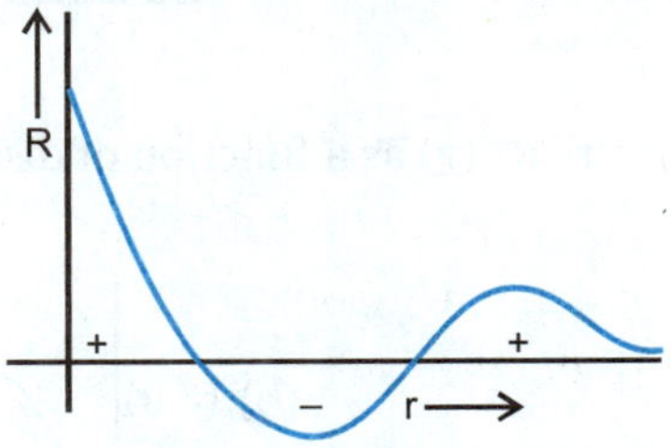

(a) 1*s* (b) 2*s*

(c) 3*s* (d) 2*p*

180. The following graph between probability density (ψ^2) and distance from the nucleus (r) represents :

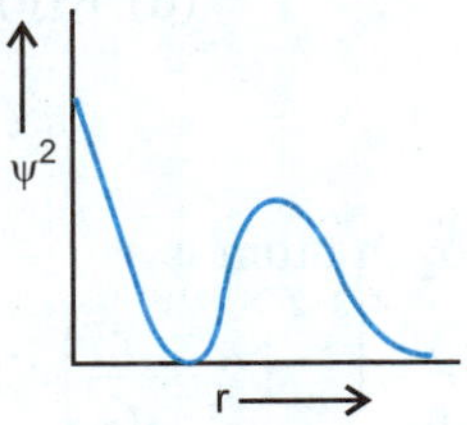

(a) 2*s* (b) 3*s*

(c) 1*s* (d) 2*p*

181. Which of the following graph is incorrectly labelled?

(a) 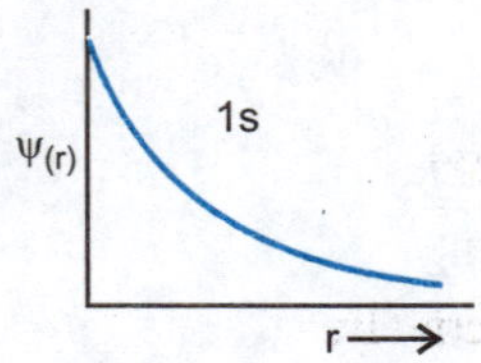

(b) 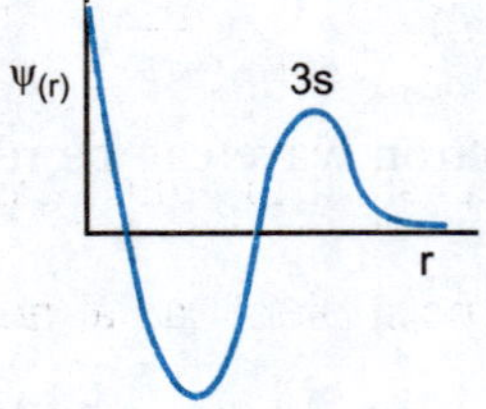

(c) 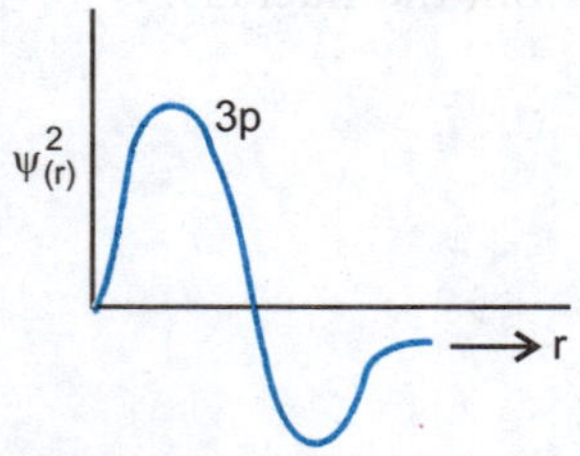

(d) 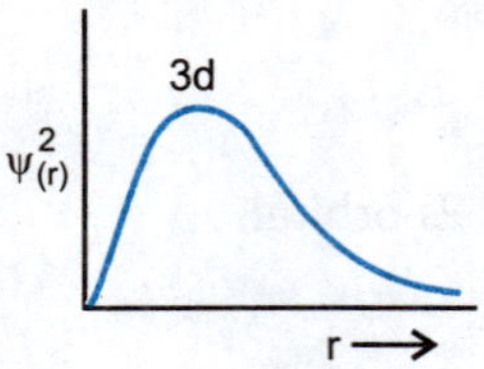

182. Which of the following graph is correct for 2*p* orbital?

(a) 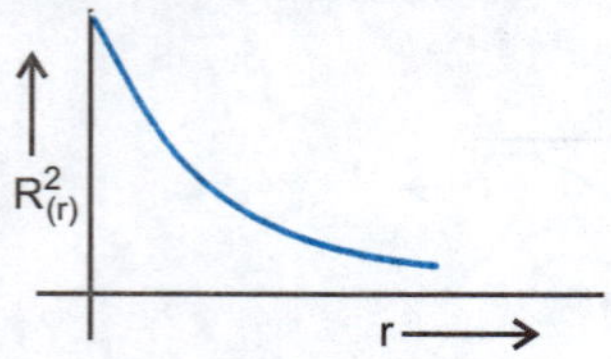

(b) 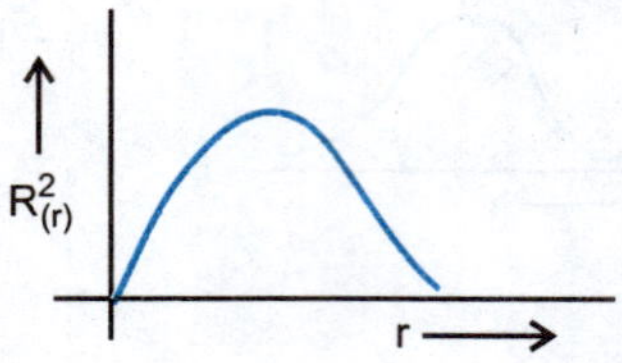

(c) 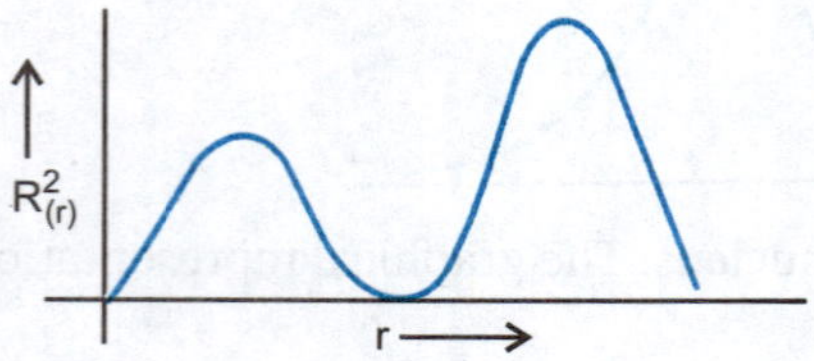

(d) 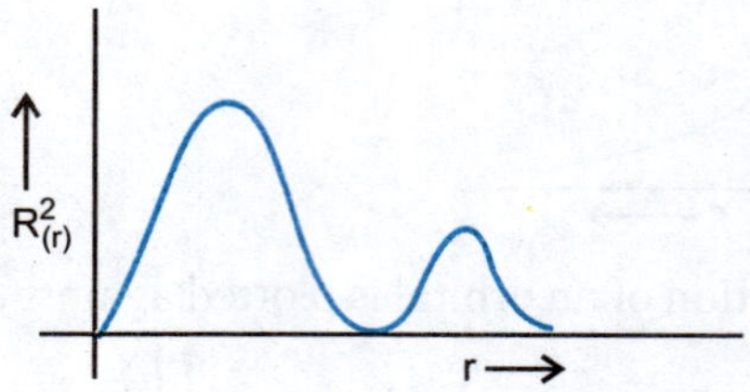

183. The variation of radial probability density $R^2(r)$ as a function of distance r of the electron from the nucleus for 3*p*-orbital :

(a) 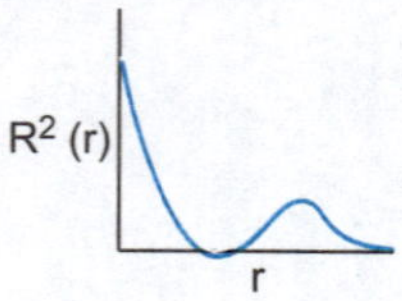

(b) 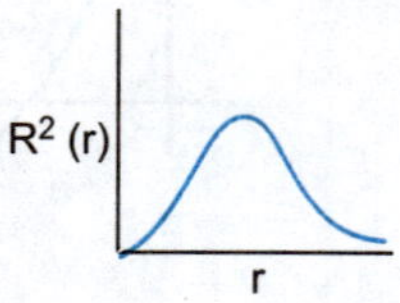

(c) 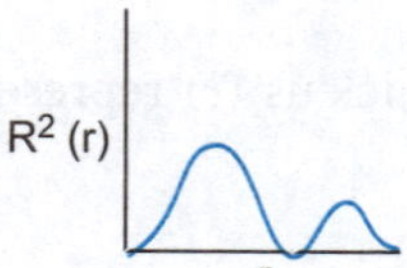

(d) 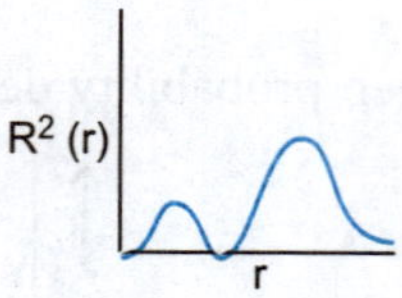

184. The Schrodinger wave equation for hydrogen atom is

$$\Psi_{2s} = \frac{1}{4\sqrt{2\pi}}\left(\frac{1}{a_0}\right)^{3/2}\left(2 - \frac{r}{a_0}\right)e^{-r/a_0}$$

where a_0 is Bohr's radius. If the radial node in 2*s* be at r_0, then r_0 would be equal to :

(a) $\frac{a_0}{2}$ (b) $2a_0$

(c) $\sqrt{2}\, a_0$ (d) $\frac{a_0}{\sqrt{2}}$

185. The Schrodinger wave equation for hydrogen atom is

$$\Psi\,(\text{radial}) = \frac{1}{16\sqrt{4}}\left(\frac{Z}{a_0}\right)^{3/2}\left[(\sigma - 1)\,(\sigma^2 - 8\sigma + 12)\right] e^{-\sigma/2}$$

where a_0 and Z are the constant in which answer can be expressed and $\sigma = \frac{2Zr}{a_0}$ minimum and maximum position of radial nodes from nucleus are ......respectively.

(a) $\frac{a_0}{Z}, \frac{3a_0}{Z}$

(b) $\frac{a_0}{2Z}, \frac{a_0}{Z}$

(c) $\frac{a_0}{2Z}, \frac{3a_0}{Z}$

(d) $\frac{a_0}{2Z}, \frac{4a_0}{Z}$

Level-2

1. Potential energy of electron present in He^+ is :

(a) $\dfrac{e^2}{2\pi\varepsilon_0 r}$ (b) $\dfrac{3e^2}{4\pi\varepsilon_0 r}$

(c) $\dfrac{-2e^2}{4\pi\varepsilon_0 r}$ (d) $\dfrac{-e^2}{4\pi\varepsilon_0 r^2}$

2. A single electron in an ion has ionization energy equal to 217.6 eV. What is the total number of neutrons present in one ion of it?

(a) 2 (b) 4

(c) 5 (d) 9

3. For a hypothetical hydrogen like atom, the potential energy of the system is given by $U(r) = \dfrac{-Ke^2}{r^3}$, where r is the distance between the two particles. If Bohr's model of quantization of angular momentum is applicable then velocity of particle is given by :

(a) $v = \dfrac{n^2h^3}{Ke^2 8\pi^3 m^2}$ (b) $v = \dfrac{n^3h^3}{8Ke^2\pi^3 m^2}$

(c) $v = \dfrac{n^3h^3}{24Ke^2\pi^3 m^2}$ (d) $v = \dfrac{n^2h^3}{24Ke^2\pi^3 m^2}$

4. A small particle of mass m moves in such a way that P.E. $= -\dfrac{1}{2}mkr^2$, where k is a constant and r is the distance of the particle from origin. Assuming Bohr's model of quantization of angular momentum and circular orbit, r is directly proportional to :

(a) n^2 (b) n

(c) $\sqrt{n}$ (d) none of these

5. A beam of specific kind of particles of velocity 2.1×10^7 m/s is scattered by a gold ($Z = 79$) nuclei. Find out specific charge (charge/mass) of this particle if the distance of closest approach is 2.5×10^{-14} m.

(a) 4.84×10^7 C/kg (b) 4.84×10^{-7} C/kg

(c) 2.42×10^7 C/kg (d) 3×10^{-12} C/kg

6. What is the angular velocity (ω) of an electron occupying second orbit of Li^{2+} ion?

(a) $\dfrac{8\pi^3me^4}{h^3}K^2$ (b) $\dfrac{8\pi^3me^4}{9h^3}K^2$

(c) $\dfrac{64}{9} \times \dfrac{\pi^3me^4}{h^3}K^2$ (d) $\dfrac{9\pi^3me^4}{h^3}K^2$

7. The ratio of the radius difference between 4th and 3rd orbit of H-atom and that of Li^{2+} ion is :

(a) 1 : 1 (b) 3 : 1

(c) 3 : 4 (d) 9 : 1

8. The velocity of an electron in excited state of H-atom is 1.093×10^6 m/s. What is the circumference of this orbit?

(a) 3.32×10^{-10} m (b) 6.64×10^{-10} m

(c) 13.30×10^{-10} m (d) 13.28×10^{-8} m

9. The angular momentum of an electron in a Bohr's orbit of He^+ is 3.1652×10^{-34} kg-m^2/sec. What is the wave number in terms of Rydberg constant (R) of the spectral line emitted when an electron falls from this level to the first excited state. [Use $h = 6.626 \times 10^{-34}$ J·s]

(a) $3R$ (b) $\frac{5R}{9}$

(c) $\frac{3R}{4}$ (d) $\frac{8R}{9}$

10. If radiation corresponding to second line of "Balmer series" of Li^{2+} ion, knocked out electron from first excited state of H-atom, then kinetic energy of ejected electron would be:

(a) 2.55 eV (b) 4.25 eV

(c) 11.25 eV (d) 19.55 eV

11. When an electron makes a transition from $(n+1)$ state to nth state, the frequency of emitted radiations is related to n according to $(n >> 1)$:

(a) $\nu = \frac{2cRZ^2}{n^3}$ (b) $\nu = \frac{cRZ^2}{n^4}$

(c) $\nu = \frac{cRZ^2}{n^2}$ (d) $\nu = \frac{2cRZ^2}{n^2}$

12. In a collection of H-atoms, all the electrons jump from $n = 5$ to ground level finally (directly or indirectly), without emitting any line in Balmer series. The number of possible different radiations is :

(a) 10 (b) 8

(c) 7 (d) 6

13. An electron is allowed to move freely in a closed cubic box of length of side 10 cm. The uncertainty in its velocity will be :

(a) 3.35×10^{-4} m sec^{-1} (b) 5.8×10^{-4} m sec^{-1}

(c) 4×10^{-5} m sec^{-1} (d) 4×10^{-6} m sec^{-1}

14. An element undergoes a reaction as shown :

$X + 2e^- \longrightarrow X^{2-}$, energy released $= 30.87$ eV/atom. If the energy released, is used to dissociate 4 g of H_2 molecules, equally into H^+ and H*, where H* is excited state of H atoms where the electron travels in orbit whose circumference equal to four times its de Broglie's wavelength. Determine the least moles of X that would be required :

Given : I.E. of H = 13.6 eV/atom, bond energy of $H_2 = 4.526$ eV/molecule.

(a) 1 (b) 2

(c) 3 (d) 4

15. If the energy of H-atom in the ground state is $-E$, the velocity of photo-electron emitted when a photon having energy E_p strikes a stationary Li^{2+} ion in ground state, is given by :

(a) $v = \sqrt{\frac{2(E_p - E)}{m}}$ (b) $v = \sqrt{\frac{2(E_p + 9E)}{m}}$

(c) $v = \sqrt{\frac{2(E_p - 9E)}{m}}$ (d) $v = \sqrt{\frac{2(E_p - 3E)}{m}}$

16. At which temperature will the translational kinetic energy of H-atom equal to that for H-atom of first line Lyman transition? (Given $N_A = 6 \times 10^{23}$)

(a) 780 K (b) 1.32×10^5 K

(c) 7.84×10^4 K (d) 1000 K

17. For a 3s-orbital, value of ψ is given by following relation :

$$\Psi(3s) = \frac{1}{9\sqrt{3}}\left(\frac{1}{a_0}\right)^{3/2}(6 - 6\sigma + \sigma^2)\, e^{-\sigma/2}; \quad \text{where } \sigma = \frac{2r \cdot Z}{3a_0}$$

What is the maximum radial distance of node from nucleus ?

(a) $\frac{(3+\sqrt{3})\, a_0}{Z}$ (b) $\frac{a_0}{Z}$

(c) $\frac{3}{2}\frac{(3+\sqrt{3})\, a_0}{Z}$ (d) $\frac{2a_0}{Z}$

18. Monochromatic radiation of specific wavelength is incident on H-atoms in ground state. H-atoms absorb energy and emit subsequently radiations of six different wavelength. Find wavelength of incident radiations:

(a) 9.75 nm (b) 50 nm

(c) 85.8 nm (d) 97.25 nm

19. The energy of a I, II and III energy levels of a certain atom are E, $\frac{4E}{3}$ and $2E$ respectively. A photon of wavelength λ is emitted during a transition from III to I. What will be the wavelength of emission for transition II to I?

(a) $\frac{\lambda}{2}$ (b) λ

(c) 2λ (d) 3λ

20. Calculate the minimum and maximum number of electrons which may have magnetic quantum number, $m = +1$ and spin quantum number, $s = -\frac{1}{2}$ in chromium (Cr) :

(a) 0, 1 (b) 1, 2

(c) 4, 6 (d) 2, 3

21. An electron in a hydrogen atom in its ground state absorbs 1.5 times as much energy as the minimum required for it to escape from the atom. What is the velocity of the emitted electron?

(a) 1.54×10^6 m/s (b) 1.54×10^8 m/s

(c) 1.54×10^3 m/s (d) 1.54×10^4 m/s

22. In a measurement of quantum efficiency of photosynthesis in green plants, it was found that 10 quanta of red light of wavelength 6850 Å were needed to release one molecule of O_2. The average energy storage in this process for 1 mole O_2 evolved is 112 kcal.

What is the energy conversion efficiency in this experiment?

Given : 1 cal = 4.18 J; $N_A = 6 \times 10^{23}$; $h = 6.63 \times 10^{-34}$ J.s

(a) 23.5 (b) 26.9

(c) 66.34 (d) 73.1

23. A hydrogen like species (atomic number Z) is present in a higher excited state of quantum number n. This excited atom can make a transition to the first excited state by successive emission of two photons of energies 10.20 eV and 17.0 eV respectively. Alternatively, the atom from the same excited state can make a transition to the second excited state by successive emission of two photons of energy 4.25 eV and 5.95 eV respectively. Determine the value of Z.

(a) 1 (b) 2

(c) 3 (d) 4

24. H-atom is exposed to electromagnetic radiation of $\lambda = 1025.6$ Å and excited atom gives out induced radiations. What is the minimum wavelength of these induced radiations?

(a) 102.6 nm (b) 12.09 nm

(c) 121.6 nm (d) 810.8 nm

25. If the lowest energy X-rays have $\lambda = 3.055 \times 10^{-8}$ m, estimate the minimum difference in energy between two Bohr's orbits such that an electronic transition would correspond to the emission of an X-ray. Assuming that the electrons in other shells exert no influence, at what Z (minimum) would a transition from the second energy level to the first result in the emission of an X-ray?

(a) 1 (b) 2

(c) 3 (d) 4

26. An α-particle having kinetic energy 5 MeV falls on a Cu-foil. The shortest distance from the nucleus of Cu to which α-particle reaches is (Atomic no. of Cu = 29, $K = 9 \times 10^9 \text{ Nm}^2/\text{C}^2$)

(a) 2.35×10^{-13} m (b) 1.67×10^{-14} m

(c) 5.98×10^{-15} m (d) None of these

27. In the graph between $\sqrt{v}$ and Z for the Mosley's equation $\sqrt{v} = a(Z - b)$, the intercept OX is -1 on $\sqrt{v}$ axis.

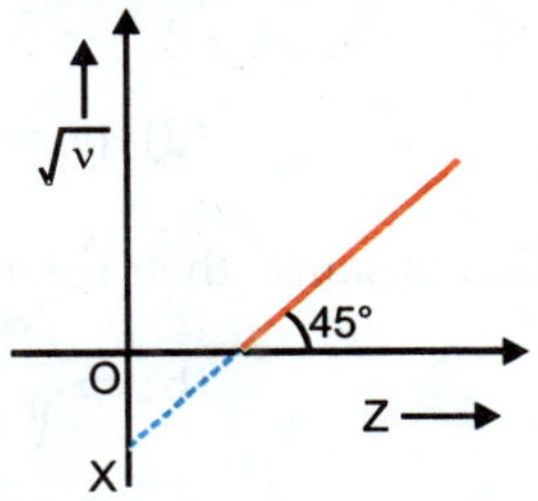

What is the frequency when atomic number (Z) is 51 ?

(a) 50 s^{-1} (b) 100 s^{-1}

(c) 2500 s^{-1} (d) None of these

28. Balmer gave an equation for wavelength of visible region of H-spectrum as $\lambda = \dfrac{Kn^2}{n^2 - 4}$.

Where n = principal quantum number of energy level, K = constant in terms of R (Rydberg constant). The value of K in terms of R is:

(a) R (b) $\dfrac{R}{2}$

(c) $\dfrac{4}{R}$ (d) $\dfrac{5}{R}$

29. The energy of separation of an electron in a hydrogen like atom in excited state is 3.4 eV. The de-Broglie wave length (in Å) associated with the electron is:
(Given radius of first orbit of H-atom is 0.53Å)

(a) 3.33 (b) 6.66

(c) 13.31 (d) None of these

30. If I excitation energy for the H-like (hypothetical) sample is 24 eV, then binding energy in III excited state is:

(a) 2 eV (b) 3 eV

(c) 4 eV (d) 5 eV

Level-3

Passage-1

Werner Heisenberg considered the limits of how precisely we can measure the properties of an electron or other microscopic particle. He determined that there is a fundamental limit to how closely we can measure both position and momentum. The more accurately we measure the momentum of a particle, the less accurately we can determine its position. The converse is also true. This is summed up in what we now call the Heisenberg uncertainty principle.

The equation is $\Delta x \cdot \Delta (mv) \geq \frac{h}{4\pi}$

The uncertainty in the position or in the momentum of a macroscopic object like a baseball is too small to observe. However, the mass of microscopic object such as an electron is small enough for the uncertainty to be relatively large and significant.

1. If the uncertainties in position and momentum are equal, the uncertainty in the velocity is :

(a) $\sqrt{\frac{h}{\pi}}$ (b) $\sqrt{\frac{h}{2\pi}}$

(c) $\frac{1}{2m}\sqrt{\frac{h}{\pi}}$ (d) none of these

2. If the uncertainty in velocity and position is same, then the uncertainty in momentum will be :

(a) $\sqrt{\frac{hm}{4\pi}}$ (b) $m\sqrt{\frac{h}{4\pi}}$

(c) $\sqrt{\frac{h}{4\pi m}}$ (d) $\frac{1}{m}\sqrt{\frac{h}{4\pi}}$

3. What would be the minimum uncertainty in de-Broglie wavelength of a moving electron accelerated by potential difference of 6 volt and whose uncertainty in position is $\frac{7}{22}$ nm?

(a) 6.25 Å (b) 6 Å

(c) 0.625 Å (d) 0.3125 Å

Passage-2

One of the fundamental laws of physics is that matter is most stable with the lowest possible energy. Thus, the electron in a hydrogen atom usually moves in the $n = 1$ orbit, the orbit in which it has the lowest energy. When the electron is in this lowest energy orbit, the atom is said to be in its ground electronic state. If the atom receives energy from an outside source, it is possible for the electron to move to an orbit with a higher n value, in which case the atoms is in an excited state with a higher energy.

The law of conservation of energy says that we cannot create or destroy energy. Thus, if a certain amount of external energy is required to excite an electron from one energy level to another, then that same amount of energy will be liberated when the electron returns to its initial state.

Lyman series is observed when the electron returns to the lowest orbit while Balmer series is formed when the electron returns to second orbit. Similarly, Paschen, Brackett and Pfund series are formed when electrons returns to the third, fourth and fifth orbits from higher energy orbits respectively.

When electrons return from n_2 to n_1 state, the number of lines in the spectrum will equal to

$$\frac{(n_2 - n_1)(n_2 - n_1 + 1)}{2}$$

If the electron comes back from energy level having energy E_2 to energy level having energy E_1, then the difference may be expressed in terms of energy of photon as :

$$E_2 - E_1 = \Delta E,\ \Delta E \Rightarrow \frac{hc}{\lambda}$$

Since, h and c are constants, ΔE corresponds to definite energy. Thus, each transition from one energy level to another will produce a radiation of definite wavelength. This is actually observed as a line in the spectrum of hydrogen atom.

Wave number of a spectral line is given by the formula

$$\bar{\nu} = R\left(\frac{1}{n_1^2} - \frac{1}{n_2^2}\right).$$

where R is a Rydberg's constant ($R = 1.1 \times 10^7\ m^{-1}$)

1. If the wavelength of series limit of Lyman series for He^+ ion is x Å, then what will be the wavelength of series limit of Balmer series for Li^{2+} ion?

(a) $\frac{9x}{4}$ Å
(b) $\frac{16x}{9}$ Å
(c) $\frac{5x}{4}$ Å
(d) $\frac{4x}{7}$ Å

2. The emission spectra is observed by the consequence of transition of electrons from higher energy state to ground state of He^+ ion. Six different photons are observed during the emission spectra, then what will be the minimum wavelength during the transition?

(a) $\frac{4}{27R_H}$
(b) $\frac{4}{15R_H}$
(c) $\lambda = \frac{15}{16R_H}$
(d) $\frac{16}{15R_H}$

3. What transition in the hydrogen spectrum would have the same wavelength as Balmer transition, $n = 4$ to $n = 2$ in the He^+ spectrum?

(a) $n = 3$ to $n = 1$
(b) $n = 3$ to $n = 2$
(c) $n = 4$ to $n = 1$
(d) $n = 2$ to $n = 1$

4. An electron in H-atom in M-shell on de-excitation to ground state gives maximum .... spectrum lines.

(a) 10
(b) 6
(c) 2
(d) 1

Passage-3

If hydrogen atoms (in the ground state) are passed through an homogeneous magnetic field, the beam is split into two parts. This interaction with the magnetic field shows that the atoms must have magnetic moment. However, the moment cannot be due to the orbital angular momentum since $l = 0$. Hence one must assume existence of intrinsic angular momentum, which as the experiment shows, has only two permitted orientations.

Spin of the electron produce angular momentum equal to $S = \sqrt{s(s+1)}\frac{h}{2\pi}$ where $S = +\frac{1}{2}$.

Total spin of an atom $= +\frac{n}{2}$ or $-\frac{n}{2}$

where n is the number of unpaired electrons.

The substance which contain species with unpaired electrons in their orbitals behave as paramagnetic substances. The paramagnetism is expressed in terms of magnetic moment.

The magnetic moment of an atom

$$\mu_s \sqrt{s(s+1)}\frac{eh}{2\pi mc} = \sqrt{\frac{n}{2}\left(\frac{n}{2}+1\right)}\frac{eh}{2\pi mc} \qquad s=\frac{n}{2}$$

$$\Rightarrow \qquad \mu_s = \sqrt{n(n+2)} \text{ B.M.}$$

n = number of unpaired electrons

1. B.M. (Bohr magneton) $= \dfrac{eh}{4\pi mc}$

If magnetic moment is zero the substances is diamagnetic.

1. Which of the following ion has lowest magnetic moment ?

(a) Fe^{2+} (b) Mn^{2+}

(c) Cr^{3+} (d) V^{3+}

2. If an ion of $_{25}Mn$ has a magnetic moment of 3.873 B.M. Then oxidation state of Mn in ion is :

(a) +2 (b) +3

(c) +4 (d) +5

Passage-4

Ozone in the upper atmosphere absorbs ultraviolet radiation which induces the following chemical reaction

$$O_3(g) \longrightarrow O_2(g) + O(g)$$

O_2 produced in the above photochemical dissociation undergoes further dissociation into one normal oxygen atom (O) and more energetic oxygen atom O*.

$$O_2(g) \longrightarrow O + O^*$$

(O*) has 1 eV more energy than (O) and normal dissociation energy of O_2 is 480 kJ mol^{-1}.

[1 eV/ Photon = 96 kJ mol^{-1}]

1. What is the maximum wavelength effective for the photochemical dissociation of O_2 molecule

(a) 2440 Å (b) 2066.67 Å

(c) 1000 Å (d) 155 Å

2. If dissociation of O_3 into O_2 and O requires 400kJ mol^{-1} and O_2 produced in this reaction is further dissociated to O and O* then the total energy required to for the dissociation of O_3 into O and O* is:

(a) 1168kJ/mol (b) 976kJ/mol

(c) 880kJ/mol (d) None of these

Passage-5

The existence of negatively charged particle in an atom was shown by J.J. Thomson as a result of the studies of the passage of electricity through gases at extremely low pressures known as discharge tube experiments. When a high voltage of the order of 10,000 volts or more was impressed across the electrodes, some sort of invisible rays moved from the negative electrode to the positive electrode these rays are called as cathode rays.

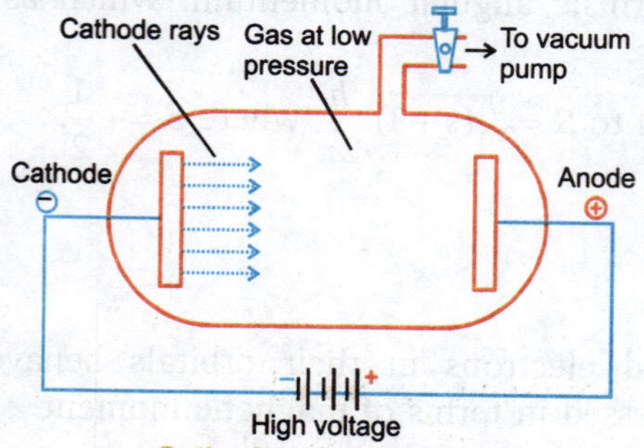

Cathode rays experiment

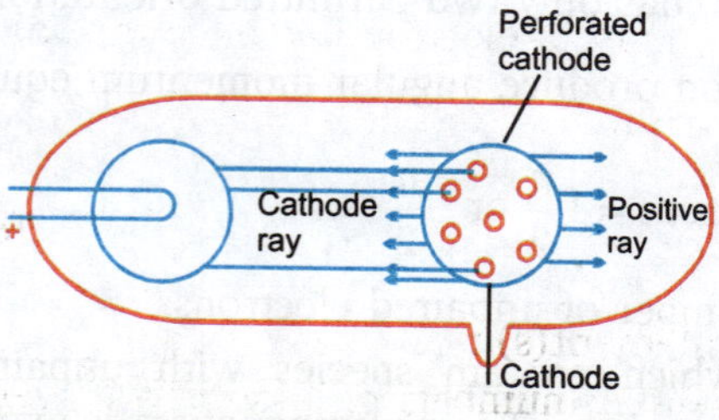

Anode rays experiment

Cathode rays travel in straight path in absence of electrical and magnetic field. Cathode rays consist of material part and charged particles? Cathode rays produce **X-rays** and light is emitted when they strike on ZnS screen. Cathode rays penetrate through thin sheets of aluminium and other metals. They affect the photographic plate and passes heating effect when they strike on metal foil. The ratio of charge to mass *i.e.* charge/mass is same for all the cathode rays irrespective of the gas used in the tube.

The existence of positively charged particles in an atom was shown by E. Goldstein. He repeated the same discharge tube experiments by using a perforated cathode. It was observed that when a high potential difference was applied between the electrodes, not only cathode rays were produced but also a new type of rays were produced simultaneously from anode moving towards cathode and passed through the holes or canal of the cathode. These termed as canal rays or anode rays.

These rays travel in straight lines and consist of positively charged particles. These rays have kinetic energy and produces heating effect also. The e/m ratio of these rays is smaller than that of electrons. Unlike cathode rays, their e/m value is is dependent upon the nature of the gas taken in the tube. These rays produce flashes of light on ZnS screen and can pass throughs thin metal foils. They can produce physical and chemical changes and are capable to produce ionisation in gases.

1. For cathode rays the value of e/m:

(a) is independent of the nature of the cathode and the gas filled in the discharge tube

(b) is constant

(c) is -1.7588×10^8 coulombs/g

(d) all of the above are correct

2. Which is not true with respect to cathode rays ?

(a) A stream of electrons (b) Charged particles

(c) Move with same speed as that of light (d) Can be deflected by the electric field

3. Select the incorrect statement.

(a) Cathode rays has charge only and no mass

(b) Anode rays are deflected by electrical and magnetic field

(c) Canal rays is named for beam of positive charged particle

(d) Anode rays particles carrying positive charge

One or More Answer is/are Correct

1. Select the correct statement(s) :

(a) The phenomena of diffraction of light can only be explained by assuming that light behaves as waves

(b) de-Broglie postulate the dual character existed with matter

(c) In his atomic model Bohr considered electron as a particle

(d) Wave nature of electrons was proved when diffraction rings were observed photographically when a stream of protons was passed through a metal foil

2. The angular momentum of electron can have the value(s) :

(a) $0.5\frac{h}{\pi}$ (b) $\frac{h}{\pi}$

(c) $\frac{h}{0.5\,\pi}$ (d) $2.5\frac{h}{2\pi}$

3. Select incorrect statement(s) :

(a) Only three quantum numbers n, l and m are needed to define an orbital

(b) Four quantum numbers are needed for complete description of an electron

(c) Two quantum numbers n and l are needed to identify subshell and shape of orbital
(d) Splitting of spectrum lines in presence of electric field is known as Zeeman effect

4. Select the correct statement(s):
(a) An electron near the nucleus is attracted by the nucleus and has a low potential energy
(b) According to Bohr's theory, an electron continuously radiate energy if it stays in one orbit
(c) Bohr's model could not explain the spectra of multielectron atoms
(d) Bohr's model was the first atomic model based on quantisation of energy

5. Choose the correct statement(s) :
(a) The shape of an atomic orbital depends upon azimuthal quantum number
(b) The orientation of an atomic orbital depends upon the magnetic quantum number
(c) The energy of an electron in an atomic orbital of multi-electron atom depends upon principal quantum number only
(d) The number of degenerate atomic orbitals of one type depends upon the value of spin quantum number

6. For radial probability curves, which of the following is/are correct?
(a) The number of maxima in 2s orbital are two
(b) The number of spherical or radial nodes is equal to $n - l - 1$
(c) The number of angular nodes are 'l'
(d) $3d_z^2$ has 3 angular nodes

7. Select the correct statement(s):
(a) Lower value of quantum number l indicates that there is a higher probability of finding the 3s electron close to the nucleus than those of 3p and 3d electrons
(b) Energy of 3s orbital is less than the 3p and 3d orbitals for multi-electronic species
(c) At the node, the value of the radial function changes from positive to negative
(d) The radial function depends upon the quantum numbers n and l

8. Choose the incorrect statement(s):
(a) For a particular orbital in hydrogen atom, the wave function may have negative value
(b) Radial probability distribution function may have zero value but can never have negative value
(c) $3d_{x^2-y^2}$ orbital has two angular nodes and one radial node
(d) yz and xz planes are nodal planes for d_{xy} orbital

9. Select the correct statement(s) :
(a) Heisenberg's principle is applicable to stationary electron
(b) Pauli's exclusion principle is not applicable to photons
(c) For an electron the product of velocity and principal quantum number will be independent to principal quantum number
(d) Quantum numbers l and m determine the value of angular wave function

10. Choose the correct statements among the following :
(a) A node is a point in space where the wave-function Ψ has zero amplitude
(b) The number of maxima (peaks) in radial probability distribution function is $(n - l)$
(c) Radial probability density is $4\pi r^2 R_{n,l}^2(r)$
(d) Ψ^2 represents probability of finding electron at certain point in space

11. Select the correct statement(s) regarding $3p_y$ orbital :
(a) Total number of nodes are 2
(b) Number of maxima in the curve $4\pi r^2 R^2$ vs r are two
(c) Quantum number n, l and m for an orbital may be 3, 1, –1 respectively
(d) The magnetic quantum number may have a positive value

12. Select the correct statement(s) :
(a) In wave mechanical model, the energy of electron in the orbital remains constant
(b) d_{xy} orbital is lies in yz plane
(c) Nodal planes are yz and xy in $d_{x^2-y^2}$ orbital
(d) Rest mass of photon is zero and increases with its velocity

13. Hydrogen has :
(a) half filled subshell
(b) half filled shell
(c) one electron in valence shell
(d) half filled orbital

14. Select incorrect statement(s) .
(a) If the value of $l = 0$, the electron distribution is spherical
(b) The shape of the orbital is given by magnetic quantum number
(c) Angular momentum of $1s, 2s, 3s$ orbit electrons are equal
(d) In an atom, all the electrons travel with the same velocity

15. The radial distribution functions $[P(r)]$ is used to determine the most probable radius, which is used to find the electron in a given orbital. $\frac{dP(r)}{dr}$ for 1s-orbital of hydrogen like atom having atomic number Z, is

$$\frac{dP}{dr} = \frac{4Z^3}{a_0^3}\left(2r - \frac{2Zr^2}{a_0}\right)e^{-2Zr/a_0} :$$

Then which of the following statements is/are correct?
(a) At the point of maximum value of radial distribution function $\frac{dP(r)}{dr} = 0$; one antinode is present
(b) Most probable radius of Li^{2+} is $\frac{a_0}{3}$ pm
(c) Most probable radius of He^+ is $\frac{a_0}{2}$ pm
(d) Most probable radius of hydrogen atom is a_0 pm

16. Select the correct statement(s) :
(a) An orbital with $l = 0$ is symmetrical about the nucleus
(b) An orbital with $l = 1$ is spherically symmetrical about the nucleus
(c) $3d_{xy}$ is spherically symmetrical about the z-axis
(d) All are correct

17. Select the correct statement(s) :
(a) Radial function $[R(r)]$ is a part of wave function which depends upon quantum number n only
(b) Angular function depends only on the direction, and is independent to the distance from the nucleus
(c) $\Psi^2(r, \theta, \phi)$ is the probability density of finding the electron at a particular point in space
(d) Radial distribution function $(4\pi r^2 R^2)$ gives the probability of the electron being present at a distance r from the nucleus

18. Which is/are correct graph?

(a)
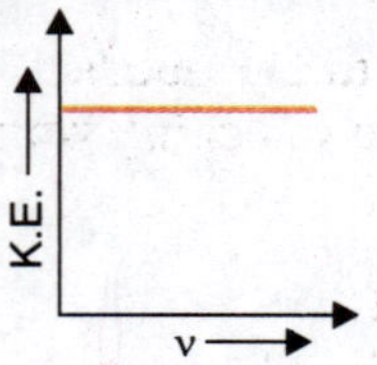

(b)
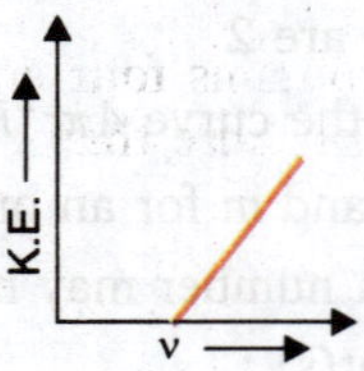

(c)
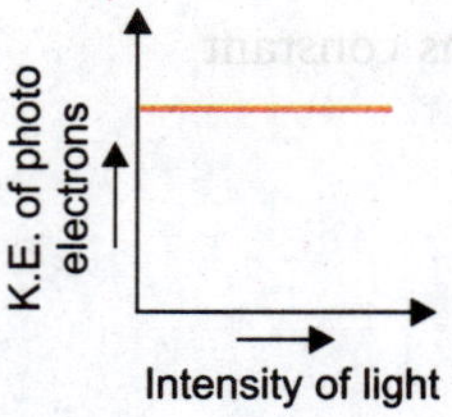

(d)
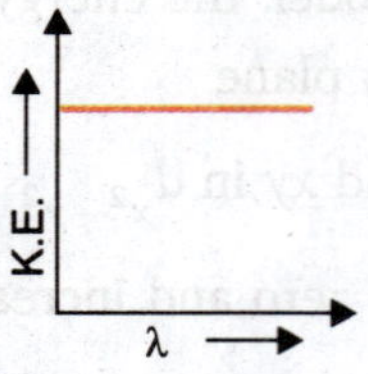

19. Select the correct curve(s) :

If v = Velocity of electron in Bohr's orbit

r = Radius of electron in Bohr's orbit

P.E. = Potential energy of electron in Bohr's orbit

K.E. = Kinetic energy of electron in Bohr's orbit

(a)
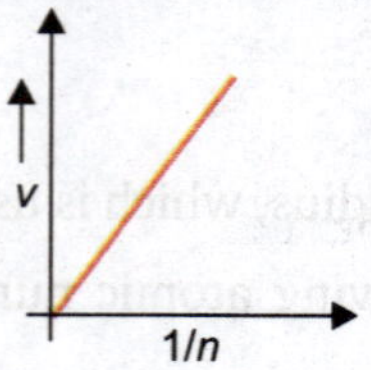

(b)
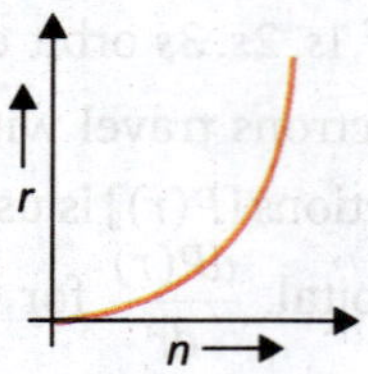

(c)
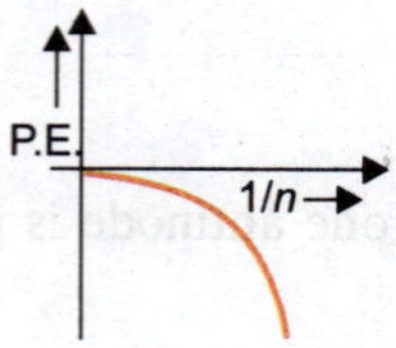

(d)
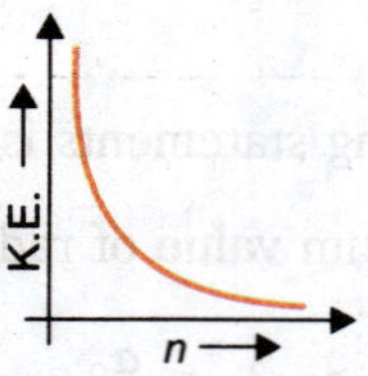

20. Select the correct set (s) of quantum numbers

(a) $n = 3, l = 0, m_l = -1$ (b) $n = 3, l = 3, m_l = -2$

(c) $n = 3, l = 2, m_l = -2$ (d) $n = 3, l = 1, m_l = 0$

21. Which is/are correct statement?

(a) Number of subshell present in M-shell = 3

(b) Number of orbitals present in N-shell = 16

(c) Cu^+ (z = 29) is paramagnetic

(d) Zeeman effect explains splitting of spectral lines in magnetic field.

22. In a sample of H-atoms electrons are de-excited from 4th excited state to ground state. Which is/are correct statement?

(a) No line observed in P-fund series.

(b) Total ten lines observed in spectrum.

(c) 4 line in UV-region and 3 line in visible region observed.

(d) One line observed in Brackett series.

Match the Column

Column-I and Column-II contains four entries each. Entries of Column-I are to be matched with some entries of Column-II. One or more than one entries of Column-I may have the matching with the same entries of Column-II.

1.

Column-I	Column-II
(A) Electron	(P) Negative charge
(B) Proton	(Q) Positive charge
(C) Neutron	(R) 1.6×10^{-19} C
(D) Positron	(S) Chargeless

2.

Column-I	Column-II
(A) Thomson model of atom	(P) Electrons are present in extra nuclear region
(B) Rutherford model of atom	(Q) Electron in the atom is described as wave
(C) Bohr model of atom	(R) Positive charge is accumulated in the nucleus
(D) Schrodinger model of hydrogen atom	(S) Uniform sphere of positive charge with embedded electrons

3.

Column-I	Column-II
(A) Atomic theory of matter	(P) Rutherford scattering experiment
(B) Quantization of charge	(Q) Muliken's oil drop experiment
(C) Quantization of electronic energy level	(R) Atomic spectra
(D) Size of nucleus	(S) Law of multiple proportions

4.

Column-I	Column-II
(A) $\frac{\text{K.E.}}{\text{P.E.}}$	(P) 2
(B) P.E. + 2 K.E.	(Q) $-\frac{1}{2}$
(C) $\frac{\text{P.E.}}{\text{T.E.}}$	(R) –1
(D) $\frac{\text{K.E.}}{\text{T.E.}}$	(S) 0

5.

Column-I	Column-II
(A) Lyman series	(P) Visible region
(B) Humphery series	(Q) Ultraviolet region
(C) Paschen series	(R) Infrared region
(D) Balmer series	(S) Far infrared region

6. In case of hydrogen spectrum wave number is given by

$$\bar{\nu} = R_H \left[\frac{1}{n_1^2} - \frac{1}{n_2^2}\right] \quad \text{where } n_1 < n_2$$

Column-I		Column-II
(A) Lyman series	(P)	$n_2 = 2$
(B) Balmer series	(Q)	$n_2 = 3$
(C) Pfund series	(R)	$n_2 = 6$
(D) Brackett series	(S)	$n_2 = 5$

7.

Column-I (Shell)		Column-II (Value of l)
(A) 2nd	(P)	1
(B) 3rd	(Q)	2
(C) 4th	(R)	3
(D) 1st	(S)	0

8. In Bohr's model, for a unielectronic atom, following symbols are used
$r_{n,Z} \rightarrow$ Radius of nth orbit with atomic number Z;
$U_{n,Z} \rightarrow$ Potential energy of electron ; $K_{n,Z} \rightarrow$ Kinetic energy of electron ;
$V_{n,Z} \rightarrow$ Velocity of electron ; $T_{n,Z} \rightarrow$ Time period of revolution

Column-I		Column-II
(A) $U_{1,2} : K_{1,1}$	(P)	1 : 8
(B) $r_{2,1} : r_{1,2}$	(Q)	–8 : 1
(C) $V_{1,3} : V_{3,1}$	(R)	9 : 1
(D) $T_{1,2} : T_{2,2}$	(S)	8 : 1

9.

Column-I		Column-II
(A) The radial node of 5s atomic orbital is	(P)	1
(B) The angular node of $3d_{yz}$ atomic orbital is	(Q)	4
(C) The sum of angular node and radial node of $4d_{xy}$ atomic orbital	(R)	2
(D) The angular node of 3p atomic orbital is	(S)	3

10.

Column-I		Column-II
(A) The d-orbital which has two angular nodes	(P)	$3d_{x^2-y^2}$
(B) The d-orbital with two nodal surfaces form cones	(Q)	$3d_{z^2}$
(C) The orbital without angular node	(R)	$4f$
(D) The orbital which has three angular nodes	(S)	$3s$

11.

	Column-I		Column-II
(A)	Orbital angular momentum of an electron	(P)	$\sqrt{s(s+1)}\,\dfrac{h}{2\pi}$
(B)	Angular momentum of an electron in an orbit	(Q)	$\sqrt{n(n+2)}$
(C)	Spin angular momentum of an electron	(R)	$\dfrac{nh}{2\pi}$
(D)	Magnetic moment of atom	(S)	$\sqrt{l(l+1)}\,\dfrac{h}{2\pi}$

12.

	Column-I		Column-II
(A)	Number of orbitals in the nth shell	(P)	$2(2l+1)$
(B)	Maximum number of electrons in a subshell	(Q)	n
(C)	Number of subshells in n^{th} shell	(R)	$2l+1$
(D)	Number of orbitals in a subshell	(S)	n^2

13.

	Column-I		Column-II
(A)	2s	(P)	$n=4, l=2, m=0$
(B)	$2p_z$	(Q)	$n=4, l=2, m=-2$ or $+2$
(C)	$4d_{x^2-y^2}$	(R)	$n=2, l=1, m=0$
(D)	$4d_{z^2}$	(S)	$n=2, l=0, m=0$

14.

	Column-I		Column-II
(A)	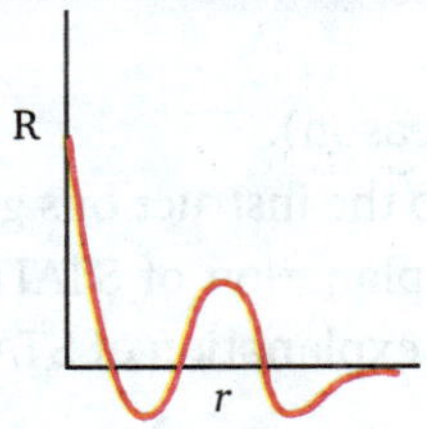	(P)	$4s$
(B)	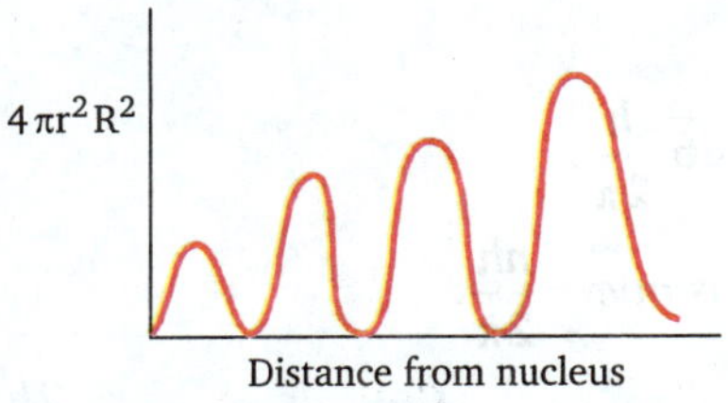	(Q)	$5p_y$
(C)	Angular probability depends upon θ and ϕ	(R)	$3s$
(D)	Atleast one angular node is present	(S)	$6d_{xy}$

15.

	Column-I		Column-II
(A)	3*s*	(P)	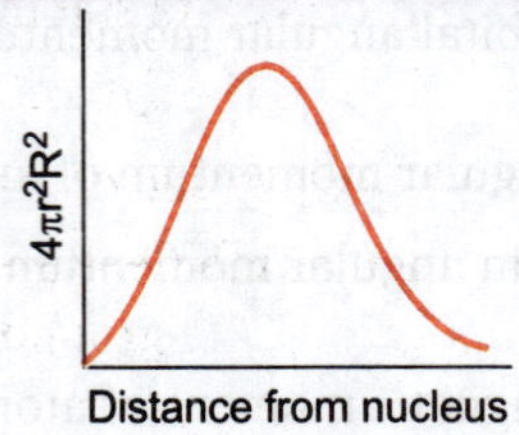
(B)	3*p*	(Q)	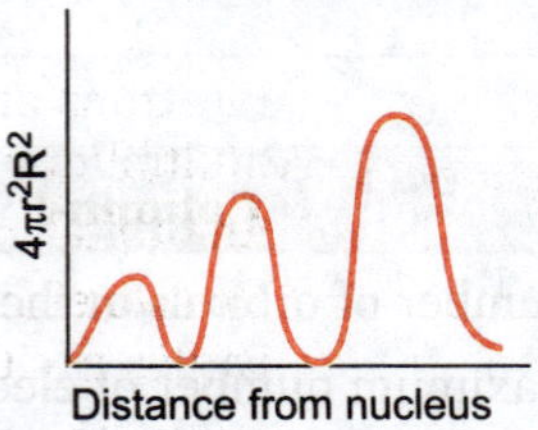
(C)	3*d*	(R)	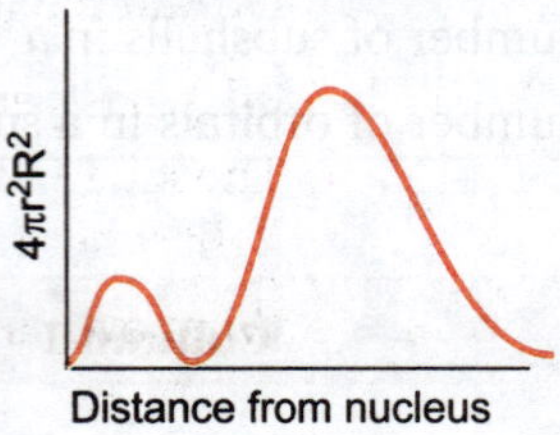
(D)	2*p*	(S)	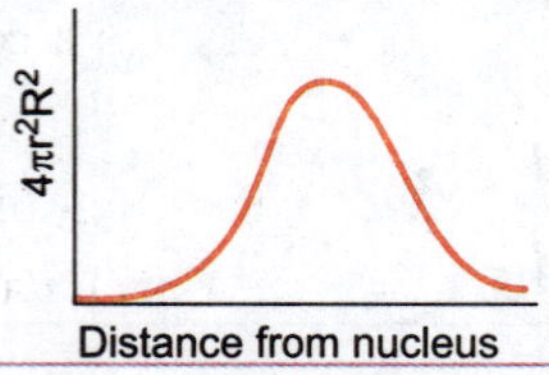

Assertion-Reason Type Questions

Each question contains STATEMENT-1 (Assertion) and STATEMENT-2 (Reason).
Examine the statements carefully and mark the correct answer according to the instructions given below :
(A) If both the statement are TRUE and STATEMENT-2 is the correct explanation of STATEMENT-1
(B) If both the statements are TRUE but STATEMENT-2 is NOT the correct explanation of STATEMENT-1
(C) If STATEMENT-1 is TRUE and STATEMENT-2 is FALSE
(D) If STATEMENT-1 is FALSE and STATEMENT-2 is TRUE

1. **STATEMENT-1 :** The angular momentum of *d*-orbitals is $\sqrt{6}\,\frac{h}{2\pi}$.
   STATEMENT-2 : Angular momentum of electron in orbit is $mvr = \frac{nh}{2\pi}$.
2. **STATEMENT-1 :** Angular momentum of the electron in the orbit which has four subshell is $\frac{2h}{\pi}$.
   STATEMENT-2 : Angular momentum of electron is quantized.
3. **STATEMENT-1 :** Line emission spectra useful in the study of atomic structure.
   STATEMENT-2 : Each element has a unique line emission spectrum.
4. **STATEMENT-1 :** Emitted radiation will fall in visible range when an electron jump from $n = 4$ to $n = 2$ in H-atom.
   STATEMENT-2 : Balmer series radiations belong to visible range for hydrogen atom only.

5. **STATEMENT-1 :** Half-filled and fully-filled degenerate orbitals are more stable.

STATEMENT-2 : Extra stability is due to the symmetrical distribution of electrons and exchange energy.

6. **STATEMENT-1 :** The ground state configuration of Cr is $3d^5 4s^1$.

STATEMENT-2 : A set of half-filled orbitals containing one electrons each with their spin parallel provides extra stability.

7. **STATEMENT-1 :** The ground state electronic configuration of nitrogen is

[↑↓] [↑↓] [↑ | ↑ | ↑]

STATEMENT-2 : Electrons are filled in orbitals as per aufbau principle, Hund's rule of maximum spin multiplicity and Pauli's principle.

8. **STATEMENT-1 :** An orbital cannot have more than two electrons and they must have opposite spins.

STATEMENT-2 : No two electrons in an atom can have same set of all the four quantum numbers as per Pauli's exclusion principle.

9. **STATEMENT-1 :** Orbital having *xz* plane as node may be $3d_{xy}$.

STATEMENT-2 : $3d_{xy}$ has zero radial node.

10. **STATEMENT-1 :** The kinetic energy of photo-electrons increases with increase in frequency of incident light where $\nu > \nu_0$.

STATEMENT-2 : Whenever intensity of light is increased the number of photo-electron ejected always increases.

11. **STATEMENT-1 :** Cu^{2+} is a coloured ion.

STATEMENT-2 : Every ion with unpaired electron is coloured.

12. **STATEMENT-1 :** For $n = 3$, l may be 0,1 and 2 and m may be 0; 0 ± 1 ; 0 ± 1 and ± 2

STATEMENT-2 : For each value of n, there are 0 to $(n - 1)$ possible values of l; and for each value of l, there are 0 to $\pm l$ values of m.

Level-4

Subjective Problems

1. Given $r_{n+1} - r_{n-1} = 2r_n$, where r_n, r_{n-1}, r_{n+1} are Bohr radius for hydrogen atom in n^{th}, $(n-1)^{th}$ and $(n+1)^{th}$ shell respectively. Calculate the value of n.

2. The energy of separation of an electron is 30.6 eV moving in an orbit of Li^{+2}. Find out the number of waves made by the electron in one complete revolution in the orbit.

3. Calculate the number of waves made by a Bohr electron in one complete revolution in n^{th} orbit of H-atom, if ratio of de-Broglie wavelength associated with electron moving in n^{th} orbit and 2nd orbit is 1.5.

4. A certain dye absorbs lights of $\lambda = 400$ nm and then fluorescence light of wavelength 500 nm. Assuming that under given condition 40% of the absorbed energy is re-emitted as fluorescence, calculate the ratio of quanta absorbed to number of quanta emitted out.

5. A photon of energy 4.5 eV strikes on a metal surface of work function 3.0 eV. If uncertainty in position is $\frac{25}{4\pi}$ Å, find the uncertainty in measurement of de-Broglie wavelength (in Å).

6. Find out the difference in number of angular nodes and number of radial nodes in the orbital to which last electron of chromium present.

7. What is the total number of radial and angular nodes present in $5f$ orbital ?

8. Infrared lamps are used in restaurants to keep the food warm. The infrared radiation is strongly absorbed by water, raising its temperature and that of the food. If the wavelength of infrared radiation is assumed to be 1500 nm, and the number of quanta of infrared radiation produced per second by an infrared lamp (that consumes energy at the rate of 100 W and is 12% efficient only) is $(x \times 10^{19})$, then the value of x is :

(Given : $h = 6.625 \times 10^{-34}$ J-s)

9. When an electron makes transition from $(n+1)$ state to n state the wavelength of emitted radiations is related to n $(n >>> 1)$ according to $\lambda \propto n^x$.

What is the value of x ?

10. For 3s orbital of hydrogen atom, the normalised wave function is

$$\psi_{3s} = \frac{1}{81\sqrt{3\pi}}\left(\frac{1}{a_0}\right)^{3/2}\left[27 - \frac{18r}{a_0} + \frac{2r^2}{a_0^2}\right]e^{\frac{-r}{3a_0}}$$

If distance between the radial nodes is d, calculate the value of $\frac{d}{1.73\,a_0}$.

11. Find the separation between two electrons (in Å) in vacuum, if electrostatic potential energy between these electrons is 7.67×10^{-19} J.

[Given: $e = 1.6 \times 10^{-19}$ C ; $\epsilon_0 = 8.85 \times 10^{-12}$ J^{-1} $C^2 m^{-1}$, $\pi = 3.14$]

12. An α- particle moving with velocity $\frac{1}{30}$ th times of velocity of light. If uncertainty in position is $\frac{3.31}{\pi}$ pm, then minimum uncertainty in kinetic energy is $y \times 10^{-16}$ J. Calculate the value of y.

13. In a sample of excited hydrogen atoms electrons make transition from $n = 2$ to $n = 1$. Emitted quanta strike on a metal of work function (ϕ) 4.2eV.

Calculate the wavelength (in Å) associated with ejected electrons having maximum kinetic energy.

14. For 1s orbital of Hydrogen atom radial wave function is given as:

$$R(r)=\frac{1}{\sqrt{\pi}}\left(\frac{1}{a_0}\right)^{3/2} e^{-r/a_0} \qquad (\text{where } a_0 = 0.529\text{Å})$$

The ratio of radial probability density of finding electron at $r=a_0$ to the radial probability density of finding electron at the nucleus is given as $(x.e^{-y})$. Calculate the value of $(x+y)$.

15. Calculate the value of A.

$A=\frac{E_{1,2}}{2E_{2,1}}$ where $E_{n,Z}$ = Energy of electron in n^{th} orbit, Z= atomic number of hydrogen like species.

Numerical Problems

(If the numerical value has more than two decimal places, truncate/round-off the value to TWO decimal places.)

1. The radius of an orbit of H-atom is 0.2116 nm. Kinetic energy (in eV) of electron present in this orbit is:

2. What is the value of $\frac{x}{y}$; if x = velocity of electron in second Bohr's orbit of hydrogen atom and y = velocity of electron in sixth Bohr's orbit of He^+.

3. An electron in a hydrogen atom in its ground state absorbs 1.5 times as much energy as the minimum required for it to escape from the atom.
If the velocity of the emitted electron is $y \times 10^6$ m/sec then find the value of y?

4. The excitation energy of an electron moving in an orbit in H-atom is 10.2 eV. The number of waves made by electron in excited state is:

5. Calculate change in potential energy (only magnitude) of an electron, when electron jump take place corresponding to α line in U.V. region of H-atom. (Give your answer in eV)

6. In a discharge tube, there are only two hydrogen atoms. If the electrons in both atoms are de-exciting from 4th orbit. The maximum number of spectral lines of different λ should respectively be:

7. Consider the following graph between stopping potential (V_0) and incident frequency (ν).

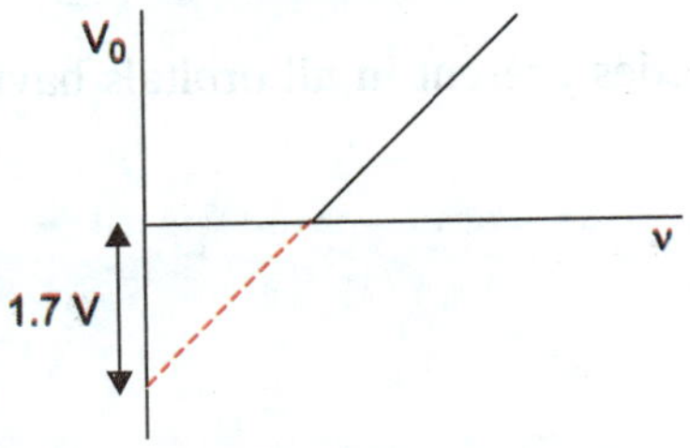

If radiation of wavelength 310 nm incident on metal surface then calculate value of stopping potential (in volt). $(hc = 1240\text{ eV-nm})$

8. The wave motion of an electron in an Bohr's orbit of hydrogen atom is as shown in following figure.

Its de-Broglie wavelength (in Å) is:

9. The wavelength of electrons accelerated from rest through a potential difference of 40 kV is $x \times 10^{-12}$ m. The value of x is ________. (Nearest integer)

Given: Mass of electron $= 9.1 \times 10^{-31}$ kg

Charge on an electron $= 1.6 \times 10^{-19}$ C

Planck's constant $= 6.63 \times 10^{-34}$ Js

10. A source of monochromatic radiation of wavelength 400 nm provides 1000 J of energy in 10 seconds. When this radiation falls on the surface of sodium, $x \times 10^{20}$ electrons are ejected per second. Assume that wavelength 400 nm is sufficient for ejection of electron from the surface of sodium metal. The value of x is _____. (Nearest integer) ($h = 6.626 \times 10^{-34}$ Js)

11. A proton and a Li^{3+} nucleus are accelerated by the same potential. If λ_{Li} and λ_P denote the de-Broglie wavelengths of Li^{3+} and proton respectively, then the value of $\frac{\lambda_{Li^{3+}}}{\lambda_P}$ is $x \times 10^{-1}$.

The value of x is________. (Rounded off to the nearest integer)

(Mass of Li^{3+} = 8.3 mass of proton)

12. A metal surface is exposed to 500 nm radiation. The threshold frequency of the metal for photoelectric current is 4.3×10^{14} Hz. The velocity of ejected electron is ________ $\times 10^5$ ms^{-1}.

(Nearest integer)

(Use: $h = 6.63 \times 10^{-34}$ Js, $m_e = 9.0 \times 10^{-31}$ kg)

13. If number of orbitals present in the highest energy subshell of a shell are 5 then the maximum number of electrons having $s = +1/2$ in this energy shell, will be :

14. Calculate total no. of nodes (radial & angular) in orbital represented by wave function $\Psi_{5, 1, 1}$.

15. Given graph is possible for various orbitals present in different subshell.

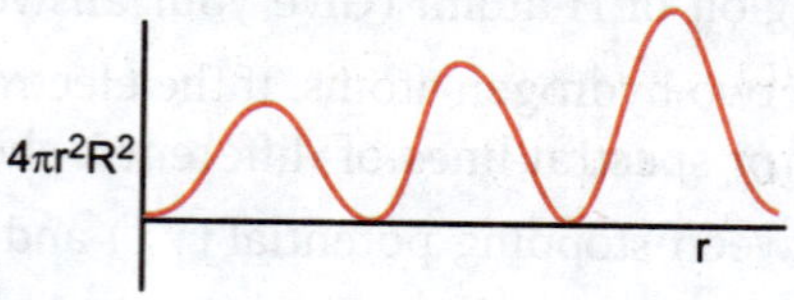

Calculate total number of radial nodes present in all orbitals having $(n + l) \leq 7$.

Answers

Level-1

1. (b)	2. (c)	3. (d)	4. (b)	5. (c)	6. (b)	7. (d)	8. (c)	9. (a)	10. (d)
11. (d)	12. (b)	13. (c)	14. (c)	15. (b)	16. (d)	17. (b)	18. (a)	19. (d)	20. (b)
21. (d)	22. (c)	23. (d)	24. (a)	25. (d)	26. (b)	27. (a)	28. (d)	29. (a)	30. (d)
31. (c)	32. (d)	33. (d)	34. (b)	35. (b)	36. (b)	37. (b)	38. (c)	39. (c)	40. (c)
41. (d)	42. (d)	43. (d)	44. (b)	45. (d)	46. (c)	47. (b)	48. (d)	49. (d)	50. (b)
51. (a)	52. (b)	53. (b)	54. (c)	55. (c)	56. (a)	57. (c)	58. (b)	59. (c)	60. (a)
61. (d)	62. (c)	63. (b)	64. (b)	65. (c)	66. (d)	67. (c)	68. (b)	69. (c)	70. (c)
71. (b)	72. (b)	73. (d)	74. (c)	75. (c)	76. (b)	77. (a)	78. (a)	79. (a)	80. (a)
81. (c)	82. (b)	83. (a)	84. (d)	85. (b)	86. (b)	87. (b)	88. (a)	89. (d)	90. (c)
91. (b)	92. (a)	93. (d)	94. (b)	95. (d)	96. (a)	97. (c)	98. (c)	99. (a)	100. (c)
101. (a)	102. (c)	103. (c)	104. (c)	105. (d)	106. (a)	107. (a)	108. (c)	109. (b)	110. (d)
111. (c)	112. (d)	113. (c)	114. (a)	115. (c)	116. (a)	117. (d)	118. (d)	119. (d)	120. (d)
121. (b)	122. (c)	123. (d)	124. (c)	125. (d)	126. (c)	127. (d)	128. (d)	129. (d)	130. (c)
131. (c)	132. (d)	133. (b)	134. (a)	135. (b)	136. (c)	137. (d)	138. (c)	139. (a)	140. (a)
141. (d)	142. (b)	143. (d)	144. (b)	145. (d)	146. (c)	147. (a)	148. (a)	149. (d)	150. (a)
151. (d)	152. (c)	153. (c)	154. (b)	155. (a)	156. (b)	157. (b)	158. (d)	159. (b)	160. (c)
161. (d)	162. (d)	163. (a)	164. (c)	165. (a)	166. (a)	167. (d)	168. (a)	169. (a)	170. (b)
171. (d)	172. (d)	173. (c)	174. (b)	175. (d)	176. (a)	177. (a)	178. (b)	179. (c)	180. (a)
181. (c)	182. (b)	183. (c)	184. (b)	185. (c)					

Level-2

1. (c)	2. (c)	3. (c)	4. (c)	5. (a)	6. (d)	7. (b)	8. (c)	9. (b)	10. (d)
11. (a)	12. (d)	13. (a)	14. (b)	15. (c)	16. (c)	17. (c)	18. (d)	19. (d)	20. (d)
21. (a)	22. (b)	23. (c)	24. (a)	25. (b)	26. (b)	27. (c)	28. (c)	29. (b)	30. (a)

Level-3

Passage–1	1.	(c)	2.	(a)	3.	(c)		
Passage–2	1.	(b)	2.	(b)	3.	(d)	4.	(c)
Passage–3	1.	(d)	2.	(c)				
Passage–4	1.	(b)	2.	(b)				
Passage–5	1.	(d)	2.	(c)	3.	(a)		

One or More Answers is/are Correct

1. (a,b,c)	2. (a,b,c)	3. (d)	4. (a,c,d)	5. (a,b,d)	6. (a,b,c)
7. (a,b,c,d)	8. (c)	9. (b,c,d)	10. (a,b,d)	11. (a,b,c,d)	12. (a,d)
13. (a,b,c,d)	14. (b,c,d)	15. (a,b,c,d)	16. (a)	17. (b,c,d)	18. (b,c)
19. (a,b,c,d)	20. (c,d)	21. (a,b,d)	22. (a,b,c,d)		

Match the Column

1. $A \rightarrow P, R$;	$B \rightarrow Q, R$;	$C \rightarrow S$;	$D \rightarrow Q, R$
2. $A \rightarrow S$;	$B \rightarrow P, R$;	$C \rightarrow P, R$;	$D \rightarrow P, Q, R$
3. $A \rightarrow S$;	$B \rightarrow Q$;	$C \rightarrow R$;	$D \rightarrow P$
4. $A \rightarrow Q$;	$B \rightarrow S$;	$C \rightarrow P$;	$D \rightarrow R$
5. $A \rightarrow Q$;	$B \rightarrow S$;	$C \rightarrow R$;	$D \rightarrow P$
6. $A \rightarrow P, Q, R, S$;	$B \rightarrow Q, R, S$;	$C \rightarrow R$;	$D \rightarrow R, S$
7. $A \rightarrow P, S$;	$B \rightarrow P, Q, S$;	$C \rightarrow P, Q, R, S$;	$D \rightarrow S$
8. $A \rightarrow Q$;	$B \rightarrow S$;	$C \rightarrow R$;	$D \rightarrow P$
9. $A \rightarrow Q$;	$B \rightarrow R$;	$C \rightarrow S$;	$D \rightarrow P$
10. $A \rightarrow P; Q$	$B \rightarrow Q$;	$C \rightarrow S$;	$D \rightarrow R$
11. $A \rightarrow S$;	$B \rightarrow R$;	$C \rightarrow P$;	$D \rightarrow Q$
12. $A \rightarrow S$;	$B \rightarrow P$;	$C \rightarrow Q$;	$D \rightarrow R$
13. $A \rightarrow S$;	$B \rightarrow R$;	$C \rightarrow Q$;	$D \rightarrow P$
14. $A \rightarrow P$;	$B \rightarrow P, Q, S$;	$C \rightarrow Q, S$;	$D \rightarrow Q, S$
15. $A \rightarrow Q$;	$B \rightarrow R$;	$C \rightarrow S$;	$D \rightarrow P$

Assertion-Reason Type Questions

1. (B)	2. (B)	3. (A)	4. (A)	5. (A)	6. (A)	7. (A)	8. (A)	9. (B)	10. (C)
11. (C)	12. (A)								

Level-4

Subjective Problems

1. 2	2. 2	3. 3	4. 2	5. 4	6. 2	7. 4	8. 9	9. 3	10. 3
11. 3	12. 5	13. 5	14. 3	15. 8					

Numerical Problems

1. 3.40	2. 1.50	3. 1.55	4. 2.00	5. 20.40	6. 4.00	7. 2.30	8. 9.97
9. 6.00	10. 2.00	11. 2.00	12. 5.00	13. 9.00	14. 4.00	15. 18.00	

Hints & Solutions

Level 1

1. (b) This statement is Gay-Lussac's law of gaseous volumes.

2. (c) Travel perpendicular to both electric and magnetic fields.

3. (d) Charge/mass for $n = 0$, for $\alpha = \frac{2}{4}$,

for $p = \frac{1}{1}$, for $e^- = \frac{1}{1/1837}$

6. (b) $r_{atom} \approx 10^{-10}$m

$r_{nucleus} \approx 10^{-5}$m

$$\frac{V_{Atom}}{V_{Nucleus}} = \frac{4/3\pi(10^{-10})^3}{4/3\,\pi(10^{-15})^3} = 10^{15}$$

9. (a) Permissible value of orbit angular momentum of $e^- = n\left(\frac{h}{2\pi}\right)$

$n = 1, 2, 3, \ldots\ldots$

for $n = 3$, orbit angular momentum

$$= 1.5\left(\frac{h}{\pi}\right)$$

10. (d) $r \propto \frac{n^2}{Z}$, $\frac{r_1}{r_2} = \frac{(1)^2}{(3)^2}$

$r_2 = 9r_1$

11. (d) $\frac{r_1}{r_2} = \left(\frac{n_1}{n_2}\right)^2 = \frac{4}{9}$

12. (b) r_1 of H-atom $= 0.529$ Å r_n

(n like atom) $= \frac{n^2}{Z} \times r_1$ (H-atom)

r_n of Be^{3+} $\Rightarrow \frac{n^2}{Z} \times r_1$ (H-atom)

$= 0.529$ Å ($Z = 4$ for Be^{3+})

$$\Rightarrow \frac{n^2}{Z} \times 0.529 = 0.529 = n^2 = Z$$

$$\Rightarrow \qquad n^2 = 4, n = 2$$

13. (c) $r_1 = 0.529$ Å; $r_{4(X)} = r_1 \times \frac{n^2}{Z}$;

$$r_{4(X)} \Rightarrow \frac{0.529 \times (4)^2}{Z}; \; Z = 16$$

14. (c) $r = 0.529\,\frac{n^2}{Z}$ Å;

$$r_4 - r_3 = 0.529\left(\frac{16}{2} - \frac{9}{2}\right) \text{Å} = 1.851 \times 10^{-10} \text{ m}$$

15. (b) $v = 2.188 \times 10^6\left(\frac{Z}{n}\right)$ ms^{-1}

$$= 2.188 \times 10^6\left(\frac{3}{5}\right) \text{ms}^{-1}$$

17. (b) $v \propto \frac{Z}{n}$

18. (a) $\because \quad \frac{mv^2}{r} = \frac{1}{4\pi\varepsilon_0} \cdot \frac{Ze \cdot e}{r^2}$

$\therefore \quad v^2 = \frac{e^2}{4\pi\varepsilon_0 rm}$ (For hydrogen $Z = 1$)

$$v^2 = \frac{e^2}{4\pi\varepsilon_0 rm} \Rightarrow v = \frac{e}{\sqrt{4\pi\varepsilon_0 rm}}$$

19. (d) Frequency of revolution

$$= \frac{\text{velocity in second orbit } (v_2)}{2\pi r_2}$$

$$= \frac{1.082 \times 10^6 \text{ ms}^{-1}}{2 \times \pi \times (2.12 \times 10^{-10}) \text{ m}} = 8.2 \times 10^{14} \text{ s}^{-1}$$

20. (b) $T \propto \frac{n^3}{Z^2}$; $\frac{T_1}{T_2} = \frac{n_1^3}{Z_1^2} \times \frac{Z_2^2}{n_2^3}$

$$= \frac{2^3}{1} \times \frac{2^2}{3^3} = \frac{32}{27}$$

21. (d) $T \propto \frac{n^3}{Z^2}$; $\frac{T_{1,2}}{T_{2,1}} = \frac{1}{4} \times \frac{1}{8} = \frac{1}{32}$

22. (c) $v \propto \frac{Z}{n}$; $r \propto \frac{n^2}{Z}$;

frequency of revolution $= \frac{v_n}{2\pi r_n}$;

Coulombic force of attraction $= \frac{Ze^2}{(4\pi\varepsilon_0)\,r^2}$

23. (d) KE $= \frac{1}{2}\frac{Ze^2}{r}$

24. (a) $E_n = \frac{1}{2}$ P.E. $= -\frac{6.8}{2} = -3.4$ eV

$\because$ $E_n \Rightarrow \frac{-13.6}{n^2} = -3.4$

$n = 2$ or first excited state

25. (d) Energy of N-shell $= \frac{-13.6 \times (4)^2}{(4)^2} = -13.6$ eV

$\therefore$ P.E. $= 2 \times E \Rightarrow 2 \times -13.6 = -27.2$ eV

26. (b) Total energy of third shell $= \frac{-13.6}{3^2}$

$= -1.51$ eV

K.E. = – Total energy $\Rightarrow$ 1.51 eV

P.E. = 2 × T.E. = –3.02 eV

27. (a) Change in P.E. $= -\frac{2x}{4} + (2x) \Rightarrow \frac{3}{2}x$

30. (d) E_H in first orbit $= \frac{-19.6 \times 10^{-18}}{4}$ J

$E_{Be^{3+}}$ in second orbit

$= -\left(\frac{19.6 \times 10^{-18}}{4}\right) \times \frac{16}{4} = -19.6 \times 10^{-18}$ J

31. (c) $\Delta E = 13.6\, Z^2 \left(\frac{1}{n_1^2} - \frac{1}{n_2^2}\right)$ eV atom^{-1}

For the ionization of Li^{2+} ($Z = 3$) from first excited state, $n_1 = 2$ and $n_2 = \infty$.

Hence, IP $= \Delta E = 13.6 \times 3^2 \times \left(\frac{1}{2^2} - \frac{1}{\infty^2}\right)$

$= 30.6$ eV

32. (d) For Be^{3+} $E_\infty - E_2 = +13.6\frac{z^2}{n^2}$

$= 13.6 \times \frac{4^2}{2^2} = 54.4$ eV

33. (d) $E \propto -\frac{Z^2}{n^2}$

$\therefore$ $\frac{E_{1,1}}{E_{3,4}} = \frac{1}{1} \times \frac{9}{4^2} = \frac{9}{16}$

34. (b) $\Delta E = 13.6 \times 9\left(\frac{1}{2^2} - \frac{1}{3^2}\right)$

$= 13.6 \times 9 \times \frac{5}{4 \times 9} = 17$eV

35. (b) $\frac{E_1}{E_2} = \left(\frac{Z_1}{Z_2}\right)^2 = \left(\frac{1}{2}\right)^2$

$E_2 = 4E_1 = -13.64$ eV

36. (b) $\because$ $E_n \propto -\frac{Z^2}{n^2} \Rightarrow E_n \propto -Z^2$

38. (c) When an electron of charge e and mass m is accelerated with a potential difference of V volts K.E. $= eV$

$\Rightarrow$ $\frac{1}{2}mv^2 = eV$ or $v^2 = \frac{2\,eV}{m}$

$\Rightarrow$ $v = \sqrt{\frac{2\,eV}{m}}$

39. (c) Frequency order

γ-radiation > X-rays > U-V > Green light > Infrared > Microwave > Radio waves.

40. (c) $E = h\nu = 6.62 \times 10^{-34} \times 4.2 \times 10^{14}$

$= 2.8 \times 10^{-19}$ J

41. (d) $E = \frac{1240}{\lambda_{nm}} = \frac{1240}{91} = 13.6$ eV

42. (d) $E = n\frac{hc}{\lambda}$

$E = \frac{2 \times 10^{-17}}{1.6 \times 10^{-19}}$ eV $= n \times \frac{1240}{595.2} \times \frac{\text{eV.nm}}{\text{nm}}$

$\Rightarrow n = 60$

43. (d) $E = \frac{hc}{\lambda}$; $E\lambda$ = constant

44. (b) $E \propto \frac{1}{\lambda} \Rightarrow \frac{E_1}{E_2} = \frac{\lambda_2}{\lambda_1} = \frac{16000\ \text{Å}}{8000\ \text{Å}} = 2$

$\therefore$ $E_1 = 2E_2$

45. (d) $\frac{E_1}{E_2} = \frac{\bar{\nu}_1}{\bar{\nu}_2} = \frac{\bar{\nu}_1}{\bar{\nu}_2} = \frac{2}{3}$

46. (c) $E = \frac{nhc}{\lambda} = nhc\bar{\nu}$ $\left(\because \bar{\nu} = \frac{1}{\lambda}\right)$

$\therefore$ $10 = nhc\bar{\nu}$ or $n = \frac{10}{hc\bar{\nu}} = \frac{10}{hcx}$

52. (b) $K\,L\,M\,N$..........> energy increases.

53. (b) $\Delta E = 13.6Z^2\left[\frac{1}{n^2} - \frac{1}{(n+1)^2}\right]$

55. (c) $\Delta E = 13.6 \times (2)^2\left[\frac{1}{4} - \frac{1}{36}\right] \times \frac{1.6 \times 10^{-19} \times 6.022 \times 10^{23}}{1000}$

$= 1.16 \times 10^3$ kJ/mol

56. (a) For the transition $n = 1$ to $n = 2$, the energy change, ΔE is positive, *i.e.*, energy is absorbed. For the transition $n = 5$ to $n = 1$, ΔE is negative, *i.e.*, energy s released.

57. (c) $\frac{1}{\lambda} = RZ^2\left[\frac{1}{n_1^2} - \frac{1}{n_2^2}\right] = R \times 3^2\left[\frac{1}{3^2} - \frac{1}{\infty^2}\right] = R$

$\therefore$ $\lambda = \frac{1}{R}$

58. (b) $\frac{1}{\lambda} = RZ^2\left[\frac{1}{n_1^2} - \frac{1}{n_2^2}\right] = R \times 2^2\left[\frac{1}{1^2} - \frac{1}{2^2}\right] = 3R$

$\therefore$ $\lambda = \frac{1}{3R}$

59. (c) For hydrogen atom

$\frac{1}{\lambda} = R_H\left[\frac{1}{n_1^2} - \frac{1}{n_2^2}\right]$;

$\lambda = 2170$ nm $= 2170 \times 10^{-9}$ m;

$R_H = 1.09677 \times 10^7$ m^{-1}

$\therefore \frac{10^9}{2170} = 1.09677 \times 10^7 \left[\frac{1}{n_1^2} - \frac{1}{7^2}\right]; n_1 = 4$

So, electron transition from $n = 7$ to $n = 4$ will produce infrared light of wavelength 2170 nm.

60. (a) Total number of spectral lines given by

$\frac{1}{2}[n-1] \times n = 15; \therefore n = 6$

Thus, electron are excited upto 6th energy level from ground state. Therefore,

$\frac{1}{\lambda} = R_H\left[\frac{1}{1^2} - \frac{1}{n^2}\right] = 109737 \times \frac{35}{36};$

$\lambda = 9.373 \times 10^{-6}$ cm $= 937.3$ Å

61. (d) $\Delta E \propto \frac{1}{\lambda}$ for $\lambda_{max}, \Delta E_{min}$

62. (c) $\lambda = \frac{91.2}{9\left[\frac{1}{(2)^2} - \frac{1}{(3)^2}\right]} = 91.2 \times \frac{36}{9 \times 5} = 73$ nm

63. (b) $n_1 + n_2 = 4; \; n_2 - n_1 = 2; \therefore n_1 = 1, \; n_2 = 3$

$\bar{\nu} = R(2)^2\left[\frac{1}{1^2} - \frac{1}{3^2}\right] = \frac{32R}{9}$

64. (b) $n_1 = 2$

$n_2 = n$

$\bar{\nu} = R(1)^2\left[\frac{1}{(2)^2} - \frac{1}{n^2}\right] = R\left[\frac{n^2 - 2^2}{4n^2}\right]$

65. (c) Max. no. of radiation of different λ emitted

$= \frac{n(n-1)}{2} = \frac{4 \times 3}{2} = 6$

66. (d) $n_1 = 2$

$n_2 = 6$

Total number of photon emitted

$= \frac{(n_2 - n_1)(n_2 - n_1 + 1)}{2}$

$= \frac{(6-2)(6-2+1)}{2} = 10$

68. (b) $h\nu = h\nu_0 + \text{KE}$

69. (c) $h\nu = h\nu_0 + \text{KE}$; $\text{KE} = h\nu - h\nu_0$

70. (c) As per Einstein's equation of photoelectric effect

$h\nu = h\nu_0 + \text{K.E.}$

$\therefore \frac{1}{2}mv^2 = h\nu - h\nu_0 = \frac{hc}{\lambda} - \frac{hc}{\lambda_0}$

$v^2 = \frac{2hc}{m}\left(\frac{1}{\lambda} - \frac{1}{\lambda_0}\right);$

$v = \left[\frac{2hc}{m}\left(\frac{1}{\lambda} - \frac{1}{\lambda_0}\right)\right]^{1/2}$

$\Rightarrow \left[\frac{2hc}{m}\left(\frac{\lambda_0 - \lambda}{\lambda\lambda_0}\right)\right]^{1/2}$

71. (b) $E_{photon} = 10.2$ eV, KE $= 10.2 - 8.7 = 1.5$ eV

72. (b) $\because \quad E = \frac{hc}{\lambda}$

$6.63 \times 10^{-19} = \frac{6.63 \times 10^{-34} \times 3 \times 10^8}{\lambda}$

$\lambda = 3 \times 10^{-7}$ m $= 300 \times 10^{-9}$ m $= 300$ nm

74. (c) $\frac{hc}{\lambda} = 1 + \phi$ ...(1)

$3 \times \frac{hc}{\lambda} = 4 + \phi$ ...(2)

from, eq. (1) and (2) $\phi = 0.5$ eV

75. (c) $\frac{1}{2}mv^2 = \frac{1240 \text{ eV nm}}{31 \text{ nm}} = 40 - 12.8$ eV

$= 27.2$ eV

$= \frac{1}{2} \times 9.1 \times 10^{-31} \times v^2 = 27.2 \times 1.6 \times 10^{-19}$

$v = 3.09 \times 10^6$ m/s

76. (b) $h\nu = h\nu_0 + eV_0$

$V_0 = \frac{h\nu}{e} - \frac{h\nu_0}{e}$

77. (a) $h\nu = h\nu_0 + eV_0; \; eV_0 = h\nu - h\nu_0$ or

$V_0 = \frac{h}{e}\nu - \frac{h}{e}\nu_0; \;$ slope$_1 = \frac{h}{e}$

Similarly, $h\nu = h\nu_0 + K_{max}$

or $K_{max} = h\nu - h\nu_0;$

slope$_2 = h$, $\frac{\text{slope}_2}{\text{slope}_1} = \frac{h}{h/e} = e$

80. (a) We know that,

$\lambda = \frac{h}{mv} = \frac{6.625 \times 10^{-34}}{1 \times 10^{-3} \times 100} = 6.625 \times 10^{-33}$m

$(\because 1\text{g} = 1 \times 10^{-3}$ kg$)$

81. (c) $\lambda \propto \frac{1}{m}$

82. (b) $\lambda = \frac{h}{mv} = \frac{6.63 \times 10^{-34}}{10^{-6} \times 10} = 6.63 \times 10^{-29}$ m

83. (a) $\lambda = \frac{h}{mv} \Rightarrow v = \frac{h}{m\lambda}$

$= \frac{6.62 \times 10^{-34}}{1.67 \times 10^{-27} \times 0.150 \times 10^{-9}} = 26.42 \times 10^2$

$= 2.64 \times 10^3$ m/s

84. (d) $\lambda = \frac{h}{p}$

85. (b) $\lambda = \frac{h}{\sqrt{2qVm}}$

86. (b) $\because \quad \frac{1}{\lambda} = R\left(\frac{1}{n_1^2} - \frac{1}{n_2^2}\right); \; n_1 = 1, n_2 = ?;$

$\frac{1}{\lambda} = R\left(\frac{1}{1} - \frac{1}{n_2^2}\right) \Rightarrow n_2^2 = \frac{R\lambda}{R\lambda - 1}$

$\Rightarrow \quad n_2 = \sqrt{\frac{\lambda R}{\lambda R - 1}}$

87. (b) $E = E_1 + E_2$; $\dfrac{hc}{\lambda} = \dfrac{hc}{\lambda_1} + \dfrac{hc}{\lambda_2}$

$$\Rightarrow \frac{hc}{\lambda} = hc\left(\frac{\lambda_2+\lambda_1}{\lambda_1\lambda_2}\right); \quad \lambda = \frac{\lambda_1\lambda_2}{\lambda_1+\lambda_2}$$

88. (a) $\dfrac{1}{400} = \dfrac{1}{\lambda_1} + \dfrac{1}{\lambda_2}$ ....(1)

$$\frac{1}{400} = \frac{1}{\lambda_1} + \frac{4}{\lambda_1} = \frac{5}{\lambda_1} \Rightarrow \lambda_1 = 2000 \text{ nm}$$

$$\lambda_2 = 500 \text{ nm}$$

89. (d) $\lambda_p = \dfrac{h}{\sqrt{2\,qV\,m_p}}$

$$\lambda_{Be^{3+}} = \frac{h}{\sqrt{2\times 3\,eV\,m_{Be^{3+}}}} = \frac{h}{\sqrt{2\times 3\,eV\times 9m_p}}$$

Hence, $\dfrac{\lambda_{Be^{3+}}}{\lambda_p} = \sqrt{\dfrac{2\,eV\,m_p}{2\times 3\,eV\times 9m_p}} = \dfrac{1}{3\sqrt{3}}$

90. (c) $\lambda = \dfrac{h}{\sqrt{2\,eV\,m}} = \dfrac{1.23}{\sqrt{V}}\text{nm}$

91. (b) $\lambda = \dfrac{h}{m_e x} = \dfrac{h}{m_p V} = \dfrac{h}{1840\,m_e V}$ $[m_p = 1840\,m_e]$

Hence, $V = \dfrac{x}{1840}$

92. (a) $E = (mC)\cdot C$ or $P = \dfrac{E}{C}$

$$= \frac{6\times 10^6 \times 1.6\times 10^{-19}}{3\times 10^8}$$

$$= 3.2\times 10^{-21} \text{ kg-m/s}$$

96. (a) $\Delta x = \Delta p$

$$\Delta p^2 = \frac{h}{4\pi}; \quad m^2\Delta v^2 = \frac{h}{4\pi}; \quad \Delta v = \frac{1}{2m}\sqrt{\frac{h}{\pi}}$$

97. (c) $\Delta x = \dfrac{h}{4\pi\Delta p} = \dfrac{h}{4\pi\times m\Delta v}$

$$\approx \frac{6.63\times 10^{-34}}{4\times 3.14\times 9.1\times 10^{-31}\times 1} = 58\times 10^{-6}\text{ m}$$

$(\because \Delta v = 1\text{ m/s})$

$\therefore \quad \Delta x = 58\ \mu\text{m}$

98. (c) $\Delta x\cdot\Delta p = \dfrac{h}{4\pi}$

$$\Delta x\cdot m\Delta v = \frac{h}{4\pi}$$

$$\Delta v = 5\times 10^6\times\frac{0.02}{100}$$

$$\Delta v = 1000\text{ m/s}$$

$$\Delta x = \frac{h}{4\pi\times m\cdot\Delta v}$$

$$= \frac{6.63\times 10^{-34}}{4\times 3.14\times 9.1\times 10^{-31}\times 1000\text{ m/s}}$$

$$= 5.8\times 10^{-8}\text{ S} = 58\times 10^{-9}\text{ S}$$

99. (a) $m = 10^{-10}\text{ g} \Rightarrow 10^{-13}\text{ kg}$;

$$\Delta v = \frac{0.0001}{100}\times 10^{-6}\times 10^{-2} = 10^{-14}\text{ m sec}^{-1}$$

$$= \Delta x\cdot\Delta p = \frac{h}{4\pi}$$

$$\Rightarrow \quad \Delta x = \frac{h}{4\pi\Delta p} = \frac{h}{4\pi\cdot m\Delta v};$$

$$\Delta x = \frac{6.62\times 10^{-34}}{4\times 3.14\times 10^{-13}\times 10^{-14}}$$

$$\Delta x = \frac{6.62}{12.56}\times\frac{10^{-34}}{10^{-27}}; \quad \Delta x = 5.2\times 10^{-8}\text{ m}$$

101. (a) n, l and m are related to size, shape and orientation respectively.

102. (c) Azimuthal quantum number defines angular momentum.

103. (c) m = orientation of orbital.

104. (c) When $l = 3$ then

$$m = -3, -2, -1, 0, +1, +2, +3$$

105. (d) $|m| \not> l$

108. (c) Maximum number of electrons

$$= 2n^2 = 2\times 4^2 = 32 \qquad (\because n = 4)$$

109. (b) $5f$ subshell

110. (d) $l = 3$ means f-subshell, maximum number of electron in f-subshell = 14.

111. (c) For f subshell $l = 3$; $\therefore$ g subshell $l = 4$

for principal shell, $l = 4, n = 5$

total no. of orbital in shell $= n^2 = 5^2 = 25$

112. (d) g-subhsell $l = 4$, $2(2l+1) = 18e^-$

113. (c) $3d$ subshell 5 orbital exist.

114. (a) Pauli's exclusion principle

115. (c) $1s^2 2s^2$ $\boxed{2p^6}$ $3s^2$ $\boxed{3p^6}$

⥮	⥮	⥮		⥮	⥯	⥮
0	+1	−1		0	+1	−1

117. (d) $n = 2, l = 2$ is not possible.

119. (d) $3d$-subshell contains maximum 10 electrons.

122. (c) Maximum number of electrons with same spin is equal to maximum number of orbitals, *i.e.*, $(2l+1)$.

124. (c) For p-orbital, $l = 1$ means dumb-bell shape.

125. (d) Orbital angular momentum $= \sqrt{l(l+1)}\,\dfrac{h}{2\pi}$

$l = 1$ for p-orbital.

126. (c) I $\quad \mu_l = \sqrt{l(l+1)}\,\dfrac{h}{2\pi} = \sqrt{12}\cdot\dfrac{h}{2\pi} = \sqrt{3}\cdot\dfrac{h}{\pi}$

II $\quad \mu_l = \sqrt{l(l+1)}\,\dfrac{h}{2\pi}$

$$l = 0$$

128. (d) 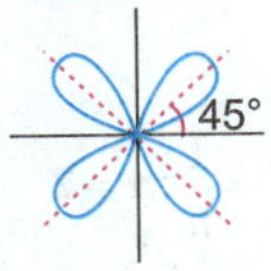

132. (d) No. of spherical nodes $= n - l - 1$;
for *s, p, d, f* values of $l = 0, 1, 2, 3$ respectively.

135. (b) Total no. of nodes $= (n-1)$

136. (c) Degenerate orbitals having equal energy.

137. (d) $E_{2p_x} = E_{2p_y} = E_{2p_z}$

138. (c) For H-atom, energy of 2*s*-orbital = energy of 2*p*-orbital so degeneracy is $1 + 3 = 4$

139. (a) For multielectron species (H^-) first excited state means e^- will transfer from 1*s* to 2*s* subshell & 2*s* subshell has only one orbital so degeneracy is 1.

152. (c) $P(33) = [Ar]\,4s^2 4p^3$

$n = 3$

153. (c) Unpaired electron of Cl atom is $3p^1$ for which $n = 3$,

Cl(17) $3s^2, 3p^5$ | ↑↓ | ↓↑ | ↑ |

$l = 1, m = -1, 0$ or $+1$ and $s = \pm\frac{1}{2}$

154. (b) $_{11}Na = 1s^2, 2s^2 2p^6, 3s^1$

$n = 3,\ l = 0,\ m = 0,\ s = +1/2$

157. (b) 4*f* has the highest energy among 3*d*, 4*f*, 4*p*, 5*s* orbitals.

158. (d) Orbitals are 4*s*, 2*s*, 3*p* and 3*d*. Out of these 3*d* has highest energy.

159. (b) 4*d*-orbital have higher energy in given data.

160. (c)

	$n + l$
e_1	3
e_2	4
e_3	5
e_4	4

$e_3 > e_2 > e_4 > e_1$

161. (d) 4*s*-orbital

162. (d) $[Ne] = 1s^2 2s^2, 2p^6$

163. (a) 3*d*-subshell filled with 5 electrons (half-filled) is more stable than that filled with 4 electrons. 1, 4*s*-electron jump into 3*d*-subshell for more stability.

169. (a)

Mn(25)	$[Ar]3d^5, 4s^2$	$n = 5$
Ti(22)	$[Ar]3d^2, 4s^2$	$n = 2$
V(23)	$[Ar]3d^3, 4s^2$	$n = 3$
Al(13)	$[Ne]3s^2, 3p^1$	$n = 1$

170. (b) Fe(III)—[Ar] $3d^5$; unpaired electrons = 5;

magnetic moment $= \sqrt{5(5+2)}$ BM

Co(II)—[Ar] $3d^7$; unpaired electrons = 3;

magnetic moment $= \sqrt{3 \times (3+2)}$ BM

171. (d) Given $\mu = \sqrt{n(n+2)} = 1.73$ BM

(where *n* is number of unpaired electrons)

$\therefore n = 1;\quad {}_{23}V = 1s^2 2s^2 2p^6 3s^2 3p^6 4s^2 3d^3$

| 1 | 1 | 1 | | | | ⇅ |

$3d^3$ $\quad$ $4s^2$

for $n = 1$ it must release 4 electrons, first two from 4*s*-orbital and then next two electrons from 3*d*-orbital. So, $x = 4$.

173. (c) ψ is related to only three quantum number $(n, l\ \&\ m)$.

175. (d) Permissible value of $n > l$.

178. (b) All *s* orbitals graph start from non-zero value.

No. of radial node $= n - l - 1 = 1$

When wave function change it's sign. indicates position of radial node(s).

179. (c) Radial node $(n - l - 1) = 2$

$l = 0,\ n = 3$

180. (a) Graph start from non-zero value, so *s*-orbital and 1 radial node is 3*s*.

181. (c) ψ^2 never be negative.

182. (b) For 2*p* orbital,

Radial node $= 2 - 1 - 1 = 0$

and $R_{(r)}$ vs. *r* start from origin.

183. (c) Radial node in 3*p* orbital $= n - l - 1$

$= 3 - 1 - 1 = 1$

All *p, d, f* orbital graph start from zero and height of peak (s) decreases when *r* increases.

184. (b) For radial node $\Psi_{2s} = 0,\ 2 - \frac{r}{a_0} = 0$

$\boxed{r = 2a_0}$

185. (c) Probability of finding e^- is zero implies that

$\Psi^2 = 0$ or $\Psi = 0$

$\Rightarrow \quad (\sigma - 1) = 0 \ \Rightarrow\ \sigma = 1$

or $\quad r_1 = \frac{a_0}{2z}$ or $(\sigma^2 - 8\sigma + 12) = 0$

and $\quad (\sigma - 6)(\sigma - 2) = 0$

$\sigma = 6,\ r = \frac{3a_0}{z}$

and $\quad \sigma = 2,\ r = \frac{a_0}{z}$

$r_2 = \frac{3a_0}{z}$

Level 2

1. (c) P.E. $= \frac{1}{4\pi\varepsilon_0}\frac{(+Ze)(-e)}{r}$

$= \frac{1}{4\pi\varepsilon_0}\frac{(+2e)(-e)}{r} = -\frac{e^2}{2\pi\varepsilon_0 r}$

2. (c) Ionization energy :

$-217.6 = -13.6 \times \frac{Z^2}{1^2}; \quad Z = 4$ m

So, it is ${}^{9}_{4}Be^{3+}$; no. of neutrons $9 - 4 = 5$

3. (c) $\frac{d[U(r)]}{dr} = \frac{3Ke^2}{r^4} \Rightarrow$ Magnitude of the force

$\therefore \quad \frac{3Ke^2}{r^4} = \frac{mv^2}{r}$

and we know $mvr = \frac{nh}{2\pi}$ or $r = \frac{nh}{2\pi m \cdot v}$,

$3Ke^2 \times \frac{8\pi^3m^3v^3}{n^3h^3} = mv^2, \quad v = \frac{n^3h^3}{24Ke^2\pi^3m^2}$

4. (c) $-\frac{1}{2} \times$ P.E. = K.E

$= -\frac{1}{2}\left(-\frac{1}{2}mkr^2\right) = \frac{1}{2}mv^2, \quad mvr = \frac{nh}{2\pi},$

$v^2 = \frac{n^2h^2}{4\pi^2m^2r^2}; \quad r^4 = \frac{n^2h^2}{2\pi^2m^2k}$

or $\quad r \propto \sqrt{n}$

5. (a) $\frac{1}{2}mv^2 = \frac{k(q_1)q_2}{r} \Rightarrow \frac{q_2}{m} = \frac{r.v^2}{2k.q_1.Z}$

$\frac{q_2}{m} = \frac{2.5 \times 10^{-14} \times (2.1 \times 10^7)^2}{2 \times 9 \times 10^9 \times 79 \times 1.6 \times 10^{-19}}$

$\Rightarrow \quad 4.84 \times 10^7$ coulomb/kg

6. (d) $v = r\omega$ where $r_n = \frac{n^2h^2}{4\pi^2me^2Z \cdot K}$

and $\quad v_n = \frac{2\pi \cdot Z \cdot e^2 \cdot K}{n \cdot h}$

$\therefore \quad \frac{2\pi Ze^2 \cdot K}{n \cdot h} = \frac{n^2h^2}{4\pi^2me^4Z \cdot K} \times \omega;$

$\omega = \frac{8\pi^3me^4 \cdot Z^2 \cdot K^2}{n^3 \cdot h^3}$

$= \frac{9\pi^3me^4 \cdot K^2}{h^3} \qquad (\because n = 2 \text{ and } Z = 3)$

7. (b) $r_n \propto \frac{v^2}{Z}$; for H, $r_4 - r_3 = 0.529(16 - 9)$

$\Rightarrow 0.529 \times 7$ Å

$r_4 - r_3$ for $Li^{2+} \Rightarrow 0.529\left(\frac{16}{3} - \frac{9}{3}\right)$

$\Rightarrow 0.529 \times \frac{7}{3}$ so ratio $\frac{7}{7/3} = 3:1$

8. (c) $v_n = 2.186 \times 10^6 \frac{Z}{n}$

$\Rightarrow 1.093 \times 10^6 = 2.186 \times 10^6 \times \frac{1}{n}; n = 2$

from Bohr theory we know $2\pi r = n\lambda$

$= 2\lambda, \quad$ where $\lambda = \frac{h}{mv}$

or $r = 0.529\frac{n^2}{Z} \Rightarrow 0.529 \times 4$ Å

$\therefore$ Circumference of the orbit

$\Rightarrow 2 \times \frac{22}{7} \times 0.529 \times 4 \times 10^{-10}$

$\Rightarrow 13.30 \times 10^{-10}$ m

9. (b) Angular momentum $= \frac{nh}{2\pi}$

$3.1652 \times 10^{-34} = \frac{n \times 6.626 \times 10^{-34}}{2\pi};$

$n = 3$

$\therefore \quad \bar{v} = R \cdot Z^2 \cdot \left(\frac{1}{n_1^2} - \frac{1}{n_2^2}\right);$

$\bar{v} = R \cdot 2^2\left(\frac{1}{2^2} - \frac{1}{3^2}\right) \Rightarrow \frac{5R}{9}$

10. (d) Energy of photon corresponding to second line of Balmer series for Li^{2+} ion

$= (13.6) \times (3)^2\left[\frac{1}{2^2} - \frac{1}{4^2}\right]$

$= 13.6 \times \frac{27}{16}$

Energy needed to eject electron from $n = 2$ level in H-atom;

$= 13.6 \times 1^2 \times \left[\frac{1}{2^2} - \frac{1}{\infty^2}\right] \Rightarrow \frac{13.6}{4}$

K.E. of ejected electron

$= 13.6 \times \frac{9 \times 3}{16} - \frac{13.6}{4} = 13.6 \times \left(\frac{27 - 4}{16}\right)$

$\Rightarrow \quad 19.55$ eV

11. (a) $\frac{1}{\lambda} = RZ^2\left(\frac{1}{n_1^2} - \frac{1}{n_2^2}\right)$, where $n_1 = n, n_2 = n + 1$

$\therefore \quad \frac{1}{\lambda} = RZ^2\left(\frac{1}{n^2} - \frac{1}{(n+1)^2}\right)$

$\Rightarrow \quad \frac{1}{\lambda} = \left(\frac{2n+1}{n^2(n+1)^2}\right)RZ^2$

Since, $\quad n >> 1;$

Therefore, $2n + 1 \approx 2n$

and $\quad (n+1)^2 \approx n^2$

$\therefore \quad \frac{1}{\lambda} = RZ^2\left(\frac{2n}{n^2 \cdot n^2}\right)$

$$\Rightarrow \frac{v}{c} = \frac{2RZ^2}{n^3} \quad \text{or} \quad v = \frac{2cRZ^2}{n^3}$$

12. (d)

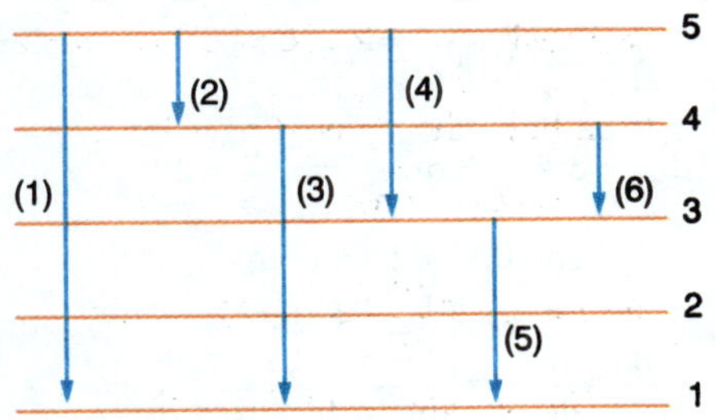

Total radiations are = 6

13. (a) If a is side of cube, then $\Delta x = a\sqrt{3}$

$\therefore \quad \Delta x = 10\sqrt{3}\text{ cm} = 10\sqrt{3}\times 10^{-2}\text{m}$

$$\Delta x \cdot \Delta p = \frac{h}{4\pi}; \quad \Delta x \cdot m \cdot \Delta v = \frac{h}{4\pi}$$

$$\Delta v = \frac{h}{4\pi m \cdot \Delta x}$$

$$= \frac{6.63\times 10^{-34}}{4\times 3.14\times 9.1\times 10^{-31}\times 10\times\sqrt{3}\times 10^{-2}}$$

$\therefore \Delta v \approx 3.34\times 10^{-4}\text{ ms}^{-1}$

14. (b) $2\pi r = 4\lambda; \quad n = 4$

Total energy required + total energy released = 0

$$2\times 4.526\times N_A + 2\times 13.6\times N_A + 2\times 13.6 \times\left(1-\frac{1}{16}\right)\times N_A - 30.87\times x\times N_A = 0$$

$x = 2 \quad \therefore$ moles of X required = 2

15. (c) Work function for $Li^{2+} = 9E$.

$$E_p = w + \frac{1}{2}mv^2; \quad E_p = 9E + \frac{1}{2}mv^2$$

$$v = \sqrt{\frac{2(E_p - 9E)}{m}}$$

16. (c) $E = hc\bar{v} \quad \Rightarrow \quad 1.63\times 10^{-18}\text{ J}$

where $\bar{v} = R(1)^2\left(\frac{1}{1^2}-\frac{1}{2^2}\right) \Rightarrow \frac{3}{4}R$

Translational K.E. of H-atom $= \frac{3}{2}\times\frac{R}{N_A}\times T$

$$\frac{3}{2}\times\frac{8.314}{6\times 10^{23}}\times T = 1.63\times 10^{-18}$$

$$T = 7.84\times 10^4\text{ K}$$

17. (c) Radial node occurs where probability of finding e^- is zero.

$\therefore \quad \psi^2 = 0 \quad \text{or} \quad \psi = 0$

$\therefore \quad 6 - 6\sigma + \sigma^2 = 0; \quad \sigma = 3\pm\sqrt{3}$

For max. distance $r = \frac{3}{2}\frac{(3+\sqrt{3})a_0}{Z}$

18. (d) $\frac{n(n-1)}{2} = 6; n = 4,$

$n = 4 \qquad E_4 = -0.85\text{ eV}$

$n = 1 \qquad E_1 = -13.6\text{ eV}$

$\therefore \qquad \Delta E = 12.75\text{ eV}$

$$12.75\text{ eV} = \frac{1240\text{ eV-nm}}{\lambda}$$

$$\lambda = 97.25\text{ nm}$$

19. (d) For II to I transition

$$\Delta E = \frac{4E}{3} - E = \frac{hc}{\lambda_{II\to I}}; \quad \frac{E}{3} = \frac{hc}{\lambda_{II-I}}$$

For III to I transition

$$\Delta E = 2E - E = \frac{hc}{\lambda} \quad \text{or} \quad E = \frac{hc}{\lambda}$$

$$\therefore \quad \frac{hc}{3\times\lambda} = \frac{hc}{\lambda_{II-I}} \quad \lambda_{II-I} = 3\lambda$$

20. (d)

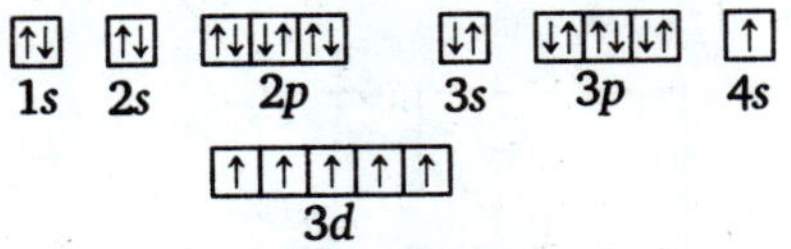

Out of 6 electrons in $2p$ and $3p$ must have one electron with $m = +1$ and $s = -\frac{1}{2}$ but in $3d$-subshell an orbital having $m = +1$ may have spin quantum no. $-\frac{1}{2}$ or $+\frac{1}{2}$.

Therefore, minimum and maximum possible values are 2 and 3 respectively.

21. (a) Energy absorbed $= 13.6\times 1.5 = 20.4$ eV out of this 6.8 eV is converted to K.E.

$6.8\text{ eV} \quad \Rightarrow \quad 6.8\times 1.6\times 10^{-19}$ J;

$$6.8\times 1.6\times 10^{-19} = \text{K.E.} \Rightarrow \left(\frac{1}{2}\right)mv^2$$

$$v = \sqrt{\frac{2KE}{m}} = \sqrt{\frac{2\times 1.088\times 10^{-18}}{9.1\times 10^{-31}}}$$

$$= 1.54\times 10^6\text{ m/s}$$

22. (b) $E = \frac{hc}{\lambda} = 2.9\times 10^{-19}$ J

Total energy of 10 quanta

$\Rightarrow 10\times 2.9\times 10^{-19} \Rightarrow 29\times 10^{-19}$ J

Energy stored for process

$$= \frac{112\times 4.18\times 10^3}{6\times 10^{23}} = 7.80\times 10^{-19}\text{ J}$$

% efficiency $= \frac{7.8\times 10^{-19}}{29\times 10^{-19}}\times 100 \Rightarrow 26.9\%$

23. (c) Total energy emitted by photo-electron

$= 10.2 + 17 = 27.20$ eV

Since, E_1 = Photon of energy emitted through the transition

$n = n \quad \text{to } n = 2 \quad \Rightarrow \quad \frac{hc}{\lambda_1} = 27.20\text{ eV}$

We have $\frac{1}{\lambda_1} = R_H\cdot Z^2\left(\frac{1}{2^2}-\frac{1}{n^2}\right)$

or $\frac{hc}{\lambda_1} = (hc)R_H\cdot Z^2\left(\frac{1}{2^2}-\frac{1}{n^2}\right)$

$\therefore\ 27.20 = (hc)\,R_H\,Z^2\left(\dfrac{1}{4}-\dfrac{1}{n^2}\right)$ ...(1)

Similarly, total energy liberated during transition of electron from $n = n$ to $n = 3$ is $E_2 = \dfrac{hc}{\lambda_2}$

$= (4.25 + 5.95) = 10.20$ eV

$\therefore\ 10.20 = (hc)\,R_H\,Z^2\left(\dfrac{1}{9}-\dfrac{1}{n^2}\right)$ ...(2)

Dividing Eq. (1) by (2), we get $n = 6$ and putting $n = 6$ in Eq. (1) or (2), we get, $Z = 3$.

24. (a) $\Delta E = \dfrac{hc}{\lambda} \Rightarrow \dfrac{1240\text{ eV-nm}}{1025.6\times 10^{-10}\times 10^{9}}$

$\Delta E = 12.09$ eV

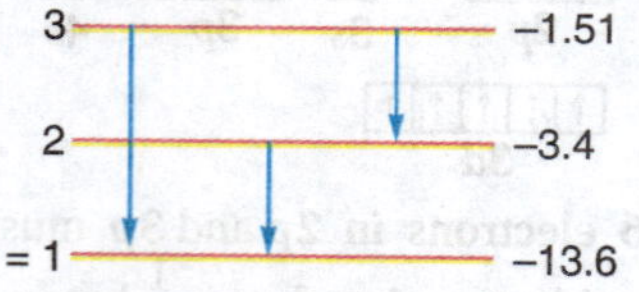

$\Delta E = 12.09$; $\therefore$ n = 3

In three different radiations, minimum wavelength for $3 \rightarrow 1$ transition

$\lambda_{3-1} = \dfrac{hc}{\Delta E} \Rightarrow \dfrac{1240\text{ eV-nm}}{12.09\text{ eV}} \approx 102.6$ nm

25. (b) $\Delta E = \dfrac{hc}{\lambda} = \dfrac{(6.63\times 10^{-34}\text{ J-s})\,(3.00\times 10^{8}\text{ m/s})}{3.055\times 10^{-8}\text{ m}}$

$= 6.52\times 10^{-18}$ J

$\Delta E_H = \dfrac{3}{4}(2.176\times 10^{-18}\text{ J})$

$= 1.63\times 10^{-18}$ J; $\Delta E = \Delta E_H (Z^2)$

$Z^2 = \dfrac{\Delta E}{\Delta E_H} = \dfrac{(6.52\times 10^{-18})}{(1.63\times 10^{-18})} = 4;$

$Z = 2$ (helium)

26. (b) K.E. $= \dfrac{K.Ze.2e}{r}$

$r = \dfrac{9\times 10^9\times 29\times 2\times (1.6\times 10^{-19})^2}{5\times 1.6\times 10^{-19}\times 10^6}$

$r = \dfrac{9\times 10^9\times 29\times 2\times 1.6\times 10^{-19}}{5\times 10^6}$

$= 1.67\times 10^{-14}$ m

27. (c) $\sqrt{v} = aZ - ab$

$ab = 1,\ a = \tan 45° = 1$

$\sqrt{v} = 51 - 1 = 50$

$v = 50^2 = 2500\text{ s}^{-1}$

28. (c) $\dfrac{1}{\lambda} = R\left[\dfrac{1}{2^2}-\dfrac{1}{n^2}\right] = R\left[\dfrac{n^2-4}{4n^2}\right]$

$\lambda = \dfrac{4}{R}\times\dfrac{n^2}{n^2-4}$ ...(1)

Given: $\lambda = k\times\dfrac{n^2}{n^2-4}$ ...(2)

Comparing equation (1) and (2) we have

$K = \dfrac{4}{R}$

29. (b) $2\pi r_n = n\lambda \Rightarrow 2\pi\times 0.53\dfrac{n^2}{z} = n\lambda$

$\lambda = 2\pi\times 0.53\times\dfrac{n}{z}$ ...(1)

$E_{\text{sep}} = 3.4 = 13.6\dfrac{z^2}{n^2} \Rightarrow \dfrac{n}{z} = 2$

$\lambda = 2p\times 0.53\times 2 = 6.66$ Å

30. (a) $\dfrac{\Delta E}{E_{4\text{th}}} = \dfrac{24}{E_{4\text{th}}} = \dfrac{\left[\dfrac{1}{1}-\dfrac{1}{4}\right]}{\left[\dfrac{1}{16}\right]} = \dfrac{3/4}{1/16}$

$E_{4\text{th}} = 2$ eV.

Level 3

Passage-1

1. (c) $\Delta x\,\Delta p = \dfrac{h}{4\pi} \Rightarrow \Delta p^2 = \dfrac{h}{4\pi}$

$\Rightarrow m^2\Delta v^2 = \dfrac{h}{4\pi} \Rightarrow \Delta v = \dfrac{1}{2m}\sqrt{\dfrac{h}{\pi}}$

2. (a) $\Delta x = \sqrt{\dfrac{h}{4\pi m}}$; $\Delta x\,\Delta p = \dfrac{h}{4\pi}$

$\sqrt{\dfrac{h}{4\pi m}}\,\Delta p = \dfrac{h}{4\pi},\ \Delta p = \sqrt{\dfrac{mh}{4\pi}}$

3. (c) $\lambda_{\text{D.B.}} = \sqrt{\dfrac{150}{6}}$ Å $= 5$ Å

and $\Delta x\cdot\Delta p \geq \dfrac{h}{4\pi}$; $p = \dfrac{h}{\lambda}$ or $\Delta p = \dfrac{h}{\lambda^2}\Delta\lambda$

$\Rightarrow \Delta x\cdot\dfrac{h}{\lambda^2}\times\Delta\lambda \geq \dfrac{h}{4\pi}$

$\Rightarrow \dfrac{1}{\pi}\times\dfrac{10^{-9}}{\lambda^2}\times\Delta\lambda > \dfrac{1}{4\pi} \Rightarrow \Delta\lambda \geq \dfrac{2.5}{4}\times 10^{-10}$

$\Delta\lambda \geq 0.625$ Å

Passage-2

1. (b) $\because \dfrac{1}{\lambda} = R_H\times Z^2\left[1-\dfrac{1}{n^2}\right]$ for Lyman's series

For He^+ ion series limit $n = \infty$;

$\dfrac{1}{\lambda_1} = R_H\times 2^2\left[1-\dfrac{1}{\infty^2}\right] \Rightarrow \dfrac{1}{\lambda_1} = R_H\times 4$

Similarly, Balmer series limit for Li^{2+} ion

$\dfrac{1}{\lambda_2} = R_H\times 3^2 = 9R_H \Rightarrow 9\times\dfrac{1}{4\lambda} \Rightarrow \lambda_2 = \dfrac{4x}{9}$

2. (b) $6 = (n_2 - n_1)\dfrac{(n_2 - n_1 + 1)}{2}$; $n_2 = 4,\ n_1 = 1$

$\therefore \dfrac{1}{\lambda} = R_H\times 2^2\left[1-\dfrac{1}{16}\right]$

$\Rightarrow R\times\dfrac{15}{4}$; $\lambda = \dfrac{4}{15R_H}$

3. (d) $\frac{1}{\lambda} = R_{H \cdot Z^2}\left(\frac{1}{n_1^2} - \frac{1}{n_2^2}\right)$

$$\frac{1}{\lambda_{He^+}} = R_H \times 2^2\left(\frac{1}{2^2} - \frac{1}{4^2}\right)$$

$$= R_H\left(\frac{1}{1^2} - \frac{1}{2^2}\right) = \frac{1}{\lambda_H}$$

for $n_2 = 2$ to $n_1 = 1$

Passage-3

1. (d) Magnetic moment $\mu = \sqrt{n(n+2)}$

(A)	Fe^{2+}	$3d^6\,4s^0$	$n = 4$
(B)	Mn^{2+}	$3d^5\,4s^0$	$n = 5$
(C)	Cr^{3+}	$3d^3\,4s^0$	$n = 3$
(D)	V^{3+}	$3d^2\,4s^0$	$n = 2$

Smaller n, smaller μ

2. (c) $\mu = \sqrt{n(n+2)} = 3.873$

$n = 3$

Mn	$3d^5$	$4s^2$
Mn^{4+}	$3d^3$	$4s^0$

Passage-4

1. (b) Total energy required for dissociation of O_2 molecule and then assigning (O^*) 1eV more energy than (O)$\Rightarrow$ (480+96) $\Rightarrow$ 576 kJ/mol

Maximum wavelength $E = \frac{hc}{\lambda}$;

$$\lambda_{maximum} = \frac{hc}{E_{min}};\ \lambda_{max}(\text{Å}) = \frac{12400}{E_{min}(\text{eV})}$$

$$= \frac{12400}{6} = 2066.67\text{Å}$$

2. (b) $O_3(g) \longrightarrow O_2(g) + O$ ....(i)

$O_2(g) \longrightarrow O + O^*$ ...(ii)

Energy required for (i) reaction is = 400 kJ/mol

Normal dissociation of O_2 required = 480 kJ/mol (given)

each (O^*) has 1 eV more energy than (0) and given 1 eV/photon = 96 kJ/mol

So total energy required for the dissociation of O_3 into O and O^* is 400 + 480 + 96 = 976 kJ/mol

One or More Answers is/are Correct

2. (a, b, c)

$$mvr = n\frac{h}{2\pi}$$

n = integer value = 1,2,3,4, ...

15. (a,b,c,d) At the point of maximum value of RDF

$$\frac{dP}{dr} = 0$$

$$\left(2r - \frac{2Zr^2}{a_0}\right) = 0;\quad r = \frac{a_0}{Z}$$

where $Z = 3$ for Li^{2+} and $Z = 2$ for the He^+;

$Z = 1$ for hydrogen.

18. (b, c)

$h\nu = W + KE$

Kinetic energy of electrons does not depend on intensity of light $\lambda = \frac{h}{\sqrt{2KE \cdot m}}$

19. (a, b, c, d)

(a) $v \propto \frac{Z}{n}$ (b) $r \propto \frac{n^2}{Z}$

(c) P.E. $\propto -\frac{Z^2}{n^2}$ (d) K.E. $\propto \frac{Z^2}{n^2}$

20. (c, d)

Given n, $l = 0$ to $n - 1$

$m_l = -l \ldots\ldots 0 \ldots\ldots +l$

21. (a, b, d)

(a) No. of subshell $= n$, $n = 3$

(b) No. of orbital $= 2n^2$, $n = 4$

(c) $Cu^+ - 3d^{10}4s^0$ diamagnetic

22. (a, b, c, d)

5 − 1 5 − 2 5 − 3 $\underbrace{5-4}_{\text{Brackett}}$

4 − 1 4 − 2 $\underbrace{4-3}_{\text{Paschen}}$

3 − 1 $\underbrace{3-2}_{\text{Balmer}}$

$\underbrace{2-1}_{\text{Lyman}}$

Match the Column

13. (a) s-orbital $\because$ $r = 0$, $\psi \neq 0$ and 3 radial nodes $\Rightarrow 4s$

(b) 3 radial nodes (s, p, d) $\Rightarrow$ $4s$, $5p_x$ $6d_{xy}$

(c) Angular probability is dependent of θ and ϕ for $5p_y$, $6d_{xy}$

(d) At least one angular node $\Rightarrow$ $5p_x(1); 6d_{xy}(2)$

Level 4

Subjective Problems

1. $r_{n+1} - r_{n-1} = 2r_n$

$$0.529\frac{(n+1)^2}{Z} - 0.529\frac{(n-1)^2}{Z} = 2 \times 0.529\frac{n^2}{Z}$$

$$n^2 + 2n + 1 - [n^2 - 2n + 1] = 2n^2$$

$$4n = 2n^2$$

$$n = 2$$

2. $E = 30.6 = 13.6\frac{(3)^2}{(n)^2}$

$n = 2$ = no. of wave in one complete evolution in the orbit.

3. $\lambda \propto n$

$\frac{\lambda}{\lambda_{2nd}}$ orbit $= 1.5 = \frac{n}{2}$

$n = 3$ = no. of wave in one complete revolution in the orbit.

4. $E_{abs} \times \frac{40}{100} = E_{Emitted.}$

$$n_{ab} \times \frac{hc}{400} \times \frac{40}{100} = n_{em} \times \frac{hc}{500}$$

$$\frac{n_{ab}}{n_{em}} = \frac{400}{500} \times \frac{100}{40} = 2$$

5. $E_{inc} = \phi + KE$

$4.5 = 3.0 + KE$

$KE = 1.5\,\text{eV}$

$$\lambda = \sqrt{\frac{150}{1.5}} = 10\,\text{Å}$$

$$\Delta P \cdot \Delta x = \frac{\lambda}{4\pi}$$

$$\frac{h}{\lambda^2} \times \Delta\lambda \cdot \Delta x = \frac{h}{4\pi}$$

$$\frac{\Delta\lambda}{(10 \times 10^{-10})^2} \times \frac{25}{4\pi} \times 10^{-10} = \frac{1}{4\pi}$$

$\Delta\lambda = 4\,\text{Å}$

6. $Cr \longrightarrow 3d^5 4s^1$

No. of angular nodes $= 2$

No. of radial nodes $= 3 - 2 - 1 = 0$

Angular node – radial node $= 2$

7. Total node $=$ Radial node + angular node

$= n - 1 - 1 + 1 = n - 1$

$5f = 5 - 1 = 4$

8. $$P = \frac{n\frac{hc}{\lambda}}{t}$$

$$\frac{100 \times 12}{100} = (x \times 10^{19}) \times \frac{6.63 \times 10^{-34} \times 3 \times 10^8}{1500 \times 10^{-9} \times 1}$$

$x = 9$

9. $$\frac{1}{\lambda} = RZ^2\left(\frac{1}{n^2} - \frac{1}{(n+1)^2}\right)$$

$$\frac{1}{\lambda} = RZ^2 \frac{(n+1)^2 - n^2}{n^2(n+1)^2}$$

$$\frac{1}{\lambda} = RZ^2 \frac{(n+1+n)(n+1-n)}{n^2(n+1)^2}$$

$$\frac{1}{\lambda} = RZ^2 \frac{(2n+1)}{n^2(n+1)^2}$$

$\lambda \propto n^3$

$x = 3$

10. For radial node

$$\psi_{3s} = 0 = \frac{1}{81\sqrt{3\pi}}\left(\frac{1}{a_0}\right)^{3/2}\left[27 - \frac{18r}{a_0} + \frac{2r^2}{a_0^2}\right]e^{-r/3a_0}$$

$$27 - \frac{18r}{a_0} + \frac{2r^2}{a_0^2} = 0$$

$$r = -\frac{-(18a_0) \pm \sqrt{(-18a_0)^2 - 4 \times 2 \times 27a_0^2}}{2 \times 2}$$

$$= \frac{18a_0 \pm 10.39\,a_0}{4}$$

distance between to radial node (d)

$$\frac{d}{1.73a_0} = \frac{2 \times 10.39a_0}{4 \times 1.73a_0}$$

$d = 3$

11. $$7.67 \times 10^{-19} = \frac{(1.6 \times 10^{-19})^2}{4 \times 3.14 \times 8.85 \times 10^{-12} \times r}$$

$\Rightarrow \quad r = 3.00 \times 10^{-10}\,\text{m} = 3\text{Å}$

12. $$d(KE) = mv\,dv = mv\frac{h}{4\pi m\Delta x}$$

$$= \frac{3 \times 10^8}{3} \times \frac{6.62 \times 10^{-34}}{4 \times \pi \times \frac{3.31}{\pi} \times 10^{-12}}$$

$= 5 \times 10^{-16}\,\text{J}$

13. $E_{in} = 10.2\,\text{eV}$

$\phi = 4.2\,\text{eV}$

$KE_{max} = 10.2 - 4.2 = 6\,\text{eV}$

$$\therefore \quad \lambda e^- = \sqrt{\frac{150}{6}}\text{Å} = 5\text{Å}$$

14. $$\frac{\text{Radial probability density at } r = a_0}{\text{Radial probability density at } r = 0} = \frac{R^2(a_0)}{R^2(0)}$$

For 1s orbital: $$R_{(r)} = \frac{1}{\sqrt{\pi}}\left(\frac{1}{a_0}\right)^{3/2} e^{\frac{r}{a_0}}$$

$$\Rightarrow \quad \frac{R^2(a_0)}{R^2(0)} = \frac{(1/\pi a_0^3)e^{-2r/a_0}}{(1/\pi a_0^3)e^0} = e^{-2}$$

15. $$A = \frac{E_{1,2}}{2E_{21}} = \frac{-13.6 \times 2^2 \times 2^2}{2 \times 1^2 \times (-13.6) \times 1^2} = \frac{16}{2} = 8$$

Numerical Problems

1. $$r_n = 0.529 \times \frac{n^2}{z} = 2.116$$

$\Rightarrow \quad n = 2$

$$E_n = -\frac{13.6\,z^2}{n^2} = -\frac{13.6 \times 1^2}{2^2} = -3.4\,\text{eV}$$

$KE = -E_n = +3.4\,\text{eV}$

2. $$\frac{x}{y} = \frac{2.188 \times 10^6 \times \frac{1}{2}}{2.188 \times 10^6 \times \frac{2}{6}} = 1.50$$

3. K.E. of emitted electron $= 0.5 \times 13.6\,\text{eV}$

Now, K.E. $= \frac{1}{2}mv^2$

or, $6.8 \times 1.6 \times 10^{-19} = \frac{1}{2} \times 9.1 \times 10^{-31} \times v^2$

$\Rightarrow \quad v = 1.55 \times 10^6$ m/s

4. Excitation energy $= 13.6 \times 1^2 \left(\frac{1}{1^2} - \frac{1}{n^2}\right) = 10.2$

$\frac{1}{n^2} = 0.25 \quad \Rightarrow \quad n = 2$

No. of waves $= n = 2$

5.

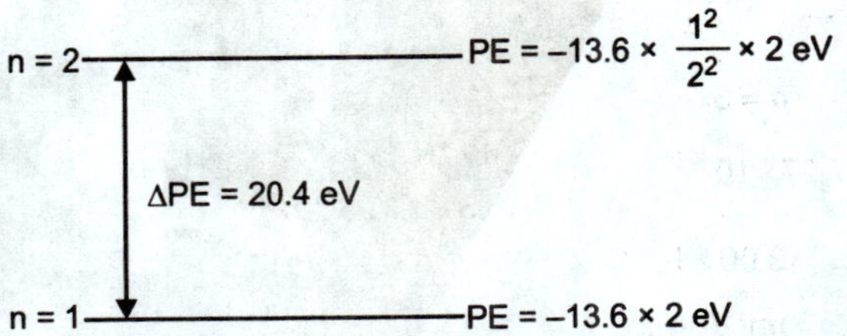

6. For maximum spectral lines, any two possibility may be considered.

Hence, max. no. of spectral lines = 4

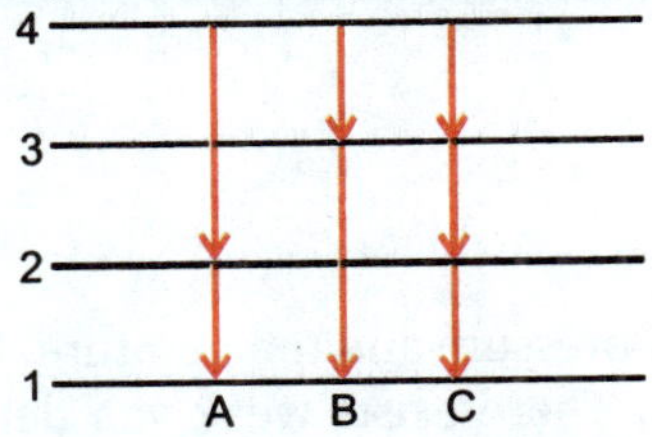

7. From given graph work function $= 1.7$ eV

$KE_{max} = \frac{1240}{310} - 1.7 = 2.3$ eV

Stopping potential is 2.3 V

8. According to diagram e^- in 3rd shell

$\therefore$ TE in 3rd shell $= -1.51$

$\lambda = \sqrt{\frac{150}{1.51}} = 9.97$ Å

9. $\lambda = \sqrt{\frac{150}{V}}$Å; $\lambda = \sqrt{\frac{150}{40 \times 1000}}$Å $= 6.1 \times 10^{-12}$ M

10. $E = n\left(\frac{hc}{\lambda}\right)$

$\frac{1000}{10} = n\left(\frac{6.626 \times 10^{-34} \times 3 \times 10^8}{400 \times 10^{-9}}\right); n = 2$

11. $\lambda = \frac{h}{\sqrt{2mqV}}$

$\frac{\lambda_{Li^{3+}}}{\lambda_P} = \sqrt{\frac{m_P \cdot q_P \cdot V_P}{M_{Li^{3+}} \cdot q_{Li^{3+}} \cdot V_{Li^{3+}}}}$

$= \sqrt{\frac{1}{8.3} \times \frac{1}{3}} = 0.2 = 2 \times 10^{-1}$

12. Max. K.E. of ejected $e^- = h\nu - h\nu_0 = h\left[\frac{c}{\lambda} - \nu_0\right]$

$\frac{1}{2} \times 9 \times 10^{-31} \times v^2 = 6.63 \times 10^{-34}\left[\frac{3 \times 10^8}{500 \times 10^{-9}} - 4.3 \times 10^{14}\right]$

$= 6.63 \times 10^{-34} \times 1.7 \times 10^{14}$

$v = 5 \times 10^5$ ms^{-1}

13. $2l + 1 = 5 \quad \Rightarrow l = 2 \quad \Rightarrow n = 3$

3s	3p	3d	
1	3	5	$e^- = 9$

14. $\Psi_{n,l,m} = \Psi_{5,1,1}; n = 5; l = 1; m = 1$

Given orbital is $5p$.

Total no. of nodes $= n - 1 = 5 - 1 = 4$

15. Given graph is possible for $3s, 4p$ and $5d$ orbitals.

Total number of radial nodes present in $3s, 4p$ (3 orbitals) and $5d$ (5 orbitals) $= 9 \times 2 = 18$

CHAPTER 3

Gaseous State

Gaseous State

This state is characterized by sensitivity of volume change with change of pressure and temperature. It is due to large distance between molecules as compared to their own dimensions. There exists weak van der Waal's forces. (but no force in ideal state).

Ideal Gas: A gas with no intermolecular attractions & having very negligible volume occupied by molecules when compared with volume of gas is termed as ideal gas. An ideal gas is a hypothetical gas and none of the gases known is ideal. These gases are called real gases and they may behave like an ideal gas under specific conditions.

Parameter associated with the ideal gas : P, V, T, n

where

P represents pressure exerted by the gas molecules on the walls of the container assuming negligible intermolecular attractions,

V represents free volume available for motion (equal to the volume of the container),

T represents absolute temperature, n represents no. of moles.

- **Conversion factors**

Pressure : atm $= 1.013 \times 10^5$ Pa $= 760$ mm of Hg $= 760$ torr $= 1.013$ bar

Volume : $1\,\text{L} = 1\,\text{dm}^3 = 10^{-3}\,\text{m}^3 = 1000\,\text{mL} = 1000\,\text{cm}^3$

Temperature : $T_k = T_C^\circ + 273 = \frac{5}{9}T_F^\circ + 255.22$

Instruments for Pressure Measurement

- **Barometer:** The instrument used for the measurement of atmospheric pressure is called a barometer.

$$P = \frac{A \times h \times d \times g}{A} = d \times g \times h$$

where d = density of fluid

h = vertical height

g = acceleration due to gravity

- **Manometer:** The instrument used for the measurement of the pressure of a gas is called a manometer. Manometer are of two types
  (i) Open end (ii) Closed end

Experimental Gas Laws

I.	**Boyle's law**	$V \propto \frac{1}{P}$	(T & n constant)	$P_1V_1 = P_2V_2$
II.	**Charle's law**	$V \propto T$	(P & n constant)	$\frac{V_1}{T_1} = \frac{V_2}{T_2}$
III.	**Gay Lussac's law**	$P \propto T$	(V & n constant)	$\frac{P_1}{T_1} = \frac{P_2}{T_2}$
IV.	**Avogadro's law**	$V \propto n$	(T & P constant)	$\frac{V_1}{n_1} = \frac{V_2}{n_2}$

Combined Gas law : $\frac{P_1V_1}{T_1} = \frac{P_2V_2}{T_2}$

Equation of State: $PV = nRT$

$PV = \frac{w}{M}RT$ R is Universal Gas constant,

$R = 0.0821$ atm. L/K. mol $= 8.314$ joule/K. mol $\approx$ 2cal/ K.mol

density of gas $(d) = \frac{PM}{RT}$

- **Dalton's law of partial pressure:**

$P_{total} = P_A + P_B + ...$

P_A, P_B are partial pressures; P_A = mole fraction$_A$ × Total pressure

and mole % of gas in mixture $= \frac{\text{Partial pressure}}{\text{Total pressure}} \times 100$

$P_{\text{wet gas}} = P_{\text{dry gas}} + P_{H_2O\ \text{vapour}}$ (aqueous tension)

- **Amagat's law:** At constant temperature and pressure, the total volume of a mixture of gases is equal to the sum of the partial volumes of the constituent gases.

- **Relative humidity:**

$$\%\ \text{Relative humidity} = \frac{\text{Partial pressure of } H_2O(g)}{\text{Vapour pressure of } H_2O\,(l)} \times 100$$

- **Graham's law of Diffusion or Effusion:**

$$r \propto \frac{1}{\sqrt{d}} \text{ or } r \propto \frac{1}{\sqrt{M}} \quad r \propto \frac{P}{\sqrt{M}}$$ [For gases effusing at different pressures]

r is rate of diffusion of any gas, d is density at same temperature.

$$\frac{r_1}{r_2} = \sqrt{\frac{d_2}{d_1}} = \sqrt{\frac{M_2}{M_1}}\ ;\ \frac{V_1/t_1}{V_2/t_2} = \sqrt{\frac{M_2}{M_1}} = \frac{\text{Pressure drop I}}{\text{Pressure drop II}}$$

$$r = \frac{\text{moles diffused}}{\text{time taken}} = \frac{\text{distance travelled in a narrow tube}}{\text{time taken}}$$

- **Isotopic Separation Factor:**

After x steps of separation, isotopic separation factor $(f) = \frac{(n_1/n_2)_{\text{final}}}{(n_1/n_2)_{\text{initial}}} = \left(\sqrt{\frac{M_2}{M_1}}\right)^x$

or
$$x = \frac{2\log f}{\log\left(\frac{M_2}{M_1}\right)}$$

Kinetic Theory of Gases

$$PV = \frac{1}{3} m N u^2 = \frac{1}{3} M u^2 \text{ (For 1 mole)}$$

Types of speeds:

- Mean square speed, $u^2 = \frac{u_1^2 + u_2^2 + \ldots + u_N^2}{N}$
- Root mean square speed, $u = \sqrt{\frac{3RT}{M}} = \sqrt{\frac{3PV}{nM}} = \sqrt{\frac{3P}{d}}$
- Average speed $= \frac{u_1 + u_2 + \ldots + u_N}{N} = \sqrt{\frac{8RT}{\pi M}} = \bar{u}$
- Most probable speed $= \sqrt{\frac{2RT}{M}} = u_{\text{mps}}$

Relationship between three types of speeds

- Most probable: average : r.m.s. $= \sqrt{2} : \sqrt{\frac{8}{\pi}} : \sqrt{3} = 1:1.13:1.22$
- $u_{\text{rms}} > u_{\text{av}} > u_{\text{mp}}$
- Average kinetic energy of a single molecule $= \frac{3}{2} \cdot \frac{R}{N} \cdot T = \frac{3}{2} kT$

where, k = Boltzman constant $= 1.3806 \times 10^{-16}$ erg deg^{-1}

- Total kinetic energy for **one mole** of a gas $= \frac{3}{2} RT$
- Kinetic energy of ***n* moles** of a gas $= n \times \frac{3}{2} RT$

Maxwells speed distribution Law:

$$dN_u = 4\pi N\left(\frac{M}{2\pi RT}\right)^{3/2} \exp\left(-Mu^2/2RT\right) u^2 du$$

$$= 4\pi N\left(\frac{m}{2\pi kT}\right)^{3/2} \exp\left(-mu^2/2kT\right) u^2 du$$

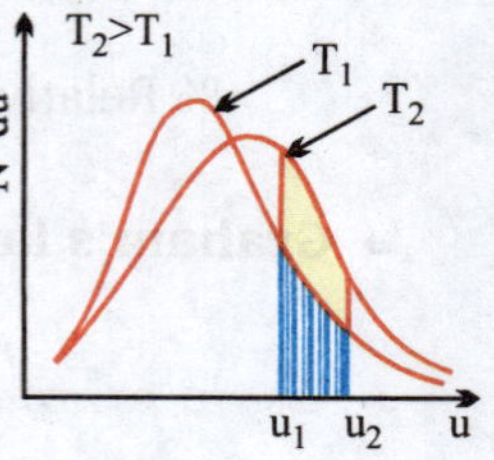

Real Gases

Real gases: Gases which do not obey perfect gas laws over a wide range of pressure and temperature are called real or non-ideal gases. These deviations are due to the interactions of molecules with each other and volume occupied by gas molecules. The repulsive forces between molecules assist expansion and attractive forces assist compression.

» **Deviation from ideal behaviour**

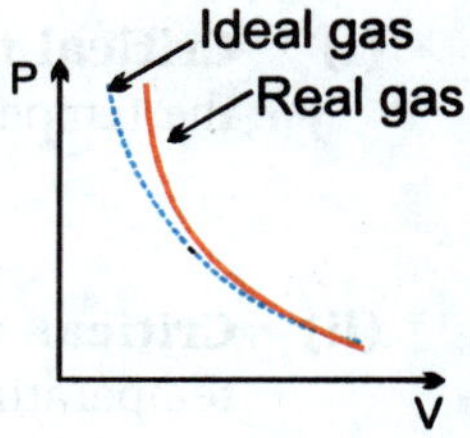

The curve for the real gas has a tendency to coincide with that of an ideal gas at low pressures when the volume is large. At higher pressures, however deviations are observed.

» **van der Waals equation of state**

$$\left(P+\frac{an^2}{V^2}\right)(V-nb)=nRT$$

a, b are van der Waals constants; different for each gas

unit of $a \longrightarrow$ atm L^2 mol^{-2} ; S.I. unit $\longrightarrow$ Pa m^6 mol^{-2}
unit of $b \longrightarrow$ L mol^{-1} ; S.I. unit $\longrightarrow$ m^3 mol^{-1}

greater the value of 'a' more easily the gas is liquefiable;
greater the value of 'b' greater the molecular size,

» **Compressibility factor**

$$Z=\frac{PV}{nRT}=\frac{\text{observed molar volume}}{\text{ideal molar volume}}$$

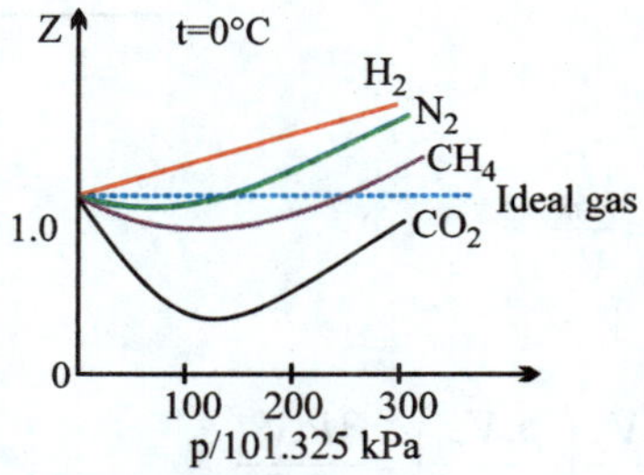

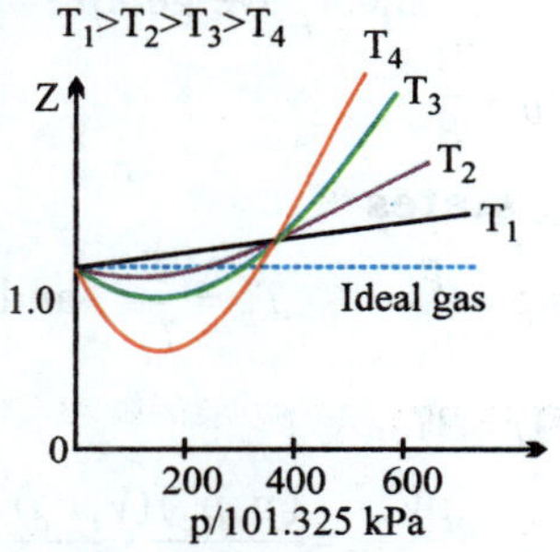

» **Interpretation of deviation using van der Waals equation**

(i) At low pressure $Z=\frac{PV}{RT}=1-\frac{a}{VRT}$

(ii) At high pressure $Z=\frac{PV}{RT}=1+\frac{Pb}{RT}$

(iii) At extremely low pressure $Z=\frac{PV}{RT}=1$; $PV=nRT$

» **Virial equation of state for 1 mole of gas**

$$Z=\frac{PV}{RT}=1+B\frac{1}{V}+C\frac{1}{V^2}+D\frac{1}{V^3}+\ldots.$$

$B=$ second virial coefficient $=b-\frac{a}{RT}$
$C=$ third virial coefficient $=b^2$
(gas dependent)

» **Boyle temperature (T_B):** The temperature at which a real gas obeys Boyle's law (*i.e.*, behaves as an ideal gas) in low pressure range.

$$T_B=\frac{a}{bR}$$

» **Inversion temperature:** The temperature at which a real gas show no Joule-Thomson effect is called its inversion temperature (T_i)

$$T_i=2T_B=\frac{2a}{bR}$$

In Joule-Thomson's effect, all gases except H_2 and He show cooling effect while H_2 and He show heating effect.

» **Critical constants**

(i) Critical temperature (T_c) : T_c is the maximum temperature at which a gas can be liquefied, *i.e.*, the temperature above which a liquid can not exist.

$$T_c = \frac{8a}{27Rb}$$

(ii) Critical pressure (P_c) : P_c is the minimum pressure required to cause liquefaction at the temperature T_c.

$$P_c = \frac{a}{27b^2}$$

(iii) Critical volume (V_c) : V_c is the volume occupied by one mole of gas at critical temperature and critical pressure.

$$V_c = 3b$$

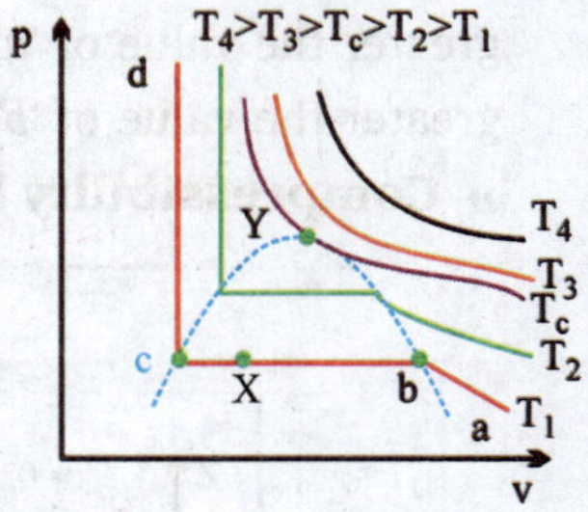

(iv) Compressibility factor under critical state of a gas

$$Z_c = \frac{P_cV_c}{RT_c}$$

Now substituting the values of P_c, V_c and T_c we get

$$Z_c = \frac{3}{8}$$

» **The law of corresponding states**

$$p_r = \frac{p}{p_c}, \quad T_r = \frac{T}{T_c} \quad \text{and } V_r = \frac{V_m}{V_c}$$

$$(p_r + 3/V_r^2)(3V_r - 1) = 8T_r$$

$$Z = \frac{pV_m}{RT} = \frac{(p_rp_c)(V_rV_c)}{R(T_rT_c)} = \frac{p_cV_c}{T_c}\left(\frac{p_rV_r}{T_r}\right) = \frac{3P_rV_r}{8T_r}$$

Level-1

1. Which one of the following statements is not correct about the three states of matter *i.e.,* solid, liquid and gaseous ?

(a) Molecules of a solid possess least energy whereas those of a gas possess highest energy

(b) The density of solids is highest whereas that of gases is lowest

(c) Gases and liquids possess definite volumes

(d) Molecules of a solids possess vibratory motion

2. Which of the following curve does not represent Boyle's law?

(a) 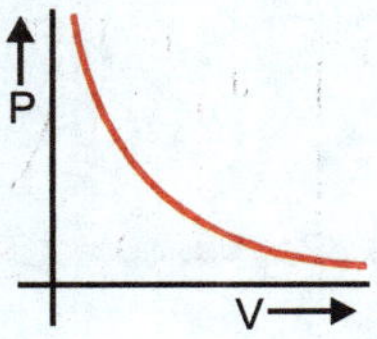

(b) 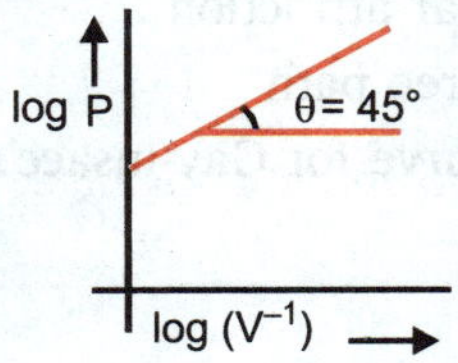

(c) 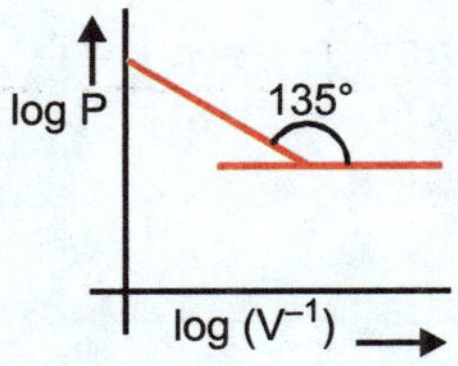

(d) 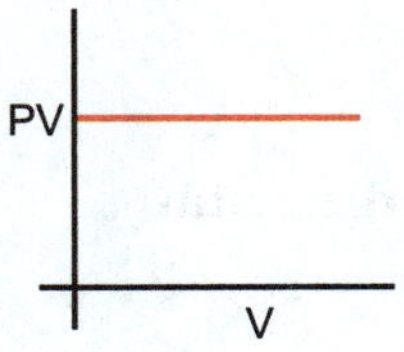

3. A certain sample of gas has a volume of 0.2 litre measured at 1 atm pressure and 0°C. At the same pressure but at 273°C, its volume will be:

(a) 0.4 litre (b) 0.8 litre

(c) 27.8 litres (d) 55.6 litres

4. Among the following curves, which is not according to Charle's law ?

(a) 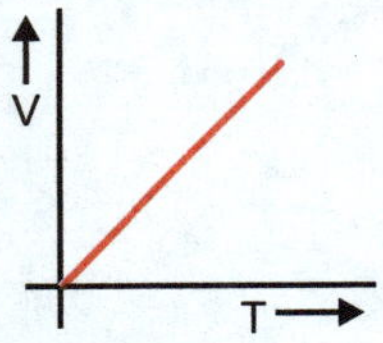

(b) 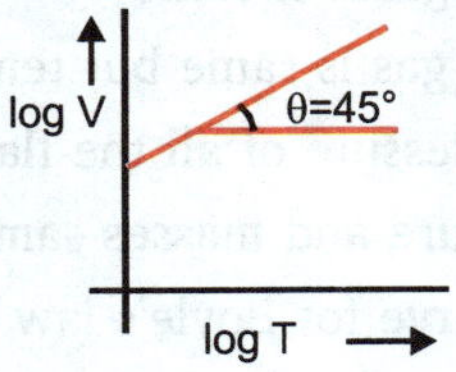

(c) 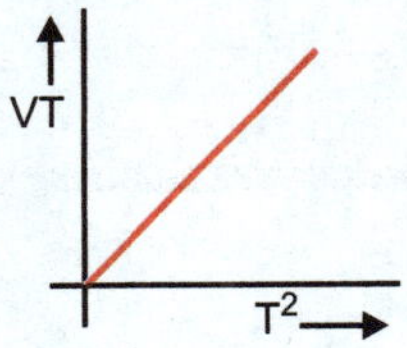

(d) 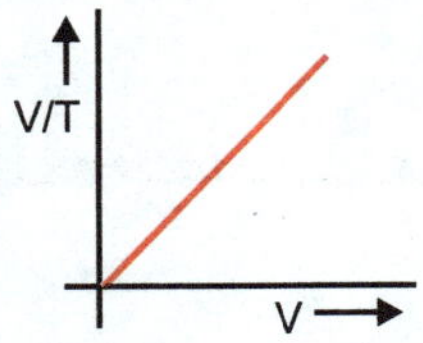

5. Initial temperature of an ideal gas is 75°C. At what temperature, the sample of neon gas would be heated to double its pressure, if the initial volume of gas is reduced by 15%?

(a) 319°C (b) 592°C

(c) 128°C (d) 60°C

6. Which is correct curve for Charle's law, when the curve is plotted at 0.821 atm pressure for 10 mole ideal gas?

(a) 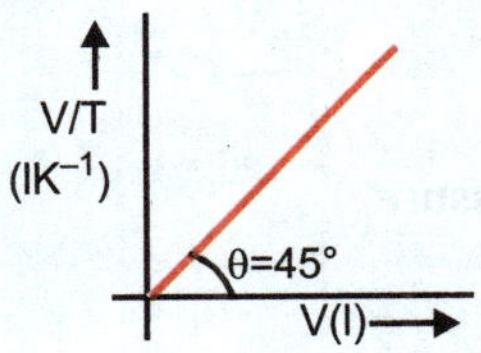

(b) 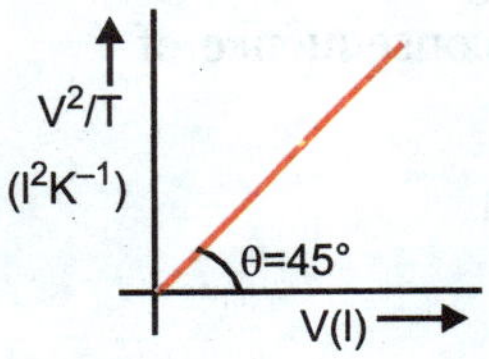

(c) 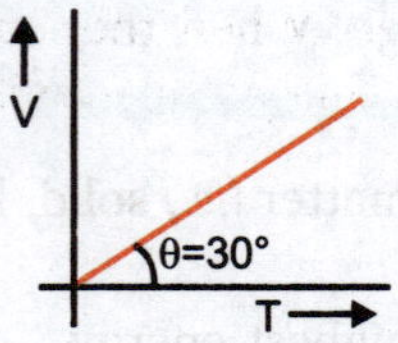

(d) 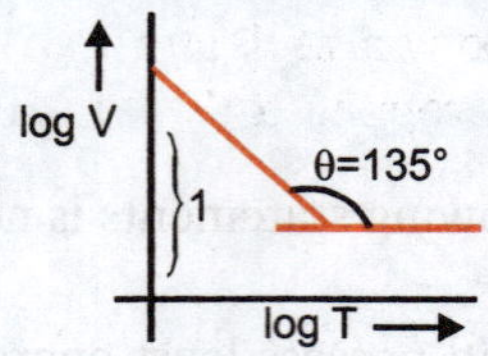

7. At constant volume, for a fixed number of moles of a gas, the pressure of the gas increases with increase in temperature due to:

(a) increase in the average molecular speed
(b) decrease in rate of collision amongst molecules
(c) increase in molecular attraction
(d) decrease in mean free path

8. Which in not correct curve for Gay-lusacc's law ?

(a) 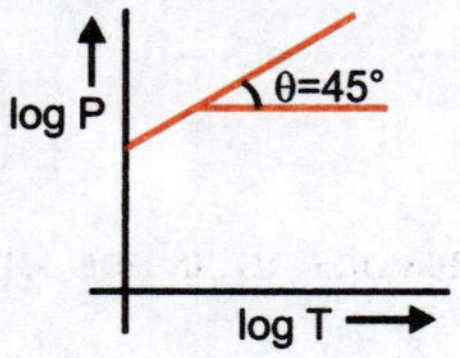

(b) 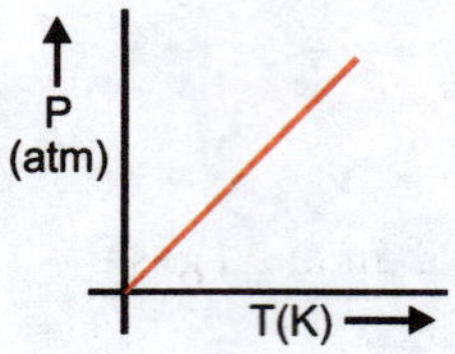

(c) 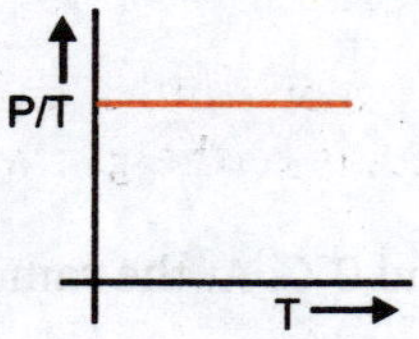

(d) 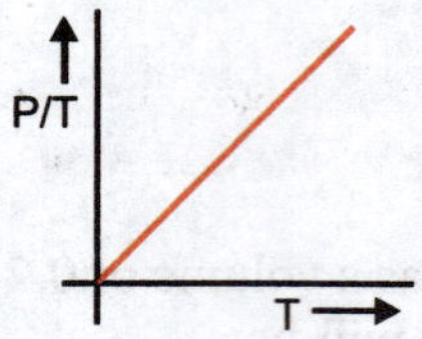

9. Three flasks of equal volumes contain CH_4, CO_2 and Cl_2 gases respectively. They will contain equal number of molecules if :

(a) the mass of all the gases is same
(b) the mass of all the gas is same but temperature is different
(c) temperature and pressure of all the flasks are same
(d) temperature, pressure and masses same in the flasks

10. Which is **incorrect** curve for Boyle's law ?

(a) 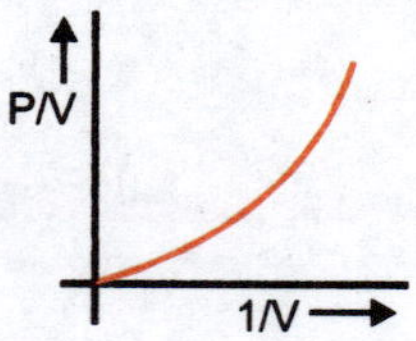

(b) 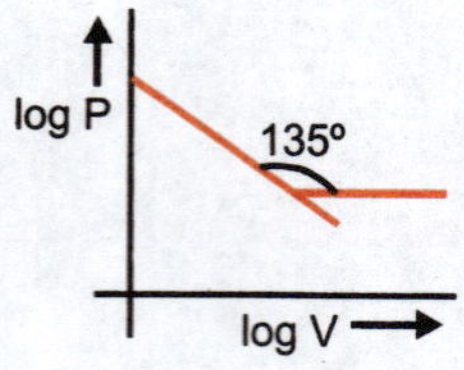

(c) 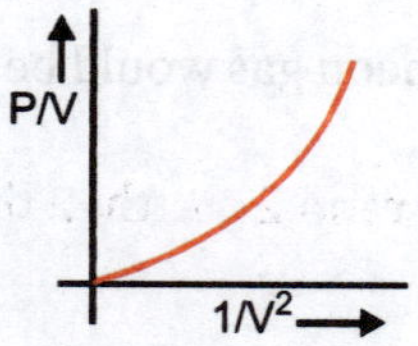

(d) 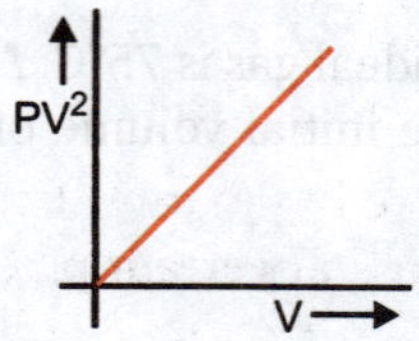

11. "Equal volumes of all gases at the same temperature and pressure contain equal number of particles." This statement is a direct consequence of :

(a) Avogadro's law
(b) Charle's law
(c) ideal gas equation
(d) law of partial pressure

12. A 2.24L cylinder of oxygen at 1 atm and 273 K is found to develop a leakage. When the leakage was plugged the pressure dropped to 570 mm of Hg. The number of moles of gas that escaped will be :
(a) 0.025 (b) 0.050
(c) 0.075 (d) 0.09

13. Which of the following curve is correct for an ideal gas ?

(a)
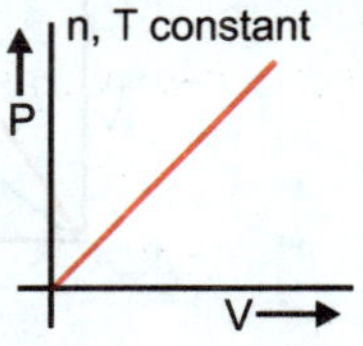

(b)
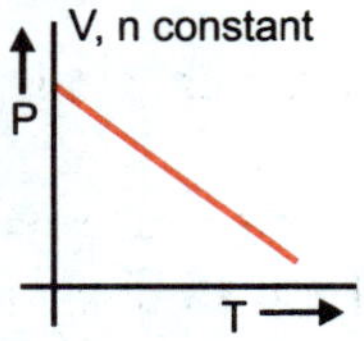

(c)
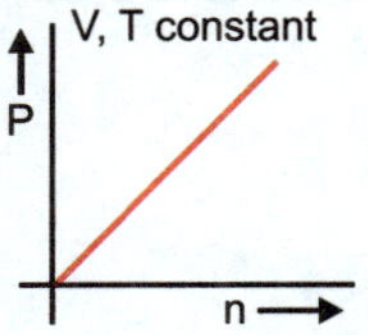

(d)
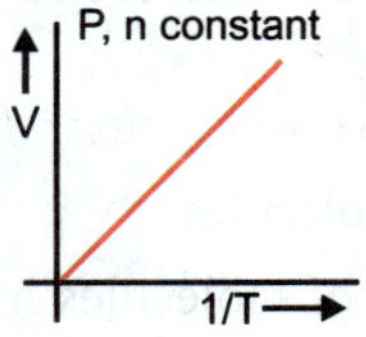

14. In the equation of state of an ideal gas $PV = nRT$, the value of the universal gas constant is not correct :
(a) 8.314 $JK^{-1}mol^{-1}$ (b) 0.0821 atm L $mol^{-1}K^{-1}$
(c) 0.8314 bar L $mol^{-1}K^{-1}$ (d) 2 cal $mol^{-1}K^{-1}$

15. At 0°C and one atm pressure, a gas occupies 100 cc. If the pressure is increased to one and a half-time and temperature is increased by one-third of absolute temperature, then final volume of the gas will be:
(a) 80 cc (b) 88.9 cc
(c) 66.7 cc (d) 100 cc

16. 10 g of a gas at 1 atm and 273 K occupies 5 litres. The temperature at which the volume becomes double for the same mass of gas at the same pressure is:
(a) 273 K (b) –273°C
(c) 273°C (d) 546°C

17. Which of the following curve does not represent Gay-lusacc's law?

(a)
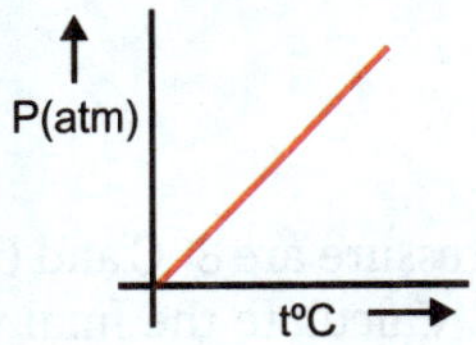

(b)
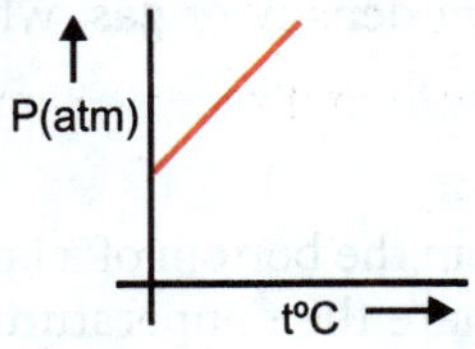

(c)
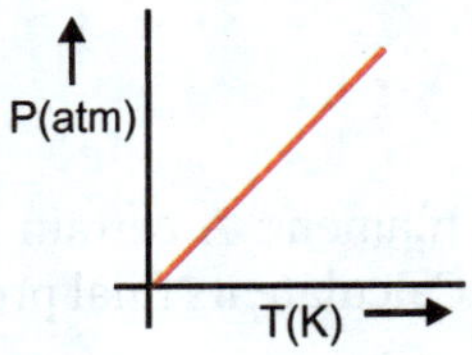

(d)
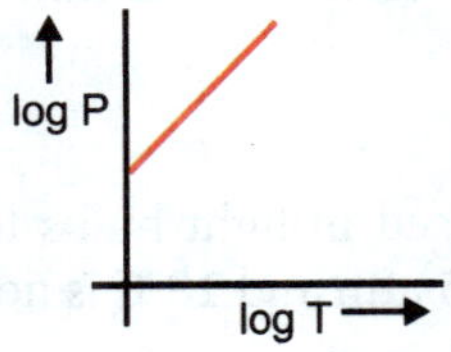

18. If densities of two gases are in the ratio 1 : 2 and their temperatures are in the ratio 2 : 1, then the ratio of their respective molar mass at certain pressure is:
(a) 1 : 1 (b) 1 : 2
(c) 2 : 1 (d) 4 : 1

19. Two separate bulbs contain ideal gases *A* and *B*. The density of gas *A* is twice that of gas *B*. The molecular mass of *A* is half that of gas *B*. The two gases are at the same temperature. The ratio of the pressure of *A* to that of gas *B* is :
(a) 2 (b) 1/2
(c) 4 (d) 1/4

20. Volume of the air that will be expelled from a vessel of 300 cm^3 when it is heated from 27°C to 37°C at the constant pressure will be:
(a) 310 cm^3 (b) 290 cm^3
(c) 10 cm^3 (d) 37 cm^3

21. *V versus T* curves at constant pressure P_1 and P_2 for an ideal gas are shown in fig. Which is correct?

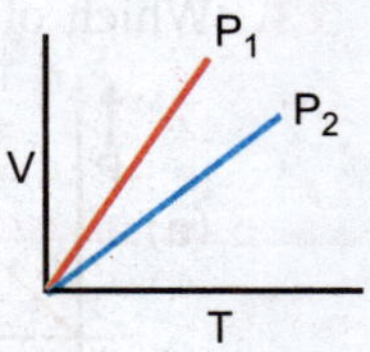

(a) $P_1 > P_2$ (b) $P_1 < P_2$
(c) $P_1 = P_2$ (d) All of these

22. Two flasks *A* and *B* of 500 mL each are respectively filled with O_2 and SO_2 at 300 K and 1 atm. pressure. The flasks will contain:
(a) the same number of atoms
(b) the same number of molecules
(c) more number of moles of molecules in flask *A* as compared to flask *B*
(d) the same amount of gases

23. 2.8 g of a gas at 1atm and 273K occupies a volume of 2.24 litres. The gas can not be:
(a) O_2 (b) CO
(c) N_2 (d) C_2H_4

24. Five grams each of the following gases at 87°C and 750 mm pressure are taken. Which of them will have the least volume ?
(a) HF (b) HCl
(c) HBr (d) HI

25. At what pressure a quantity of gas will occupy a volume of 60 mL, if it occupies a volume of 100 mL at a pressure of 720 mm (while temperature is constant) :
(a) 700 mm (b) 800 mm
(c) 100 mm (d) 1200 mm

26. At 1 atm and 273 K the density of gas, whose molecular weight is 45, is:
(a) 44.8 g/L (b) 11.4 g/L
(c) 2 g/L (d) 3 g/L

27. A small bubble rises from the bottom of a lake, where the temperature and pressure are 8° C and 6.0 atm, to the water's surface, where the temperature is 25°C and pressure is 1.0 atm. Calculate the final volume of the bubble if its initial volume was 2 mL.
(a) 14 mL (b) 12.72 mL
(c) 11.31 mL (d) 15 mL

28. Argon is an inert gas used in light bulbs to retard the vaporization of the filament. A certain light-bulb containing argon at 1.25 atm and 18°C is heated to 85°C at constant volume. Calculate its final pressure.
(a) 1.53 atm (b) 1.25 atm
(c) 1.35 atm (d) 2 atm

29. Calculate the volume of O_2 at 1 atm and 273 K required for the complete combustion of 2.64 L of acetylene (C_2H_2) at 1 atm and 273 K. $2C_2H_2(g) + 5O_2(g) \longrightarrow 4CO_2(g) + 2H_2O(l)$
(a) 3.6 L (b) 1.056 L
(c) 6.6 L (d) 10 L

30. The density of O_2 (*g*) is maximum at :
(a) STP (b) 273 K and 2 atm
(c) 546 K and 1 atm (d) 546 K and 2 atm

31. At 27°C a sample of ammonia gas exerts a pressure of 5.3 atm. What is the pressure when the volume of the gas is reduced to one-tenth of the original value at the same temperature?

(a) 0.53 atm (b) 5.3 atm

(c) 53 atm (d) None of these

32. A certain amount of gas at 25°C and at a pressure of 0.80 atm is kept in a glass vessel. Suppose that the vessel can withstand a pressure of 2.0 atm. How high can you raise the temperature of the gas without bursting the vessel?

(a) 745°C (b) 472°C

(c) 500°C (d) None of these

33. Which one of these graphs for an ideal gas, the arrow indication is incorrectly marked ?

(a)

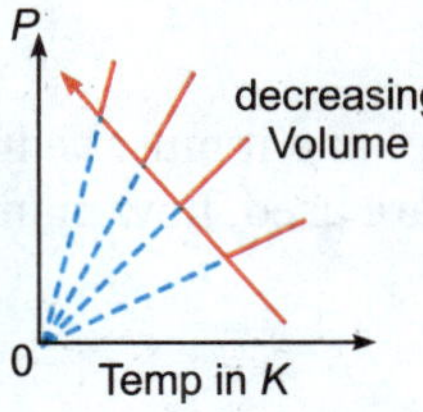

(b)

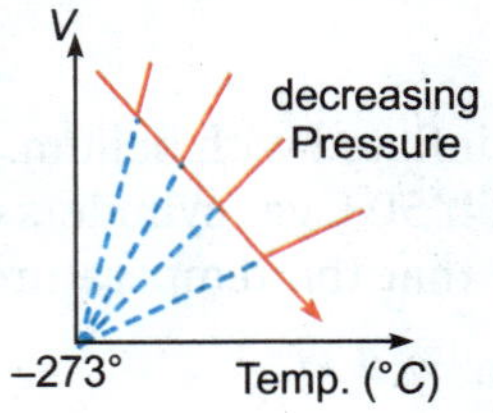

(c)

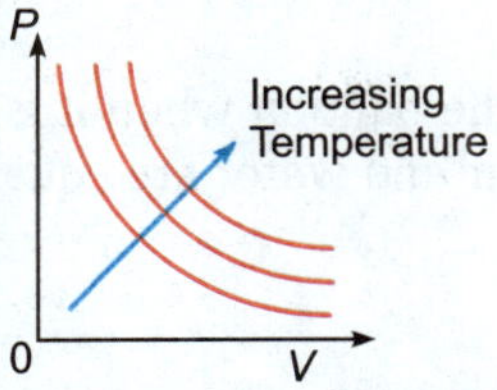

(d)

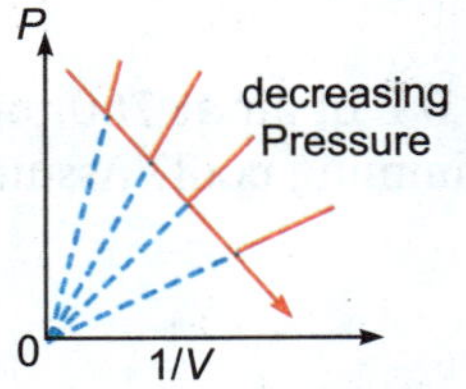

34. The pressure of sodium vapour in a 1.0 L container is 10 torr at 1000°C. How many atoms are in the container?

(a) 9.7×10^{17} (b) 7.6×10^{19}

(c) 4.2×10^{17} (d) 9.7×10^{19}

35. An ideal gaseous mixture of ethane (C_2H_6) and ethene (C_2H_4) occupies 28 litre at 1 atm and 273 K. The mixture reacts completely with 128 gm O_2 to produce CO_2 and H_2O. Mole fraction at C_2H_4 in the mixture is :

(a) 0.6 (b) 0.4

(c) 0.5 (d) 0.8

36. A certain hydrate has the formula $MgSO_4.xH_2O$. A quantity of 54.2 g of the compound is heated in an oven to drive off the water. If the steam generated exerts a pressure of 24.8 atm in a 2.0 L container at 120°C, calculate *x*.

(a) 2 (b) 5

(c) 6 (d) 7

37. Nitrogen forms several gaseous oxides. One of them has a density of 1.33 g/L measured at 764 mmHg and 150°C. Write the formula of the compound.

(a) NO (b) N_2O

(c) NO_2 (d) N_2O_5

38. Starting out on a trip into the mountains, you inflate the tires on your automobile to a recommended pressure of 3.21×10^5 Pa on a day when the temperature is $-5.0°$ C. You drive to the beach, where the temperature is 28.0°C. Assume that the volume of the tire has increased by 3%. What is the final pressure in the tyres?

(a) 350 Pa (b) 3500 Pa

(c) 3.5×10^5 Pa (d) None of these

39. A compressed cylinder of gas contains 1.50×10^3 g of N_2 gas at a pressure of 2.0×10^7 Pa and a temperature of 17.1°C. What volume of gas has been released into the atmosphere if the final pressure in the cylinder is 1.80×10^5 Pa? Assume ideal behaviour and that the gas temperature is unchanged.

(a) 1264 L (b) 126 L

(c) 12600 L (d) 45 L

40. The density of liquid gallium at 30°C is 6.095 g/mL. Because of its wide liquid range (30 to 2400°C), gallium is used as a barometer fluid at high temperature. What height (in cm) of gallium will be observed on a day when the mercury barometer reads 740 torr?
(The density of mercury is 13.6 g/mL.).

(a) 322 (b) 285

(c) 165 (d) 210

41. A weather balloon is inflated with helium. The balloon has a volume of 100 m^3 and it must be inflated to a pressure of 0.10 atm. If 50 L gas cylinders of helium at a pressure of 100 atm are used, how many cylinders are needed? Assume that the temperature is constant.

(a) 2 (b) 3

(c) 4 (d) 1

42. A balloon contains 14.0 L of air at 760 torr. What will be the volume of the balloon when it is taken to a depth of 10 ft. in a swimming pool? Assume that the temperature of the air and water are equal. (density : Hg = 13.6 g/mL.)

(a) 11.0 (b) 11.3

(c) 10 (d) 10.8

43. A 0.50 L container is occupied by nitrogen at a pressure of 800 torr and a temperature of 0°C. The container can only withstand a pressure of 3.0 atm. What is the highest temperature (°C) to which the container may be heated?

(a) 505 (b) 450

(c) 625 (d) 560

44. Equal volumes of oxygen gas and a second gas weigh 1.00 and 2.375 grams respectively under the same experimental conditions. Which of the following is the unknown gas?

(a) NO (b) SO_2

(c) CS_2 (d) CO

45. A high altitude balloon contains 6.81 g of helium in 1.16×10^4 L at – 23°C. Assuming ideal gas behaviour, how many grams of helium would have to be added to increase the pressure to 4.0×10^{-3} atm?

(a) 1.27 g (b) 1.58 g

(c) 2.68 g (d) 2.23 g

46. A 4.40 g piece of solid CO_2 (dry ice) is allowed to sublime in a balloon. The final volume of the balloon is 1.00 L at 300 K. What is the pressure (atm) of the gas?

(a) 0.122 (b) 2.46

(c) 122 (d) 24.6

47. For a closed (not rigid) container containing $n = 10$ moles of an ideal gas fitted with movable, frictionless, weightless piston operating such that pressure of gas remains constant at 0.821 atm. Which graph represents correct variation of $\log V$ *vs* $\log T$ where V is in litre and T in kelvin.

(a) 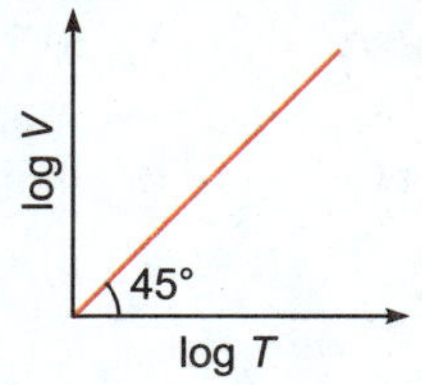

(b) 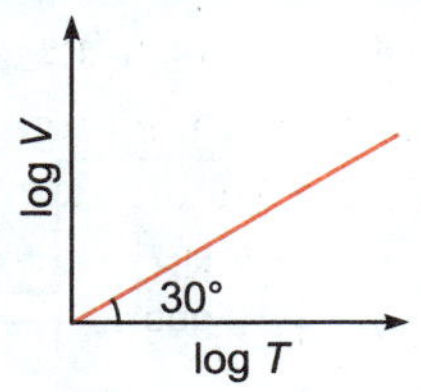

(c) 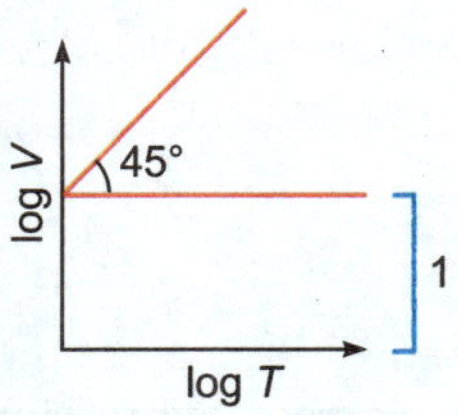

(d) 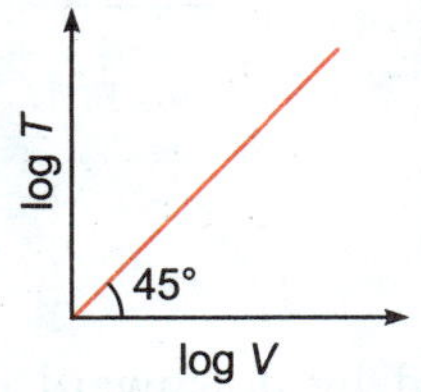

48. The intercept on y-axis and slope of curve plotted between P/T vs. T

(For an ideal gas having 10 moles in a closed rigid container of volume 8.21 L. (P = Pressure in atm and T = Temp. in K, $\log_{10} 2 = 0.30$)) are respectively :

(a) 0.01, 0 (b) 0.1, 0
(c) 0.1, 1 (d) 10, 1

49. The density of gas A is twice that to B at the same temperature. The molecular weight of gas B is twice that of A. The ratio of pressure of gas A and B will be :

(a) 1 : 6 (b) 1 : 1
(c) 4 : 1 (d) 1 : 4

50. Two inflated ballons I and II (thin skin) having volume 600 mL and 1500 mL at 300 K are taken as shown in diagram. If maximum volume of inner and outer balloons are 800 mL and 1800 mL respectively then find the balloon which will burst first on gradual heating.

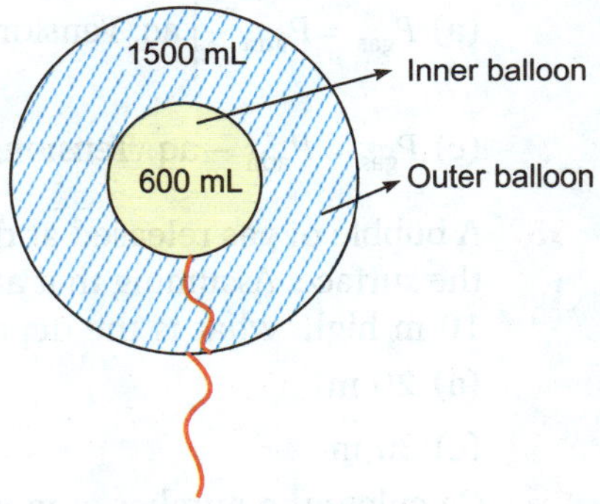

(a) inner balloon
(b) outer balloon
(c) both simultaneously
(d) unpredictable

51. An open flask containing air is heated from 300 K to 500 K. What percentage of air will be escaped to the atmosphere, if pressure is keeping constant?

(a) 80 (b) 40
(c) 60 (d) 20

52. The value of universal gas constant R depends on :

(a) temperature of gas (b) volume of gas
(c) number of moles of gas (d) units of volume and pressure

53. A manometer attached to a flask contains with ammonia gas have no difference in mercury level initially as shown in diagram. After sparking into the flask, ammonia is partially dissociated as $2NH_3(g) \longrightarrow N_2(g) + 3H_2(g)$ now it have difference of 6 cm in mercury level in two columns, what is partial pressure of $H_2(g)$ at equilibrium?

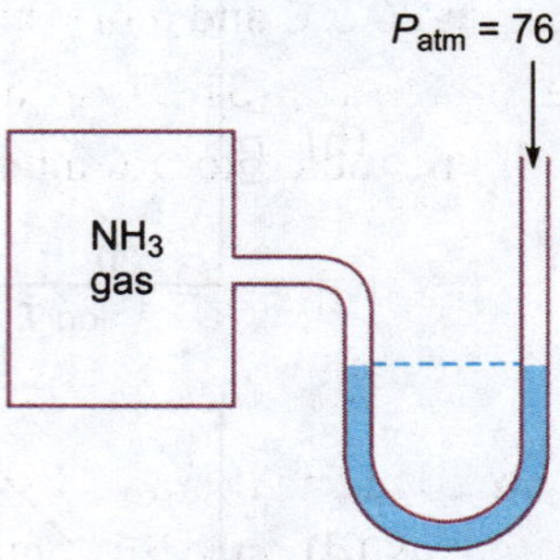

(a) 9 cm Hg (b) 18 cm Hg

(c) 27 cm Hg (d) None of these

54. An ideal gas is collected by downward displacement of water. Select the correct expression for P_{gas} according to the diagram [$d_{Hg} = 13.6\, g/cm^3$] :

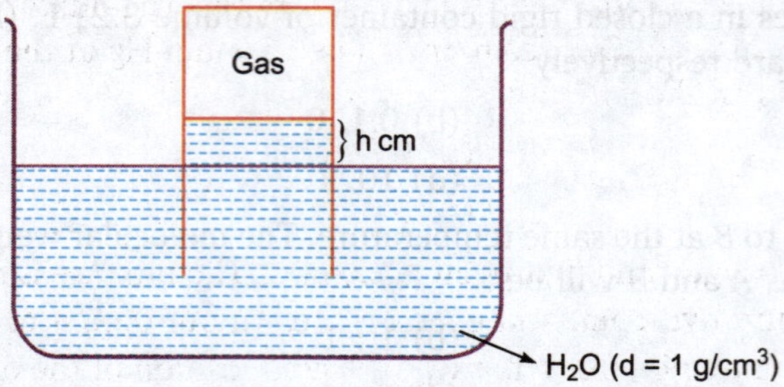

(a) $P_{gas} = P_{atm} - \left[\text{aq. Tension} + \frac{h}{13.6}\right]$ (b) $P_{gas} = P_{atm} - hdg$

(c) $P_{gas} = P_{atm} - \text{aq. Tension} + \frac{h}{13.6}$ (d) none of these

55. A bubble of gas released at the bottom of a lake increases to four times its original volume when it reaches the surface. Assuming that atmospheric pressure is equivalent to the pressure exerted by a column of water 10 m high, what is the depth of the lake?

(a) 20 m (b) 10 m

(c) 30 m (d) 40 m

56. Calculate the number of moles of gas present in the container of volume 10 L at 300 K, if the manometer containing glycerine shows 5 m difference in level as shown diagram.

(Given : $d_{glycerin} = 2.72$ g/mL; $d_{mercury} = 13.6$ g/mL)

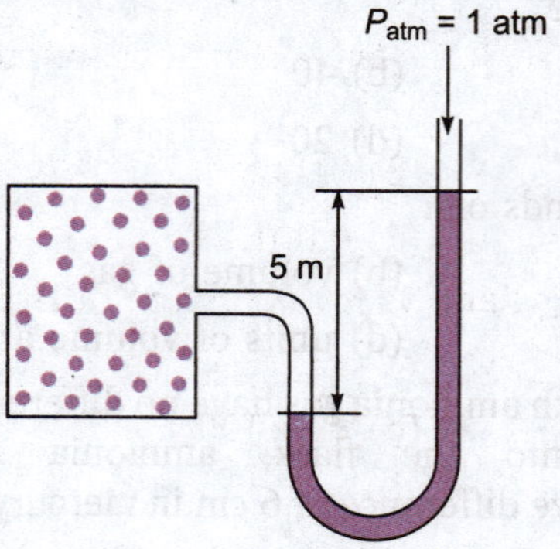

(a) 0.94 mole (b) 0.49 mole

(c) 0.64 mole (d) none of these

57. A rigid vessel of volume 0.50 m^3 containing H_2 at 20.5°C and a pressure of 611×10^3 Pa is connected to a second rigid vessel of volume 0.75 m^3 containing Ar at 31.2°C at a pressure of 433×10^3 Pa. A valve separating the two vessels is opened and both are cooled to a temperature of 14.5°C. What is the final pressure in the vessels?

(a) 2×10^5 (b) 3.22×10^5 Pa

(c) 4840 Pa (d) 4.84×10^5 Pa

58. Two glass bulbs *A* and *B* at same temperature are connected by a very small tube having a stop-cock. Bulb *A* has a volume of 100 cm^3 and contained the gas while bulb *B* is empty. On opening the stop-cock, the pressure fell down to 20%. The volume of the bulb *B* is :

(a) 100 cm^3 (b) 200 cm^3

(c) 250 cm^3 (d) 400 cm^3

59. A mixture of C_2H_2 and C_3H_8 occupied a certain volume at 80 mm Hg. The mixture was completely burnt to CO_2 and $H_2O(l)$. When the pressure of CO_2 was found to be 230 mm Hg at the same temperature and volume, the mole fraction of C_3H_8 in the mixture is :

(a) 0.125 (b) 0.875

(c) 0.6 (d) 0.8

60. The total pressure of a mixture of oxygen and hydrogen is 1.0 atm. The mixture is ignited and the water is removed. The remaining gas is pure hydrogen and exerts a pressure of 0.40 atm when measured at the same values of *T* and *V* as the original mixture. What was the composition of the original mixture in mole fraction ?

(a) $x_{O_2} = 0.2; x_{H_2} = 0.8$ (b) $x_{O_2} = 0.4; x_{H_2} = 0.6$

(c) $x_{O_2} = 0.6; x_{H_2} = 0.4$ (d) $x_{O_2} = 0.8; x_{H_2} = 0.2$

61. Two closed vessel *A* and *B* of equal volume of 8.21 L are connected by a narrow tube of negligible volume with open valve. The left hand side container is found to contain 3 mole CO_2 and 2 mole of He at 400 K. What is the partial pressure of He in vessel *B* at 400 K?

(a) 2.4 atm (b) 8 atm

(c) 12 atm (d) None of these

62. At STP, a container has 1 mole of He, 2 mole Ne, 3 mole O_2 and 4 mole N_2. Without changing total pressure if 2 mole of O_2 is removed. The partial pressure of O_2 will be decreased by :

(a) 26% (b) 40%

(c) 58.33% (d) 66.66%

63. 821 mL N_2 (*g*) was collected over liquid water at 300 K and 1 atm. If vapour pressure of H_2O is 30 torr then moles of N_2 (*g*) in moist gas mixture is :

(a) 0.39 (b) 0.032

(c) 0.96 (d) 0.0013

64. Let *p* and p_s be the partial pressure of H_2O (*g*) and vapour pressure of H_2O (*l*) respectively. Then the % relative humidity is given by:

(a) $\frac{p_s + p}{p_s} \times 100$ (b) $\frac{p}{p_s} \times 100$

(c) $\frac{p_s}{p} \times 100$ (d) $(p + p_s) \times 100$

65. The vapour pressure of water at 80°C is 355 mm of Hg. 1 L vessel contains O_2 at 80°C, which is saturated with water and the total pressure being 760 mm of Hg. The contents of the vessel were pumped into 0.3 L vessel at the same temperature. What is the partial pressure of O_2 ?

(a) 1350 Hg (b) 2178.3 Hg

(c) 121.5 Hg (d) 355 Hg

66. Which of the following gaseous mixture does not follow Dalton's law of partial pressure?

(a) SO_2 and Cl_2 (b) CO_2 and N_2

(c) CO and CO_2 (d) CO and N_2

67. Equal masses of methane and oxygen are mixed in an empty container at 25°C. The fraction of the total pressure exerted by oxygen is:

(a) $\frac{2}{3}$ (b) $\frac{1}{3} \times \frac{273}{298}$

(c) $\frac{1}{3}$ (d) $\frac{1}{2}$

68. A box of 1 L capacity is divided into two equal compartments by a thin partition which are filled with 2g H_2 and 16 g CH_4 respectively. The pressure in each compartment is recorded as *P* atm. The total pressure when partition is removed will be:

(a) *P* (b) 2*P*

(c) *P*/2 (d) *P*/4

69. If 10^{-4} dm^3 of water is introduced into a 1.0 dm^3 flask at 300 K, how many moles of water are in the vapour phase when equilibrium is established?
(Given : Vapour pressure of H_2O at 300 K is 3170 Pa; $R = 8.314\ JK^{-1}\ mol^{-1}$)

(a) 1.27×10^{-3} mole (b) 5.56×10^{-3} mole

(c) 1.53×10^{-2} mole (d) 4.46×10^{-2} mole

70. At room temperature Dalton's law of partial pressure is not applicable to :

(a) H_2 and N_2 mixture (b) H_2 and Cl_2 mixture

(c) H_2 and CO_2 mixture (d) none of these

71. 56 g of nitrogen and 96 g of oxygen are mixed isothermaly and at a total pressure of 10 atm. The partial pressures of oxygen and nitrogen (in atm) are respectively :

(a) 4, 6 (b) 5, 5

(c) 2, 8 (d) 6, 4

72. The closed containers of the same capacity and at the same temperature are filled with 44 g of H_2 in one and 44 g of CO_2 in the other. If the pressure of carbon dioxide in the second container is 1 atm, then pressure of hydrogen in the first container would be :

(a) 1 atm (b) 10 atm

(c) 22 atm (d) 44 atm

73. A jar contains a gas and a few drops of water. The pressure in the jar is 830 mm of Hg. The temperature of the jar is reduced by 1%. The vapour pressure of water at two temperatures are 30 and 25 mm of Hg. Calculate the new pressure in jar.

(a) 792 mm of Hg (b) 817 mm of Hg

(c) 800 mm of Hg (d) 840 mm of Hg

74. O_2 and SO_2 gases are filled in ratio of 1 : 3 by mass in a closed container of 3 L at temperature of 27°C. The partial pressure of O_2 is 0.60 atm, the concentration of SO_2 would be

(a) 0.36 mol/L (b) 0.036 mol/L

(c) 3.6 mol/L (d) 36 mol/L

75. A gaseous mixture contains three gases *A*, *B* and *C* with a total number of moles of 10 and total pressure of 10 atm. The partial pressure of *A* and *B* are 3 atm and 1 atm respectively. If *C* has molecular weight of 2 g/mol. then, the weight of *C* present in the mixture will be :

(a) 8 g (b) 12 g
(c) 3 g (d) 6 g

76. A rigid container contains 5 mole H_2 gas at some pressure and temperature. The gas has been allowed to escape by simple process from the container due to which pressure of the gas becomes half of its initial pressure and temperature become $(2/3)^{rd}$ of its initial. The mass of gas remaining is :
(a) 7.5 g (b) 1.5 g
(c) 2.5 g (d) 3.5 g

77. Pressure of 1 g ideal gas X at 300 K is 2 atm. When 2 g of another gas Y is introduced in the same vessel at same temperature, the pressure become 3 atm. The correct relationship between molar mass of X and Y is :
(a) $M_Y = 2M_X$ (b) $M_Y = 4M_X$
(c) $M_X = 4M_Y$ (d) None of these

78. Dry ice is solid carbon dioxide. A 0.050 g sample of dry ice is placed in an evacuated 4.6 L vessel at 30°C. Calculate the pressure inside the vessel after all the dry ice has been converted to CO_2 gas.
(a) 6.14 atm (b) 0.614 atm
(c) 0.0614 atm (d) 6.14×10^{-3} atm

79. A mixture of helium and neon gases is collected over water at 28.0°C and 745 mmHg. If the partial pressure of helium is 368 mmHg, what is the partial pressure of neon?
(Vapour pressure of water at 28° C = 28.3 mmHg)
(a) 348.7 mmHg (b) 377 mmHg
(c) 384.7 mmHg (d) none of these

80. Consider the following apparatus. Calculate the partial pressure of helium after the opening valve. The temperature remains constant at 16°C.

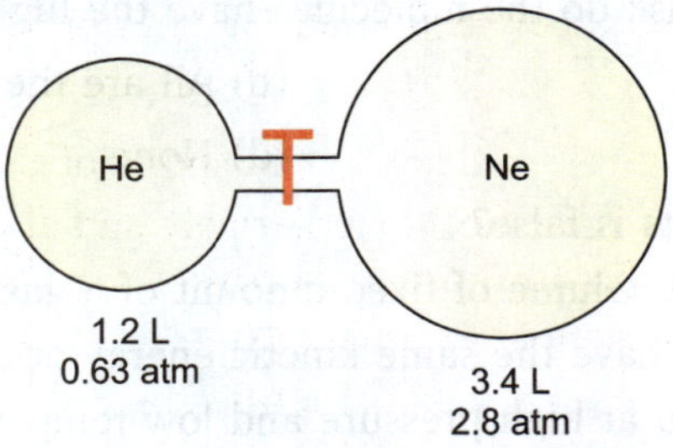

(a) 0.164 atm (b) 1.64 atm
(c) 0.328 atm (d) 1 atm

81. Oxygen gas generated by the decomposition of potassium chlorate is collected over water. The volume of oxygen collected at 24°C and atmospheric pressure of 760 mmHg is 128 mL. Calculate the mass of oxygen gas obtained. The vapour pressure of the water vapour at 24°C is 22.4 mm Hg.
(a) 1.36 g (b) 1.52 g
(c) 0.163 g (d) 1.63 g

82. The quantity $\frac{PV}{k_B T}$ represents the (k_B : Boltzmann constant)
(a) number of particles of the gas (b) mass of the gas
(c) number of moles of the gas (d) translation energy of the gas

83. Which of the following statements about kinetic energy (K.E.) is true?
(a) All objects moving with the same velocity have the same K.E.
(b) The K.E. of a body will quadruple if its velocity doubles
(c) As the velocity of a body increases, its K.E. decreases
(d) The K.E. of a body is independent of its mass

84. The Ne atom has 10 times the mass of H_2 molecule. Which of the following statements is true?
I. At 25°C both of them have the same kinetic energy.
II. Ten moles of H_2 would have the same volume as 1 mole of Ne at same temp. and pressure.
III. One mole of Ne exerts the same pressure as one mole of H_2 at STP.
IV. A H_2 molecule travels 10 times faster than Ne atom at same temperature.
V. At STP, one litre of Ne has 10 times the density of 1 litre of H_2.
(a) II, IV, V (b) I, III, V
(c) I, II, III (d) I, II

85. Which of the following is NOT a postulate of the kinetic molecular theory of gases?
(a) The gas molecules possess a volume that is negligibly small compared to the container
(b) The pressure and volume of a gas are inversely related
(c) Gases consist of discrete particles that are in constant chaotic motion
(d) The average kinetic energy of the molecules is directly proportional to the absolute temperature

86. Which one of the following relationships when graphed does not give a straight line for helium gas?
I. K.E. and T at constant pressure and volume
II. P vs. V at constant temperature for a constant mass
III. V vs. $1/T$ at constant pressure for a constant mass
(a) II (b) II and III
(c) III (d) I

87. Consider three one-litre flasks labeled A, B and C filled with the gases NO, NO_2, and N_2O, respectively, each at 1 atm and 273 K. In which flask do the molecules have the highest average kinetic energy?
(a) Flask C (b) All are the same
(c) Flask A (d) None

88. Which of the following statements is false?
(a) The product of pressure and volume of fixed amount of a gas is independent of temperature.
(b) Molecules of different gases have the same kinetic energy at a given temperature.
(c) The gas equation is not valid at high pressure and low temperature.
(d) The gas constant per molecule is known as Boltzmann's constant.

89. Which is not correct in terms of kinetic theory of gases?
(a) Gaseous particles are considered as point mass.
(b) The gaseous molecules are in random motion.
(c) When gaseous molecules collide, they lose energy.
(d) When the gas is heated , the molecules moves faster.

90. Two flasks A and B have equal volumes. A is maintained at 300 K and B at 600 K. A contains H_2 gas, B has an equal mass of CO_2 gas. Find the ratio of total K.E. of gases in flask A to that of B.
(a) 1 : 2 (b) 11 : 1
(c) 33 : 2 (d) 55 : 7

91. Kinetic energy and pressure of a gas of unit volume are related as :
(a) $P = \frac{2}{3}E$ (b) $P = \frac{3}{2}E$
(c) $P = \frac{E}{2}$ (d) $P = 2E$

92. Two flask A and B of equal volumes maintained at temperature 300 K and 700 K contain equal mass of He(g) and $N_2(g)$ respectively. What is the ratio of translational kinetic energy of gas in flask A to that of flask B?

(a) 1 : 3 (b) 3 : 1
(c) 3 : 49 (d) None of these

93. Which of the following change is observed occurs when a substance X is converted from liquid to vapour phase at the standard boiling point?
I. Potential energy of the system decreases
II. The distance between molecules increases
III. The average kinetic energy of the molecules in both phases are equal

(a) I only (b) II only
(c) III only (d) II and III only

94. A mixture of Ne and Ar kept in a closed vessel at 250 K has a total K.E. = 3 kJ. The total mass of Ne and Ar is 30 g. Find mass % of Ne in gaseous mixture at 250 K.

(a) 61.63 (b) 38.37
(c) 50% (d) 28.3

95. In two vessels of 1 litre each at the same temperature 1g of H_2 and 1g of CH_4 are taken. For these gases:

(a) V_{rms} values will be same (b) Kinetic energy per mol will be same
(c) Total kinetic energy will same (d) Pressure will be same

96. Four particles have speed 2,3,4 and 5 cm/s respectively. Their rms speed is:

(a) 3.5 cm/s (b) (27/2) cm/s
(c) $\sqrt{54}$ cm/s (d) $(\sqrt{54}/2)$ cm/s

97. A gaseous mixture contains 4 molecules with a velocity of 6 cm sec^{-1}, 5 molecules with a velocity of 2 cm sec^{-1} and 10 molecules with a velocity of 3 cm sec^{-1}. What is the rms speed of the gas:

(a) 2.5 cm sec^{-1} (b) 1.9 cm sec^{-1}
(c) 3.6 cm sec^{-1} (d) 4.6 cm sec^{-1}

98. The ratio between the root mean square velocity of H_2 at 50 K and that of O_2 at 800 K is:

(a) 4 (b) 2
(c) 1 (d) 1/4

99. If $C_1, C_2, C_3 \ldots$ represent the speeds of $n_1, n_2, n_3 \ldots$ molecules respectively, then the root mean square speed will be:

(a) $\sqrt{\dfrac{n_1C_1^2 + n_2C_2^2 + n_3C_3^2 + \ldots}{n_1 + n_2 + n_3 + \ldots}}$ (b) $\sqrt{\dfrac{(n_1 + n_2 + n_3 + \ldots)^2}{n_1C_1^2 + n_2C_2^2 + n_3C_3^2 + \ldots}}$

(c) $\sqrt{\dfrac{(n_1C_1)}{n_1} + \dfrac{(n_2C_2)}{n_2} + \dfrac{(n_3C_3)}{n_3}}$ (d) $\sqrt{\dfrac{(n_1C_1 + n_2C_2 + n_2C_3 + \ldots)^2}{n_1 + n_2 + n_3 + \ldots}}$

100. The root mean square speed of hydrogen is $\sqrt{5}$ times than that of nitrogen. If T is the temperature of the gas, then :

(a) $T_{H_2} = T_{N_2}$ (b) $T_{H_2} > T_{N_2}$
(c) $T_{H_2} < T_{N_2}$ (d) $T_{H_2} = \sqrt{7}\, T_{N_2}$

101. At a definite temperature (T), the distribution of speeds is given by the curve. In the curve points A, B and C indicates the speeds corresponding to :

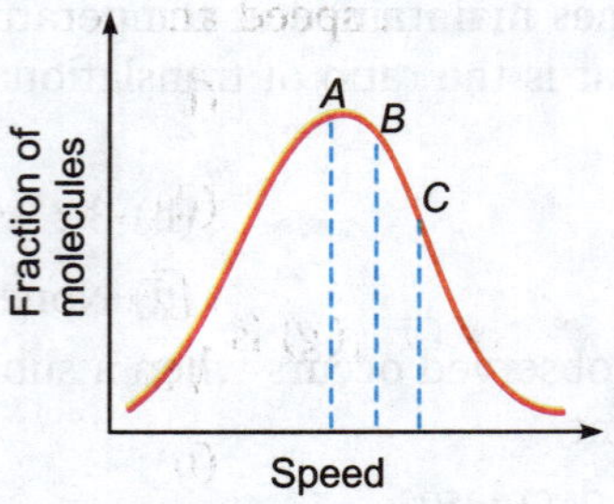

(a) most probable, average and root mean square speeds
(b) average, root mean square and most probable speeds
(c) root mean square, average and most probable speeds
(d) most probable, root mean square and average speeds

102. The most probable speed of 8 g of H_2 is 200 ms^{-1}. Average kinetic energy (neglect rotational and vibrational energy) of H_2 gas is :
(a) 480 J (b) 240 J
(c) 120 J (d) none of these

103. At what temperature will average speed of the molecules of the second member of the series C_nH_{2n} be the same to that of Cl_2 at 627°C?
(a) 259.4 K (b) 400 K
(c) 532.4 K (d) None of these

104. If u_{rms} of a gas is $30\,R^{1/2}$ ms^{-1} at 27°C then the molar mass of gas is :
(a) 0.02 kg/mol (b) 0.001 kg/mol
(c) 0.003 kg/mol (d) 1 kg/mol

105. 6×10^{22} gas molecules each of mass 10^{-24} kg are taken in a vessel of 10 litre. What is the pressure exerted by gas molecules? The root mean square speed of gas molecules is 100 m/s.
(a) 20 Pa (b) 2×10^4 Pa
(c) 2×10^5 Pa (d) 2×10^7 Pa

106. At what temperature will most probable speed of the molecules of the second member of homologous series C_nH_{2n-2} be the same as that of SO_2 at 527°C.
(a) 500°C (b) 727°C
(c) 227°C (d) None of these

107. The root mean square speed of 8 g of He is 300 ms^{-1}. Total kinetic energy of He gas is :
(a) 120 J (b) 240 J
(c) 360 J (d) None of these

108. If T_1, T_2 and T_3 are the temperatures at which the u_{rms}, $u_{average}$, u_{mp} of oxygen gas are all equal to 1500 m/s then the correct statement is :
(a) $T_1 > T_2 > T_3$ (b) $T_1 < T_2 < T_3$
(c) $T_1 = T_2 = T_3$ (d) None of these

109. The density of a gas filled in electric lamp is 0.75 kg/m^3. After the lamp has been switched on, the pressure in it increases from 4×10^4 Pa to 9×10^4 Pa. What is increases in u_{rms} ?
(a) 100 (b) 200
(c) 300 (d) None of these

110. The root mean square speed of an ideal gas at constant pressure varies with density (d) as :
(a) d^2 (b) d
(c) $\sqrt{d}$ (d) $1/\sqrt{d}$

111. The ratio among most probable speed, mean speed and root mean square speed is given by

(a) $1 : 2 : 3$ (b) $1 : \sqrt{2} : \sqrt{3}$

(c) $\sqrt{2} : \sqrt{3} : \sqrt{8/\pi}$ (d) $\sqrt{2} : \sqrt{8/\pi} : \sqrt{3}$

112. The average speed at temperature T°C of $CH_4(g)$ is $\sqrt{\frac{28}{88}} \times 10^3\ ms^{-1}$. What is the value of T ?

(a) 240.55°C (b) – 32.45°C

(c) 3000°C (d) – 24.055°C

113. At what temperature most probable speed of SO_2 molecules have the same value as root mean square speed of O_2 molecules at 300 K?

(a) 150 K (b) 600 K

(c) 750 K (d) 900 K

114. The rms speed of N_2 gas molecules is u. If the temperature is doubled and the nitrogen molecules dissociate into nitrogen atoms, the rms speed becomes:

(a) $u/2$ (b) $2u$

(c) $4u$ (d) $14u$

115. The rate of diffusion of a gas is proportional to:

(a) $\frac{P}{\sqrt{M}}$ (b) $\frac{P}{M}$

(c) $\sqrt{\frac{P}{M}}$ (d) $\frac{\sqrt{P}}{M}$

116. At constant volume and temperature conditions, the rates of diffusion r_A and r_B of gases A and B having densities ρ_A and ρ_B are related by the expression :

(a) $r_A = r_B.(\rho_A/\rho_B)^2$ (b) $r_A = r_B(\rho_B/\rho_A)^{1/2}$

(c) $r_A = (r_B.\rho_B/\rho_A)^{1/2}$ (d) $r_A = r_B(\rho_A/\rho_B)^{1/2}$

117. What is the ratio of diffusion rate of oxygen to hydrogen?

(a) 1 : 4 (b) 4 : 1

(c) 1 : 8 (d) 8 : 1

118. The molecular weight of a gas which diffuses through a porous plug at $1/6^{th}$ of the speed of hydrogen under identical conditions is:

(a) 27 (b) 72

(c) 36 (d) 48

119. X mL of H_2 gas effuses through a hole in a container in 5 seconds. The time taken for the effusion of the same volume of the gas specified below under identical condition is:

(a) 10 sec : He (b) 20 sec : O_2

(c) 25 sec : CO (d) 55 sec : CO_2

120. At identical temperature and pressure, the rate of diffusion of hydrogen gas is $3\sqrt{3}$ times that of a hydrocarbon having molecular formula C_nH_{2n-2}. What is the value of n ?

(a) 1 (b) 4

(c) 3 (d) 8

121. Calculate relative rate of effusion of O_2 to CH_4 from a container containing O_2 and CH_4 in 3 : 2 mass ratio.

(a) $\frac{3\sqrt{2}}{4}$ (b) $\frac{3}{4\sqrt{2}}$

(c) $\frac{3}{\sqrt{2}}$ (d) none of these

122. Calculate relative rates of effusion of SO_2 to CH_4, in the mixture obtained by effusing out a mixture with initial molar ratio $\frac{n_{SO_2}}{n_{CH_4}} = \frac{8}{1}$ for three effusing steps.

(a) 2 : 1 (b) 1 : 4

(c) 1 : 2 (d) none of these

123. A gaseous mixture containing He, CH_4 and SO_2 was allowed to effuse through a fine hole. If mixture contain He, CH_4 and SO_2 in 1 : 2 : 3 mole ratio, calculate the molar ratio of gases effusing out initially.

(a) 2 : 2 : 3 (b) 6 : 6 : 1

(c) $\sqrt{2} : \sqrt{2} : 3$ (d) 4 : 4 : 3

124. 80 mL of O_2 takes 2 minute to pass through the hole. What volume of SO_2 will pass through the hole in 3 minute?

(a) $\frac{120}{\sqrt{2}}$ mL (b) $120 \times \sqrt{2}$ mL

(c) $\frac{12}{\sqrt{2}}$ mL (d) None of these

125. When two cotton plugs, one moistened with ammonia and the other with hydrochloric acid, are simulataneously inserted into opposite ends of a glass tube 87.0 cm long, a white ring of NH_4Cl forms where gaseous NH_3 and gaseous HCl first come into contact.

$$NH_3(g) + HCl(g) \longrightarrow NH_4Cl(s)$$

At what distance from the ammonia-moistened plug does this occur?

(a) 51.7 cm from NH_3 end (b) 51.7 cm from HCl end

(c) 43.5 at mid point (d) None of these

126. Dimethyl ether decomposes as

$$CH_3OCH_3(g) \longrightarrow CH_4(g) + CO(g) + H_2(g)$$

When CH_3OCH_3 decomposes to 20% extent under certain conditions, what is the ratio of diffusion of pure CH_3OCH_3 with methane?

(a) 0.59 : 1 (b) 1.18 : 1

(c) 2.36 : 1 (d) 1.77 : 1

127. 4 gm of sulphur dioxide gas diffuses from a container in 8 min. Mass of helium gas diffusing from the same container over the same time interval is :

(a) 0.5 g (b) 1 g

(c) 2 g (d) None of these

128. Under identical conditions of pressure and temperature, 4 L of gaseous mixture (H_2 and CH_4) effuses through a hole in 5 min whereas 4 L of a gas X of molecular mass 36 takes to 10 min to effuse through the same hole. The mole ratio of $H_2 : CH_4$ in the mixture is :

(a) 1 : 2 (b) 2 : 1

(c) 2 : 3 (d) 1 : 1

129. According to the law of equipartition of energy, the energy associated with each degree of freedom is :

(a) $\frac{1}{3} K_B T$ (b) $\frac{1}{2} K_B T$

(c) $K_B T$ (d) $\frac{3}{2} K_B T$

130. Calculate γ (ratio of C_p and C_v) for triatomic linear gas at high temperature. Assume that the contribution of vibrational degree of freedom is 75% :

(a) 1.222 (b) 1.121
(c) 1.18 (d) 1.33

131. If one mole each of a monoatomic and diatomic gases are mixed at low temperature then C_p/C_v ratio for the mixture is :

(a) 1.40 (b) 1.428
(c) 1.5 (d) 1.33

132. If one mole of a mono-atomic gas ($\gamma = 5/3$) is mixed with one mole of a diatomic gas ($\gamma = 7/5$), the value of γ for the mixture is :

(a) 1.4 (b) 1.5
(c) 1.53 (d) 3.07

133. Select the correct statement :

(a) Internal energy of a real gas at a given temperature increases as the volume increases
(b) Internal energy of an ideal gas at given temperature increase as the volume increases
(c) Internal energy of an ideal gas molecules is not a function of temperature
(d) The internal energy of a real gas at a constant temperature is independent of change in volume

134. Which gas shows real behaviour?

(a) 16 g O_2 at 1 atm & 273 K occupies 11.2 L
(b) 1 g H_2 in 0.5 L flask exerts pressure of 24.63 atm at 300 K
(c) 1 mole NH_3 at 300 K and 1 atm occupies volume 22.4 L
(d) 5.6 L of CO_2 at 1 atm & 273 K is equal to 11 g

135. Consider a real gas placed in a container. If the intermolecular attractions are supposed to disappear suddenly which of the following would happen?

(a) Pressure decreases (b) Pressure increases
(c) Pressure remains unchanged (d) Gas collapses

136. The pressure of a real gas is less than the pressure of an ideal gas because of :

(a) increase in number of collisions (b) finite size of molecule
(c) increase in KE of molecules (d) intermolecular forces of attraction

137. A gas behaves like an ideal gas at :

(a) high pressure and low temperature (b) low pressure and high temperature
(c) high pressure and high temperature (d) low pressure and low temperature

138. If temperature and volume are same, the pressure of a gas obeying van der Waal's equation is :

(a) less than that of an ideal gas (b) more than that of an ideal gas
(c) same as that of an ideal gas (d) none of these

139. 1 mole of each of X_1, X_2, X_3 with van der Waal's constants a (in atm L^3 mol^{-2}) 1.0, 3.8, 2.1 respectively is kept separately in three different vessels of equal volume at identical temperature. Their pressures are observed to P_1, P_2, and P_3 respectively. On the basis of this data alone, select the correct option (neglect the effect of 'b'):

(a) $P_1 < P_2 < P_3$ (b) $P_2 < P_1 < P_3$
(c) $P_2 < P_3 < P_1$ (d) $P_1 = P_2 = P_3$

140. A gas obeys the equation of state $P(V - b) = RT$ (The parameter b is a constant). The slope for an isochore will be :

(a) negative (b) zero
(c) $R/(V - b)$ (d) R/P

141. van der Waals constant b of helium is 24 mL mol^{-1}. Find molecular diameter of helium.

(a) 1.335×10^{-10} cm
(b) 1.335×10^{-8} cm
(c) 2.67×10^{-8} cm
(d) 4.34×10^{-8} cm

142. Which of the following can be most readily liquefied?
(Given: value of 'a' for $NH_3 = 4.17$ atm $.L^3mol^{-2}$, $CO_2 = 3.59$ atm $.L^3mol^{-2}$, $SO_2 = 6.71$ atm $.L^3mol^{-2}$, $Cl_2 = 6.49$ atm $.L^3mol^{-2}$)

(a) NH_3
(b) Cl_2
(c) SO_2
(d) CO_2

143. For which of the following gases should the correction for the molecular volume be largest : CO, CO_2, NH_3 or SF_6 ?

(a) CO
(b) CO_2
(c) NH_3
(d) SF_6

144. Under which of the following sets of conditions is a real gas expected to deviate from ideal behaviour?
(I) High pressure, small volume
(II) High temperature, low pressure
(III) Low temperature, high pressure

(a) only I
(b) only II
(c) only III
(d) I and III both

145. For a certain gas which deviates positive from ideal behaviour, a plot between P/ρ vs P was found to be non-linear. The intercept on y-axis will be :

(a) $\frac{RT}{M}$
(b) $\frac{M}{RT}$
(c) $\frac{MZ}{RT}$
(d) $\frac{R}{TM}$

146. At low pressure, the van der Waal's equation become :

(a) $PV_m = RT$
(b) $P(V_m - b) = RT$
(c) $\left(P + \frac{a}{V_m^2}\right)V_m = RT$
(d) $P = \frac{RT}{V_m} + \frac{a}{V_m^2}$

147. At low pressure, if $RT = 2\sqrt{a \cdot P}$, then the volume occupied by a real gas is :

(a) $\frac{2RT}{P}$
(b) $\frac{2P}{RT}$
(c) $\frac{RT}{2P}$
(d) $\frac{RT}{P}$

148. For a gas deviation from ideal behaviour is maximum at :

(a) 0°C and 1.0 atm
(b) 100°C and 2.0 atm
(c) – 13°C and 1.0 atm
(d) – 13°C and 2.0 atm

149. At low pressures, van der Waals' equation is written as $\left(P + \frac{a}{V^2}\right)V = RT$. The compressibility factor is then equal to:

(a) $\left(1 - \frac{a}{RTV}\right)$
(b) $\left(1 - \frac{RTV}{a}\right)$
(c) $\left(1 + \frac{a}{RTV}\right)$
(d) $\left(1 + \frac{RTV}{a}\right)$

150. The compressibility factor for a real gas at high pressure is :

(a) 1 (b) $1+\frac{Pb}{RT}$

(c) $1-\frac{Pb}{RT}$ (d) $1+\frac{RT}{Pb}$

151. The compressibility of a gas is greater than unity at 1 atm and 273 K. Therefore :

(a) $V_m > 22.4$ L (b) $V_m < 22.4$ L

(c) $V_m = 22.4$ L (d) $V_m = 44.8$ L

152. At 273 K temperature and 9 atm pressure, the compressibility for a gas is 0.9. The volume of 1 milli-mole of gas at this temperature and pressure is :

(a) 2.24 litre (b) 0.020 mL

(c) 2.24 mL (d) 22.4 mL

153. The compressibility factor for nitrogen at 330 K and 800 atm is 1.90 and at 570 K and 200 atm is 1.10. A certain mass of N_2 occupies a volume of 1 dm^3 at 330 K and 800 atm. Calculate volume occupied by same quantity of N_2 gas at 570 K and 200 atm :

(a) 1 L (b) 2 L

(c) 3 L (d) 4 L

154. Consider the equation $Z=\frac{PV}{RT}$. Which of the following statements is correct?

(a) When $Z > 1$, real gases are easier to compress than the ideal gas

(b) When $Z = 1$, real gases get compressed easily

(c) When $Z > 1$, real gases are difficult to compress

(d) When $Z = 1$, real gases are difficult to compress

155. What is the compressibility factor (Z) for 0.02 mole of a van der Waals' gas at pressure of 0.1 atm. Assume the size of gas molecules is negligible.

Given : $RT = 20$ L atm mol^{-1} and $a = 1000$ atm L^2 mol^{-2}

(a) 2 (b) 1

(c) 0.02 (d) 0.5

156. Consider the following statements. If the van der Waal' s parameters of two gases are given as

	a/dm^6 bar mol^{-2}	b/dm^3 mol^{-1}
Gas A	6.5	0.055
Gas B	2	0.01

then :

1. Critical volume of $A <$ Critical volume of B
2. Critical pressure $A >$ Critical pressure of B
3. Critical temperature of $A >$ Critical temperature of B

Which of the above statements is/are incorrect?

(a) 1 alone (b) 1 and 2

(c) 1, 2 and 3 (d) 2 and 3

157. The van der Waals parameters for gases *W, X, Y* and *Z* are

Gas	a (atm L^2 mol^{-2})	b (L mol^{-1})
W	4.0	0.027
X	8.0	0.030
Y	6.0	0.032
Z	12.0	0.027

Which one of these gases has the highest critical temperature?
(a) W (b) X
(c) Y (d) Z

158. Pressure remaining the constant, the volume of a given mass of an ideal gas increases for every degree centigrade rise in temperature by definite fraction of its volume at:
(a) 0°C (b) its critical temperature
(c) absolute zero (d) its Boyle temperature

159. The critical temperature of a substance is:
(a) the temperature above which the substance undergoes decomposition
(b) the temperature above which a substance can exist only in gaseous state
(c) boiling point of the substance
(d) all are wrong

160. The excluded volume of a gas will be larger, if $\frac{T_C}{P_C}$ is :
(a) small (b) large
(c) equal to 1 (d) less than unity

161. Select incorrect statement :
(a) we can condense vapour simply by applying pressure
(b) to liquefy a gas one must lower the temperature below T_C and also apply pressure
(c) at T_C, there is no distinction between liquid and vapour state hence density of the liquid is nearly equal to density of the vapour
(d) However great the pressure applied, a gas cannot be liquified below it's critical temp.

162. Inversion temperature $\left(T_i = \frac{2a}{Rb}\right)$ is defined as the temperature above which if a gas is expanded adiabatically it gets warmed up but if temperature of gas is lower than T_i then it will cool down. What will happen to a gas if it is adiabatically expanded at 600 K if its Boyle's temperature is 290 K?
(a) Heating (b) Cooling
(c) Constant (d) None

163. The temperature at which the second virial coefficient of real gas is zero is called :
(a) Critical temperature (b) Triple point
(c) Boiling point (d) Boyle's temperature

164. The van der Waals' equation for one mole may be expressed as

$$V_m^3 - \left(b + \frac{RT}{p}\right)V_m^2 + \frac{aV_m}{p} - \frac{ab}{p} = 0$$

where V_m is the molar volume of the gas. Which of the following is incorrect?
(a) For a temperature less than T_C, V has three real roots

(b) For a temperature less than T_C, V has three imaginary roots

(c) For a temperature equal to T_C all three roots of V are real and identical

(d) On increasing the temp. ($T < T_C$), the three roots become closer to one another

165. For a gas obeying the van der Waals' equation, at the critical point

(a) both $\left(\frac{\partial P}{\partial V}\right)_T$ and $\left(\frac{\partial^2 P}{\partial V^2}\right)_T$ are zero

(b) only $\left(\frac{\partial^2 P}{\partial V^2}\right)_T$ is not zero

(c) $\left(\frac{\partial P}{\partial V}\right)_T$ is zero but $\left(\frac{\partial^2 P}{\partial V^2}\right)_T$ is non-zero

(d) $\left(\frac{\partial P}{\partial V}\right)_T$ is non-zero but $\left(\frac{\partial^2 P}{\partial V^2}\right)_T$ is zero

166. However great the pressure, a gas cannot be liquified above its :

(a) Boyle temperature (b) Inversion temperature

(c) Critical temperature (d) Room temperature

167. The temperature at which real gases obey the ideal gas laws over a wide range of low pressure is called:

(a) Critical temperature (b) Inversion temperature

(c) Boyle temperature (d) Reduced temperature

168. The correct order of temperature of a real gas is :

(I) Boyle's temperature

(II) Critical temperature

(III) Inversion temperature

(a) III > I > II (b) I > II > III

(c) II > I > III (d) I > III > II

169. For an ideal gas, the critical temperature is:

(a) 0°C (b) 0 K

(c) 298 K (d) Undefined

170. The PV–P isotherms of 1 mole of different real gases at 546 K, if the $P \to 0$ then $PV \to$?

(a) 0 L atm (b) 11.2 L atm

(c) 22.4 L atm (d) 44.8 L atm

171. The van der Waals' equation of law of corresponding states for 1 mole of gas is :

(a) $\left(P_r + \frac{3}{V_r^2}\right)(3V_r - 1) = 8\,T_r$

(b) $\left(P_r - \frac{3}{V_r^2}\right)(3V_r - 1) = 8\,T_r$

(c) $\left(P_r + \frac{3}{V_r^2}\right)(3V_r + 1) = 8\,\pi T_r$

(d) $\left(P_r + \frac{3}{V_r^2}\right)(3V_r + 1) = 8$

172. Calculate the volume occupied by 16 gram O_2 at 300 K and 8.31 MPa if $\frac{P_c V_c}{RT_c} = 3/8$ and $\frac{P_r V_r}{T_r} = 2.21$ (Given : $R = 8.314$ J/K- mol)

(a) 125.31 mL (b) 124.31 mL

(c) 248.62 mL (d) none of these

173. 6 litre H_2O is placed in a closed room of volume 827 L at the temperature of 300 K. If vapour pressure of liquid water is 22.8 mm of Hg at 300 K and its density is 1 g/cm^3:

[Given : $R = 0.0821$ atm. L mol^{-1} K^{-1}, Assuming volume of liquid water to be constant]

Column-I		Column-II	
(P)	Mass of H_2O in gaseous form (in g)	**(1)**	6
(Q)	Moles of H_2O in gaseous state	**(2)**	18
(R)	Approximate mass of water left in liquid state (in kg)	**(3)**	3
(S)	Total number of moles of all atoms in vapour form	**(4)**	1

Codes :

	P	Q	R	S
(a)	1	2	4	3
(b)	4	3	2	1
(c)	2	4	1	3
(d)	1	2	3	4

174. Match the items of columns I and II.

Column-I		Column-II	
(P)	Z for ideal gas behaviour	**(1)**	3/8
(Q)	Z for real gas at low pressure	**(2)**	$\left(1+\frac{Pb}{RT}\right)$
(R)	Z for real gas at high pressure	**(3)**	1
(S)	Z for critical state	**(4)**	$\left(1-\frac{a}{RTV}\right)$

Codes :

	P	Q	R	S
(a)	1	2	4	3
(b)	3	4	2	1
(c)	2	1	4	3
(d)	1	2	3	4

175.

List-I (Conditions for real gas)		List-II	
(P)	If force of attraction among gas particles are negligible	**(1)**	$PV_m = RT$
(Q)	At 1 atm and 273 K	**(2)**	$PV_m = RT - \frac{a}{V_m}$
(R)	If the volume of gas particles is negligible	**(3)**	$\left(P+\frac{a}{V_m^2}\right)(V_m - b) = RT$
(S)	At low pressure and high temperature	**(4)**	$PV_m = RT + Pb$

Codes :

	P	Q	R	S
(a)	4	1	3	2
(b)	4	3	2	1
(c)	2	1	4	3
(d)	1	2	3	4

Level-2

1. A graph is plotted between P (atm) vs $t°C$ for 10 mol of an ideal gas as follows: Then slope of curve and volume of container (L) respectively, are:

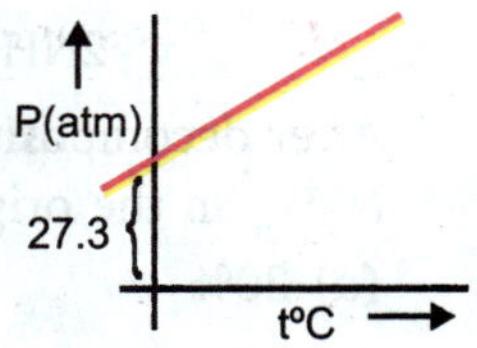

(a) 0.1, 8.21
(b) 8.21, 0.1
(c) 27.3, 8.21
(d) 8.21, 27.3

2. For two samples of ideal gases A and B curves are plotted n vs V (volume of container) at 16.42 atm pressure. Then temperature of A and B respectively are:

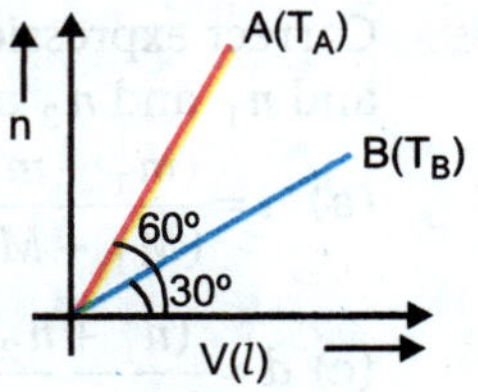

(a) $\frac{200}{\sqrt{3}}$ K, $200\sqrt{3}$ K
(b) $\frac{200}{\sqrt{3}}$ °C, $(200\sqrt{3})$ °C
(c) $200\sqrt{3}$ K, $\frac{200}{\sqrt{3}}$ K
(d) 200 K, $\frac{\sqrt{3}}{200}$ K

3. Consider the composite system, which is held at 300 K, shown in the following figure. Assuming ideal gas behavior, calculate the total pressure if the barriers separating the compartments are removed. Assume that the volume of the barriers is negligible. (Given : $R = 0.082$ atm L/mol.K)

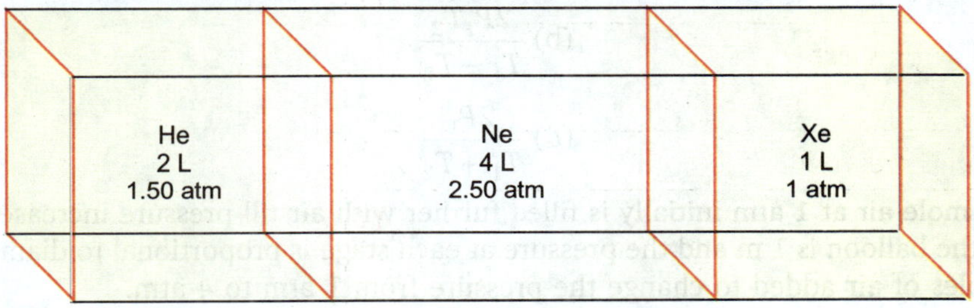

(a) 1 atm
(b) 2 atm
(c) 2.3 atm
(d) 3.2 atm

4. 11 moles of N_2 and 12 moles of H_2 mixture reacted in 20 litre vessel at 800 K. After equilibrium was reached, 6 mole of H_2 was present. 3.58 litre of liquid water is injected in equilibrium mixture and resultant gaseous mixture suddenly cooled to 300K. What is the final pressure of gaseous mixture? Neglect vapour pressure of liquid solution. Assume (i) all NH_3 dissolved in water (ii) no change in volume of liquid (iii) At 300 K no reaction takes place between N_2 and H_2

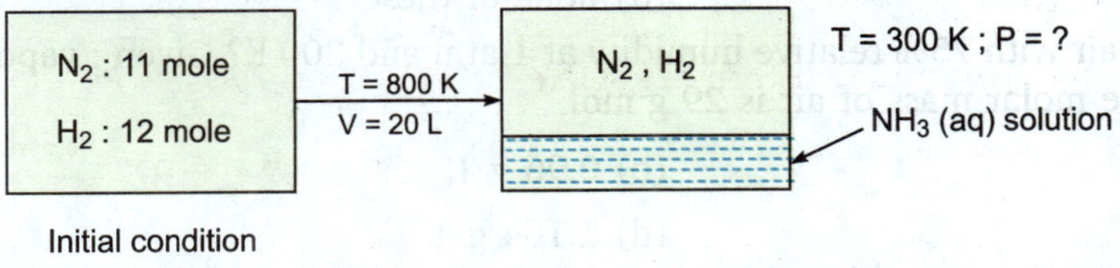

(a) 18.47 atm
(b) 60 atm
(c) 22.5 atm
(d) 45 atm

5. Two vessels connected by a valve of negligible volume. One container (I) has 2.8 g of N_2 at temperature T_1(K). The other container (II) is completely evacuated. The container (I) is heated to T_2 (K) while container (II) is maintained at $T_2/3$ (K). volume of vessel (I) is half that of vessel (II). If the valve is opened then what is the mass ratio of N_2 in both vessel (W_I/W_{II})?

(a) 1 : 2
(b) 1 : 3
(c) 1 : 6
(d) 3 : 1

6. A mixture of $NH_3(g)$ and $N_2H_4(g)$ is placed in a sealed container at 300 K. The total pressure is 0.5 atm. The container is heated to 1200 K at which time both substances decompose completely according to the equations

$$2NH_3(g) \longrightarrow N_2(g) + 3H_2(g) \text{ and } N_2H_4(g) \longrightarrow N_2(g) + 2H_2(g)$$

After decomposition is complete, the total pressure at 1200 K is found to be 4.5 atm. Find the mole% of N_2H_4 in the original mixture :

(a) 20% (b) 25%

(c) 50% (d) 75%

7. Correct expression for density of an ideal gas mixture of two gases 1 and 2, where m_1 and m_2 are masses and n_1 and n_2 are moles and M_1 and M_2 are molar masses.

(a) $d = \frac{(m_1 + m_2)}{(M_1 + M_2)}$ (b) $d = \frac{(m_1 + m_2)}{(n_1 + n_2)} \frac{P}{RT}$

(c) $d = \frac{(n_1 + n_2)}{(m_1 + m_2)} \times \frac{P}{RT}$ (d) None of these

8. Two closed vessel A and B of equal volume containing air at pressure P_1 and temperature T_1 are connected to each other through a narrow open tube. If the temperature of one is now maintained at T_1 and other at T_2 (where $T_1 > T_2$) then that what will be the final pressure?

(a) $\frac{T_1}{2P_1T_2}$ (b) $\frac{2P_1T_2}{T_1 + T_2}$

(c) $\frac{2P_1T_1}{T_1 - T_2}$ (d) $\frac{2P_1}{T_1 + T_2}$

9. A balloon containing 1 mole air at 1 atm initially is filled further with air till pressure increases to 4 atm. The initial diameter of the balloon is 1 m and the pressure at each stage is proportional to diameter of the balloon. How many moles of air added to change the pressure from 1 atm to 4 atm.

(a) 80 (b) 257

(c) 255 (d) 256

10. If Pd vs. P (where P denotes pressure in atm and d denotes density in gm/L) is plotted for He gas (assume ideal) at a particular temperature. If $\left[\frac{d}{dP}(Pd)\right]_{P = 8.21 \text{ atm}} = 5$, then the temperature will be

(a) 160 K (b) 320 K

(c) 80 K (d) none of these

11. What is the density of wet air with 75% relative humidity at 1 atm and 300 K? Given : vapour pressure of H_2O is 30 torr and average molar mass of air is 29 g mol^{-1}.

(a) 1.614 g/L (b) 0.96 g/L

(c) 1.06 g/L (d) 1.164 g/L

12. 7 moles of a tetra-atomic non-linear gas 'A' at 10 atm and T K are mixed with 6 moles of another gas B at $\frac{T}{3}$ K and 5 atm in a closed, rigid vessel without energy transfer with surroundings. If final temperature of mixture was $\frac{5T}{6}$ K, then gas B is? (Assuming all modes of energy are active)

(a) monoatomic (b) diatomic

(c) triatomic (d) tetra atomic

13. Three closed rigid vessels, A, B and C, which initially contain three different gases at different temperatures are connected by tube of negligible volume, without any energy transfer with surroundings. The vessel A contain 2 mole Ne gas, at 300 K, vessel 'B' contain 2 mole SO_2 gas at 400 K and vessel 'C' contain 3 mole CO_2 gas at temperature 500 K. What is the final pressure (in atm) attained by gases when all valves of connecting three vessels are opened and additional 15.6 kcal heat supplied to vessels through valve. The volume of A, B and C vessel is 2, 2 and 3 litre respectively

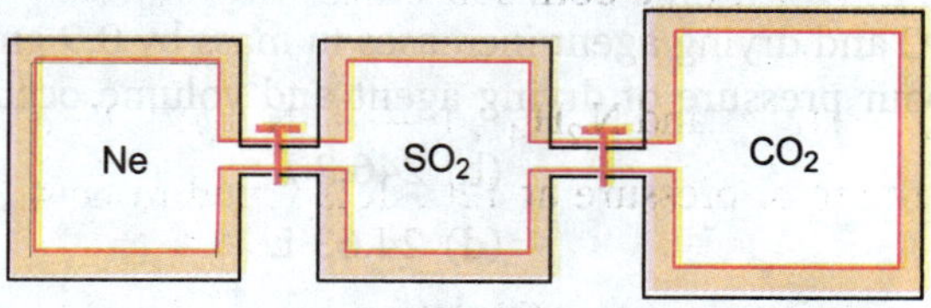

Given : $R = 2$ calorie/mol-K; C_v (Ne) $= 3/2\,R$, C_v (CO) $= 5/2\ R$ and C_v $(SO_2) = 3\,R$

(a) 73.89 atm (b) 67.31 atm

(c) 80 atm (d) none of these

14. Gas molecules each of mass 10^{-26} kg are taken in a container of volume $1\,dm^3$. The root mean square speed of gas molecules is $1\,km\,sec^{-1}$. What is the temperature of gas molecules.

(Given : $N_A = 6 \times 10^{23}$; $R = 8$ J/mol.K)

(a) 298 K (b) 25 K

(c) 250 K (d) 2500 K

15. A balloon of diameter 21 meter weighs 100 kg. Calculate its pay-load, if it is filled with He at 1.0 atm and 27°C. Density of air is $1.2\,kg\,m^{-3}$. (Given : $R = 0.082\,L\,atm\,K^{-1}mol^{-1}$)

(a) 4952.42 kg (b) 4932.42 kg

(c) 493.242 kg (d) none of these

16. A given volume of ozonised oxygen (containing 60% oxygen by volume) required 220 sec to effuse while an equal volume of oxygen took 200 sec only under identical conditions. If density of O_2 is 1.6 g/L then find density of O_3.

(a) 1.936 g/L (b) 2.16 g/L

(c) 3.28 g/L (d) 2.24 g/L

17. If 250 mL of N_2 over water at 30°C and a total pressure of 740 torr is mixed with 300 mL of Ne over water at 25°C and a total pressure of 780 torr, what will be the total pressure if the mixture is in a 500 mL vessel over water at 35°C.

(Given : Vapour pressure (Aqueous tension) of H_2O at 25°C, 30°C and 35°C are 23.8, 31.8 and 42.2 torr respectively. Assume volume of $H_2O(l)$ is negligible in final vessel)

(a) 760 torr (b) 828.4 torr

(c) 807.6 torr (d) 870.6 torr

18. A bulb of constant volume is attached to a manometer tube open at other end as shown in figure. The manometer is filled with a liquid of density $(1/3)^{rd}$ that of mercury. Initially h was 228 cm.

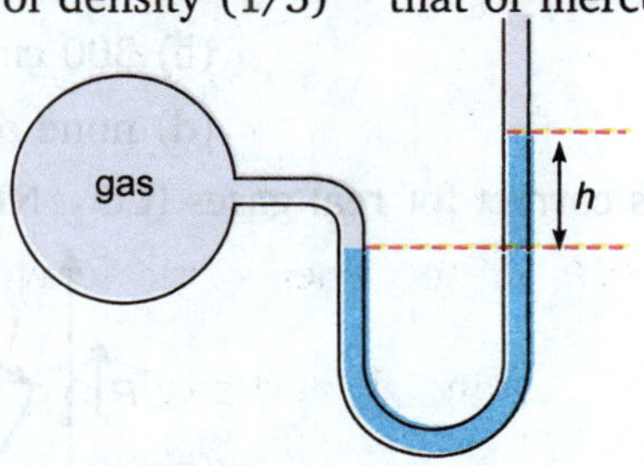

Through a small hole in the bulb gas leaked assuming pressure decreases as $\frac{dP}{dt} = -kP$.

If value of h is 114 cm after 14 minutes, what is the value of k (in $hour^{-1}$)?

[Use : $\ln(4/3) = 0.28$ and density of Hg $= 13.6$ g/mL.].

(a) 0.6 (b) 1.2

(c) 2.4 (d) none of these

19. A mixture of nitrogen and water vapours is admitted to a flask at 760 torr which contains a sufficient solid drying agent. After long time the pressure attained a steady value of 722 torr.

If the experiment is done at 27°C and drying agent increases in mass by 0.9 gm, what is the volume of the flask? Neglect any possible vapour pressure of drying agent and volume occupied by drying agent.

(a) 443.34 L (b) 246.3 L

(c) 12.315 L (d) 24.63 L

20. At room temperature following reaction goes to completion

$$2AB(g) + B_2(g) \longrightarrow 2AB_2(s)$$

AB_2 is solid with negligible vapour pressure below 0°C. At 300 K, the AB in the smaller flask exerts a pressure of 3 atm and in the larger flask B_2 exerts a pressure of 1 atm at 400 K when they are separated out by a close valve. The gases are mixed by opening the stop cock and after the end of the reaction the flask are cooled to 250 K

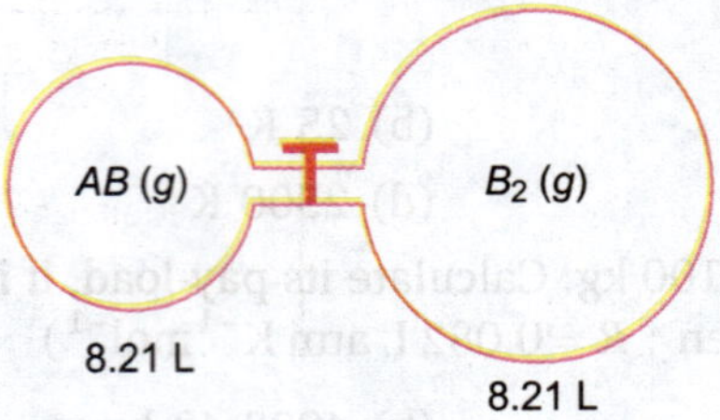

The final pressure is :

(a) 0.156 atm (b) 0.3125 atm

(c) 0.625 atm (d) 3.2 atm

21. A vessel of uniform cross-section of length 500 cm as shown in figure is divided in two parts by a weightless and frictionless piston one part contains 5 moles of He(g) and other part 2 moles of $H_2(g)$ and 4 moles of $O_2(g)$ added at the same temperature and pressure in which reaction takes place finally vessel cooled to 300 K and 1 atm. What is the length of He compartment?

(Assume volume of piston and vol. of $H_2O(l)$ formed are negligible)

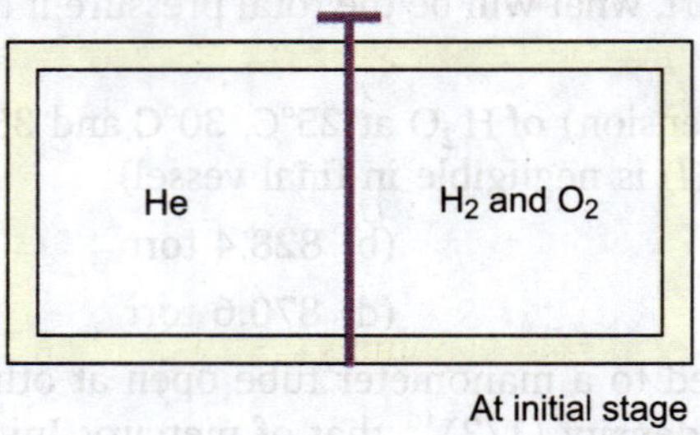

At initial stage

(a) 187.5 cm (b) 300 cm

(c) 312.5 cm (d) none of these

22. Which of the following graph is correct for real gases (CO_2, NH_3 etc.) at 0°C ?

(c) 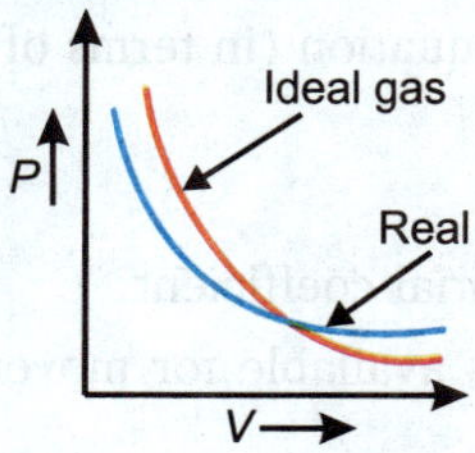

(d) 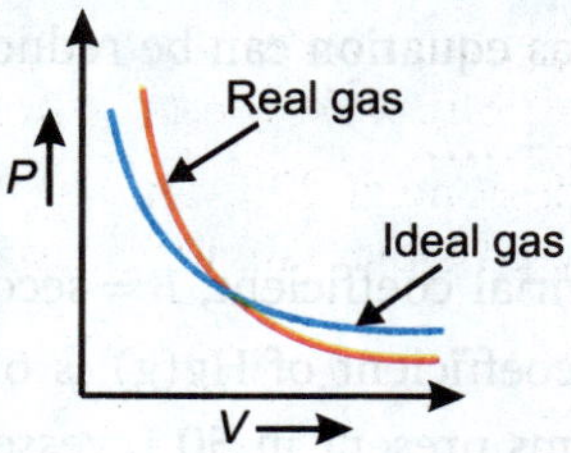

23. Which of the following graph is correct for He(*g*) at 0°C ?

(a) 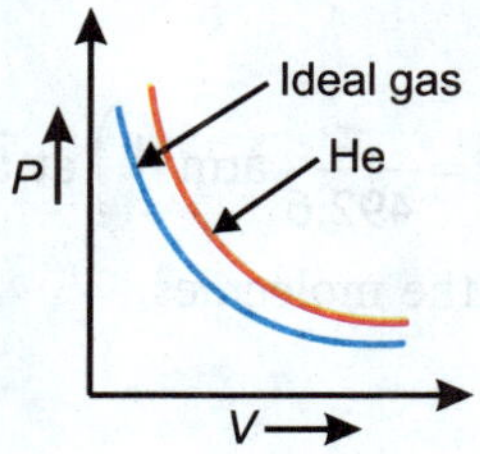

(b) 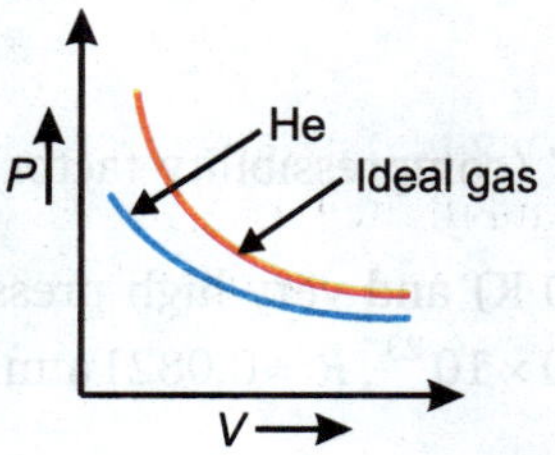

(c) 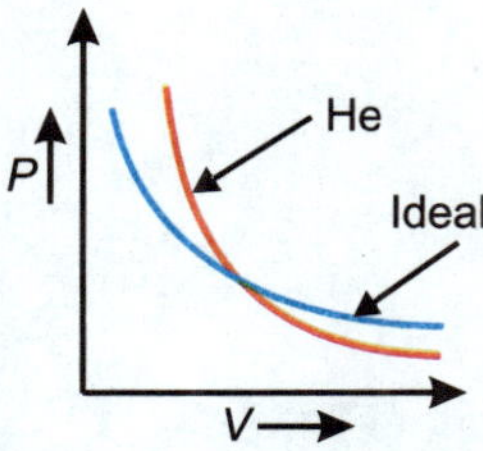

(d) 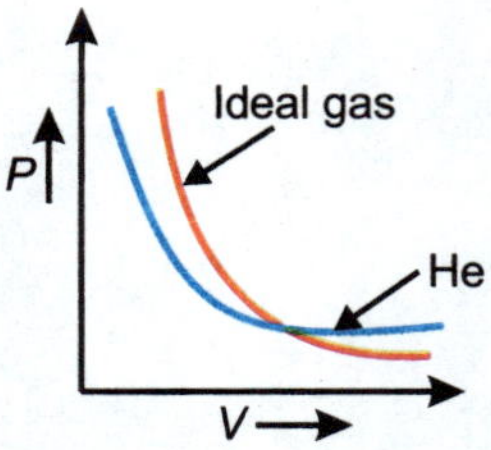

24. The equation of state for a real gas is $P(V - nb) = nRT$, where b and R are constants. If the pressure and temperature are such that molar volume occupied by real gas $= 11b$, what is the value of compressibility factor of gas in this condition?

(a) 1.20 (b) 1.69
(c) 1.10 (d) 1.0

25. For a real gas behave ideally, the molar volume (V_m) may be:

(a) $\frac{b}{aP}$ (b) $\frac{a}{Pb}$
(c) $\frac{8a}{27\,Pb}$ (d) $\frac{27\,a}{8\,Pb}$

26. For a real gas (mol. mass = 60) if density at critical point is 0.80 g/cm^3 and its $T_c = \frac{4 \times 10^5}{821}$ K, then van der Waals' constant a (in atm L^2 mol^{-2}) is

(a) 0.3375 (b) 3.375
(c) 1.68 (d) 0.025

27. The van der Waals' constant 'b' of a gas is $4\pi \times 10^{-4}$ L/mol. How near can the centres of the two molecules approach each other? [Use : $N_A = 6 \times 10^{23}$]

(a) 10^{-7} m (b) 10^{-10} m
(c) 5×10^{-11} m (d) 5×10^{-9} m

28. The density of vapour of a substance (X) at 1 atm pressure and 500 K is 0.8 kg/m^3. The vapour effuses through a small hole at a rate of 4/5 times slower than oxygen under the same condition. What is the compressibility factor (Z) of the vapour?

(a) 0.974 (b) 1.35
(c) 1.52 (d) 1.22

29. van der Waal's gas equation can be reduced to virial equation and virial equation (in terms of volume) is $Z = A + \frac{B}{V_m} + \frac{C}{V_m^2} + \ldots\ldots$

where A = first virial coefficient, B = second virial coefficient, C = third virial coefficient.

The third virial coefficient of Hg(g) is 625 $(\text{cm}^2/\text{mol})^2$. What volume is available for movement of 10 moles He(g) atoms present in 50 L vessel?

(a) 49.75 L
(b) 49.25 L
(c) 25 L
(d) 50 L

30. If the slope of 'Z' (compressibility factor) vs. 'P' curve is constant $\left(\text{slope} = \frac{\pi}{492.6}\ \text{atm}^{-1}\right)$ at a particular temperature (300 K) and very high pressure, then calculate diameter of the molecules.
(Given : $N_A = 6.0 \times 10^{23}$, $R = 0.0821\ \text{atm. L mol}^{-1}\ \text{K}^{-1}$)

(a) 7.5 Å
(b) 5 Å
(c) 2.5 Å
(d) 1.25 Å

Level-3

Passage-1

van der Waals' equation for calculating the pressure of a non ideal gas is

$$\left(P + \frac{an^2}{V^2}\right)(V - nb) = n\,RT$$

van der Waals' suggested that the pressure exerted by an ideal gas, P_{ideal}, is related to the experimentally measured pressure, P_{real} by the equation

$$\underset{\substack{\uparrow \\ \text{observed} \\ \text{pressure}}}{P_{\text{ideal}} = P_{\text{real}}} \quad + \underset{\substack{\uparrow \\ \text{correction} \\ \text{term}}}{\frac{an^2}{V^2}}$$

Constant **'a'** is measure of intermolecular interaction between gaseous molecules that gives rise to nonideal behavior. It depends upon how frequently any two molecules approach each other closely. Another correction concerns the volume occupied by the gas molecules. In the ideal gas equation, V represents the volume of the container. However, each molecule does occupy a finite, although small, intrinsic volume, so the effective volume of the gas becomes $(V - nb)$, where n is the number of moles of the gas and b is a constant. The term nb represents the volume occupied by gas particles present in n moles of the gas.

Having taken into account the corrections for pressure and volume, we can rewrite the ideal gas equation as follows :

$$\underset{\text{corrected pressure}}{\left(P + \frac{an^2}{V^2}\right)} \quad \underset{\text{corrected volume}}{(V - nb)} = nRT$$

1. At relatively high pressures, the van der Waals' equation of state reduces to

(a) $PV = RT - a/V_m$ (b) $PV = aRT/V_m^2$

(c) $P = RT - a/V_m^2$ (d) $PV_m = RT + Pb$

2. For non-zero value of force of attraction between gas molecules at large volume, gas equation will be :

(a) $PV = nRT - \dfrac{n^2a}{V}$ (b) $PV = nRT + nbP$

(c) $P = \dfrac{nRT}{V - b}$ (d) $PV = nRT$

3. The van der Waals' constant 'a' for CO_2 gas is greater than that of H_2 gas. Its mean that the

(a) strength of van der Waals' force of CO_2 gas is less than that of H_2 gas

(b) strength of van der Waals' force of CO_2 gas is equal to that of H_2 gas

(c) CO_2 gas can be more easily liquified

(d) H_2 gas can be more easily liquified

4. Using van der Waals' equation, find the constant 'a' (in atm L^2 mol^{-2}) when two moles of a gas confined in 4 L flask exerts a pressure of 11.0 atmospheres at a temperature of 300 K. The value of b is 0.05 L mol^{-1}. ($R = 0.082$ atm.L/K mol)

(a) 2.62 (b) 2.64

(c) 6.24 (d) 6.46

Passage-2

Collision cross-section is an area of an imaginary sphere of radius σ around the molecule within which the centre of another molecule cannot penetrate.

The volume swept by a single molecule in unit time is

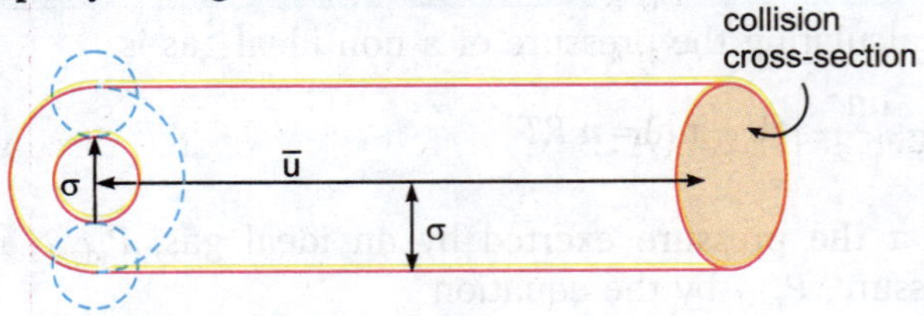

$V = (\pi\sigma^2)\bar{u}$ where $\bar{u}$ is the average speed

If N^* is the number of molecules per unit volume, then the number of molecules within the volume V is

$$N = VN^* = (\pi\sigma^2\bar{u})N^*$$

Hence, the number of collision made by a single molecule in unit time will be

$$Z = N = (\pi\sigma^2\bar{u})N^*$$

In order to account for the movements of all molecules, we must consider the average velocity along the line of centres of two colliding molecules instead of the average velocity of a single molecule. If it is assumed that, on an average, molecules collide while approaching each other perpendicularly, then the average velocity along their centres is $\sqrt{2}\,\bar{u}$ as shown below.

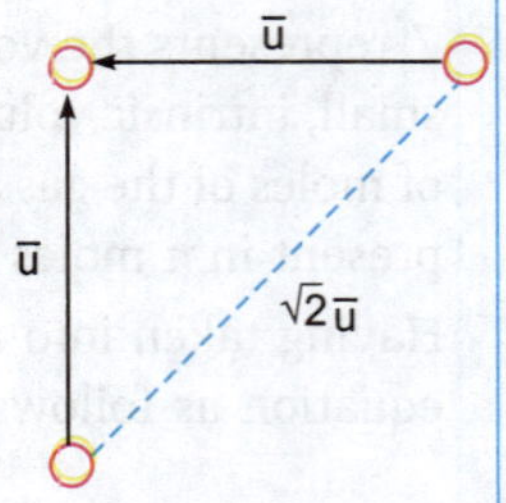

Number of collision made by a single molecule with other molecules per unit time is given by

$$Z_1 = \pi\sigma^2(\bar{u}_{rel})N^* = \sqrt{2}\,\pi\sigma^2\,\bar{u}N^*$$

The total number of bimolecular collisions Z_{11} per unit volume per unit time is given by

$$Z_{11} = \frac{1}{2}(Z_1N^*) \text{ or } Z_{11} = \frac{1}{2}(\sqrt{2}\pi\sigma^2\bar{u}N^*)\,N^* = \frac{1}{\sqrt{2}}\,\pi\sigma^2\bar{u}N^{*2}$$

If the collision involve two unlike molecules then the number of collisions Z_{12} per unit volume per unit time is given as

$$Z_{12} = \pi\sigma_{12}^2\left(\sqrt{\frac{8kT}{\pi\mu}}\right)N_1N_2$$

where N_1 and N_2 are the number of molecules per unit volume of the two types of molecules, σ_{12} is the average diameter of the two molecules and μ is the reduced mass.

The mean free path is the average distance travelled by a molecule between two successive collisions. We can express it as follows :

$$\lambda = \frac{\text{Average distance travelled per unit time}}{\text{No. of collisions made by a single molecule per unit time}} = \frac{\bar{u}}{Z_1}$$

or $$\lambda = \frac{\bar{u}}{\sqrt{2}\,\pi\sigma^2\bar{u}N^*} \Rightarrow \frac{1}{\sqrt{2}\,\pi\sigma^2N^*}$$

1. For a given gas the mean free path at a particular pressure is :
(a) independent to temperature
(b) decreases with rise in temperature
(c) increases with rise in temperature
(d) directly proportional to T^2

2-3 Three ideal gas samples in separate equal volume containers are taken and following data is given :

	Pressure	Temperature	Mean free paths	Mol. wt.
Gas *A*	1 atm	1600 K	0.16 nm	20
Gas *B*	2 atm	200 K	0.16 nm	40
Gas *C*	4 atm	400 K	0.04 nm	80

2. Calculate ratio of collision frequencies (Z_{11}) $(A:B:C)$ of following for the three gases.

(a) 1 : 2 : 4 (b) 4 : 2 : 1

(c) 1 : 4 : 16 (d) 16 : 4 :1

3. Calculate number of collision by one molecule per sec (Z_1).

(a) 4 : 1 : 4 (b) 1 : 4 : 4

(c) 4 : 3 : 2 (d) 1 : 2 : 4

Passage-3

A manometer contains a liquid of density 5.44 g/cm^3 is attached to a flask containing gas 'A' as follows

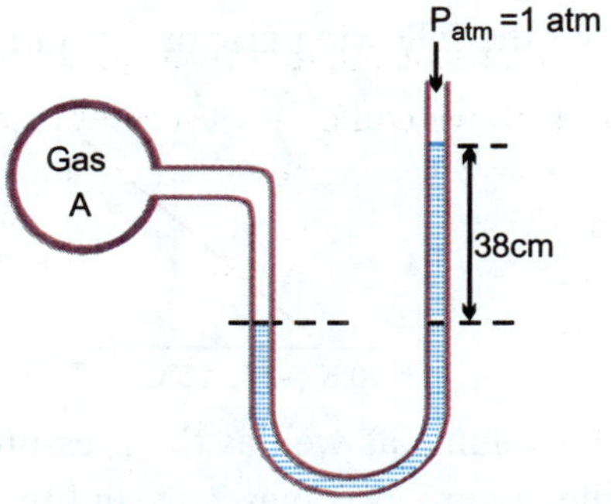

1. If the same liquid is used in barometer to measure the atmospheric pressure, then what will be the length of the liquid column, which exerts pressure equal to 1 atm ? (density of Hg = 13.6 g/cm^3)

(a) 190 cm (b) 76 cm

(c) 30.4 cm (d) 266 cm

2. The initial pressure of gas *A* in the flask is:

(a) 1.5 atm (b) 1 atm

(c) 1.3 atm (d) 1.2 atm

3. If gas *A* undergoes 30% trimerisation $[3A(g) \rightleftharpoons A_3(g)]$ then the difference in height of the liquid level in two columns is:

(a) 38 cm (b) 7.6 cm

(c) 3.04 cm (d) 15.1 cm

One or More Answer is/are Correct

1. Which of the following curves represent(s) Boyle's law?

(a) 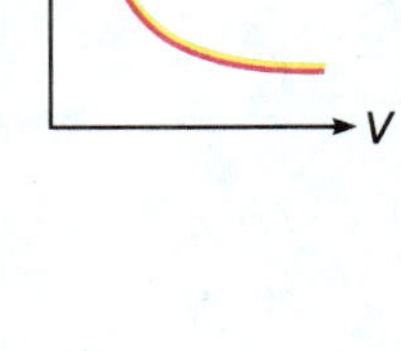

(b) 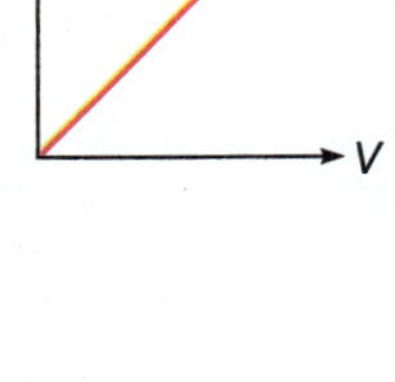

(c) 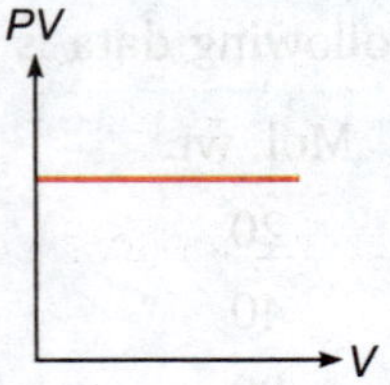

(d) 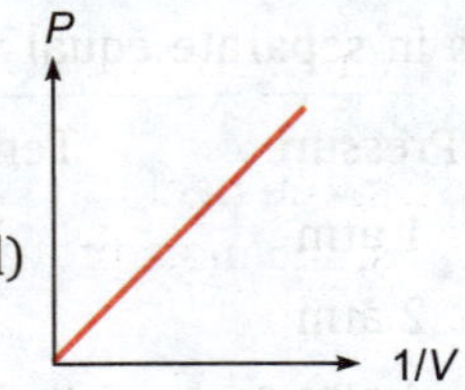

2. If a gas expands at constant temperature:
(a) the pressure decreases
(b) the kinetic energy of the molecules remains the same
(c) the K.E. of the molecules decrease
(d) the number of molecules of the gas increase

3. Which of the following statements are correct?
(a) It is not possible to compress a gas at a temperature below T_C
(b) At a temperature below T_C, the molecules are close enough for the attractive forces to act, and condensation occurs
(c) No condensation takes place above T_C
(d) Due to higher kinetic energy of the gas molecules above T_C, it is considered as super critical fluid

4. What conclusion would you draw from the following graphs for an ideal gas?

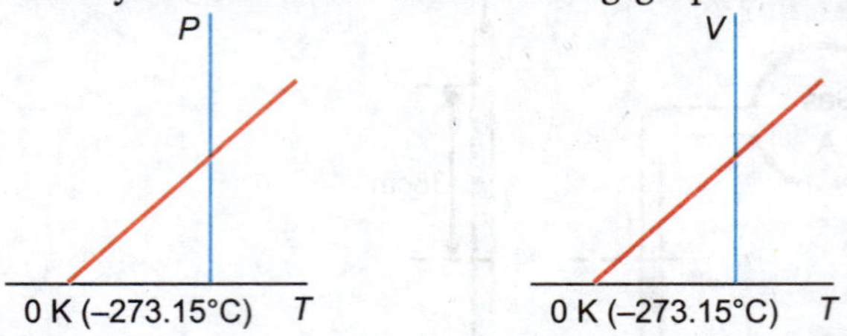

(a) As the temperature is reduced, the volume as well as the pressure increase
(b) As the temperature is reduced, the volume becomes zero and the pressure reaches infinity
(c) As the temperature is reduced, the pressure decrease
(d) A point is reached where, theoretically, the volume become zero

5. At Boyle temperature:
(a) the effects of the repulsive and attractive intermolecular forces just offset each other
(b) the repulsive intermolecular forces are stronger than the attractive intermolecular forces
(c) the repulsive intermolecular forces are weaker than the attractive intermolecular forces
(d) $b - \frac{a}{RT} > 0$

6. Indicate the correct statement for equal volumes of $N_2(g)$ and $CO_2(g)$ at 25°C and 1 atm.
(a) The average translational K.E. per molecule is the same for N_2 and CO_2
(b) The rms speed is same for both N_2 and CO_2
(c) The density of N_2 is less than that of CO_2
(d) The total translational K.E. of both N_2 and CO_2 is the same

7. Which of the following is correct for critical temperature?
(a) It is the highest temperature at which liquid and vapour can coexist
(b) Beyond this temperature, a gas cannot be liquefied by compression.
(c) At this temperature, the gas and the liquid phases have different critical densities
(d) All are correct

8. Consider the following statement regarding Maxwell's distribution of speeds. The correct statement(s) is/are :

(a) as temperature increases, the peak (maxima) of a curve is shifted towards right side

(b) as temperature increases, the most probable speed of molecules increases but fraction of molecules of very high speed increases

(c) the area under the curve at all the temperatures is the same

(d) the fractions of molecules having different speeds are different at a given temperature

9. If a gas expands at a constant pressure by providing heat :

(a) the temperature increases

(b) the kinetic energy of the gaseous molecules remains same

(c) the kinetic energy of gaseous molecules decreases

(d) the number of molecules of the gas decreases

10. Select the incorrect statement(s) :

(a) The product of pressure and volume of fixed amount of a gas is independent of temperature

(b) The value of universal gas constant depends upon temperature, volume and number of gaseous molecules

(c) The gas constant also known as Boltzmann's constant

(d) The average kinetic energy of molecules depends only on temperature

11. Following represents the Maxwell distribution curve for an ideal gas at two temperature T_1 and T_2. Which of the following option(s) is/are true?

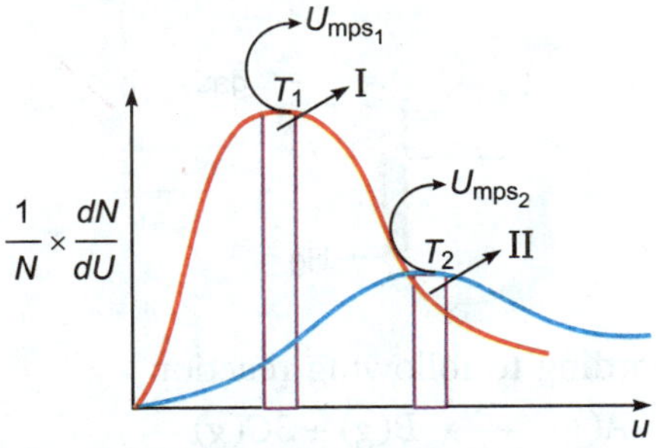

(a) Total area under the two curves is independent to moles of gas

(b) u_{mp} decreases as temperature decreases

(c) $T_1 > T_2$ and hence higher the temperature, sharper the curve

(d) The fraction of molecules having speed u_{mp} decreases as temperature increases

12. Two container each containing liquid water are connected as shown in diagram.

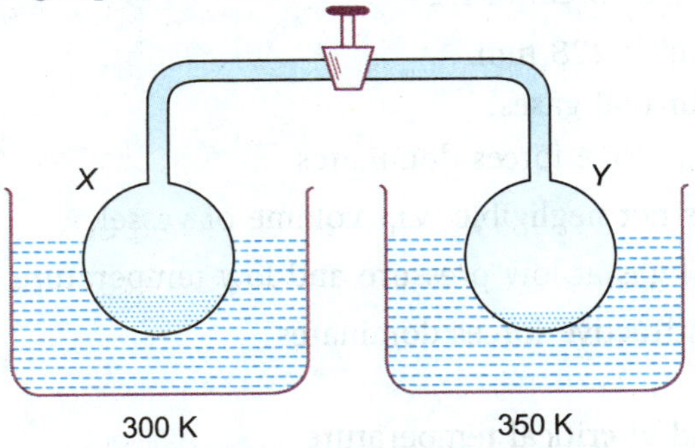

Given that vapour pressure of $H_2O(l)$ at 300 K and 350 K are 22 torr and 40 torr, select correct statement(s):

(a) The final pressure in each container if valve is opened while keeping the containers at the given temperature is 22 torr

(b) The final pressure in each container if valve is opened while keeping the containers at the given temperature is 40 torr

(c) Mass of $H_2O(l)$ is decreased in container X

(d) Mass of $H_2O(l)$ is decreased in container Y

13. Select the correct statement(s) :

(a) At Boyle's temperature a real gas behaves like an ideal gas at low pressure

(b) Above critical conditions, a real gas is called super critical fluid

(c) For hydrogen gas repulsive forces domainates over attractive forces at all temperatures

(d) At high pressure van der Waals' constant repulsive forces domainates over attractive forces

14. Select the correct statement :

(a) The value of compressibility factor 'Z' for H_2 gas is greater than one at room temperature and pressure

(b) The real gas behaves as an ideal gas at Boyle's temperature in low pressure reason

(c) For a real gas following van der Waals' equation of state, the expression of critical temperature is $\frac{8a}{27R.b}$

(d) At low pressure, the compressibility factor 'Z' $= 1 + \frac{P.b}{RT}$ for a van der Waals' gas.

15. A open ended mercury manometer is used to measure the pressure exerted by a trapped gas as shown in the figure. Initially manometer shows no difference in mercury level in both columns as shown in diagram.

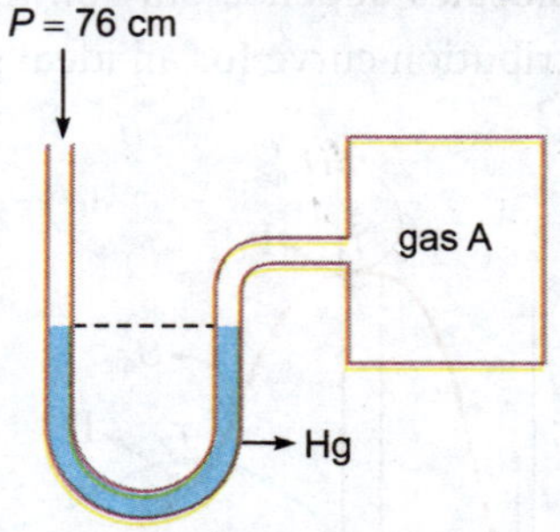

After sparking 'A' dissociates according to following reaction

$$A(g) \longrightarrow B(g) + 3C(g)$$

If pressure of Gas "A" decreases to 0.9 atm, then (Assume temperature to be constant and is 300 K)

(a) total pressure increases to 1.3 atm

(b) total pressure increases by 0.3 atm

(c) total pressure increases by 22.3 cm of Hg

(d) difference in mercury level is 228 mm.

16. Select incorrect statements for real gases:

(a) In low pressure region repulsive forces dominates

(b) Volume of gas particles is not negligible w.r.t volume of vessel

(c) Gases behaves as an ideal gas at low pressure and low temperature

(d) In high pressure region attractive forces dominates

17. Select correct statements:

(a) A real gas can be liquified at critical temperature

(b) Critical pressure is the maximum pressure at which a substance is present in its liquid state at T_C

(c) Ideal gas can be liquified below T_C

(d) Critical volume is the molar volume of substance in gaseous state at T_C and P_C

18. Which is/are correct for real gases?

(a) $\lim_{P \to 0} (PV_m) =$ constant at constant high temperature

(b) $\lim_{V_m \to 0} (PV_m) =$ constant at constant low temperature

(c) $\lim_{P\to 0}\left(\frac{PV_m}{RT}\right)=1$ at high temperature

(d) $\lim_{V\to 0}\left(\frac{PV_m}{RT}\right)=R$

19. Select incorrect statement(s)
(a) At very low pressure real gases show minimum deviation from ideal behaviour.
(b) The compressibility factor for an ideal gas is zero.
(c) At Boyle temperature real gas behave as ideal gas in high pressure region.
(d) Real gas show maximum deviation at high pressure and low temperature.

20. If an ideal gas is heated at constant pressure:
(a) the volume increases
(b) the mass of gas remains same
(c) the kinetic energy of the molecules increases
(d) attraction forces between gas particles increases

Match the Column

Column-I and **Column-II** contains four entries each. Entries of Column-I are to be matched with some entries of Column-II. One or more than one entries of Column-I may have the matching with the same entries of Column-II.

1.

	Column I		Column II
(A)	Boyle's temperature	(P)	a/Rb
(B)	Inversion temperature	(Q)	$8a/27Rb$
(C)	Critical temperature	(R)	$2a/Rb$
(D)	Critical pressure	(S)	$a/27b^2$

2.

	Column I		Column II
(A)	Root mean square speed	(P)	$\frac{3}{2}RT$
(B)	Most probable speed	(Q)	$\sqrt{\frac{8P}{\pi d}}$
(C)	Average speed	(R)	$\sqrt{\frac{3P}{d}}$
(D)	K.E. of gas mol^{-1}	(S)	$\sqrt{\frac{2RT}{M}}$

3.

	Column I		Column II
(A)	If force of attraction among the gas molecules be negligible	(P)	$\left(P+\frac{a}{V^2}\right)(V-b)=RT$
(B)	If the volume of the gas molecules be negligible w.r.t. volume of vessel	(Q)	$PV=RT-\frac{a}{V}$
(C)	At STP	(R)	$PV=RT+Pb$
(D)	At low pressure and at high temperature	(S)	$PV=RT$

4. Match gases under specified conditions listed in Column I with their properties/laws in Column II.

Column I		Column II	
(A)	Hydrogen gas ($P = 200$ atm, $T = 273$K)	(P)	Compressibility factor $\neq 1$
(B)	Hydrogen gas ($P \sim 0, T = 273$ K)	(Q)	Attractive forces are dominant
(C)	CO_2 ($P = 1$ atm, $T = 273$ K)	(R)	$PV = nRT$
(D)	Real gas with very large molar volume	(S)	$P(V - nb) = nRT$

5.

Column I		Column II	
(A)	$\frac{1}{V^2}$ $vs \cdot P$ for ideal gas at constant T and n.	(P)	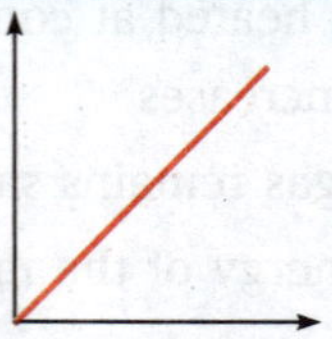
(B)	V $vs \cdot \frac{1}{T}$ for ideal gas at constant P and n	(Q)	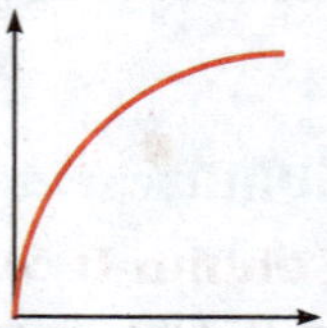
(C)	PT $vs \cdot T^2$ for ideal gas at constant V and n.	(R)	
(D)	V $vs \cdot \frac{1}{P^2}$ for ideal gas at constant T and n.	(S)	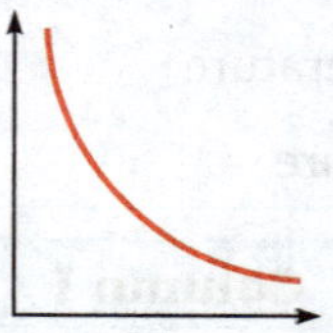

6. Match the correct column from list-1 to list-2 on the basis of following Andrew's isotherm of real gas.

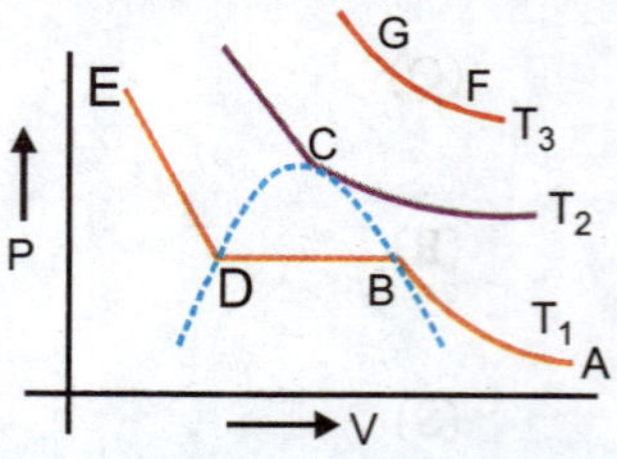

Column-I		Column-II	
(A)	Substance exist in both liquid and gas state	(P)	At AB part
(B)	Only liquid state exist	(Q)	At BD part
(C)	Substance exist in gas state only	(R)	At DE part
(D)	Real gas is called super critical fluid	(S)	At point C
		(T)	At GF curve

7.

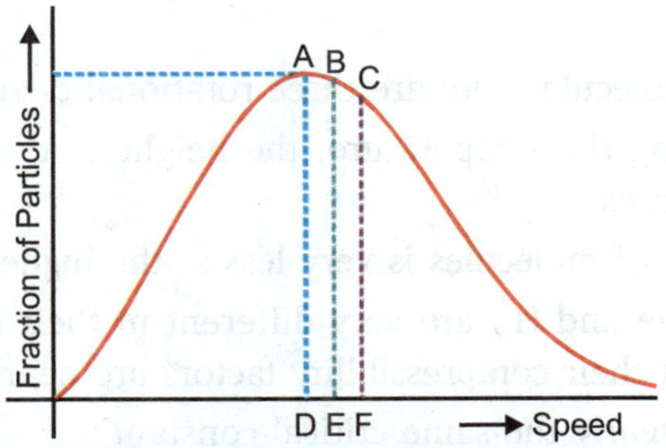

Column-I		Column-II
(A) Speed at point 'A' is	(P)	$\sqrt{\frac{8RT}{\pi M}}$
(B) Speed at point 'B' is	(Q)	$\sqrt{\frac{3RT}{M}}$
(C) Speed at point 'C' is	(R)	U_{rms}
(D) The speed possessed by maximum fraction of the gas particles	(S)	$\sqrt{\frac{2RT}{M}}$

Assertion-Reason Type Questions

Each question contains STATEMENT-1 (Assertion) and STATEMENT-2 (Reason).

Examine the statements carefully and mark the correct answer according to the instructions given below

(A) If both the statement are TRUE and STATEMENT-2 is the correct explanation of STATEMENT-1

(B) If both the statement are TRUE but STATEMENT-2 is NOT the correct explanation of STATEMENT-1

(C) If STATEMENT-1 is TRUE and STATEMENT-2 is FALSE

(D) If STATEMENT-1 is FALSE and STATEMENT-2 is TRUE

1. **STATEMENT-1 :** The heat absorbed during the isothermal expansion of an ideal gas against vacuum is zero.

STATEMENT-2 : The volume occupied by the molecules of an ideal gas is zero.

2. **STATEMENT-1 :** A lighter gas diffuses more rapidly than a heavier gas.

STATEMENT-2 : At a given temperature, the rate of diffusion of a gas is inversely proportional to the density.

3. **STATEMENT-1 :** The value of the van der Waals' constant 'a' is larger for ammonia than for nitrogen.

STATEMENT-2 : Hydrogen bonding is present in ammonia.

4. **STATEMENT-1 :** Helium shows only positive deviation from ideal behaviour at room temperature.

STATEMENT-2 : Helium is an inert gas.

5. **STATEMENT-1 :** CH_4, CO_2 has value of Z (compressibility factor) less than one at 0°C.

STATEMENT-2 : $Z < 1$ is due to the attractive forces dominate among the molecules.

6. **STATEMENT-1 :** Liquefaction of an ideal gas is not possible.

STATEMENT-2 : There are no intermolecular attractive forces in an ideal gas.

7. STATEMENT-1 : The average translational kinetic energy per molecule of the gas per degree of freedom is 1/2 KT.

STATEMENT-2 : For every molecule there are three rotational degree of freedom.

8. STATEMENT-1 : On increasing the temperature, the height of the peak of the Maxwell distribution curve decreases.

STATEMENT-2 : The fraction of molecules is very less at the higher speeds.

9. STATEMENT-1 : The gases He and H_2 are very different in their behaviour at any temperature and pressure but their compressibility factors are nearly the same at the critical point.

STATEMENT-2 : They have nearly the same critical constant.

10. STATEMENT-1 : Most probable speed is the speed possessed by maximum fraction of molecules at the same temperature.

STATEMENT-2 : On collision, more and more molecules acquire higher speed at the same temperature.

11. STATEMENT-1 : 1 mole of H_2 and O_2 each occupy 22.7 L of volume at 0°C and 1 bar pressure, considering ideal behaviour.

STATEMENT-2 : Molar volume for all ideal gases at the same temperature and pressure are equal.

12. STATEMENT-1 : Reacting gases react to form a new gas having pressure equal to the sum of their partial pressure.

STATEMENT-2 : Pressure exerted by a mixture of non-reacting gases present in a container is equal to the sum of their partial pressures.

13. STATEMENT-1 : $1/4^{th}$ of the initial mole of the air is expelled, if air present in an open vessel is heated from 27°C to 127°C.

STATEMENT-2 : Rate of diffusion of a gas is inversely proportional to the square root of its molecular mass.

14. STATEMENT-1 : Compressibility factor for hydrogen varies with pressure with positive slope at all pressures at 0°C.

STATEMENT-2 : Even at low pressures, repulsive forces dominate in hydrogen gas at 0°C.

15. STATEMENT-1 : Wet air is heavier than dry air.

STATEMENT-2 : The density of dry air is less than the density of water at 1 atm and 273K.

Level-4

Subjective Problems

1. Collapsible balloon is inflated to a volume of 10 L at a pressure of 1 atm. When the balloon is immersed to the bottom of a lake, its volume reduce to 1.25 L. Assuming atmospheric pressure to be equivalent to 10 m column of water and no change in temperature, if the depth of the lake is $x \times 10^3$ cm, what is the value of x?
2. A gaseous mixture containing equal moles of H_2, O_2 and He is subjected to series of effusion steps. The composition (by moles) of effused mixture after 4 effusion steps is $x:1:y$ respectively. Then find the value of $\left(\frac{x}{y}\right)$.
3. One mole of a gas changed from its initial state (15L, 2 atm) to final state (4L, 10 atm) reversibly. If this change can be represented by a straight line in $P-V$ curve maximum temperature (approximate), the gas attained is $x \times 10^2$K. Then find the value of x.
4. An empty LPG cylinder weighs 14.8 kg. When full, it weighs 29.0 kg and shows a pressure of 3.47 atm. In the course of use at ambient temperature, the mass of the cylinder is reduced to 23.0 kg. The final pressure inside the cylinder is__________atm. (Nearest integer)
   (Assume LPG to be an ideal gas)
5. 1 mole of a diatomic gas present in 10 L vessel at certain temperature exert a pressure of 0.96 atm. Under similar conditions an ideal gas exerted 1.0 atm pressure. If volume of gas molecule is negligible, then find the value of van der Waals' constant "a" (in atm L^2/mol^2).
6. The graph of compressibility factor (Z) *vs.* P for one mole of a real gas is shown in following diagram. The graph is plotted at constant temperature 273 K. If the slope of graph at very high pressure $\left(\frac{dZ}{dP}\right)$ is $\left(\frac{1}{2.8}\right)$ atm^{-1}, then calculate volume of one mole of real gas molecules
   (in L /mol)
   Given : $N_A = 6 \times 10^{23}$ and $R = \frac{22.4}{273}$ L atm $K^{-1} mol^{-1}$

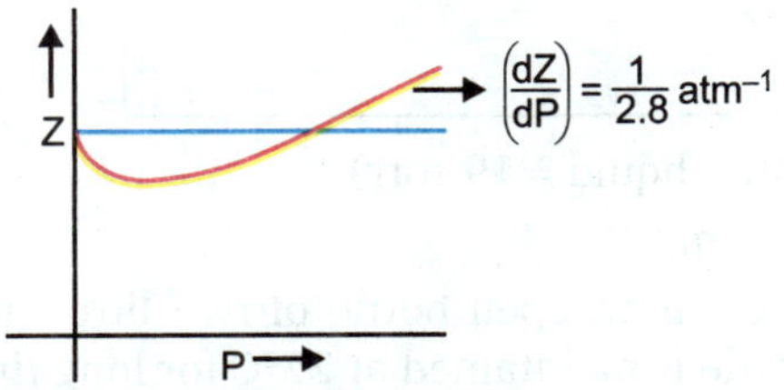

7. Under the identical conditions of temperature, the density of a gas X is two times to that of gas Y while molecular mass of gas Y is three times that of X. Calculate the ratio of pressure of X and Y.
8. The time taken for a certain volume of a certain gas to diffuse through a small hole was 2 min. Under similar conditions an equal volume of oxygen took 5.65 min to pass. What is the molecular mass of gas (in amu) ?
9. Excess $F_2(g)$ reacts at 150° C and 1.0 atm pressure with $Br_2(g)$ to give a compound BrF_n. If 423 mL of $Br_2(g)$ at the same temperature and pressure produced 4.2 g of BrF_n, what is n ? [Atomic mass Br = 80, F = 19]
10. A certain gas obeys $P(V_m - b) = RT$. The value of $\left(\frac{\partial Z}{\partial P}\right)_T$ is $\frac{xb}{RT}$. The value of x is ________.
    (Z: compressibility factor)

11. Air is trapped in a horizontal glass tube by 36 cm mercury column as shown below :

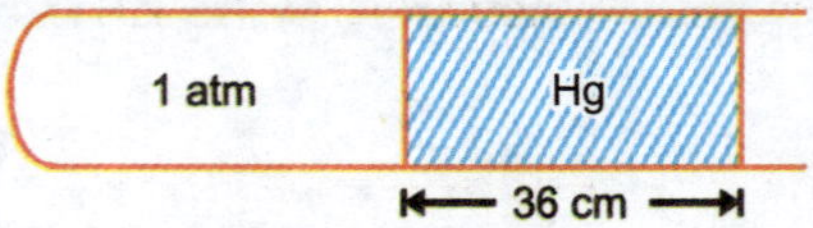

If the tube is held vertical keeping the open end up, length of air column shrink to 19 cm. What is the length (in cm) by which the mercury column shifts down?

12. A flask containing air at 107°C and 722 mm of Hg is cooled to 100 K and 760 mm of Hg. If density in the initial condition 1 g/cm^3, then what is the final density (g/cm^3)?

13. If an ideal gas at 100 K is heated to 109 K in a rigid container, the pressure increases by X%. What is the value of X?

14. The van der Waal's constants for a gas are $a = 3.6$ atm L^2 mol^{-2}, $b = 0.6$ L mol^{-1}. If $R = 0.08$ L atm K^{-1} mol^{-1} and the Boyle's temperature (K) is T_B of this gas, then what is the value of $\frac{T_B}{15}$?

15. A flask has 10 gas particles, out of which four particles are moving at 7 ms^{-1} and the remaining are moving at the same speed of 'X' ms^{-1}. If the r.m.s. of the gas is 5 ms^{-1}, what is the value of X ?

Numerical Problems

(If the numerical value has more than two decimal places, truncate/round-off the value to TWO decimal places.)

1. A rigid vessel is filled with a mixture of 1 mole $C_2H_6(g)$ with stoichiometric amount of $O_2(g)$ at a total pressure of one atm at 27°C. The gas mixture is exploded, what would be the final pressure (in atm) of the products at 127°C?

2. An LPG cylinder contains gas at a pressure of 300 kPa at 27°C. The cylinder can withstand the pressure of 1.2×10^6 Pa. The room in which the cylinder is kept catches fire. The minimum temperature at which the bursting of cylinder will take place is ______ °C. (Nearest integer)

3. Given

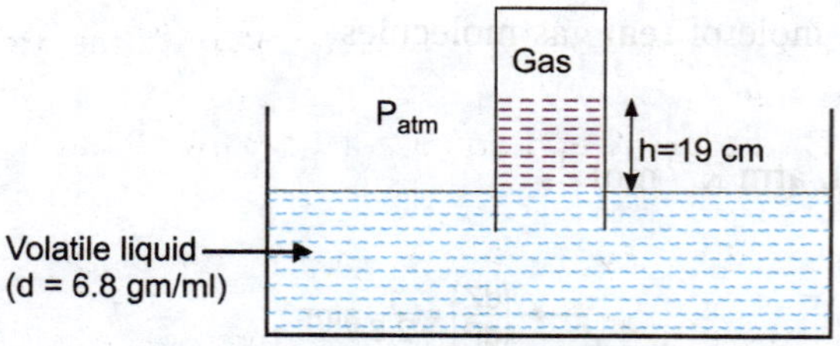

(Given : Vapour pressure of volatile liquid = 19 torr)

The pressure of dry gas is _____ atm.

4. A quantity of 88 g dry ice is taken in an open bottle of two litre capacity and the bottle is then closed properly. If the temperature of bottle is maintained at 27°C for long time, in which all the CO_2 is sublimed, then the pressure in bottle (in atm) is _____.

5. 0.32 gm of $CH_4(g)$ is diffused through a pin hole of rigid vessel in 20 sec., if all the $SO_2(g)$ at same pressure and temperature that will be diffused through same pin hole in 40 sec. in separate experiment is collected in a closed vessel of volume 2.463 L over liquid water at 27°C, then calculate total pressure in container (in torr).

(Given : Vapour pressure of water at 27°C is 28 mm of Hg. Consider $SO_2(g)$ is water insoluble and volume of water is negligible w.r.t. volume of vessel)

6. If initial rate of effusion ratio $\left(\frac{r_{CH_4}}{r_{He}}\right)$ for CH_4 and He is 2 : 1, at certain temperature, then mole fraction of CH_4 in initial mixture is ______ .

7. The speed corresponding to a gas "X" (MW = 50) corresponding to the maxima of the curve given below at a given temperature is 200 m/sec. Then kinetic energy (in kJ) of 300 gram of gas "X" will be :

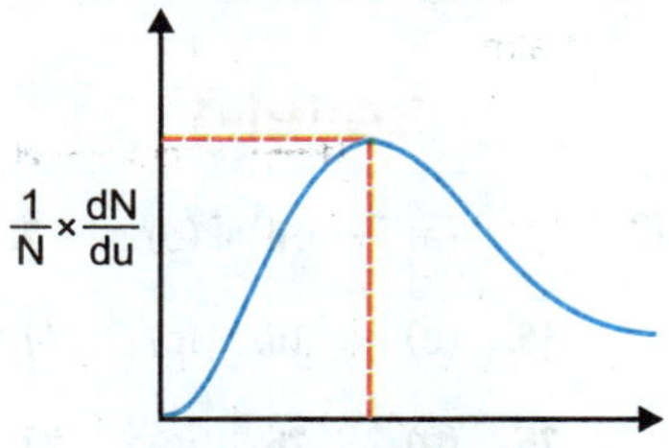

8. If the density of $H_2(g)$ at 20 bar is 4.0 g dm^{-3} then the most probable speed of H_2 molecules (in km/sec) is:

9. Initially bulb "A" contained oxygen gas at 27° C and 950 mm of Hg and bulb "B" contained neon gas at 27° C and 900 mm of Hg. These bulbs are connected by a narrow tube of negligible volume equipped with a stopcock and gases were allowed to mix-up freely. The pressure obtained in the combined system was found to be 910 mm of Hg.

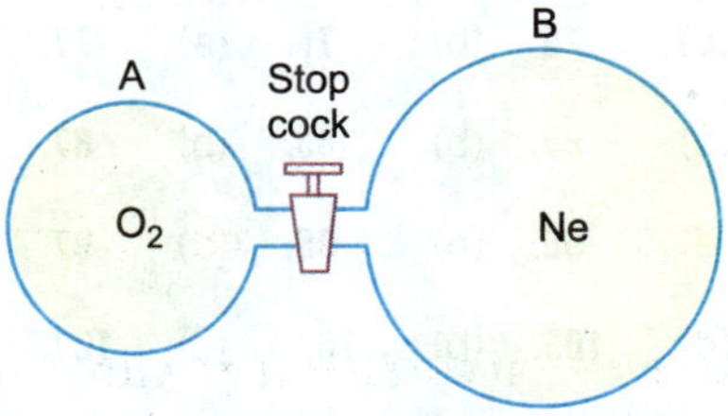

If volume of bulb B was measured to be 10 L, then find the volume of oxygen gas present initially in bulb "A".

10. One mole of a gas occupies 21.6 L at 27°C. The compressibility factor of gas at this temperature is 0.90. If $b = 0.6\,L\,mol^{-1}$, then what is the value of van der Waals' constant a (in atm L^2/mol^2) for the gas ? (Given : $R = 0.08$ L-atm/K-mol)

11. The critical temperature (T_C) and critical pressure of a van der Waals' gas are 160 K and 82.1 atm respectively. The critical volume (in L/mol) is :

12. For a van der Waals' gas, the T_C and P_C are 227°C and 200 atm respectively. Calculate volume (in L) of 40 moles of gas. (Given : $R = 0.080$ atm L/mol K)

13. At moderate pressure, the compressibility factor for a gas is given as $Z = 1 + 0.25P - \frac{100}{T} \cdot P$, where P is in bar and T is in kelvin. What is the Boyle's temperature of the gas (in K) ?

14. The van der Waals' constants a and b for a real gas ($M_{gas} = 82.1$ g/mol) are 4.105 atm-L^2/mol^2 and 0.2 L/mol respectively. Calculate density of gas (in g/L) at 5 atm and 250 K.

15. For real gas we can write the virial equation as

$$PV_m = RT\left[1 + \frac{B}{V_m} + \frac{C}{V_m^2} + \ldots\right]$$

where B and C are virial coefficients. At 300 K for 1 mole real gas value of $B = 0.004\,L\,mol^{-1}$ and $C = 0.0036\,L^2\,mol^{-2}$. Calculate the temperature (in K) of real gas above which it is not liquified on applying pressure.

(Given : $R = \frac{1}{12}$ L atm mol^{-1} K^{-1})

Answers

Level-1

1. (c)	2. (c)	3. (a)	4. (d)	5. (a)	6. (b)	7. (a)	8. (d)	9. (c)	10. (c)
11. (a)	12. (a)	13. (c)	14. (c)	15. (b)	16. (c)	17. (a)	18. (a)	19. (c)	20. (c)
21. (b)	22. (b)	23. (a)	24. (d)	25. (d)	26. (c)	27. (b)	28. (a)	29. (c)	30. (b)
31. (c)	32. (b)	33. (b)	34. (b)	35. (a)	36. (d)	37. (c)	38. (c)	39. (a)	40. (c)
41. (a)	42. (d)	43. (a)	44. (c)	45. (d)	46. (b)	47. (a)	48. (b)	49. (c)	50. (b)
51. (b)	52. (d)	53. (a)	54. (a)	55. (c)	56. (a)	57. (d)	58. (d)	59. (b)	60. (a)
61. (b)	62. (c)	63. (b)	64. (b)	65. (a)	66. (a)	67. (c)	68. (a)	69. (a)	70. (b)
71. (d)	72. (c)	73. (b)	74. (b)	75. (b)	76. (a)	77. (b)	78. (d)	79. (a)	80. (a)
81. (c)	82. (a)	83. (b)	84. (b)	85. (b)	86. (b)	87. (b)	88. (a)	89. (c)	90. (b)
91. (a)	92. (b)	93. (d)	94. (d)	95. (b)	96. (d)	97. (c)	98. (c)	99. (a)	100. (c)
101. (a)	102. (b)	103. (c)	104. (d)	105. (b)	106. (c)	107. (c)	108. (b)	109. (b)	110. (d)
111. (d)	112. (b)	113. (d)	114. (b)	115. (a)	116. (b)	117. (a)	118. (b)	119. (b)	120. (b)
121. (b)	122. (c)	123. (d)	124. (a)	125. (a)	126. (c)	127. (b)	128. (d)	129. (b)	130. (c)
131. (c)	132. (b)	133. (a)	134. (c)	135. (b)	136. (d)	137. (b)	138. (a)	139. (c)	140. (c)
141. (c)	142. (c)	143. (d)	144. (d)	145. (a)	146. (c)	147. (c)	148. (d)	149. (a)	150. (b)
151. (a)	152. (c)	153. (d)	154. (c)	155. (d)	156. (c)	157. (d)	158. (a)	159. (b)	160. (b)
161. (d)	162. (a)	163. (d)	164. (b)	165. (a)	166. (c)	167. (c)	168. (a)	169. (d)	170. (d)
171. (a)	172. (b)	173. (c)	174. (b)	175. (b)					

Level-2

1. (a)	2. (a)	3. (b)	4. (c)	5. (c)	6. (b)	7. (b)	8. (b)	9. (c)	10. (a)
11. (d)	12. (b)	13. (a)	14. (c)	15. (b)	16. (d)	17. (d)	18. (b)	19. (d)	20. (c)
21. (c)	22. (d)	23. (a)	24. (c)	25. (b)	26. (b)	27. (b)	28. (c)	29. (a)	30. (b)

Level-3

Passage–1	1. (d)	2. (a)	3. (c)	4. (d)
Passage–2	1. (c)	2. (c)	3. (a)	
Passage–3	1. (a)	2. (d)	3. (b)	

One or More Answers is/are correct

1.	(a,b,c,d)	2.	(a,b)	3.	(b,c,d)	4.	(c,d)	5.	(a)	6.	(a,c,d)	7.	(a,b)	8.	(a,b,c,d)
9.	(a)	10.	(a,b,c)	11.	(a,b,d)	12.	(a,d)	13.	(a,b,d)	14.	(a,b,c)	15.	(a,b,d)	16.	(a,b,c,d)
17.	(a,d)	18.	(a,c)	19.	(b,c)	20.	(a,b,c)								

Match the Column

1. A → P; B → R; C → Q; D → S
2. A → R; B → S; C → Q; D → P
3. A → R; B → Q; C → P; D → S
4. A → P, S; B → R; C → P, Q; D → R
5. A → R; B → S; C → P; D → Q
6. A → Q, S; B → R; C → P, T; D → T
7. A → S; B → P; C → Q, R; D → S

Assertion-Reason Type Questions

1. (B)	2. (C)	3. (A)	4. (B)	5. (A)	6. (A)	7. (C)	8. (B)	9. (C)	10. (C)
11. (A)	12. (D)	13. (B)	14. (A)	15. (D)					

Level-4

Subjective Problems

1. 7	2. 4	3. 7	4. 2	5. 4	6. 2	7. 6	8. 4	9. 5	10. 1
11. 9	12. 4	13. 9	14. 5	15. 3					

Numerical Problems

1. 1.48	2. 927.00	3. 0.85	4. 25.13	5. 180	6. 0.80	7. 9.00	8. 1.00
9. 2.5	10. 66.65	11. 0.06	12. 0.25	13. 400.00	14. 20.00	15. 82.96	

Hints & Solutions

Level 1

2. (c) Boyle's Law

Temperature and mass = constant

$PV = K$

$\log P = \log K - \log V$

3. (a) $\frac{V_1}{T_1} = \frac{V_2}{T_2}$; pressure and mass = constant

$\frac{0.2}{273} = \frac{V_2}{546}$

$V_2 = 0.4\,\text{L}$

4. (d) Charle's law

Pressure and mass = constant

$\frac{V}{T} = \text{constant} = K$

$V = KT$

$\log V = \log K + \log T$

5. (a) $\frac{P_1V_1}{T_1} = \frac{P_2V_2}{T_2}$

$\frac{P \times V}{348} = \frac{(2P) \times (0.85\,V)}{T_2}$

$T_2 = 592\,\text{K} = 319°\text{C}$

6. (b) $V = KT$

(i) $\frac{V}{T} = K$ (ii) $VT = KT^2$

(iii) $\log V = \log K + \log T$

8. (d) V = constant, mass = constant

$\frac{P}{T} = \text{constant}$

$P = KT$

$\log P = \log K + \log T$

10. (c) Boyle's law $PV = K$

$\frac{P}{V} = \frac{K}{V^2}$ ...(i)

$PV^2 = KV$ ...(ii)

$\log P = \log K - \log V$ ...(iii)

12. (a) $P \propto n$, initial mole $n_1 = \frac{PV}{RT} = 0.1$ mole

$\frac{P_1}{P_2} = \frac{n_1}{n_2}$

$\frac{1}{0.75} = \frac{0.1}{n_2} \Rightarrow n_2 = 0.075$

Escape mole = 0.025

13. (c) For ideal gas, $PV = nRT$

15. (b) $\frac{P_1V_1}{T_1} = \frac{P_2V_2}{T_2}$

$\frac{P \times 0.1}{T} = \frac{1.5\,P \times V_2}{\frac{4T}{3}}$

$V_2 = 88.9\,\text{mL}$

16. (c) $\frac{V_1}{T_1} = \frac{V_2}{T_2}$

17. (a) $P = KT$

$P = K(t°\text{C} + 273)$

18. (a) $PM = dRT$

$M \propto dT$

$\frac{M_1}{M_2} = \frac{d_1T_1}{d_2T_2} = \frac{1 \times 2}{2 \times 1} = \frac{1}{1}$

19. (c) $PM = dRT$

$\frac{P_A}{P_B} = \frac{d_A}{d_B} \times \frac{M_B}{M_A} = \frac{4}{1}$

20. (c) $V \propto T$

Because temperature ↑ by 10 K

expelled air 10 cm^3

21. (b)

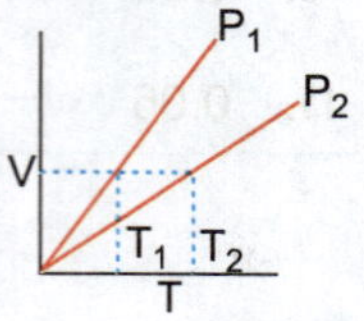

$PV = nRT$

n, V = constant

$P \propto T$

$T_2 > T_1$

$P_2 > P_1$

22. (b) $PV = \frac{W}{M}RT$

24. (d) $PV = \frac{W}{M}RT$

$V \propto \frac{1}{M}$

25. (d) $P_1V_1 = P_2V_2$

26. (c) $d = \frac{PM}{RT}$

27. (b) The moles of the gas in the bubble remains constant, so that $n_1 = n_2$. To calculate the final volume, V_2,

$$V_2 = V_1 \times \frac{P_1}{P_2} \times \frac{T_2}{T_1}$$

$$= 2.0 \text{ mL} \times \frac{6.0 \text{ atm}}{1.0 \text{ atm}} \times \frac{298 \text{ K}}{281 \text{ K}}$$

$$= 12.72 \text{ mL}$$

29. (c) $n_{C_2H_2} = \frac{PV}{RT} = \frac{1 \times 2.64}{0.0821 \times 273} = 0.118$ mole

$n_{O_2} = \frac{5}{2} n_{C_2H_2} = \frac{5}{2} \times 0.118 = 0.294$

$V = \frac{nRT}{P} = 6.6 \text{ L}$

30. (b) $d = \frac{PM}{dRT}$

31. (c) $P_1V_1 = P_2V_2$

$V_1 = V$, $V_2 = V/10$

32. (b) $\frac{P_1}{T_1} = \frac{P_2}{T_2}$

33. (b)

(a) 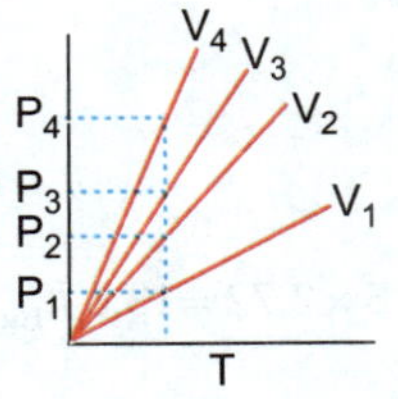

$P \propto 1/V$

$V_1 > V_2 > V_3 > V_4$

$PV = nRT$

(b) 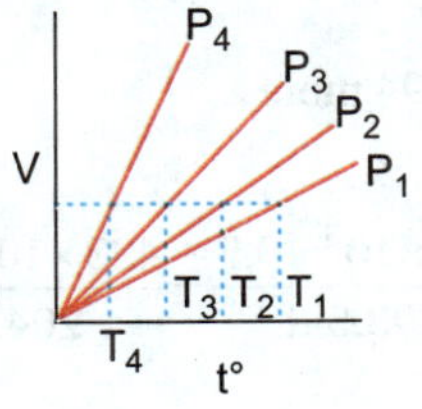

$P \propto T$

$P \propto T$

$P_1 > P_2 > P_3 > P_4$

34. (b) $PV = nRT$

$n = \frac{10 \times 1}{76 \times 0.0821 \times 1273} = 1.26 \times 10^{-3}$

No. of Na atoms $= 1.26 \times 10^{-3} \times 6.02 \times 10^{23}$

$= 7.6 \times 10^{19}$

35. (a) Moles of C_2H_6 and C_2H_4 are a and b respectively.

$a + b = \frac{28}{22.4} = 1.25$

$$\underset{a}{C_2H_6(g)} + \underset{3.5a}{3.5\,O_2(g)} \longrightarrow 2\,CO_2(g) + 3H_2O(l)$$

$$\underset{b}{C_2H_4(g)} + \underset{3b}{3O_2(g)} \longrightarrow 2\,CO_2(g) + 2H_2O(l)$$

Moles of $O_2 = 3.5a + 3b = 4$

$a + b = 1.25$

$a = 0.5$

$b = 0.75$

36. (d) $n_{H_2O} = \frac{PV}{RT} = 1.54$

$\frac{n_{MgSO_4 \times H_2O}}{n_{H_2O}} = \frac{1}{x} = \frac{54.2}{(120 + 18x) \times 1.54}$

$x = 7$

37. (c) $PM = dRT$

$M = 46$

38. (c) Because the number of moles is constant.

$\frac{P_iV_i}{T_i} = \frac{P_fV_f}{T_f}$; $P_f = \frac{P_iV_iT_f}{V_fT_i}$

$P_f = \frac{P_iV_iT_f}{V_fT_i}$

$= 3.21 \times 10^5 \text{ Pa} \times \frac{V_i}{1.03\,V_i} \times \frac{(273 + 28.0)}{(273 - 5.00)}$

$= 3.50 \times 10^5$ Pa

39. (a) $n_{N_2} = 53.57$ mole

$\frac{P_1}{P_2} = \frac{n_1}{n_2}$

$n_2 = \frac{1.8}{200} \times 53.57 = 0.48213$

n_{N_2} escape $= 53.087$

volume $= \frac{nRT}{p} = 1264$ L

40. (c) $h_{Gallium} = \frac{74 \times 13.6}{6.095} = 165.1$

41. (a) Volume of He at 0.1 atm

$P_1V_1 = P_2V_2$

$50 \times 100 = V_2 \times 1$

$V_2 = 5000$ L

Volume available for balloon

$= 5000 - 50 = 4950$ L at 1 atm

Volume of balloon $= 100 \text{ m}^3$

$= 100000 \text{ L} = 49500$ L at 0.1 atm

So, 2 cylinder required.

42. (d) $P_1V_1 = P_2V_2$, $P_2 = \dfrac{76+30.40\times10}{13.6}$

$$V_2 = \frac{76\times14}{98.41} = 10.8\,\text{L}$$

43. (a) $\dfrac{P_1}{T_1} = \dfrac{P_2}{T_2}$

$$T_2 = 3\times\frac{76}{80}\times273$$

44. (c) $n_1 = n_2$

$$\frac{1}{32} = \frac{2.375}{M_2}$$

$$M_2 = 76$$

45. (d) $P_i = \dfrac{6.81}{4}\times\dfrac{0.0821\times250}{1.1\times104}$

$$= 3.012\times10^{-3}$$

$$P_f = 4\times10^{-3}$$

$$\Delta P\uparrow = 1\times10^{-3}$$

$$n_{\text{He-added}} = \frac{10^{-3}\times1.16\times10^{4}}{0.0821\times250}$$

$$wt = n_{\text{He}}\times4 = 2.24\,\text{gm}.$$

46. (b) $P = \dfrac{nRT}{V} = 2.46$ atm

47. (a) $V = \dfrac{nRT}{P} = \dfrac{10\times0.0821\times T}{0.821}$

$$V = T$$

$$\log V = \log T$$

slope = 1

$\theta = 45°$

48. (b) Intercept on y-axis

$$= \log_{10}\frac{nR}{V} = \log_{10}\frac{10\times0.0821}{8.21} = -1.0$$

$\dfrac{P}{T}$ v/s curve $\dfrac{P}{T} = \dfrac{nR}{V}$

$$\text{Intercept} = \frac{nR}{V}$$

$$= \frac{10\times0.0821}{8.21} = 0.1,\quad \text{slope} = 0$$

49. (c) $d_A = 2d_B$; $3M_A = M_B$; $PM = dRT$

$$= \frac{P_A}{P_B}\times\frac{M_A}{M_B} = \frac{d_A}{d_B}\times\frac{RT}{RT}$$

$$= \frac{P_A}{P_B}\times\frac{1}{2} = 2$$

$$\frac{P_A}{P_B} = \frac{4}{1}$$

50. (b) Case I—Suppose inner balloon burst first

$$\frac{600}{300} = \frac{800}{T_2} \Rightarrow T_2 = 400\text{ K}$$

Case II—Suppose outer balloon burst first

$$\frac{1500}{300} = \frac{1800}{T_2};\quad T_2 = 360\text{ K}$$

51. (b) $T_1 = 300\,\text{K}$, $T_2 = 500\,\text{K}$

At constant P and V

$$n_1T_1 = n_2T_2$$

$$n_1\times300 = n_2\times500$$

$$n_2 = 0.6\,n_1;$$

Moles escaped $= n_1 - 0.6n_1 = 0.4\,n_1$

% of air escaped $= \dfrac{0.4\,n_1}{n_1}\times100 = 40$

52. (d) $R = \dfrac{PV}{nT}$

53. (a)

	$2NH_3(g)$	$\longrightarrow$	$N_2(g)$	$+\,3H_2(g)$
Before sparking	76		0	0
After sparking at eqm	$76-2x$		x	$3x$

Increase in pressure $2x = 6$; $x = 3$ cm Hg

Partial pressure of $H_2 = 3\times3 = 9$ cm Hg

54. (a) $P_{\text{Atm}} = \dfrac{h}{13.6} + P_{\text{Gas}} + aq$ Tension

$$P_{\text{Gas}} = P_{\text{Atm}} - \left(\text{aq tension} + \frac{h}{13.6}\right)$$

55. (c)

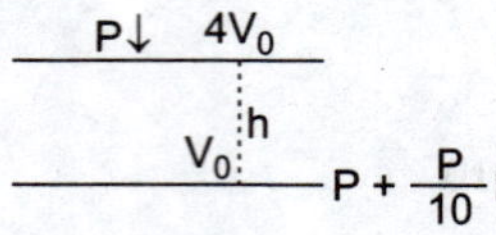

$$P_1V_1 = P_2V_2$$

$$P\times4\,V = P\left(1+\frac{h}{10}\right)V$$

$$4 = 1+\frac{h}{10}$$

$$h = 30\,\text{m}$$

56. (a) $h_g d_g = h_{\text{Hg}}d_{\text{Hg}}$; $5\times2.72 = 13.6\,h_{\text{Hg}}$

$$h_{\text{Hg}} = 1\,\text{m}$$

$$P_{\text{gas}} = 0.76+1 = 1.76\,\text{m of Hg}$$

$$= 1760\text{ mm of Hg}$$

$$\frac{1760}{760}\times10 = n\times R\times300$$

$$n = 0.94\text{ mole}$$

57. (d) $n_T = n_1 + n_2$

$$\frac{P\times1.25}{R\times287.5} = \frac{611\times10^3\times0.5}{293.5\,R} + \frac{433\times10^3\times0.75}{204.2\,R}$$

$$P = 4.84\times10^5\,\text{Pa}$$

58. (d)

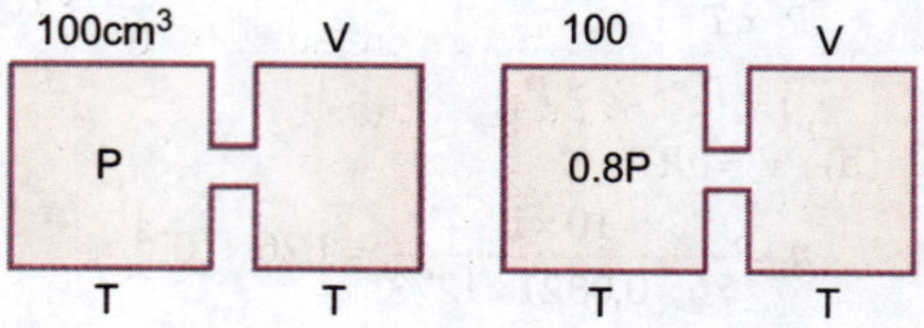

$$\frac{0.2P\times(100+V)}{RT}=\frac{100P}{RT}$$

$$0.2(100+V)=100$$

$$V=400\,cm^3$$

59. (b) Moles of C_2H_2 and C_3H_8 are a and b

$$\underset{a}{C_2H_2}+2.5\,O_2\longrightarrow \underset{2a}{2CO_2}+H_2O$$

$$\underset{b}{C_3H_8}+5O_2\longrightarrow \underset{3b}{3CO_2}+4H_2O$$

$$2a+3b\propto 230$$

$$a+b\propto 80$$

$$b\propto 70$$

$$a\propto 10$$

$$\text{mole fraction}=\frac{7}{8}=0.875$$

60. (a) $p\times n$

$$2H_2(g)+O_2(g)\longrightarrow 2H_2O(g)$$

$$a \qquad b$$

$$a-2b \qquad b-b$$

$$a+b=1$$

$$a-2b=0.4$$

$$3b=0.6$$

$$b=0.2$$

$$a=0.8$$

61. (b) $P_1=\frac{5\times RT}{V}$; p.pr. of He $=\frac{2}{5}\times\left(\frac{5RT}{V}\right)$

$$=\frac{2\times0.0821\times400}{8.21}=8\text{ atm}$$

62. (c) $P_{O_2}=\frac{3}{10}\times P_T$;

After removing 2 mole of O_2,

$$P'_{O_2}=\frac{1}{8}\times P_T$$

Decrease in partial pressure of O_2

$$=\frac{\frac{3P_T}{10}-\frac{P_T}{8}}{\frac{3P_T}{10}}\times100$$

$$=58.33$$

63. (b) $P_{N_2}=760-30=730$ torr.

$$n_{N_2}=\frac{730\times0.821}{760\times0.0821\times300}=0.032$$

64. (b) % Relative humidity

$$=\frac{\text{Partial pressure of }H_2O(g)}{\text{Vapour pressur of }H_2O}\times100$$

at constant temperature

65. (a) Volume of $O_2=1$ L

and $p_{O_2}=760-355=405$ mm Hg

Since, the temperature is constant, using Boyle's law we get

$$1\text{ L}\times405\text{ mm}\equiv0.3\text{ L}\times p'_{O_2}$$

$$\therefore\quad p'_{O_2}=1350\text{ mm Hg}$$

66. (a) Dalton's law is applicable for a non-reacting gaseous mixture.

$$SO_2+Cl_2\longrightarrow SO_2Cl_2$$

67. (c) $n_{CH_4}=\frac{x}{16}\qquad n_{O_2}=\frac{x}{32}$

mole fraction $O_2=1/3$

68. (a) $n_T=n_1+n_2\qquad n_1=\frac{P\times1}{R\times T}=1$

$$\frac{P'\times2L}{RT}=2\qquad P=RT$$

$$P'=RT=P$$

69. (a) $n=\frac{PV}{RT}=\frac{3170\times10^{-3}}{8.314\times300}=1.27\times10^{-3}$ mole.

70. (b) Dalton's Law is applicable on a non-reacting gas mixture.

71. (d) mole $N_2=\frac{56}{28}=2$, mole $O_2=\frac{96}{32}=3$

$$P_{N_2}=\frac{2}{5}\times10=4\text{ atm}$$

$$P_{O_2}=\frac{3}{5}\times10=6\text{ atm}$$

72. (c) $P\propto n$

$$\frac{P_{H_2}}{P_{CO_2}}=\frac{n_{H_2}}{n_{CO_2}}=\frac{22}{1}$$

$$P_{H_2}=22\text{ atm}$$

73. (b) $P_{gas}=P_{dry\ gas}+P_{moisture}$ at T K

or $\quad P_{dry}=830-30=800$

Now at $T_2=0.99\,T_1$;

at constant volume $\frac{P_1}{T_1}=\frac{P_2}{T_2}$

$$P_{dry}=\frac{800\times0.99\,T}{T}=792\text{ mm}$$

$$\therefore\quad P_{gas}=P_{dry}+P_{moisture}$$

$$=792+25=817\text{ mm}$$

74. (b) We have $P_{O_2}=X_{O_2}\times P_{total}$

$$\therefore\quad 0.60=\frac{n_{O_2}}{n_{O_2}+n_{SO_2}}\times P_{total}$$

Let no. of moles of O_2 are x;

$$0.60=\frac{x/32}{x/32+3x/64}\times P_{total}$$

$$\therefore\ 0.60=0.4\times P_{total}$$

$$\therefore\quad P_{total}=\frac{0.60}{0.4}=1.5\text{ atm}$$

$$\because\ P_{O_2}+P_{SO_2}=1.5\text{ atm}$$

$$\Rightarrow P_{SO_2}=1.5-P_{O_2}=1.5-0.60=0.9\text{ atm}$$

$\therefore$ Concentration of $SO_2 = \dfrac{0.9}{0.0821 \times 300}$

$= 0.036$

75. (b) Given $P = 10$ atm,

total numbers of moles; $n_A + n_B + n_C = 10$

$P_A = 3$ atm, $P_B = 1$ atm

$\because \quad P_A = x_A \times P_{(total)} = \dfrac{n_A}{n_A + n_B + n_C} \times 10$

$= \dfrac{n_A}{10} \times 10, n_A = 3$

Similarly, $P_B = x_B \times P_{(total)}$

So, $\quad n_B = 1$

$\therefore \quad n_C = 10 - (n_A + n_B) = 10 - 4 = 6$

Mass of C $= 6 \times 2 = 12$ g

76. (a) $\quad PV = \dfrac{10}{M} RT \quad \ldots(1)$

Let x g of the gas remain in the flask when final pressure $\dfrac{P}{2}, \dfrac{2T}{3}$

$\therefore \quad \left(\dfrac{P}{2}\right) V = \left(\dfrac{x}{M}\right) R \left(\dfrac{2T}{3}\right) \quad \ldots(2)$

$(1)/(2) \Rightarrow \quad 2 = \dfrac{10}{x} \times \dfrac{3}{2}$

$x = \dfrac{3 \times 10}{4} = 7.5$ g

77. (b) $\dfrac{P_x}{P_y} = \dfrac{n_x}{n_y}$

$\dfrac{2}{1} = \dfrac{1/m_x}{2/m_y}$

$m_y = 4m_x$

78. (d) $P = \dfrac{nRT}{V}$

$P = \dfrac{0.05 \times 0.0821 \times 303}{44 \times 4.6} = 6.14 \times 10^{-3}$ atm

79. (a) $P_T = P_{He} + P_{Ne} + \text{V.P } H_2O$

$745 = 368 + P_{Ne} + 28.3$

$P_{Ne} = 348.7$ mm of Hg

80. (a)

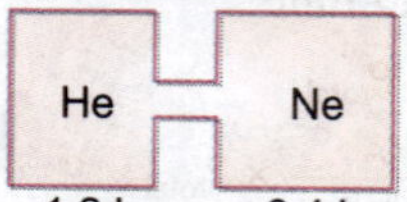

1.2 L   3.4 L
0.63 atm   2.8 atm

$n_{He} = \dfrac{1.2 \times 0.63}{0.082 \times 289}, \quad n_{Ne} = \dfrac{3.4 \times 2.8}{0.0821 \times 289}$

$= 0.032 \qquad = 0.4$

$P_f = \dfrac{0.431 \times 0.0821 \times 289}{4.6} = 2.223$

$P_{He} = \dfrac{0.032}{0.432} \times 2.223 = 0.164$ atm

81. (c) From the total pressure and the vapour pressure of water we can calculate the partial pressure of O_2.

$P_{O_2} = P_T - P_{H_2O}$

$= 760 - 22.4$

$= 737.6$ mm Hg

From the ideal gas equation we write

$m = \dfrac{PVM}{RT}$

$= \dfrac{(0.974 \text{ atm})(0.128 \text{ L})(32.0 \text{ g/mol})}{(0.0821 \text{ L atm/K mol})(273 + 24) \text{ K}}$

$= 0.163$ g

82. (a) $PV = nRT \qquad R = K \times N_A$

$PV = n \times N_A KT$

$n \times N_A = \dfrac{PV}{KT} =$ no. of particles

83. (b) $KE = \dfrac{1}{2} Mv^2$

84. (b) K.E is temperature dependent

same moles would have same volume at same P and T

$PV = nRT$

$d = \dfrac{PM}{RT}$

85. (b) According to Boyle's law at constant temperature and mass

$PV = K$

$P \propto \dfrac{1}{V}$

86. (b) $KE_T = \dfrac{3}{2} nRT$ at constant P and V

$PV = K$ at constant T and mass

$V = KT$ at constant P and mass

87. (b) $KE_{avg} = \dfrac{3}{2} KT$

88. (a) $PV = nRT$

$PV \propto T \qquad$ at given mass.

89. (c) According to kinetic theory of gases collision between molecules are perfectly elastic.

90. (b) $KE_T = \dfrac{3}{2} nRT = \dfrac{3}{2} \dfrac{W}{M} RT$

$KE \propto \dfrac{T}{M}$

$\dfrac{KE_1}{KE_2} = \dfrac{T_1}{M_1} \dfrac{M_2}{T_2} = \dfrac{300}{2} \times \dfrac{44}{600} = \dfrac{11}{1}$

91. (a) $PV = \dfrac{1}{3} mu^2$

$P \times 1 = \dfrac{2}{3} \times \dfrac{1}{2} mu^2$

$P = \dfrac{2}{3} E.$

92. (b) Moles ratio $= \frac{n_{He}}{n_{N_2}} = \frac{w/4}{w/28} = \frac{7}{1}$

Ratio of translational K.E. $= \frac{n_{He} \cdot T_{He}}{n_{N_2} \cdot T_{N_2}}$

$$= \frac{7}{1} \times \frac{300}{700} = 3:1$$

93. (d) Due to phase transfer distance between particle changes, because temperature remains constant, so *KE* remains constant.

94. (d) $K = \frac{3}{2} n_T RT \Rightarrow n_T = \frac{3000 \times 2}{3 \times 250 \times 8.314}$

$$3 \times 10^3 = \frac{3}{2} n_T \times 8.314 \times 250$$

$$\frac{x}{20} + \frac{30-x}{40} = \frac{2}{8.314}$$

$$\%\text{Ne} = 28.3$$

95. (b) $KE = \frac{3}{2} RT$

$$U_{rms} = \sqrt{\frac{3RT}{M}}$$

96. (d) $U_{rms} = \sqrt{\frac{U_1^2 + U_2^2 + U_3^2 + U_4^2}{4}}$

$$= \sqrt{\frac{54}{4}} = \frac{\sqrt{54}}{2}$$

97. (c) $U_{ms} = \sqrt{\frac{4 \times (6)^2 + 5(2)^2 + 10(3)^2}{19}}$

$$= 3.6 \text{ cm/sec.}$$

98. (c) $U_{ms} = \sqrt{\frac{3RT}{Mw}}$

$$\frac{U_{H_2}}{U_{O_2}} = \sqrt{\frac{50}{2} \times \frac{32}{800}} = 1$$

100. (c) $U_{rms} = \sqrt{\frac{3RT}{M}} \quad \frac{(U_{rms})_{H_2}}{(U_{rms})_{N_2}} = \sqrt{\frac{T_{H_2}}{M_{H_2}} \times \frac{M_{N_2}}{T_{N_2}}};$

$$(U_{rms})_{H_2} = \sqrt{5}\,(U_{rms})_{N_2}$$

$$\therefore \frac{(U_{rms})_{H_2}}{(U_{rms})_{N_2}} = \sqrt{5} = \sqrt{\frac{T_{H_2}}{T_{N_2}} \times \frac{28}{2}}$$

$$= \frac{\sqrt{5}}{1} = \sqrt{\frac{T_{H_2}}{T_{N_2}} \times 14}$$

$$= 5 = \frac{T_{H_2}}{T_{N_2}} \times 14$$

$$T_{N_2} \times 5 = T_{H_2} \times 14$$

$$\therefore T_{N_2} > T_{H_2}$$

101. (a) $U_{rms} = \sqrt{\frac{2RT}{M}} \quad U_{avg} = \sqrt{\frac{8}{\pi} \frac{RT}{M}} \quad U_{mp} = \sqrt{\frac{2RT}{M}}$

U_{rms}	: U_{avg}	: U_{mp}
1.224	1.128	1

$$U_{rms} > U_{avg} > U_{mp}$$

102. (b) $200 = \sqrt{\frac{2RT}{2 \times 10^{-3}}} = RT = 40$

Average K.E. $= \frac{3}{2} nRT = \frac{3}{2} \times \frac{8}{2} \times 40$

$$= 240 \text{ J}$$

103. (c) Second member of C_nH_{2n} series

$$= C_3H_6 = 42$$

$$= \sqrt{\frac{8RT_1}{\pi M_1}} = \sqrt{\frac{8RT_2}{\pi M_2}} = \frac{900}{71} = \frac{T_2}{42}$$

$$T_2 = 532.4 \text{ K}$$

104. (d) $U_{rms} = \sqrt{\frac{3RT}{M}}$

105. (b) $P = \frac{1}{3} \frac{nmc^2}{V}$

$$= \frac{1}{3} \times \frac{6 \times 10^{22} \times 10^{-24} \times (100)^2}{10 \times 10^{-3}}$$

$$= 2 \times 10^4 \text{ Pa}$$

106. (c) Second member is C_3H_4;

$$\sqrt{\frac{2RT_1}{M_1}} = \sqrt{\frac{2RT_2}{M_2}}$$

$$T_1 = T_2 \left(\frac{M_1}{M_2}\right) = 800 \left(\frac{40}{64}\right) \text{K}$$

$$= 500 \text{ K} \quad \text{or} \quad 227°\text{C}$$

107. (c) $300 = \sqrt{\frac{3RT}{4 \times 10^{-3}}}; \quad RT = 120$

Total K.E. of He gas $= \frac{3}{2} nRT$

$$= \frac{3}{2} \times \frac{8}{4} \times 120 \text{ J}$$

$$= 360 \text{ J}$$

108. (b) $U_{mp} = \sqrt{\frac{2RT}{M}}$

$$U_{avg} = \sqrt{\frac{8}{\pi} \frac{RT_2}{M}}$$

$$U_{rms} = \sqrt{\frac{RT_1}{M}}$$

$$T_3 > T_2 > T_1$$

109. (b) $U_1 = \sqrt{\frac{3P_1}{d_1}}$

$$\therefore \Delta U_{rms} = \sqrt{\frac{3}{d}}\,(\sqrt{P_2} - \sqrt{P_1})$$

$$= \sqrt{\frac{3}{0.75}}\,(300 - 200)$$

$$= \sqrt{4} \times 100 = 200$$

110. (d) Rate of diffusion $\propto \dfrac{1}{\sqrt{\text{density}}}$

at constant P and T

$d = \dfrac{PM}{RT}$, $\quad d \propto M$

111. (d) $\alpha : v : u = \sqrt{2} : \sqrt{\dfrac{8}{\pi}} : \sqrt{3}$

112. (b) $\sqrt{\dfrac{8RT}{\pi M}} = \sqrt{\dfrac{28}{88}} = \sqrt{\dfrac{7}{22}}$

$$\frac{8 \times 8.314 \times T}{\pi \times 16 \times 10^{-3}} = \frac{7}{22} \times 10^6$$

$$T = \frac{1000 \times 2}{8.314} = 240.55 \text{ K}$$

$$T\,°\text{C} = 240.55 - 273 = -32.45\,°\text{C}$$

113. (d) $\sqrt{\dfrac{2RT}{M_{SO_2}}} = \sqrt{\dfrac{3R \times T_1}{M_{O_2}}} \Rightarrow \dfrac{2 \times T}{64} = \dfrac{3 \times 300}{32}$

$$T = 900 \text{ K}$$

114. (b) $U_{rms} = \sqrt{\dfrac{3RT}{M}}$

115. (a) Rate of diffusion $\propto \dfrac{1}{\sqrt{M}} \propto P$

$\therefore$ Rate of diffusion $\propto \dfrac{P}{\sqrt{M}}$

116. (b) According to Grahm's law

$r \propto \dfrac{1}{\sqrt{d}}$, $\quad \dfrac{r_1}{r_2} = \sqrt{\dfrac{d_2}{d_1}}$ at constant P and T

117. (a) $\dfrac{r_1}{r_2} = \sqrt{\dfrac{M_2}{M_1}}$ at constant P and T

118. (b) $\dfrac{r_1}{r_{H_2}} = \sqrt{\dfrac{2}{M_1}} = \dfrac{1}{6}$

$$M = 72$$

119. (b) $\dfrac{r_1}{r_2} = \sqrt{\dfrac{M_2}{M_1}}$

$$\frac{V_1}{t_1} \times \frac{t_2}{V_2} = \sqrt{\frac{M_2}{M_1}}$$

120. (b) $\dfrac{r_{H_2}}{r_{H.C}} = \sqrt{\dfrac{M}{2}} = 3\sqrt{3}$

$$\frac{M}{2} = 9 \times 3$$

$$M = 54$$

$$n = 4$$

121. (b) $\dfrac{r_{O_2}}{r_{CH_4}} = \dfrac{n_{O_2}}{n_{CH_4}} \cdot \sqrt{\dfrac{M_{CH_4}}{M_{O_2}}}$

$$= \frac{3}{2} \times \frac{16}{32} \times \sqrt{\frac{16}{32}} = \frac{3}{4\sqrt{2}}$$

122. (c) $\dfrac{n'_{SO_2}/n'_{CH_4}}{8/1} = \left(\dfrac{16}{64}\right)^{3/2} = \dfrac{1}{1}$;

$$\frac{r_{SO_2}}{r_{CH_4}} = \frac{1}{1} \cdot \sqrt{\frac{16}{64}} = \frac{1}{2}$$

123. (d) $\dfrac{n'_{He}}{n'_{CH_4}} = \dfrac{1}{2}\sqrt{\dfrac{16}{4}} = \dfrac{1}{1}$

$$\frac{n'_{He}}{n'_{SO_2}} = \frac{1}{3}\sqrt{\frac{64}{4}} = \frac{4}{3}$$

$$n'_{He} : n'_{CH_4} : n'_{SO_2} = 4 : 4 : 3$$

124. (a) $\dfrac{r_1}{r_2} = \dfrac{V_1/t_1}{V_2/t_2} = \dfrac{V_1 \times t_2}{V_2 \times t_1} = \sqrt{\dfrac{M_2}{M_1}}$;

$$\frac{80 \times 3}{V_2 \times 2} = \sqrt{\frac{64}{32}} = \sqrt{2}; \quad V_2 = \frac{120}{\sqrt{2}}$$

125. (a) $\dfrac{r_{NH_3}}{r_{HCl}} = \sqrt{\dfrac{36.5}{17}} = \dfrac{x}{87.5 - x}$

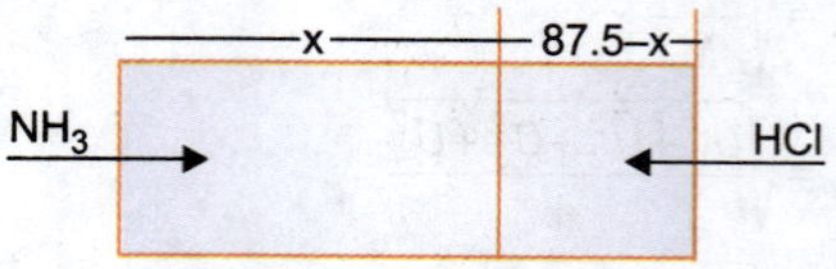

$$x = 51.7 \text{ cm}$$

126. (c) $\dfrac{r_{CH_3OCH_3}}{r_{CH_4}} = \sqrt{\dfrac{M_{CH_4}}{M_{CH_3OCH_3}}} \times \dfrac{P_{CH_3OCH_3}}{P_{CH_4}}$

$$= \sqrt{\frac{16}{46}} \times \frac{0.8}{0.2} = 2.36 : 1$$

127. (b) $\dfrac{n_1 \cdot t_2}{n_2 \cdot t_1} = \sqrt{\dfrac{M_2}{M_1}}$; $\quad \dfrac{w_1}{M_1} \times \dfrac{M_2}{w_2} = \sqrt{\dfrac{M_2}{M_1}}$

$$\frac{w_1}{w_2} = \sqrt{\frac{M_1}{M_2}}; \quad \frac{w_1}{4} = \sqrt{\frac{4}{64}}$$

$$\Rightarrow \quad w_1 = 1 \text{ gm}$$

128. (d) $\dfrac{r_{mix}}{r_x} = \dfrac{4/5}{4/10} = 2 = \sqrt{\dfrac{M_x}{M_{Mix}}}$

$$\Rightarrow M_{mix} = 9$$

$$M_{mix} = M_{H_2} X_{H_2} + M_{CH_4} X_{CH_4}$$

$$= 2X_{H_2} + 16(1 - X_{H_2}) = 9$$

$$\Rightarrow \quad X_{H_2} = 0.5$$

129. (b) Energy associated with each degree of freedom is $\dfrac{1}{2} k_B T$

130. (c) $C_v = \dfrac{3}{2}R + R + 4R \times \dfrac{3}{4} = \dfrac{11}{2}R$;

$$C_p = \frac{11}{2}R + R = \frac{13}{2}R$$

$$\frac{C_p}{C_v} = \frac{13}{11} = 1.18$$

131. (c) $C_v = \dfrac{\left(\frac{3}{2}R + \frac{5}{2}R\right)}{2} = 2R$

$C_p = \dfrac{\left(\frac{5}{2}R + \frac{7}{2}R\right)}{2} = 3R$

$\dfrac{C_p}{C_v} = 1.5$

132. (b) $\left.\begin{aligned} C_v &= \frac{3}{2}RT \\ C_p &= \frac{5}{2}RT \end{aligned}\right\}$ for monoatomic gas;

$\left.\begin{aligned} C_v &= \frac{5}{2}2RT \\ C_p &= \frac{7}{2}RT \end{aligned}\right\}$ for diatomic gas

Thus, for mixture of 1 mole each,

$C_v = \dfrac{\frac{3}{2}RT + \frac{5}{2}RT}{2}$

and $C_p = \dfrac{\frac{5}{2}RT + \frac{7}{2}RT}{2}$

Thus, $\dfrac{C_p}{C_v} = \dfrac{3RT}{2RT} = 1.5$

133. (a) Internal energy of a substance is a function of temperature and volume (except ideal gas). For ideal gas it is only function of temperature.

134. (c) For ideal gas $PV = nRT$

volume of 1 mole gas at 1 atm pressure and 273 K is 22.4 L and at 300 K and 1 atm pressure volume = 24.63 L.

135. (b) If attraction are disappear than pressure of real gas ↑

136. (d) $P_{\text{Ideal}} = P_{\text{Real}} + \dfrac{n^2a}{V^2}$

137. (b) Real gas behaves like an ideal gas at low pressure and high temperature at this condition $Z = 1$.

138. (a) $P_{\text{Ideal}} = P_r + \dfrac{n^2a}{V^2}$

139. (c) $P_{\text{Real}} = P_i - \dfrac{n^2a}{V^2}$

$a \uparrow P_r \downarrow$

$P_2 < P_3 < P_1$

140. (c) $P(V-b) = RT$; $P = \dfrac{RT}{(V-b)}$;

$P = \left(\dfrac{R}{(V-b)}\right)T + 0$

141. (c) $b = 4 \times$ volume occupied by molecules of 1 mol of a gas,

or $\quad b = 4N_A\left(\dfrac{4}{3}\pi r^3\right)$

or $\quad r = \left(\dfrac{3b}{16N_A\pi}\right)^{1/3}$

$= \left[\dfrac{3 \times 24}{16 \times 6.023 \times 10^{23} \times 3.14}\right]^{1/3}$ cm

$= 1.355 \times 10^{-8}$ cm

Now, molecular diameter, $d = 2r$

$= 2.67 \times 10^{-8}$ cm

142. (c) Liquifaction $\propto$ I.M.A.F. $\propto a$

143. (d) $V_i = V - V_{ex}$

$V_{ex} = n.b$

$b \propto$ size

144. (d) A real gas can behave like an ideal gas at low pressure and high temperature.

145. (a) $P(V - nb) = nRT$

$PV - Pnb = nRT \qquad d = \dfrac{\text{Mass}}{\text{Volume}}$

$PV = nRT + Pnb \qquad n = \dfrac{\text{Mass}}{M}$

$P\dfrac{\text{Mass}}{d} = nRT + Pnb$

$\dfrac{P}{d} = \dfrac{n}{\text{Mass}}RT + \dfrac{Pnb}{\text{Mass}}$

$\dfrac{P}{d} = \dfrac{RT}{M} + \dfrac{Pb}{M}$

146. (c) $\left(P + \dfrac{a}{V_m^2}\right)(V_m - b) = RT$

147. (c) At low pressure,

$\left(P + \dfrac{a}{V^2}\right)(V) = RT$

i.e., $\quad PV^2 - RTV + a = 0$

$V = \dfrac{RT \pm \sqrt{R^2T^2 - 4Pa}}{2P} = \dfrac{RT}{2P}$

$(\because\ 4a \cdot P = R^2T^2)$

148. (d) Real gas shows ideal behaviour at low pressure and high temperature.

149. (a) $\dfrac{PV}{RT} = Z \quad \therefore \quad Z = 1 - \dfrac{a}{VRT}$

150. (b) van der Waals' equation for one mole of a real gas is

$\left(P + \dfrac{a}{V^2}\right)(V - b) = RT$

or $\quad PV = RT + Pb + \dfrac{ab}{V^2} - \dfrac{a}{V}$

At high pressures, the van der Waals' gas equation reduces to $PV = RT + Pb$

or $\dfrac{PV}{RT} = 1 + \dfrac{Pb}{RT}$

$\therefore$ Compressibility factor $Z = \dfrac{PV}{RT} = 1 + \dfrac{Pb}{RT}$

151. (a) $\dfrac{PV}{RT} > 1$ RT constant

$PV > 1$ at 1 atm

$V_m > 22.4$ L

152. (c) $Z = \dfrac{PV}{nRT} = \dfrac{V}{n} = \dfrac{0.9 \times 0.0821 \times 273}{9}$

$= 2.24$ litre/mol

$\therefore$ Volume of 1 milli-mole of gas $= 2.24$ mL

153. (d) $Z = \dfrac{PV}{nRT}$; $1.90 = \dfrac{1 \times 800}{n \times R \times 330}$;

$n = \dfrac{1 \times 800}{1.90 \times R \times 330}$

$Z = 1.10 = \dfrac{V \times 200}{n \times R \times 570}$;

$1.10 = \dfrac{V \times 200 \times 1.90 \times R \times 330}{800 \times R \times 570}$

$V = 4$ L

154. (c) $Z > 1$ real gas shows positive deviation from ideal behaviour when repulsive forces dominate.

155. (d) $\left(0.1 + \dfrac{1000 \times (0.02)^2}{V^2}\right) V = 20 \times 0.02$

$= 0.1\,V^2 - 0.4\,V + 0.4 = 0$

$= V^2 - 4V + 4 = 0$

$\Rightarrow$ $V = 2$ L

$Z = \dfrac{PV}{nRT} = \dfrac{0.1 \times 2}{20 \times 0.02} = 0.5$

156. (c) $T_C = \dfrac{8a}{27\,Rb}$ $P_C = \dfrac{a}{27\,b^2}$ $V_C = 3b$

158. (a) $V = K(t°C + 273)$

1°C $V_1 = K(273 + 1) = K \times 273 + K$

$V = (K \times 273)$ $t = 0°C$

2°C $V_2 = K(273 + 2) = K \times 273 + 2K$

159. (b) It is a temperature above which a gas can not be liquified.

160. (b) $\dfrac{T_C}{P_C} = \dfrac{8a \times 27b^2}{27\,Rb \times a} = \dfrac{8b}{R}$

$V_{\text{Excluded}} = nb$

161. (d) $T_{\text{Liquifaction}} \propto \dfrac{1}{\text{I.M.A.F}} \propto \dfrac{1}{a} \propto \dfrac{1}{T_C}$

Critical temperature is a temperature above which a gas can't be liquify.

162. (a) Because $T > T_i$ and if gas expanded adiabatically then $T \uparrow$

163. (d) At low pressure $Z = 1 + \left(b - \dfrac{a}{RT}\right)\dfrac{1}{V_m}$ at Boyles temperature $b - \dfrac{a}{RT} = 0$

164. (b) Below critical temperature a gas can be liquified. So V_m has three real roots.

165. (a) At critical temperature slope of pressure volume curve is maximum and it is zero.

166. (c) Above critical temperature a gas can't be liquified.

167. (c) At Boyle's temperature and low pressure $Z = 1$.

At low pressure $Z = 1 + \left(b - \dfrac{a}{RT}\right)\dfrac{1}{V_m}$

At $T_B = a/Rb$, $Z = 1$

168. (a) $T_C = \dfrac{8a}{27\,Rb}$, $T_b = \dfrac{a}{Rb}$, $T_i = \dfrac{2a}{Rb}$

$T_i > T_b > T_c$

169. (d) No. liquification of ideal gas.

170. (d) When $P \to 0$ real gases behave as ideal gases so $PV = nRT = 1 \times 0.0821 \times 546 = 44.8$ L atm

171. (a) The van der Waals' equation for 1 mole of gas is

$$\left(P + \frac{a}{V^2}\right)(V - b) = RT$$

if we put, $P_r = \dfrac{P}{P_C}$, $V_r = \dfrac{V_m}{V_C}$ and $T_r = \dfrac{T}{T_C}$;

$\left(\because P_C = \dfrac{a}{27b^2}, V_C = 3b \text{ and } T_C = \dfrac{8a}{27Rb}\right)$

$$\left(P_r\left(\frac{a^2}{27b^2}\right) + \frac{a}{V_r^2\,(3b)^2}\right)(V_r(3b) - b) = RT_r\left(\frac{8a}{27Rb}\right)$$

This equation is called van der Waals' equation of law of corresponding states.

172. (b) $\dfrac{P_r V_r}{T_r} = 2.21$

$\dfrac{P}{P_C} \times \dfrac{V}{V_C} \times \dfrac{T_C}{T} = 2.21$

$\dfrac{PV}{T} \times \dfrac{T_C}{P_C V_C} = 2.21$

$= 0.1243$ L $= 124.31$ mL

Level 2

2. (a) $n = \left(\dfrac{P}{RT}\right)V$

Slope of line $= \dfrac{P}{RT} = \tan\theta$

$\Rightarrow$ $T = \dfrac{P}{R\tan\theta} = \dfrac{16.42}{0.0821\tan\theta}$

For B, $T = 200\sqrt{3}$ K

For A, $T = \dfrac{200}{\sqrt{3}}$ K

3. (b) The number of moles of He, Ne and Xe is given by

$$n_{He} = \frac{PV}{RT} = \frac{3}{RT}; \quad n_{Ne} = \frac{PV}{RT} = \frac{10}{RT};$$

$$n_{Xe} = \frac{PV}{RT} = \frac{1}{RT}$$

$$n = n_{He} + n_{Ne} + n_{Xe} = \frac{(3+10+1)}{RT}$$

The total pressure is given by

$$P = \frac{(n_{He} + n_{Ne} + n_{Xe})RT}{V}$$

$$P_{total} = \frac{(3+10+1)}{RT} \times \frac{RT}{7} = 2 \text{ atm}$$

4. (c) $N_2 + 3H_2 \rightleftharpoons 2NH_3$

Initial moles 11 12 0

at equilibrium 9 6 4

moles of N_2 and H_2 present at equilibrium

$= 15$

after addition of water

$NH_3(g) + H_2O(l) \longrightarrow NH_4OH(l)$

Volume of vessel available for gaseous mixture of N_2 and $H_2 = 20 - 3.58$

$\Rightarrow$ 16.42 L

Pressure exerted by gaseous mixture at

$$300 \text{ K} = \frac{15 \times 0.821 \times 300}{16.42} \Rightarrow 22.5 \text{ atm}$$

5. (c) I II

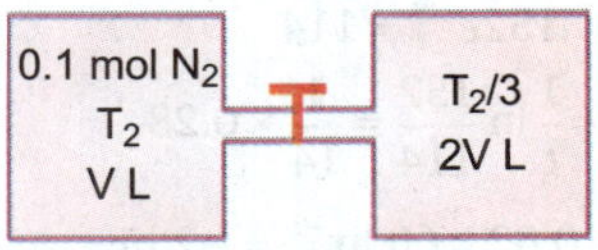

Let x mole of N_2 present into vessel II and P is final pressure of N_2

$P(2V) = xR(T_2/3)$ and $P(V) = (0.1 - x)RT_2$

$$\Rightarrow \quad 2 = \frac{x}{3(0.1-x)}$$

$\Rightarrow \quad x = 0.6/7$ mole;

$$\frac{0.6}{7} \times 28 \Rightarrow 2.4 \text{ g } N_2$$

II has 2.4 g N_2 and I has 0.4 g of N_2;

$$\frac{W_I}{W_{II}} = \frac{0.4}{2.4} \Rightarrow 1:6$$

6. (b) Let initial mixture contains n_1 and n_2 moles of NH_3 and N_2H_4 respectively

Total moles of gases originally present

$= n_1 + n_2$

Total moles of gases after decomposition of gases $= 2n_1 + 3n_2$

$0.5 \times V = (n_1 + n_2) R \times 300$

$4.5 \times V = (2n_1 + 3n_2) R \times 1200$

$$\frac{2n_1 + 3n_2}{n_1 + n_2} = \frac{9}{4}$$

$$\frac{n_1}{n_2} = \frac{1}{3}$$

$$\frac{n_2}{n_1 + n_2} \times 100 = 25\%$$

8. (b) Let $T_1 > T_2$; final pressure will be same, Let x mole transfer from A to B vessel.

$\therefore \quad P_A V = (n - x)RT$

and $\quad P_A V = (n + x)RT_2$

$$\therefore \quad x = \frac{n(T_1 - T_2)}{T_1 + T_2};$$

finally $P_1 \times 2V = 2nRT_1; V = \dfrac{nRT_1}{P_1}$

$$\therefore \quad P_A \times \frac{nRT_1}{P_1} = \left(n - \frac{n(T_1 - T_2)}{(T_1 + T_2)}\right)RT_1$$

$$P_A = \frac{2P_1T_2}{T_1 + T_2}$$

9. (c) $P \propto d$; $P = kd$ and $k = \dfrac{1 \text{ atm}}{1 \text{ metre}}$

$$PV = nRT; \; kd\left(\frac{1}{6}\pi d^3\right) = nRT;$$

$$\frac{d_1^4}{d_2^4} = \frac{n_1}{n_2}; \; \frac{1}{4^4} = n_1/n_2; \; n_2 = 256$$

no. of moles added $= 256 - 1 = 255$

10. (a) $PM = dRT$

$$Pd = P^2\left(\frac{M}{RT}\right)$$

$$\frac{d(Pd)}{dP} = \frac{2PM}{RT} = 5$$

$$\frac{0.2 \times 8.21 \times 4}{0.0821 \times T} = 5$$

$$T = 160 \text{ K}$$

11. (d) % Relative humidity

$$= \frac{\text{Partial pressure of } H_2O}{\text{Vapour pressure of } H_2O} \times 100$$

$$75 = \frac{P_{H_2O}}{30} \times 100$$

$P_{H_2O} = 22.5$ torr;

% of H_2O vapour in air

$$= \frac{(22.5)}{760} \times 100 = 2.96$$

molar mass of wet air

$$= \frac{29 \times 97.04 + 2.96 \times 18}{100}$$

$$= \frac{2814.16 + 53.28}{100} = 28.67$$

density of wet air $= \frac{PM}{RT} = \frac{1 \times 28.67}{0.0821 \times 300}$

$= 1.164\ g/L$

12. (b) C_v of gas $A = \frac{3}{2}R + \frac{3}{2}R + 6R = 9R$

Let C_v of gas $B = X$

Heat lost by gas $A = 7 \times 9R \times \left(T - \frac{5T}{6}\right)$

Heat gained by gas $B = 6X\left(\frac{5T}{6} - \frac{T}{3}\right)$

$\Rightarrow \quad 63R\left(\frac{T}{6}\right) = 6X\left(\frac{T}{2}\right) = X = \frac{7}{2}R;$

$\therefore$ gas is diatomic

13. (a) Total heat received by gas = heat provided

$2 \times \frac{3}{2}R(T_f - 300) + 2 \times 3R(T_f - 400)$

$+ 3 \times \frac{5}{2}R(T_f - 500) = 15600$

$T_f = 900\ K$

$P_f = \frac{900 \times 0.0821 \times 7}{7}$

$P_f = 73.89$ atm

14. (c) $1000 = \sqrt{\frac{3RT}{M}}$; $10^3 = \left[\frac{3 \times 8.314 \times T}{10^{-26} \times 6 \times 10^{23}}\right]^{1/2}$

$T = \frac{10^6 \times 10^{-26} \times 6 \times 10^{23}}{3 \times 8} = 250\ K$

15. (b) Mass of balloon $= 100$ kg $= 10 \times 10^4$ g

Volume of balloon $= \frac{4}{3}\pi r^3$

$= \frac{4}{3} \times \frac{22}{7} \times \left(\frac{21}{2}\right)^3$

$= 4851\ m^3 = 4851 \times 10^3\ L$

Mass of gas (He) in balloon $= \frac{PVM}{RT}$

$= \frac{1 \times 4851 \times 10^3 \times 4}{0.082 \times 300}$

$= 78.878 \times 10^4$

$\therefore$ Total mass of gas and balloon

$= 78.878 \times 10^4 + 10 \times 10^4$

$= 88.878 \times 10^4$ g or 888.78 kg

Mass of air displaced $= 1.2 \times 4851$

$= 5821.2$ kg

$\therefore$ Pay load = mass of air displaced

− mass of (balloon + gas)

$\therefore$ Pay load $= 5821.2 - 888.78 = 4932.42$ kg

16. (d) Let V mL of gas effused

$\frac{V/220}{V/200} = \sqrt{\frac{d_{O_2}}{d_{mix}}} \Rightarrow d_{mix} 1.6 \times (1.1)^2 = 1.936\ g/L$

Let density of ozone is d; In 100 volume ozonised oxygen, 60% O_2 and 40% by volume O_3 is present

$\therefore$ Mass of mixture = mass of ozone

+ mass of oxygen

$100 \times 1.936 = 40 \times d + 60 \times 1.6$

density of O_3 is 2.44 g/L

17. (d) $n_{N_2} = \frac{\left(\frac{708.2}{760} \times 0.25\right)}{0.0821 \times 303} = 9.36 \times 10^{-3}$

$n_{O_2} = \frac{\left(\frac{756.2}{760}\right) \times 0.3}{(0.0821) \times 298}$

$= 0.0122$

n_{total} moles $= 0.02156$

pressure in final vessel $= P$

$\frac{(n_{total})RT}{V} = \frac{0.02156 \times 00821 \times 308}{0.5}$

$P = 1.09$ atm or 828.4 torr

$P_{total} = P_{(O_2 + N_2)} +$ V.pr. of H_2O

$= 828.4 + 42.2 = 870.6$ torr

18. (b) $P = P_0 e^{-kt}$; $P_0 = \frac{228}{3} + 76 = 152$ cm Hg

At $t = 14$ min; $P = \frac{114}{3} + 76 = 114$ cm Hg

$\Rightarrow \quad 152 e^{-kt} = 114$

$\Rightarrow \quad k = \frac{1}{t} \ln \frac{152}{114} = \frac{1}{14} \times 0.28$

$= 0.02 \times 60\ hr^{-1} = 1.2\ hr^{-1}$

19. (d) $P_{H_2O} = 760 - 722 = \frac{38}{760}$ atm; $n_{H_2O} = \frac{0.9}{18}$

$V = \frac{0.9}{18} \times \frac{0.821 \times 300}{38} \times 760$; $V = 24.63$ L

20. (c) $n_{AB} = \frac{3 \times 8.21}{0.0821 \times 300} = 1$; $n_{B_2} = \frac{1 \times 8.21}{0.0821 \times 400} = 0.25$

$2AB(g) + B_2(g) \longrightarrow 2AB_2(s)$

	$2AB(g)$	$B_2(g)$
Initial moles	1	0.25
After reaction	0.5	0

$P_{AB} \times (8.21 + 8.21) = 0.5 \times 0.0821 \times 250$

$P_{AB} = 0.625$ atm

21. (c) Moles of O_2 remaining after reaction $= 4 - 1 = 3$

$V \propto n$ is due to same T and P in each compartment after reaction.

$\therefore$ Length of He compartment

$\frac{5}{5+3} \times 500 = 312.5$ cm

24. (c) $Z = \frac{PV_m}{RT} = \frac{V_m}{V_m - b} = \frac{11b}{11-b} = \frac{11}{10} = 1.10$

25. (b) $PV_m = RT_B$

$PV_m = R\frac{a}{Rb}$

$V_m = \frac{a}{bP}$

26. (b) $V_c = \frac{60}{0.80} = 75 \text{ cm}^3 \text{ mol}^{-1}$;

$b = \frac{V_c}{3} = 25 \text{ cm}^3 \text{ mol}^{-1} = 0.025 \text{ L mol}^{-1}$

$\therefore \quad T_c = \frac{8a}{27Rb}$

$\frac{4 \times 10^5}{821} = \frac{8 \times a}{27 \times 0.0821 \times 0.025}$

$\Rightarrow \quad a = 3.375$

27. (b) $b = 4 \times \frac{4}{3} \times \pi r^3 \times N_A$;

$4 \times \pi \times 10^{-4} \times 1000$

$= 4 \times \frac{4}{3} \times \pi \times r^3 \times 6 \times 10^{23}$

$r = 5 \times 10^{-9}$ cm

Distance of closest approach $= 2r$

$= 10^{-8}$ cm or 10^{-10} m

28. (c) $\frac{r_x}{r_{O_2}} = \sqrt{\frac{M_{O_2}}{M_x}} = \left(\frac{4}{5}\right)^2 = \frac{32}{M_x} = M_x = 50$

$d_x = 0.80 \text{ kg/m}^3$,

$V_m = \frac{1000}{800} \times 50 = 62.5$ L

$Z = \frac{PV_m}{RT} = \frac{1 \times 62.5}{0.0821 \times 500} = 1.52$

29. (a) $b = 25 \text{ cm}^3/\text{mol}$

$b = 0.025$ L/mol

$V_{Real} = V_i - nb = 50 - 10 \times 0.025 = 49.75$ L

30. (b) $Z = 1 + \frac{Pb}{RT}$ at high pressure

slope $= \frac{b}{RT} = \frac{\pi}{492.6}$

$\Rightarrow b = \frac{\pi}{492.6} \times 0.0821 \times 300$ and

$b = \frac{4}{3}\pi r^3 \times 4 N_A$

$= \frac{\pi}{492.6} \times 0.0821 \times 300 \times 10^{-3}$

$\Rightarrow \quad r = 2.5$ Å or $d = 5$ Å

Level 3

Passage-1

1. (d) $\left(p + \frac{a}{V_m^2}\right)(V_m - b) = RT$

$P\uparrow\uparrow \Rightarrow V_m \downarrow\downarrow \qquad P + \frac{a}{V_m^2} \approx p$

$P(V_m - b) = RT$

$PV_m = RT + Pb$

2. (a) $V\uparrow\uparrow \Rightarrow P\downarrow\downarrow$

$V - nb \approx V$

$\left(P + \frac{n^2a}{V^2}\right)V = nRT$

$PV + \frac{n^2a}{V} = nRT$

$PV = nRT - \frac{n^2a}{V}$

3. (c) van der Waal's constant 'a' $\propto$ I.M.A.F.

So CO_2 gas can be more easily liquified.

4. (d) $\left(P + \frac{n^2a}{V^2}\right)(V - nb) = nRT$

Passage-2

2 & 3. Let σ_A, σ_B and σ_C be collision diameters of A, B and C

$\lambda = \frac{kT}{\sqrt{2}\pi\sigma^2 P}$, i.e., $\sigma \propto \sqrt{\frac{T}{\lambda P}}$

$\therefore \quad \sigma_A : \sigma_B : \sigma_C$

$= \sqrt{\frac{1600}{0.16 \times 1}} : \sqrt{\frac{200}{0.16 \times 2}} : \sqrt{\frac{400}{0.04 \times 4}}$

$= 4 : 1 : 2$

2. (c) $Z_{11} \propto \frac{\sigma^2 P^2}{\sqrt{m}T^{3/2}}$

$\Rightarrow A : B : C = \frac{16 \times 1}{\sqrt{20} \times (1600)^{3/2}} :$

$\frac{1 \times 4}{\sqrt{40} \times (200)^{3/2}} : \frac{4 \times 16}{\sqrt{80} \times (400)^{3/2}}$

$= 1 : 4 : 16$

3. (a) $Z_1 \propto \frac{(\sigma^2)P}{\sqrt{MT}}$

$\Rightarrow A : B : C = \frac{0.6 \times 1}{\sqrt{20 \times 1600}} : \frac{1 \times 2}{\sqrt{40 \times 200}}$

$: \frac{4 \times 4}{\sqrt{80 \times 400}}$

$= 4 : 1 : 4$

Passage-3

1. (a) $h_L d_L = h_{Hg} d_{Hg}$

$$h_L = \frac{76 \times 13.6}{5.44} = 190 \text{ cm}$$

2. (d) $P_{Gas} = P_{Atm} + P_L = 1 + \frac{38}{190} = 1.2$

3. (b)

	$3A(g)$	$\rightleftharpoons A_3g$
$t=0$	1.2 atm	$A_3(g)$
$t = t_{eq.}$	$1.2-0.36$	$\frac{1}{3}(0.36) = 0.12$ atm

$\therefore$ $P_T = 1.2 - 0.36 + 0.12 = 0.96$ atm

$\therefore$ Pressure difference in column

$= 1 - 0.96 = 0.04$ atm

$\therefore$ The difference in height of the liquid level in two columns $= 0.04 \times 190 = 7.6$ cm

One or More Answers is/are Correct

1. (a, b, c, d)

For Boyle's law

$T =$ constant

$PV = K =$ constant

2. (a, b)

$PV = nRT$

$T =$ constant

$PV =$ constant

$V\uparrow \Rightarrow P\downarrow$

KE is a function of temperature.

3. (b, c, d)

T_c is a temperature above which a gas can not be liquified at this point, slope of $P-V$ curve is maximum and attraction force is dominating at this point. Both liquid vapour are exist in equilibrium.

4. (c, d)

at constant volume	at constant P
$P \propto T$	$V \propto T$
$P = KT$	$V = KT$
$P = K(t°C + 273.15)$	$V = K(t°C + 273.15)$

5. (a)

T_B is a temperature at which gas behave like an ideal gas mean attractive and repulsing forces just offset each other.

6. (a, c, d)

$$KE = \frac{3}{2} RT \text{ (1 mole)} U_{rms} = \sqrt{\frac{3RT}{M_w}}$$

$$d = \frac{PM_w}{RT}$$

7. (a, b)

Refer. Q.3

8. (a, b, c, d)

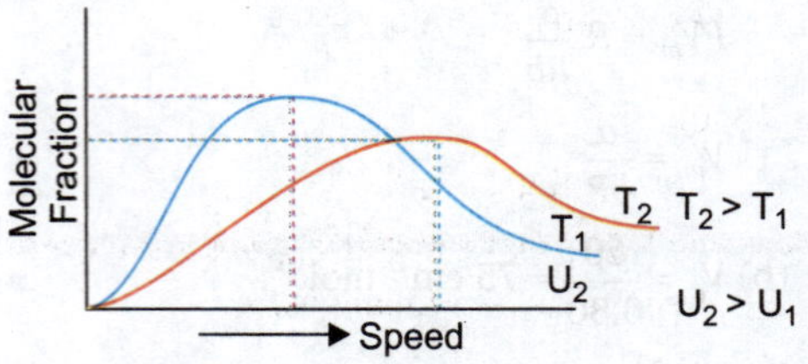

9. (a) Refer Q.2

10. (a, b, c)

$PV = nRT$

R is constant and independent of P and T

$R = K \times N_A$

$K =$ Boltzmann's constant

11. (a,b,d)

(A) $\because$ area under the curve gives fraction of molecules and total area is constant.

(B) u_{mp} decreases with decrease in temperature.

(C) T_2 is higher temperature

(D) As seen from graph; $\therefore$ *A, B, D*

12. (a, d)

Vapour pressure of liquid is only and only temperature dependent.

13. (a, b, d)

For H_2 gas IMAF ≈ 0

$Z > 1$ $\quad Z = 1 + \frac{Pb}{RT}$

at low pressure $Z < 1$ $\quad Z = 1 - \frac{a}{RTV_m}$

Real gas behave like an ideal gas at low pressure and high temperature.

14. (a, b, c)

$T_C = \frac{8a}{27Rb}$ Refer. Q.14

15. (a,b,d)

$P_T = (1 + 3x) = 1 + 3 \times 0.1 = 1.3$ atm.

$\Delta P = 0.2$ atm or 76×0.3 cm of Hg

or 760×0.3 mm of Hg

16. (a, b, c, d)

Refer. Q. No. 13

17. (a, d)

Critical pressure is the minimum pressure at which substance convert into liquid state at critical point.

18. (a,c)

Low P and high temperature gas behaves as an ideal gas.

$\therefore$ $PV = \text{constant}$ and $\dfrac{PV_m}{RT} = 1$.

19. (b, c)

For ideal gas $Z = 1$

at low pressure and Boyle's temperature real gas behave like an ideal gas.

20. (a, b, c)

$V \propto T$ at constant P

$KE = \dfrac{3}{2}RT$ (1 mole)

■ Match the Column

5. $y = \dfrac{1}{V^2}$ or $\sqrt{y} = \dfrac{1}{V}$

$P = x$ and $P = \dfrac{\text{constant}}{V}$

(A) $x = (k)\sqrt{y} \Rightarrow y = k^1 x^2$

(B) $V = kT;\ \ y = V \,\&\, \dfrac{1}{T} = x \ \ \therefore y = \dfrac{k}{x}$

(C) $P = kT;\ \ PT = kT^2$ or $y = kx$

(D) $v = \dfrac{c}{p} \Rightarrow y = c\sqrt{x};\ y^2 = cx$

Level 4

■ Subjective Problems

1. Refer Q.No. 59 Level-1

2. For H_2 and He.

$$f^{4th} = \frac{x/y}{1} = \left(\sqrt{\frac{M_{He}}{M_{H_2}}}\right)^4$$

$$\frac{x}{y} = (\sqrt{2})^4 = 4$$

3. Equation of line

$$P - 2 = \frac{10-2}{4-15}(V - 15)$$

$$P - 2 = -\frac{8}{11}(V - 15)$$

$$P = 2 - \frac{8V}{11} + \frac{15 \times 8}{11}$$

$$P = \left(\frac{142}{11} - \frac{8V}{11}\right)$$

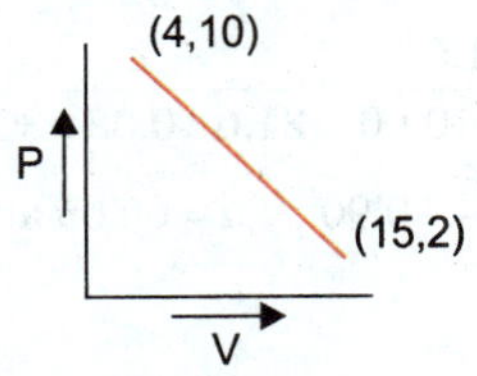

$$f(T) = \frac{1}{nR}\left(\frac{142V}{11} - \frac{8V^2}{11}\right)$$

$$\frac{d(F(T))}{dV} = \frac{1}{nR}\left(\frac{142}{11} - \frac{8V^2}{11}\right) = 0$$

$$V = \frac{142}{11 \times 16} = 8.875$$

$$P = \frac{142}{11} - \frac{8}{11} \times \frac{8.875}{16} = \frac{71}{11}$$

$$T_{Max} = \frac{PV}{nR} = \frac{71}{11} \times \frac{142}{16 \times 0.0821} = 700$$

$$= 7 \times 10^2\,\text{K}$$

4. Initial wt. of butane $= 29 - 14.8 = 14.2$ kg

Final wt. of butane $= 23 - 14.8 = 8.2$ kg

$$\frac{P_1}{n_1} = \frac{P_2}{n_2} \text{ or } \frac{P_1}{w_1} = \frac{P_2}{w_2}$$

$$\frac{3.47}{14.2} = \frac{P_2}{8.2} \ ; \ P_2 = 2.00 \text{ atm}$$

5. $P + \dfrac{n^2 a}{V^2} = P_i$

$$\frac{n^2 a}{V^2} = 0.04$$

$$a = 4$$

6. $Z = 1 + \dfrac{Pb}{RT}$ high pressure

$$\frac{dZ}{dP} = \frac{b}{RT} = \frac{1}{2.8}$$

$$b = \frac{RT}{2.8} = \frac{22.4}{2.8} = 4 \times \left(N_A \times \frac{4}{3}\pi R^3\right)$$

$$\left(N_A \times \frac{4}{3}\pi R^3\right) = \text{Volume of 1 mole gas}$$

$$= \frac{5.6}{2.8} = 2$$

7. $P = \dfrac{dRT}{M_w};\ \ \dfrac{p_x}{p_y} = \dfrac{d_x}{d_y} \times \dfrac{M_y}{M_x} = 6$

8. $\dfrac{r_x}{r_{O_2}} = \sqrt{\dfrac{M_{O_2}}{M_x}}$

$$\frac{V}{2} \times \frac{5.65}{V} = \sqrt{\frac{32}{M}}$$

$$8 = \frac{32}{M}$$

$$M = 4$$

9. $n_{Br_2} = \dfrac{1 \times 0.423}{0.0821 \times 423} = \dfrac{1}{82.1}$ mole

POAC on 'Br'

$$\frac{1}{82.1} \times 2 = \frac{4.2}{80 + 19n} \times 1$$

$$n \approx 5$$

10. $Z = 1 + \frac{Pb}{RT}$; $\left(\frac{\partial Z}{\partial P}\right)_T = \frac{b}{RT}$; $\therefore x = 1.00$

11. $P_f = 1 + \frac{36}{76} = \frac{112}{76}$ atm. Final height = 19 cm

$P_i = 1$ atm, initial length = h_i cm

$\therefore$ Boyle's law $\quad P_iV_i = P_fV_f$

$$1 \times h_iA = \frac{112}{76} \times 19A$$

$$h_i = 28 \text{ cm}$$

$\therefore$ The length by which the Hg column shifts down $= h_i - h_f$

$$= 28 - 19 = 9 \text{ cm}$$

12. $P_i = 722$ mm $\quad P_f = 760$ mm

$T_i = 107 + 273 = 380$ K $\quad T_f = 100$ K

$d_i = 1 \text{ g/cm}^3 \quad d_f = ?$

$$\frac{d_iT_iR}{P_i} = \frac{d_fT_fR}{P_f}$$

$$\Rightarrow d_f = \left(\frac{P_f}{P_i}\right)\left(\frac{T_i}{T_f}\right)d_i = \left(\frac{760}{722}\right)\left(\frac{380}{100}\right) \times 1 \text{ g/cm}^3$$

$$= 4 \text{ g/cm}^3$$

13. V, n constant.

$$\frac{P_i}{T_i} = \frac{P_f}{T_f} \Rightarrow P_f = \frac{T_f}{T_i}P_i = \left(\frac{109}{100}\right)P_i$$

$$\Rightarrow P_{\text{increases}}\ \Delta P = P_f - P_i = \frac{9}{100}P_i$$

$\therefore$ % Pressure increases

$$= \frac{\Delta P_f}{P_i} \times 100 = \frac{9P_i}{100P_i} \times 100\%$$

$$X\% = 9\%$$

$$X = 9$$

14. $T_B = \frac{a}{Rb} = \frac{3.6}{0.08 \times 0.6} = 75 \text{ K}, \frac{75}{15} = 5 \text{ K}$

15. 3 ms^{-1}

$$u_{\text{rms}} = \sqrt{\frac{u_1^2N_1 + u_2^2N_2}{N_1 + N_2}} \Rightarrow u_{\text{rms}}^2 = \frac{u_1^2N_1 + u_2^2N_2}{N_1 + N_2}$$

$$25 = \frac{4 \times 7^2 + u_2^2}{10} \Rightarrow u_2 = \sqrt{\frac{54}{6}} = 3 \text{ ms}^{-1}$$

Numerical Problems

1. $C_2H_6(g) + \frac{7}{2}O_2(g) \longrightarrow 2CO_2(g) + 3H_2O(g)$

1 mole $\quad$ 3.5 $\quad$ 2 $\quad$ 3

$$\frac{P_1}{n_1T_1} = \frac{P_2}{n_2T_2} ; \quad \frac{1}{4.5 \times 300} = \frac{P_2}{5 \times 400} ; P_2 = 1.48 \text{ atm}$$

2. $\frac{P_1}{T_1} = \frac{P_2}{T_2}$; $\quad \frac{300 \times 10^3}{300} = \frac{1.2 \times 10^6}{T_2}$;

$$T_2 = 1200 \text{ K} = 927°\text{C}$$

3. $1 = P_{\text{Gas}} + \frac{19 \times 6.8}{13.6 \times 76} + \frac{19}{760}$

$$P_{\text{Gas}} = 0.85 \text{ atm}$$

4. $P = P_{CO_2} + P_{\text{Air}} = \frac{2 \times 0.0821 \times 300}{2} + 1$

$$= 25.13 \text{ atm}$$

5. $\frac{n_1}{20} \times \frac{40}{n_2} = \sqrt{\frac{64}{16}}$; $\quad \frac{0.02 \times 2}{n_2} = 2$; $n_2 = 0.02$ mole

$$P_{SO_2} = \frac{nRT}{V} = \frac{0.02 \times 24.63}{2.463} = 0.2 \text{ atm}$$

$$= 0.2 \times 760 = 152 \text{ mm of Hg}$$

$$P_T = 152 + 28 = 180 \text{ torr}$$

6. $\frac{r_{CH_4}}{r_{He}} = \frac{n_{CH_4}}{n_{He}}\sqrt{\frac{M_{He}}{M_{CH_4}}}$

$$\frac{2}{1} = \frac{n_{CH_4}}{n_{He}}\sqrt{\frac{4}{16}} = \frac{n_{CH_4}}{n_{He}} \times \frac{1}{2}$$

$$\frac{n_{CH_4}}{n_{He}} = 4$$

$\therefore$ Mole fraction of $CH_4 = \frac{4}{5} = 0.8$

7. The velocity corresponding to the maxima is the most probable velocity which is given by the expression.

$$C_{mp} = \sqrt{\frac{2RT}{M}}$$

$$RT = \frac{C_{mp}^2M}{2} = \frac{(200)^2 \times 50 \times 10^{-3}}{2}$$

$$= 40000 \times 25 \times 10^{-3} \text{ J mol}^{-1} = 40 \times 25 \text{ J mol}^{-1}$$

No. of moles $= \frac{300}{50} = 6$;

$$E = \frac{3}{2}nRT = \frac{3}{2} \times 6 \times 40 \times 25 = 9 \times 10^3 \text{ J} = 9 \text{ kJ}$$

8. $u_{mp} = \sqrt{\frac{2RT}{M}} = \sqrt{\frac{2P}{d}} = \sqrt{\frac{2 \times 20 \times 10^5}{4}}$

$$= 1000 \text{ ms}^{-1} = 1 \text{ km/sec}$$

$(d = 0.4 \text{ g/dm}^3 = 0.4 \text{ kg/m}^3)$

9. $n_T = n_1 + n_2$

$$\frac{910 \times (10 + V)}{RT} = \frac{950 \times V}{RT} + \frac{900 \times 10}{RT}$$

$$V = 2.5 \text{ L}$$

10. $Z = \left(\frac{V_m}{V_m - b}\right) - \frac{a}{V_mRT}$

$$0.90 = \frac{21.6}{21.6 - 0.60} - \frac{a}{21.6 \times 0.08 \times 300}$$

$$\frac{a}{518.4} = \frac{21.6}{21} - 0.90 ; \quad a = 66.65 \text{ atm L}^2/\text{mol}^2$$

11. $\dfrac{T_C}{P_C} = \dfrac{8b}{R}$

$\dfrac{160}{82.1} \times \dfrac{0.0821}{8} = b$; $b = 0.02$ L/mol

$V_C = 3b = 0.06$ L/mol

12. $P_C = \dfrac{a}{27b^2}$; $T_C = \dfrac{8a}{27\,Rb}$

$\dfrac{200}{500} = \dfrac{R}{8b}$; $b = \dfrac{2.5\,R}{8} = \dfrac{2.5 \times 0.08}{8} = 0.025$

$b = 4 \times$ volume of 1 mole gas molecules

Volume of 40 moles gas molecules $= \dfrac{b}{4} \times 40$

$= \dfrac{0.025}{4} \times 40 = 0.25$ L

13. $\dfrac{dZ}{dP} = 0$, $0.25 = \dfrac{100}{T}$, $T = 400$ K

14. $T_B = \dfrac{a}{Rb} = \dfrac{4.105}{0.0821 \times 0.2} = 250$ K ;

So this gas behave ideally at 250 K.

Density of gas, $d = \dfrac{PM}{RT}$;

$d = \dfrac{5 \times 82.1}{0.0821 \times 250} = 20$ g/L

15. Second virial coefficients, $B = b - \dfrac{a}{RT} = 0.004$

Third virial coefficients, C

$= b^2 = 0.0036 \Rightarrow b = 0.06$

at $T = 300$ K, $0.06 - \dfrac{a \times 12}{300} = 0.004 \Rightarrow a = 1.4$

$T_C = \dfrac{8}{27} \times \dfrac{a}{Rb} = \dfrac{8}{27} \times \dfrac{1.4 \times 12}{0.06} = 82.96$ K

CHAPTER 4

Thermodynamics

- The subject of **Thermodynamics** deals basically with the interaction of one body with another in terms of quantities of heat & work.
- A **System** is defined as that macroscopic part of the universe which is at the moment under investigation.
- **Surroundings** are the rest of the universe outside the system.
- **Boundary :** The actual or imaginary surface that separates the system from the surroundings.

System

- **Open System** can exchange matter as well as energy with surroundings
- **Closed System** can exchange energy but not matter with surroundings
- **Isolated System** cannot exchange matter as well as energy with surroundings

Thermodynamic Properties

- **Intensive Properties** do not depend on the amount of matter present in system
  Examples : Pressure, Temperature, Density, Melting point, Boiling point etc.
- **Extensive Properties** depend on the amount of matter present in the system.
  Examples : Mass, Volume, Heat Capacity, Enthalpy, Entropy etc.

Thermodynamic Process : The change of a system from one state to another state.

Thermodynamic Processes

Isothermal	Isobaric	Isochoric	Adiabatic	Polytropic
temperature remains constant throughout the process *i.e.*, : $dT=0$	pressure remains constant throughout the process *i.e.*, : $dP=0$	volume remains constant throughout the process *i.e.*, : $dV=0$	system does not exchange heat with surrounding *i.e.*, : $dq=0$	PV^n = constant

- **Cyclic Process :** System undergoes a number of different processes and finally returns to its initial state.
- **REVERSIBLE PROCESS (QUASI-STATIC) & IRREVERSIBLE PROCESS :**
  A process which is carried out so slowly that the system and the surroundings are nearly in equilibrium is known as a **reversible Process.** If this condition does not hold good, the process is said to be, **irreversible.**
  In a reversible process the driving force is infinitesimally larger than the opposing force. If the driving force is made infinitely smaller than opposing force, the system can be brought back without producing any permanent change.
- **All natural processes are irreversible process.**
  Spontaneous Process : Proceeds on its own *i.e.*, without any external help.
- **INTERNAL ENERGY (*U*) :** The sum of all microscopic forms of energy.
  $U = U_{\text{Translational}} + U_{\text{Rotational}} + U_{\text{Vibrational}} + U_{\text{Bonding}} + U_{\text{Electronic}} + \ldots\ldots$
- U is a state function & an extensive property.
- $\Delta U = U_{\text{Final}} - U_{\text{Initial}}$; $\Delta U = q_v$
- $dU = \left(\frac{\partial U}{\partial T}\right)_v dT + \left(\frac{\partial U}{\partial V}\right)_T dV$, $dU = nC_{Vm}(T_2 - T_1)$ at constant volume

Enthalpy

Chemical reactions are generally carried out at constant pressure (atmospheric pressure) so it has been found useful to define a new state function called Enthalpy (H) as :

- $H = U + PV$; $\Delta H = \Delta U + \Delta(PV)$; at constant pressure $\Delta H = \Delta U + P\Delta V$; combining with first law. $\Delta H = q_p$
- The difference between ΔH & ΔU becomes significant only when gases are involved (insignificant in solids and liquids)

$$\Delta H = \Delta U + \Delta(PV) = \Delta U + nR\,\Delta T$$
$$\Delta H = \Delta U + (\Delta n_g)RT$$

- For a given system consider; $H = f(T, P)$; $dH = \left(\frac{\partial H}{\partial T}\right)_P dT + \left(\frac{\partial H}{\partial P}\right)_T dP$
- For isobaric process : $dP = 0$; $dH = \left(\frac{\partial H}{\partial T}\right)_P dT$; $dH = C_P.dT$; $\Delta H = \int C_P.dT$
- For an ideal gas $H = f(T)$ only
  $\left(\frac{\partial H}{\partial P}\right)_T = 0$ or $dH = C_P.dT$; $\Delta H = \int C_P.dT$; for n moles of ideal gas $\Delta H = \int nC_P\,dT$

Kirchoff's Equation

- For chemical reactions $\int d\Delta_r H = \int \Delta_r C_p.dT$

- $\Delta_r H_{T_2} - \Delta_r H_{T_1} = \Delta_r C_p (T_2 - T_1)$
- Similarly $\Delta_r U_{T_2} - \Delta_r U_{T_1} = \Delta_r C_v (T_2 - T_1)$

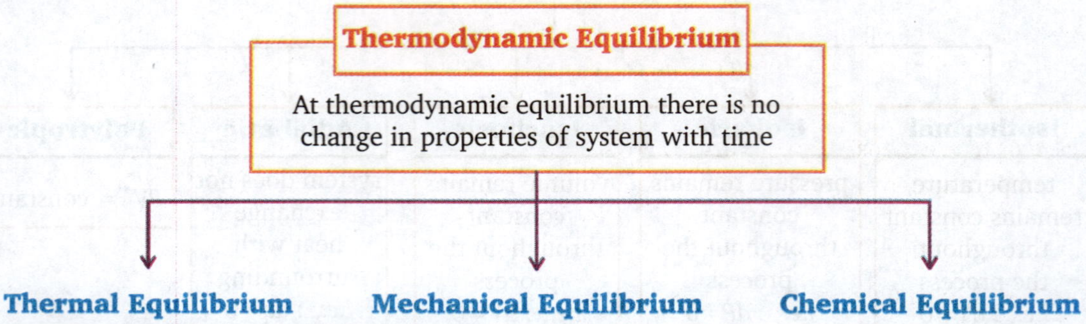

- **ZEROTH LAW OF THERMODYNAMICS**
  Two systems in thermal equilibrium with a third system, are also in thermal equilibrium with each other.
- **FIRST LAW OF THERMODYNAMICS**
  It is law of conservation of energy. Mathematically for **closed system** at rest in absence of external fields this law is written as :

$$\Delta U = q + w$$

Where ΔU is change in internal energy of the system, q is the transfer of heat from surroundings to the system and w is the work involved (either done on the system or by the system).

- **Sign convention according to IUPAC :**

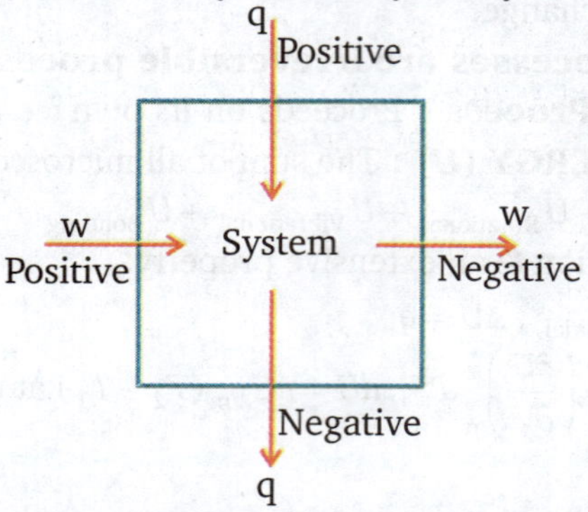

- **HEAT :** The energy that flow into or out of a system because of a difference in temperature between the system and its surrounding.
- **TYPES OF WORK :** Two types of work normally come across in chemistry.
  1. Mechanical work (*e.g.*, *PV* work) **2.** Electrical work (Non *PV* work)
- *Mechanical work* is involved when a system changes its volume in the presence of an externally applied pressure (*i.e.*, pressure volume work). It is especially important in system containing gases

$$dw = -P_{ext} \cdot dV \; ; \quad w = -\int_{V_1}^{V_2} P_{ext.} \, dV$$

- **CALCULATION OF WORK DONE IN VARIOUS PROCESS**
- **Isochoric process:**
  Since $dV = 0$ so $w = 0$
  from 1st law $\Delta U = q_v + w = q_v$
- **Isobaric process:**
  $w = -P_{ext}(V_2 - V_1)$
  & $\Delta H = q_p$
- **Isothermal process:**
  Since $dT = 0$ so $dU = 0$ for an ideal gas
  from I[st] law $q + w = 0$ or $q = -w$

If process is reversible

$$w = -nRT\ln\left(\frac{V_2}{V_1}\right) = -nRT\ln\left(\frac{P_1}{P_2}\right) \quad \text{or} \quad w = -2.303\,nRT\log\frac{V_2}{V_1} = -2.303\,nRT\log\frac{P_1}{P_2}$$

If process is irreversible

$$w = -P_{ext}\left(\frac{nRT}{P_2} - \frac{nRT}{P_1}\right)$$

In case of free expansion $p_{ext} = 0$ so $w = 0$

- **Adiabatic process :** $q = 0$

$$\Delta U = w = nC_v(T_2 - T_1) = \frac{P_2V_2 - P_1V_1}{\gamma - 1}, \text{ where } \gamma = \frac{C_P}{C_V}$$

If process is reversible $T_1V_1^{\gamma-1} = T_2V_2^{\gamma-1}$

If process is irreversible $\frac{T_2 - T_1}{\gamma - 1} = -P_{ext}\cdot\left(\frac{T_2}{P_2} - \frac{T_1}{P_1}\right)$

- **Cyclic process:**

Work done = Area enclosed in PV-diagram
For clockwise it is –ve
For anti-clockwise it is +ve

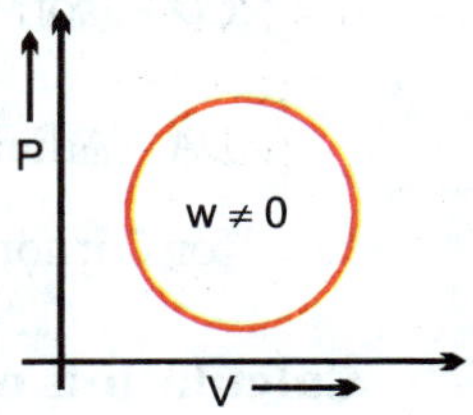

- Electrical work $(w = V.I.t)$ is a type of non PV-work.

Heat Capacity : Between any two temperatures, the quantity of heat required to raise the temperature of the system from the lower to the higher temperature divided by the temperature difference.

- $C = dq/dT$.
- $q_v = \Delta U$
- $C_v = \left(\frac{q}{T_2 - T_1}\right)_v = \left(\frac{\Delta U}{T_2 - T_1}\right)_v$ or $C_v = (\partial U/\partial T)_v$
- $C_P = \left(\frac{\partial H}{\partial T}\right)_P$
- $C_{p,m} - C_{v,m} = R$ (for ideal gas)
- $C_{p,m}$ and $C_{v,m}$ depends on temperature even for an ideal gas. $(C = a + bT + cT^2)$
- $\gamma_{mix} = \frac{n_1C_{P_1} + n_2C_{P_2} + \ldots\ldots}{n_1C_{v_1} + n_2C_{v_2} + \ldots\ldots}$

Reversible Adiabatic process involving ideal gas

- $T_1V_1^{\gamma-1} = T_2V_2^{\gamma-1}$
- $TV^{\gamma-1} = \text{constant}$
- $PV^{\gamma} = \text{constant}$
- $T^{\gamma}P^{1-\gamma} = \text{constant}$
- **Molar Specific heat (C_m) of ideal gas in any Reversible Polytropic Process :**

$$C_m = C_{v,m} + \frac{R}{1-x} = \frac{R}{\gamma - 1} + \frac{R}{1-x}$$

Second Law of Thermodynamics

Statements :

(i) No cyclic engine is possible which take heat from one single source and in a cycle completely convert it into work without producing any change in surroundings.

(ii) In an irreversible process, entropy of universe increases but it remains constant : in a reversible process

$\Delta S_{sys} + \Delta S_{surr} = 0$ for reversible process

$\Delta S_{sys} + \Delta S_{surr} > 0$ for irreversible process

$\Delta S_{sys} + \Delta S_{surr} \geq 0$ (In general)

(iii) Efficiency of Carnot engine working reversibly is maximum.

Carnot cycle

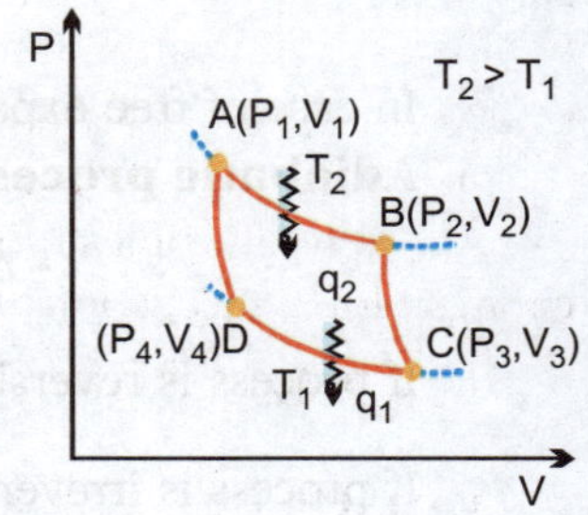

AB – Isothermal Reversible Expansion $w_{AB} = -nRT_2 \ln \dfrac{V_2}{V_1}$

BC – Adiabatic Reversible Expansion $w_{BC} = C_V(T_1 - T_2)$

CD – Isothermal Reversible Expansion $w_{CD} = -nRT_1 \ln\left(\dfrac{V_4}{V_3}\right) T_2 > T_1$

DA – Adiabatic Reversible Expansion $w_{DA} = C_V(T_2 - T_1)$

for Carnot efficiency (η) calculation $= \dfrac{-w_{\text{Total}}}{q_2} = \dfrac{T_2 - T_1}{T_2} = \dfrac{q_2 - q_1}{q_2}$

Calculation of Entropy change (ΔS or ΔS_{sys})

Ideal gas in any Reversible/Irreversible process from (P_1, V_1, T_1) to (P_2, V_2, T_2)

- $\Delta S = nC_v \ln \dfrac{T_2}{T_1} + nR \ln \dfrac{V_2}{V_1} = nC_p \ln \dfrac{T_2}{T_1} + nR \ln \dfrac{P_1}{P_2}$
- Reversible & irreversible isothermal expansion / contraction of an ideal gas

$$\Delta S = nR \ln \frac{V_2}{V_1}$$

- Isobaric heating or cooling : $\Delta S = nC_P \ln\left(\dfrac{T_2}{T_1}\right)$
- Isochoric heating or cooling : $\Delta S = nC_V \ln\left(\dfrac{T_2}{T_1}\right)$
- Adiabatic process :
- $\Delta S = nC_V \ln \dfrac{T_2}{T_1} + nR \ln \dfrac{V_2}{V_1}$ for irreversible process

$\Delta S = 0$ for reversible adiabatic compression and expansion.

- General reversible heating or cooling of any substance

$$\Delta S = \int_{T_1}^{T_2} \frac{dq_{\text{rev}}}{T}$$

- Reversible Phase Transformation

(a) $\Delta S_{\text{Fusion}} = \Delta H_{\text{Fusion}} / T_m$, T_m is normal melting point (K)

(b) $\Delta S_{\text{vap}} = \Delta H_{\text{vap}} / T_B$, T_B is normal boiling point (K)

(c) $\Delta S_{\text{sub}} = \Delta H_{\text{sub}} / T_{\text{sub}}$, T_{sub} is sublimation temperature (K)

(d) $\Delta S_{\text{Trans}} = \Delta H_{\text{Trans}} / T_{\text{Trans}}$, T_{Trans} is the phase transition temperature (K)

- For chemical reaction

$$\Delta_r S° = \Sigma \nu_P \Delta_f S° \text{ (products)} - \Sigma \nu_R \Delta_f S° \text{ (reactants)}$$

Third Law of Thermodynamics

"At absolute zero temperature, the entropy of a perfectly crystalline substance is taken as zero", which means that at absolute zero every crystalline solid is in a state of perfect order and its entropy should be zero.

By virtue of the third law, the absolute value of entropy (unlike absolute value of enthalpy) for any pure substance can be calculated at room temperature.

$$S_T - S_{0K} = \int_0^T \frac{q_{rev}}{T}, \text{ since } S_{0K} = 0 \text{ ; } S_T = \int_0^T \frac{q_{rev}}{T}$$

Absolute entropies of various substances have been tabulated and these value are used to calculate entropy changes for the reactions by the formula;

$$\Delta S^\circ = \Sigma \nu_P S_m^\circ \text{ (products) } - \Sigma \nu_R S_m^\circ \text{ (reactants)}$$

Variation of ΔS_r with temperature & pressure : $\int dS_r = \int \frac{\Delta_r C_{p,m} \cdot dT}{T}$

$$\Delta_r S_{T_2} - \Delta_r S_{T_1} = \Delta_r C_{p,m} \ln \frac{T_2}{T_1}$$

Gibbs Free Energy (G) and Spontaneity

- A new thermodynamic state function G, the Gibbs free energy is defined as :
  $G = H - TS$ or $\Delta G = \Delta H - T\Delta S$ (at constant temperature and pressure)
- $\Delta G < 0$ for spontaneous process
- $\Delta G = 0$ for equilibrium.

Criteria for Spontaneity

For a feasible process $\Delta S_T = \Delta S_{sys} + \Delta S_{surr.} > 0$

In absence of non PV work at constant T and P for a feasible process $\Delta G < 0$.

Standard Free Energy Change (ΔG°)

The free energy change for a process at a specified temperature in which the reactants in their standard state are converted to the products in their standard state. It is denoted by ΔG°.

- **Standard states :**

  (a) Standard temperature (any specified temperature)

  (b) For gases/solid/liquid: Standard pressure $P^\circ = 1$ bar

  (c) For ion/substance in solution: Standard concentration $(C^\circ) = 1M$

Like the standard enthalpy of formation of an element "the standard free energy of formation of an element in its standard state is zero". And so;

- $\Delta_r G^\circ = \sum \nu_P \Delta_f G^\circ \text{ (products) } - \sum \nu_R \Delta_f G^\circ \text{ (reactants)}$

A Thermodynamic Relationship

For any reversible reaction : $\Delta G = \Delta G^\circ + RT \ln Q$.

where Q is reaction quotient.

At equilibrium $\Delta G = 0$, $Q = K$

$\Delta G^\circ = -RT \ln K$, $\Delta G^\circ = -2.303\, RT \log K_{eq.}$

Effect of Temperature : Van't Hoff Equation

(a) $\dfrac{d(\ln K)}{dT} = \dfrac{\Delta H^\circ}{RT^2}$ (b) $\dfrac{d(\ln K)}{d\left(\frac{1}{T}\right)} = -\dfrac{\Delta H^\circ}{R}$

$\Delta G^\circ = \Delta H^\circ - T\Delta S^\circ$ $\ln K = \dfrac{-\Delta H^\circ}{RT} + \dfrac{\Delta S^\circ}{R}$

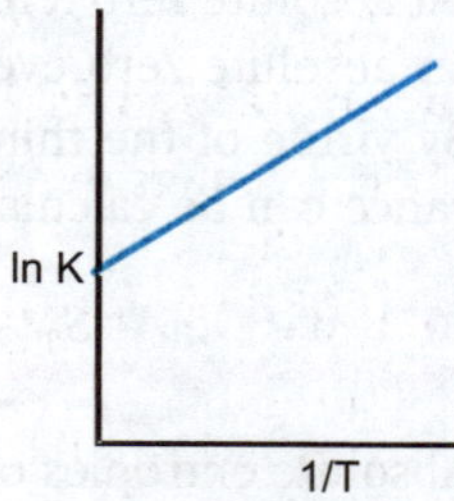

If $\Delta H^\circ < 0$

Integrated form $\ln\left(\dfrac{K_2}{K_1}\right) = -\dfrac{\Delta H^\circ}{R}\left(\dfrac{1}{T_2} - \dfrac{1}{T_1}\right)$

- **RELATIONSHIP BETWEEN ΔG & $w_{non\text{-}PV}$**

  $dU = dq + dw_{PV} + dw_{non\text{-}PV}$

  for reversible process at constant T & P

  $dU + pdV - TdS = dw_{non\text{-}PV}$

  $dH - TdS = dw_{non\text{-}PV}$

  $(dG_{system})_{T,P} = dw_{non\text{-}PV}$

  $-(dG_{system})_{T,P} = (dw_{non\text{-}PV})_{by\ system}$

 Means free energy change for a process is equal to the maximum possible work that can be derived from the process *i.e.*,

- $\Delta G^\circ = w_{max}$ (for a reversible change at constant pressure and temperature)
- In case of a galvanic cell. $\Delta G = -nFE_{cell}$, where E_{cell} = e.m.f. of the cell ; F = Faraday constant and n = number of electrons being transferred in the chemical process
- So $\Delta G^\circ = -nF E^\circ_{cell}$, where E°_{Cell} is the standard cell potential.

THERMOCHEMISTRY

- **Enthalpy of Reaction :**
  The amount of heat evolved or absorbed when the number of moles of reactants react completely to give the products as given by the balanced chemical equation.
- **Enthalpy of formation :**
  The change in heat enthalpy when one mole of a substance is formed directly from its constituent elements at standard states.
  (i) For elements, $\Delta_f H^\circ = 0$; *e.g.*, $\Delta_f H^\circ(\text{Zn, S}) = 0$, $\Delta_f H^\circ(Br_2, l) = 0$, $\Delta_r H^\circ(O_2, g) = 0$
  (ii) If element exists in two or more allotropic forms then for most stable state $\Delta_f H^\circ = 0$
  e.g., $\Delta_f H^\circ[\text{C}(s, \text{graphite})] = 0$, $\Delta_f H^\circ\ [\text{S}(s, \text{rhombic})] = 0$
- **Enthalpy of combustion :**
  The change in enthalpy when one mole of a substance is completely burnt in oxygen.
- **Enthalpy of Transition :**
  The enthalpy change when one mole of one allotropic form changes to another.
  For example : C(graphite) $\longrightarrow$ C(diamond) ; $\Delta H^\circ_{trs} = 1.90\ \text{kJ mol}^{-1}$.
- **Enthalpy of fusion :**
  The change in enthalpy when one mole of a solid is completely converted into liquid state at its melting point.
- **Enthalpy of vaporisation :**
  The change in enthalpy when one mole of a liquid is completely converted into its vapours at its boiling point.
- **Enthalpy of sublimation :**
  The change in enthalpy when one mole of a solid is directly converted into vapours at sublimation temperature.

- **Enthalpy of solution :**

The change in enthalpy when one mole of a substance is dissolved in excess of water so that further dilution does not involve any heat change.

- **Enthalpy of Dilution :**

The change in enthalpy when a solution containing 1 mole of solute is diluted from one concentration to another.

- **Enthalpy of Hydration : (For anhydrous salt)**

The enthalpy change when 1 mole an hydrous salt combines with the requisite amount of water to form a new hydrated stable salt. For example, the hydration of anhydrous cupric sulphate is represented by

$$CuSO_4(s) + 5H_2O(l) \longrightarrow CuSO_4 \cdot 5H_2O(s)$$

- **Enthalpy of neutralisation :**

Enthalpy of neutralisation is the change in enthalpy when one gram equivalent of an acid is completely neutralised by one gram equivalent of a base in dilute solutions.

$$\text{Strong acid} + \text{Strong base} \longrightarrow \text{Salt} + \text{Water}, \Delta H^{\circ}_{neut} \approx -56\,\text{kJ mol}^{-1}$$

- **Enthalpy of formation of ions :**

The enthalpy change when one mole of aqueous ions is obtained from element in its standard state as.

$$\frac{1}{2}Cl_2(g) + aq \longrightarrow Cl^-(aq)\,;\Delta H^{\circ}_r = \Delta H^{\circ}_f(Cl^-, aq)$$

By convention, the standard enthalpy of formation of $H^+(aq)$ is taken to be zero.

- **Hess's law of constant heat summation :**

The total heat change in a particular reaction is the same whether the reaction takes place in one step or in a number of steps.

e.g., (i) $A \xrightarrow{\Delta H} B$

(ii) $A \xrightarrow{\Delta H_1} C \xrightarrow{\Delta H_2} B$

According to Hess's Law $\Delta H = \Delta H_1 + \Delta H_2$.

- **Enthalpy of atomisation :**

The change in enthalpy when one mole of substance converts into gaseous atoms.

- **Bond enthalpies :**

The average of enthalpies required to dissociate the bond present in different gaseous compounds into free atoms or radicals in the gaseous state.

The bond enthalpy given for any particular pair of atoms is the average value of the dissociation enthalpies of the bond for a number of molecules in which the pair of atoms appears.

- **Estimation of Enthalpy of a reaction from bond Enthalpies :**

For the gaseous reaction; $C_2H_4(g) + HCl(g) \longrightarrow C_2H_5Cl(g)$

$$\Delta H = \begin{pmatrix}\text{Energy required to}\\ \text{break reactants into}\\ \text{gaseous atoms}\end{pmatrix} + \begin{pmatrix}\text{Energy released to}\\ \text{from products from the}\\ \text{gaseous atoms}\end{pmatrix}$$

- **Resonance Enthalpy :**

$$\Delta H^{\circ}_{Resonance} = \Delta H^{\circ}_{f,\,experimental} - \Delta H^{\circ}_{f,\,calculated}$$
$$= \Delta H^{\circ}_{c,\,calculated} - \Delta H^{\circ}_{c,\,experimental}$$

Level-1

1. Out of molar entropy (I), specific volume (II), heat capacity (III), volume (IV), extensive properties are :
   (a) I, II (b) I, II, IV
   (c) II, III (d) III, IV
2. Out of internal energy (I), boiling point (II), pH (III) and E.M.F. of the cell (IV) intensive properties are :
   (a) I, II (b) II, III, IV
   (c) I, III, IV (d) All of these
3. Thermodynamic equilibrium involves
   (a) chemical equilibrium (b) mechanical equilibrium
   (c) thermal equilibrium (d) all the above simultaneously
4. Which has maximum internal energy at 290 K?
   (a) Neon gas (b) Nitrogen gas
   (c) Ozone gas (d) Equal for all
5. A 10 g piece of iron ($C = 0.45$ J/g°C) at 100°C is dropped into 25 g of water ($C = 4.2$ J/g°C) at 27°C. Find temperature of the iron and water system at thermal equilibrium.
   (a) 30°C (b) 33°C
   (c) 40°C (d) None of these
6. When freezing of a liquid takes place in a system it:
   (a) may have $q > 0$ or $q < 0$ depending on the liquid
   (b) is represented by $q > 0$
   (c) is represented by $q < 0$
   (d) has $q = 0$
7. Mechanical work is specially important in systems that contain
   (a) gas-liquid (b) liquid-liquid
   (c) solid-solid (d) amalgam
8. Determine which of the following reactions taking place at constant pressure represents system that do work on the surrounding environment
   I. $Ag^+(aq) + Cl^-(aq) \longrightarrow AgCl(s)$ II. $NH_4Cl(s) \longrightarrow NH_3(g) + HCl(g)$
   III. $2NH_3(g) \longrightarrow N_2(g) + 3H_2(g)$
   (a) I (b) III
   (c) II and III (d) I and II
9. Determine which of the following reactions taking place at constant pressure represent surrounding that do work on the system environment
   I. $4NH_3(g) + 7O_2(g) \longrightarrow 4NO_2(g) + 6H_2O(g)$
   II. $CO(g) + 2H_2(g) \longrightarrow CH_3OH(l)$
   III. $C(s, \text{graphite}) + H_2O(g) \longrightarrow CO(g) + H_2(g)$
   IV. $H_2O(s) \longrightarrow H_2O(l)$
   (a) III, IV (b) II and III
   (c) II, IV (d) I and II, IV
10. A sample of liquid in a thermally insulated container (a calorimeter) is stirred for 2 hr. by a mechanical linkage to a motor in the surrounding. For this process :
    (a) $w < 0;\ q = 0;\ \Delta U = 0$ (b) $w > 0;\ q > 0;\ \Delta U > 0$
    (c) $w < 0;\ q > 0;\ \Delta U = 0$ (d) $w > 0;\ q = 0;\ \Delta U > 0$
11. A system undergoes a process in which $\Delta U = +300$ J while absorbing 400 J of heat energy and undergoing an expansion against 0.5 bar. What is the change in the volume (in L)?
    (a) 4 (b) 5
    (c) 2 (d) 3

12. An ideal gas expands against a constant external pressure of 2.0 atmosphere from 20 litre to 40 litre and absorbs 10 kJ of heat from surrounding. What is the change in internal energy of the system? (Given : 1 atm-litre = 101.3 J)

(a) 4052 J (b) 5948 J
(c) 14052 J (d) 9940 J

13. One mole of an ideal gas at 25°C expands in volume from 1.0 L to 4.0 L at constant temperature. What work (in J) is done if the gas expands against vacuum ($P_{external} = 0$)?

(a) -4.0×10^2 (b) -3.0×10^2
(c) -1.0×10^2 (d) Zero

14. At 25°C, a 0.01 mole sample of a gas is compressed from 4.0 L to 1.0 L at constant temperature. What is the work done for this process if the external pressure is 4.0 bar?

(a) 1.6×10^3 J (b) 8.0×10^2 J
(c) 4.0×10^2 J (d) 1.2×10^3 J

15. Calculate the work done (in J) when 4.5 g of H_2O_2 reacts against a pressure of 1.0 atm at 25°C

$$2H_2O_2(l) \longrightarrow O_2(g) + 2H_2O(l)$$

(a) -1.63×10^2 (b) 4.5×10^2
(c) 3.2×10^2 (d) -6.1×10^2

16. 2 mole of zinc is dissolved in HCl at 25°C. The work done in open vessel is :

(a) -2.477 kJ (b) -4.955 kJ
(c) 0.0489 kJ (d) None

17. Temperature of 1 mole of a gas is increased by 2°C at constant pressure. Work done by the gas is :

(a) R (b) $2R$
(c) $R/2$ (d) $3R$

18. A sample of an ideal gas is expanded $1\,m^3$ to $3\,m^3$ in a reversible process for which $P = KV^2$, with $K = 6\,bar/m^6$. Work done by the gas is :

(a) 5200 kJ (b) 15600 kJ
(c) 52 kJ (d) 5267.6 kJ

19. A given mass of gas expands reversibly from the state A to the state B by three paths 1, 2 and 3 as shown in the figure. If w_1, w_2 and w_3 respectively be the work done by the gas along three paths then:

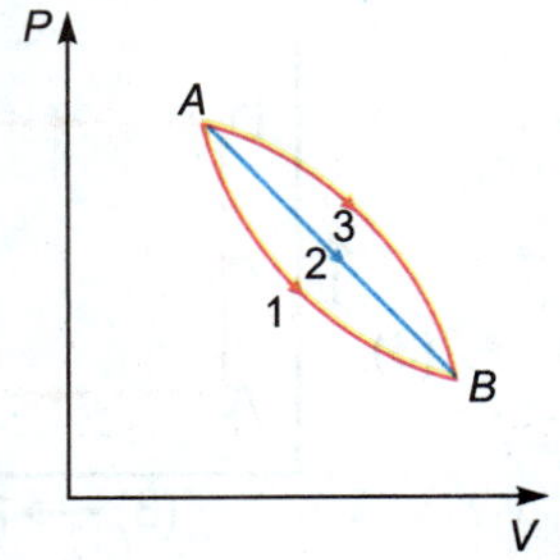

(a) $w_1 > w_2 > w_3$
(b) $w_1 < w_2 < w_3$
(c) $w_1 = w_2 = w_3$
(d) $w_2 < w_3 < w_1$

20. Heat absorbed by a system in going through a cyclic process shown in figure is :

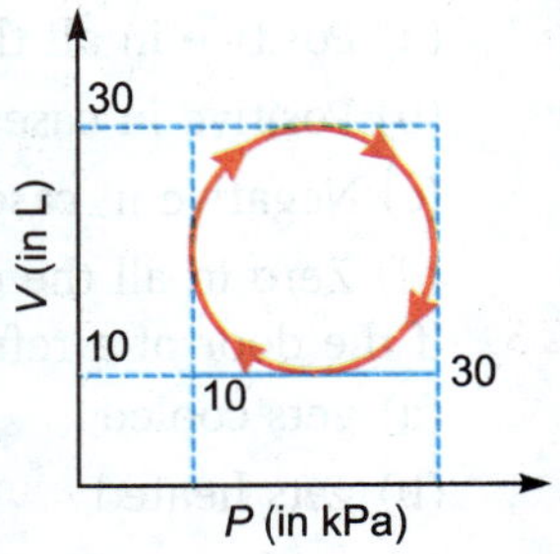

(a) $10^7 \pi$ J (b) $10^6 \pi$ J
(c) $10^2 \pi$ J (d) $10^4 \pi$ J

21. In the cyclic process shown in P–V diagram, the magnitude of the work done is:

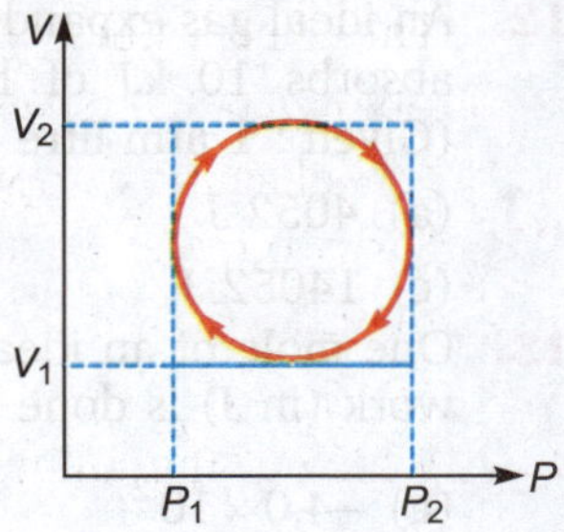

(a) $\pi\left(\dfrac{P_2 - P_1}{2}\right)^2$ (b) $\pi\left(\dfrac{V_2 - V_1}{2}\right)^2$

(c) $\dfrac{\pi}{4}(P_2 - P_1)(V_2 - V_1)$ (d) $\pi(V_2 - V_1)^2$

22. An ideal gas is taken around the cycle $ABCA$ as shown in P–V diagram. The net work done during the cycle is equal to :

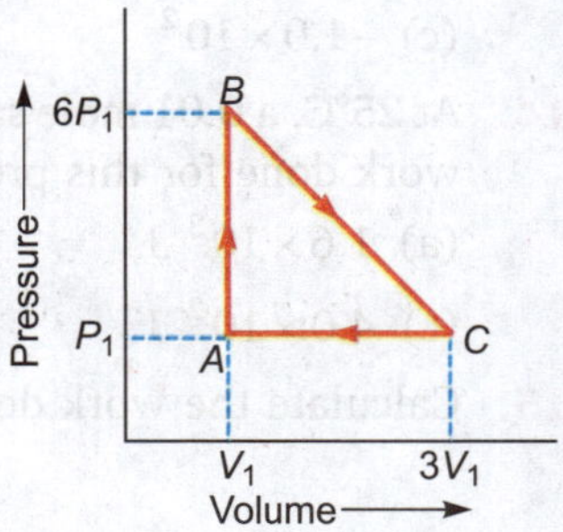

(a) $12P_1V_1$ (b) $6P_1V_1$

(c) $5P_1V_1$ (d) P_1V_1

23. An ideal gas is at pressure P and temperature T in a box, which is kept in vacuum with in a large container. The wall of the box is punctured. What happens as the gas occupies entire container?

(a) It's temperature falls (b) Its temperature rises

(c) Its temperature remains the same (d) Unpredictable

24. In diagrams (1 to 4), variation of volume with changing pressure is shown. A gas is taken along the path $ABCD$. The change in internal energy of the gas will be :

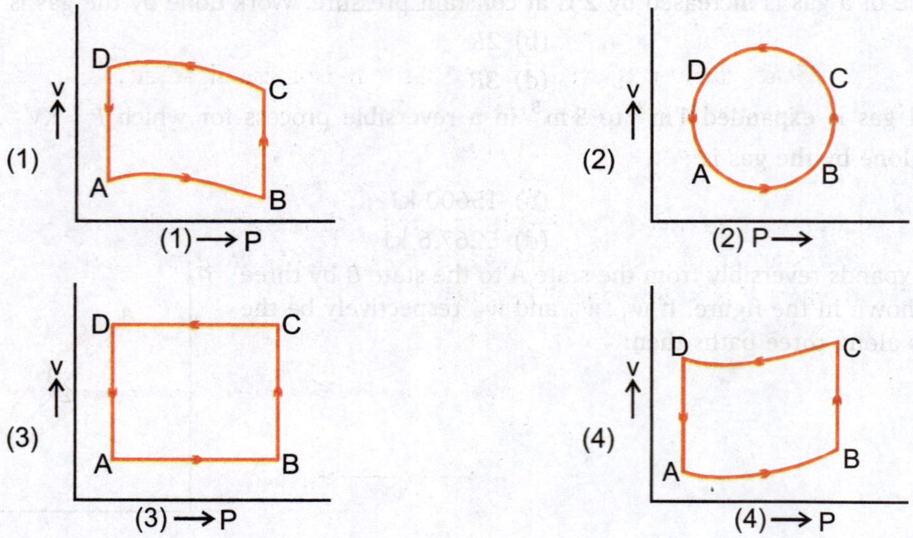

(a) Positive in all the cases (1) to (4)

(b) Positive in cases (1), (2), (3) but zero in case (4)

(c) Negative in cases (1), (2), (3) but zero in case (4)

(d) Zero in all the cases

25. If the door of a refrigerator is kept open, the room in which the refrigerator is kept

(a) gets cooled

(b) gets heated

(c) neither gets cooled nor gets heated

(d) gets cooled or heated depending on the initial temperature of the room

26. The temperature of an ideal gas increases in an :

(a) adiabatic expansion
(b) isothermal expansion
(c) adiabatic compression
(d) isothermal compression

27. For two mole of an ideal gas :

(a) $C_v - C_p = R$
(b) $C_p - C_v = 2R$
(c) $C_p - C_v = R$
(d) $C_v - C_p = 2R$

28. Which of the following expressions is true for an ideal gas ?

(a) $\left(\frac{\partial V}{\partial T}\right)_P = 0$
(b) $\left(\frac{\partial P}{\partial T}\right)_V = 0$
(c) $\left(\frac{\partial U}{\partial V}\right)_T = 0$
(d) $\left(\frac{\partial U}{\partial T}\right)_V = 0$

29. Liquefied oxygen at 1 atmosphere is heated from 50 K to 300 K by supplying heat at a constant rate. The graph that correctly shows the relationship between temperature and time is :

(a) 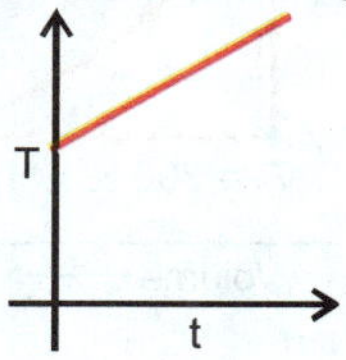

(b) 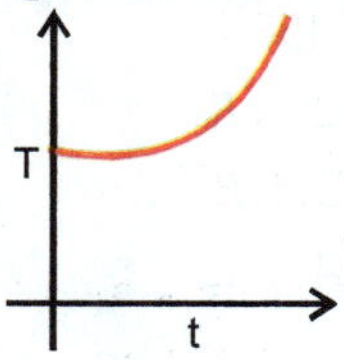

(c) 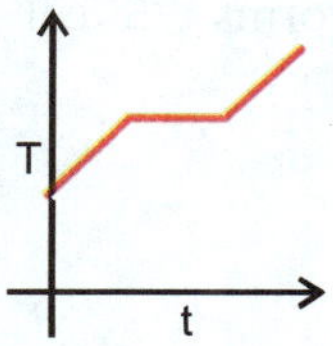

(d) 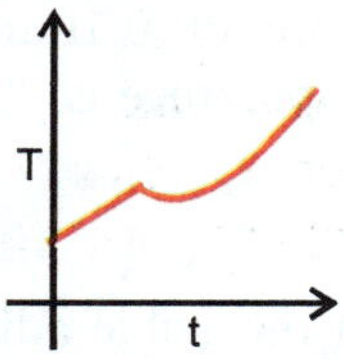

30. If w_1, w_2, w_3 and w_4 for an ideal gas are magnitude of work done in isothermal, adiabatic, isobaric and isochoric reversible expansion processes, the correct order will be :

(a) $w_1 > w_2 > w_3 > w_4$
(b) $w_3 > w_2 > w_1 > w_4$
(c) $w_3 > w_2 > w_4 > w_1$
(d) $w_3 > w_1 > w_2 > w_4$

31. For a closed container containing 100 mole of an ideal gas fitted with movable, frictionless, weightless piston operating such that pressure of gas remains constant at 8.21 atm. Which graph represents correct variation of log V *vs.* log T where V is in litre and T in kelvin?

(a) 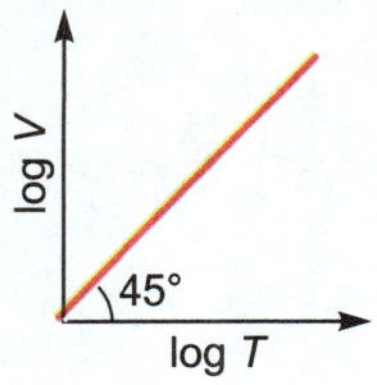

(b) 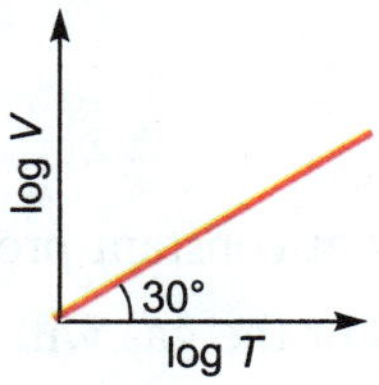

(c) 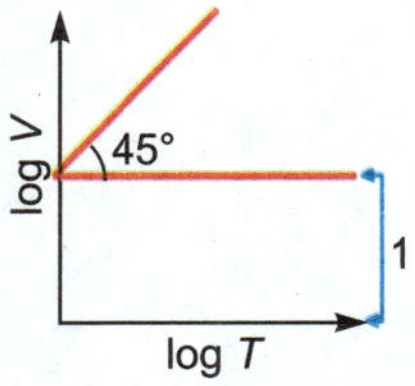

(d) 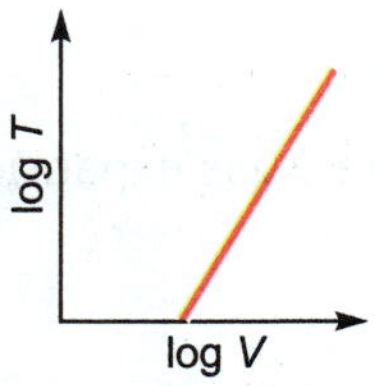

32. A gas expands against a variable pressure given by $P = \frac{20}{V}$ (where P in atm and V in L). During expansion from volume of 1 litre to 10 litre, the gas undergoes a change in internal energy of 400 J. How much heat is absorbed by the gas during expansion?

(a) 46 J (b) 4660 J
(c) 5065.8 J (d) 4260 J

33. 2 mole of an ideal gas at 27°C expands isothermally and reversibly from a volume of 4 litre to 40 litre. The work done (in kJ) by the gas is :
(a) $w = -28.72$ kJ (b) $w = -11.488$ kJ
(c) $w = -5.736$ kJ (d) $w = -4.988$ kJ

34. 10 mole of ideal gas expand isothermally and reversibly from a pressure of 10 atm to 1atm at 300 K. What is the largest mass which can lifted through a height of 100 meter?
(a) 31842 kg (b) 58.55 kg
(c) 342.58 kg (d) None of these

35. A heat engine carries one mole of an ideal mono-atomic gas around the cycle as shown in the figure. Select the correct option:

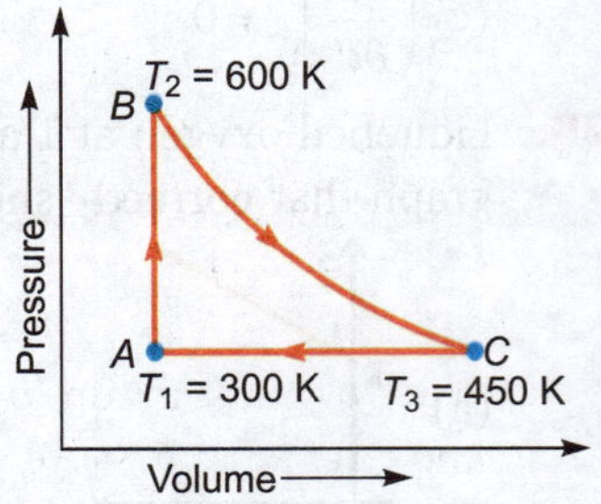

(a) $q_{AB} = 450\,R$ and $q_{CA} = -450\,R$
(b) $q_{AB} = 450\,R$ and $q_{CA} = -225\,R$
(c) $q_{AB} = 450\,R$ and $q_{CA} = -375\,R$
(d) $q_{AB} = 375\,R$ and $q_{CA} = -450\,R$

36. What is the final temperature of 0.10 mole monoatomic ideal gas that performs 75 cal of work adiabatically if the initial temperature is 227°C? (use $R = 2$ cal/K-mol)
(a) 250 K (b) 300 K
(c) 350 K (d) 750 K

37. The work done by the gas in reversible adiabatic expansion process is :
(a) $\dfrac{P_2V_2 - P_1V_1}{\gamma - 1}$ (b) $\dfrac{nR(T_1 - T_2)}{\gamma - 1}$
(c) $\dfrac{P_2V_2 - P_1V_1}{\gamma}$ (d) None of these

38. During an adiabatic process, the pressure of gas is found to be proportional to the cube of its absolute temperature. The ratio of $(C_{p,m}/C_{v,m})$ for gas is :
(a) $\dfrac{3}{2}$ (b) $\dfrac{5}{3}$
(c) $\dfrac{7}{2}$ (d) $\dfrac{4}{3}$

39. A gas expands adiabatically at constant pressure such that $T \propto V^{-1/2}$
The value of $\gamma\,(C_{p,m}/C_{v,m})$ of the gas will be :
(a) 1.30 (b) 1.50
(c) 1.70 (d) 2

40. For a reversible adiabatic ideal gas expansion $\dfrac{dP}{P}$ is equal to :
(a) $\gamma \dfrac{dV}{V}$ (b) $-\gamma \dfrac{dV}{V}$
(c) $\left(\dfrac{\gamma}{\gamma - 1}\right)\dfrac{dV}{V}$ (d) $\dfrac{dV}{V}$

41. *P–V* plot for two gases (assuming ideal)during adiabatic processes are given in the Fig. Plot *A* and plot *B* should correspond respectively to:

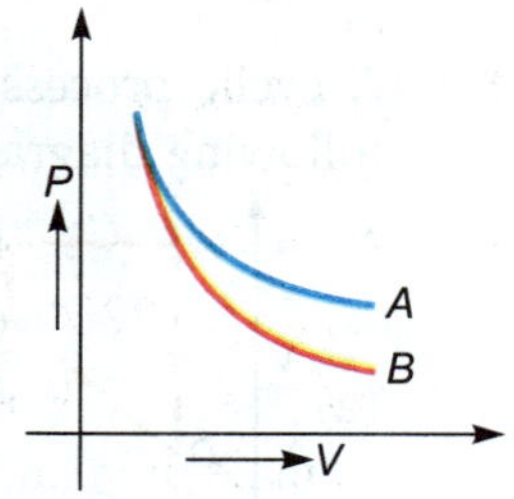

(a) He and H_2 (b) H_2 and He
(c) He and Ne (d) H_2 and Cl_2

42. Calculate the final temperature of a monoatomic ideal gas that is compressed reversible and adiabatically from 16 L to 2 L at 300 K :
(a) 600 K (b) 1044.6 K
(c) 1200 K (d) 2400 K

43. 5 mole of an ideal gas expand isothermally and irreversibly from a pressure of 10 atm to 1 atm against a constant external pressure of 1 atm. w_{irr} at 300 K is :
(a) −15.921 kJ (b) −11.224 kJ
(c) −110.83 kJ (d) None of these

44. With what minimum pressure (in kPa), a given volume of an ideal gas ($C_{p,m} = 7/2\,R$), originally at 400 K and 100 kPa pressure can be compressed irreversibly adiabatically in order to raise its temperature to 600 K:
(a) 362.5 kPa (b) 275 kPa
(c) 437.5 kPa (d) 550 kPa

45. The work done in adiabatic compression of 2 mole of an ideal monoatomic gas against constant external pressure of 2 atm starting from initial pressure of 1 atm and initial temperature of 300 K (R = 2 cal/mol-degree)
(a) 360 cal (b) 720 cal
(c) 800 cal (d) 1000 cal

46. One mole of an ideal gas $\left(C_{v,m} = \frac{5}{2}R\right)$ at 300 K and 5 atm is expanded adiabatically to a final pressure of 2 atm against a constant pressure of 2 atm. Final temperature of the gas is :
(a) 270 K (b) 273 K
(c) 248.5 K (d) 200 K

47. 10 litre of a non linear polyatomic ideal gas at 127°C and 2 atm pressure is suddenly released to 1 atm pressure and the gas expanded adiabatically against constant external pressure. The final temperature and volume of the gas respectively are.
(a) $T = 350\,K; V = 17.5\,L$ (b) $T = 300\,K; V = 15\,L$
(c) $T = 250\,K; V = 12.5\,L$ (d) None of these

48. Calculate average molar heat capacity at constant volume of gaseous mixture contained 2 mole of each of two ideal gases $A\left(C_{v,m} = \frac{3}{2}R\right)$ and $B\left(C_{v,m} = \frac{5}{2}R\right)$:
(a) R (b) $2R$
(c) $3R$ (d) $8R$

49. 0.5 mole each of two ideal gases $A\left(C_{v,m} = \frac{5}{2}R\right)$ and B ($C_{v,m} = 3R$) are taken in a container and expanded reversibly and adiabatically, during this process temperature of gaseous mixture decreased from 350 K to 250 K. Find ΔH (in cal/mol) for the process :
(a) $-100\,R$ (b) $-137.5\,R$
(c) $-375\,R$ (d) None of these

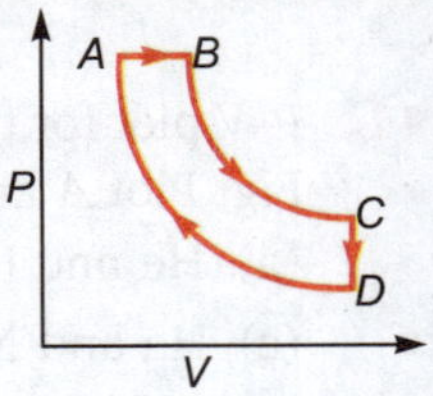

50. A cyclic process $ABCD$ is shown in P–V diagram for an ideal gas. Which of the following diagram represents the same process?

(a) 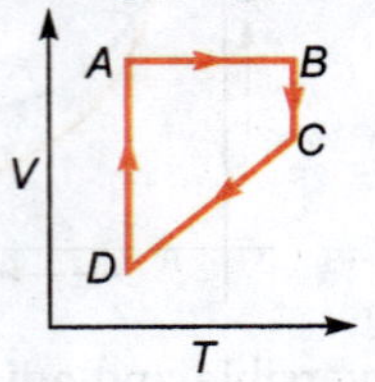

(b) 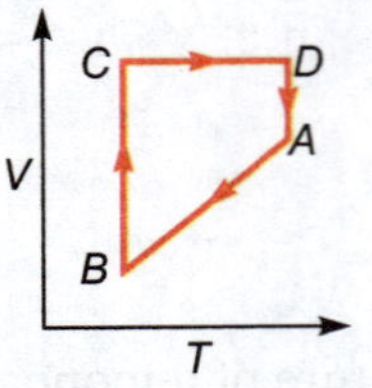

(c) 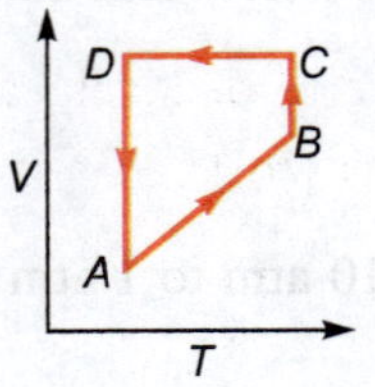

(d) 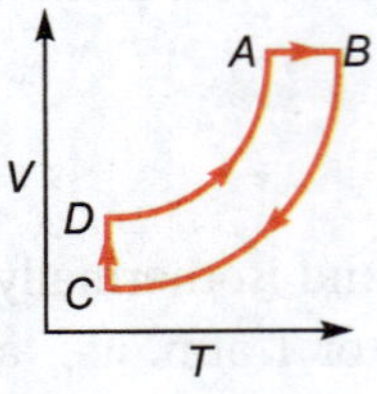

51. 36 mL of pure water takes 100 sec to evaporate from a vessel and heater connected to an electric source which delivers 806 watt. The $\Delta H_{\text{vaporization}}$ of H_2O is :

(a) 40.3 kJ/mol (b) 43.2 kJ/mol
(c) 4.03 kJ/mol (d) None of these

52. For the reaction : $PCl_5(g) \longrightarrow PCl_3(g) + Cl_2(g)$:

(a) $\Delta H = \Delta U$ (b) $\Delta H > \Delta U$
(c) $\Delta H < \Delta U$ (d) None of the above

53. Consider the reaction at 300 K

$$H_2(g) + Cl_2(g) \longrightarrow 2HCl(g); \quad \Delta H° = -185 \text{ kJ}$$

If 2 mole of H_2 completely react with 2 mole of Cl_2 to form HCl, what is $\Delta U°$ for this reaction?

(a) 0 (b) –185 kJ
(c) 370 kJ (d) None of these

54. Which of the indicated relationship is correct for the following exothermic reaction carried out at constant pressure?

$$CO(g) + 3H_2(g) \longrightarrow CH_4(g) + H_2O(g)$$

(a) $\Delta U = \Delta H$ (b) $\Delta U > \Delta H$
(c) $w < 0$ (d) $q > 0$

55. One mole of an ideal gas undergoes a change of state (2.0 atm, 3.0 L) to (2.0 atm, 7.0 L) with a change in internal energy $(\Delta U) = 30$ L-atm. The change in enthalpy (ΔH) of the process in L-atm :

(a) 22 (b) 38
(c) 25 (d) None of these

56. What is the change in internal energy when a gas contracts from 377 mL to 177 mL under a constant pressure of 1520 torr, while at the same time being cooled by removing 124 J heat?

(a) 40.52 J (b) –83.48 J
(c) –248 J (d) None of these

57. For the real gases reaction $2CO(g) + O_2(g) \longrightarrow 2CO_2(g); \Delta H = -560$ kJ. In 10 litre rigid vessel at 500 K the initial pressure is 70 bar and after the reaction it becomes 40 bar. The change in internal energy is :

(a) –557 kJ (b) –530 kJ
(c) –563 kJ (d) None of these

58. One mole of a non-ideal gas undergoes a change of state from (1.0 atm, 3.0 L, 200 K) to (4.0 atm, 5.0 L, 250 K) with a change in internal energy $(\Delta U) = 40$ L-atm. The change in enthalpy of the process in L-atm :

(a) 43 (b) 57
(c) 42 (d) None of these

59. Consider the reaction at 300 K

$$C_6H_6(l) + \frac{15}{2}O_2(g) \longrightarrow 6\,CO_2(g) + 3H_2O(l);\ \Delta H = -3271\,\text{kJ}$$

What is ΔU for the combustion of 1.5 mole of benzene at 27°C?

(a) -3267.25 kJ (b) -4900.88 kJ
(c) -4906.5 kJ (d) -3274.75 kJ

60. For the reaction; $FeCO_3(s) \longrightarrow FeO(s) + CO_2(g)$; $\Delta H = 82.8$ kJ at 25°C, what is (ΔE or ΔU) at 25°C?

(a) 82.8 kJ (b) 80.32 kJ
(c) –2394.77 kJ (d) 85.28 kJ

61. At 5×10^5 bar pressure density of diamond and graphite are 3 g/cc and 2 g/cc respectively, at certain temperature 'T'. Find the value of $\Delta U - \Delta H$ for the conversion of 1 mole of graphite to 1 mole of diamond at temperature 'T' :

(a) 100 kJ/mol (b) 50 kJ/mol
(c) –100 kJ/mol (d) None of these

62. Predict which of the following reaction(s) has a positive entropy change?

I. $Ag^+(aq) + Cl^-(aq) \longrightarrow AgCl(s)$
II. $NH_4Cl(s) \longrightarrow NH_3(g) + HCl(g)$
III. $2NH_3(g) \longrightarrow N_2(g) + 3H_2(g)$

(a) I and II (b) III
(c) II and III (d) II

63. Predict which of the following reaction(s) has a negative entropy change?

I. $CH_4(g) + 2O_2(g) \longrightarrow CO_2(g) + 2H_2O(l)$
II. $NH_3(g) + HCl(g) \longrightarrow NH_4Cl(s)$
III. $2KClO_4(s) \longrightarrow 2KClO_3(s) + O_2(g)$

(a) III (b) II
(c) I and II (d) I

64. Which of the following reactions is associated with the most negative change in entropy?

(a) $2SO_2(g) + O_2(g) \longrightarrow 2SO_3(g)$ (b) $C_2H_4(g) + H_2(g) \longrightarrow C_2H_6(g)$
(c) $C(s, \text{graphite}) + O_2(g) \longrightarrow CO_2(g)$ (d) $3C_2H_2(g) \longrightarrow C_6H_6(l)$

65. When two mole of an ideal gas $\left(C_{p,m} = \frac{5}{2}R\right)$ heated from 300 K to 600 K at constant pressure, the change in entropy of gas (ΔS) is :

(a) $\frac{3}{2}R\ln 2$ (b) $-\frac{3}{2}R\ln 2$
(c) $5R\ln 2$ (d) $\frac{5}{2}R\ln 2$

66. Which of the following expression for entropy change of an irreversible process ?

(a) $dS > \frac{dq}{T}$ (b) $dS = \frac{dq}{T}$
(c) $dS < \frac{dq}{T}$ (d) $dS = \frac{dU}{T}$

67. Which of the following expressions is known as Clausius inequality ?

(a) $\oint \frac{dq}{T} \le 0$ (b) $\oint \frac{ds}{T} = 0$
(c) $\oint \frac{T}{dq} \le 0$ (d) $\oint \frac{dq}{T} \ge 0$

68. In problem 65, calculate ΔS_{gas} if process is carried out at constant volume :

(a) $5R \ln 2$ (b) $\frac{3}{2} R \ln 2$

(c) $3 R \ln 2$ (d) $-3 R \ln 2$

69. If one mole of an ideal gas $\left(C_{p,m} = \frac{5}{2} R\right)$ is expanded isothermally at 300 K until it's volume is tripled, then change in entropy of gas is :

(a) zero (b) infinity

(c) $\frac{5}{2} R \ln 3$ (d) $R \ln 3$

70. In previous problem if expansion is carried out freely ($P_{ext} = 0$), then ΔS is :

(a) zero (b) infinity

(c) $R \ln 3$ (d) None

71. When one mole of an ideal gas is compressed to half of its initial volume and simultaneously heated to twice its initial temperature, the change in entropy of gas (ΔS) is :

(a) $C_{p,m} \ln 2$ (b) $C_{v,m} \ln 2$

(c) $R \ln 2$ (d) $(C_{v,m} - R) \ln 2$

72. What is the change in entropy when 2.5 mole of water is heated from 27°C to 87°C? Assume that the heat capacity is constant. ($C_{p,m}$ $(H_2O) = 4.2$ J/g-K and $\ln(1.2) = 0.18$)

(a) 16.6 J/K (b) 9 J/K

(c) 34.02 J/K (d) 1.89 J/K

73. Calculate standard entropy change in the reaction

$$Fe_2O_3(s) + 3H_2(g) \longrightarrow 2Fe(s) + 3H_2O(l)$$

Given : S°_m $(Fe_2O_3, S) = 87.4$, S°_m $(Fe, S) = 27.3$,

S°_m $(H_2, g) = 130.7$, S°_m $(H_2O, l) = 69.9$ JK^{-1} mol^{-1}

(a) -212.5 JK^{-1} mol^{-1} (b) -215.2 JK^{-1} mol^{-1}

(c) -120.9 JK^{-1} mol^{-1} (d) None of these

74. Calculate the entropy change (J/mol K) of the given reaction. The molar entropies [J/K-mol] are given in brackets after each substance.

$$2PbS(s)\ [91.2] + 3O_2(g)\ [205.1] \longrightarrow 2PbO(s)\ [66.5] + 2SO_2(g)\ [248.2]$$

(a) -113.5 (b) -168.3

(c) $+72.5$ (d) -149.2

75. Given $\Delta_r S^\circ = -266$ and the listed [S°_m values] calculate S° for $Fe_3O_4(s)$:

$$4Fe_3O_4(s)\ [......] + O_2(g)\ [205] \longrightarrow 6Fe_2O_3(s)\ [87]$$

(a) $+111.1$ (b) $+122.4$

(c) 145.75 (d) 248.25

76. The entropy change for a phase transformation is :

(a) $\frac{\Delta U}{\gamma + dT}$ (b) $\frac{\Delta T}{\Delta H}$

(c) $\frac{\Delta H}{T}$ (d) $\frac{\Delta H + \Delta G}{T}$

77. What is the melting point of benzene if $\Delta H_{fusion} = 9.95$ kJ/mol and $\Delta S_{fusion} = 35.7$ J/K-mol?

(a) 278.7°C (b) 278.7 K

(c) 300 K (d) 298 K

78. ΔS for freezing of 10 g of $H_2O(l)$ (enthalpy of fusion is 80 cal/g) at 0°C and 1 atm is :

(a) 12.25 J/K (b) -0.244 J/K

(c) -2.93 J/K (d) -12.25 J/K

79. Chloroform has $\Delta H_{vaporization} = 29.2$ kJ/mol and boils at 61.2°C. What is the value of $\Delta S_{vaporization}$ for chloroform?
(a) 87.3 J/mol-K (b) 477.1 J/mol-K
(c) – 87.3 J/mol-K (d) – 477.1 J/mol-K

80. The entropy of vaporization of benzene is 85 JK^{-1} mol^{-1}. When 117g benzene vaporizes at it's normal boiling point, the entropy change of surrounding is :
(a) –85 JK^{-1} (b) -85×1.5 JK^{-1}
(c) 85×1.5 JK^{-1} (d) None of these

81. Identify the correct statement regarding entropy
(a) At absolute zero temperature, the entropy of perfectly crystalline substances is +ve
(b) At absolute zero temperature entropy of perfectly crystalline substance is taken to be zero
(c) At 0°C the entropy of a perfectly crystalline substance is taken to be zero
(d) At absolute zero temperature, the entropy of all crystalline substances is taken to be zero

82. Calculate ΔS for following process :
$$\underset{\text{at 100 K}}{X(s)} \longrightarrow \underset{\text{at 200 K}}{X(l)}$$
Given : Melting point of $X_{(s)} = 100$ K; $\Delta H_{Fusion} = 20$ kJ/mol; $C_{p,m}(X,l) = 10$ J/mol.K
(a) 26.93 J/K (b) 206.93 J/K
(c) 203 J/K (d) 206.93 kJ/K

83. For a perfectly crystalline solid $C_{p,m} = aT^3$, where a is constant. If $C_{p,m}$ is 0.42 J/K mol at 10 K, molar entropy at 20 K is :
(a) 0.42 J/K mol (b) 0.14 J/K mol
(c) 1.12 J/K mol (d) zero

84. Consider the following spontaneous reaction $3X_2(g) \rightarrow 2X_3(g)$. What are the sign of $\Delta H, \Delta S$ and ΔG for the reaction?
(a) +ve, +ve, +ve (b) +ve, – ve, – ve
(c) –ve, + ve, – ve (d) –ve, – ve, – ve

85. For the reaction $2H(g) \longrightarrow H_2(g)$, the sign of ΔH and ΔS respectively are :
(a) +, – (b) +, +
(c) –, + (d) –, –

86. Consider the following reaction.
$$C_6H_6(l) + \frac{15}{2}O_2(g) \longrightarrow 6CO_2(g) + 3H_2O(g)$$
signs of $\Delta H, \Delta S$ and ΔG for the above reaction will be
(a) +,–,+ (b) –,+,–
(c) –,+,+ (d) +,+,–

87. Consider the following reaction at temperature T :
$$CH_2{=}CH_2(g) + Cl_2(g) \longrightarrow ClCH_2CH_2Cl(g)$$
$$\Delta_r H° = -217.5 \text{ kJ/mol}, \quad \Delta_r S° = -233.9 \text{ J/K-mol}$$
Reaction is supported by :
(a) entropy (b) enthalpy
(c) both (a) & (b) (d) neither

88. For a process to be spontaneous at constant T and P :
(a) $(\Delta G)_{system}$ must be negative (b) $(\Delta G)_{system}$ must be positive
(c) $(\Delta S)_{system}$ must be positive (d) $(\Delta S)_{system}$ must be negative

89. For a reaction to occur spontaneously :

(a) ΔS must be negative (b) $(-\Delta H + T\,\Delta S)$ must be positive

(c) $\Delta H + T\Delta S$ must be negative (d) ΔH must be negative

90. Which of the following conditions regarding a chemical process ensures its spontaneity at all temperature?

(a) $\Delta H > 0, \Delta S < 0$ (b) $\Delta H < 0, \Delta S > 0$

(c) $\Delta H < 0, \Delta S < 0$ (d) $\Delta H > 0, \Delta S < 0$

91. The free energy change $\Delta G = 0$, when

(a) the system is at equilibrium

(b) catalyst is added

(c) reactants are initially mixed thoroughly

(d) the reactants are completely consumed

92. Which of the following conditions will always lead to a non-spontaneous change?

(a) ΔH and ΔS both +ve (b) ΔH is –ve and ΔS is +ve

(c) ΔH and ΔS both –ve (d) ΔH is +ve and ΔS is –ve

93. Suppose that a reaction has $\Delta H = -40$ kJ and $\Delta S = -50$ J/K. At what temperature range will it change from spontaneous to non-spontaneous?

(a) 0.8 K to 1 K (b) 799 K to 800 K

(c) 800 K to 801 K (d) 799 K to 801 K

94. For isothermal expansion in case of an ideal gas :

(a) $\Delta G = \Delta S$ (b) $\Delta G = \Delta H$

(c) $\Delta G = -T.\Delta S$ (d) None of these

95. What is the normal boiling point of mercury?

Given : ΔH_f° (Hg, l) = 0; S° (Hg, l) = 77.4 J/K-mol

ΔH_f° (Hg, g) = 60.8 kJ/mol; S° (Hg, g) = 174.4 J/K-mol

(a) 624.8 K (b) 626.8 K

(c) 636.8 K (d) None of these

96. 18 gm of ice is converted into water at 0°C and 1 atm. The entropies of $H_2O(s)$ and $H_2O(l)$ are 38.2 and 60 J/mol K respectively. The enthalpy change for this conversion is :

(a) 5951.4 J/mol (b) 595.14 J/mol

(c) –5951.4 J/mol (d) None of these

97. Using the listed [$\Delta G^\circ{}_f$ values] calculate ΔG° for the reaction :

$3H_2S(g)$ [–33.6] + $2HNO_3(l)$ [–80.6] $\longrightarrow$ $2NO(g)$ [+86.6] + $4H_2O(l)$ [–237.1] + $3S(s)$ [0.0]

(a) –513.2 (b) –1037.0

(c) +433.4 (d) +225.0

98. From the following ΔH° and ΔS° values, predict which of reactions I, II, and III would be spontaneous at 25°C.

	ΔH° (kJ)	ΔS° (J/K)
I.	+10.5	+30
II.	+1.8	–113
III.	–126	+84

(a) III (b) I

(c) II and III (d) I and II

99. Calculate ΔH_f° for UBr_4 from the ΔG° of reaction and the S° values at 298 K.

$U(s) + 2Br_2(l) \longrightarrow UBr_4(s)$; $\Delta G^\circ = -788.6$ kJ; S° (J/K-mol) 50.3, 152.3, 242.6

(a) –822.1 kJ/mol (b) –841.2 kJ/mol
(c) –775.6 kJ/mol (d) –804.3 kJ/mol

100. The entropies of $H_2(g)$ and $H(g)$ are 130.6 and 114.6 J mol^{-1} K^{-1} respectively at 298 K. Using the data given below calculate the bond energy of H_2 (in kJ/mol) :

$$H_2(g) \longrightarrow 2H(g);\ \Delta G° = 406.6\ kJ$$

(a) 377.2 (b) 436.0
(c) 425.5 (d) 430.5

101. Consider the ΔG_f° and ΔH_f° (kJ/mol) for the following oxides. Which oxide can be most easily decomposed to form the metal and oxygen gas?

(a) ZnO ($\Delta G° = -318.4, \Delta H° = -348.3$) (b) Cu_2O ($\Delta G° = -146.0, \Delta H° = -168.8$)
(c) HgO ($\Delta G° = -58.5, \Delta H° = -90.8$) (d) PbO ($\Delta G° = -187.9, \Delta H° = -217.3$)

102. Which of the following option is correct ?

(a) $\left[\frac{\partial \ln K_p}{\partial T}\right] = \frac{\Delta H°}{RT^2}$ (b) $\frac{\partial \ln k}{\partial T} = \frac{E_a}{RT^2}$
(c) $\left[\frac{\partial \ln K_c}{\partial T}\right] = \frac{\Delta U°}{RT^2}$ (d) All of these

103. Calculate $\Delta G°$ (kJ/mol) at 127°C for a reaction with $K_{\text{equilibrium}} = 10^5$:

(a) –38.294 (b) –16.628
(c) –9.16 (d) None of these

104. When reaction is carried out at standard states then at the equilibrium :

(a) $\Delta H° = 0$ (b) $\Delta S° = 0$
(c) equilibrium constant $K = 0$ (d) equilibrium constant $K = 1$

105. At 25°C, $\Delta G°$ for the process $H_2O(l) \rightleftharpoons H_2O(g)$ is 8.6 kJ. The vapour pressure of water at this temperature, is nearly :

(a) 24 torr (b) 285 torr
(c) 32.17 torr (d) 100 torr

106. For the auto-ionization of water at 25°C, $H_2O(l) \rightleftharpoons H^+(aq) + OH^-(aq)$ equilibrium constant is 10^{-14}. What is $\Delta G°$ for the process?

(a) $\simeq 8 \times 10^4$ J (b) $\simeq 3.5 \times 10^4$ J
(c) $\simeq 10^4$ J (d) None of these

107. The molar entropies of HI(g), H(g) and I(g) at 298 K are 206.5, 114.6, and 180.7 J mol^{-1} K^{-1} respectively. Using the $\Delta G°$ given below, calculate the bond energy of HI.

$$HI(g) \longrightarrow H(g) + I(g);\quad \Delta G° = 271.8\ kJ$$

(a) 282.4 kJ mol^{-1} (b) 298.3 kJ mol^{-1}
(c) 290.1 kJ mol^{-1} (d) 315.4 kJ mol^{-1}

108. Hess's law states that :

(a) the standard enthalpy of an overall reaction is the sum of the enthalpy changes in individual reactions.
(b) enthalpy of formation of a compound is same as the enthalpy of decomposition of the compound into constituent elements, but with opposite sign.
(c) at constant temperature the pressure of a gas is inversely proportional to its volume
(d) the mass of a gas dissolved per litre of a solvent is proportional to the pressure of the gas in equilibrium with the solution

109. An imaginary reaction $X \longrightarrow Y$ takes place in three steps

$X \longrightarrow A, \Delta H = -q_1$; $B \longrightarrow A, \Delta H = -q_2$; $B \longrightarrow Y, \Delta H = -q_3$

If Hess' law is applicable, then the heat of the reaction ($X \rightarrow Y$) is :

(a) $q_1 - q_2 + q_3$ (b) $q_2 - q_3 - q_1$

(c) $q_1 - q_2 - q_3$ (d) $q_3 - q_2 - q_1$

110. The enthalpy change for a reaction does not depend upon :

(a) the physical states of reactants and products

(b) use of different reactants for the same product

(c) the number of intermediate reaction steps

(d) the differences in initial or final temperatures of involved substances

111. The standard enthalpy of formation of gaseous H_2O at 298 K is –241.82 kJ/mol. Calculate $\Delta H°$ at 373 K given the following values of the molar heat capacities at constant pressure :

$H_2O(g) = 33.58\ JK^{-1}\ mol^{-1}$; $H_2(g) = 29.84\ JK^{-1}\ mol^{-1}$; $O_2(g) = 29.37\ JK^{-1}\ mol^{-1}$

Assume that the heat capacities are independent of temperature :

(a) –242.6 kJ/mol (b) – 485.2 kJ/mol

(c) –121.3 kJ/mol (d) – 286.4 kJ/mol

112. Which of the following value of ΔH_f° represent that the product is least stable ?

(a) $-94.0\ kcal\ mol^{-1}$ (b) $-231.6\ kcal\ mol^{-1}$

(c) $+21.4\ kcal\ mol^{-1}$ (d) $+64.8\ kcal\ mol^{-1}$

113. For which of the following substances the enthalpy of formation in the standard state is zero ?

(a) Sucrose (b) Ethanol

(c) Aluminium (d) Calcium chloride

114. Calculate the standard enthalpy of reaction for the following reaction using the listed enthalpies of reaction:

$$3\,Co(s) + 2O_2(g) \longrightarrow Co_3O_4(s)$$

$$2\,Co(s) + O_2(g) \longrightarrow 2\,CoO(s);\ \Delta H_1^\circ = -475.8\ kJ$$

$$6CoO(s) + O_2(g) \longrightarrow 2\,Co_3O_4(s);\ \Delta H_2^\circ = -355.0\ kJ$$

(a) –891.2 kJ (b) –120.8 kJ

(c) +891.2 kJ (d) –830.8 kJ

115. From given following equations and $\Delta H°$ values, determine the enthalpy of reaction at 298 K for the reaction :

$$C_2H_4(g) + 6F_2(g) \longrightarrow 2\,CF_4(g) + 4HF(g)$$

$$H_2(g) + F_2(g) \longrightarrow 2HF(g);\ \Delta H_1^\circ = -537\ kJ$$

$$C(s) + 2F_2(g) \longrightarrow CF_4(g);\ \Delta H_2^\circ = -680\ kJ$$

$$2C(s) + 2H_2(g) \longrightarrow C_2H_4(g);\ \Delta H_3^\circ = 52\ kJ$$

(a) –1165 (b) –2486

(c) +1165 (d) +2486

116. Given :

C(diamond) + $O_2 \longrightarrow CO_2$; $\Delta H = -395\ kJ$

C(graphite) + $O_2 \longrightarrow CO_2$; $\Delta H = -393\ kJ$

The enthalpy of formation of diamond from graphite is

(a) +2.0 kJ (b) –1.5 kJ

(c) –788 kJ (d) 788 kJ

117. Which of the following equations represents a reaction that provides the enthalpy of formation of CH_3Cl?

(a) $C(s) + HCl(g) + H_2(g) \longrightarrow CH_3Cl(g)$
(b) $C(s) + 3H(g) + Cl(g) \longrightarrow CH_3Cl(g)$
(c) $C(s) + 3/2H_2(g) + 1/2Cl_2(g) \longrightarrow CH_3Cl(g)$
(d) $CH_4(g) + Cl_2(g) \longrightarrow CH_3Cl(g) + HCl(g)$

118. Use the given standard enthalpies of formation (in kJ/mol) to determine the enthalpy of reaction of the following reaction :

$$NH_3(g) + 3F_2(g) \longrightarrow NF_3(g) + 3HF(g)$$

$\Delta H_f^\circ (NH_3, g) = -46.2$; $\Delta H_f^\circ (NF_3, g) = -113.0$; $\Delta H_f^\circ (HF, g) = -269.0$

(a) –335.8 kJ/mol
(b) –873.8 kJ/mol
(c) –697.2 kJ/mol
(d) –890.4 kJ/mol

119. The standard enthalpy of formation of octane (C_8H_{18}) is –250 kJ/mol. Calculate the enthalpy of combustion of C_8H_{18}. The enthalpy of formation of $CO_2(g)$ and $H_2O(l)$ are –394 kJ/mol and –286 kJ/mol respectively :

(a) – 5200 kJ/mol
(b) – 5726 kJ/mol
(c) – 5476 kJ/mol
(d) – 5310 kJ/mol

120. Determine the enthalpy of formation of $B_2H_6(g)$ in kJ/mol of the following reaction :

$$B_2H_6(g) + 3O_2(g) \longrightarrow B_2O_3(s) + 3H_2O(g);$$

Given : $\Delta_r H^\circ = -1941$ kJ/mol; $\Delta H_f^\circ (B_2O_3, s) = -1273$ kJ/mol; $\Delta H_f^\circ (H_2O, g) = -241.8$ kJ/mol

(a) –75.6
(b) +75.6
(c) –57.4
(d) –28.4

121. Consider the following reactions :

$$C(s) + O_2(g) \longrightarrow CO_2(g) + x \text{ kJ}$$
$$CO(g) + \tfrac{1}{2}O_2(g) \longrightarrow CO_2(g) + y \text{ kJ}$$

The heat of formation of CO(g) is :

(a) $-(x + y)$ kJ/mol
(b) $(x - y)$ kJ/mol
(c) $(y - x)$ kJ/mol
(d) None of these

122. If $\Delta_f H^\circ (C_2H_4)$ and $\Delta_f H^\circ (C_2H_6)$ are x_1 and x_2 kcal mol^{-1}, then heat of hydrogenation of C_2H_4 is :

(a) $x_1 + x_2$
(b) $x_1 - x_2$
(c) $x_2 - x_1$
(d) $x_1 + 2x_2$

123. What amount of energy (kJ) is released in the combustion of 5.8 g of $C_4H_{10}(g)$?

$$2C_4H_{10}(g) + 13O_2(g) \longrightarrow 8CO_2(g) + 10H_2O(l); \quad \Delta H^\circ = -5756 \text{ kJ}$$

(a) 575.6
(b) 287.8
(c) 182
(d) 57.56

124. The enthalpy of the reaction forming PbO according to the following equation is 438 kJ. What heat energy (kJ) is released in formation of 22.3 g PbO(s)?
(Atomic masses : Pb = 207, O = 16.0)

$$2Pb(s) + O_2(g) \longrightarrow 2PbO(s)$$

(a) 21.9
(b) 28.7
(c) 14.6
(d) 34.2

125. The fat, $C_{57}H_{104}O_6(s)$, is metabolized via the following reaction. Given the enthalpies of formation, calculate the $\Delta_r H^\circ$ (kJ) when 1.0 g of this fat reacts.

$$C_{57}H_{104}O_6(s) + 80O_2(g) \longrightarrow 57CO_2(g) + 52H_2O(l)$$

$\Delta_f H^\circ (C_{57}H_{104}O_6, s) = -70870$ kJ/mol; $\Delta_f H^\circ (H_2O, l) = -285.8$ kJ/mol;
$\Delta_f H^\circ (CO_2, g) = -393.5$ kJ/mol

(a) $\Delta H^\circ = +37.98$ (b) $\Delta H^\circ = -40.4$
(c) $\Delta H^\circ = -33.4$ (d) $\Delta H^\circ = -30.2$

126. The $\Delta_f H^\circ$ of $NH_3(g)$ is $-46\ \text{kJ mol}^{-1}$. The ΔH (in kJ mol^{-1}) of the reaction, $2NH_3(g) \longrightarrow N_2(g) + 3H_2(g)$ is :

(a) 46 (b) −46
(c) 92 (d) −92

127. Consider the following reaction:

$$H_2O(l) \longrightarrow H_2O(g);\ \Delta H_1 = 44\ \text{kJ}$$
$$2CH_3OH(l) + 3O_2(g) \longrightarrow 4H_2O(l) + 2CO_2(g);\ \Delta H_2 = -1453\ \text{kJ}$$

What is the value of ΔH for second reaction if water vapour instead of liquid water is formed as product?

(a) −1409 kJ (b) −1629 kJ
(c) −1277 kJ (d) None of these

128. The standard enthalpy change for the following reaction is 436.4 kJ :

$$H_2(g) \longrightarrow H(g) + H(g)$$

What is the $\Delta_f H^\circ$ of atomic hydrogen (H)?

(a) 872.8 kJ/mol (b) 218.2 kJ/mol
(c) −218.2 kJ/mol (d) −436.9 kJ/mol

129. Determine enthalpy of formation for $H_2O_2(l)$, using listed enthalpies of reaction :

$$N_2H_4(l) + 2H_2O_2(l) \longrightarrow N_2(g) + 4H_2O(l);\quad \Delta_r H_1^\circ = -818\ \text{kJ/mol}$$
$$N_2H_4(l) + O_2(g) \longrightarrow N_2(g) + 2H_2O(l);\quad \Delta_r H_2^\circ = -622\ \text{kJ/mol}$$
$$H_2(g) + 1/2\,O_2(g) \longrightarrow H_2O(l);\quad \Delta_r H_3^\circ = -285\ \text{kJ/mol}$$

(a) −383 kJ/mol (b) −187 kJ/mol
(c) −498 kJ/mol (d) None of these

130. Heat of combustion of ethanol at constant pressure and at temperature T K is found to be $-q\ \text{J mol}^{-1}$. Hence, heat of combustion (in J mol^{-1}) of ethanol at the same temperature at constant volume will be :

(a) $RT - q$ (b) $-(q + RT)$
(c) $q - RT$ (d) $q + RT$

131. Stearic acid $[CH_3(CH_2)_{16}CO_2H]$ is a fatty acid, the part of fat that stores most of the energy. 1.0 g of stearic acid was burned in a bomb calorimeter. The bomb had a heat capacity of 652 J/°C. If the temperature of 500 g water ($c = 4.18$ J/g °C) increase from 25.0 to 39.3°C, how much heat was released when the stearic acid was burned?
[Given $C_p (H_2O) = 4.18$ J/g°C]

(a) 39.21 kJ (b) 29.91 kJ
(c) 108 kJ (d) 9.32 kJ

132. Gasoline has an enthalpy of combustion 24000 kJ/ gallon. When gasoline burns in an automobile engine, approximately 30% of the energy released is used to produce mechanical work. The remainder is lost as heat transfer to the engine's cooling system. As a start on estimating how much heat transfer is required, calculate what mass of water could be heated from 25°C to 75°C by the combustion of 1.0 gallon of gasoline in an automobile?
(Given : $C (H_2O) = 4.18$ J/g°C)

(a) 34.45 kg (b) 80.383 kg
(c) 22 kg (d) 224 kg

133. A 0.05 L sample of 0.2 *M* aqueous hydrochloric acid is added to 0.05 L of 0.2 *M* aqueous ammonia in a calorimeter. Heat capacity of entire calorimeter system is 480 J/K. The temperature increase is 1.09 K. Calculate $\Delta_r H^\circ$ in kJ/mol for the following reaction :

$$HCl(aq.) + NH_3(aq.) \longrightarrow NH_4Cl(aq.)$$

(a) –52.32 (b) –61.1
(c) –55.8 (d) –58.2

134. A coffee cup calorimeter initially contains 125 g of water, at a temperature of 24.2°C. 8 g of ammonium nitrate (NH_4NO_3), also at 24.2°C, is added to the water, and the final temperature is 18.2°C. What is the heat of solution of ammonium nitrate in kJ/mol? The specific heat capacity of the solution is 4.2 J/°C g.

(a) 33.51 kJ/mol (b) 39.5 kJ/mol
(c) 32.2 kJ/mol (d) 37.3 kJ/mol

135. Read following statement(s) carefully and select the right option :

(I) The enthalpy of solution of $CaCl_2 \cdot 6H_2O$ in a large volume of water is endothermic to the extent of 3.5 kcal/ mol. If $\Delta H = -23.2$ kcal for the reaction,

$$CaCl_2(s) + 6H_2O(l) \longrightarrow CaCl_2.6H_2O(s)$$

then heat of solution of $CaCl_2$ (anhydrous) in a large volume of water is – 19.7 kcal/mol

(II) For the reaction $2Cl(g) \rightarrow Cl_2(g)$; the sign of ΔH and ΔS are negative.

(a) statement I and II both are wrong (b) both are correct
(c) only I is correct (d) only II is correct

136. If the enthalpy of formation and enthalpy of solution of HCl(*g*) are –92.3 kJ/mol and –75.14 kJ/mol respectively then find enthalpy of formation of $Cl^-(aq)$:

(a) –17.16 kJ/mol (b) –167.44 kJ/mol
(c) 17.16 kJ/mol (d) None of these

137. At 25°C, when 1 mole of $MgSO_4$ was dissolved in water, the heat evolved was found to be 91.2 kJ. One mole of $MgSO_4 \cdot 7H_2O$ on dissolution gives a solution of the same composition accompanied by an absorption of 13.8 kJ. The enthalpy of hydration, *i.e.*, ΔH_h for the reaction

$$MgSO_4(s) + 7H_2O(l) \longrightarrow MgSO_4 \cdot 7H_2O(s) \text{ is :}$$

(a) –105 kJ/mol (b) –77.4 kJ/mol
(c) 105 kJ/mol (d) None of these

138. The enthalpies of neutralization of a weak base *A*OH and a strong base *B*OH by HCl are –12250 cal/mol and –13000 cal/mol respectively. When one mole of HCl is added to a solution containing 1 mole of *A*OH and 1 mole of *B*OH, the enthalpy change was –12500 cal/ mol. In what ratio is the acid distribution between *A*OH and *B*OH?

(a) 2 : 1 (b) 2 : 3
(c) 1 : 2 (d) None of these

139. The magnitude of enthalpy change for neutralization of the reaction ;

$$Mg(OH)_2(aq) + 2HCl(aq) \longrightarrow MgCl_2(aq) + 2H_2O(l)$$

(Given : $H^+(aq) + OH^-(aq) \longrightarrow H_2O(l)$; –57 kJ/mol) will be:

(a) less that $57\,kJ\,mol^{-1}$ (b) $+57\,kJ\,mol^{-1}$
(c) greater than $57\,kJ\,mol^{-1}$ (d) None of these

140. Enthalpy of neutralization of HCl by NaOH is –55.84 kJ/mol and by NH_4OH is –51.34 kJ/mol. The enthalpy of ionization of NH_4OH is :

(a) 107.18 kJ/mol (b) 4.5 kJ/mol
(c) –4.5 kJ/mol (d) None of these

141. Which of the following reaction is endothermic ?

(a) $CaCO_3 \longrightarrow CaO + CO_2$ (b) $Fe + S \longrightarrow FeS$

(c) $NaOH + HCl \longrightarrow NaCl + H_2O$ (d) $CH_4 + 2O_2 \longrightarrow CO_2 + 2H_2O$

142. Which of the following is not correct ?

(a) Dissolution of a salt in excess of water may be endothermic process

(b) Neutralisation is always exothermic

(c) The absolute value of enthalpy (H) can be determined by calorimeter

(d) The heat of reaction at constant volume is denoted by ΔU

143. Substance $A_2B(g)$ can undergoes decomposition to form two set of products :

$$A_2B(g) \longrightarrow A_2(g) + B(g); \quad \Delta H° = 40 \text{ kJ/mol}$$

$$A_2B(g) \longrightarrow A(g) + AB(g); \quad \Delta H° = 50 \text{ kJ/mol}$$

If the molar ratio of $A_2(g)$ to $A(g)$ is 5 : 3 in a set of product gases, then the energy involved in the decomposition of 1 mole of $A_2B(g)$ is :

(a) 48.75 kJ/mol (b) 43.75 kJ/mol

(c) 46.25 kJ/mol (d) None of these

144. Boron can undergo the following reactions with the given enthalpy changes :

$$2B(s) + \frac{3}{2}O_2(g) \longrightarrow B_2O_3(s); \quad \Delta H = -1260 \text{ kJ}$$

$$2B(s) + 3H_2(g) \longrightarrow B_2H_6(g); \quad \Delta H = 30 \text{ kJ}$$

Assume no other reactions are occurring.

If in a container (operating at constant pressure) which is isolated from the surrounding, mixture of H_2 (gas) and O_2 (gas) are passed over excess of B(*s*), then calculate the molar ratio ($O_2 : H_2$) so that temperature of the container do not change :

(a) 15 : 3 (b) 42 : 1

(c) 1 : 42 (d) 1 : 84

145. The enthalpy change for the following reaction is 368 kJ. Calculate the average O—F bond energy.

$$OF_2(g) \longrightarrow O(g) + 2F(g)$$

(a) 184 kJ/mol (b) 368 kJ/mol

(c) 536 kJ/mol (d) 736 kJ/mol

146. The enthalpy change for the reaction, $C_2H_6(g) \longrightarrow 2C(g) + 6H(g)$ is X kJ. The bond energy of C—H bond is :

(a) $\frac{X}{2}$ (b) $\frac{X}{3}$

(c) $\frac{X}{6}$ (d) data insufficient

147. The table given below lists the bond dissociation energy (E_{diss}) for single covalent bonds formed between C and atoms A, B, D, E.

Bond	E_{diss} (kcal mol^{-1})
C — A	240
C — B	382
C — D	276
C — E	486

Which of the atoms has smallest size ?

(a) D (b) E

(c) A (d) D

148. Calculate P—Cl bond enthalpy

Given : $\Delta_f H\,(PCl_3, g) = 306$ kJ/mol; $\Delta H_{atomization}\,(P, s) = 314$ kJ/mol;

$\Delta_f H\,(Cl, g) = 121$ kJ/mol

(a) 123.66 kJ/mol (b) 371 kJ/mol

(c) 19 kJ/mol (d) None of these

149. Calculate the enthalpy for the following reaction using the given bond energies (kJ/mol) :

(C—H = 414; H—O = 463; H—Cl = 431, C—Cl = 326; C—O = 335)

$$CH_3\text{—}OH(g) + HCl(g) \longrightarrow CH_3\text{—}Cl(g) + H_2O(g)$$

(a) –23 kJ/mol (b) –42 kJ/mol

(c) –59 kJ/mol (d) –511 kJ/mol

150. Based on the values of B.E. given, $\Delta_f H°$ of $N_2H_4(g)$ is :

Given : N—N = 159 kJ mol^{-1}; H—H = 436 kJ mol^{-1}

N≡N = 941 kJ mol^{-1}; N—H = 398 kJ mol^{-1}

(a) 711 kJ mol^{-1} (b) 62 kJ mol^{-1}

(c) –98 kJ mol^{-1} (d) –711 kJ mol^{-1}

Level-2

1. A heating coil is immersed in a 100 g sample of $H_2O\,(l)$ at 1 bar and 100°C in a closed vessel. In this heating process, 60% of the liquid is converted in to gaseous form at constant pressure of 1 bar. Densities of liquid and gaseous water under these conditions are 1000 kg/m^3 and 0.60 kg/m^3 respectively. Magnitude of the work done for the process is :

(a) 4997 J (b) 4970 J
(c) 9994 J (d) None of these

2. A rigid and insulated tank of 3 m^3 volume is divided into two compartments. One compartment of volume of 2 m^3 contains an ideal gas at 0.8314 MPa and 400 K while the second compartment of volume 1 m^3 contains the same gas at 8.314 MPa and 500 K. If the partition between the two compartments is ruptured, the final temperature of the gas is :

(a) 420 K (b) 450 K
(c) 480 K (d) None of these

3. What is the value of change in internal energy at 1 atm in the process?

$$H_2O\,(l, 323\text{ K}) \longrightarrow H_2O\,(g, 423\text{ K})$$

Given : $C_{v,m}\,(H_2O, l) = 75.0\text{ JK}^{-1}\text{ mol}^{-1}$; $C_{p,m}\,(H_2O, g) = 33.314\text{ JK}^{-1}\text{ mol}^{-1}$

ΔH_{vap} at 373 K = 40.7 kJ/mol

(a) 42.91 kJ/mol (b) 43086 kJ/mol
(c) 42.6 kJ/mol (d) 49.6 kJ/mol

4. For an ideal gas $\frac{C_{p,m}}{C_{v,m}} = \gamma$. The molecular mass of the gas is M, its specific heat capacity at constant volume is :

(a) $\frac{R}{M(\gamma - 1)}$ (b) $\frac{M}{R(\gamma - 1)}$
(c) $\frac{\gamma RM}{\gamma - 1}$ (d) $\frac{\gamma R}{M(\gamma - 1)}$

5. 1 mole of an ideal gas $A\,(C_{v,m} = 3R)$ and 2 mole of an ideal gas B are $\left(C_{v,m} = \frac{3}{2}R\right)$ taken in a container and expanded reversible and adiabatically from 1 litre to 4 litre starting from initial temperature of 320 K. ΔE or ΔU for the process is :

(a) $-240\,R$ (b) $240\,R$
(c) $480\,R$ (d) $-960\,R$

6. Calculate the work done by the system in an irreversible (single step) adiabatic expansion of 2 mole of a polyatomic gas ($\gamma = 4/3$) from 300 K and pressure 10 atm to 1 atm :

(a) $-227\,R$ (b) $-205\,R$
(c) $-405\,R$ (d) None of these

7. One mole of an ideal gas is carried through the reversible cyclic process as shown in figure. The max. temperature attained by the gas during the cycle :

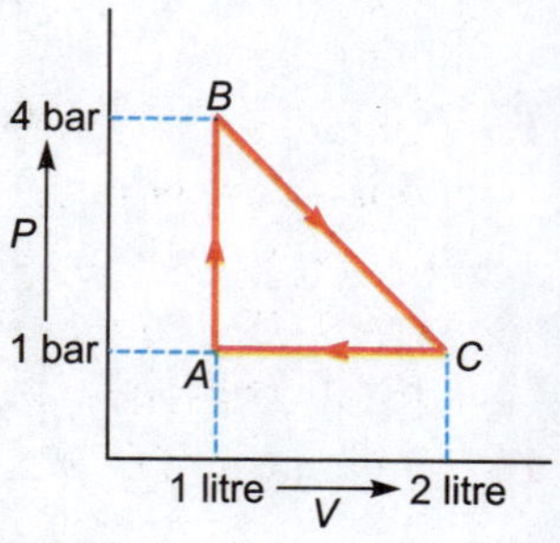

(a) $\frac{7}{6R}$ (b) $\frac{12}{49R}$
(c) $\frac{49}{12R}$ (d) None of these

8. Two moles of a triatomic linear gas (neglect vibration degree of freedom) are taken through a reversible process ideal starting from A as shown in figure. The volume ratio $\frac{V_B}{V_A} = 4$. If the temperature at A is $-73°$C, then:

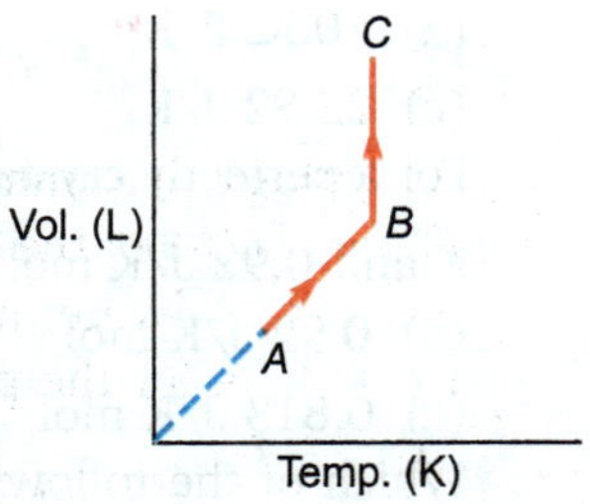

(i) Work done by the gas in AB process is :

(a) 6.16 kJ (b) 308.3 kJ

(c) 9.97 kJ (d) 0 J

(ii) Total enthalpy change in both steps is :

(a) 3000 R (b) 4200 R

(c) 2100 R (d) 0

9. A gas $\left(C_{v,m} = \frac{5}{2}R\right)$ behaving ideally was allowed to expand reversibly and adiabatically from 1 litre to 32 litre. Its initial temperature was 327°C. The molar enthalpy change (in J/mol) for the process is :

(a) $-1125\,R$ (b) -675

(c) $-1575\,R$ (d) None of these

10. Two mole of an ideal gas is heated at constant pressure of one atmosphere from 27°C to 127°C. If $C_{v,m} = 20 + 10^{-2}\,T\ \text{JK}^{-1}\cdot\text{mol}^{-1}$, then q and ΔU for the process are respectively :

(a) 6362.8 J, 4700 J (b) 3037.2 J, 4700 J

(c) 7062.8 J, 5400 J (d) 3181.4 J, 2350 J

11. 10 mole of an ideal gas is heated at constant pressure of one atmosphere from 27°C to 127°C. If $C_{v,m} = 21.686 + 10^{-3}\,T\ (\text{JK}^{-1}\cdot\text{mol}^{-1})$, then ΔH for the process is :

(a) 3000 J (b) 3350 J

(c) 3700 J (d) 30350 J

12. For polytropic process $PV^n = \text{constant}$, molar heat capacity (C_m) of an ideal gas is given by :

(a) $C_{v,m} + \frac{R}{(n-1)}$ (b) $C_{v,m} + \frac{R}{(1-n)}$

(c) $C_{v,m} + R$ (d) $C_{p,m} + \frac{R}{(n-1)}$

13. 2 mole of an ideal monoatomic gas undergoes a reversible process for which $PV^2 = C$. The gas is expanded from initial volume of 1 L to final volume of 3 L starting from initial temperature of 300 K. Find ΔH for the process :

(a) $-600\,R$ (b) $-1000\,R$

(c) $-3000\,R$ (d) None of these

14. Calculate ΔS for 3 mole of a diatomic ideal gas which is heated and compressed from 298 K and 1 bar to 596 K and 4 bar: [Given: $C_{v,m}$ (gas) $= \frac{5}{2}R$; $\ln(2) = 0.70$; $R = 2\ \text{cal K}^{-1}\ \text{mol}^{-1}$]

(a) $-14.7\ \text{cal K}^{-1}$ (b) $+14.7\ \text{cal K}^{-1}$

(c) $-4.9\ \text{cal K}^{-1}$ (d) $6.3\ \text{cal K}^{-1}$

15. One mole of an ideal monoatomic gas at 27°C is subjected to a reversible isoentropic compression until final temperature reaches to 327°C. If the initial pressure was 1.0 atm then find the value of ($\ln P_2$): (Given : $\ln 2 = 0.7$)

(a) 1.75 atm (b) 0.176 atm

(c) 1.0395 atm (d) 2.0 atm

16. Two mole of an ideal gas is expanded irreversibly and isothermally at 37°C until its volume is doubled and 3.41 kJ heat is absorbed from surrounding. ΔS_{total} (system + surrounding) is :

(a) –0.52 J/K (b) 0.52 J/K
(c) 22.52 J/K (d) 0

17. For a perfectly crystalline solid $C_{p,m} = aT^3 + bT$, where a and b are constant. If $C_{p,m}$ is 0.40 J/K mol at 10 K and 0.92 J/K mol at 20 K, then molar entropy at 20 K is :
(a) 0.92 J/K mol (b) 8.66 J/K mol
(c) 0.813 J/K mol (d) None of these

18. Which of the following statement(s) is/are correct?
Statement (i) : The entropy in isolated system with P–V work only, is always maximized at equilibrium.
Statement (ii) : It is possible for the entropy of a closed system to decrease substantially in an irreversible process.
Statement (iii) : Entropy can be created but not destroyed.
Statement (iv) : ΔS_{system} is zero for reversible process in an isolated system.
(a) Statement i, ii, iii (b) Statement ii, iv
(c) Statement i, ii, iv (d) All of these

19. Combustion of sucrose is used by aerobic organisms for providing energy for the life sustaining processes. If all the capturing of energy from the reaction is done through electrical process (non P–V work) then calculate maximum available energy which can be captured by combustion of 34.2 gm of sucrose
Given : $\Delta H_{combustion}$ (sucrose) $= -6000$ kJ mol^{-1}
$\Delta S_{combustion} = 180$ J/K-mol and body temperature is 300 K
(a) 600 kJ (b) 594.6 kJ
(c) 5.4 kJ (d) 605.4 kJ

20. For the hypothetical reaction

$$A_2(g) + B_2(g) \rightleftharpoons 2AB(g)$$

$\Delta_r G°$ and $\Delta_r S°$ are 20 kJ/mol and –20 JK^{-1} mol^{-1} respectively at 200 K.
If $\Delta_r C_p$ is 20 JK^{-1} mol^{-1} then $\Delta_r H°$ at 400 K is :
(a) 20 kJ/mol (b) 7.98 kJ/mol
(c) 28 kJ/mol (d) None of these

21. Calculate $\Delta_f G°$ for (NH_4Cl, s) at 310 K.
Given : $\Delta_f H°(NH_4Cl, s) = -314.5$ kJ/mol; $\Delta_r C_p = 0$
$S°_{N_2(g)} = 192$ JK^{-1} mol^{-1}; $S°_{H_2(g)} = 130.5$ JK^{-1} mol^{-1};
$S°_{Cl_2(g)} = 233$ JK^{-1} mol^{-1}; $S°_{NH_4Cl(s)} = 99.5$ JK^{-1} mol^{-1}
All given data at 300 K.
(a) –198.56 kJ/mol (b) –426.7 kJ/mol
(c) –202.3 kJ/mol (d) None of these

22. Using listed informations, calculate $\Delta_r G°$ (in kJ/mol) at 27°C

$$Co_3O_4(s) + 4CO(g) \longrightarrow 3Co(s) + 4CO_2(g)$$

Given : At 300 K $\Delta H°_f$ (kJ/mol) –891, –110.5, 0.0, –393.5
$S°$ (J/K-mol) 102.5, 197.7, 30.0, 213.7
(a) –214.8 (b) –195.0
(c) –200.3 (d) –256.45

23. Fixed mass of an ideal gas contained in a 24.63 L sealed rigid vessel at 1 atm is heated from –73° C to 27° C. Calculate change in Gibb's energy if entropy of gas is a function of temperature as $S = 2 + 10^{-2}\,T$ (J/K):
(Use 1 atm L = 0.1 kJ)
(a) 1231.5 J (b) 1281.5 J
(c) 781.5 J (d) 0

24. The molar heat capacities at constant pressure (assumed constant with respect to temperature) of A, B and C are in ratio of 3 : 1.5 : 2.0. The enthalpy change for the exothermic reaction $A + 2B \longrightarrow 3C$ at 300 K and 310 K is ΔH_{300} and ΔH_{310} respectively then :

(a) $\Delta H_{300} > \Delta H_{310}$
(b) $\Delta H_{300} < \Delta H_{310}$
(c) $\Delta H_{300} = \Delta H_{310}$
(d) if $T_2 > T_1$ then $\Delta H_{310} > \Delta H_{300}$ and if $T_2 < T_1$ then $\Delta H_{310} < \Delta H_{300}$

25. Determine $\Delta U°$ at 300 K for the following reaction using the listed enthalpies of reaction :

$$4CO(g) + 8H_2(g) \longrightarrow 3CH_4(g) + CO_2(g) + 2H_2O(l)$$
$$C\,(graphite) + 1/2\,O_2(g) \longrightarrow CO(g);\ \Delta H_1° = -110.5\,kJ$$
$$CO(g) + 1/2\,O_2(g) \longrightarrow CO_2(g);\ \Delta H_2° = -282.9\,kJ$$
$$H_2(g) + 1/2\,O_2(g) \longrightarrow H_2O(l);\ \Delta H_3° = -285.8\,kJ$$
$$C\,(graphite) + 2H_2(g) \longrightarrow CH_4(g);\ \Delta H_4° = -74.8\,kJ$$

(a) –653.5 kJ (b) –686.2 kJ
(c) –747.4 kJ (d) None of these

26. Calculate $\Delta_f H°$ (in kJ/mol) for Cr_2O_3 from the $\Delta_r G°$ and the $S°$ values provided at 27°C

$$4Cr(s) + 3O_2(g) \longrightarrow 2Cr_2O_3(s);\quad \Delta_r G° = -2093.4\ kJ/mol$$
$$S°\ (J/K\ mol) :\ S°(Cr, s) = 24;\quad S°(O_2, g) = 205;\quad S°(Cr_2O_3, s) = 81$$

(a) –2258.1 kJ/mol (b) –1129.05 kJ/mol
(c) –964.35 kJ/mol (d) None of these

27. Calculate the heat produced (in kJ) when 224 gm of CaO is completely converted to $CaCO_3$ by reaction with CO_2 at 27°C in a container of fixed volume.

Given : $\Delta H_f^\circ(CaCO_3, s) = -1207$ kJ/mol; $\Delta H_f^\circ(CaO, s) = -635$ kJ/mol

$\Delta H_f^\circ(CO_2, g) = -394$ kJ/mol; [Use $R = 8.3\ JK^{-1}\ mol^{-1}$]

(a) 702.04 kJ (b) 721.96 kJ
(c) 712 kJ (d) 721 kJ

28. When 1.0 g of oxalic acid ($H_2C_2O_4$) is burned in a bomb calorimeter whose heat capacity is 8.75 kJ/K, the temperature increases by 0.312 K. The enthalpy of combustion of oxalic acid at 27°C is :

(a) –245.7 kJ/mol (b) –244.452 kJ/mol
(c) –241.5 kJ/mol (d) None of these

29. Enthalpy of neutralization of H_3PO_3 acid is –106.68 kJ/mol using NaOH. Enthalpy of neutralization of HCl by NaOH is –55.84 kJ/mol. Calculate $\Delta H_{ionization}$ of H_3PO_3 into its ions :

(a) 50.84 kJ/mol (b) 5 kJ/mol
(c) 2.5 kJ/mol (d) None of these

30. The enthalpy of neutralization of a weak monoprotic acid (HA) in 1 M solution with a strong base is –55.95 kJ/mol. If the unionized acid required 1.4 kJ/mol heat for its complete ionization and enthalpy of neutralization of the strong monobasic acid with a strong monoacidic base is –57.3 kJ/mol, what is the % ionization of the weak acid in molar solution ?

(a) 1% (b) 3.57%
(c) 35.7% (d) 10%

31. Determine C—C and C—H bond enthalpy (in kJ/mol)

Given : $\Delta_f H°(C_2H_6, g) = -85$ kJ/mol, $\Delta_f H°(C_3H_8, g) = -104$ kJ/mol

$\Delta_{sub} H°(C, s) = 718$ kJ/mol, B.E. (H—H) = 436 kJ/mol

(a) 414, 345 (b) 345, 414
(c) 287, 404.5 (d) None of these

32. Consider the following data : $\Delta_f H°(N_2H_4, l) = 50$ kJ/mol, $\Delta_f H°(NH_3, g) = -46$ kJ/mol
B.E. (N—H) = 393 kJ/mol and B.E. (H—H) = 436 kJ/mol,
$\Delta_{vap} H(N_2H_4, l) = 18$ kJ/mol
The N—N bond energy in N_2H_4 is :
(a) 226 kJ/mol (b) 154 kJ/mol
(c) 190 kJ/mol (d) None of these

33. What is the bond enthalpy of Xe—F bond?

$$XeF_4(g) \longrightarrow Xe^+(g) + F^-(g) + F_2(g) + F(g)\,;\quad \Delta_r H = 292 \text{ kcal/mol}$$

Given : Ionization energy of Xe = 279 kcal/mol
B.E. (F—F) = 38 kcal/mol, Electron affinity of F = 85 kcal/mol
(a) 24 kcal/mol (b) 34 kcal/mol
(c) 8.5 kcal/mol (d) None of these

34. If enthalpy of hydrogenation of $C_6H_6(l)$ into $C_6H_{12}(l)$ is –205 kJ/mol and resonance energy of $C_6H_6(l)$ is – 152 kJ/ mol then enthalpy of hydrogenation of 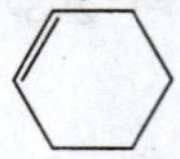 is
Assume ΔH_{vap} of $C_6H_6(l), C_6H_8(l), C_6H_{12}(l)$ all are equal :
(a) –535.5 kJ/mol (b) –238 kJ/mol
(c) –357 kJ/mol (d) – 119 kJ/mol

35. The enthalpy of combustion of propane (C_3H_8) gas in terms of given data is :

Bond energy (kJ/mol)				
ε_{C-H}	$\varepsilon_{O=O}$	$\varepsilon_{C=O}$	ε_{O-H}	ε_{C-C}
$+x_1$	$+x_2$	$+x_3$	$+x_4$	x_5

Resonance energy of CO_2 is $-z$ kJ/mol and $\Delta H_{vaporization}$ $[H_2O(l)]$ is y (kJ/mol)
(a) $8x_1 + 2x_5 + 5x_2 - 6x_3 - 8x_4 - 4y - 3z$ (b) $6x_1 + x_5 + 5x_2 - 3x_3 - 4x_4 - 4y - 3z$
(c) $8x_1 + 2x_5 + 5x_2 - 6x_3 - 8x_4 - y - z$ (d) $8x_1 + x_5 + 5x_2 - 6x_3 - 8x_4 - 4y + 3z$

36. The efficiency of an ideal gas with adiabatic exponent 'γ' for the shown cyclic process would be

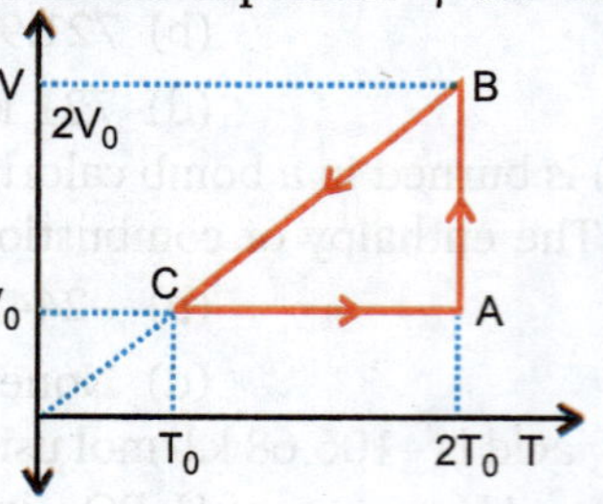

(a) $\dfrac{(\gamma-1)(2\ln 2-1)}{1+(\gamma-1)2\ln 2}$ (b) $\dfrac{(\gamma-1)(1-2\ln 2)}{(\gamma-1)2\ln 2-1}$

(c) $\dfrac{(2\ln 2+1)(\gamma-1)}{(\gamma-1)2\ln 2+1}$ (d) $\dfrac{(2\ln 2-1)}{\gamma/(\gamma-1)}$

37. Initially one mole of ideal gas $\left(C_v = \dfrac{5}{2}R\right)$ at 1.0 atm and 300 K is put through the following cycle :

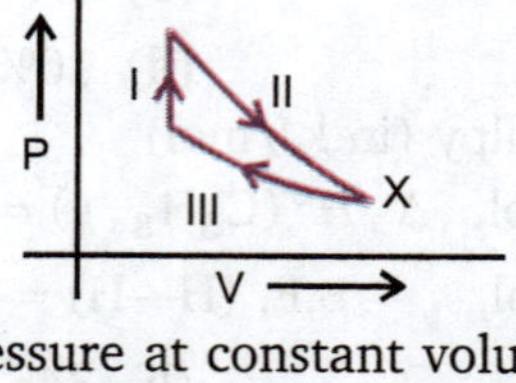

Step-I : Heating to twice its initial pressure at constant volume.
Step-II : Adiabatic expansion to its initial temperature.

Step-III : Isothermal compression back to 1.00 atm.
What is the volume at state X ?

(a) 40.4 L (b) 65.0 L
(c) 139 L (d) 4.35 L

38. What is the enthalpy of neutralization of HF against a strong base ?
Given : $H^+(aq) + OH^-(aq) \longrightarrow H_2O(l)$; $\Delta_r H° = -56$ kJ/mol
$\Delta_s H°(HF, aq) = -329$ kJ/mol ; $\Delta_s H°(H_2O, l) = -285$ kJ/mol
$\Delta_s H°(F^-, aq) = -320$ kJ/mol

(a) −17 kJ/mol (b) −38 kJ/mol
(c) −47 kJ/mol (d) −43 kJ/mol

39. The molar heat capacities of Iodine vapour and solid are 7.8 and 14 cal/mol respectively. If enthalpy of sublimation of iodine is 6096 cal/mol at 200°C, then what is ΔU (internal energy change) at 250°C in cal/mol

(a) 5360 (b) 4740
(c) 6406 (d) None of these

40. For the reaction taking place at certain temperature $NH_2COONH_4(s) \rightleftharpoons 2NH_3(g) + CO_2(g)$, if equilibrium pressure is $3X$ bar then $\Delta_r G°$ would be

(a) $-RT \ln 9 - 3RT \ln X$ (b) $RT \ln 4 - 3RT \ln X$
(c) $-3RT \ln X$ (d) None of these

Level-3

Passage-1

The first law of thermodynamics for a closed system is $dU = dq + dw$, where $dw = dw_{pv} + dw_{non\text{-}pv}$. The most common type of $w_{non\text{-}pv}$ is electrical work. As per IUPAC convention work done on the system is positive.

1. A system generates 50 J of electrical energy, has 150 J of pressure-volume work done on it by the surroundings while releasing 300 J of heat energy. What is the change in the internal energy of the system?

(a) –500 (b) –100
(c) –300 (d) –200

2. A system generates 50 J of electrical energy and delivers 150 J of pressure-volume work against the surroundings while releasing 300 J of heat energy. What is the change in the internal energy of the system?

(a) –100 (b) –400
(c) –300 (d) –500

Passage-2

If the boundary of system moves by an infinitesimal amount, the work involved is given by $dw = -P_{ext}\, dV$

for irreversible process $w = -P_{ext}\, \Delta V$ (where $\Delta V = V_f - V_i$)

for reversible process $P_{ext} = P_{int} \pm dP \simeq P_{int}$

so for reversible isothermal process $w = -nRT \ln \dfrac{V_f}{V_i}$

2 mole of an ideal gas undergoes isothermal compression along three different paths :

(i) reversible compression from $P_i = 2$ bar and $V_i = 8$ L to $P_f = 20$ bar

(ii) a single stage compression against a constant external pressure of 20 bar, and

(iii) a two stage compression consisting initially of compression against a constant external pressure of 10 bar until $P_{gas} = P_{ext}$, followed by compression against a constant pressure of 20 bar until $P_{gas} = P_{ext}$.

1. Work done (in bar-L) on the gas in reversible isothermal compression is :

(a) 9.212 (b) 36.848
(c) 18.424 (d) None of these

2. Work done on the gas in single stage compression is :

(a) 36 (b) 72
(c) 144 (d) None of these

3. Total work done on the gas in two stage compression is :

(a) 40 (b) 80
(c) 160 (d) None of these

4. Order of magnitude of work is :

(a) $w_1 > w_2 > w_3$ (b) $w_3 > w_2 > w_1$
(c) $w_2 > w_3 > w_1$ (d) $w_1 = w_2 = w_3$

Passage-3

Standard Gibb's energy of reaction ($\Delta_r G°$) at a certain temperature can be computed as $\Delta_r G° = \Delta_r H° - T.\Delta_r S°$ and the change in the value of $\Delta_r H°$ and $\Delta_r S°$ for a reaction with temperature can be computed as follows :

$$\Delta_r H°_{T_2} - \Delta_r H°_{T_1} = \Delta_r C°_p (T_2 - T_1)$$

$$\Delta_r S°_{T_2} - \Delta_r S°_{T_1} = \Delta_r C°_p \ln\left(\frac{T_2}{T_1}\right)$$

$$\Delta_r G° = \Delta_r H° - T.\Delta_r S°$$

and by $\Delta_r G° = -RT \ln K_{eq}$.

Consider the following reaction : $CO(g) + 2H_2(g) \rightleftharpoons CH_3OH(g)$

Given : $\Delta_f H° (CH_3OH, g) = -201$ kJ/mol; $\Delta_f H° (CO, g) = -114$ kJ/mol

$S° (CH_3OH, g) = 240$ J/K-mol ; $S° (H_2, g) = 29$ JK^{-1} mol^{-1}

$S° (CO, g) = 198$ J/mol-K ; $C°_{p,m} (H_2) = 28.8$ J/mol-K

$C°_{p,m} (CO) = 29.4$ J/mol-K; $C°_{p,m} (CH_3OH) = 44$ J/mol-K

and $\ln\left(\frac{320}{300}\right) = 0.06$, all data at 300 K

1. $\Delta_r S°$ at 300 K for the reaction is :
(a) 152.6 J/K-mol (b) 181.6 J/K-mol
(c) –16 J/K-mol (d) None of these

2. $\Delta_r H°$ at 300 K for the reaction is :
(a) –87 kJ/mol (b) 87 kJ/mol
(c) –315 kJ/mol (d) –288 kJ/mol

3. $\Delta_r S°$ at 320 K is :
(a) 155.18 J/mol-K (b) 150.02 J/mol-K
(c) 172 J/mol-K (d) None of these

4. $\Delta_r H°$ at 320 K is :
(a) –288.86 kJ/mol (b) –289.1 kJ/mol
(c) –87.86 kJ/mol (d) None of these

5. $\Delta_r G°$ at 320 K is :
(a) – 48295.2 kJ/mol (b) – 240.85 kJ/mol
(c) 240.85 kJ/mol (d) – 81.91 kJ/mol

Passage-4

Enthalpy of neutralization is defined as the enthalpy change when 1 mole of acid/base is completely neutralized by base/acid in dilute solution.

For strong acid and strong base neutralization net chemical change is

$$H^+(aq) + OH^-(aq) \longrightarrow H_2O(l); \quad \Delta_r H° = -55.84 \text{ kJ/mol}$$

$\Delta H°_{\text{ionization}}$ of aqueous solution of strong acid and strong base is zero.

When a dilute solution of a weak acid or base is neutralized, the enthalpy of neutralization is some what less because of the absorption of heat in the ionization of the weak acid or base,

for weak acid/base

$$\Delta H°_{\text{neutralization}} = \Delta H°_{\text{ionization}} + \Delta_r H° \ (H^+ + OH^- \rightarrow H_2O)$$

1. If enthalpy of neutralization of CH_3COOH by HCl is –49.86 kJ/mol then enthalpy of ionization of CH_3COOH is :
   (a) 5.98 kJ/mol (b) –5.98 kJ/mol
   (c) 105.7 kJ/mol (d) None of these
2. What is $\Delta H°$ for complete neutralization of strong diacidic base $A(OH)_2$ by HNO_3?
   (a) –55.84 kJ/mol (b) –111.68 kJ/mol
   (c) 55.84 kJ/mol (d) None of these
3. Under the same conditions how many mL of 0.1 *M* NaOH and 0.05 *M* H_2A (strong diprotic acid) solution should be mixed for a total volume of 100 mL produce the highest rise in temperature :
   (a) 25 : 75 (b) 50 : 50
   (c) 75 : 25 (d) 66.66 : 33.33

Passage-5

Gibbs Helmholtz equation relates the enthalpy, entropy and free energy change of the process at constant pressure and temperature as

$$\Delta G = \Delta H - T\Delta S \quad \text{(at constant P, T)}$$

In General the magnitude of ΔH does not change much with the change in temperature but the term $T\Delta S$ changes appreciably. Hence in some process spontaneity is very much dependent on temperature and such processes are generally known as entropy driven process.

1. For the reaction at 298 K; $A_2B_4 \longrightarrow 2AB_2$
   $\Delta H = 2\,\text{kJ}$ and $\Delta S = 20\,\text{J/K}$ at constant *P* and *T*, the reaction will be
   (a) spontaneous and entropy driven (b) spontaneous and enthalpy driven
   (c) non-spontaneous (d) at equilibrium
2. When $CaCO_3$ is heated to a high temperature it decomposes into CaO and CO_2, however it is quite stable at room temperature. It can be explained by the fact that
   (a) $\Delta_r H$ dominates the term $T\Delta S$ at high temperature
   (b) the term $T\Delta S$ dominates the $\Delta_r H$ at high temperature
   (c) at high temperature both $\Delta_r S$ and $\Delta_r H$ becomes negative
   (d) thermodynamics can not say anything about spontaneity
3. The Dissolution of $CaCl_2 \cdot 6H_2O$ in a large volume of water is endothermic to the extent of 3.5 kcal mol^{-1} and ΔH for the reaction is $-23.2\,\text{kcal mol}^{-1}$.
   $$CaCl_2(s) + 6H_2O(l) \longrightarrow CaCl_2 \cdot 6H_2O(s)$$
   Select the correct statement :
   (a) $\Delta H_{Solution}$ for anhydrous $CaCl_2$ is –19.7 kcal/mol and the process is enthalpy driven
   (b) $\Delta H_{Solution}$ for anhydrous $CaCl_2$ is –19.7 kcal/mol and the process is entropy driven
   (c) Dissolution of $CaCl_2 \cdot 6H_2O$ in water is enthalpy driven process
   (d) The $\Delta_r S$ the reaction $CaCl_2(s) + 6H_2O(l) \longrightarrow CaCl_2 \cdot 6H_2O(s)$ is negative

One or More Answer is/are Correct

1. Identify the intensive quantities from the following :
   (a) Enthalpy (b) Temperature
   (c) Pressure (d) Mass
2. Identify the extensive quantities from the following :
   (a) Gibb's energy (b) Entropy
   (c) Refractive index (d) Specific heat

3. Identify the state functions from the following :

(a) Heat (b) Work

(c) Enthalpy (d) Enthalpy change

4. Which of the following statements is/are correct as per IUPAC sign convention?

(a) The work done by the system on the surrounding is negative

(b) The work done by the surrounding on the system is positive

(c) The heat absorbed by the system from the surrounding is positive

(d) The heat absorbed by the surrounding from the system is positive

5. In an isothermal irreversible expansion of an ideal gas as per IUPAC sign convention :

(a) $\Delta U = 0$ (b) $\Delta H = 0$

(c) $w = -nRT \ln \frac{P_1}{P_2}$ (d) $w = -q$

6. In reversible isothermal expansion of an ideal gas :

(a) $w = 0$ (b) $U_1 = U_2$

(c) $H_1 = H_2$ (d) $q = nRT \ln \frac{V_2}{V_1}$

7. An adiabatic process is that process in which :

(a) energy is transferred as heat

(b) no energy is transfer red as heat

(c) $\Delta U = w$

(d) the temp. of gas increases in a reversible adiabatic compression

8. In adiabatic process, the work involved during expansion or compression of an ideal gas is given by :

(a) $nC_v \Delta T$ (b) $\frac{nR}{\gamma - 1}(T_2 - T_1)$

(c) $-nRP_{ext}\left[\frac{T_2P_1 - T_1P_2}{P_1P_2}\right]$ (d) $-2.303\, RT \log \frac{V_2}{V_1}$

9. One mole of an ideal gas is subjected to a two step reversible process (A—B and B—C). The pressure at A and C is same. Mark the correct statement(s) :

(a) Work involved in the path AB is zero

(b) In the path AB work will be done on the gas by the surrounding

(c) Volume of gas at $C = 3 \times$ volume of gas at A

(d) Volume of gas at B is 16.42 litres

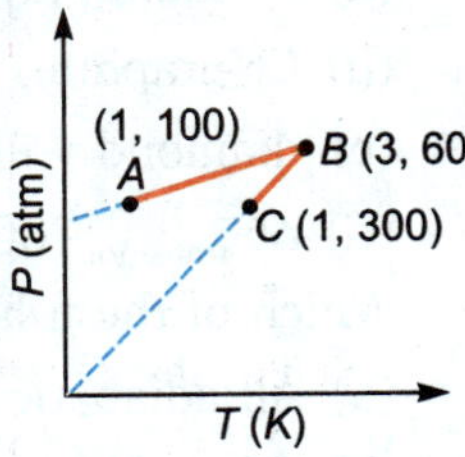

10. Assume ideal gas behaviour for all the gases considered and neglect vibrational degrees of freedom. Separate equimolar samples of Ne, O_2, CO_2 and SO_2 were subjected to a two process as mentioned. Initially all are at same state of temperature and pressure.

Step I ⟶ All undergo reversible adiabatic expansion to attain same final volume, which is double the original volume thereby causing the decreases in their temperature.

Step II ⟶ After step I all are given appropriate amount of heat isochorically to restore the original temperature.

Mark the correct option(s) :

(a) Due to step I only, the decrease in temperature will be maximum for Ne

(b) During step II, heat given will be minimum for SO_2

(c) There will be no change in internal energy for any of the gas after both the steps of process are competed

(d) The P–V graph of O_2 and CO_2 will be same

11. Which of the following is/are correct?

(a) $\Delta H = \Delta U + \Delta (PV)$ when P and V both changes

(b) $\Delta H = \Delta U + P\Delta V$ when pressure is constant

(c) $\Delta H = \Delta U + V\Delta P$ when volume is constant

(d) $\Delta H = \Delta U + P\Delta V + V\Delta P$ when P and V both changes

12. $\Delta H < \Delta U$ for the reaction(s) :

(a) $N_2(g) + 3H_2(g) \longrightarrow 2NH_3(g)$

(b) $Ag_2O(s) \longrightarrow 2Ag(s) + \frac{1}{2}O_2(g)$

(c) $CO(g) + \frac{1}{2}O_2(g) \longrightarrow CO_2(g)$

(d) $C(s) + O_2(g) \longrightarrow CO_2(g)$

13. Which of the following conditions may lead to a non-spontaneous change?

(a) ΔH and ΔS both + ve

(b) $\Delta H = -$ ve; $\Delta S = +$ ve

(c) $\Delta H = +$ ve; $\Delta S = -$ ve

(d) $\Delta H = -$ ve; $\Delta S = -$ ve

14. For a process to be spontaneous :

(a) $(\Delta G_{system})_{T,\,p} = 0$

(b) $\Delta S_{system} + \Delta S_{surrounding} > 0$

(c) $\Delta S_{system} + \Delta S_{surrounding} < 0$

(d) $(\Delta G_{system})_{T,\,p} < 0$

15. The normal boiling point of a liquid 'X' is 400 K. Which of the following statement is true about the process $X(l) \longrightarrow X(g)$?

(a) at 400 K and 1 atm pressure $\Delta G = 0$

(b) at 400 K and 2 atm pressure $\Delta G = +$ ve

(c) at 400 K and 0.1 atm pressure $\Delta G = -$ ve

(d) at 410 K and 1 atm pressure $\Delta G = +$ ve

16. When ice melts at 1°C :

(a) an increase in entropy

(b) a decrease in enthalpy

(c) a decrease in free energy

(d) process is spontaneous

17. The value of $\Delta H_{transition}$ of C (graphite) $\longrightarrow$ C (diamond) is 1.9 kJ/mol at 25°C. Entropy of graphite is higher than entropy of diamond. This implies that :

(a) C (diamond) is more thermodynamically stable than C (graphite) at 25°C

(b) C (graphite) is more thermodynamically stable than C (diamond) at 25°C

(c) diamond will provide more heat on complete combustion at 25°C

(d) $\Delta G_{transition}$ of C (diamond) $\longrightarrow$ C (graphite) is –ve

18. Which of the following statement(s) is/are false?

(a) All adiabatic processes are isoentropic (or isentropic) processes

(b) When $(\Delta G_{system})_{T,\,p} < 0$; the reaction must be exothermic

(c) $dG = VdP - SdT$ is applicable for closed system, both PV and non-PV work

(d) The heat of vaporisation of water at 100°C is 40.6 kJ/mol. When 9 gm of water vapour condenses to liquid at 100°C and 1 atm, then $\Delta S_{system} = 54.42$ J/K

19. Which of the following statement(s) is/are true?

(a) $\Delta U = 0$ for combustion of $C_2H_6(g)$ in a sealed rigid adiabatic container

(b) $\Delta_f H°$ (S, monoclinic) $\neq 0$

(c) If dissociation energy of $CH_4(g)$ is 1656 kJ/mol and $C_2H_6(g)$ is 2812 kJ/mol, then value of C—C bond energy will be 328 kJ/mol

(d) If $\Delta H_f(H_2O, g) = -242$ kJ/mol; $\Delta H_{vap}(H_2O, l) = 44$ kJ/mol then, $\Delta_f H°(OH^-, aq.)$ will be –142 kJ/mol

20. From the following data, mark the option(s) where ΔH is correctly written for the given reaction.

Given : $H^+(aq) + OH^-(aq) \longrightarrow H_2O(l)$; $\Delta H = -57.3$ kJ

$\Delta H_{\text{solution}}$ of $HA(g) = -70.7$ kJ/mol

$\Delta H_{\text{solution}}$ of $BOH(g) = 20$ kJ/mol

$\Delta H_{\text{ionization}}$ of $HA = 15$ kJ/mol and BOH is a strong base.

Reaction	$\Delta_r H$ (kJ/mol)
(a) $HA(aq) + BOH(aq) \longrightarrow BA(aq) + H_2O$	–42.3
(b) $HA(g) + BOH(g) \longrightarrow BA(aq) + H_2O$	–93
(c) $HA(g) \longrightarrow H^+(aq) + A^-(aq)$	–55.7
(d) $B^+(aq) + OH^-(aq) \longrightarrow BOH(aq)$	–20

21. Select correct statement(s)

(a) An adiabatic system can exchange energy with its surroundings.

(b) A thermodynamic property which is intensive is additive.

(c) Work done may be zero in a cyclic process.

(d) For a simple compressible substance, the relation $dq - P.dv = 0$ is true for any cycle involving mechanical work only.

22. For an isolated system, the entropy :

(a) either increases or remains constant

(b) either decreases or remains constant

(c) can never decrease

(d) can never increase

23. The normal boiling point of a liquid X is 400 K. ΔH_{vap} at normal boiling point is 40 kJ/mol. Select correct statement(s) :

(a) $\Delta S_{\text{Vaporisation}} < 100$ J/mol.K at 400 K and 2 atm

(b) $\Delta S_{\text{Vaporisation}} < 10$ J/mol.K at 400 K and 1 atm

(c) $\Delta G_{\text{Vaporisation}} < 0$ at 410 K and 1 atm

(d) $\Delta U = 43.32$ kJ/mol.K at 400 K and 1 atm

24. Select incorrect statement(s)

(a) A closed system with all adiabatic boundaries must be an isolated system

(b) Total heat exchange in a cyclic process may be zero

(c) Entropy of a closed system is maximum at equilibrium

(d) Molar Gibb's Energy is an extensive property

25. Select correct statement(s) for the reaction $H_2O(g) + CO(g) \longrightarrow H_2(g) + CO_2(g)$

Substance	$CO(g)$	$CO_2(g)$	$H_2O(g)$	$H_2(g)$
$\Delta_f H^\circ_{400}$ (kcal mol^{-1})	–25	–95	–55	0
S°_{400} (cal mol^{-1}K^{-1})	45	50	40	30

(a) Reaction is enthalpy driven

(b) Reaction is entropy driven

(c) Reaction is spontaneous at 400 K

(d) Reaction is non-spontaneous at 400 K

Match the Column

Column-I and **Column-II** contains four entries each. Entries of Column-I are to be matched with some entries of column-II. One or more than one entries of Column-I may have the matching with the same entries of Column-II.

1.

Column-I	Column-II
(A) Reversible cooling of an ideal gas at constant volume	(P) $w = 0; q < 0; \Delta U < 0$
(B) Reversible isothermal expansion of an ideal gas	(Q) $w < 0; q > 0; \Delta U > 0$
(C) Adiabatic expansion of non-ideal gas into vacuum	(R) $w = 0; q = 0; \Delta U = 0$
(D) Reversible melting of sulphur at normal melting point	(S) $w < 0; q > 0; \Delta U = 0$

2.

Column-I	Column-II
(A) Adiabatic process	(P) $q = 0$
(B) Isothermal process	(Q) $\Delta H = 0$
(C) Isoenthalpic process	(R) $\Delta T = 0$
(D) Isoentropic process	(S) $\Delta S = 0$

3.

Column-I	Column-II
(A) Reversible isothermal expansion of an ideal gas	(P) $w = -2.303\, nRT \log\left(\frac{V_2}{V_1}\right)$
(B) Reversible adiabatic compression of an ideal gas	(Q) $PV^\gamma = \text{constant}$
(C) Irreversible adiabatic expansion of an ideal gas	(R) $w = \frac{nR}{(\gamma - 1)}(T_2 - T_1)$
(D) Irreversible isothermal compression of an ideal gas	(S) $\Delta H = 0$

4.

Column-I	Column-II
(A) $PCl_5(g) \rightleftharpoons PCl_3(g) + Cl_2(g)$ in a closed system	(P) $\Delta H < \Delta U$
(B) $2HI(g) \rightleftharpoons H_2(g) + I_2(g)$ in a closed system	(Q) $\Delta H = \Delta U \neq 0$
(C) $N_2(g) + 3H_2(g) \rightleftharpoons 2NH_3(g)$ in a closed system	(R) $\Delta H > \Delta U$
(D) $CO(g) + \frac{1}{2}O_2(g) \rightleftharpoons CO_2(g)$ in an isolated system	(S) $\Delta U = 0$

5.

Column-I (Process)	Column-II (Entropy Change)
(A) Reversible isothermal compression of an ideal gas	(P) $\Delta S_{\text{system}} > 0$
(B) Isothermal free expansion ($P_{\text{ext}} = 0$) of an ideal gas	(Q) $\Delta S_{\text{system}} < 0$
(C) Reversible adiabatic expansion of an ideal gas	(R) $\Delta S_{\text{system}} = 0$
(D) Reversible ideal gas expansion	(S) Information insufficient

6.

Column-I (Process)	Column-II (Entropy change)
(A) Reversible isothermal ideal gas expansion	(P) $\Delta S_{surrounding} = 0$
(B) Reversible adiabatic ideal gas compression	(Q) $\Delta S_{surrounding} < 0$
(C) Adiabatic free expansion ($P_{ext} = 0$) of an ideal gas	(R) $\Delta S_{surrounding} > 0$
(D) Irreversible isothermal ideal gas compression	(S) $\Delta S_{system} = 0$

7.

Column-I	Column-II
(A) $(\Delta G_{system})_{T,P} = 0$	(P) Process is in equilibrium
(B) $\Delta S_{system} + \Delta S_{surrounding} > 0$	(Q) Process is nonspontaneous
(C) $\Delta S_{system} + \Delta S_{surrounding} < 0$	(R) Process is spontaneous
(D) $(\Delta G_{system})_{T,P} > 0$	(S) System is unable to do useful work

8.

Column-I	Column-II
(A) $(\Delta G_{system})_{T,P}$	(P) $nR \ln\left(\frac{V_2}{V_1}\right)$
(B) Work done in reversible isothermal ideal gas expansion	(Q) $nRT \ln\left(\frac{P_2}{P_1}\right)$
(C) ΔG for reversible isothermal expansion of an ideal gas	(R) $-nFE$
(D) ΔS_{gas} for isothermal expansion of an ideal gas	(S) $nR \ln\left(\frac{P_1}{P_2}\right)$

9.

Column-I	Column-II
(A) Heating of an ideal gas at constant pressure	(P) $\Delta H = nC_{p,m} \Delta T \neq 0$
(B) Compression of liquid at constant temperature	(Q) $\Delta U = 0$
(C) Reversible process for an ideal gas at constant temperature	(R) $\Delta G = V \Delta P$
(D) Adiabatic free expansion of an ideal gas	(S) $\Delta G = nRT \ln\left(\frac{P_2}{P_1}\right)$

10.

Column-I (Sign of ΔH and ΔS respectively)	Column-II (Nature of reaction)
(A) – & –	(P) Spontaneous only at low temperature
(B) – & +	(Q) Spontaneous only at high temperature
(C) + & +	(R) Spontaneous at all temperature
(D) + & –	(S) Non-spontaneous at all temperature

11.

Column-I	Column-II
(A) Reversible adiabatic compression	(P) $\Delta S_{system} > 0$
(B) Reversible vaporisation of liquid	(Q) $\Delta S_{system} < 0$
(C) $2N(g) \rightarrow N_2(g)$	(R) $\Delta S_{surrounding} < 0$
(D) $MgCO_3(s) \xrightarrow{\Delta} MgO(s) + CO_2(g)$	(S) $\Delta S_{surrounding} = 0$

12.

Column-I	Column-II
(A) $H^+(aq)$	(P) $\Delta_f H^\circ = 0$
(B) $H(g)$	(Q) $\Delta_f H^\circ \neq 0$
(C) $H_2(g)$	(R) $\Delta_f G^\circ = 0$
(D) C (s, diamond)	(S) $\Delta_f S^\circ < 0$

13.

Column-I (Partial derivative)	Column-II (Thermodynamic variable)
(A) $\left(\frac{\partial U}{\partial T}\right)_V$	(P) C_P
(B) $\left(\frac{\partial H}{\partial T}\right)_P$	(Q) C_V
(C) $\left(\frac{\partial G}{\partial T}\right)_P$	(R) $-S$
(D) $\left(\frac{\partial G}{\partial P}\right)_T$	(S) V

14.

Column-I	Column-II
(A) $C(s, \text{graphite}) + O_2(g) \rightarrow CO_2(g)$	(P) $\Delta H^\circ_{combustion}$
(B) $C(s, \text{graphite}) \rightarrow C(g)$	(Q) $\Delta H^\circ_{formation}$
(C) $CO(g) + \frac{1}{2}O_2(g) \rightarrow CO_2(g)$	(R) $\Delta H^\circ_{atomization}$
(D) $CH_4(g) \rightarrow C(g) + 4H(g)$	(S) $\Delta H^\circ_{sublimation}$

15.

Column-I	Column-II
(A) $O_2(g)$	(P) $\Delta_f H^\circ = +\text{ve}$; $\Delta_f S^\circ = +\text{ve}$
(B) $O_3(g)$	(Q) $\Delta_f H^\circ = -\text{ve}$; $\Delta_f S^\circ = -\text{ve}$
(C) $Br_2(g)$	(R) $\Delta_f H^\circ = +\text{ve}$; $\Delta_f S^\circ = -\text{ve}$
(D) $H_2O(l)$	(S) $\Delta_f H^\circ = 0$; $\Delta_f S^\circ = 0$

Assertion-Reason Type Questions

Each question contains STATEMENT-1 (Assertion) and STATEMENT-2 (Reason).
Examine the statements carefully and mark the correct answer according to the instructions given below :
(A) If both the statements are TRUE and STATEMENT-2 is the correct explanation of STATEMENT-1
(B) If both the statements are TRUE but STATEMENT-2 is NOT the correct explanation of STATEMENT-1
(C) If STATEMENT-1 is TRUE and STATEMENT-2 is FALSE
(D) If STATEMENT-1 is FALSE and STATEMENT-2 is TRUE

1. **STATEMENT-1 :** Heat and work are "definite quantities".
   STATEMENT-2 : Heat and work are not properties of a system. Their values depend on the path of the process and vary accordingly.
2. **STATEMENT-1 :** There is no change in internal energy for an ideal gas at constant temperature.
   STATEMENT-2 : Internal energy of an ideal gas is a function of temperature only.
3. **STATEMENT-1 :** The heat absorbed during the isothermal expansion of an ideal gas against vacuum is zero.
   STATEMENT-2 : The volume occupied by the molecules of an ideal gas is zero.
4. **STATEMENT-1 :** ΔH and ΔU are the same for the reaction, $N_2(g) + O_2(g) \rightleftharpoons 2NO(g)$,
   STATEMENT-2 : All reactants and products are gases where all gases are ideal.
5. **STATEMENT-1 :** The magnitude of the work involved in an isothermal expansion is greater than that involved in an adiabatic expansion.
   STATEMENT-2 : P–V curve (P on y-axis and V on x-axis) decrease more rapidly for reversible adiabatic expansion compared to reversible isothermal expansion starting from same initial state.
6. **STATEMENT-1 :** Entropy change in reversible adiabatic expansion of an ideal gas is zero.
   STATEMENT-2 : The increase in entropy due to volume increase just compensate the decrease in entropy due to fall in temperature.
7. **STATEMENT-1 :** There cannot be chemical equilibrium in an open system.
   STATEMENT-2 : There is no fixed mass in an open system.
8. **STATEMENT-1 :** The standard free energy changes of all spontaneously occuring reactions are negative.
   STATEMENT-2 : The standard free energies change of the elements in their standard states at 1 bar and 298 K are taken as zero.
9. **STATEMENT-1 :** Enthalpy and entropy of any elementary substance in the standard states are taken as zero.
   STATEMENT-2 : At absolute zero, particles of the perfectly crystalline substance become completely motionless.
10. **STATEMENT-1 :** A reaction which is spontaneous and accompanied by decrease of randomness must be exothermic.
    STATEMENT-2 : All exothermic reactions are accompanied by decrease of randomness.
11. **STATEMENT-1 :** Many endothermic reactions that are not spontaneous at room temperatures become spontaneous at high temperature.
    STATEMENT-2 : ΔH° of the endothermic reaction may increases with increase in temperature.
12. **STATEMENT-1 :** Decrease of free energy during the process under constant temperature and pressure provides a measure of its spontaneity.
    STATEMENT-2 : A spontaneous change must have +ve sign of ΔS_{system}.
13. **STATEMENT-1 :** $|\Delta_f H|$ of $(H_2O, l) > |\Delta_f H|$ of (H_2O, g)
    STATEMENT-2 : $\Delta H_{condensation}$ is negative.

14. STATEMENT-1 : All combustion reactions are exothermic.

STATEMENT-2 : Enthalpies of products are greater than enthalpies of reactants ($\Sigma\, \nu_p \Delta_f H(P) > \Sigma\, \nu_R \Delta_f H(R)$)

15. STATEMENT-1 : Enthalpy of neutralization of CH_3COOH by NaOH is less than that of HCl by NaOH.

STATEMENT-2 : Enthalpy of neutralization of CH_3COOH is less because of the absorption of heat in the ionization process.

16. STATEMENT-1 : Internal energy of a real gas may change during expansion at constant temperature.

STATEMENT-2 : Internal energy of a real gas is a function of temperature and volume.

17. STATEMENT-1 : Work is a state function which is expressed in joule.

STATEMENT-2 : Work appears only at the boundary of the system.

18. STATEMENT-1 : The expansion of a gas into an evacuated space takes place non-spontaneously.

STATEMENT-2 : A process in which all steps cannot be retraced by themselves is called a spontaneous process.

Level-4

Subjective Problems

1. A perfect gas undergoes a reversible adiabatic expansion from (300 K, 200 atm) to (90 K, 10 atm). Find the atomicity of gas.
2. 5 mole of an ideal gas at temp. T are compressed isothermally from 12 atm. to 24 atm. Calculate the value of $10r$
   Where, $r = \dfrac{\text{Work done along reversible process}}{\text{Work done along single step irreversible process}}$ (Given : $\ln 2 = 0.7$)
3. A diatomic ideal gas is expanded according to PV^3 = constant, under very high temperature (Assume vibration mode active). Calculate the molar heat capacity of gas (in cal / mol K) in this process.
4. A heat engine is operating between 500 K to 300 K and it absorbs 10 kcal of heat from 500 K reservoir reversibly per cycle. Calculate the work done (in kcal) per cycle.
5. In a carnot cycle involving ideal non-linear triatomic gas, if during adiabatic expansion volume increases from 2L to 16 L and heat absorbed during isothermal expansion is 8 kcal, then calculate magnitude of work done by carnot engine (in kcal).
6. Molar heat capacities at constant pressure for A, B and C are 3, 1.5 and 2 J/K mol. The enthalpy of reaction and entropy of reaction, $A + 2B \longrightarrow 3C$ are 20 kJ/mol and 20 J/K mol at 300 K. Calculate ΔG (in kJ / mol) for the reaction ,
   $$\frac{1}{2}A + B \longrightarrow \frac{3}{2}C$$
7. Standard molar enthalpy of combustion of glucose is –2880 kJ. If only 25% of energy is available for muscular work and 1.0 km walk consumes 90 kJ of energy, what maximum distance (in km) a person can walk after eating 90 g of glucose.
8. Given $C_2H_2(g) + H_2(g) \longrightarrow C_2H_4(g)$: $\Delta H^\circ = -175\,\text{kJ mol}^{-1}$
   $\Delta H^\circ_{f(C_2H_4,g)} = 50\,\text{kJ mol}^{-1}$; $\Delta H^\circ_{f(H_2O,l)} = -280\,\text{kJ mol}^{-1}$; $\Delta H^\circ_{f(CO_2 g)} = -390\,\text{kJ mol}^{-1}$
   If ΔH° is enthalpy of combustion (in kJ mol^{-1}) of $C_2H_2(g)$, then calculate the value of $\left|\dfrac{\Delta H^\circ}{257}\right|$.
9. The integral enthalpies of solution of anhydrous $CuSO_4(s)$ and hydrated $CuSO_4 \cdot 5H_2O(s)$ are –70 kJ per mol and 10 kJ per mol respectively. Determine the magnitude of enthalpy of hydration of 0.1 mole anhydrous $CuSO_4(s)$ as
   $CuSO_4(s) + 5H_2O(l) \longrightarrow CuSO_4 \cdot 5H_2O(s)$
10. If enthalpy of neutralisation of HCl by NaOH is $-57\,\text{kJ mol}^{-1}$ and with NH_4OH is $-50\,\text{kJ mol}^{-1}$. Calculate enthalpy of ionisation of $NH_4OH(aq)$.
11. Lattice energy of NaCl(s) is $-790\,\text{kJ mol}^{-1}$ and enthalpy of hydration is $-785\,\text{kJ mol}^{-1}$. Calculate enthalpy of solution of NaCl(s).
12. x g sample of NH_4NO_3 is decomposed in a Bomb calorimeter. The temperature of calorimeter increase by 4° C. The heat capacity of the system is 1.25 kJ/° C. Calculate the value of x.
   Given molar heat of decomposition of NH_4NO_3 at constant volume is $400\,\text{kJ mol}^{-1}$.
13. A heat engine operating between 227°C and 77°C absorbs 10 kcal of heat from the 227°C reservoir reversibly per cycle. Calculate total work done (in kcal) in two cycles.
14. Calculate work done in chemical reaction (in kcal)
   $A(s) + 3B(g) \longrightarrow C(l)$ at 227°C at 1 atm in closed vessel.

15. One mole ideal monoatomic gas is heated according to path AB and AC. If temperature of state B and state C are equal.

Calculate $\frac{q_{AC}}{q_{AB}} \times 10$.

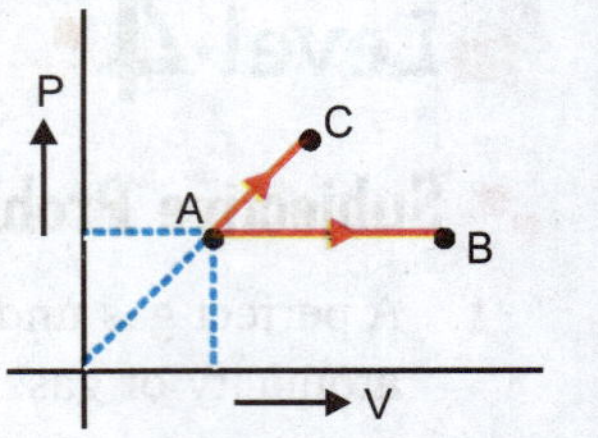

Numerical Problems

(If the numerical value has more than two decimal places, truncate/round-off the value to TWO decimal places.)

1. Five moles of an ideal gas at 293 K is expanded isothermally from an initial pressure of 2.1 MPa to 1.3 MPa against at constant external pressure 4.3 MPa. The heat transferred in this process is ______ kJ mol^{-1}. (Rounded-off to the nearest integer)
(Given : Use $R = 8.314\ \text{J mol}^{-1}\ \text{K}^{-1}$)

2. A gas expands from $3\ \text{dm}^3$ to $5\ \text{dm}^3$ against a constant pressure of 3 bar. The work done during expansion is used to heat 5 mole of water of temperature 290 K. Calculate final temperature of water.
Specific heat of water $= 4.2\ \text{J g}^{-1}\ \text{K}^{-1}$.

3. A real gas is subjected to an adiabatic process from (4 bar, 40 L) to (6 bar, 25L) against a constant pressure in a signal step. Calculate the enthalpy change for the process (in bar litre).

4. If the enthalpy change (in kcal) for 2 moles of an ideal gas $\left(\gamma = \frac{7}{5}\right)$ under following process is $-x$ kcal, then find the value of x.

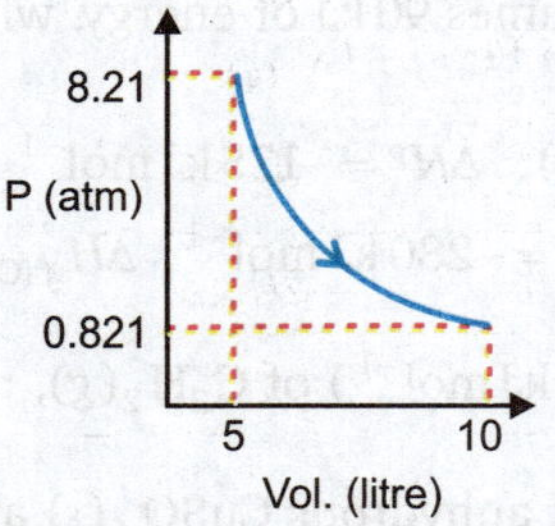

Given : $R = 2$ cal/mole-K, $R = 0.0821$ L-atm/mole-K

5. A monoatomic ideal gas expands from 1 bar, 1L until it's volume becomes 8 times, adiabatically reversibly. Calculate work done ($|w|$) in J.

6. Two moles of ideal gas undergoes following cyclic process:
(I) Isochoric heating from $(P_1V_1T_1)$ to double temperature.
(II) Isobaric expansion to double volume.
(III) Linear expansion (on PV curve) to $(P_1, 8V_1)$
(IV) Isobaric compression to initial state.
Calculate the magnitude of work done (in kcal), if initial temperature of the gas is 400 K ?
(Given : $R = 2$ cal/mol-K)

7. 2 moles of an ideal gas is involved in the process AB as shown in the diagram.
What is $|\Delta S_{\text{System}}|$ (in cal/K)?

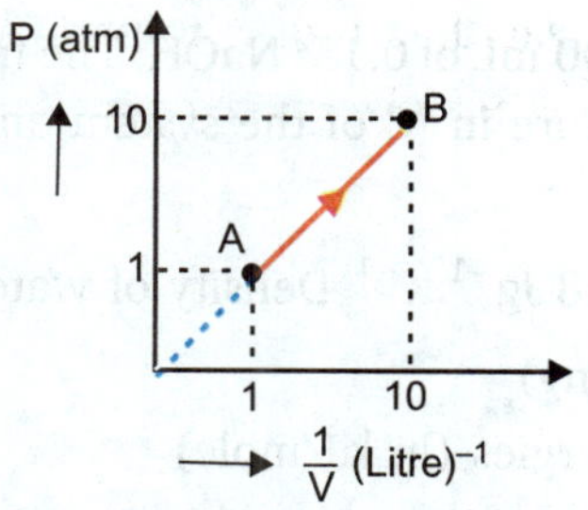

8. 1.0 mole of an ideal gas, initially present in a 2.00 L insulated cylinder at 300K is allowed to expand against vacuum to 8.00 L. Determine $\Delta S_{\text{Universe}}$ for the process in cal/K.
(Given : $\ln 2 = 0.7$)

9. Two blocks of the same metal having same mass and at temperature 200 K and 400 K respectively, are brought in contact with each other and allowed to attain thermal equilibrium at constant pressure. Calculate the change in entropy (ΔS) for this process (in J/K.mol).
(Given : $C_{p,m}(s) = 20$ J/K.mol, $\ln 2 = 0.7$, $\ln 3 = 1.1$)

10. Suppose molar internal energy of a substance over a limited temperature range could be expressed as $E_m(T) = bT + cT^2$. ($b = 16\ \text{JK}^{-1}\ \text{mol}^{-1}$ and $c = 6 \times 10^{-3}\ \text{JK}^{-2}\ \text{mol}^{-1}$)
Determine change in entropy when 1.0 mol of gas is heated from 300 K to 600 K at constant volume.
(Given : $\ln 2 = 0.7$)

11. Calculate $\Delta S^\circ_{\text{Universe}}$ for the reaction at 300 K (in J/K · mol)

$$A(g) + 3B(g) \longrightarrow 2C(g)\ ,\ \Delta_r H^\circ_{300} = -99\ \text{kJ/mol}$$

Given : $S^\circ_m(A) = 20$ J/K · mol ; $S^\circ_m(B) = 50$ J/K · mol ; $S^\circ_m(C) = 70$ J/K · mol

12. Data given for the following reaction is as follows :

$$FeO_{(s)} + C_{(graphite)} \longrightarrow Fe_{(s)} + CO_{(g)}$$

Substance	$\Delta_f H^\circ$ (kJmol^{-1})	ΔS° (Jmol^{-1} K^{-1})
$FeO_{(s)}$	–266.3	57.49
$C_{(Graphite)}$	0	5.74
$Fe_{(s)}$	0	27.28
$CO_{(g)}$	–110.5	197.6

The minimum temperature in K at which the reaction becomes spontaneous is ______ .

13. For the reaction $2CuO(s) \rightleftharpoons 2Cu(s) + O_2(g)$; $K_p = 10^{-48}$ bar at 300 K.
What is the magnitude of $\Delta_f G^\circ(CuO, s)$ in kcal/mol?
(Given : $R = 2$ cal mol^{-1} K^{-1})

14. Calculate the heat produced when 11.2 g of CaO is completely converted to $CaCO_3$ by reaction with CO_2 at 27°C in a container of fixed volume.
Given : $\Delta H^\circ_f[CaCO_3(s)] = -1207$ kJ/mol ; $\Delta H^\circ_f[CaO(s)] = -635$ kJ/mol
$\Delta H^\circ_f[CO_2, g] = -394$ kJ/mol ; [Use $R = 8.3\ \text{J} \cdot \text{K}^{-1}\ \text{mol}^{-1}$]

15. A solution of 6.3 g of haemoglobin (molar mass = 64,000 g/mol) in 25 mL of solution shows a temperature rise of 0.03°C for complete oxygenation. Each mole of haemoglobin binds 4 moles of oxygen.
If the heat capacity per mL of the solution is 4.2 J/K-mL, then the amount of heat released per mole of oxygen bound (in kJ) is ______.

16. 200 mL of 0.2 *M* HCl is mixed with 300 mL of 0.1*M* NaOH. The molar heat of neutralization of this reaction is –57.1kJ. The increase in temperature in °C of the system on mixing is $x \times 10^{-2}$.

The value of x is.(Nearest integer)

(Given : Specific heat of water $= 4.18\ \text{J g}^{-1}\ \text{K}^{-1}$ Density of water $= 1.00\ \text{g cm}^{-3}$)

(Assume no volume change on mixing)

17. Consider following data of bond energies (in kJ/mole)

C — H = 415, C — C = 348, C ═ C = 615, H — Cl = 432, C — Cl = 328

Determine the magnitude of enthalpy change of following gaseous reaction (in kJ/mol).

$C_2H_4 + HCl \longrightarrow \underset{\displaystyle H}{\underset{|}{H_2C}} - \underset{\displaystyle Cl}{\underset{|}{CH_2}}$

18. For the gaseous reaction

$$C_2H_6 \longrightarrow C_2H_4 + H_2$$

the reaction enthalpy $\Delta_r H =$ .... kJ mol^{-1}. (Round off to the Nearest Integer).

(Given : Bond enthalpies in kJ mol^{-1}: C — C:347, C ═ C : 611; C — H : 414, H — H : 436)

19. Given bond energy (in kJ mol^{-1})

Bond	H — H	O ═ O	O — H	O — O
Bond energy	435	498	464	142

Enthalpy of vaporisation of $H_2O_2(l) = 52$ kJ mol^{-1}.

Calculate the magnitude of enthalpy of formation of $H_2O_2(l)$ (in kJ mol^{-1})

20. The Born-Haber cycle for KCl is evaluated with the following data

$\Delta_f H°$ for KCl $= -436.7\ \text{kJ mol}^{-1}$; $\Delta_{\text{Sub}} H°$ for $K = 89.2\ \text{kJ mol}^{-1}$;

$\Delta_{\text{Ionization}} H°$ for $K = 419.0\ \text{kJ mol}^{-1}$; $\Delta_{\text{Electron gain}} H°$ for Cl(g) $= -348.6\ \text{kJ mol}^{-1}$;

$\Delta_{\text{Bond}} H°$ for $Cl_2 = 243.0\ \text{kJ mol}^{-1}$

The magnitude of lattice enthalpy of KCl in kJ mol^{-1} is ______. (Nearest integer)

Answers

Level-1

1.	(d)	**2.**	(b)	**3.**	(d)	**4.**	(c)	**5.**	(a)	**6.**	(c)	**7.**	(a)	**8.**	(c)	**9.**	(d)	**10.**	(d)
11.	(c)	**12.**	(b)	**13.**	(d)	**14.**	(d)	**15.**	(a)	**16.**	(b)	**17.**	(b)	**18.**	(a)	**19.**	(b)	**20.**	(c)
21.	(c)	**22.**	(c)	**23.**	(c)	**24.**	(d)	**25.**	(b)	**26.**	(c)	**27.**	(b)	**28.**	(c)	**29.**	(c)	**30.**	(d)
31.	(a)	**32.**	(c)	**33.**	(b)	**34.**	(b)	**35.**	(c)	**36.**	(a)	**37.**	(a)	**38.**	(a)	**39.**	(b)	**40.**	(b)
41.	(b)	**42.**	(c)	**43.**	(b)	**44.**	(b)	**45.**	(b)	**46.**	(c)	**47.**	(a)	**48.**	(b)	**49.**	(c)	**50.**	(c)
51.	(a)	**52.**	(b)	**53.**	(d)	**54.**	(b)	**55.**	(b)	**56.**	(b)	**57.**	(b)	**58.**	(b)	**59.**	(b)	**60.**	(b)
61.	(a)	**62.**	(c)	**63.**	(c)	**64.**	(d)	**65.**	(c)	**66.**	(a)	**67.**	(a)	**68.**	(c)	**69.**	(d)	**70.**	(c)
71.	(d)	**72.**	(c)	**73.**	(b)	**74.**	(b)	**75.**	(c)	**76.**	(c)	**77.**	(b)	**78.**	(d)	**79.**	(a)	**80.**	(b)
81.	(b)	**82.**	(b)	**83.**	(c)	**84.**	(d)	**85.**	(d)	**86.**	(b)	**87.**	(b)	**88.**	(a)	**89.**	(b)	**90.**	(b)
91.	(a)	**92.**	(d)	**93.**	(d)	**94.**	(c)	**95.**	(b)	**96.**	(a)	**97.**	(a)	**98.**	(a)	**99.**	(a)	**100.**	(b)
101.	(c)	**102.**	(d)	**103.**	(a)	**104.**	(d)	**105.**	(a)	**106.**	(a)	**107.**	(b)	**108.**	(a)	**109.**	(b)	**110.**	(c)
111.	(a)	**112.**	(d)	**113.**	(c)	**114.**	(a)	**115.**	(b)	**116.**	(a)	**117.**	(c)	**118.**	(b)	**119.**	(c)	**120.**	(c)
121.	(c)	**122.**	(c)	**123.**	(b)	**124.**	(a)	**125.**	(a)	**126.**	(c)	**127.**	(c)	**128.**	(b)	**129.**	(b)	**130.**	(a)
131.	(a)	**132.**	(b)	**133.**	(a)	**134.**	(a)	**135.**	(b)	**136.**	(b)	**137.**	(a)	**138.**	(a)	**139.**	(c)	**140.**	(b)
141.	(a)	**142.**	(c)	**143.**	(b)	**144.**	(d)	**145.**	(a)	**146.**	(d)	**147.**	(b)	**148.**	(a)	**149.**	(a)	**150.**	(b)

Level-2

1.	(c)	**2.**	(c)	**3.**	(c)	**4.**	(a)	**5.**	(d)	**6.**	(c)	**7.**	(c)	**8.**	i (c), ii (b)	**9.**	(c)	**10.**	(a)
11.	(d)	**12.**	(b)	**13.**	(b)	**14.**	(d)	**15.**	(a)	**16.**	(b)	**17.**	(c)	**18.**	(d)	**19.**	(d)	**20.**	(a)
21.	(a)	**22.**	(d)	**23.**	(c)	**24.**	(c)	**25.**	(d)	**26.**	(b)	**27.**	(a)	**28.**	(d)	**29.**	(b)	**30.**	(b)
31.	(b)	**32.**	(c)	**33.**	(b)	**34.**	(d)	**35.**	(a)	**36.**	(a)	**37.**	(c)	**38.**	(c)	**39.**	(b)	**40.**	(d)

Level-3

Passage–1	**1.**	(d)	**2.**	(d)						
Passage–2	**1.**	(b)	**2.**	(c)	**3.**	(b)	**4.**	(c)		
Passage–3	**1.**	(c)	**2.**	(a)	**3.**	(d)	**4.**	(c)	**5.**	(d)
Passage–4	**1.**	(a)	**2.**	(b)	**3.**	(b)				
Passage–5	**1.**	(a)	**2.**	(b)	**3.**	(a)				

One or More Answers is/are Correct

1. (b,c)	2. (a,b)	3. (c)	4. (a,b,c)	5. (a,b,d)
6. (b,c,d)	7. (b,c,d)	8. (a,b,c)	9. (c,d)	10. (a,c,d)
11. (a,b,c)	12. (a,c)	13. (a,c,d)	14. (b,d)	15. (a,b,c)
16. (a,c,d)	17. (b,c,d)	18. (a,b,c,d)	19. (a,b,c)	20. (a,b,c)
21. (a,c,d)	22. (a,c)	23. (a,c)	24. (a,c,d)	25. (a,c)

Match the Column

1. $A \rightarrow P$; $B \rightarrow S$; $C \rightarrow R$; $D \rightarrow Q$
2. $A \rightarrow P$; $B \rightarrow R$; $C \rightarrow Q$; $D \rightarrow S$
3. $A \rightarrow P, S$; $B \rightarrow Q, R$; $C \rightarrow R$; $D \rightarrow S$
4. $A \rightarrow R$; $B \rightarrow Q$; $C \rightarrow P$; $D \rightarrow S$
5. $A \rightarrow Q$; $B \rightarrow P$; $C \rightarrow R$; $D \rightarrow S$
6. $A \rightarrow Q$; $B \rightarrow P, S$; $C \rightarrow P$; $D \rightarrow R$
7. $A \rightarrow P, S$; $B \rightarrow R$; $C \rightarrow Q, S$; $D \rightarrow Q, S$
8. $A \rightarrow R$; $B \rightarrow Q$; $C \rightarrow Q$; $D \rightarrow P, S$
9. $A \rightarrow P$; $B \rightarrow R$; $C \rightarrow Q, S$; $D \rightarrow Q, S$
10. $A \rightarrow P$; $B \rightarrow R$; $C \rightarrow Q$; $D \rightarrow S$
11. $A \rightarrow S$; $B \rightarrow P, R$; $C \rightarrow Q$; $D \rightarrow P, R$
12. $A \rightarrow P, R$; $B \rightarrow Q$; $C \rightarrow P, R$; $D \rightarrow Q, S$
13. $A \rightarrow Q$; $B \rightarrow P$; $C \rightarrow R$; $D \rightarrow S$
14. $A \rightarrow P, Q$; $B \rightarrow Q, R, S$; $C \rightarrow P$; $D \rightarrow R$
15. $A \rightarrow S$; $B \rightarrow R$; $C \rightarrow P$; $D \rightarrow Q$

Assertion-Reason Type Questions

1. (D)	2. (A)	3. (B)	4. (B)	5. (A)	6. (A)	7. (A)	8. (B)	9. (D)	10. (C)
11. (B)	12. (C)	13. (A)	14. (C)	15. (A)	16. (A)	17. (D)	18. (D)		

Level-4

Subjective Problems

1. 1	2. 7	3. 6	4. 4	5. 4	6. 7	7. 4	8. 5	9. 8	10. 7
11. 5	12. 1	13. 6	14. 3	15. 8					

Numerical Problems

1. 15.00	2. 291.58	3. 80.00	4. 2.80	5. 112.50	6. 6.40	7. 9.212	8. 2.80
9. 2.00	10. 14.80	11. 300	12. 964.00	13. 33.16	14. 35.10	15. 8.00	16. 82.00
17. 44.00	18. 128.00	19. 189.00	20. 718.00				

Hints & Solutions

Level 1

1. (d) Heat capacity and volume are extensive properties.
2. (b) Boiling point, pH and emf are intensive property.
3. (d) If system is subjected to chemical equilibrium, mechanical equilibrium and thermal equilibrium then system is consider in thermodynamic equilibrium.
4. (c) Ozone have maximum internal energy $E=\frac{3RT}{2}+\frac{3RT}{2}+(3N-6)RT$. For non-linear molecule

   $E=\frac{3RT}{2}+\frac{2RT}{2}+(3N-5)RT$

   For linear molecule.
5. (a) Heat lost by iron + Heat gained by water $=0$

   $10\times 0.45\,(T-373)+25\times 4.2\times (T-300)=0$

   $T=303$ K or 30°C
6. (c) In freezing heat released so $q<0$
7. (a) $w=-\int_{V_1}^{V_2} PdV$
8. (c) $W=-\Delta n_g RT=-\text{ve}$

   II. $\Delta n_g=2$, III. $\Delta n_g=2$
9. (d) $w=-P_{ext}.\Delta V=-\Delta n_g RT$ and Δn_g is –ve for I and III

   also $\Delta V=-\text{ve}$ for IV
10. (d) $q=0$, $\Delta U=q+w$

    $w>0$ so $\Delta U>0$
11. (c) $\Delta U=q+w$

    $300=400-0.5\times\Delta V\times 100$

    $\Delta V=2\text{L}$
12. (b) $\Delta U=q+w$

    $=10\times 1000-2\times(20)\times 101.3=5948$ J
13. (d) $w=-\int PdV=0$
14. (d) $w=-4\int_1^4 dV=+4\times 3\times 100,\ w=+1.2\times 10^3$ J
15. (a) $w=-\Delta n_g RT=\frac{-4.5}{34\times 2}\times 8.314\times 298,$

    $w=-163.9$ J
16. (b) $w=-P.\Delta V=-n_{H_2}$

    $RT=-\frac{2\times 8.314\times 298}{1000}=-4.955$ kJ
17. (b) $w=-nR(T_2-T_1),\ w=-1R(2)=-2R$
18. (a) $\int dw=-\int P.dV$

    $\Rightarrow\ w_{rev}=-\int 6.V^2dV$

    $=-6\left[\frac{V_2^3}{3}-\frac{V_1^3}{3}\right]$ bar.m^3;

    $w=-5200$ kJ

    so, work done by the gas is 5200 kJ.
19. (b) $w=-$ Area of P-V curve
20. (c) For cyclic process : $dU=0$

    $\therefore\ -w=q$

    Net work done by system = Area of the circle
21. (c) Process direction in V–P diagram is clockwise so process direction in P–V diagram is anti-clockwise.

    Net work done by system = Area

    $=\pi\times\frac{(P_2-P_1)}{2}.\frac{(V_2-V_1)}{2}$
22. (c) Work done in the cyclic process

    = Area bounded $(ABCA)=5P_1V_1$
23. (c) $w=-P\Delta V=0$

    $\Delta U=q+w=0$

    Temperature remains constant.
24. (d) Zero

    Internal energy is state function.
27. (b) $C_P-C_V=nR=2R$
28. (c) No intramolecular force act between gaseous molecule.
29. (c)

30. (d) $w_3>w_1>w_2>w_4$

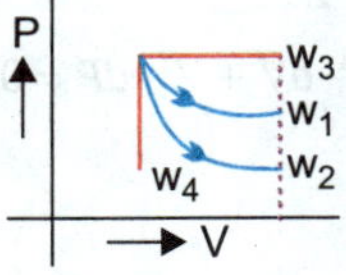

31. (a) $V = \frac{nRT}{P} = \frac{100 \times 0.0821 \times T}{8.21} = T$

$\therefore \quad \log V = \log T$

32. (c) $\int dw = \int -P.dV$

$$\Rightarrow \quad w = -\int 20 \cdot \frac{dV}{V} = -20 \ln \frac{V_2}{V_1}$$

$w = -46.06$ L atm $= -4665.8$ J

$\Delta U = q + w \Rightarrow 400 = q - 4665.8$

$q = 5065.8$ J

33. (b) $w = -nRT \ln \frac{V_2}{V_1}$

$$w = -2 \times 8.314 \times 10^{-3} \times 300 \ln \frac{40}{4},$$

$w = -11.488$ J

34. (b) $w = -nRT \ln \frac{P_1}{P_2}$

$$= -10 \times 8.314 \times 300 \ln \frac{10}{1}$$

$= -57441.42$ J

$w = -mgh$

$\therefore \quad m \times 9.81 \times 100 = 57441.42$

$m = 58.55$ kg

35. (c) At constant volume $\Delta U = q_v = nC_{vm} \Delta T$

at constant pressure $\Delta H = q_p = nC_{p,m} \Delta T$

36. (a) $\Delta U = w$

$$\Rightarrow \quad -75 = 0.1 \times \frac{3}{2} \times 2\, [T_2 - 500]$$

$T_2 = 250$ K

37. (a) $\Delta U = w = \frac{P_2V_1 - P_1V_1}{\gamma - 1}$

38. (a) $T^{\frac{\gamma}{1-\gamma}}.P = \text{constant}$ or $P \propto T^{\frac{\gamma}{\gamma-1}}$

$\because \quad P \propto T^3$

$\therefore \quad \frac{\gamma}{\gamma - 1} = 3; \quad \gamma = \frac{3}{2}$

39. (b) $T \propto \frac{1}{\sqrt{V}}; \quad TV^{1/2} = \text{constant}$

For adiabatic process,

$TV^{\gamma-1} = \text{constant}$

$\therefore \quad \gamma - 1 = \frac{1}{2}, \quad \gamma = \frac{3}{2}$

40. (b) $PV^{\gamma} = \text{constant}, P \cdot \gamma V^{\gamma-1} \cdot dV + V^{\gamma} \cdot dP = 0$

$$\therefore \quad \frac{dP}{P} = -\gamma \cdot \frac{dV}{V}$$

41. (b) $PV = \text{constant}$ for isothermal process

$PV^{\gamma} = \text{constant}$ for adiabatic process so more is the value of γ, more is decrease in pressure.

42. (c) $T.V^{\gamma-1} = \text{constant}$

$$\therefore \quad \frac{T_2}{T_1} = \left(\frac{V_1}{V_2}\right)^{\gamma-1}$$

$$\Rightarrow \quad T_2 = 300 \times \left(\frac{16}{2}\right)^{\frac{5}{3}-1}$$

$= 1200$ K

43. (b) $\int dw = -\int P_{ext} \cdot dV$

$$\therefore \quad w_{irr} = -P_{ext}\left[\frac{nRT}{P_2} - \frac{nRT}{P_1}\right]$$

$$w_{irr} = -1 \times (5 \times 8.314 \times 300) \times \left[1 - \frac{1}{10}\right]$$

$w_{irr} = -11.224$ kJ

44. (b) For an adiabatic irreversible compression, $\Delta U = w$

$\therefore \quad nC_{v,m}(T_2 - T_1) = -P_{ext}(V_2 - V_1)$

here $P_{ext} = P_2$

$$nC_{v,m}(T_2 - T_1) = -P_2\left[\frac{nRT_2}{P_2} - \frac{nRT_1}{P_1}\right]$$

$$\Rightarrow \quad \frac{5}{2} R (T_2 - T_1) = -P_2 \times R \left[\frac{T_2}{P_2} - \frac{T_1}{P_1}\right]$$

$\Rightarrow \quad P_2 = 275$ kPa

45. (b) $w = \Delta U = n\int_{T_1}^{T_2} C_{vm}\, dT$

$$w = 2 \times \frac{R}{\gamma - 1}(T_2 - T_1)$$

$$\frac{T_2 - T_1}{\gamma - 1} = -P_{ext}\left(\frac{T_2}{P_2} - \frac{T_1}{P_1}\right)$$

$$\frac{T_2 - 300}{\frac{5}{3} - 1} = -2\left(\frac{T_2}{2} - \frac{300}{1}\right)$$

$1.5\, T_2 - 450 = -T_2 + 600 \Rightarrow T_2 = 420$ K

$$w = 2 \times \frac{2}{\frac{5}{3} - 1}(420 - 300) = 720$$

46. (c) $\because \quad q = 0 \qquad \therefore \quad \Delta U = w$

$$\Rightarrow nC_{v,m}(T_2 - T_1) = -P_{ext}\left[\frac{nRT_2}{P_2} - \frac{nRT_1}{P_1}\right]$$

$$\therefore \quad C_{v,m}[T_2 - T_1] = P_{ext} \cdot R\left[\frac{T_1}{P_1} - \frac{T_2}{P_2}\right]$$

$$\Rightarrow \frac{5}{2} R\, [T_2 - 300] = 2 \times R \left[\frac{300}{5} - \frac{T_2}{2}\right]$$

$\Rightarrow \quad T_2 = 248.5$ K

47. (a) $\Delta U = w$

$$n \times \frac{R}{\gamma - 1}(T_2 - T_1) = -P_{ext}\left(\frac{nRT_2}{P_2} - \frac{nRT_1}{P_1}\right)$$

$$3(T_2 - T_1) = -1\left(\frac{T_2}{1} - \frac{T_1}{2}\right]$$

$$3T_2 - 3T_1 = -T_2 + \frac{T_1}{2}$$

$$4T_2 = \frac{7T_1}{2}$$

$$T_2 = \frac{7T_1}{8} = \frac{7 \times 400}{8} = 350\,K$$

$$\frac{P_1V_1}{T_1} = \frac{P_2V_2}{T_2} \Rightarrow \frac{2 \times 10}{400} = \frac{1 \times V_2}{350} \Rightarrow V_2 = 17.5\,L$$

48. (b) Average $C_{v,m} = \frac{n_1C_{v,m_1} + n_2C_{v,m_2}}{n_1 + n_2}$

$$= \frac{2 \times \frac{3}{2}R + 2 \times \frac{5}{2}R}{2 + 2} = 2R$$

49. (c) $\Delta H = (n_1C_{p,m_1} + n_2C_{p,m_2})\Delta T$

$$= \left(0.5 \times \frac{7}{2}R + 0.5 \times 4R\right)(-100)$$

$$= -375\,R$$

50. (c) AB : Isobaric expansion

BC : Isothermal expansion

CD : Isochoric

DA : Isothermal compression

51. (a) 1 watt = 1 J/sec

Total heat supplied for 36 mL H_2O

$= 806 \times 100$

$= 80600$ J

$$\Delta H_{vap} = \frac{80600}{36} \times 18$$

$= 40300$ J or 40.3 kJ/mol

52. (b) $\Delta H = \Delta U + \Delta n_gRT$, $\Delta n_g = 1$, $\Delta H > \Delta U$

53. (d) $\Delta n_g = 0 \therefore \Delta H^\circ = \Delta U^\circ$

For 2 mole $\Delta U^\circ = -370$ kJ

54. (b) $\Delta H = \Delta U + \Delta n_gRT$, $\Delta n_g = -2$, $\Delta H < \Delta U$

$w = -\Delta n_gRT$, $w > 0$, $q = -ve$

55. (b) At constant pressure, $\Delta H = \Delta U + P \cdot \Delta V$

$= 30 + 2 \times 4$

$= 38$ atm-L

56. (b) $\Delta H = \Delta U + P \cdot \Delta V$

$$\Rightarrow -124 = \Delta U + \frac{1520}{760} \times (177 - 377) \times 10^{-3} \times 101.3$$

$\Delta U = -83.48$ J

57. (b) At constant volume $\Delta H = \Delta U + V\Delta P$

$\Rightarrow -560 = \Delta U + 10 \times (-30) \times 0.1$

$\Delta U = -530$ kJ

58. (b) When both P and V are changing

$\Delta H = \Delta U + \Delta(PV)$

$= \Delta U + (P_2V_2 - P_1V_1)$

$\Delta H = 40 + (20 - 3)$

$= 57$ L-atm

59. (b) For combustion of 1 mole of benzene

$\Delta n_g = -1.5$

$\Delta H = \Delta U + \Delta n_gRT$

$$\Rightarrow -3271 = \Delta U - \frac{1.5 \times 8.314 \times 300}{1000}$$

$\Rightarrow \Delta U = -3267.25$ kJ

For 1.5 mole of combustion of benzene

$\Delta U = -3267.25 \times 1.5$

$= -4900.88$ kJ

60. (b) $\Delta H = \Delta E + \Delta n_gRT$

$$\Rightarrow 82.8 = \Delta U + \frac{1 \times 8.314 \times 298}{1000}$$

$\Rightarrow \Delta U = 80.32$

61. (a) C (graphite) $\longrightarrow$ C (diamond)

$\Delta H = \Delta U + P \cdot \Delta V$

$$V_m\,(\text{diamond}) = \frac{12}{3}\text{ mL}$$

$$V_m\,(\text{graphite}) = \frac{12}{2}\text{ mL}$$

$$\Delta H - \Delta U = (500 \times 10^3 \times 10^5\,N/m^2)\left(\frac{12}{3} - \frac{12}{2}\right) \times 10^{-6}$$

$= -100$ kJ/mol

$\Delta U - \Delta H = +100$ kJ/mol

62. (c) Δn_g is +ve

63. (b) $\Delta n_g = -ve$, $\Delta S = -ve$

64. (d) $\Delta n_g = -ve$, $\Delta S = -ve$

(a) $\Delta n_g = -1$ (b) $\Delta n_g = -1$

(c) $\Delta n_g = 0$ (d) $\Delta n_g = -3$

65. (c) $\Delta S = nC_{p,m} \ln \frac{T_2}{T_1}$

$$= 2 \times \frac{5}{2} R \ln \frac{600}{300}$$

$= 5\,R \ln 2$

68. (c) $\Delta S_{gas} = nC_{v,m} \ln \frac{T_2}{T_1}$

$$= 2 \times \left(\frac{5}{2} - 1\right) R \ln 2$$

$= 3\,R \ln 2$

69. (d) $\Delta S = nR \ln \frac{V_2}{V_1} = 1R \ln \frac{3V}{V} = R \ln 3$

70. (c) ΔS will be same because entropy is a state function.

71. (d) $\Delta S = nC_{v,m} \ln \frac{T_2}{T_1} + nR \ln \frac{V_2}{V_1}$

$= C_{v,m} \ln 2 + R \ln \left(\frac{1}{2}\right)$

$\Delta S = (C_{v,m} - R) \ln 2$

72. (c) $\Delta S = nC_{p,m} \ln \frac{T_2}{T_1}$

$= 2.5 \times 18 \times 4.2 \ln \left(\frac{360}{300}\right)$

$= 34.02$ J/K

73. (b) $\Delta_r S° = \Sigma\, v_p S°$ (Products) $- \Sigma\, v_R S°$ (Reactants)

$= (2 \times 27.3 + 3 \times 69.9) - (87.4 + 3 \times 130.7)$

$= -215.2\ \text{JK}^{-1}\ \text{mol}^{-1}$

74. (b) $\Delta_r S = (2 \times 248.2) + (2 \times 66.5)$

$-(2 \times 91.2) - 3(205.1)$

$= -168.3$ J

75. (c) $-266 = 6 \times 87 - 205 - 4x$

$x = 145.75$

76. (c) $\Delta S = \frac{\Delta H}{T}$

77. (b) at M.P. $\Delta S_{\text{fusion}} = \frac{\Delta H_{\text{fusion}}}{T}$

$\Rightarrow \quad T = \frac{\Delta H}{\Delta S} = \frac{9.95 \times 1000}{35.7} = 278.7$ K

78. (d) $\Delta S_{\text{freezing}} = -\frac{\Delta H_{\text{fusion}}}{T}$

$\Rightarrow \quad \Delta S = -\frac{80 \times 10}{273} \times 4.18$

$= -12.25$ J/K

79. (a) $\Delta S = \frac{\Delta H}{T_{\text{B.P.}}}$

80. (b) $n_{C_6H_6} = \frac{117}{78} = 1.5$

$\Delta S_{\text{system}} = 1.5 \times 85$ J/K

$\therefore \quad \Delta S_{\text{surrounding}} = -1.5 \times 85$ J/K

81. (b) According to 3rd law of thermodynamics at absolute zero temperature entropy of perfectly crystalline substance is taken to be zero.

82. (b)

$X(s) \xrightarrow{\Delta S} X(l)$

$\downarrow \Delta S_1 \qquad \nearrow \Delta S_2$

$X(l)$

$\Delta S = \frac{20 \times 1000}{100} + 1 \times \int_{100}^{200} C_{pm} \frac{dT}{T}$

$\Delta S = 200 + 10 \ln 2 = 200 + 10 \times 0.693,$

$\Delta S = 206.93\ \text{JK}^{-1}$

83. (c) $0.42 = a(10)^3 \quad \Rightarrow \quad a = 0.42 \times 10^{-3}$

$S_m = \int_0^{20} \frac{C_{p,m}}{T} dT$

$= \int_0^{20} aT^2 dT = \frac{a}{3}[20^3 - 0]$

$= 1.12$ J/K-mol

84. (d) For spontaneous reaction $\Delta G = -$ve

$\Delta S = -$ve ($\Delta n_g = -$ve)

$\Delta H = -$ve

$\Delta G = \Delta H - T\Delta S$

85. (d) $\Delta H = -$ve (H—H bond formed)

$\Delta S = -$ve ($\Delta n_g = -1$)

86. (b) $\Delta H = -$ve (Combustion reaction)

$\Delta S = +$ve ($\Delta n_g = +$ve)

$\Delta G = \Delta H - T\Delta S = -$ve

87. (b) For $\Delta_r G° -$ve, $\Delta_r H°$ should be $-$ve and $\Delta S°$ should be $+$ve

88. (a) For spontaneous reaction

$(\Delta G_{\text{System}})_{T,P} = -$ve

89. (b) $(\Delta G)_{\text{system}} = \Delta H - T \cdot \Delta S$

and $\quad \Delta H - T \cdot \Delta S < 0$

90. (b) $\Delta G = \Delta H - T\Delta S, \Delta H = -$ve , $\quad \Delta S = +$ve

91. (a) At equilibrium $\Delta G = 0$

92. (d) $\Delta G = \Delta H - T\Delta S = +$ve

for spontaneous change $\Delta H > 0, \Delta S < 0$

93. (d) $\Delta G_{799} = -40 \times 1000 - 799 \times (-50),$

$\Delta G_{799} = -50$ J spontaneous

$\Delta G_{801} = -40 \times 1000 - 801(-50)$

$\Delta G_{801} = +50$ J non-spontaneous

94. (c) For ideal gas isothermal expansion $\Delta H = 0$

$\therefore \quad \Delta G = -T \cdot \Delta S$

where $\quad \Delta S = nR \ln \left(\frac{V_2}{V_1}\right)$

95. (b) $Hg(l) \rightleftharpoons Hg(g),$

$\Delta_r S° = 174.4 - 77.4 = 97$ J/K-mol

$\because \quad \Delta G° = \Delta H° - T \cdot \Delta S° = 0$

$T = \frac{\Delta H°}{\Delta S°}$

$= \frac{60.8 \times 1000}{97} = 626.8$ K

96. (a) At equilibrium $\therefore \quad \Delta H = T \cdot \Delta S$

$H_2O(s) \longrightarrow H_2O(l)$

$\Delta S = S_{H_2O(l)} - S_{H_2O(s)} = 21.8$ J/mol-K

$\Delta H = 273 \times (21.8) = 5951.4$ J/mol

97. (a) $\Delta G^\circ = 2\times 86.6 + 4\times(-237.1)$
$+3\times 0 - 3(-33.6)$
$-2(-80.6) = -513.2$

98. (a) $\Delta G = \Delta H - T\Delta S$
$\Delta G_{I} = 10.5\times 1000 - 298\times 30 = 1560$
$\Delta G_{II} = 1.8\times 1000 + 298\times 113 = 35474$
$\Delta G_{IV} = -126\times 1000 - 298\times 84 = -ve$
III is spontaneous

99. (a) $\Delta G = \Delta H - T\Delta S$
$\Delta S = 242.6 - 50.3 - (2\times 152.3), \Delta S = -112.3$
$\Delta G = -788.6\times 1000 = \Delta H + 298\times 112.3$
$\Delta H = -788.6\times 1000 - 298\times 112.3$
$\Delta H = -822.1 \text{ kJ mol}^{-1}$

100. (b) $\Delta G = \Delta H - T\Delta S$
$406.6\times 1000 = \Delta H - 298(2\times 114.6 - 130.6)$
$\Delta H = 406.6\times 1000 + 298(98.6)$
$\Delta H = 436 \text{ kJ mol}^{-1}$

101. (c) $\Delta G^\circ = -RT\ln K$; $K\uparrow$ stability of reactant $\downarrow$

102. (d) $\frac{\partial}{\partial T}\ln K_p = \frac{\Delta H}{RT^2}$
$\frac{\partial}{\partial T}\ln K = \frac{E_a}{RT^2}$

103. (a) $\Delta G^\circ = -RT\ln K = -8.314\times 10^{-3}\times 400\ln 10^5$
$\Delta G^\circ = -38.294 \text{ kJ}$

104. (d) At standard state on equilibrium $\Delta G^\circ = 0$
$\Delta G^\circ = 0 = -RT\ln K \Rightarrow K = 1$

105. (a) At equilibrium $\Delta G^\circ = -RT\ln K_{eq}$
where $K_{eq} = (P_{H_2O}/P^\circ)$ and $P^\circ = 1$ bar

106. (a) $\Delta G^\circ = -RT\ln K_w$
$= -8.314\times 298\times 2.303(-14)$
$\simeq 80000$ J/mol

107. (b) $\Delta S^\circ = -206.5 + 114.6 + 180.7 = 88.8$
$\Delta G^\circ = \Delta H^\circ - T\Delta S^\circ$
$\Delta H^\circ = 271.8 + 298\times 88.8\times 10^{-3}$
$\Delta H^\circ = 298.3$

108. (a) According to Hess's law the standard enthalpy of overall reaction is sum of the enthalpy change in individual reaction.

109. (b) $X \longrightarrow A + q_1$
$q_2 + A \longrightarrow B$
$B \longrightarrow Y + q_3$
$q_2 + X \longrightarrow Y + q_1 + q_3, \quad \Delta H = q_2 - q_1 - q_3$

110. (c) The change in enthalpy does not depend upon number of intermediate steps.

111. (a) $\Delta_r C_p^\circ = -10.945$ for
$H_2(g) + \frac{1}{2}O_2(g) \rightarrow H_2O(g)$
According to Kirchhoff's law
$\Delta_r H^\circ_{T_2} - \Delta_r H^\circ_{T_1} = \Delta_r C^\circ_p[T_2 - T_1]$
$\Delta_r H^\circ_{T_2} - (241.82) = [-10.94 \text{ J}(373 - 298)/100]$
$\Rightarrow \Delta_r H_{T_2} = -242.6$ kJ/mol

112. (d) $\Delta_f H\uparrow$ stability $\downarrow$

113. (c) $(\Delta_f H)_{Al(s)} = 0$

114. (a) $\Delta H^\circ = \frac{3}{2}\times\Delta H_1^\circ + \frac{\Delta H_2^\circ}{2}$
$= -713.7 - 177.5$
$= -891.2$ kJ

115. (b) $\Delta H^\circ = 2\times\Delta H_1^\circ + 2\times\Delta H_2^\circ - \Delta H_3^\circ$

116. (a) $C(\text{graphite}) \longrightarrow C(\text{diamond})$
$\Delta H = 395 - 393 = 2$

117. (c) $C(s) + \frac{3}{2}H_2(g) + \frac{1}{2}Cl_2(g) \longrightarrow CH_3Cl(g)$

118. (b) $\Delta_r H = (\Delta_f H)_P - (\Delta_f H)_R$
$\Delta_r H = -113 + 3\times(-269) + 46.2 = -873.8$ kJ

119. (c) $C_8H_{18}(g) + \frac{25}{2}O_2(g) \rightarrow 8CO_2(g) + 9H_2O(l)$
$\Delta_r H^\circ = 8\times(-394) + 9\times(-286) - (-250)$
$= -5476$ kJ/mol

120. (c) $-1941 = -1273 + 3(-241.8) - (\Delta_f H)B_2H_6$
$(\Delta_f H)_{B_2H_6} = -1273 - 725.4 + 1941 = -57.4$

121. (c) $C(s) + O_2(g) \longrightarrow CO_2(g)$;
$\Delta_r H_1 = -x$ kJ/mol ...(1)
$CO(g) + 1/2\,O_2(g) \longrightarrow CO_2(g)$
$\Delta_r H_2 = -y$ kJ/mol ...(2)
Equation (1) − (2)
$C(s) + \frac{1}{2}O_2(g) \longrightarrow CO(g)$
$\Delta_f H = (y - x)$ kJ/mol

122. (c) $C_2H_4(g) + H_2(g) \longrightarrow C_2H_6(g)$
$\Delta_r H = x_2 - x_1$

123. (b) $Q = \frac{5.8}{58}\times\frac{5756}{2} = 287.8$ kJ

124. (a) $Q = \frac{1}{2}\times 438\times\frac{22.3}{223} = 21.9$

125. (a) $\Delta_r H^\circ = 57\times(-393.5) + 52\times(-285.8) - (-70870)$
$= +33578.9$ kJ/g
$= +\frac{33578.9}{884} = +37.98$ kJ/g

126. (c) $\Delta_r H = 46\times 2 = 92$

127. (c) $2CH_3OH + 3O_2(g) \longrightarrow 4H_2O(l) + 2CO_2(g)$ +1453
$4\times 44 + 4H_2O(l) \longrightarrow 4H_2O(g)$
$2CH_3OH + 3O_2(g) \longrightarrow 4H_2O(g) + 2CO_2(g)$ +1277
$\Delta H = -1277$

128. (b) $(\Delta_f H) = \frac{436.4}{2} = 218.2$ kJ/mol

129. (b) For $H_2(g) + O_2(g) \longrightarrow H_2O_2(l)$

$$\Delta_f H^\circ (H_2O_2, l) = \Delta_r H_3{}^\circ + \frac{\Delta_r H_2^\circ}{2} - \frac{\Delta_r H_1^\circ}{2}$$

130. (a) $C_2H_5OH(l) + 3O_2(g) \longrightarrow 2CO_2(g) + 3H_2O(l)$

$\Delta n_g = 2 - 3 = -1$

$\Delta H = \Delta U + \Delta n_g RT$

$\Delta U = \Delta H + RT = RT - q$

131. (a) $-q_{reaction} = q_{bomb} + q_{water}$

$q_{reaction} = (C\text{ (bomb)} + (m_{water} \times c))\, \Delta T$

$= (652 + 500 \times 4.18) \times 14.3$

$= 39210$ J or 39.21 kJ

132. (b) $q = m \times c \times \Delta T,\ m = q/(c \times \Delta T)$

$= (24 \times 10^6 \times 0.7)/(4.18 \times 50)$

$= 80383$ g or 80.383 kg

133. (a) m mole of acid $= 0.05 \times 0.2 = 0.01$

$$\Delta_r H^\circ = -\frac{480 \times 1.09}{0.01 \times 1000} = -52.32 \text{ kJ/mol}$$

134. (a) $\Delta H_{solution} = \frac{(125 + 8) \times 4.2 \times 6}{8} \times 80$

$= 33516$ J/mol or 33.51 kJ/mol

135. (b) (I) $CaCl_2(s) + aq \longrightarrow CaCl_2(aq) + q$

$CaCl_2(aq) + 6H_2O(l) \longrightarrow CaCl_2 \cdot 6H_2O(s) + 23.2$

$CaCl_2(s) + 6H_2O \longrightarrow CaCl_2 \cdot 6H_2O(s) + q + 23.2$

$3.5 = q + 23.2$

$q = -19.7$

(II) $2\,Cl(g) \longrightarrow Cl_2(g)$

$\Delta H = -$ve, $\Delta S = -$ve.

136. (b) $HCl(g) + aq \longrightarrow H^+(aq.) + Cl^-(aq.)$;

$\Delta_r H = -75.14$

$-75.14 = \Delta_f H\,(H^+, aq) + \Delta_f H\,(Cl^-, g) - \Delta_f H\,(HCl, g)$

$\because \quad \Delta_f H\,(H^+, aq) = 0$

$\Delta_f H\,(Cl^-, aq) = -75.14 - 92.3$

$= -167.44$ kJ/mol

137. (a) Given that

$MgSO_4(s) + nH_2O \longrightarrow MgSO_4\,(nH_2O)$;

$\Delta_r H_1 = -91.2$ kJ/mol ...(i)

$MgSO_4 \cdot 7H_2O(s) + (n - 7)H_2O \longrightarrow MgSO_4\,(nH_2O)$

$\Delta_r H_2 = 13.8$ kJ/mol ...(ii)

Equation (i) – (ii)

or $\Delta H_{hyd} = \Delta_r H_1 - \Delta_r H_2$

$= -91.2$ kJ/mol $- 13.8$ kJ/mol

$= -105$ kJ/mol

138. (a) $-12250x - 13000\,(1 - x) = -12500$

$750x = 500 \Rightarrow x = 2/3$ and $y = 1/3$

So, $\frac{x}{y} = \frac{2}{1}$

139. (c) |Enthalpy of change| > 57

$2H^+$ participate in reaction.

140. (b) $\Delta H_{neut.}$

$= \Delta H_{ionization} + \Delta_r H\,(H^+ + OH^- \rightarrow H_2O)$

$\Rightarrow \quad -51.34 = x - 55.84$

So, $\Delta H_{ionization}$ or $x = 4.5$ kJ/mol

141. (a) $CaCO_3(s) + \text{Heat} \longrightarrow CaO(s) + CO_2(g)$

142. (c) By calorimetry we can determine change in enthalpy.

143. (b) $\Delta_r H = \frac{5}{8} \times 40 + \frac{3}{8} \times 50 = 43.75$ kJ/mol

144. (d) No. of moles of O_2 required to supplied 30 kJ heat to second reaction

$= \frac{30}{1260} \times \frac{3}{2} = \frac{1}{28}$

So $n_{O_2} : n_{H_2} = \frac{1}{28} : 3$ or $1 : 84$

145. (a) $(O—F) = \frac{368}{2} = 184$

146. (d) C—C bond enthalpy is not given.

147. (b) B.E.↑ size↓

148. (a) $P(s) + \frac{3}{2}Cl_2(g) \longrightarrow PCl_3(g)$

$306 = (314 + 3 \times 121) - [\text{B.E. (P—Cl)} \times 3]$;

B.E. (P—Cl) $= 123.66$ kJ/mol

149. (a) $CH_3OH(g) + HCl(g) \longrightarrow CH_3Cl(g) + H_2O(g)$

$\Delta_r H = (\text{B.E.})_R - (\text{B.E.})_P$

$= 3(C—H) + (C—O) + (O—H) + (H—Cl)$

$-3(C—H) - (C—Cl)) - 2(O—H)$

$= 335 + 463 + 431 - 326 - 2 \times 463 = -23$ kJ

150. (b) $N_2(g) + 2H_2(g) \longrightarrow N_2H_4(g)$

$\Delta_f H\,(N_2H_4, g)$

$= (941 + 2 \times 436) - (159 + 4 \times 398)$

$= 1813 - 1751 = 62$ kJ mol^{-1}

Level 2

1. (c) $w = -P_{ext}(V_f - V_i)$

$$= -10^5\left(\frac{60\times10^{-3}}{0.60} + \frac{40\times10^{-3}}{1000} - \frac{100\times10^{-3}}{1000}\right)$$

$$= -10^5(100\times10^{-3} + 0.04\times10^{-3} - 0.1\times10^{-3})$$

$$|w| = 9994\text{ J}$$

2. (c) Mole of the gas in the first compartment

$$n_1 = \frac{P_1V_1}{RT_1} = -\frac{0.8314\times10^6\times2}{8.314\times400} = 500$$

Similarly, $n_2 = 2000$

The tank is rigid and insulated hence $w = 0$ and $q = 0$ therefore $\Delta U = 0$

Let T_f and P_f denote the final temperature and pressure respectively

$$\Delta U = n_1C_{V,m}[T_f - T_1] + n_2C_{V,m}[T_f - T_2] = 0$$

$$500(T_f - 400) + 2000(T_f - 500) = 0$$

$$T_f = 480\text{ K}$$

3. (c)

$$H_2O(l, 323\text{ K}) \xrightarrow{\Delta U_1} H_2O(l, 373\text{ K})$$
$$\downarrow \Delta U_2$$
$$H_2O(g, 323\text{ K}) \xleftarrow{\Delta U_3} H_2O(g, 373\text{ K})$$

$$C_{V,m}(H_2O, g) = 33.314 - 8.314 = 25\text{ J/K mol}$$

$$\Delta U_2 = \Delta H_2 - \Delta n_g RT = 37.6$$

$$\Delta U_{total} = \Delta U_1 + \Delta U_2 + \Delta U_3 = C_{v,m}(l)\cdot\Delta T + \Delta V_{vap.} + C_{v,m}(g)\,\Delta T$$

$$= \frac{75\times50}{1000} + 37.6 + \frac{25\times50}{1000}$$

$$= 42.6\text{ kJ/mol}$$

4. (a) $\because \frac{C_{p,m}}{C_{V,m}} = \gamma$ and $C_{p,m} - C_{V,m} = R$

$$\therefore\ C_{V,m} = \frac{R}{\gamma - 1}$$

$$C_{V,m} = \frac{C_V}{n} \text{ and } C_V = m\cdot c_V$$

$$\therefore\ \frac{R}{\gamma-1} = \frac{m\cdot c_V}{m}\times M$$

$$\therefore\ c_V = \frac{R}{(\gamma-1)M}$$

5. (d) Av. $C_{V,m} = \frac{n_1C_{V,m_1} + n_2C_{V,m_2}}{n_1 + n_2} = 2R$

for adiabatic process $dU = dW$

$$\frac{dT}{T} = -\frac{R}{C_{V,m}}\left(\frac{dV}{V}\right)$$

$$n_1C_{V,m_1}\,dT + n_2C_{V,m_2}\,dT = -(n_1RT + n_2RT)\times\frac{dV}{V}$$

$$\ln\frac{T_2}{T_1} = -\frac{1}{2}\ln\left(\frac{V_2}{V_1}\right)$$

$$\Rightarrow\ T_2 = 320\times\left(\frac{1}{4}\right)^{1/2} = 160\text{ K}$$

$$\Delta U = (n_1C_{V,m_1} + n_2C_{V,m_2})\,\Delta T = -960\,R$$

6. (c) $2\times C_{V,m}(T_2 - T_1) = -1\left(\frac{2RT_2}{1} - \frac{2RT_1}{10}\right)$

$$\gamma = \frac{4}{3} = \frac{C_{p,m}}{C_{V,m}}$$

and $C_{p,m} = C_{V,m} + R \qquad \therefore\ C_{V,m} = 3R;$

$$3R(T_2 - 300) = -RT_2 + \frac{R\times300}{10}$$

$$T_2 = 232.5\text{ K}$$

$$w = \Delta U = 2\times3R(300 - 232.5)$$

$$\Rightarrow\ w = -405\,R$$

7. (c) Max. temp. attained by gas in between B to C

According to equation of straight line

$$\frac{P-4}{1-4} = \frac{V-1}{2-1}$$

$$\Rightarrow\ P - 4 = -3V + 3$$

$$\Rightarrow\ P = 7 - 3V$$

For 1 mole gas

$$\frac{RT}{V} = 7 - 3V;\ RT = 7V - 3V^2 \qquad \ldots(1)$$

$$R\frac{dT}{dV} = 7 - 6V = 0$$

$V = \frac{7}{6}$ put in Eq. (1)

$$RT = \left(7 - 3\times\frac{7}{6}\right)\times\frac{7}{6}$$

$$\Rightarrow\ T = \frac{49}{12R}$$

8. (i) (c) $w = -P\cdot\Delta V = -nR\Delta T = -2\times8.314\times600 = -9.97\text{ kJ}$

(ii) (b) $\Delta H_{total} = \Delta H_{AB} + \Delta H_{BC} = nC_{p,m}\,\Delta T + 0$

$$= 2\times\frac{7}{2}\times R\times(800 - 200)$$

$$= 4200\,R$$

9. (c) $\frac{T_2}{T_1} = \left(\frac{V_1}{V_2}\right)^{\gamma-1}$

$$T_2 = T_1\cdot\left(\frac{1}{32}\right)^{\frac{7}{5}-1} = 600\cdot\left(\frac{1}{2^5}\right)^{\frac{2}{5}}$$

$$= 600(0.5)^2 = 150\text{ K}$$

$$\Delta H_m = \frac{7}{2}R\times(150 - 600) = -1575\,R$$

10. (a) $w = -nR\,\Delta T = -2\times8.314\times100 = -1662.8\text{ J}$

$$\Delta U = n\int C_{v,m}\,dT$$

$$= 2\times\int(20 + 10^{-2}T)\,dT$$

$$= 2\times 20\times (T_2 - T_1) + 2\times 10^{-2}\times \frac{(T_2^2 - T_1^2)}{2}$$

$$= 4700 \text{ J}$$

$$4700 = q - 1662.8$$

$$\therefore \quad q = 6362.8 \text{ J}$$

11. (d) $C_{p,m} = (21.686 + 8.314 + 10^{-3}T)$

$$= 30 + 10^{-3}T$$

$$\int \Delta H = \int nC_{p,m}\,\Delta T$$

$$= 10\times \int (30 + 10^{-3}T)\cdot dT$$

$$\Delta H = 10\times 30\,(T_2 - T_1) + \frac{10^{-3}}{2}\times [T_2^2 - T_1^2]\times 10$$

$$= 30350 \text{ J}$$

12. (b) $dU = dq + dw$

$$nC_{v,m}\cdot dT = nC_m\cdot dT - P\cdot dV$$

$$C_m = C_{v,m} + \frac{P\cdot dV}{n\cdot dT} \quad \text{...(1)}$$

$$PV^n = K \text{ and } PV = nRT$$

$$\therefore \quad KV^{1-n} = nRT$$

$$K(1-n)V^{-n}\cdot dV = nRdT$$

$$\frac{dV}{dT} = \frac{nR}{K(1-n)V^{-n}} \quad \text{...(2)}$$

from Eqs. (1) and (2)

$$C_m = C_{v,m} + \frac{R}{(1-n)}$$

13. (b) $P_2V_2^2 = P_1V_1^2$

$$\frac{P_2}{P_1} = \left(\frac{V_1}{V_2}\right)^2$$

Now, $\frac{T_2}{T_1} = \frac{P_2V_2}{P_1V_1} = \frac{V_1}{V_2} = \frac{1}{3}$

$$\therefore \quad T_2 = \frac{300}{3} = 100 \text{ K}$$

$$\Delta H = nC_{p,m}\,\Delta T$$

$$= 2\times \frac{5}{2}R\times (-200 \text{ K}) = 1000 \text{ R}$$

14. (d) $\Delta S = nC_{p,m}\ln\frac{T_2}{T_1} + nR\ln\frac{P_1}{P_2}$

$$= 3\times \frac{7}{2}R\ln\left(\frac{596}{298}\right) + 3R\ln\frac{1}{4}$$

$$= -6.3 \text{ cal K}^{-1}$$

15. (a) For isoentropic process $\Delta S_{\text{system}} = 0$

$$\therefore \quad nC_{p,m}\ln\frac{T_2}{T_1} + nR\ln\frac{P_1}{P_2} = 0$$

$$\Rightarrow \quad \ln(P_2) = \frac{5}{2}\times \ln\left(\frac{600}{300}\right)$$

$$= 1.75 \text{ atm}$$

16. (b) $\Delta S_{\text{system}} = nR\ln\frac{V_2}{V_1}$

$$= 2\times R\times \ln 2$$

$$= 11.52 \text{ J/K}$$

$$\Delta S_{\text{surrounding}} = -\frac{3.41\times 1000}{310}$$

$$= -11 \text{ J/K}$$

$$\Delta S_{\text{total}} = +11.52 - 11$$

$$= +0.52 \text{ J/K}$$

17. (c) $0.40 = aT_1^3 + bT_1$

$$0.40 = a\times (1000) + b\times 10$$

$$0.4 = 1000a + 10b \quad \text{...(1)}$$

$$0.92 = aT_2^3 + bT_2$$

$$\Rightarrow \quad 0.92 = a\times 8000 + 20b \quad \text{...(2)}$$

from Eqs. (1) and (2)

$$a = 2\times 10^{-5}, \quad b = 0.038$$

$$S_m = \int \frac{aT^3 + bT}{T}\cdot dT$$

$$= \frac{a\,[T_2^3 - T_1^3]}{3} + b\,[T_2 - T_1]$$

$$= 0.813 \text{ J/K-mol}$$

19. (d) No. of moles of sucrose $= \frac{34.2}{342} = 0.1$

$-(\Delta G)_{T,P}$ = useful work done by the system

$$-\Delta G = -\Delta H + T\cdot \Delta S$$

$$= +(6000\times 0.1) + \frac{180\times 0.1\times 300}{1000}$$

$$= 605.4 \text{ kJ}$$

20. (a) $\Delta G^\circ_{200} = \Delta H^\circ_{200} - T\,\Delta S^\circ_{200}$

$$\Delta H^\circ_{200} = 20 - 4 = 16 \text{ kJ/mol}$$

$$\Delta H^\circ_{T_2} = \Delta H^\circ_{T_1} + \Delta C_P\,[T_2 - T_1]$$

$$\Delta H^\circ_{400} = \Delta H^\circ_{200} + \frac{20\times 200}{1000} \text{ kJ/mol}$$

$$= 16 + 4 = 20 \text{ kJ/mol}$$

21. (a) $\Delta_f S^\circ (NH_4Cl, s)$ at 300 K

$$= S^\circ_{NH_4Cl(s)} - \left[\frac{1}{2}S^\circ_{N_2} + 2\,S^\circ_{H_2} + \frac{1}{2}S^\circ_{Cl_2}\right]$$

$$= -374 \text{ JK}^{-1}\text{ mol}^{-1}$$

$$\because \quad \Delta_r C_P = 0$$

$$\therefore \quad \Delta_f S^\circ_{310} = \Delta_f S^\circ_{300}$$

$$= -374 \text{ JK}^{-1}\text{ mol}^{-1}$$

$$\Delta_f H^\circ_{310} = \Delta_f H^\circ_{300} = -314.5$$

$$\Delta_f G^\circ_{310} = \Delta_f H^\circ - 310\,\Delta S^\circ$$

$$= -314.5 - \frac{310(-374)}{1000}$$

$$= -198.56 \text{ kJ/mol}$$

22. (d) $\Delta_r H^\circ = 3\cdot \Delta_f H^\circ (CO, s) + 4\,\Delta_f H^\circ (CO_2, g) - \Delta_f H^\circ (Co_3O_4, s) - 4\cdot \Delta_f H^\circ (CO, g)$

$$\Delta_r H^\circ = -241 \text{ kJ/mol}$$

$\Delta_r S° = 3 \times 30 + 4 \times 213.7 - 102.5 - 4 \times 197.7$
$= 51.5$ J/K-mol

$\therefore \quad \Delta_r G° = \Delta_r H° - T \cdot \Delta_r S°$
$= -241 - \dfrac{300 \times 51.5}{1000}$
$= -256.45$ kJ/mol

23. (c) At constant volume, $\dfrac{P_1}{T_1} = \dfrac{P_2}{T_2}$

$\Rightarrow \quad P_2 = 1 \times \dfrac{300}{200} = \dfrac{3}{2}$

and $\quad V_1 = 24.63$ L

for single phase

$\because \quad dG = Vdp - S\,dT$

$\Delta G = V \cdot \Delta P - \int (2 + 10^{-2} T) \cdot dT$

$= 1231.5 - 200 - \dfrac{10^{-2} \times 50{,}000}{2}$

$= 781.5$ J

24. (c) $\because \Delta_r C_p = 0, \quad \therefore \Delta H_{300} = \Delta H_{310}$

25. (d) $\Delta H° = -3\,\Delta H_1° + \Delta H_2° + 2\,\Delta H_3° + 3\,\Delta H_4°$
$= -747.4$ kJ

$\Delta H° = \Delta U° + \Delta n_g RT$; where $\Delta n_g = -8$

$-747.4 = \Delta U° - \dfrac{8 \times 8.314 \times 300}{1000}$

$\Delta U° = -727.44$ kJ

26. (b) $\Delta_r G° = \Delta_r H° - T \times \Delta_r S°$

$\Delta_r S° = 2 \times 81 - 4 \times 24 - 3 \times 205$ J/mol

$\therefore \quad \Delta_r H° = -2258.1$ kJ/mol

$\Delta_r H° = 2 \times \Delta_f H° (Cr_2O_3, s)$

$\therefore \quad \Delta_f H° (Cr_2O_3, s) = -\dfrac{2258.1}{2}$

$= -1129.05$ kJ/mol

27. (a) $CaO(s) + CO_2(g) \longrightarrow CaCO_3(s)$

$\Delta_f H° = \Delta H_f° (CaCO_3) - \Delta H_f° (CaO) - \Delta H_f° (CO_2)$
$= -1207 - (-635) - (-394)$
$= -178$ kJ/mol

$\therefore \Delta U = \Delta H - \Delta n_g RT$

$\Delta U = -178 - \dfrac{(-1) \times 8.3 \times 300}{1000}$

$= -175.51$ kJ

$n_{CaO} = \dfrac{224}{56} = 4$

$\therefore \quad q_v = n \cdot \Delta_r U = 4 \times (-175.51)$
$= -702.04$ kJ

28. (d) $H_2C_2O_4(l) + \frac{1}{2} O_2(g) \rightarrow H_2O(l) + 2\,CO_2(g)$;

$\Delta n_g = 3/2$

$\Delta U_c = -\dfrac{0.312 \times 8.75}{1} \times 90$

$= -245.7$ kJ/mol

$\Delta H = \Delta U + \Delta n_g RT$

$= -245.7 + \frac{3}{2} \times \dfrac{8.314 \times 300}{1000}$

$= -241.95$ kJ/mol

29. (b) $H_3PO_3 \longrightarrow 2H^+ + HPO_3^{2-}$; $\Delta_r H = ?$

$2H^+ + 2\,OH^- \longrightarrow 2H_2O$;

$\Delta_r H = -55.84 \times 2 = -111.68$

$-106.68 = \Delta_{ion} H - 55.84 \times 2$

$\Delta_{ion}\, H = 5$ kJ/mol

30. (b) $HA \longrightarrow H^+ + A^-$; $\Delta_r H = 1.4$ kJ/mol

$\Delta H_{neutralization} = \Delta H_{ionization} + \Delta_r H$
$(H^+ + OH^- \rightarrow H_2O)$

$-55.95 = \Delta H_{ionization} - 57.3$

$\Delta H_{ionization}$ for 1 M HA = 1.35 kJ/mol

% heat utilized by 1 M acid for ionization

$= \dfrac{1.35}{1.4} \times 100 = 96.43\%$

so, acid is $100 - 96.43 = 3.57\%$ ionized

31. (b) $2\,C(s) + 3H_2(g) \longrightarrow C_2H_6(g)$

$$\Delta_f H° = \begin{bmatrix} 2 \times \Delta_{sub} H\,(C, s) \\ 3 \times B.E.\,(H—H) \end{bmatrix} - \begin{bmatrix} B.E.\,(C—C) \\ + 6 \times B.E.\,(C—H) \end{bmatrix}$$

$-85 = 2 \times 718 + 3 \times 436 - (x + 6y)$

$x + 6y = 2829 \quad ...(1)$

Similarly, for $C_3H_8(g)$

$2x + 8y = 4002 \quad ...(2)$

from Eqs. (1) and (2)

$x = 345$ kJ/mol; $y = 414$ kJ/mol

32. (c) $\dfrac{1}{2} N_2(g) + \dfrac{3}{2} H_2(g) \longrightarrow NH_3(g)$

Let B.E. of N≡N is x

$-46 = \dfrac{x}{2} + \dfrac{3}{2} \times 436 - 3 \times 393 \Rightarrow x = 958$

$N_2H_4(l) \longrightarrow N_2(g) + 2H_2(g)$;

$\Delta_r H = -50$ kJ/mol

$$\Delta_r H = \begin{bmatrix} \Delta_{vap} H\,(N_2H_4, l) \\ + 4 \times B.E.\,(N—H) \\ + B.E.\,(N—N) \end{bmatrix} - \begin{pmatrix} B.E.\,(N \equiv N) \\ + 2\,B.E.\,(H—H) \end{pmatrix}$$

$-50 = (18 + 4 \times 393 + y) - (958 + 2 \times 436)$

$-50 = (1590 + y) - (1830)$

B.E. (N—N) or $y = 190$ kJ/mol

33. (b) $\Delta_r H$ = [Heat supplied] − [Heat evolved]

$292 = [4x + 279] - [38 + 85]$

$\Rightarrow \quad x = 34$ kcal/mol

34. (d) Theoretical heat of hydrogenation of benzene = (Actual heat of hydrogenation) + (Resonance energy)

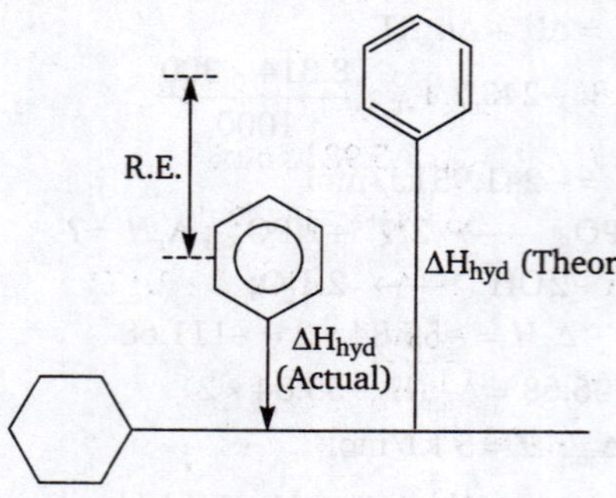

$= -205 - 152 = -357$

Enthalpy of hydrogenation of (cyclohexene)

$= \frac{-357}{3} = -$ **119 kJ/mol**

35. (a) $C_3H_8(g) + 5\,O_2(g) \rightarrow 3\,CO_2(g) + 4H_2O(l)$

$$\Delta_c H = \begin{bmatrix} 8 \times \text{B.E.}(\text{C—H}) \\ +2 \times \text{B.E.}(\text{C—C}) \\ +5 \times \text{B.E.}(\text{O═O}) \end{bmatrix} - \begin{bmatrix} 6 \times \text{B.E.}(\text{C═O}) \\ +8 \times \text{B.E.}(\text{O—H}) \\ +3 \times |\text{R.E.}| \text{ of } CO_2 \\ +4 \times \Delta_{vap} H(H_2O) \end{bmatrix}$$

36. (a)

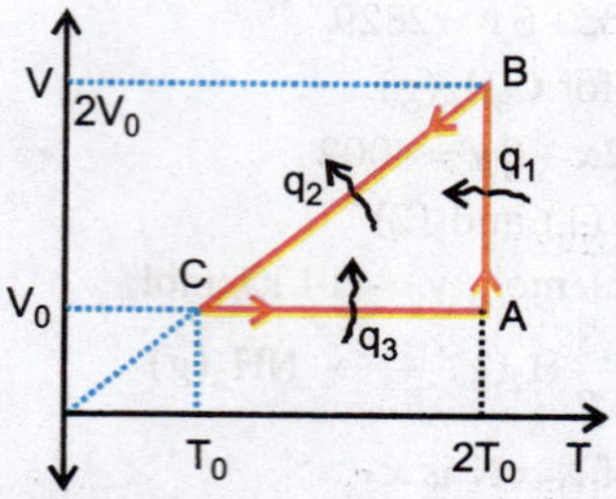

AB process

$0 = q_1 + w_1$

$= q_1 - nR(2T_0)\ln 2$

BC process $\Delta U_2 = q_2 + w_2$

$$\frac{nR}{(\gamma-1)}(T_0 - 2T_0) = q_2 - \left(\frac{nRT_0}{V_0}\right)\cdot(V_0 - 2V_0)$$

CA process

$\Delta U_3 = q_3 + w_3$

$$\frac{nR}{(\gamma-1)}(2T_0 - T_0) = q_3 + 0$$

$$\text{Efficiency} = \frac{\text{Total work done}}{\text{Total heat absorbed}} = \frac{w_1 + w_2}{q_1 + q_3}$$

$$= \frac{(-2RT_0 \ln 2) + (nRT_0)}{(2RT_0 \ln 2) + \left(\frac{nRT_0}{\gamma - 1}\right)}$$

37. (c) $V_1 = 24.63\,\text{L}$ $\quad T_1 = 300\,\text{K}\,; T_2 = 600\,\text{K}$

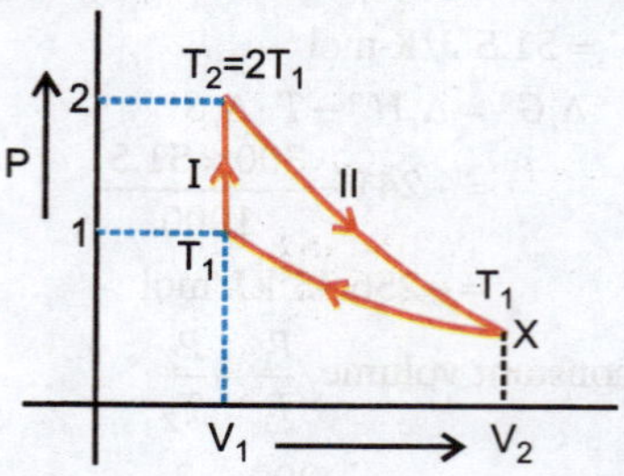

For path (II)

$$T_1 V_1^{\gamma-1} = T_2 V_2^{\gamma-1}$$

$$\frac{T_1}{T_2} = \left(\frac{V_2}{V_1}\right)^{\gamma-1}$$

$$\frac{600}{300} = \left(\frac{V_2}{V_1}\right)^{\frac{7}{5}-1}$$

$$(2)^{5/2} = \frac{V_2}{V_1}$$

$$V_2 = 4\sqrt{2} \times 24.63 = 139.3\,\text{L} \approx 139\,\text{L}$$

38. (c) $\Delta_r H° = \Delta_r H°(H_2O, l) - \Delta_r H°(OH^-, aq) = -229$

$-56 = -285 - \Delta_r H°(OH^-, aq);$

$\Delta_r H°(OH^-, aq) = -229$

$HF(aq) + OH^-(aq) \longrightarrow F^-(aq) + H_2O(l)$

$\Delta.\overset{\circ}{H}_{neut} = (-320 - 285) - (-329 - 229)$

$= -47$ kJ/mol

39. (b) $I_{2(s)} \longrightarrow I_2(g)$

$\Delta H_2 = \Delta H_1 + \Delta C_P(T_2 - T_1)$

$\Delta C_P = C_P(I_{2,g}) - C_P(I_{2,s}) = -6.2$ cal/mole

$\Delta H_2 = 6096 - 6.2(50) = 5786$ cal/mole

$\Delta H = \Delta U + 1 \times 2 \times 523$

$5786 = \Delta U + 1 \times 2 \times 523$

$\Delta U = 5786 - 1046 = 4740$ cal

40. (d) $\Delta G° = -RT \ln K_p\,;\; K_p = (2x)^2 X = 4X^3$

$\Delta G° = -RT \ln(4X^3)$

$\Delta G° = -RT \ln 4 - 3RT \ln X$

Level 3

Passage-1

1. (d) $\Delta U = q + w$

$\Delta U = -300 + (-50 + 150) = -200$

2. (d) $\Delta U = -300 + (-50 - 150) = -500$

Passage-2

1. (b) $PV = nRT \Rightarrow 2 \times 8 = 2 \times 0.080 \times T$

$T = 100$ K

$$w_{rev} = -2.303 \times n \times R \times T \log \frac{P_1}{P_2}$$

$$= -2.303 \times 2 \times 0.08 \times 100 \times \log\left(\frac{1}{10}\right)$$

$= 36.848$ bar-L

2. (c) $w_{irr} = -P_{ext}(V_2 - V_1)$

$$= -20\left(\frac{nRT}{P_2} - \frac{nRT}{P_1}\right)$$

$= 144$ bar-L

3. (b) $w_{irr\,(total)} = w_1 + w_2$

$$= -10\left(\frac{nRT}{10} - \frac{nRT}{2}\right)$$

$$= -20\left(\frac{nRT}{20} - \frac{nRT}{10}\right)$$

$= 5 \times nRT = 80$ bar-L

4. (c) In isothermal compression is

$w_{\text{One-stage}} > w_{\text{Two-stage}} > w_{\text{Reversible}}$

Passage-3

1. (c) $\Delta_r S = 240 - 198 - 29 \times 2 = -16$

2. (a) $\Delta_r H = -201 + 114 = -87$

3. (d) $\Delta_r S = -16 + \int_{300}^{320} \Delta_r C_p \frac{dT}{T}$

$$\Delta_r S = -16 + (44 - 29.4 - 2 \times 28.8) \ln \frac{320}{300}$$

$$\Delta_r S = -16 + (-43) \ln \frac{320}{300} = -18.78$$

4. (c) $\Delta_r H = -87 + \int_{300}^{320} \Delta_r C_P \, dT$

$= -87 + C - 4.31 \times 70 \times 10^{-3}$

$= -87.86$ kJ/mol

5. (d) $\Delta_r S° = S°_{CH_3OH} — S°_{CO} - 2\,S°_{H_2} = -$ J/K-mol

$\Delta_r H° = \Delta_f H°(CH_3OH) - \Delta_f H°(CO)$
$-2 \cdot \Delta_f H°(H_2)$

$= -87$ kJ/mol

$\Delta_r S°_{320} - \Delta_r S°_{300} = \Delta_r C_p [T_2 - T_1]$

where $\Delta_r C_p^\circ = 44 - 29.4 - 2 \times 28.8$

$= -43$ J/K-mol

$$\Delta_r S°_{320} = -16 + (-43) \ln \frac{320}{320}$$

$= -18.58$

$\Delta_r H°_{320} = \Delta_r H°_{300} + \Delta_r C_p^\circ [T_2 - T_1]$

$$= -87 + \frac{(-43) \times 20}{1000}$$

$= -87.86$ kJ/mol

$\Delta_r G°_{320} = \Delta_r H°_{320} - T \cdot \Delta_r S°_{320}$

$$= -87.86 - \frac{320 \times (-18.58)}{1000}$$

$= -81.91$ kJ/mol

Passage-4

1. (a) $-49.86 = (\Delta_{Ion} H)_{CH_3COOH} - 55.84$

$(\Delta_{Ion} H)_{CH_3COOH} = 5.98 \text{ kJ mol}^{-1}$

2. (b) $\Delta_r H = 2 \times (-55.84) = -111.68$ kJ

3. (b) $2NaOH + H_2A \longrightarrow Na_2A + 2H_2O$

$\frac{0.1 V_1}{1000}$ $\quad \frac{0.05 V_2}{1000}$

$V_1 + V_2 = 100$ ...(1)

$$2 \times \frac{0.05\, V_2}{1000} = \frac{0.1\, V_1}{1000}$$

$\Rightarrow V_1 = V_2$

$V_1 = V_2 = 50$ mL

Passage-5

1. (a) $(\Delta G)_{PT} = 2000 - (20 \times 298)$

$= -3960$ J/mol

2. (b) $CaCO_3 \longrightarrow CaO + CO_2 \quad \Delta H$ +ve

Reaction becomes spontaneous at high temperature because $T\Delta S$ dominates over ΔH_{rxn}.

3. (a)

$CaCl_2(s) + aq + 6H_2O \longrightarrow CaCl_2(aq) + x$

$CaCl_2(aq) \longrightarrow CaCl_2 \cdot 6H_2O(s) + aq + 3.5$

$CaCl_2(s) + 6H_2O \longrightarrow CaCl_2 \cdot 6H_2O(s) + x + 3.5$

$x + 3.5 = 23.2$

$x = 19.7$

$\Delta H = -x = -19.7$

enthalpy driven

One or More Answers is/are Correct

1. (b,c)

Intensive properties : Temperature, pressure.

2. (a, b)

Extensive properties : Gibb's energy, Entropy.

3. (c) State function : Enthalpy

4. (a, b, c)

Work done by system is negative.
Work done by surrounding is positive.
Heat absorbed by system is positive.
Heat absorbed by surrounding is negative.

5. (a, b, d)

$\Delta U = n\int C_{vm} dT \qquad \Delta H = \int C_{pm} dT$

$\Delta U = q + w$

6. (b,c,d) $\Delta U = 0 \quad \therefore \; U_2 - U_1 = 0$

Similarly, $\Delta H = 0$

$$q = -w = nRT \ln \frac{V_2}{V_1}$$

7. (b, c, d)

No heat transfer

$\Delta U = q + w = 0 + w = w$

$\Delta U = w$

$T_2 + T_1 + \frac{w}{nC_{vm}}$

$w = +ve$

$T_2 > T_1$

8. (a, b, c)

$\Delta U - q + w = 0 + w = w$

$w = n\int C_{vm} dT$

$w = \frac{nR}{\gamma - 1}(T_2 - T_1)$

$w = -nR\, P_{ext}\left(\frac{T_2}{P_2} - \frac{T_1}{P_1}\right)$

9. (c,d)

At $V_A = \frac{1 \times R \times 100}{1} = 100\, R$

$V_B = \frac{1 \times R \times 600}{3} = \mathbf{200\, R}$

$\because$ $V_B > V_A$ so expansion of gas takes place

$V_B = 200 \times 0.0821 = 16.42$ L

10. (a,c,d)

First step is adiabatic $(q = 0)$ so $\Delta U_1 = w_1$

Second step is isochoric $(w = 0)$

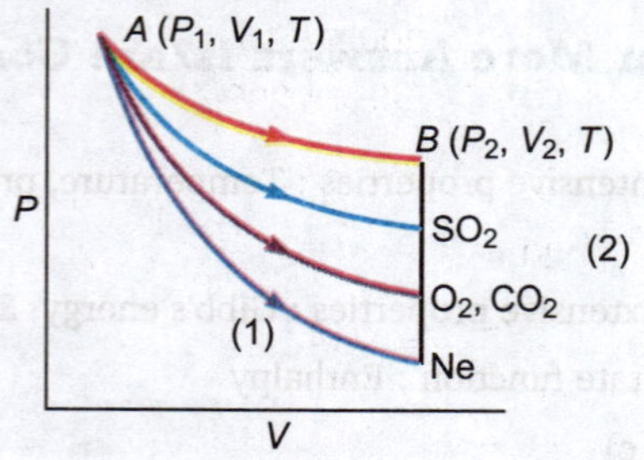

So, $\Delta U_2 = q_2$

$\because$ initial and final temp. are same

$\therefore$ $\Delta U_{total} = \Delta U_1 + \Delta U_2 = 0$

or $w_1 + q_2 = 0$

Max. work done by the gas, SO_2 is (area) under the curve

so, SO_2 absorbed

$\because$ $\gamma_{SO_2} < \gamma_{CO_2} = \gamma_{O_2} < \gamma_{Ne}$

so, max. decrease in temp. of Ne due to step 1.

11. (a,b,c)

$\Delta H = \Delta U + P \cdot \Delta V + V \cdot \Delta P + \Delta P \cdot \Delta V$

is correct relation.

12. (a, c)

$\Delta H = \Delta U + \Delta n_g RT$

(a) $\Delta n_g = -2$ $\quad \Delta U > \Delta H$

(b) $\Delta n_g = +\frac{1}{2}$ $\quad \Delta U > \Delta H$

(c) $\Delta n_g = -\frac{1}{2}$ $\quad \Delta U > \Delta H$

(d) $\Delta n_g = 0$ $\quad \Delta U = \Delta H$

13. (a, c, d)

$\Delta G = \Delta H - T\Delta S$

(a) $\Delta H = +ve, \Delta S = +ve$ at low temperature $\Delta G = +ve$

(b) $\Delta H = -ve, \Delta S = +ve\,; \Delta G = -ve$ at any temperature.

(c) $\Delta H = +ve\,,\ \Delta S = -ve,\ \Delta G = +ve$ at any temperature

(d) $\Delta H = -ve\,,\ \Delta S = -ve,\ \Delta G = +ve$ at high temperature

14. (b, d)

For spontaneous process

$\Delta S_{System} + \Delta S_{Surrounding} > 0$

$(\Delta G_{System})_{T,P} < 0$

15. (a, b, c)

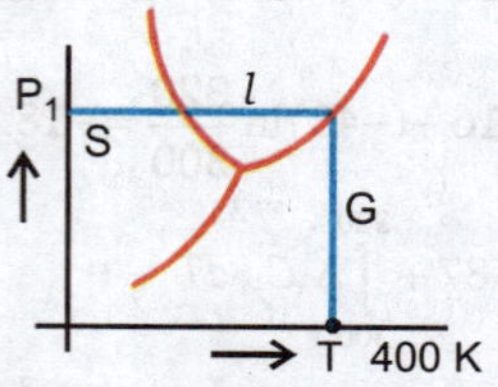

$x(l) \rightleftharpoons x(g) \quad T_b^0 = 400\,K$

(a) At 400 K and 1 atm pressure, $\Delta G = 0$

(b) at 2 atm and 400 K, $\Delta G = +ve$

(c) at 400 K and 0.1 atm pressure, $\Delta G = -ve$

(d) at 410 K and 1 atm pressure, $\Delta G = -ve$

16. (a, c, d)

Entropy ↑ intermolecular force ↓

$\Delta H > 0$

Spontaneous process $\Delta G < 0$

17. (b, c, d)

$C(graphite) \longrightarrow C(diamond)$

$\Delta G = \Delta H - T\Delta S$

$= 1.9 - 298(-ve) = +ve$

Graphite is more stable than diamond thermodynamically

$\Delta_r H = (\Delta_C H)_G - (\Delta_C H)_D\,, (\Delta_C H)_G = -x$

$(\Delta_C H)_D = -y$

$1.9 = -x + y \Rightarrow \quad y = 1.9 + x$

Diamond provides more heat on complete combustion in comparison of graphite.

18. (a, b, c, d)

(a) Reversible adiabatic process is iso-entropic process

(b) When $(\Delta G_{System})_{T,P} < 0$ then process may be endothermic / exothermic

(c) $dG = VdP - SdT$, only PV work consider

(d) $\Delta S = -\frac{9}{18} \times \frac{40.6 \times 1000}{373} = -54.42$

19. (a, b, c)

(a) $\Delta E = q + W = 0 + 0 = 0$

(b) $(\Delta_F H^\circ)(S, \text{rhombic}) = 0$

(c) $(C-C) + (C-H) = 2812$

$(C-C) = 2812 - 6 \times \frac{1656}{4} = 328$

(d) $H_2O(l) \longrightarrow H_2O(g)$

$44 = -(\Delta_F H)_{H_2O(l)} - 242$

$(\Delta_F H)_{H_2O(l)} = -286$

$H^+(aq) + OH^-(aq) \longrightarrow H_2O(l)$

$-57.3 = -286 - 4(\Delta_f H)_{OH^-}(aq)$

$(\Delta_f H)_{OH^-} = -228.7$

20. (a, b, c)

(a) $\Delta_r H = 15 + 0 - 57.3 = -42.3$ kJ

(b)

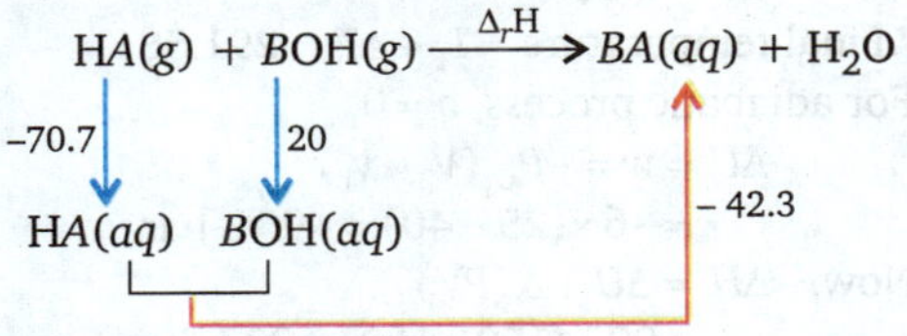

(c)

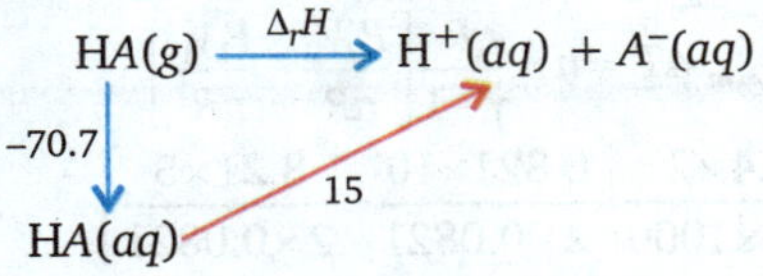

$\Delta_r H = -70.7 + 15 = -55.7$

(d) $B^+(aq) + OH^-(aq) \longrightarrow BOH(aq)$

$\Delta_r H = 0$

21. (a,c,d)

Work is a form of energy exchange between system and surrounding in adiabatic process.

Intensive property is not additive.

In cyclic process work may be zero.

For cyclic process

$dU = 0 = dq + dw \Rightarrow dq - Pdv = 0$

22. (a,c,)

Entropy of isolated system increases or remains constant but it can never decreases.

23. (a,c)

$\Delta S_{vap} = \frac{\Delta H}{\text{B.P.}} = \frac{40 \times 1000}{400} = 100$

(a) $P = 2$ atm B.P. > 400 K

$P \uparrow$ B.P $\uparrow$ $\Delta S_{vap} < 100$

(c) $\Delta G = \Delta H - T\Delta S$

$= 40 \times 1000 - 410 \times 100 = -ve$

$\Delta G < 0$

(d) $\Delta H = \Delta U + \Delta n_g RT$

$\Delta O = \Delta U + 1 \times 8.31 \times 10^{-3} \times 400$

$\Delta U = 36.676$ kJ/mol

24. (a, c, d)

(a) In isolated system no change possible

(b) $\Delta U = q + w = 0 \Rightarrow q = -w$

w may be zero so q may be zero.

(c) Entropy of a closed system is not maximum at equilibrium.

(d) Intensive properties.

25. (a, c)

$\Delta H = -ve$ Enthalpy driven

$\Delta S = +ve$ Entropy driven

$\Delta G = -ve$ Spontaneous process

$\Delta G = +ve$ Non spontaneous process

Level 4

Subjective Problems

1. For reversible adiabatic process

$P_1^{1-\gamma} \cdot T_1^\gamma = P_2^{1-\gamma} \cdot T_2^\gamma$

$\frac{T_1}{T_2} = \left(\frac{P_2}{P_1}\right)^{\frac{1-\gamma}{\gamma}}$; $\frac{300}{90} = \left(\frac{10}{200}\right)^{\frac{1-\gamma}{\gamma}}$;

$\frac{10}{3} = \left(\frac{1}{20}\right)^{\frac{1-\gamma}{\gamma}}$; $\log\frac{10}{3} = \frac{1-\gamma}{x} \cdot \log\frac{1}{20}$

$\frac{1-\gamma}{\gamma} = \frac{0.5229}{-1.301} = -0.4$

$\frac{1}{\gamma} = 0.6 \Rightarrow \gamma = 1.66$

Atomicity = 1

2. $r = \frac{-nRT \ln\frac{12}{24}}{-P_{ext} nRT\left(\frac{1}{P_2} - \frac{1}{P_1}\right)}$

$r = \frac{-0.7}{24\left(\frac{1}{24} - \frac{1}{12}\right)} = \frac{0.7 \times 24}{24(1)}$

$10r = 10 \times 0.7 = 7$

3. $C_m = C_{VM} + \frac{R}{1-x} = C_{VM} + \frac{R}{1-3}$

$C_m = \frac{3R}{2} + \frac{2R}{2} + R - \frac{R}{2} = 3R = 6$

4. $\eta = \frac{w}{10} = 1 - \frac{300}{500} \Rightarrow w = 4\,\text{kcal}$

5. $\eta = \frac{|w|}{8}$ ...(i)

$\eta = 1 - \frac{T_1}{T_2}$

$T_2 \times 2^{\gamma-1} = T_1 \times 16^{\gamma-1} \Rightarrow \frac{T_1}{T_2} = \frac{1}{2}$

$\frac{|w|}{8} = 1 - \frac{1}{2} = \frac{1}{2} \qquad \Rightarrow |w| = \frac{1}{2} \times 8 = 4$

6. $\frac{1}{2}A + B \longrightarrow \frac{3}{2}C$

$\Delta G = \Delta H - T\Delta S$

$= (10 \times 1000 - 300 \times 10) \times 10^{-3}$

$\Delta G = 7\,\text{kJ/mol}$

7. Energy available $= \frac{25}{100} \times 2880 \times \frac{1}{2} = 360$

Minimum distance $= \frac{1}{90} \times 360 = 4\,\text{km}$

8. $C_2H_2(g) + H_2(g) \longrightarrow C_2H_4(g)$

$-175 = -(\Delta_f H)_{C_2H_2} + 50$

$\Rightarrow (\Delta_f H)_{C_2H_2} = 225$

$C_2H_2 + \frac{5}{2}O_2 \longrightarrow 2CO_2 + H_2O$

$\Delta H^\circ = 2 \times (-390) - 280 - 225 = -1285$

$\left|\frac{\Delta H^\circ}{257}\right| = \left|-\frac{1285}{257}\right| = 5$

9. $5H_2O + CuSO_4(s) + aq \longrightarrow CuSO_4(aq) + 70$

$CuSO_4(aq) \longrightarrow CuSO_4 \cdot 5H_2O(s) + aq + 10$

$CuSO_4(s) + 5H_2O \longrightarrow CuSO_4 \cdot 5H_2O + 80$

$|\Delta H| = 80$

For 0.1 mol : $80 \times 0.1\,\text{mol} : 80 \times 0.1 = 8\,\text{kJ}$

10. $-50 = (\Delta_{\text{Ion}}H)_{NH_4OH} - 57$

$(\Delta_{\text{Ion}}H)_{NH_4OH} = 7\,\text{kJ}$

11.

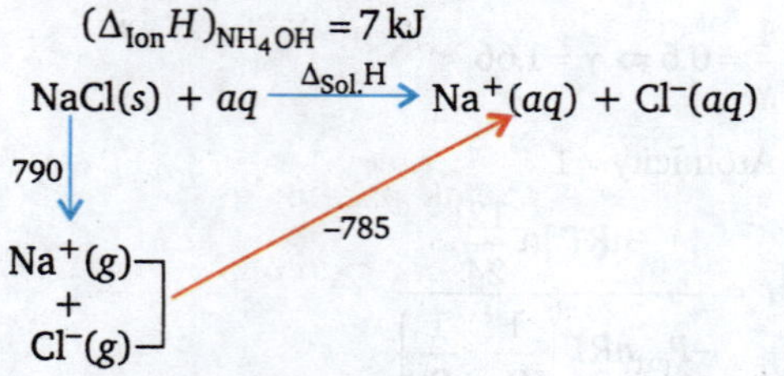

$\Delta_{\text{Sol.}}H = 790 - 785 = 5$

12. $Q_v = 1.25 \times 4 = 5$

$Q_v = 5 = \frac{x}{80} \times 400$

$\Rightarrow x = 1$

13. $\eta = \frac{T_2 - T_1}{T_2} = \frac{|-w_{\text{Total}}|}{q_2} \Rightarrow \frac{500 - 350}{500} = \frac{|-w_{\text{Total}}|}{10}$

Work done in one cycle = 3

Work done in two cycles $= 3 \times 2 = 6$

14. $W = -\Delta n_g RT = \frac{-(-3) \times 2 \times 500}{1000} = 3$

15. Process AC = polytropic process ($P = KV$)

Molar Heat capacity $c_m = c_V + R/2 = 2R$

Process AB = Isobaric

$c_m = c_p = 5R/2$

$$\frac{q_{AC}}{q_{AB}} = \frac{\int_{T_A}^{T_C} nC_m \cdot dT}{\int_{T_A}^{T_B} n \cdot C_{p,m} \cdot dT} = \frac{2R}{\frac{5}{2}R} = 0.8 \; :$$

$\frac{q_{AC}}{q_{AB}} \times 10 = 0.8 \times 10 = 8$

Numerical Problems

2. Since work is done against constant P and thus, irreversible

$\Delta V = 5 - 3 = 2\,\text{dm}^3 = 2$ litre ; $P = 3$ atm

$\therefore \; w = -P \cdot \Delta V = -3 \times 2$ litre atm $= -600$ J

Now this work is used up in heating water

$\therefore \quad w = n \times C \times \Delta T$

$600 = 5 \times 4.2 \times 18 \times \Delta T$

$\therefore \quad \Delta T = 1.58$

$\therefore$ Final temperature $= T_1 + \Delta T = 291.58$ K

3. For adiabatic process, $q = 0$

$\therefore \quad \Delta U = w = -P_{ext}(V_2 - V_1)$

$= -6 \times (25 - 40) = +90$ L-bar

Now, $\Delta H = \Delta U + \Delta(PV)$

$= 90 + (150 - 160) = 80$ L-bar

4. $\Delta H = nC_{p,m}\,\Delta T = n \times \frac{\gamma R}{\gamma - 1}\left[\frac{P_2V_2}{nR} - \frac{P_1V_1}{nR}\right]$

$= 2 \times \frac{1.4 \times 2}{0.4 \times 1000}\left[\frac{0.821 \times 10}{2 \times 0.0821} - \frac{8.21 \times 5}{2 \times 0.0821}\right]$

$= -2.8$ kcal, $x = 2.80$

5. $P_1V_1^\gamma = P_2V_2^\gamma \Rightarrow 1 \times 1^{5/3} = P_2 \times (8)^{5/3} \Rightarrow P_2 = \frac{1}{32}$ bar

$w = \frac{P_2V_2 - P_1V_1}{\gamma - 1} = \frac{\left(\frac{1}{32}\right) \times 8 - 1 \times 1}{\frac{2}{3}} = -\frac{9}{8}$ bar L

$= -\frac{9}{8} \times 100 = -112.5$ J, $|w| = 112.50$ J

6.

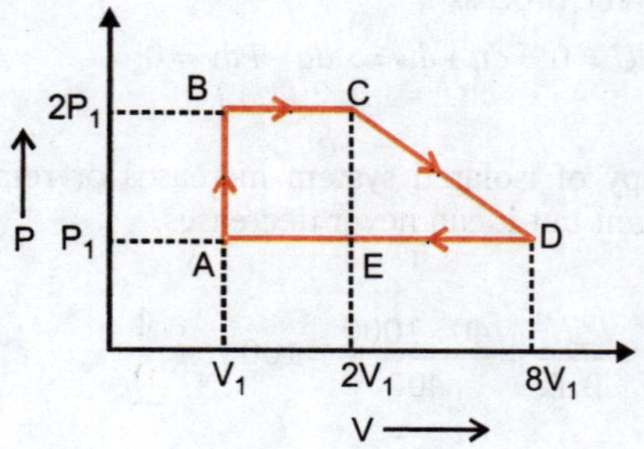

$|w| =$ Area under PV curve

$|w| =$ Area of $ABCEA + ECDE$

$$= P_1V_1 + \frac{1}{2} \times 6V_1P_1 = 4P_1V_1$$

$$= 4nRT_1 = 4 \times 2 \times 2 \times 400 \text{ cal} = 6400 \text{ cal}$$

$|w| = 6.40$ kcal

7. $\Delta S = nR \ln \frac{V_2}{V_1}$ (Isothermal process)

$$= 2.303 \times 2 \times R \log \frac{V_2}{V_1}$$

$$= 2.303 \times 2 \times 2 \times \log \frac{0.1}{1}$$

$$= -4.606 \times 2 \text{ cal/K}$$

$|\Delta S_{\text{System}}| = 9.212$ cal/K

8. This is adiabatic free expansion

$w = 0$; $q = 0$ so, $\Delta U = 0$

i.e., $T_1 = T_2 = 300$ K

$$\Delta S_{\text{sys.}} = nR \ln \frac{V_2}{V_1} = 1 \times R \times \ln \frac{8}{2} = R \ln 4$$

$$= 2 \times 1.4 = 2.8 \text{ cal/K}$$

$$\Delta S_{\text{surr.}} = \frac{-q_{\text{sys}}}{T} = 0$$

$$\Delta S_{\text{univ.}} = \Delta S_{\text{sys.}} + \Delta S_{\text{surr.}} = 2.80 \text{ cal/K}$$

9. $\Delta U = 0$

$\Delta U_1 + \Delta U_2 = 0$

$nC_{vm}(T_f - T_1) + nC_{vm}(T_f - T_2) = 0$

$$T_f = \frac{T_1 + T_2}{2} \quad \ldots\text{(i)}$$

$\Delta S = \Delta S_1 + \Delta S_2$

$$\Delta S = nC_{p,m} \ln \frac{T_f}{T_1} + nC_{p,m} \ln \frac{T_f}{T_2}$$

$$\Delta S = nC_{P,m} \ln \frac{T_f^2}{T_1 \times T_2} \quad \ldots\text{(ii)}$$

From (i) and (ii)

$$\Delta S = nC_{p,m} \ln \frac{(T_1 + T_2)^2}{4(T_1T_2)} = 2$$

10. $dS = \frac{dq}{T} = C_v \frac{dT}{T} + \frac{p}{T} dV \quad \left[\left(\frac{\partial U}{\partial T}\right)_v = C_v = b + 2cT\right]$

$$= (b + 2cT)\frac{dT}{T} + 0 \quad [\because dV = 0]$$

$$\int dS = \int \left(\frac{b}{T} + 2c\right) dT$$

$$\Rightarrow \Delta S = b \ln \frac{T_2}{T_1} + 2c\Delta T = b \ln 2 + 600\,c$$

$$= 16 \times 0.7 + 600 \times 6 \times 10^{-3}$$

$$= 14.80 \text{ JK}^{-1} \text{ mol}^{-1}$$

11. $\Delta_r S° = 2 \times 70 - 20 - 50 \times 3 = -30$ J/K $\cdot$ mol

$$\Delta_r G° = \Delta_r H° - T \cdot \Delta_r S° = -99 - \frac{300 \times (-30)}{1000}$$

$$= -99 + 9 = -90 \text{ kJ/mol}$$

$$\Delta G° = -T \cdot \Delta S°_{\text{Total}}$$

$$-90 \text{ kJ/mol} = -300 \cdot \Delta S°_{\text{Total}}$$

$$-\frac{90 \times 1000}{-300} = \Delta S°_{\text{Total}}$$

$$\Delta S°_{\text{Total}} = 300 \text{ J/k} \cdot \text{mol}$$

12. $\Delta_r H° = -110.5 - (-266.3) = 155.8$

$\Delta_r S° = 27.28 + 197.6 - 57.49 - 5.74$

$\Delta_r S° = 161.65$; at eq^{m}. $\Delta_r G° = 0$

$155.8 \times 1000 = T \times 161.65$

$$T = 963.8 \approx 964 \text{ K}$$

13. $Cu(s) + \frac{1}{2}O_2(g) \rightleftharpoons CuO(s)$;

$$K_p = \sqrt{10^{+48}} = 10^{24}$$

$$\Delta_r G° = \Delta_f G°(\text{CuO}, s) = -RT \ln K_p$$

$$= -2 \times 300 \times 2.303 \log(10^{24})$$

$$= -2 \times 300 \times 2.303 \times 24 \text{ cal/mol}$$

$$= -33.16 \text{ kcal/mol}$$

$$|\Delta_r G°| = 33.16 \text{ kcal/mol}$$

14. $CaO(s) + CO_2(g) \longrightarrow CaCO_3(s)$

$$\Delta_r H° = \Delta H_f°(CaCO_3) - \Delta H_f°(CaO) - \Delta H_f°(CO_2)$$

$$= -1207 - (-635) - (-394)$$

$$= -178 \text{ kJ/mol}$$

$$\therefore \Delta E_r° = \Delta H_r° - \Delta n_g RT$$

$$\Delta E_r° = -178 - \frac{(-1) \times 8.3 \times 300}{1000} = -175.51 \text{ kJ}$$

$$n_{\text{CaO}} = \frac{11.2}{56} = 0.2$$

$$\therefore q_v = n \cdot \Delta_r E = 0.2 \times (-175.51)$$

$$= -35.10 \text{ kJ}$$

15. Heat released by $\frac{6.3}{64000}$ mole haemoglobin

$$= 25 \times 4.2 \times 0.03 = 3.15 \text{ J}$$

$\therefore$Heat released per mole haemoglobin

$$= \frac{3.15 \times 64000}{6.3} = 32000 \text{ J}$$

$\therefore$ Heat released per mole

$$O_2 = \frac{32000}{4} = 8000 \text{ J} = 8.00 \text{ kJ}$$

16. $\Delta H_{\text{Neutralization}} = -\frac{M.C.\Delta T}{\text{Mol of NaOH}}$

$$57.1 \times 1000 = \frac{500 \times 4.18 \times \Delta T}{300 \times 0.1 \times 10^{-3}}$$

$$\Delta T = 0.8196 = 81.96 \times 10^{-2}$$

$$= 82 \times 10^{-2}$$

17. $\Delta H = BE$ data for reactant $-BE$ data for product

$= [4 \times 415 + 1 \times 615 + 1 \times 432]$

$-[5 \times 415 + 1 \times 348 + 1 \times 328] = -44$ kJ/mole

18.

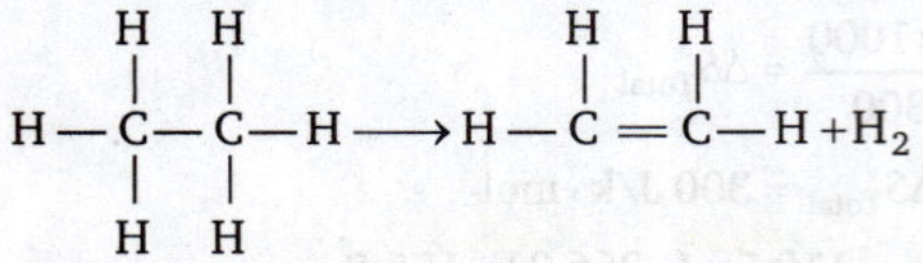

$\Delta_r H = \text{B.E.}(C-C) + \text{B.E.}(C-H) \times 2 - \text{B.E.}(C=C) - \text{B.E.}(H-H)$

$= 347 + 44 \times 2 - 611 - 436 = 128$ kJ/mol

19. $H_2(g) + O_2(g) \longrightarrow H_2O_2(l)$

$\downarrow BE_{H-H}$ $\quad \downarrow BE_{O=O}$ $\quad \downarrow \Delta H_{vap.}$

$2H(g) \quad 2O(g) \qquad H_2O_2(g)$

$\downarrow 2BE_{O-H} + BE_{O-O}$

$2H(g) + 2O(g)$

$\Delta H_f = (435 + 498) - (52 + 2 \times 464 + 142)$

$= -189$ kJ/mol, $|\Delta H_f| = 189.00$ kJ/mol

20. $K(s) + \frac{1}{2}Cl_2(g) \longrightarrow KCl(s)$

$89.2 + 419 + \frac{1}{2} \times 243 - 348.6 = -436.7 + \text{L.E.}$

$\text{L.E.} = -717.8$ kJ/mol

$|\text{L.E.}| = 717.8 \approx 718$ kJ/mol

CHAPTER 5

Chemical Equilibrium

Types of Reactions

	Reversible	Irreversible
(i)	A reaction in which not only the reactants react to form products but also the products react to form reactants under the same conditions is called reversible reaction.	A reaction which cannot take place in the reverse direction is called an irreversible reaction
(ii)	A few examples are given below $CaCO_3(s) \rightleftharpoons CaO(s) + CO_2(g)$ $N_2O_4(g) \rightleftharpoons 2NO_2(g)$	A few examples are given below $AgNO_3(aq) + NaCl(aq) \rightarrow AgCl(s) + NaNO_3(aq)$ $BaCl_2(aq) + Na_2SO_4(aq) \rightarrow BaSO_4(s) + 2NaCl(aq)$

Characteristics of Equilibrium State

- A reaction (or a process) is said to be in equilibrium when the rate of forward reaction (process) becomes equal to the rate of backward reaction (process).
- At equilibrium, the concentration of each of the reactants and products become constant.
- Chemical equilibrium is dynamic in nature and not static *i.e.*, even after equilibrium is attained, the forward as well as the backward reaction take place but at equal speeds.
- A chemical equilibrium can be established only if none of the products is allowed to escape out.
- Chemical equilibrium can be attained from either direction.
- When reaction attains equilibrium at certain temperature and pressure, $\Delta_r G = 0$
- Catalyst increases rate of reaction in forward as well as backward direction upto same extent.

Law of Mass Action

It was put forward by Guldberg and Waage. It sates that the rate at which a substance reacts is directly proportional to its active mass raised with suitable powers and hence the rate at which substances react together is directly proportional to the product of their active masses. Active mass means molar concentration.

Law of Chemical Equilibrium

For the reaction $aA(aq) + bB(aq) \rightleftharpoons xX(aq) + yY(aq)$

$$K_c = \frac{[X]^x [Y]^y}{[A]^a [B]^b}$$

Where K_c is called equilibrium constant at constant temperature.

Types of Equilibrium Constants

K_c = Equilibrium constant in terms of molar concentration $= \frac{[X]^x [Y]^y}{[A]^a [B]^b}$

K_p = Equilibrium constant in terms of pressure $= \frac{p_X^x \cdot p_Y^y}{p_A^a \cdot p_B^b}$

K_x = Equilibrium constant in terms of mole fraction $= \frac{(X_x)^x (X_y)^y}{(X_A)^a (X_B)^b}$

Unit of Equilibrium Constant

Unit of $K_p = (\text{atm})^{\Delta n_g}$ (where Δn_g = change in gas mole of reaction)

Unit of $K_c = M^{\Delta n_g}$

Unit of K_x = Unitless

Relationship between K_p and K_c: $K_p = K_c (RT)^{\Delta n_g}$

where R = universal gas constant, T = temperature

Relationship between K_p and K_x: $K_p = K_x (P)^{\Delta n_g}$

where P = equilibrium pressure in container.

Types of Chemical Equilibria

- Homogeneous equilibrium. When in an equilibrium reaction, all the reactants and products are present in the same phase (*i.e.* gaseous or liquid), it is called homogeneous equilibrium.

$PCl_5(g) \rightleftharpoons PCl_3(g) + Cl_2(g)$; $CH_3COOH(l) + C_2H_5OH(l) \rightleftharpoons CH_3COOC_2H_5(l) + H_2O(l)$ etc.

- Heterogeneous equilibrium. When in an equilibrium reaction, the reactants and the products are present in two or more than two phases, it is called a heterogeneous equilibrium.

$PbCl_2(s) \rightleftharpoons Pb^{2+}(aq) + 2Cl^-(aq)$; $CaO(s) + CO_2(g) \rightleftharpoons CaCO_3(s)$ etc.

Expression of K for Equilibrium Reaction

- **Homogeneous equilibria:** If reactants and products are present in same phase.

Example: (i) $NH_3(aq) + H_2O(l) \rightleftharpoons NH_4^+(aq) + OH^-(aq)$

(ii) $3O_2(g) \rightleftharpoons 2O_3(g)$

(iii) $C_3H_8(g) + 5O_2(g) \rightleftharpoons 3CO_2(g) + 4H_2O(g)$

Equilibrium constant expression for them are

(i) $K_c = \frac{[NH_4^+(aq)][OH^-(aq)]}{[NH_3(aq)]}$; [] represents concentration in mol/ litre at equilibrium.

(ii) $K_c = \frac{[O_3]^2}{[O_2]^3}$; $K_p = \frac{P_{O_3}^2}{P_{O_2}^3}$ (iii) $K_C = \frac{[CO_2]^3[H_2O]^4}{[C_3H_8][O_2]^5}$; $K_p = \frac{P_{CO_2}^3 \cdot P_{H_2O}^4}{P_{C_3H_8} \cdot P_{O_2}^5}$

Note: Equilibrium constant for gaseous homogeneous equilibrium can be expressed in two ways K_p and K_C. This means value of equilibrium constant depends upon choice of standard state in which concentration of reactants and products are expressed.

Heterogeneous equilibria

If reactants and products are present in two or more phases, the equilibria describing them is called heterogeneous equilibrium.

Example: (i) $AgI(s) \rightleftharpoons Ag^+(aq) + I^-(aq)$

(ii) $CaO(s) + CO_2(g) \rightleftharpoons CaCO_3(s)$

(iii) $Br_2(l) \rightleftharpoons Br_2(g)$

Equilibrium expression for them can be written as

(i) $K_C = [Ag^+(aq)][I^-(aq)]$ (ii) $K_P = \dfrac{1}{P_{CO_2}}$; $K_C = \dfrac{1}{[CO_2(g)]}$

(iii) $K_P = P_{Br_2}(g)$; $K_C = [Br_2(g)]$

Predicting the Extent of a Reaction

High value of equilibrium constant indicates that product(s) concentration is high and its low value indicates that concentration of the product(s) in equilibrium mixture is low.
Large values of K_P or K_C (larger than about 10^3), favour the products strongly. For intermediate values of K (approximately in the range of 10^{-3} to 10^3), the concentrations of reactants and products are comparable. Small values of equilibrium constant (smaller than 10^{-3}), favour the reactants strongly.

Predicting the Direction of the Reaction

For this purpose, we calculate the reaction quotient, Q. The reaction quotient is defined in the same way as the equilibrium constant (with molar concentrations to give Q_C, or with partial pressure to give Q_P) at any stage of reaction. For a general reaction

$$aA + bB \rightleftharpoons cC + dD$$

$$Q_C = \frac{[C]^c[D]^d}{[A]^a[B]^d}$$

Then, if $Q_C > K_C$, the reaction will proceed towards direction of reactants.
If $Q_C < K_c$, the reaction will move towards direction of the products.
If $Q_C = K_c$, the reaction mixture is at equilibrium

Important Point about Equilibrium Constant

- If the reaction is reversed, the value of equilibrium constant is inversed *e.g.*,

$$CH_3COOH(aq) + C_2H_5OH(aq) \rightleftharpoons CH_3COOC_2H_5(aq) + H_2O(l), \text{ Equilibrium constant} = K$$

then for $CH_3COOC_2H_5(aq) + H_2O(l) \rightleftharpoons CH_3COOH(aq) + C_2H_5OH(aq)$, Equilibrium constant $= 1/K$

- If the reaction is divided by 2, equilibrium constant is the square root of the original *e.g.*,

for $N_2(g) + 3H_2(g) \rightleftharpoons 2NH_3(g)$, Equilibrium constant $= K$

then for $\frac{1}{2}N_2(g) + \frac{3}{2}H_2(g) \rightleftharpoons NH_3(g)$, Equilibrium constant $= \sqrt{K}$

If the reaction is multiplied by 2, equilibrium constant is the square of the original.

- If the reaction is written in two steps, equilibrium constant is equal to the product of the equilibrium constants of the step reactions *e. g.*, if for

$$N_2(g) + 2O_2(g) \rightleftharpoons 2NO_2(g), \text{ Equilibrium constant} = K$$

and for the same reaction taking place in steps *i.e.*,

$$N_2(g) + O_2(g) \rightleftharpoons 2NO(g), \text{ Equilibrium constant} = K_1$$

and $$NO(g) + O_2(g) \rightleftharpoons 2NO_2(g), \text{ Equilibrium constant} = K_2$$

then $$K = K_1 \times K_2$$

- Effect of temperature: According to van't Hoff equation, $\log \dfrac{K_2}{K_1} = \dfrac{\Delta H}{2.303R}\left(\dfrac{T_2 - T_1}{T_1 T_2}\right)$

where K_1 and K_2 are the equilibrium constants at temperatures T_1 and T_2 respectively and ΔH is the molar enthalpy change in the temperature range T_1 to T_2.
For exothermic reaction, as temperature increases K decreases.
For endothermic reaction, as temperature increases K increases.

Relationship Between Degree of Dissociation (α) and Vapour Density

- For dissociation reaction: $A_n(g) \rightleftharpoons nA(g)$

$$\text{Degree of dissociation } (\alpha) = \frac{\text{No. of moles dissociated}}{\text{Total no. of moles taken}}$$

$$= \frac{D - d}{d(n-1)} = \frac{M_t - M_0}{M_0(n-1)}$$

where M_t = theoretical (calculated) molecular mass
M_0 = observed (experimental) molecular mass
D = theoretical vapour density, d = observed vapour density.
n = number of moles of product formed from 1 mole reactant

$$\text{Density of gas mixture} = \frac{PM}{RT}$$

- For polymerization reaction

$$nA(g) \rightleftharpoons A_n(g), \text{ where } n \geq 2$$

$$\alpha = \frac{D - d}{d\left(\frac{1}{n} - 1\right)} = \frac{M_t - M_0}{M_0\left(\frac{1}{n} - 1\right)}$$

Le Chatelier's Principle

"If a system in equilibrium is subjected to a change of concentration, temperature or pressure, the equilibrium shifts in a direction so as to undo the effect of the change imposed."

- Effect of change in concentration on equilibrium

As we add or remove reactant (or product) the ratio of new concentrations become 'Q' (reaction quotient)
$Q < K$: equilibrium will shift in forward direction.
$Q > K$: equilibrium will shift in backward direction.

- Effect of change in pressure

If a system in equilibrium consists of gases, then the concentrations of all the components can be altered by changing the pressure. When the pressure on the system is increased, the volume decreases proportionately. The equilibrium will shift in the direction in which there is decrease in number of moles *i.e.*, towards the direction in which there is decrease in volume.

- **Effect of change in pressure on melting point:** There are two types of solids:

(i) Solids whose volume decreases on melting, *e.g.*, ice, diamond, carborundum, magnesium nitride and quartz.

Solid (higher volume) $\rightleftharpoons$ Liquid (lower volume)

The process of melting is facilitated at high pressure, thus melting point is lowered.

(ii) Solid whose volume increase on melting, *e.g.*, Fe, Cu, Ag, Au, etc.

Solid (lower volume) $\rightleftharpoons$ Liquid (higher volume)

In this case the process of melting becomes difficult at high pressure, thus melting point becomes high.

- **Solubility of substances :** When solid substances are dissolved in water, either heat is evolved (exothermic) or heat is absorbed (endothermic).

  (i) For endothermic solubility process, solubility increase with increase in temperature.

  (ii) For exothermic solubility process, decrease with increase in temperature.

- **Solubility of gases in liquids :** When a gas dissolves in liquid, there is decrease in volume. Thus increase of pressure will favour the dissolution of gas in liquid.

- **Effect of temperature:** For endothermic reaction as temperature increases reaction shift in forward direction. For exothermic reaction as temperature increases reaction shift in backward direction.

- **Addition of inert gas**

  (i) For reactions in which $n_p = n_r$ there is no effect of adding an inert gas at constant volume or at constant pressure on the equilibrium.

  (ii) For reactions in which $n_p \neq n_r$, there is no effect of adding inert gas on the equilibrium at constant volume but at constant pressure, equilibrium shifts towards larger mole side.

Applying Le Chatelier's principle, the favourable conditions for the decomposition of NH_3 by Haber's process

i.e., $$N_2(g) + 3H_2(g) \rightleftharpoons 2NH_3(g), \quad \Delta H = -92.5 \text{ kJ}$$

are (i) High temperature (ii) Low pressure (iii) Removal of N_2 and H_2 (iv) Addition of inert gas at constant pressure.

Physical Equilibrium

Only physical changes are involved.

Example : (i) Solid liquid equilibria

$$H_2O(s) \rightleftharpoons H_2O(l)$$

At equilibrium: Net rate of conversion of ice into water = Net rate of conversion of water into ice.

For $H_2O(l)$ normal freezing point is 273 K.

(ii) Liquid vapour equilibria :

$$H_2O(l) \rightleftharpoons H_2O(g)$$

Equilibrium is characterized by constant value of vapour pressure of $H_2O(l)$ at certain temperature.

For $H_2O(l)$ normal boiling point is 373 K.

Level-1

1. A reversible reaction is one which :
(a) proceeds in one direction
(b) proceeds in both directions
(c) proceeds spontaneously
(d) all the statements are wrong

2. The equilibrium constant K_c for the reaction

$$P_4(g) \rightleftharpoons 2P_2(g)$$

is 1.4 at 400°C. Suppose that 3 moles of $P_4(g)$ and 2 moles of $P_2(g)$ are mixed in 2 litre container at 400°C. What is the value of reaction quotient (Q_c)?
(a) $\frac{3}{2}$
(b) $\frac{2}{3}$
(c) 1
(d) None of these

3. In a chemical reaction equilibrium is established when :
(a) opposing reaction ceases
(b) concentrations of reactants and product are equal
(c) velocity of opposing reaction is the same as that of forward reaction
(d) reaction ceases to generate heat

4. The equilibrium constant for a reaction is K, and the reaction quotient is Q. For a particular reaction mixture, the ratio $\frac{K}{Q}$ is 0.33. This means that :
(a) the reaction mixture will equilibrate to form more reactant species
(b) the reaction mixture will equilibrate to form more product species
(c) the equilibrium ratio of reactant to product concentrations will be 3
(d) the equilibrium ratio of reactant to product concentrations will be 0.33

5. Consider the reaction $2SO_2(g) + O_2(g) \rightleftharpoons 2SO_3(g)$ for which $K_c = 278\ M^{-1}$. 0.001 mole of each of the reagents $SO_2(g), O_2(g)$ and $SO_3(g)$ are mixed in a 1.0 L flask. Determine the reaction quotient of the system and the spontaneous direction of the system :
(a) $Q_c = 1000$; the equilibrium shifts to the right
(b) $Q_c = 1000$; the equilibrium shifts to the left
(c) $Q_c = 0.001$; the equilibrium shifts to the left
(d) $Q_c = 0.001$; the equilibrium shifts to the right

6. In Q. No. 5, if the mixture of gases was allowed to come to equilibrium. The volume of the reaction vessel was then rapidly increased by a factor of two. As a result of the change the reaction quotient (Q_c) would :
(a) increase because of the pressure decrease
(b) decrease because of the pressure decrease
(c) remain the same because the equilibrium constant is independent of volume
(d) increase because the reaction is endothermic

7. For the reaction $A(g) + 3B(g) \rightleftharpoons 2C(g)$ at 27°C, 2 moles of A, 4 moles of B and 6 moles of C are present in 2 litre vessel. If K_c for the reaction is 1.2, the reaction will proceed in :
(a) forward direction
(b) backward direction
(c) neither direction
(d) none of these

8. For a reversible gaseous reaction $N_2 + 3H_2 \rightleftharpoons 2NH_3$ at equilibrium, if some moles of H_2 are replaced by same number of moles of T_2 (T is tritium, isotope of H and assume isotopes do not have different chemical properties) without affecting other parameters, then :
(a) the sample of ammonia obtained after sometime will be radioactive.
(b) moles of N_2 after the change will be different as compared to moles of N_2 present before the change
(c) the value of K_p or K_c will change
(d) the average molecular mass of new equilibrium will be same as that of old equilibrium

9. For the synthesis of ammonia by the reaction $N_2 + 3H_2 \rightleftharpoons 2NH_3$ in the Haber's process, the attainment of equilibrium is correctly predicted by the curve

(a)
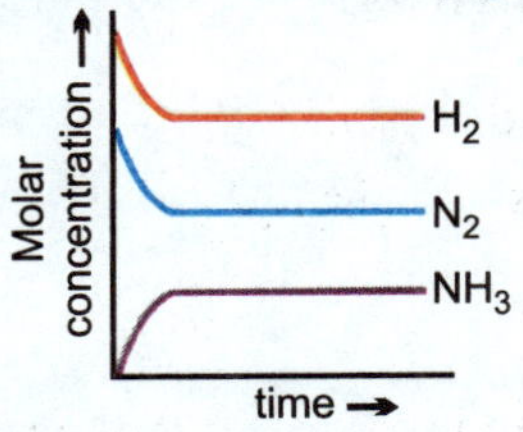

(b)
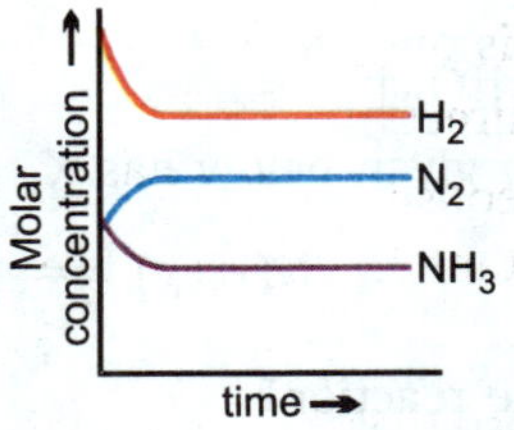

(c)
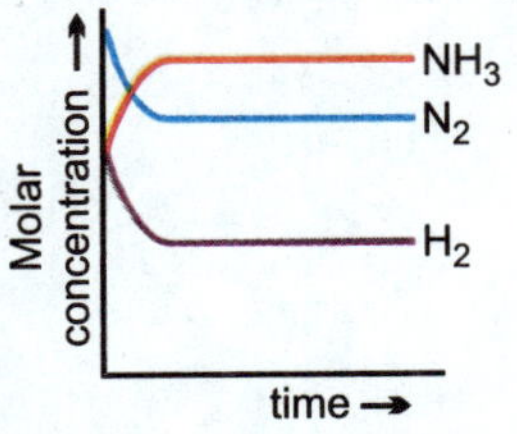

(d)
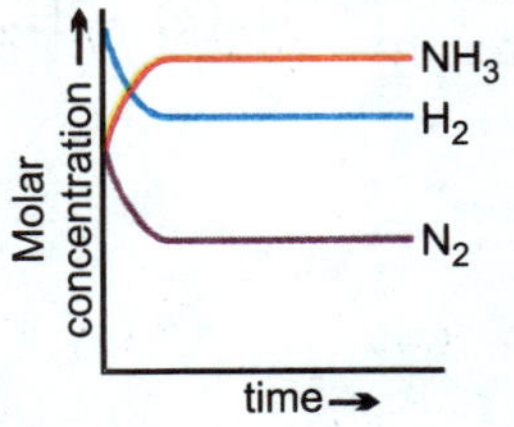

10. The figure shows the change in concentration of species *A* and *B* as a function of time.

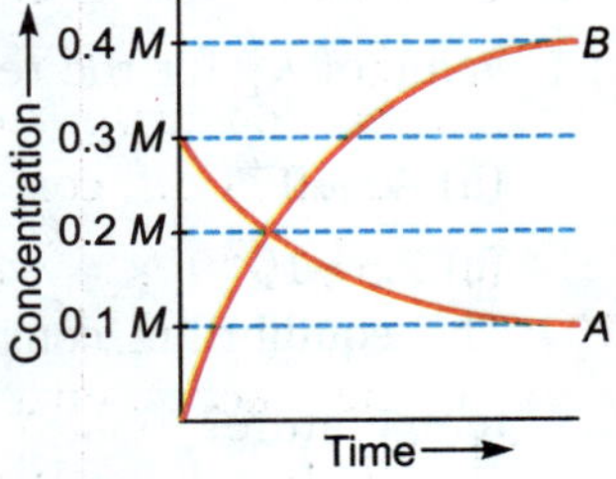

The equilibrium constant K_c for the reaction $A(g) \rightleftharpoons 2B(g)$ is:

(a) $K_c > 1$ (b) $K < 1$

(c) $K = 1$ (d) data insufficient

11. Attainment of the equilibrium $A(g) \rightleftharpoons 2C(g) + B(g)$ gave the following graph. Find the correct option. (% dissociation = fraction dissociated ×100)

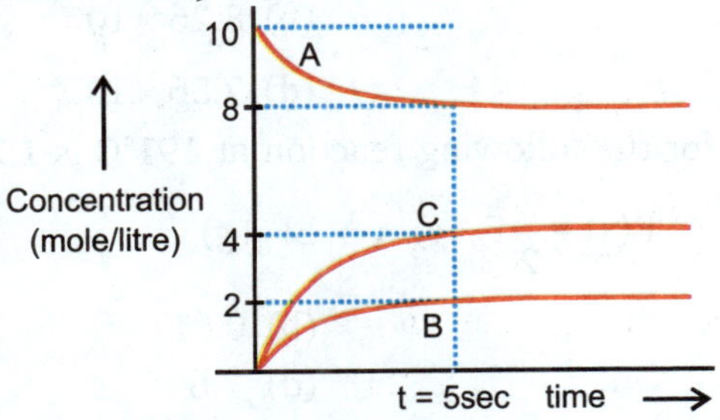

(a) At $t = 5$ sec equilibrium has been reached and $K_c = 40\ (\text{mol/litre})^2$

(b) At $t = 5$ sec equilibrium has been reached and % dissociation of *A* is 20%

(c) At $t = 5$ sec equilibrium has been reached and % dissociation of *A* is 30%

(d) None of these

12. Using molar concentrations, what is the unit of K_c for the reaction?

$$CH_3OH(g) \rightleftharpoons CO(g) + 2H_2(g)$$

(a) M^{-2} (b) M^2

(c) M^{-1} (d) M

13. What is the unit of K_p for the reaction?

$$CS_2(g) + 4H_2(g) \rightleftharpoons CH_4(g) + 2H_2S(g)$$

(a) atm (b) atm^{-2}

(c) atm^2 (d) atm^{-1}

14. What is the equilibrium expression for the reaction $P_4(s) + 5O_2(g) \rightleftharpoons P_4O_{10}(s)$

(a) $K_c = [O_2]^5$ (b) $K_c = [P_4O_{10}] / 5[P_4][O_2]$

(c) $K_c = [P_4O_{10}] / [P_4][O_2]^5$ (d) $K_c = 1/[O_2]^5$

15. At 527°C, the reaction given below has $K_c = 4$

$$NH_3(g) \rightleftharpoons \frac{1}{2}N_2(g) + \frac{3}{2}H_2(g)$$

What is the K_p for the reaction?

$$N_2(g) + 3H_2(g) \rightleftharpoons 2NH_3(g)$$

(a) $16 \times (800\,R)^2$ (b) $\left(\frac{800R}{4}\right)^{-2}$

(c) $\left(\frac{1}{4 \times 800R}\right)^2$ (d) None of these

16. The equilibrium constant for the reaction $N_2(g) + O_2(g) \rightleftharpoons 2NO(g)$ at temperature (T) is 4×10^{-4}. The value of K_c for the reaction $NO(g) \rightleftharpoons \frac{1}{2}N_2(g) + \frac{1}{2}O_2(g)$ at the same temperature is :

(a) 4×10^{-4} (b) 50

(c) 2.5×10^2 (d) 0.02

17. The equilibrium constant K_c for the following reaction at 842°C is 7.90×10^{-3}. What is K_p at same temperature?

$$\frac{1}{2}F_2(g) \rightleftharpoons F(g)$$

(a) 8.64×10^{-5} (b) 8.26×10^{-4}

(c) 7.90×10^{-2} (d) 7.56×10^{-2}

18. The equilibrium constant K_p for the following reaction at 191°C is 1.24. What is K_c?

$$B(s) + \frac{3}{2}F_2(g) \rightleftharpoons BF_3(g)$$

(a) 6.7 (b) 0.61

(c) 8.30 (d) 7.6

19. For the equilibrium $SO_2Cl_2(g) \rightleftharpoons SO_2(g) + Cl_2(g)$, what is the temperature at which $\frac{K_p\,(\text{atm})}{K_c\,(\text{M})} = 3$?

(a) 0.027 K (b) 0.36 K

(c) 36.54 K (d) 273 K

20. For the reversible reaction,

$N_2(g) + 3H_2(g) \rightleftharpoons 2NH_3(g)$

at 500°C, the value of K_p is 1.44×10^{-5} when partial pressure is measured in atmospheres. The corresponding value of K_c with concentration in mole litre^{-1}, is:

(a) $1.44 \times 10^{-5}/(0.082 \times 500)^{-2}$ (b) $1.44 \times 10^{-5}/(8.314 \times 773)^{-2}$

(c) $1.44 \times 10^{-5}/(0.082 \times 773)^2$ (d) $1.44 \times 10^{-5}/(0.082 \times 773)^{-2}$

21. For the reaction $CO(g) + Cl_2(g) \rightleftharpoons COCl_2(g)$ the value of $\frac{K_c}{K_p}$ is equal to :

(a) $\sqrt{RT}$ (b) RT

(c) $\frac{1}{RT}$ (d) 1.0

22. The concentration of a pure solid or liquid phase is not included in the expression of equilibrium constant because :

(a) density of solid and liquid are independent of their quantities.
(b) solids and liquids react slowly.
(c) solids and liquids at equilibrium do not interact with gaseous phase.
(d) the molecules of solids and liquids cannot migrate to the gaseous phase.

23. A catalyst is a substance which :

(a) increases the equilibrium concentration of the product.
(b) changes the equilibrium constant of the reaction.
(c) shortens the time to reach equilibrium.
(d) supplies energy to the reaction.

24. What will be the effect on the equilibrium constant on increasing temperature, if the reaction neither absorbs heat nor releases heat?

(a) Equilibrium constant will remain constant.
(b) Equilibrium constant will decrease.
(c) Equilibrium constant will increase.
(d) Can not be predicted.

25. The equilibrium constant for the reaction $N_2(g) + O_2(g) \rightleftharpoons 2NO(g)$ is 4×10^{-4} at 200 K. In presence of a catalyst, equilibrium is attained ten times faster. Therefore, the equilibrium constant in presence of the catalyst at 200 K is :

(a) 40×10^{-4} (b) 4×10^{-4}
(c) 4×10^{-3} (d) difficult to compute without more data

26. For the reaction $H_2(g) + I_2(g) \rightleftharpoons 2HI(g)$, the equilibrium constant changes with :

(a) total pressure (b) catalyst
(c) concentration of H_2 and I_2 (d) temperature

27. Consider the reactions

(i) $2CO(g) + 2H_2O(g) \rightleftharpoons 2CO_2(g) + 2H_2(g)$; Eqm. Constant $= K_1$
(ii) $CH_4(g) + H_2O(g) \rightleftharpoons CO(g) + 3H_2(g)$; Eqm. Constant $= K_2$
(iii) $CH_4(g) + 2H_2O(g) \rightleftharpoons CO_2(g) + 4H_2(g)$; Eqm. Constant $= K_3$

Which of the following relation is correct?

(a) $K_3 = \frac{K_1}{K_2}$ (b) $K_3 = \frac{K_1^2}{K_2^2}$
(c) $K_3 = K_1 K_2$ (d) $K_3 = \sqrt{K_1} \cdot K_2$

28. For the reaction $2NO_2(g) + \frac{1}{2}O_2(g) \rightleftharpoons N_2O_5(g)$, if the equilibrium constant is K_p, then the equilibrium constant for the reaction $2N_2O_5(g) \rightleftharpoons 4NO_2(g) + O_2(g)$ would be :

(a) K_p^2 (b) $\frac{2}{K_p}$
(c) $\frac{1}{K_p^2}$ (d) $\frac{1}{\sqrt{K_p}}$

29. The equilibrium constant (K_c) for the reaction

$$2HCl(g) \rightleftharpoons H_2(g) + Cl_2(g)$$

is 4×10^{-34} at 25°C. What is the equilibrium constant for the reaction?

$$\frac{1}{2}H_2(g) + \frac{1}{2}Cl_2(g) \rightleftharpoons HCl(g)$$

(a) 2×10^{-17} (b) 2.5×10^{33}
(c) 5×10^{6} (d) None of these

30. At a certain temperature, the following reactions have the equilibrium constants as shown below:

$$S(s) + O_2(g) \rightleftharpoons SO_2(g); \quad K_c = 5 \times 10^{52}$$

$$2S(s) + 3O_2(g) \rightleftharpoons 2SO_3(g); \quad K_c = 10^{29}$$

What is the equilibrium constant K_c for the reaction at the same temperature ?

$$2SO_2(g) + O_2(g) \rightleftharpoons 2SO_3(g)$$

(a) 2.5×10^{76} (b) 4×10^{23}
(c) 4×10^{-77} (d) None of these

31. Consider the following gaseous equilibria given below :

(I) $N_2 + 3H_2 \rightleftharpoons 2NH_3$; Eqm. Constant $= K_1$
(II) $N_2 + O_2 \rightleftharpoons 2NO$; Eqm. Constant $= K_2$
(III) $H_2 + \frac{1}{2}O_2 \rightleftharpoons H_2O$; Eqm. Constant $= K_3$

The equilibrium constant for the reaction, $2NH_3 + \frac{5}{2}O_2 \rightleftharpoons 2NO + 3H_2O$ in terms of K_1, K_2 and K_3 will be :

(a) $K_1K_2K_3$ (b) $\frac{K_1K_2}{K_3}$
(c) $\frac{K_1K_3^2}{K_2}$ (d) $\frac{K_2K_3^3}{K_1}$

32. In the reaction $X(g) + Y(g) \rightleftharpoons 2Z(g)$, 2 mole of X, 1 mole of Y and 1 mole of Z are placed in a 10 litre vessel and allowed to reach equilibrium. If final concentration of Z is 0.2 M, then K_c for the given reaction is :

(a) 1.60 (b) $\frac{80}{3}$
(c) $\frac{16}{3}$ (d) None of these

33. An equilibrium mixture of the reaction $2H_2S(g) \rightleftharpoons 2H_2(g) + S_2(g)$ had 0.5 mole H_2S, 0.10 mole H_2 and 0.4 mole S_2 in one litre vessel. The value of equilibrium constant (K) in mol litre^{-1} is :

(a) 0.004 (b) 0.008
(c) 0.016 (d) 0.160

34. Given $[CS_2] = 0.120\,M$, $[H_2] = 0.10\,M$, $[H_2S] = 0.20\,M$ and $[CH_4] = 8.40 \times 10^{-5}\,M$ for the following reaction at 900°C, at eq. Calculate the equilibrium constant (K_c).

$$CS_2(g) + 4H_2(g) \rightleftharpoons CH_4(g) + 2H_2S(g)$$

(a) 0.0120 (b) 0.0980
(c) 0.280 (d) 0.120

35. The equilibrium constant for the following reaction is 10.5 at 500 K. A system at equilibrium has $[CO] = 0.250\,M$ and $[H_2] = 0.120\,M$. What is the $[CH_3OH]$?

$$CO(g) + 2H_2(g) \rightleftharpoons CH_3OH(g)$$

(a) 0.03618 (b) 0.0435
(c) 0.546 (d) 0.0499

36. When sulphur (in the form of S_8) is heated at temperature T, at equilibrium, the pressure of S_8 falls by 30% from 1.0 atm, because $S_8(g)$ is partially converted into $S_2(g)$.
Find the value of K_p for this reaction.
(a) 2.96 (b) 6.14
(c) 204.8 (d) None of these

37. 9.2 grams of $N_2O_4(g)$ is taken in a closed one litre vessel and heated till the following equilibrium is reached $N_2O_4(g) \rightleftharpoons 2NO_2(g)$
At equilibrium, 50% $N_2O_4(g)$ is dissociated. What is the equilibrium constant (in mol litre $^{-1}$) (molecular mass of $N_2O_4 = 92$)
(a) 0.1 (b) 0.4
(c) 0.2 (d) 2

38. Two moles of NH_3 when put into a previously evacuated vessel (one litre), partially dissociated into N_2 and H_2. If at equilibrium one mole of NH_3 is present, the equilibrium constant is :
(a) 3/4 $mol^2 litre^{-2}$ (b) 27/64 $mol^2 litre^{-2}$
(c) 27/32 $mol^2 litre^{-2}$ (d) 27/16 $mol^2 litre^{-2}$

39. In the presence of excess of anhydrous $SrCl_2$, the amount of water taken up is governed by $K_p = 10^{12}$ atm^{-4} for the following reaction at 273 K

$$SrCl_2 \cdot 2H_2O(s) + 4H_2O(g) \rightleftharpoons SrCl_2 \cdot 6H_2O(s)$$

What is equilibrium vapour pressure (in torr) of water in a closed vessel that contains $SrCl_2 \cdot 2H_2O(s)$?
(a) 0.001 torr (b) 10^3 torr
(c) 0.76 torr (d) 1.31 torr

40. $CuSO_4 \cdot 5H_2O(s) \rightleftharpoons CuSO_4 \cdot 3H_2O(s) + 2H_2O(g)$; $K_p = 4 \times 10^{-4}$ atm^2. If the vapour pressure of water is 38 torr then percentage of relative humidity is : (Assume all data at constant temperature)
(a) 4 (b) 10
(c) 40 (d) None of these

41. $NH_4HS(s) \rightleftharpoons NH_3(g) + H_2S(g)$
The equilibrium pressure at 25°C is 0.660 atm. What is K_p for the reaction?
(a) 0.109 (b) 0.218
(c) 1.89 (d) 2.18

42. For the reaction $2A(g) \rightleftharpoons B(g) + 3C(g)$, at a given temperature, $K_c = 16$. What must be the volume of the flask, if a mixture of 2 mole each of A, B and C exist in equilibrium?
(a) $\frac{1}{4}$ (b) $\frac{1}{2}$
(c) 1 (d) None of these

43. One mole of pure ethyl alcohol was treated with one mole of pure acetic acid at 25°C. One-third of the acid changes into ester at equilibrium. The equilibrium constant for the reaction will be :
(a) $\frac{1}{4}$ (b) 2
(c) 3 (d) 4

44. $I_2(aq) + I^-(aq) \rightleftharpoons I_3^-(aq)$. We started with 1 mole of I_2 and 0.5 mole of I^- in one litre flask. After equilibrium is reached, excess of $AgNO_3$ gave 0.25 mole of yellow precipitate. Equilibrium constant is :
(a) 1.33 (b) 2.66
(c) 2.0 (d) 3.0

45. At 87°C, the following equilibrium is established.

$$H_2(g) + S(s) \rightleftharpoons H_2S(g); \; K_c = 0.08$$

If 0.3 mole hydrogen and 2 mole sulphur are heated to 87°C in a 2 L vessel, what will be the concentration of H_2S at equilibrium?

(a) 0.011 *M* (b) 0.022 *M*
(c) 0.044 *M* (d) 0.08 *M*

46. In the equilibrium $2SO_2(g) + O_2(g) \rightleftharpoons 2SO_3(g)$, the partial pressure of SO_2, O_2 and SO_3 are 0.662, 0.10 and 0.331 atm respectively. What should be the partial pressure of oxygen so that the equilibrium concentrations of SO_2 and SO_3 are equal?

(a) 0.4 atm (b) 1.0 atm
(c) 0.8 atm (d) 0.25 atm

47. When heated, ammonium carbamate decomposes as follows :

$$NH_4COONH_2(s) \rightleftharpoons 2NH_3(g) + CO_2(g)$$

At a certain temperature, the equilibrium pressure of the system is 0.318 atm. K_p for the reaction is :

(a) 0.128 (b) 0.426
(c) 4.76×10^{-3} (d) None of these

48. In a system $A(s) \rightleftharpoons 2B(g) + 3C(g)$, if the concentration of C at equilibrium is increased by a factor of 2, it will cause the equilibrium concentration of B to change to :

(a) two times the original value (b) one half of its original value
(c) $2\sqrt{2}$ times to the original value (d) $\frac{1}{2\sqrt{2}}$ times the original value

49. For $A + B \rightleftharpoons C + D$, if initially the concentration of A and B are both equal but at equilibrium concentration of D will be twice of that of A then what will be the equilibrium constant of reaction ?

(a) $\frac{4}{9}$ (b) $\frac{9}{4}$
(c) $\frac{1}{9}$ (d) 4

50. The equilibrium constant K_c for the reaction $SO_2(g) + NO_2(g) \rightleftharpoons SO_3(g) + NO(g)$ is 16. If 1 mole of each of all the four gases is taken in 1 dm^3 vessel, the equilibrium concentration of NO would be :

(a) 0.4 *M* (b) 0.6 *M*
(c) 1.4 *M* (d) 1.6 *M*

51. On increasing the temperature, the rate of a reaction :

(a) always increases
(b) always decreases
(c) first increases and then decreases
(d) may increase or decrease depending upon the nature of the reaction

52. A catalyst increases the rate of a reaction by :

(a) increasing the activation energy of the reaction
(b) increasing the value of rate constants (k_f and k_b)
(c) increasing the enthalpy change of the reaction
(d) decreasing the enthalpy change of the reaction

53. At a certain temperature, only 50% HI is dissociated at equilibrium in the following reaction :

$$2HI(g) \rightleftharpoons H_2(g) + I_2(g)$$

The equilibrium constant for this reaction is :

(a) 0.25 (b) 1.0
(c) 3.0 (d) 0.5

54. The equilibrium constant K_p for the reaction

$$H_2(g) + CO_2(g) \rightleftharpoons H_2O(g) + CO(g)$$

is 4.0 at 1660°C. Initially 0.80 mole H_2 and 0.80 mole CO_2 are injected into a 5.0 litre flask. What is the equilibrium concentration of $CO_2(g)$?

(a) 0.533 *M* (b) 0.0534 *M*
(c) 0.535 *M* (d) None of these

55. At 273 K and 1 atm, 10 litre of N_2O_4 decomposes to NO_2 according to equation

$$N_2O_4(g) \rightleftharpoons 2NO_2(g)$$

What is degree of dissociation (α) when the original volume is 25% less than that of existing volume?

(a) 0.25 (b) 0.33
(c) 0.66 (d) 0.5

56. The equilibrium constant for the reaction $CO(g) + H_2O(g) \rightleftharpoons CO_2(g) + H_2(g)$ is 5. How many moles of CO_2 must be added to 1 litre container already containing 3 moles each of CO and H_2O to make 2M equilibrium concentration of CO?

(a) 15 (b) 19
(c) 5 (d) 20

57. A nitrogen-hydrogen mixture initially in the molar ratio of 1 : 3 reached equilibrium to form ammonia when 25% of the N_2 and H_2 had reacted. If the total pressure of the system at equilibrium is 21 atm, the partial pressure of ammonia at the equilibrium was :

(a) 4.5 atm (b) 3.0 atm
(c) 2.0 atm (d) 1.5 atm

58. Ammonia under a pressure of 15 atm at 27°C is heated to 347°C in a closed vessel in the presence of a catalyst. Under the conditions, NH_3 is partially decomposed according to the equation, $2NH_3 \rightleftharpoons N_2 + 3H_2$ the vessel is such that the volume remains effectively constant where as pressure increases to 50 atm. Calculate the percentage of NH_3 actually decomposed :

(a) 65% (b) 61.3%
(c) 62.5% (d) 64%

59. 0.1 mole of $N_2O_4(g)$ was sealed in a tube under one atmospheric conditions at 25°C. Calculate the number of moles of $NO_2(g)$ present, if the equilibrium $N_2O_4(g) \rightleftharpoons 2NO_2(g)$ ($K_p = 0.14$) is reached after some time :

(a) 1.8×10^2 (b) 2.8×10^2
(c) 0.034 (d) 2.8×10^{-2}

60. 5 moles of SO_2 and 5 moles of O_2 are allowed to react. At equilibrium, it was found that 60% of SO_2 is used up. If the pressure of the equilibrium mixture is one atmosphere, the partial pressure of O_2 is:

(a) 0.52 atm (b) 0.21 atm
(c) 0.41 atm (d) 0.82 atm

61. $N_2(g) + 3H_2(g) \rightleftharpoons 2NH_3(g)$

For the reaction initially the mole ratio was 1 : 3 of $N_2 : H_2$. At equilibrium 50% of each has reacted. If the equilibrium pressure is *P*, the partial pressure of NH_3 at equilibrium is :

(a) $\frac{P}{3}$ (b) $\frac{P}{4}$
(c) $\frac{P}{6}$ (d) $\frac{P}{8}$

62. 2.0 mole of PCl_5 were introduced in a vessel of 5.0 L capacity of a particular temperature. At equilibrium, PCl_5 was found to be 35% dissociated into PCl_3 and Cl_2. The value of K_c for the reaction

$$PCl_3(g) + Cl_2(g) \rightleftharpoons PCl_5(g)$$

(a) 1.89 (b) 0.377
(c) 1.33 (d) 13.3

63. At certain temperature compound $AB_2(g)$ dissociates according to the reaction

$$2AB_2(g) \rightleftharpoons 2AB(g) + B_2(g)$$

With degree of dissociation α, which is small compared with unity, the expression of K_p, in terms of α and initial pressure P is :

(a) $P\frac{\alpha^3}{2}$ (b) $\frac{P\alpha^2}{3}$
(c) $P\frac{\alpha^3}{3}$ (d) $\frac{P\alpha^2}{2}$

64. For the reaction

$H_2(g) + CO_2(g) \rightleftharpoons CO(g) + H_2O(g)$, if the initial concentration of $[H_2] = [CO_2] = 1M$ and x moles/litre of hydrogen is consumed at equilibrium, the correct expression of K_p is :

(a) $\frac{x^2}{(1-x)^2}$ (b) $\frac{(1+x)^2}{(1-x)^2}$
(c) $\frac{x^2}{(2+x)^2}$ (d) $\frac{x^2}{1-x^2}$

65. If D_T and D_O are the theoretical and observed vapour densities at a definite temperature and α be the degree of dissociation of a substance, then, α in the terms of D_O, D_T and n (number of moles of products formed from 1 mole reactant) is calculated by the formula :

(a) $\alpha = \frac{D_O - D_T}{(1-n)\,D_T}$ (b) $\alpha = \frac{D_T - D_O}{(n-1)\,D_T}$
(c) $\alpha = \frac{D_T - D_O}{(n-1)\,D_O}$ (d) $\alpha = \frac{D - D_T}{(n-1)\,D_T}$

66. For the dissociation of PCl_5 into PCl_3 and Cl_2 in gaseous phase reaction, if d is the observed vapour density and D the theoretical vapour density with 'α' as degree of dissociation, variation of D/d with 'α' is given by?

(a) 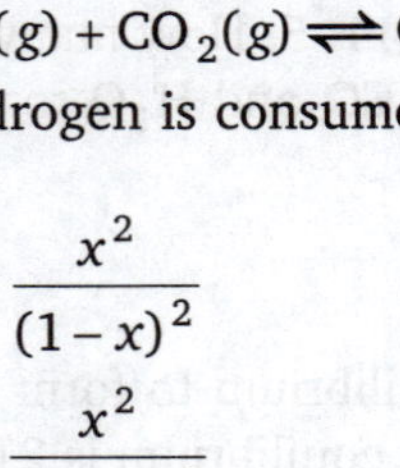

(b) 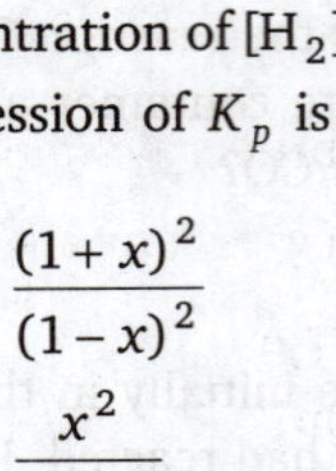

(c) 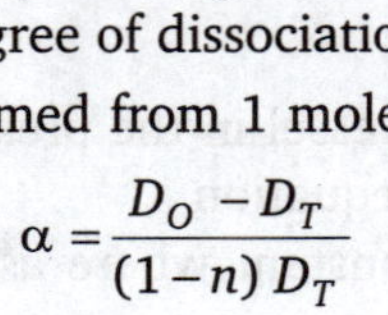

(d) None of these

67. At 27°C and 1 atm pressure, N_2O_4 is 20% dissociation into NO_2. What is the density of equilibrium mixture of N_2O_4 and NO_2 at 27°C and 1 atm?

(a) 3.11 g/litre (b) 2.11 g/litre
(c) 4.5 g/litre (d) None of these

68. $COCl_2$ gas dissociates according to the equation, $COCl_2(g) \rightleftharpoons CO(g) + Cl_2(g)$. When heated to 700 K the density of the gas mixture at 1.16 atm and at equilibrium is 1.16 g/litre. The degree of dissociation of $COCl_2$ at 700 K is :

(a) 0.28 (b) 0.50
(c) 0.72 (d) 0.42

69. The degree of dissociation of I_2 molecule of 1000°C and under atmospheric pressure is 40% by volume. If the dissociation is reduced to 20% at the same temp., total equilibrium pressure on the gas is :

(a) 1.57 atm (b) 2.57 atm
(c) 3.57 atm (d) 4.57 atm

70. Determine the value of equilibrium constant (K_c) for the reaction

$$A_2(g) + B_2(g) \rightleftharpoons 2AB(g)$$

if 10 moles of A_2, 15 moles of B_2 and 5 moles of AB are placed in a 2 litre vessel and allowed to come to equilibrium. The final concentration of AB is 7.5 M :

(a) 4.5 (b) 1.5
(c) 0.6 (d) None of these

71. At 87°C, the following equilibrium is established

$$H_2(g) + S(s) \rightleftharpoons H_2S(g); \quad K_p = 7 \times 10^{-2}$$

If 0.50 mole of hydrogen and 1.0 mole of sulphur are heated to 87° C in 1.0 L vessel, what will be the partial pressure of H_2S at equilibrium?

(a) 0.966 atm (b) 1.38 atm
(c) 0.0327 atm (d) 1 atm

72. Pure PCl_5 is introduced into an evacuated chamber and comes to equilibrium at 247°C and 2.0 atm. The equilibrium gaseous mixture contains 40% chlorine by volume.
Calculate K_p at 247°C for the reaction

$$PCl_5(g) \rightleftharpoons PCl_3(g) + Cl_2(g)$$

(a) 0.625 atm (b) 4 atm
(c) 1.6 atm (d) None of these

73. For the reaction

$$SnO_2(s) + 2H_2(g) \rightleftharpoons 2H_2O(g) + Sn(l)$$

calculate K_p at 900 K, where the equilibrium steam-hydrogen mixture was 45% H_2 by volume :

(a) 1.49 (b) 1.22
(c) 0.67 (d) None of these

74. For the reaction $XCO_3(s) \rightleftharpoons XO(s) + CO_2(g)$, $K_p = 1.642$ atm at 727°C. If 4 moles of $XCO_3(s)$ was put into a 50 litre container and heated to 727°C.
What mole percent of the XCO_3 remains unreacted at equilibrium?

(a) 20 (b) 25
(c) 50 (d) None of these

75. $Fe_2O_3(s)$ may be converted to Fe by the reaction

$$Fe_2O_3(s) + 3H_2(g) \rightleftharpoons 2Fe(s) + 3H_2O(g),$$

for which $K_c = 8$ at temp. 720°C.
What percentage of the H_2 remains unreacted after the reaction has come to equilibrium?

(a) $\simeq$22% (b) $\simeq$34%
(c) $\simeq$66% (d) $\simeq$78%

76. $AB_3(g)$ is dissociates as $AB_3(g) \rightleftharpoons AB_2(g) + \frac{1}{2}B_2(g)$.

When the initial pressure of AB_3 is 800 torr and the total pressure developed at equilibrium is 900 torr, what fraction of $AB_3(g)$ is dissociated?

(a) 10% (b) 20%
(c) 25% (d) 30%

77. At 1000 K, a sample of pure NO_2 gas decomposes as :

$$2NO_2(g) \rightleftharpoons 2NO(g) + O_2(g)$$

The equilibrium constant K_p is 156.25 atm. Analysis shows that the partial pressure of O_2 is 0.25 atm at equilibrium. The partial pressure of NO_2 at equilibrium is :

(a) 0.01 (b) 0.02
(c) 0.04 (d) None of these

78. Pure nitrosyl chloride (NOCl) gas was heated to 240°C in a 1.0 L container. At equilibrium the total pressure was 1.0 atm and the NOCl pressure was 0.64 atm. What would be the value of K_p?

(a) 1.02 atm (b) 16.875×10^{-3} atm
(c) 16×10^{-2} atm (d) None of these

79. At a certain temperature the equilibrium constant K_c is 0.25 for the reaction

$$A_2(g) + B_2(g) \rightleftharpoons C_2(g) + D_2(g)$$

If we take 1 mole of each of the four gases in a 10 litre container, what would be equilibrium concentration of $A_2(g)$?

(a) 0.331 *M* (b) 0.033 *M*
(c) 0.133 *M* (d) 1.33 *M*

80. At 200°C PCl_5 dissociates as follows :

$$PCl_5(g) \rightleftharpoons PCl_3(g) + Cl_2(g)$$

It was found that the equilibrium vapours are 62 times as heavy as hydrogen. The % degree of dissociation of PCl_5 at 200°C is nearly :

(a) 10% (b) 42%
(c) 50% (d) 68%

81. For the dissociation reaction $N_2O_4(g) \rightleftharpoons 2NO_2(g)$, the degree of dissociation (α) in terms of K_p and total equilibrium pressure P is :

(a) $\alpha = \sqrt{\dfrac{4P + K_P}{K_P}}$ (b) $\alpha = \sqrt{\dfrac{K_P}{4P + K_P}}$

(c) $\alpha = \sqrt{\dfrac{K_P}{4P}}$ (d) None of these

82. Consider the following equilibrium

$$N_2O_4(g) \rightleftharpoons 2NO_2(g)$$

Then the select the correct graph, which shows the variation in concentrations of NO_2 against concentrations of N_2O_4:

(a) 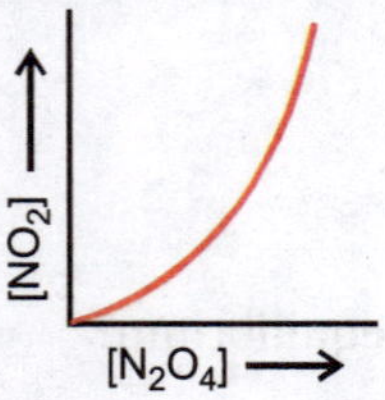

(b) 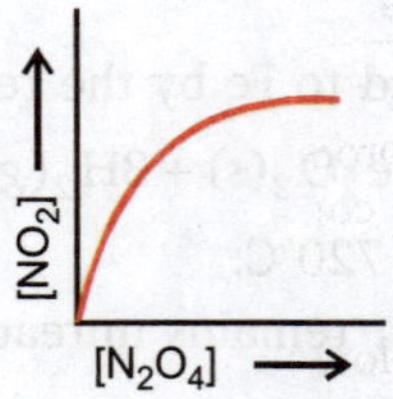

(c) 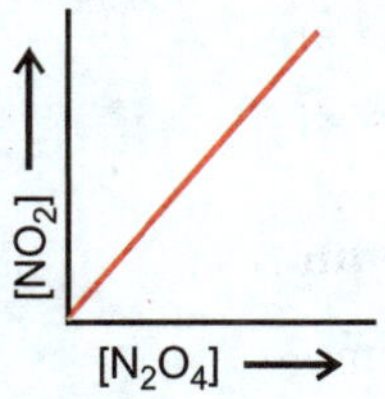
(d) 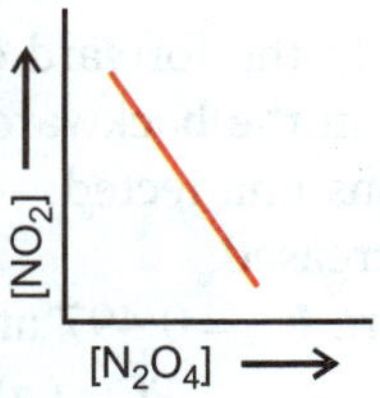

83. The vapour pressure of mercury is 0.002 mm Hg at 27°C. K_c for the process $Hg(l) \rightleftharpoons Hg(g)$ is:

(a) 0.002 (b) 8.12×10^{-5}

(c) 6.48×10^{-5} (d) 1.068×10^{-7}

84. Calculate the equilibrium constant (K_c) for the reaction given below, if at equilibrium mixture contains 5.0 mole of A_2, 3 mole of B_2 and 2 mole of AB_2 at 8.21 atm and 300 K

$$A_2(g) + 2B_2(g) \rightleftharpoons 2AB_2(g) + \text{Heat}$$

(a) 1.333 (b) 2.66

(c) 20 (d) None of these

85. For the reaction (1) and (2)

$$A(g) \rightleftharpoons B(g) + C(g) \quad \ldots(1)$$

$$X(g) \rightleftharpoons 2Y(g) \quad \ldots(2)$$

Given, $K_{P_1} : K_{P_2} = 9:1$

If the degree of dissociation of $A(g)$ and $X(g)$ be same then the total pressure at equilibrium (1) and (2) are in the ratio :

(a) 3 : 1 (b) 36 : 1

(c) 1 : 1 (d) 0.5 : 1

86. Given the following reaction at equilibrium, $N_2(g) + 3H_2(g) \rightleftharpoons 2NH_3(g)$. At constant pressure if some inert gas added in to the system, then:

(a) More $NH_3(g)$ is produced (b) Less $NH_3(g)$ is produced

(c) No affect on the equilibrium (d) K_p of the reaction is decreased

87. In which of the following equilibrium, change in volume of the system does not alter the number of moles :

(a) $N_2(g) + O_2(g) \rightleftharpoons 2NO(g)$ (b) $PCl_5(g) \rightleftharpoons PCl_3(g) + Cl_2(g)$

(c) $N_2(g) + 3H_2(g) \rightleftharpoons 2NH_3(g)$ (d) $SO_2Cl_2(g) \rightleftharpoons SO_2(g) + Cl_2(g)$

88. For the reaction

$$N_2(g) + 3H_2(g) \rightleftharpoons 2NH_3(g); \quad \Delta H = -93.6 \text{ kJ mol}^{-1}$$

The number of moles of H_2 at equilibrium will increase if :

(a) volume is increased (b) volume is decreased

(c) argon gas is added at constant volume (d) NH_3 is removed

89. The volume of the reaction vessel containing an equilibrium mixture is increased in the following reaction

$$SO_2Cl_2(g) \rightleftharpoons SO_2(g) + Cl_2(g)$$

When equilibrium is re-established :

(a) the amount of $Cl_2(g)$ remains unchanged

(b) the amount of $Cl_2(g)$ increases

(c) the amount of $SO_2Cl_2(g)$ increases

(d) the amount of $SO_2(g)$ decreases

90. Some inert gas is added at constant volume to the following reaction at equilibrium

$$NH_4HS(s) \rightleftharpoons NH_3(g) + H_2S(g)$$

Predict the effect of adding the inert gas :

(a) The equilibrium shifts in the forward direction
(b) The equilibrium shifts in the backward direction
(c) The equilibrium remains unaffected
(d) The value of K_p is increased

91. Consider the reaction where $K_p = 0.497$ at 500 K

$$PCl_5(g) \rightleftharpoons PCl_3(g) + Cl_2(g)$$

If the three gases are mixed in a rigid container so that the partial pressure of each gas is initially 1 atm, then which is correct observation ?
(a) More PCl_5 will be produced
(b) More PCl_3 will be produced
(c) Equilibrium will be established when 50% reaction is complete
(d) None of the above

92. The preparation of $SO_3(g)$ by reaction $SO_2(g) + \frac{1}{2}O_2(g) \rightleftharpoons SO_3(g)$ is an exothermic reaction. If the preparation follows the following temperature-pressure relationship for its % yield, then for temperatures T_1, T_2 and T_3. The correct option is :

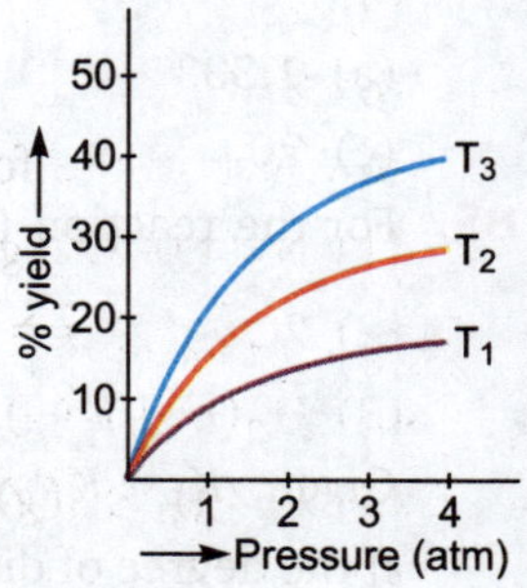

(a) $T_3 > T_2 > T_1$
(b) $T_1 > T_2 > T_3$
(c) $T_1 = T_2 = T_3$
(d) Nothing could be predicted about temperature through given information

93. An equilibrium mixture at 700 K of $0.05\,M\,N_2(g)$, $0.3\,M\,H_2(g)$ and $0.2\,M\,NH_3(g)$ is present in a container. Now if this equilibrium is disturbed by adding $N_2(g)$ and equilibrium concentration of NH_3 is 0.24 M then which of the following graphs represents the above situation more appropriately:

(a)
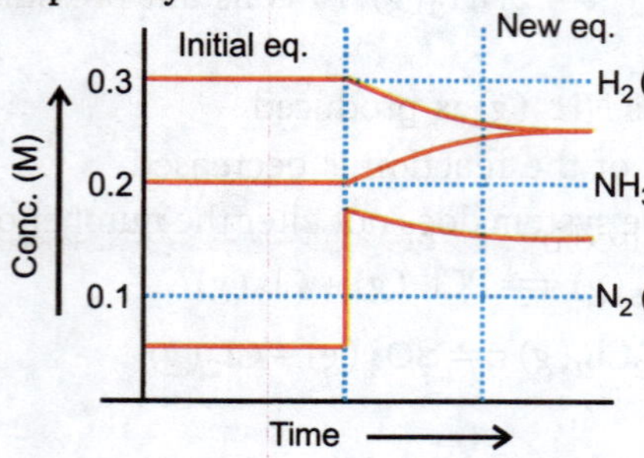

(b)
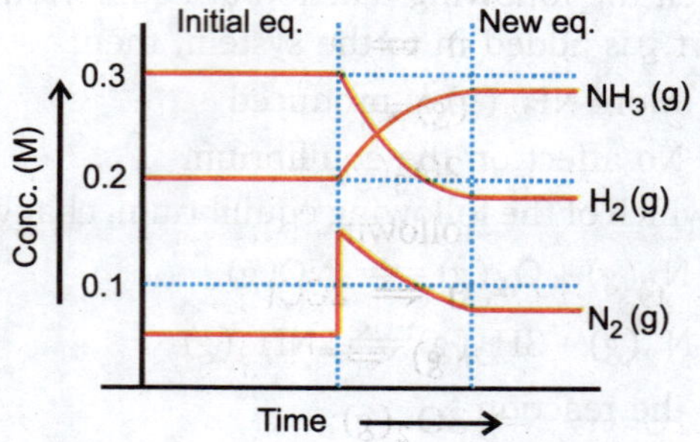

(c)
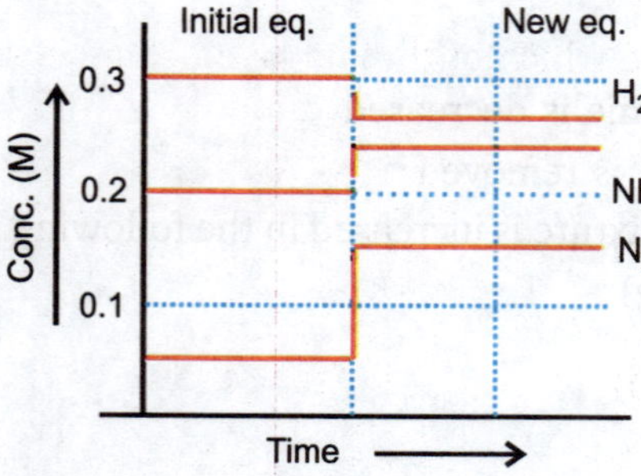

(d)
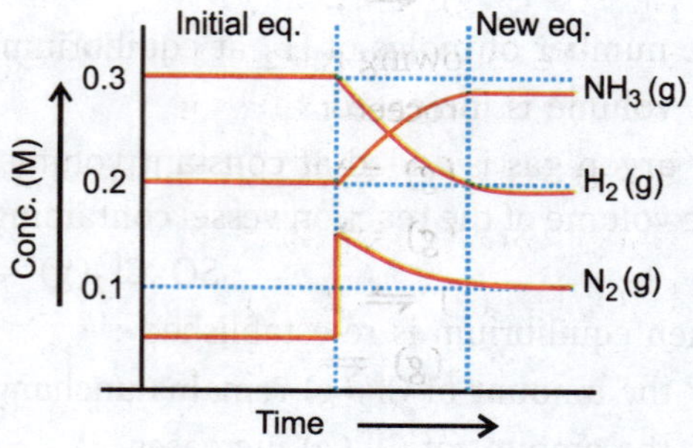

94. In a vessel containing N_2, H_2 and NH_3 at equilibrium, some helium gas is introduced so that total pressure increase while temperature and volume remain constant. According to Le Chatelier's principle, the dissociation of NH_3 :
(a) increases
(b) decreases
(c) remains unaltered
(d) changes unpredictably

95. Le-Chatelier principle is not applicable to :

(a) $H_2(g) + I_2(g) \rightleftharpoons 2HI(g)$
(b) $Fe(s) + S(s) \rightleftharpoons FeS(s)$
(c) $N_2(g) + 3H_2(g) \rightleftharpoons 2NH_3(g)$
(d) $N_2(g) + O_2(g) \rightleftharpoons 2NO(g)$

96. Consider the following reactions. In which case the formation of product is favoured by decrease in pressure?

(1) $CO_2(g) + C(s) \rightleftharpoons 2CO(g)$; $\Delta H^\circ = +172.5$ kJ
(2) $N_2(g) + 3H_2(g) \rightleftharpoons 2NH_3(g)$; $\Delta H^\circ = -91.8$ kJ
(3) $N_2(g) + O_2(g) \rightleftharpoons 2NO(g)$; $\Delta H^\circ = 181$ kJ
(4) $2H_2O(g) \rightleftharpoons 2H_2(g) + O_2(g)$; $\Delta H^\circ = 484.6$ kJ

(a) 2, 3
(b) 3, 4
(c) 2, 4
(d) 1, 4

97. In which of the following reactions, the formation of product is favoured by decrease in temperature?

(1) $N_2(g) + O_2(g) \rightleftharpoons 2NO(g)$; $\Delta H^\circ = 181$
(2) $2CO_2(g) \rightleftharpoons 2CO(g) + O_2(g)$; $\Delta H^\circ = 566$
(3) $H_2(g) + I_2(g) \rightleftharpoons 2HI(g)$; $\Delta H^\circ = -9.4$
(4) $H_2(g) + F_2(g) \rightleftharpoons 2HF(g)$; $\Delta H^\circ = -541$

(a) 1, 2
(b) 2 only
(c) 1, 2, 3
(d) 3, 4

98. For which of the following reactions is product formation favoured by low pressure and high temperature?

(a) $H_2(g) + I_2(g) \rightleftharpoons 2HI(g)$; $\Delta H^\circ = -9.4$ kJ
(b) $CO_2(g) + C(s) \rightleftharpoons 2CO(g)$; $\Delta H^\circ = 172.5$ kJ
(c) $CO(g) + 2H_2(g) \rightleftharpoons CH_3OH(l)$; $\Delta H^\circ = -21.7$ kJ
(d) $3O_2(g) \rightleftharpoons 2O_3(g)$; $\Delta H^\circ = 285$ kJ

99. For which of the following reaction is product formation favoured by low pressure and low temperature?

(a) $CO_2(g) + C(s) \rightleftharpoons 2CO(g)$; $\Delta H^\circ = 172.5$ kJ
(b) $CO(g) + 2H_2(g) \rightleftharpoons CH_3OH$; $\Delta H^\circ = -21.7$ kJ
(c) $2O_3(g) \rightleftharpoons 3O_2(g)$; $\Delta H^\circ = -285$ kJ
(d) $H_2(g) + F_2(g) \rightleftharpoons 2HF(g)$; $\Delta H^\circ = -541$ kJ

100. Consider the following reactions at equilibrium and determine which of the indicated changes will cause the reaction to proceed to the right.

(1) $CO(g) + 3H_2(g) \rightleftharpoons CH_4(g) + H_2O(g)$ (add CH_4)
(2) $N_2(g) + 3H_2(g) \rightleftharpoons 2NH_3(g)$ (remove NH_3)
(3) $H_2(g) + F_2(g) \rightleftharpoons 2HF(g)$ (add F_2)
(4) $BaO(s) + SO_3(g) \rightleftharpoons BaSO_4(s)$ (add BaO)

(a) 2, 3
(b) 1, 4
(c) 2, 4
(d) 2, 3, 4

101. If the pressure in a reaction vessel for the following reaction is increased by decreasing the volume, what will happen to the concentrations of CO and CO_2?

$$H_2O(g) + CO(g) \rightleftharpoons H_2(g) + CO_2(g) + \text{Heat}$$

(a) both the [CO] and [CO_2] will decrease
(b) neither the [CO] nor the [CO_2] will change
(c) the [CO] will decrease and the [CO_2] will increase
(d) both the [CO] and [CO_2] will increase

102. Consider the following reaction and determine which of the conditions will shift the equilibrium position to the right?

$$4NH_3(g) + 5O_2(g) \rightleftharpoons 4NO(g) + 6H_2O(g) + \text{Heat}$$

(a) increasing the temperature
(b) increasing the pressure
(c) adding a catalyst
(d) none of the above is correct

103. The conversion of ozone into oxygen is exothermic. Under what conditions is ozone the most stable?

$$2O_3(g) \rightleftharpoons 3O_2(g)$$

(a) At low pressure and low temperature
(b) At high pressure and high temperature
(c) At high pressure and low temperature
(d) At low pressure and high temperature

104. A system at equilibrium is described by the equation of fixed temperature *T*.

$$SO_2Cl_2(g) \rightleftharpoons SO_2(g) + Cl_2(g)$$

What effect will be the effect on equilibrium, if total pressure is increased by reducing volume?

(a) Concentration of $SO_2Cl_2(g)$ increases
(b) Concentration of $SO_2(g)$ increases
(c) Concentration of $Cl_2(g)$ increases
(d) Concentration of all gases increases

105. The reaction $2NO_2(g) \rightleftharpoons N_2O_4(g)$ is an exothermic equilibrium. This means that :

(a) equilibration of this gas mixture will be slower at high temperature
(b) a mole of N_2O_4 will occupy twice the volume of a mole of NO_2 at the same temperature
(c) the equilibrium will move to the right if an equilibrium mixture is cooled
(d) the position of equilibrium will move to the left with increasing gas pressure

106. Densities of diamond and graphite are 3.5 and 2.3 g/mL

$$\text{C (diamond)} \rightleftharpoons \text{C (graphite)}; \quad \Delta_r H = -1.9 \text{ kJ/mol}$$

favourable conditions for the formation of graphite are :

(a) high pressure and low temperature
(b) low pressure and high temperature
(c) high pressure and high temperature
(d) low pressure and low temperature

107. For an equilibrium $H_2O(s) \rightleftharpoons H_2O(l)$, which of the following statements is true?

(a) The pressure changes do not affect the equilibrium
(b) More of ice melts if pressure on the system is increased
(c) More of liquid freezes if pressure on the system is increased
(d) The pressure changes may increase or decrease the degree of advancement of the process

108. A pressure cooker reduces cooking time for food because:

(a) the higher pressure inside the cooker crushes the food material
(b) cooking involves chemical changes helped by a rise in temperature
(c) heat is more evenly distributed in the cooking space
(d) boiling point of water involved in cooking is increased

109. The vapour pressure of a liquid in a closed container depends on :

(1) temperature of liquid
(2) quantity of liquid
(3) surface area of the liquid

(a) 1 only
(b) 2 only
(c) 1 and 3 only
(d) 1, 2 and 3

110. The pressure on a sample of water at its triple point is reduced while the temperature is held constant. Which phases changes are favoured?

(I) melting of ice

(II) sublimation of ice

(III) vaporization of liquid water

(a) I only (b) III only

(c) II only (d) II and III

111. An exothermic reaction is represented by the graph :

(a)
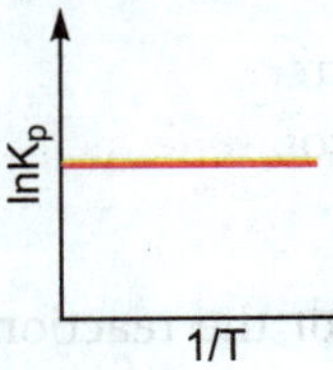

(b)
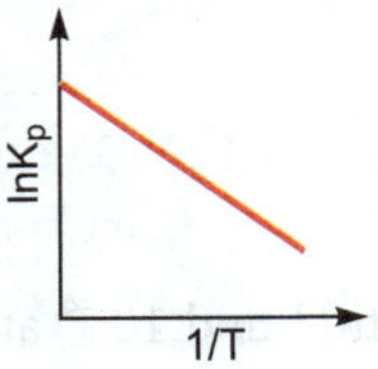

(c)
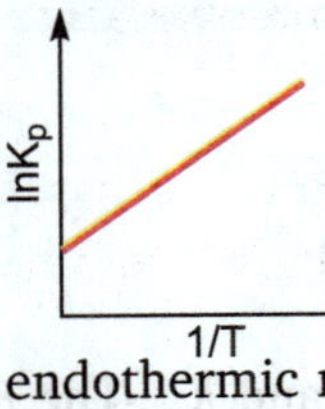

(d)
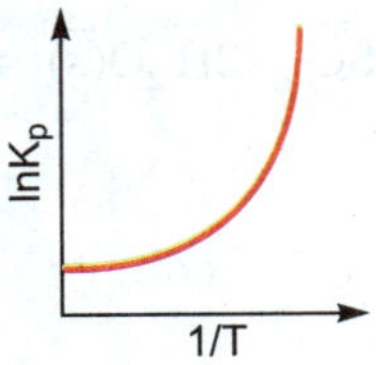

112. An endothermic reaction is represented by the graph :

(a)
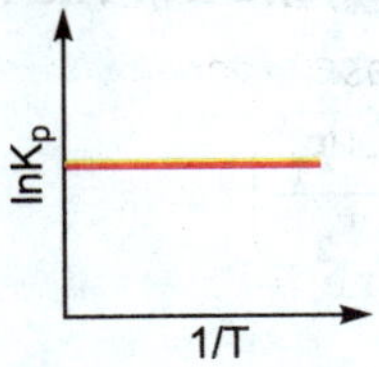

(b)
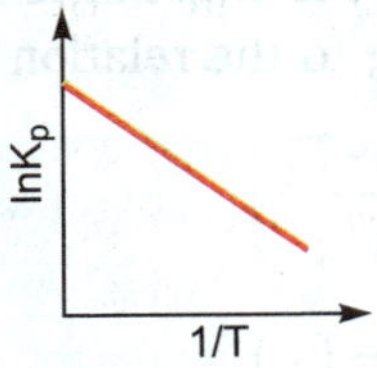

(c)
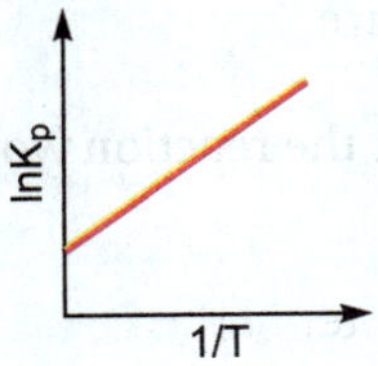

(d)
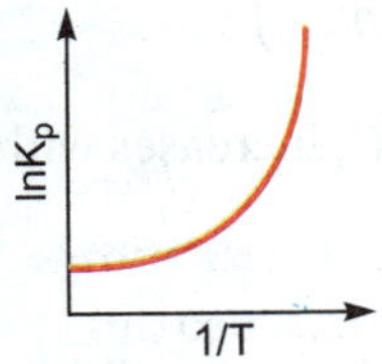

113. A schematic plot of ln K_{eq} *versus* inverse of temperature for a reaction is shown below

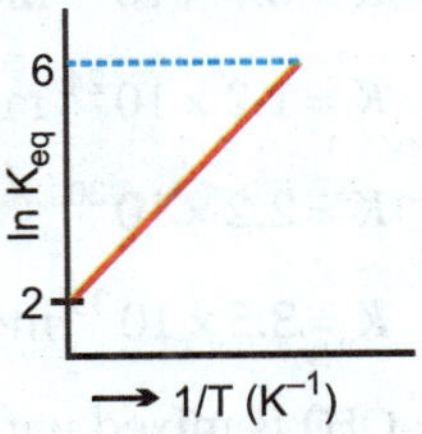

the reaction must be :

(a) Exothermic

(b) Endothermic

(c) One with negligible enthalpy change

(d) Highly spontaneous at ordinary temperature

114. What is the correct relationship between free energy change and equilibrium constant of a reaction?

(a) $\Delta G° = RT \ln K$ (b) $\Delta G° = -RT \ln K$

(c) $\Delta G = RT \ln K$ (d) $\Delta G = -RT \ln K$

115. For the chemical equilibrium $CaCO_3(s) \rightleftharpoons CaO(s) + CO_2(g), \Delta_r H°$ can be determined from which one of the following plots ?

(a) 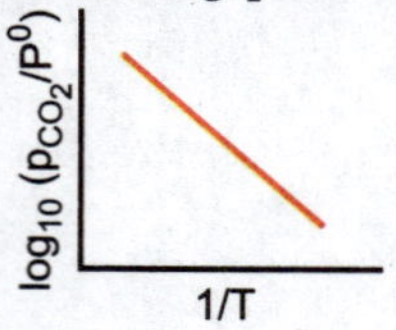

(b) 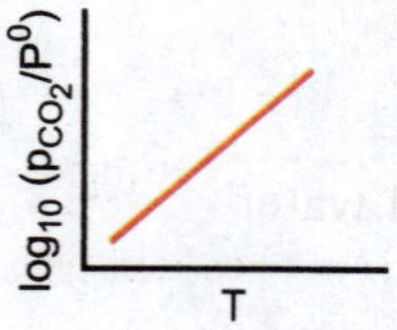

(c) 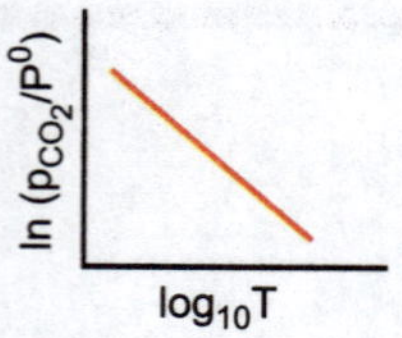

(d) 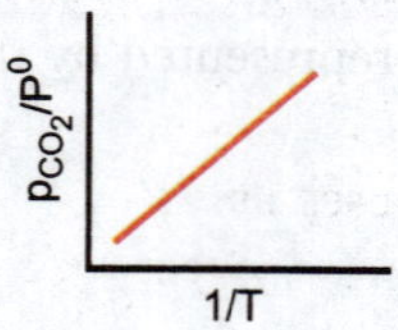

116. K_p has the value of 10^{-6} atm^3 and 10^{-4} atm^3 at 298 K and 323 K respectively for the reaction

$$CuSO_4 \cdot 3H_2O(s) \rightleftharpoons CuSO_4(s) + 3H_2O(g)$$

$\Delta_r H°$ for the reaction is :

(a) 7.7 kJ/mol
(b) −147.41 kJ/mol
(c) 147.41 kJ/mol
(d) None of these

117. van't Hoff's equation shows the effect of temperature on equilibrium constants K_c and K_p. The K_p varies with temperature according to the relation :

(a) $\log \dfrac{K_{p_2}}{K_{p_1}} = \dfrac{\Delta H°}{2.303\,R}\left(\dfrac{T_1 - T_2}{T_1 T_2}\right)$

(b) $\log \dfrac{K_{p_2}}{K_{p_1}} = \dfrac{\Delta H°}{2.303\,R}\left(\dfrac{T_2 - T_1}{T_1 T_2}\right)$

(c) $\log \dfrac{K_{p_2}}{K_{p_1}} = \dfrac{\Delta E°}{2.303\,R}\left(\dfrac{T_1 - T_2}{T_1 T_2}\right)$

(d) None of these

118. For a reaction, the value of K_p increases with increase in temperature. The ΔH for the reaction would be :

(a) positive
(b) negative
(c) zero
(d) cannot be predicted

119. The most stable oxides of nitrogen will be :

(a) $2NO_2(g) \rightleftharpoons N_2(g) + 2O_2(g)$; $K = 6.7 \times 10^{16}$ mol L^{-1}

(b) $2N_2O_5(g) \rightleftharpoons 2N_2(g) + 5O_2(g)$; $K = 1.2 \times 10^{24}$ mol^5 L^{-5}

(c) $2NO(g) \rightleftharpoons N_2(g) + O_2(g)$; $K = 2.2 \times 10^{30}$

(d) $2N_2O(g) \rightleftharpoons 2N_2(g) + O_2(g)$; $K = 3.5 \times 10^{33}$ mol L^{-1}

120. When 1 mole of pure ethyl alcohol (C_2H_5OH) is mixed with 1 mole of acetic acid at 25°C, the equilibrium mixture contains 2/3 mole each of ester and water

$$C_2H_5OH(l) + CH_3COOH(l) \rightleftharpoons CH_3COOC_2H_5(l) + H_2O(l)$$

The $\Delta G°$ for the reaction at 298 K is :

(a) 3435 J
(b) 4 J
(c) −3435 J
(d) Zero

121. The value of $\Delta G°$ for a reaction in aqueous phase having $K_c = 1$, would be :

(a) $-RT$
(b) −1
(c) 0
(d) $+RT$

122. A plot of Gibbs energy of a reaction mixture against the extent of the reaction is :

(a) minimum at equilibrium
(b) zero at equilibrium
(c) maximum at equilibrium
(d) None of these

123. For the reaction at 300 K

$$A(g) \rightleftharpoons V(g) + S(g)$$

$$\Delta_r H^\circ = -30 \text{ kJ/mol}, \ \Delta_r S^\circ = -0.1 \text{ kJ.K}^{-1}.\text{mol}^{-1}$$

What is the value of equilibrium constant?

(a) 0
(b) 1
(c) 10
(d) None of these

124. Solid $Ca(HCO_3)_2$ decomposes as

$Ca(HCO_3)_2(s) \rightleftharpoons CaCO_3(s) + CO_2(g) + H_2O(g)$

If the total pressure is 0.2 bar at 420 K, what is the standard free energy change for the given reaction $(\Delta_r G^\circ)$?

(a) 840 kJ/mol
(b) 3.86 kJ/mol
(c) 6.98 kJ/mol
(d) 16.083 kJ/mol

125. The standard free energy change of a reaction is $\Delta G^\circ = -115$ kJ mol^{-1} at 298 K. Calculate the value of $\log_{10} K_p$ ($R = 8.314$ JK^{-1} mol^{-1})

(a) 20.16
(b) 2.303
(c) 2.016
(d) 13.83

Level-2

1. The following equilibrium constants were determined at 1120 K :

$$2CO(g) \rightleftharpoons C(s) + CO_2(g); \quad K_{P_1} = 10^{-14} \text{ atm}^{-1}$$

$$CO(g) + Cl_2(g) \rightleftharpoons COCl_2(g); \quad K_{P_2} = 6 \times 10^{-3} \text{ atm}^{-1}$$

What is the equilibrium constant K_c for the following reaction at 1120 K :

$$C(s) + CO_2(g) + 2Cl_2(g) \rightleftharpoons 2COCl_2(g)$$

(a) $3.31 \times 10^{11}\ M^{-1}$ (b) $5.5 \times 10^{10}\ M^{-1}$
(c) $5.51 \times 10^{6}\ M^{-1}$ (d) None of these

2. One mole of $N_2(g)$ is mixed with 2 moles of $H_2(g)$ in a 4 litre vessel. If 50% of $N_2(g)$ is converted to $NH_3(g)$ by the following reaction :

$$N_2(g) + 3H_2(g) \rightleftharpoons 2NH_3(g)$$

What will be the value of K_c for the following equilibrium?

$$NH_3(g) \rightleftharpoons \frac{1}{2}N_2(g) + \frac{3}{2}H_2(g)$$

(a) 256 (b) 16
(c) $\frac{1}{16}$ (d) None of these

3.

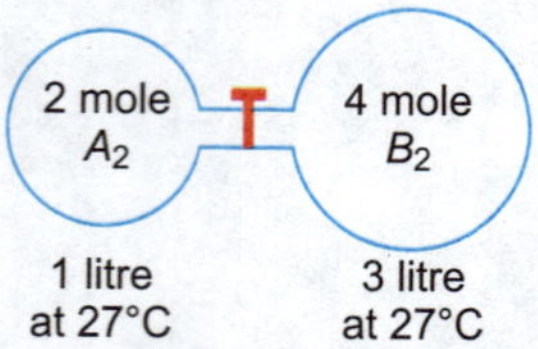

The gas A_2 in the left flask allowed to react with gas B_2 present in right flask as $A_2(g) + B_2(g) \rightleftharpoons 2AB(g)$; $K_c = 4$ at 27°C. What is the concentration of AB when equilibrium is established?

(a) 1.33 M (b) 2.66 M
(c) 0.66 M (d) 0.33 M

4. Assume that the decomposition of HNO_3 can be represented by the following equation

$$4HNO_3(g) \rightleftharpoons 4NO_2(g) + 2H_2O(g) + O_2(g)$$

and the reaction approaches equilibrium at 400 K temperature and 30 atm pressure. At equilibrium partial pressure of HNO_3 is 2 atm.

Calculate K_c in $(mol/L)^3$ at 400 K :

(Use : $R = 0.08$ atm-L/mol-K)

(a) 4 (b) 8
(c) 16 (d) 32

5. For the equilibrium

$$LiCl \cdot 3NH_3(s) \rightleftharpoons LiCl \cdot NH_3(s) + 2NH_3(g);$$

$K_p = 9 \text{ atm}^2$ at 37°C. A 5 litre vessel contains 0.1 mole of $LiCl \cdot NH_3$. How many moles of NH_3 should be added to the flask at this temperature to derive the backward reaction for completion?

Use : $R = 0.082$ atm-L/mol K

(a) 0.2 (b) 0.59
(c) 0.69 (d) 0.79

6. Ammonium carbamate dissociates as

$$NH_2COONH_4(s) \rightleftharpoons 2NH_3(g) + CO_2(g)$$

In a closed vessel containing ammonium carbamate in equilibrium, ammonia is added such that partial pressure of NH_3 now equals to the original total pressure. Calculate the ratio of partial pressure of CO_2 now to the original partial pressure of CO_2 :

(a) 4 (b) 9

(c) $\frac{4}{9}$ (d) $\frac{2}{9}$

7. For the reaction $C_2H_6(g) \rightleftharpoons C_2H_4(g) + H_2(g)$

K_p is 5×10^{-2} atm. Calculate the mole per cent of $C_2H_6(g)$ at equilibrium if pure C_2H_6 at 1 atm is passed over a suitable catalyst at 900 K :

(a) 20 (b) 33.33

(c) 66.66 (d) None of these

8. $2NOBr(g) \rightleftharpoons 2NO(g) + Br_2(g)$. If nitrosyl bromide (NOBr) is 40% dissociated at certain temp. and a total pressure of 0.30 atm. K_p for the reaction $2NO(g) + Br_2(g) \rightleftharpoons 2NOBr(g)$ is :

(a) 45 (b) 25

(c) 0.022 (d) 0.25

9. Consider the partial decomposition of A as

$$2A(g) \rightleftharpoons 2B(g) + C(g)$$

At equilibrium 700 mL gaseous mixture contains 100 mL of gas C at 10 atm and 300 K. What is the value of K_p for the reaction?

(a) $\frac{40}{7}$ (b) $\frac{1}{28}$

(c) $\frac{10}{28}$ (d) $\frac{28}{10}$

10. At a certain temperature and 2 atm pressure equilibrium constant (K_p) is 25 for the reaction

$$SO_2(g) + NO_2(g) \rightleftharpoons SO_3(g) + NO(g)$$

Initially if we take 2 moles of each of the four gases and 2 moles of inert gas, what would be the equilibrium partial pressure of NO_2?

(a) 1.33 atm (b) 0.1665 atm

(c) 0.133 atm (d) None of these

11. 0.020 g of selenium vapour at equilibrium occupying a volume of 2.463 mL at 1 atm and 27°C. The selenium is in a state of equilibrium according to reaction

$$3Se_2(g) \rightleftharpoons Se_6(g)$$

What is the degree of association of selenium?

(At. mass of Se = 79)

(a) 0.205 (b) 0.315

(c) 0.14 (d) None of these

12. Determine the degree of association (polymerization) for the reaction in aqueous solution, if observed (mean) molar mass of HCHO and $C_6H_{12}O_6$ is 150 :

$$6HCHO \rightleftharpoons C_6H_{12}O_6$$

(a) 0.50 (b) 0.833

(c) 0.90 (d) 0.96

13. A reaction system in equilibrium according to reaction $2SO_2(g) + O_2(g) \rightleftharpoons 2SO_3(g)$ in one litre vessel at a given temperature was found to be 0.12 mole each of SO_2 and SO_3 and 5 mole of O_2. In another vessel of one litre contains 32 g of SO_2 at the same temperature. What mass of O_2 must be added to this vessel in order that at equilibrium 20% of SO_2 is oxidized to SO_3?

(a) 0.4125 g (b) 11.6 g

(c) 1.6 g (d) None of these

14. The equilibrium constant K_p for the following reaction is 4.5

$$N_2O_4(g) \rightleftharpoons 2NO_2(g)$$

What would be the average molar mass (in g/mol) of an equilibrium mixture of N_2O_4 and NO_2 formed by the dissociation of pure N_2O_4 at a total pressure of 2 atm?

(a) 69 (b) 57.5
(c) 80.5 (d) 85.5

15. A flask containing 0.5 atm pressure of $A_2(g)$, some solid AB added into flask which undergoes dissociation according to $2AB(s) \rightleftharpoons A_2(g) + B_2(g)$ $K_p = 0.06\ \text{atm}^2$

The total pressure (in atm) at equilibrium is :

(a) 0.70 (b) 0.6
(c) 0.10 (d) None of these

16. A vessel of 250 litre was filled with 0.01 mole of Sb_2S_3 and 0.01 mole of H_2 to attain the equilibrium at 440°C as

$$Sb_2S_3(s) + 3H_2(g) \rightleftharpoons 2Sb(s) + 3H_2S(g)$$

After equilibrium, the H_2S formed was analysed by dissolved it in water and treating with excess of Pb^{2+} to give 1.19 g of PbS as precipitate. What is the value of K_c at 440°C?

(a) 1 (b) 2
(c) 4 (d) 8

17. For the reaction $2A(g) + B(g) \rightleftharpoons C(g) + D(g)$; $K_c = 10^{12}$. If the initial moles of A, B, C and D are 2, 1, 7 and 3 moles respectively in a one litre vessel, what is the equilibrium concentration of A?

(a) 4×10^{-4} (b) 2×10^{-4}
(c) 10^{-4} (d) 8×10^{-4}

18. The equilibrium constant for the following reaction in aqueous solution is 0.90.

$$H_3BO_3 + \text{glycerin} \rightleftharpoons (H_3BO_3 - \text{glycerin})$$

How many moles of glycerin should be added per litre of 0.10 M H_3BO_3 so that 80% of the H_3BO_3 is converted to the boric-acid-glycerin complex?

(a) 4.44 (b) 4.52
(c) 3.6 (d) 0.08

19. Rate of diffusion of ozonized oxygen is $0.4\sqrt{5}$ times that of pure oxygen. What is the per cent degree of association of oxygen assuming pure O_2 in the sample initially?

(a) 20 (b) 40
(c) 60 (d) None of these

20. One mole of SO_3 was placed in a two litre vessel at a certain temperature. The following equilibrium was established in the vessel

$$2SO_3(g) \rightleftharpoons 2SO_2(g) + O_2(g)$$

The equilibrium mixture reacts with 0.2 mole $KMnO_4$ in acidic medium. Hence, K_c is :

(a) 0.50 (b) 0.25
(c) 0.125 (d) None of these

21. At 800°C, the following equilibrium is established as

$$F_2(g) \rightleftharpoons 2F(g)$$

The composition of equilibrium may be determined by measuring the rate of effusion of the mixture through a pin hole. It is found that at 800°C and 1 atm mixture effuses 1.6 times as fast as SO_2 effuses under the similar conditions. (At. mass of F = 19). What is the value of K_p (in atm)?

(a) 0.315 (b) 0.685
(c) 0.46 (d) 1.49

22. The equilibrium constant for the ionization of $RNH_2(g)$ in water as

$$RNH_2(g) + H_2O(l) \rightleftharpoons RNH_3^+(aq) + OH^-(aq)$$

is 8×10^{-6} at 25°C. Find the pH of a solution at equilibrium when pressure of $RNH_2(g)$ is 0.5 bar :

(a) ≈ 12.3 (b) ≈ 11.3

(c) ≈ 11.45 (d) None

23. Calculate $\Delta_r G$ for the reaction at 27°C

$$H_2(g) + 2Ag^+(aq) \rightleftharpoons 2Ag(s) + 2H^+(aq)$$

Given : $P_{H_2} = 0.5$ bar; $[Ag^+] = 10^{-5}\ M$;

$[H^+] = 10^{-3}\ M$; $\Delta_r G° [Ag^+(aq)] = 77.1$ kJ/mol

(a) −154.2 kJ/mol (b) −178.9 kJ/mol

(c) −129.5 kJ/mol (d) None of these

24. When N_2O_5 is heated at certain temperature, it dissociates as $N_2O_5(g) \rightleftharpoons N_2O_3(g) + O_2(g)$; $K_c = 2.5$. At the same time N_2O_3 also decomposes as : $N_2O_3(g) \rightleftharpoons N_2O(g) + O_2(g)$. If initially 4.0 moles of N_2O_5 are taken in 1.0 litre flask and allowed to dissociate. Concentration of O_2 at equilibrium is 2.5 *M*. Equilibrium concentration of N_2O_5 is :

(a) 1.0 *M* (b) 1.5 *M*

(c) 2.166 *M* (d) 1.846 *M*

25. Two solid compounds *X* and *Y* dissociates at a certain temperature as follows

$$X(s) \rightleftharpoons A(g) + 2B(g);\ K_{P_1} = 9 \times 10^{-3}\ \text{atm}^3$$

$$Y(s) \rightleftharpoons 2B(g) + C(g);\ K_{P_2} = 4.5 \times 10^{-3}\ \text{atm}^3$$

The total pressure of gases over a mixture of *X* and *Y* is :

(a) 4.5 atm (b) 0.45 atm

(c) 0.6 atm (d) None of these

Level-3

Passage-1

For a gaseous reaction

$$aA(g) + bB(g) \rightleftharpoons cC(g) + dD(g)$$

equilibrium constants K_c, K_p and K_x are represented by the following relations

$$K_c = \frac{[C]^c\,[D]^d}{[A]^a\,[B]^b};\quad K_p = \frac{p_C^c \cdot p_D^d}{p_A^a \cdot p_B^b} \text{ and } K_x = \frac{x_C^c \cdot x_D^d}{x_A^a \cdot x_B^b}$$

where $[A]$ represents molar concentration of A, p_A represents partial pressure of A and P represents total pressure, x_A represents mole fraction of A

1. On the basis of above work-up, select the write option

(a) $K_p = K_c(RT)^{\Delta n_g}; K_x = K_p(RT)^{\Delta n_g}$
(b) $K_c = K_p(RT)^{-\Delta n_g}; K_p = K_x P^{\Delta n_g}$
(c) $K_c = K_x P^{\Delta n_g}; K_p = K_x P^{\Delta n_g}$
(d) $K_c = K_p(RT)^{-\Delta n_g}; K_x = K_p(RT)^{\Delta n_g}$

2. For the reaction $SO_2Cl_2(g) \rightleftharpoons SO_2(g) + Cl_2(g)$, $K_p > K_x$ is obtained at :

(a) 0.5 atm
(b) 0.8 atm
(c) 1 atm
(d) 2 atm

3. For the following equilibrium relation between K_c and K_x (in terms of mole fraction) is

$$PCl_3(g) + Cl_2(g) \rightleftharpoons PCl_5(g)$$

(a) $K_c = K_x(RT)^{-1}$
(b) $K_c = K_x(RT)$
(c) $K_c = K_x\left(\frac{RT}{P}\right)$
(d) $K_c = K_x\left(\frac{P}{RT}\right)$

Passage-2

Variation of equilibrium constant K with temperature is given by van't Hoff equation

$$\ln K = \frac{\Delta_r S^\circ}{R} - \frac{\Delta_r H^\circ}{RT}$$

from this equation, $\Delta_r H^\circ$ can be evaluated if equilibrium constants K_1 and K_2 at two temperature T_1 and T_2 are known.

$$\log\left(\frac{K_2}{K_1}\right) = \frac{\Delta_r H^\circ}{2.303\,R}\left[\frac{1}{T_1} - \frac{1}{T_2}\right]$$

1. For an isomerization $X(g) \rightleftharpoons Y(g)$, the temperature dependency of equilibrium constant is given by :

$$\ln K = 2 - \frac{1000}{T}$$

The value of $\Delta_r S^\circ$ at 300 K is :

(a) $2R$
(b) $\frac{2}{R}$
(c) $1000\,R$
(d) None of these

2. Select the correct statement :

(a) Value of K_{eq} always increases with increasing temperature
(b) For exothermic reaction value of K_{eq} increases with decreasing in temperature
(c) For endothermic reaction value of K_{eq} increases with decreasing in temperature
(d) For exothermic reaction slope is ($\log K$ Vs. $1/T$) negative

3. Variation of $\log_{10} K$ with $\frac{1}{T}$ is shown by the following graph in which straight line is at 45° hence $\Delta H°$ is :

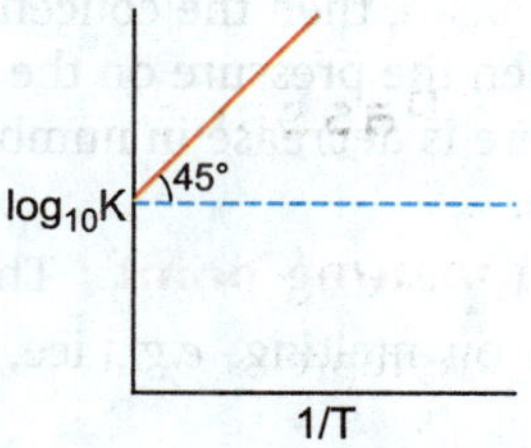

(a) –4.606 J/mol (b) –19.147 J/mol
(c) 8.314 J/mol (d) 10 J/mol

4. The equilibrium constant K_p for the following reaction is 1 at 27°C and 4 at 47°C.

$$A(g) \rightleftharpoons B(g) + C(g)$$

For the reaction calculate enthalpy change for the

$B(g) + C(g) \rightleftharpoons A(g)$ (Given : $R = 2$ cal/mol-K)

(a) – 13.31 kcal/mol (b) 13.31 kcal/mol
(c) – 19.2 kcal/mol (d) 55.63 kcal/mol

Passage-3

N_2O_3 is an unstable oxide of nitrogen and it decomposes into $NO(g)$ and $NO_2(g)$ where $NO_2(g)$ is further dimerise into N_2O_4 as

(i) $N_2O_3(g) \rightleftharpoons NO_2(g) + NO(g)$; $K_{P1} = 2.5$ bar

(ii) $2NO_2(g) \rightleftharpoons N_2O_4(g)$; K_{P2}

A flask is initially filled with pure $N_2O_3(g)$ having pressure 2 bar and equilibria was established. At equilibrium partial pressure of $NO(g)$ was found to be 1.5 bar.

1. The equilibrium partial pressure of $N_2O_3(g)$ is:

(a) 0.5 bar (b) 1.0 bar
(c) 1.5 bar (d) 0.1 bar

2. The equilibrium partial pressure of $NO_2(g)$ is:

(a) 0.066 bar (b) 0.133 bar
(c) 0.423 bar (d) 0.83 bar

3. The value of K_{P2} is:

(a) 0.16 bar^{-1} (b) 0.32 bar^{-1}
(c) 0.48 bar^{-1} (d) 0.64 bar^{-1}

Passage-4

If a system at equilibrium is subjected to a change of any one of the factors such as concentration, pressure or temperature, the system adjusts itself in such a way so as to minimise the effect of that change.

Effect of change in concentration on equilibrium:

As we add or remove reactant (or product) the ratio of equilibrium concentration become 'Q' (reaction quotient) and depending upon.

$Q < K$: equilibrium will shift in forward direction

$Q > K$: equilibrium will shift in backward direction

Effect of change in pressure :

If a system in equilibrium consists of gases, then the concentrations of all the components can be altered by changing the pressure. When the pressure on the system is increased, then equilibrium will shift in the direction in which there is decrease in number of moles *i.e.,* towards the direction in which there is decrease in volume.

Effect of change in pressure on melting point : There are two types of solids :

(a) Solids whose volume decreases on melting, *e.g.,* ice, diamond, carborundum, magnesium nitride and quartz.
Solid (higher volume) $\rightleftharpoons$ Liquid (lower volume)
The process of melting is facilitated at high pressure, thus melting point is lowered.

(b) Solid whose volume increase on melting, *e.g.,* Fe, Cu, Ag, Au, etc.
Solid (lower volume) $\rightleftharpoons$ Liquid (higher volume)
In this case the process of melting becomes difficult at high pressure, thus melting point becomes high.

(c) **Solubility of substances :** When solid substance are dissolved in water, either heat is evolved.
For endothermic solubility process solubility increase with increase in temperature.
For exothermic solubility decrease with increase in temperature.

(d) **Solubility of gases in liquids :** When a gas dissolves in liquid, there is decrease in volume. Thus increase of pressure will favour the dissolution of gas in liquid.
Effect of temperature : For endothermic reaction as temperature increases reaction shift in forward direction. For exothermic reaction as temperature increases reaction shift in backward direction.

1. A 'X'(g) solute when dissolved in water heat is evolved. Then solubility of 'X' will increase :
(a) high temperature, low pressure (b) low temperature, high pressure
(c) high temperature, high pressure (d) low temperature, low pressure

2. $Fe(l) \rightleftharpoons Fe(s)$
Above equilibrium is favoured at :
(a) high pressure, low temperature (b) high pressure, high temperature
(c) low pressure, high temperature (d) low pressure, low temperature

3. For the reaction
$N_2(g) + O_2(g) \rightleftharpoons 2NO(g)$
If pressure is increased by reducing the volume of the container then :
(a) total pressure at equilibrium will remain same
(b) concentration of all the component at equilibrium will change
(c) concentration of all the component at equilibrium will remain same
(d) equilibrium will shift in the backward direction

One or More Answer is/are Correct

1. A catalyst :
(a) increases the average kinetic energy of reacting molecules
(b) decreases the activation energy
(c) can alters the reaction mechanism
(d) can change the equilibrium composition of mixture

2. Which of the following is correct about the chemical equilibrium?
(a) $(\Delta G)_{T,P} = 0$
(b) Equilibrium constant is independent of initial concentration of reactants

(c) Catalyst has no effect on equilibrium state
(d) Reaction stops at equilibrium

3. For the reaction

$$AB_2(g) \rightleftharpoons AB(g) + B(g)$$

If α is negligiable w.r.t. 1 then degree of dissociation (α) of AB_2 is proportional to :

(a) $\frac{1}{P}$ (b) $\frac{1}{V}$

(c) $\frac{1}{\sqrt{P}}$ (d) $\sqrt{V}$

4. Consider the reactions given below. In which cases will the reaction proceed toward right by increasing the pressure?

(a) $4HCl(g) + O_2(g) \longrightarrow 2Cl_2(g) + 2H_2O(g)$

(b) $Cl_2(g) + H_2O(g) \longrightarrow 2HCl(g) + \frac{1}{2}O_2(g)$

(c) $CO_2(g) + 4H_2(g) \longrightarrow CH_4(g) + 2H_2O(g)$

(d) $N_2(g) + O_2(g) \longrightarrow 2NO(g)$

5. Ammonia is a weak base that reacts with water according to the equation

$$NH_3(aq) + H_2O(l) \rightleftharpoons NH_4^+(aq) + OH^-(aq)$$

Select the correct option(s) that can increase the moles of ammonium ion in water:

(a) Addition of HCl (b) Addition of NaOH
(c) Addition of NH_4Cl (d) Addition of H_2O

6. Consider the reaction $2CO(g) + O_2(g) \rightleftharpoons 2CO_2(g) + \text{Heat}$

Under what conditions shift is undeterminable?

(a) Addition of O_2 and decrease in volume
(b) Addition of CO and removal of CO_2 at constant volume
(c) Increase in temperature and decrease in volume
(d) Addition of CO and increase in temperature at constant volume

7. What will be the effect of addition of catalyst at constant temperature?

(a) The equilibrium constant will remain constant
(b) ΔH of the reaction will remain constant
(c) k_f and k_b will increase upto same extent
(d) equilibrium composition will change

8. For the reaction $PCl_5(g) \rightleftharpoons PCl_3(g) + Cl_2(g)$, the forward reaction at constant temperature is favoured by:

(a) introducing an inert gas at constant volume
(b) introducing chlorine gas at constant volume
(c) introducing an inert gas at constant pressure
(d) increasing the volume of the container

9. For the reaction : $Cl_2(g) + 3F_2(g) \rightleftharpoons 2ClF_3(g)$; $\Delta H = -329$ kJ, dissociation of $ClF_3(g)$ will be favoured by :

(a) increasing the temperature
(b) increasing the volume of the container
(c) adding of F_2 gas
(d) adding of inert gas at constant pressure

10. Increase in the pressure for the following equilibrium results in the:

$$H_2O(l) \rightleftharpoons H_2O(g)$$

(a) formation of more $H_2O(l)$
(b) formation of more $H_2O(g)$
(c) increase in b.p. of $H_2O(l)$
(d) decrease in b.p. of $H_2O(l)$

11. Heating a II group metal carbonate leads to decomposition as :

$$BaCO_3(s) \rightleftharpoons BaO(s) + CO_2(g)$$

Equilibrium will shift left

(a) by addition of BaO(s)
(b) by addition of $CO_2(g)$
(c) by decreasing the temperature
(d) by decreasing the volume of the vessel

12. $N_2(g)$ and $H_2(g)$ are allowed to react in a closed vessel at given temp. and pressure for the formation of $NH_3(g)$ [$N_2(g) + 3H_2(g) \rightleftharpoons 2NH_3(g) + 22.4$ kcal]. If He(g) is added at equilibrium at constant pressure then which is/are correct ?

(a) Concentration of $N_2(g)$, $H_2(g)$ and $NH_3(g)$ decrease.
(b) Moles of $NH_3(g)$ decreases.
(c) The extent of cooling depends on amount of He(g) added.
(d) Concentration of N_2 and H_2 increases and concentration of NH_3 decreases.

Match the Column

Column-I and **Column-II** contains four entries each. Entries of Column-I are to be matched with some entries of Column-II. One or more than one entries of Column-I may have the matching with the same entries of Column-II.

1.

Column-I	Column-II
(A) $CaCO_3(s) \rightleftharpoons CaO(s) + CO_2(g)$	(P) $K_p > K_c$ above room temperature
(B) $CO(g) + Cl_2(g) \rightleftharpoons COCl_2(g)$	(Q) $K_p = K_c$ above room temperature
(C) $H_2(g) + I_2(g) \rightleftharpoons 2HI(g)$	(R) $K_p < K_c$ above room temperature
(D) $HCl(g) \rightleftharpoons H^+(aq) + Cl^-(aq)$	(S) K_p and K_c not defined

2.

Column-I	Column-II
(A) $3O_2(g) \rightleftharpoons 2O_3(g)$	(P) no unit
(B) $SO_2(g) + \frac{1}{2}O_2(g) \rightleftharpoons SO_3(g)$	(Q) atm$^{-1/2}$
(C) $2HF(g) \rightleftharpoons H_2(g) + F_2(g)$	(R) atm^{-1}
(D) $CO(g) + 3H_2(g) \rightleftharpoons CH_4(g) + H_2O(g)$	(S) atm^{-2}

3.

Column-I (Reaction) (If initial concentration of reactant = 1 *M*)	Column-II (If α is negligiable w.r.t. 1)
(A) $2X(g) \rightleftharpoons Y(g) + Z(g)$	(P) $\alpha = 2 \times \sqrt{K_c}$
(B) $X(g) \rightleftharpoons Y(g) + Z(g)$	(Q) $\alpha = 3 \times \sqrt{K_c}$
(C) $3X(g) \rightleftharpoons Y(g) + Z(g)$	(R) $\alpha = (2K_c)^{1/3}$
(D) $2X(g) \rightleftharpoons Y(g) + 2Z(g)$	(S) $\alpha = \sqrt{K_c}$

4.

Column-I	Column-II
(A) $N_2(g) + 3H_2(g) \rightleftharpoons 2NH_3(g); \Delta H = -\text{ve}$	(P) K increases with increase in temperature
(B) $N_2(g) + O_2(g) \rightleftharpoons 2NO(g);\ \Delta H = +\text{ve}$	(Q) K decreases with increase in temperature
(C) $A(g) + B(g) \rightleftharpoons 2C(g) + D(g); \Delta H = +\text{ve}$	(R) Pressure has no effect
(D) $PCl_5(g) \rightleftharpoons PCl_3(g) + Cl_2(g);\ \Delta H = +\text{ve}$	(S) Moles of product increase due to addition of inert gas at constant pressure

5.

Column-I	Column-II
(A) $\dfrac{K_{10+T\,°C}}{K_{T\,°C}} = 2$	(P) Endothermic
(B) $\dfrac{K_{10+T\,°C}}{K_{T\,°C}} = \dfrac{1}{2}$	(Q) Not affected by pressure
(C) $A(g) + B(g) \rightleftharpoons C(g)$	(R) Exothermic
(D) $X(s) + Y(g) \rightleftharpoons Z(g)$	(S) Affected by volume

6.

Column-I	Column-II
(A) Pressure increased in $2NO(g) \rightleftharpoons N_2(g) + O_2(g)$	(P) Equilibrium shifted in forward direction
(B) Pressure increased in $CH_4(g) + H_2O(g) \rightleftharpoons CO(g) + 3H_2(g)$	(Q) Equilibrium shifted in backward direction
(C) Temp. increased and pressure increased $3O_2(g) \rightleftharpoons 2O_3(g);\quad \Delta H = 285\ \text{kJ}$	(R) Equilibrium remains unaffected
(D) Pressure decreased and moles of N_2 increased $N_2(g) + 2O_2(g) \rightleftharpoons 2NO_2(g);\quad \Delta H = 66.4\ \text{kJ}$	(S) Theoretically we cannot predict

7.

Column-I (Reactions)	Column-II (Equilibrium States)
(A) (A)  (g) $\rightleftharpoons$ 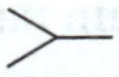 (g) n-butane isobutane	(P) 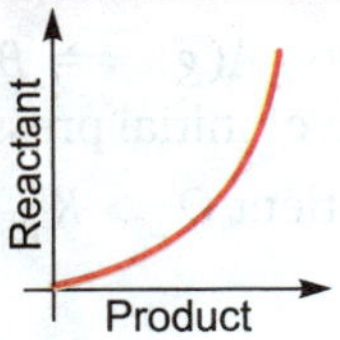
(B) $I_2(s) \rightleftharpoons I_2(aq)$	(Q) 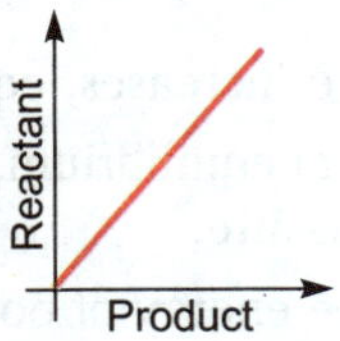

(C) $N_2O_4(g) \rightleftharpoons 2NO_2(g)$	(R) 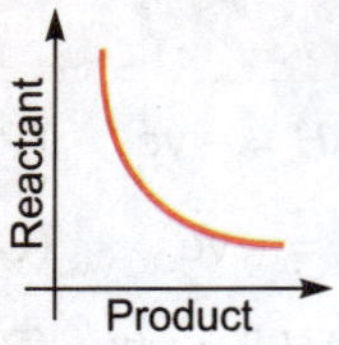
(D) $AgI(s) \rightleftharpoons Ag^+(aq) + I^-(aq)$	(S) 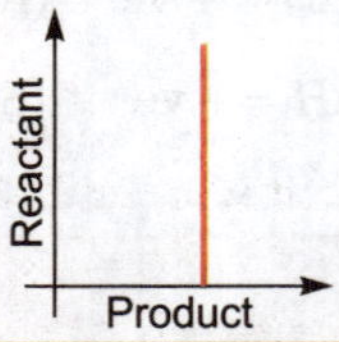

Assertion-Reason Type Questions

Each question contains STATEMENT-1 (Assertion) and STATEMENT-2 (Reason).
Examine the statements carefully and mark the correct answer according to the instructions given below :
(A) If both the statements are TRUE and STATEMENT-2 is the correct explanation of STATEMENT-1
(B) If both the statements are TRUE but STATEMENT-2 is NOT the correct explanation of STATEMENT-1
(C) If STATEMENT-1 is TRUE and STATEMENT-2 is FALSE
(D) If STATEMENT-1 is FALSE and STATEMENT-2 is TRUE

1. STATEMENT-1 : The endothermic reactions are favoured at lower temperature and the exothermic reactions are favoured at higher temperature.
STATEMENT-2 : When a system in equilibrium is disturbed by changing the temperature, it will tend to adjust itself so as to overcome the effect of change.

2. STATEMENT-1 : The melting point of ice decreases with increase of pressure.
STATEMENT-2 : Concentration in volume take place during melting of ice.

3. STATEMENT-1 : The equilibrium of $A(g) \rightleftharpoons B(g) + C(g)$ is not affected by changing the volume.
STATEMENT-2 : K_c for the reaction does not depend on volume of the container.

4. STATEMENT-1 : For a chemical reaction at initial stage rate of forward reaction (r_f) is greater than rate of reversed reaction (r_b)
STATEMENT-2 : When $r_f = r_b$, chemical reaction is at equilibrium.

5. STATEMENT-1 : For the reaction $A(g) \rightleftharpoons B(g) + C(g), K_p = 1$ atm. If we start with equal moles of all gases at 9 atm of initial pressure, then at equilibrium partial pressure of A increases.
STATEMENT-2 : Reaction quotient $Q_p > K_p$ hence equilibrium shifts in backward direction.

6. STATEMENT-1 : The gas phase reaction $PCl_3(g) + Cl_2(g) \rightleftharpoons PCl_5(g)$ shifts to the right on increasing pressure.
STATEMENT-2 : When pressure increases, equilibrium shifts towards more number of moles.

7. STATEMENT-1 : For a reaction at equilibrium, the Gibb's free energy of reaction is minimum at constant temp. and pressure.
STATEMENT-2 : The Gibb's free energy of both reactants and products increases and become equal at equilibrium.

8. STATEMENT-1 : The physical equilibrium is not static but dynamic in nature.
STATEMENT-2 : The physical equilibrium is a state in which two opposing process are proceeding at the same rate.

9. STATEMENT-1 : Equilibrium constant for the reverse reaction is the inverse of the equilibrium constant for the reaction in the forward direction.

STATEMENT-2 : Equilibrium constant depends upon the way in which the reaction is written.

10. STATEMENT-1 : If $Q_p < K_p$ reaction moves in direction of reactants.

STATEMENT-2 : Reaction quotient is defined in the same way as equilibrium constant at any stage of the reaction.

11. STATEMENT-1 : For the reaction $H_2(g) + I_2(g) \rightleftharpoons 2HI(g)$ if the volume of vessel is reduced to half of its original volume, equilibrium concentration of all gases will be doubled.

STATEMENT-2 : According to Le-Chatelier's principle, reaction shifts in a direction that tends to minimized the effect of the stress.

12. STATEMENT-1 : The equilibrium constant of the exothermic reaction at high temperature decreases.

STATEMENT-2 : Since $\ln \frac{K_2}{K_1} = \frac{\Delta H^\circ}{R}\left[\frac{1}{T_1} - \frac{1}{T_2}\right]$ and for exothermic reaction,

$\Delta H^\circ = -$ ve and thereby; $\frac{K_2}{K_1} < 1$

13. STATEMENT-1 : For the reaction at certain temperature

$$A(g) + B(g) \rightleftharpoons C(g)$$

there will be no effect by addition of inert gas at constant volume.

STATEMENT-2 : Molar concentration of all gases remains constant.

14. STATEMENT-1 : For the physical equilibrium $H_2O(s) \rightleftharpoons H_2O(l)$ on increasing temperature and increasing pressure more water will form.

STATEMENT-2 : Since forward reaction is endothermic in nature and volume of water is greater than that of the volume of ice.

15. STATEMENT-1 : The catalyst does not alter the equilibrium constant.

STATEMENT-2 : Because for the catalysed reaction and uncatalysed reaction ΔH remains same and equilibrium constant depends on ΔH.

Level-4

Subjective Problems

1. If 50% of CO_2 converts to CO at the following equilibrium :

$$\frac{1}{2}C(s) + \frac{1}{2}CO_2(g) \rightleftharpoons CO(g)$$

and the equilibrium pressure is 12 atm. Calculate K_p.

2. Calculate partial pressure of B at equilibrium in the following equilibrium

$$A(s) \rightleftharpoons B(g) + 2C(g);\ K_p = 32\ \text{atm}^3.$$

3. In a gaseous reaction $A + 2B \rightleftharpoons 2C + D$ the initial concentration of B was 1.5 times that of A. At equilibrium the concentrations of A and D were equal. Calculate the equilibrium constant K_C.

4. For the reaction $A(g) \rightleftharpoons B(g);\ K_C = 10$

$$B(g) \rightleftharpoons C(g);\ K_C = 2$$

$$C(g) \rightleftharpoons D(g);\ K_C = 0.01$$

Calculate K_C for the reaction $D(g) \rightleftharpoons A(g)$.

5. 5 litre vessel contains 2 moles of each of gases A and B at equilibrium. If 1 mole each of A and B are removed. Calculate K_C for the reaction $A(g) \rightleftharpoons B(g)$.

6. Calculate K_p (in atm^2) for the reaction $A(g) \rightleftharpoons B(s) + 2C(g)$; $K_C = 0.2\,M^2$ at 305 K.

7. A mixture of 3 moles of SO_2, 4 moles of NO_2, 1 mole of SO_3 and 4 moles of NO is placed in a 2.0 L vessel. $SO_2(g) + NO_2(g) \rightleftharpoons SO_3(g) + NO(g)$. At equilibrium, the vessel is found to contain 1 mole of SO_2. Calculate the value of K_C.

8. The density of an equilibrium mixture of N_2O_4 and NO_2 at 1 atm and 373.5 K is 2.0 g/L. Calculate K_C for the reaction $N_2O_4(g) \rightleftharpoons 2NO_2(g)$, if initial concentration of $N_2O_4 = 1M$.

9. If chemical equilibrium is attained at standard states then what is the value of $\Delta G°$?

10. Calculate the equilibrium concentration ratio of C to A if equimolar ratio of A and B were allowed to come to equilibrium at 300 K.

$$A(g) + B(g) \rightleftharpoons C(g) + D(g)\ ;\ \Delta G° = -830\ \text{cal}.$$

11. A definite amount of solid NH_4HS is placed in a flask already containing ammonia gas at a certain temperature and 0.1 atm pressure. NH_4HS decomposes to give NH_3 and H_2S and at equilibrium total pressure in flask is 1.1 atm. If the equilibrium constant K_p for the reaction $NH_4HS(s) \rightleftharpoons NH_3(g) + H_2S(g)$ is represented as $z \times 10^{-1}$ then find the value of z.

12. The gaseous reaction : $A(g)+nB(g) \rightleftharpoons mC(g)$ is represented by following curves

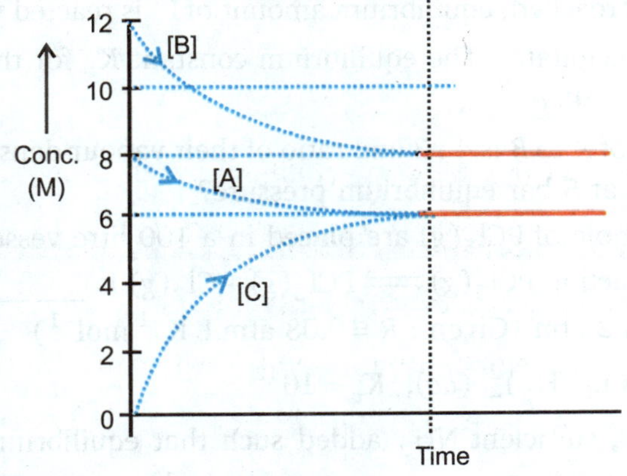

What is the value of $n+m$?

Numerical Problems

(If the numerical value has more than two decimal places, truncate/round-off the value to TWO decimal places.)

1. At 727°C, some amount of $CaCO_3(s)$ decomposes in a 5.0 L closed rigid vessel as following,

$$CaCO_3 \rightleftharpoons CaO(s)+CO_2(g)$$

If K_p for reaction at 727°C is 1.642 atm then the equilibrium amount of CaO(s) formed is :

2. One mole of PCl_5 in a closed rigid vessel at 2 atm and 300 K temperature is heated upto 600 K temperature. So that equilibrium is achieved and degree of dissociation of PCl_5 is 50%. Calculate total pressure (in atm) at equilibrium.

3. Consider the reaction $2A(g) \rightleftharpoons A_2(g)$, in an experiment 150 g of A at 546 K in 22.4 L vessel dimerised to the extent of 50%. Calculate the pressure (in atm) developed in vessel at equilibrium. (Given : $M_A=50$ g/mol)

4. The reaction rate for the reaction : $[PtCl_4]^{2-}+H_2O \rightleftharpoons [Pt(H_2O)Cl_3]^- + Cl^-$ was measured as a function of concentrations of different species. It was observed that

$$\frac{-d[PtCl_4]^{2-}}{dt}=4.8\times10^{-5}[PtCl_4]^{2-}-2.4\times10^{-3}[Pt(H_2O)Cl_3]^-[Cl^-]$$

where square brackets are used to denote molar concentrations. The equilibrium constant K_c for the reverse reaction is _____.

5. At high temperature the following equilibria exist in a mixture of carbon, oxygen and their compounds:

$$C(s)+O_2(g) \rightleftharpoons CO_2(g) \quad ; \quad K_1=10^{-2}$$
$$2C(s)+O_2(g) \rightleftharpoons 2CO(g) \quad ; \quad K_2=10^{-3}$$
$$C(s)+CO_2(g) \rightleftharpoons 2CO(g) \quad ; \quad K_3=x$$
$$2CO(g)+O_2(g) \rightleftharpoons 2CO_2(g) \quad ; \quad K_4=y$$

What will be the ratio of x and y?

6. At 1990 K and 1 atm pressure, there are equal number of Cl_2 molecules and Cl atoms in the reaction mixture. The value of K_p for the reaction $Cl_2(g) \rightleftharpoons 2Cl(g)$ under the above conditions is $x\times10^{-1}$. The value of x is _____.

7. $AB_3(g)$ is dissociate as, $AB_3(g) \rightleftharpoons AB_2(g)+\frac{1}{2}B_2(g)$

When the initial pressure of AB_3 is 800 torr and total pressure developed at equilibrium is 1000 torr. Calculate vapour density of equilibrium mixture.
(Given : Atomic mass of $A=15$ and $B=20$)

8. $I_2 + I^- \rightleftharpoons I_3^-$. This reaction is set up in aqueous medium. We start with 1 mol of I_2 and 0.5 mol of I^- in 10 L flask. After equilibrium is reached, equilibrium amount of I^- is reacted with excess of $AgNO_3$ separately gave 0.25 mol of yellow precipitate. The equilibrium constant K_c for the reaction is :

9. For the reaction : $A(g) \rightleftharpoons 2B(g)$

If at equilibrium mass ratio of A to B is 4 : 3 and ratio of their vapour density is 2 : 1 then what is the value of equilibrium constant K_p at 5 bar equilibrium pressure?

10. One mole of $N_2(g)$ and 3 mole of $PCl_5(g)$ are placed in a 100 litre vessel heated to 227°C. The K_P of the following reaction $PCl_5(g) \rightleftharpoons PCl_3(g) + Cl_2(g)$ is _______ . The equilibrium pressure is 2 atm (Given : $R = 0.08$ atm.L K^{-1} mol^{-1})

11. $Cu^{2+}(aq) + 4NH_3(aq) \rightleftharpoons Cu(NH_3)_4^{2+}(aq)$, $K_c = 10^{12}$

In a solution of 0.5 M Cu^{2+}, sufficient NH_3 added such that equilibrium conc. of NH_3 is 0.5 M and if equilibrium concentration of Cu^{2+} is represented as $y \times 10^{-12}$ M, then find the value of y.

12. For the reaction $A + 2B \rightleftharpoons C + D$, the equilibrium constant $K_c = 10^9$. If the initial moles of A, B, C and D are 1.5, 1, 3.5 and 0.5 moles respectively in a 10 litre vessel, if the equilibrium conc. of B is $y \times 10^{-5}$ M then the value of y is _____.

13. A mixture of $A(g)$, $B(g)$ and $C(g)$ at equilibrium has a average molecular weight 80. According to the reaction

$$A(g) \rightleftharpoons B(g) + C(g)$$

The degree of dissociation for $A(g)$ is :

(Given : mol. wt. of $A = 100$; mol. wt. of $B = 60$; mol. wt. of $C = 40$)

14. $A(s) \rightleftharpoons 2B(g) + C(g)$

The above equilibrium was established by initially taking $A(s)$ only. At equilibrium, B is removed so that its partial pressure at new equilibrium becomes (1/3)rd of original total pressure. Find the ratio of total pressure at new equilibrium and at initial equilibrium.

15. The decomposition of calcium sulphate at 1625 K,

$$2CaSO_4(s) \rightleftharpoons 2CaO(s) + 2SO_2(g) + O_2(g); \quad K_p = 3.2 \times 10^{-6} \text{ atm}^3$$

A sample of $CaSO_4(s)$ is introduced in a reaction vessel that is filled with air at 1.0 atm pressure at 1625 K, the partial pressure of $SO_2(g)$ is $y \times 10^{-3}$ atm. The value of y is ____.

(Assume air contains 20% by volume of O_2)

Answers

Level-1

1. (b)	2. (b)	3. (c)	4. (a)	5. (b)	6. (a)	7. (a)	8. (a)	9. (a)	10. (a)
11. (b)	12. (b)	13. (b)	14. (d)	15. (c)	16. (b)	17. (d)	18. (d)	19. (c)	20. (d)
21. (b)	22. (a)	23. (c)	24. (a)	25. (b)	26. (d)	27. (d)	28. (c)	29. (d)	30. (c)
31. (d)	32. (c)	33. (c)	34. (c)	35. (a)	36. (a)	37. (c)	38. (d)	39. (c)	40. (c)
41. (a)	42. (b)	43. (a)	44. (a)	45. (a)	46. (a)	47. (c)	48. (d)	49. (d)	50. (d)
51. (a)	52. (b)	53. (a)	54. (b)	55. (b)	56. (b)	57. (b)	58. (b)	59. (c)	60. (c)
61. (a)	62. (d)	63. (a)	64. (a)	65. (c)	66. (a)	67. (a)	68. (c)	69. (d)	70. (a)
71. (a)	72. (c)	73. (a)	74. (d)	75. (b)	76. (c)	77. (b)	78. (b)	79. (c)	80. (d)
81. (b)	82. (b)	83. (d)	84. (b)	85. (b)	86. (b)	87. (a)	88. (a)	89. (b)	90. (c)
91. (a)	92. (b)	93. (a)	94. (c)	95. (b)	96. (d)	97. (d)	98. (b)	99. (c)	100. (a)
101. (d)	102. (d)	103. (b)	104. (d)	105. (c)	106. (d)	107. (b)	108. (d)	109. (a)	110. (d)
111. (c)	112. (b)	113. (a)	114. (b)	115. (a)	116. (c)	117. (b)	118. (a)	119. (a)	120. (c)
121. (c)	122. (a)	123. (b)	124. (d)	125. (a)					

Level-2

1. (a)	2. (c)	3. (c)	4. (d)	5. (d)	6. (c)	7. (c)	8. (a)	9. (c)	10. (c)
11. (b)	12. (d)	13. (b)	14. (b)	15. (a)	16. (a)	17. (a)	18. (b)	19. (c)	20. (c)
21. (d)	22. (b)	23. (c)	24. (d)	25. (b)					

Level-3

Passage–1:	1. (b)	2. (d)	3. (c)	
Passage–2:	1. (a)	2. (b)	3. (b)	4. (a)
Passage–3:	1. (a)	2. (d)	3. (c)	
Passage–4:	1. (b)	2. (a)	3. (b)	

One or More Answers is/are correct

1. (b,c)	2. (a,b,c)	3. (c,d)	4. (a,c)	5. (a,c,d)	6. (c,d)	7. (a,b,c)	8. (c,d)
9. (a,b,d)	10. (a,c)	11. (b,c,d)	12. (a,b,c)				

Match the Column

1. $A \rightarrow P$; $B \rightarrow R$; $C \rightarrow Q$; $D \rightarrow S$
2. $A \rightarrow R$; $B \rightarrow Q$; $C \rightarrow P$; $D \rightarrow S$
3. $A \rightarrow P$; $B \rightarrow S$; $C \rightarrow Q$; $D \rightarrow R$
4. $A \rightarrow Q$; $B \rightarrow P, R$; $C \rightarrow P, S$; $D \rightarrow P, S$
5. $A \rightarrow P$; $B \rightarrow R$; $C \rightarrow S$; $D \rightarrow Q$
6. $A \rightarrow R$; $B \rightarrow Q$; $C \rightarrow P$; $D \rightarrow S$
7. $A \rightarrow Q$; $B \rightarrow S$; $C \rightarrow P$; $D \rightarrow S$

Assertion-Reason Type Questions

1. (D)	2. (A)	3. (D)	4. (B)	5. (A)	6. (C)	7. (C)	8. (A)	9. (A)	10. (D)
11. (B)	12. (A)	13. (A)	14. (C)	15. (A)					

Level-4

Subjective Problems

1. 4	2. 2	3. 4	4. 5	5. 1	6. 5	7. 9	8. 2	9. 0	10. 2
11. 3	12. 5								

Numerical Problems

1. 5.60	2. 6.00	3. 4.50	4. 50.00	5. 1.00	6. 5.00	7. 30.00	8. 13.33
9. 4.50	10. 0.20	11. 8.00	12. 2.00	13. 0.25	14. 1.67	15. 4.00	

Hints & Solutions

Level 1

2. (b) $Q_C = \frac{[P_2(g)]^2}{[P_4(g)]} = \frac{(1)^2}{(3/2)} = \frac{2}{3}$

4. (a) $\because \frac{K}{Q} = 0.33$

$\Rightarrow K < Q$

i.e., reaction moves in backward direction to achieve equilibrium.

5-6. (b,a)

$$Q_C = \frac{[SO_3(g)]^2}{[SO_2(g)]^2[O_2(g)]^1} = 1000 > K_c$$

i.e., reaction moves backward

also if volume increases pressure decreases

$\Rightarrow Q_c \uparrow$

7. (a) $Q_C = \frac{\left(\frac{6}{2}\right)^2}{\left(\frac{2}{2}\right)\left(\frac{4}{2}\right)^3} = \frac{9}{8}$

$Q_c < K_c$ so reaction will proceed in forward direction.

8. (a) Since D_2 has similar chemical properties as H_2 so upon mixing, NH_2D and NHD_2 are formed.

9. (a) $[N_2]$ and $[H_2]$ start from non-zero value, decrease & then become constant.

$[NH_3]$ start from zero, increase & then become constant.

10. (a) From graph, $[A]_{eq} = 0.1\,M, [B]_{eq} = 0.4\,M$

$$K_c = \frac{[B]^2}{[A]^1} = 1.6$$

11. (b)

	$A(g)$	$\rightleftharpoons$	$2C(g)$	+	$B(g)$
$t = 0$	$10M$				
t_{eq}	$(10 - 10\alpha)$		$2(10\alpha)$		10α

From graph

$[A]_{eq} = 10(1-\alpha) = 8$

$\Rightarrow \alpha = 0.2$

$$K_c = \frac{[C]^2[B]^1}{[A]^1} = \frac{4^2 \times 2}{8}$$

$= 4\,(\text{mol/L})^2$

12. (b) $K_c = \frac{[CO]^1[H_2]^2}{[CH_3OH]^1}$

So units $\longrightarrow M^2$

13. (b) $K_p \longrightarrow (\text{atm})^{\Delta n_g}$

$= (\text{atm})^{-2}$

14. (d) Activity of solid component is unit.

15. (c) If, $K_c = \frac{[N_2]^{1/2}[H_2]^{3/2}}{[NH_3]^1} = 4$

then $\frac{[NH_3]^2}{[N_2][H_2]^3} = (4)^{-2} = \frac{1}{16}$

also $K_p = K_c \cdot (RT)^{\Delta n_g}$

16. (b) For, $N_2 + O_2 \rightleftharpoons 2NO, \Delta n_g = 0$

$\Rightarrow K_c = K_p = K_x = 4 \times 10^{-4}$

So for $NO \rightleftharpoons \frac{1}{2}N_2 + \frac{1}{2}O_2$

$K_c = (4 \times 10^{-4})^{-1/2}$

17-21. (d, d,c, d, b)

$K_p = K_c \cdot (RT)^{\Delta n_g}$

24. (a) $\ln \frac{K_2}{K_1} = \frac{\Delta H}{R}\left(\frac{1}{T_1} - \frac{1}{T_2}\right)$

25-26. (b,d)

Equilibrium constant is temperature dependant only.

27. (d) $CO(g) + H_2O(g) \rightleftharpoons CO_2(g) + H_2(g);\ \sqrt{K_1}$

$CH_4(g) + H_2O(g) \rightleftharpoons CO(g) + 3H_2(g);\ K_2$

Addition

$CH_4(g) + 2H_2O(g) \rightleftharpoons CO_2(g) + 4 \cdot H_2(g);$ $(\sqrt{K_1} \cdot K_2)$

$\Rightarrow K_3 = \sqrt{K_1} \cdot K_2$

28-29. (c,d) Similar to Q no. 16

30-31. (c,d) Similar to Q. no. 27

32. (c)

	$X(g) + Y(g) \rightleftharpoons 2Z(g)$		
$t = 0$	$0.2M$	$0.1M$	$0.1M$
t_{eq}	$(0.2 - x)$	$(0.1 - x)$	$(0.1 + 2x)$

$\because \quad [Z] = (0.1 + 2x) = 0.2$

$\Rightarrow \quad x = 0.05$

Now, $K_c = \frac{[Z]^2}{[X]^1[Y]^1}$

33. (c) $K_c = \frac{[H_2]^2[S_2]}{[H_2S]}$

34. (c) $K_c = \frac{[CH_4][H_2S]^2}{[CS_2][H_2]^4}$

35. (a) $K_c = \frac{[CH_3OH]}{[CO][H_2]^2}$

36. (a) $K_p = \frac{P^4_{S_2(g)}}{P_{S_8(g)}} = \frac{(4 \times 0.3)^4}{0.70} = 2.96$

37. (c) $N_2O_4(g) \rightleftharpoons 2NO_2(g)$

$t = 0 \quad 0.1M$

$t_{eq.} \quad 0.1(1-\alpha) \quad 2(0.1\alpha)$

$\because \alpha = 0.5$

$K_c = \frac{[NO_2]^2}{[N_2O_4]}$

38. (d) $2NH_3(g) \rightleftharpoons N_2(g) + 3H_2(g)$

$t = 0 \quad 2 \text{ moles}$

$t_{eq.} \quad 2-2x \quad x \quad 3x$

Given : $(2-2x) = 1$

$\Rightarrow \quad x = 0.5$

Now $K_c = \frac{[N_2][H_2]^3}{[NH_3]^2}$

39. (c) $K_p = \frac{1}{P^4_{H_2O(g)}}$

$\Rightarrow P_{H_2O(g)} = \left(\frac{1}{K_p}\right)^{1/4}$

$\Rightarrow (10^{-12})^{1/4} = 10^{-3}$ atm

$\therefore K_p = 10^{-3} \times 760 = 0.76$ torr

40. (c) $P_{H_2O} = 2 \times 10^{-2}$ atm

$R.H. = \frac{2 \times 10^{-2} \times 760}{38} \times 100$

$= 40\%$

41. (a) $NH_4HS(s) \rightleftharpoons NH_3(g) + H_2S(g)$

$t = 0 \quad a \text{ moles}$

$t_{eq.} \quad (a-x) \quad x \quad x$

$\because P_{Gas} = X_{Gas} \cdot P_{Total}$

$\Rightarrow P_{NH_3} = \frac{1}{2} \times 0.66 = 0.33$ atm

$P_{H_2S} = \frac{1}{2} \times 0.66 = 0.33$ atm

Now $K_p = (P_{NH_3})^1 (P_{H_2S})^1$

42. (b) $K_c = \frac{n_B n_C^3}{n_A^2} \times \frac{1}{V^2}$

$\Rightarrow \quad 16 = \frac{2 \times 2^3}{2^2 \times V^2}$

$\Rightarrow \quad V = \frac{1}{2}$

43. (a) $C_2H_5OH(l) + CH_3COOH(l) \rightleftharpoons CH_3COOC_2H_5(l) + H_2O(l)$

$t = 0 \quad 1 \text{ mole} \quad 1 \text{ mole} \quad - \quad -$

$t_{eq.} \quad (1-x) \quad (1-x) \quad (x) \quad (x)$

$x = \frac{1}{3}$

$K_c = \frac{[CH_3COOC_2H_5][H_2O]}{[C_2H_5OH][CH_3COOH]}$

44. (a) $I_2(aq.) + I^-(aq.) \rightleftharpoons I_3^-(aq.)$

$t = 0 \quad 1 \text{ mole} \quad 0.5 \text{ mole} \quad -$

$t_{eq.} \quad (1-x) \quad (0.5-x) \quad (x)$

$AgNO_3$ reacts with I^- and forms precipitate.

i.e., $AgNO_3 + I^- \longrightarrow AgI(s) + NO_3^-(aq.)$

$\Rightarrow (0.5 - x) = 0.25$

$\Rightarrow x = 0.25$

$K_c = \frac{[I_3^-]}{[I_2][I^-]}$

45. (a) $K_c = \frac{[H_2S(g)]}{[H_2(g)]}$

$\Rightarrow 8 \times 10^{-2} = \frac{x}{0.3 - x}$

$\Rightarrow \quad x = 0.022$

$[H_2S(g)] = \frac{0.022}{2}$

$= 0.011M$

46. (a) $K_p = \frac{(P_{SO_3})^2}{(P_{SO_2})^2(P_{O_2})^1}$ will be same in both cases

47. (c) $P_{total} = 3P$

$\Rightarrow \quad P = \frac{0.318}{3} = 0.106$

$\therefore \quad K_p = 4P^3 = 4.76 \times 10^{-3}$

48. (d) $K_c = [B]^2[C]^3$

and K_c remains constant.

49. (d) $A + B \rightleftharpoons C + D$

$t = 0 \quad a \quad a$

$t_{eq.} \quad (a-x) \quad (a-x) \quad (x) \quad (x)$

$\because [D] = 2[A]$

$\Rightarrow x = 2(a-x)$

$\Rightarrow x = 2a/3$

Now $K_c = \frac{[C][D]}{[A][B]}$

50. (d) $Q = 1 < K_{eq.}$

reaction moves forward

i. e., $SO_2 + NO_2 \rightleftharpoons SO_3 + NO$

	SO_2	NO_2	SO_3	NO
$t=0$	1	1	1	1
$t_{eq.}$	$(1-x)$	$(1-x)$	$(1+x)$	$(1+x)$

$$K_c = \frac{[SO_3][NO]}{[SO_2][NO_2]} = 16$$

$$4 = \frac{1+x}{1-x}$$

$$x = 0.6 \quad [NO] = 1 + x = 1.6\,M$$

53. (a) $2HI(g) \rightleftharpoons H_2(g) + I_2(g)$

	HI	H_2	I_2
$t=0$	aM		
$t_{eq.}$	$(a - a\alpha)$	$\frac{a\alpha}{2}$	$\frac{a\alpha}{2}$

Given : $\alpha = 0.5$

So $K_{eq} = \frac{[H_2][I_2]}{[HI]^2}$

54. (b) $H_2(g) + CO_2(g) \rightleftharpoons H_2O(g) + CO(g)$

	H_2	CO_2	H_2O	CO
moles at eqm	$0.8-x$	$0.8-x$	x	x
conc. at eqm	$\frac{0.8-x}{5}$	$\left(\frac{0.8-x}{5}\right)$	$\left(\frac{x}{5}\right)$	$\left(\frac{x}{5}\right)$

$\because \ \Delta n_g = 0$

$$\therefore \quad K_c = \frac{\left(\frac{x}{5}\right)^2}{\left(\frac{0.8-x}{5}\right)^2}$$

$$\Rightarrow \quad 2 = \frac{x}{0.8-x}$$

$$\Rightarrow \quad x = 0.533$$

$$[CO_2(g)] = \frac{0.8 - 0.533}{5}$$

55. (b) For ideal gas mole % ≡ volume %

$N_2O_4(g) \rightleftharpoons 2NO_2(g)$

	N_2O_4	NO_2
Initial moles	a	0
at eqm	$a(1-\alpha)$	$2a\alpha$

As per given original volume

$$= \frac{75}{100} \times \text{Volume at eqm}$$

at constant T and P : $V \propto n$

$$\therefore \quad a = 0.75 \times a(1+\alpha)$$

$$\Rightarrow \quad \alpha = 0.33$$

56. (b) $CO + H_2O \rightleftharpoons CO_2 + H_2$

	CO	H_2O	CO_2	H_2
Initial conc.	3	3	x	0
At eqm	2	2	$(x+1)$	1

$$5 = \frac{(x+1)}{4} = 20 = x + 1$$

$$\Rightarrow \quad x = 19$$

57. (b) $N_2(g) + 3H_2(g) \rightleftharpoons 2NH_3(g)$

	N_2	H_2	NH_3
$t=0$	a mole	$3a$	
$t_{eq.}$	$(a-x)$	$(3a-3x)$	$(2x)$

$x = a/4$

$t_{eq.}$ $\left(\frac{3a}{4}\right)$mole $\left(\frac{9a}{4}\right)$mole $\left(\frac{a}{2}\right)$mole

$$P_{Gas} = X_{Gas} \cdot P_{Total}$$

58. (b) $2NH_3(g) \rightleftharpoons N_2(g) + 3H_2(g)$

	NH_3	N_2	H_2
$t=0$	a		
$t_{aq.}$	$(a - a\alpha)$	$\frac{a\alpha}{2}$	$\frac{3a\alpha}{2}$

$\because \ PV = nRT$

at $t = 0$; $15 \times V = a.R.300$ ...(1)

at $t_{eq.}$; $50 \times V = (a + a\alpha).R.620$ ...(2)

$$(1) \div (2); \frac{3}{10} = \frac{300}{(1+\alpha).620}$$

59. (c) $N_2O_4(g) \rightleftharpoons 2NO_2(g)$

	N_2O_4	NO_2
$t=0$	0.1 mole	–
$t = eq.$	$(0.1-x)$	$(2x)$

$$K_p = \frac{(P_{NO_2})^2}{(P_{N_2O_4})^1}$$

$$0.14 = \frac{\left(\frac{2x}{0.1+x} \times 1\right)^2}{\left(\frac{0.1-x}{0.1+x} \times 1\right)^1}$$

60. (c) $2SO_2 + O_2 \rightleftharpoons 2SO_3$

	SO_2	O_2	SO_3
$t=0$	5 mole	5 mole	
$t_{eq.}$	$5-2x$	$5-x$	$2x$

Given : $(2x) = \frac{60}{100}(5)$

$x = 1.5$

Now $P_{Gas} = X_{Gas} \cdot P_{Total}$

61. (a) Similar to Q. no. 60.

62. (d) $PCl_5 \rightleftharpoons PCl_3 + Cl_2$

	PCl_5	PCl_3	Cl_2
$t=0$	$0.4M$	(x)	(x)
$t_{eq.}$	$(0.4-x)$		

$$x = \frac{35}{100}(0.4) = 0.14$$

Now $K_c = \frac{[PCl_3][Cl_2]}{[PCl_5]} = \frac{0.14 \times 0.14}{0.26}$

$$K'_c = \frac{1}{K_c} = 13.3$$

63. (a) $2AB_2(g) \rightleftharpoons 2AB(g) + B_2(g)$

	AB_2	AB	B_2
$t=0$	P		
$t_{eq.}$	$P - P\alpha$	$P\alpha$	$\frac{P\alpha}{2}$

$$K_p = \frac{(P_{AB})^2(P_{B_2})^1}{(P_{AB_2})^2}$$

64. (a) $H_2(g) + CO_2(g) \rightleftharpoons CO(g) + H_2O(g)$

	H_2	CO_2	CO	H_2O
$t=0$	$1M$	$1M$		
$t_{eq.}$	$(1-x)$	$(1-x)$	(x)	(x)

$\because \ \Delta n_g = 0$

$$\Rightarrow \quad K_p = K_c = \frac{x^2}{(1-x)^2}$$

65. (c) $A_n(g) \rightleftharpoons n.A(g)$

$t=0$ a mole

$t_{eq.}$ $(a-a\alpha)$ $na\alpha$

$$M_0 = \frac{a.M_T}{(a-a\alpha+na\alpha)}$$

$$\Rightarrow \alpha = \frac{M_T - M_0}{(n-1)\cdot M_0}$$

$\because$ $D = M/2$

$$\Rightarrow \alpha = \frac{D_T - D_0}{(n-1)\cdot D_0}$$

66. (a) For $PCl_5(g) \rightleftharpoons PCl_3(g) + Cl_2(g)$

$n = 2$

$$\alpha = \frac{D-d}{(2-1)\cdot d}, \quad \alpha = \frac{D}{d} - 1$$

If $\left(\frac{D}{d}\right) \longrightarrow$ at x-axis and $(\alpha) \longrightarrow$ at y-axis

then it is a straight line positive slope and negative intercept.

67. (a)
$$\alpha = \frac{M_T - M_O}{(n-1)M_O}$$

$$\Rightarrow 0.2 = \frac{92 - M_O}{M_O}$$

$$\Rightarrow M_O = 76.66$$

$$d_{mixture} = \frac{PM_{mixture}}{RT}$$

$$= \frac{1 \times 76.66}{0.821 \times 300} = 3.11 \text{ g/litre}$$

68. (c)
$$M_O = \frac{dRT}{P} = 57.47$$

$$\alpha = \frac{99 - 57.47}{57.47} = 0.72$$

69. (d) $I_2(g) \rightleftharpoons 2I(g)$

$t=0$ a mole

$t_{eq.}$ $(a-a\alpha)$ $(2a\alpha)$

$$P_{Gas} = x_{Gas} \cdot P_{Total}$$

$$K_p = \frac{\left(\frac{2\alpha}{1+\alpha}\cdot P_{Total}\right)^2}{\left(\frac{1-\alpha}{1+\alpha}\cdot P_{Total}\right)^1}$$

K_p will be same in both cases.

70. (a) $A_2(g) + B_2(g) \rightleftharpoons 2AB(g)$

	A_2	B_2	AB
Initial conc.	10/2	15/2	5/2
at eqm	$5-x$	$7.5-x$	$2.5+2x$

$\because$ $2.5 + 2x = 7.5$

$\Rightarrow$ $x = 2.5$

$$K_c = \frac{(7.5)^2}{2.5 \times 5} = 4.5$$

71. (a) $H_2(g) + S(s) \rightleftharpoons H_2S(g)$

Conc. at eqm $0.5-x$ — x

$$K_c = \frac{[H_2S]}{[H_2]}$$

$$\Rightarrow 7 \times 10^{-2} = \frac{x}{0.5 - x}$$

$$\Rightarrow x = 0.0327$$

$$P_{H_2S} = \left(\frac{n_{H_2S}}{V}\right) RT$$

$$= 0.0327 \times 0.0821 \times 360 = 0.966$$

72. (c) The gaseous mixture contains 40% Cl_2 and 40% PCl_3, since they are produced in 1 : 1 mole ratio. The PCl_5 % is 20.

For ideal gases mole % $\equiv$ volume %

$$P_{Cl_2} = P_{PCl_3}$$

$$\Rightarrow 2 \times 0.40 = 0.80 \text{ atm}$$

$$P_{PCl_5} = 2 \times 0.2 = 0.40 \text{ atm}$$

$$\therefore K_p = \frac{P_{PCl_3} \cdot P_{Cl_2}}{P_{PCl_5}}$$

$$= \frac{0.80 \times 0.80}{0.40} = 1.6 \text{ atm}$$

73. (a) For an ideal gas mole % $\equiv$ volume %

$$K_p = \frac{P^2_{H_2O}}{P^2_{H_2}} = \frac{(0.55)^2}{(0.45)^2} = 1.49$$

74. (d) Moles of CO_2 present at equilibrium

$$= \frac{1.642 \times 50}{0.0821 \times 1000} = 1$$

Mole % of XCO_3 decomposed

$$= \frac{1}{4} \times 100 = 25\%$$

Hence, 75% remains undecomposed.

75. (b) Let initial moles of $H_2(g)$ is 1

$Fe_2O_3(s) + 3H_2(g) \rightleftharpoons 2Fe(s) + 3H_2O(g)$

at eqm — $1-3x$ — $3x$

$$K_c = \frac{\left(\frac{3x}{V}\right)^3}{\left(\frac{1-3x}{V}\right)^3}$$

$$\Rightarrow 8 = \left(\frac{3x}{1-3x}\right)^3$$

$$\Rightarrow x = 0.22$$

% of H_2 unreacted

$$= \frac{1 - 3 \times 0.22}{1} \times 100 = 34$$

76. (c) $AB_3(g) \rightleftharpoons AB_2(g) + \frac{1}{2}B_2(g)$

At eqm $800-x$ x $x/2$

$$800 - x + x + \frac{x}{2} = 900 \Rightarrow x = 200$$

$$\% \text{ dissociation} = \frac{200}{800} \times 100 = 25$$

77. (b) Let P is initial pressure of NO_2

$$2NO_2(g) \rightleftharpoons 2NO(g) + O_2(g)$$

At eqm $P - 2x$ $\quad 2x \quad x$

as per given $x = 0.25$

$$K_p = \frac{(2x)^2 (x)}{(P - 2x)^2}$$

$$\Rightarrow \quad 156.25 = \frac{(0.5)^2 (0.25)}{P^2_{NO_2}}$$

$$\Rightarrow \quad P_{NO_2} = 0.02$$

78. (b) $\quad 2NOCl(g) \rightleftharpoons 2NO(g) + Cl_2(g)$

At eqm $P - 2x$ $\quad 2x \quad x$

$P + x = 1; \quad P - 2x = 0.64, \quad x = 0.12$

$$K_p = \frac{(0.24)^2 (0.12)}{(0.64)^2}$$

$$= 16.875 \times 10^{-3} \text{ atm}$$

79. (c) $Q_c = \dfrac{1 \times 1}{1 \times 1} = 1$

$\because$ $Q_c > K_c$ so reaction will proceed in backward direction

$$A_2(g) + B_2(g) \rightleftharpoons C_2(g) + D_2(g)$$

Conc eqm $\frac{1+x}{10} \quad \frac{1+x}{10} \quad \frac{1-x}{10} \quad \frac{1-x}{10}$

$$0.25 = \frac{\left(\frac{1-x}{10}\right)^2}{\left(\frac{1+x}{10}\right)^2} \Rightarrow x = 0.333$$

$$[A_2(g)] = \frac{1+x}{10} = \frac{1.333}{10} = 0.133$$

80. (d) Given : $M_{Mix} = 62 \cdot M_{H_2} = 124$

For $PCl_5 \rightleftharpoons PCl_3 + Cl_2$

$$\alpha = \frac{(M)_{PCl_5} - (M)_{Mix}}{(2-1) \cdot (M)_{Mix}}$$

81. (b) Let initial mole of N_2O_4 is 1

$$N_2O_4(g) \rightleftharpoons 2NO_2(g)$$

Mole at eqm $\quad 1 - \alpha \quad 2\alpha$

Total moles at eqm $= 1 + \alpha$

$$K_p = \frac{P^2_{NO_2}}{P_{N_2O_4}} = \frac{\left(\frac{2\alpha}{(1+\alpha)} \times P\right)^2}{\frac{(1-\alpha)}{(1+\alpha)} \times P} = \frac{4\alpha^2 P}{1 - \alpha^2}$$

$$K_p - K_p\alpha^2 = 4\alpha^2 P \Rightarrow \alpha^2(4P + K_p) = K_p$$

$$\alpha = \sqrt{\frac{K_p}{K_p + 4P}}$$

82. (b) $K_c = \dfrac{[NO_2]^2}{[N_2O_4]}$

If $[NO_2] \rightarrow$ at y-axis

$[N_2O_4] \rightarrow$ at x-axis

$$\Rightarrow \quad y = Kc \cdot x^2$$

Parabolic curve

83. (d) $K_c = [Hg(g)]; P = CRT$

$$C = \left(\frac{0.002}{760}\right) \times \frac{1}{(0.0821 \times 300)}$$

$$= 1.068 \times 10^{-7} \ M$$

84. (b) $K_c = \dfrac{n^2_{AB_2}}{n_{A_2} \cdot n^2_{B_2}} \times V = \dfrac{n^2_{AB_2}}{n_{A_2} \cdot n^2_{B_2}} \times \left(\dfrac{n_{total} RT}{P}\right)$

$$= \frac{2^2}{5 \times 3^2} \times \left(\frac{10 \times 0.0821 \times 300}{8.21}\right)$$

85. (b) $\quad A(g) \rightleftharpoons B(g) + C(g)$

$t = 0 \quad a$ mole

$t_{eq.} \quad (a - a\alpha) \quad (a\alpha) \quad (a\alpha)$

if $\quad p_{Total} = P_1$

then, $\quad K_{p_1} = \dfrac{\left(\frac{\alpha}{1+\alpha} p_1\right)\left(\frac{\alpha}{1+\alpha} \cdot p_1\right)}{\left(\frac{1-\alpha}{1+\alpha} \cdot p_1\right)}$

also $\quad X(g) \rightleftharpoons 2Y(g)$

$t = 0 \quad b$ mole

$t_{eq.} \quad (b - b\alpha) \quad (2b\alpha)$

if $p_{Total} = p_2$

$$K_{p_2} = \frac{\left(\frac{2\alpha}{1+\alpha} \cdot p_2\right)^2}{\left(\frac{1-\alpha}{1+\alpha} \cdot p_2\right)^1}$$

Given : $\dfrac{K_{p_1}}{K_{p_2}} = \dfrac{1}{9}$

86. (b) Addition of inert gas at constant pressure will increase volume so reaction moves in the direction having more gaseous moles.

87. (a) Volume change has no effect for $\Delta n_g = 0$.

88, 89 (a, b)

Volume increase leads the reaction to move towards more gaseous moles.

90. (c) Addition of inert gas at constant volume has no effect.

91. (a) $Q = 1 \text{ atm} > K_p$

92. (b) For exothermic reaction yield $\uparrow$ on temperature $\downarrow$

i.e., $\quad T_1 > T_2 > T_3$

93. (a) $\quad N_2(g) + 3H_2(g) \rightleftharpoons 2NH_3(g)$

$(t_{eq})_1 \quad 0.05M \quad 0.3M \quad 0.2M \ ; \quad K_c = \dfrac{800}{27}$

$(t_{eq})_2 \quad y \quad 0.3 - 3x \quad 0.2 + 2x = 0.24$

$x = 0.02 \ M \qquad [H_2] = 0.3 - 3x = 0.24 \ M$

$$K_c = \frac{800}{27} = \frac{(0.24)^2}{y(0.24)^3} \quad ; \quad y = 0.14 \ M$$

So graph (a) is correct.

94. (c) Similar to Q. No. 90.

95. (b) Le-Chatelier's principle is not applicable in solid phase reaction.

96. (d) When pressure decreases Eq. shift toward more gaseous moles.

97. (d) When temperature decreases Eq. shift toward exothermic direction.

98-99 (b,c)

Forward reaction must be exothermic with more gaseous moles at product side.

100. (a) Solid phase not involved in equilibrium constant/reaction quotient, addition of solid not disturbed position of equilibrium.

101. (d) When volume decreases Eq. shift toward less gaseous moles.

103. (b) For O_3 to be stable the reaction shouldn't be favourable in forward direction.

104. (d) Similar to Q. No. 101.

105. (c) Temperature ↓ favours exothermic reaction.

106. (d) Graphite has lesser density

$\Rightarrow$ more volume

$\Rightarrow$ low pressure

also for exothermic $\Rightarrow$ low temperature

107. (b) Molar volume of $H_2O(s)$ is more than $H_2O(l)$ so on increasing pressure $\Rightarrow$ formation of $H_2O(l)$.

110. (d) Pressure ↓ $\Rightarrow$ towards more volume.

111. (c) $\because \quad \ln K_p = \left(-\frac{\Delta H^0}{R}\right)\left(\frac{1}{T}\right)+\left(\frac{\Delta S^0}{R}\right)$

112. (b) For endothermic; $\Delta H > 0$, so slope is negative.

113. (a) For exothermic; $\Delta H < 0$, so slope is positive.

115. (a) For dissociation reaction; $\Delta H > 0$

116. (c) $\ln \frac{K_{P_2}}{K_{P_1}} = \frac{\Delta H^\circ}{R}\left[\frac{1}{T_1} - \frac{1}{T_2}\right]$

$$= \ln\left(\frac{10^{-4}}{10^{-6}}\right) = \frac{\Delta H^\circ}{8.314}\left(\frac{25}{298 \times 323}\right)$$

$$\Delta_r H^\circ = 147.41 \text{ kJ/mol}$$

117-118 (b, a)

See Q. No. 116

119. (a) The value of K_{eq} is the measure of extent of reaction. For most stable oxide $\Rightarrow K_{eq}$ should be minimum.

120-125. (c, c, a, b, d, a)

$\Delta G^\circ = -RT \ln K_{eq}$

$\Delta G^\circ = (\Delta H^\circ - T\Delta S^\circ) = -RT \ln K_{eq}$

123. (b) $\Delta_r G^\circ = \Delta_r H^\circ - T \cdot \Delta_r S^\circ$

$= -30 - 300 \times 0.1 = 0$

$\Delta_r G^\circ = -2.303 \, RT \log K$

$K = 1$

124. (d) $P_{total} = 0.2$ bar

At equilibrium $P_{CO_2} = P_{H_2O} = 0.1$ bar

$K_p = (0.1)^2 = 0.01$

$\Delta_r G^\circ = -RT \ln K_p$

$= -8.314 \times 420 \ln (0.01)$

$= 16083.6$ J/mol or 16.083 kJ/mol

Level 2

1. (a) $C(s) + CO_2(g) \rightleftharpoons 2CO(g)$; ...(1)

$K_{P_1} = 10^{14}$ atm

$2CO(g) + 2Cl_2(g) \rightleftharpoons 2COCl_2(g)$; ...(2)

$K_{P_2} = (6 \times 10^{-3})^2 \text{ atm}^{-2}$

Add (1) and (2)

$C(s) + CO_2(g) + 2Cl_2(g) \rightleftharpoons 2COCl_2(g)$;

$K_p = 10^{14} \times 36 \times 10^{-6}$

$= 36 \times 10^8$

For given reaction $\Delta n_g = -1$

$\therefore \quad K_c = K_p(RT)$

$= 36 \times 10^8 \times 0.0821 \times 1120$

$K_c = 3.31 \times 10^{11} \, M^{-1}$

2. (c)

	$N_2(g)$	+ $3H_2(g)$	$\rightleftharpoons$ $2NH_3(g)$
Initial moles	1	2	0
at eqm	$1-x$	$2-3x$	$2x$

where $x = 0.5$

$$K_c = \frac{(1/4)^2}{\left(\frac{0.5}{4}\right)\left(\frac{0.5}{4}\right)^3} = 256$$

equilibrium constant for the reaction

$$NH_3(g) \rightleftharpoons \frac{1}{2}N_2(g) + \frac{3}{2}H_2(g)$$

$$K_c^1 = \frac{1}{\sqrt{K_c}} = \frac{1}{16}$$

3. (c)

	$A_2(g)$ +	$B_2(g)$ $\rightleftharpoons$	$2AB(g)$
Moles at eqm	$2-x$	$4-x$	$2x$

$$K_c = \frac{4x^2}{(2-x)(4-x)}$$

$\Rightarrow \quad x = \frac{32}{24} = 1.33$ mole

$[AB(g)] = \frac{2 \times 1.33}{4} = 0.66 \, M$

4. (d) $P_{total} = P_{HNO_3} + P_{NO_2} + P_{H_2O} + P_{O_2}$

$\because \quad P_{NO_2} = 4P_{O_2}$ and $P_{H_2O} = 2P_{O_2}$

$\therefore \quad P_{total} = P_{HNO_3} + 7P_{O_2}$

$\Rightarrow \quad 30 - 2 = P_{O_2} \times 7$

$\Rightarrow \quad P_{O_2} = \frac{28}{7} = 4$

$$K_p = \frac{P_{NO_2}^4 \cdot P_{H_2O} \cdot P_{O_2}}{P_{HNO_3}^4}$$

$$= \frac{(4 \times 4)^4 \times (2 \times 4)^2 \times 4}{2^4} = 2^{20}$$

$K_p = K_c (RT)^{\Delta n_g} = K_c (0.08 \times 400)^3$

$$\Rightarrow \quad K_c = \frac{2^{20}}{(32)^3} = 32$$

5. (d) $LiCl \cdot 3NH_3(s) \rightleftharpoons LiCl \cdot NH_3(s) + 2NH_3(g)$;

$$[K_p = 9 \text{ atm}^2]$$

Therefore,

$$LiCl \cdot NH_3(s) + 2NH_3(g) \rightleftharpoons LiCl \cdot 3NH_3(s);$$

Initial moles	0.1	a	0
Final moles at eqm	0	$(a-0.2)$	0.1

$$\left[K'_p = \frac{1}{9} (\text{atm})^{-2}\right]$$

Let initial moles of NH_3 are a

At eqm $K'_p = \frac{1}{(P'_{NH_3})^2}$ or $\frac{1}{9} = \frac{1}{(P'_{NH_3})^2}$

$\therefore \quad P'_{NH_3} = 3$ atm

$\because \quad PV = nRT \Rightarrow 3 \times 5 = n \times 0.0820 \times 310$

$\therefore \quad n = 0.59$ *i.e.*, $(a - 0.2) = 0.59$

$\therefore$ initial moles of $NH_3 = 0.79$

6. (c) $NH_2COONH_4(s) \rightleftharpoons 2NH_3(g) + CO_2(g)$

Let partial pressure at equilibrium of CO_2 be P, then $P_{NH_3} = 2P$ and total pressure at equilibrium $= 3P$

$$K_p = (2P)^2 \times P = 4P^3 \quad \ldots(1)$$

If NH_3 is added and the pressure of NH_3 after addition at equilibrium is $3P$

$$K_p = 4P^3 = (P'_{NH_3})^2 \times P'_{CO_2}$$

$$\Rightarrow \quad P'_{CO_2} = \frac{4}{9} P$$

$$\therefore \text{ ratio of } \frac{P'_{CO_2}}{P_{CO_2}} = \frac{4}{9}$$

7. (c) $C_2H_6(g) \rightleftharpoons C_2H_4(g) + H_2(g)$

At eqm	$1-x$	x	x

$$K_p = \frac{P_{C_2H_4} \cdot P_{H_2}}{P_{C_2H_6}} \Rightarrow \frac{x^2}{1-x} = 5 \times 10^{-2}$$

$$x^2 + 0.05\,x - 0.05 = 0$$

$$x = \frac{-0.05 + \sqrt{(0.05)^2 + 4 \times 0.05}}{2} = 0.20 \text{ atm}$$

Partial pressure of C_2H_6 = mole fraction × total pressure

$\Rightarrow \quad 0.80 = \text{mole fraction} \times 1.2$

$$\therefore \text{ mole \% of } C_2H_6 = \frac{0.8}{1.2} \times 100 = 66.66$$

8. (a) $2NOBr(g) \rightleftharpoons 2NO(g) + Br_2(g)$

Initial pressure	P	0	0
at eqm	$P-2x$	$2x$	x

where $2x = 0.40\ P$

$\Rightarrow \quad x = 0.20\ P$;

$\Rightarrow \quad 1.20\ P = 0.30$

$\Rightarrow \quad P = 0.25$ atm

$$K_p = \frac{P_{NO}^2 \cdot P_{Br_2}}{P_{NOBr}^2}$$

$$= \frac{(0.4P)^2 (0.2P)}{(0.6P)^2} = 0.0222$$

$\therefore K_p$ for the reaction

$$2NO(g) + Br_2(g) \rightleftharpoons 2NOBr(g)$$

is $\frac{1}{0.0222} = 45$

9. (c) $\because$ Initially only A is present so at eqm B and C should be present in 2 : 1

$$2A(g) \rightleftharpoons 2B(g) + C(g)$$

At eqm	400 mL	200 mL	100 mL

for ideal gases volume % ≡ mole %

$$K_p = \frac{\left(\frac{200}{700} \times 10\right)^2 \left(\frac{100}{700} \times 10\right)}{\left(\frac{400}{700} \times 10\right)^2} = \frac{10}{28}$$

10. (c) $SO_2(g) + NO_2(g) \rightleftharpoons SO_3(g) + NO(g)$

Initial moles	2	2	2	2
at eqm	$2-x$	$2-x$	$2+x$	$2+x$

$(\because\ Q_p < K_p)$

Total no. of moles of gases at equilibrium

$$= 8 + 2 = 10$$

$$K_p = \frac{P_{SO_3} \cdot P_{NO}}{P_{SO_2} \cdot P_{NO_2}}$$

$$\Rightarrow \quad 25 = \frac{\left(\frac{2+x}{10} \times P\right)^2}{\left(\frac{2-x}{10} \times P\right)^2}$$

$$\Rightarrow \quad 5 = \frac{2+x}{2-x}$$

$$\Rightarrow \quad x = 1.33$$

Partial pressure of $NO_2 = \frac{2-x}{10} \times P_{total}$

$$= \frac{0.666}{10} \times 2$$

$$= 0.133 \text{ atm}$$

11. (b) $3\,Se_2(g) \rightleftharpoons Se_6(g)$

moles at eqm	$a(1-\alpha)$	$\frac{a\alpha}{3}$

$$\frac{\text{Observed no. of moles at equilibrium}}{\text{Initial no. of moles of reactant}} = \frac{M_T}{M_O},$$

$$\Rightarrow \quad \frac{a(1-\alpha) + \frac{a\alpha}{3}}{a} = \frac{M_T}{M_O} = \alpha = 0.315$$

where molar mass of Se_2

$$(M_T) = 79 \times 2 = 158$$

and molar mass of mixture

$$(M_O) = \frac{wRT}{PV} = \frac{0.02 \times 24.63}{1 \times 2.463 \times 10^{-3}} = 200$$

12. (d) $6HCHO \rightleftharpoons C_6H_{12}O_6$

Conc. at eqm	$C(1-\alpha)$	$\frac{C\alpha}{6}$

$$\frac{\text{Observed molar concentration}}{\text{Initial molar concentration}} = \frac{M_T}{M_O}$$

M_T = Theoretical molar mass of HCHO

$$\frac{C(1-\alpha)+\frac{C\alpha}{6}}{C} = \frac{M_T}{M_O} = \frac{30}{150}$$

$\Rightarrow \quad \alpha = 0.96$

13. (b) $2SO_2(g) + O_2(g) \rightleftharpoons 2SO_3(g)$

$$K_c = \frac{(0.12)^2}{(0.12)^2 \times 5} = 0.2$$

Another vessel

$$2SO_2(g) + O_2(g) \rightleftharpoons 2SO_3(g)$$

moles at eqm $\quad 0.5-2x \quad y-x \quad 2x$

as per given $2x = \frac{20}{100} \times 0.5 = 0.1$

$$K_c = \frac{(0.1)^2}{(0.4)^2(y-0.05)} = 0.20$$

$y = 0.3625$ mole

$\therefore$ mass of O_2 added $= 11.6$ g

14. (b) Let initial mole of $N_2O_4(g)$ is one

$$N_2O_4(g) \rightleftharpoons 2NO_2(g)$$

Initial moles $\quad 1 \quad 0$

At equilibrium $\quad 1-\alpha \quad 2\alpha$

Total no. of moles at equilibrium $= 1+\alpha$

$$P_{N_2O_4} = \frac{1-\alpha}{1+\alpha} \times P$$

$$P_{NO_2} = \frac{2\alpha}{(1+\alpha)} \times P$$

Hence, $\quad K_p = \frac{P^2_{NO_2}}{P_{N_2O_4}} = \frac{4\alpha^2}{(1-\alpha^2)} \times P$

$\therefore \quad 4.5 = \frac{4\alpha^2}{(1-\alpha^2)} \times 2$

$\Rightarrow \quad \alpha = 0.6$

Mole fraction of N_2O_4:

$$X_{N_2O_4} = \frac{1-\alpha}{1+\alpha} = 0.25$$

$\therefore \quad X_{NO_2} = 0.75$

Average molar mass of mixture

$$= 0.25 \times 92 + 0.75 \times 46 = 57.5$$

15. (a) $\quad 2AB(s) \rightleftharpoons A_2(g) + B_2(g)$

$\quad 0.5+x \quad x$

$$K_p = P_{A_2} \cdot P_{B_2}$$

$\Rightarrow \quad 0.06 = (0.5+x)\,x$

$\quad = x^2 + 0.5x - 0.06$

$\Rightarrow \quad x = 0.1$

$P_{\text{total}} = P_{A_2} + P_{B_2} = 0.6 + 0.1 = 0.70$ atm

16. (a) $Sb_2S_3(s) + 3H_2(g) \longrightarrow 2Sb(s) + 3H_2S(g)$

$\quad 0.01-x \quad 0.01-3x \quad 2x \quad 3x$

where $3x = 0.005$

$$H_2S + Pb^{2+} \longrightarrow PbS + 2H^+$$

no. of moles of PbS formed

$$= \frac{1.19}{238} = 0.005 \text{ mole}$$

At eqm $[H_2] = \left[\frac{0.005}{250}\right] = \left(\frac{.01-.005}{250}\right)$;

$$K_c = \left(\frac{0.005}{0.005}\right)^3 = 1$$

17. (a) $\quad 2A(g) + B(g) \rightleftharpoons C(g) + D(g)$

Initial moles $\quad 1 \quad 1 \quad 7 \quad 3$

at eqm $\quad 2-2x \quad 1-x \quad 7+x \quad 3+x$

Due to very high value of K_c we can assume that reactant is almost converted into products so

$1-x = y; \quad 2-2x = 2y \Rightarrow \quad x \simeq 1$

$$10^{12} = \frac{8 \times 4}{(2y)^2 y}$$

$\Rightarrow \quad y^3 = 8 \times 10^{-12}$

$\therefore$ equilibrium concentration of

$$A = 2y = 4 \times 10^{-4}$$

18. (b) $K_c = \frac{[\text{complex}]}{[H_3BO_3]\,[\text{glycerine}]} = 0.90$

$$\Rightarrow \quad \frac{[\text{complex}]}{[H_3BO_3]} = \frac{80}{20}$$

$\therefore \quad [\text{glycerine}] = \frac{80}{20 \times 0.9} = 4.44\ M$

$$H_3BO_3 + \text{glycerine} \rightleftharpoons (H_3BO_3 - \text{glycerine})$$

At eq^m $\quad 0.1-x \quad a-x \quad x$

$$x = 0.1 \times \frac{80}{100} = 0.08;$$

$$a - x = 4.44$$

$\Rightarrow \quad a = 4.44 + 0.08 = 4.52\ M$

or Initial moles = 4.52

19. (c) $\frac{r_{\text{mix}}}{r_{O_2}} = \sqrt{\frac{32}{M_{\text{mix}}}} = 0.4\sqrt{5}$

$\Rightarrow \quad M_{\text{mix}} = 40$ g/mol

$$O_2(g) \rightleftharpoons \frac{2}{3}O_3(g)$$

Initial moles $\quad 1 \quad 0$

at eqm $\quad 1-\alpha \quad \frac{2}{3}\alpha$

$$\frac{32}{40} = \frac{1-\alpha+\frac{2}{3}\alpha}{1}$$

$\Rightarrow \quad \alpha = 0.6$

$\therefore \quad \%\,\alpha = 60$

20. (c) $2SO_3(g) \rightleftharpoons 2SO_2(g) + O_2(g)$

moles at eqm $1-2x$ $\quad 2x \quad x$

Only SO_2 (O. No. = 4) will oxidized

so equivalent of SO_2 = equivalent of $KMnO_4$

$$2x \times 2 = 0.2 \times 5$$

$$\Rightarrow \quad 2x = 0.5$$

$$K_c = \frac{\left(\frac{0.5}{2}\right)^2\left(\frac{0.25}{2}\right)}{\left(\frac{0.5}{2}\right)^2} = 0.125$$

21. (d) $$\frac{r_{mix}}{r_{SO_2}} = \sqrt{\frac{M_{SO_2}}{M_{mix}}}$$

$$\Rightarrow \quad 2.56 = \frac{64}{M_{mix}}$$

$$\Rightarrow \quad M_{mixture} = 25$$

Let mole fraction of F_2 is x

$$25 = \frac{38 \times x + (1-x) \times 19}{1}$$

$$x = 0.315;$$

$$\Rightarrow \quad K_p = \frac{P_F^2}{P_{F_2}} = \frac{(0.685\,P)^2}{(0.315\,P)} \simeq 1.49\text{atm}$$

22. (b) $8 \times 10^{-6} = \frac{x^2}{0.5}$

$$\Rightarrow \quad x = 2 \times 10^{-3}$$

$\therefore$ $\quad$ pOH = 2.7

So, $\quad$ pH = 11.3

23. (c) $\Delta_r G^\circ = 0 - 77.1 \times 2 = -154.2$ kJ/mol

$$Q = \frac{[H^+]^2}{P_{H_2} \cdot [Ag^+]^2}$$

$$= \frac{10^{-6}}{0.5 \times (10^{-10})} = 2 \times 10^4$$

$$\Delta G = \Delta_r G^\circ + RT \ln Q$$

$$\Delta_r G = -154.2 + \frac{8.314 \times 300 \ln(2 \times 10^4)}{1000}$$

$$= -129.5 \text{ kJ/mol}$$

24. (d) $N_2O_5 \rightleftharpoons N_2O_3 + O_2$

$4-x \quad\quad x-y \quad\quad x+y$

$N_2O_3 \rightleftharpoons N_2O + O_2$

$x-y \quad\quad y \quad\quad y+x$

$\because$ $[O_2] = x + y = 2.5$

and $\quad 2.5 = \frac{(x+y)(x-y)}{4-x}$

$\therefore$ $\quad x = 2.166$

$[N_2O_5] = 4 - x = 1.846$

25. (b) Let x is partial pressure of A and y is partial pressure of C when both equilibrium simultaneously established in a vessel

$X(s) \rightleftharpoons A(g) + 2B(g)$

$x \quad\quad (2x+2y)$;

$$K_{P_1} = P_A \cdot P_{B(\text{total})}^2$$

$Y(s) \rightleftharpoons C(g) + 2B(g)$

$y \quad\quad (2y+2x)$;

$$K_{P_2} = P_C \cdot P_{B(\text{total})}^2$$

$$\frac{K_{P_1}}{K_{P_2}} = \frac{x}{y} \quad \Rightarrow \quad x = 2y$$

$$K_{P_1} = x(2x+2y)^2$$

$\Rightarrow \quad x = 0.1$ atm;

$\therefore \quad y = 0.05$ atm

Total pressure of gases $= P_A + P_B + P_C$

$= 3(x+y)$

$= 0.45$ atm.

Level 3

Passage-2 :

1,2,3. (a, b, b)

$$\ln K_p = \left(-\frac{\Delta H^\circ}{R}\right)\left(\frac{1}{T}\right) + \left(\frac{\Delta S^\circ}{R}\right)$$

4. (a) $\ln\frac{K_2}{K_1} = \frac{\Delta H}{R}\left(\frac{1}{T_1} - \frac{1}{T_2}\right)$

Passage-3

$N_2O_3(g) \rightleftharpoons NO_2(g) + NO(g)$; $K_{P_1} = 2.5$ bar

$2NO_2(g) \rightleftharpoons N_2O_4(g)$; K_{P_2}

$x - y \quad\quad y/2$

$P_{NO} = x = 1.5$ b

$$\therefore \quad K_{P_1} = \frac{P_{NO} \cdot P_{NO_2}}{P_{N_2O_3}}$$

$$2.5 = \frac{x(x-y)}{2-x} = \frac{1.5\,(1.5-y)}{2-1.5} = 4.5 - 3y$$

$$\Rightarrow \quad y = \frac{2}{3}\text{ bar}$$

(1) $P_{N_2O_3} = 2 - x = 2 - 0.5 = 0.5$ b

(2) $P_{NO_2} = x - y = 1.5 - \frac{2}{3} = \frac{2.5}{3}$ b

(3) $K_{P_2} = \frac{y}{2(x-y)^2} = \frac{2/3}{2\left(\frac{2.5}{3}\right)^2} = 0.48\text{ bar}^{-1}$

Passage-4 :

1. (b) For $X(g) \xrightleftharpoons{H_2O(l)} X_{(aq.)}$; $\Delta H < 0$

For more solubility ⇒ forward shift

$\Rightarrow$ low temperature (exothermic)

$\Rightarrow$ high pressure (less volume)

2. (a) For $Fe(l) \rightleftharpoons Fe(s)$

i.e., $\Delta H < 0 \Rightarrow$ low temperature

less volume $\Rightarrow$ high pressure

One or More Answers is/are Correct

6. (c, d) Le Chatelier's principle is not quantitative. If both stress would cause the same direction of shift, the shift is determinable. If the two stresses would cause shifts in opposite directions, no deduction is possible.

Level 4

Subjective Problems

1. $\frac{1}{2}C(s)+\frac{1}{2}CO_2(g) \rightleftharpoons CO(g)$

$t=0$ $\quad P$ atm

$t_{eq.}$ $\quad (P - P/2) \quad P$

$P/2+P=12$

$P=8$ atm

$$K_p=\frac{(p_{CO})^1}{(p_{CO_2})^{1/2}}$$

2. $A(s) \rightleftharpoons B(g)+2C(g)$

$t=0 \quad a$

$t_{eq.} \quad a-p \quad p \quad 2p$

$K_p=(P_B)^1(P_C)^2$

$32=(P)^1(2P)^2$

$\Rightarrow P=2$ atm

3. $A + 2B \rightleftharpoons 2C + D$

$t=0 \quad 2a \quad 3a$

$t_{eq} \quad (2a-x) \; (3a-2x) \quad (2x) \quad (x)$

Given : $(2a-x)=x$

$\Rightarrow \quad x=a$

$$K_c=\frac{[C]^2[D]^2}{[A]^1[B]^2}$$

4. $D \rightleftharpoons C;\; K_C=\left(\frac{1}{0.01}\right)$

$B \rightleftharpoons A;\; K_C=\left(\frac{1}{10}\right)$

$C \rightleftharpoons B;\; K_C=\left(\frac{1}{2}\right)$

Addition

$D \rightleftharpoons A;\; K_C=\left(\frac{1}{0.01}\times\frac{1}{10}\times\frac{1}{2}\right)$

5. $A(g) \rightleftharpoons B(g)$

$(t_{eq.})_1 \quad 0.4M \quad 0.4M$

$K_C=1$

$\because K_C=$ remains constant

6. $K_P=k_C(RT)^{\Delta n_g}$

7. $SO_2 + NO_2 \rightleftharpoons SO_3 + NO$

$t=0$ $\quad$ 3 mole $\quad 4 \quad 1 \quad 4$

$t_{eq.} \quad (3-x) \quad (4-x) \quad (1+x) \quad (4+x)$

Given : $(3-x)=1$

$\Rightarrow x=2$

$$K_C=\frac{[SO_3][NO]}{[SO_2][NO_2]}$$

8. For mixture

$P.M_{Mix}=d.RT$ and for $N_2O_4 \rightleftharpoons 2NO_2$

$$\alpha=\frac{(M)_{N_2O_4}-(M)_{Mix}}{(2-1)\cdot(M)_{Mix}}$$

9. $\Delta G°=-RT\ln K_{eq}$

At standard states all concentration and partial pressure are unity.

$\Rightarrow K_{eq}=1$

10. $\Delta G°=-RT\ln K_{eq}$

$-830=-2\times300\times2303\log_{10}K_{eq}$

$\Rightarrow K_{eq}=4$

Now

$A + B \rightleftharpoons C + D$

$t=0 \quad a \quad a$

$t_{eq.} \quad (a-x) \quad (a-x) \quad (x) \quad (x)$

$$K_C=\frac{[C][D]}{[A][B]}$$

$\Rightarrow x=2a/3$

11. $NH_4HS(s) \rightleftharpoons NH_3(g) + H_2S(g)$

At eqm. $\quad - \quad (0.1+P) \quad P \quad ;$

As per given $0.1+2P=1.1;\quad P=0.5$

$K_P=P_{NH_3}\cdot P_{H_2S}=(0.1+P)\times P$

$=0.6\times0.5=0.3=3\times10^{-1}=3$

12. $A + nB \rightleftharpoons mC$

$t=0 \quad 8M \quad 12M$

$t=t_{eq.} \quad 8-x \quad 12-nx \quad mx$

Given: $[A]_{eq}=6M$

$8-x=6$

$x=2M$

$[B]_{eq}=8M$

$12-nx=8$

$n(2)=4$

$n=2$

$[C]_{eq}=6M$

$mx=6$

$m(2)=6$

$m=3$

$\therefore \quad m+n=3+2=5$

Numerical Problems

1. $PV = nRT$

$$n = \frac{1.642 \times 5}{1000 \times 0.0821}$$

Mole of $CO_2 = 0.1$

Mole of $CaO(s) = 0.1$

Wt. of $CaO = 5.6\,g$

2. $PCl_5 \rightleftharpoons PCl_3 + Cl_2$

1	0	0
1 – 0.5	0.5	0.5

$$\frac{P_1}{P_2} = \frac{n_1 T_1}{n_2 T_2}$$

$$\Rightarrow \quad P_2 = 2 \times \frac{1.5 \times 600}{1 \times 300} = 6$$

3. $2A \rightleftharpoons A_2$;

Moles at eqm. $3 - x$ $\quad x/2$

As per given $x = 1.5$, $n_{Total} = 3 - \frac{x}{2} = 2.25$

$$P = \frac{n_{Total}RT}{V} = \frac{2.25 \times 44.8}{22.4} = 4.50 \text{ atm}$$

4. For given reaction, $K'_C = \frac{k_f}{k_b} = \frac{4.8 \times 10^{-5}}{2.4 \times 10^{-3}} = \frac{1}{50}$

For reverse reaction $K_C = \frac{1}{K'_C} = 50.00$

5. $2C(s) + O_2(g) \rightleftharpoons 2CO(g)$; K_2

$CO_2(g) \rightleftharpoons C(s) + O_2(g)$; $\frac{1}{K_1}$

$C(s) + CO_2(g) \rightleftharpoons 2CO(g)$; K_3

$$K_3 = \frac{K_2}{K_1} = \frac{10^{-3}}{10^{-2}} = 10^{-1}$$

$C(s) + O_2(g) \rightleftharpoons CO_2(g)$; K_1

$2CO(g) \rightleftharpoons C(s) + CO_2(g)$; $\frac{1}{K_3}$

$2CO(g) + O_2(g) \rightleftharpoons 2CO_2(g)$; K_4

$$K_4 = \frac{K_1}{K_3} = \frac{10^{-2}}{10^{-1}} = 10^{-1}$$

$$\frac{x}{y} = 1$$

6. Partial pressure of both gas are equal due to equal molecules

$P_{Total} = P_{Cl_2} + P_{Cl}$;

$1 = 2P_{Cl_2}$; $P_{Cl_2} = P_{Cl} = 0.5$

$$K_P = \frac{P_{Cl}^2}{P_{Cl_2}} = \frac{(0.5)^2}{0.5} = 0.5$$

$K_P = 5 \times 10^{-1}$

$\therefore \quad x = 5$

7. $AB_3(g) \rightleftharpoons AB_2(g) + \frac{1}{2}B_2(g)$

$t = 0$	800 torr	0	0
$t = t_{eq.}$	$(800 - P)$	P	$\frac{P}{2}$

$\therefore$ At equilibrium total pressure = 1000

$$800 + \frac{P}{2} = 1000$$

$\therefore \quad P = 400$ torr

Degree of dissociation of $AB_3(g) = \frac{400}{800} = 0.5$

$$\text{V.D.} = \frac{M_{Avg}}{2} = \frac{15 + 20 \times 3}{2(1 + 0.25)} = 30$$

8. $Ag^+(aq) + I^-(aq) \longrightarrow AgI(s)$

$I_2(aq) + I^-(aq) \rightleftharpoons I_3^-(aq)$

Mole at eqm. $(1 - x)$ $\quad (0.5 - x)$ $\quad x$

Moles of I^- at eqm. = Moles of $AgNO_3$ reacted = Moles of AgI formed

Moles of AgI formed $= 0.5 - x = 0.25$, $x = 0.25$

$$K_C = \frac{\left(\frac{0.25}{10}\right)}{\left(\frac{0.75}{10}\right) \times \left(\frac{0.25}{10}\right)} = 13.33$$

9. $A \rightleftharpoons 2B$

Moles : n_A $\quad n_B$

$$\frac{4}{2} : \frac{3}{1}, \quad \frac{n_A}{n_B} = \frac{2}{3}$$

$P_A = \frac{2}{5} \times 5 = 2$ bar ; $\quad P_B = \frac{3}{5} \times 5 = 3$ bar

$$K_P = \frac{3^3}{2} = 4.5 \text{ bar}$$

10. $PCl_5(g) \rightleftharpoons PCl_3(g) + Cl_2(g)$

Mole at $t = 0$	3	0	0
Mole at eqm.	$(3 - x)$	x	x

$\therefore$ Total mole present at equilibrium $= 4 + x$

Given, total pressure at equilibrium = 2

Now, $PV = nRT$

$$2 \times 100 = (4 + x) \times 0.08 \times 500$$

$\therefore \quad x = 1$

$$\therefore K_P = \left[\frac{n_{PCl_3} \times n_{Cl_2}}{n_{PCl_5}}\right] \times \left[\frac{P}{\Sigma n}\right]^{+1} = \frac{x^2}{(3 - x)} \times \left[\frac{2}{4 + x}\right]$$

$K_P = 0.2$

11. K_C is very high, almost Cu^{2+} converted into complex so $[Cu(NH_3)_4^{2+}] = 0.5\,M$

$$K_C = \frac{[Cu(NH_3)_4^{2+}]}{[Cu^{2+}][NH_3]^4}$$

$$10^{12} = \frac{0.5}{y \times (0.5)^4}$$

$$y = 8 \times 10^{-12}$$

12. $A + 2B \rightleftharpoons C + D$

	A	$2B$	C	D
Initial	1.5	1	3.5	0.5
At eqm.	$1.5 - x$	$1 - 2x$	$3.5 + x$	$0.5 + x$
	$= 1$	$= y$	$= 4$	$= 1$

$$10^9 = \frac{4 \times 1}{1 \times (y)^2} \times 10;\ y = 2 \times 10^{-4}$$

$\therefore \quad [B] = 2 \times 10^{-5} M, \quad y = 2$

13. $A(g) \rightleftharpoons B(g) + C(g)$

$$\alpha = \frac{M_T - M_{avg}}{M_{avg}(n-1)} = \frac{100 - 80}{80(2-1)}$$

$$\alpha = \frac{1}{4} = 0.25$$

14. $A(s) \rightleftharpoons 2B(g) + C(g)$

eq. 1 $\quad 2P \quad P \quad P_{T_1} = 3P$

$$K_p = (2P)^2 P = 4P^3$$

eq. 2 $\quad P \quad P_c \quad P_{T_2} = P + P_c$

$$K_p = (P)^2 P_c = 4P^3$$

$\therefore \quad P_c = 4P$

$$\frac{P_{T_2}}{P_{T_1}} = \frac{5P}{3P} = \frac{5}{3} = 1.66 \approx 1.67$$

15. Air pressure = 1 atm; $\therefore P_{O_2} = 0.2$ atm

$$2CaSO_4(s) \rightleftharpoons 2CaO(s) + \underset{2y}{2SO_2(g)} + \underset{(y+0.2)}{O_2(g)}$$

$$K_P = 3.2 \times 10^{-6} = (2y)^2(y + 0.2)$$

$\because K_P <<< 1 \quad \therefore y <<< 0.2 \quad \therefore y + 0.2 \simeq 0.2$

$$3.2 \times 10^{-6} = (4y^2)0.2$$

$$y = 2 \times 10^{-3}$$

$\therefore \quad P_{SO_2} = 2y = 4 \times 10^{-3}$ atm

CHAPTER 6

Ionic Equilibrium

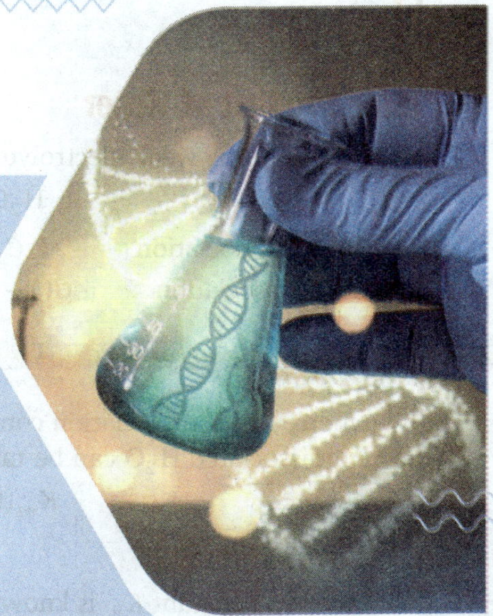

Arrhenius Theory

When dissolved in water, the substances which releases

(i) H^+ ions are called acids *e.g.* HNO_3, $HClO_4$, CH_3COOH,

(ii) OH^- ions are called bases, *e.g.* NaOH, KOH, NH_4OH

Bronsted - Lowry Protonic Concept : acid is H^+ donor and base is H^+ acceptor

Conjugate Acid and Base

To get conjugate acid of a given species add H^+ to it. *e.g.* conjugate acid of N_2H_4 is $N_2H_5^+$.

To get conjugate base of any species subtract H^+ from it. *e.g.* Conjugate base of NH_3 is NH_2^-.

HCl is an acid while its conjugate ion Cl^- is a base.

HCO_3^- can donate proton as well as can accept proton hence amphiprotic.

Lewis Concept : Acids are substances which accept a lone pair of electrons to form a coordinate bond and bases are the substances which donate a lone pair of electrons to form a coordinate bond.

- Lewis-acid is lone pair acceptor ; Lewis-base is lone-pair donor.

Ostwald dilution law: Applicable on weak electrolytes (weak acids/bases)

If α is negligiable with respect to 1 ($\alpha < 0.05$), than for mono basic acid and mono acidic base.

$$\therefore \qquad K_C = C\alpha^2 \qquad \text{or} \qquad \alpha = \sqrt{\left(\frac{K_C}{C}\right)}$$

So $\qquad \alpha \propto \dfrac{1}{\sqrt{C}} \qquad$ or $\qquad \alpha \propto \sqrt{V} \qquad$ where, V is the volume of solution

at infinite dilution α reaches its maximum value, unity.

Presence of other Solute (Common Ion Effect)

Degree of dissociation of a weak electrolyte is suppressed by the addition of a substance having an ion common to weak electrolyte is known as common ion effect.

Ionic Product of Water

Pure water is weak electrolyte and dissociated as:

$$H_2O \rightleftharpoons H^+ + OH^-$$

	H_2O	H^+	OH^-
Before dissociation	C	0	0
After dissociation	$C(1-\alpha)$	$C\alpha$	$C\alpha$

$$K_{eq.} = \frac{[H^+][OH^-]}{[H_2O]}$$

where, K_{eq} is dissociation constant or equilibrium constant of water. Since, dissociation of water is too less and undissociated H_2O can be taken as constant, therefore

$$K_{eq}[H_2O] \rightleftharpoons [H^+][OH^-]$$

$$K_w = [H^+][OH^-]$$

The new constant K_w is known as ionic product of water. The numerical value of K_w increases considerably with temperature. It is 1.0×10^{-14} at 25°C. The variation of ionic product of water with temperature is given by

$$\ln\frac{K_{w_2}}{K_{w_1}} = \frac{\Delta_r H}{R}\frac{[T_2 - T_1]}{T_1 T_2}$$

pH Concept

S. Sorenson used a new term pH to express the hydrogen ion concentration.

$$pH = -\log[H^+] \quad \text{or} \quad pH = -\log[H_3O^+] \quad \text{or} \quad [H^+] = 10^{-pH}$$

For simplicity hydrogen ions are simply written as H^+, with the understanding that H^+ ions in solution are always solvated.

- Ionic product (auto dissociation constant) of water
  $K_w = 1 \times 10^{-14} = [H_3O^+][OH^-]$ at 298 K
- K_w increases with increase in T
- $pH + pOH = pK_w = 14$ (at 298 K)
- For weak acid and its conjugate base

$$K_a K_b = K_w$$

or $$pK_a + pK_b = pK_w = 14 \quad \text{(at 298 K)}$$

pH calculation of weak acid (HA)

$$HA(aq) \rightleftharpoons H^+(aq) + A^-(aq)$$

	HA	H^+	A^-
Initial conc.	C	–	–
Conc. at eq.	$C(1-\alpha)$	$C\alpha$	$C\alpha$

$$K_a = \frac{C\alpha^2}{1-\alpha}$$

if $\alpha << 1$; $K_a \approx C\alpha^2$

$$[H^+] = C\alpha = \sqrt{K_a C}$$

$$\therefore \quad pH = \frac{1}{2}(pK_a - \log C)$$

α = Degree of dissociation

K_a = dissociation constant of weak acid

pH calculation of weak base (BOH)

$$BOH(aq) \rightleftharpoons B^+(aq) + OH^-(aq)$$

	BOH	B^+	OH^-
Initial conc.	C		
Conc. at eq.	$C(1-\alpha)$	$C\alpha$	$C\alpha$

$$K_b = \frac{C\alpha^2}{1-\alpha}$$

if $\alpha << 1$; $K_b \approx C\alpha^2$

$$[OH^-] = C\alpha = \sqrt{K_b C}$$

$$\therefore \quad pOH = \frac{1}{2}(pK_b - \log C)$$

α = Degree of dissociation

K_b = dissociation constant of weak base

pH Calculation of mixture of two weak acid HA and HB :

$[H^+] = \sqrt{K_{a_1}C_1 + K_{a_2}C_2}$; if α_1 and α_2 of both acids are negligiable w.r.t. unity.

K_{a_1} and K_{a_2} are dissociation constant of HA and HB respectively.

C_1 and C_2 are concentration of HA and HB respectively.

pH calculation of mixture of two weak base *AOH* and *BOH*.

$[OH^-] = \sqrt{K_{b_1}C_1 + K_{b_2}C_2}$; if α_1 and α_2 of both bases are negligiable w.r.t. unity.

K_{b_1} and K_{b_2} are dissociation constant of AOH and BOH respectively.

C_1 and C_2 are concentration of AOH and BOH respectively.

Salt Hydrolysis

- Hydrolysis is the interaction of cation/anion/both ions of the salt with H_2O
- Salt of strong acid (SA) and strong base (SB) is not hydrolysed ; solution is neutral, pH = 7 at 25°C.
- Solution of salt of SA and weak base (WB) is acidic due to hydrolysis of cation

$$h = \text{degree of hydrolysis} = \sqrt{\frac{K_w}{K_b C}} = \sqrt{\frac{K_h}{C}}, \quad K_h \text{ (hydrolysis constant)} = \frac{Ch^2}{(1-h)}$$

$$pH = 7 - \frac{1}{2}(pK_b + \log C) \qquad \text{At } 25°C$$

- Solution of salt of weak acid (WA) and SB is alkaline due to hydrolysis of anion

$$h = \sqrt{\frac{K_w}{K_a C}} = \sqrt{\frac{K_h}{C}}$$

$$pH = 7 + \frac{1}{2}(pK_a + \log C) \qquad \text{At } 25°C$$

- Solution of salt of WA and WB is neutral $(K_a = K_b)$, acidic $(K_a > K_b)$ and basic $(K_b > K_a)$ due to hydrolysis of both

$$h = \sqrt{\frac{K_w}{K_a K_b}} = \sqrt{K_h} \quad ; \quad pH = 7 + \frac{1}{2}(pK_a - pK_b) \qquad \text{At } 25°C$$

- For amphiprotic salt like HS^-, HCO_3^-, $pH = \frac{pK_1 + pK_2}{2}$ or $[H^+] = \sqrt{K_1 K_2}$

(K_1 = first dissociation constant, K_2 = second dissociation constant of acid)

Buffer Solutions are the solutions whose pH does not change significantly on adding a small quantity of strong base or strong acid.

In general buffer solutions are simple (salts of weak acid and weak base), or mixed buffer.

Mixed buffer are of two types :

(A) Acidic buffer solutions : weak acid with its conjugate base :

made by mixing a weak acid with its conjugate base. *e.g.*, CH_3COOH with CH_3COONa, $NaHCO_3$ and H_2CO_3, H_3PO_4 and NaH_2PO_4.

(B) Basic buffer solutions : weak base with its conjugate acid :

$NH_3(aq)$ with $NH_4Cl(aq)$, $RNH_2(aq.)$ and $RNH_3Cl(aq.)$ etc,

If K_a for acid (or K_b for base) is not too high, we may write :

Henderson's Equation

For acidic buffer solution : $pH = pK_a + \log$ ([salt]/[acid])

For basic buffer solution : $pOH = pK_b + \log$ ([salt]/[base])

- For good buffer capacity, [salt] : [acid] ratio should be as close to one as possible. In such a case, $pH = pK_a$. (This also is the case at midpoint of titration)

- Buffer capacity $= \dfrac{\text{No. of moles of acid (or base) added per litre of buffer solution}}{\text{Change in pH}}$

Maximum Buffer Capacity

It can be proved that maximum buffer capacity is achieved when the salt and acid or base concentration is equal.

Indicators

Indicator is a substance which indicates the point of equivalence in a titration by undergoing a change in its colour. They are weak acids or weak bases.

Theory of Indicators : The ionized and unionized forms of indicators have different colours. If 90 % or more of a particular form (ionised or unionised) is present, then its colour can be distinctly seen. In general, for an indicator which is weak acid, $HIn \rightleftharpoons H^+ + In^-$, the ratio of ionized to unionized form can be determined from

$$pH = pK_a + \log \frac{[In^-]}{[HIn]}$$

So, for detectable colour change, $pH = pK_a \pm 1$

This roughly gives the range of indicators. Ranges for some popular indicators are

Indicators	pH range	Colour	
		Acid medium	Basic medium
Methyl orange	3.1 – 4.4	red	yellow
Methyl red	4.2 – 6.3	red	yellow
Phenolphathlene	8.3 – 10	colourless	pink

Equivalence point: The point at which exactly equivalent amounts of titrate and titrant have been mixed.

Acid Base Titration: For choosing a suitable indicator titration curves are of great help. In a titration curve, change in pH is plotted against the volume of alkali to a given acid. Four cases arise.

- **Strong acid vs strong base:** The curve is almost vertical over the pH range 3.5-10. This abrupt change corresponds to equivalence point. Any indicator suitable.
- **Weak acid vs strong base:** Final solution is basic at equivalence point. Vertical region (not so sharp) lies in pH range 6.5-10. So, phenolphathlene is suitable.
- **Strong acid vs weak base:** Final solution acidic. Vertical point in pH range 3.8-7.2. Methyl red or methyl orange suitable.
- **Weak acid vs weak base:** No sharp change in pH. No suitable indicator.

Note : *At midpoint of titration, $pH = pK_a$, thus by pH measurements, K_a for weak acids (or K_b for weak bases) can be determined.*

Polyprotic acids and bases : K_2, K_3 etc. can be safely neglected. If $K_1 >> K_2 >> K_3$ and only K_1 plays a significant role in pH calculation.

Solubility (S) and Solubility Product (K_{sp})

This is generally used for sparingly soluble salts,

Let the salt $A_x B_y$ is present in water. Let the solubility of A_xB_y, is S

$$A_xB_y(s) \rightleftharpoons xA^{y+}(aq) + yB^{x-}(aq)$$

$$- \qquad\qquad xS \qquad\qquad yS$$

$\therefore \qquad K_{sp} = (xS)^x \cdot (yS)^y = x^x y^y \cdot (S)^{x+y}$

e.g., For sparingly soluble salts (*e.g.*, $Ag_2C_2O_4$) an equilibrium which exists as

$$Ag_2C_2O_4(s) \rightleftharpoons 2Ag^+(aq.) + C_2O_4^{2-}(aq.)$$

Then $\qquad K_{sp} = [Ag^+]^2[C_2O_4^{2-}]$

Ionic product (Q): It is defined as product of concentration of all the ions of a sparingly soluble salt at time '*t*'.

- **Case-1 :** If $Q < K_{sp}$ then solution is unsaturated no precipitation takes place.
- **Case-2 :** If $Q = K_{sp}$ then solution is saturated.
- **Case-3 :** If $Q > K_{sp}$ solution is super saturated, precipitation takes place.

Common ion effect : Suppression of dissociation by adding an ion common with dissociated products. *e.g.*, Solubility of AgCl decreases in presence of NaCl (Cl^- ion common) OR in presence of $AgNO_3$(Ag^+ ion common)

	$AgCl(s) \rightleftharpoons$	$Ag^+(aq)$	+ $Cl^-(aq)$
Solubility in water at eq.:	–	S	S
Solubility in 0.01*M* NaCl at eq.:	–	S'	$(S'+0.01)$

$$K_{sp} = [Ag^+][Cl^-]_{Total}$$

$$K_{sp} = S'(S'+0.01) \qquad \text{where} \quad S' < S$$

Simultaneous solubility : When two sparingly soluble salts are present in water with atleast a common ion, then solubility of both salts decreases *e.g.*, AgCl and AgBr, $SrSO_4$ and $BaSO_4$

e.g. Calculate simultaneous solubility of AgCl and AgBr in the solution.

If simultaneous solubility of AgCl and AgBr are x and y respectively.

	$AgCl(s) \rightleftharpoons$	$Ag^+(aq.)$	+ $Cl^-(qq.)$		$AgBr(s) \rightleftharpoons$	$Ag^+(aq.)$	+ $Br^-(qq.)$
At eq.	–	$(x+y)$	x	At eq.	–	$(x+y)$	y

$K_{sp_1}(AgCl) = [Ag^+]_{Total}[Cl^-] \qquad K_{sp_2}(AgBr) = [Ag^+]_{Total}[Br^-]$

$K_{sp_1} = (x+y)x \qquad K_{sp_2} = (x+y)y$

Effect of complex formation and hydrolysis on solubility : Solubility increases due to complex formation and hydrolysis of anion (cation/anion)

Solubility of AgBr in aq. sol. of NH_3.

Let solubility of AgBr is x

	$AgBr(s) \rightleftharpoons$	$Ag^+(aq)$	$+ Br^-(qq)$	; K_{sp}
At eq.	–	$(x-y)$	x	
	$Ag^+(aq)$	$+ 2NH_3(aq) \rightleftharpoons$	$Ag(NH_3)_2^+(aq)$	; K_f
At eq.	$(x-y)$	$(z-2y)$	y	
Overall	$AgBr(s) + 2NH_3(aq) \rightleftharpoons$		$Ag(NH_3)_2^+(aq) + Br^-(aq)$	
Reaction at eq.	–	$(z-2y)$	y	x

Due to very large value of K_f we can assume $y \approx x$

$$K_{sp} \cdot K_f = \frac{x^2}{(z-2x)^2}$$

Level-1

1. Morphine ($C_{17}H_{19}NO_3$), which is used medically to relieve pain is a base. What is its conjugate acid?
(a) $C_{17}H_{18}NO_3^+$ (b) $C_{17}H_{18}NO_3$
(c) $C_{17}H_{20}NO_3^-$ (d) $C_{17}H_{20}NO_3^+$

2. The conjugate base of $H_2PO_4^-$ is :
(a) H_3PO_4 (b) $H_2PO_4^-$
(c) HPO_4^{2-} (d) PO_4^{3-}

3. The strongest Bronsted base in the following anion is:
(a) CN^- (b) Cl^-
(c) I^- (d) Br^-

4. Which salt can furnish H^+ in its aqueous solution?
(a) NaH_2PO_2 (b) Na_2HPO_3
(c) Na_2HPO_4 (d) All of these

5. Which is the set of amphiprotic species?
(a) $H_3O^+, HPO_4^{2-}, HCO_3^-$ (b) $H_2O, HPO_3^{2-}, H_2PO_2^-$
(c) $H_2PO_4^-, H_2PO_3^-, H_2O$ (d) All of these

6. The K_a values for HPO_4^{2-} and HSO_3^- are 4.8×10^{-13} and 6.3×10^{-8} respectively. Therefore, it follows the HPO_4^{2-} is a .... acid than HSO_3^- and PO_4^{3-} is a .... base than SO_3^{2-}.
(a) weaker, stronger (b) stronger, weaker
(c) weaker, weaker (d) stronger, stronger

7. Given the following K_a values, determine which species is the strongest base ?
$HSO_4^- = 1.2 \times 10^{-2}$, $H_2PO_4^- = 6.3 \times 10^{-8}$, $HCO_3^- = 4.7 \times 10^{-11}$
(a) CO_3^{2-} (b) H_2SO_4
(c) SO_4^{2-} (d) HPO_4^{2-}

8. Given that K_w for water is 10^{-13} M^2 at 62°C, compute the sum of pOH and pH for a neutral aqueous solution at 62°C :
(a) 7.0 (b) 13.30
(c) 14.0 (d) 13.0

9. The value of the ion product constant for water, (K_w) at 60°C is 9.6×10^{-14} M^2. What is the $[H_3O^+]$ of a neutral aqueous solution at 60°C and an aqueous solution with a pH = 7.0 at 60°C are respectively?
(a) 3.1×10^{-8}, acidic (b) 3.1×10^{-7}, neutral
(c) 3.1×10^{-8}, basic (d) 3.1×10^{-7}, basic

10. For pure water :
(a) pH increases while pOH decreases with rise in temperature
(b) pH decreases while pOH increases with rise in temperature
(c) both pH and pOH decreases with rise in temperature
(d) both pH and pOH increases with rise in temperature

11. A beer has a pH of 4.30. What is the $[H_3O^+]$?
(a) 3.0×10^{-4} (b) 2.0×10^{-4}
(c) 2.0×10^{-5} (d) 5.0×10^{-5}

12. The hydrogen ion concentration of the oceans is about $2 \times 10^{-9} M$. What is the pH?

(a) 8.85 (b) 9.3

(c) 7.85 (d) 8.7

13. The hydroxide ion concentration of a wine is $8 \times 10^{-11} M$. What is the pH of the wine?

(a) 2.10 (b) 2.9

(c) 3.9 (d) 4.9

14. The pH of a solution is 5. To this solution acid was added so that its pH value becomes 2.0. The increase in H^+ concentration is :

(a) 100 times (b) 5 times

(c) 2.5 times (d) 1000 times

15. A solution has a pH = 9. It is 1000 times more basic than the original solution. What was the pH of the original solution?

(a) 12 (b) 6

(c) 9 (d) 10

16. Equal volumes of two HCl solutions of pH = 3 and pH = 5 were mixed. What is the pH of the resulting solution?

(a) 3.5 (b) 4.0

(c) 4.5 (d) 3.3

17. pOH of 0.002 M HNO_3 is :

(a) $11 + \log 2$ (b) $11 - \log 2$

(c) $-3 + \log 2$ (d) None of these

18. Number of equivalents of HCl present in 100 mL of its solution whose pH is 4 :

(a) 10^{-4} (b) 10^{-3}

(c) 10^{-2} (d) 10^{-5}

19. To a 10 mL of 10^{-3} N H_2SO_4 solution water has been added to make the total volume of one litre. Its pOH would be :

(a) 3 (b) 12

(c) 9 (d) 5

20. The pH of a solution of H_2SO_4 is 1. Assuming complete ionisation, find the molarity of H_2SO_4 solution :

(a) 0.1 (b) 0.2

(c) 0.05 (d) 2.0

21. pH of a strong diprotic acid (H_2A) at concentrations :

(i) 10^{-4} M, (ii) 10^{-4} N

are respectively :

(a) 3.7 and 4.0 (b) 4 and 3.7

(c) 4 and 4 (d) 3.7 and 3.7

22. Calcium hydroxide is a strong base. Compute $[Ca^{2+}]$ and $[OH^-]$ for a solution that is prepared by dissolving 0.60 g of $Ca(OH)_2$ in enough water to make a 1500 mL of solution.

[Atomic mass : Ca = 40, O = 16, H = 1]

(a) $5.4 \times 10^{-3}, 9.1 \times 10^{-13}$ (b) $5.4 \times 10^{-3}, 1.08 \times 10^{-2}$

(c) $5.4 \times 10^{-3}, 5.4 \times 10^{-3}$ (d) $8.1 \times 10^{-3}, 8.1 \times 10^{-3}$

23. pH of 10^{-6} M HCl(*aq.*) is :

(a) just less than 6 (b) exactly equal to 6

(c) just greater than 6 (d) just less than 7

24. 10^{-5} *M* NaOH solution at 25°C is diluted 1000 times. The pH of the resultant solution will :
(a) be equal to 8
(b) lie between 7 and 8
(c) lie between 6 and 7
(d) remain unchanged

25. 4.0 g of NaOH and 4.9 g of H_2SO_4 are dissolved in water and volume is made upto 250 mL. The pH of this solution is:
(a) 7.0
(b) 1.0
(c) 2.0
(d) 12.0

26. A 25.0 mL sample of 0.10 *M* HCl is titrated with 0.10 *M* NaOH. What is the pH of the solution at the points where 24.9 and 25.1 mL of NaOH have been added?
(a) 3.70, 10.70
(b) 3.30, 10.30
(c) 3.70, 10.30
(d) 3.0, 11.0

27. What is the pH of a solution in which 25.0 mL of 0.1 *M* NaOH is added to 25 mL of 0.08 *M* HCl and final solution is diluted to 500 mL?
(a) 3
(b) 11
(c) 12
(d) 13

28. What is the pH of a solution in which 10.0 mL of 0.010 *M* $Sr(OH)_2$ is added to 10.0 mL of 0.010 *M* HCl?
(a) 2.30
(b) 1.50
(c) 11.70
(d) 7.00

29. At 90°C, pure water has $[H^+] = 10^{-6}$ *M*. If 100 mL of 0.2 *M* HCl is added to 200 mL of 0.1 *M* KOH at 90°C then pH of the resulting solution will be :
(a) 5
(b) 6
(c) 7
(d) None of these

30. What change will occur for the following reaction if the hypochlorous acid solution is diluted from 0.1 to 0.01 *M*?

$$HOCl(aq.) + H_2O(l) \rightleftharpoons OCl^-(aq.) + H_3O^+(aq.)$$

(a) a decrease in the fraction of acid ionized
(b) an increase in the fraction of acid ionized
(c) no change in the fraction of acid ionized
(d) we can not predict

31. Given K_a values of 5.76×10^{-10} and 4.8×10^{-10} for NH_4^+ and HCN respectively. What is the equilibrium constant for the following reaction?

$$NH_4^+(aq.) + CN^-(aq.) \rightleftharpoons NH_3(aq.) + HCN(aq.)$$

(a) 0.83
(b) 1.2
(c) 8.0×10^{-11}
(d) 27.6×10^{-10}

32. Which is the strongest acid (pK_a value is given)?
(a) HCOOH [3.77]
(b) C_6H_5COOH [4.22]
(c) CH_3COOH [4.7]
(d) CH_3CH_2COOH [4.88]

33. Given : Enthalpy of ionization of two acids :
$\Delta H°$ (HCN) = 45.2 kJ mol^{-1}
$\Delta H°$ (CH_3COOH) = 2.1 kJ mol^{-1}
Which relationship for the two acids is true ?
(a) $pK_a(HCN) = pK_a(CH_3COOH)$
(b) $pK_a(HCN) > pK_a(CH_3COOH)$
(c) $pK_a(HCN) < pK_a(CH_3COOH)$
(d) $pK_a(HCN) = \frac{45.2}{2.1} pK_a(CH_3COOH)$

34. What is the hydronium ion concentration of a 0.25 *M* HA solution? ($K_a = 4 \times 10^{-8}$)

(a) 10^{-4} (b) 10^{-5}

(c) 10^{-7} (d) 10^{-10}

35. What is the percent dissociation (α) of a 0.01 *M* HA solution? ($K_a = 10^{-4}$)

(a) 9.5% (b) 1%

(c) 10.5% (d) 17%

36. Given the two concentration of HCN ($K_a = 10^{-9}$) are 0.1 *M* and 0.001 *M* respectively. What will be the ratio of degree of dissociation?

(a) 1 (b) 0.1

(c) 0.003 (d) 0.01

37. A 0.10 *M* solution of HF is 8.0% dissociated What is the K_a?

(a) 6.4×10^{-4} (b) 8.8×10^{-4}

(c) 6.95×10^{-4} (d) 7.6×10^{-4}

38. A weak base *M*OH of 0.1 *N* concentration shows a pH value of 9. What is the percentage degree of dissociation of the base?

(a) 0.01% (b) 0.001%

(c) 0.1% (d) 0.02%

39. 0.01 *M* HA (*aq.*) is 2% dissociated; $[OH^-]$ of solution is :

(a) 2×10^{-4} (b) 10^{-8}

(c) 5×10^{-11} (d) 5×10^{-12}

40. If degree of dissociation is 0.01 of decimolar solution of weak acid HA then pK_a of acid is :

(a) 2 (b) 3

(c) 5 (d) 7

41. What concentration of $HCOO^-$ is present in a solution of 0.01 *M* HCOOH ($K_a = 1.8 \times 10^{-4}$) and 0.01 *M* HCl?

(a) 1.8×10^{-3} (b) 10^{-2}

(c) 1.8×10^{-4} (d) 10^{-4}

42. Choose the correct code

	Column-I		Column-II
(P)	pK_b of X^- (K_a of $HX = 10^{-6}$)	**(1)**	6.9
(Q)	pH of 10^{-8} *M* HCl	**(2)**	8
(R)	pH of 10^{-2} *M* acetic acid solution (Take K_a of acetic acid $= 1.6 \times 10^{-5}$)	**(3)**	10.7
(S)	pOH of a solution obtained by mixing equal volumes of solution with pH 3 and 5.	**(4)**	3.4

Codes :

	P	Q	R	S
(a)	1	2	4	3
(b)	4	3	2	1
(c)	2	1	4	3
(d)	1	2	3	4

43. How much water must be added to 300 mL of 0.2 *M* solution of CH_3COOH ($K_a = 1.8 \times 10^{-5}$) for the D.O.D. (α) of the acid to double?

(a) 600 mL (b) 900 mL

(c) 1200 mL (d) 1500 mL

44. What is $[NH_4^+]$ in a solution that contain 0.02 *M* NH_3 ($K_b = 1.8 \times 10^{-5}$) and 0.01 *M* KOH?

(a) 9×10^{-6} (b) 1.8×10^{-5}

(c) 3.6×10^{-5} (d) None of these

45. A hand book states that the solubility of $RNH_2(g)$ in water at 1 atm and 0°C is 22.41 volumes of $RNH_2(g)$ per volume of water. (pK_b of $RNH_2 = 4$) Find the max. pOH that can be attained by dissolving RNH_2 in water :

(a) 1 (b) 2

(c) 4 (d) 6

46. The $[H^+]$ of a resulting solution that is 0.01 *M* acetic acid ($K_a = 1.8 \times 10^{-5}$) and 0.01 *M* in benzoic acid ($K_a = 6.3 \times 10^{-5}$) :

(a) 9×10^{-4} (b) 81×10^{-4}

(c) 9×10^{-5} (d) 2.8×10^{-3}

47. 6.0 g weak acid HA (mol. mass = 60 g/mol.) is dissolved in water and formed 10 m^3 solution. If $K_{a(HA)} = 10^{-9}$, then pOH of solution is : [Given : log 4 = 0.6]

(a) 6.7 (b) greater than 6.7 and less than 7.0

(c) greater than 7.0 and less than 7.3 (d) greater than 7.3

48. Carbonic acid (H_2CO_3), a diprotic acid has $K_{a_1} = 4.0 \times 10^{-7}$ and $K_{a_2} = 7.0 \times 10^{-11}$. What is the $[HCO_3^-]$ of a 0.025 *M* solution of carbonic acid?

(a) 7.8×10^{-3} (b) 6.6×10^{-4}

(c) 10^{-10} (d) 1.0×10^{-4}

49. Carbonic acid (H_2CO_3), a diprotic acid has $K_{a_1} = 4.0 \times 10^{-7}$ and $K_{a_2} = 7.0 \times 10^{-11}$. What is the $[CO_3^{2-}]$ of a 0.025 *M* solution of carbonic acid?

(a) 5.5×10^{-9} (b) 5.5×10^{-8}

(c) 7.0×10^{-9} (d) 7.0×10^{-11}

50. Selenious acid (H_2SeO_3), a diprotic acid has $K_{a_1} = 3.0 \times 10^{-3}$ and $K_{a_2} = 5.0 \times 10^{-8}$. What is the $[OH^-]$ of a 0.30 *M* solution of selenious acid?

(a) 2.85×10^{-3} (b) 5.0×10^{-6}

(c) 3.5×10^{-12} (d) 3.5×10^{-13}

51. Which of the hydrated species may exist?

I : $H_5O_2^+$ II : H_3O^+ III : $H_3O_2^-$ IV : $H_7O_3^+$

(a) II only (b) I and II

(c) I, II and IV (d) I, II, III and IV

52. Consider the following salts. Which one(s) when dissolved in water will produce an acidic solution?

1. NH_4Cl 2. $KHSO_4$ 3. NaCN 4. KNO_3

(a) 2 and 3 (b) 1 and 2

(c) only 3 (d) 2 and 4

53. Consider the following salts. Which one(s) when dissolved in water will produce a basic solution?

1. $RbClO_4$ 2. $NaNO_2$ 3. NH_4Cl 4. NaCl

(a) 1 and 3
(b) only 2
(c) 1 and 2
(d) 3 and 4

54. At 25°C dissociation constants of acid HA and base BOH in aqueous solution are same. The pH of 0.01 M solution of HA is 5. The pOH of 10^{-4} M solution of BOH at the same temperature is:
(a) 3.5
(b) 4
(c) 6
(d) None of these

55. Which of the following solutions has the highest pH?
(a) 0.2 M $HClO_4$
(b) 0.20 M CH_3COOH
(c) 0.020 M HCl
(d) 0.2 M NaCl

56. pH of solutions of four sodium salts NaW, NaX, NaY and NaZ were found to be 7.0, 9.0, 10.0 and 11.0 respectively. If each solution has concentration 0.1 M, the weakest acid is :
(a) HW
(b) HX
(c) HY
(d) HZ

57. The pH values 0.1 M solution of HCOONa (I), HCOOH (II), CH_3COONH_4 (III), NaOH (IV), HCl (V), will be in the order :
(a) IV > III > I > II > V
(b) IV > I > III > II > V
(c) II > III > I > IV > V
(d) V > II > III > I > IV

58. pH of an aqueous NaCl solution at 50°C should be :
(a) 7
(b) > 7
(c) < 7
(d) 0

59. Upon hydrolysis of sodium carbonate, the reaction takes place between:
(a) Na^+ and water
(b) Na^+ and OH^-
(c) CO_3^{2-} and water
(d) CO_3^{2-} and H^+

60. The solution of blue vitrol in water is acidic because:
(a) $CuSO_4$ reacts with water
(b) Cu^{2+} reacts with water
(c) SO_4^{2-} reacts with water
(d) $CuSO_4$ removes OH^- ions from water

61. 1 mL of 0.1 N HCl is added to 999 mL solution of NaCl. The pH of the resulting solution will be :
(a) 7
(b) 4
(c) 2
(d) 1

62. If a salt of strong acid and weak base hydrolyses appreciably ($\alpha = 0.1$), which of the following formula is to be used to calculate degree of hydrolysis 'α'?
(a) $\alpha = \sqrt{\dfrac{K_w}{K_a \cdot a}}$
(b) $\alpha = \sqrt{\dfrac{K_w}{K_b \cdot a}}$
(c) $\alpha = \sqrt{\dfrac{K_w}{K_a \cdot K_b}}$
(d) None of these

63. The correct formula to calculate the hydroxyl ion concentration of an aqueous solution of NH_4NO_3 is :
(a) $\sqrt{\dfrac{C \times K_w}{K_b}}$
(b) $\sqrt{\dfrac{K_w \times K_b}{C}}$
(c) $\sqrt{\dfrac{C \times K_w}{K_a}}$
(d) $\sqrt{\dfrac{K_a \times K_w}{C}}$

64. $[H^+] = \sqrt{\dfrac{K_w K_a}{C}}$ is suitable for
(a) NaCl, NH_4Cl
(b) CH_3COONa , NaCN
(c) CH_3COONa, $(NH_4)_2SO_4$
(d) CH_3COONH_4, $(NH_4)_2CO_3$

65. What is the hydrolysis constant of the OCl^- ion? The ionization constant of HOCl is 3.0×10^{-8}.

(a) 3.33×10^{-8} (b) 3.33×10^{-7}
(c) 3.0×10^{-7} (d) 3.33×10^{-6}

66. What is the pH of a 0.10 *M* $C_6H_5O^-$ solution? The K_a of C_6H_5OH is 1.0×10^{-10}.

(a) 10.51 (b) 11.04
(c) 11.50 (d) 12

67. Calculate the $[OH^-]$ in 0.01 *M* aqueous solution of NaOCN (K_b for $OCN^- = 10^{-10}$) :

(a) 10^{-6} *M* (b) 10^{-7} *M*
(c) 10^{-8} *M* (d) None of these

68. What is the ionization constant of an acid if the hydronium ion concentration of a 0.40 *M* solution is 1.40×10^{-4} *M*?

(a) 1.96×10^{-8} (b) 1.22×10^{-9}
(c) 4.90×10^{-8} (d) 1.40×10^{-6}

69. The degree of hydrolysis of 0.1 *M* RNH_3Cl solution is 1.0%. If the concentration of RNH_3Cl is made 0.4 *M*, what is the new degree of hydrolysis (in percentage)?

(a) 0.01 (b) 0.001
(c) 0.2 (d) 0.5

70. % hydrolysis of 0.1*M* CH_3COONH_4, when $K_a(CH_3COOH) = K_b(NH_4OH) = 1.8 \times 10^{-5}$ is:

(a) 0.55 (b) 7.63
(c) 0.55×10^{-2} (d) 7.63×10^{-3}

71. The enthalpy of neutralisation of four acids *HA*, *HB*, *HC* and *HD* with NaOH are −13, −12, −11, −10 kcal/mol. Which salt has maximum degree of hydrolysis?

(a) 1*M* Na*A* (b) 1*M* Na*B*
(c) 1*M* Na*C* (d) 1*M* Na*D*

72. Calculate $[H^+]$ at neutralization point between titration of 0.1 *M*, 25 mL of weak acid *HA* ($K_{a(HA)} = 10^{-5}$) with 0.05 *M* NaOH solution :

(a) 3×10^{-9} (b) 1.732×10^{-9}
(c) 8 (d) 10

73. When a salt of weak acid and weak base is dissolved in water at 25°C, the pH of the resulting solution will always :

(a) be 7 (b) be greater than 7
(c) be less than 7 (d) depend upon K_a and K_b values

74. What will be the pH of an aqueous solution of 1.0 *M* ammonium formate?
Given : $pK_a = 3.8$ and $pK_b = 4.8$

(a) 7.5 (b) 3.4
(c) 6.5 (d) 10.2

75. What will be the pH and % α (degree of hydrolysis) respectively for the salt *BA* of 0.1 *M* concentration?
Given : K_a for $HA = 10^{-6}$ and K_b for $BOH = 10^{-6}$

(a) 5, 1% (b) 7, 10%
(c) 9, 0.01% (d) 7, 0.01%

76. The percentage degree of hydrolysis of a salt of weak acid (HA) and weak base (*B*OH) in its 0.1 *M* solution is found to be 10%. If the molarity of the solution is 0.05 *M*, the percentage hydrolysis of the salt should be :

(a) 5% (b) 10%
(c) 20% (d) None of these

77. What is the hydronium ion concentration of a 0.02 M solution of Cu^{2+} solution of copper(II) perchlorate? The acidity constant of the following reaction is 5×10^{-9}.

$$Cu^{2+}(aq.) + 2H_2O(l) \rightleftharpoons Cu(OH)^+(aq.) + H_3O^+(aq.)$$

(a) 1×10^{-5} (b) 7×10^{-4}
(c) 5×10^{-4} (d) 1×10^{-4}

78. What is the acidity constant for the following reaction given that the hydronium ion concentration of a 0.04 M solution of Ni^{2+} solution of nickel(II) perchlorate is 4.5×10^{-6}?

$$Ni^{2+}(aq.) + 2H_2O(l) \rightleftharpoons Ni(OH)^+(aq.) + H_3O^+(aq.)$$

(a) 2×10^{-12} (b) 4×10^{-6}
(c) 5×10^{-12} (d) 5×10^{-10}

79. Calculate the pH at 25°C of a solution that is 0.10 M in $Fe(NO_3)_3$. The acid dissociation constant for the reaction given below is 1.0×10^{-3}.

$$[Fe(H_2O)_6]^{3+} + H_2O(l) \rightleftharpoons H_3O^+(aq.) + [Fe(H_2O)_5(OH)]^{2+}$$

(a) 2.00 (b) 2.02
(c) 2.30 (d) 2.50

80. Approximate pH of 0.01 M NaHA is calculated by :
($K_{a_1} = 10^{-6}$ and $K_{a_2} = 10^{-8}$ are ionization constants of H_2A)

(a) $pH = 7 + \frac{pK_{a_1}}{2} + \frac{\log C}{2}$ (b) $pH = 7 - \frac{pK_{a_1}}{2} - \frac{\log C}{2}$
(c) $pH = \frac{pK_{a_1} + pK_{a_2}}{2}$ (d) None of these

81. H_3PO_4 is a weak triprotic acid; approximate pH of 0.1 M $Na_2HPO_4(aq.)$ is calculated by :

(a) $\frac{1}{2}[pK_{a_1} + pK_{a_2}]$ (b) $\frac{1}{2}[pK_{a_2} + pK_{a_3}]$
(c) $\frac{1}{2}[pK_{a_1} + pK_{a_3}]$ (d) $pK_{a_1} + pK_{a_2}$

82. Which of the following is a buffer solution?
(a) 500 mL of 0.1 N CH_3COOH + 500 mL of 0.1 N NaOH
(b) 500 mL of 0.1 N CH_3COOH + 500 mL of 0.1 N HCl
(c) 500 mL of 0.1 N CH_3COOH + 500 mL of 0.2 N NaOH
(d) 500 mL of 0.2 N CH_3COOH + 500 mL of 0.1 N NaOH

83. If 20 mL of 0.1 M NaOH is added to 30 mL of 0.2 M CH_3COOH ($pK_a = 4.74$), the pH of the resulting solution is :

(a) 4.44 (b) 9.56
(c) 8.96 (d) 9.26

84. $H_2CO_3 + NaHCO_3$ found in blood helps in maintaining pH of the blood close to 7.4. An excess of acid entering the blood stream is removed by:

(a) HCO_3^- (b) H_2CO_3
(c) H^+ ion (d) CO_3^{2-} ion

85. 100 mL of 0.02 M benzoic acid ($pK_a = 4.20$) is titrated using 0.02 M NaOH, pH after 50 mL and 100 mL of NaOH have been added are:
(a) 3.50, 7 (b) 4.2, 7
(c) 4.2, 8.1 (d) 4.2, 8.25

86. What is the pH of a solution of 0.28 M acid and 0.84 M of its conjugate base if the ionization constant of acid is 4×10^{-4}?
(a) 3.88 (b) 3.34
(c) 7 (d) 10.12

87. The toxic compound 2,4-dinitrophenol has $K_a = 10^{-4}$. In an experiment, a buffer solution of 2,4-dinitrophenol was prepared with the pH adjusted to 5. Calculate the ratio of the concentrations of the dissociated ion to the undissociated acid :
(a) 0.01 (b) 0.1
(c) 10 (d) 100

88. Equilibrium constant for the following reaction is 1×10^{-9} :

$$C_5H_5N(aq.) + H_2O(l) \rightleftharpoons C_5H_5NH^+(aq.) + OH^-(aq.)$$

Determine the moles of pyridinium chloride ($C_5H_5N.HCl$) that should be added to 500 mL solution of 0.4 M pyridine (C_5H_5N) to obtain a buffer solution of pH = 5 :
(a) 0.1 mole (b) 0.2 mole
(c) 0.3 mole (d) 0.4 mole

89. Which one of the following mixture does not act as a buffer solution?
(a) Ammonium chloride and ammonium hydroxide
(b) Sodium phosphate & disodium hydrogen phosphate
(c) Sodium propionate and propionic acid
(d) Sod. acetate and sodium propionate

90. The acid dissociation constant of uric acid is $K_a = 4.0 \times 10^{-6}$ M. The pH of a sample of urine is 6.0. What is the ratio of concentration of urate ion to uric acid in the urine?
(a) 2.0 (b) 4.0
(c) 6.0 (d) 0.25

91. CH_3NH_2 (0.12 mole, $pK_b = 3.3$) is added to 0.08 moles of HCl and the solution is diluted to one litre, resulting pH of solution is :
(a) 10.7 (b) 3.6
(c) 10.4 (d) 11.3

92. An aqueous solution at room temperature contains 0.1 M NH_4Cl and 0.01M NH_4OH ($pK_b = 5$), the pH of the solution is :
(a) 7.5 (b) 6.8
(c) 6.5 (d) 8.0

93. Which of the following will decrease with dilution at a given temperature?
(a) pH of 10^{-3} M acetic acid solution
(b) pH of 10^{-3} M aniline solution
(c) degree of dissociation of 10^{-3} M acetic acid
(d) degree of dissociation of 10^{-3} M aniline solution

94. A 1L solution contains 0.2 M NH_4OH and 0.2 M NH_4Cl. If 1.0 mL of 0.001 M HCl is added to it, what will be the $[OH^-]$ of the resulting solution ($K_b = 2 \times 10^{-5}$)
(a) 2×10^{-5} (b) 5×10^{-10}
(c) 2×10^{-3} (d) None of these

95. 0.1 M formic acid solution is titrated against 0.1 M NaOH solution. What would be the difference in pH between 1/5 and 4/5 stages of neutralization of acid ?

(a) 2 log 3/4 (b) 2 log 1/5

(c) log 1/3 (d) 2 log 4

96. In the neutralization process of H_3PO_4 and NaOH, the number of buffers formed will be:

(a) 3 (b) 1

(c) 2 (d) 4

97. A buffer solution is made up of acetic acid [$pK_a = 5$] having conc. = 1.5 M and sodium acetate having conc. = 0.15 M. What is the number of OH^- ions present in 1 litre solution?

(a) $10^{-10}\ N_A$ (b) $10^{-4}\ N_A$

(c) $10^{-3}\ N_A$ (d) $10^{-6}\ N_A$

98. The pH of a solution of 0.10 M CH_3COOH increases when which of the following substances is added?

(a) $NaHSO_4$ (b) $HClO_4$

(c) KNO_3 (d) K_2CO_3

99. H^+ ion concentration of water does not change by adding :

(a) CH_3COONa (b) $NaNO_3$

(c) NaCN (d) Na_2CO_3

100. pK_a of NH_4^+ is 9.26. Hence, effective range for $NH_4OH - NH_4Cl$ buffer is about pH :

(a) 8.26 to 10.26 (b) 4.74 to 5.74

(c) 3.74 to 5.74 (d) 8.26 to 9.26

101. 1.0 L solution is prepared by mixing 61 g benzoic acid ($pK_a = 4.2$) with 72 g of sodium benzoate and then 300 mL 1.0 M HBr solution was added. The pH of final solution is:

(a) 3.6 (b) 3.8

(c) 4.2 (d) 4.8

102. The pH of a solution containing 0.4 M HCO_3^- and 0.2 M CO_3^{2-} is :

[K_{a_1} $(H_2CO_3) = 4 \times 10^{-7}$; K_{a_2} $(HCO_3^-) = 4 \times 10^{-11}$]

(a) 10.4 (b) 10.1

(c) 6.1 (d) 10.7

103. The pH of the resultant solution of 20 mL of 0.1 M H_3PO_4 and 20 mL of 0.1 M Na_3PO_4 is :

(a) $pK_{a_1} + \log 2$ (b) pK_{a_1}

(c) pK_{a_2} (d) $\dfrac{pK_{a_1} + pK_{a_2}}{2}$

104. Which one of the following curves represents the graph of pH during the titration of NaOH and HCl($aq.$)?

(a) 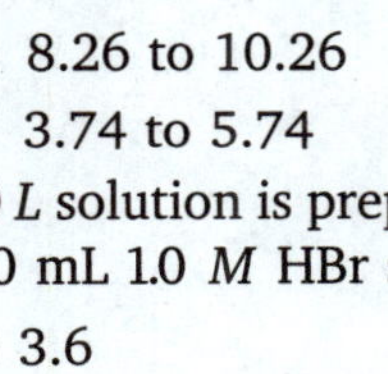

(b) 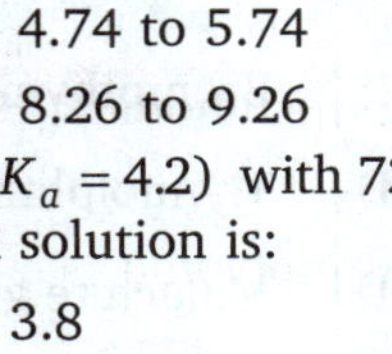

(c) 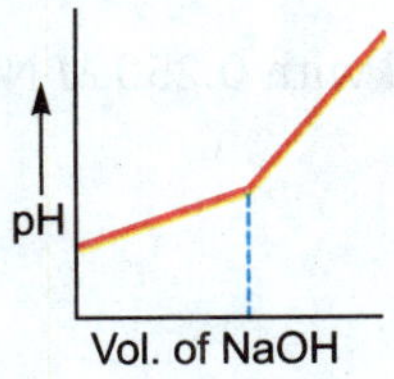

(d) 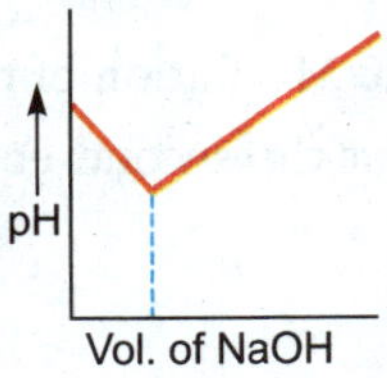

105. When 100 mL of 0.1 *M* NaCN solution is titrated with 0.1 *M* HCl solution the variation of pH of solution with volume of HCl added will be :

(a) 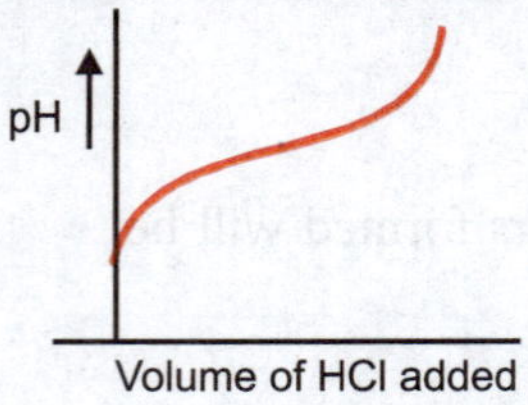

(b) 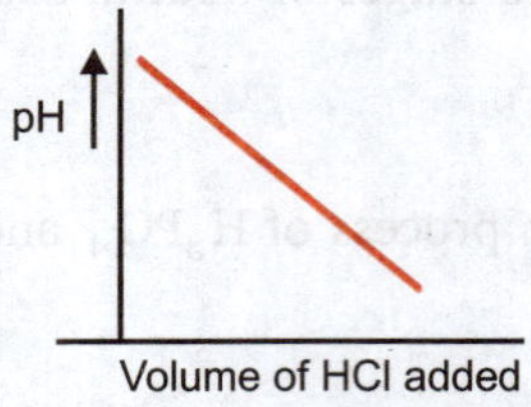

(c) 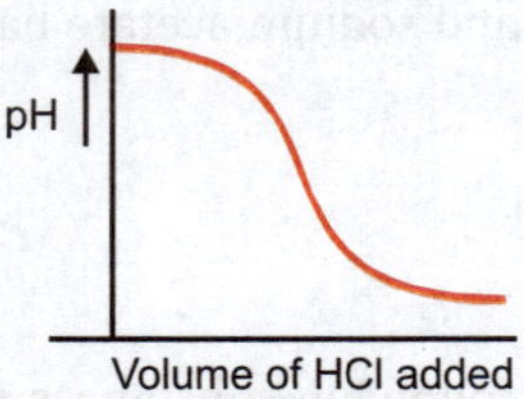

(d) 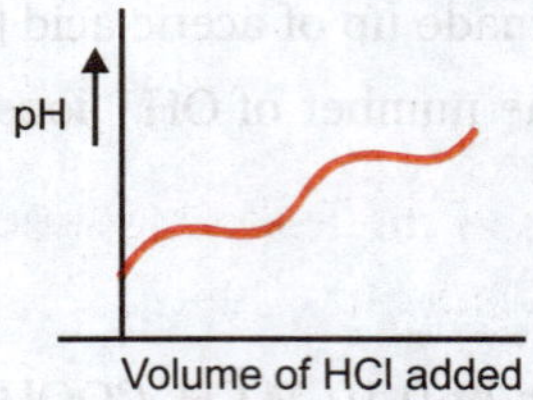

106. The best indicator for the detection of end point in titration of a weak acid and a strong base is :

(a) Methyl orange (3.1 to 4.4) (b) Methyl red (4.2 to 6.3)
(c) Bromothymol blue (6 to 7.6) (d) Phenolphthalein (8.2 to 10)

107. Select the best indicator from the given table for titration of 20 mL of 0.02 *M* CH_3COOH with 0.02 *M* NaOH. Given pK_a (CH_3COOH) = 4.74

	Indicator	pH range
(I)	Bromothymol blue	6.0–7.6
(II)	Thymolphthalein	9.3–10.5
(III)	Malachite green	11.4–13
(IV)	M-Cresol purple	7.4–9.0

(a) I (b) II
(c) III (d) IV

108. Bromothymol blue is an indicator with a K_a value of 6×10^{-5}. What % of this indicator is in its basic form at a pH of 5 ?

(a) 40 (b) 85.7
(c) 14.3 (d) 60

109. An acid-base indicator has $K_{HIn} = 3.0 \times 10^{-5}$. The acid form of the indicator is red and the basic form is blue. The change in $[H^+]$ required to change the indicator from 75% red to 75% blue is :

(a) 8×10^{-5} *M* (b) 9×10^{-5} *M*
(c) 1×10^{-5} *M* (d) 3×10^{-4} *M*

110. An acid-base indicator which is a weak acid has a pK_a value = 5.45. At what concentration ratio of sodium acetate to acetic acid would the indicator show a colour half-way between those of its acid and conjugate base forms ? [pK_a of acetic acid = 4.75]

(a) 4 : 1 (b) 6 : 1
(c) 5 : 1 (d) 3 : 1

111. A 20.0 mL sample of a 0.20 *M* solution of the weak diprotic acid H_2A is titrated with 0.250 *M* NaOH. The concentration of solution at the second equivalent point is :

(a) 0.10 *M* NaHA (b) 0.153 *M* Na_2A
(c) 0.10 *M* Na_2A (d) 0.0769 *M* Na_2A

112. During the titration of a weak diprotic acid (H_2A) against a strong base (NaOH), the pH of the solution half-way to the first equivalent point and that at the first equivalent point are given respectively by :

(a) pK_{a_1} and $pK_{a_1} + pK_{a_2}$

(b) $\sqrt{K_{a_1} C}$ and $\dfrac{pK_{a_1} + pK_{a_2}}{2}$

(c) pK_{a_1} and $\dfrac{pK_{a_1} + pK_{a_2}}{2}$

(d) pK_{a_1} and pK_{a_2}

113. In which of the following cases is the solution of AgCl unsaturated?

(a) $[Ag^+][Cl^-] < K_{sp}$

(b) $[Ag^+][Cl^-] > K_{sp}$

(c) $[Ag^+][Cl^-] = K_{sp}$

(d) $[Ag^+][Cl^-] \leq K_{sp}$

114. When equal volumes of the following solutions are mixed, the precipitation of AgCl ($K_{sp} = 1.8 \times 10^{-10}$) will occur with :

(a) 10^{-4} M (Ag^+) and 10^{-4} M (Cl^-)

(b) 10^{-5} M (Ag^+) and 10^{-5} M (Cl^-)

(c) 10^{-5} M (Ag^+) and 10^{-6} M (Cl^-)

(d) 10^{-10} M (Ag^+) and 10^{-10} M (Cl^-)

115. Choose the correct set of True/False for following statements:

(i) Silver chloride is more soluble in very concentrated sodium chloride solution than in pure water.

(ii) The pH of a buffer solution does not change on addition of small amount of an acid or a base.

(iii) Addition of NH_4Cl does not affect the pH of a solution of NH_4OH

(iv) Degree of hydrolysis of ammonium acetate does not depend upon the concentration of ammonium acetate solution.

(v) A mixture of acetic acid and sodium acetate can act as buffer solution.

(a) TTFTT

(b) FTTTF

(c) TFTFT

(d) FFTTT

116. A 1 litre solution containing NH_4Cl and NH_4OH has hydroxide ion concentration of 10^{-6} mol/litre. Which of the following hydroxides could be precipitated when the solution is added to 1 litre solution of 0.1 M metal ions?

(I) $Ba(OH)_2$ ($K_{sp} = 5 \times 10^{-3}$)

(II) $Ni(OH)_2$ ($K_{sp} = 1.6 \times 10^{-16}$)

(III) $Mn(OH)_2$ ($K_{sp} = 2 \times 10^{-13}$)

(IV) $Fe(OH)_2$ ($K_{sp} = 8 \times 10^{-16}$)

(a) I, II, IV

(b) IV

(c) II and IV

(d) II, III, IV

117. 150 mL of 0.0008 M ammonium sulphate is mixed with 50 mL of 0.04 M calcium nitrate. The ionic product of $CaSO_4$ will be : ($K_{sp} = 2.4 \times 10^{-5}$ for $CaSO_4$)

(a) $< K_{sp}$

(b) $> K_{sp}$

(c) $\approx K_{sp}$

(d) None of these

118. In a saturated solution of AgCl, NaCl is added gradually. The concentration of Ag^+ is plotted against the concentration of Cl^-. The graph appears as :

(a) 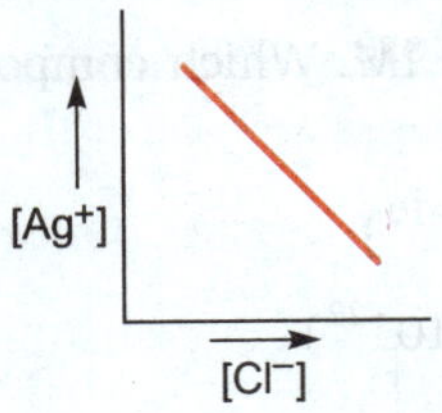

(b) 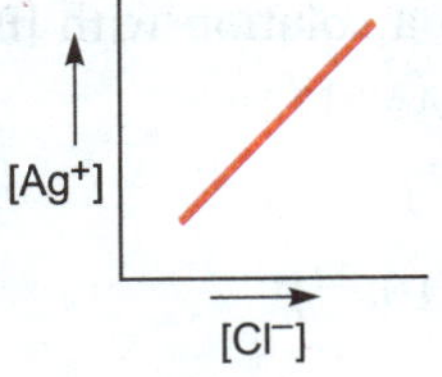

(c) $[Ag^+]$ 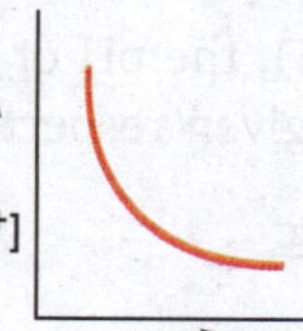

(d) $[Ag^+]$ 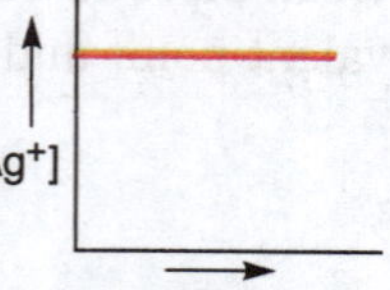

119. K_{sp} of AgCl is 1×10^{-10}. Its solubility in $0.1M$ KNO_3 will be :

(a) 10^{-5} moles/litre (b) $> 10^{-5}$ moles/litre
(c) $< 10^{-5}$ moles/litre (d) None of these

120. 50 mL of a solution containing 10^{-3} mole of Ag^+ is mixed with 50 mL of a 0.1 M HCl solution. How much Ag^+ remains in solution? (K_{sp} of AgCl $= 1.0 \times 10^{-10}$)

(a) 2.5×10^{-9} (b) 2.5×10^{-7}
(c) 2.5×10^{-8} (d) 2.5×10^{-10}

121. The solubility of a sparingly soluble salt A_xB_y in water is 'S' moles per litre. The solubility product has the value :

(a) S^2 (b) $x^y y^x \cdot S^{x+y}$
(c) $x^x y^y S^{x+y}$ (d) S^{x+y}

122. What is the molarity of a saturated solution of $CaCO_3$? ($K_{sp} = 2.8 \times 10^{-9}$)

(a) 2.6×10^{-5} (b) 2.8×10^{-9}
(c) 5.2×10^{-5} (d) 5.6×10^{-9}

123. K_{sp} of $Zr_3(PO_4)_4$ in terms of solubility (S) is :

(a) $108\,S^7$ (b) $4\,S^3$
(c) $6912\,S^7$ (d) None of these

124. The solubility of electrolytes MX_1, MX_2 and MX_3 is 1×10^{-3} moles per litre. Hence their respective solubility products are :

(a) $10^{-6}, 4 \times 10^{-9}, 27 \times 10^{-12}$ (b) $10^{-9}, 4 \times 10^{-9}, 32 \times 10^{-12}$
(c) $10^{-9}, 8 \times 10^{-8}, 32 \times 10^{-12}$ (d) None of these

125. A saturated solution of $Ca_3(PO_4)_2$ has $[Ca^{2+}] = 2 \times 10^{-8}\,M$ and $[PO_4^{3-}] = 1.6 \times 10^{-5}\,M$ K_{sp} of $Ca_3(PO_4)_2$ is:

(a) 3.2×10^{-13} (b) 3.2×10^{-34}
(c) 2.048×10^{-33} (d) None of these

126. Which of the following is most soluble in water?

(a) $Ba_3(PO_4)_2$ ($K_{sp} = 6 \times 10^{-39}$) (b) ZnS ($K_{sp} = 7 \times 10^{-16}$)
(c) $Fe(OH)_3$ ($K_{sp} = 6 \times 10^{-38}$) (d) $Ag_3(PO_4)$ ($K_{sp} = 1.8 \times 10^{-18}$)

127. Silver ions are added to a solution with $[Br^-] = [Cl^-] = [CO_3^{2-}] = [AsO_4^{3-}] = 0.1M$. Which compound will precipitate with lowest $[Ag^+]$?

(a) AgBr ($K_{sp} = 5 \times 10^{-13}$) (b) AgCl ($K_{sp} = 1.8 \times 10^{-10}$)
(c) Ag_2CO_3 ($K_{sp} = 8.1 \times 10^{-12}$) (d) Ag_3AsO_4 ($K_{sp} = 1 \times 10^{-22}$)

128. The solubility product of different sparingly soluble salts are given as under :

S. No.	Formula Type	Solubility product
1.	AB	4.0×10^{-20}
2.	A_2B	3.2×10^{-11}
3.	AB_3	2.7×10^{-31}

The correct increasing order of solubility is :

(a) 1, 3, 2 (b) 2, 1, 3
(c) 1, 2, 3 (d) 3, 1, 2

129. If K_{sp} for $HgSO_4$ is 6.4×10^{-5}, then solubility of this substance in mole per m^3 is :

(a) 8×10^{-3} (b) 6.4×10^{-5}
(c) 8×10^{-6} (d) None of these

130. The solubility of $Ba_3(AsO_4)_2$ (formula mass $= 690$) is 6.9×10^{-2} g/100 mL. What is the K_{sp} ?

(a) 1.08×10^{-11} (b) 1.08×10^{-13}
(c) 1.0×10^{-15} (d) 6.0×10^{-13}

131. The solubility of $AgBrO_3$ (formula mass $= 236$) is 0.0072 g in 1000 mL. What is the K_{sp} ?

(a) 2.2×10^{-8} (b) 3.0×10^{-10}
(c) 3.0×10^{-5} (d) 9.3×10^{-10}

132. The solubility of PbF_2 (formula mass $= 245$) is 0.46 g/L. What is the solubility product?

(a) 1.1×10^{-10} (b) 2.6×10^{-8}
(c) 1.1×10^{-7} (d) 6.8×10^{-9}

133. How many grams of MgC_2O_4 (formula mass $= 112$) will dissolve in 1.5 L of water? ($K_{sp} = 8.1 \times 10^{-5}$)

(a) 1.0 (b) 1.29
(c) 1.512 (d) 4.65

134. What is the molarity of F^- ions in a saturated solution of BaF_2? ($K_{sp} = 1.0 \times 10^{-6}$)

(a) 1.0×10^{-2} (b) 1.0×10^{-3}
(c) 1.26×10^{-2} (d) 6.3×10^{-3}

135. What is the molarity of F^- in a saturated solution of InF_3? ($K_{sp} = 7.9 \times 10^{-10}$)

(a) 2.3×10^{-3} (b) 8.3×10^{-3}
(c) 1.0×10^{-3} (d) 7.0×10^{-3}

136. What is the pH of a saturated solution of $Cu(OH)_2$? ($K_{sp} = 2.6 \times 10^{-19}$)

(a) 6.1 (b) 7.30
(c) 8.42 (d) 7.90

137. The solubility product of AgCl is $10^{-10}\ M^2$. The minimum volume (in m^3) of water required to dissolve 14.35 mg of AgCl is approximately :

(a) 0.01 (b) 0.1
(c) 100 (d) 10

138. What is the molar solubility of $Fe(OH)_2$ ($K_{sp} = 8.0 \times 10^{-16}$) at pH 13.0?

(a) 8.0×10^{-18} (b) 8.0×10^{-15}
(c) 8.0×10^{-17} (d) 8.0×10^{-14}

139. What is the minimum pH necessary to cause a precipitate of $Pb(OH)_2$ ($K_{sp} = 1.2 \times 10^{-5}$) to form in a 0.12 M $PbCl_2$ solution?

(a) 12.4 (b) 10.8

(c) 12.0 (d) 11.1

140. Which of the following would increase the solubility of $Pb(OH)_2$?

(a) Add hydrochloric acid

(b) Add a solution of $Pb(NO_3)_2$

(c) Add a solution of NaOH

(d) None of the above–the solubility of a compound is constant at constant temperature

141. What is the molar solubility of Ag_2CO_3 ($K_{sp} = 4 \times 10^{-13}$) in 0.1 M Na_2CO_3 solution?

(a) 10^{-6} (b) 10^{-7}

(c) 2×10^{-6} (d) 2×10^{-7}

142. What is the concentration of Pb^{2+} when $PbSO_4$ ($K_{sp} = 1.8 \times 10^{-8}$) begins to precipitate from a solution that is 0.0045 M in SO_4^{2-}?

(a) 4.0×10^{-8} M (b) 1.0×10^{-6} M

(c) 2.0×10^{-8} M (d) 4.0×10^{-6} M

143. What is the concentration of Ba^{2+} when BaF_2 ($K_{sp} = 1.0 \times 10^{-6}$) begins to precipitate from a solution that is 0.30 M F^-?

(a) 9.0×10^{-7} (b) 3.3×10^{-5}

(c) 1.1×10^{-5} (d) 3.0×10^{-7}

144. Solubility of AgCl in 0.2 M NaCl is x and that in 0.1 M $AgNO_3$ is y then which of the following is correct?

(a) $x = y$ (b) $x > y$

(c) $x < y$ (d) We cannot predict

145. What is the molarity of $Fe(CN)_6^{4-}$ in a saturated solution of $Ag_4[Fe(CN)_6]$? ($K_{sp} = 1.6 \times 10^{-41}$)

(a) 1.6×10^{-8} (b) 5.2×10^{-8}

(c) 2.0×10^{-8} (d) 2.3×10^{-9}

146. At 25°C, K_{sp} for $PbBr_2$ is equal to 8×10^{-5}. If the salt is 80% dissociated, what is the solubility of $PbBr_2$ in mol/litre?

(a) $\left[\dfrac{10^{-4}}{1.6 \times 1.6}\right]^{1/3}$ (b) $\left[\dfrac{10^{-5}}{1.6 \times 1.6}\right]^{1/3}$

(c) $\left[\dfrac{10^{-4}}{0.8 \times 0.8}\right]^{1/3}$ (d) $\left[\dfrac{10^{-5}}{1.6 \times 1.6}\right]^{1/2}$

147. What is the molar solubility of $Mn(OH)_2$ ($K_{sp} = 4.5 \times 10^{-14}$) in a buffer solution containing equal amounts of NH_4^+ and NH_3 ($K_b = 1.8 \times 10^{-5}$)?

(a) 3.0×10^{-4} (b) 1.38×10^{-4}

(c) 1.38×10^{-3} (d) 7.3×10^{-4}

148. Find moles of NH_4Cl required to prevent $Mg(OH)_2$ from precipitating in a litre of solution which contains 0.02 mole of NH_3 and 0.001 mole of Mg^{2+} ions.

Given : K_b $(NH_3) = 10^{-5}$; K_{sp} $[Mg(OH)_2] = 10^{-11}$.

(a) 10^{-4} (b) 2×10^{-3}

(c) 0.02 (d) 0.1

149. What mass of AgI will dissolve in 1.0 L of 1.0 M NH_3? Neglect change in conc. of NH_3.

[Given : K_{sp} (AgI) $= 1.5 \times 10^{-16}$; K_f $[Ag(NH_3)_2^+] = 1.6 \times 10^7$]; (At. mass Ag = 108; I = 127)

(a) 4.9×10^{-5} g (b) 0.0056 g

(c) 0.035 g (d) 0.011 g

150. Consider the following statement and select correct option :

(I) K_{sp} of $Fe(OH)_3$ in aqueous solution is 3.8×10^{-38} at 298 K. The concentration of Fe^{3+} will increase when $[H^+]$ ion concentration decreases.

(II) In a mixture of NH_4Cl and NH_4OH in water, a further amount of NH_4Cl is added. The pH of the mixture will decreases.

(III) An aqueous solution of each of the following salts (NH_4I, HCOOK) will be basic, acidic respectively.

(a) only I is correct (b) only II is correct

(c) only III is correct (d) II and III are correct

Level-2

1. Equilibrium constants of T_2O (T or ${}^3_1H$ is an isotope of ${}^1_1H$) and H_2O are different at 298 K. Let at 298 K pure T_2O has pT (like pH) is 7.62. The pT of a solution prepared by adding 10 mL of 0.2 *M* TCl to 15 mL of 0.25 *M* NaOT is :
(a) $2 - \log 7$ (b) $14 + \log 7$
(c) $13.24 - \log 7$ (d) $13.24 + \log 7$

2. Liquid NH_3 dissociation to a slight extent. At a certain temp. its self dissociation constant $K_{SDC\,(NH_3)} = 10^{-30}$. The number of NH_4^+ ions are present per 100 cm^3 of pure liquid are :
(a) 10^{-15} (b) 6.022×10^8
(c) 6.022×10^7 (d) 6.022×10^6

3. To what volume of 10 litre of 0.5 *M* CH_3COOH ($K_a = 1.8 \times 10^{-5}$) be diluted in order to double the hydroxide ion concentration :
(a) 20 L (b) 30 L
(c) 40 L (d) None of these

4. 20 mL of 0.1 *M* weak acid HA ($K_a = 10^{-5}$) is mixed with solution of 10 mL of 0.3 *M* HCl and 10 mL of 0.1 *M* NaOH. Find the value of $[A^-]/([HA]+[A^-])$ in the resulting solution :
(a) 2×10^{-4} (b) 2×10^{-5}
(c) 2×10^{-3} (d) 0.05

5. What concentration of FCH_2COOH ($K_a = 2.6 \times 10^{-3}$) is initially needed so that $[H^+] = 2 \times 10^{-3}$ *M* ?
(a) 2×10^{-3} *M* (b) 2.6×10^{-3} *M*
(c) 5.2×10^{-3} *M* (d) 3.53×10^{-3} *M*

6. Calculate the ratio of $[HCOO^-]$ and $[F^-]$ in a mixture of 0.2 *M* HCOOH ($K_a = 2 \times 10^{-4}$) and 0.1 *M* HF ($K_a = 6.6 \times 10^{-4}$) :
(a) 1 : 6.6 (b) 1 : 3.3
(c) 2 : 3.3 (d) 3.3 : 2

7. If first dissociation of $X(OH)_3$ is 100% where as second dissociation is 50% and third dissociation is negligible then the pH of 4×10^{-3} *M* $X(OH)_3$ is :
(a) 11.78 (b) 10.78
(c) 2.5 (d) 2.22

8. H_3A is a weak triprotic acid ($K_{a_1} = 10^{-5}$, $K_{a_2} = 10^{-9}$, $K_{a_3} = 10^{-13}$)

What is the value of pX of 0.1 *M* H_3A (*aq.*) solution? where $pX = -\log X$ and $X = \dfrac{[A^{3-}]}{[HA^{2-}]}$
(a) 7 (b) 8
(c) 9 (d) 10

9. Calcium lactate is a salt of weak organic acid and strong base represented as $Ca(LaC)_2$. A saturated solution of $Ca(LaC)_2$ contains 0.6 mole in 2 litre solution. pOH of solution is 5.60. If 90% dissociation of the salt takes place then what is pK_a of lactic acid?
(a) $2.8 - \log(0.54)$ (b) $2.8 + \log(0.54)$
(c) $2.8 + \log(0.27)$ (d) None of these

10. What is the concentration of $CH_3COOH(aq.)$ in a solution prepared by dissolving 0.01 mole of $NH_4^+CH_3COO^-$ in 1 L H_2O? [$K_{a(CH_3COOH)} = 1.8 \times 10^{-5}$; $K_{b(NH_4OH)} = 1.8 \times 10^{-5}$]

(a) 5.55×10^{-5} (b) 0.10

(c) 6.4×10^{-4} (d) 5.55×10^{-3}

11. K_a for the reaction; $Fe^{3+}(aq.) + H_2O(l) \rightleftharpoons Fe(OH)^{2+}(aq.) + H_3O^+(aq.)$ is 6.5×10^{-3}. What is the max. pH value which could be used so that at least 80% of the total iron (III) in a dilute solution exists as Fe^{3+}?

(a) 2 (b) 2.41

(c) 2.79 (d) 1.59

12. $Fe(OH)_2$ is diacidic base has $K_{b_1} = 10^{-4}$ and $K_{b_2} = 2.5 \times 10^{-6}$

What is the concentration of $Fe(OH)_2$ in 0.1 *M* $Fe(NO_3)_2$ solution?

(a) 4×10^{-9} (b) 2.5×10^{-6}

(c) 10^{-10} (d) 10^{-14}

13. How many gm of solid KOH must be added to 100 mL of a buffer solution to make the pH of solution 6.0, if it is 0.1 *M* each w.r.t. acid HA and salt K*A*.

[Given : pK_a (HA) = 5]

(a) 0.458 (b) 0.327

(c) 5.19 (d) None of these

14. Fixed volume of 0.1 *M* benzoic acid ($pK_a = 4.2$) solution is added into 0.2 *M* sodium benzoate solution and formed a 300 mL, resultant acidic buffer solution. If pH of this buffer solution is 4.5 then find added volume of benzoic acid :

(a) 100 mL (b) 150 mL

(c) 200 mL (d) None of these

15. A 1.025 g sample containing a weak acid HX (mol. mass = 82) is dissolved in 60 mL water and titrated with 0.25 *M* NaOH. When half of the acid was neutralised the pH was found to be 5.0 and at the equivalence point the pH is 9.0. Calculate mass percentage of HX in sample :

(a) 50% (b) 75%

(c) 80% (d) None of these

16. Which of the following expression for % dissociation of a monoacidic base (*B*OH) in aqueous solution at appreciable concentration is not correct?

(a) $100 \times \sqrt{\frac{K_b}{c}}$ (b) $\frac{1}{1 + 10^{(pK_b - pOH)}}$

(c) $\frac{K_w[H^+]}{K_b + K_w}$ (d) $\frac{K_b}{K_b + [OH^-]}$

17. A solution of weak acid HA was titrated with base NaOH. The equivalent point was reached when 40 mL of 0.1 *M* NaOH has been added. Now 20 mL of 0.1 *M* HCl were added to titrated solution, the pH was found to be 5.0. What will be the pH of the solution obtained by mixing 20 mL of 0.2 *M* NaOH and 20 mL of 0.2 *M* HA?

(a) 7 (b) 9

(c) 11 (d) None of these

18. A buffer solution 0.04 M in Na_2HPO_4 and 0.02 M in Na_3PO_4 is prepared. The electrolytic oxidation of 1.0 milli-mole of the organic compound $RNHOH$ is carried out in 100 mL of the buffer. The reaction is

$$RNHOH + H_2O \longrightarrow RNO_2 + 4H^+ + 4e^-$$

The approximate pH of solution after the oxidation is complete is :
[Given : for H_3PO_4, $pK_{a_1} = 2.2$; $pK_{a_2} = 7.20$; $pK_{a_3} = 12$]
(a) 6.90 (b) 7.20
(c) 7.5 (d) None of these

19. When a 20 mL of 0.08 M weak base BOH is titrated with 0.08 M HCl, the pH of the solution at the end point is 5. What will be the pOH if 10 mL of 0.04 M NaOH is added to the resulting solution?
[Given : $\log 2 = 0.30$ and $\log 3 = 0.48$]
(a) 5.40 (b) 5.88
(c) 4.92 (d) None of these

20. Calculate approximate pH of the resultant solution formed by titration of 25 mL of 0.04 M Na_2CO_3 with 50 mL of 0.025 M HCl. [Given : $pK_{a_1} = 6.4$ and $pK_{a_2} = 10.3$ for H_2CO_3]
(a) 5.92 (b) 6.88
(c) 6.4 (d) 5.88

21. In the titration of a solution of a weak acid HA and NaOH, the pH is 5.0 after 10 mL of NaOH solution has been added and 5.60 after 20 mL NaOH has been added.
What is the value of pK_a for HA?
(a) 5.15 (b) 5.3
(c) 5.6 (d) None of these

22. 50 mL of 0.05 M Na_2CO_3 is titrated against 0.1 M HCl. On adding 40 mL of HCl, pH of the solution will be
[Given : For H_2CO_3, $pK_{a_1} = 6.35$, $pK_{a_2} = 10.33$; $\log 3 = 0.477$, $\log 2 = 0.30$]
(a) 6.35 (b) 6.526
(c) 8.34 (d) 6.173

23. 10 mL of 0.1 M tribasic acid H_3A is titrated with 0.1 M NaOH solution. What is the ratio of $\frac{[H_3A]}{[A^{3-}]}$ at 2nd equivalence point? [Given : $K_{a_1} = 10^{-3}$, $K_{a_2} = 10^{-8}$, $K_{a_3} = 10^{-12}$]
(a) $\simeq 10^{-4}$ (b) $\simeq 10^{+4}$
(c) $\simeq 10^{-7}$ (d) $\simeq 10^{+6}$

24. A_3B_2 is a sparingly soluble salt of molar mass M (g mol^{-1}) and solubility x g $litre^{-1}$. The ratio of the molar concentration of B^{3-} to the solubility product of the salt is :
(a) $108\,\frac{x^5}{M^5}$ (b) $\frac{1}{108}\,\frac{M^4}{x^4}$
(c) $\frac{1}{54}\,\frac{M^4}{x^4}$ (d) None of these

25. A solution is 0.10 M $Ba(NO_3)_2$ and 0.10 M $Sr(NO_3)_2$. If solid Na_2CrO_4 is added to the solution, what is $[Ba^{2+}]$ when $SrCrO_4$ begins to precipitate?
[$K_{sp}(BaCrO_4) = 1.2 \times 10^{-10}$; $K_{sp}(SrCrO_4) = 3.5 \times 10^{-5}$]
(a) 7.4×10^{-7} (b) 2.0×10^{-7}
(c) 6.1×10^{-7} (d) 3.4×10^{-7}

26. A solution is 0.01 M KI and 0.1 M KCl. If solid $AgNO_3$ is added to the solution, what is the $[I^-]$ when AgCl begins to precipitate?
$[K_{sp}\ (AgI) = 1.5 \times 10^{-16};\ K_{sp}\ (AgCl) = 1.8 \times 10^{-10}]$

(a) 3.5×10^{-7}
(b) 6.1×10^{-8}
(c) 2.2×10^{-7}
(d) 8.3×10^{-8}

27. A solution of 0.1 M in Cl^- and 10^{-4} M CrO_4^{-2}. If solid $AgNO_3$ is gradually added to this solution, what will be the concentration of Cl^- when Ag_2CrO_4 begins to precipitate?

$$[K_{sp}\ (AgCl) = 10^{-10}\ M^2;\ K_{sp}\ (Ag_2CrO_4) = 10^{-12}\ M^3]$$

(a) $10^{-6}\ M$
(b) $10^{-4}\ M$
(c) $10^{-5}\ M$
(d) $10^{-9}\ M$

28. If 500 mL of 0.4 M $AgNO_3$ is mixed with 500 mL of 2 M NH_3 solution then what is the concentration of $Ag(NH_3)^+$ in solution?
Given : $K_{f_1}\ [Ag(NH_3)]^+ = 10^3;\ K_{f_2}\ [Ag(NH_3)_2^+] = 10^4$

(a) $3.33 \times 10^{-7}\ M$
(b) $3 \cdot 33 \times 10^{-5}\ M$
(c) $3 \times 10^{-4}\ M$
(d) $10^{-7}\ M$

29. The simultaneous solubility of AgCN ($K_{sp} = 2.5 \times 10^{-16}$) and AgCl ($K_{sp} = 1.6 \times 10^{-10}$) in 1.0 M $NH_3(aq.)$ are respectively : [Given : $K_f\ [Ag(NH_3)_2^+] = 10^7$]

(a) $0.037, 5.78 \times 10^{-8}$
(b) $5.78 \times 10^{-8}, 0.037$
(c) $0.04, 6.25 \times 10^{-8}$
(d) $1.58 \times 10^{-3}, 1.26 \times 10^{-5}$

30. There exist an equilibrium between solid $SrSO_4$ and Sr^{2+} and SO_4^{2-} ion in aqueous medium. The possible equilibrium states are shown in figure as thick line. Now, if equilibrium is disturbed by addition of (a) $Sr(NO_3)_2$ and (b) K_2SO_4 and dotted line represent approach of system towards equilibrium. Match the columns given below :

(iii)
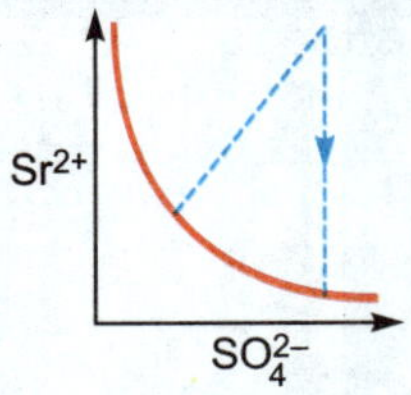

(iv)
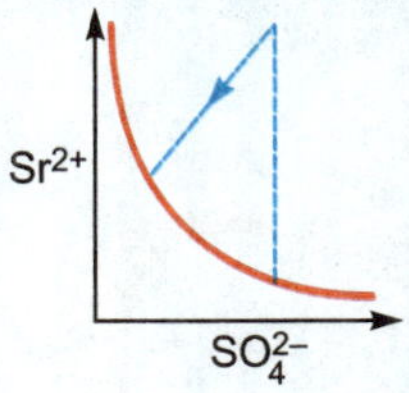

(v)
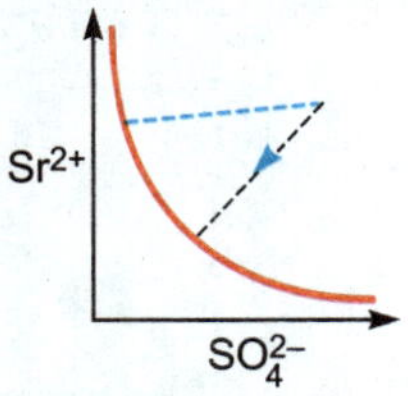

(vi)
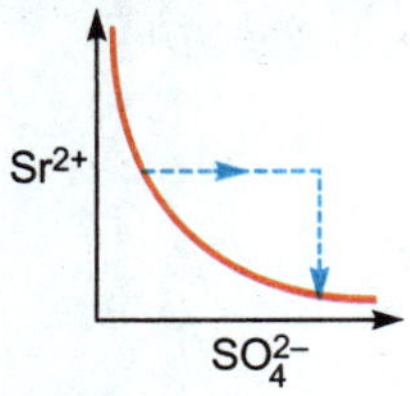

(I) addition of $Sr(NO_3)_2$
(II) addition of K_2SO_4

(a) (I) (iii), (II) (iv)
(b) (I) (iv), (II) (v)
(c) (I) (vi), (II) (v)
(d) (I) (iv), (II) (vi)

31. Solubility of AgCN is maximum in :

(a) acidic buffer solution
(b) basic buffer solution
(c) in pure water
(d) equal in all solution

32. $AgBr(s) + 2S_2O_3^{2-}(aq.) \rightleftharpoons Ag(S_2O_3)_2^{3-}(aq.) + Br^-(aq.)$
[Using : $K_{sp}\ (AgBr) = 5 \times 10^{-13}$ $\quad K_f\ (Ag(S_2O_3)_2^{3-}) = 5 \times 10^{13}$]

What is the molar solubility of AgBr in 0.1 M $Na_2S_2O_3$?

(a) 0.5 M
(b) 0.45 M
(c) 0.045 M
(d) None of these

33. What is $[Ag^+]$ in a solution made by dissolving both Ag_2CrO_4 and $Ag_2C_2O_4$ until saturation is reached with respect to both salts?

$$[K_{sp}\,(Ag_2C_2O_4) = 2\times10^{-11}\ ,\quad K_{sp}\,(Ag_2CrO_4) = 2\times10^{-12}]$$

(a) 2.80×10^{-4} (b) 7.6×10^{-5}

(c) 6.63×10^{-6} (d) 3.52×10^{-4}

34. What is the minimum pH required to prevent the precipitation of ZnS in a solution that is 0.01 *M* $ZnCl_2$ and saturated with 0.10 *M* H_2S?

[Given : $K_{sp} = 10^{-21}$, $K_{a_1}\times K_{a_2} = 10^{-20}$]

(a) 0 (b) 1

(c) 2 (d) 4

35. The $Al(OH)_3$ is involved in the following two equilibria,

$$Al(OH)_3(s) \rightleftharpoons Al^{3+}(aq.) + 3OH^-(aq.);\ K_{sp}$$

$$Al(OH)_3(s) + OH^-(aq.) \rightleftharpoons Al(OH)_4^-(aq.);\ K_c$$

Which of the following relationship is correct at which solubility is minimum?

(a) $[OH^-] = \left(\dfrac{K_{sp}}{K_c}\right)^{1/3}$ (b) $[OH^-] = \left(\dfrac{K_c}{K_{sp}}\right)^{1/4}$

(c) $[OH^-] = \sqrt{\left(\dfrac{K_{sp}}{K_c}\right)^{1/4}}$ (d) None of these

Level-3

Passage-1

Solution of a weak acid and its anion (that is, its conjugate base) or of a base and its common cation are buffered. When we add a small amount of acid or base to any one of them, the pH of solution changes very little. pH of buffer solution can be computed as

for acidic buffer : $pH = pK_a + \log \dfrac{[\text{Conjugate base}]}{[\text{Acid}]}$

for basic buffer : $pOH = pK_b + \log \dfrac{[\text{Conjugate acid}]}{[\text{Base}]}$

It is generally accepted that a solution has useful buffer capacity (pH change resisting power) provided that the value of [salt or conjugate base]/[acid] for acidic buffer lies within the range of 1 : 10 to 10 : 1. Buffer capacity is maximum when [conjugate base] = [acid]

1. One litre of an aqueous solution contains 0.15 mole of CH_3COOH ($pK_a = 4.8$) and 0.15 mole of CH_3COONa. After the addition of 0.05 mole of solid NaOH to this solution, the pH will be :

(a) 4.5 (b) 4.8
(c) 5.1 (d) 5.4

2. Calculate the pH of a solution made by adding 0.01 mole of HCl in 100 mL of a solution which is 0.2 *M* in NH_3 ($pK_b = 4.74$) and 0.3 *M* in NH_4^+ :

(Assuming no change in volume)

(a) 5.34 (b) 8.66
(c) 7.46 (d) None of these

3. Useful buffer range of weak acid *HA* ($K_a = 10^{-5}$) is :

(a) 5 to 7 (b) 4 to 6
(c) 3 to 5 (d) None of these

4. Select correct statement :

(a) When we add small amount of NaOH in acidic buffer solution, pOH of solution is increases
(b) When we add small amount of NaOH in basic buffer solution, pH of solution is increases
(c) When we add small amount of water in acidic buffer solution, pH of solution is decreases
(d) When 100 mL of 0.2 *M* CH_3COOH react with 200 mL of 0.1 *M* NaOH buffer solution is formed

Passage-2

Hydrolysis is an acid-base reaction of a cation or anion or both ions of a salt with water. Resultant solution of hydrolysis may be acidic, basic or neutral. The anion A^- which is a weaker base than OH^- and which has its conjugate acid *HA* stronger than water but weaker than H_3O^+ shows the phenomenon of hydrolysis Ex. : CH_3COO^-, CN^-, NO_2^-, etc.

The cation B^+ which is a weaker acid than H_3O^+ and which has its conjugate base *BOH* stronger than water but weaker than OH^- shows the phenomenon of hydrolysis Ex. : NH_4^+, $C_6H_5NH^+$, $N_2H_5^+$ etc.

The hydrolysis constant of anion and cation are given by

$$A^-(aq.) + H_2O(l) \rightleftharpoons HA(aq.) + OH^-(aq.)$$

$$K_h = \frac{K_w}{K_a} \Rightarrow \frac{[HA(aq.)][OH^-(aq.)]}{[A^-(aq.)]}$$

$$B^+(aq.) + H_2O(l) \rightleftharpoons BOH(aq.) + H^+(aq.)$$

$$K_h = \frac{K_w}{K_b} \Rightarrow \frac{[BOH(aq.)][H^+(aq.)]}{[B^+(aq.)]}$$

1. Which of the following statement is true?
(a) Weaker the acid, greater will be hydrolysis of its anion
(b) Weaker the base, greater will be hydrolysis of its cation
(c) Both (a) and (b)
(d) None of these

2. Select the correct statement :
(a) KCl undergoes hydrolysis
(b) $K_h = K_b(A^-)$ and $K_h = K_a(B^+)$
(c) 0.1 *M* solution of NaCN is acidic
(d) resultant solution of equal volume of 0.1 *M* NH_3 and 0.1 *M* HCl is basic

3. When pure ammonium chloride is dissolved in pure water, the pH of the resulting solution is not 7. This is because :
(a) ammonium ions accept protons from water molecules leaving free OH^- ions in solution
(b) ammonium ions donate protons to water molecules forming H_3O^+ ions in solution
(c) ammonium ions combine with water molecule to give the weak base, ammonium hydroxide
(d) chloride ion made the solution acidic

4. Calculate percentage degree of hydrolysis in a 0.1 *M* solution of CH_3COONa.
(K_a of $CH_3COOH = 10^{-5}$)
(a) 0.1 (b) 0.01
(c) 10^{-4} (d) None of these

Passage-3

Acid-base indicators are either weak organic acids or weak organic bases. Indicator change colour in dilute solution when the hydronium ion concentration reaches a particular value. For example, phenolphthalein is a colourless substance in any aqueous solution with a pH less than 8.3. In between the pH range 8.3 to 10, transition of colour (colourless to pink) takes place and if pH of solution is greater than 10 solution is dark pink. Considering an acid indicator HIn, the equilibrium involving it and its conjugate base In^- can be represented as :

$$\underset{\text{acidic form}}{HIn} \rightleftharpoons H^+ + \underset{\text{basic form}}{In^-}$$

pH of solution can be computed as :

$$pH = pK_{In} + \log\frac{[In^-]}{[HIn]}$$

In general, transition of colour takes place in between the pH range $pK_{In} \pm 1$.

1. An indicator is a weak acid and pH range is 4.0 to 6.0. If indicator in 50% ionized in a given solution then what is the ionization constant of the acid?
(a) 10^{-4} (b) 10^{-5}
(c) 10^{-6} (d) None of these

2. Select the correct statement(s) :
(a) At midway in the transition of an acidic indicator, $pH = pK_{In}$
(b) Methyl orange (3.1 to 4.4) is a suitable indicator for titration of weak acid and strong base
(c) Bromothymol blue (6.0 to 7.6) is a good indicator for titration of HCl and NaOH
(d) Thymol blue (1.2 to 2.8) is a very good indicator for titration of 100 mL of 0.1 *M* NH_4OH ($pK_b = 4.74$) and 0.1 *M* HCl

3. Following is the titration curve of two acids HA and HB (5 milli-moles each) titrated against strong base NaOH (0.1 M)

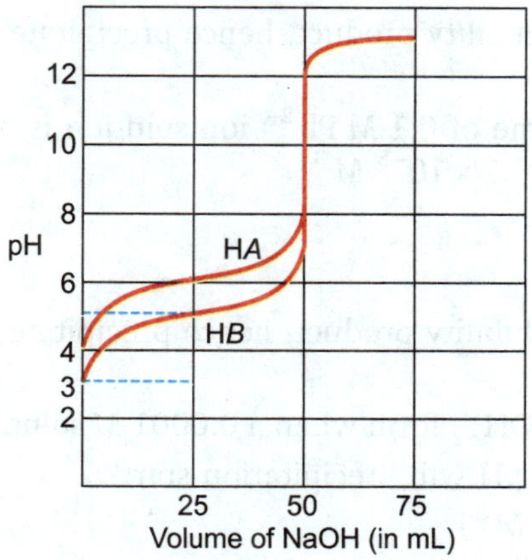

What is pK_a for HB acid ?

(a) 3 (b) 4

(c) 5 (d) 6

4. What is the equilibrium constant for the reaction :

$$HB(aq.) + NaA(aq.) \rightleftharpoons HA(aq.) + NaB(aq.)?$$

(a) 10 (b) 0.1

(c) 10^{-7} (d) 10^{-11}

5. Calculate the pH at equivalent point when HB is titrated with NaOH.

(a) 8.75 (b) 8.85

(c) 9.0 (d) None of these

6. Which of the following indicator is most suitable for titration of HB with strong base :

(a) Phenolphthalein (8.3 – 10) (b) Bromothymol blue (6 – 7.6)

(c) Methyl red (4.2 – 6.3) (d) Malachite green (11.4 – 13)

Passage-4

Consider a saturated solution of silver chloride that is in contact with solid silver chloride. The solubility equilibrium can be represented as

$$AgCl(s) \rightleftharpoons Ag^+(aq.) + Cl^-(aq.); \quad K_{sp} = [Ag^+(aq.)][Cl^-(aq.)]$$

Where K_{sp} is called the solubility product constant or simply the solubility product. In general, the solubility product of a compound is the product of the molar concentrations of the constituent ions, each raised to the power of its stoichiometric coefficient in the equilibrium equation.

For concentrations of ions that do not necessarily correspond to equilibrium conditions we use the reaction quotient (Q) which is called the ion or ionic product (Q), to predict whether a precipitate will form. Note that Q has the same form as K_{sp}.

The possible relationships between Q and K_{sp} are

$Q < K_{sp}$ Unsaturated solution

$Q = K_{sp}$ Saturated solution

$Q > K_{sp}$ Supersaturated solution; precipitate will form

1. Will a precipitate form if 50 cm^3 of 0.01 M $AgNO_3$ and 50 cm^3 of 2×10^{-5} M NaCl are mixed?
[Given : K_{sp} (AgCl) $= 10^{-10}$ M^2]

(a) Yes
(b) No
(c) Ionic product is less than solubility product, hence precipitate will form
(d) Data insufficient

2. Will a precipitate form if 1 volume of 0.1 M Pb^{2+} ion solution is mixed with 3 volume of 0.3 M Cl^- ion solution? [Given : K_{sp} $(PbCl_2) = 1.7 \times 10^{-5}$ M^3]
(a) Yes
(b) No
(c) Ionic product is less than solubility product, hence precipitate will form
(d) Data insufficient

3. At 25°C, will a precipitate of $Mg(OH)_2$ form when a 0.0001 M solution of $Mg(NO_3)_2$ is adjusted to a pH of 9.0? At what minimum value of pH will precipitation start?
[Given : K_{sp} $(Mg(OH)_2) = 10^{-11}$ M^3]
(a) No, pH = 3.5
(b) No, pH = 10.5
(c) No, pH = 6.0
(d) Yes, pH = 8.5

4. Determine the molar solubility of MgF_2 from its solubility product $K_{sp} = 4 \times 10^{-9}$:
(a) 10^{-3}
(b) 6.32×10^{-5}
(c) 2×10^{-5}
(d) None of these

5. The solubility product of ferric hydroxide in aqueous solution is 6×10^{-38} at 298 K. The solubility of Fe^{3+} ion will increase when the :
(a) pH is increased
(b) pH is 7.0
(c) pH is decreased
(d) saturated solution is exposed to the atmosphere

One or More Answer is/are Correct

1. Which is/are wrong statement(s)?
(a) Arrhenius acids are also Bronsted acids but all Arrhenius bases are not Bronsted bases
(b) All Lewis bases are Bronsted bases
(c) All Bronsted acids are Lewis acids
(d) Conjugate base of a strong acid is weak

2. Which of the following are conjugate acid-base pairs ?
(a) HCO_3^-, CO_3^{2-}
(b) $C_6H_5\overset{+}{N}H_3, C_6H_5NH_2$
(c) $H_2C_2O_4, HC_2O_4^-$
(d) OH^-, H^+

3. Which are the set of amphiprotic species?
(a) $H_2O, H_2PO_4^-, HPO_4^{2-}$
(b) HPO_4^{2-}, HCO_3^-
(c) $H_2PO_2^-, H_2PO_3^-, HC_2O_4^-$
(d) $HPO_3^{2-}, H_2O, CO_3^{2-}$

4. Which of the following statements is/are not correct?
(a) A substance which can provide OH^- in aqueous medium is a base
(b) A substance which can accept a pair of electron is a base
(c) A substance which can accept a proton in aqueous medium is a base
(d) A substance which can donate a pair of electron is a base

5. If degree of ionization (α) of a weak electrolyte AB is very less then α is :
(a) directly proportional to the square root of volume of solution

(b) inversely proportional to the dilution
(c) inversely proportional to the square root of concentration
(d) directly proportional to concentration

6. Factor influencing the degree of ionization of a weak electrolyte is :
(a) dilution (b) temperature
(c) presence of other ions (d) nature of solvent

7. Which of the following statement(s) is/are correct about the ionic product of water?
(a) K_i (ionization constant of water) $< K_w$ (ionic product of water)
(b) $pK_i > pK_w$
(c) At 25°C, $K_i = 1.8 \times 10^{-14}$
(d) Ionic product of water at 10°C is 10^{-14}

8. Which among the following statement is/are correct?
(a) $pH = -\log_{10} [H_3O^+]$ for dilute solution
(b) pH of H_2O decreases with increase of temperature
(c) pH can not more than 14
(d) If a solution is diluted ten times, its pH always increases by 1

9. If concentration of two weak acids are different and D.O.D. (α) are very less then their relative strength can be compared by :
(a) $\frac{[H^+]_1}{[H^+]_2}$ (b) $\frac{\alpha_1}{\alpha_2}$
(c) $\frac{C_1\alpha_1}{C_2\alpha_2}$ (d) $\frac{K_{a_1}C_1}{K_{a_2}C_2}$

10. If concentration of two weak bases are same and D.O.D. (α) are very less then their relative strength can be compared by :
(a) $\frac{[OH^-]_1}{[OH^-]_2}$ (b) $\frac{K_{b_1}}{K_{b_2}}$
(c) $\frac{\alpha_1}{\alpha_2}$ (d) $\frac{\sqrt{K_{b_1}}}{\sqrt{K_{b_2}}}$

11. Which of the following expressions is/are true?
(a) $[H^+] = [OH^-] = \sqrt{K_w}$ for a neutral solution
(b) $[OH^-] < \sqrt{K_w}$ for an acidic solution
(c) $pH + pOH = 14$ at all temperature
(d) $[OH^-] = 10^{-7}\ M$ for a neutral solution at 25°C

12. If K_{a_1}, K_{a_2} and K_{a_3} be the first, second and third dissociation constant of H_3PO_4 and $K_{a_1} >> K_{a_2} >> K_{a_3}$ which is/are correct :
(a) $[H^+] \simeq \sqrt{K_{a_1}[H_3PO_4]}$ (b) $[H^+] \simeq [HPO_4^{2-}]$
(c) $K_{a_2} \simeq [HPO_4^{2-}]$ (d) $[HPO_4^{2-}] = [PO_4^{3-}]$

13. Which of the following mixtures constitute a buffer?
(a) HCOOH + HCOONa (b) $Na_2CO_3 + NaHCO_3$
(c) NaCl + HCl (d) $NH_4Cl + (NH_4)_2SO_4$

14. Which of the following mixtures can act as a buffer?
(a) NaOH + HCOONa (1 : 1 molar ratio) (b) HCOOH + NaOH (2 : 1 molar ratio)
(c) NH_4Cl + NaOH (2 : 1 molar ratio) (d) HCOOH + NaOH (1 : 1 molar ratio)

15. Which of the following will function as buffer?

(a) NaCl + NaOH (b) Borax + boric acid
(c) $NaH_2PO_4 + Na_2HPO_4$ (d) $NH_4Cl + NH_4OH$

16. Which of the following statements is/are correct?

(a) The conjugate acid of NH_2^- is NH_3
(b) Solubility product constant increases with increase in concentration of ions
(c) On diluting a buffer solution pH change is negligible
(d) In alkaline buffer solution, if some HCl is added, its $[OH^-]$ will increase

17. Degree of hydrolysis (α) for a salt of strong acid and weak base is :

(a) independent of dilution (b) increases with dilution
(c) increases with decrease in K_b (d) increases with increase in temperature

18. The compound whose 0.1 *M* solution is acidic :
[pK_a of HCOOH = 3.75, pK_b of NH_4OH = 4.74]

(a) Ammonium formate (b) Ammonium sulphate
(c) Ammonium chloride (d) Sodium formate

19. Formic acid is a weak acid and hydrochloric acid is a strong acid. It shows that the :

(a) $[OH^-]$ of 0.01 *M* HCl (*aq.*) will be less than that of 0.01 *M* HCOOH (*aq.*)
(b) solution containing 0.1 *M* NaOH (*aq.*) and 0.1 *M* HCOONa (*aq.*) is a buffer solution
(c) pH of 10^{-9} *M* HCl (*aq.*) will be approximately 7 at 25°C
(d) pH of a solution formed by mixing equimolar quantities of HCOOH and HCl will be less than that of a similar solution formed from HCOOH and HCOONa

20. If you have a saturated solution of CaF_2, then :

(a) $[Ca^{2+}] = (K_{sp}/4)^{1/3}$ (b) $2 \times [Ca^{2+}] = [F^-]$
(c) $[Ca^{2+}] = 2[F^-]$ (d) $[Ca^{2+}] = \sqrt{K_{sp}}$

21. Which is/are correct statement(s) about the solubility of AgCl(*s*).?
Given : $K_{sp(AgCl)} = 10^{-10}$; $K_f[Ag(NH_3)_2^+] = 10^8$.

(a) Solubility of AgCl in pure water is 10^{-5} gm/litre
(b) Solubility of AgCl in 2 *M* KBr is 10^{-5} mol/litre
(c) Solubility of AgCl in 2 *M* $AgNO_3$ is 5×10^{-11} *M*
(d) Solubility of AgCl in 2*M* NH_3 is 0.166 *M*

22. H_2A is a weak diprotic acid. If the pH of 0.1*M* H_2A solution is 3 and concentration of A^{2-} is 10^{-12} at 25°C. Select correct statement(s)

(a) $[H^+]_{total} \approx [H^+]$ from first step of ionization of acid H_2A
(b) Concentration of OH^- in solution is 10^{-3} *M*
(c) The value of K_{a_1} is nearly 10^{-5}
(d) $pK_{a_2} - pK_{a_1} = 9$

23. Which is/are correct statement(s)?

(a) CH_3COONH_4 have greater degree of hydrolysis in 0.2 *M* solution in comparison of 0.4 *M* solution.
(b) Anions which are weaker base than OH^-, do not hydrolyse
(c) The CH_3COO^- have greater degree of hydrolysis in comparision of $HCOO^-$ when their salt solution have equal conc.
(d) SO_4^{2-} neglegible hydrolysis but HSO_4^- does not undergo hydrolysis

24. 0.01M NH_4Cl (*aq*) solution at 25°C has:

(a) $[Cl^-(aq)] < 10^{-2} M$ (b) $[NH_4^+(aq)] < 10^{-2} M$

(c) pOH > 7 (d) $[H^+] > 10^{-7}$ M

25. In an acidic indicator HIn has ionization constant is 10^{-8}. The acid form of indicator is yellow and alkaline form is red. Which is correct statement ?

(Given : log2 = 0.3, log3 = 0.48)

(a) The pH range of indicator is 7 to 9

(b) Change in pH is 0.96 when 75% yellow colour change to 75% red colour

(c) This indicator is suitable for the titration of strong acid *vs.* strong base

(d) pH of indicator is 8.3 when ratio of acid form to alkaline form is 2.

Match the Column

Column-I and **Column-II** contains four entries each. Entries of Column-I are to be matched with some entries of Column-II. One or more than one entries of Column-I may have the matching with the same entries of Column-II.

1.

Column-I	Column-II
(A) HCl	(P) Bronsted base
(B) NH_3	(Q) Bronsted acid
(C) H_2O	(R) Arrhenius acid
(D) CN^-	(S) Lewis base in adduct displacement reaction

2.

Column-I	Column-II
(A) Conjugate acid-base pair	(P) Bronsted-Lowry concept
(B) Acid-base adduct	(Q) Lewis concept
(C) An acid-base reaction	(R) Arrhenius concept
(D) Proton donation	(S) $K_a \cdot K_b = K_w$

3.

Column-I	Column-II
(A) $Fe(NO_3)_2(aq.)$	(P) Only cationic hydrolysis
(B) $KClO_4(aq.)$	(Q) Only anionic hydrolysis
(C) $HCOONa(aq.)$	(R) Both cationic as well as anionic hydrolysis
(D) $NH_4CN(aq.)$	(S) No hydrolysis

4.

Column-I	Column-II
(A) Salt of weak acid and weak base	(P) $pH = 1/2\,[pK_w + pK_a + \log C]$
(B) Salt of weak acid and strong base	(Q) $pH = 1/2\,[pK_w + pK_a - pK_b]$
(C) Salt of strong acid and strong base	(R) $pH = 1/2\,[pK_w - pK_b - \log C]$
(D) Salt of strong acid and weak base	(S) $pH = 1/2\,[pK_w]$

5.

Column-I	Column-II
(A) Salt of weak acid and weak base ($pK_a = pK_b$)	(P) pH of solution at 25°C less than 7
(B) Salt of weak acid and strong base	(Q) pH of solution at 25°C greater than 7
(C) Salt of strong acid and strong base	(R) pH of solution at 25°C equal to 7
(D) Salt of strong acid and weak base	(S) pH cannot be find until the value of K_a / K_b is given

6.

Column-I	Column-II
(A) pH of 0.1 M HA ($pK_a = 5$) and 0.01 M NaA	(P) 4
(B) pH of 0.1 M BOH ($pK_b = 6$) and 0.1 M BCl	(Q) 7
(C) pH of 0.1 M salt of weak acid ($pK_a = 5$) and weak base ($pK_b = 7$)	(R) 6
(D) pH of 500 litre of 0.02 M HNO_3 and 500 litre 0.01 M $Sr(OH)_2$	(S) 8

7.

Column-I	Column-II
(A) CH_3COOH ($pK_a = 4.74, 0.1\ M$) + CH_3COONa (0.1 M)	(P) Acidic buffer at its maximum capacity
(B) CH_3COOH (0.1 M) + HCl (0.1 M)	(Q) Buffer solution
(C) CH_3COOH ($pK_a = 4.74$, 0.1 M) + NH_4OH ($pK_b = 4.74, 0.1\ M$)	(R) pH < 7 at 25°C
(D) CH_3COONa (300 mL of 0.1 M) + HCl (100 mL of 0.1 M)	(S) pH = 7 at 25°C

8.

Column-I	Column-II
(A) Titration of a strong acid with strong base	(P) Methyl orange (3.1 – 4.4)
(B) Titration of weak acid with strong base	(Q) Methyl red (4.2 – 6.3)
(C) Titration of strong acid with weak base	(R) Phenolphthalein (8.3 – 10)
(D) Titration of weak acid with weak base	(S) No general indicator is suitable

9. When we titrate sodium carbonate solution (in beaker) with hydrochloric acid

Column-I	Column-II
(A) At the start of titration	(P) Buffer solution of HCO_3^- and CO_3^{2-}
(B) Before the first equivalent point	(Q) Buffer solution of H_2CO_3 and HCO_3^-
(C) At the first equivalent point	(R) Amphiprotic anion, $pH = 1/2\,(pK_{a_1} + pK_{a_2})$
(D) Between the first and second equivalent points	(S) Hydrolysis of CO_3^{2-}

10.

Column-I	Column-II
(A) Mercurous iodide	(P) $108\,S^5$
(B) Aluminium phosphate	(Q) $4\,S^3$
(C) Calcium phosphate	(R) S^2
(D) Zirconium phosphate	(S) $6912\,S^7$

Assertion-Reason Type Questions

Each question contains STATEMENT-1 (Assertion) and STATEMENT-2 (Reason).
Examine the statements carefully and mark the correct answer according to the instructions given below :
(A) If both the statements are TRUE and STATEMENT-2 is the correct explanation of STATEMENT-1
(B) If both the statements are TRUE but STATEMENT-2 is NOT the correct explanation of STATEMENT-1
(C) If STATEMENT-1 is TRUE and STATEMENT-2 is FALSE
(D) If STATEMENT-1 is FALSE and STATEMENT-2 is TRUE

1. STATEMENT-1 : All strong monoprotic acid with same concentration in dilute solution show same pH.
STATEMENT-2 : Water shows levelling effect.

2. STATEMENT-1 : If water is heated to 50°C then pH will increase.
STATEMENT-2 : K_w increases with increase in temperature.

3. STATEMENT-1 : Addition of HCl(*aq.*) to HCOOH(*aq.*) decrease the dissociation of HCOOH(*aq.*)
STATEMENT-2 : Due to common ion effect of H^+, dissociation of HCOOH decrease.

4. STATEMENT-1 : pH of 10^{-7} M HCl is less than 7 at 25°C.
STATEMENT-2 : At very low concentration of HCl, contribution of H^+ from water is considerable.

5. STATEMENT-1 : The dissociation constants of weak diprotic acid are in the order of $K_{a_1} > K_{a_2}$.
STATEMENT-2 : Removal of H^+ from anion is difficult as compared to neutral molecule.

6. STATEMENT-1 : When 0.1 M weak diprotic acid H_2A dissociates with its dissociation constants $K_{a_1} = 10^{-3}$ and $K_{a_2} = 10^{-8}$, then $[A^{2-}]$ is almost equal to $10^{-3}\,M$.
STATEMENT-2 : Since $K_{a_2} << K_{a_1}$ for 0.1 M H_2A, so $[A^{2-}]$ is negligible w.r.t. $[HA^-]$.

7. STATEMENT-1 : pH value of acidic buffer solution change, if buffer solution is diluted upto very larger extent.
STATEMENT-2 : $[H^+]$ decrease due to change in concentration as well as α and decrease in concentration is more as compared to increase in α.

8. STATEMENT-1 : In a titration of weak acid with strong base, the pH at the half equivalence point is pK_a.
STATEMENT-2 : At half equivalence point, it will form acidic buffer at its maximum capacity where [acid] = [salt].

9. STATEMENT-1 : In the titration of Na_2CO_3 with HCl using methyl orange indicator, the volume of the acid required at the equivalence point is twice that of the acid required using phenolphthalein as indicator.

STATEMENT-2 : Two moles of HCl are required for the complete neutralisation of one mole of Na_2CO_3.

10. STATEMENT-1 : In the acid-base titration involving strong base and weak acid, methyl red can be used as an indicator.

STATEMENT-2 : Methyl red changes its colour in the pH range 4.2 to 6.3.

11. STATEMENT-1 : Sparingly soluble salts AB and XY_2 with the same solubility product, will have different solubility.

STATEMENT-2 : Solubility of sparingly soluble salt depend upon solubility product.

12. STATEMENT-1 : Solubility product of BaF_2 will increase on dilution.

STATEMENT-2 : Solubility of BaF_2 will change on changing temperature.

13. STATEMENT-1 : Solubility of sparingly soluble salt decreases due to common ion effect.

STATEMENT-2 : Solubility product constant does not depend on common ion effect.

14. STATEMENT-1 : Solubility of AgCl in $NH_3(aq.)$ is greater than in pure water.

STATEMENT-2 : When AgCl dissolve in $NH_3(aq.)$, complex ion formation $[Ag(NH_3)_2]^+$ takes place and solubility equilibria of AgCl shifted in forward direction.

15. STATEMENT-1 : Solubility of AgCN in acidic solution is greater than that in pure water.

STATEMENT-2 : Solubility equilibria of AgCN in acidic solution is shifted in forward direction due to formation of HCN.

Level-4

Subjective Problems

1. Calculate pOH of 0.1M aq. solution of weak base BOH ($K_b = 10^{-7}$) at 25°C.
2. pH of 0.01M aq. solution of HA is 4. Find the value of pK_a of HA at 25°C.
3. Calculate approximate pH of 10^{-10} M NaOH at 25°C.
4. Calculate pH of a resultant solution of 25 mL of 0.1M HCl, 50 mL of 0.02 M HNO_3 and 25 mL of 0.1M NaOH.
5. Calculate pH of a resultant solution of 0.1M HA ($K_a = 10^{-6}$) and 0.45 M HB ($K_a = 2 \times 10^{-6}$) at 25°C.
6. 0.16 g N_2H_4 ($K_b = 4 \times 10^{-6}$) are dissolved in water and the total volume made up to 500 mL. Calculate the percentage of N_2H_4 that has reacted with water in this solution
7. Calculate pH of a buffer solution that contains 0.1M NH_4OH ($K_b = 10^{-5}$) and 0.1M NH_4Cl.
8. Calculate the ratio of sodium formate and formic acid ($K_a = 2 \times 10^{-4}$) in a buffer solution of pH = 4.3.
9. What is the pOH of 0.1M KB (salt of weak acid and strong base) at 25°C ? (Given : pK_b of $B^- = 7$)
10. A certain weak acid has $K_a = 10^{-5}$. If the equilibrium constant for its reaction with a strong base is represented as 1×10^y then find the value of y.
11. If solubility of AgCl in 0.2 M solution of $AgNO_3$ is represented as $y \times 10^{-10}$ then find the value of y. (Given : $K_{sp(AgCl)} = 10^{-10}$)
12. When one litre of a saturated solution of $PbCl_2$ (mol. mass = 278) is evaporated, the residue is found to weight 2.78 g. If K_{sp} of $PbCl_2$ is represented as $y \times 10^{-6}$ then find the value of y.
13. A solution is saturated in $SrCO_3$ and SrF_2. The CO_3^{2-} was found to be 10^{-3} mol/L. If the concentration of F^- in solution is represented as $y \times 10^{-2}$ M then what is the value of y? [Given : $K_{sp}(SrCO_3) = 2.5 \times 10^{-10}$; $K_{sp}(SrF_2) = 10^{-10}$]

14. 10 mL of H_2A (weak diprotic acid) solution is titrated against 0.1M NaOH. pH of the solution is plotted against volume of strong base added and following observation is made.

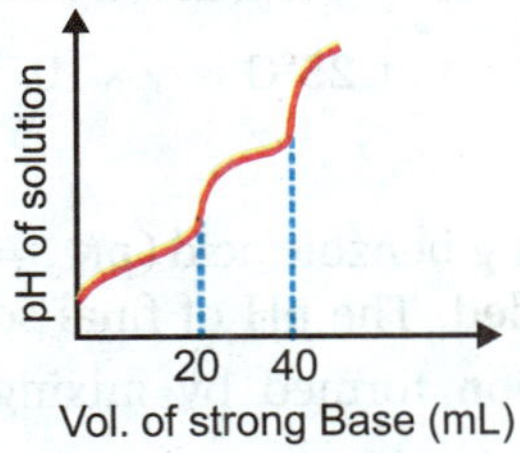

If pH of the solution at first equivalence point is pH_1 and at second equivalence point is pH_2. Calculate the value of $(pH_2 - pH_1)$ at 25°C

Given: For H_2A, $pK_{a_1} = 4.6$ and $pK_{a_2} = 8, \log 25 = 1.4$

15. Amongst the following, the total number of compounds whose aqueous solution turns red litmus paper blue is:

NaCN	KCl	CH_3COONH_4	NaH_2PO_4	$ZnCl_2$	Na_3PO_4
$Fe(NO_3)_3$	Na_2CO_3	NH_4Cl	$NaHCO_3$	$Na_2C_2O_4$	Na_2HPO_4

Given:

Acid	Ka_1	Ka_2	Ka_3
H_3PO_4	10^{-3}	10^{-8}	10^{-12}
H_2CO_3	10^{-6}	10^{-11}	–
$H_2C_2O_4$	10^{-2}	10^{-5}	–

Numerical Problems

(If the numerical value has more than two decimal places, truncate/round-off the value to TWO decimal places.)

1. Equal volume of 10^{-2} M each HCl(aq) and $Ca(OH)_2(aq)$ solutions are mixed at 90°C. Calculate pH of resultant solution. (Given : $\log 2 = 0.3$ and K_w at 90° C $= 10^{-12}$).

2. A student is mixed small amount of NaOH into a vessel containing 200 L of water, which increases the pH from 7 to 9. How many milli grams of H_2SO_4 must be added further to bring the pH back to 7?

3. At 298 K, one gram sample containing $CH_3COOH (K_a = 10^{-5})$ and some inert impurity is dissolved in water and the solution is made upto 100 mL. If pH of resulting solution is 3 then % purity of sample is____.

4. An aqueous solution contains 0.01M RNH_2 $(K_b = 10^{-6})$ and 10^{-3} M KOH.

If concentration of RNH_3^+ is represented as 10^{-z} then the value of z is ____ .

5. What is the pOH of final solution prepared by mixing 50 mL, 0.2 M NaX(aq) and 50 mL, 0.2 M NaY(aq) solutions.

[Given : $K_a(HX) = 5 \times 10^{-9}$; $K_a(HY) = 1.25 \times 10^{-9}$]

6. After solid $SrCO_3$ was dissolved in a basic buffer solution and pH of buffer solution maintained 8.6, equilibrated with a buffer solution at pH 8.6, the solution was found to have $[Sr^{2+}] = 2.0 \times 10^{-4}$ M, if the K_{sp} of $SrCO_3$ is $y \times 10^{-8}$. Find the value of y?

(Given : K_{a_2} for $H_2CO_3 = 5.0 \times 10^{-11}$, $\log 2 = 0.3$)

7. Calculate pH of 10^{-4} M $RNH_3Cl(aq)$ solution at 62°C.

[Given : $K_a(RNH_3^+) = 10^{-7}$; $K_w = 10^{-13}$]

8. If the degree of hydrolysis of 0.05 M $(NH_4)_2SO_4$ at 298 K is 10^{-x} then find the value of x.

[Given : $pK_a(NH_4^+) = 9$]

9. What is the percentage degree of hydrolysis of a mixture of 0.02 M aniline and 0.02 M acetic acid ?
[Given : At 298 K, $K_b(C_6H_5NH_2) = 5 \times 10^{-10}$, $K_a(CH_3COOH) = 2 \times 10^{-5}$]

10. If $[OH^-]$ of $0.01M$ $C_6H_5ONa\,(aq)$ solution at 25°C is $y \times 10^{-4}$ then the value of y is ________.
[Given : $K_a(C_6H_5OH) = 10^{-10}$]

11. 1.0 L solution is prepared by mixing 61 g benzoic acid ($pK_a = 4.2$) with 72 g of sodium benzoate and then 300 mL of 1.0 M HBr solution was added. The pH of final solution is ______.

12. What is the pH of a resultant solution formed by mixing of equal volume of 0.6 M $RNH_2(aq)$ and 0.2 M $HNO_3(aq)$ at 25°C?
[Given : $\log 2 = 0.3 \log 3 = 0.48; K_b(RNH_2) = 10^{-6}$]

13. When 200 mL of 0.2 M HCN is titrated with 0.2 M KOH solution at neutralisation (equivalence) point. What is the pH of the solution at 25°C ?
[Given : $K_a(HCN) = 10^{-9}$]

14. $0.1M$ of $BOH(aq)$ is titrated against $0.1M$ $HCl(aq)$, if pOH at half neutralisation point of BOH is 5 then calculate pH at $\left(\frac{3}{4}\right)^{th}$ neutralisation point of BOH.

(Given : $\log 2 = 0.3; \log 3 = 0.48$)

15. Calculate pOH of 0.02 M K_2HA (A potassium salt of triprotic weak acid H_3A) at 83°C. If K_{a_1}, K_{a_2} and K_{a_3} for H_3A are $10^{-4}, 10^{-6}, 10^{-10}$ respectively and $K_w = 10^{-12}$ at the same temperature.

16. A solution of 0.3 M, 100 mL H_3PO_4 and 0.3 M, 100 mL of Na_3PO_4 are mixed then calculate pH of resulting solution ($pK_{a_1} = 2.9$, $pK_{a_2} = 7.5$, $pK_{a_3} = 12.5$ for H_3PO_4)

17. Consider the titration of 40.0 mL of a 0.05 M solution of the protonated form of the amino acid methionine (H_2A^+) with 0.1M NaOH.
Calculate the pH after addition of 20.0 mL of base.

$H_3\overset{+}{N}CHCOOH$
|
$CH_2CH_2SCH_3$

For Methionine cation (H_2A^+): $K_{a_1} = 4 \times 10^{-3}$; $K_{a_2} = 8 \times 10^{-10}$

18. An amount of 0.10 moles of AgCl(s) is added to one litre of water. Next, the crystals of NaBr are added until 80% of the AgCl is converted to AgBr(s). If the $[Br^-]$ at this point is $y \times 10^{-4}$ then the value of y is ______.
(Given : K_{sp} of AgCl $= 10^{-10}$ and K_{sp} of AgBr $= 4 \times 10^{-13}$)

19. Calculate the concentration of CH_3COO^- ion in a solution prepared by adding 0.1 mole of $CH_3COOAg(s)$ in 1 L of 0.1M HCl solution.
[Given : $K_{sp}(AgCl) = 10^{-10}$; $K_{sp}(CH_3COOAg) = 10^{-8}$; $K_a(CH_3COOH) = 10^{-5}$]

20. If concentration of free CN^- must be maintained in a solution that is 1.0 M $AgNO_3\,(aq)$ and 0.1M NaCl(aq) to prevent AgCl from precipitating is represented by 0.5×10^{-z} then find the value of z.
[Given : $K_{sp}(AgCl) = 10^{-10}$; $K_f(Ag(CN)_2^-) = 4 \times 10^{17}$]

21. 500 mL of saturated solution of $Ca(OH)_2$ is mixed with equal volume of aqueous NaOH solution is added such that resultant solution has concentration of OH^- ion is 0.04 M.
Calculate the mass of $Ca(OH)_2$ precipitated (in mg)
[Given : $K_{sp}(Ca(OH)_2) = 4 \times 10^{-6}$]

Answers

Level 1

1. (d)	2. (c)	3. (a)	4. (c)	5. (c)	6. (a)	7. (a)	8. (d)	9. (d)	10. (c)
11. (d)	12. (d)	13. (c)	14. (d)	15. (b)	16. (d)	17. (a)	18. (d)	19. (c)	20. (c)
21. (a)	22. (b)	23. (a)	24. (b)	25. (a)	26. (c)	27. (b)	28. (c)	29. (b)	30. (b)
31. (b)	32. (a)	33. (b)	34. (a)	35. (a)	36. (b)	37. (c)	38. (a)	39. (c)	40. (c)
41. (c)	42. (c)	43. (b)	44. (c)	45. (b)	46. (a)	47. (c)	48. (d)	49. (d)	50. (d)
51. (d)	52. (b)	53. (b)	54. (c)	55. (d)	56. (d)	57. (b)	58. (c)	59. (c)	60. (b)
61. (b)	62. (d)	63. (b)	64. (b)	65. (b)	66. (c)	67. (a)	68. (c)	69. (d)	70. (a)
71. (d)	72. (b)	73. (d)	74. (c)	75. (b)	76. (b)	77. (a)	78. (d)	79. (b)	80. (c)
81. (b)	82. (d)	83. (a)	84. (a)	85. (c)	86. (a)	87. (c)	88. (b)	89. (d)	90. (b)
91. (c)	92. (d)	93. (b)	94. (a)	95. (d)	96. (a)	97. (a)	98. (d)	99. (b)	100. (a)
101. (a)	102. (b)	103. (c)	104. (b)	105. (c)	106. (d)	107. (d)	108. (b)	109. (a)	110. (c)
111. (d)	112. (c)	113. (a)	114. (a)	115. (a)	116. (c)	117. (a)	118. (c)	119. (a)	120. (a)
121. (c)	122. (c)	123. (c)	124. (a)	125. (c)	126. (d)	127. (a)	128. (a)	129. (d)	130. (b)
131. (d)	132. (b)	133. (c)	134. (c)	135. (d)	136. (d)	137. (a)	138. (d)	139. (c)	140. (a)
141. (a)	142. (d)	143. (c)	144. (c)	145. (d)	146. (a)	147. (b)	148. (b)	149. (d)	150. (b)

Level 2

1. (d)	2. (c)	3. (c)	4. (a)	5. (d)	6. (c)	7. (a)	8. (d)	9. (a)	10. (a)
11. (d)	12. (c)	13. (a)	14. (b)	15. (c)	16. (c)	17. (b)	18. (c)	19. (b)	20. (b)
21. (b)	22. (d)	23. (c)	24. (c)	25. (d)	26. (d)	27. (a)	28. (b)	29. (a)	30. (b)
31. (a)	32. (c)	33. (d)	34. (b)	35. (d)					

Level 3

Passage–1:	1. (c)	2. (b)	3. (b)	4. (b)		
Passage–2:	1. (c)	2. (b)	3. (b)	4. (b)		
Passage–3:	1. (b)	2. (a,c)	3. (c)	4. (a)	5. (b)	6. (a)
Passage–4:	1. (a)	2. (a)	3. (b)	4. (a)	5. (c)	

One or More Answer is/are Correct

1. (b,c)	**2.** (a,b,c)	**3.** (a,b)	**4.** (b)	**5.** (a,c)	**6.** (a,b,c,d)	**7.** (a,b)	**8.** (a,b)
9. (a,c)	**10.** (a,c,d)	**11.** (a,b,d)	**12.** (a,c)	**13.** (a,b)	**14.** (b, c)	**15.** (b, c, d)	**16.** (a, c)
17. (b, c, d)	**18.** (a, b, c)	**19.** (a, c, d)	**20.** (a, b)	**21.** (b,c,d)	**22.** (a,c)	**23.** (c,d)	**24.** (b,c,d)
25. (a,b,c)							

Match the Column

1.	A → Q, R;	B → P, S;	C → P Q, S;	D → P, S
2.	A → P, S;	B → Q;	C → P, Q, R;	D → P, R
3.	A → P;	B → S;	C → Q;	D → R
4.	A → Q;	B → P;	C → S;	D → R
5.	A → R;	B → Q, S;	C → R;	D → P, S
6.	A → P;	B → S;	C → R;	D → Q
7.	A → P, Q,R;	B → R;	C → Q, S;	D → Q, R
8.	A → P, Q, R;	B → R;	C → P, Q;	D → S
9.	A→ S;	B→ P;	C→ R;	D→ Q
10.	A→ Q;	B→ R;	C→ P;	D→ S

Assertion-Reason Type Questions

1. (A)	**2.** (D)	**3.** (A)	**4.** (A)	**5.** (A)	**6.** (D)	**7.** (A)	**8.** (A)	**9.** (B)	**10.** (D)
11. (B)	**12.** (D)	**13.** (B)	**14.** (A)	**15.** (A)					

Level 4

Subjective Problems

1. 4	**2.** 6	**3.** 7	**4.** 2	**5.** 3	**6.** 2	**7.** 9	**8.** 4	**9.** 4	**10.** 9
11. 5	**12.** 4	**13.** 2	**14.** 4	**15.** 6					

Numerical Problems

1. 9.70	**2.** 98.00	**3.** 60.00	**4.** 5.00	**5.** 3.00	**6.** 0.08	**7.** 5.50	**8.** 4.00
9. 50.00	**10.** 9.50	**11.** 3.60	**12.** 8.30	**13.** 11.00	**14.** 8.52	**15.** 4.00	**16.** 7.50
17. 5.75	**18.** 3.20	**19.** 4.00	**20.** 4.00	**21.** 185			

Hints & Solutions

Level 1

1,2 (d, c)

$HX \rightleftharpoons H^+ + X^-$

If, $HX \rightarrow$ acid

then, $X^- \rightarrow$ conjugate base

3. (a) Among HF, HCl, HBr and HI,

HF is weakest acid

So, F^- is strongest base.

4. (c) H_3PO_2 and H_3PO_3 are mono and dibasic acids. Hence, NaH_2PO_2 and Na_2HPO_3 are normal salts, while Na_2HPO_4 is acid-salt as one proton is there in the compound.

7. (a) For a conjugate acid-base pair,

$K_a \cdot K_b = K_w$

8. (d) $pH + pOH = pK_w$

9. (d) In neutral solution

$[H^+] = \sqrt{K_w}$

10. (c) As temperature ↑, $[H^+]$ and $[OH^-]$ both↑

11-12-14. (d, d, d)

$pH = -\log_{10}[H^+]$

13. (c) $[H^+] = \dfrac{K_w}{[OH^-]}$ and $pH = -\log_{10}[H^+]$

15. (b) $[H^+] = 10^{-9}$

If it is 1000 times more basic then,

$[H^+]_{\text{Original}} = 10^{-6} M$

16. (d) When equal volumes are taken, the concentration becomes half.

$\therefore \quad [H^+] = \dfrac{1}{2}(10^{-3} + 10^{-5})$

$= 505 \times 10^{-6}$

$\therefore \quad pH = 3.3$

17. (a) $[H^+] = 0.002\ M$

$pH = \log 2 \times 10^{-3}$

$= 3 - \log 2$

$\therefore \quad pOH = 11 + \log 2$

18. (d) $pH = 4, \quad [H^+] = 10^{-4}$ mol/litre

Equivalent of HCl in 100 mL solution

$= \dfrac{10^{-4} \times 100}{1000} = 10^{-5}$

19. (c) $N_1V_1 = N_2V_2$

$10^{-3} \times 10 = N_2 \times 1000$

$\Rightarrow \quad N_2 = 10^{-5}$

$\therefore \quad pH = 5$

and $\quad pOH = 14 - 5 = 9$

20. (c) $pH = 1 \Rightarrow [H^+] = 0.1 M$

For $H_2SO_4(aq.)$

$[H^+] = 2 \cdot [\text{Acid}]$

21. (a) $H_2A(aq) \xrightarrow{100\%} 2H^+(aq) + A^{2-}(aq)$

$pH = -\log [H^+]$

22. (b) $[Ca(OH)_2(aq.)] = \dfrac{0.60}{74} \times \dfrac{1000}{1500}$

$= 5.40 \times 10^{-3}$

$Ca(OH)_2(aq.) \longrightarrow Ca^{2+}(aq.) + 2OH^-(aq.)$

$[Ca^{2+}] = 5.40 \times 10^{-3}$

$[OH^-] = 2 \times 5.40 \times 10^{-3} = 1.08 \times 10^{-2}$

23-24. (a, b)

At these concentration consider ionization of water also.

25. (a)

	H_2SO_4	+ 2NaOH	$\longrightarrow Na_2SO_4$	$+ 2H_2O$
	0.05 mole	0.1 mole		
	–	–	0.05	0.1

Aqueous solution of *SASB* type salt

i.e., $pH = 7$

26. (c) HCl + NaOH $\longrightarrow$ NaCl + H_2O

(i) 2.5 m mole 2.49 mole

$[H^+] = \dfrac{0.01}{50} = 2 \times 10^{-4} M$

$pH = 4 - \log 2 = 3.7$

(ii) 2.5 m mole 2.51 M mole

$[OH^-] = \dfrac{0.01}{50} = 2 \times 10^{-4} M$

$pOH = 3.7$

27. (b) Milli-equivalents of HCl $= 25 \times 0.08 = 2$

Milli-equivalents of NaOH $= 2.5$

remaining $[OH^-] = \frac{2.5-2}{500} = 10^{-3}$

$\therefore$ $[H^+] = 10^{-11}$ or pH = 11

28. (c) $H^+ + OH^- \longrightarrow H_2O$

0.1 m mole   0.2 m mole

LR

$[OH^-]_{Rem.} = \frac{(0.2-0.1)}{10+10} M$

29. (b) 20 meq. of HCl reacts completely with 20 meq. of KOH. Hence, the final solution is neutral. Thus, at 90°C, pH = 6.

30. (b) In aqueous reaction, upon dilution, the reaction moves towards more equation moles.

31. (b) $NH_4^+(aq) \rightleftharpoons NH_3(aq) + H^+(aq)$: K_1

$HCN(aq) \rightleftharpoons H^+(aq) + CN^-(aq)$; K_2

$NH_4^+(aq) + CN^-(aq) \rightleftharpoons NH_3(aq) + HCN(aq)$

$K_{eq} = \frac{K_1}{K_2}$

32. (a) $\because$ Strength of acid $\propto \frac{1}{pK_a}$

$\therefore$ Formic acid will be the strongest acid.

33. (b) More $\Delta H_{ion} \Rightarrow$ weaker is the acid.

34. (a) $\alpha = \sqrt{\frac{K_a}{C}} < 0.05$

i.e., $[H^+] = \sqrt{K_a \cdot C}$

35. (a) $HA(aq.) \rightleftharpoons H^+(aq.) + A^-(aq.)$

at equilibrium $C(1-\alpha)$ $C\alpha$ $C\alpha$

$K_a = \frac{C\alpha^2}{1-\alpha}$

$\Rightarrow 10^{-4} = \frac{(0.01)\alpha^2}{1-\alpha}$

$\Rightarrow \alpha = 0.095$

$\Rightarrow \% \alpha = 9.5$

36. (b) $\alpha = \sqrt{\frac{K_a}{C}}$

37. (c) $\because$ $\alpha > 0.05$

$\Rightarrow K_a = \frac{C\alpha^2}{1-\alpha}$

38. (a) pH = 9; $[OH^-] = 10^{-5}$; $C\alpha = 10^{-5}$

$\therefore \alpha = 10^{-4}$

% dissociation $= 10^{-4} \times 100 = 0.01\%$

39. (c) $[H^+] = C \cdot \alpha$ and $[OH^-] = \frac{K_w}{[H^+]}$

40. (c) α is negligible w.r.t. 1 so $K_a = C\alpha^2$

$\Rightarrow 0.1 \times (0.01)^2 \Rightarrow 10^{-5}$

$\therefore$ $pK_a = 5$

41. (c) $HCOOH \rightleftharpoons H^+ + HCOO^-$

$C-x$ $x+0.01$ x

$K_a = \frac{(x+0.01)x}{0.01-x}$

neglect x w.r.t. 0.01 due to common ion effect

$K_a = x$

$\therefore$ $x \simeq [HCOO^-] = 1.8 \times 10^{-4}$

42. (c) (P) $(pK_a)_{HX} + (pK_b)_{X^-} = pK_w$

(Q) Very dilute solution

(R) $pH = \frac{1}{2} pK_a - \frac{1}{2} \log_{10} C$

(S) $[H^+]_1 = 10^{-3} M$, $[H]_2^+ = 10^{-5} M$

$[H^+]_{Final} = \frac{M_1V_1 + M_2V_2}{V_1 + V_2}$

43. (b) $\because$ α is negligible w.r.t. 1

$K_a = C_1\alpha_1^2 = C_2\alpha_2^2$

$\therefore$ $C_2 = C_1 \left(\frac{\alpha_1}{\alpha_2}\right)^2$

$= 0.2 \times \frac{1}{4} = 0.05$

$C_1V_1 = C_2V_2$

$\Rightarrow 300 \times 0.2 = 0.05 \times V_2$

$\Rightarrow V_2 = 1200$ mL

Volume of H_2O added $= 1200 - 300$

$= 900$ mL

44. (c) Due to low value of K_b and common ion effect we can neglect x w.r.t. 0.01 when $x = c\alpha$

$K_b = \frac{x \times 0.01}{0.02}$

$\therefore$ $x = [NH_4^+] = 3.6 \times 10^{-5}$

45. (b) $n_{RNH_2(g)} = \frac{1 \times 22.41}{0.0821 \times 273} = 1$

$[RNH_2(aq.)] = \frac{n_{RNH_2}}{\text{Volume of solution}} = 1$

α is negligible w.r.t. 1 so

$pOH = \frac{1}{2}[pK_b - \log C] = 2$

46. (a) α is negligible w.r.t. 1 for both acid so

$[H^+] = \sqrt{K_{a_1}C_1 + K_{a_2}C_2}$

$= \sqrt{1.8 \times 10^{-5} \times 0.01 + 6.3 \times 10^{-5} \times 0.01}$

$[H^+] = \sqrt{81 \times 10^{-8}} = 9 \times 10^{-4}$

47. (c) $[HA] = \frac{\left(\frac{6}{60}\right)}{10000} \Rightarrow 10^{-5} M$

$\alpha = \sqrt{\frac{10^{-9}}{10^{-5}}} \Rightarrow 0.01$ (negligible w.r.t.1)

$\therefore$ $[H^+] = \sqrt{10^{-9} \times 10^{-5}} \Rightarrow \sqrt{10^{-14}}$

$\Rightarrow \quad 10^{-7}M, \quad \because [H^+] < 10^{-6}M$

So we should consider $[H^+]$ from H_2O

$\Rightarrow \quad [H^+]_{total} = 2 \times 10^{-7}$ without considering common ion effect pH = 6.7, so pOH = 7.3

but due to C.I. effect H^+ from H_2O will be less

so pH > 6.7

$\therefore$ pOH of solution is exist between 7.0 and 7.3.

48. (d) and **49.** (d)

$$K_{a_1} = \frac{[H^+]_{total}[HCO_3^-]}{[H_2CO_3]}$$

$\because K_{a_1} >> K_{a_2}$ and concentration is appreciable so we can neglect α w.r.t. 1

$\therefore \quad [H^+]_{total} \simeq [HCO_3^-];$

$[HCO_3^-] = \sqrt{4 \times 10^{-7} \times 0.025} \Rightarrow 10^{-4}$

and $\quad [CO_3^{2-}] \simeq K_{a_2}$

50. (d) $[H^+]_{total} \simeq [H^+]$ from first step

$\therefore K_{a_1} = \dfrac{C\alpha^2}{(1-\alpha)}$; α is not negligible w.r.t. 1 so after solving quadratic equation $\alpha = 0.095$

$\therefore \quad [OH^-] = \dfrac{10^{-14}}{0.095 \times 0.3}$

$\Rightarrow \quad 3.5 \times 10^{-13}$

52. (b) Salt of *SAWB* type.

53. (b) Salt of *WASB* type.

54. (c) $\quad pH = \frac{1}{2} pK_a - \frac{1}{2} \log C$

$\Rightarrow \quad 5 = \frac{1}{2} pK_a + \frac{1}{2} \times 2$

$pK_a = 8 \quad \Rightarrow \quad pK_b = 8;$

$pOH = \frac{1}{2} pK_b - \frac{1}{2} \log C = 4 + 2 = 6$

56. (d) All are *WASB* salts,

So, $[OH^-] = \sqrt{\dfrac{K_w \cdot C}{K_a}}$

For weakest acid, K_a should be minimum.

$\Rightarrow$ $[OH^-]$ minimum

$\Rightarrow$ $[H^+]$ maximum

$\Rightarrow$ pH minimum

57. (b) I $\rightarrow$ *WASB* salt

II $\rightarrow$ *WA*

III $\rightarrow$ *WAWB* salt

IV $\rightarrow$ *SB*

V $\rightarrow$ SA

58. (c) As temperature $\uparrow$ $pKw \downarrow$

$\Rightarrow$ pH also $\downarrow$

61. (b) $[H^+]_{final} = \dfrac{1 \times 0.1}{(1+999)} = 10^{-4}M$

$\Rightarrow \quad pH = 4$

65. (b) Similar to Q. No.7

66. (c) $(K_b)_{C_6H_5O^-} = \dfrac{K_w}{10^{-10}}$

$[OH^-] = \sqrt{K_b \cdot c}$

67. (a) $K_{a(HOCN)} = \dfrac{10^{-14}}{10^{-10}} = 10^{-4}$

$K_h = K_{b(OCN^-)} = 10^{-10}$

$\alpha = \sqrt{\dfrac{K_h}{C}} = \sqrt{\dfrac{10^{-10}}{0.01}} = 10^{-4}$

$\because \alpha << 0.1; \therefore K_h = c\alpha^2$

$[OH^-] = c\alpha; [OH^-] = 0.01 \times 10^{-4} = 10^{-6}M$

68. (c) $[H^+] = C \cdot \alpha$

$\alpha < 0.05 \quad \Rightarrow K_a = \dfrac{C \cdot \alpha^2}{1-\alpha}$

69. (d) $\dfrac{\alpha_1}{\alpha_2} = \sqrt{\dfrac{C_2}{C_1}} \Rightarrow \dfrac{0.01}{\alpha_2} = \sqrt{\dfrac{0.4}{0.1}}$

$\alpha_2 = 0.005 \quad$ or $\quad \% \alpha_2 = 0.5$

70. (a) For *WAWB* salt, $= \sqrt{\dfrac{K_w}{K_a \cdot K_b}} = \dfrac{\alpha}{1-C} = \sqrt{K_h}$

71. (d) $\alpha = \sqrt{\dfrac{K_w}{K_a C}}$

|Enthalpy of neutralisation| $\downarrow K_a \downarrow h \uparrow$

72. (b) At neutralization point:

m-mole of acid = m-mole of base

$0.1 \times 25 = 0.05 \times V_{NaOH} \Rightarrow V_{NaOH} = 50$ mL

$[NaA] = \dfrac{25 \times 0.1}{25+50} = \dfrac{2.5}{75}$

$[H^+] = \sqrt{\dfrac{K_w \cdot K_a}{C}} = 1.732 \times 10^{-9}$

74. (c) $pH = \frac{1}{2}(pK_w + pK_a - pK_b) = 6.5$

75. (b) $\because \quad pH = \frac{1}{2} pK_w + \frac{1}{2} pK_a - \frac{1}{2} pK_b = 7$

and $\alpha = \sqrt{\dfrac{K_w}{K_a K_b}} = \sqrt{\dfrac{10^{-14}}{10^{-12}}} = 0.1$

or $\% \quad \alpha = 10$

76. (b) α does not depend on concentration in this case.

77. (a) $\alpha = \sqrt{\dfrac{K_h}{C}}; \quad [H_3O^+] = C\alpha$

$= \sqrt{K_h \times C} = \sqrt{0.02 \times 5 \times 10^{-9}} = 10^{-5}$

78. (d)

$$Ni^{2+}(aq.) + H_2O(l) \rightleftharpoons Ni(OH)^+(aq.) + H^+(aq.)$$

$t = 0$	$0.04M$		
$t_{eq.}$	$(0.04 - x)$	(x)	(x)

$x = 4.5 \times 10^{-6}$

$K_a = \dfrac{x^2}{(0.04 - x)}$

79. (b) α is not negligible w.r.t. 1 so after solving quadratic equation $\alpha = 0.095$

$[H^+] = C\alpha = 0.0095; \quad pH = 2.02$

80. (c) For amphiprotic salt,

81. (b) $pH = \frac{pK_{a_2} + pK_{a_3}}{2}$

82. (d) $CH_3COOH + NaOH \longrightarrow CH_3COONa + H_2O$

	CH_3COOH	$NaOH$	CH_3COONa	H_2O
	100	50	—	—
	50	—	50	—

$$pH = pK_a + \log\frac{[CH_3COO^-]}{[CH_3COOH]} = pK_a$$

83. (a)

$CH_3COOH + NaOH \longrightarrow CH_3COONa + H_2O$

	CH_3COOH	$NaOH$	CH_3COONa	H_2O
initial milli-moles	6	2	—	—
after reaction	4	—	2	—

$$pH = pK_a + \log\frac{[CH_3COO^-]}{[CH_3COOH]} = 4.44$$

85. (c)

	HX	+ NaOH	⟶ NaX	+ H_2O
(I)	2 m mole	1 m mole	–	–
	(2 – 1)	–	1	

ABS

$$\Rightarrow pH = pK_a + \log\frac{[Salt]}{[Acid]}$$

	HX	+ NaOH	⟶ NaX	+ H_2O
(II)	2 m mole	2 m mole	–	–
	2	2	2	

$[NaX] = 0.01\,M$

$$pH = 7 + \frac{1}{2}[pK_a + \log C]$$

86,87, 90 (a, c, b)
For *ABS*

$$pH = pK_a + \log\frac{[Salt]}{[Acid]}$$

88. (b) For *BBS*

$$pOH = pK_b + \log\frac{[Salt]}{[Base]}$$

91. (c) $CH_3NH_2 + HCl \longrightarrow CH_3NH_3Cl$

	CH_3NH_2	HCl	CH_3NH_3Cl
initial moles	0.12	0.08	
after reaction	0.04	—	0.08

$$pOH = pK_b + \log\frac{[CH_3NH_3^+]}{[CH_3NH_2]}$$

$$= 3.3 + \log\frac{0.08}{0.04} = 3.6$$

$\therefore$ pH = 10.4

92. (d) $pOH = pK_b + \log\frac{[NH_4^+]}{[NH_4OH]} = 6$

$\therefore$ pH = 14 – 6 = 8.0

94. (a) *BBS*

$NH_4OH \rightleftharpoons NH_4^+ + OH^-$

Addition of HCl takes equilibrium to right

NH_4OH	+ H^+ ⟶	$NH_4^+ + H_2O$
0.2 mole	10^{-6}	0.2
$(0.2 - 10^{-6})$	–	$(0.2 + 10^{-6})$

$$[OH^-] = K_b \cdot \frac{[Base]}{[Salt]}$$

95. (d) $HCOOH + NaOH \longrightarrow HCOONa + H_2O$

	HCOOH	NaOH	HCOONa
(I)	a	$a/5$	
	$(a - a/5)$	–	$(a/5)$

$$pH_1 = pK_a + \log\frac{[HCOOH]}{[HCOONa]}$$

	HCOOH	NaOH	HCOONa
(II)	a	$4a/5$	
	$(a - 4a/5)$		$4a/5$

$$pH_2 = pKa + \log\frac{[HCOOH]}{[HCOONa]}$$

$$\Delta pH = pH_2 - pH_1$$

97. (a) Henderson equation is

$$pH = pK_a + \log\frac{[salt]}{[acid]} = 4$$

$\therefore$ pOH = 10, $[OH^-] = 10^{-10}$

Number of OH^- ions $= 10^{-10}\,N_A$;

where N_A = Avogadro's number.

98. (d) pH↑ upon addition of base.

99. (b) SASB salt.

100. (a) $pK_b(NH_4OH) = 14 - pK_a(NH_4^+) = 4.74$

So, buffer range,

$$pK_b \pm 1$$

101. (a) Mol of $C_6H_5COOH = \frac{61}{122} = 0.5$;

Mol of $C_6H_5COONa = \frac{72}{144} = 0.5$

$$pH = pK_a + \log\frac{(C_6H_5COONa)}{(C_6H_5COOH)};$$

$$pH = 4.2 + \log\frac{0.2}{0.8} = 3.6$$

102. (b) $pH = pK_{a_2} + \log\frac{[CO_3^{2-}]}{[HCO_3^-]}$

$$= (11 - \log 4) + \log\left(\frac{0.2}{0.4}\right) = 10.1$$

103. (c) $H_3PO_4 + PO_4^{3-} \longrightarrow H_2PO_4^- + HPO_4^{2-}$

	H_3PO_4	PO_4^{3-}	$H_2PO_4^-$	HPO_4^{2-}
initial milli moles	20 × 0.1	20 × 0.1		
after reaction	—	—	2	2

Buffer solution of $H_2PO_4^-$ (acid) and HPO_4^{2-} (conjugate base) is formed;

$$pH = pK_{a_2} + \log\frac{[HPO_4^{2-}]}{[H_2PO_4^-]} = pK_{a_2}$$

106. (d) At eq. point, *WASB* salt forms means base solution So, *CTR* > 7

107. (d) At equivalent point

$$pH = \frac{1}{2}[pK_w + pK_a + \log C]$$

$$= \frac{1}{2}[14 + 4.74 - 2] = 8.37$$

Where $[\text{Salt}] = \dfrac{\text{milli-moles of acid}}{\text{Total volume}}$

$$= \frac{20 \times 0.02}{20 + 20} = 0.01$$

For best indicator pH at equivalent point should lie between colour transition range of indicator.

108. (b) $\text{HIn} \rightleftharpoons \text{H}^+ + \text{In}^-$

Acid Basic

$$Ka = \frac{[\text{H}^+][\text{In}^-]}{[\text{HIn}]}$$

$$6 \times 10^{-5} = \frac{10^{-5}(x)}{(100 - x)}$$

109. (a) $\text{HIn} \rightleftharpoons \text{In}^- + \text{H}^+; \quad K_{\text{In}} = \dfrac{[\text{H}^+][\text{In}^-]}{[\text{HIn}]}$

red blue

Case I : $3 \times 10^{-5} = \dfrac{[\text{H}^+]_1 \times 0.25}{0.75}$

Case II : $3 \times 10^{-5} = \dfrac{[\text{H}^+]_2 \times 0.75}{0.25}$

$[\text{H}^+]_1 = 9 \times 10^{-5}$ $[\text{H}^+]_2 = 10^{-5}$

Change in $[\text{H}^+] = 8 \times 10^{-5}$

110. (c) $\text{pH} \Rightarrow \text{pK}_{\text{In}} + \log \dfrac{[\text{In}^-]}{[\text{HIn}]} = pK_a + \log \dfrac{[\text{Salt}]}{[\text{Acid}]}$

111. (d) $\text{H}_2A + 2\text{NaOH} \longrightarrow \text{Na}_2A + \text{H}_2\text{O}$

For second eq. pt

$(n)_{\text{NaOH}} = 2(n)_{\text{H}_2A}$ $\quad [\text{Na}_2A] = \dfrac{0.4}{(32 + 250)}$

$0.25 \times V = 2(0.2 \times 20)$

$\Rightarrow \quad V = 32 \text{ mL}$

112. (c) $\text{H}_2A + \text{NaOH} \longrightarrow \text{NaHA} + \text{H}_2\text{O}$

(I) a $\quad a/2$

$(a - a/2)$ $\quad -$ $\quad a/2$

WA + its conjugate $\Rightarrow ABS$

$$\text{pH} = pKa_1 + \log \frac{[\text{NaHA}]}{[\text{H}_2A]} = \frac{[\text{NaHA}]}{[\text{H}_2A]} = pKa_1$$

(II) a $\quad a$

$-$ $\quad -$ $\quad (a)$

aq. solution of amphiprotic salt

$$\text{pH} = \frac{pK_{a_1} + pK_{a_2}}{2}$$

114. (a) When equal volume mixed concentration become half also, for ppt,

$$[\text{Ag}^+][\text{Cl}^-] > K_{sp}$$

116. (c) When 1 litre each solution are mixed

$[\text{OH}^-] = 10^{-6} M$ (Buffer solution)

$\text{M}^{n+} = 0.05 M$

for $Q = [0.5][10^{-6}]^2 = 5 \times 10^{-14}$

$Q > K_{sp}$ for Fe^{2+} and Ni^{2+}

So Fe(OH)_2 and Ni(OH)_2 are ppt.

117. (a) For $[\text{Ca}^{+2}]$, $M_1V_1 = M_2V_2$

$0.04 \times 50 = M_2 \times 200 \Rightarrow M_2 = 0.01$

Similarly, for $[\text{SO}_4^{-2}]$

$0.0008 \times 150 = M_2 \times 200$

$\Rightarrow \quad M_2 = 0.0006$

ionic product $(Q) = [\text{Ca}^{+2}] \times [\text{SO}_4^{-2}]$

$= 6 \times 10^{-6}$ $\quad$ So $Q < K_{sp}$

118. (c) For AgCl

$K_{sp} = [\text{Ag}^+][\text{Cl}^-]$

$y \quad x$

$\Rightarrow \quad xy = \text{constant}$

i.e., Rectangular hyperbolic

119. (a) KNO_3 does not effect the equilibrium.

120. (a) $[\text{Ag}^+] = \dfrac{10^{-3}}{50 + 50} \times 1000 = 0.01$

$[\text{Cl}^-] = \dfrac{0.1 \times 50}{50 + 50} = 0.05$

$Q = [\text{Ag}^+][\text{Cl}^-] = 5 \times 10^{-4}$

$\because \quad Q > K_{sp}$

$\text{Ag}^+(aq.) + \text{Cl}^-(aq.) \longrightarrow \text{AgCl}(s)$

at given condition $\quad 0.01 \quad 0.05$

after precipitation $\quad x \quad \simeq 0.04$

$K_{sp} = [\text{Ag}^+][\text{Cl}^-]; \quad 10^{-10} = x \times 0.04$

$x = [\text{Ag}^+] = 2.5 \times 10^{-9}$

121. (c) $\quad A_xB_y(s) \rightleftharpoons x \cdot A^{y+}_{(aq.)} + y \cdot B^{x-}_{(aq.)}$

$t = 0 \quad a$

$t_{eq.} \quad (a - s) \quad (x \cdot s) \quad (x \cdot y)$

$k_{sp} = [A^{y+}]^x[B^{x-}]^y$

$= (x \cdot s)^x (y \cdot s)^y$

122. (c) Molarity of saturated solution

$\Rightarrow$ Solubility

123. (c) $\quad \text{Zr}_3(\text{PO}_4)_4(s) \rightleftharpoons 3 \cdot \text{Zr}^{4+}_{(aq.)} + 4\text{PO}^{3-}_{4(aq.)}$

$t = 0 \quad a$

$t_{eq.} \quad (a - s) \quad 3s \quad 4s$

$K_{sp} = [\text{Zr}^{4+}]^3[\text{PO}_4^{3-}]^4$

124. (a) The corresponding values of solubility products of the given electrolytes are :

$MX_1 \quad MX_2 \quad MX_3$

$S^2 = 10^{-6}, \quad 4S^3 = 4 \times 10^{-9}, \quad 27S^4 = 27 \times 10^{-12}$

125. (c) $K_{sp} = [\text{Ca}^{2+}]^3[\text{PO}_4^{3-}]^2$

126. (d) For most soluble salt, solubility should be maximum.

127. (a) (A) $[\text{Ag}^+] = \dfrac{K_{sp}}{[\text{Br}^-]} = 5 \times 10^{-12} M$

(B) $[\text{Ag}^+] = \dfrac{K_{sp}}{[\text{Cl}^-]} = 1.8 \times 10^{-7} M$

(C) $[Ag^+] = \sqrt{\dfrac{K_{sp}}{[CO_3^{2-}]}} = 9\times10^{-6}\ M$

(D) $[Ag^+] = \left(\dfrac{K_{sp}}{[AsO_4^{3-}]}\right)^{1/3} = 10^{-7}\ M$

128. (a) Solubility of $AB = \sqrt{K_{sp}} = 2\times10^{-10}$

Solubility of $A_2B = \sqrt[3]{\dfrac{K_{sp}}{4}} = 2\times10^{-4}$

Solubility of $AB_3 = \left[\dfrac{K_{sp}}{27}\right]^{1/4} = 10^{-8}$

129. (d) $HgSO_4(s) \rightleftharpoons Hg^{2+}(aq.) + SO_4^{2-}(aq.)$

$S = \sqrt{6.4\times10^{-5}} = 8\times10^{-3}$ mol/L

$= 8$ mol/m^3

130. (b) $s = \dfrac{6.9\times10^{-2}}{690}$ mol/100 mL $= 10^{-3}$ mol/L

For $Ba_3(AsO_4)_2$, $K_{sp} = 108\,s^5$

131-132-133. (d, b, c)

Similar to Q.No. 130

134. (c) $BaF_2(s) \rightleftharpoons Ba^{2+}(aq.) + 2F^-(aq.)$

$S \qquad 2S$

$K_{sp} = 4S^3;\ S = \left(\dfrac{10^{-6}}{4}\right)^{1/3} = 0.63\times10^{-2}$

Molarity of $F^- = 2S = 2\times0.63\times10^{-2}$

$= 1.26\times10^{-2}$

135. (d) $InF_3(s) \rightleftharpoons In^{3+}(aq) + 3F^-(aq)$

$t = eq \qquad s \qquad 3s$

$K_{sp} = (s)^1(3s)^3$

136. (d) $Cu(OH)_2(s) \rightleftharpoons Cu^{2+}(aq.) + 2OH^-(aq.)$

$S \qquad 2S$

$K_{sp} = 4S^3$

$\therefore \quad S = 4\times10^{-7}$

$[OH^-] = 2S = 8\times10^{-7}$

$pOH = 6.1 \quad \therefore\ pH = 7.90$

137. (a) $[Ag^+]$ or $S = \sqrt{K_{sp}} = 10^{-5}\ M$

$10^{-5} = \dfrac{\dfrac{14.35\times10^{-3}}{143.5}}{V\text{ (in litre)}}$

$V = 10$ litre or 0.01 m^3

138. (d) $Fe(OH)_2(s) \rightleftharpoons Fe^{2+}(aq.) + 2OH^-(aq.)$

$t_{eq.} \qquad s \qquad 10^{-1}M$

$K_{sp} = [Fe^{2+}][OH^-]^2$

$8\times10^{-16} = (s)^1(10^{-1})^2$

139. (c) To precipitate

$[Pb^{2+}][OH^-]^2 \ge K_{sp}$

$[OH^-]_{Min} = \sqrt{\dfrac{K_{sp}}{[Pb^{2+}]}}$

141. (a) Let solubility of Ag_2CO_3 in presence of Na_2CO_3 is x,

$Ag_2CO_3(s) \rightleftharpoons 2Ag^+(aq.) + CO_3^{2-}(aq.)$

$2x \qquad (x+0.1)\simeq 0.1$

$K_{sp} = [Ag^+]^2[CO_3^{2-}]$

$\Rightarrow \quad 4\times10^{-13} = (2x)^2\times0.1$

$\Rightarrow \quad x = 10^{-6}$

142-143. (d, c)

Similar to Q.No. 139

144. (c) More common ion $\Rightarrow$ more $\downarrow$ in solubility.

145. (d) $K_{sp} = 256\cdot s^5$

146. (a) $PbBr_2(s) \rightleftharpoons Pb^{+2}(aq.) + 2Br^-(aq.)$

$0.8\,S \qquad 2\times0.8\times S$

$K_{sp} = [Pb^{+2}][Br^-]^2$

$\Rightarrow \quad 8\times10^{-5} = (0.8\,S)(1.6\,S)^2$

$\Rightarrow \quad S = \left[\dfrac{10^{-4}}{1.6\times1.6}\right]^{1/3}$

147. (b) For basic buffer solution

$pOH = pK_b + \log\dfrac{[NH_4^+]}{[NH_3]} = pK_b$

$\therefore \quad [OH^-] = 1.8\times10^{-5}$

$K_{sp} = [Mn^{2+}][OH^-]^2$

$[Mn^{2+}] = 1.38\times10^{-4}\ M$

148. (b) $[OH^-] = \sqrt{\dfrac{K_{sp}}{[Mg^{2+}]}} = \sqrt{\dfrac{10^{-11}}{10^{-3}}} = 10^{-4}$

$pOH = pK_b + \log\dfrac{[NH_4Cl]}{[NH_3]}$

$\Rightarrow \quad 4 = 5 + \log\dfrac{[NH_4Cl]}{[NH_3]}$

$\Rightarrow \quad 0.10 = \dfrac{[NH_4Cl]}{[NH_3]}$

moles of NH_4Cl required $= 0.1\times0.02$

$= 2\times10^{-3}$

149. (d) Due to higher value of K_f, mostly Ag^+ converted into complex.

$AgI(s) \rightleftharpoons Ag^+(aq.) + I^-(aq.);\ K_{sp}$

$Ag^+(aq.) + 2NH_3(aq.) \rightleftharpoons Ag(NH_3)_2^+(aq.);\ K_f$

overall reaction is

$AgI(s) + 2NH_3(aq.) \rightleftharpoons Ag(NH_3)_2^+(aq.) + I^-(aq.)$

$K_{eq} = K_{sp}\cdot K_f$

$K_{sp}\cdot K_f = \dfrac{x^2}{1^2}$

$\therefore \quad x = 4.9\times10^{-5}$ mol/litre

So mass of AgI required is

$= 4.9\times10^{-5}\times235 = 0.011$ g

Level 2

1. (d) $TCl + NaOT \longrightarrow NaCl + T_2O$

Initial milli-moles: 2, 3.75

milli-moles of remaining NaOT = 1.75

$$\therefore \quad [OT^-] = \frac{1.75}{25} = 7 \times 10^{-2} \text{ moles}$$

$$pOT = 2 - \log 7$$

$$pT + pOT = 7.62 \times 2$$

$$\therefore \quad pT = 15.24 - 2 + \log 7$$

$$= 13.24 + \log 7$$

2. (c) $2NH_3 \rightleftharpoons NH_4^+ + NH_2^-; \quad [NH_4^+] = \sqrt{K_{SIC}}$

$$K_{SIC} = [NH_4^+][NH_2^-]$$

$$\therefore \quad [NH_4^+] = 10^{-15}\,M$$

$\because$ Number of NH_4^+ ions in 1000 cm^3

$$= 10^{-15} \times 6.022 \times 10^{23}$$

$\therefore$ Number of NH_4^+ ions in 100 cm^3

$$= 10^{-15} \times 6.022 \times 10^{23} \times \frac{100}{1000}$$

$$= 6.022 \times 10^7$$

3. (c) α is very less so $[H^+] = \sqrt{K_a \times C} = 3 \times 10^{-3}$

$$\text{Now } [H^+] = \frac{3 \times 10^{-3}}{2} = 1.5 \times 10^{-3}$$

$$(1.5 \times 10^{-3}) = \sqrt{K_a \cdot C_2}$$

$$\Rightarrow \quad C_2 = 0.125$$

milli-moles of acid remain constant

$$C_1V_1 = C_2V_2$$

$$10 \times 0.5 = 0.125 \times V_2$$

$$\Rightarrow \quad V_2 = 40 \text{ litre}$$

4. (a) $[H^+] = \dfrac{10 \times 0.3 - 10 \times 0.1}{10 + 10 + 20} = 0.05$

$$[HA] = \frac{20 \times 0.1}{20 + 20} = 0.05$$

$$\underset{0.05 - x}{HA} \rightleftharpoons \underset{0.05 + x}{H^+} + \underset{x}{A^-}$$

Due to common ion effect neglect x w.r.t. 0.05

$$K_a = \frac{(0.05 + x) \cdot x}{(0.05 - x)} \simeq x$$

$$\therefore \quad x = 10^{-5}$$

$$\therefore \quad \frac{[A^-]}{[HA] + [A^-]} = \frac{x}{x + 0.05} = 2 \times 10^{-4}$$

5. (d) $FCH_2COOH(aq.) \rightleftharpoons FCH_2COO^-(aq.) + H^+(aq.)$

at eqm. $C(1 - \alpha)$ $\quad C\alpha \quad C\alpha$

$$K_a = \frac{[H^+][FCH_2COO^-]}{[FCH_2COOH]}$$

$$\Rightarrow \quad 2.6 \times 10^{-3} = \frac{(2 \times 10^{-3})^2}{[FCH_2COOH]}$$

at eqm. $[FCH_2COOH] = 1.53 \times 10^{-3}$

$$\text{Total concentration} = 2 \times 10^{-3} + 1.53 \times 10^{-3}$$

$$= 3.53 \times 10^{-3}$$

6. (c) $\underset{C_1 - x}{HCOOH(aq.)} + H_2O(l) \rightleftharpoons \underset{x}{HCOO^-(aq.)} + \underset{x + y}{H_3O^+(aq.)}$

$$\underset{C_2 - y}{HF(aq.)} + H_2O(l) \rightleftharpoons \underset{y}{F^-(aq.)} + \underset{x + y}{H_3O^+(aq.)}$$

$\because$ α is very less for both acid so

$$K_a(HCOOH) = \frac{x(x + y)}{C_1};$$

$$K_a(HF) = \frac{y(x + y)}{C_2}$$

$$\therefore \quad \frac{x}{y} = \frac{K_a(HCOOH)\,C_1}{K_a(HF)\,C_2}$$

$$\frac{[HCOO^-]}{[F^-]} = \frac{x}{y}$$

$$= \frac{2 \times 10^{-4} \times 0.2}{6.6 \times 10^{-4} \times 0.1} = \frac{2}{3.3}$$

7. (a) First dissociation

$$X(OH)_3 \longrightarrow X(OH)_2^+ + OH^-$$

Second dissociation :

$$X(OH)_2^+ \longrightarrow X(OH)^{2+} + OH^-$$

$$\text{Total } [OH^-] = 4 \times 10^{-3} + 2 \times 10^{-3} = 6 \times 10^{-3}$$

$$pOH = 3 - \log 6 \simeq 2.22$$

$$\therefore \quad pH = 11.78$$

8. (d) α is negligible w.r.t. 1

$[H^+]$ mainly from first step

$$[H^+] = \sqrt{K_{a_1} \times C} = \sqrt{10^{-5} \times 0.1} = 10^{-3}$$

$$\therefore \quad K_{a_3} = \frac{[H^+][A^{3-}]}{[HA^{2-}]} = 10^{-13} = \frac{10^{-3}[A^{3-}]}{[HA^{2-}]}$$

$$X = \frac{[A^{3-}]}{[HA^{2-}]} = 10^{-10};$$

$$\therefore \quad pX = 10$$

9. (a) $Ca(LaC)_2 \longrightarrow Ca^{2+} + 2LaC^-$

Initial concentration $\quad 0.6/2 = 0.3\,M$

After dissociation $\quad [LaC^-] = 2 \times 0.3 \times 0.9$

$$= 0.54\,M$$

$$pH = \frac{1}{2}[pK_w + pK_a + \log(LaC^-)]$$

$$14 - 5.6 = \frac{1}{2}[14 + pK_a + \log(0.54)]$$

$$pK_a = 2.8 - \log(0.54)$$

10. (a) $CH_3COO^-(aq.) + NH_4^+(aq.) + H_2O(l) \rightleftharpoons$

at equilibrium $C(1-\alpha)$ $C(1-\alpha)$

$CH_3COOH(aq.) + NH_4OH(aq.)$

$C\alpha$ $C\alpha$

$$K_h = \frac{K_w}{K_a K_b} = \frac{\alpha^2}{(1-\alpha)^2}$$

$$\frac{10^{-14}}{(1.8 \times 10^{-5})^2} = \frac{\alpha^2}{(1-\alpha)^2}$$

$$\frac{10^{-2}}{1.8} = \frac{\alpha}{1-\alpha}; \alpha = 5.55 \times 10^{-3}$$

$$[CH_3COOH] = C\alpha$$

$$= 0.01 \times 0.0055 = 5.55 \times 10^{-5}$$

11. (d) $Fe^{3+}(aq.) + H_2O(l) \rightleftharpoons Fe(OH)^{2+}(aq.) + H_3O^+(aq.)$

$$K_a = \frac{[Fe(OH)^{2+}][H_3O^+]}{[Fe^{3+}]}$$

$$\Rightarrow 6.5 \times 10^{-3} = \frac{0.20}{0.80} \times [H_3O^+]$$

$$\Rightarrow [H_3O^+] = 2.6 \times 10^{-2}; \; pH = 1.59$$

12. (c) Due to larger difference between K_{h_1} and K_{h_2} and common ion effect we can assume that y is negligible w.r.t. x

$Fe^{2+} + H_2O \rightleftharpoons Fe(OH)^+ + H^+$

at equilibrium $C - x$ $x - y$ $x + y$

$Fe(OH)^+ + H_2O \rightleftharpoons Fe(OH)_2 + H^+$

at equilibrium $x - y$ y $y + x$

$$K_{h_2} = \frac{K_w}{K_{b_1}} = \frac{10^{-14}}{10^{-4}} = 10^{-10}$$

$\because$ $x + y \approx x - y$

$\therefore$ $K_{h_2} = y \Rightarrow Fe(OH)_2 = 10^{-10}$

13. (a) Let x milli-moles of KOH is added

$$6 = 5 + \log\left[\frac{s+x}{a-x}\right]; \; \frac{s+x}{a-x} = 10$$

or $\frac{10+x}{10-x} = 10 \Rightarrow x = 8.18$

$\therefore$ Mass $= 8.18 \times 10^{-3} \times 56 = 0.458$ gm

14. (b) $pH = pK_a + \log\frac{[C_6H_5COO^-]}{[C_6H_5COOH]}$

$$\therefore \frac{[C_6H_5COO^-]}{[C_6H_5COOH]} = 2$$

Let volume of acid is V mL

$$\frac{0.2 \times (300 - V)}{0.1 \times V} = 2 \Rightarrow V = 150 \text{ mL}$$

15. (c) When half acid is neutralized $pH = pK_a$ and at the equivalent point

$$pH = \frac{1}{2}[pK_w + pK_a + \log C]$$

$$= 9 = \frac{1}{2}[14 + 5 + \log C] = C = 0.1$$

Let V mL of NaOH is used in titration milli eq. of NaOH $= 0.25 \times V =$ milli eq. of salt formed

$$\therefore \quad \frac{0.25 \times V}{V + 60} = 0.1 = V = 40 \text{ mL}$$

milli equivalent or milli-moles of acid $= 10$

Mass of acid $= 10 \times 10^{-3} \times 82 = 0.82$ gm

Mass % of acid $= \frac{0.82}{1.025} \times 100 = 80\%$

16. (c) $BOH(aq.) \rightleftharpoons B^+(aq.) + OH^-(aq.)$

At equilibrium : $c(1-\alpha)$ $c\alpha$ $c\alpha$

$$K_b = \frac{[B^+][OH^-]}{[BOH]} \Rightarrow \frac{c\alpha^2}{(1-\alpha)};$$

neglect α w.r.t. 1 as concentration is appreciable, %

$$\alpha = 100 \times \sqrt{\frac{K_b}{c}}$$

Total dissolved base present in solution as BOH and B^+

So $\alpha = \frac{[B^+]}{[B^+] + [BOH]}$

$$\Rightarrow \frac{1}{1 + \frac{[BOH]}{[B^+]}} = \frac{1}{1 + \frac{[OH^-]}{K_b}}$$

$$\Rightarrow \frac{K_b}{K_b + [OH^-]} \Rightarrow \frac{K_b \cdot [H^+]}{K_b[H^+] + K_w}$$

also $pOH = -\log[OH^-]; [OH^-] = 10^{-pOH};$

$$K_b = 10^{-pK_b}$$

$$\alpha = \frac{1}{1 + \frac{10^{-pOH}}{10^{-pK_b}}} \Rightarrow \frac{1}{1 + 10^{(pK_b - pOH)}}$$

17. (b) $HA + NaOH \longrightarrow NaA + H_2O$

milli-moles of salt NaA or $A^- = 40 \times 0.1 = 4$

Now $A^- + H^+ \longrightarrow HA$

	A^-	H^+	HA
Initial milli-moles	4	2	
Final milli-moles	2	—	2

Acidic buffer solution is formed and $[A^-] = [HA]$

$$pH = pK_a + \log \frac{[A^-]}{[HA]} \quad \Rightarrow \quad pK_a = 5$$

Now $\quad HA + NaOH \longrightarrow NaA + H_2O$

hydrolysis of A^- will takes place

$$[NaA] = \frac{\text{milli moles of acid}}{\text{Total volume}} = \frac{20 \times 0.2}{20 + 20} = 0.1$$

$$pH = \frac{1}{2}(pK_w + pK_a + \log C)$$

$$= \frac{1}{2}[14 + 5 - 1] = 9$$

18. (c) Moles of H^+ produced $= 4 \times 10^{-3}$

$\therefore \quad [H^+] = \dfrac{4 \times 10^{-3}}{0.1} = 0.04\ M$

$$PO_4^{3-}(aq.) + H^+(aq.) \longrightarrow HPO_4^{2-}(aq.)$$

	PO_4^{3-}	H^+	HPO_4^{2-}
Initial milli-moles	0.02	0.04	
Final milli-moles	—	0.02	0.02

$$HPO_4^{2-}(aq.) + H^+(aq.) \longrightarrow H_2PO_4^-(aq.)$$

	HPO_4^{2-}	H^+	$H_2PO_4^-$
Initial milli-moles	0.06	0.02	—
Final milli-moles	0.04	—	0.02

Finally buffer solution of $H_2PO_4^-$ (acid) and HPO_4^{2-} (conjugate base) is formed

$$pH = pK_{a_2} + \log \frac{[HPO_4^{2-}]}{[H_2PO_4^-]} = 7.2 + \log \frac{0.04}{0.02} = 7.5$$

19. (b) $\quad BOH + HCl \longrightarrow BCl + H_2O$

at equilibrium point $\quad N_1V_1 = N_2V_2; \quad V_2 = 20$ mL

$\therefore \quad [BCl] = \dfrac{20 \times 0.08}{20 + 20} = 0.04$

$$pH = \frac{1}{2}[pK_w - pK_b - \log(C)]$$

$\therefore \quad pK_b = 5.4$

$$B^+ + OH^- \longrightarrow BOH; \text{ Basic Buffer is formed}$$

	B^+	OH^-	BOH
Initial milli-moles	1.6	0.4	
Final milli-moles	1.2	—	0.4

$$pOH = pK_b + \log \frac{[B^+]}{[BOH]}$$

$$= 5.4 + \log\left(\frac{1.2}{0.4}\right) = 5.4 + 0.48$$

$$pOH = 5.88$$

20. (b) $\quad CO_3^{2-} + H^+ \longrightarrow HCO_3^-$ acidic buffer is formed.

	CO_3^{2-}	H^+	HCO_3^-
Initial milli-moles	25×0.04	50×0.025	1
Final milli-moles	—	0.25	

$$HCO_3^- + H^+ \longrightarrow H_2CO_3$$

	HCO_3^-	H^+	H_2CO_3
Initial milli-moles	1	0.25	
Final milli-moles	0.75	—	0.25

$$pH = pK_{a_1} + \log \frac{[HCO_3^-]}{[H_2CO_3]} = 6.4 + \log\left(\frac{0.75}{0.25}\right) = 6.88$$

21. (b) Let a is initial milli-moles of HA and molarity of NaOH solution is x

$$HA + NaOH \longrightarrow NaA + H_2O$$

In both case solution is acidic, so acidic buffer solution is formed.

$$5 = pK_a + \log \frac{10x}{a - 10x}$$

$$\Rightarrow \quad 5.60 = pK_a + \log \frac{20x}{a - 20x}$$

$$\Rightarrow \quad 0.60 = \log \frac{20x}{(a - 20x)} \times \frac{(a - 10x)}{10x}$$

$$\Rightarrow \quad 4 = \frac{2(a - 10x)}{(a - 20x)} \Rightarrow a = 30x$$

$$\therefore \quad 5 = pK_a + \log \frac{10x}{20x}$$

or $\quad pK_a = 5.3$

22. (d)

	CO_3^{2-}	+ H^+	$\longrightarrow$ HCO_3^-
Initial milli-moles	50×0.05	40×0.1	—
Final milli-moles	—	1.5	2.5

	HCO_3^-	+ H^+	$\longrightarrow$ H_2CO_3
Initial milli-moles	2.5	1.5	—
Final milli-moles	1	—	1.5

$$pH = pK_{a_1} + \log \frac{[HCO_3^-]}{[H_2CO_3]} = 6.173$$

23. (c) At second equivalence point the only species present at appreciable concentration is HA^{2-}

So, $\quad pH = \frac{pK_{a_2} + pK_{a_3}}{2} = \frac{8 + 12}{2} = 10$

$$[H^+] = 10^{-10}$$

and $\quad \frac{[H_3A]}{[A^{3-}]} = \frac{[H^+]^3}{K_{a_1}K_{a_2}K_{a_3}} = 10^{-7}$

24. (c) Solubility $\quad S = \frac{x}{M}$

$$A_3B_2(s) \rightleftharpoons 3A^{2-}(aq.) + 2B^{3-}(aq.)$$

$$\qquad\qquad 3S \qquad\quad 2S$$

$$K_{sp} = (3S)^3(2S)^2 = 108S^5$$

$$\frac{[B^{3-}]}{K_{sp}} = \frac{2S}{108S^5} = \frac{1}{54}\frac{M^4}{x^4}$$

25. (d) $\quad K_{sp}(SrCrO_4) = [Sr^{2+}][CrO_4^{2-}]$

$$[CrO_4^{2-}] = \frac{3.5 \times 10^{-5}}{0.1} = 3.5 \times 10^{-4}$$

$K_{sp}(BaCrO_4) = [Ba^2][CrO_4^{2-}]$

$[CrO_4^{2-}]_{total} \simeq [CrO_4^{2-}]$ from $SrCrO_4$

$$[Ba^{2+}] = \frac{1.2 \times 10^{-10}}{3.5 \times 10^{-4}} = 3.4 \times 10^{-7}$$

27. (a) $[Ag^+]$ required for precipitation of AgCl

$$[Ag^+] = \frac{K_{sp}(AgCl)}{[Cl^-]} = 10^{-9}$$

$[Ag^+]$ required for precipitation of Ag_2CrO_4

$$= \sqrt{\frac{K_{sp}\,(Ag_2CrO_4)}{[CrO_4^{2-}]}} = \sqrt{\frac{10^{-12}}{10^{-4}}} = 10^{-4}$$

At this point, the concentration of Cl^- ion in the solution can be calculated from K_{sp} (AgCl)

$$\therefore \qquad [Cl^-] = \frac{K_{sp}\,(AgCl)}{[Ag^+]} = \frac{10^{-10}}{10^{-4}} = 10^{-6}\ M$$

28. (b) After mixing $[Ag^+] = 0.2\ M; [NH_3] = 1\ M$

Due to very high value of K_f; Ag^+ mainly converted into complex

$$Ag^+(aq.) + 2NH_3(aq.) \rightleftharpoons Ag(NH_3)_2^+(aq.)$$

Initial concentration	0.2	1	
At equilibrium	x	0.6	$\simeq 0.2$

$$Ag(NH_3)_2^+(aq.) \rightleftharpoons Ag(NH_3)^+(aq.) + NH_3(aq.)$$

At equilibrium	$0.2 - y$	y	$0.6 + y$
	$\simeq 0.2$		$\simeq 0.6$

$$\frac{1}{K_{f_2}} = \frac{y \times 0.6}{0.2} \Rightarrow \frac{0.1}{10^4};\ y = [Ag(NH_3)^+] \Rightarrow 3.33 \times 10^{-5} M$$

29. (a) $$AgCl(s) + 2NH_3(aq.) \rightleftharpoons Ag(NH_3)_2^+(aq.) + Cl^-(aq.)$$

$$K_1 = 1.6 \times 10^{-10} \times 10^7 = 1.6 \times 10^{-3}$$

$$= \frac{[Ag(NH_3)_2^+]\,[Cl^-]}{[NH_3]^2}$$

$$AgCN(s) + 2NH_3(aq.) \rightleftharpoons Ag(NH_3)_2^+(aq.) + CN^-(aq.)$$

$$K_2 = 2.5 \times 10^{-16} \times 10^7 = 2.5 \times 10^{-9}$$

$$= \frac{[Ag(NH_3)_2^+]\,[CN^-]}{[NH_3]^2}$$

$$\frac{[Cl^-]}{[CN^-]} = \frac{1.6 \times 10^{-3}}{2.5 \times 10^{-9}} = 6.4 \times 10^5$$

$$K_1 = \frac{x^2}{(1-2x)^2} \Rightarrow \frac{x}{1-2x} = 0.04$$

$$x = 0.037$$

$$\therefore \qquad [CN^-] = \frac{0.037}{6.4 \times 10^5} = 5.78 \times 10^{-8}$$

32. (c) $$AgBr(s) \rightleftharpoons Ag^+(aq.) + Br^-(aq.);\ K_{sp}$$

$$Ag^+(aq.) + 2S_2O_3^{2-}(aq.) \rightleftharpoons Ag(S_2O_3)_2^{3-}(aq.);\ K_f$$

$$AgBr(s) + 2S_2O_3^{2-}(aq.) \rightleftharpoons Ag(S_2O_3)_2^{3-}(aq.) + Br^-(aq.);\ K_{sp} \cdot K_f = 25$$

$$0.1 - 2x \qquad\qquad x \qquad\qquad x$$

Value K_f is very higher so we can assume almost Ag^+ converts into complex

$$2.5 = \frac{x^2}{(0.1-2x)^2} \quad \text{or} \quad 5 = \frac{x}{0.1-2x}$$

$$\Rightarrow \quad x = 0.045\ M$$

33. (d)

$$Ag_2CrO_4(s) \rightleftharpoons 2Ag^+(aq.) + CrO_4^{2-}(aq.)$$
$$2x + 2y \qquad x$$
$$Ag_2C_2O_4(s) \rightleftharpoons 2Ag^+(aq.) + C_2O_4^{2-}(aq.)$$
$$2y + 2x \qquad y$$

$$\frac{K_{sp_1}}{K_{sp_2}} = \frac{x}{y} = \frac{2\times10^{-12}}{2\times10^{-11}} \Rightarrow \frac{x}{y} = 0.1$$

$$2\times10^{-11} = (2x+2y)^2 \cdot y$$
$$2\times10^{-11} = 4.84y^3$$
$$y = 1.6\times10^{-4};$$
$$\therefore \quad x = 0.16\times10^{-4}$$

$$\text{Total } [Ag^+] = 2x + 2y$$
$$= (2\times0.16 + 2\times1.6)\times10^{-4}$$
$$= 3.52\times10^{-4}$$

34. (b) $K_{sp} = [Zn^{2+}][S^{2-}]$

$$[S^{2-}] = \frac{10^{-21}}{0.01} = 10^{-19}$$

$$\text{for } K_{a_1}\cdot K_{a_2} = \frac{[H^+]^2[S^{2-}]}{[H_2S]}$$

$$10^{-20} = \frac{[H^+]^2\times10^{-19}}{0.1} \Rightarrow [H^+] = 0.1$$

or $\quad pH = 1$

35. (d) Dissolved $Al(OH)_3$ present in solution as $Al^{3+}(aq.)$ as well as $[Al(OH)_4]^-(aq.)$

$$\therefore \quad S = [Al^{3+}(aq.)] + [Al(OH)_4(aq.)]$$

$$S = \frac{K_{sp}}{[OH^-]^3} + K_c[OH^-]$$

for minimum solubility

$$\frac{dS}{d(OH^-)} = 0$$

or $$-\frac{K_{sp}\times3}{[OH^-]^4} + K_c = 0$$

$$[OH^-] = \left(\frac{3K_{sp}}{K_c}\right)^{1/4}$$

Level 3

Passage-1

2. (b) $[HCl] = \dfrac{0.01}{0.1} = 0.1\,M$

$$pOH = 4.74 + \log\left(\frac{0.3+0.1}{0.2-0.1}\right) = 5.34$$

$$\therefore \quad pH = 14 - 5.34 = 8.66$$

Passage-3

3. (c) When acid is half neutralized $pH = pK_a$

$$\therefore \quad pK_a = 5$$

4. (a) $K_{eq} = \dfrac{K_a(HB)}{K_a(HA)} = \dfrac{10^{-5}}{10^{-6}} = 10$

5. (b) When $V_{NaOH} = 0$ mL; $pH = 3$

or $\quad [H^+] = 0.001$

$$K_{a(HB)} = \frac{C\alpha^2}{(1-\alpha)} = \frac{(0.001)\alpha}{1-\alpha} = 10^{-5}$$

$$\alpha = 0.01$$

$$C = \frac{0.001}{0.01} = 0.1$$

$$C = \frac{n_{HB}}{V}$$

$$0.1 = \frac{5}{V};\ V = 50 \text{ mL}$$

$$\therefore \quad [NaB] = \frac{\text{milli-moles of acid}}{\text{total volume}} = \frac{5}{50+50} = 0.05$$

$$pH = \frac{1}{2}[pK_w + pK_a + \log C]$$

$$= \frac{1}{2}[14 + 5 + \log(0.05)] = 8.85$$

6. (a) For best indicator, $pH \simeq pK_{In}$ of an indicator as well as pH transition range of the indicator must coincide with the steep portion of the titration curve.

One or More Answers is/are Correct

21. (b,c,d)

(a) $S = \sqrt{K_{sp}} = 10^{-5}$ mol lit^{-1}

(b) $s = 10^{-5}$ mol lit^{-1}

(c) $AgCl(s) \rightleftharpoons Ag^+ + Cl^-$

$$S+2 \qquad S$$

$$10^{-10} = (S+2)S \Rightarrow S = \frac{10^{-10}}{2} = 5\times10^{-11}$$

(d) $$AgCl(s) \rightleftharpoons [Ag(NH_3)_2]^+ + Cl^-$$

$$Ag^+ + 2NH_3 \rightleftharpoons [Ag(NH_3)_2]^2 + Cl^-$$

$$AgCl(s) + 2NH_3 \rightleftharpoons [Ag(NH_3)_2]^+ + Cl^-$$

$$K = \frac{[Ag(NH_3)_2]^+[Cl^-]}{[NH_3]^2} = K_{sp}\times K_f$$

$$\frac{S^2}{(2-2S)^2} = 10^{+8}\times10^{-10} = 10^{-2}$$

$$S = 0.166\,M$$

22. (a,c) $H_2A \rightleftharpoons H^+ + HA^-$; $\because$ y is very less so $x+y \approx x$

$x-y \approx x$

At equilibrium $c-x \quad x+y \quad x-y$ $\quad K_{a_2} = y = 10^{-12}$; $pK_{a_2} = 12$

$HA^- \rightleftharpoons H^+ + A^{2-}$ $\quad \therefore \; K_{a_1} = \frac{x^2}{c-x}$

At equilibrium $x-y \quad x+y \quad y$

$pK_{a_2} - pK_{a_1} = 12 - 5 = 7$ $\quad pH = \frac{1}{2}[pK_{a_1} - \log c]$

$3 = \frac{1}{2}[pK_{a_1} + 1] \qquad pK_{a_1} = 5$

25. (a,b,c)

$$pH_1 = 8 \log \frac{25}{75} = 8 + \log \frac{1}{3}$$

$$pH_2 = 8 + \log \frac{75}{25} = 8 + \log 3$$

$$\text{Change in pH} = \log 3 - \log \frac{1}{3} = 2 \log 3$$

Level 4

Subjective Problems

1. $[OH^-] = \sqrt{K_b \cdot c}$

2. $[H^+] = 10^{-4} M = C\alpha$

$\because \; \alpha < 0.05 \quad \Rightarrow K_a = \frac{C\alpha^2}{1-\alpha}$

3. Consider ionisation of water.

4.
$$H^+ + OH^- \longrightarrow H_2O$$

(2.5 + 1) mole (2.5) m mole

L.R.

$$[H^+]_{Final} = \frac{(3.5-2.5)}{25+50+25} = 10^{-3} M$$

5. $[H^+]_{Final} = \sqrt{K_{a_1} \cdot c_1 + K_{a_2} \cdot c_2}$

6. $C = \frac{0.16/32}{500/1000} M$

$\alpha = \sqrt{K_b \cdot c}$

7. $pOH = pK_b + \log \frac{[NH_4Cl]}{[NH_4OH]}$

8. $pH = pK_a + \log \frac{[Salt]}{[Acid]}$

9. For *WASB* salt

$$[OH^-] = \sqrt{\frac{K_w \cdot c}{ka}}$$

10.
$$HX(aq.) \rightleftharpoons H^+(aq.) + X^-(aq.); K_a = 10^{-5}$$
$$H^+(aq.) + OH^-(aq.) \rightleftharpoons H_2O(l); Kw = 10^{-14}$$

Addition

$$HX(aq.) + OH^-(aq.) \rightleftharpoons H_2O(l) + X^-(aq.)$$

$$K_{eq.} = Ka \times Kw$$

11. $AgCl(s) \rightleftharpoons \underset{0.2M}{Ag^+(aq)} + \underset{s}{Cl^-(aq)}$

$K_{sp} = [Ag^+][Cl^-]$

12. Upon evaporation of saturated solution, all water will be escaped and only dissolved $PbCl_2$ will be left.

i.e., Solubility $= \frac{2.78/278}{1} = 10^{-2}$ mol/L

for $PbCl_2$ $K_{sp} = 4s^2$

$y \times 10^{-6} = 4(10^{-2})^3$

$\Rightarrow \quad y = 4$

13. Let's assume simultaneous solubility of $SrCO_3$ as $x\,M$ while SrF_2 as $y\,M$

$SrCO_3(s) \rightleftharpoons Sr^{2+}(aq) + CO_3^{2-}(aq)$

At eq. $(x+y)M \times M$; $K_{sp_1} = (x+y)x$

$SrF_2(s) \rightleftharpoons Sr^{2+}(aq) + 2F^-(aq)$;

$2.5 \times 10^{-10} = (x+y)x \quad \ldots(1)$

At eq. $(y+x)M\; 2y\,M$; $K_{sp_2} = (x+y)(2y)^2$

$10^{-10} = 4y^2(x+y) \quad \ldots(2)$

$2.5 = \frac{x}{4y^2}$

$10y^2 = x = 10^{-3}$

$y^2 = 10^{-4}$

$y = 10^{-2}M$

$\therefore \quad [F^-] = 2y = 2 \times 10^{-2}\,M$

14. $H_2CO_3 + NaOH \longrightarrow NaHCO_3 + H_2O$

$pH_1 = \frac{1}{2}(pK_{a_1} + pK_{a_2}) = \frac{1}{2}(4.6 + 8) = 6.3$

$N_1V_1 = N_2V_2$

$N_1 \times 10 = 0.1 \times 20$

$N_1 = 0.2$

$NaHCO_3 + NaOH \longrightarrow Na_2CO_3 + H_2O$

$c = \frac{2m\text{ mole}}{50\text{ mL}} = \frac{1}{25}M$

$pH_2 - pH_1 = 10.3 - 6.3 = 4$

15. NaCN; Na_3PO_4; Na_2CO_3; $NaHCO_3$; $Na_2C_2O_4$; Na_2HPO_4

Numerical Problems

1. $2HCl + Ca(OH)_2 \longrightarrow CaCl_2 + 2H_2O$

mol: $10^{-2}V$ $\quad 10^{-2}V$

$0 \quad \frac{10^{-2}V}{2} \quad - \quad -$

$[Ca(OH)_2] = \frac{\frac{10^{-2}V}{2}}{2V} = \frac{10^{-2}}{4}$

So $[OH^-] = \frac{10^{-2}}{2}M$

$pOH = 2 + \log 2 = 2.3$

$pH = 12 - 2.3 = 9.7$

2. When $pH = 9$

$\Rightarrow \quad [OH^-] = 10^{-5}\,M$

So, $10^{-5} = \frac{(n)_{NaOH}}{200\,L}$

Moles of NaOH added $= 2 \times 10^{-3}$ mol for pH back to 7

$2NaOH + H_2SO_4 \longrightarrow Na_2SO_4 + 2H_2O$

For 2 m mole of NaOH, 1 m mole of H_2SO_4 required

$= 1 \times 10^{-3} \times 98$ g

$\approx 0.98 = 98$ mg

3. $[H^+] = \sqrt{K_a \times C}$

$10^{-6} = 10^{-5} \times C = 0.01$

$m_{CH_3COOH} = 0.01 \times 60$

% purity $= \frac{0.01 \times 60}{1} \times 100 = 60\%$

4. $CH_3NH_2 + H_2O \rightleftharpoons CH_3\overset{+}{N}H_3 + OH^-$

$0.01 \qquad\qquad 0 \qquad 0$

$0.01 - x \qquad\qquad x \qquad x + 0.001$

$10^{-6} = \frac{x(x + 0.001)}{0.01 - x}$

$x = 10^{-5}$; $z = 5$

5. $[OH^-] = \sqrt{K_{h_1}C_1 + K_{h_2}C_2} = 10^{-3}$

(where $C_1 = 0.1$ and $C_2 = 0.1$;

$K_{h_1} = \frac{K_w}{K_{a(HX)}}, K_{h_2} = \frac{K_w}{K_{a(HY)}})$

$pOH = 3$

6. $SrCO_3(s) \rightleftharpoons Sr^{2+} + CO_3^{2-}$

$2 \times 10^{-4}\,M \qquad 2 \times 10^{-4} - x$

$CO_3^{2-} + H_2O \rightleftharpoons HCO_3^- + OH^-$

$(2 \times 10^{-4} - x)M \qquad xM \qquad 4 \times 10^{-6}\,M$

$K_h = \frac{K_w}{K_{a_2}}; \frac{10^{-14}}{5 \times 10^{-11}} = \frac{x \times 4 \times 10^{-6}}{(2 \times 10^{-4} - x)}$

$x = 1.96 \times 10^{-4}$

$\therefore \quad K_{sp} = (2 \times 10^{-4}) \times (2 \times 10^{-4} - x)$

$= 2 \times 10^{-4} \times (2 \times 10^{-4} - 1.96 \times 10^{-4})$

$= 0.08 \times 10^{-8}$

7. $pH = \frac{1}{2}[pK_w - pK_b - \log C]$

$pH = \frac{1}{2}[13 - 6 + 4] = 5.5$

8. $\alpha = \sqrt{\dfrac{K_w}{K_b \times c}} = \sqrt{\dfrac{10^{-14}}{10^{-5} \times 0.1}} = 10^{-4}$; $x = 4$

9. $K_h = \dfrac{K_w}{K_a \cdot K_b} = \dfrac{10^{-14}}{5 \times 10^{-10} \times 2 \times 10^{-5}} = 1$

$\sqrt{K_h} = \dfrac{\alpha}{1-\alpha}$; $1 = \dfrac{\alpha}{1-\alpha}$; $\alpha = 0.5$

$\% \alpha = 0.5 \times 100 = 50$

10. $K_h = \dfrac{K_w}{K_a} = 10^{-4}$

$10^{-4} = \dfrac{C\alpha^2}{(1-\alpha)}$; $10^{-4} - 10^{-4}\alpha = 0.01\alpha^2$;

$\alpha^2 + 10^{-2}\alpha - 10^{-2} = 0$

$\alpha = \dfrac{-0.01 \pm \sqrt{10^{-4} + 4 \times 10^{-2}}}{2}$

$C_6H_5O^-(aq) + H_2O(l) \rightleftharpoons C_6H_5OH(aq) + OH^-(aq)$

$\alpha = \dfrac{-0.01 + 0.2}{2}$

$\alpha = 0.095$

$[OH^-] = C\alpha = 0.01 \times 0.095 = 9.5 \times 10^{-4}$

11. Mol of $C_6H_5COOH = \dfrac{61}{122} = 0.5$

Mol of $C_6H_5COONa = \dfrac{72}{144} = 0.5$

$pH = pK_a + \log \dfrac{(C_6H_5COONa)}{(C_6H_5COOH)}$

$pH = 4.2 + \log \dfrac{0.2}{0.8} = 3.6$

12. $RNH_2(aq) + HNO_3(aq) \longrightarrow RNH_3^+(aq) + H_2O(l)$

M.moles: $0.6 \times V$ $\quad 0.2 \times V$

$0.4 \times V$ $\quad -$ $\quad 0.2 \times V$

$[RNH_2] = \dfrac{0.4 \times V}{2V} = 0.2\,M$

$[RNH_3^+] = \dfrac{0.2 \times V}{2V} = 0.1\,M$

$pOH = pK_b + \log \dfrac{[RNH_3^+]}{[RNH_2]} = 6 + \log\left(\dfrac{0.1\,V}{0.2\,V}\right) = 5.7$

$pH = 14 - 5.7 = 8.3$

13. At eq. point : $V \times 0.2 = 200 \times 0.2$

$V_{KOH} = 200\,mL$

$HCN + KOH \longrightarrow KCN + H_2O$

$[KCN] = \dfrac{40}{400} = 0.1$

$pH = \dfrac{1}{2}[pK_w + pK_a + \log C]$

$pH = 11$

14. At half equivalent point

$pOH = pK_b = 5$

At $\dfrac{3}{4}$ equivalent point

$BOH + HCl \longrightarrow BCl + H_2O$

$\dfrac{1}{4}$ $\quad -$ $\quad \dfrac{3}{4}$

$pOH = pK_b + \log \dfrac{3/4}{1/4} = 5 + \log 3 = 5.48$

$pH = 14 - 5.48 = 8.52$

15. $pH = \dfrac{pK_{a_2} + pK_{a_3}}{2} = \dfrac{6 + 10}{2} = 8$

$pOH = 12 - 8 = 4$

16. $H_3PO_4 + Na_3PO_4 \longrightarrow NaH_2PO_4 + Na_2HPO_4$

$0.3M, 100\,mL$ $\quad 0.3M, 100\,mL$

$-$ $\quad -$ $\quad 30$ $\quad 30$

Resultant solution is a buffer solution

$pH = pK_{a_2} + \log \dfrac{[Na_2HPO_4]}{[NaH_2PO_4]}$

$pH = pK_{a_2} = 7.50$

17. $pH = \dfrac{1}{2}(2.4 + 9.1) = 5.75$

18. $AgCl(s) + Br^-(aq) \rightleftharpoons AgBr(s) + Cl^-(aq)$

Mole: $0.1 - 0.08$ $\quad x$ $\quad 0.08$ $\quad 0.08$

$K_{eq} = \dfrac{[Cl^-]}{[Br^-]} = \dfrac{K_{sp}(AgCl)}{K_{sp}(AgBr)}$

$\dfrac{0.08}{[Br^-]} = \dfrac{10^{-10}}{4 \times 10^{-13}}$

$\therefore$ $[Br^-] = 3.2 \times 10^{-4}\,M$

$y = 3.20$

19. $CH_3COOAg(s) + H^+(aq) + Cl^-(aq) \rightleftharpoons$

Moles: $(0.1 - x)$ $\quad (0.1 - x) \approx y$ $\quad (0.1 - x) \approx y$

$CH_3COOH(aq) + AgCl(s)$

$x \approx 0.1$

$K_C = \dfrac{K_{sp}(CH_3COOAg)}{K_{sp}(AgCl)} \times \dfrac{1}{K_{a(CH_3COOH)}}$

$= \dfrac{10^{-8}}{10^{-10} \times 10^{-5}} = 10^7$

$\because$ K_C is very high so almost completion of reaction takes place so $x \approx 0.1$.

$K_C = \dfrac{[CH_3COOH(aq)]}{y^2}$;

$10^7 = \dfrac{0.1}{y^2}$; $y = [H^+] = 10^{-4}\,M$

$pH = 4.00$

20. $K_{sp}(AgCl) = [Ag^+][Cl^-]$; $10^{-10} = [Ag^+] \times 0.1$;

$[Ag^+] = 10^{-9}$ M

$$Ag^+(aq) + 2CN^-(aq) \rightleftharpoons Ag(CN)_2^-(aq)$$

$1 - x = y \qquad (z - 2x) \qquad x \approx 1$

$$K_f = \frac{1}{y[CN^-]^2}$$

$$4 \times 10^{17} = \frac{1}{(10^{-9})[CN^-]^2}$$

$$[CN^-]^2 = \frac{1}{4 \times 10^8} = 0.25 \times 10^{-8}$$

$$[CN^-] = 0.5 \times 10^{-4}$$

$\therefore \qquad z = 4$

21. $K_{sp}(Ca(OH)_2) = 4s^3 = 4 \times 10^{-6}$

$s = 0.01\,M$

Mole of $Ca^{2+} = \dfrac{0.01 \times 500}{1000} = 0.005$

After mixing $[Ca^{2+}] = \dfrac{0.01 \times 500}{1000} = 0.005$

$K_{sp} = [Ca^{2+}]_{Left}[OH^-]^2$

$4 \times 10^{-6} = [Ca^{2+}]_{Left}(0.04)^2$

$[Ca^{2+}]_{Left} = 2.5 \times 10^{-3}$ M

Moles of $Ca(OH)_2$ precipitated

$= [Ca^{2+}]_{Initial} - [Ca^{2+}]_{Left}$

$= 0.005 - 2.5 \times 10^{-3} = 2.5 \times 10^{-3}$

Wt. of $Ca(OH)_2$ precipitated $= 2.5 \times 10^{-3} \times 74$

$= 0.185$ g $= 185$ mg

CHAPTER 7

Redox Reactions

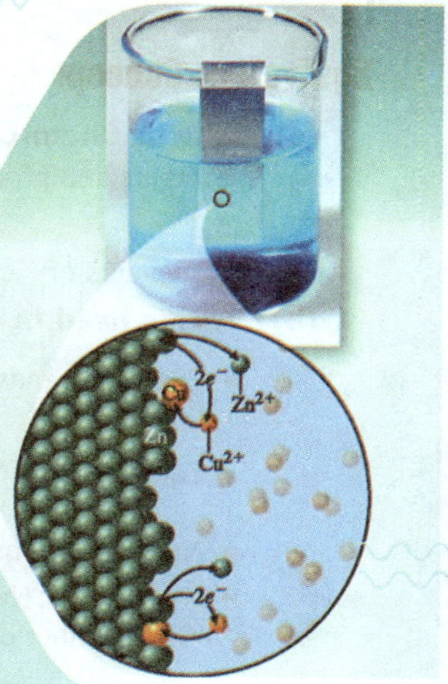

Redox Processes

(a) Oxidation Number : It is the charge (real or imaginary) which an atom appears to have when it is in combination. It may be a whole no. or fractional. An element may have different values of oxidation number depending . It depends on nature of compound in which it is present. There are some operational rules to determine oxidation number.

(b) Definition of Oxidation and Reduction :

- **Oxidation :** Addition of oxygen , removal of hydrogen , addition of electronegative element, removal of electropositive element , loss of electrons , increase in oxidation number (de-electronation).
- **Reduction :** Removal of oxygen, addition of hydrogen, removal of electronegative element, addition of electropositive element, gain of electrons, decrease in oxid. no. (electronation).
- **Redox Reaction :** A reaction in which oxidation & reduction occur simultaneously.

(c) Agents :

- **Oxidising Agents (Oxidants, Oxidisers) :** They oxidise others, themselves are reduced & gain electrons. *e.g.,* O_2,O_3,HNO_3,MnO_2,H_2O_2, halogens, $KMnO_4$,$K_2Cr_2O_7$,KIO_3,$Cl(SO_4)_3$, $FeCl_3$, NaOCl, hydrogen ions.(Atoms present in their higher oxidation state.)
- **Reducing Agents (Reductants or Reducers) :** They reduce others, themselves get oxidised & lose electrons. Also called reductants or reducers . H_2 molecular form is weak reducing agent but Nascent hydrogen is powerful. *e. g.*, C, CO, H_2S, SO_2, $SnCl_2$, Sodium thio Sulphate ($Na_2S_2O_3$), Al, Na, CaH_2, $NaBH_4$, $LiAlH_4$. (Atoms present in their lower oxidation state.)

  Both Oxidising & Reducing Agents : SO_2, H_2O_2, O_3, NO_2, etc.

(d) Balancing of redox reactions :

- Ion - electron method
- Oxidation number method

[Concept involved that in any chemical reaction e^- cannot be produced so no. of e^-s in L.H.S. & R.H.S. should be same]

Equivalent Concept

(a) Law of Chemical equivalence : It states that in any chemical reaction the equivalents of all the reactants and products must be same.

$$2A + 3B \longrightarrow 4C$$

Equivalents of 'A' = Equivalents of 'B' = Equivalents of 'C'

(b) Terms used in equivalent concept :

- Equivalent mass of $A = \dfrac{\text{Molar mass of } A}{\text{Valency factor or } n \text{ factor}}$
- Equivalents of 'A' $= \dfrac{\text{Weight of '}A\text{' (in g)}}{\text{Equivalent weight of '}A\text{'}}$
- Numbers of equivalents of 'A' = no. of moles of 'A'$\times n$-factor
- **Normality (*N*) :** For solutions concentration term **normality (*N*)** is used, which can be defined as "The number of equivalent of solute present in one litre (1000 mL) solution".

$$N = \frac{\text{Number of equivalents of solute}}{\text{Volume of solution (in L)}}$$

$$N = \frac{W(g) \times 1000}{E \times V \text{ (in mL)}}$$

$$N = \text{Molarity} \times \text{Valence factor}$$

milli-equivalents of solute $= N \times V$ (in mL) $= M \times V$ (in mL) $\times n$ factor

(c) Valence factor (*n*-factor) calculation : *n*-factor here we mean a conversion factor by which we divide molar mass of substance to get equivalent mass and it depends on nature of substance which vary from one condition to another condition. We can divide *n*-factor calculations in two category.

- **In case of non-redox reaction.**

  n-factor = mole of charge displaced per mole of specie

- **In case of redox reaction.**

  n-factor of oxidizing or reducing agent = mole of electrons gain or lost per mol of specie.

(d) Volumetric analysis (Titration) : Titration is a procedure for determining the concentration of a solution by allowing a carefully measured volume to react with a standard solution of another substance, whose concentration is known.

- **Primary standard :** A substance available in a pure form or state of known purity which is used in standardizing a solution.
- **Standardization :** The process by which the concentration of a solution is accurately ascertained.
- **Standard solution :** A solution whose concentration has been accurately determined.
- **Titrant :** The reagent (a standard solution) which is added from a buret to react with the analyte.
- **Titrate :** This mainly involve titrations based chemistry. It can be divided into two major category.

  (I) Non-redox reactions (II) Redox reactions

(e) Type of reactions :

(i) Non-redox reactions : This involve following kind of titrations :

1. Acid-Base titrations
2. Double indicator acid-base titration
3. Precipitation titration
4. Back titration

(ii) Redox reactions : This involve following kind of titrations :

1. Iodimetry titrations

2. Iodometry titrations

3. Back titration

(f) Titrations :

(I) Non-redox titrations

(1) Acid-Base titration : To find out strength or concentration of unknown acid or base it is titrated against base or acid of known strength. At the equivalence point we can know amount of acid or base used and then with the help of law of equivalents we can find strength of unknown.

milliequivalent of acid at equivalence point = milliequivalent of base at equivalence point

(2) Double indicator acid-base titration : In the acid-base titration the equivalence point is known with the help of indicator which changes its colour at the end point. In the titration of polyacidic base or polybasic acid there are more than one end point. Some times one indicator is not able to give colour change at every end point. So to find out each end point we have to use more than one indicator. For example in the titration of Na_2CO_3 against HCl there are two end points.

$$Na_2CO_3 + HCl \longrightarrow NaHCO_3 + NaCl$$
$$NaHCO_3 + HCl \longrightarrow H_2CO_3 + NaCl$$

When we use phenolphthalein in the above titration it changes its colour at first end point when $NaHCO_3$ is formed and with it we can not know second end point. Similarly with methyl orange it changes its colour at second end point only and we can not know first end point. It is because all indicator changes colour on the basis of pH of medium. So in titration of $NaHCO_3$, $KHCO_3$ against acid phenolphthalein can not be used.

Titration	**Indicator**	**pH Range**	***n*-factor**
Na_2CO_3 against acid	Phenolphthalein	8.3 – 10	1
Na_2CO_3	Methyl orange	3.1 – 4.4	2

Note : When we carry out dilution of solution, milliequivalent, equivalent, milli mole or mole of substance does not change because they represent amount of substance, however molar concentration may change.

(3) Precipitation titration : In ionic reaction we can know strength of unknown solution of salt by titrating it against a reagent with which it can form precipitate. For example NaCl strength can be known by titrating it against $AgNO_3$ solution with which it form white ppt. of AgCl.

milliequivalent of NaCl at equivalence point = milliequivalent of $AgNO_3$ used = milliequivalent of AgCl formed

(II) Redox Titrations

At equivalence point,

milliequivalent of Oxidizing agent used = milliequivalent of reducing agent reacted.

S.No.	Reagent	Half Reaction	*n*-factor of reagent
1.	FAS (Mohr's salt) $[FeSO_4(NH_4)_2SO_4 \cdot 6H_2O]$	$Fe^{2+} \rightarrow Fe^{3+} + e^-$	1
2.	MnO_4^- (Permanganate ion) (in acidic medium)	$MnO_4^- + 8H^+ + 5e^- \rightarrow Mn^{2+} + 4H_2O$	5
3.	MnO_4^- (in basic medium)	$Mn^{7+} + e^- \rightarrow Mn^{6+}$	1
4.	MnO_4^- (in mild basic or neutral medium)	$Mn^{7+} + e^- \rightarrow Mn^{4+}$	3

5.	$Cr_2O_7^{2-}$ (dichromate ion)	$Cr_2O_7^{2-} + 14H^+ + 6e^- \rightarrow 2Cr^{3+} + 7H_2O$	6
6.	$C_2O_4^{2-}$ (Oxalate ion)	$C_2O_4^{2-} \rightarrow 2CO_2 + 2e^-$	2
7.	As_2O_3	$As_2O_3 + 5H_2O \rightarrow 2AsO_4^{3-} + 10H^+ + 4e^-$	4
8.	$CaOCl_2$ (Bleaching powder)	$CaOCl_2 + H_2O + 2KI \rightarrow Ca(OH)_2 + I_2 + 2KCl$	2
9.	MnO_2	$MnO_2 + 4HCl\,(Conc.) \xrightarrow{\Delta} MnCl_2 + Cl_2 + 2H_2O$	2
10.	IO_3^-	$IO_3^- + 5I^- + 6H^+ \rightarrow 3I_2 + 3H_2O$	5
11.	H_2O_2 (act as oxidizing agent)	$H_2O_2 + 2e^- \rightarrow 2H_2O$	2
12.	H_2O_2 (act as reducing agent)	$H_2O_2 \rightarrow O_2 + 2e^-$	2
13.	H_2O_2 (disproportion)	$H_2O_2 \rightarrow H_2O + 1/2\,O_2$	1
14.	Cl_2 (disproportion)	$3Cl_2 + 6OH^-$ (strong) $\rightarrow ClO_3^- + 5Cl^- + 3H_2O$	5/3
15.	H_2S (in acidic medium)	$H_2S + I_2 \rightarrow S + 2I^- + 2H^+$	2
16.	Sn^{2+} (in acidic medium)	$Sn^{2+} + I_2 \rightarrow Sn^{4+} + 2I^-$	2
17.	N_2H_4	$N_2H_4 \rightarrow N_2 + 4H^+ + 4e^-$	4
18.	SO_3^{2-} (in acidic medium)	$SO_3^{2-} + H_2O \rightarrow SO_4^{2-} + 2e^- + 2H^+$	2
19.	$Na_2S_2O_3$ (Sodium thiosulphate or Hypo)	$2Na_2S_2O_3 + I_2 \rightarrow Na_2S_4O_6 + 2NaI$	1
20.	I_2	$I_2 + 2e^- \rightarrow 2I^-$	2

(1) Iodimetry Titration : In such titrations iodine solution is used as an oxidant and iodine is directly titrated against a reducing agent. This type of titrations are used for the determination of strength of reducing agents like sulphides, arsenides, thiosulphates etc., by titrating them against a standard solution of iodine.

This type of titration involves free iodine, here iodine solution is treated with known sodium thiosulphate solution.

$$I_2 + 2Na_2S_2O_3 \longrightarrow 2NaI + Na_2S_4O_6$$

Equivalents of I_2 = Equivalents of $Na_2S_2O_3$ used

(2) Iodometry Titration : It is an indirect method of estimation of iodine. In this titration an oxidizing agent is reacted with excess $KI_{(aq.)}$ solution and the liberated iodine is treated with a standard solution of a reducing agent added from a burette. Here a neutral or an acidic solution of an oxidizing agent is used and the amount of liberated I_2 is equal to the equivalents of this oxidizing agent.

These titrations are used to determine the concentration of $K_2Cr_2O_7$, $KMnO_4$, $CuSO_4$, Ferric ions, H_2O_2 etc.

These titrations are carried out in following two steps :

- **Step-1 :** Oxidizing agent (X) + KI (excess) $\longrightarrow I_2$ + reduced state of oxidant

  Equivalents of (X) = Equivalents of I_2

- **Step-2 :** Liberated $I_2 + 2Na_2S_2O_3 \longrightarrow 2NaI + Na_2S_4O_6$

  Equivalents of I_2 = Equivalents of $Na_2S_2O_3$ used

(g) Back titration : Back titration is used in volumetric analysis to find out excess of reagent added by titrating it with suitable reagent. It is also used to find out percentage purity of sample.

For example in acid-base titration suppose we have added excess base $[B(OH)_q]$ in acid (H_nA) solution. To find excess base it is back titrated with another acid (H_mB) of known strength.

Equivalent of base = equivalent of $(H_nA + H_mB)$

$\Rightarrow$ $q \times$ mole of base taken $= n \times$ mole of H_nA reacted $+ m \times$ mole of H_mB reacted

Hardness of Water

(a) Definition of hard water : Hard water is having soluble salts of calcium and magnesium ions.

(b) Degree of hardness : Degree of hardness defined as number of parts by mass of $CaCO_3$ (or its equivalent quantities of other substance) present in million parts of mass of water.

$$\text{Hardness of water} = \left[\frac{\text{Mass of } CaCO_3}{\text{Mass of water}} \times 10^6\right] \text{ppm}$$

Level-1

1. Phosphorous has the oxidation state of +1 in:

(a) Orthophosphoric acid (b) Phosphorous acid

(c) Hypophosphorous acid (d) Metaphosphoric acid

2. The oxidation state(s) of Cl in $CaOCl_2$ (bleaching powder) is/are:

(a) +1 only (b) –1 only

(c) +1 and –1 (d) none of these

3. The oxidation number of sulphur in S_8, S_2F_2, H_2S and H_2SO_4 respectively are:

(a) $0, +1, -2$ and 6 (b) $+2, 0, +2$ and 6

(c) $0, +1, +2$ and 4 (d) $-2, 0, +2$ and 6

4. When SO_2 is passed into an acidified potassium dichromate solution, the oxidation numbers of sulphur and chromium in the final products respectively are:

(a) +6, +6 (b) +6, +3

(c) 0, +3 (d) +2, +3

5. The oxidation number of nitrogen atoms in NH_4NO_3 are:

(a) +3, +3 (b) +3, –3

(c) $-3, +5$ (d) $-5, +3$

6. The oxidation states of S-atoms in Caro's and Marshell's acids are:

(a) $+6, +6$ (b) $+6, +4$

(c) $+6, -6$ (d) $+4, +6$

7. The oxidation number of oxygen in KO_3, Na_2O_2 is :

(a) 3, 2 (b) 1, 0

(c) 0, 1 (d) –0.33, –1

8. The oxidation number of phosphorus in $Ba(H_2PO_2)_2$ is :

(a) –1 (b) +1

(c) +2 (d) +3

9. Oxidation state of Co in $[Co(NH_3)_5Cl]Cl_2$ is :

(a) +3 (b) +1

(c) +2 (d) 0

10. In which of the following the oxidation number of oxygen has been arranged in increasing order :

(a) $OF_2 < KO_2 < BaO_2 < O_3$ (b) $BaO_2 < KO_2 < O_3 < OF_2$

(c) $BaO_2 < O_3 < OF_2 < KO_2$ (d) $KO_2 < OF_2 < O_3 < BaO_2$

11. Which of the following has been arranged in order of increasing oxidation number of nitrogen?

(a) $NH_3 < N_2O_5 < NO < N_2$ (b) $NO_2^+ < NO_3^- < NO_2^- < N_3^-$

(c) $NH_4^+ < N_2H_4 < NH_2OH < N_2O$ (d) $NO_2 < NaN_3 < NH_4^+ < N_2O$

12. Which ordering of compounds is according to the decreasing order of the oxidation state of nitrogen ?

(a) HNO_3, NO, NH_4Cl, N_2 (b) HNO_3, NO, N_2, NH_4Cl

(c) HNO_3, NH_4Cl, NO, N_2 (d) NO, HNO_3, NH_4Cl, N_2

13. The oxidation states of sulphur in the anions SO_3^{2-}, $S_2O_4^{2-}$ and $S_2O_6^{2-}$ follow the order:

(a) $S_2O_6^{2-} < S_2O_4^{2-} < SO_3^{2-}$

(b) $S_2O_4^{2-} < SO_3^{2-} < S_2O_6^{2-}$

(c) $SO_3^{2-} < S_2O_4^{2-} < S_2O_6^{2-}$

(d) $S_2O_4^{2} < S_2O_6^{2-} < SO_3^{2-}$

14. When $K_2Cr_2O_7$ is converted to K_2CrO_4, the change in the oxidation state of chromium is :

(a) 0

(b) 6

(c) 4

(d) 3

15. 2 mole of N_2H_4 loses 16 mole of electron is being converted to a new compound *X*. Assuming that all of the N appears in the new compound. What is the oxidation state of 'N' in *X* ?

(a) -1

(b) -2

(c) $+2$

(d) $+4$

16. *A* compound contains atoms *A*, *B* and *C*. The oxidation number of *A* is +2, of *B* is +5 and of *C* is –2. The possible formula of the compound is :

(a) ABC_2

(b) $B_2(AC_3)_2$

(c) $A_3(BC_4)_2$

(d) $A_3(B_4C)_2$

17. If it is known that in $Fe_{0.96}O$, Fe is present in +2 and +3 oxidation state. What is the mole fraction of Fe^{2+} in the compound ?

(a) 12/25

(b) 25/12

(c) 1/12

(d) 11/12

18. In Fe(II)–MnO_4^- titration, HNO_3, is not used because:

(a) it oxidises Mn^{2+}

(b) it reduces MnO_4^-

(c) it oxidises Fe^{2+}

(d) it reduces Fe^{3+} formed

19. Which species are oxidized and reduced in the reaction ?

$FeC_2O_4 + KMnO_4 \longrightarrow Fe^{3+} + CO_2 + Mn^{2+}$

(a) Oxidised : Fe, C ; Reduced : Mn

(b) Oxidised : Fe ; Reduced : Mn

(c) Reduced : Fe, Mn ; Oxidised : C

(d) Reduced : C ; Oxidised : Mn, Fe

20. In which of the following reactions, H_2O_2 is acting as a reducing agent ?

(a) $SO_2 + H_2O_2 \longrightarrow H_2SO_4$

(b) $2KI + H_2O_2 \longrightarrow 2KOH + I_2$

(c) $PbS + 4H_2O_2 \longrightarrow PbSO_4 + 4H_2O$

(d) $Ag_2O + H_2O_2 \longrightarrow 2Ag + H_2O + O_2$

21. Following reaction describes the reaction of iron with oxygen:

$$4Fe + 3O_2 \longrightarrow 4Fe^{3+} + 6O^{2-}$$

Which one of the following statement is incorrect ?

(a) This is an example of a redox reaction

(b) Metallic iron is reduced to Fe^{3+}

(c) Fe^{3+} is an oxidising agent

(d) Metallic iron is a reducing agent

22. Select which of the following is non-redox reaction :

(a) $CH_4(g) + O_2(g) \longrightarrow CO_2(g) + H_2O(l)$

(b) $Sn(s) + NO_3^-(aq) + H^+(aq) \longrightarrow H_2SnO_3(aq) + NO(g) + H_2O(l)$

(c) $Mg(s) + HCl(aq) \longrightarrow MgCl_2(aq) + H_2(g)$

(d) $Cu(NO_3)_2(aq) + Na_2S(aq) \longrightarrow CuS(s) + 2NaNO_3(aq)$

23. The redox reaction among the following is :

(a) combination of dinitrogen with dioxygen at 2000 K

(b) reaction of $[Co(H_2O)_6]Cl_3$ with $AgNO_3$

(c) reaction of H_2SO_4 with NaOH

(d) formation of ozone from atmospheric oxygen in the presence of sunlight

24. Which reaction involves neither oxidation nor reduction?

(a) $S_2O_3^{2-} \longrightarrow S_4O_6^{2-}$ (b) $Cr \longrightarrow CrCl_3$

(c) $Na \longrightarrow Na^+$ (d) $CrO_4^{2-} \longrightarrow Cr_2O_7^{2-}$

25. An example of a disproportionation reaction is :

(a) $2NaBr + Cl_2 \longrightarrow 2NaCl + Br_2$

(b) $2KMnO_4 \longrightarrow K_2MnO_4 + MnO_2 + O_2$

(c) $2MnO_4^- + 10I^- + 16H^+ \longrightarrow 2Mn^{2+} + 5I_2 + 8H_2O$

(d) $2CuBr \longrightarrow CuBr_2 + Cu$

26. Which of the following species can show disproportionation reaction?

(a) $Cr_2O_7^{2-}$ (b) MnO_4^-

(c) P_4 (d) F_2

27. $Cr_2O_7^{2-} + X \xrightarrow{H^+} Cr^{3+} +$ oxidised product of $X + H_2O$. Here X in the above reaction can not be :

(a) $C_2O_4^{2-}$ (b) Fe^{2+}

(c) SO_4^{2-} (d) S^{2-}

28. (1) $I_2 + H_2O_2 + 2OH^- \longrightarrow 2I^- + 2H_2O + O_2$

(2) $H_2O_2 + HOCl \longrightarrow Cl^- + H_3O^+ + O_2$

(a) H_2O_2 is acting as oxidising agent in both the reaction

(b) H_2O_2 is acting as reducing agent in both the reaction

(c) H_2O_2 is acting as oxidising agent in reaction (1) and as reducing agent in reaction (2)

(d) H_2O_2 is acting as reducing agent in reaction (1) and as oxidising agent in reaction (2)

29. Which reaction does not represent auto redox or disproportionation ?

(a) $Cl_2 + OH^- \longrightarrow Cl^- + ClO_3^- + H_2O$ (b) $2H_2O_2 \longrightarrow H_2O + O_2$

(c) $2Cu^+ \longrightarrow Cu^{2+} + Cu$ (d) $(NH_4)_2Cr_2O_7 \longrightarrow N_2 + Cr_2O_3 + 4H_2O$

30. In the reaction, $P + NaOH \longrightarrow PH_3 + NaH_2PO_2$

(a) P is oxidized only (b) P is reduced only

(c) P is oxidized as well as reduced (d) Na is reduced

31. Which of the following is a redox-reaction ?

(a) $2Na[Ag(CN)_2] + Zn \longrightarrow Na_2[Zn(CN)_4] + 2Ag$

(b) $BaO_2 + H_2SO_4 \longrightarrow BaSO_4 + H_2O_2$

(c) $N_2O_5 + H_2O \longrightarrow 2HNO_3$

(d) $AgNO_3 + KI \longrightarrow AgI + KNO_3$

32. Balance the following equation and choose the quantity which is the sum of the coefficients of the products :

$$\ldots\ldots CS_2 + \ldots\ldots Cl_2 \longrightarrow CCl_4 + \ldots\ldots S_2Cl_2$$

(a) 5 (b) 3

(c) 6 (d) 2

33. Balance the following equation and choose the quantity which is the sum of the coefficients of reactants and products:

$$....PtCl_4 + ...XeF_2 \longrightarrow PtF_6 + ...ClF + ...Xe$$

(a) 16
(b) 13
(c) 18
(d) 12

34. For the redox reaction, $MnO_4^- + C_2O_4^{2-} + H^+ \longrightarrow Mn^{2+} + CO_2 + H_2O$ the correct coefficients of the reactants for the balanced reaction are respectively MnO_4^-, $C_2O_4^{2-}$, H^+:

(a) 2, 5, 16
(b) 16, 3, 12
(c) 15, 16, 12
(d) 2, 16, 5

35. In following redox reaction,

$$xFe^{2+} + yCr_2O_7^{2-} + 14H^+ \longrightarrow 6Fe^{3+} + 2Cr^{3+} + zH_2O$$

The value of x, y, z are:

(a) $x = 6, y = 1, z = 7$
(b) $x = 3, y = 6, z = 8$
(c) $x = 6, y = 2, z = 7$
(d) $x = 4, y = 6, z = 7$

36. Balance the following reaction and find the value of a, b and g reapectively:

$$aKMnO_4 + bH_2O_2 + 6H_2SO_4 \longrightarrow dMnSO_4 + eK_2SO_4 + fO_2 + gH_2O$$

(a) 2, 5, 8
(b) 4, 10, 16
(c) 10, 4, 16
(d) 8, 5, 2

37. In the ionic equation $BiO_3^- + xH^+ + ye^- \longrightarrow Bi^{3+} + H_2O$, the value of $(x + y)$ is :

(a) 8
(b) 6
(c) 4
(d) None of these

38. In a balanced reaction, $H_2SO_4 + xHI \longrightarrow H_2S + yI_2 + zH_2O$, the values of x, y, z are :

(a) $x = 3, y = 5, z = 2$
(b) $x = 4, y = 8, z = 5$
(c) $x = 8, y = 4, z = 4$
(d) $x = 5, y = 3, z = 4$

39. $S_8 + bOH^- \longrightarrow cS^{2-} + dS_2O_3^{2-} + 6H_2O$. Find the value of b.

(a) 2
(b) 8
(c) 10
(d) 12

40. For the redox reaction $xP_4 + yHNO_3 \longrightarrow H_3PO_4 + NO_2 + H_2O$:

(a) $x = 1, y = 5$
(b) $x = 1, y = 10$
(c) $x = 1, y = 20$
(d) $x = 1, y = 15$

41. $xI_2 + 12OH^- \longrightarrow yI^- + 2IO_3^- + zH_2O$, here x, y and z are respectively :

(a) 4, 1, 6
(b) 6, 10, 6
(c) 8, 2, 6
(d) None of these

42. In the balanced chemical reaction $IO_3^- + aI^- + bH^+ \longrightarrow cH_2O + dI_2$, a, b, c and d respectively, correspond to :

(a) 5, 6, 3, 3
(b) 5, 3, 6, 3
(c) 3, 5, 3, 6
(d) 5, 6, 5, 5

43. In the reaction of oxalate with permanganate in acidic medium, the number of electron(s) involved in producing one molecule of CO_2 is :

(a) 10
(b) 5
(c) 2
(d) 1

44. $xMnO_4^- + SO_3^{2-} + H_2O \longrightarrow MnO_2 + ySO_4^{2-} + OH^-$. Find the $\frac{x}{y}$.

(a) $\frac{1}{2}$ (b) $\frac{2}{3}$

(c) $\frac{3}{2}$ (d) $\frac{2}{5}$

45. The equivalent weight of $MnSO_4$ is half its molecular weight when it is converted to :

(a) Mn_2O_3 (b) MnO_2

(c) MnO_4^- (d) MnO_4^{2-}

46. The equivalent weight of NaH_2PO_4 (molecular weight M) can be :

(a) $\frac{M}{2}$ as acid (b) $\frac{M}{1}$ as base

(c) Both (a) and (b) (d) neither (a) and (b)

47. Acetic acid on chlorination yields trichloroacetic acid. Its equivalent weight will be :

(a) 60 (b) 40

(c) 20 (d) 10

48. In the reaction, $Pb + PbO_2 + H_2SO_4 \longrightarrow PbSO_4 + H_2O$, the equivalent weight of H_2SO_4 is :

(a) M (b) $M/2$

(c) $2M$ (d) $M/4$

49. 0.1 mole H_3PO_x is completely neutralised by 5.6 g KOH then the true statement is:

(a) $x = 3$ and given acid is dibasic

(b) $x = 4$ and given acid has no P—H linkage

(c) $x = 2$ and given acid does not form acid salt

(d) all of these

50. When potassium permanganate is titrated against ferrous ammonium sulphate in acidic medium, the equivalent weight of potassium permanganate is:

(a) $\frac{\text{molecular weight}}{3}$ (b) $\frac{\text{molecular weight}}{5}$

(c) $\frac{\text{molecular weight}}{2}$ (d) $\frac{\text{molecular weight}}{10}$

51. Equivalent weight of FeS_2 in the half reaction, $FeS_2 \longrightarrow Fe_2O_3 + SO_2$ is :

(a) $M/10$ (b) $M/11$

(c) $M/6$ (d) $M/1$

52. The equivalent weight of HCl in the given reaction is:

$K_2Cr_2O_7 + 14HCl \longrightarrow 2KCl + 2CrCl_3 + 3Cl_2 + H_2O$

(a) 16.25 (b) 36.5

(c) 73 (d) 85.1

53. Equivalent weight of H_3PO_2 when it disproportionate into PH_3 and H_3PO_3 is:

(a) M (b) $M/2$

(c) $M/4$ (d) $3M/4$

54. In the following reaction, $As_2S_3 + H^+ + NO_3^- \longrightarrow NO + H_2O + AsO_4^{3-} + SO_4^{2-}$

the equivalent weight of As_2S_3 is related to its molecular weight by :

(a) $M/2$ (b) $M/4$

(c) $M/24$ (d) $M/28$

55. Sulphur forms the chlorides S_2Cl_2 and SCl_2. The equivalent mass of sulphur in SCl_2 is :

(a) 8 g/mol (b) 16 g/mol

(c) 64.8 g/mol (d) 32 g/mol

56. The equivalent weight of an element is 4. Its chloride has a vapour density 59.25. Then, the valency of the elements is :

(a) 4 (b) 3

(c) 2 (d) 1

57. 6×10^{-3} mole $K_2Cr_2O_7$ reacts completely with 9×10^{-3} mole X^{n+} to give XO_3^- and Cr^{3+}. The value of n is :

(a) 1 (b) 2

(c) 3 (d) None of these

58. What weight of $H_2C_2O_4 \cdot 2H_2O$ (mol. wt. = 126) should be dissolved in water to prepare 250 mL of centinormal solution which act as a reducing agent ?

(a) 0.63 g (b) 0.1575 g

(c) 0.126 g (d) 0.875 g

59. The equivalent wt. of the salt, $KHC_2O_4 \cdot H_2C_2O_4 \cdot 4H_2O$ when it act as reducing agent is :

(a) $\frac{\text{Mol. wt}}{1}$ (b) $\frac{\text{Mol. wt}}{2}$

(c) $\frac{\text{Mol. wt}}{3}$ (d) $\frac{\text{Mol. wt}}{4}$

60. The equivalent weight of divalent metal is W. The molecular weight of its chloride is :

(a) W + 35.5 (b) W + 71

(c) 2W + 71 (d) 2W + 35.5

61. When BrO_3^- ion reacts with Br^- in acid medium, Br_2 is liberated. The equivalent weight of Br_2 in this reaction is:

(a) $\frac{5M}{8}$ (b) $\frac{5M}{3}$

(c) $\frac{3M}{5}$ (d) $\frac{4M}{6}$

62. If m_A gram of a metal A displaces m_B gram of another metal B from its salt solution and if the equivalent weights are E_A and E_B respectively then equivalent weight of A can be expressed as:

(a) $E_A = \frac{m_A}{m_B} \times E_B$ (b) $E_A = \frac{m_A \times m_B}{E_B}$

(c) $E_A = \frac{m_B}{m_A} \times E_B$ (d) $E_A = \sqrt{\frac{m_A}{m_B} \times E_B}$

63. Hydrazine reacts with KIO_3 in presence of HCl as

$$N_2H_4 + IO_3^- + 2H^+ + Cl^- \longrightarrow ICl + N_2 + 3H_2O$$

The equivalent masses of N_2H_4 and KIO_3 respectively are:

(a) 8 and 53.5 (b) 16 and 53.5

(c) 8 and 35.6 (d) 8 and 87

64. In the reaction, $2CuSO_4 + 4KI \longrightarrow Cu_2I_2 + 2K_2SO_4 + I_2$, if M is the molar mass of $CuSO_4$ then it's equivalent weight is :

(a) M (b) $\frac{M}{2}$

(c) $2M$ (d) $4M$

65. In the reaction, $2CuSO_4 + 4KI \longrightarrow Cu_2I_2 + 2K_2SO_4 + I_2$, if M is the molar mass of KI then it's equivalent weight is :

(a) M (b) $\frac{M}{2}$

(c) $2M$ (d) $4M$

66. Find the number of moles of $KMnO_4$ needed to oxidise one mole Cu_2S in acidic medium. Here Cu_2S is oxidised into Cu^{2+} and SO_2.

(a) 1.6 (b) 0.4

(c) 1.2 (d) 1

67. An oxide, X_2O_3 is oxidized to XO_4^- by $Cr_2O_7^{2-}$ in acid medium. The number of moles of X_2O_3 oxidized per mole of $Cr_2O_7^{2-}$ is :

(a) 2/3 (b) 3/4

(c) 3/2 (d) 3

68. H_2O_2 react with $K[Cr(OH)_4]$ to give CrO_4^{2-} and H_2O. How many moles of H_2O_2 will be required to react completely with 1.5 mole of $K[Cr(OH)_4]$ in basic medium?

(a) 1.25 (b) 2.25

(c) 3 (d) 4.5

69. What will be the normality of a solution obtained by mixing 0.45 N and 0.60 N NaOH in the ratio 2 : 1 by volume ?

(a) 0.4 N (b) 0.5 N

(c) 1.05 N (d) 0.15 N

70. A solution containing 2.7×10^{-3} mole of A^{2+} ions required 1.6×10^{-3} mole of MnO_4^- for the oxidation of A^{2+} to AO_3^- the medium used is :

(a) neutral (b) acidic

(c) strong basic (d) none of these

71. If equal volume of $1M$ $KMnO_4$ and $1M$ $K_2Cr_2O_7$ solutions are allowed to oxidise Fe(II) to Fe(III) in acidic medium, then Fe(II) oxidise will be:

(a) More by $KMnO_4$ (b) More by $K_2Cr_2O_7$

(c) Equal in both cases (d) Can't be determined

72. Number of moles of MnO_4^- required to oxidizes one mole of ferrous oxalate completely in acidic medium will be:

(a) 7.5 moles (b) 0.2 moles

(c) 0.6 moles (d) 0.4 moles

73. How many grams of I_2 are present in a solution which requires 40 mL of 0.11 N $Na_2S_2O_3$ to react with it? (I = 127)

$$S_2O_3^{2-} + I_2 \longrightarrow S_4O_6^{2-} + 2I^-$$

(a) 12.7 g (b) 0.558 g

(c) 25.4 g (d) 11.4 g

74. H_2O_2 is used as bleaching reagent because on dissociation it gives oxygen ($2H_2O_2 \longrightarrow 2H_2O + O_2$). If "Chachi ji" used H_2O_2 solution to bleach her hair at 1 atm and 273 K. She has a H_2O_2 solution labelled as '5.6 V' then what normality of such solution must she required to bleach her hair ?

(a) 0.25 *N* (b) 0.5 *N*

(c) 0.75 *N* (d) 1 *N*

75. A volume of 100 mL of H_2O_2 is oxidized by 100 mL of 1 *M*-$KMnO_4$ in acidic medium. A volume of 100 mL of same H_2O_2 is oxidized by '*V*' mL of 1 *M*-$KMnO_4$ in mild basic medium. The value of '*V*' is :

(a) 500 (b) 166.67

(c) 100 (d) 33.33

76. A solution of H_2O_2 is titrated with a solution of $KMnO_4$. The reaction is

$$2MnO_4^- + 5H_2O_2 + 6H^+ \longrightarrow 2Mn^{2+} + 5O_2 + 8H_2O$$

It requires 50 mL of 0.1 *M*-$KMnO_4$ to oxidize 10 mL of H_2O_2.

The strength of H_2O_2 solution is :

(a) 4.25% (*w*/v) (b) 2.125% (*w*/v)

(c) 0.85% (*w*/v) (d) 8.5% (*w*/v)

77. 1.25 g of a solid dibasic acid is completely neutralised by 25 mL of 0.25 molar $Ba(OH)_2$ solution. Molecular mass of the acid is :

(a) 100 (b) 150

(c) 120 (d) 200

78. 10 mL of *N*-HCl, 20 mL of *N*/2 H_2SO_4 and 30 mL of *N*/3 HNO_3 are mixed together and volume made to one litre. The normality of H^+ in the resulting solution is :

(a) 3*N*/100 (b) *N*/10

(c) *N*/20 (d) *N*/40

79. 0.45 g of an acid of mol. wt. 90 was neutralised by 20 mL of 0.54 *N* caustic potash (KOH). The basicity of acid is :

(a) 1 (b) 2

(c) 3 (d) 4

80. A 3.4 g sample of H_2O_2 solution containing *x*% H_2O_2 by mass requires *x* mL of a $KMnO_4$ solution for complete oxidation under acidic condition. The molarity of $KMnO_4$ solution is :

(a) 1 (b) 0.5

(c) 0.4 (d) 0.2

81. What volume of $O_2(g)$ measured at 1 atm and 273 K will be formed by action of 100 mL of 0.5 *N* $KMnO_4$ on hydrogen peroxide in an acid solution ? The skeleton equation for the reaction is

$$KMnO_4 + H_2SO_4 + H_2O_2 \longrightarrow K_2SO_4 + MnSO_4 + O_2 + H_2O$$

(a) 0.12 litre (b) 0.028 litre

(c) 0.56 litre (d) 1.12 litre

82. $KMnO_4$ reacts with oxalic acid according to the equation

$$2MnO_4^- + 5C_2O_4^{2-} + 16H^+ \longrightarrow 2Mn^{2+} + 10CO_2 + 8H_2O$$

Here, 20 mL of 0.1 *M* $KMnO_4$ is equivalent to :

(a) 120 mL of 0.25 *M* $H_2C_2O_4$ (b) 150 mL of 0.10 *M* $H_2C_2O_4$

(c) 25 mL of 0.20 *M* $H_2C_2O_4$ (d) 50 mL of 0.20 *M* $H_2C_2O_4$

83. Ratio of moles of Fe (II) oxidised by equal volumes of equimolar $KMnO_4$ and $K_2Cr_2O_7$ solutions in acidic medium will be:

(a) 5 : 3 (b) 1 : 1

(c) 1 : 2 (d) 5 : 6

84. The weight of a mixture containing HCl and H_2SO_4 is 0.1 g. On treatment with an excess of an $AgNO_3$ solution, reacted with this acid mixture gives 0.1435 g of AgCl. Weight % of the H_2SO_4 is mixture is :

(a) 36.5 (b) 63.5

(c) 50 (d) None of these

85. The equivalent mass of crystalline oxalic acid ($H_2C_2O_4 \cdot 2H_2O$) is equal to the equivalent weight of:

(a) Na_2CO_3 (b) HNO_3

(c) KOH (d) CH_3COOH

86. The NH_3 evolved due to complete conversion of N from 1.12 g sample of protein was absorbed in 45 mL of 0.4 *N* HNO_3. The excess acid required 20 mL of 0.1 *N* NaOH. The % N in the sample is:

(a) 8 (b) 16

(c) 20 (d) 25

87. Find out % of oxalate ion in a given sample of an alkali metal oxalate salt, 0.30 g of it is dissolved in 100 mL water required 90 mL of centimolar $KMnO_4$ solution in acidic medium.

(a) 66% (b) 55%

(c) 44% (d) 6.6%

88. A 0.5 g sample of KH_2PO_4 is titrated with 0.1 *M* NaOH. The volume of base required to do this is 25.0 mL. The reaction is represented as

$$H_2PO_4^- + OH^- \longrightarrow HPO_4^{2-} + H_2O$$

The percentage purity of KH_2PO_4 is : (K = 39, P = 31)

(a) 68% (b) 34%

(c) 85% (d) 51%

89. 6.0 gm of metal carbonate converted into 4 gm of metal sulphide. The equivalent mass of metal is :

(a) 6 (b) 12

(c) 18 (d) 24

90. What volume of HNO_3 (sp. gravity 1.05 g mL^{-1} containing 12.6 (*w/W*) of HNO_3) that reduce into NO is required to oxidise iron 1 g $FeSO_4 \cdot 7H_2O$ in acid medium is:

(a) 70 mL (b) 0.57 mL

(c) 80 mL (d) 0.65 mL

91. The total volume of 0.1 *M* $KMnO_4$ solution that are needed to oxidize 100 mg each of ferrous oxalate and ferrous sulphate in a mixture in acidic medium is:

(a) 1.096 mL (b) 1.32 mL

(c) 5.48 mL (d) none of these

92. When 2.5 g of a sample of Mohr's salt reacts completely with 50 mL of $\frac{N}{10}$ $KMnO_4$ solution. The % purity of the sample of Mohr's salt is :

(a) 78.4 (b) 70

(c) 37 (d) 40

93. 4 mole of a mixture of Mohr's salt and $Fe_2(SO_4)_3$ requires 500 mL of 1 M $K_2Cr_2O_7$ for complete oxidation in acidic medium. The mole % of the Mohr's salt in the mixture is:

(a) 25 (b) 50
(c) 60 (d) 75

94. The equivalent weight of a metal is double than of oxygen. How many times is the equivalent weight of it's oxide than the equivalent weight of the metal ?

(a) 1.5 (b) 2
(c) 3 (d) 4

95. Calculate the mass of anhydrous oxalic acid, which can be oxidised to $CO_2(g)$ by 100 mL of an MnO_4^- solution, 10 mL of which is capable of oxidising 50 mL of $1N$ I^- to I_2.

(a) 45 g (b) 22.5 g
(c) 30 g (d) 12.25 g

96. Normality of a solution in which 1.11 g $CaCl_2$ is dissolved in 250 mL solution is :

(a) 0.4 N (b) 0.16 N
(c) 0.04 N (d) 0.08 N

97. 50 mL of 0.5 M oxalic acid is needed to neutralize 25 mL of sodium hydroxide solution. The amount of NaOH of the given sodium hydroxide solution is :

(a) 8 g (b) 4 g
(c) 1 g (d) 2 g

98. Calculate volume of 0.1 M dibasic acid, which required for complete titration with $\frac{N}{2}$, 50 mL base.

(a) 125 mL (b) 50 mL
(c) 75 mL (d) 100 mL

99. What volume of 0.1 N NaOH required to neutralise 10 mL of 0.1 N phosphinic (H_3PO_2) acid is:

(a) 2 mL (b) 5 mL
(c) 10 mL (d) 20 mL

100. How many milli-litre of 0.5 N $SnCl_2$ solution will reduce 600 mL of 0.1 N $HgCl_2$ to Hg_2Cl_2?

(a) 120 mL (b) 60 mL
(c) 30 mL (d) 240 mL

101. 5 mole of equimolar mixture of ferric oxalate and ferous oxalate require x mole of $KMnO_4$. In acidic medium for complete oxidation, x is:

(a) 4.0 (b) 4.5
(c) 8.0 (d) 0.9

102. If 200 mL of 0.1 M KHC_2O_4, 300 mL of 0.2 M $H_2C_2O_4$ and 500 mL of 0.3 M NaCl is neutralized by 140 mL of $Ca(OH)_2$. What is its molarity?

(a) 1 (b) 0.5
(c) 1.5 (d) 2

103. 100 mL, $2M$ $KMnO_4$ oxidises 250 mL of $NaHC_2O_4$ in acidic medium. If 50 mL of same $NaHC_2O_4$ requires 25 mL of HCl solution then find molarity of HCl solution.

(a) 2 M (b) 3 M
(c) 4 M (d) None of these

104. The number of moles of $KMnO_4$ needed to oxidise a mixture of 1 mol each of $FeSO_4$, FeC_2O_4 completely in acidic medium is :

(a) 0.2 (b) 0.4
(c) 0.8 (d) 1

105. 54 g of a metal is required 48 g of oxygen gas to completely convert in it's oxide then equivalent mass of metal is :

(a) 9 gram (b) 18 gram
(c) 27 gram (d) 32 gram

106. $S_2O_3^{2-}$ ion is oxidized by $S_2O_8^{2-}$ ion, the products are $S_4O_6^{2-}$ and SO_4^{2-} ions. What volume of 0.25 *M* thiosulphate solution would be needed to reduce 1 g of $K_2S_2O_8$? (K = 39, S = 32)

(a) 29.63 mL (b) 32.69 mL
(c) 36.92 mL (d) 62.93 mL

107. Manganese ions (Mn^{2+}) can be oxidized by persulphate ions ($S_2O_8^{2-}$) according to the following equation : $Mn^{2+} + S_2O_8^{2-} \longrightarrow SO_4^{2-} + MnO_4^-$ (Acidic medium). How many moles of $S_2O_8^{2-}$ are required to oxidize 1 mole of Mn^{2+} ?

(a) 2.5 (b) 2.0
(c) 1.0 (d) 0.4

108. V_1 mL of permanganate solution of molarity M_1 reacts exactly with V_2 mL of ferrous sulphate solution in acidic medium of molarity M_2, then :

(a) $V_1M_1 = V_2M_2$ (b) $5V_1M_1 = V_2M_2$
(c) $V_1M_1 = 5V_1M_2$ (d) None of these

109. What volume (in mL) of 0.02 *M* $K_2Cr_2O_7$ solution required to react with 0.288 g of ferrous oxalate in acidic medium?
(Molar mass of Fe = 56 g mol^{-1})

(a) 16.66 (b) 50
(c) 150 (d) 300

110. 10 mL of H_2O_2 solution (volume strength '*X*') required 10 mL of $\left(\frac{1}{0.56}\right)$ *N* $KMnO_4$ solution in acidic medium, hence, '*X*' is :

(a) 0.56 (b) 5.6
(c) 0.1 (d) 10

111. A 20.0 mL solution containing 0.2 g impure H_2O_2 reacts completely with 0.316 g of $KMnO_4$ in acid solution. The purity of H_2O_2 (in %) is :

(a) 8.5 (b) 25
(c) 75 (d) 85

112. 180 g of iron pyrite (FeS_2) is heated strongly in air. The SO_2 thus produced is passed through acidified semimolar (*M*/2) $K_2Cr_2O_7$ solution. The volume (in L) of acidified $K_2Cr_2O_7$ used in redox titration is :
(Given; Fe = 56, S = 32)

(a) 1 L (b) 2 L
(c) 2.5 L (d) 4 L

113. 32 g of a sample of $FeSO_4 \cdot 7H_2O$ were dissolved in dilute sulphuric acid and water and it's volume was made up to 1 litre, 25 mL of this solution required 20 mL of 0.02 M $KMnO_4$ solution for complete oxidation. Calculate the weight % of $FeSO_4 \cdot 7H_2O$ in the sample.

(a) 34.75 (b) 69.5
(c) 89.5 (d) None of these

114. A mixture containing 0.05 mole of $K_2Cr_2O_7$ and 0.02 mole of $KMnO_4$ was treated with excess of KI in acidic medium. The liberated iodine required 1.0 L of $Na_2S_2O_3$ solution for titration. Concentration of $Na_2S_2O_3$ solution was:

(a) 0.40 mol L^{-1} (b) 0.20 mol L^{-1}
(c) 0.25 mol L^{-1} (d) 0.30 mol L^{-1}

115. If a g is the mass of $NaHC_2O_4$ required to neutralize 100 mL of 0.2 M NaOH and b g that required to reduce 100 mL of 0.02 M $KMnO_4$ in acidic medium, then:

(a) $a = b$ (b) $2a = b$
(c) $a = 2b$ (d) None of these

116. 2 mole, equimolar mixture of $Na_2C_2O_4$ and $H_2C_2O_4$ required V_1 L of 0.1 M $KMnO_4$ in acidic medium for complete oxidation. The same amount of the mixture required V_2 L of 0.2 M NaOH for neutralization. The ratio of V_1 to V_2 is:

(a) 1 : 2 (b) 2 : 1
(c) 4 : 5 (d) 5 : 4

117. A mixture of $NaHC_2O_4$ and $KHC_2O_4 \cdot H_2C_2O_4$ required equal volumes of 0.2 N $KMnO_4$ and 0.12 N NaOH separately. What is the molar ratio of $NaHC_2O_4$ and $KHC_2O_4 \cdot H_2C_2O_4$ in the mixture ?

(a) 6 : 1 (b) 1 : 6
(c) 1 : 3 (d) 3 : 1

118. In an iodometric estimation, the following reactions occur

$2Cu^{2+} + 4I^- \longrightarrow Cu_2I_2 + I_2; \quad I_2 + 2Na_2S_2O_3 \longrightarrow 2NaI + Na_2S_4O_6$

0.12 mole of $CuSO_4$ was added to excess of KI solution and the liberated iodine required 120 mL of hypo. The molarity of hypo solution was:

(a) 2 (b) 0.20
(c) 0.1 (d) 1.0

119. What volume of 0.40 M-$Na_2S_2O_3$ would be required to react with the I_2 liberated by adding excess of KI to 50 mL of 0.20 M $CuSO_4$?

(a) 12.5 mL (b) 25 mL
(c) 50 mL (d) 2.5 mL

120. A solution containing sodium hydroxide and sodium carbonate required 20 mL of $N/20$ HCl solution when titrated with phenolphthalein as an indicator. But the same amount of the solution when titrated with methyl orange as an indicator required 25 mL of the same acid. The amount of sodium hydroxide present in the solution is :

(a) 0.03 g (b) 0.053 g
(c) 0.0265 g (d) None of these

121. 0.63 g of a metal carbonate is mixed with 250 mL of 0.1 N HCl solution. 50 mL of 0.2 N NaOH solution was required to neutralise excess of the acid. The equivalent weight of the metal carbonate is :

(a) 63 g (b) 42 g
(c) 23.4 g (d) None of these

122. 10 L of hard water required 5.6 g of lime for removing hardness. Hence temporary hardness in ppm of $CaCO_3$ is:

(a) 1000 (b) 2000

(c) 100 (d) 1

123. The hardness of a water sample (in terms of equivalents of $CaCO_3$) containing 10^{-3} M $CaSO_4$ is : (molar mass of $CaSO_4 = 136$ g mol^{-1})

(a) 10 ppm (b) 90 ppm

(c) 50 ppm (d) 100 ppm

124. 1 L of pond water contains 20 mg of Ca^{2+} and 12 mg of Mg^{2+} ions. What is the volume of a 2 N Na_2CO_3 solution required to soften 5000 L of pond water?

(a) 500 L (b) 50 L

(c) 5 L (d) None of these

125. 100 mL of a water sample contains 0.81 g of calcium bicarbonate and 0.73 g of magnesium bicarbonate. The hardness of this water sample expressed in terms of equivalents of $CaCO_3$ is : (molar mass of calcium bicarbonate is 162 g mol^{-1} and magnesium bicarbonate is 146 g mol^{-1})

(a) 10,000 ppm (b) 100 ppm

(c) 1,000 ppm (d) 5,000 ppm

Level-2

1. During developing of an exposed camera film, one step involves in the following reaction :

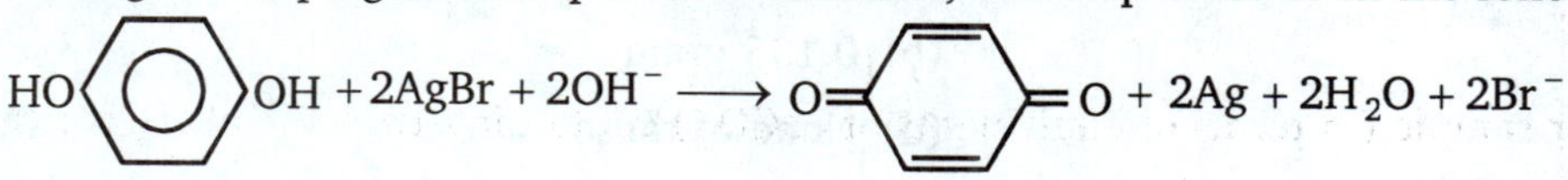

(Quinol)

Which of the following best describes the role of quinol ?

(a) It acts as an acid
(b) It acts as reducing agent
(c) It acts as oxidant
(d) It acts as a base

2. 100 cm^3 of a solution of an acid (Molar mass = 98) containing 29.4 g of the acid per litre were completely neutralized by 90.0 cm^3 of aq. NaOH containing 20 g of NaOH per 500 cm^3. The basicity of the acid is:

(a) 3
(b) 2
(c) 1
(d) data insufficient

3. 20 mL of 0.1 *M* solution of compound $Na_2CO_3 \cdot NaHCO_3 \cdot 2H_2O$ is titrated against 0.05 *M* HCl. *x* mL of HCl is used when phenolphthalein is used as an indicator and *y* mL of HCl is used when methyl orange is the indicator in two separate titrations. Hence $(y - x)$ is:

(a) 40 mL
(b) 80 mL
(c) 120 mL
(d) None of these

4. A sample containing $HAsO_2$ (mol. mass = 108) and weighing 3.78 g is dissolved and diluted to 250 mL in a volumetric flask. A 50 mL sample (aliquot) is withdrawn with a pipet and titrated with 35 mL of 0.05 *M* solution of I_2. Calculate the percentage $HAsO_2$ in the sample:

(a) 25%
(b) 20%
(c) 10%
(d) none of these

5. A mixture of FeO and Fe_2O_3 is completely reacted with 100 mL of 0.25 *M* acidified $KMnO_4$ solution. The resultant solution was then treated with Zn dust which converted Fe^{3+} of the solution to Fe^{2+}. The Fe^{2+} required 1000 mL of 0.10 *M* $K_2Cr_2O_7$ solution. Find out the weight % Fe_2O_3 in the mixture.

(a) 80.85
(b) 19.15
(c) 50
(d) 89.41

6. To a 10 mL, 1 *M* aqueous solution of Br_2, excess of NaOH is added so that all Br_2 is disproportionated to Br^- and BrO_3^-. The resulting solution is free from Br^-, by extraction and excess of OH^- neutralised by acidifying the solution. The resulting solution is sufficient to react with 2 g of impure CaC_2O_4 (*M* = 128 g/mol) sample. The % purity of oxalate sample is:

(a) 85.3%
(b) 12.5%
(c) 90%
(d) 64%

7. 0.10 g of a sample containing $CuCO_3$ and some inert impurity was dissolved in dilute sulphuric acid and volume made up to 50 mL. This solution was added into 50 mL of 0.04 *M* KI solution where copper precipitates as CuI and I^- is oxidized into I_3^-. A 10 mL portion of this solution is taken for analysis, filtered and made up free I_3^- and then treated with excess of acidic permanganate solution. Liberated iodine required 20 mL of 2.5 *mM* sodium thiosulphate solution to reach the end point.
Determine mass percentage of $CuCO_3$ in the original sample.

(a) 7.41
(b) 74.1
(c) 61.75
(d) None of these

8. 20 mL of 0.2 M NaOH(aq) solution is mixed with 35 mL of 0.1M NaOH(aq) solution and the resultant solution is diluted to 100 mL. 40 mL of this diluted solution reacted with 10% impure sample of oxalic acid ($H_2C_2O_4$). The mass of impure sample is:

(a) 0.15 gram (b) 0.135 gram
(c) 0.59 gram (d) None of these

9. An impure sample of sodium oxalate ($Na_2C_2O_4$) weighing 0.20 g is dissolved in aqueous solution of H_2SO_4 and solution is titrated at 70°C, requiring 45 mL of 0.02 M $KMnO_4$ solution. The end point is overrun, and back titration in carried out with 10 mL of 0.1 M oxalic acid solution. Find the % purity of $Na_2C_2O_4$ in sample:

(a) 75 (b) 83.75
(c) 90.25 (d) None of these

10. 0.5 g mixture of $K_2Cr_2O_7$ and $KMnO_4$ was treated with excess of KI in acidic medium. Iodine liberated required 150 cm^3 of 0.10 N solution of thiosulphate solution for titration.
Find the percentage of $K_2Cr_2O_7$ in the mixture:

(a) 14.64 (b) 34.2
(c) 65.69 (d) 50

11. A 150 mL of solution of I_2 is divided into two unequal parts. I part reacts with hypo solution in acidic medium. 15 mL of 0.4 M hypo was consumed. II part was added with 100 mL of 0.3 M NaOH solution. Residual base required 10 mL of 0.3 M H_2SO_4 solution for complete neutralization. What was the initial concentration of I_2 ?

(a) 0.08 M (b) 0.1 M
(c) 0.2 M (d) None of these

12. A mixture of H_2SO_4 and $H_2C_2O_4$ (oxalic acid) and some inert impurity weighing 3.185 g was dissolved in water and the solution made up to 1 litre. 10 mL of this solution required 3 mL of 0.1 N NaOH for complete neutralization. In another experiment 100 mL of the same solution in hot condition required 4 mL of 0.02 M $KMnO_4$ solution for complete reaction. The mass % of H_2SO_4 in the mixture was:

(a) 40 (b) 50
(c) 60 (d) 80

13. The concentration of an oxalic acid solution is 'x' mol $litre^{-1}$. 40 mL of this solution reacts with 16 mL of 0.05 M acidified $KMnO_4$. What is the pH of 'x' M oxalic acid solution ? (Assume that oxalic acid dissociates completely.)

(a) 1.3 (b) 1.699
(c) 1 (d) 2

14. RH_2 (ion exchange resin) can replace Ca^{2+} ions in hard water as $RH_2 + Ca^{2+} \longrightarrow RCa + 2H^+$. If 1 L of hard water after passing through RH_2 has pH = 3 then hardness in parts per million of Ca^{2+} is:

(a) 20 (b) 10
(c) 40 (d) 100

15. 15g of $Ba(MnO_4)_2$ sample containing inert impurity is completely reacted with 100 mL of 16.8 V. H_2O_2 at 1 atm and 273 K in acidic medium then what will be the % purity of $Ba(MnO_4)_2$ in the sample (At mass of Ba = 137, Mn = 55)

(a) 50% (b) 10%
(c) 75% (d) 25%

16. A 0.25 g sample of pyrolusite (impure MnO_2) is titrated with 1.26 g hydrated oxalic acid ($H_2C_2O_4 \cdot 2H_2O$) in an acidic medium. The excess of oxalic acid required 30 mL of 0.1*M* $KMnO_4(aq.)$ solution. What percentage by mass of MnO_2 in pyrolusite sample?

(a) 97 (b) 87

(c) 75.5 (d) 21.75

17. 10.5 gram of a sample of $MgCO_3$ that is 80% pure was strongly heated and the all CO_2 liberated was absorbed in 1000 mL of 0.3 *M* NaOH. How much mL of 0.5 *M* HCl would be required to react with resultant solution to reach the end point indicated by phenolphthalein?

(a) 200 mL (b) 300 mL

(c) 400 mL (d) None of these

18. An impure sample of $Ba(OH)_2$ (mol. wt. = 171) of mass 1.5 g was allowed to react with 80 mL of 0.20 *M* HCl(*aq.*). When the excess acid was titrated against NaOH, 20 mL of NaOH (*aq.*) was required. 10 mL of the same NaOH (*aq.*) required 30 mL of the 0.1*M* HCl(*aq.*) in a separate titration. What is the percentage purity of the $Ba(OH)_2$ sample?

(a) 85.5 (b) 57

(c) 42.75 (d) None of these

19. V_1 mL of NaOH of normality X and V_2 mL of $Ca(OH)_2$ of normality Y are mixed together. The mixture is completely neutralized by 100 mL of 0.18 *N* HCl. If $V_1 = \frac{V_2}{2}$ and $X = 4Y$. The fraction of acid is neutralized by $Ca(OH)_2$ is:

(a) 0.33 (b) 0.15

(c) 0.66 (d) 0.45

20. A 6.4 g impure sample of Cu dissolved in excess of H_2SO_4 . Further it treated with excess of KI. The liberated I_2 titrated with 100 mL of 0.2 *M* hypo solution. Calculate % mass composition of Cu in original sample [Atomic mass of Cu = 64]:

(a) 5 (b) 10

(c) 15 (d) 20

Level-3

Passage-1

The strength of H_2O_2 is expressed in several ways like molarity, normality, % (*w/V*), volume strength, etc. The strength of "10 V" means 1 volume of H_2O_2 on decomposition gives 10 volumes of oxygen at 1 atm and 273 K or 1 litre of H_2O_2 gives 10 litre of O_2 at 1 atm and 273 K The decomposition of H_2O_2 is shown as under :

$$H_2O_2(aq) \longrightarrow H_2O(l) + \frac{1}{2}O_2(g)$$

H_2O_2 can acts as oxidising as well as reducing agent. As oxidizing agent H_2O_2 is converted into H_2O and as reducing agent H_2O_2 is converted into O_2. For both cases its *n*-factor is 2.

$\therefore$ Normality of H_2O_2 solution $= 2 \times$ Molarity of H_2O_2 solution

1. What is the molarity of "11.2 *V*" H_2O_2?

(a) 1 *M* (b) 2 *M*

(c) 5.6 *M* (d) 11.2 *M*

2. What is the percentage strength (% *w/V*) of "11.2 *V*" H_2O_2?

(a) 1.7 (b) 3.4

(c) 34 (d) None of these

3. 20 mL of H_2O_2 solution is reacted with 80 mL of 0.05 *M* $KMnO_4$ in acidic medium then what is the volume strength of H_2O_2?

(a) 2.8 (b) 5.6

(c) 11.2 (d) None of these

4. 40 g $Ba(MnO_4)_2$ (mol. mass = 375) sample containing some inert impurities in acidic medium completely reacts with 125 mL of "33.6 *V*" of H_2O_2. What is the percentage purity of the sample?

(a) 28.12% (b) 70.31%

(c) 85% (d) None of these

Passage-2

A water is said to be a soft water if it produces sufficient foam with the soap and water that does not produce foam with soap is known as hard water. Hardness has been classified into two types (i) Temporary hardness (ii) Permanent hardness.

Temporary hardness is due to presence of calcium and magnesium bicarbonate. It is simply removed by boiling as

$$Ca(HCO_3)_2 \xrightarrow{\Delta} CaCO_3\downarrow + CO_2\uparrow + H_2O$$

$$Mg(HCO_3)_2 \xrightarrow{\Delta} MgCO_3\downarrow + CO_2\uparrow + H_2O$$

Temporary hardness can also be removed by addition of slaked lime, $Ca(OH)_2$

$$Ca(HCO_3)_2 + Ca(OH)_2 \longrightarrow 2CaCO_3\downarrow + 2H_2O$$

Permanent hardness is due to presence of sulphates and chlorides of Ca, Mg, etc. It is removed by washing soda as

$$CaCl_2 + Na_2CO_3 \longrightarrow CaCO_3\downarrow + 2NaCl$$

$$CaSO_4 + Na_2CO_3 \longrightarrow CaCO_3\downarrow + Na_2SO_4$$

Permanent hardness also removed by ion exchange resin process as

$$2RH + Ca^{2+} \longrightarrow R_2Ca + 2H^+$$

$$2ROH + SO_4^{2-} \longrightarrow R_2SO_4 + 2OH^-$$

The degree of hardness of water is measured in terms of ppm of $CaCO_3$. 100 ppm means 100 g of $CaCO_3$ is present in 10^6 g of H_2O. If any other water sample which contain 120 ppm of $MgSO_4$, hardness in terms of $CaCO_3$ is equal to = 100 ppm.

1. One litre of a sample of hard water ($d = 1$ g/mL) contains 136 mg of $CaSO_4$ and 190 mg of $MgCl_2$. What is the total hardness of water in terms of $CaCO_3$?
(a) 100 ppm (b) 200 ppm
(c) 300 ppm (d) None of these

2. What is the mass of $Ca(OH)_2$ required for 10 litre of water remove temporary hardness of 100 ppm due to $Ca(HCO_3)_2$?
(a) 1.62 g (b) 0.74 g
(c) 7.4 g (d) None of these

3. A 200 g sample of hard water is passed through the column of cation exchange resin, in which H^+ is exchanged by Ca^{2+}. The outlet water of column required 50 mL of 0.1 M NaOH for complete neutralization. What is the hardness of Ca^{2+} ion in ppm?
(a) 250 ppm (b) 500 ppm
(c) 750 ppm (d) 1000 ppm

Passage-3

$$\text{Equivalent mass} = \frac{\text{Molecular mass / Atomic mass}}{n\text{-factor}}$$

n-factor is very important in redox as well as non-redox reactions. With the help of n-factor we can predict the molar ratio of the reactant species taking part in reactions. The reciprocal of n-factor's ratio of the reactants is the molar ratio of the reactants.

In general n-factor of acid/base is number of moles of H^+/OH^- furnished per mole of acid/base.

n-factor of a reactant is number of moles of electrons lost or gained per mole of reactant.

Example 1 :

1. In acidic medium : $KMnO_4\ (n = 5) \longrightarrow Mn^{2+}$
2. In neutral medium : $KMnO_4\ (n = 3) \longrightarrow Mn^{2+}$
3. In basic medium : $KMnO_4\ (n = 1) \longrightarrow Mn^{6+}$

Example 2 : $FeC_2O_4 \longrightarrow Fe^{3+} + 2CO_2$

Total number of moles of e^- lost by 1 mole of FeC_2O_4

$$= 1 + 1 \times 2 \Rightarrow 3$$

$\therefore$ n-factor of $FeC_2O_4 = 3$

1. n-factor of $Ba(MnO_4)_2$ in acidic medium is :
(a) 2 (b) 6
(c) 10 (d) None of these

2. Consider the following reaction.
$$H_3PO_2 + NaOH \longrightarrow NaH_2PO_2 + H_2O$$
What is the equivalent mass of H_3PO_2? (mol. wt. is M)

(a) M (b) $M/2$

(c) $M/3$ (d) None of these

3. For the reaction, $Fe_{0.95}O$ (molar mass = M) $\longrightarrow Fe_2O_3$, what is the eq. mass of $Fe_{0.95}O$?

(a) $\frac{M}{0.85}$ (b) $\frac{M}{0.95}$

(c) $\frac{M}{0.8075}$ (d) None of these

4. In the reaction, $xVO + yFe_2O_3 \longrightarrow FeO + V_2O_5$, what is the value of x and y respectively?

(a) 1, 1 (b) 2, 3

(c) 3, 2 (d) None of these

One or More Answer is/are Correct

1. Which of the following statements regarding the compound A_xB_y is/are correct?

(a) 1 mole of A_xB_y contains 1 mole of A and 1 mole B

(b) 1 equivalent of A_xB_y contains 1 equivalent of A and 1 equivalent of B

(c) 1 mole of A_xB_y contains x moles of A and y moles of B

(d) equivalent mass of A_xB_y = equivalent mass of A + equivalent mass of B

2. If 1 mole of H_3PO_4 reacts with 1 mole of $X(OH)_2$ as shown below:

$H_3PO_4 + X(OH)_2 \longrightarrow XHPO_4 + 2H_2O$ then :

(a) the equivalent mass of base is $\frac{\text{mol. mass}}{2}$

(b) the eq. mass of H_3PO_4 is $\frac{98}{3}$

(c) the resulting solution requires 1 mole NaOH for complete neutralization

(d) minimum 1 mole of $X(OH)_2$ is required for complete neutralization of $XHPO_4$

3. In acidic medium dichromate ion oxidizes stannous ion as:

$$xSn^{2+} + yCr_2O_7^{2-} + zH^+ \longrightarrow aSn^{4+} + bCr^{3+} + cH_2O$$

(a) the value of $x : y$ is 1 : 3 (b) the value of $x + y + z$ is 18

(c) $a : b$ is 3 : 2 (d) the value of $z - c$ is 7

4. When a equimolar mixture of Cu_2S and CuS is titrated with $Ba(MnO_4)_2$ in acidic medium, the final products contain Cu^{2+}, SO_2 and Mn^{2+}. If the mol. mass of Cu_2S, CuS and $Ba(MnO_4)_2$ are M_1, M_2 and M_3 respectively then :

(a) eq. mass of Cu_2S is $\frac{M_1}{8}$

(b) eq. mass of CuS is $\frac{M_2}{6}$

(c) eq. mass of $Ba(MnO_4)_2$ is $\frac{M_3}{5}$

(d) Cu_2S and CuS both have same equivalents in mixture

5. Which is/are incorrect statement ?

(a) Equivalent mass of $H_2PO_3^-$ is 40.5.

(b) Eq. mass of $H_2PO_4^-$ may be equal to molar mass or less than molar mass because it depends on the reaction.

(c) $KMnO_4$ has maximum eq. mass in acidic medium.

(d) Oxidation state of H in MgH_2 is greater than in H_2O_2.

6. 10 mL of $NaHC_2O_4$ is oxidised by 20 mL of 0.1 N acidic $KMnO_4$. 20 mL of same $NaHC_2O_4$ is neutralised by following solution(s):

(a) 10 mL of 0.1 M KOH

(b) 10 mL of 0.2 N NaOH

(c) 10 mL of 0.1 M $Ca(OH)_2$

(d) 10 mL of 0.1 N $Ca(OH)_2$

7. For the reaction "p"HNO_2 + "q"$KMnO_4$ + "r"$H_2SO_4 \longrightarrow$ "s"HNO_3 + "t"$MnSO_4$ + "u"K_2SO_4 + $3H_2O$. Select correct statement(s).

(a) H_2SO_4 is reducing agent

(b) HNO_2 is reducing agent

(c) $p + q + r = 10$

(d) Eq. wt. of HNO_3 in reaction is (molar mass)/2

8. A 0.25 g sample of pyrolusite (impure MnO_2) is titrated with 1.26 g hydrated oxalic acid ($H_2C_2O_4 \cdot 2H_2O$) in an acidic medium. The excess of oxalic acid required 30 mL of 0.1 M $KMnO_4(aq.)$ solution. ($Mn \rightarrow 55$ g/mol).

Which is/are correct statement(s) ?

(a) 0.25×10^{-2} moles of MnO_2 present in sample

(b) % of MnO_2 in pyrolusite sample is 87

(c) % of impurities in pyrolusite sample is 6.5

(d) Eq. mass (E) of MnO_2 is 43.5 g in above titration

9. 1.5 mole Na_2CO_3, 1 mole $NaHCO_3$ and 0.5 mole NaOH dissolved in water to prepare 500 mL solution. 50 mL of this solution was first titrated with $\frac{N}{10}$ HCl with phenolphthalein indicator, consume V_1 volume. After this MeOH was added in 20 mL of titrated solution and V_2 volume of same HCl was again required for next end point. Select correct option(s).

(a) Volume of HCl used with MeOH indicator is twice with phenolphthalein indicator.

(b) Volume of HCl used with phenolphthalein indicator is twice with MeOH indicator.

(c) HCl can neutralizes Na_2CO_3, $NaHCO_3$ and NaOH completely with phenolphthalein indicator.

(d) HCl can naturalizes Na_2CO_3, $NaHCO_3$ and NaOH completely with MeOH indicator.

10. One mole of FeC_2O_4 is separately oxidised by $KMnO_4$ and $K_2Cr_2O_7$ in acid medium into Fe^{3+} ions and CO_2 molecules. Which statements are true?

(a) The mole ratio of $KMnO_4$ and $K_2Cr_2O_7$ required is 6 : 5.

(b) The ratio of number of equivalents of $KMnO_4$ and $K_2Cr_2O_7$ required is 1 : 1.

(c) Equivalent weight of FeC_2O_4 is $\frac{1}{3}$rd of its formula weight.

(d) FeC_2O_4 undergoes disproportionation.

Match the Column

1.

Column-I	Column-II (*n*-factor)
(A) When Bi_2S_3 converted into Bi^{5+} and S	(P) 18
(B) When $Al_2(Cr_2O_7)_3$ reduced into Cr^{3+} in acidic medium	(Q) 11
(C) When FeS_2 converted into Fe_2O_3 and SO_2	(R) 2
(D) When $Mn(NO_3)_2$ converted into MnO_4^{2-} and NO	(S) 10

2.

Column-I	Column-II
(A) Eq. mass $= \dfrac{\text{Molecular mass}}{33}$	(P) When CrI_3 oxidises into $Cr_2O_7^{2-}$ and IO_4^-
(B) Eq. mass $= \dfrac{\text{Molecular mass}}{27}$	(Q) When $Fe(SCN)_2$ oxidises into Fe^{3+}, SO_4^{2-}, CO_3^{2-} and NO_3^-
(C) Eq. mass $= \dfrac{\text{Molecular mass}}{28}$	(R) When NH_4SCN oxidizes into SO_4^{2-}, CO_3^{2-} and NO_3^-
(D) Eq. mass $= \dfrac{\text{Molecular mass}}{24}$	(S) When As_2S_3 oxidises into AsO_3^- and SO_4^{2-}

3.

Column-I	Column-II
(A) $\underline{P_2H_4} \longrightarrow PH_3 + P_4H_2$	(P) $E = \dfrac{3M}{4}$
(B) $\underline{I_2} \longrightarrow I^- + IO_3^-$	(Q) $E = \dfrac{3M}{5}$
(C) $MnO_4^- + Mn^{2+} + H_2O \longrightarrow \underline{Mn_3O_4} + H^+$	(R) $E = \dfrac{15M}{26}$
(D) $\underline{H_3PO_2} \longrightarrow PH_3 + H_3PO_3$	(S) $E = \dfrac{5M}{6}$

4. A sample of raw material contain $NaNO_3$. It contains some $NaIO_3$ also. The $NaIO_3$ can be used as a source of iodine, produced in the following reactions:

$$IO_3^- + HSO_3^- \longrightarrow I^- + SO_4^{2-} \quad \text{...(1)}$$

$$I^- + IO_3^- \longrightarrow I_2 + H_2O \quad \text{...(2)}$$

One litre of sample solution containing 396 g of $NaIO_3$ is treated with stoichiometric quantity of $NaHSO_3$. Now a substantial amount of same solution is added to reaction mixture to bring about the reaction (2).

Column-I	Column-II
(A) *n*-factor of IO_3^- in reaction (2)	(P) 6
(B) Number of moles of HSO_3^- used in reaction (1)	(Q) 1.2
(C) Moles of I_2 produced	(R) 2
(D) Equivalents of IO_3^- used in reaction (2)	(S) 5

5.

Column-I	Column-II (Type of Redox Reaction)
(A) $Br_2 + OH^- \longrightarrow BrO_3^- + Br^-$	(P) Intermolecular
(B) $FeCl_2 + KMnO_4 + HCl \longrightarrow MnCl_2 + FeCl_3$	(Q) Intramolecular
(C) $Ag_2O \longrightarrow Ag + \frac{1}{2}O_2$	(R) Disproportion
(D) $NH_4NO_3 \longrightarrow N_2O + H_2O$	(S) Comproportion

Assertion-Reason Type Questions

Each question contains STATEMENT-1 (Assertion) and STATEMENT-2 (Reason).
Examine the statements carefully and mark the correct answer according to the instructions given below

(A) If both the statement are TRUE and STATEMENT-2 is the correct explanation of STATEMENT-1

(B) If both the statement are TRUE but STATEMENT-2 is NOT the correct explanation of STATEMENT-1

(C) If STATEMENT-1 is TRUE and STATEMENT-2 is FALSE

(D) If STATEMENT-1 is FALSE and STATEMENT-2 is TRUE

1. **STATEMENT-1 :** The oxidation state of S in $H_2S_2O_8$ is 6.
   STATEMENT-2 : Maximum oxidation state of S is 6.

2. **STATEMENT-1 :** 0.1 M $H_3PO_3(aq)$ solution has normality equal to 0.3 N when completely reacted with NaOH.
   STATEMENT-2 : H_3PO_3 is a dibasic acid.

3. **STATEMENT-1 :** MnO_2 can act as an oxidizing agent as well as reducing agent.
   STATEMENT-2 : Oxidation state of Mn in MnO_2 lies between highest and lowest oxidation state.

4. **STATEMENT-1 :** Equivalent volume of H_2 is 11.2 L at 1 atm and 273 K.
   STATEMENT-2 : 1/2 mole H_2 has produced when 1 mole of $H^+(aq)$ accepted 1 mole of e^-.

5. **STATEMENT-1 :** For the reaction in titration $Na_2CO_3 + HCl \longrightarrow NaCl + NaHCO_3$, the suitable indicator is phenolphthalein.
   STATEMENT-2 : Phenolphthalein provide its colour in acidic medium.

6. **STATEMENT-1 :** $[Fe(CN)_6]^{4-} \longrightarrow Fe^{3+} + CO_2 + NO_3^-$, the equivalent mass of reactant is 3.74.
   STATEMENT-2 : Equivalent mass of reactant $= \frac{\text{Mol. mass}}{61}$.

7. **STATEMENT-1 :** In the balanced redox reaction,
   $$xAs_2S_3 + yNO_3^- + 4H_2O \longrightarrow aAsO_4^{3-} + bNO + cSO_4^{2-} + 8H^+$$
   the n-factor of As_2S_3 and NO_3^- is 28 and 3 respectively.
   STATEMENT-2 : Molar ratio is reciprocal of n-factor's ratio so $x : y$ is 3 : 28.

8. **STATEMENT-1 :** In the given reaction, $NaOH + H_3PO_4 \longrightarrow NaH_2PO_4 + H_2O$ equivalent mass of H_3PO_4 is $M/3$
   STATEMENT-2 : H_3PO_4 is tribasic acid.

9. **STATEMENT-1 :** In CrO_5 oxidation number of Cr is +6.
   STATEMENT-2 : CrO_5 has butterfly structure in which peroxide bonds are present.

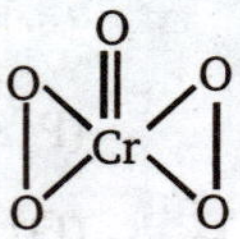

10. **STATEMENT-1 :** $I_2 \rightarrow IO_3^- + I^-$, is an example of a disproportionation reaction.

STATEMENT-2 : Oxidation number of I can vary from –1 to +7.

11. **STATEMENT-1 :** Fluorine exists only in –1 oxidation state in compounds.

STATEMENT-2 : Fluorine has $2s^2 2p^5$ configuration.

12. **STATEMENT-1 :** H_2SO_4 can not act as reducing agent.

STATEMENT-2 : Sulphur can not increase its oxidation number beyond +6.

Level-4

Subjective Problems

1. 0.7875 g of crystalline barium hydroxide is dissolved in water. For the neutralization of this solution 20 mL of $N/4$ HNO_3 is required. How many moles of water of crystallization are present in one mole of this base ? (Given : Atomic mass Ba = 137, O = 16, N = 14, H = 1)
2. 2.0 g of polybasic organic acid (Molecular mass = 600) required 100 mL of a $\frac{M}{6}$ NaOH solution for complete neutralisation. Find the basicity of acid.
3. A mixture contains 1.0 mole each of NaOH, Na_2CO_3 and $NaHCO_3$. When half of mixture is titrated with HCl, it required x mole of HCl in presence of phenolphthalein. In another experiment, half of mixture required y mole of same HCl in presence of methyl orange. Find the value of $(x + y)$.
4. When BrO_3^- ion reacts with Br^- ion in acidic medium, Br_2 is liberated. Calculate the ratio of molecular mass and equivalent mass of $KBrO_3$.
5. A volume of 12.5 mL of 0.05 M SeO_2 reacts with 25 mL of 0.1 M $CrSO_4$ which is oxidised to Cr^{3+}. To what oxidation state was the selenium converted by the reaction ?
6. A 0.276 g impure sample of copper ore is dissolved and Cu^{2+} is titrated with KI solution. I_2 liberated required 40 mL of 0.1 M $Na_2S_2O_3$ solution for titration. What is the % of impurities in the ore ?
7. A sample of 28 mL of $H_2O_2\,(aq)$ solution required 10 mL of 0.1 M $KMnO_4\,(aq)$ solution for complete reaction in acidic medium. What is the volume strength of H_2O_2 ?
8. For the redox reaction given below, what is the value of $\frac{x}{z}$?

   $$xNO_3^- + y\,As_2S_3 + zH_2O \longrightarrow \text{---}\,AsO_4^{3-} + \text{---}\,NO + \text{---}\,SO_4^{2-} + \text{---}\,H^+.$$
9. On heating 0.220 g of a metallic oxide in presence of hydrogen, 0.045 g of water is formed. If the equivalent mass of the metal is E, then what is the value of $E/9$?
10. 10 g mixture of $K_2Cr_2O_7$ and $KMnO_4$ was treated with excess of KI in acidic medium. Iodine liberated required 100 cm^3 of 2.2 N sodium thiosulphate solution for titration. If the mass percent of $KMnO_4$ in the mixture Z, then what is the value of $2Z/5$?
11. In an ore the only oxidisable material is Sn^{2+}. This ore is titrated with a dichromate solution containing 2.5 g $K_2Cr_2O_7$ in 0.50 litre. A 0.40 g of sample of the ore required 10.0 cm^3 of the titrant to reach equivalent point. If the percentage of tin in ore is x, then what is the value of $x/5$? (K = 39.1, Cr = 52, Sn = 118.7)
12. $Mn^{2+}(aq.)$ can be determined by titration with $MnO_4^-(aq.)$

   $$Mn^{2+} + MnO_4^- + OH^- \longrightarrow MnO_2 + H_2O \text{ (unbalanced)}$$

   A 30.0 mL sample of $Mn^{2+}(aq.)$ requires 40.0 mL of 0.5 M $KMnO_4(aq.)$ for its titration.

   What is the molarity of $Mn^{2+}(aq.)$?
13. If Cu_2S is treated with acidic solution of $KMnO_4$, it is oxidised completely to Cu^{2+} and SO_2 is liberated. Calculate mole of $KMnO_4$ required for complete oxidation of 1.25 mole of Cu_2S.
14. 20 mL of 0.2 M NaOH$(aq.)$ solution is mixed with 35 mL of 0.1 M NaOH$(aq.)$ solution and the resultant solution is diluted to 100 mL. 40 mL of this diluted solution reacted with 0.15 gram impure sample of oxalic acid $(H_2C_2O_4)$. If % of impurity in sample is x then find the value of $\frac{x}{2}$.
15. Equimolar mixture of $NaHCO_3$ and Na_2CO_3 is titrated against 0.2 N HCl in presence of phenolphthalein indicator, if volume of HCl required for this titration is 20 mL. Resulting solution is further titrated with 1 N NaOH solution. Calculate volume in mL of NaOH required for this titration.

Numerical Problems

(If the numerical value has more than two decimal places, truncate/round-off the value to TWO decimal places.)

1. 10.0 mL of 0.05 M $KMnO_4$ solution was consumed in a titration with 10.0 mL of given oxalic acid dihydrate solution. The strength of given oxalic acid solution is ______ $\times 10^{-2}$ g/L.
(Round off to the Nearest Integer).

2. 0.8 g of a metal oxide MO were dissolved in excess dilute sulphuric acid and $Ba(MnO_4)_2$ of strength 0.2 M are reacted. Volume of $Ba(MnO_4)_2$ required to reach the equivalence point is 8 mL and given that M^{2+} oxidizes into M^{4+} then atomic weight of metal M would be:

3. x mole of $KHC_2O_4 \cdot 3H_2C_2O_4$ is oxidised completely by 40 mL of 1.0 M $KMnO_4$ in acidic medium and y mole of $KHC_2O_4 \cdot H_2C_2O_4$ is neutralised completely by 30 mL of 1.0 M of $Ca(OH)_2$ then the ratio of x and y is :

4. In the Kjeldahl method for estimation of nitrogen present in a soil sample, ammonia evolved with 1 g sample neutralise by 10 mL of 1 M H_2SO_4. The excess of acid was neutralised by 5 mL of 0.5 M $Ca(OH)_2$. The percentage by mass of nitrogen in the soil is:

5. What volume of (in mL) 0.82 M $KMnO_4$ solution is needed for complete reaction with 34.32 g $Fe_{0.94}O_{1.0}$, in acidic medium? (Fe = 56)

6. A sample of a metal carbonate MCO_3 was neutralized by 10 mL of 0.1 N HCl and the resulting chloride gave 0.0517 g of phosphate, $M_3(PO_4)_2$. The equivalent weight of M is ______. (Nearest integer)

7. In order to oxidise a mixture two mole of each of FeC_2O_4, $Fe_2(C_2O_4)_3$ and $FeSO_4$ in acidic medium, the number of moles of $KMnO_4$ required is :

8. An aqueous solution containing 1.07 g KIO_3 and excess of KI is acidified with HCl. Liberated I_2 consumed hypo solution. If the molarity of hypo is 2/5 M, then the volume of $Na_2S_2O_3$ (in mL) to reach at the equivalence point is ______. [Given : At. mass K = 39, I = 127]

9. 2.175 g of pyrolusite ore was treated with 50 mL of 0.5 M oxalic acid and H_2SO_4 in excess. The oxalic acid left was raised to 250 mL in flask by dilution. 25 mL of this solution was titrated with 30 mL of 0.02 M $KMnO_4$ solution. Find out wt. % of MnO_2 in sample.
Given : $MnO_2 + H_2C_2O_4 + H_2SO_4 \longrightarrow MnSO_4 + 2CO_2 + 2H_2O$

10. 0.4 g mixture of NaOH, Na_2CO_3 and some inert impurities was first titrated with $\frac{N}{10}$ HCl using phenolphthalein as an indicator, 17.5 mL of HCl was required at the end point. After this methyl orange was added and titrated. 1.5 mL of same HCl was required for the next end point. The weight percentage of Na_2CO_3 in the mixture is ______. (Rounded-off to the nearest integer)

11. Calculate the equivalent weight of HCHO in disproportionation reaction in which HCHO disproportionate to $HCOO^-$ and CH_3OH in presence of OH^-.

12. In acidic medium 200 mL 0.2 M $Ba(MnO_4)_2$ is titrated with 50 mL 2.27 V H_2O_2 solution. The excess of $Ba(MnO_4)_2$ is back titrated with $\frac{10}{3}$ M FeC_2O_4 solution in acidic medium. Calculate volume (in mL) of FeC_2O_4 solution used.

13. 1 L of equimolar mixture of $Fe_2(C_2O_4)_3$ and $H_2C_2O_4$ require 12 mole of $K_2Cr_2O_7$ for complete titration. 2L of same equimolar mixture reacts completely with n mole of NaOH. Calculate the value of n.

14. Pure $Na_2C_2O_4$ and $KHC_2O_4 \cdot H_2C_2O_4$ are mixed in such a ratio that each gram of the mixture reacts with equal volumes of 0.1 M $KMnO_4$ and 0.1 M NaOH. Find mole ratio of $Na_2C_2O_4$ and $KHC_2O_4 \cdot H_2C_2O_4$ in mixture.

15. In 10 litre water sample 146 mg $Mg(HCO_3)_2$, 324 mg $Ca(HCO_3)_2$, 272 mg $CaSO_4$, 190 mg $MgCl_2$, 117 mg NaCl present. Calculate the temporary hardness of water (in ppm).

Answers

Level-1

1. (c)	2. (c)	3. (a)	4. (b)	5. (c)	6. (a)	7. (d)	8. (b)	9. (a)	10. (b)
11. (c)	12. (b)	13. (b)	14. (a)	15. (c)	16. (c)	17. (d)	18. (c)	19. (a)	20. (d)
21. (b)	22. (d)	23. (a)	24. (d)	25. (d)	26. (c)	27. (c)	28. (b)	29. (d)	30. (c)
31. (a)	32. (d)	33. (a)	34. (a)	35. (a)	36. (b)	37. (a)	38. (c)	39. (d)	40. (c)
41. (b)	42. (a)	43. (d)	44. (b)	45. (b)	46. (c)	47. (d)	48. (a)	49. (c)	50. (b)
51. (b)	52. (d)	53. (d)	54. (d)	55. (b)	56. (b)	57. (a)	58. (b)	59. (d)	60. (c)
61. (c)	62. (a)	63. (a)	64. (a)	65. (c)	66. (a)	67. (b)	68. (b)	69. (b)	70. (b)
71. (b)	72. (c)	73. (b)	74. (b)	75. (b)	76. (a)	77. (d)	78. (a)	79. (b)	80. (c)
81. (c)	82. (c)	83. (d)	84. (b)	85. (b)	86. (c)	87. (a)	88. (a)	89. (b)	90. (b)
91. (c)	92. (a)	93. (d)	94. (a)	95. (b)	96. (d)	97. (d)	98. (a)	99. (c)	100. (a)
101. (b)	102. (b)	103. (c)	104. (c)	105. (a)	106. (a)	107. (a)	108. (b)	109. (b)	110. (d)
111. (d)	112. (b)	113. (b)	114. (a)	115. (d)	116. (c)	117. (d)	118. (d)	119. (b)	120. (a)
121. (b)	122. (a)	123. (d)	124. (c)	125. (a)					

Level-2

1. (b)	2. (a)	3. (b)	4. (a)	5. (a)	6. (d)	7. (b)	8. (a)	9. (b)	10. (a)
11. (b)	12. (a)	13. (c)	14. (a)	15. (c)	16. (b)	17. (c)	18. (b)	19. (a)	20. (d)

Level-3

Passage–1	1.	(a)	2.	(b)	3.	(b)	4.	(b)
Passage–2	1.	(c)	2.	(b)	3.	(b)		
Passage–3	1.	(c)	2.	(a)	3.	(a)	4.	(b)

One or More Answers is/are correct

1. (b,c,d)	2. (a,c)	3. (b,c,d)	4. (a,b)	5. (a,c,d)	6. (b,c)	7. (b,c,d)	8. (a,b,d)
9. (b,d)	10. (a,b,c)						

Match the Column

1.	A→ S;	B→ P;	C→ Q;	D→ R
2.	A→ Q;	B→ P;	C→ S;	D→ R
3.	A→ S;	B→ Q;	C→ R;	D→ P
4.	A→ S;	B→ P;	C→ Q;	D→ R
5.	A→ P, R;	B→ P;	C→ Q;	D→ Q, S

Assertion-Reason Type Questions

1. (B)	2. (D)	3. (A)	4. (A)	5. (C)	6. (D)	7. (A)	8. (D)	9. (A)	10. (B)
11. (B)	12. (A)								

Level-4

Subjective Problems

1. 8	2. 5	3. 3	4. 5	5. 0	6. 8	7. 1	8. 7	9. 4	10. 6
11. 3	12. 1	13. 2	14. 5	15. 8					

Numerical Problems

1. 1575.00	2. 84.00	3. 1.25	4. 21.00	5. 100.00	6. 20.00	7. 4.00	8. 75.00
9. 40.00	10. 5.30	11. 30.00	12. 38.00	13. 36.00	14. 05.50	15. 30.00	

Hints & Solutions

Level 1

1. (c) Hypophosphorous Acid —H_3PO_2

2. (c) $\overset{+2}{Ca}$ with $\overset{-2(+1)}{OCl}$ and $\underset{(-1)}{Cl}$

3. (a) S_8^0, $S_2^{+1}F_2$, H_2S^{-2}, $H_2SO_4^{+6}$

4. (b) The redox reaction is :

$Cr_2O_7^{2-} + 3SO_2 + 2H^+ \longrightarrow 2Cr^{3+} + 3SO_4^{2-} + H_2O$

5. (c) $\overset{-3}{N}H_4\overset{+5}{N}O_3$

6. (a) Caro's acid $= H_2SO_5$

= Peroxomono-sulphuric acid

Marshell's acid $= H_2S_2O_8$

= Peroxodi-sulphuric acid

The oxidation state in both the acids is +6 each because oxidation state cannot be greater than the number of valence electrons.

7. (d) $KO_3^{-1/3}$, $Na_2O_2^{-1}$

8. (b) $Ba(H_2PO_2)_2 \longrightarrow Ba^{2+} + H_2\overset{+1}{P}O_2^-$

9. (a) $x + 5(0) - 1 - 2 = 0$

$x = +3$

10. (b) $BaO_2^{-1} < KO_2^{-1/2} < O_3^0 < \overset{+2}{O}F_2$

11. (c) O.No. of N is −3, −2, −1, +1 respectively.

12. (b) $H\overset{+5}{N}O_3 > \overset{+2}{N}O > \overset{0}{N_2} > \overset{-3}{N}H_4Cl$

13. (b) $\overset{a}{S_2}O_4^{2-}$: $2a + 4\times(-2) = -2$

$\therefore$ $a = +3$

$\overset{a}{S}O_3^{2-}$: $a + 3\times(-2) = -2$

$\therefore$ $a = +4$

$\overset{a}{S_2}O_6^{2-}$: $2a + 6\times(-2) = -2$

$\therefore$ $a = +5$

$\therefore$ $S_2O_4^{2-} < SO_3^{2-} < S_2O_6^{2-}$

14. (a) $\overset{+6}{Cr_2}O_7^{2-} \longrightarrow \overset{+6}{Cr}O_4^{2-}$

15. (c) 1 mole N_2H_4 loses 8 mole e^-; 1 mole N loses 4 mole of e^-

$\therefore$ New oxidation state of N is $-2 + 4 \Rightarrow 2$

16. (c) Compound should be electrically neutral.

17. (d) $Fe_{0.96}O$: Fe^{2+} (x), Fe^{3+} $(0.96 - x)$

→ Charge balance $2x + 3(0.96 - x) - 2 = 0$,

$x = \frac{11}{12}$

18. (c) $Fe^{2+} + MnO_4^- + H^+ \longrightarrow Fe^{3+} + Mn^{2+} + H_2O$

19. (a) Both Fe and C are oxidised and MnO_4^- is reduced.

20. (d) H_2O_2 if self be oxidised to O_2.

22. (d) No change in oxidation no. of any element.

23. (a) $N_2 + O_2 \longrightarrow 2NO$ is a redox reaction, other reactions are precipitation neutralization and photochemical reactions respectively.

24. (d) O.No. of Cr remains constant.

25. (d) Cu^+ converted into Cu & Cu^{2+}.

26. (c) $4P_4 + 12OH^- + 12H_2O \longrightarrow 12H_2PO_2^- + 4PH_3$

Cr & Mn already present in max. O.No., oxidation of F_2 is not possible.

27. (c) S in SO_4^{2-} is already present in highest O.No. so oxidation of S is not possible.

28. (b) (1)

$I_2 + \underset{(-1)}{H_2O_2} + 2OH^- \longrightarrow 2I^- + 2H_2O + \underset{(0)}{O_2}$

(2) $\underset{(-1)}{H_2O_2} + HOCl \longrightarrow Cl^- + H_3O^+ + \underset{(0)}{O_2}$

30. (c) $\overset{0}{P} + NaOH \longrightarrow \overset{-3}{P}H_3 + NaH_2\overset{+1}{P}O_2$

In above reaction, the ox. no. of P increases from zero to +1 in NaH_2PO_2 and decrease zero to −3 in PH_3. Thus P is oxidised as well as reduced and it is redox reaction.

32. (d) $CS_2 + 3Cl_2 \longrightarrow CCl_4 + S_2Cl_2$

33. (a) Balanced reaction is

$$PtCl_4 + 5XeF_2 \longrightarrow PtF_6 + 4ClF + 5Xe$$

∴ Sum of stoichiometric coefficients

$$= 1 + 5 + 1 + 4 + 5 = 16$$

34. (a) The balanced redox reaction is

$$2MnO_4^- + 5C_2O_4^{2-} + 16H^+ \longrightarrow 2Mn^{2+} + 10CO_2 + 8H_2O$$

35. (a)

$$(Fe^{2+} \longrightarrow Fe^{3+} + 1e^-) \times 6$$

$$6e^- + 14H^+ + Cr_2O_7^{2-} \longrightarrow 2Cr^{3+} + 7H_2O$$

$$6Fe^{2+} + 14H^+ + Cr_2O_7^{2-} \longrightarrow 2Cr^{3+} + 6Fe^{3+} + 7H_2O$$

$x = 6, \quad y = 1, \quad z = 7$

36. (b) $KMnO_4 + H_2O_2 + H_2SO_4 \longrightarrow MnSO_4 + K_2SO_4 + O_2 + H_2O$

$n_f = 5 \quad n_f = 2$

$$2KMnO_4 + 5H_2O_2 + 3H_2SO_4 \longrightarrow 2MnSO_4 + K_2SO_4 + 5O_2 + 8H_2O$$

Reaction is multiply by 2 (due to coefficient of H_2SO_4 is required 6)

$$4KMnO_4 + 10H_2O_2 + 6H_2SO_4 \longrightarrow 4MnSO_4 + 2K_2SO_4 + 10O_2 + 16H_2O$$

37. (a) $BiO_3^- + 6H^+ + 2e^- \longrightarrow Bi^{3+} + 3H_2O,$

$x + y = 6 + 2 = 8$

38. (c) $H_2\overset{+6}{S}O_4 + 8H\overset{-1}{I} \longrightarrow H_2S + 4\overset{0}{I_2} + 4H_2O$

39. (d) $S_8 + 12OH^- \longrightarrow 4S^{2-} + 2S_2O_3^{2-} + 6H_2O$

40. (c) $P_4 + 20HNO_3 \rightarrow 4H_3PO_4 + 20NO_2 + 4H_2O$

41. (b) $6I_2 + 12OH^- \longrightarrow 10I^- + 2IO_3^- + 6H_2O$

42. (a) $IO_3^- + 5I^- + 6H^+ \longrightarrow 3I_2 + 3H_2O$

43. (d) $C_2O_4^{2-} \longrightarrow 2CO_2 + 2e^-$, No. of electrons for producing 1 mole of $CO_2 = 1$.

44. (b) $2MnO_4^- + 3SO_3^{2-} + H_2O \longrightarrow 2MnO_2 + 3SO_4^{2-} + 2OH^-$

45. (b) $\underline{\overset{+2}{Mn}}SO_4^- \longrightarrow \underline{\overset{+4}{Mn}}O_2 + 2e^-,$

$$E_{MnSO_4} = \frac{M_{MnSO_4}}{2}$$

46. (c) As Base :

$n_f = 1$ $NaH_2PO_4 + H_2O \rightleftharpoons H_3PO_4 + OH^-Na^+$

As Acid :

$n_f = 2$; $H_2PO_4 \rightleftharpoons 2H^+ + PO_4^{3-}$

47. (d) $CH_3COOH + 3Cl^- \rightarrow CCl_3COOH + 3H^+ + 6e^-$

$$\therefore E_{CH_3COOH} = \frac{60}{6} = 10$$

48. (a) $Pb + PbO_2 + 2H_2SO_4 \longrightarrow 2PbSO_4 + 2H_2O$

The reaction is balanced by the loss and gain of 2 moles of electron per mole of Pb and hence,

$$E_{H_2SO_4} = \frac{2M}{2} = M$$

49. (c) If $x = 4$

$$H_3PO_4 + 3KOH \longrightarrow Na_3PO_4 + 3H_2$$

0.1 mole → 0.3 mole

If $x = 3$

$$H_3PO_3 + 2KOH \longrightarrow Na_2PO_3 + 2H_2O$$

0.1 mole → 0.2 mole

If $x = 2$

$$H_3PO_2 + KOH \longrightarrow NaPO_2 + H_2O$$

0.1 mole → 0.1 mole

So, if $x = 2$, 5.6 g KOH will be required.

50. (b) In acidic medium $MnO_4^- \longrightarrow Mn^{2+}$

$$n_f = 5, \quad E = \frac{M}{5}$$

51. (b) $n = |1 + 5 \times 2|$

52. (d) 14 mole HCl^- loses 6 mole e^-;

∴ 1 mole HCl loses $\frac{6}{14}$ mole e^-

$$\therefore \text{eq. wt. of HCl} = \frac{M}{\left(\frac{6}{14}\right)}$$

$$\Rightarrow \frac{36.5 \times 14}{6} = 85.1$$

53. (d) $n\text{-factor} = \frac{n_1 \times n_2}{n_1 + n_2} = \frac{4 \times 2}{4 + 2} = \frac{4}{3}$

$$\text{Eq. wt.} = \frac{\text{mol wt.}}{n\text{-factor}} = \frac{M \times 3}{4}$$

54. (d) $n\text{-factor} = |2 \times 2 + 8 \times 3| \Rightarrow 28$

55. (b) In $SnCl_2$

71 g chlorine combines with 32 g S

So, 35.5 g chlorine will combine with = 16 g

So, Eq. mass of S = 16 g

As the mass which combines with 35.5 g chlorine is equivalent mass.

56. (b) $E_{Element} = 4$, Formula of chloride MCl_x

Eq. mass of chloride $= E_{Element} + E_{Chlorine}$

$$\frac{M.M.}{n_f} = E_{Element} + E_{Chlorine}$$

$$\frac{59.25 \times 2}{x} = 4 + 35.5$$

$$x = 3$$

57. (a) $K_2Cr_2O_7 + X^{n+} \longrightarrow X^{5+}O_3^- + Cr^{3+}$

$$6 \times 10^{-3} \times 6 = (5 - n) \times 9 \times 10^{-3} \Rightarrow n = 1$$

58. (b) Normality $= \frac{1}{100}$ (Centinormal)

Number of eq. = Normality × Volume (in L)

$$\text{Number of eq.} = \frac{1}{100} \times \frac{250}{1000} = \frac{1}{400}$$

$\frac{\text{Mass}}{\text{Molar mass}} \times n_f \times \frac{1}{400}$

As R.A. $C_2O_4^{2-} \longrightarrow CO_2$

$\frac{w}{126} \times 2 = \frac{1}{400}$

$w = 0.1575$ g

59. (d) As R.A.

$\overset{+3}{KHC_2O_4} \cdot \overset{+3}{H_2C_2O_4} \cdot 4H_2 \longrightarrow \overset{+4}{CO_2}$

$n_f = 1 \times 4 = 4$

$E = \frac{M}{4}$

60. (c) $E_{\text{Chloride}} = E_{\text{Element}} + E_{\text{Chlorine}}$

$E_{\text{Chloride}} = w + 35.5$

$\therefore$ Metal is divalent, n_f for metal chloride = 2

$E = \frac{M}{2} \Rightarrow M = 2(w + 35.5) = 2w + 71$

61. (c) $2BrO_3^- + 12H^+ + 10Br^- \longrightarrow 6Br_2 + 6H_2O$

10 mole e^- required for formation of 6 moles of Br_2

$\therefore$ n-factor of $Br_2 = \frac{10}{6} = \frac{5}{3}$

eq. wt. $= \frac{\text{mol. wt.}}{n} = \frac{m}{5/3} = \frac{3M}{5}$

62. (a) Eq. of A = Eq. of B

$\frac{mA}{E_A} = \frac{mB}{E_B}$

63. (a) O.N. of N in N_2H_4 is –2 which changes to 0 in N_2. Hence, equivalent mass of

$N_2H_4 = \frac{\text{molar mass}}{2 \times 2}$

O.N. of iodine changes from +5 in IO_3^- to +1 in ICl. Hence, equivalent mass of

$IO_3^- = \frac{\text{molar mass}}{4}$

64. (a) n-factor of $CuSO_4$

$= \frac{\text{Moles of } e^- \text{ involved in redox}}{\text{Stoichiometric coefficient}}$

$= \frac{2}{2} = 1$

$\therefore \quad E = \frac{M}{n\text{-factor}} = M$

65. (c) n-factor of KI

$= \frac{\text{Moles of } e^- \text{ involved in redox}}{\text{Stoichiometric coefficient}}$

$= \frac{2}{4} = 0.5$

$\therefore \quad E = \frac{M}{n\text{-factor}} = \frac{M}{0.5} = 2M$

66. (a) No. of equivalent of $KMnO_4$ = No. of equivalent of Cu_2S

$x \times 5 = 1 \times 8 \Rightarrow x = 1.6$

$KMnO_4 \longrightarrow Mn^{2+}; Cu_2S \longrightarrow Cu^{2+} + SO_2$

$(n = 5) \qquad n = 2 + 6 = 8$

67. (b) No. of equivalent of X_2O_3 = No. of equivalent of $Cr_2O_7^{2-}$

or $n \times 8 = 1 \times 6, \; n_{X_2O_3} = \frac{3}{4}$

68. (b) M.Eq^n of H_2O_2 = M.Eq^n of $K[Cr(OH)_4]$;

$X \times 2 = 1.5 \times 3 \Rightarrow x = 2.25$

69. (b) $N_R = \frac{N_1V_1 + N_2V_2}{V_1 + V_2} = \frac{2 \times 0.45 + 0.6 \times 1}{3} = 0.5$

70. (b) Eq. of A^{2+} = Eq. of MnO_4^-

$2.7 \times 10^{-3} \times 3 = 1.6 \times 10^{-3} \times n_f$

$\Rightarrow n_f = 5 \Rightarrow$ Acidic medium.

71. (b) $5e^- + KMnO_4 \longrightarrow Mn^{2+}$

$6e^- + K_2Cr_2O_7 \longrightarrow 2Cr^{3+} + 6e^-$

$Fe^{2+} \longrightarrow Fe^{3+} + e^-$

5 mole of Fe^{2+} oxidize by 1 mole of $KMnO_4$

6 mole of Fe^{3+} oxidize by 1 mole of $K_2Cr_2O_7$

72. (c) $\overset{+2}{Fe}\overset{+3}{C_2}O_4 + \overset{+7}{Mn}O_4^- \longrightarrow \overset{+3}{Fe} + \overset{+4}{C}O_2 + Mn^{2+}$

$n_f = 3 \qquad n_f = 5$

$1 \times 3 = n_{MnO_4^-} \times 5 \Rightarrow n_{MnO_4^-} = 0.6$ mole

73. (b) Eq. of I_2 = Eq. of $S_2O_3^{2-}$

$\frac{w}{254} \times 2 = \frac{40}{1000} \times 0.11 \Rightarrow w = 0.558$ g

74. (b) At 1 atm and 273 K

$M = \frac{\text{Volume strength of } H_2O_2}{11.2} = \frac{5.6}{11.2} = 0.5$

In this disproportionation reaction

n-factor of H_2O_2 is 1 $\left(n = \frac{2 \times 2}{2 + 2} = 1\right)$

$N = M \times n\text{-factor} = 0.5 \times 1 = 0.5$

75. (b) $\frac{100 \times 1}{1000} \times 5 = \frac{V \times 1}{1000} \times 3 \Rightarrow V = 166.67$ mL

76. (a) $n_{eq}(KMnO_4) = n_{eq}H_2O_2$

or $\frac{50 \times 0.1}{1000} \times 5 = \frac{w}{34} \times 2 \Rightarrow w = 0.425$ g

$\therefore$ Strength $= \frac{0.425}{10} \times 100 = 4.25\%$ (w/V)

77. (d) Eq. of acid = Eq. of base

$$\frac{1.25}{M}\times 2 = 25\times 0.25\times 2$$

$$M = 200$$

78. (a) $N_{R(H^+)} = \frac{\left|10\times 1 + 20\times\frac{1}{2}\times 30\times\frac{1}{3}\right|}{1000} = \frac{3N}{100}$

79. (b) Eq. of Acid = Eq. of base

$$\frac{0.45}{90}\times n_f = \frac{20\times 0.54}{1000} \Rightarrow n_f = 2$$

80. (c) Eq. mass of $H_2O_2 = \frac{34}{2} = 17$

$$\text{Eq. of } H_2O_2 = \frac{3.4\times x}{100\times 17}$$

$$= \text{Eq. of } KMnO_4 = x\times N\times 10^{-3} \Rightarrow N = 2$$

$$\text{Molarity of } KMnO_4 = \frac{2}{5} = 0.4$$

81. (c) 50 m-eq. $KMnO_4$ = 10 m-mole $KMnO_4$

or 25 m-mole of $O_2 = 25\times 22.4\times 10^{-3}$

$= 0.56$ L

82. (c) Eq. of $KMnO_4 = 20\times 0.1\times 5 = 10$

Eq. of oxalic acid $= 25\times 0.2\times 2 = 10$

83. (d) $Fe^{2+} + MnO_4^- \longrightarrow Fe^{3+} + Mn^{2+}$

$n_f = 1$ $\quad n_f = 5$

Moles of $Fe^{2+}\times 1$ by $KMnO_4$

$= \text{moles of } MnO_4^-\times 5$ ...(1)

$Fe^{2+} + Cr_2O_7^{2-} \longrightarrow Fe^{3+} + Cr^{3+}$

$n_f = 1$ $\quad n_f = 6$

Moles of $Fe^{2+}\times 1$ by $K_2Cr_2O_7$

$= \text{moles of } MnO_4^-\times 6$ ...(2)

(1) + (2)

$$\frac{n_1}{n_2} = \frac{5}{6}$$

84. (b) Moles of AgCl = moles of $AgNO_3$

$$= \text{moles of HCl} = \frac{0.1435}{143.5} = 10^{-3}$$

wt. of HCl = 0.0365

wt. of $H_2SO_4 = 0.0635$

$$\text{wt. \% of } H_2SO_4 = \frac{0.0635\times 100}{0.1} = 63.5$$

85. (b) $H_2\overset{+3}{C_2}O_4\cdot 2H_2O \longrightarrow \overset{+4}{C}O_2$

$n_f = 2$

$$E = \frac{126}{2} = 63$$

$$HNO_3 (n_f = 1) \quad E = \frac{63}{1} = 63$$

86. (c) milli-equivalent of NH_3 reacted with HNO_3

$= 45\times 0.4 - 20\times 0.1 = 16$

$$\therefore \frac{W}{17}\times 1000 = 16; \quad W_{NH_3} = 0.272 \text{ g};$$

$$\text{wt. of N} = 0.272\times\frac{14}{17} = 0.224$$

$$\text{\% N in the sample} = \frac{0.224}{1.12}\times 100 = 20\%$$

87. (a) $1000\times\frac{W}{24+64}\times 2 = 90\times\frac{1}{100}\times 5;$

$W = 0.198$ g

% of oxalate ion in a given sample

$$= \frac{0.198}{0.3}\times 100 = 66$$

88. (a) $n_{eq}\ (KH_2PO_4) = n_{eq}(OH^-)$

or $\frac{w}{136}\times 1 = \frac{25\times 0.1}{1000} \Rightarrow w = 0.34$ g

$$\therefore \text{ \% Purity} = \frac{0.34}{0.5}\times 100 = 68\%$$

89. (b) $\frac{6}{E_M + 30} = \frac{4}{E_M + 16}$

$$1.5E_M + 24 = E_M + 30$$

90. (b) Let volume of HNO_3 is V then

$$V\times 1.05\times\frac{12.6}{100}\times\frac{3}{63} = \frac{1}{278}\times 1$$

$V = 0.57$ mL

91. (c) m-eq. of $KMnO_4$ = m-eq. of FeC_2O_4 + m-eq. of $FeSO_4$

$$(0.1\times 5)\times V = \frac{100}{\frac{144}{3}} + \frac{100}{\frac{152}{1}};$$

$V = 5.48$ mL

92. (a) meq. $FeSO_4(NH_4)_2SO_4\cdot 6H_2O$ = meq of $KMnO_4$

$(n = 1)$

$$\frac{W}{392}\times 1\times 1000 = 0.1\times 50; \quad W = 1.96 \text{ g}$$

Hence, % purity of Mohr's salt

$$= \frac{1.96}{2.5}\times 100 = 78.4\%$$

93. (d) $Cr_2O_7^{2-} + 6Fe^{2+}$ $(n = 1)$ (Mohr's salt) $+ 14H^+ \longrightarrow 2Cr^{3+} + 6Fe^{3+} + 7H_2O$

Equivalent of Fe^{2+} = moles of Mohr's salt

= equivalent of $K_2Cr_2O_7$

$= 500\times 10^{-3}\times 6\times 1 = 3.0$

Hence, mole per cent of Mohr's salt

$$= \frac{3}{4}\times 100 = 75$$

94. (a) Eq. mass of metal $= 2\times 8 = 16$

Eq. mass of oxide $= E_{Element} + E_{Oxygen}$

$= 16 + 8 = 24$

$$\frac{E_{Oxide}}{E_{Metal}} = \frac{24}{16} = \frac{3}{2}$$

95. (b) Let normality of $KMnO_4$ solution is N

$\therefore \ N \times 10 = 50 \times 1 \ \Rightarrow \ N = 5$

m-eq. of MnO_4^- = m-eq. of oxalic acid

$$= N_1V_1 = \frac{W}{E} \times 1000$$

$$5 \times 100 = \frac{W}{(90/2)} \times 1000 \Rightarrow W = 22.5 \text{ g}$$

96. (d)

$$\text{Normality } (N) = \frac{\text{No. of Eqn. of } CaCl_2}{\text{Volume of solution (in mL)}} \times 1000$$

$$= \frac{1.11}{111/2} \times \frac{1000}{250} = 0.08\,N$$

97. (d) $50 \times 0.5 \times 2 = \frac{W}{40} \times 1000 \Rightarrow W = 2\text{ g}$

98. (a) No. of milli equivalent of acid = No. of milli equivalent of base

$0.1 \times 2 \times V = (1/2) \times 50, V = 125$ mL

99. (c) M. Eq^n of NaOH = M. Eq^n of H_3PO_2;

$0.1 \times V = 10 \times 0.1 \Rightarrow V = 10$ mL

100. (a) $\overset{+2}{Sn}Cl_2 + \overset{+2}{Hg}Cl_2 \longrightarrow \overset{+1}{Hg}_2Cl_2 + \overset{+4}{Sn}$

$n_f = 2 \quad n_f = 1$

$0.5 \times V(\text{mL}) = 600 \times 0.1$

$V(\text{mL}) = 120$ mL

101. (b) Eq. of $KMnO_4$ = Eq. of FeC_2O_4 + Eq. of $Fe_2(C_2O_4)_3$

$x \times 5 = 2.5 \times 3 + 2.5 \times 6$

$\Rightarrow x = 4.5$ mole

102. (b) Eq. of $Ca(OH)_2$ = Eq. of KHC_2O_4 + Eq. of $H_2C_2O_4$

$140 \times M \times 2 = 200 \times 0.1 \times 1 + 300 \times 0.2 \times 2$

$\Rightarrow \quad M = 0.5$

103. (c) gm eq. of $KMnO_4$ = gm eq. of $NaHC_2O_4$

$100 \times 2 \times 5 = 250 \times M_f \times 2$

$M_f = 2$

gm eq. of $NaHC_2O_4$ = gm eq. of HCl

$2 \times 50 \times 1 = 25 \times M_{HCl} \times 1$

$M_{HCl} = 4M$

104. (c) No. of equivalent of $KMnO_4$ = No. of equivalent of $(FeSO_4 + FeC_2O_4)$

$5 \times n = 1 \times 1 + 1 \times 3, \ n = 0.8$

105. (a) No. of equivalent of metal = No. of equivalent of oxygen ($E = M/n$-factor)

$54/E = 48/8, \ E = 9$ gram

106. (a) $n_{eq}(S_2O_3^{2-}) = n_{eq}(K_2S_2O_8)$

or $\frac{V \times 0.25}{1000} \times 1 = \frac{1}{270} \times 2 \Rightarrow V = 29.63$ mL

107. (a) No. of equivalent of Mn^{2+} = No. of equivalent of $S_2O_8^{2-}$

1×5 = mole of $S_2O_8^{2-} \times 2$,

mole of $S_2O_8^{2-} = 2.5$

108. (b) M. Eq^n of $KMnO_4$ = M. Eq^n of $FeSO_4$

$M_1 \times 5 \times V_1 = M_2 \times V_2 \times 1 \Rightarrow 5M_1V_1 = M_2V_2$

109. (b) $V \times 0.02 \times 6 = \frac{0.288}{(56+88)} \times 3 \times 1000$

$\Rightarrow \quad V = 50$ mL

110. (d) $H_2O_2 + KMnO_4 \longrightarrow O_2 + Mn^{2+}$

$n_f = 2 \quad n_f = 5$

Equivalent of H_2O_2 = Equivalent of $KMnO_4$

$$\frac{10}{1000} \times M \times 2 = \frac{10}{1000} \times \frac{1}{0.56}$$

$$M_{H_2O_2} = \frac{1}{1.12} = \frac{\text{V.S.}}{11.2}$$

$$\therefore \ \text{Volume strength} = \frac{11.2}{1.12} = 10$$

111. (d) $\frac{W_{H_2O_2}}{34} \times 2 = \frac{0.316}{158} \times 5 \Rightarrow W_{H_2O_2} = 0.17$ gram

$$\% \text{ purity} = \frac{W_{H_2O_2}}{W_{sample}} \times 100 = \frac{0.17}{0.2} \times 100 = 85$$

112. (b) $\underset{1.5}{FeS_2} \longrightarrow \underset{3\text{ mole}}{2SO_2}$

$Cr_2O_7^{2-} + SO_2 \longrightarrow Cr^{3+} + SO_4^{2-}$

No. of equivalent of $(n_{eq})_{K_2Cr_2O_7}$ = No. of equivalent of $(n_{eq.})_{SO_2}$

$0.5 \times V \times 6 = 3 \times 2 \Rightarrow V = 2L$

113. (b) m-eq. of $FeSO_4 \cdot 7H_2O$ in 1 litre

$$= 20 \times 0.02 \times 5 \times \left(\frac{1000}{25}\right) = 80$$

$$\therefore \quad \frac{W}{278} \times 1 \times 1000 = 80 \Rightarrow W = 22.24$$

weight % of $FeSO_4 \cdot 7H_2O$ in given sample

$$= \frac{22.24}{32} \times 100 = 69.5$$

114. (a) eq. wt. of $K_2Cr_2O_7 = \frac{\text{molar mass}}{6}$

eq. wt. of $KMnO_4 = \frac{\text{molar mass}}{5}$

eq. of $Na_2S_2O_3$ = eq. of I_2 liberated

= eq. of $KMnO_4$ + eq. of $K_2Cr_2O_7$

or $N \times 1 = 0.02 \times 5 + 0.05 \times 6$

$\Rightarrow \quad N = 0.4$ or $M = 0.4$

115. (d) eq. wt. of $NaHC_2O_4$ as an acid = 112

eq. of $NaHC_2O_4$ = eq. of NaOH

$= 100 \times 10^{-3} \times 0.2$

$= 0.02$

$\Rightarrow \quad a = 0.02 \times 112 = 2.24$ g

eq. wt. of $NaHC_2O_4$ as reductant

$(C_2O_4^{2-} \longrightarrow 2CO_2 + 2e^-) = \frac{112}{2} = 56$

eq. of $NaHC_2O_4$ as reductant = eq. of $KMnO_4 = 100 \times 10^{-2} \times 0.02 \times 5 = 0.01$

$\Rightarrow \quad b = 0.01 \times 56 = 0.56$ g

Hence, $a : b = \frac{2.24}{0.56} = 4 : 1$ or $a = 4b$

116. (c) eq. of $Na_2C_2O_4$ + eq. of $H_2C_2O_4$

$= $ eq. of $KMnO_4 = V \times 0.1 \times 5$

$2 + 2 = 0.5V; \quad V = 8L$

eq. of $H_2C_2O_4$ = eq. of NaOH

$\Rightarrow \quad 1 \times 2 = 0.2 \times V_2; \quad V_2 = 10$ L

Hence, $V_1 : V_2 = 8 : 10 = 4 : 5$

117. (d) Let the moles of $NaHC_2O_4$ be 'a' and that of $KHC_2O_4 \cdot H_2C_2O_4$ be 'b' and volumes of each consumed be V litre.

In redox :

when $KMnO_4$ used : $0.2 \times V = 2a + 4b$ ...(1)

In acid-base neutralization :

when NaOH used: $0.12 \times V = a + 3b$ ...(2)

from (1) and (2) $a = 0.06\ V$

$b = 0.02\ V$ or $a : b = 3 : 1$

118. (d) Moles of pure $CuSO_4 = 0.12 =$ moles of $Na_2S_2O_3$ reacted with I_2

Moles of $Na_2S_2O_3 = 120 \times 10^{-3} \times M$

$= 0.12$

$M = 1.0$

119. (b) M. Eq^n of $Na_2S_2O_3 = M$. Eq^n of $CuSO_4$

$\Rightarrow \quad 0.4 \times V = 50 \times 0.20 \Rightarrow V = 25$ mL

120. (a) NaOH + Na_2CO_3

a milli-mole — *b* milli-mole

$a + b = 20 \times \frac{1}{20}$ ...(i)

(in presence of phenolphthalein)

$a + 2b = 25 \times \frac{1}{20}$ ...(ii)

(in presence of Methyl orange)

By solving (i) & (ii), $a = 0.75$ milli-mole;

$b = 0.25$ milli-mole

Mass of NaOH $= \frac{1}{1000} \times 0.75 \times 40 = 0.03$ g

121. (b) $\frac{W}{E} \times 1000 = 250 \times 0.1 - 50 \times 0.2$

$\frac{0.63}{E} \times 1000 = 15 \Rightarrow E = 42$

122. (a) $Ca(HCO_3)_2$ + $CaO \longrightarrow CaCO_3 + H_2O$

162 g — 56 g

16.2 g — 5.6 g

162 g $Ca(HCO_3)_2 = 100$ g $CaCO_3$

16.2 g $Ca(HCO_3)_2 = 10$ g $CaCO_3$

Hardness $= \frac{\text{Mass of } CaCO_3}{\text{Mass of water}} \times 10^6$

$= \frac{10}{10^4} \times 10^6 = 1000$ ppm

123. (d) $10^{-3}\ M\ CaSO_4 = 10^{-3}\ M\ CaCO_3$ because n-factor of both is same,

mass of 1 L water = 1000 gram

Hardness of water (in ppm)

$= \frac{W_{CaCO_3}\text{ (in g)}}{W_{\text{water sample}}\text{(in g)}} \times 10^6$

$= \frac{10^{-3} \times 1 \times 100 \times 10^6}{1000} = 100$

124. (c) Milli-equivalents of Ca^{2+}

+ milli-equivalents of Mg^{2+}

= milli-equivalents of Na_2CO_3

$\frac{20}{20} + \frac{12}{12} = V \times 2$

$\therefore \quad V = 1$ mL for 1 L of tap water

For 5000 L of pond water

$V = 5000$ mL or 5 L

125. (a) Moles of $Ca(HCO_3)_2 = 0.81/162 = 5 \times 10^{-3}$

= Moles of $CaCO_3$

Moles of $Mg(HCO_3)_2 = 0.73/146 = 5 \times 10^{-3}$

= Moles of $CaCO_3$

Hardness of water (in ppm)

$= \frac{W_{CaCO_3}\text{ (in g)}}{W_{\text{water sample}}\text{(in g)}} \times 10^6$

$= \frac{(5 \times 10^{-3} + 5 \times 10^{-3}) \times 100}{100} \times 10^6$

$= 10000$ ppm

Level 2

2. (a) m-eq. of acid = m-eq. of base

$\therefore \quad N_1V_1 = N_2V_2$

$$\frac{29.4}{\left(\frac{98}{n}\right)} \times 100 = 90 \times \left(\frac{20}{40} \times \frac{1000}{500}\right)$$

$n = 3$

3. (b) When phenolphthalein is used as an indicator :

$0.05\ x = 20 \times 0.1 \times 1; x = 40$ mL

When methyl orange is used as an indicator:

$0.05\ y = 20 \times 0.1 \times 3$

$y = 120$ mL

$\therefore \quad y - x = 80$ mL

4. (a) $I_2 + AsO_2^- + 2H_2O \longrightarrow HAsO_4^{2-} + 2I^- + 3H^+$

m-eq. of $HAsO_2$ (in 50 mL) = m-eq. of I_2

$= 35 \times 0.05 \times 2 = 3.5$

m-eq. of $HAsO_2$ in 250 mL $= 3.5 \times \dfrac{250}{50}$

$= 17.5$

mass of $HAsO_2$ in sample

$= \dfrac{17.5}{2} \times (108) \times 10^{-3} = 0.945$ g

% of $HAsO_2$ in the sample $= \dfrac{0.945}{3.78} \times 100$

$= 25\%$

5. (a) m-eq. of FeO = m-eq. of $KMnO_4$

$= 0.25 \times 5 \times 100$

m-mole of FeO $(n = 1) = \dfrac{0.25 \times 100 \times 5}{1}$

$= 125$

total m-eq. or m-mole. of Fe^{2+}

$= 1000 \times 0.1 \times 6 = 600$

(from FeO and Fe_2O_3 after reaction with Zn dust)

m-mole of Fe^{2+} from $Fe_2O_3 = 600 - 125$

$= 475$

m-mole of $Fe_2O_3 = \dfrac{475}{2}$

Mass of FeO $= \dfrac{125 \times (56 + 16)}{1000}$ g $= 9$ g

Mass of $Fe_2O_3 = \dfrac{475}{2} \times \dfrac{160}{1000} = 38$ g

% $Fe_2O_3 = \dfrac{38}{38 + 9} \times 100 = 80.85\%$

6. (d) $3Br_2 + 6OH^- \longrightarrow 5Br^- + BrO_3^- + 3H_2O$ ...(1)

$6H^+ + BrO_3^- + 3C_2O_4^{2-} \longrightarrow Br^- + 6CO_2 + 3H_2O$ ...(2)

10 m-mole of Br_2 produce (10/3) m-mole BrO_3^- and m-moles of

CaC_2O_4 required $= \dfrac{10}{3} \times 3 = 10$

$\therefore$ Mass of $CaC_2O_4 = 10 \times 10^{-3} \times 128$

% purity $= \dfrac{10 \times 10^{-3} \times 128}{2} \times 100 = 64$

7. (b) m-moles of KI $= 50 \times 0.04 = 2$

m-equivalent of $Na_2S_2O_3$ reacted with 10 mL solution $= 20 \times 2.5 \times 10^{-3} = 50 \times 10^{-3}$

m-equivalent of I^- left unreacted in 100 mL solution

$= 50 \times 10^{-3} \times \dfrac{100}{10} = 0.5$

m-moles of I^- which reacts with $CuCO_3 = 2 - 0.5$

$= 1.5$

$\because \quad 2Cu^{2+} + 5I^- \longrightarrow 2CuI + I_3^-$

m-moles of $Cu^{2+} = \dfrac{2}{5} \times 1.5 = 0.6$

weight of $CuCO_3 = 0.6 \times 10^{-3} \times 123.5$

$= 0.0741$

weight % of $CuCO_3 = \dfrac{0.0741}{0.1} \times 100$

$= 74.1$

8. (a) M_{NaOH} resultant

$= \dfrac{20 \times 0.2 + 35 \times 0.1}{100} = 0.075\ M$

Milli-equivalent of NaOH = milli-equivalent of $H_2C_2O_4$

Let mass of impure sample is x gram

$40 \times 0.075 = \dfrac{x \times 0.90}{90} \times 2 \times 1000$

$x = 0.15$ gram

9. (b) m-eq. of $Na_2C_2O_4$ = m-eq. of $KMnO_4$ reacted

total m-eq. of $KMnO_4$– excess m-eq. of $KMnO_4$ reacted with $H_2C_2O_4$

$= 45 \times 0.02 \times 5 - 10 \times 0.1 \times 2 = 2.5$

$1000 \times \dfrac{W}{134} \times 2 = 2.5$

$W_{Na_2C_2O_4} = 0.1675$ g

% purity of $Na_2C_2O_4$ in sample

$= \dfrac{0.1675}{0.2} \times 100 = 83.75$

10. (a) Reactions of $K_2Cr_2O_7$ and $KMnO_4$ with KI may be given as :

$K_2Cr_2O_7 + 7H_2SO_4 + 6KI \longrightarrow 4K_2SO_4 + Cr_2(SO_4)_3 + 7H_2O + 3I_2$

$2KMnO_4 + 8H_2SO_4 + 10KI \longrightarrow 6K_2SO_4 + 2MnSO_4 + 5I_2$

Thus, equivalent mass of $K_2Cr_2O_7 = \dfrac{294}{6} = 49$

equivalent mass of $KMnO_4 = \dfrac{158}{5} = 31.6$

$\because$ m-eq. of $K_2Cr_2O_7$ + m-eq. of $KMnO_4$

= m-eq. of I_2 = m-eq. of hypo.

Let the mass of $K_2Cr_2O_7 = x$ g;

$\therefore$ mass of $KMnO_4 = (0.5 - x)$ g

$\dfrac{x}{49} + \dfrac{(0.5 - x)}{31.6} = 150 \times 0.1 \times 10^{-3}$

or $\quad x = 0.0732$

$\therefore$ % of $K_2Cr_2O_7 = \dfrac{0.0732}{0.5} \times 100$

$\Rightarrow 14.64$

11. (b) $I_2 + 2Na_2S_2O_3 \longrightarrow 2NaI + Na_2S_4O_6$ ...(i)

m-moles of $Na_2S_2O_3$ consumed

$= 15 \times 0.4 = 6$ m-mole

m-moles of I_2 consumed = 3 m-mole

$3I_2 + 6NaOH \longrightarrow 5NaI + NaIO_3 + 3H_2O$...(ii)

m-mol of I_2 reacted with NaOH are

$= \frac{30 - 2 \times 3}{2} = 12$ m-mole

Total m-mole of I_2 consumed in reaction

(i) and (ii) = 3 + 12 = 15 m-mole

Molarity of $I_2 = \frac{15}{150} = 0.1\ M$

12. (a) Let x and y are milli-moles H_2SO_4 and $H_2C_2O_4$ present in the given mixture reacts with base

m-eq. of acids = m-eq. of base;

$(x + y) \times 2 = 3 \times 0.1 \times \frac{1000}{10}$ or $x + y = 15$

In second experiment only $KMnO_4$ reacted with $H_2C_2O_4$ in which Mn^{7+} converted into Mn^{2+} and $C_2O_4^{2-}$ converted into CO_2

$\therefore$ milli-equivalent of $H_2C_2O_4$

= milli-equivalent of $KMnO_4$

$y \times 2 = 4 \times 0.02 \times 5 \times \frac{1000}{100}$

$y = 2$

$\therefore$ $x = 13$, mass of $H_2SO_4 = 13 \times 10^{-3} \times 98$

$= 1.274$ g

wt. % of H_2SO_4 in sample $= \frac{1.274}{3.185} \times 100$

= 40%

13. (c) $KMnO_4 + C_2O_4^{2-} \xrightarrow{H^+} Mn^{2+} + CO_2$

$n_f = 5$ $\quad n_f = 2$

16 mL $\quad$ 40 mL

0.05 M $\quad$ XM

m. eq. of $C_2O_4^{2-}$ = m.eq. of $KMnO_4$

$2(40)X = 5 \times 16 \times 0.05$

$X = 0.05M$

$\therefore$ $[H^+] = 2X = 0.1\ M$

pH = 1

14. (a) $[H^+] = 10^{-3}\ M$

So $[Ca^{2+}] = \frac{1}{2} \times 10^{-3}\ M$

Mass of $Ca^{2+} = \frac{10^{-3}}{2} \times 40$

So, mass of Ca^{2+} ions in 10^6 mL hard water

$= \frac{\frac{1}{2} \times 10^{-3} \times 40 \times 10^6}{10^3} = 20$

15. (c) Equivalent of $Ba(MnO_4)_2$ = Equivalent of H_2O_2

$$\left(\frac{x}{375}\right) \times 10 = \left(\frac{16.8}{5.6}\right) \times \frac{100}{1000}$$

$x = 11.25$

% purity $= \frac{11.25}{15} \times 100 = 75\%$

16. (b) $MnO_2 + H_2C_2O_4 \cdot 2H_2O \longrightarrow Mn^{2+} + CO_2$

$n_f = 2$ $\quad n_f = 2$

a mol $\quad \frac{1.26}{126} = 10^{-2}$ mol

Equiv. of $H_2C_2O_4 \cdot 2H_2O$ = Equiv. of (MnO_2 pure + $KMnO_4$)

$10^{-2} \times 2 = (a \times 2) + \frac{30}{1000} \times 0.1 \times 5$

$2a = 2 \times 10^{-2} - 1.5 \times 10^{-2} = 0.5 \times 10^{-2}$

$\therefore$ $a = 0.25 \times 10^{-2}$ mol

% of $MnO_2 = \frac{0.25 \times 10^{-2} \times 87}{0.25} \times 100\% = 87\%$

17. (c) Moles of $MgCO_3$ = Moles of CO_2 produced

$= \frac{10.5 \times 0.8}{84} = 0.1$

Moles of NaOH = 1 × 0.3 = 0.3

	CO_2 +	$2NaOH \longrightarrow$	Na_2CO_3 +	$2H_2O$
Initial moles	0.1	0.3	–	–
Final moles	–	0.1	0.1	–

HCl will react with Na_2CO_3 and NaOH

$Na_2CO_3 + HCl \xrightarrow{HPh} NaHCO_3 + NaCl$

Moles of HCl = Moles of Na_2CO_3 + Moles of NaOH

$V \times 0.5 = 0.1 + 0.1$

$V_{HCl} = 0.4$ L or 400 mL

18. (b) $Ba(OH)_2 + 2HCl \longrightarrow BaCl_2 + 2H_2O$

$HCl + NaOH \longrightarrow NaCl + H_2O$

m. moles of $Ba(OH)_2 = \frac{1}{2} \times$ m. moles of HCl reacted

$= \frac{1}{2} \times 10 = 5$

m. moles of HCl reacted $= 80 \times 0.2 - 20 \times M_{NaOH}$

Where $M_{NaOH} \times 10 = 30 \times 0.1$

$M = 0.3$

m. moles of HCl reacted $= 80 \times 0.2 - 20 \times 0.3 = 10$

Wt. of $Ba(OH)_2 = 5 \times 10^{-3} \times 171$

$= 0.855$

% of $Ba(OH)_2$ in sample $= \frac{0.855}{1.5} \times 100 = 57$

19. (a) milli equivalent + milli equivalent = milli equivalent of

of NaOH of $Ca(OH)_2$ HCl

$$V_1X + V_2Y = 100 \times 0.18$$

$$\left(\frac{V_2}{2}\right)\times(4Y) + V_2Y = 18$$

$$2V_2Y + V_2Y = 18$$

$$V_2Y = 6$$

$$\text{fraction} = \frac{6}{18} = \frac{1}{3} = 0.33$$

20. (d) $CuSO_4 + KI \longrightarrow I_2 + K_2SO_4 + Cu_2I_2$

$I_2 + Na_2S_2O_3 \longrightarrow Na_2S_4O_6 + NaI$

No. of Eq. $CuSO_4$ = No. of Eq. I_2 = No. of Eq. hypo

$$\text{mole } CuSO_4 \times 1 = \text{mole } I_2 \times 2 = \frac{1 \times 0.2 \times 100}{1000}$$

$\therefore$ mole of Cu in impure sample = 0.02

mass of Cu in impure sample = $0.02 \times 64 = 1.28$

$$\therefore \quad \% \text{ mass of Cu} = \frac{1.28}{6.4} \times 100 = 20$$

Level 3

Passage-1

1. (a) 1 litre of $H_2O_2(aq)$ provide 11.2 litre of O_2 at 1 atm and 273 K

$$\text{moles of } O_2 = \frac{11.2}{22.4} = 0.5$$

$n_{H_2O_2}$ required $= 0.5 \times 2$

$$M_{H_2O_2} = \frac{n_{H_2O_2}}{V_{\text{solution}}} = 1\ M$$

2. (b) Strength in percentage means how many g H_2O_2 present per 100 mL

$\because$ $M \Rightarrow$ 1 and mol. mass of $H_2O_2 = 34$

$\therefore$ 34 g H_2O_2 present per litre of solution or 3.4 g H_2O_2 present per 100 mL of solution.

3. (b) m-eq. of H_2O_2 = m-eq. of $KMnO_4$

$$20 \times N = 0.05 \times 5 \times 80 \Rightarrow N = 1$$

$$N = \frac{\text{volume strength of } H_2O_2}{5.6}$$

$\Rightarrow$ volume strength of $H_2O_2 = 5.6$

4. (b) m-eq. of $Ba(MnO_4)_2$ = m-eq. of H_2O_2

$$\left(\because M = \frac{33.6}{11.2} \Rightarrow 3\right)$$

$$\frac{w}{375} \times 10 \times 1000 = 3 \times 125 \times 2; \quad w = 28.125$$

$$\% \text{ purity} = \frac{w}{40} \times 100$$

$$= \frac{28.125}{40} \times 100 = 70.31$$

Passage-2

1. (c) 136 mg of $CaSO_4 \equiv 100$ mg of $CaCO_3$;

190 mg of $MgCl_2 \equiv 2 \times 100$ mg of $CaCO_3$

1 litre or 1000 g water contain 300 mg of $CaCO_3$

10^6 g water contain

$$\frac{300 \times 10^{-3}}{1000} \times 10^6 = 300 \text{ ppm}$$

2. (b) 10^6 g water contain 100 g $CaCO_3$

10 L or 10^4 g water contain

$$\frac{100}{10^6} \times 10^4 = 1 \text{ g of } CaCO_3$$

100 g $CaCO_3 \equiv 162$ g of $Ca(HCO_3)_2$

or 1 g $CaCO_3 \equiv 1.62$ g of $Ca(HCO_3)_2$

or 0.01 mole

$$Ca(HCO_3)_2 + Ca(OH)_2 \longrightarrow 2CaCO_3\downarrow + 2H_2O$$

moles of $Ca(OH)_2$ required = 0.01

or mass of $Ca(OH)_2$ required = 0.74 g

3. (b) m-moles of H^+ present in outlet water

= m-moles of $OH^- = 50 \times 0.1 = 5$

m-moles of Ca^{2+} present in hard water $= \frac{5}{2}$

= 2.5; (1 Ca^{2+} replaced by $2H^+$)

no. of mg of Ca^{2+} ion $= 2.5 \times 40 = 100$

200 g sample hard water contain 100 mg of Ca^{2+}

$\therefore$ 10^6 g sample hard water contain

$$\frac{100}{200} \times 10^6 \times 10^{-3} = 500 \text{ ppm}$$

Passage-3

1. (c) n-factor $= 5 \times 2 = 10$

2. (a) H_3PO_2 is a monobasic acid $\therefore$ n-factor = 1

3. (a) $n\text{-factor} = \left(3 - \frac{2}{0.95}\right) \times 0.95 = 0.85$

$$\therefore \quad E = \frac{M}{0.85}$$

4. (b) n-factor of VO = 3; $Fe_2O_3 = 1 \times 2 = 2$;

$\therefore$ x and y are 2 and 3

One or More Answers is/are Correct

1. (b, c, d)

$$xA + yB \longrightarrow A_xB_y$$

As per equivalent concept

Number of equivalent of A = No. of equivalent of B = No. of equivalent of A_xB_y.

2. (a, c)

$$\underset{n\text{-factor}=2}{H_3PO_4} + \underset{n\text{-factor}=2}{X(OH)_2} \longrightarrow XHPO_4 + 2H_2O$$

(A) Equivalent weight of base $= \frac{\text{Mol. wt}}{2}$

(B) Equivalent weight of $H_3PO_4 = \frac{98}{2}$

(C) $\underset{n\text{-factor}=1}{XHPO_4} + \underset{n\text{-factor}=1}{NaOH} \longrightarrow$

Moles of NaOH × 1 = moles of $XHPO_4$ × 1

Moles of NaOH = 1 × 1 = 1

(D)

$$2XHPO_4 + X(OH)_2 \longrightarrow X_3(PO_4)_2 + 2H_2O$$

Moles 1 $\frac{1}{2}$

Moles of $X(OH)_2$ required $= \frac{1}{2}$

3. (b, c, d)

$$xSn^{2+} + yCr_2O_7^{2-} + zH^+ \longrightarrow aSn^{4+} + bCr^{3+} + cH_2O$$

Upon balancing the redox-reaction

$$3Sn^{2+} + Cr_2O_7^{2-} + 14H^+ \longrightarrow 3Sn^{4+} + 2Cr^{3+} + 7H_2O$$

4. (a, b)

Let $n_{Cu_2S} = a$; $n_{CuS} = a$

$$\underset{n\text{-factor}=8}{Cu_2S} + \underset{6}{CuS} + \underset{10}{Ba(MnO_4)_2} + H^+ \longrightarrow Cu^{2+} + SO_2 + Mn^{2+}$$

5. (a, c, d)

Equivalent weight $= \frac{\text{Molecular weight}}{n\text{-factor}}$

(A) n-factor of H_3PO_3 as acid = 1

as base = 1

In both case, eq. wt. of $H_3PO_3 = \frac{81}{1} \Rightarrow 81$

(B) n-factor of $H_2PO_4^-$ as acid = 1 or 2

as base = 1

(C) In acidic medium

Eq. wt. of $KMnO_4 = \frac{\text{Molecular weight}}{5}$

In basic medium

Eq. wt. of $KMnO_4 = \frac{\text{Molecular weight}}{1}$

In neutral/slightly basic medium

Eq. wt. of $KMnO_4 = \frac{\text{Molecular weight}}{3}$

(D) In MgH_2, O.S. of H = −1

In H_2O_2, O.S. of H = +1

6. (b, c)

m. eqn. of $NaHC_2O_4$ = m. eqn. of $KMnO_4$

$10 \times N = 20 \times 0.1$

$N = 0.2$ or $M = 0.1$

In acid base neutralization : m. eqn. of acid = m. eqn. of base ($Ba(OH)_2$)

$$20 \times 0.1 \times 1 = 10 \times 0.1 \times 2$$

7. (b, c, d)

Balance reaction : $5HNO_2 + 2KMnO_4 + 3H_2SO_4 \longrightarrow 5HNO_3 + 2MnSO_4 + K_2SO_4 + 3H_2O$

8. (a, b, d)

$$MnO_2 + H_2C_2O_4 \cdot 2H_2O \longrightarrow Mn^{2+} + CO_2$$

$n_f = 2$ $n_f = 2$

a mol $\frac{1.26}{126} = 10^{-2}$ mole

Equiv. of $H_2C_2O_4 \cdot 2H_2O$ = Equiv. of (MnO_2 pure + $KMnO_4$)

$$10^{-2} \times 2 = (a \times 2) + \frac{30}{1000} \times 0.1 \times 5$$

$$2a = 2 \times 10^{-2} - 1.5 \times 10^{-2} = 0.5 \times 10^{-2}$$

$\therefore$ $a = 0.25 \times 10^{-2}$ mole

$$\% \text{ of } MnO_2 = \frac{0.25 \times 10^{-2} \times 87}{0.25} \times 100\% = 87\%$$

9. (b, d)

Na_2CO_3 $NaHCO_3$ $NaOH$

HPh indicator

g-eq. of Na_2CO_3 + g-eq. of NaOH = g-eq. of HCl

$$(3 \times 1 + 1 \times 1)\frac{50}{1000} = \frac{N}{10} \times V_1 \; ; \; V_1 = 2\,L$$

Further MeOH indicator

g-eq. of $NaHCO_3$ = g-eq. of HCl

$$(3 \times 1 + 2 \times 1)\frac{20}{1000} = \frac{N}{10} \times V_2 \; ; V_2 = 1\,L$$

10. (a, b, c)

With $KMnO_4$

$$FeC_2O_4 + MnO_4^- \longrightarrow Fe^{3+} + CO_2 + Mn^{2+}$$

$n_f = 3$ $n_f = 5$

Number of equivalents of MnO_4^- = number of equivalents of FeC_2O_4

5 × number of mole of MnO_4^- = 3 × number of mole of FeC_2O_4

Number of moles of $MnO_4^- = \frac{3}{5} \times 1$

With $K_2Cr_2O_7$

$$FeC_2O_4 + Cr_2O_7^- \longrightarrow Fe^{3+} + CO_2 + Cr^{3+}$$

$n_f = 3$ $n_f = 6$

Number of equivalents of $Cr_2O_7^-$ = number of equivalents of FeC_2O_4

$6 \times$ number of mole of $Cr_2O_7^- = 3 \times$ number of mole of FeC_2O_4

Number of moles of $Cr_2O_7^- = \frac{3}{6} \times 1$

Hence, mole ratio of $KMnO_4$ and $K_2Cr_2O_7 = 6:5$

Since, n-factor for FeC_2O_4 is same in both reactions so number of equivalents will also be same hence equivalents of $KMnO_4$ and $K_2Cr_2O_7$ will be same

Equivalent weight of

$$FeC_2O_4 = \frac{\text{Formula weight}}{n\text{-factor}} = \frac{\text{Formula weight}}{3}$$

Match the Column

1. (A) $\overset{+3}{Bi_2}\overset{-2}{S_3} \longrightarrow Bi^{5+} + \overset{0}{S}$

n-factor $= (2 \times 2) + (2 \times 3) = 10$

(B) $Al_2(\overset{+6}{Cr_2}O_7)_3 \longrightarrow Cr^{3+}$

n-factor $= (3 \times 6) = 18$

(C) $\overset{+2}{Fe}\overset{-1}{S_2} \longrightarrow \overset{+3}{Fe_2}O_3 + \overset{+4}{S}O_2$

n-factor $= (1 \times 1) + (5 \times 2) = 11$

(D) $\overset{+2}{Mn}(\overset{+5}{N}O_3)_2 \longrightarrow \overset{+6}{Mn}O_4^{2-} + \overset{+2}{N}O$

n-factor $= |(4 \times 1) - (3 \times 2)| = 2$

2. (P) $\overset{+3}{Cr}\overset{-1}{I_3} \longrightarrow \overset{+6}{Cr_2}O_7^{2-} + \overset{+7}{I}O_4^-$

n-factor $= (3 \times 1) + (8 \times 3) = 27$

(Q) $\overset{+2}{Fe}(\overset{-2}{S}\overset{+4}{C}\overset{-3}{N})_2 \longrightarrow \overset{+6}{Fe^{3+}} + \overset{+4}{S}O_4^{2-} + \overset{+5}{C}O_3^{2-} + NO_3^-$

n-factor $= (1 \times 1) + (8 \times 2) + (8 \times 2) = 33$

(R) $\overset{-3}{N}H_4\overset{-2}{S}\overset{+4}{C}\overset{-3}{N} \longrightarrow \overset{+6}{S}O_4^{2-} + \overset{+4}{C}O_3^{2-} + \overset{+5}{N}O_3^-$

n-factor $= (8 \times 2) + (8 \times 1) = 24$

(S) $\overset{+3}{AS_2}\overset{-2}{S_3} \longrightarrow \overset{+5}{As}O_3 + \overset{+6}{S}O_4^{2-}$

n-factor $= (2 \times 2) + (8 \times 3) = 28$

4. Moles of $NaIO_3 = \frac{396}{198} = 2$

(1) $\overset{+5}{I}O_3^- + 3H\overset{+4}{S}O_3^- \longrightarrow \overset{-1}{I^-} + 3\overset{+6}{S}O_4^{2-} + 3H^+$

n-factor $\quad \frac{6}{1} \quad \frac{6}{3} \quad \frac{6}{1} \quad \frac{6}{3} \quad \frac{6}{3}$

(2) $5I^- + IO_3^- \xrightarrow{6H^+} 3I_2 + 3H_2O$

n-factor $= \frac{5}{5} = 1, \frac{5}{1} \quad \frac{5}{3} \quad \frac{5}{3}$

(A) n-factor of IO_3^- in reaction (2) $= 5$

(B) Moles of HSO_3^- used in (1) reaction $= 3 \times$ moles of $IO_3^- = 3 \times 2 = 6$

(C) Moles of I_2 produced $=$ moles of I^- produced $\times \frac{3}{5} = 2 \times \frac{3}{5} = 1.2$

(D) Equivalents of IO_3^- used in reaction (2)

$=$ Equivalents of I^-

$= 2 \times 1 = 2$

5. (A) $\overset{0}{Br_2} + OH^- \longrightarrow \overset{+5}{Br}O_3^- + \overset{-1}{Br^-}$

Interamolecular, Disproportionation Redox Reaction

(B) $\overset{+2}{Fe}Cl_2 + K\overset{+7}{Mn}O_4 + HCl \longrightarrow \overset{+2}{Mn}Cl_2 + \overset{-3}{Fe}Cl_3$

Interamolecular Redox Reaction

(C) $\overset{+1}{Ag_2}\overset{-2}{O} \longrightarrow \overset{0}{Ag} + \frac{1}{2}\overset{0}{O_2}$

Interamolecular Redox

(D) $\overset{-3}{N}H_4\overset{+5}{N}O_3 \longrightarrow \overset{+1}{N_2}O + H_2O$

Interamolecular comproportionation Redox Reaction

Assertion-Reason Type Question

2. $N = M \times n$-factor

and n-factor of $H_3PO_3 = 2$ (max).

6. $[\overset{+2}{Fe}(\overset{+2}{C}\overset{-3}{N})_6]^{4-} \longrightarrow \overset{+3}{Fe^{3+}} + \overset{+4}{C}O_2 + \overset{+5}{N}O_3^-$

n-factor $= (1 \times 1) + (2 \times 6) + (8 \times 6) = 61$

7. $\overset{+3}{As_2}\overset{-2}{S_3} + \overset{+5}{N}O_3^- \longrightarrow \overset{+5}{As}O_4^{3-} + \overset{+2}{N}O + \overset{+6}{S}O_4^{2-}$

n-factor $= (2 \times 2) + (8 \times 3)$ $\quad n-f = (3 \times 1)$

$= 28 \qquad = 3$

8. $NaOH + H_3PO_4 \longrightarrow NaH_2PO_4 + H_2O$

n-factor $= 1 \qquad 1$

Eq. wt of $H_3PO_4 = \frac{M}{1}$

10. $\overset{0}{I_2} \longrightarrow \overset{+5}{I}O_3^- + \overset{-1}{I^-}$

Level 4

Subjective Problems

1. $Ba(OH)_2 \cdot xH_2O(aq) + HNO_3(aq) \longrightarrow Ba(NO_3)_2 + H_2O(l)$

Eq. of $Ba(OH)_2 \cdot xH_2O =$ eq. of HNO_3

$$\frac{0.7875}{(171 + 18x)} \times 2 = \frac{1}{4} \times \frac{20}{1000}$$

$x = 8$

2. n-factor of acid $=$ Basicity (n)

Eq. of acid $=$ Eq. of base

$$\frac{2}{600} \times n = \frac{1}{6} \times \frac{100}{1000} \times 1$$

$n = 5$

3. In HPh Eq. of NaOH + Eq. of Na_2CO_3 = Eq. of HCl

$(0.5)(1)+(0.5)(1)=(x)(1)$

$x=1$

In MeOH Eq. of NaOH + Eq. of Na_2CO_3 + Eq. of $NaHCO_3$ = Eq. of HCl

$(0.5)(1)+(0.5)(2)+(0.5)(1)=(y)(1)$

$y=2$

$x+y=3$

4. $\overset{+5}{BrO_3^-}+\overset{-1}{Br^-}+H^+ \longrightarrow \overset{0}{Br_2}(l)$

n-factor = 5

$$E_{KBrO_3}=\frac{M_{KBrO_3}}{5}$$

5. $\overset{+4}{SeO_2}+\overset{+2}{CrSO_4} \longrightarrow Se^x + Cr^{3+}$

Eq. of SeO_2 = Eq. of $CrSO_4$

$(12.5)(0.05)(4-x)=(25)(0.1)(1)$

$(4-x)=4$

$x=0$

6. $\underset{n\text{-factor}=1}{Cu^{2+}}+KI \longrightarrow \underset{n\text{-factor}=2}{I_2}+Cu^+$

$\underset{n\text{-factor}=2}{I_2}+\underset{n\text{-factor}=1}{Na_2S_2O_3} \longrightarrow NaI+Na_2S_4O_6$

Eq. of Cu^{2+} = Eq. of I_2 produced = Eq. of I_2 reacted = Eq. of $Na_2S_2O_3$.

$$(x)(1)=\left(\frac{40}{1000}\right)(0.1)(1)$$

$x=4\times10^{-3}$ mol

Mass of pure Cu $=4\times10^{-3}\times63.5=0.254$ g

$$\%\ \text{Purity}=\frac{0.254}{0.276}\times100=92\%$$

% impurity = 8%

7. Eq. of H_2O_2 = Eq. of $KMnO_4$

$(25\times M\times2)=(10\times0.1\times5)$

$M=0.089$

Volume strength $=M\times11.2=1$

8. Balanced redox reaction

$$28NO_3^-+3As_2S_3+4H_2O \longrightarrow 6AsO_4^{3-}+28NO+9SO_4^{2-}+8H^+$$

9. $\underset{0.220\text{g}}{M_2O_x}$ (Metallic oxide) $+H_2 \longrightarrow M(s)+\underset{0.045\text{g}}{H_2O(l)}$

$\Rightarrow$ *POAC* on O

$$\Rightarrow \left(\frac{0.220}{2M+16x}\right)x=\left(\frac{0.045}{18}\times1\right)$$

$\Rightarrow 3.96x=0.09M+0.72x$

$$\Rightarrow E=\frac{M}{x}=36$$

$$\frac{E}{9}=4$$

10. Let mass of $K_2Cr_2O_7=a$ g

Mass of $KMnO_4=(10-a)$ g

Eq. of $K_2Cr_2O_7$ + Eq. of $KMnO_4$ = Eq. of $Na_2S_2O_3$

$$\left(\frac{9}{294}\times6\right)+\left(\frac{10-a}{158}\times5\right)=(2.2)(0.1)$$

$a\approx8.5$ g

Mass % of $KMnO_4=Z=15$

11. $a=5\times10^{-4}$

% of Sn^{2+} in sample

$$(x)=\frac{5\times10^{-4}\times11.8}{0.4}\times100$$

$x\simeq15$

$$\therefore \frac{x}{5}=\frac{15}{5}=3$$

12. Let molarity = x mol/L

$25x=m$ mole of Mn^{2+}

$3Mn^{2+}+2MnO_4^-+OH^- \longrightarrow 5MnO_2+H_2O$

m-mole of $KMnO_4$ used $=0.5\times40=20$

$$m\text{-mole of } Mn^{2+}=20\times\frac{3}{2}=30\ \text{m-mole}$$

$$\text{Conc. of } Mn^{2+}=\frac{30}{30}=1M$$

13. $Cu_2S + KMnO_4 \longrightarrow Cu^{2+}+SO_2+Mn^{2+}$

$nf=8 \quad nf=5$

Equivalent of $KMnO_4$ = Equivalent of Cu_2S

$5\times n=8\times1.25$

$n=2$

14. $$M_{NaOH}\ \text{resultant}=\frac{20\times0.2+35\times0.1}{100}=0.075\ M$$

Milli-equivalent of NaOH = milli-equivalent of $H_2C_2O_4$

$$40\times0.075=\frac{x}{90}\times2\times1000$$

$x=0.135$ gram

$$\%\ \text{of } H_2C_2O_4 \text{ in a sample}=\frac{x}{0.15}\times100=90\ ;\ \%\ \text{impurity}$$

$(x)=10;$

$$\therefore \frac{x}{2}=5$$

15. $\underset{a\text{ mole}}{Na_2CO_3}+\underset{a\text{ mole}}{NaHCO_3}$

Eq. Na_2CO_3 = Eq. HCl

$$a\times1=0.2\times\frac{20}{1000}=0.004$$

Eq. $NaHCO_3$ = Eq. NaOH

$$0.008 \times 1 = 1 \times \frac{V_{mL}}{1000}$$

$$V_{mL} = 8 \text{ mL}$$

■ Numerical Problems

1. m. eqn. of $KMnO_4$ = m. eqn. of oxalic acid

$$10 \times 0.05 \times 5 = 10 \times M \times 2; \quad M = 0.125,$$

Strength of $H_2C_2O_4 \cdot 2H_2O = 0.125 \times 126$

$$= 15.75 \text{ g/L}$$

2. $\overset{+2}{M}O + Ba(MnO_4)_2 \xrightarrow{H^+} Mn^{2+} + M^{4+}$

$$\frac{0.8}{m+16} \times 2 = \frac{0.2 \times 8}{1000} \times 10;$$

where m is atomic mass of metal

$$m + 16 = 100$$

$$m = 84$$

3. When $KHC_2O_4 \cdot 3H_2C_2O_4$ is reacted as reducing agent

$$5 \times \frac{40 \times 1.0}{1000} = 8x \qquad \ldots\text{(i)}$$

When $KHC_2O_4 \cdot 3H_2C_2O_4$ is reacted as acid

$$2 \times 1 \times \frac{30}{1000} = 1 \times 3y \qquad \ldots\text{(ii)}$$

(i) ÷ (ii)

$$\frac{8x}{3y} = \frac{20}{6} \Rightarrow \frac{x}{y} = \frac{5}{4} = 1.25$$

4. m. eqn. of NH_3 + m. eqn. of $Ca(OH)_2$ = m. eqn. of H_2SO_4

$$n_{NH_3} \times 1 + 0.5 \times \frac{5}{1000} \times 2 = 1 \times \frac{10}{1000} \times 2$$

$$n_{NH_3} = n_N = 0.015$$

$$W_{NH_3} = 0.015 \times 14 = 0.21;$$

$$\text{Mass \% of } N = \frac{0.21}{1.0} \times 100 = 21.00$$

5. m. eqn. of $KMnO_4$ = m. eqn. of $Fe_{0.94}O$;

$$Fe_{0.94}O \longrightarrow Fe^{3+}$$

$$0.82 \times 5 \times V \text{ (in L)} = \frac{34.32}{(56 \times 0.94 + 16)} \times 0.82$$

$$V_{KMnO_4} = 0.1 \text{ L} = 100 \text{ mL}$$

6. m. eqn. of HCl = m. eqn. of $M_3(PO_4)_2$

$$10 \times 0.1 = \left(\frac{0.0517}{E + 31.66}\right) \times 1000$$

where $E_{PO_4^{3-}} = \frac{95}{3} = 31.66$

$$E = 20.04 \approx 20.00$$

7. $2 \times 3 + 6 \times 2 + 2 \times 1 = 5 \times n_{KMnO_4}$

$$n_{KMnO_4} = \frac{20}{5} = 4$$

8. $KIO_3 + 5KI + 6HCl \longrightarrow 3I_2 + 6KCl + 3H_2O$

$$I_2 + 2Na_2S_2O_3 \longrightarrow 2NaI + Na_2S_4O_6$$

Moles of $Na_2S_2O_3$ = 6 × moles of KIO_3

$$n_{hypo} = \frac{2}{5} \times V(L) = \frac{1.07}{214} \times 3 \times 2$$

$$V = 0.075 \text{ L} = 75 \text{ mL}$$

9. Total m. eqn of oxalic acid $= 50 \times 0.5 \times 2 = 50$

m. eqn of $KMnO_4$ (in 25 mL) $= 30 \times 0.025 \times 5$

m. eqn of $KMnO_4$ reacted with 250 mL oxalic acid left

$$= 30 \times 0.02 \times 5 \times \left(\frac{250}{25}\right) = 30$$

m. eqn. of MnO_2 reacted with oxalic acid

$$= 50 - 30 = 20$$

$$\frac{W}{87} \times 2 \times 1000 = 20$$

$$W = 0.87 \text{ g}$$

$$\text{\% of } MnO_2 = \frac{0.87}{2.175} \times 100 = 40\%$$

10. Let m. moles of NaOH = x,

m. moles of $Na_2CO_3 = y$

When used HPh : m. moles of NaOH + m. moles of Na_2CO_3 = m. moles of HCl

$$x + y = 17.5 \times \frac{1}{10}$$

$$x + y = 1.75$$

After end point when MeOH is added : $NaHCO_3$ converted into H_2CO_3

$$\therefore \quad y = 2.0 \times \frac{1}{10}$$

$$y = 0.2$$

wt. of $Na_2CO_3 = 0.2 \times 106 \times 10^{-3} = 0.0212$ g

wt. % of Na_2CO_3 in sample

$$= \frac{0.0212}{0.4} \times 100 = 5.30\%$$

11. $2\,\overset{0}{H}CHO \xrightarrow{OH^-} H\overset{+2}{C}OO^- + \overset{-2}{C}H_3OH$ ($-2e^-$ to $HCOO^-$; $+2e^-$ to CH_3OH)

$$nf = \frac{2}{2} = 1 \quad ; \quad \text{Eq. wt.} = \frac{\text{mol wt.}}{nf} = \frac{30}{1}$$

12. $MH_2O_2 = \frac{2.27}{11.35} = 0.2$

$$x_{MnO_4^-} = x_{H_2O_2} + x_{FeC_2O_4}$$

$$\frac{5 \times 2 \times 200 \times 0.2}{1000} = \frac{2 \times 50 \times 0.2}{1000} + \frac{3 \times 10 \times V}{3 \times 1000}$$

$$V = 38 \text{ mL}$$

13. $Fe_2(C_2O_4)_3 : x$ mole ; $H_2C_2O_4$: x mole

$n_{C_2O_4^{2-}} = 3x + x = 4x$

Eq^n. of $C_2O_4^{2-}$ = Eq^n. of $K_2Cr_2O_7$

$$2 \times 4x = 6 \times 12$$

$$x = 9$$

Eq^n. of NaOH = Eq^n. of $H_2C_2O_4$

$$1 \times n = 2 \times 18$$

$$n = 36$$

14. Let moles of $Na_2C_2O_4 = x, KHC_2O_4 \cdot H_2C_2O_4 = y$

$x \times 3 = 0.1 \times V$ ...(i)

$y \times 2 + x \times 4 = 0.1 \times V \times 5$ ...(ii)

$$\frac{y}{x} \times 2 + \frac{x}{x} \times 4 = \frac{x}{x} \times 3 \times 5$$

$$\frac{y}{x} \times 2 + 4 = 15$$

$$\frac{y}{x} = \frac{11}{2}$$

15. Temporary hardness is due to $Mg(HCO_3)_2$ and $Ca(HCO_3)_2$

2×10^{-3} mole $Ca(HCO_3)_2$

Mole of $CaCO_3$ produced = moles of $Mg(HCO_3)_2$ + moles of $Ca(HCO_3)_2$

Mole of $CaCO_3$ produced $= 1 \times 10^{-3} + 2 \times 10^{-3} = 3 \times 10^{-3}$

Wt. of $CaCO_3 = 3 \times 10^{-3} \times 100 = 0.3$ g

$$\text{Hardness} = \frac{0.3}{1000} \times 10^6 = 30 \text{ ppm}$$

CHAPTER 8

Chemical Kinetics & Nuclear Chemistry

Rate of Appearance or Disappearance of a Substance C

(i) Average $\left(\pm\frac{\Delta[c]}{\Delta t}\right)$ (ii) Instantaneous $\left(\pm\frac{d[c]}{dt}\right)$

- **Expressions of the rate:**

For a general reaction: $aA + bB \longrightarrow cC + dD$,

Rate of disappearance of $A = -\frac{d[A]}{dt}$; Rate of disappearance of $B = -\frac{d[B]}{dt}$

Rate of appearance of $C = \frac{d[C]}{dt}$ & Rate of appearance of $D = \frac{d[D]}{dt}$

The positive sign shows that concentrations of C and D increases with time and the negative sign is indicating that concentrations of A and B decrease with time.

Instantaneous rate of reaction : $-\frac{1}{a}\frac{d[A]}{dt} = -\frac{1}{b}\frac{d[B]}{dt} = \frac{1}{c}\frac{d[C]}{dt} = \frac{1}{d}\frac{d[D]}{dt}$

Average rate of reaction: $-\frac{1}{a}\frac{\Delta[A]}{\Delta t} = -\frac{1}{b}\frac{\Delta[B]}{\Delta t} = \frac{1}{c}\frac{\Delta[C]}{\Delta t} = \frac{1}{d}\frac{\Delta[D]}{\Delta t}$

$$\text{Rate of reaction (ROR)} = \frac{\text{Rate of disappearance of reactant}}{\text{Stoichiometry coefficient of reactant}}$$

$$= \frac{\text{Rate of appearance of product}}{\text{Stoichiometry coefficient of product}}$$

- **Units of Rate of Reaction:** $= \text{mol L}^{-1}\text{s}^{-1}$ or $\text{mol L}^{-1}\text{ min}^{-1}$ (concentration time^{-1}).

Rate Law and Rate Constant

$$\text{Rate} \propto [A]^a.[B]^b \text{ or Rate} = k[A]^a[B]^b$$

The constant of proportionality, k is known as the rate constant (specific reaction rate) and may be defined as the rate at unit concentrations of the reactants.

- k depends on the temperature and is independent of the concentration of the reactants.
- At a fixed temperature, k is a constant and is a characteristic of the reaction. Larger value of k indicates fast reaction and small k indicates slow reactions.

Molecularity

It is always a whole number (not zero) and never a fraction. Its value does not exceed 3 & it has no meaning for a complex reaction.

Order of Reaction

$$aA + bB \longrightarrow cC$$

$$\text{Rate of reaction} = k[A]^m[B]^n$$

Order of reaction $= m + n$ & the order w.r.t. A, B are m, n respectively.

- The order of reaction is obtained from the experimentally determined rate and may be zero, integral or a fraction. In a multi-step complex reaction, the order of the reaction can be determined with the help of slowest step, which is called rate determining step.
- In elementary reaction stoichiometric coefficient of reactants is equal to order of reaction :$m + n = a + b$.

Order	Differential rate law	Integrated rate law	Straight line plot	Half life	Units of k	Example
0	$-d[R]/dt = k$	$kt = [R]_0 - [R]$	$[R]_0$ vs t	$[R]_0/2k$	conc time^{-1} or mol $L^{-1}s^{-1}$	$H_2(g)+Cl_2(g) \xrightarrow{h\nu} 2HCl(g)$ $2HI(g) \xrightarrow[\text{surface}]{Au} H_2(g)+I_2(g)$
1	$-d[R]/dt = k[R]$	$[R] = [R]_0e^{-kt}$ $kt = \ln\{[R]_0/[R]\}$	ln $[R]$ vs t	$(\ln 2)/k$ or $\frac{0.693}{k}$	time^{-1} or s^{-1}	Decomposition of H_2O_2 in aqueous solution Radioactive disintegration
2	$-d[R]/dt = k[R]^2$	$kt = \frac{1}{[R]} - \frac{1}{[R]_0}$	$\frac{1}{[R]}$ vs t	$\frac{1}{k[R]_0}$	$mol^{-1}L\ s^{-1}$	$CH_3COOC_2H_5 + NaOH \downarrow CH_3COONa + C_2H_5OH$ $C_2H_4 + H_2 \xrightarrow{100°C} C_2H_6$

Important Graphical Representation

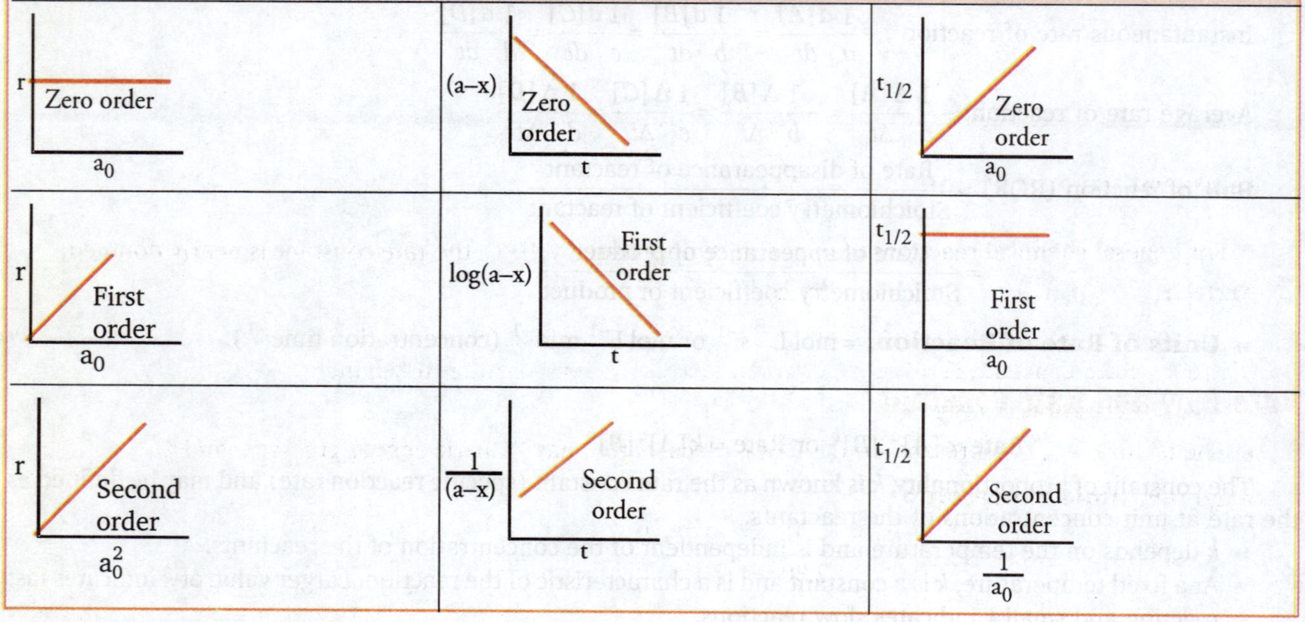

where a_0 = Initial Concentration of reactant, r = rate of reaction

- Some useful relationships between times for different fractions of reaction of first order to complete

$$t_{3/4} \text{ or } t_{75\%} = 2t_{1/2}$$
$$t_{87.5\%} = 3t_{1/2}$$
$$t_{99.9\%} \cong 10t_{1/2}$$

- Amount of the substance left after n half-lives $= \dfrac{A_0}{2^n}$

n^{th} Order Reaction

$A \rightarrow$ Product

$$k \cdot t = \frac{1}{n-1}\left\{\frac{1}{(a-x)^{n-1}} - \frac{1}{a^{n-1}}\right\} \qquad [n \neq 1, n = \text{order}]$$

$$t_{1/2} = \frac{1}{k(n-1)}\left[\frac{2^{n-1}-1}{a^{n-1}}\right]$$

- **Reversible reaction** $A \underset{k_b}{\overset{k_f}{\rightleftharpoons}} B$

$$(k_f + k_b) = \frac{1}{t}\ln\left(\frac{x_{eq}}{x_{eq}-x}\right) \quad ; \quad \text{At equilibrium } k_{eq} = \frac{k_f}{k_b} = \frac{[B]}{[A]} = \frac{x_{eq}}{a - x_{eq}}$$

Parallel/Side Reactions

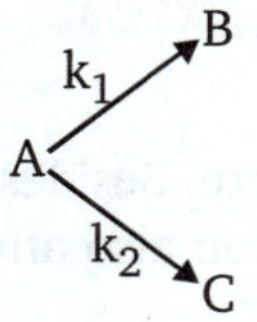

$$; \ (k_1 + k_2)t = \ln\frac{[A]_0}{[A]_t} \quad ; \quad \frac{[B]}{[C]} = \frac{k_1}{k_2}$$

$$\% \text{ Yield of } B = \frac{k_1}{k_1 + k_2} \times 100 \text{ and } \% \text{ Yield of } C = \frac{k_2}{k_1 + k_2} \times 100$$

Series/Consecutive Reactions

$$A \longrightarrow B \longrightarrow C$$

$$t_{max} = \frac{1}{k_1 - k_2}\ln\left(\frac{k_1}{k_2}\right)$$

Temperature coefficient (μ) $= \dfrac{\boldsymbol{k_{T+10}}}{\boldsymbol{k_T}}$

- For general chemical reactions with rise in temperature by 10°C, the rate constant is nearly doubled.

Arrhenius Equation

$$k = Ae^{-E_a/RT}$$

where k = Rate constant, A = Pre-exponential factor, T = Temperature in kelvin,

E_a = Activation Energy

- The factor $e^{-E_a/RT}$ represents fraction of molecules that have kinetic energy greater than E_a
- **Logarithmic expression**

$$\log_{10}\left(\frac{k_2}{k_1}\right) = \frac{E_a}{2.303R}\left(\frac{1}{T_1} - \frac{1}{T_2}\right)$$

- Energy profile for exothermic process & endothermic process:

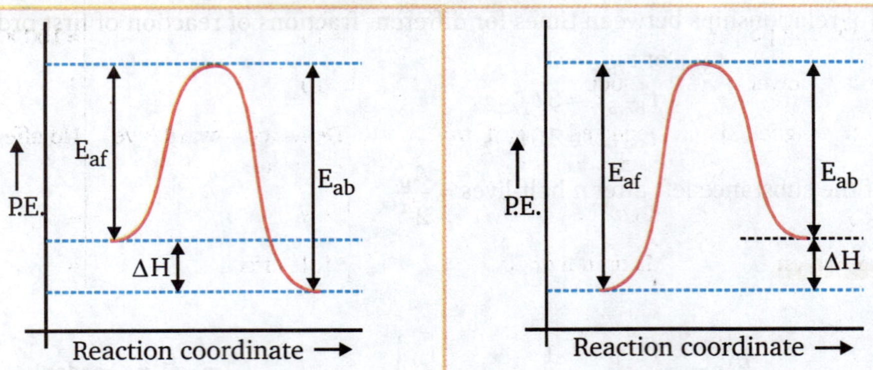

$\Delta H = E_{af} - E_{ab} = -ve$ ($\because E_{af} < E_{ab}$) — Exothermic process

$\Delta H = E_{af} - E_{ab} = +ve$ ($\because E_{af} > E_{ab}$) — Endothermic process

- Threshold Energy = Activation Energy + Energy possessed by reactant molecules.
- **Activation energy diagram for two step endothermic reaction:**

(A) (i) I step-Slow (ii) II step-Fast

(B) (i) Rapid equilibrium (ii) Slow

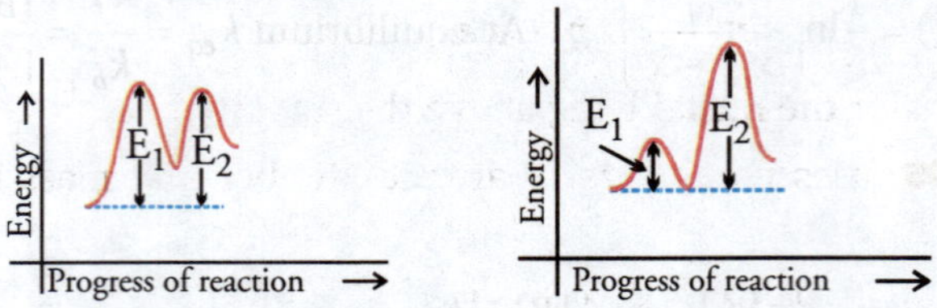

Factors affecting reaction rates : Concentration of reactants and Reaction temperature. Besides these, presence of catalyst and surface area (if a reactant or a catalyst is a solid) exposure to radiation also affect the reaction rates.

Nuclear Chemistry

Nuclear chemistry is the study of the properties and reactions of atomic nuclei. In nuclear reaction only the nuclides (nuclide : the nucleus of a specific isotope) participates and overall (mass + energy) remains conserved.

Radioactivity

The property of a nucleus emitting radiations like α, β and γ is known as radioactivity and the substance possessing the property is called a radioactive substance. The emission of these particles takes place because of the instability of the nucleus. Radioactivity is a property of nucleus.

Properties of α, β-particles and γ-rays.

Properties	Alpha	Beta	Gamma
Nature	Fast moving He nuclei (He^{2+})	Fast moving electrons	High energy electromagnetic
Representation	${}_2^4He$ or α	${}_{-1}^{0}e$ or ${}_{-1}^{0}\beta$	γ or ${}_0^0\gamma$
Charge	2 unit (+ve)	1 unit (–ve)	No charge
Mass	6.65×10^{-24} g/particle	9.11×10^{-28} g/particle	0
Velocity	1/10 of light (min.)	nearly velocity of light	Same as light waves (max.)

Properties	Alpha	Beta	Gamma
Relative penetrating power	1	≈ 100	≈ 10000
Relative ionising power	≈ 10000	≈ 100	1
Effect of electromagnetic field	Deflected toward –ve electrode	Deflected toward +ve electrode	No effect
Kinetic energy	High	Low	–
Effect on ZnS Screen	Maximum effect	Little effect	–

Types of Radioactive Decay

1. **α-decay** causes decrease of atomic number of 2 units and mass number by 4 units, *e.g.*,

$$^{215}_{84}Po \longrightarrow {}^{211}_{82}Pb + {}^{4}_{2}He$$

All nucleides with atomic number greater than 83 are beyond the band of stability and are radioactive.

2. **β-decay** causes increase of atomic number by 1 unit and no change in mass number, *e.g.*,

$$^{228}_{88}Ra \longrightarrow {}^{228}_{89}Ac + {}^{0}_{-1}e\ ; \quad {}^{14}_{6}C \longrightarrow {}^{14}_{7}N + {}^{0}_{-1}e$$

A neutron is converted to proton in this process.

$$^{1}_{0}n \longrightarrow {}^{1}_{1}p + {}^{0}_{-1}e$$

Such emission occurs for the nuclei lying above the stability belt.

3. **γ-ray** emission ($^{0}_{0}\gamma$) causes no change in atomic number and mass number, since it represents the energy lost, *e.g.*,

$$^{60m}_{28}Ni \longrightarrow {}^{60}_{28}Ni + {}^{0}_{0}\gamma$$

4. **Positron Emission :** Causes decrease the atomic number by 1 unit, *e.g.*,

$$^{15}_{8}O \longrightarrow {}^{15}_{7}N + {}^{0}_{+1}e\ ; \quad {}^{118}_{54}Xe \longrightarrow {}^{0}_{+1}e + {}^{118}_{53}I$$

A proton is converted to neutron and positron in this process.

$$^{1}_{1}p \longrightarrow {}^{1}_{0}n + {}^{0}_{1}e$$

Such emission occurs for the nuclei lying below the stability belt.

- Positron ($^{0}_{+1}e$) is a particle having the same mass as an electron, but positively charged.

5. **Electron-capture :**

Causes decrease the atomic number by 1 unit, *e.g.*,

$$^{204}_{84}Po + {}^{0}_{-1}e \longrightarrow {}^{204}_{83}Bi\ ; \quad {}^{40}_{19}K + {}^{0}_{-1}e \longrightarrow {}^{40}_{18}Ar$$

A proton is converted to neutron in this process.

$$^{1}_{1}p + {}^{0}_{-1}e \longrightarrow {}^{1}_{0}n$$

Electron capture occurs with the nuclei lying below the stability belt, in which an electron from the K-shell is captured by the nucleus.

Group Displacement Law (Soddy-Fajans)

In an α-particle emission, the resulting element has a mass number less by four units and atomic number less by two units and so lies two places to the left in the periodic table.

In a β-particle emission the resulting element has the same mass number but has an atomic number greater by one unit and so lies one place to the right in the periodic table.

Neutron/proton ratio and stability zone

- For atomic number < 20, most stable nuclei have *n: p* ratio nearly 1 : 1 (except H & Ar)
- For *n/p* ratio > 1.5, nucleus is unstable. Largest stable nucleus is $^{208}_{82}Pb$ for which *n/p* ratio is 1.53.
- For atomic number > 82, there are no stable nuclei.

Note: Heaviest stable nuclide is $^{208}_{82}Pb$ $t_{1/2}$ for $_{83}Bi = 1.9 \times 10^{19}$ yr., which is α-emitter reference. Pierre De Marcillae exp det.....

nature, 422, 876-878, 2003

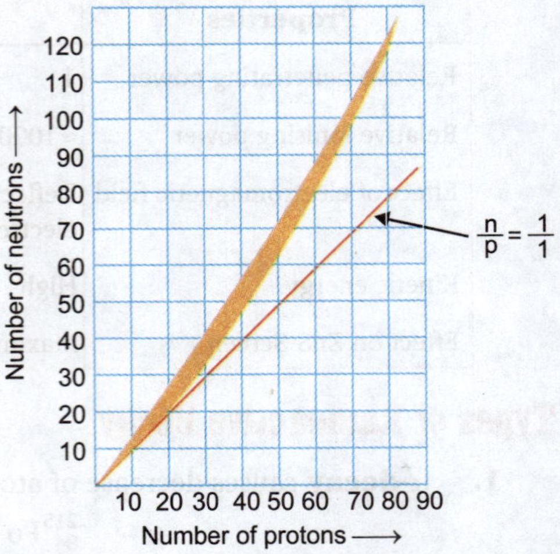

Magic numbers and nuclear stability

Nuclei with 2, 8, 20, 28, 50, 82 or 126 protons or neutrons are unusually stable and have a larger number of stable isotopes than neighboring nuclei in the periodic table. These numbers are called magic numbers. They are supposed to represent completely filled nuclear shells of energy levels.

- Nuclei with magic number of protons as well as neutrons have notably high stabilities. [*e.g.*, ^4_2He, $^{16}_8O$, $^{40}_{20}Ca$ and $^{208}_{82}Pb$].

Even Odd Theory

Most naturally occurring nuclides have even numbers of neutrons and even numbers of protons. 165 such stable nuclei are known. There exist 55 known nuclei with even number of protons and odd number of neutrons, and 50 known stable nuclei with odd number of protons and even number of neutrons. On the other hand, the number of known stable nuclei having odd numbers of both neutrons and protons is only 5.

Artificial nuclear reactions

The first artificial transmutation was carried out by Rutherford in 1919 who bombarded nitrogen gas with alpha particles and obtained hydrogen and oxygen.

$$^{14}_7N + ^4_2He \longrightarrow ^{17}_8O + ^1_1H$$

1. Alpha particle induced or (α, *n*) type reactions :

$$^9_4Be + ^4_2He \longrightarrow ^{12}_6C + ^1_0n$$

Since α-particle is used and a neutron is produced, the reaction may be termed as (α,*n*) reaction. In another α-bombardment nuclear reaction, the isotope ($^{30}_{15}P$) produced is itself radioactive.

$$^{27}_{13}Al + ^4_2He \longrightarrow ^{30}_{15}P + ^1_0n$$

2. Deuteron-induced or (*D*, α) type reaction :

$$^{16}_8O + ^2_1H \longrightarrow ^{14}_7N + ^4_2He$$

3. Proton-induced or (*p*, γ) reaction :

$$^{14}_7N + ^1_1H \longrightarrow ^{15}_8O + \gamma$$

4. Neutron-induced or (*n*, γ) reaction :

$$^{31}_{15}P + ^1_0n \longrightarrow ^{32}_{15}P + \gamma$$

Radioactive decay

- Radioactive decay is a first order process. The rate of nuclear decay is determined from measurements of the activity (*A*) of a sample.

Hence $$-\frac{dN}{dt} = \lambda N \quad \text{or} \quad N = N_0 e^{-\lambda t}$$

where N = number of radioactive nuclei at any time t ; N_0 = number of radioactive nuclei at $t = 0$; λ = decay constant.

S.I. units : The SI unit is becquerel (Bq).

1 disintegration per second $(dps) = 1Bq$.

Other units : 1 curie (Ci) $= 3.7 \times 10^{10}$ *dps*, 1 Rutherford (*Rd*) $= 10^6$ *dps*.

- **Specific activity :** Activity per unit mass of radioactive sample (dps/g)
- **Half life ($t_{1/2}$) :** The time taken by half the nuclei (originally present) to decay. $t_{1/2} = 0.693/\lambda$

Note : After n half-lives have passed, activity is reduced to $\frac{1}{2^n}$ of its initial value.

- **Average life (t_{av}) :** $t_{av} = 1/\lambda = 1.44\, t_{1/2}$

Radioactive equilibrium

Among the members of a decay chain, the state which prevails when the ratios between the activities of successive members remain constant. (This is not an equilibrium in the strict sense since radioactive decay is an irreversible process).

Secular equilibrium

Radioactive equilibrium where the half life of the intermediate isotope is so long that the change of its activity can be ignored during the period of interest and all activities remain constant.

$$A \xrightarrow{\lambda_A} B \xrightarrow{\lambda_B} C$$

Number of nuclei of B is max. at t_{max} ; $\quad t_{max} = \frac{1}{(\lambda_1 - \lambda_2)} \ln\left(\frac{\lambda_1}{\lambda_2}\right)$

$$\frac{dN_B}{dt} = \lambda_A N_A - \lambda_B N_B.$$

Secular equilibrium occurs when $\frac{dN_B}{dt} = 0$ or $\frac{N_B}{N_A} = \frac{\lambda_A}{\lambda_B}$

- Parallel decay $A \begin{cases} \xrightarrow{\lambda_1} B \\ \xrightarrow{\lambda_2} C \end{cases}$

$$\% \text{ of } B = \left(\frac{\lambda_1}{\lambda_1 + \lambda_2}\right) \times 100$$

$$\% \text{ of } C = \left(\frac{\lambda_2}{\lambda_1 + \lambda_2}\right) \times 100$$

Applications of Radioactivity :

- **Age of Minerals and Rocks :** We can determine the age of the rock by measuring the ratio of $^{238}_{92}U$ and $^{206}_{82}Pb$.

$$^{238}_{92}U \longrightarrow \ ^{206}_{82}Pb + 8\,^{4}_{2}He + 6\,^{0}_{-1}e$$

At time t $\quad$ w gram $\quad$ y gram

Age of Rock, $\quad t = \frac{1}{\lambda} \ln\left(\frac{w_0}{w}\right)$ where $w_0 = w + x$,

w_0 = Initial mass of $^{238}_{92}U$; x = mass of $^{238}_{92}U$ disintegrated. $x = \frac{238}{206} \times y$

- **Radiocarbon Dating :** $^{14}_{6}C$ is used to date historical artifacts.

Age of wood artifacts , $t = \frac{1}{\lambda} \ln\left(\frac{A_0}{A}\right)$ (where $t_{1/2}$ of $^{14}_{6}C = 5770$ years)

A_0 = Activity of fresh wood of plant/tree. ; A = Activity of old wood.

- Uses in Medicines and other Areas

Nuclear Fission

In a nuclear fission reaction, a heavy nucleus splits up into two main fragments of lighter nuclei and several neutrons.

Of the three natural isotopes of uranium ($^{238}_{92}U$, $^{235}_{92}U$ and $^{234}_{92}U$) the $^{235}_{92}U$ nucleus undergoes nuclear fission when bombarded with slow neutrons. The $^{236}_{92}U$ formed breaks up in several different ways, for example :

$$^{235}_{92}U + ^{1}_{0}n \longrightarrow ^{236}_{92}U \longrightarrow ^{140}_{56}Ba + ^{93}_{36}Kr + 3\,^{1}_{0}n$$

$$^{235}_{92}U + ^{1}_{0}n \longrightarrow ^{236}_{92}U \longrightarrow ^{144}_{54}Xe + ^{90}_{38}Sr + 2\,^{1}_{0}n$$

$$^{235}_{92}U + ^{1}_{0}n \longrightarrow ^{236}_{92}U \longrightarrow ^{144}_{54}Cs + ^{90}_{37}Rb + 2\,^{1}_{0}n$$

- The key to the liberation of energy in the nuclear fission reaction is the production of two or more neutrons per reaction initiated by one neutron.
- Atom bomb is based on nuclear fission.

A nuclear fission chain reaction has three general steps :

1. Initiation : The reaction of a single atom is needed to start the chain reaction Fission of $^{235}_{92}U$ is initiated by the absorption of a neutron.

2. Propagation : This part of the process repeats itself over and over, with each step yielding more product. The fission of $^{236}_{92}U$ releases neutrons that initiate the fission of other uranium atoms.

3. Termination : Eventually, the chain will end. Termination could occur if the reactant ($^{235}_{92}U$) is used up, or if the neutrons that continue the chain escape from the sample without being captured by $^{235}_{92}U$.

Nuclear Fusion

Just as the fission of heavy nuclei is accompanied by mass losses resulting into the liberation of large amounts of energy, the fusion of light nuclei is accompanied by mass losses and the evolution of large quantities of energy. Some such reactions and energy release in each process are shown below,

Fusion reaction	Mass loss	Energy released (kJ mol^{-1})
$^{2}_{1}H + ^{2}_{1}H \longrightarrow ^{4}_{2}He$	0.026	2.3×10^{9}
$^{2}_{1}H + ^{3}_{1}H \longrightarrow ^{4}_{2}He + ^{1}_{0}n$	0.018	1.79×10^{9}

Compared with fission reactions, fusion reactions have the advantage that large amounts of highly radioactive nuclides are not obtained as by-products which may pose problem of safe storage.

- Very high temperatures is required for a fusion reaction.
- Hydrogen bomb is based on nuclear fusion.
- **Einstein's mass-energy equation :** Energy changes in nuclear reactions can determined by using Einstein's mass-energy equation $E = mc^2$

Where E is the energy equivalent of mass m and c is the velocity of light.

For a change of 1 amu (atomic mass unit), the corresponding energy change, $E \approx 931 \times 10^6$ eV ≈ 931MeV, *i.e.*, a mass of 1 amu is equivalent to 931 MeV of energy.

Mass Defect

It has invariably been found that the actual mass of an isotope of an element is less than the sum of masses of the protons, neutrons and electrons present in it. This difference is called mass defect. The mass defect is nothing, but the loss of mass during the formation of the nucleus of the isotope.

- **Mass defect in nuclear reaction :**

  Δm = mass of nuclei of reactants – mass of nuclei of products.

  Energy liberated in nuclear reaction : $\Delta E = \Delta m . c^2$.

Mass defect in an isotope formation :

Let m_p, m_n and m_e are the respective masses of proton, neutron and electron. Then, the calculated mass of this isotope,

$M' = Z m_p + Z m_e + (A - Z) m_n = Z m_H + (A - Z) m_n$

$(\because m_p + m_e = \text{Mass of hydrogen atom} = m_n)$

Let M = Actual atomic mass as determined experimentally.

then, Δm = Mass defect $= M' - M$

Binding Energy

Loss of mass during the formation of the nucleus from nucleons is converted into energy. The release of energy imparts stability to the nucleus. The energy released when constituent nucleons combine to form a nucleus, is called binding energy of the nucleus. In other words, energy equal to binding energy will be needed to break up the nucleus into its constituent nucleons. Consequently, the greater the binding energy, the more stable is the nucleus.

Binding Energy and Nuclear Stability

$$\text{B.E. per nucleon} = \frac{\text{B.E.}}{\text{No. of nucleons}}$$

Binding energy per nucleon is a direct indicator of its nuclear stability. Higher the binding energy per nucleon of an isotope, greater is its nuclear stability.

Level-1

1. The differential rate law equation for the elementary reaction $A+2B \xrightarrow{k} 3C$, is :

(a) $-\frac{d[A]}{dt}=-\frac{d[B]}{dt}=\frac{d[C]}{dt}=k[A][B]^2$

(b) $-\frac{d[A]}{dt}=-\frac{1}{2}\frac{d[B]}{dt}=\frac{1}{3}\frac{d[C]}{dt}=k[A]^2[B]$

(c) $-\frac{d[A]}{dt}=-\frac{1}{2}\frac{d[B]}{dt}=\frac{1}{3}\frac{d[C]}{dt}=k[A][B]^2$

(d) None of these

2. The rate of a reaction is expressed in different ways as follows:

$$+\frac{1}{2}\frac{d[C]}{dt}=-\frac{1}{3}\frac{d[D]}{dt}=+\frac{1}{4}\frac{d[A]}{dt}=-\frac{d[B]}{dt}$$

The reaction is:

(a) $4A+B \longrightarrow 2C+3D$ (b) $B+3D \longrightarrow 4A+2C$

(c) $A+B \longrightarrow C+D$ (d) $B+D \longrightarrow A+C$

3. In the reaction, $A+2B \longrightarrow 6C+2D$, if the initial rate $\frac{d[A]}{dt}$ at $t=0$ is 2.6×10^{-2}M sec^{-1}, what will be the value of $-\frac{d[B]}{dt}$ at $t=0$?

(a) $8.5\times10^{-2}\,M\,\text{sec}^{-1}$ (b) $2.5\times10^{-2}\,M\,\text{sec}^{-1}$

(c) $5.2\times10^{-2}\,M\,\text{sec}^{-1}$ (d) $7.5\times10^{-2}\,M\,\text{sec}^{-1}$

4. For the reaction $2A \longrightarrow B+3C$; if $-\frac{d[A]}{dt}=k_1[A]^2$; $\frac{d[B]}{dt}=k_2[A]^2$; $\frac{d[C]}{dt}=k_3[A]^2$

the correct relation between k_1, k_2 and k_3 is :

(a) $k_1=k_2=k_3$ (b) $2k_1=k_2=3k_2$

(c) $4k_1=k_2=3k_2$ (d) $\frac{k_1}{2}=k_2=\frac{k_3}{3}$

5. The rate constant of nth order has units :

(a) litre^{1-n} mol^{1-n} sec^{-1} (b) Mol^{1-n} litre^{1-n} sec

(c) Mol$^{1-n^2}$ litre$^{n^2}$ sec^{-1} (d) Mole^{1-n} litre^{n-1} sec^{-1}

6. Which of the following statement is incorrect?

(a) Unit of rate of disappearance is Ms^{-1}

(b) Unit of rate of reaction is Ms^{-1}

(c) Unit of rate constant k depends upon order

(d) Unit of k for first order reaction is Ms^{-1}

7. Which of the following relation is correct for k_f and k_b in an equilibrium process that contains equal moles of reactants and products.

(a) $k_f=k_b$ (b) $k_f>k_b$

(c) $k_f<k_b$ (d) we cannot predict

8. Listed in the table are forward and reverse rate constants for the reaction

$$2NO(g) \rightleftharpoons N_2(g)+O_2(g)$$

Temperature (K)	$k_f(M^{-1}s^{-1})$	$k_b\ (M^{-1}s^{-1})$
1400	0.29	1.1×10^{-6}
1500	1.3	1.4×10^{-5}

Select the correct statement :

(a) Reaction is exothermic and value of equilibrium constant (K_{eq}) at 1400 K is 3.79×10^{-6}

(b) Reaction is endothermic and value of K_{eq} at 1400 K is 2.63×10^{5}

(c) Reaction is exothermic and value of K_{eq} at 1400 K is 2.63×10^{5}

(d) Reaction is endothermic and value of K_{eq} at 1500 K is 9.28×10^{4}

9. The rate constant of a reaction depends on

(a) temperature

(b) pressure

(c) extent of reaction

(d) initial concentration of the reactant

10. In the following reaction, how is the rate of appearance of the underlined product related to the rate of disappearance of the underlined reactant

$$\underline{BrO_3^-(aq)} + 5Br^-(aq) + 6H^+(aq) \longrightarrow \underline{3Br_2(l)} + 3H_2O(aq)$$

(a) $-\dfrac{d[BrO_3^-]}{dt} = \dfrac{d[Br_2]}{dt}$

(b) $-\dfrac{1}{3}\dfrac{d[BrO_3^-]}{dt} = \dfrac{d[Br_2]}{dt}$

(c) $\dfrac{-d[BrO_3^-]}{dt} = \dfrac{1}{3}\dfrac{d[Br_2]}{dt}$

(d) None of these

11. Consider a reaction $A(g) \xrightarrow{k=0.1\,M\,min^{-1}} 2B(g)$. If initial concentration of A is 0.5M then select correct graph.

(a)

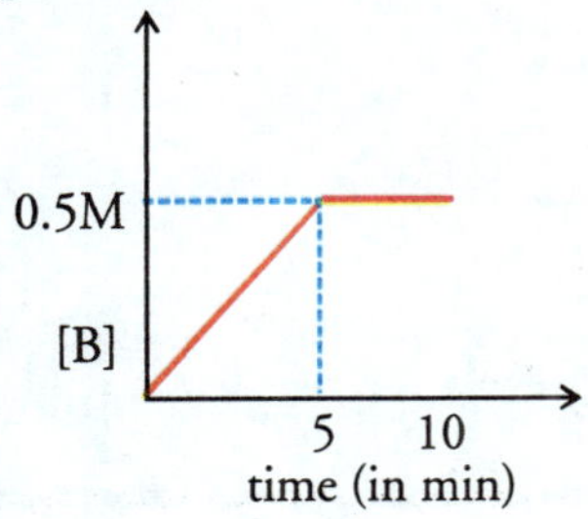

(b)

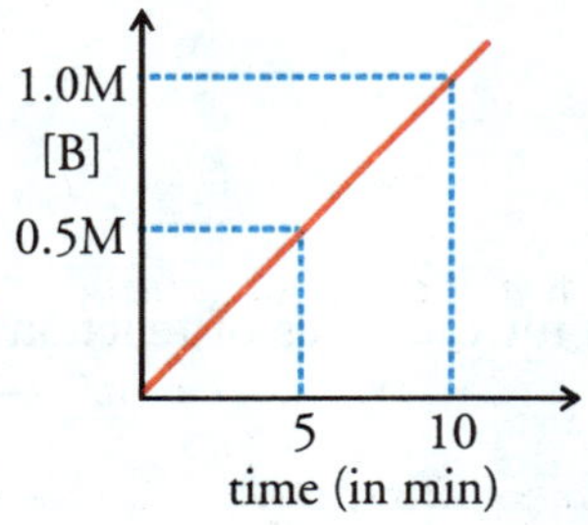

(c)

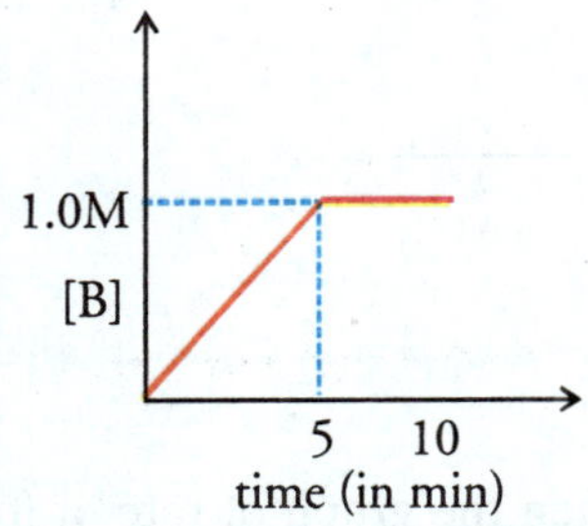

(d)

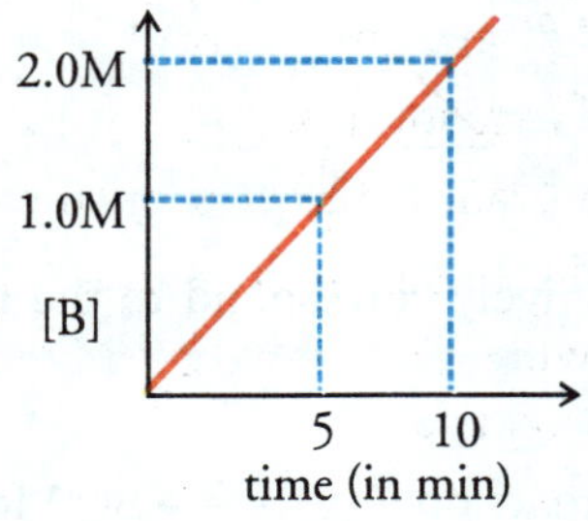

12. Which of the following statements is incorrect?

(a) A second order reaction must be a bimolecular elementary reaction

(b) A bimolecular elementary reaction must be a second order reaction

(c) Zero order reaction must be a complex reaction

(d) First order reaction may be complex or elementary reaction

13. The molecularity of a complex reaction given below is :

$$2N_2O_5(g) \longrightarrow 4NO_2(g) + O_2(g)$$

(a) 1 (b) 2

(c) 3 (d) has no meaning

14. Decomposition of NH_4NO_2 (*aq*) into N_2 (*g*) and $2H_2O$ (*l*) is first order reaction. Which of the following graph is correct?

(a)
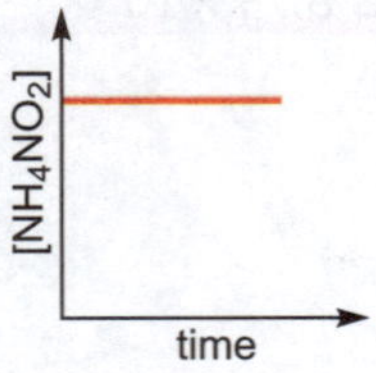

(b)
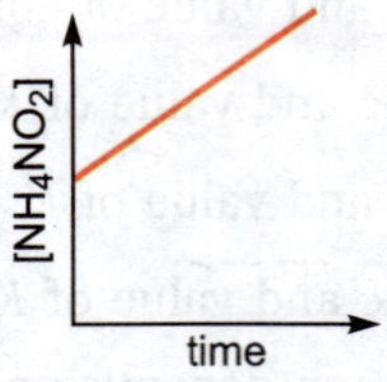

(c)
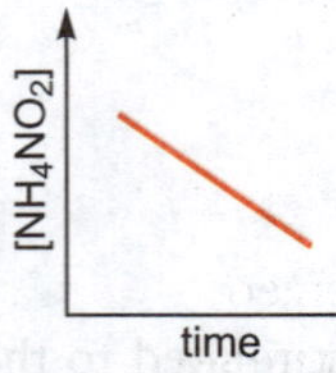

(d)
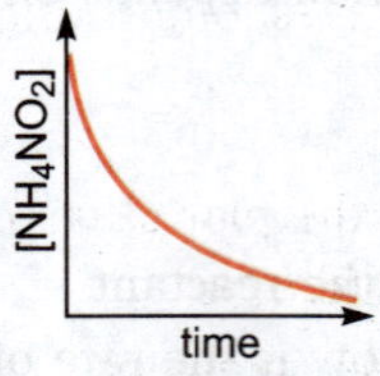

15. Decomposition of HI (*g*) on Gold surface is zero order reaction. Initially few moles of H_2 are present in container then which of the following graph is correct?

(a)
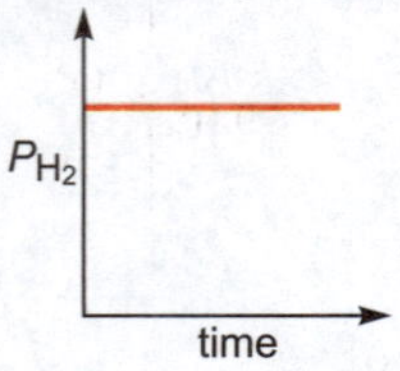

(b)
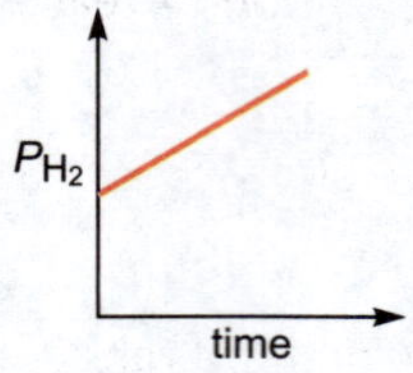

(c)
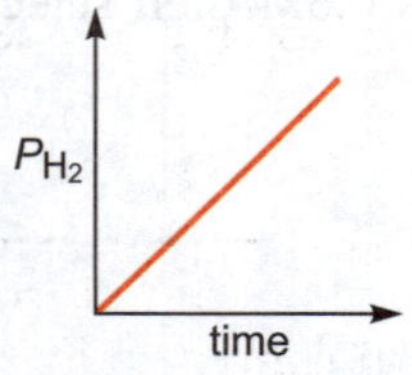

(d)
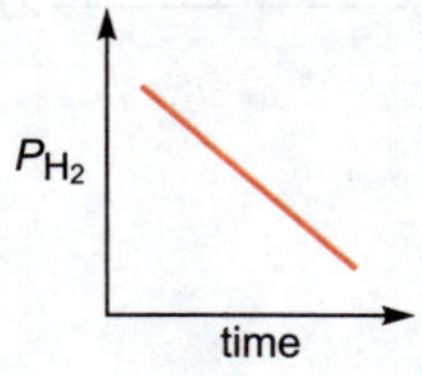

16. Consider the plots for the types of reaction

$$nA \longrightarrow B + C$$

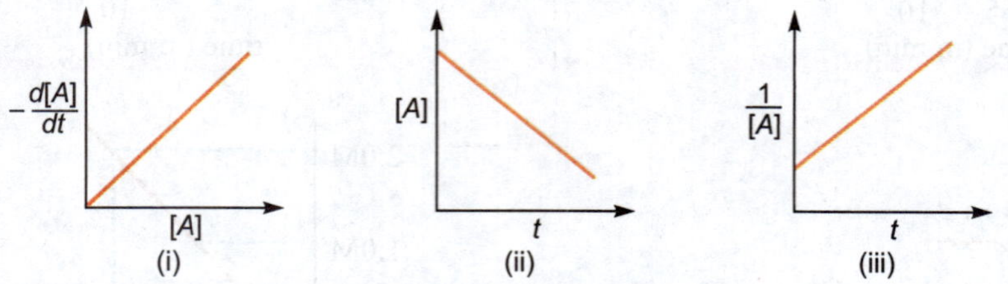

These plots respectively correspond to the reaction orders :

(a) 0, 2, 1 (b) 0, 1, 2

(c) 1, 1, 2 (d) 1,0, 2

17. If decomposition reaction $A(g) \longrightarrow B(g)$ follows first order kinetics then the graph of rate of formation (R) of B against time t will be:

(a)
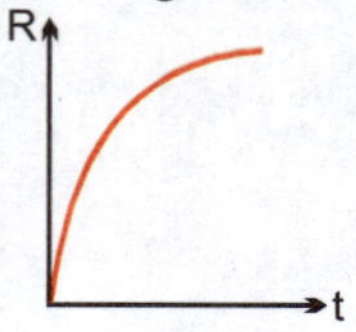

(b)
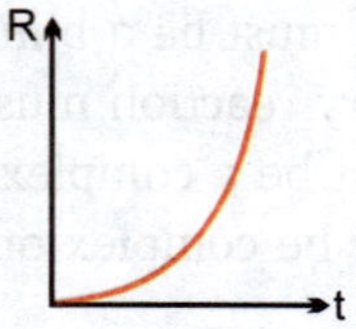

(c) 

(d) 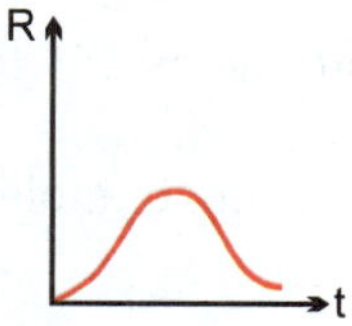

18. Consider the plots, given below, for the types of reaction

$$nA \longrightarrow B + C$$

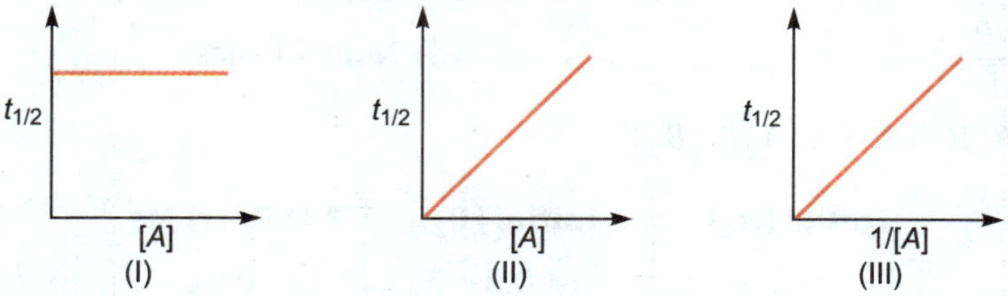

These plots respectively correspond to the reaction orders :

(a) 0, 1, 2
(b) 1, 2, 0
(c) 1, 0, 2
(d) None of these

19. For a zero order reaction, the plot of conc. $(a - x)$ vs time is linear with

(a) + ve slope and zero intercept
(b) – ve slope and zero intercept
(c) + ve slope and non-zero intercept
(d) – ve slope and non-zero intercept

20. What will be the order of reaction for a chemical change having log $t_{1/2}$ *vs* log *a*?

(where a = initial concentration of reactant; $t_{1/2}$ = half-life)

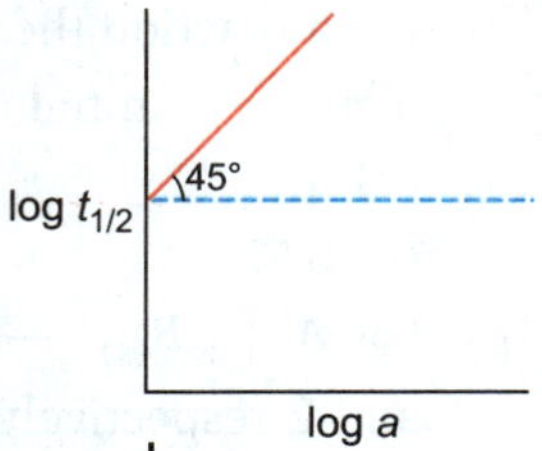

(a) Zero order
(b) First order
(c) Second order
(d) None of these

21. A graph between log $t_{1/2}$ and log a (abscissa), a being the initial concentration of A in the reaction For reaction $A \longrightarrow$ Product, the rate law is :

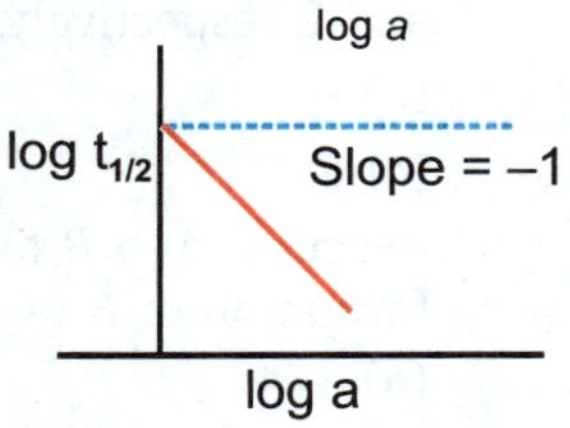

(a) $\frac{-d[A]}{dt} = K$
(b) $\frac{-d[A]}{dt} = K[A]$
(c) $\frac{-d[A]}{dt} = K[A]^2$
(d) $\frac{-d[A]}{dt} = K[A]^3$

22. For the reaction $A \longrightarrow B$, for which graph between half life $(t_{1/2})$ and initial concentration (a) of the reactant is as given below

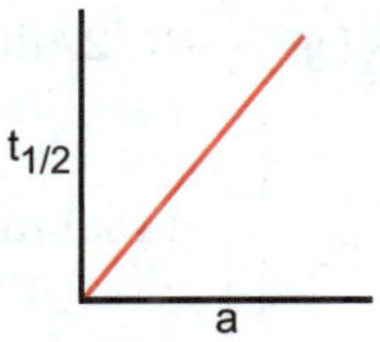

Hence graph between $-\frac{d[A]}{dt}$ and time will be:

(a) 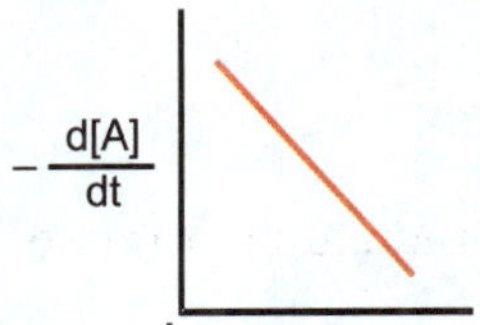

(b) 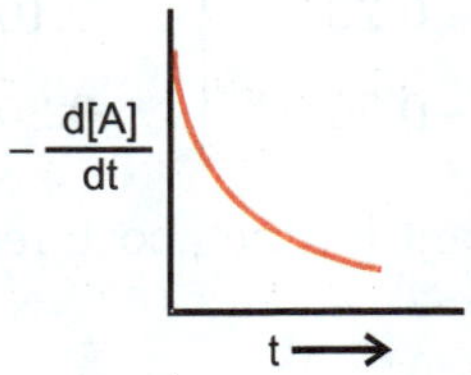

(c) 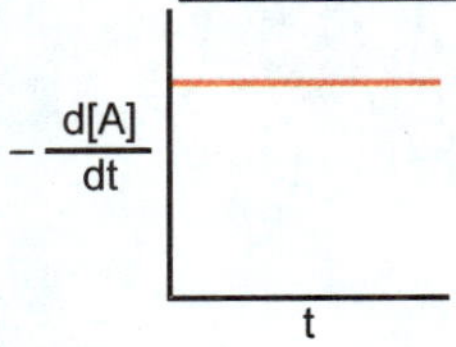

(d) 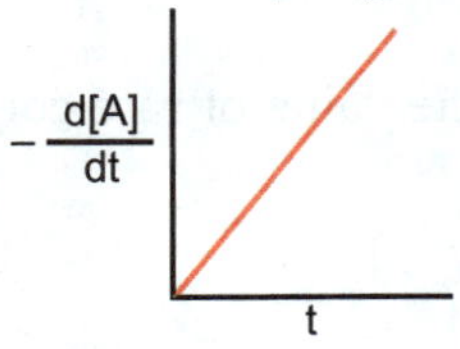

23. For the ideal gaseous reaction, the rate is generally expressed in terms of $\frac{dP}{dt}$ instead of $\frac{dC}{dt}$ or $\frac{dn}{dt}$ (where $C = \frac{n}{V}$ is concentration and n the no. of moles). What is the relation among these three expressions if T and V are constant?

(a) $\frac{dC}{dt} = \frac{dn}{dt} = \frac{dP}{dt}$
(b) $\frac{dC}{dt} = \frac{1}{V}\frac{dn}{dt} = \frac{1}{RT}\left(\frac{dP}{dt}\right)$
(c) $RT\frac{dC}{dt} = \frac{dn}{dt} = \frac{dP}{dt}$
(d) None of these

24. $A_2 + B_2 \longrightarrow 2AB$; R.O.R. $= k[A_2]^a [B_2]^b$

Initial $[A_2]$	Initial $[B_2]$	R.O.R. (r) Ms^{-1}
0.2	0.2	0.04
0.1	0.4	0.04
0.2	0.4	0.08

Order of reaction with respect to A_2 and B_2 are respectively :

(a) $a = 1, b = 1$
(b) $a = 2, b = 0$
(c) $a = 2, b = 1$
(d) None

25. For a reaction the initial rate is given as: $R_0 = k[A_0]^2[B_0]$. By what factor, the initial rate of reaction will increase if initial concentration of A is taken 1.5 times and of B is tripled?

(a) 4.5
(b) 2.25
(c) 6.75
(d) None of these

26. For $A_{(s)} + B_{(s)} \longrightarrow C_{(s)}$; rate $= k[A]^{1/2}[B]^2$, if initial concentration of A and B are increased by factors 4 and 2 respectively, then the initial rate is changed by the factor:

(a) 4
(b) 6
(c) 8
(d) None of these

27. Reaction $A \rightarrow B$ follows second order kinetics. Doubling the concentration of A will increase the rate of formation of B by a factor of :

(a) 1/4
(b) 1/2
(c) 2
(d) 4

28. The reaction of A_2 and B_2 follows the equation

$$A_2(g) + B_2(g) \longrightarrow 2AB(g)$$

The following data were observed

$[A_2]_0$	$[B_2]_0$	Initial rate of appearance of $AB(g)$ (in Ms^{-1})
0.10	0.10	2.5×10^{-4}
0.20	0.10	5×10^{-4}
0.20	0.20	10×10^{-4}

The value of rate constant for the above reaction is:

(a) 2.5×10^{-4}
(b) 2.5×10^{-2}
(c) 1.25×10^{-2}
(d) None of these

29. Above given question, the value of rate constant for appearance of $AB(g)$ is :

(a) 2.5×10^{-4}
(b) 2.5×10^{-2}
(c) 1.25×10^{-2}
(d) None of these

30. The following data pertain to reaction between A and B

S. No	[A] mol L^{-1}	[B] mol L^{-1}	Rate mol L^{-1} sec^{-1}
I	1×10^{-2}	2×10^{-2}	2×10^{-4}
II	2×10^{-2}	2×10^{-2}	4×10^{-4}
III	2×10^{-2}	4×10^{-2}	8×10^{-4}

Which of the following inference(s) can be drawn from the above data

(a) Rate constant of the reaction 10^{-4}

(b) Rate law of the reaction is $k[A][B]$

(c) Rate of reaction increases four times on doubling the concentration of both the reactant, Select the correct answer

codes

(a) a, b and c (b) a and b

(c) b and c (d) c alone

31. The unit of rate constant of elementary reaction depends upon the :

(a) temperature of the reaction (b) concentration of reactants

(c) activation energy of the reaction (d) molecularity of the reaction

32. Select the rate law that corresponds to the data shown for the reaction $A+B \longrightarrow C$

Exp.	[A]	[B]	Rate
1.	0.012	0.035	0.10
2.	0.024	0.070	0.80
3.	0.024	0.035	0.10
4.	0.012	0.070	0.80

(a) Rate= $k[B]^3$ (b) Rate= $k[B]^4$

(c) Rate $k=[A][B]^3$ (d) Rate= $k[A]^2[B]^2$

33. An elementary reaction between A and B is a second order reaction. Which of the following rate equations must be correct?

(a) $r=k[A]^2[B]^0$ (b) $r=k[A]^{3/2}[B]^{1/2}$

(c) $r=k[A]^0[B]^2$ (d) $r=k[A][B]$

34. If 'a' is the initial concentration of the reactant, the half-life period of the reaction of nth order is inversely proportional to :

(a) a^{n-1} (b) a^n

(c) a^{1-n} (d) a^{n+1}

35. Which of the following expressions is correct for zero order and first order respectively [where a is initial concentration]?

(a) $t_{1/2} \propto a; t_{1/2} \propto \frac{1}{a}$ (b) $t_{1/2} \propto a; t_{1/2} \propto a^0$

(c) $t_{1/2} \propto a^0; t_{1/2} \propto a$ (d) $t_{1/2} \propto a; t_{1/2} \propto \frac{1}{a^2}$

36. The unit of rate constant of zero order and first order chemical reactions are respectively :

(a) mol L^{-1} s^{-1}, mol L^{-1} s^{-1} (b) s^{-1}, mol L^{-1} s^{-1}

(c) mol L^{-1} s^{-1}, s^{-1} (d) None of these

37. The units of rate of reaction and rate constant are same for a :

(a) zero order reaction (b) first order reaction

(c) second order reaction (d) third order reaction

38. $CH_3COOC_2H_5(aq) + H_2O(l) \xrightarrow{H^+(aq)} CH_3COOH(aq) + C_2H_5OH(aq)$. What type of reaction is this?

(a) Unimolecular elementary
(b) Pseudo first order
(c) Zero order
(d) Second order

39. When ethyl acetate was hydrolysed in presence of 0.1 M HCl, the rate constant was found to be $5.4 \times 10^{-5}\ s^{-1}$. But in presence of 0.1 $M\ H_2SO_4$ the rate constant was found to be $6.25 \times 10^{-5}\ s^{-1}$. Thus it may be concluded that :

(a) H_2SO_4 furnishes more H^+ than HCl
(b) H_2SO_4 furnishes less H^+ than HCl
(c) both have the same strength
(d) will depend on concentration of ethyl acetate

40. For an elementary reaction $2A + B \longrightarrow A_2B$ if the volume of vessel is quickly reduced to half of its original volume then rate or reaction will

(a) remain unchanged
(b) increase four times
(c) increase eight times
(d) decrease eight times

41. In the reaction $A \rightarrow B + C$, rate constant is 0.001 Ms^{-1}. If we start with 1 M of A then conc. of A and B after 10 minutes are respectively :

(a) 0.5 M, 0.5 M
(b) 0.6 M, 0.4 M
(c) 0.4 M, 0.6 M
(d) none of these

42. For a reaction $A \xrightarrow{k_r = 0.6 M\ min^{-1}} 2B$

starting with 1 M of 'A' only, concentration of B (in M) after 100 sec. and 200 sec. is respectively ?

(a) 2 and 4
(b) 1 and 2
(c) 2 and 3
(d) None of these

43. Half-life ($t_{1/2}$) and completion time (T) of the above reaction ($A \longrightarrow 2B$) are :

(a) 500 min, 750 min
(b) 500 sec, 750 sec
(c) 50 sec, 100 sec
(d) None of these

44.

Column I	Column II
P. Zero order reaction	**1.** $t_{1/2} \propto \frac{1}{[A]_0}$
Q. First order reaction	**2.** $t_{100\%} = [A]_0 / k$
R. Second order reactions	**3.** Involves at least two reactants
S. Pseudo unimolecular reaction	**4.** $[A] = [A]_0 e^{-kt}$

Code:

	P	Q	R	S
(a)	2	1	4	2
(b)	2	4	1	3
(c)	2	1	3	4
(d)	3	2	1	4

45. For the zero order reaction $A \rightarrow B + C$; initial concentration of A is 0.1 M. If $[A] = 0.08\ M$ after 10 minutes, then its half-life and completion time are respectively :

(a) 10 min; 20 min
(b) 2×10^{-3} min; 4×10^{-3} min
(c) 25 min, 50 min
(d) 250 min, 500 min

46. For an elementary reaction, $X(g) \rightarrow Y(g) + Z(g)$ the half life period is 10 min. In what period of time would the concentration of X be reduced to 10% of original concentration ?

(a) 20 Min. (b) 33 Min
(c) 15 Min (d) 25 Min

47. In the presence of an acid, the initial concentration of cane sugar was reduced from 0.20 to 0.10 molar in 5 hours and from 0.2 to 0.05 molar in 10 hours. The reaction is of :
(a) Zero order (b) First order
(c) Second order (d) Third order

48. A first order reaction is 75% completed in 100 minutes. How long time will it take for its 87.5% completion?
(a) 125 min (b) 150 min
(c) 175 min (d) 200 min

49. The rate constant for a first order reaction whose half life is 480 sec, is :
(a) 1.44×10^{-3} sec^{-1} (b) $1.44 \times$ sec^{-1}
(c) 0.72×10^{-3} sec^{-1} (d) 2.88×10^{-3} sec^{-1}

50. Rate constant $k = 2.303$ min^{-1} for a particular reaction. The initial concentration of the reactant is 1 mol/litre then rate of reaction after 1 minute is :
(a) 2.303 M min^{-1} (b) 0.2303 M min^{-1}
(c) 0.1 M min^{-1} (d) none of these

51. For the reaction $3A(g) \xrightarrow{k} B(g) + C(g)$, k is 10^{-4} L/mol.min.

If $[A] = 0.5M$ then the value of $-\frac{d[A]}{dt}$ (in Ms^{-1}) is:
(a) 7.5×10^{-5} (b) 3×10^{-4}
(c) 2.5×10^{-5} (d) none of these

52. 99% of a first order reaction was completed in 32 minutes when 99.9% of the reaction will complete :
(a) 50 min (b) 46 min
(c) 48 min (d) 49 min

53. Which of the following represents the expression for $\frac{3}{4}$th life of first order reaction
(a) $\frac{k}{2.303} \log 4/3$ (b) $\frac{2.303}{k} \log 3/4$
(c) $\frac{2.303}{k} \log 4$ (d) $\frac{2.303}{k} \log 3$

54. Consider the following first order competing reactions:

$X \xrightarrow{k_1} A + B$ and $Y \xrightarrow{k_2} C + D$

If 50% of the reaction of X was completed when 96% of the reaction of Y was completed, the ratio of their rate constants (k_2 / k_1) is:
(a) 4.06 (b) 0.215
(c) 1.1 (d) 4.65

55. For the reactions
(i) $A \xrightarrow{k_I} P$
(ii) $B \xrightarrow{K_{II}} Q$, following observation is made.

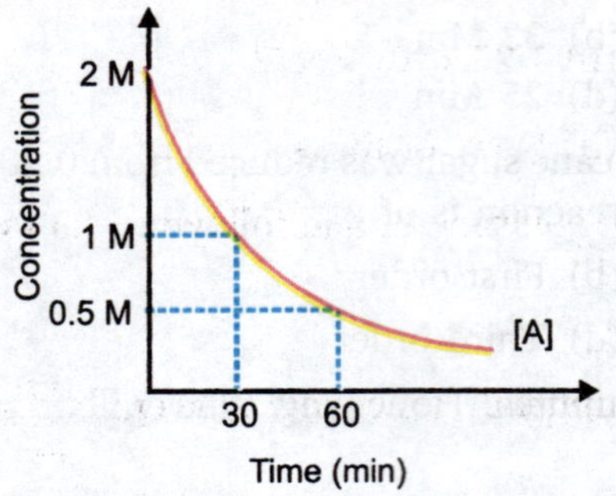

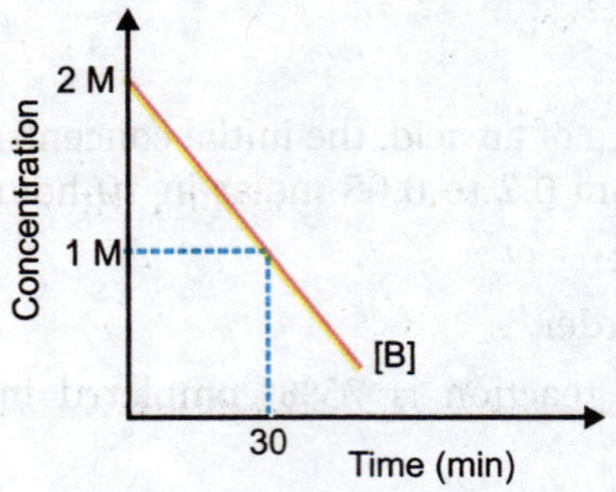

Calculate $\dfrac{k_I}{K_{II}}$, where k_I and k_{II} and rate constant for the respective reaction.

(a) 2.303 (b) 1

(c) 0.36 (d) 0.693

56. The decomposition of N_2O_5 in carbon tetrachloride was followed by measuring the volume of O_2 gas evolved : $2N_2O_5\,(CCl_4) \longrightarrow 2N_2O_4\,(CCl_4) + O_2(g)$. The maximum volume of O_2 gas obtained was 100 cm^3. In 500 minutes, 90 cm^3 of O_2 were evolved. The first order rate constant (in min^{-1}) for the disappearance of N_2O_5 is :

(a) $\dfrac{2.303}{500}$ (b) $\dfrac{2.303}{500}\log\dfrac{100}{90}$

(c) $\dfrac{2.303}{500}\log\dfrac{90}{100}$ (d) $\dfrac{100}{10\times 500}$

57. For a first order homogeneous gaseous reaction $A \longrightarrow 3B$, if pressure after time t was P_T and after completion of reaction, pressure was P_∞ then select correct relation

(a) $k = \dfrac{1}{t}\ln\left(\dfrac{P_\infty}{3(P_\infty - P_t)}\right)$ (b) $k = \dfrac{1}{t}\ln\left(\dfrac{2P_\infty}{(P_\infty - P_T)}\right)$

(c) $k = \dfrac{1}{t}\ln\left(\dfrac{3P_\infty}{2P_\infty - P_t}\right)$ (d) $k = \dfrac{1}{t}\ln\left(\dfrac{2P_\infty}{3(P_\infty - P_T)}\right)$

58. The half-life of first order decomposition of NH_4NO_3 is 2.10 hr at 288 K temperature
$NH_4NO_3(aq) \longrightarrow N_2O(g) + 2H_2O(l)$. If 6.2 g of NH_4NO_3 is allowed to decompose, the time required for NH_4NO_3 to decompose 90% is :

(a) 6.978 hr (b) 0.319

(c) 0.319 hr (d) None of these

59. For a first order homogeneous gaseous reaction, $A \longrightarrow 2B + C$
then initial pressure was P_i while total pressure after time 't' was P_t. The right expression for the rate constants k in terms of P_i, P_t and t is :

(a) $k = \dfrac{2.303}{t}\log\left(\dfrac{2P_i}{3P_i - P_t}\right)$ (b) $k = \dfrac{2.303}{t}\log\left(\dfrac{2P_i}{2P_t - P_i}\right)$

(c) $k = \dfrac{2.303}{t}\log\left(\dfrac{P_i}{P_i - P_t}\right)$ (d) none of these

60. The decomposition of azo methane, at certain temperature according to the equation
$(CH_3)_2N_2 \longrightarrow C_2H_6 + N_2$ is a first order reaction.
After 40 minutes from the start, the total pressure developed is found to be 350 mm Hg in place of initial pressure 200 mm Hg of azo methane. The value of rate constant k is :

(a) $2.88 \times 10^{-4} \sec^{-1}$ (b) $1.25 \times 10^{-4} \sec^{-1}$

(c) $5.77 \times 10^{-4} \sec^{-1}$ (d) None of these

61. The hydrolysis of sucrose was studied with the help of polarimeter and following data were collected time (min.) : 0 70 ∞

observed rotation (degrees) : 44 16.5 –11

The time taken when the reaction mixture will be optically inactive ? (Given: ln 2=0.7, ln 3=1.1, ln 5=1.6)

(a) 16 min. (b) 69.47 min.

(c) 160 min. (d) none of these

62. For a particular reaction with initial conc. of the reactants as a_1 and a_2, the half-life period are t_1 and t_2 respectively. The order of the reaction (n) is given by :

(a) $n = 1 + \dfrac{\log(t_2/t_1)}{\log(a_2/a_1)}$ (b) $n = \dfrac{\log(t_1/t_2)}{\log(a_2/a_1)}$

(c) $n = 1 + \dfrac{\log(t_1/t_2)}{\log(a_2/a_1)}$ (d) none of these

63. The value of $\dfrac{t_{0.875}}{t_{0.50}}$ for n^{th} order reaction is

(a) $2^{(2n-2)}$ (b) $2^{(2n-2)-1}$

(c) $\dfrac{8^{n-1}-1}{2^{n-1}-1}$ (d) None of these

64. $A \longrightarrow B$ first order reaction A is optical active and B is optically inactive. A series of experiment were conducted on a solution of A

Time	0	60 min	∞
optical rotation	82°	77°	2°

Assume some impurity is present calculate the optical rotation after 5 hours.
(Given ln 1.066 = 0.064, $e^{0.16} = 1.17$)

(a) 60 (b) 30

(c) 20 (d) 120

65. At 300 K the half-life of a sample of a gaseous compound initially at 1 atm is 100 sec. When the pressure is 0.5 atm the half-life is 50 sec. The order of reaction is :

(a) 0 (b) 1

(c) 2 (d) 3

66. A decomposes as $A(g) \begin{cases} \xrightarrow{k_1 = 2\times10^{-3}\ s^{-1}} 2B(g) \\ \xrightarrow[k_2 = 1\times10^{-3}\ s^{-1}]{} C(g) \end{cases}$. The rate of appearance of B, taking $2\,M$ concentration of A, is equal to :

(a) $2 \times 10^{-3}\ M\ s^{-1}$ (b) $4 \times 10^{-3}\ M\ s^{-1}$

(c) $8 \times 10^{-3}\ M\ s^{-1}$ (d) None of these

67. For an endothermic reaction where ΔH represents the enthalpy of reaction in kJ/mol, the minimum value for the energy of activation will be :

(a) less than ΔH (b) more than ΔH

(c) equal to ΔH (d) zero

68. The activation energy of the reaction, $A + B \longrightarrow C + D + 38$ kcal is 20 kcal. What would be the activation energy of the following reaction.

$C + D \rightarrow A + B$

(a) 20 kcal (b) −20 kcal

(c) 18 kcal (d) 58 kcal

69. When the activation energies of the forward and backward reactions are equal, then:

(a) $\Delta U = 0, \Delta S = 0$ (b) $\Delta U = 0; \Delta G = 0$

(c) $\Delta S = 0; \Delta G = 0$ (d) only $\Delta U = 0$

70. For an exothermic chemical process occurring in two steps as follows

(i) $A + B \longrightarrow X$ (slow) (ii) $X \longrightarrow AB$ (fast)

The progress of reaction can be best described by:

(a)
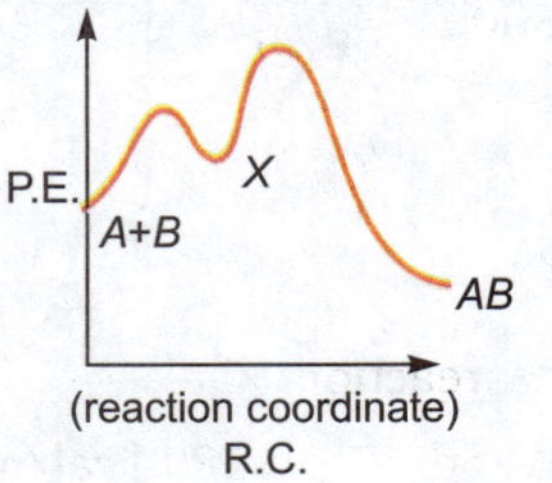

(b)
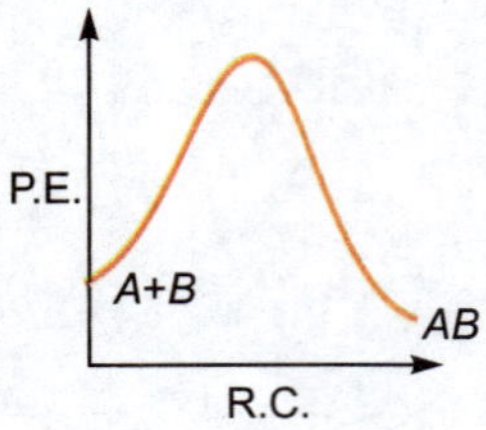

(c)
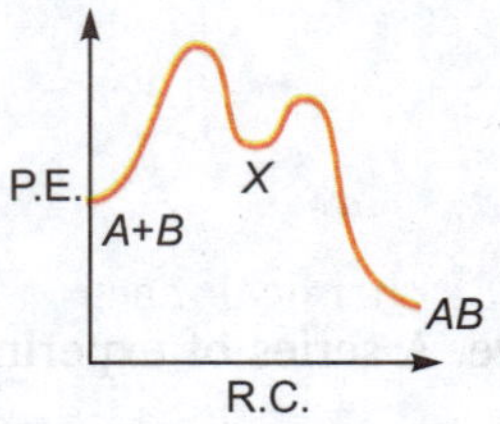

(d)
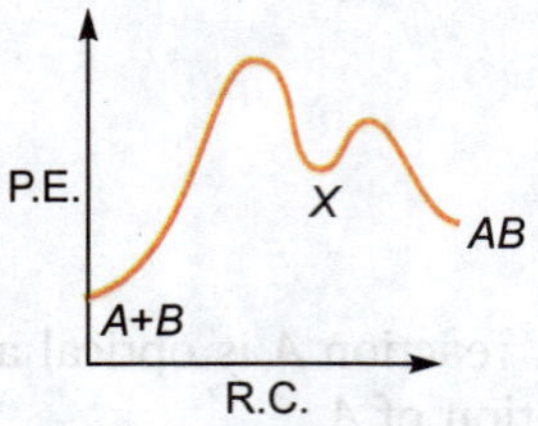

71. Select the correct diagram for an endothermic reaction that proceeds through two steps, with the second step is rate determining :

(a)
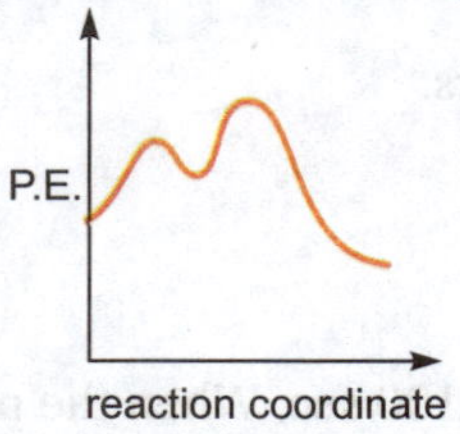

(b)
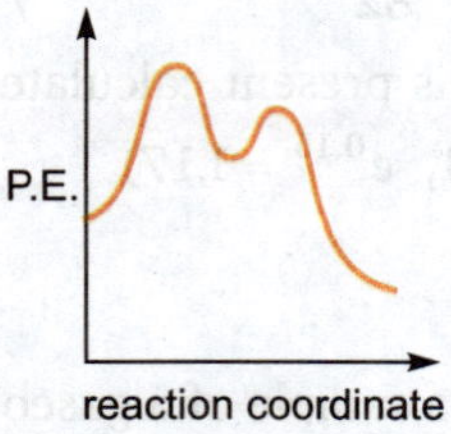

(c)
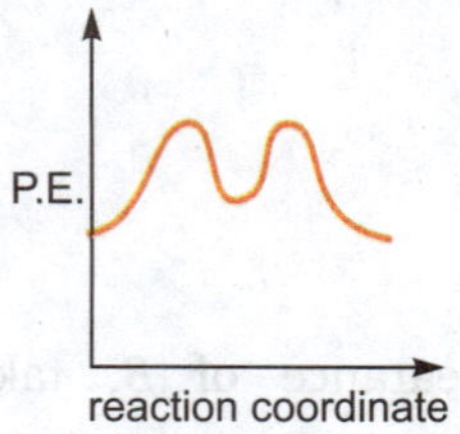

(d)
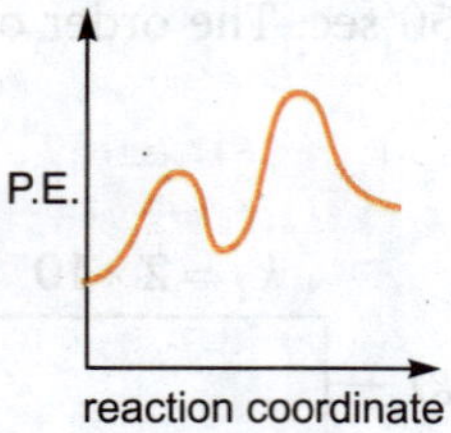

72. $\dfrac{k_{35^\circ}}{k_{34^\circ}} > 1$, this means that

(a) rate increases with the rise in temperature

(b) rate decreases with rise in temperature

(c) rate does not change with rise in temperature

(d) None of the above

73. The plot of ln k *versus* $1/T$ is linear with slope of:

(a) $-E_a/R$
(b) E_a/R
(c) $E_a/2.303\ R$
(d) $-E_a/2.303\ R$

74. Rate constant for a chemical reaction taking place at 500K is expressed as $k = A.\ e^{-1000}$
The activation energy of the reaction is:

(a) 100 cal/mol
(b) 1000 kcal/mol
(c) 10^4 kcal/mol
(d) 10^6 kcal/mol

75. For a complex reaction $A \xrightarrow{k}$ products

$$E_{a_1} = 180\ \text{kJ/mol};\ E_{a_2} = 80\ \text{kJ/mol};\ E_{a_3} = 50\ \text{kJ/mol}$$

Overall rate constant k is related to individual rate constant by the equation $k = \left(\frac{k_1 . k_2}{k_3}\right)^{2/3}$. Activation energy (kJ/mol) for the overall reaction is :

(a) 100
(b) 43.44
(c) 150
(d) 140

76. For reaction $A \longrightarrow B$, the rate constant $k_1 = A_1(e^{-E_{a_1}/RT})$ and for the reaction $X \longrightarrow Y$, the rate constant $k_2 = A_2(e^{-E_{a_2}/RT})$. If $A_1 = 10^9$, $A_2 = 10^{10}$ and $E_{a_1} = 1200$ cal/mol and $E_{a_2} = 1800$ cal/mol, then the temperature at which $k_1 = k_2$ is: (Given; $R = 2$ cal/K-mol)

(a) 300K
(b) 300×2.303K
(c) $\frac{300}{2.303} K$
(d) None of these

77. The activation energies of the forward and backward reactions in the case of a chemical reaction are 30.5 and 45.4 kJ/mol respectively. The reaction is :

(a) exothermic
(b) endothermic
(c) neither exothermic nor endothermic
(d) independent of temperature

78. A reaction rate constant is given by : $k = 1.2 \times 10^{14} e^{\frac{-25000}{RT}}$ sec^{-1}. It means :

(a) log k *versus* log T will give a straight line with a slope as 25000
(b) log k *versus* log T will give a straight line with a slope as –25000
(c) log k *versus* T will give a straight line with a slope as –25000
(d) log k *versus* $1/T$ will give a straight line

79. The temperature coefficient of a reaction is :

(a) the rate constant
(b) the rate constant at a fixed temperature
(c) the ratio of rate constant at two temperature
(d) the ratio of rate constant differing by 10°C preferably k_{308}/k_{298}

80. Which graph shows zero activation energy?

(a) 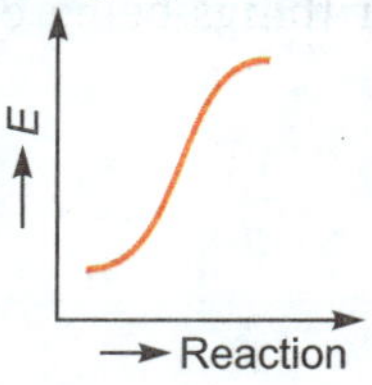

(b) 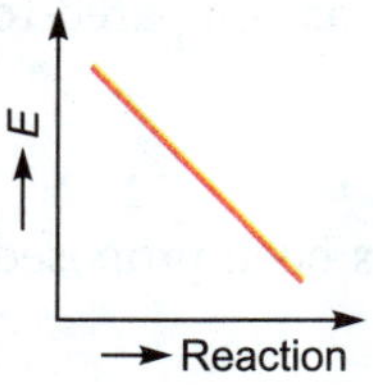

(c) 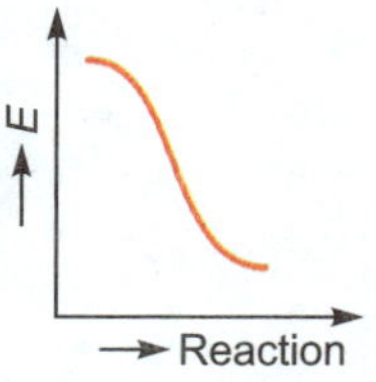

(d) 

81. A homogeneous catalytic reaction takes place through the three alternative plots A, B and C shown in the given figure. Which one of the following indicates the relative ease with which the reaction can take place ?

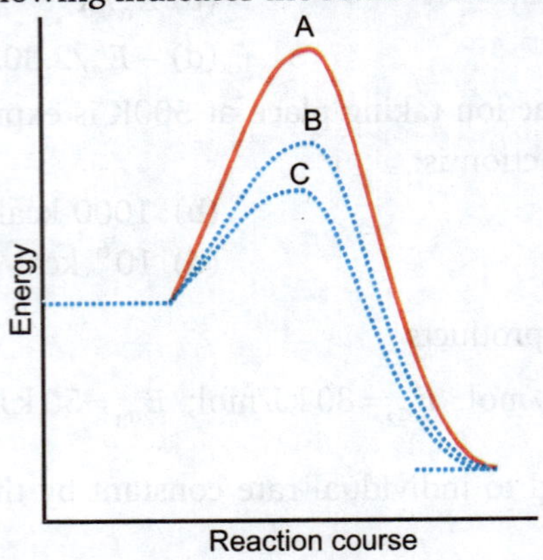

(a) $A > B > C$ (b) $C > B > A$
(c) $A > C > B$ (d) $A = B = C$

82. A first order reaction is 50% completed in 20 minutes at 27°C and in 5 minutes at 47°C. The energy of activation of the reaction is :
(a) 43.85 kJ/mol (b) 55.14 kJ/mol
(c) 11.97 kJ/mol (d) 6.65 kJ/mol

83. The rate of a reaction gets doubled when temperature increases by 27°C to 37°C. By what factor will it change for the temperature range 17°C to 27°C.
(a) 1.81 (b) 1.71
(c) 2.1 (d) 2.41

84. Which of the following explains the increase of the reaction rate by catalyst ?
(a) Catalyst decreases the rate of backward reaction so that the rate of forward reaction increases
(b) Catalyst provides extra energy to reacting molecules so that they may reduce effective collisions
(c) Catalyst provides an alternative path of lower activation energy to the reactants
(d) Catalyst increases the number of collisions between the reacting molecules.

85. Collision theory is satisfactory for:
(a) First order reactions (b) Zero order reactions
(c) Bimolecular elementary reactions (d) Any order reactions

86. For the first order reaction $A \longrightarrow B + C$, carried out at 27°C. If 3.8×10^{-16}% of the reactant molecules exists in the activated state, the E_a (activation energy) of the reaction is :
(a) 12 kJ/mol (b) 831.4 kJ/mol
(c) 100 kJ/mol (d) 88.57 kJ/mol

87. A catalyst lowers the activation energy for a certain reaction from 83.314 to 75 kJ mol^{-1} at 500 K. What will be the rate of reaction as compared to uncatalysed reaction? Assume other things being equal.
(a) Double (b) 28 times
(c) 7.38 times (d) 7.38×10^3 times

88. A following mechanism has been proposed for a reaction
$2A + B \rightarrow D + E$
$A + B \rightarrow C + D$ (slow)
$A + C \rightarrow E$ (fast)
The rate law expression for the reaction by RDS method is:
(a) $r = k[A]^2[B]$ (b) $r = k[A][B]$
(c) $r = k[A]^2$ (d) $r = k[A][C]$

89. A hypothetical reaction $A_2 + B_2 \longrightarrow 2AB$ follows the mechanism as given below

$A_2 \rightleftharpoons A + A$ (fast)

$A + B_2 \rightarrow AB + B$ (slow)

$A + B \rightarrow AB$ (fast)

The order of the over all reaction by RDS method is

(a) 2 (b) 1

(c) $\frac{3}{2}$ (d) 0

90. Chemical reaction occurs as a result of collisions between reacting molecules. Therefore, the reaction rate is given by

(a) total number of collisions occurring in a unit volume per second

(b) fraction of molecules which possess energy less than the threshold energy

(c) total number of effective collisions which have enough activation energy

(d) none of the above

91. Radioactivity is affected by :

(a) temperature (b) pressure

(c) electric and magnetic field (d) none of these

92. The radiation from naturally occurring radioactive substance as seen after deflection by a magnetic field in one direction are :

(a) α-rays (b) β-rays

(c) both α and β rays (d) either α or β-rays

93. In the radioactive decay

${}_ZX^A \rightarrow {}_{Z+1}Y^A \rightarrow \underset{\text{high energy}}{{}_{Z-1}Z^{A-4}} \rightarrow \underset{\text{low energy}}{{}_{Z-1}Z^{A-4}}$ the sequence of the radiation emitted is :

(a) α, β, γ (b) γ, α, β

(c) β, γ, α (d) β, α, γ

94. A radioactive nuclide emitts γ-rays due to the :

(a) emission of an electron from its orbital

(b) nuclear energy transition from a higher state to a lower state

(c) presence of less neutrons than protons

(d) presence of more neutrons than protons

95. Consider the following decay ${}^A_ZX \rightarrow {}^{A}_{Z+1}Y + {}^{0}_{-1}e$, X is unstable because :

(a) its nucleus has excess energy (b) $\frac{n}{p}$ ratio is high

(c) $\frac{n}{p}$ ratio is low (d) none of these

96. Consider the following decay ${}^A_ZX \rightarrow {}_{Z-1}^{A}Y + {}^{0}_{+1}e\,(\beta^+)$ X is unstable because :

(a) its nucleus has excess energy (b) $\frac{n}{p}$ ratio is high

(c) $\frac{n}{p}$ ratio is low (d) none of these

97. During α-decay :

(a) $\frac{n}{p}$ ratio decreases (b) $\frac{n}{p}$ ratio increases

(c) $\frac{n}{p}$ ratio remains constant (d) $\frac{n}{p}$ ratio may increase or decrease

98. Which of the following processes causes the emission of X-ray?
(a) α-emission (b) β-emission
(c) β^+ (Positron) emission (d) electron capture

99. Which of the following processes result in an increase in the atomic number of a nuclide?
(a) α-emission (b) electron capture
(c) γ-emission (d) β-(Beta)emission

100. .... is produced when a positron and an electron collide.
(a) *X*-ray (b) Neutron
(c) γ-radiation (d) Neutrino

101. ${}^{165}_{67}Ho$ is stable isotope. ${}^{150}_{67}Ho$ is expected to disintegrate by:
(a) α-emission (b) β-emission
(c) positron emission (d) γ-emission

102. ${}^{1}_{1}H$ is a stable isotope. ${}^{3}_{1}H$ is expected to disintegrated by :
(a) α-emission (b) β-emission
(c) positron emission (d) proton emission

103. Emission of β-particle is equivalent to :
(a) increase of one proton only (b) decrease of one neutron only
(c) both (a) and (b) (d) none of these

104. Atoms ${}^{A}_{7}X$, ${}^{B}_{8}Y$ and ${}^{17}_{9}Z$ are such that ${}_{8}Y$ is an isobar of ${}_{7}X$ and atom ${}^{17}_{9}Z$ is isotone of ${}_{8}Y$. Mass no. of *X* and no. of neutrons in *Y* are respectively :
(a) 8, 8 (b) 17, 7
(c) 9, 8 (d) 16, 8

105. ${}^{234}_{90}Th$ disintegrates to give ${}^{206}_{82}Pb$ as the final product. Total no. of α and β particles emitted out during this process are :
(a) 6 (b) 7
(c) 8 (d) 13

106. An isotone of ${}^{76}_{32}Ge$ is :
(a) ${}^{77}_{32}Ge$ (b) ${}^{77}_{33}As$
(c) ${}^{77}_{34}Se$ (d) ${}^{77}_{36}Se$

107. Pair of isobar is :
(a) ${}^{13}_{6}C$, ${}^{13}_{7}N$ (b) ${}^{13}_{6}C$, ${}^{14}_{7}N$
(c) ${}^{14}_{6}C$; ${}^{15}_{8}N$ (d) none of these

108. Isodiaphers are atoms having:
(a) n/p same (b) p/n same
(c) $(n-p)$ same (d) $(n-p)$ different

109. The 'Group displacement law' was given by:
(a) Bacqueral (b) Rutherford
(c) Madam Curie (d) Soddy and Fajan

110. ${}^{7}_{3}Li + {}^{1}_{1}p \longrightarrow X$; Identify *X* if reaction is (p, α) type.
(a) ${}^{8}_{4}Be$ (b) ${}^{4}_{2}He$
(c) ${}^{0}_{0}\gamma$ (d) none of these

111. Identify reaction type:

$$^{27}_{13}Al + ^{2}_{1}H \longrightarrow ^{28}_{13}Al + ^{1}_{1}H$$

(a) (d, p) (b) (p, p)
(c) (p, d) (d) none of these

112. $^{27}_{13}Al + ^{1}_{1}p \longrightarrow X + ^{0}_{0}\gamma$; Identify X if reaction is (p, γ) type artificial radioactive reaction.

(a) $^{28}_{13}Al$ (b) $^{27}_{14}Si$
(c) $^{28}_{14}Si$ (d) none of these

113. The number of neutrons accompanying in the formation of $^{139}_{54}Xe$ and $^{94}_{38}Sr$ from the absorption of slow neutron by $^{235}_{92}U$ followed by nuclear fission is :

(a) 0 (b) 1
(c) 2 (d) 3

114. What will be the product of reaction $^{255}_{101}Md(\alpha, 2n)$?

(a) $^{256}_{103}Lr$ (b) $^{257}_{102}No$
(c) $^{257}_{103}Lr$ (d) $^{205}_{82}Pb$

115. Complete the following nuclear equation by supplying the symbol for the other product of the fission :

$$^{235}_{92}U + ^{1}_{0}n \longrightarrow ^{94}_{38}Sr + \ldots\ldots\ldots + 2^{1}_{0}n$$

(a) $^{139}_{54}Xe$ (b) $^{140}_{54}Xe$
(c) $^{104}_{64}Gd$ (d) none of these

116. $^{235}_{92}U + ^{1}_{0}n \longrightarrow ^{139}_{56}Ba + ^{94}_{36}Kr + 3^{1}_{0}n + 200MeV$

Total energy released (in MeV) after 5th stage of fission is :

(a) 48600 (b) 16200
(c) 24200 (d) None of these

117. Proton bombardment on Th^{230} followed by emission of two alpha particles will produce :

(a) Rn^{232} (b) Ra^{233}
(c) Fr^{223} (d) Fr^{222}

118. $X \longrightarrow ^{206}_{82}Pb + ^{4}_{2}He$. In this reaction predict the position of group of X :

(a) II B (b) IV B
(c) VI A (d) VI B

119. $_{90}Th$ is a member of III group. After losing α-particle it forms a new element belonging to :

(a) I group (b) II group
(c) III group (d) IV group

120. Alpha decay of $^{238}_{92}U$ forms $^{234}_{90}Th$. What kind of decay from $^{234}_{90}Th$ produces $^{234}_{89}Ac$?

(a) α (b) β
(c) β^+ (positron) (d) γ-emission

121. $^{214}_{83}Bi$ decays to A by α-emission. A then decays to B by beta emission, which further decays to C by another beta emission. Element C decays to D by still another beta emission, and D decays by α-emission to form a stable isotope E. What is element E?

(a) $^{207}_{81}Tl$ (b) $^{206}_{80}Hg$
(c) $^{206}_{79}Au$ (d) $^{206}_{82}Pb$

122. The activity of a radioactive nuclide (X^{100}) is 6.023 curie at a certain time 't'. If its disintegration constant is $3.7 \times 10^4\ s^{-1}$ the mass of X after t sec is :

(a) 6.022×10^6 g (b) 10^{-13} g
(c) 10^{-15} g (d) 10^{-17} g

123. Activity of a radioactive substance is A_1 at time t_1 and A_2 at time $t_2 (t_2 > t_1)$, then the ratio of $\frac{A_2}{A_1}$ is :
(a) $e^{\lambda(t_2+t_1)}$ (b) $e^{\lambda(t_1-t_2)}$
(c) $e^{-\lambda(t_1+t_2)}$ (d) $\frac{t_2}{t_1}$

124. The half-life of $^{14}_{6}C$ is 5730 year. What fraction of its original C^{14} would left after 22920 year of storage?
(a) 0.50 (b) 0.25
(c) 0.125 (d) 0.0625

125. A radioactive sample had an initial activity of 56 dpm. After 69.3 minutes, it was found to have an activity of 28 dpm. Find the number of atoms in a sample having an activity of 100 dpm.
(a) 693 (b) 100
(c) 1000 (d) 10,000

126. A radioactive sample has initial activity of 28 dpm 30 minutes later its activity 14 dpm. How many atoms of nuclide were present initially?
(a) 2800 (b) 1212
(c) 528 (d) 2802

127. The value of decay constant of Co^{60} is 2.5×10^{-7} min^{-1}. The activity of 2.0 g of the sample is nearly :
(a) 5×10^5 dpm (b) 2.5×10^{10} dpm
(c) 5×10^{15} dpm (d) 10^{10} dpm

128. Half-life $(t_{1/2})$ for a radioactive decay is 6930 sec. The time required to fall the rate of decay to $\left(\frac{1}{100}\right)^{th}$ of its initial value is :
(a) 69.3 sec (b) 20,000 sec
(c) 46060 sec (d) none of these

129. A sample of radioactive substance is found 90% of its initial amount after one day. What % of the original sample can be found after 3 days?
(a) 81 (b) 72.9
(c) 25 (d) 65.61

130. If time t is required for a radioactive substance to become one third of its initial amount, what fraction would be left after 0.5 t?
(a) $\frac{1}{2}$ (b) $\frac{1}{\sqrt{3}}$
(c) $\frac{1}{3}$ (d) $\sqrt{\frac{2}{3}}$

131. The present activity of the hair of Egyption mummy is 1.75 dpm. $t_{1/2}$ of $^{14}_{6}C$ is 5770 year and disintegration rate of fresh sample of C^{14} is 14 dpm. Find out age of mummy.
(a) 23080 year (b) 138480 year
(c) 11998.3 year (d) 17310 year

132. The amount of $^{14}_{6}C$ isotope in a piece of wood is found to one fourth (1/4) of that present in a fresh piece of wood. Calculate the age of the piece of wood ($t_{1/2}$ of $^{14}_{6}C = 5770$ years)
(a) 7999 year (b) 11540 year
(c) 16320 year (d) 23080 year

133. A radioactive element undergoing decay is left 20% of its initial weight after certain period of time t. How many such periods should elapse from the start for the 50% of the element to be left over?

(a) 3 (b) 4
(c) 5 (d) None of these

134. In a sample of wood, the reading of a counter is 32 dpm and in a fresh sample of tree it is 122dpm. Due to error counter gives the reading 2 dpm in absence of ^{14}C. Half life of ^{14}C is 5770 years.
The approximate age (in years) of wood sample is:

(a) 7997.2 (b) 57570
(c) 11,540 (d) 15140

135. A 0.50g sample of rock was found to have 2.5×10^{-6} mol of $^{40}_{19}K$ ($t_{1/2} = 1.3 \times 10^9$ yr) and 7.5×10^{-6} mol of $^{40}_{20}$Ca. How old is the rock ?

(a) 6.5×10^8 yr (b) 1.3×10^9 yr
(c) 2.6×10^9 yr (d) 5.2×10^9 yr

136. Indium-112 is radioactive and has a very short half-life ($t_{1/2} = 14$ min). Its decay constant and average life are respectively :

(a) 0.0495 min^{-1}, 9.7 min (b) 0.495 min^{-1}, 20.2 min
(c) 9.7 min^{-1}, 20.2 min (d) 0.0495 min^{-1}, 20.2 min

137. The half-life of a radioactive element is 100 minutes. The time interval between the stages to 50% and 87.5% decay will be :

(a) 100 min (b) 50 min
(c) 200 min (d) 25 min

138. The half-life of Tc^{99} is 6.0 hr. The total residual activity in a patient 30 hr after receiving an injection containing Tc^{99} must be more than 0.01 μC_i. What is the maximum activity (in μC_i) that the sample injected can have?

(a) 0.16 (b) 0.32
(c) 0.64 (d) 0.08

139. A pure radio-chemical preparation was observed to disintegrate at the rate of 2140 counts/minutes at 12.35 P.M. At 3.55 P.M. of the same day, the disintegration rate of the sample was only 535 count/minutes. What is the half-life of the material?

(a) 50 min (b) 100 min
(c) 200 min (d) None of these

140. A certain radioactive isotope $^{A}_{Z}X$ ($t_{1/2} = 100$ days) decays to $^{A-8}_{Z-2}Y$. If 1 mole of $^{A}_{Z}X$ is kept in sealed container, how much He gas will accumulate at 1 atm and 273 K in 200 days?

(a) 11.2 litre (b) 33.6 litre
(c) 22.4 litre (d) 44.8 litre

141. A radioactive substance decay 25% in 10 minute. If at start there are 4×10^{20} atoms present. After how much time will the number of atoms be reduced to 10^{20} atoms? (given $\ln 3 = 1.098$)

(a) 10.98 min (b) 21.97 min
(c) 48.19 min (d) None of these

142. The time of decay for a nuclear reaction is given by $t = 4\,t_{1/2}$. The relation between the mean life (T) and time of decay (t) is given by :

(a) $2\,T \ln 2$ (b) $4\,T \ln 2$
(c) $2\,T^4 \ln 2$ (d) $\frac{1}{T^2} \ln 2$

143. Two radio isotopes A and B of atomic mass X and Y are mixed in equal amount by mass. After 20 days, their mass ratio is found to be 1 : 4. Isotope A has a half-life of 1 day. The half-life of isotope B is :

(a) $1.11\,\dfrac{Y}{X}$ day (b) $0.11\,\dfrac{X}{Y}$ day

(c) 0.6237 day (d) 1.10 day

144. Two radioactive nuclides A and B have half-lives 50 min and 10 min respectively. A fresh sample contains the nuclide of B to be eight times that of A. How much time should elapse so that the number of nuclides of A becomes double of B?

(a) 30 (b) 40

(c) 50 (d) 100

145. A radioactive nuclide is produced at a constant rate of α per second. It's decay constant is λ. If N_0 be the no. of nuclei at time $t = 0$, then max. no. of nuclei possible are :

(a) N_0 (b) α/λ

(c) $N_0 + \dfrac{\alpha}{\lambda}$ (d) $\dfrac{\lambda}{\sigma} + N_0 s$

146. An analysis of the rock shows that the relative number of Sr^{87} and Rb^{87} ($t_{1/2} = 4.7 \times 10^{10}$ year) atoms is 0.05. What is the age of the rock? Assume all the Sr^{87} have been formed from Rb^{87} only.

(a) 7.62×10^9 year (b) 1.43×10^9 year

(c) 3.28×10^9 year (d) 4.32×10^8 year

147. There are two radio nuclei A and B. A is a α-emitter and B is β-emitter. Their disintegration constant are in the ratio of 1 : 2. What should be the number of atoms ratio of A and B at time $t = 0$, so that initially probability of getting of α and β-particles are same.

(a) 2 : 1 (b) 4 : 1

(c) 1 : 2 (d) 1 : 4

148. A radioactive substance (parent) decays to its daughter element. The age of radioactive substance (t) is related to the daughter (d)/parent (p) ratio by the equation :

(a) $t = \dfrac{1}{\lambda}\ln\left(1 + \dfrac{p}{d}\right)$ (b) $t = \dfrac{1}{\lambda}\ln\left(1 + \dfrac{d}{p}\right)$

(c) $t = \dfrac{1}{\lambda}\ln\left(\dfrac{d}{p}\right)$ (d) $t = \dfrac{1}{\lambda}\ln\left(\dfrac{p}{d}\right)$

149. Ac^{227} has a half-life of 22 years. The decays follows two parallel paths

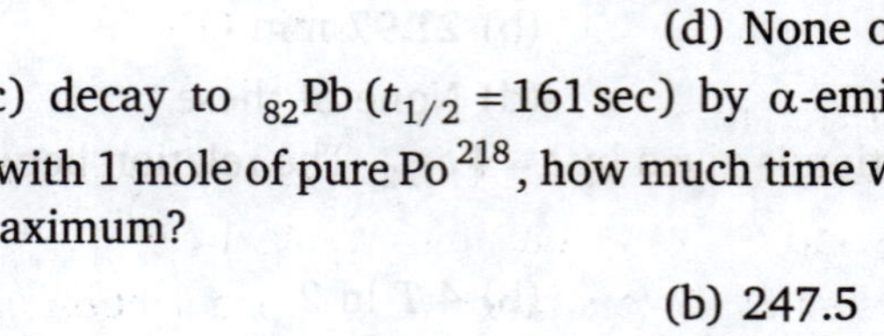

What are the decay constants (λ) for Th and Fr respectively?

(a) 0.03087, 0.00063 (b) 0.00063, 0.03087

(c) 0.02, 0.98 (d) None of these

150. ${}^{218}_{84}Po$ ($t_{1/2} = 183$ sec) decay to ${}_{82}Pb$ ($t_{1/2} = 161$ sec) by α-emission, while Pb^{214} is a β-emitter. In an experiment starting with 1 mole of pure Po^{218}, how much time would be required for the number of nuclei of ${}^{214}_{82}Pb$ to reach maximum?

(a) 147.5 (b) 247.5

(c) 182 (d) 304

Level-2

1. The forward rate constant for the elementary reversible gaseous reaction

$$C_2H_6 \rightleftharpoons 2CH_3 \text{ is } 1.57 \times 10^{-3}\ s^{-1} \text{ at } 100\ K$$

What is the rate constant for the backward reaction at this temperature if 10^{-4} moles of CH_3 and 10 moles of C_2H_6 are present in a 10 litre vessel at equilibrium.

(a) $1.57 \times 10^{9}\ L\ mol^{-1}\ s^{-1}$ (b) $1.57 \times 10^{10}\ L\ mol^{-1}\ s^{-1}$

(c) $1.57 \times 10^{11}\ L\ mol^{-1}\ s^{-1}$ (d) $1.57 \times 10^{7}\ L\ mol^{-1}\ s^{-1}$

2. For a hypothetical reaction,

$A + 3B \longrightarrow P$ $\quad \Delta H = -2x$ kJ/mole of A

& $M \longrightarrow 2Q + R$ $\quad \Delta H = +x$ kJ/mole of M

These reactions are carried simultaneously in a reactor such that temperature is not changing. If rate of disappearance of B is $y\,M\ sec^{-1}$ then rate of formation (in $M\ sec^{-1}$) of Q is:

(a) $\frac{2}{3}y$ (b) $\frac{3}{2}y$

(c) $\frac{4}{3}y$ (d) $\frac{3}{4}y$

3. The kinetic data for the given reaction $A(g) + 2B(g) \xrightarrow{k} C(g)$ is provided in the following table for three experiments at 300 K

Ex. No.	[A/M]	[B/M]	Initial rate ($M\ sec^{-1}$)
1.	0.01	0.01	6.930×10^{-6}
2.	0.02	0.01	1.386×10^{-5}
3.	0.02	0.02	1.386×10^{-5}

In another experiment starting with initial concentration of 0.5 and 1 M respectively for A and B at 300 K, find the rate of reaction after 50 minutes from the starting of experiment (in M/sec).

(a) 6.93×10^{-4} (b) 0.25×10^{-7}

(c) 4.33×10^{-5} (d) 3.46×10^{-4}

4. The reaction $A(g) + 2B(g) \longrightarrow C(g)$ is an elementary reaction. In an experiment involving this reaction, the initial partial pressures of A and B are $P_A = 0.40$ atm and $P_B = 1.0$ atm respectively. When $P_C = 0.3$ atm, the rate of the reaction relative to the initial rate is :

(a) $\frac{1}{12}$ (b) $\frac{1}{50}$

(c) $\frac{1}{25}$ (d) none of these

5. Which of the following is incorrect statement ?

(a) Stoichiometry of a reaction tells about the order of the elementary reaction

(b) For a zero order reaction, rate and the rate constant are identical.

(c) A zero order reaction is controlled by factors other than concentration of reactants

(d) A zero order reaction is an elementary reaction

6. Two first order reaction have half-lives in the ratio 8 : 1. Calculate the ratio of time intervals $t_1 : t_2$. The time t_1 and t_2 are the time period for $\left(\frac{1}{4}\right)^{th}$ and $\left(\frac{3}{4}\right)^{th}$ completion.

(a) 1 : 0.301 (b) 0.125 : 0.602
(c) 1 : 0.602 (d) none of these

7. Reaction $A + B \longrightarrow C + D$ follows rate law, $r = k[A]^{1/2}[B]^{1/2}$. Starting with 1 M of A and B each, what is the time taken for concentration of A become 0.1 M?
[Given $k = 2.303 \times 10^{-2}$ sec^{-1}].
(a) 10 sec (b) 100 sec
(c) 1000 sec (d) 434 sec

8. For the two reactions I: $A \rightarrow B$; II: $C \rightarrow D$ following graph is obtained.

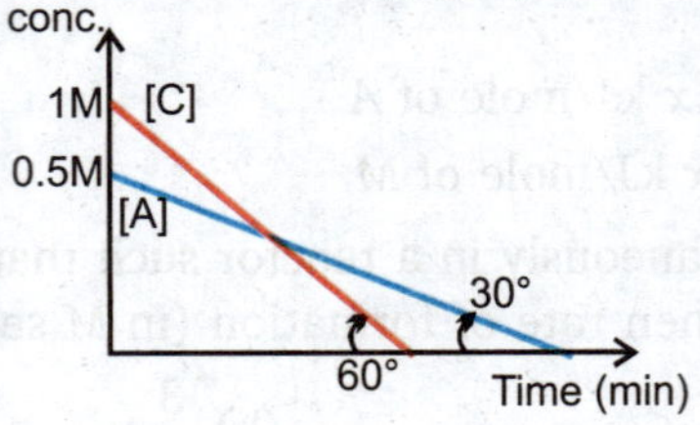

Which of the following is true:
(a) If $[B] = [A]$ then at that time $[D] = 0.75\,M$ (b) If $[C] = [A]$ then at that time $[B] > [D]$
(c) $(t_{100\%})_{\text{Reaction I}} = (t_{100\%})_{\text{Reaction II}}$ (d) $[A] = [C]$ at $t = \frac{\sqrt{3}}{2}$ min.

9. The reaction $A(g) \longrightarrow B(g) + 2C(g)$ is a first order reaction with rate constant 2.772×10^{-3} s^{-1}. Starting with 0.1 mole of A in 2 litre vessel, find the concentration of A after 250 sec when the reaction is allowed to take place at constant pressure and at 300 K ?
(a) 0.0125 M (b) 0.025 M
(c) 0.05 M (d) none of these

10. For a first order homogeneous gaseous reaction, $A \longrightarrow 2B + C$
if the total pressure after time t was P_t and after long time $(t \rightarrow \infty)$ was P_∞ then k in terms of P_t, P_∞ and t is :
(a) $k = \frac{2.303}{t} \log\left(\frac{P_\infty}{P_\infty - P_t}\right)$ (b) $k = \frac{2.303}{t} \log\left(\frac{2P_\infty}{P_\infty - P_t}\right)$
(c) $k = \frac{2.303}{t} \log\left(\frac{2P_\infty}{3(P_\infty - P_t)}\right)$ (d) none of these

11. $A(aq) \longrightarrow B(aq) + C(aq)$ is first order reaction.

Time	t	∞
moles of reagent	n_1	n_2

Reaction progress is measured with the help of titration of reagent 'R'. If all A, B and C react with reagent and have 'n' factors [n factor; eq. mass $= \frac{\text{mol. mass}}{n}$] in the ratio of 1 : 2 : 3 with the reagent, the k in terms of t, n_1 and n_2 is :
(a) $k = \frac{1}{t} \ln\left(\frac{n_2}{n_2 - n_1}\right)$ (b) $k = \frac{1}{t} \ln\left(\frac{2n_2}{n_2 - n_1}\right)$
(c) $k = \frac{1}{t} \ln\left(\frac{4n_2}{n_2 - n_1}\right)$ (d) $k = \frac{1}{t} \ln\left(\frac{4n_2}{5(n_2 - n_1)}\right)$

12. The gaseous decomposition reaction, $A(g) \longrightarrow 2B(g) + C(g)$ is observed to first order over the excess of liquid water at 25°C. It is found that after 10 minutes the total pressure of system is 188 torr and after very long time it is 388 torr. The rate constant of the reaction (in hr^{-1}) is : [Given : vapour pressure of H_2O at 25° is 28 torr ($\ln 2 = 0.7, \ln 3 = 1.1, \ln 10 = 2.3$)]

(a) 0.02 (b) 1.2
(c) 0.2 (d) none of these

13. The reaction, Sucrose $\xrightarrow{H^+}$ Glucose + Fructose, takes place at certain temperature while the volume of solution is maintained at 1 litre. At time zero the initial rotation of the mixture is 34°. After 30 minutes the total rotation of solution is 19° and after a very long time, the total rotation is –11°C. Find the time when solution was optically inactive.

(a) 135 min (b) 103.7 min
(c) 38.7 min (d) 45 min

14. A gaseous compound *A* reacts by three independent first order processes (as shown in figure) with rate constant 2×10^{-3}, 3×10^{-3} and $1.93 \times 10^{-3} sec^{-1}$ for products *B*, *C* and *D* respectively. If initially pure *A* was taken in a closed container with $P = 8$ atm, then the partial pressure of *B* (in atm) after 100 sec from starting the experiment.

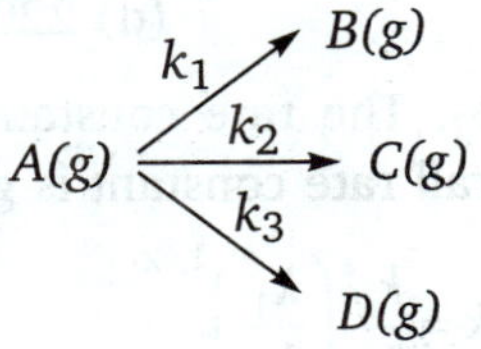

(a) 0.288 (b) 0.577
(c) 1.154 (d) none of these

15. A compound *A* dissociates by two parallel first order paths at certain temperature

$$A(g) \xrightarrow{k_1(\text{min}^{-1})} 2B(g) \qquad k_1 = 6.93 \times 10^{-3}\ \text{min}^{-1}$$

$$A(g) \xrightarrow{k_2(\text{min}^{-1})} C(g) \qquad k_2 = 6.93 \times 10^{-3}\ \text{min}^{-1}$$

The reaction is started with 1 mole of pure '*A*' in 1 litre closed container with initial pressure 2 atm. What is the pressure (in atm) developed in container after 50 minutes from start of experiment?

(a) 1.25 (b) 0.75
(c) 1.50 (d) 2.50

16. For given hypothetical elementary parallel reaction,

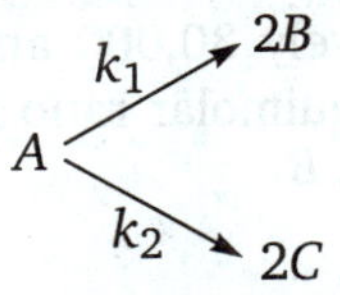

Where $\dfrac{k_1}{k_2} = \dfrac{1}{2}$

Initially only 2 moles of *A* are present. The total no. of moles of *A*, *B* and *C* at the end of 75% reaction are:

(a) 2 (b) 3
(c) 4 (d) 3.5

17. The reaction *cis*-$X \underset{k_b}{\overset{k_f}{\rightleftharpoons}}$ *trans*-X is first order in both directions. At 25°C, the equilibrium constant is 0.10 and the rate constant $k_f = 3 \times 10^{-4}\ s^{-1}$. In an experiment starting with the pure *cis*-form, how long would it take for half of the equilibrium amount of the *trans*-isomer to be formed?

(a) 150 sec (b) 200 sec
(c) 240 sec (d) 210 sec

18. A hydrogenation reaction is carried out at 500 K. If the same reaction is carried out in the presence of a catalyst at the same rate with same frequency factor, the temperature required is 400 K. What is the activation energy of the reaction, if the catalyst lowers the activation energy barrier by 16 kJ/mol?

(a) 100 kJ/mol (b) 80 kJ/mol

(c) 60 kJ/mol (d) None of these

19. Consider the reaction.

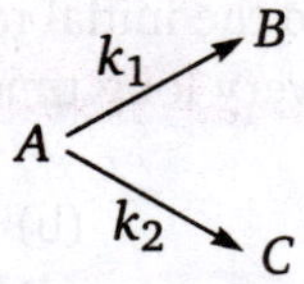

The rate constant for two parallel reactions were found to be 10^{-2} $dm^3\ mol^{-1}\ s^{-1}$ and 4×10^{-2} $dm^3\ mol^{-1}\ s^{-1}$. If the corresponding energies of activation of the parallel reaction are 100 and 120 kJ/mol respectively, what is the net energy of activation (E_a) of A?

(a) 100 kJ/mol (b) 120 kJ/mol

(c) 116 kJ/mol (d) 220 kJ/mol

20. A reaction takes place in various steps. The rate constant for first, second, third and fifth steps are k_1, k_2, k_3 and k_5 respectively. The overall rate constant is given by

$$k = \frac{k_2}{k_3}\left(\frac{k_1}{k_5}\right)^{1/2}$$

If activation energy are 40, 60, 50 and 10 kJ/mol respectively, the overall energy of activation (kJ/mol) is :

(a) 10 (b) 20

(c) 25 (d) none of these

21. For reaction $A \longrightarrow B$, the rate constant $k_1 = A_1\ e^{-E_{a_1}/(RT)}$ and for the reaction $X \longrightarrow Y$, the rate constant $k_2 = A_2 e^{-E_{a_2}/(RT)}$. If $A_1 = 10^8$, $A_2 = 10^{10}$ and $E_{a_1} = 600$ cal/mol, $E_{a_2} = 1800$ cal/mol, then the temperature at which $k_1 = k_2$ is (Given : $R = 2$ cal/K-mol)

(a) 1200 K (b) 1200×4.606 K

(c) $\frac{1200}{4.606}$ K (d) $\frac{600}{4.606}$ K

22. For first order parallel reactions k_1 and k_2 are 4 and 2 min^{-1} respectively at 300 K. If the activation energies for the formation of B and C are respectively 30,000 and 38,314 joule/mol respectively, the temperature at which B and C will be obtained in equimolar ratio is :

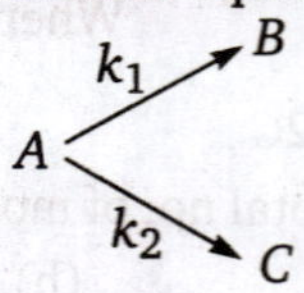

(a) 757.48 K (b) 378.74 K

(c) 600 K (d) none of these

23. In the series reaction

$A \xrightarrow{k_1} B \xrightarrow{k_2} C \xrightarrow{k_3} D$, if $k_1 > k_2 > k_3$ then the rate determing step of the reaction is

(a) $A \to B$ (b) $C \to D$

(c) $B \to C$ (d) Any step

24. The following mechanism has been proposed for the exothermic catalyzed complex reaction

$$A + B \underset{\text{fast}}{\overset{\text{slow}}{\rightleftharpoons}} IAB \xrightarrow{k_1} AB + I \xrightarrow{k_2} P + A$$

If k_1 is much smaller than k_2, the most suitable qualitative plot of potential energy (P. E.) *versus* reaction co-ordinate (R.C.) for the above reaction

(a)
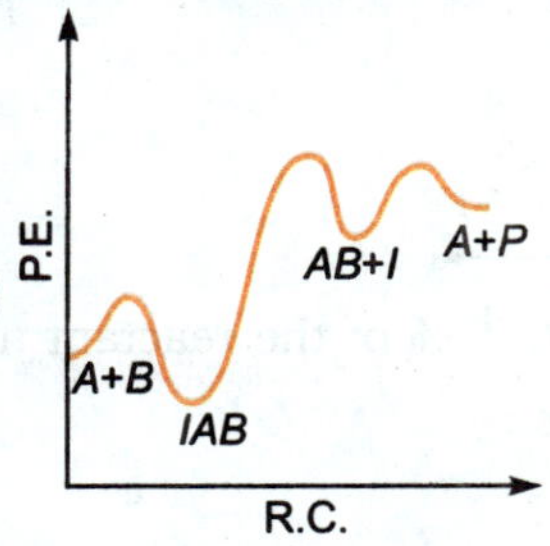

(b)
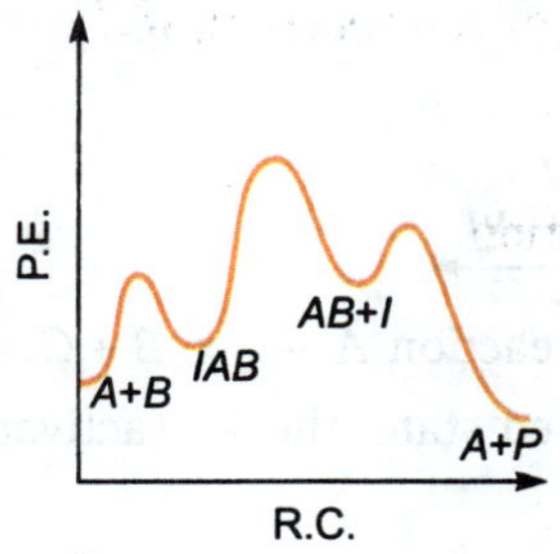

(c)
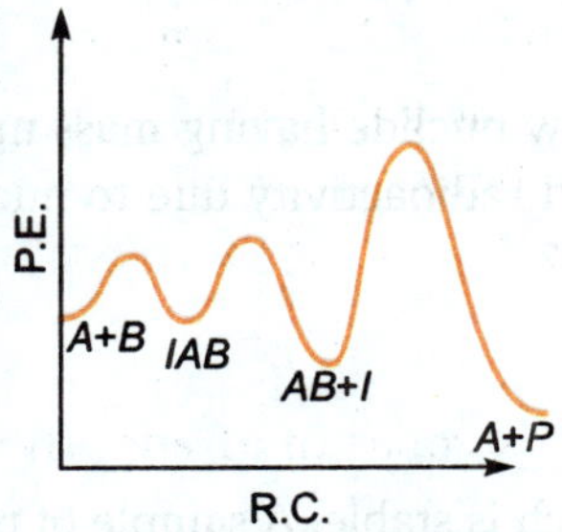

(d)
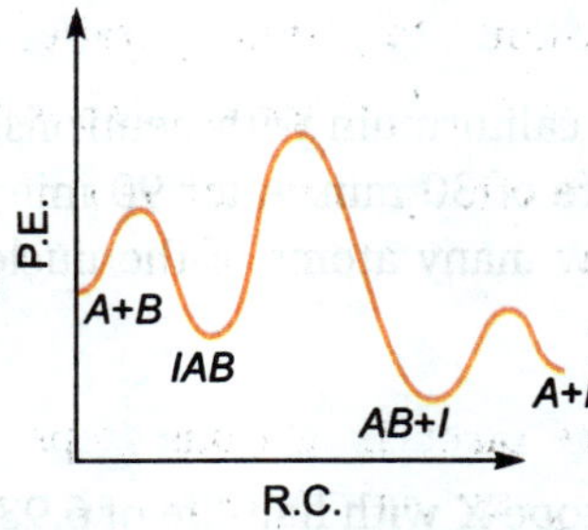

25. The mechanism of esterification in presence of acid catalyst (H_2SO_4) is proposed as follows :

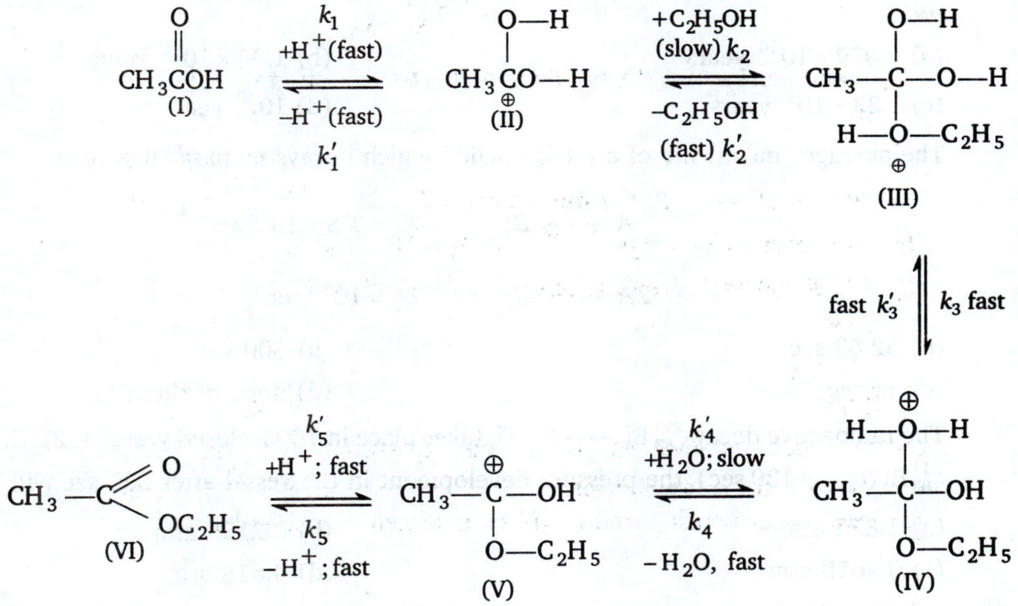

Which of the following potential energy *Vs* reaction co-ordinate diagram is consistent with given mechanism?

(a)
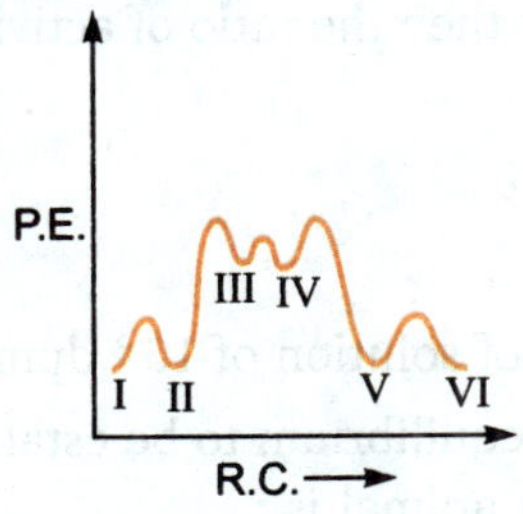

(b)
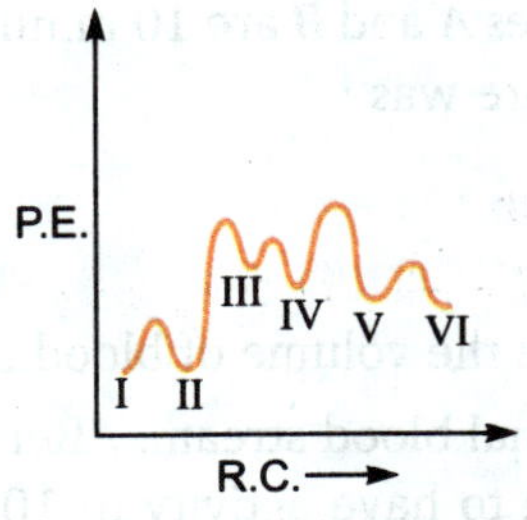

(c) 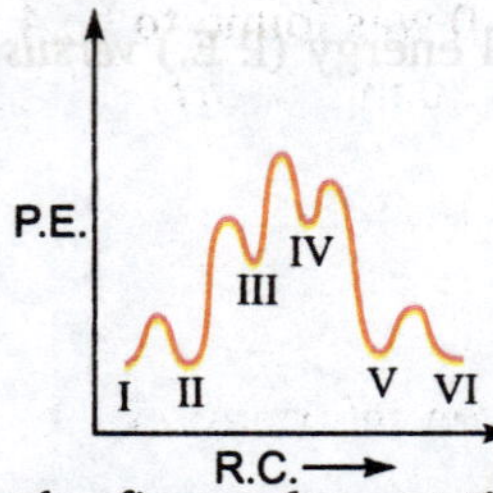

(d) 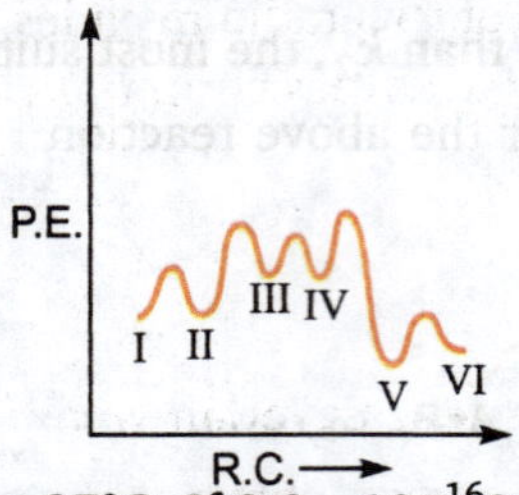

26. For the first order reaction $A \longrightarrow B + C$, carried out at 27°C. If $3.8 \times 10^{-16}\%$ of the reactant molecules exists in the activated state, the E_a (activation energy) of the reaction is :

(a) 12 kJ/mol (b) 831.4 kJ/mol
(c) 100 kJ/mol (d) 88.57 kJ/mol

27. Upon irradiating californium with neutrons, a scientist discovered a new nuclide having mass number of 250 and a half-life of 30 min. After 90 min. of irradiation, the observed radioactivity due to nuclide was 100 dis/min. How many atoms of the nucleide were prepared initially?

(a) 2.4×10^4 (b) 3.46×10^4
(c) 1900 (d) 800

28. A radioactive isotope X with half-life of 6.93×10^9 years decay to Y which is stable. A sample of rock from the moon was found to contain both the elements X and Y in the mole ratio 1 : 7. What is the age of the rock?

(a) 2.079×10^{10} years (b) 1.94×10^{10} years
(c) 1.33×10^9 years (d) 10^{10} years

29. The average (mean) life of a radio nuclide which decays by parallel path is

$$A \xrightarrow{\lambda_1} B; \qquad \lambda_1 = 1.8 \times 10^{-2} \text{ sec}^{-1}$$

$$2A \xrightarrow{\lambda_2} C; \qquad \lambda_2 = 10^{-3} \text{ sec}^{-1}$$

(a) 52.63 sec (b) 500 sec
(c) 50 sec (d) none of these

30. The radioactive decay ${}^{211}_{83}\text{Bi} \longrightarrow {}^{207}_{81}\text{Tl}$, takes place in 100 L closed vessel at 27°C. Starting with 2 moles of ${}^{211}_{83}\text{Bi}$ ($t_{1/2} = 130$ sec), the pressure development in the vessel after 520 sec will be :

(a) 1.875 atm (b) 0.2155 atm
(c) 0.4618 atm (d) 4.618 atm

31. A fresh radioactive mixture contains short lived species A and B. Both emitting α-particles initially of 8000 α-particles per minute. 20 minutes later, they emits at the rate of 3500 α-particles per minute. If the half-lives of the species A and B are 10 minutes and 500 hours respectively, then the ratio of activities of A : B in the initial mixture was :

(a) 4 : 6 (b) 6 : 4
(c) 3 : 4 (d) 3 : 1

32. In order to determine the volume of blood in an animal, a 1.0 mL sample of solution of 10^3 dpm of ${}^3_1\text{H}$ is injected into the animal blood stream. After sufficient time for circulatory equilibrium to be established, 2 mL of blood is found to have activity to 10 dpm. The volume of blood in animal is :

(a) 199 mL (b) 198 mL
(c) 200 mL (d) 20 mL

33. The ratio of activities of two radio nuclides X and Y in a mixture at time $t = 0$ was found to be 4 : 1. After two hours, the ratio of activities become 1 : 1. If the $t_{1/2}$ of radio nuclide X is 20 min then $t_{1/2}$ [in minutes] of radio nuclide Y is :

(a) 10
(b) 20
(c) 30
(d) 40

34. Find the age of an ancient Egyptian wooden article (in years) from the given information.

(i) Activity of 1 g of carbon obtained from ancient wooden article = 7 counts/min/g

(ii) Activity of 1 g carbon obtained from fresh wooden sample = 15.4 counts per min/g

(iii) Percentage increase in level of C^{14} due to nuclear explosions in past 100 years is 10%

(iv) $t_{1/2}$ of ${}^{14}_{6}C = 5770$ years

(a) 5.770×10^3
(b) 16.87×10^3
(c) 2488
(d) none of these

35. The isotopes ${}^{238}U$ and ${}^{235}U$ occur in nature in the mass ratio 140 : 1. It is assumed that initially they were found in equal mass. If half life ($t_{1/2}$) of ${}^{238}U = 4.5 \times 10^9$ and $t_{1/2}$ of ${}^{235}U = 5 \times 10^8$ year respectively then the age of earth is ($\log 7 = 0.846; \log 2 = 0.3$)

(a) 4.02×10^9 year
(b) 2.01×10^9 year
(c) 8.72×10^9 year
(d) None of these

Level-3

Passage-1

Two consecutive irreversible first order reactions can be represented by

$$A \xrightarrow{k_1} B \xrightarrow{k_2} C$$

The rate equation for A is readily integrated to obtain

$$[A]_t = [A]_0 . e^{-k_1 t}; \text{ and } [B]_t = \frac{k_1 [A]_0}{k_2 - k_1}[e^{-k_1 t} - e^{-k_2 t}]$$

1. At what time will B be present in maximum concentration?

(a) $\frac{k_1}{k_2 - k_1}$

(b) $\frac{1}{k_1 - k_2} \ln \frac{k_1}{k_2}$

(c) $\frac{1}{k_2 - k_1} \ln \frac{k_1}{k_2}$

(d) None of these

2. When $k_1 = 1\,s^{-1}$ and $k_2 = 500\,s^{-1}$; select most appropriate graph

(a)

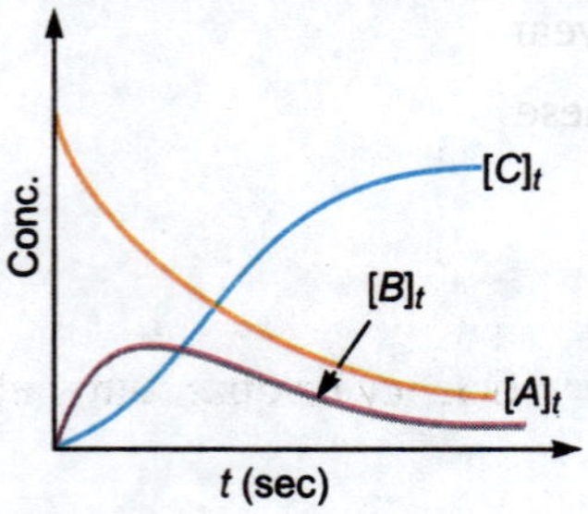

(b)

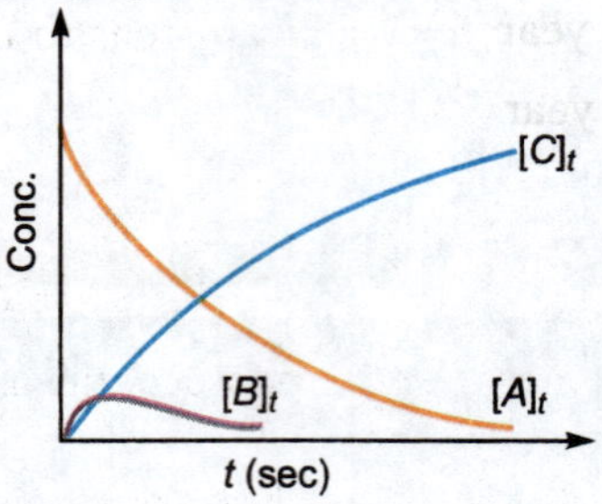

(c)

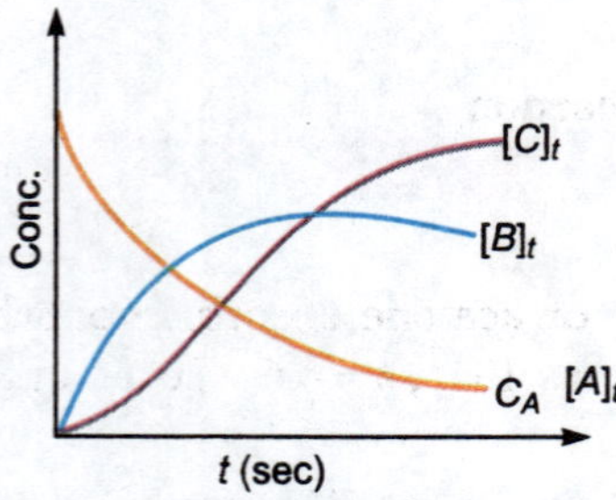

(d)

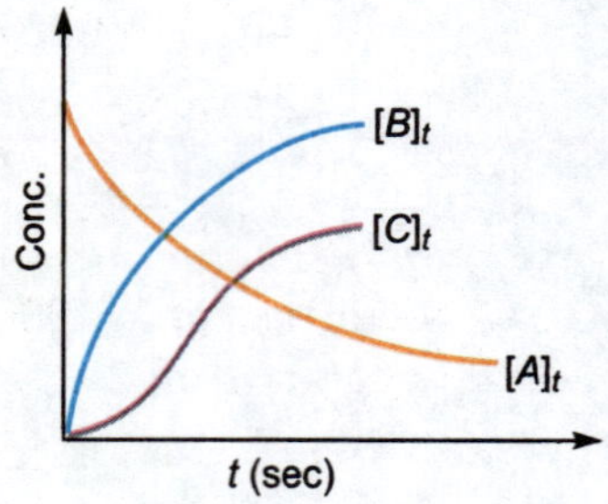

3. Select the correct statement for given reaction :

(a) A decreases linearly

(b) B rises to a max. and then constant

(c) B rises to a max. and then falls

(d) The slowest rate of increases of C occuring where B is max.

Passage-2

Arrhenius studied the effect of temperature on the rate of a reaction and postulated that rate constant varies with temperature exponentially as $k = Ae^{-E_a/RT}$. This method is generally used for finding the activation energy of a reaction. Keeping temperature constant, the effect of catalyst on the activation energy has also been studied.

1. The pre-exponential factor in the Arrhenius equation of a first order reaction has the unit :

(a) mol $L^{-1}s^{-1}$

(b) L $mol^{-1}s^{-1}$

(c) s^{-1}

(d) dimensionless

2. If x is the fraction of molecules having energy greater than E_a it will be given by :

(a) $x = -\frac{E_a}{RT}$ (b) $\ln x = -\frac{E_a}{RT}$

(c) $x = e^{E_a/RT}$ (d) Any of these

3. If the rate of reaction doubles for 10°C rise of temperature form 290K to 300K, the activation energy of the reaction will be approximately:

(a) 40 kcal mol^{-1} (b) 12 kcal mol^{-1}

(c) 60 kcal mol^{-1} (d) 70 kcal mol^{-1}

Passage-3

The rate constant of forward reaction increases twice when temperature raised from 300 K to 310 K while equilibrium constant increases four times when temperature increase from 300 K to 310 K. Assume activation energies and enthalpy of reaction remains constant in temperature range of 300 K to 310 K. (Use: $\ln 2 = 0.7$)

1. The activation energy for forward reaction is :

(a) 13.02 kcal (b) –13.02 kcal

(c) 26.04 kcal (d) –26.04 kcal

2. The activation energy for backward reaction is :

(a) 13.02 kcal (b) –13.02 kcal

(c) 26.04 kcal (d) –26.04 kcal

Passage-4

An important parameter of a photochemical reaction is the quantum efficiency or quantum yield (ϕ) which is defined as

$$\phi = \frac{\text{moles of the substance reacted}}{\text{moles of photons absorbed}}$$

Absorption of UV radiation decompose acetone according to the reaction

$$(CH_3)_2CO \xrightarrow{h\nu} C_2H_6 + CO$$

1. The quantum yield of a reaction at $\lambda = 330$ nm is 0.4. A sample of acetone absorbs monochromatic radiation at $\lambda = 330$ nm at the rate of 7.2×10^{-3} Js^{-1} (Given : $N_A = 6 \times 10^{23}$; $h = 6.6 \times 10^{-34}$ in S.I. unit). The rate of formation of CO(mol/s) is :

(a) 2×10^{-8} (b) 8×10^{-8}

(c) 8×10^{-9} (d) none of these

2. If quantum yield in 0.8 then rate of formation of C_2H_6 (mol/s) is :

(a) 2×10^{-8} (b) 1.6×10^{-9}

(c) 16×10^{-9} (d) 8×10^{-9}

Passage-5

Radioactive disintegration is a first order reaction and its rate depends only upon the nature of nucleus and does not depend upon external factors like temperature and pressure. The rate of radioactive disintegration (Activity) is represented as

$$-\frac{dN}{dt}=\lambda N$$

Where λ = decay constant; N = number of nuclei at time t; N_0 = initial no. of nuclei

The above equation after integration can be represented as

$$\lambda=\frac{2.303}{t}\log\left(\frac{N_0}{N}\right)$$

1. Half-life period of U^{237} is 2.5×10^5 years. In how much time will the amount of U^{237} remaining be only 25% of the original amount?

(a) 2.5×10^5 years (b) 1.25×10^5 years (c) 5×10^5 years (d) 10^6 years

2. Calculate the half-life period of a radioactive element which remains only 1/16 of its original amount in 4740 years :

(a) 1185 years (b) 2370 years (c) 52.5 years (d) none of these

3. What is the activity in Ci (curie) of 1.0 mole of Plutonium–239? ($t_{1/2}$ = 24,000 years)

(a) 1.49 Ci (b) 14.9 Ci (c) 5.513×10^{11} Ci (d) None of these

One or More Answer is/are Correct

1. Select the correct statement(s):

(a) Rate constants are never negative

(b) Partial orders are never negative

(c) Molecularity and order of reaction both are equal for elementary reactions

(d) Order of reaction may be change with change in practical conditions (temp. and pressure)

2. Select the correct statement(s) :

(a) The rate of reaction decreases with decrease in temperature

(b) The rate of reaction is uniform in zero order reaction

(c) The rate of reaction depends upon the surface area of the solid reactants

(d) Average and instantaneous rate of reaction defined for micro and macro-scopic time interval respectively

3. Select the correct statement(s) :

(a) The rate law of the elementary reaction; $2A \longrightarrow B + C$, must be $r = k[A]^2$

(b) The rate law for the complex reaction

$$A+B \longrightarrow C, \text{ might not be } r=k[A][B]$$

(c) If the partial orders differ from the stoichiometric coefficients in the balanced reaction, the reaction must be complex

(d) If the partial orders are equal to corresponding coefficients in the balanced reaction, the reaction must be elementary

4. Select the correct statement(s):

(a) Every substance that appears in the rate law of reaction must be a reactant or product in that reaction

(b) If we know the rate law of a reaction; we can deduce its mechanism

(c) If the reaction has rate $r = k[A][B]^{3/2}$ then reaction may be elementary

(d) A zero order reaction must be a complex reaction

5. Select the correct statement(s) :

(a) When $T \to \infty$ or $E_a \to 0$ then $k = A$

(b) A positive catalyst can change ΔH of the reaction

(c) A mixture of reactants may be thermodynamically unstable but kinetically stable

(d) A negative catalyst increases the activation energy of the reaction

6. Consider a reaction $A + B \to C$, in which both reactants are in the same phase, may be

(a) unimolecular elementary reaction

(b) Exothermic

(c) Heterogeneous

(d) Photochemical

7. In the following gaseous phase first order reaction

$$A(g) \longrightarrow 2B(g) + C(g)$$

initial pressure was found to be 400 mm of Hg and it changed to 1000 mm of Hg after 20 min. Then

(a) half life for A is 10 min

(b) rate constant is 0.0693 min^{-1}

(c) partial pressure of C at 30 min is 350 mm of Hg

(d) total pressure after 30 min is 1100 mm of Hg

8. Identify the true statement(s)

(a) A catalyst is chemically unchanged at the end of a reaction

(b) A catalyst may appear in the kinetic rate equation of the reaction

(c) A catalyst will not affect the composition of an equilibrium mixture

(d) A catalyst cannot cause a non-spontaneous ($\Delta G > 0$) reaction to proceed

9. For the reaction $A \underset{k_2 sec^{-1}}{\overset{k_1 sec^{-1}}{\rightleftharpoons}} B$ following graph is given,

$k_1 = 4 \times 10^{-2}\ sec^{-1}$. Which is/are correct statement (s) (ln 2 = 0.7, ln 8/7 = 0.14)

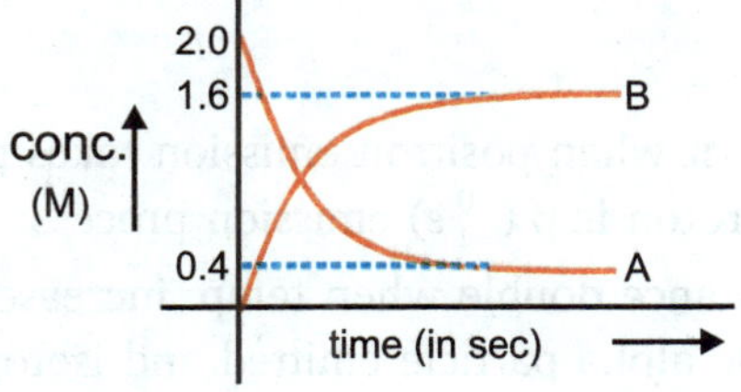

(a) Equilibrium constant is 4.0

(b) Time taken for the completion of 50% of equilibrium conc. of B is 14 sec.

(c) Time taken for the completion of 10% of initial conc. of A is 2.8 sec.

(d) Rate constant of backward reaction is $10^{-2}\ sec^{-1}$

10. Select the correct statement(s) :

(a) α-particles are simply helium atoms

(b) γ-rays travel with higher speed as compared to α-particle and have higher ionization power as compare to β-particle

(c) Loss of β-particles results in the production of isobars

(d) β-particles are considered as the best bombarding particles

11. Select the correct statement(s) :

(a) In the reaction ${}^{235}_{92}U + {}^{1}_{0}n \longrightarrow {}^{140}_{56}Ba + 2\,{}^{1}_{0}n + x$, x is ${}^{94}_{36}Kr$

(b) In the reaction $^{23}_{11}Na + z \longrightarrow ^{23}_{12}Mg + ^{1}_{0}n$, the bombarding particle z is deuteron
(c) Very large amount of energy is produced during nuclear fission and nuclear fusion
(d) In a fission reaction, a loss in mass occurs releasing a vast amount of energy

12. Select the correct statement(s):
(a) SI unit of radioactivity is becquerel (Bq)
(b) 1 Ci = 3.7×10^7 Bq
(c) $^{7}_{3}Li + ^{1}_{1}H \longrightarrow ^{4}_{2}He$ is (P, α) type reaction
(d) The half-life of a particular radioactive isotope is a characteristics constant of that isotope

13. Select the correct statement(s) :
(a) On bombarding $^{14}_{7}N$ nuclei with α-particle, the nuclei of the product formed after release of proton would be $^{17}_{8}O$
(b) Decay constant does not depend upon temperature
(c) Nuclide and its decay product after α-emission are called isodiaphers
(d) Half-life of radium is 1580 years. Its average life will be 1097.22 years

14. In the decay process

$$A \xrightarrow{-\alpha} B \xrightarrow{-\beta} C \xrightarrow{-\beta} D$$

(a) A and B are isobars
(b) A and D are isotopes
(c) B, C and D are isobars
(d) A and C are isotones

15. In electron capture (radioactive process) :
(a) a neutron is formed
(b) a proton is consumed
(c) γ-ray emission takes place
(d) X-ray emission takes place

16. Select the correct statement (s) for positron emission by unstable nucleus.
(a) X-ray emission takes place
(b) A neutron is formed
(c) $\frac{n}{p}$ of daughter nucleus increases
(d) A neutron is consumed

17. Select the correct statement (s) :
(a) Mass number remains constant when positron emission takes place
(b) One neutron converts into proton in β ($^{0}_{-1}e$) emission process
(c) Activity of a radioactive substance double when temp. increases from 300 K to 310 K
(d) Isodiaphers formed when one alpha particle emitted and isotopes formed when 2 beta particles emitted

Match the Column

Column-I and **Column-II** contains four entries each. Entries of Column-I are to be matched with some entries of Column-II. One or more than one entries of Column-I may have the matching with the same entries of Column-II.

1.

Column-I	Column-II
(A) Unit of k is always equals to	(P) 1/time
(B) Unit of k in zero order	(Q) M/time
(C) Unit of k in first order	(R) $\text{Time}^{-1}\ M^{-1}$
(D) Unit of k in second order	(S) Unit of A (pre-exponential factor)

2.

Column-I	Column-II
(A) Molecularity of a reaction	(P) 0, 1 Possible
(B) Order of reaction	(Q) 1, 2 Possible
(C) The dissociation of H_2O_2 (aq) is	(R) First order reaction
(D) $H_2(g) + Cl_2(g) \xrightarrow{h\nu} 2HCl$ is	(S) Zero order reaction

3.

Column-I (Curve)	Column-II (Slope)
(A) C vs t (abscissa) for zero order	(P) unity
(B) log C vs t (abscissa) for first order	(Q) zero
(C) $\left(\frac{-dc}{dt}\right)$ vs C for zero order	(R) $-k$
(D) $\ln\left(\frac{-dc}{dt}\right)$ vs ln C for first order	(S) $-\frac{k}{2.303}$

4.

Column-I Linear plots (with non zero slope)	Column-II (Order)
(A) $\ln\left[-\frac{d[A]}{dt}\right]$ vs. ln [A]	(P) 2
(B) $\log_e k$ vs. $\frac{1}{T}$	(Q) $\frac{1}{2}$
(C) $\log t_{1/2}$ vs. $\log [A]_0$	(R) 0
(D) $\frac{-d[A]}{dt}$ vs. $[A]^2$	(S) 1

$[A]$ = Reactant concentration at time 't'
k = rate constant
T = Absolute temperature

5.

Column-I (First order of reaction $A \rightarrow B$)	Column-II
(A) $[B]$ vs t	(P) 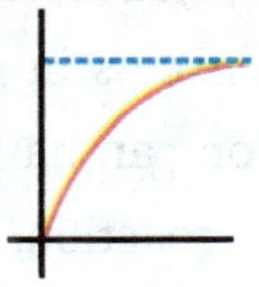
(B) $[A]$ vs t	(Q) 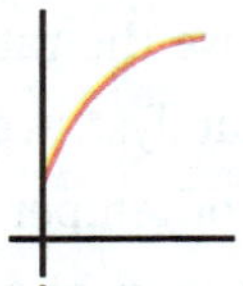
(C) $\frac{1}{[A]}$ vs t	(R) 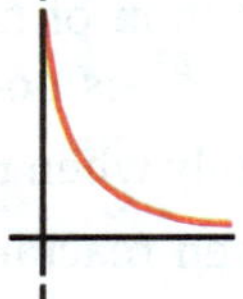
(D) ln k vs $\frac{1}{T}$	(S) 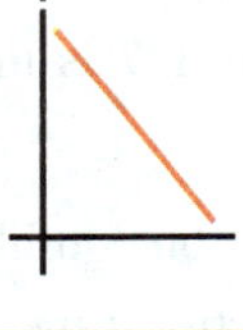

6.

Column-I	Column-II
(A) Isotones	(P) $^{234}_{91}Pa$ and $^{234}_{90}Th$
(B) Isobars	(Q) $^{12}_{6}C$ and $^{14}_{6}C$
(C) Isotopes	(R) $^{39}_{19}K$ and $^{19}_{9}F$
(D) Isodiaphers	(S) $^{39}_{18}Ar$ and $^{40}_{19}K$

7.

Column-I	Column-II
(A) α-emission	(P) Change in mass no.
(B) β-emission	(Q) No change in atomic no. and mass no.
(C) γ-emission	(R) Atomic no. decreases
(D) β^+ (Positron) emission	(S) Atomic no. increases

Assertion-Reason Type Questions

Each question contains STATEMENT-1 (Assertion) and STATEMENT-2 (Reason).
Examine the statements carefully and mark the correct answer according to the instructions given below :
(A) If both the statements are TRUE and STATEMENT-2 is the correct explanation of STATEMENT-1
(B) If both the statements are TRUE but STATEMENT-2 is NOT the correct explanation of STATEMENT-1
(C) If STATEMENT-1 is TRUE and STATEMENT-2 is FALSE
(D) If STATEMENT-1 is FALSE and STATEMENT-2 is TRUE

1. **STATEMENT-1 :** Molecularity has no meaning for a complex reaction
STATEMENT-2 : Molecularity defined only for RDS
2. **STATEMENT-1 :** An elementary reaction cannot have fractional order.
STATEMENT-2 : Stoichiometric coefficients in an elementary reaction can be fractional.
3. **STATEMENT-1 :** Concentration of reactant in zero order reaction is constant.
STATEMENT-2 : For zero order reaction $A \longrightarrow B$, successive half life of reaction decreases with the progress of the reaction.
4. **STATEMENT-1 :** Acid catalysed hydrolysis of esters is pseudo first order reaction.
STATEMENT-2 : Water is present in excess in given reaction.
5. **STATEMENT-1 :** The order of reaction can have fractional value.
STATEMENT-2 : For an elementary reaction, the partial orders are determined by the reaction stoichiometry.
6. **STATEMENT-1 :** Catalyst may increase the rate constant to a large extent.
STATEMENT-2 : By using suitable catalyst, we can significantly increase yield.
7. **STATEMENT-1 :** For each 10°C rise of temperature the k is nearly double.
STATEMENT-2 : Energy wise distribution of molecules in a gas sample is an exponential function of temperature so $e^{-E_a/RT}$ is doubled.
8. **STATEMENT-1 :** Product is formed only when the required orientation and energy conditions are met.
STATEMENT-2 : All collisions between reactants yield the desired product.
9. **STATEMENT-1 :** The plot of k *versus* $1/T$ is linear.
STATEMENT-2 : $k = A \cdot e^{-E_a/(RT)}$
10. **STATEMENT-1 :** For exothermic reaction equilibrium constant decreases with increase in temperature.
STATEMENT-2 : For exothermic reaction rate constant decreases with decrease in temperature.

11. **STATEMENT-1 :** If the activation energy of reaction is zero, temperature will have no effect on the rate constant.

STATEMENT-2 : Lower the activation energy faster is the reaction.

12. **STATEMENT-1 :** Active complex is an intermediate product.

STATEMENT-2 : Active complex is unstable with high vibrational energy.

13. **STATEMENT-1 :** The pre-exponential factor A has the same units for all reactions.

STATEMENT-2 : $e^{-E_a/RT}$ has no unit.

14. **STATEMENT-1 :** γ-rays have very high penetrating power.

STATEMENT-2 : γ-rays are electromagnetic radiations of high energy.

15. **STATEMENT-1 :** Nuclide $^{30}_{13}Al$ is less stable than $^{40}_{20}Ca$.

STATEMENT-2 : Nuclide having odd number of protons and neutrons are generally unstable.

16. **STATEMENT-1 :** Disintegration of $^{3}_{1}H$ (tritium) is accompanied by β-emission.

STATEMENT-2 : Tritium has high n/p ratio.

17. **STATEMENT-1 :** The life of radioactive object (organic origin) can found with the help of carbon dating.

STATEMENT-2 : $^{14}_{6}C$ is a α and β-emitter.

18. **STATEMENT-1 :** Neutrons are the best bombarding particles.

STATEMENT-2 : Neutrons are neutral particles.

19. **STATEMENT-1 :** Nucleus does not contain free electrons, yet it emit beta-particles

STATEMENT-2 : At high n/p ratio, one neutron supposed to give 1 proton and 1 e^-(β).

20. **STATEMENT-1 :** Rate of disintegration of thorium increases with the increase in moles of thorium.

STATEMENT-2 : Rate of disintegration does not depend upon temperature, pressure

Level-4

Subjective Problems

1. The rate of decomposition of $NH_3(g)$ at 10 atm on platinum surface is zero order. What is rate of formation (in $M\ min^{-1}$) of $H_2(g)$, if rate constant of reaction $2NH_{3(g)} \rightarrow N_2(g) + 3H_2(g)$ is $2.0\ M\ min^{-1}$?

2. $5A \longrightarrow$ Product

In above reaction, half-life period is directly proportional to initial concentration of reactant. The initial rate of reaction is $400\ mol\ lit^{-1}\ min^{-1}$.

Calculate the half-life period (in sec) when initial concentration of reactant is $200\ mol\ lit^{-1}$.

3. In an elementary reaction $A(g) + 2B(g) \longrightarrow C(g)$ the initial pressure of A and B are $P_A = 0.40$ atm and $P_B = 0.60$ atm respectively. After time T, if pressure of C is observed 0.1 atm, then find the value of $\frac{r_i \text{ (initial rate of reaction)}}{r_t \text{ (rate of reaction after time } t)}$

4. Carbon monoxide reacts with O_2 to form CO_2: $2CO(g) + O_2(g) \longrightarrow 2CO_2(g)$

Informations about this reaction are given in the table below.

[CO] mol/L	**[O_2]** mol/L	**Rate of reaction** (mol/L. min)
0.02	0.02	4×10^{-5}
0.04	0.02	1.6×10^{-4}
0.02	0.04	8×10^{-5}

What is the value for the rate constant for the reaction in properly related unit ?

5. Half-life for the zero order reaction $A(g) \longrightarrow B(g) + C(g)$ and half-life for the first order reaction $X(g) \rightarrow Y(g) + Z(g)$ are equal. If completion time for the zero order reaction is 13.86 min, then calculate the rate constant (in hr^{-1}) for the reaction $X(g) \rightarrow Y(g) + Z(g)$.

6. For any acid catalysed reaction, $A \xrightarrow{H^+} B$

half-life period is independent of concentration of A at given pH. At difinite concentration of A half -life time is 10 min at pH = 2 and half-life time is 100 min at pH = 3. If the rate law expression of reaction is $r = k[A]^x[H^+]^y$ then calculate the value of $(x + y)$.

7. For a reaction, $A \rightleftharpoons B$ equilibrium constant is 1.66 and $k_{forward} = 0.166\ hr^{-1}$.

Calculate the time (in hours) when concentration of B is 80% of its equilibrium concentration.

(Given : ln 25 = 3.20)

8. 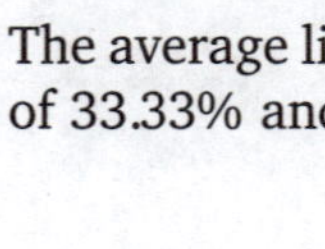

; At time $t = 0$, initial mole of A is 1.

Overall half life of the reaction is 15 days. Then calculate the number of mole of C after 45 days if the ratio of $k_1 : k_2 : k_3$ is 4 : 2 : 1

9. Iodine-131 is a radioactive isotope. If 1.0 mg of ^{131}I has an activity of 4.6×10^{12} Bq, what is the half -life of ^{131}I (in days) ?

10. The average life of a radioactive element is 7.2 min. Calculate the time interval (in min.) between the stages of 33.33% and 66.66% decay

11. A, B and C are isodiaphers while C, D and E are isobars. Calculate the difference of protons between A and E $^{206}_{82}A \longrightarrow B \longrightarrow C \longrightarrow D \longrightarrow E$

Given : Isodiaphers and isobars are formed in successive α and β-emission respectively.

12. In the given radioactive disintegration series,

$^{232}_{90}\text{Th} \longrightarrow {}^{208}_{82}\text{Pb}$

Calculate value of $(n + 2)$.

Where value of n is number of isobars formed in this series, suppose there is successive emission of β-particles.

13. In the given radioactive disintegration series

$^{235}_{92}\text{U} \longrightarrow {}^{207}_{82}\text{Pb}$

Calculate difference between number of α and number of β particles emitted in this series.

Numerical Problems

(If the numerical value has more than two decimal places, truncate/round-off the value to TWO decimal places.)

1. The reaction that occurs in a breath analyser, a device used to determine the alcohol level in a person's blood stream is

$$2K_2Cr_2O_7 + 8H_2SO_4 + 3C_2H_6O \longrightarrow 2Cr_2(SO_4)_3 + 3C_2H_4O_2 + 2K_2SO_4 + 11H_2O$$

If the rate of appearance of $Cr_2(SO_4)_3$ is 2.67 mol min^{-1} at a particular time, the rate of disappearance of C_2H_6O at the same time is .... mol min^{-1} (Nearest integer).

2. For the non-equilibrium process,

$$A + B \longrightarrow \text{Products}$$

the rate is first-order with respect to A and second order w.r.t. B.

When $[A] = 1.0\,\text{M}, [B] = 1.0\,\text{M}$, rate $= 1.0 \times 10^{-2}\,\text{Ms}^{-1}$,

When 75% each of reactant has been converted into products, rate of reaction is _____ $\times 10^{-4}\,\text{Ms}^{-1}$.

3. The following data were obtained during the first order thermal decomposition of SO_2Cl_2 at a constant volume.

$$SO_2Cl_2(g) \longrightarrow SO_2(g) + Cl_2(g)$$

Exp. No.	Time (sec)	Total pressure (atm)
1	0	0.5
2	100	0.6

The rate of reaction is _____ $\times 10^{-4}$ atm/sec when total pressure is 0.65 atm.

(Given : $\log 2 = 0.3$, $\ln x = 2.3 \log x$)

4. The reaction : $A \longrightarrow$ Product, follows first order kinetics and the progress of reaction is depicted in the figure (Given : $\ln 2 = 0.7$) :

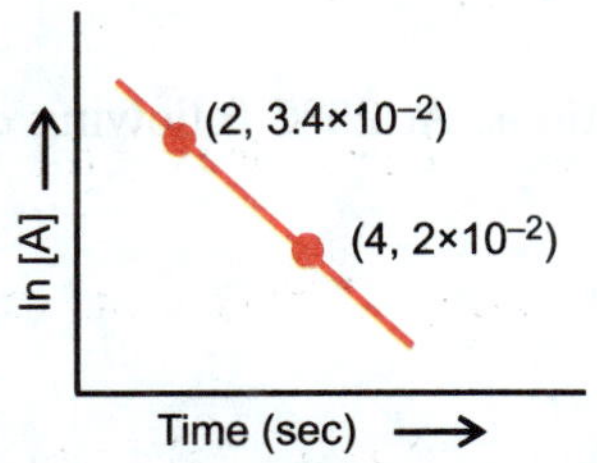

The half life of this reaction is ________ sec.

5. Sucrose hydrolysis in acid solution into glucose and fructose following first order rate law with a half-life of 3.33 h at 25°C. After 9 h, the fraction of sucrose remaining is f. The value of $\log_{10}\left(\frac{1}{f}\right)$ is ....$\times 10^{-2}$.

(Rounded off to the integer)

(Given : $\ln 10 = 2.303$, $\ln 2 = 0.693$)

6. Aqueous NH_4NO_2 decomposes according to the first order reaction :

$$NH_4NO_2(aq) \longrightarrow N_2(g) + 2H_2O(l)$$

At constant temperature and pressure, the volume of N_2 collected after 20 minutes is 20 mL and after a very long time is 40 mL. The rate constant for the reaction is ______ $\times 10^{-2}$ min^{-1}.

(Given : $\log 2 = 0.3010$)

7. Consider a first order reaction at 300 K in presence of inert gas

$$A(g) \longrightarrow B(g) + 2C(g)$$

Determine rate constant (in hr^{-1}) for the above reaction with the help of following information. ($\ln 2 = 0.7$)

Time	0	70 sec	∞
Total pressure	200	350	500

8. Consider the following gaseous reaction where initial pressure of $P(g)$ is 200 mm Hg

$$P(g) \longrightarrow 3Q(g) + R(g)$$

After time t, the pressure of system increases by 450 mm Hg.

Calculate the value of t (in second)?

(Given : Rate constant of reaction, $k = 4.606 \times 10^{-3}\ s^{-1}$, $\log 2 = 0.3$)

9. An acid-catalysed hydrolysis of ester is carried out in the presence of HCl. Small but same volume of reaction mixture were taken out at various instant and were titrated with standard NaOH solution. The following data was obtained.

Time (in min)	0	120	∞
Volume (mL) of NaOH solution needed for titration	20	55	60

The value of half-life of reaction is ______ min.

10. Consider a first order reaction

$$A \longrightarrow B + C$$

The optical rotation per mol or A, B and C are 20°, 10° and −30° after 10 min from starting total optical rotation of mixture was found to be 10° and after a long time −40°.

The value of rate constant (k) is ____ min^{-1}.

(Given : $\ln 16 = 2.8$, $\ln 10 = 2.3$)

11. Consider a reversible reaction $A \underset{k_2}{\overset{k_1}{\rightleftharpoons}} P$

which is elementary from both directions. At 25°C following curve is obtained

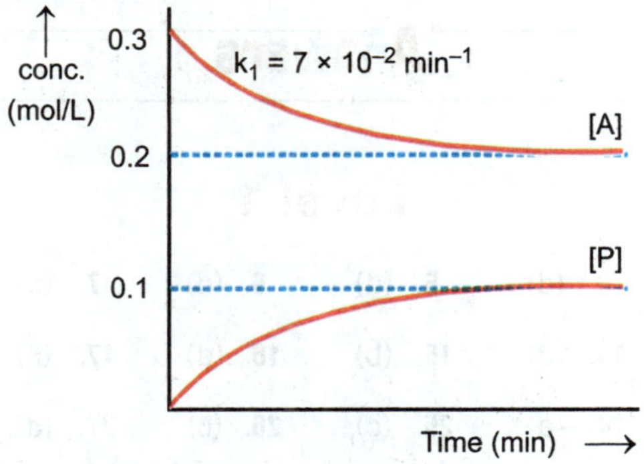

Determine the time (in min) at which 25% of A would be reacted.

(Given : $\ln 2 = 0.7$)

12. If the activation energy of a reaction is 80.9 kJ mol^{-1}, the fraction of molecules at 700 K, having enough energy to react to form product is e^{-x}. The value of x is________.

(Rounded off to the nearest integer)

(Given : $R = 8.31 JK^{-1} mol^{-1}$)

13. A first order reaction 75% completed in 120 min at 300 K temperature and in 40 min at 320 K. The activation energy of reaction is _______ kcal/mol.

(Given : $\ln 3 = 1.1$)

14. Study the following transition series.

$$A \xrightarrow{-\alpha} B \xrightarrow{-\alpha} C \xrightarrow{-\beta} D \xrightarrow{-\beta} E \xrightarrow{-\alpha} F \xrightarrow{-\beta} G \xrightarrow{-\alpha} H$$

If the mass number of 'A' is 230, then what is the mass number of 'H' ?

15. 16 g of a radioactive substance is reduced to 0.50 g after 1 hour. The average life of the radioactive substance is _______ min.

16. A mixture of two isotopes, ^{90}Y and ^{96}Y has a specific activity of $0.26\, N_A$ dps. Calculate the wt.% of ^{96}Y in mixture.

(Given : decay constant of ^{90}Y and ^{96}Y are 45 and $24\, s^{-1}$ respectively and $N_A = 6 \times 10^{23}$).

17. A radio nuclide 'A' decays simultaneously into 'B' and 'C', by α- and β-emission, respectively. The half-lives for the decay are 20 and 60 min, respectively. The time (in min) in which 93.75% of 'A' will decay is _______ .

18. A is a radioactive substance which decay into another radioactive substance B, and it also decay into stable product C as follows

$$A \xrightarrow{t_{1/2}=10 \text{ days}} B \xrightarrow{t_{1/2}=15 \text{ days}} C$$

Determine the time (in days) at which number of nucleus of B is maximum.

(Given : $\log 2 = 0.3$, $\log 3 = 0.48$, $\log 5 = 0.7$)

19. The amount of $_6C^{14}$ isotope in a piece of wood is found to one sixteenth $\left(\frac{1}{16}\right)$ of that present in a fresh piece of wood. Calculate the age ______ $\times 10^3$ years of the piece of wood.

(Given : $t_{1/2}$ of $_6C^{14} = 5770$ years)

20. An analysis of rock shows that molar ratio of Sr to Rb is 3.0. The age of the rock sample is $y \times 10^{10}$ yrs then the value of y is _______.

(Given : $^{87}_{37}Rb \longrightarrow {}^{87}_{38}Sr + {}_{-1}e^0$; Half-life period of Rb $= 5 \times 10^{10}$ years)

Answers

Level 1

1. (c)	2. (b)	3. (c)	4. (d)	5. (d)	6. (d)	7. (d)	8. (c)	9. (a)	10. (c)
11. (c)	12. (a)	13. (d)	14. (d)	15. (b)	16. (d)	17. (c)	18. (c)	19. (d)	20. (a)
21. (c)	22. (c)	23. (b)	24. (a)	25. (c)	26. (c)	27. (d)	28. (c)	29. (b)	30. (c)
31. (d)	32. (a)	33. (d)	34. (a)	35. (b)	36. (c)	37. (a)	38. (b)	39. (a)	40. (c)
41. (c)	42. (d)	43. (c)	44. (b)	45. (c)	46. (b)	47. (b)	48. (b)	49. (a)	50. (b)
51. (d)	52. (c)	53. (c)	54. (d)	55. (d)	56. (a)	57. (d)	58. (a)	59. (a)	60. (c)
61. (c)	62. (c)	63. (c)	64. (a)	65. (a)	66. (c)	67. (b)	68. (d)	69. (d)	70. (c)
71. (d)	72. (a)	73. (a)	74. (b)	75. (d)	76. (c)	77. (a)	78. (d)	79. (d)	80. (c)
81. (b)	82. (b)	83. (c)	84. (c)	85. (c)	86. (c)	87. (c)	88. (b)	89. (c)	90. (c)
91. (d)	92. (d)	93. (d)	94. (b)	95. (b)	96. (c)	97. (b)	98. (d)	99. (d)	100. (c)
101. (c)	102. (b)	103. (c)	104. (d)	105. (d)	106. (b)	107. (a)	108. (c)	109. (d)	110. (b)
111. (a)	112. (c)	113. (d)	114. (c)	115. (b)	116. (c)	117. (c)	118. (c)	119. (b)	120. (c)
121. (d)	122. (c)	123. (b)	124. (d)	125. (d)	126. (b)	127. (c)	128. (c)	129. (b)	130. (b)
131. (d)	132. (b)	133. (d)	134. (c)	135. (c)	136. (d)	137. (c)	138. (b)	139. (b)	140. (b)
141. (c)	142. (b)	143. (d)	144. (c)	145. (b)	146. (c)	147. (a)	148. (b)	149. (b)	150. (b)

Level 2

1. (d)	2. (c)	3. (c)	4. (c)	5. (d)	6. (c)	7. (b)	8. (a)	9. (a)	10. (c)
11. (d)	12. (b)	13. (b)	14. (c)	15. (d)	16. (d)	17. (d)	18. (b)	19. (c)	20. (c)
21. (d)	22. (b)	23. (b)	24. (b)	25. (a)	26. (c)	27. (b)	28. (a)	29. (c)	30. (c)
31. (d)	32. (a)	33. (c)	34. (a)	35. (a)					

Level 3

Passage–1	1. (b)	2. (b)	3. (c)
Passage–2	1. (c)	2. (b)	3. (b)
Passage–3	1. (a)	2. (b)	
Passage–4	1. (c)	2. (c)	
Passage–5	1. (c)	2. (a)	3. (b)

One or More Answers is/are Correct

1. (a,c,d)	**2.** (a,b,c)	**3.** (a,b,c)	**4.** (d)	**5.** (a,c,d)	**6.** (b,c,d)	**7.** (a,b,c,d)	**8.** (a,b,c,d)
9. (a,b,c,d)	**10.** (c)	**11.** (a,c,d)	**12.** (a,c,d)	**13.** (a,b,c)	**14.** (b,c)	**15.** (a,b,d)	**16.** (b,c)
17. (a,b)							

Match the Column

1. A → S; B → Q; C → P; D → R
2. A → Q; B → P, Q; C → R; D → S
3. A → R; B → S; C → Q; D → P
4. A → P, Q, S; B → P, Q, R, S; C → P, Q, R; D → P
5. A → P; B → R; C → Q; D → S
6. A → S; B → P; C → Q; D → R
7. A → P, R; B → S; C → Q; D → R

Assertion-Reason Type Questions

1. (C)	**2.** (C)	**3.** (D)	**4.** (A)	**5.** (B)	**6.** (C)	**7.** (A)	**8.** (C)	**9.** (D)	**10.** (B)
11. (B)	**12.** (D)	**13.** (D)	**14.** (A)	**15.** (B)	**16.** (A)	**17.** (C)	**18.** (A)	**19.** (A)	**20.** (B)

Level 4

Subjective Problems

1. 6	**2.** 3	**3.** 3	**4.** 5	**5.** 6	**6.** 2	**7.** 6	**8.** 2	**9.** 8	**10.** 5
11. 2	**12.** 7	**13.** 3							

Numerical Problems

1. 4.00	**2.** 1.56	**3.** 8.00	**4.** 100.00	**5.** 81.00	**6.** 3.47	**7.** 36.00	**8.** 300.00
9. 40.00	**10.** 0.05	**11.** 6.66	**12.** 14.0	**13.** 10.56	**14.** 214.00	**15.** 17.30	**16.** 96.00
17. 60.00	**18.** 18.00	**19.** 23.08	**20.** 10.00				

Hints & Solutions

Level 1

3. (c)

$$\text{Rate of reaction} = \frac{\text{Rate of disappearance/appearance}}{\text{Stoichiometric coefficient}}$$

$$= -\frac{d[A]}{dt} = -\frac{1}{2}\frac{d[B]}{dt} = 2.6 \times 10^{-2}$$

$$= -\frac{1}{2}\frac{d[B]}{dt}$$

$$5.2 \times 10^{-2} = -\frac{d[B]}{dt}$$

4. (d) Rate of reaction $= -\frac{1}{2}\frac{d[A]}{dt} = \frac{d[B]}{dt} = \frac{1}{3}\frac{d[C]}{dt}$

or $\frac{k_1[A]^2}{2} = \frac{k_2[A]^2}{1} = \frac{k_3[A]^3}{3}$

$$\frac{k_1}{2} = k_2 = \frac{k_3}{3}$$

5. (d) Rate of reaction $= K[\text{conc.}]^n$

$$K = \frac{\text{Rate}}{[\text{conc.}]^n} = \frac{Ms^{-1}}{M^n} = \frac{\text{mol L}^{-1}\text{ s}^{-1}}{\text{mol}^n\text{ L}^{-n}}$$

$$= \text{mol}^{1-n}\text{ L}^{n-1}\text{ s}^{-1}$$

6. (d) As explain in Q. No. 5

7. (d) $K_{eq} = \frac{k_f}{k_b} = \frac{\text{Conc. of products as per rate law}}{\text{Conc. of reactants as per rate law}}$, stoichiometry of reaction is required.

8. (c) K_{eq} at $1400\,K = \frac{0.29}{1.1 \times 10^{-6}} = 2.63 \times 10^5$; K_{eq} at $1500\,K = 9.28 \times 10^4$; Value of K_{eq} is reduces with rise in temperature so reaction is exothermic.

9. (a) Arrhenious eq^n. Rate constant $K = Ae^{-Ea/RT}$

10. (c) As explain in Q. 3

Rate of disappearance $= -\frac{d[BrO_3^-]}{dt}$

Rate of appearance $= +\frac{d[Br_2]}{dt}$

Rate of reaction $= -\frac{d[BrO_3^-]}{dt} = \frac{1}{3}\frac{d[Br_2]}{dt}$

11. (c) $x = kt$

$$t_{\text{completion}} = \frac{a}{k} = \frac{0.5}{0.1} = 5 \text{ min}$$

13. (d) Reaction mechanism required.

14. (d) $(a - x) = ae^{-kt}$ or $A_t = A_0 e^{-kt}$

where $A_t = (a - x) =$ conc. of reactant remained after time $= t$

$A_0 = a =$ Initial concentration of reactant when $t = 0$

17. (c) Rate of reaction decreases with respect to times in 1st order Rx^n.

18. (c) $t_{1/2} \propto \frac{1}{a^{n-1}}$;

where, a : initial concentration of reactant

n : order of reaction $\quad (\because n \neq 1)$

19. (d) $a - x = a - kt$

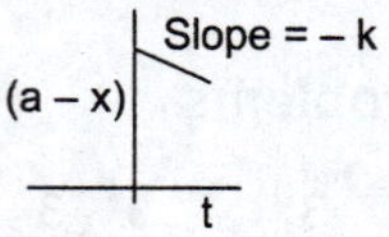

where,

a : initial conc. of reactant

x : conc. of reactant convert into product.

20. (a) $t_{1/2} = \frac{a}{2k} = k'.a$; $\log(t_{1/2}) = \log k' + \log a$

21. (c) $t_{1/2} = \frac{1}{K \cdot a}$ for 2nd order;

where a : initial conc. of reactant

22. (c) $t_{1/2} \propto a$; zero order

23. (b) $C = \frac{n}{V} = \frac{P}{RT}$

$$\therefore \frac{dC}{dt} = \frac{1}{V}\frac{dn}{dt} = \frac{1}{RT}\frac{dP}{dt}$$

24. (a) Rate of reaction $= K[A_2]^a[B_2]^b$

$0.04 = K(0.2)^a(0.2)^b$ ...(i)

$0.04 = K(0.1)^a(0.4)^b$ ...(ii)

$0.08 = K(0.2)^a(0.4)^b$ ...(iii)

$\frac{(i)}{(iii)}\quad \frac{1}{2} = \left(\frac{1}{2}\right)^b$; $\quad b = 1$

$\frac{(i)}{(ii)}\quad 1 = 2^a \times \left(\frac{1}{2}\right)^b$; $\quad a = 1$

25. (c) $R_0 = K[A_0]^2[B_0]$ ...(i)

$R_0' = K[1.5\,A_0]^2[3B_0]$

$R_0' = (1.5)^2 \times 3[A_0]^2[B_0]$ ...(ii)

From eq. (i) and (ii)

$R_0' = 6.75\,R_0$

26. (c) As explain in Q. 25

27. (d) Rate of reaction $= k[A]^2$

Rate of disappearance of $A = -\frac{d[A]}{dt}$

Rate of appearance of $B = +\frac{d[B]}{dt}$

Rate of reaction $= -\frac{d[A]}{dt} = +\frac{d[B]}{dt} = k[A]^2$

Rate of appearance of B after double the conc. of A.

$\frac{d[B]}{dt} = k[2A]^2$

28. (c) Order w.r.t. $A = 1$; order w.r.t. $B = 1$

R.O.R. $= \frac{1}{2}\frac{d(AB)}{dt} = k_r[A][B]$;

$\frac{1}{2} \times (2.5 \times 10^{-4}) = k_r\,(0.1)(0.1)$

$k_r = 1.25 \times 10^{-2}$

29. (b) Rate of appearance of $AB(g)$,

$\frac{d(AB)}{dt} = k_b[A][B]$

$\therefore \quad k_b = 2.5 \times 10^{-2}$

30. (c) As explain in Q. 24

31. (d) For elementary reaction molecularity of reaction = order of reaction.

32. (a) As explain in Q. 24

33. (d) $r = k[A]^a[B]^b$

Order of reaction $= n = a + b = 2$

In elementary reaction, partial order should be integer.

So, $a = 1$ and $b = 1$

34. (a) $t_{1/2} = \frac{1}{(n-1)k}\left[\frac{2^{n-1}-1}{a^{n-1}}\right]$

35. (b) As explain in Q. 18

36. (c) As explain in Q. 5

37. (a) Rate $= k[\text{conc.}]^0 = k$;

for zero order reaction.

38. (b) Conc. of water (solvent (l)) remains constant so rate is independent from conc. of H_2O.

39. (a) $CH_3COOC_2H_5(aq) + H_2O(l) \xrightarrow{H^+} CH_3COOH(aq) + C_2H_5OH(aq)$

$r = \frac{(k_{net})H_2SO_4}{(k_{net})HCl}$ = R.O.R. in presence of

H_2SO_4 > R.O.R. in presence of HCl

40. (c) $r_1 = k[A]^2[B]$; $r_2 = k[2A]^2[2B] = 8\,r_1$

41. (c) $[A]_t = [A] - kt = 1 - 0.001 \times 10 \times 60$

$= 0.4\,M$

$[B]_t = 0.001 \times 10 \times 60 = 0.6\,M$

42. (d) $t_{100\%} = \frac{a}{k} = \frac{1}{0.6}$ min $= 100$ sec.

Reaction will be finish in 100 sec., so concentration of B will reach maximum value $2\,M$ after 100 sec.

43. (c) $x = kt$ so $t_{1/2} = \frac{a}{2k}$ and $T = \frac{a}{k}$;

$t_{1/2} = \frac{1}{2 \times 0.001} = 50$ sec and $T = 100$ sec

45. (c) $x = 0.1 - 0.08 = 0.02\,M$;

$k = \frac{x}{t} = \frac{0.02}{10} = 2 \times 10^{-3}\,M\,\text{min}^{-1}$

$\therefore \quad t_{1/2} = \frac{[A]_0}{2K} = \frac{0.1}{2 \times 2 \times 10^{-3}} = 25$ min

$t_{\text{completion}} = 2 \times t_{1/2} = 50$ min.

46. (b) Elementary reaction,

$x(g) \longrightarrow y(g) + z(g)$

Rate $= k[x]$, first order

$k = \frac{1}{t}\ln\frac{[A_0]}{[A_t]}$

$\frac{\ln 2}{10\,\text{min}} = \frac{1}{t}\ln\frac{100}{10}$

$\frac{0.3}{10} = \frac{1}{t} \quad \Rightarrow t = 33$ min.

47. (b) $K_1 = \frac{1}{5}\ln\frac{0.2}{0.1}$; $K_2 = \frac{1}{10}\ln\frac{0.2}{0.05}$; $K_1 = K_2$.

48. (b) $K = \frac{1}{100}\ln\frac{100}{25} = \frac{1}{t}\ln\frac{100}{12.5}$

49. (a) $K = \frac{0.693}{t_{1/2}} = \frac{0.693}{480} = 1.44 \times 10^{-3}\,s^{-1}$

50. (b) $k = \frac{2.303}{t}\log\left(\frac{C_{A_0}}{C_A}\right)$;

$2.303 \times 1 = 2.303\log\left(\frac{C_{A_0}}{C_A}\right)$

$\frac{C_{A_0}}{C_A} = 10$

$\therefore \quad C_A \Rightarrow \frac{1}{10} \Rightarrow 0.1$

$\therefore$ rate after 1 min $r_1 = k \cdot C_A$

$\Rightarrow 2.303 \times 0.1 \Rightarrow 0.2303\,M\,\text{min}^{-1}$

51. (d) Reaction is second order; $k = 10^{-4}$ L/mol. min.

$$\therefore \quad -\frac{1}{3}\frac{d[A]}{dt} = k[A]^2$$

$$-\frac{d[A]}{dt} = 10^{-4} \times 3 \times (0.5)^2$$

$$= 7.5 \times 10^{-5} M\ \text{min}^{-1}$$

$$= \frac{7.5 \times 10^{-5}}{60} = 1.25 \times 10^{-6}\ \text{Ms}^{-1}$$

52. (c) $k = \frac{2.303}{t}\log\left(\frac{a}{a-x}\right)$

$$k = \frac{2.303}{32}\log\left(\frac{100}{1}\right) \Rightarrow \frac{2.303}{16}$$

$$t = \frac{2.303 \times 16}{2.303}\log\left(\frac{100}{0.1}\right) = 48 \text{ min}$$

53. (c) $t_{3/4} = \frac{1}{K}\ln\frac{100}{\frac{1}{4}\times 100}$

54. (d) $K_2 = \frac{1}{t}\ln\frac{100}{4}$ ...(i)

$K_1 = \frac{1}{t}\ln\frac{100}{50}$ ...(ii)

$$\frac{K_2}{K_1} = \frac{\ln 25}{\ln 2} = 4.65$$

55. (d) $k_I = \frac{0.693}{30}$

$$k_{II} = \frac{a}{2t_{1/2}} = \frac{2}{2 \times t_{1/2}} = \frac{1}{t_{1/2}} = \frac{1}{30}$$

$$\frac{k_I}{k_{II}} = \frac{0.693}{30 \times 1} \times 30 = 0.693$$

56. (a) $2N_2O_5(CCl_4) \longrightarrow 2N_2O_4(CCl_4) + O_2(g)$

$t = 0$	a mol		0
$t = t$	$a - x$		$\frac{x}{2}$ mol
$t = \infty$	0		$\frac{a}{2}$ mol

$a \propto 100$

$x \propto 90$

$$K = \frac{1}{t}\ln\frac{a}{a-x} = \frac{2.303}{500}$$

57. (d) $A \longrightarrow 3B$; $\quad P_T = P_0 + 2x$

After time t, $\quad P_0 - x \quad 3x$

$$x = \frac{P_T - P_0}{2} \qquad k = \frac{1}{t}\ln\left(\frac{P_0}{P_0 - x}\right)$$

After long time, 0 $\quad 3P_0$

$$x = \frac{P_T - \frac{P_\infty}{3}}{2} \qquad k = \frac{1}{t}\ln\left(\frac{\frac{P_\infty}{3}}{\frac{P_\infty}{3} - \left(\frac{3P_T - P_\infty}{6}\right)}\right)$$

$$P_\infty = 3P_O \qquad x = \frac{3P_T - P_\infty}{6} \qquad k = \frac{1}{t}\ln\left(\frac{P_\infty/3}{\frac{P_\infty}{2} - \frac{P_T}{2}}\right)$$

58. (a) $k = \frac{0.693}{2.1} = 0.33\ \text{hr}^{-1}$

Let t be the time for 90% decomposition,

so $\quad a = 100\%\ x = 90$ or $(a - x) = 10$

$$k = \frac{2.303}{t}\log\left(\frac{a}{a-x}\right)$$

$$\Rightarrow \quad t = \frac{2.303}{0.33}\log\left(\frac{100}{10}\right); \quad t = 6.978 \text{ hr}$$

$n_{NH_4NO_3}$ taken $= \frac{6.2}{62} \Rightarrow 0.1$;

as per given no. of moles of N_2O produced

$$= 0.1 \times \frac{90}{100}$$

Vol. of N_2O produced at STP

$$= 0.09 \times 22.4 = 2.016 \text{ L}$$

59. (a)

	A	$\longrightarrow$	$2B$	+	C
$t = 0$	P_i		0		0
$t = t$	$P_i - y$		$2y$		y

Total pressure $P_t = P_i - y + 2y + y = P_i + 2y$

$$\frac{P_t - P_i}{2} = y$$

$$K = \frac{2.303}{t}\log\frac{2P_i}{3P_i - P_t}$$

60. (c)

$$k = \frac{2.303}{t}\log\left(\frac{P_i}{P_t}\right); (CH_3)_2N_2 \longrightarrow C_2H_6 + N_2$$

$t = 0$ 200 ; $\quad t \quad 200 - x \quad x \quad x$

as per given $200 + x = 350 \qquad x = 150$

$$\therefore \quad k = \frac{2.303}{t}\log\left(\frac{200}{200 - 150}\right)$$

$$k = 5.77 \times 10^{-4}\ \text{sec}^{-1}$$

61. (c) $k = \frac{1}{t}\ln\left(\frac{r_\infty - r_0}{r_\infty - r_t}\right) = \frac{1}{t}\ln\left(\frac{r_0 - r_\infty}{r_t - t_\infty}\right)$

$$k = \frac{1}{70}\ln\left(\frac{44 + 11}{16.5 + 11}\right)$$

$k = \frac{\ln 2}{70}$ when solution is optically inactive $r_t = 0$

$$\therefore \quad k = \frac{1}{t}\ln\left(\frac{r_o - r_\infty}{0 - r_\infty}\right)$$

$$\frac{\ln 2}{70} = \frac{1}{t}\ln\left(\frac{44+11}{11}\right);$$

$$\frac{0.7}{70} = \frac{1}{t}\times 1.6\,;\; t = 160 \text{ min}$$

62. (c) $\frac{(t_{1/2})_1}{(t_{1/2})_2} = \left(\frac{a_2}{a_1}\right)^{n-1}$; $n = 1 + \frac{\log (t_{1/2})_1}{\log (t_{1/2})_2}$

63. (c) $kt = \frac{1}{n-1}\left[\frac{1}{A_t^{n-1}} - \frac{1}{A_0^{n-1}}\right]$

$$kt_{0.5} = \frac{1}{n-1}\left[\frac{2^{n-1}-1}{A_0^{n-1}}\right] \quad \text{...(1)}$$

$$kt_{0.875} = \frac{1}{n-1}\left[\frac{8^{n-1}}{A_0^{n-1}}\right]$$

$$\frac{t_{0.875}}{t_{0.5}} = \frac{8^{n-1}-1}{2^{n-1}-1}$$

64. (a) $kt = \ln\frac{\alpha_0}{\alpha_t}$; $k \times 60 = \ln\frac{80}{75}$

$$\Rightarrow \quad k = \frac{1}{60}\ln\frac{80}{75}$$

$$k \times 5 \times 60 = \ln\frac{80}{r_t}\,;\; 300 \times \frac{1}{60}.$$

$$\ln\frac{80}{75} = \ln\frac{80}{r_t}\,;\; \ln\frac{80}{r_t} = 0.32$$

$r_t = 58$

optical rotation observed after 5 hrs
$= 58 + 2 = 60$

65. (a) For zero order $t_{1/2} \propto [\text{conc.}]^0$

66. (c) $\frac{1}{2}\frac{dC_B}{dt} = k_1.C_A$; $\frac{dC_B}{2t} = 2k_1C_A$

$$\Rightarrow \; 2 \times 2 \times 10^{-3} \times 2 = 8 \times 10^{-3}\; M\,s^{-1}$$

67. (b)

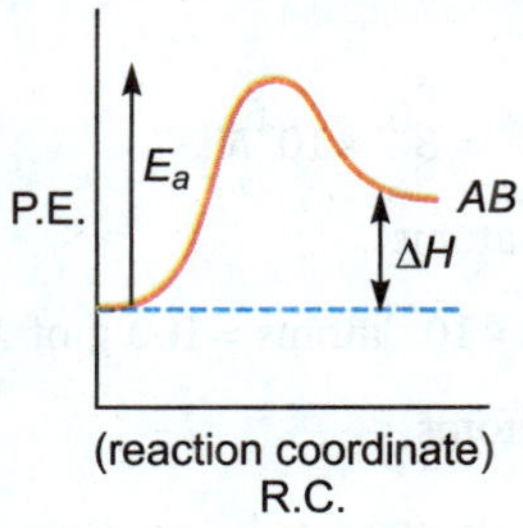

E_a is always greater than ΔH

68. (d) $Ea_2 = 58$

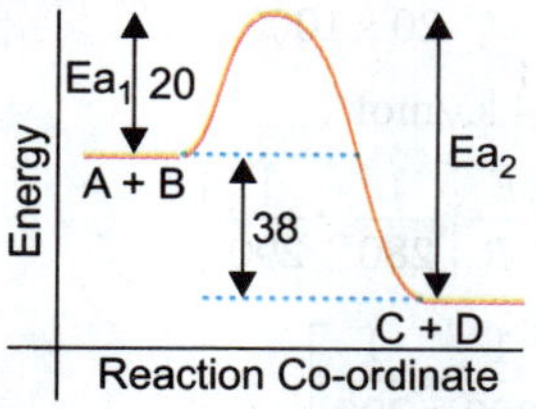

70. (c) First step is slow (require more activation energy) second step is fast (less activation energy) and overall reaction exothermic, so product energy level should be less as compared to reactants.

71. (d) Slow step require larger activation energy and product energy level is always more than reactant energy level in endothermic reaction.

72. (a) Rate of $Rx^n = K\,[\text{Conc.}]^n$

73. (a) $K = A \cdot e^{-E_a/RT}$

$$\ln K = \ln A - \frac{E_a}{RT}$$

74. (b) $K = A \cdot e^{-E_a/RT}$

Given $K = A \cdot e^{-1000}$

$$\frac{E_a}{RT} = 1000$$

75. (d) $k = \left(\frac{k_1.k_2}{k_3}\right)^{2/3}$; $E = \frac{2}{3}[E_{a_1} + E_{a_2} - E_{a_3}]$

$$\Rightarrow \; \frac{2}{3}[180 + 80 - 50] = 140 \text{ kJ/mol}$$

76. (c) $A_1 e^{-E_{a_1}/RT} = A_2 e^{-E_{a_2}/RT}$

$$\frac{A_2}{A_1} = e^{(E_{a_2} - E_{a_1})/RT}$$

$$10 = \text{Exp}\left(\frac{600}{RT}\right),\; R = 2 \text{ cal/K-mol}$$

$$\ln 10 = \frac{600}{2T}$$

$$T = \frac{300}{2.303}\text{K}$$

77. (a) Exothermic because of activation energy $E_b > E_f$

78. (d) $K = 1.2 \times 10^{14} e^{-\frac{25000}{RT}}$

$$\ln K = \ln 1.2 \times 10^{14} - \frac{25000}{RT}$$

81. (b) More activation energy, slow reaction rate.

82. (b) $k_{1(300)} = \frac{0.693}{20}$; $k_{2(320)} = \frac{0.693}{5}$

$$\ln\frac{k_{2(320)}}{k_{1(300)}} = \frac{E_a}{R}\left[\frac{1}{T_1} - \frac{1}{T_2}\right]$$

$$E_a = \frac{2.303\,RT_1T_2}{(T_2 - T_1)}\log\frac{k_2}{k_1}$$

$$= \frac{2.303\times 8.314}{20\times 1000}\times 300\times 320 \log 4$$

$$= 55.14 \text{ kJ/mol}$$

83. (c) $\left(\ln\frac{K_2}{K_1}\right)_1 = \ln 2 = \frac{E_a}{R}\left[\frac{1}{280}-\frac{1}{290}\right]$

$$\left(\ln\frac{K_2}{K_1}\right)_2 = \frac{E_a}{R}\left[\frac{1}{290}-\frac{1}{300}\right]$$

$$\frac{\ln 2}{\left(\ln\frac{K_2}{K_1}\right)_2} = \frac{\frac{10}{280\times 290}}{\frac{10}{290\times 300}}$$

$$\frac{0.693}{\ln\frac{K_2}{K_1}} = \frac{300}{280}$$

$$\frac{K_2}{K_1} = 1.9$$

84. (c) As explain in Q. 81

86. (c) $e^{-\frac{E_a}{RT}} = 3.8\times\frac{10^{-16}}{100}$

$$-\frac{E_a}{RT} = \ln 3.8\times 10^{-18}$$

$$E_a = 100 \text{ kJ/mol}$$

87. (c) $k_1 = A.e^{-\frac{83.314}{R\times 500}}$; $k_2 = A.e^{-\frac{75}{R\times 500}}$;

$$\frac{k_2}{k_1} = \exp\left(\frac{8.314\times 1000}{8.314\times 500}\right)$$

$$\Rightarrow \exp(2) \Rightarrow 7.38$$

88. (b) From slow reaction

Rate $= K[A][B]$

89. (c) Rate of reaction $= K_1[A][B_2]$ ...(i)

$$K_2 = \frac{[A]^2}{[A_2]} \quad ...(ii)$$

From Eq. (i) and (ii)

$$R = K_1\times\sqrt{K_2[A_2][B_2]} = K_1\times K_2^{1/2}[A_2]^{1/2}[B]$$

$$= K[A_2]^{1/2}[B_2]$$

$(\because K = K_1K_2^{1/2})$

91. (d) No significance of arrhenius equation in radioactivity.

92. (d) Due to charge on the α and β. Rays.

93. (d) ${}^{A}_{Z}X \longrightarrow {}^{A}_{Z+1}Y + {}^{0}_{-1}\beta$

$${}^{A}_{Z+1}Y \longrightarrow {}^{A-4}_{Z-1}Z + {}^{4}_{+2}\alpha$$

$$\underset{\text{High energy nucleous}}{{}^{A-4}_{Z-1}Z} \longrightarrow \underset{\text{Low energy nucleous}}{{}^{A-4}_{Z-1}Z} + \underset{\text{Energy}}{{}^{0}_{0}\gamma}$$

95. (b) ${}^{A}_{Z}x \longrightarrow {}^{A}_{Z+1}y + {}^{0}_{-1}e$

Neutron	$A-Z$	$A-(Z+1)$
Proton	Z	$Z+1$
$\frac{n}{p}=$	$\frac{A-Z}{Z}$	$\frac{A-Z-1}{Z+1}$

96. (c) As explain in Q. 95

104. (d) Isotones have same no. of neutron;

$$17-9 = B-8$$

$$B = 16$$

Isobars have same mass no.

$$A = B = 16$$

no. of neutrons $16-8 = 8$

105. (d) ${}^{234}_{90}\text{Th} \longrightarrow {}^{206}_{82}\text{Pb} + x{}^{4}_{2}\alpha^{+2} + y{}^{0}_{-1}\beta$

$234 = 206 + 4x = 0$

$x = 7$ α-particle

$90 = 82 + 2x - y$

$y = 6$

106. (b) Same number of neutron.

107. (a) Same mass number

108. (c) Same difference between n and p

110. (b) ${}^{7}_{3}\text{Li} + {}^{1}_{1}\text{P} \longrightarrow {}^{4}_{2}x + {}^{4}_{2}\alpha^{+2}$

112. (c) As explain Q. No. 110

113. (d) ${}_{92}\text{U}^{235} + {}_{0}n^{1} \longrightarrow {}_{54}X^{139} + {}_{38}\text{Sr}^{194} + 3{}_{0}n^{1}$

114. (c) ${}^{255}_{101}\text{Md} + {}^{4}_{2}\alpha \longrightarrow 2{}^{1}_{0}n + {}^{257}_{103}\text{Lr}$

115. (b) ${}^{235}_{92}\text{U} + {}^{1}_{0}n \longrightarrow {}^{94}_{38}\text{Sr} + {}^{140}_{54}\text{Xe} + 2{}^{1}_{0}n$

116. (c) $E_{\text{total}} = 200\frac{(3^5-1)}{3-1}$

$$= 100\times 242 = 24200 \text{ MeV}$$

117. (c) ${}^{230}_{90}\text{Th} + {}^{1}_{1}\text{P} \longrightarrow 2{}^{4}_{2}\text{He} + {}^{223}_{87}\text{Fr}$

118. (c) On emission of one α particle, formed nuclei occupied two position left in periodic table.

120. (c) On emission of positron atomic no. increases mass no. remain constant.

121. (d) ${}^{214}_{83}\text{Bi} \xrightarrow{-{}^{4}_{2}\alpha} {}^{210}_{81}\text{A} \xrightarrow{-{}^{0}_{-1}\beta} {}^{210}_{82}\text{B} \xrightarrow{-{}^{0}_{-1}\beta}$

$${}^{210}_{83}\text{C} \xrightarrow{-{}^{0}_{-1}\beta} {}^{210}_{84}\text{D} \xrightarrow{-{}^{4}_{2}\alpha} {}^{206}_{82}\text{E}$$

122. (c) $A = \lambda N$

$6.023\times 3.7\times 10^{10} = 3.7\times 10^{4}N$;

$N = 6.023\times 10^{6}$ atoms

1 mole or 6.023×10^{23} atoms $= 100$ g of X

$\therefore$ 6.023×10^{6} atoms

$$\Rightarrow \frac{100}{6.023\times 10^{23}}\times 6.023\times 10^{6} = 10^{-15} \text{ g}$$

123. (b) $A_1 = A_0e^{-\lambda t_1}$ ...(i)

$A_2 = A_0e^{-\lambda t_2}$ ...(ii)

$$\frac{A_2}{A_1} = e^{\lambda(t_1-t_2)} \quad \text{...(iii)}$$

124. (d) Four half-lives (Total time $= n \times$ half-life so, $n = 4$), hence 0.0625

125. (d) $\lambda = \frac{0.693}{69.3} = 10^{-2}\ \text{min}^{-1}$;

$$N = \frac{-\frac{dN}{dt}}{\lambda} = \frac{100}{10^{-2}} = 10{,}000$$

126. (b) $A_0 = \lambda N_0$

$$28 = \frac{\ln 2}{30} N_0$$

$$N_0 = 1212$$

129. (b) Equal fraction decay in equal periods of time, fraction of sample remaining after

3 days $\Rightarrow (0.9)^3 = 0.729$

131. (d) $\frac{N_0}{N} = \frac{A_0}{A} = 8$;

$$t = \frac{1}{\lambda}\ln\left(\frac{A_0}{A}\right) \Rightarrow t = \frac{5770}{0.693}\ln 8$$

$\Rightarrow$ 17310 year

133. (d) t for 20% left

$$\Rightarrow \quad t_1 = \frac{2.303}{\lambda}\log\frac{1}{1-0.8}$$

$$= \frac{\ln 5}{\lambda}$$

t for 50% left

$$\Rightarrow \quad t_2 = \frac{1}{\lambda}\ln 2$$

$$\frac{t_2}{t_1} = \frac{\frac{1}{\lambda}\ln 2}{\frac{1}{\lambda}\ln 5} = 0.43$$

$$t_2 = 0.43\ t_1$$

134. (c) $t = \frac{t0.5}{0.693}.\ \ln\frac{r_0}{r} = \frac{5770}{0.693}\times\ln\frac{120}{30}$;

$$t = \frac{5770}{0.693}\times 2\times 0.693 = 11{,}540 \text{ years}$$

135. (c) $\ {}^{40}_{19}\text{K} \longrightarrow {}^{40}_{20}\text{Ca} + {}^{0}_{-1}e$

At present 2.5×10^{-6} 7.5×10^{-6} mole

moles of ${}^{40}_{19}$K consumed $= 7.5\times10^{-6}$ mole

initial mole of ^{40}K was $(7.5+2.5)\times10^{-6}$ mole

$$t = \frac{1}{\lambda}\ln\left(\frac{10\times10^{-6}}{2.5\times10^{-6}}\right) = \frac{1.3\times10^9}{\ln 2}\times\ln(4)$$

$$= 1.3\times10^9\times2 = 2.6\times10^9 \text{ yr}$$

136. (d) $t_{avg} = \frac{1}{\lambda} = \frac{14}{0.693}$

137. (c) $\Delta t = |t_1 - 3t_{1/2}| = 200$

138. (b) Total time $= nt_{1/2}; n = 5; \frac{\text{Initial activity}}{2^n}$

Initial activity $= 0.01\times2^5 = 0.32\ \mu C_i$

139. (b) $\lambda t = \ln\left(\frac{A_0}{A}\right)$

$$= \frac{0.693}{t_{1/2}}\times200 = \ln\left(\frac{2140}{535}\right)$$

$$= t_{1/2} = 100 \text{ min.}$$

141. (c) $\lambda = \frac{1}{10}\ln\left(\frac{100}{100-25}\right)$

$$t = \frac{1}{\lambda}\ln\left(\frac{N_0}{N}\right)$$

$$t = \frac{10}{\ln\left(\frac{4}{3}\right)}\times\ln\left(\frac{4\times10^{20}}{10^{20}}\right)$$

$$t = 48.19$$

142. (b) $t_{1/2} = \frac{t}{4};\ t_{1/2} = T\ln 2$

so $\frac{t}{4} = T\ln 2;\ t = 4t\ln 2$

143. (d) W_0 = initial mass, W = mass. after 20 days

$$\lambda_A = \frac{2.303}{t}\log\left(\frac{W_0}{W_A}\right);$$

$$\lambda_B = \frac{2.303}{t}\log\left(\frac{W_0}{W_B}\right)$$

$$\lambda_A - \lambda_B = \frac{2.303}{t}\log\left(\frac{W_B}{W_A}\right)$$

$\therefore\ (t_{1/2})_B = 1.10$ day

144. (c) Given $(n_0)_B = 8\times(n_0)_A; (n)_A = 2\times(n)_B$

$$\lambda_A - \lambda_B = \frac{2.303}{t}\log\left[\frac{(N_A)_0}{(N_A)}\times\frac{(N_B)}{(N_B)_0}\right];$$

$$t = \frac{\ln\left(\frac{1}{16}\right)}{\frac{0.693}{50} - \frac{0.693}{10}}$$

$$= 50 \text{ min}$$

145. (b) as per given $\frac{dN}{dt} = \alpha - \lambda N$ for max. no. of nuclei;

$$\frac{dN}{dt} = 0;\ \alpha = \lambda\cdot N,\ N = \frac{\alpha}{\lambda}$$

146. (c) $\frac{\text{Atoms of Sr}}{\text{Atoms of Rb}} = 0.05$

or $\frac{\text{Atoms of (Sr + Rb)}}{\text{atoms of Rb}} = 1.05$

so, $\frac{\text{Initial no. of atoms of Rb}}{\text{Present no. of atoms of Rb}} = 1.05$

$$t = \frac{2.303}{\lambda}\log(n_0/n)$$

$$= \frac{2.303}{0.693}\times4.7\times10^{10}\log(1.05)$$

$$= 3.28\times10^9 \text{ year}$$

147. (a) $\lambda_A N_A = \lambda_B N_B$

($\because$ rate of disintegration are same)

$$\frac{N_A}{N_B} = \frac{\lambda_B}{\lambda_A} = \frac{2}{1}$$

148. (b) $N = N_0 . e^{-\lambda t}$; where N = Parent remaining (p)

and N_0 = Initial parent = Parent remaining (p) + daughter formed (d)

$$p = (p+d).e^{-\lambda t} \text{ or } \ln\frac{(p+d)}{p} = \lambda \cdot t$$

$$t = \frac{1}{\lambda}\ln\left(1+\frac{d}{p}\right)$$

149. (b) $\lambda = \lambda_1 + \lambda_2$; $\lambda = \frac{0.693}{22}$ and $\frac{\lambda_1}{\lambda_2} = \frac{2}{98}$

$\lambda_1 = 0.00063 \text{ year}^{-1}$; $\lambda_2 = 0.03087 \text{ year}^{-1}$

150. (b) ${}^{218}_{84}Po \longrightarrow {}^{214}_{82}Pb + {}^{4}_{2}He \longrightarrow {}^{214}_{83}Bi + {}^{0}_{-1}e$

Pb^{214} to reach max. no. of nuclei

$$t_{max.} = \frac{1}{\lambda_1 - \lambda_2}\ln\frac{\lambda_1}{\lambda_2} = 247.5 \text{ sec}$$

where $\lambda_1 = \frac{0.693}{183}$; $\lambda_2 = \frac{0.693}{161}$

Level 2

1. (d) $K_{eq} = \frac{k_f}{k_b} \Rightarrow \frac{[CH_3]^2}{[C_2H_6]}$

$$\therefore \quad [CH_3] = \frac{10^{-4}}{10} = 10^{-5}\, M$$

$$\frac{1.57\times10^{-3}}{k_b} = \frac{(10^{-5})^2}{1}$$

$\Rightarrow k_b = 1.57\times10^{7}\text{ L mol}^{-1}\text{ s}^{-1}$

3. (c) $r_1 = k[0.01]^a[0.01]^b = 6.93\times10^{-6}$ ...(i)

$r_2 = k[0.02]^a[0.01]^b = 1.386\times10^{-5}$ ...(ii)

$r_3 = k[00.2]^a[0.02]^b = 1.386\times10^{-5}$ ...(iii)

From data $a = 1$; $b = 0$;

overall order = 1; $k = 6.93\times10^{-4}\text{ sec}^{-1}$

$$6.93\times10^{-4} = \frac{1}{50\times60}\ln\frac{A_o}{A_t};$$

when $[A]_0 = 0.5$

$\Rightarrow \quad A_t = 0.0625$

rate of reaction $= 6.93\times10^{-4}\times0.0625$

$= 4.33\times10^{-5}\text{ Ms}^{-1}$

4. (c) $\frac{r_2}{r_1} = \frac{P_{A_2}.P_{B_2}^2}{P_{A_1}.P_{B_1}^2} = \frac{0.1\times(0.4)^2}{0.4\times1^2} = \frac{1}{25}$

6. (c)

$$t_1 = \frac{(t_{1/2})_1}{0.693}\ln\left(\frac{1}{(1-(1/4))}\right)$$

$$t_2 = \frac{(t_{1/2})_2}{0.693}\ln\left(\frac{1}{1-3/4}\right)$$

$$\frac{t_1}{t_2} = \frac{8}{1}\times\frac{\ln(4/3)}{\ln(4)} = 1 : 0.602$$

7. (b)

	A	$+ B$	$\longrightarrow C$	$+ D$
$t = 0$	1	1		
t	$1-x$	$1-x$	x	x

$$r = k[A]^{1/2}[B]^{1/2}$$

$$\Rightarrow \quad \frac{dx}{dt} = k(1-x)^{1/2}(1-x)^{1/2}$$

$$\text{or} \quad \frac{dx}{dt} = k(1-x)$$

$$\Rightarrow \quad t = \frac{1}{k}\ln\left(\frac{1}{1-x}\right);$$

$$t = \frac{2.303}{2.303\times10^{-2}}\log\left(\frac{1}{0.1}\right) = 100 \text{ sec}$$

8. (a)

(a) True $[B]_t = k_I t$; $0.25 = \frac{1}{\sqrt3}t$; $t = 0.25\sqrt3$

$[D]_t = k_{II}t = \sqrt3\times0.25\times\sqrt3 = 0.75\, M$

(b) False if $[C] = [A]$ then at that time $[B] < [D]$

(c) False $t_{100\%} = \frac{a}{k}$ (for zero order)

$$\frac{(t_{100\%})_I}{(t_{100\%})} = \frac{a_I}{a_{II}}\cdot\frac{k_{II}}{k_I} = \frac{0.5}{1}\times\frac{\sqrt3}{1/\sqrt3} = \frac{3}{2}$$

(d) $[A]_t = [A]_0 - k_I t$ or $[A]_t = 0.5 - \frac{1}{\sqrt3}t$

$[C]_t = [C]_0 - k_{II}t$ or $[C]_t = 1 - \sqrt3 t$

if $[A] = [C]_t$

i.e. $0.5 - \frac{1}{\sqrt3}t = 1 - \sqrt3 t$ or $(\sqrt3 - \frac{1}{\sqrt3})t = 0.5$

$$t = \frac{\sqrt3}{4}\text{ min.}$$

9. (a)

$$t_{1/2} = \frac{0.693}{k} = \frac{0.693}{2.772\times10^{-3}} = 250 \text{ sec}$$

Since volume is changing therefore half-life should be defined on the basis of moles, so moles of A remaining after 250 sec. Also final volume of the container

$$V_f = \frac{n_f}{n_i}\times V_i = \frac{0.05+0.15}{0.1}\times2 = 4 \text{ litre}$$

$\therefore$ Final concentration of $A = \dfrac{0.05}{4}$

$= 0.0125\ M$

10. (c) $A \longrightarrow 2B + C$

$t = 0$	P_i	0	0
t	$P_i - x$	$2x$	x
$t \to \infty$	0	$2P_i$	P_i

$P_\infty = 3P_i$ or $P_i = \dfrac{P_\infty}{3}$; $P_i + 2x = P_t$

$x = \dfrac{P_t - P_i}{2}$

As we know $k = \dfrac{2.303}{t}\log\left(\dfrac{P_i}{P_i - x}\right)$

so $k = \dfrac{2.303}{t}\log\left(\dfrac{2P_\infty}{3(P_\infty - P_t)}\right)$

11. (d) Let n is the moles of reagent 'R' when R is reacted with A at time $t = 0$

$A \longrightarrow B + C$

$t = 0$	n	0	0
at t	$n - x$	$2x$	$3x$
at $t \to \infty$	0	$2n$	$3n$

$\therefore \quad 5n = n_2 \Rightarrow n = \dfrac{n_2}{5}$

$n + 4x = n_1 \Rightarrow x = \dfrac{n_1 - n}{4}$

$k = \dfrac{2.303}{t}\log\left(\dfrac{n}{n - x}\right)$

so $\quad k = \dfrac{1}{t}\ln\left(\dfrac{4n_2}{5(n_2 - n_1)}\right)$

12. (b) $A(g) \longrightarrow 2B(g) + C(g)$

Let initial pressure	P_0	0	0
After 10 min.	$(P_0 - x)$	$2x$	x
After long time $(t \to \infty)$	0	$2P_0$	P_0

as per given $(P_0 - x) + 2x + x +$ vapour pressure of $H_2O = 188$

$P_0 + 2x = 160$ and $3P_0 + 28 = 388$

so, $\quad P_0 = 120$ and $x = 20$ torr

$k = \dfrac{1}{t}\ln\left(\dfrac{P_0}{P_0 - x}\right)$

$\Rightarrow \dfrac{1}{10}\ln\left(\dfrac{120}{100}\right) = \dfrac{1}{10} \times (\ln 4 + \ln 3 - \ln 10)$

$= 0.02\ \text{min}^{-1} = 1.2\ \text{hr}^{-1}$

13. (b) $k = \dfrac{2.303}{t}\log\left(\dfrac{r_\infty - r_0}{r_\infty - r_t}\right)$

$= \dfrac{2.303}{30}\log\left(\dfrac{-11-34}{-11-19}\right)$

$= 1.35 \times 10^{-2}\ \text{min}^{-1}$

when solution is optically inactive $r_t = 0$;

$t = \dfrac{1}{k}\ln\left(\dfrac{-45}{-11}\right) = 103.7$ min

14. (c) Overall rate constant

$= k = k_1 + k_2 + k_3 = 6.93 \times 10^{-3}$

$t_{1/2} = \dfrac{0.693}{6.93 \times 10^{-3}} = 100$ sec;

After half-life, $P_B + P_C + P_D = 4$ atm

$\dfrac{P_B}{P_B + P_C + P_D} = \dfrac{k_1}{k_1 + k_2 + k_3} = \dfrac{200}{693}$

$P_B = 4 \times \dfrac{200}{693}$

$= 1.154$ atm

15. (d) $A \xrightarrow{k_1} 2B$; $\quad A \xrightarrow{k_2} C$

$a_0 - x - y \quad 2x \quad a_0 - x - y \quad y$

$\dfrac{d[A]}{dt} = (k_1 + k_2)[A]$; $\quad \dfrac{[B]}{[C]} = \dfrac{2k_1}{k_2} = \dfrac{2x}{y}$

$= \dfrac{k_1}{k_2} = \dfrac{x}{y}$

$\because \quad k_1 = k_2 = x = y$

$(k_1 + k_2)t = \ln\dfrac{A_0}{A_t}$

$2 \times 6.93 \times 10^{-3} \times 50 = \ln\dfrac{A_0}{A_t}$

$= [A]_t = \dfrac{[A_0]}{2} = a_0 - x - y = a_0/2 \qquad (\because x = y)$

$\therefore x = a_0/4$. Total moles in container

$a_0 - x - y + 2x + y = a_0 + x = 1.25\ a_0$

% increase in moles $\Rightarrow 25$, So final pressure

$= 2 \times 1.25 = 2.5$ atm

16. (d) 1 mole of A will form 2 moles of B and C. After completion of reaction when 75% A converted into B and C then total no. of moles $= 0.5 + 2 \times 1.5 = 3.5$ moles

17. (d) $cis\text{-}X \underset{k_b}{\overset{k_f}{\rightleftharpoons}} trans\text{-}X$

Initial	a	0;	$K_{(aq)} = \dfrac{k_f}{k_b}$; $k_b = \dfrac{3 \times 10^{-4}}{0.1} = 3 \times 10^{-3}$
at time t	$a - x$	x	
at eqm	$a - x_e$	x_e	

As we know $(k_f + k_b) = \dfrac{1}{t}\ln\left(\dfrac{x_e}{x_e - x}\right)$

given $\quad x = \dfrac{x_e}{2}$

$\therefore \quad (k_f + k_b) = \dfrac{1}{2}\ln 2$

or $\quad (3 \times 10^{-3} + 3 \times 10^{-4}) = \dfrac{0.693}{t}$

$t = 210$ sec

18. (b) $k = A \cdot e^{-E_a/RT}$; $A \cdot e^{-\frac{E_a}{RT_1}} = A \cdot e^{-\frac{(E_a - x)}{RT_2}}$

$$\frac{E_a}{500} = \frac{(E_a - 16)}{400} = E_a = 80 \text{ kJ/mol}$$

19. (c) For parallel reactions

$$E_a = \frac{k_1 E_{a_1} + k_2 E_{a_2}}{k_1 + k_2}$$

$$= \frac{10^{-2} \times 100 + 4 \times 10^{-2} \times 120}{10^{-2} + 4 \times 10^{-2}}$$

$$E_a = 116 \text{ kJ/mol}$$

20. (c) $k = A \cdot e^{-E_a/(RT)}$

$\therefore$ Effective overall energy of activation

$$E_a = E_{a(2)} - E_{a(3)} + \frac{1}{2} E_{a(1)} - \frac{1}{2} E_{a(5)}$$

$$= 60 - 50 + \frac{1}{2} \times 40 - \frac{1}{2} \times 10$$

$$= 25 \text{ kJ/mol}$$

21. (d) $A_1 \cdot e^{-E_{a_1}/RT} = A_2 \cdot e^{-E_{a_2}/RT}$;

$$\frac{A_2}{A_1} = e^{(Ea_2 - Ea_1)/RT}$$

$$10^2 = \text{Exp}\left\{\frac{1200}{RT}\right\}; 2 \ln 10 = \frac{1200}{2T}$$

$$T = \left\{\frac{600}{4.606}\right\} \text{K}$$

22. (b) $\ln \frac{k_1'}{k_1} = \frac{E_1}{R}\left[\frac{1}{T_1} - \frac{1}{T_2}\right]$ ...(i)

$\ln \frac{k_2'}{k_2} = \frac{E_2}{R}\left[\frac{1}{T_1} - \frac{1}{T_2}\right]$ ...(ii)

Solving we get (ii)–(i),

$$\ln \frac{k_2'}{k_2} - \ln \frac{k_1'}{k_1} = \left(\frac{E_2 - E_1}{R}\right)\left(\frac{1}{T_1} - \frac{1}{T_2}\right)$$

(for equimolar formation of B and C, $k_2' = k_1'$)

$$\ln\left(\frac{k_1}{k_2}\right) = \left(\frac{8314}{8.314}\right)\frac{(T_2 - 300)}{300 \times T_2}$$

$$\ln 2 = (1000)\left(\frac{T_2 - 300}{300 \times T_2}\right)$$

$$T_2 = 378.74 \text{ K}$$

24. (b) $A + B \underset{\text{fast}}{\overset{\text{slow}}{\rightleftharpoons}} IAB$; So $E_{a(f)}$ is high and $E_{a(b)}$ is low.

$k_1 << k_2$; So, E_a for this step is very high and for next step is low and overall reaction is exothermic.

25. (a) Activation energy

$$E_{\text{I}} \approx E_{\text{II}} \approx E_{\text{V}} \approx E_{\text{VI}}$$

$$E_{1(f)} \approx E_{1(b)} \approx E_{2(b)} \approx E_{3(f)} \approx E_{3(b)} \approx E_{4(f)} \approx E_{5(f)} \approx E_{6(b)}$$

$E_{2(f)} \approx E_{4(b)}$ (slow steps so larger)

26. (c) $e^{-\frac{E_a}{RT}} \times 100 = 3.8 \times 10^{-16}$; $e^{-\frac{E_a}{RT}} = 3.8 \times 10^{-18}$

$$-\frac{E_a}{RT} = \ln (3.8 \times 10^{-18});$$

$R = 8.314$ and $T = 300$

$E_a = 100$ kJ/mol

27. (b) $\lambda t = \ln\left(\frac{A_0}{A}\right)$

$$= \frac{0.693}{30} \times 90 = \ln \frac{A_0}{A} = A_0 = A \times 8$$

$$\therefore \quad N_0 = \frac{A_0}{\lambda} = \frac{800}{0.693} \times 30 = 3.46 \times 10^4 \text{ atoms}$$

28. (a) at time $\underset{n_0 - x}{X} \longrightarrow \underset{x}{Y}$; $\frac{n_0 - x}{x} = \frac{1}{7}$; $n_0 = \frac{8x}{7}$;

$$\lambda = \frac{0.693}{6.93 \times 10^9} \Rightarrow 10^{-10} \text{years}^{-1}$$

$$t = \frac{2.303}{\lambda} \log\left(\frac{w_0}{w_0 - x}\right)$$

$$= \frac{2.303}{10^{-10}} \log (8)$$

$$= 2.079 \times 10^{10} \text{ years}$$

29. (c) λ_{net} or $\lambda = \lambda_1 + 2\lambda_2 = 1.8 \times 10^{-3} + 2 \times 10^{-3}$

$$= 2 \times 10^{-2}$$

Average life $(T_{\text{av}}) = \frac{1}{\lambda}$

$$= \frac{1}{2 \times 10^{-2}} = 50 \text{ sec}$$

30. (c) ${}_{83}Bi^{211} \longrightarrow {}_{81}Tl^{207} + {}_2He^4$; total time $= n \times$ half-life

moles of substance left after n halves

$$= \frac{\text{initial moles}}{2^n} = \frac{2}{2^4} = 0.125$$

moles of He produced

$$= 2 - 0.125 = 1.875$$

Pressure developed due to

$$\text{He} = \frac{1.875 \times 0.0821 \times 300}{100}$$

$$= 0.4618 \text{ atm}$$

31. (d) Let initial activities of A and B are A_0 and B_0

($\because$ after 2 half-lives of activity of A will remain $\frac{A_0}{4}$)

$$A_0 + B_0 = 8000 \text{ and also } \frac{A_0}{4} + B_0 = 3500$$

(we can assume that activity of B remains constant due to larger half-life)

so $\frac{3A_0}{4} = 4500$; $A_0 = 6000$; $B_0 = 2000$;

$$\frac{A_0}{B_0} = \frac{6000}{2000} = \frac{3}{1}$$

32. (a) No change in activity of sample during establishment of circulatory equilibrium.

Let volume of blood is V mL, so total vol. $=(V+1)$ mL after injection of sample.

2 mL sample has activity of 10 dpm, so $(V+1)$ mL sample has activity of $\frac{10}{2}\times(V+1)$

Since rate is constant so

$$\frac{10}{2}\times(V+1)=1000;\ V=199 \text{ mL}$$

33. (c) $\frac{A_{0(X)}}{A_{0(Y)}}=\frac{4}{1};\ \frac{A_x}{A_y}=1,$

$$\lambda_y-\lambda_x=\frac{1}{t}\ln\left(\frac{(A_0)_y}{(A_0)_x}\times\frac{A_x}{A_y}\right)$$

$$(\lambda_y-\lambda_x)t=\ln\left(\frac{1}{4}\right);(t_{1/2})_y=30 \text{ min.}$$

34. (a) Corrected C^{14} content :

Let initial C^{14} count $=A_0$;

$$A_0+A_0/10=15.4;\ A_0=14$$

$$t=\frac{1}{\lambda}\log\frac{14}{7}=5.770\times10^3 \text{ years};\ t=\frac{\ln 2}{\lambda}$$

$$=t_{1/2};\ t=5770 \text{ years}$$

35. (a) For U^{238}; $\lambda_1 t=\ln\left(\frac{n_0}{n_1}\right)$;

For U^{235}; $\lambda_2 t=\ln\left(\frac{n_0'}{n_2}\right)$

$$(\lambda_1-\lambda_2)t=\ln\left(\frac{n_0}{n_1}\times\frac{n_2}{n_0'}\right);$$

$$\ln 2\left[\frac{1}{0.5\times10^9}-\frac{1}{4.5\times10^9}\right]t=\ln(140)$$

$$t=4.02\times10^9 \text{ year}$$

Level 3

Passage-1

1. (b) For max. con. of B;

$$\frac{d[B]}{dt}=0;\text{ so } t_{max}=\frac{1}{k_1-k_2}\ln\frac{k_1}{k_2}$$

2. (b) $k_1 << k_2$

So rate of appearance of B is much lesser than rate of disappearance of B.

Passage-2

1. (c) $e^{-E_a/RT}$ is fraction of molecules having energy greater than activation energy (E_a), it is unit less. So A has same unit of K.

for 1st order reaction unit of K is s^{-1}.

2. (b) Answer is given in (1)

$$x=e^{-E_a/RT}$$

$$\ln x=-\frac{E_a}{RT}$$

3. (b) $\ln\frac{K_2}{K_1}=\frac{E_a}{R}\left[\frac{1}{T_1}-\frac{1}{T_2}\right]$

$$\ln 2=\frac{E_a}{2\text{ cal}}\left[\frac{1}{290}-\frac{1}{300}\right]$$

$$E_a=\frac{2.303\times0.3\times2\times290\times300}{10}\text{ cal}$$

$$\approx 12 \text{ K Cal}$$

Passage-3

1. (a) $\ln\frac{2k}{k}=\frac{E_{a_f}}{2}\left(\frac{1}{300}-\frac{1}{310}\right)=\frac{E_{a_f}\times10}{2\times300\times310}$

$$E_{a_f}=\frac{0.7\times2\times300\times310}{10}=13.02\text{ kcal}$$

2. (b) $\ln 4=\frac{\Delta H}{R}\left(\frac{1}{300}-\frac{1}{310}\right)$

$$1.4=\frac{\Delta H}{2}\times\frac{10}{310\times300}$$

$$\Delta H=2.8\times310\times30=26.04\text{ kcal}$$

$$E_{a_f}-E_{a_b}=\Delta H$$

$$E_{a_b}=E_{a_f}-\Delta H=13.02-26.04=-13.02\text{ kcal}$$

Passage-4

1. (c) $E=\frac{nhC}{\lambda}$

$$7.2\times10^{-3}\text{ J/s}=\frac{n\times6.6\times10^{-34}\times3\times10^8}{330\times10^{-9}}$$

No. of photon

$$=\frac{7.2\times10^{-3}\times330\times10^9}{6.6\times10^{-34}\times3\times10^8}=120\times10^{14}$$

$$\text{Mole of photon}=\frac{120\times10^{14}}{6\times10^{23}}=20\times10^{-9}$$

Mole of CO formed

$$=0.4\times20\times10^{-9}=8\times10^{-9}\text{ mol/s}$$

2. (c) Rate of formation of $C_2H_6=0.8\times20\times10^{-9}$

$$=16\times10^{-9}\text{ mol/s}$$

Passage-5

1. (c) $\lambda=\frac{1}{t}\ln\frac{N_0}{N}$

$$\frac{\ln 2}{t_{1/2}}=\frac{1}{t}\ln\frac{N_0}{\frac{N_0}{4}}$$

$$\frac{\ln 2}{2.5\times10^5}=\frac{1}{t}\ln 4$$

$$t=2\times2.5\times10^5=5\times10^5\text{ years.}$$

2. (a) $\frac{N}{N_0} = \left(\frac{1}{x}\right)^n$; where n = no. of halves;

$$\frac{N}{N_0} = \frac{1}{10} = \left(\frac{1}{x}\right)^n$$

$$= n = 4$$

Total time $= n \times t_{1/2}$

$$= 4740 = 4 \times t_{1/2};$$

$$t_{1/2} = 1185 \text{ years}$$

3. (b) $-\frac{dN}{dt} = \lambda \times N \Rightarrow \frac{0.693}{t_{1/2}} \times n \times N_A$

$$-\frac{dN}{dt} = \frac{0.693}{24{,}000 \times 365 \times 24 \times 60 \times 60}$$

$$\times 1 \times 6.02 \times 10^{23} \text{ dps or } 14.9 \text{ Ci}$$

One or More Answers is/are Correct

2. (A) $K = Ae^{-E_a/RT}$ as $T\downarrow$, $K\downarrow$ rate of reaction $\downarrow$

(B) Rate of reaction $= k[A]^0 = K$ = constant

(C) As surface area ↑ No. of molecules will be more to react per unit time, more surface site available for the reaction.

(D) Average rate defined for macro-scopic time and instantaneous rate defined for micro-scopic time.

3. For an elementary reaction $aA + bB \longrightarrow cC + dD$ rate law is always $r = K[A]^a[B]^b$ but not *vice versa*.

For a complex reaction $aA + bB \longrightarrow cC + dD$ rate law may or may not be $r = K[A]^a[B]^b$.

4. (A) In a rate law reactant, product or catalyst may be appeared.

(B) We can not

(C) In an elementary reaction only positive integer stoichiometric coefficient appears.

(D) A zero order reaction is always complex

5. (A) $K = Ae^{-E_a/RT}$

As $T\uparrow, K\uparrow$ so if $T \to \infty$, then $K \to A$

or as $Ea\uparrow$, $K\uparrow$ so if $E_a \to 0$, then $K \to A$

(B) Catalyst does not change ΔH of reaction.

(C) A negative catalyst decrease rate of reaction by increasing activation energy.

6. $A + B \longrightarrow C$

(A) If it is an elementary reaction then it will be bimolecular reaction.

(B) It may be exothermic eq. $NH_3(g) + HCl(g) \longrightarrow NH_4Cl(s)$

(C) It may be heterogeneous

(D) It may be photochemical reaction

7. $A(g) \longrightarrow 2B(g) + C(g)$

	$A(g)$	$2B(g)$	$C(g)$
$t = 0$	400		
$t = 20$	$400 - x$	$2x$	x

So $400 - x + 2x + x = 1000$

$$x = 300 \text{ mmHg}$$

$$K = \frac{1}{20}\ln\frac{400}{100} = \frac{2\ln 2}{20} = 0.0693 \text{ min}^{-1}$$

$$t_{1/2} = \frac{\ln 2}{K} = \frac{0.693}{0.0693} = 10 \text{ min}$$

$$\frac{30}{10} \times \ln 2 = \ln\frac{400}{400 - x}$$

$$8 = \frac{400}{400 - x} \Rightarrow 400 - x = 50$$

$$\Rightarrow x = 350 \text{ mmHg}$$

Partial pressure of C after 30 min = 350 mmHg

Total pressure after 30 min

$$= 400 + 2x = 400 + 700 = 1100 \text{ mmHg}$$

9. (a,b,c,d)

$$k = \frac{Be}{Ae} = \frac{1.6}{0.4} = 4\,;\, k = 4\,\frac{k_1}{k_2} = \frac{4 \times 10^{-2}}{k_2}$$

$$\Rightarrow k_2 = 10^{-2}$$

$$t_{0.5x_e} = \frac{1}{k + k_2}\ln\frac{x_e}{x_e - x}$$

$$= \frac{1}{5 \times 10^{-2}}\ln 2 = \frac{70}{2};\, t_{0.5x_e} = 14 \text{ min};$$

$$t_{0.1a} = \frac{1}{5 \times 10^{-2}} \times \ln\frac{x_e}{x_e - x} = \frac{1}{5 \times 10^{-2}}\ln\frac{1.6}{1.6 - 0.2}$$

$$= \frac{100}{5}\ln\frac{16}{14} = 20 \times 0.14 = 2.8$$

10. (A) α-particles are just nucleus of helium ($_2^4He^{+2}$)

(C) $_Z^AX \longrightarrow {}_{Z+1}^AY + {}_{-1}e^0$

so X and Y are isobars.

11. (A) $_{92}U^{235} + {}_0n^1 \longrightarrow {}_{56}Ba^{140} + 2\,{}_0n^1 + {}_Z^AX$

Atomic no. $\Rightarrow 92 + 0 = 56 + (2 \times 0) + Z$

$$Z = 36$$

Mass no. $\Rightarrow 235 + 1 = 140 + (2 \times 1) + A$

$$A = 236 - 172 = 94$$

So X is $_{36}^{94}Kr$

(B) $_{11}Na^{23} + {}_Z^AX \longrightarrow {}_{12}Mg^{23} + {}_0n^1$

Atomic no. $\Rightarrow 11 + Z = 12 + 0 \Rightarrow Z = 1$

Mass no. $\Rightarrow 23 + A = 23 + 1 \Rightarrow A = 1$

$X \longrightarrow {}_1^1H \longrightarrow$ Proton

(C) and (D) both are correct.

12. (A) SI unit of radioactivity is Bq

1 Bq = 1 dps

(B) 1 Ci = 3.7×10^{10} dps or Bq

13. Similar as question 11.

14. $^{M}_{Z}A \xrightarrow{-\alpha} {}^{M-4}_{Z-2}B \xrightarrow{-\beta} {}^{M-4}_{Z-1}C \xrightarrow{\beta} {}^{M-4}_{Z}D$

15. (a, b)

$^{1}_{1}P + _{-1}e^{0} \rightarrow {}^{1}_{0}n + \text{X-rays}$

17. (a,b)

$^{1}_{1}P \rightarrow {}^{1}_{0}n + {}^{e}_{+1}e$ (positron) ; $^{1}_{0}n \rightarrow {}^{1}_{1}P + {}^{0}_{-1}e$ (Beta)

Activity = λN, Decay constant λ does not depend upon temperature.

By the emission of α-particle, isodiaphers are formed.

By the emission of β-particle, isobars are formed.

Match the Column

1. (A) $K = Ae^{-E_a/RT}$, unit of K = unit of A

(B) $K = \dfrac{\text{Rate}}{(\text{Concentration})^n} = \dfrac{M/s}{(M)^n} = M^{1-n}s^{-1}$

$n = 0$ Unit of $K = Ms^{-1}$

(C) $n < 1$ Unit of $K = s^{-1}$

(D) $n = 2$ Unit of $K = M^{-1}s^{-1}$

2. (A) Molecularity of a reaction never be zero.

(B) Order of reaction may be 0, 1, 2, 3 fraction

(C) $H_2O_2 \longrightarrow H_2O + 1/2\,O_2$ if its an elementary then it is 1st order reaction.

(D) It is zero order reaction because reaction rate is depend on due to absorption of photon.

3. (A) $C_t = C_0 - Kt$ slope $= -K$

(B) $\log C_t = \log C_0 - \dfrac{K \cdot t}{2.303}$; slope $= -K/2.303$

(C) $-\dfrac{dC}{dt} = K[C]^\circ$; slope = zero

(D) $\ln\left(-\dfrac{dc}{dt}\right) = \ln K + \ln C$; slope = 1

4. (A) $-\dfrac{d[A]}{dt}$ vs $[A]$, Let the order or reaction w.r.t. A is x then for non zero slope $[A]^x$, and $x \neq 0$

(B) $K = Ae^{-E_a/RT}$

$\ln K = \ln A - \dfrac{E_a}{RT}$

It is applicable for all possible order in this slope $\neq 0$

(C) $t_{1/2} \propto [A]_0$ it is not possible for 1st order.

(D) $\dfrac{-d[A]}{dt}$ vs $[A]^2$, $[A]^2 \longrightarrow$ second order.

5. (A) Concentration of product increases w.r.t time.

(B) Concentration of reactant decreases w.r.t time.

(D) $K = Ae^{-E_a/RT}$

$\ln K = \ln A - \dfrac{E_a}{RT}$ slope $= -\dfrac{E_a}{R}$

6. (A) Isotone $\rightarrow$ atoms having same number of neutrons (*i. e.* value of mass number (A) – atomic number (Z).

(B) Isobare $\rightarrow$ Atoms has same mass no. and different atomic no.

(C) Isotope $\rightarrow$ Atoms has same atomic no. and difference mass no.

(D) Isodiaphers $\rightarrow$ Same difference between neutrons and proton.

7. (A) $^{A}_{z}X \xrightarrow{-\alpha} {}^{A-4}_{Z-2}Y$

(B) $^{A}_{z}X \xrightarrow{{}^{0}_{-1}\beta} {}^{A}_{Z+1}Y$, atomic no. increase.

(C) $^{A}_{z}X \xrightarrow{-{}^{0}_{0}\gamma} {}^{A}_{Z}Y$, no. change in mass no. and atomic no.

(D) $^{A}_{z}X \xrightarrow{-{}^{0}_{+1}\beta} {}^{A}_{Z-1}Y$, Atomic no. decrease.

Level 4

Subjective Problems

1. $2NH_3(g) \rightleftharpoons N_2(g) + 3H_2(g)$

Rate $= K[NH_3]^0$

Rate = 2

Rate $= \dfrac{1}{3} \times \dfrac{\Delta[H_2]}{\Delta t} \Rightarrow \dfrac{\Delta[H_2]}{\Delta t} = 6$

2. $t_{1/2} = \dfrac{[A_0]}{2K_A} = \dfrac{[A_0]}{2 \times 5K} = \dfrac{200 \text{ min.}}{2 \times 5 \times 400}$

$= \dfrac{60 \text{ sec.}}{20} = 3$

3. $A(g) + 2B(g) \longrightarrow C(g)$

	A	B	C
$t = 0$	0.4	0.6	
$t = t$	0.3	0.4	0.1

$r_i = KP_A^1 \cdot P_B^2$

$r_i = K \times 0.4 \times (0.6)^2$

$r_t = K(0.3)(0.4)^2$

$\dfrac{r_i}{r_t} = 3$

4. Rate of reaction $= K[CO]^x[O_2]^y$

$4 \times 10^{-5} = K(0.02)^x(0.02)^y$ ...(i)

$1.6 \times 10^{-4} = K(0.04)^x(0.02)^y$ ...(ii)

$8 \times 10^{-5} = K(0.02)^x(0.04)^y$ ...(iii)

$\dfrac{(i)}{(ii)}\ \dfrac{1}{4} = \left(\dfrac{1}{2}\right)^x$

$x = 2$

$\frac{\text{(i)}}{\text{(iii)}}\quad \frac{1}{2}=\left(\frac{1}{2}\right)^y$

$y=1$

From eq. (1)

$4\times10^{-5}=K(0.02)^2(0.02)^1$

$$K=\frac{4\times10^{-5}}{4\times10^{-4}\times2\times10^{-2}}=\frac{10}{2}=5$$

5. For zero order $2(t_{1/2})_0=t_{\text{Complete}}$;

$$(t_{1/2})_0=\frac{1.386}{2}\quad \text{...(i)}$$

For first order $(t_{1/2})_1=\frac{0.693}{k_1}$ ...(ii)

Given $(t_{1/2})_0=(t_{1/2})_1$

$\frac{13.86}{2}=\frac{0.693}{k_1}$

$k_1=0.1\,\text{min}^{-1}=0.1\times60\,\text{hr}^{-1}=6\,\text{hr}^{-1}$

6. $r=k[A]^x[H^+]^y$

Order of reaction w.r.t. $A=1$

$(x=1)$; $r=k'[H^+]^y$ $(\because k'=k[A]^1)$

$\frac{a/2}{10}=k'[10^{-2}]^y$ ...(i)

$\frac{a/2}{100}=k'[10^{-3}]^y$ ...(ii)

(i)/(ii); $y=1$;

where $a\to$ initial concentration.

7. $(k_f+k_b)t=\ln\frac{x_e}{x_{e-x}}=\ln\frac{x_e}{x_e-0.8x_e}$;

$(k_f+k_b)t=\ln\frac{x_e}{0.2x_e}=\ln 5=1.6$;

$\frac{k_f}{k_b}=K_c$, $k_b=0.1t=6$ hrs.

8. 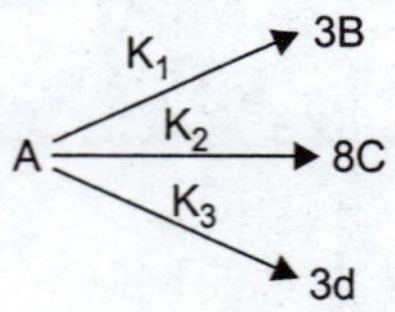

$K_1:K_2:K_3=4:2:1$

mole of A remain after 45 days

$$=\frac{N_0}{2^n}=\frac{N_0}{2^{\frac{T}{t_{1/2}}}}=\frac{1}{2^{45/15}}=\frac{1}{2^3}=\frac{1}{8}$$

moles of A convert into product $=\frac{7}{8}$ mol

moles of $[C]=\frac{K_2}{K_1+K_2+K_3}\times\frac{7}{8}\times\frac{8}{1}=2$

9. $A=\lambda N$

$$4.6\times10^{12}\text{Bq}=\frac{\ln 2}{t_{1/2}}\times\frac{1\times10^{-3}}{131}\times6\times10^{23}$$

$$t_{1/2}=\frac{0.693\times6\times10^{20}}{4.6\times10^{12}\times131\times3600\times24}=7.98\text{ day}$$

$t_{1/2}=8$

10. $\lambda=\frac{1}{t_{avg}}=\frac{1}{t}\ln\frac{[A_0]}{[A_t]}$; $\frac{1}{7.2}=\frac{1}{t_1}\ln\frac{100}{100-\frac{100}{3}}$;

$\frac{1}{7.2}=\frac{1}{t_2}\ln\frac{100}{100-\frac{200}{3}}$;

$t_2-t_1=7.2\ln 3-7.2\ln\frac{3}{2}=5$

11. ${}^{206}_{82}A\longrightarrow{}^{202}_{80}B\longrightarrow{}^{298}_{78}C\longrightarrow{}^{198}_{79}D\longrightarrow{}^{198}_{80}E.$

Proton difference in A and $E=2$

12. Number of α-particle $=\frac{232-208}{4}=6$

$Z_2=Z_1-2x+y$

(where $x=$ number of α and $y=$ number of β)

$82=90-12+y$

$y=4$

Series

${}^{A}_{Z}P\xrightarrow{\beta}{}^{A}_{Z+1}Q\xrightarrow{\beta}{}^{A}_{Z+2}R\xrightarrow{\beta}{}^{A}_{Z+3}S\xrightarrow{\beta}{}^{A}_{Z+4}T$

Total number of isobars $=5$

$n=5$, so $n+2=7$

13. Number of α-particle $=\frac{235-207}{4}=7$

$Z_2=Z_1-2x+y$

(where $x=$ number of α and $y=$ number of β)

$82=92-14+y$

$y=\mathbf{4}$

Number of α particle-number of β particle) $=\mathbf{3}$

Numerical Problems

1. R.O.R.$(r)=-\frac{1}{3}\frac{\Delta[C_2H_6O]}{\Delta t}=\frac{1}{2}\frac{\Delta[Cr_2(SO_4)_3]}{\Delta t}$

$-\frac{\Delta[C_2H_6O]}{\Delta t}=\frac{3}{2}\times2.67=4\,\text{mol min}^{-1}$

2. Rate $=k[A][B]^2$

$1\times10^{-2}=k[1][1]^2$

$k=10^{-2}$

Rate $=10^{-2}[0.25][0.25]^2$

$=1.56\times10^{-4}$

3. $SO_2Cl_2(g)\longrightarrow SO_2(g)+Cl_2(g)$

	$SO_2Cl_2(g)$	$SO_2(g)$	$Cl_2(g)$
$t=0$	0.5	0	0
$t=100$	$0.5-x$	x	x

$0.6=0.5+x\ \Rightarrow\ x=0.1$

$k\times100=\ln\frac{0.5}{0.5-0.1}=\ln\frac{10}{8}$

$k=2.3\times10^{-3}\ \text{sec}^{-1}$

Now, $0.65=0.5+y\ \Rightarrow\ y=0.15$

Rate $=kP_{SO_2Cl_2}$

Rate $=2.3\times10^{-3}(0.5-0.15)$

$=8\times10^{-4}$ atm/sec

4. For the first order process :

$$\ln[A] = \ln[A]_0 - kt$$

$$\text{Slope} = \frac{-1.4\times10^{-2}}{2}$$

$$k = 0.7\times10^{-2}\,s^{-1} = \frac{0.7}{t_{1/2}}$$

$$t_{1/2} = 100\text{ sec.}$$

5. $t = \frac{1}{k}\ln\left(\frac{n_0}{n}\right)$

$$9 = \frac{3.33}{\ln 2}\times 2.303\log\left(\frac{1}{f}\right)\text{ where } f = \frac{n_0}{n}$$

$$\log\left(\frac{1}{f}\right) = 0.8132 = 81.32\times10^{-2}$$

6. $k = \frac{2.303}{20}\log\left(\frac{40}{40-20}\right)$

$$= 3.466\times10^{-2}\,\text{min}^{-1}$$

7. $A(g) \longrightarrow B(g) + 2C(g)$

	$A(g)$	$B(g)$	$2C(g)$
At $t = 0$:	P_0	0	0
After 70 sec :	$(P_0 - x)$	x	$2x$
After long time:	0	P_0	$2P_0$

$$P_0 + P_{\text{inert}} = 200$$

$$P_0 + 2x + P_{\text{inert}} = 350$$

$$3P_0 + P_{\text{inert}} = 500$$

$$P_0 = 150;\quad x = 75$$

$$k = \frac{1}{t}\ln\left(\frac{P_0}{P_0 - x}\right)$$

$$= \frac{1}{70}\ln\left(\frac{150}{150-75}\right) = 10^{-2}\,s^{-1}$$

$$k = 60\times60\times10^{-2} = 36\,\text{hr}^{-1}$$

8. $P(g) \longrightarrow 3Q(g) + R(g)$

After time t : $(200 - x)$ $\quad 3x \quad x$

$$200 + 3x = 200 + 450$$

$$x = 150$$

$$x = 150;\quad t = \frac{1}{k}\ln\left(\frac{P_0}{P_0 - x}\right)$$

$$t = \frac{2.303}{4.606\times10^{-3}}\log\left(\frac{200}{200-150}\right)$$

$$t = 300\text{ s}$$

9. $t = \frac{t_{1/2}}{\ln 2}\cdot\ln\frac{V_\infty - V_0}{V_\infty - V_t}$

$$\Rightarrow\quad 120 = \frac{t_{1/2}}{\ln 2}\cdot\ln\frac{60-20}{60-55}$$

$$\therefore\quad t_{1/2} = 40\text{ min}$$

10. $k = \frac{2.303}{t}\cdot\log\frac{a}{a-x}$

A	$\longrightarrow$	B	+	C
a		0		0
$a - x$		x		x
0		a		a

$$20(a - x) + 10x - 30x = 10 \quad \ldots(1)$$

and $\quad 10a - 30a = -40 \quad \ldots(2)$

From (1) and (2) $a = 2$ and $x = \frac{3}{4}$

$$k = \frac{1}{10}\cdot\ln\left(\frac{2}{2-\frac{3}{4}}\right) \Rightarrow k = 0.05\text{ min}^{-1}$$

11. $k_{eq} = \frac{k_1}{k_2} = \frac{0.1}{0.2}$

$$\Rightarrow\quad k_2 = 2k_1 = 14\times10^{-2}\,\text{min}^{-1}$$

	A	$\rightleftharpoons$	P
$t = 0$	0.3		0
$t = eq.$	0.2		0.1
At time t	$0.3 - \frac{0.3}{4}$		$\frac{0.3}{4}$

$$(k_1 + k_2)t = \ln\frac{x_e}{x_e - x}$$

$$(7\times10^{-2} + 14\times10^{-2})t = \ln\frac{0.1}{0.1 - \frac{0.3}{4}}$$

$$21\times10^{-2}\times t = 1.4 \Rightarrow t = 6.66\text{ min}$$

12. $x = \frac{E_a}{RT} = \frac{80.9\times1000}{8.314\times700} = 13.9 \approx 14.00$

13. At 300 K, $t_{0.5} = \frac{120}{2} = 60$ min

At 320 K, $t_{0.875} = 1.5\times40 = 60$ min

$$\ln\frac{60}{20} = \frac{E_a}{2\times10^{-3}}\left(\frac{1}{300} - \frac{1}{320}\right)$$

$$E_a = 10.56\text{ cal/mol}$$

14. Mass no. of H $= 230 - 4\times4 = 214$

15. $\lambda = \frac{2.303}{60}\log\left(\frac{16}{0.5}\right)$

$$t_{av} = \frac{1}{\lambda} = \frac{60}{2.303\log(32)}$$

$$= \frac{60}{5\times0.693} = 17.30\text{ min}$$

16. $A = \lambda_1 N_1 + \lambda_2 N_2$

Let wt. of ^{90}Y in 1 g sample is x g

$$0.26\,N_A = 45\left(\frac{x}{90}\times N_A\right) + 24\left(\frac{1-x}{96}\times N_A\right) \qquad x = 0.04$$

wt. % of $^{90}Y = 0.04\times100 = 4$

wt. % of $^{96}Y = 100 - 4 = 96$

17. $A\,(\lambda) \xrightarrow{\lambda_1} B$, $\quad A\,(\lambda) \xrightarrow{\lambda_2} C$

or $\lambda = \lambda_1 = \lambda_2$

$$\frac{1}{t_{1/2}} = \frac{1}{(t_{1/2})_1} + \frac{1}{(t_{1/2})_2} = \frac{1}{20} + \frac{1}{60}$$

$$t_{1/2} = 15\text{ min}$$

Required time $(t) = 4\times t_{1/2} = 60.00$ min

18. $\lambda_1 = \dfrac{\ln 2}{10}$ and $\lambda_2 = \dfrac{\ln 2}{15}$

$$t_{max} = \frac{1}{(\lambda_2 - \lambda_1)} \ln\left(\frac{\lambda_2}{\lambda_1}\right)$$

$$= \frac{1}{\ln 2\left(\frac{1}{15} - \frac{1}{10}\right)} \ln\left(\frac{10}{15}\right) = \frac{\log\left(\frac{3}{2}\right) \times 30}{\log 2}$$

$$= (0.48 - 0.3) \times 100 = 18.00 \text{ days}$$

19. $\lambda t = \ln \dfrac{C^{14}_{\text{Living}}}{C^{14}_{\text{Dead}}}$

$$\frac{\ln 2}{5770} \times t = \ln \frac{C_0}{C_0/16}$$

$$t = 4 \times 5770 \text{ years} = 23080 \text{ years}$$

20.
$$^{87}_{37}\text{Rb} \longrightarrow {}^{87}_{38}\text{Sr} + {}_{-1}e^{0}$$

At present mole $\quad x \quad\quad 3x$

Initial mol of $^{87}_{37}\text{Rb} = 4x$

$$t = t_{1/2}(\text{Rb}) \times \ln 4$$

$$t = 2t_{1/2} = 2 \times 5 \times 10^{10}$$

$$= 10 \times 10^{10} \text{ years}$$

CHAPTER 9

Electrochemistry

Electrolytic Cell

- **Electrolysis :** Chemical changes in electrolyte solution by passage of electric current, resulting into deposition of metals or liberation of gases at electrodes is known as electrolysis.
- **Electrolytic Cell :** This cell converts electrical energy into chemical energy. The entire assembly except that of the external battery is known as the electrolytic cell.

Quantitative Aspect of Electrolysis

Faraday's laws of electrolysis :

- **First law of electrolysis :** Amount of substance deposited or liberated at an electrode is directly proportional to amount of charge passed (utilized) through the solution.

$$W \propto Q$$

$$W = ZQ = Z \times i \times t$$

$$\text{Current efficiency} = \frac{\text{Actual charge utilised in process}}{\text{Charge passed through battery}} \times 100$$

W = mass liberated/deposited, Q = charge in coulomb, i = current, t = time

Z = electrochemical equivalent, 1 Faraday = 96500 coulomb

- **Second law of electrolysis :** When same amount of charge is passed through different electrolytic solutions connected in series then mass of substances deposited or dissolved at anode or cathode are in ratio of their equivalent mass. *i.e.*, $W_1/W_2 = E_1/E_2$

$$\frac{W_1}{W_2} = \frac{E_1}{E_2}$$

Products of Electrolysis

S.No.	Electrolyte	Anode Product	Cathode Product
1.	NaCl (Molten) with Pt electrode	$Cl_2(g)$	Na (l)
2.	NaCl (aq) with Pt electrode	$Cl_2(g)$	$H_2(g)$
3.	$Na_2SO_4(aq)$ with Pt electrode	$O_2(g)$	$H_2(g)$

4.	$NaNO_3(aq)$ with Pt electrode	$O_2(g)$	$H_2(g)$
5.	$AgNO_3(aq)$ with Pt electrode	$O_2(g)$	Ag (s)
6.	$CuSO_4(aq)$ with inert electrode	$O_2(g)$	Cu (s)
7.	$CuSO_4(aq)$ with copper electrode	Cu dissolves	Cu (s)

Galvanic Cell

This cell converts chemical energy into electrical energy.

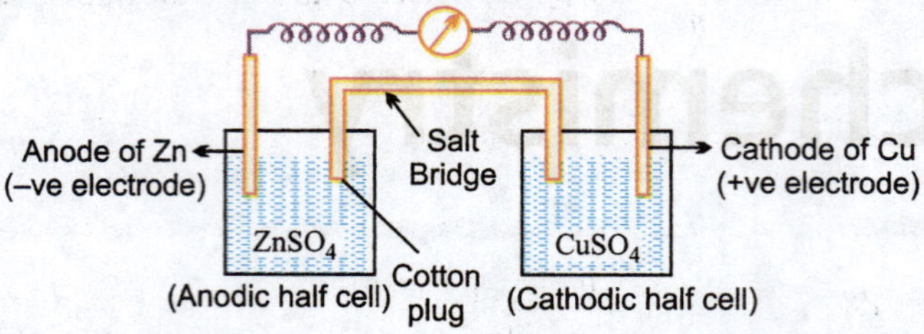

Galvanic cell is made up of two half cells *i.e.*, anodic and cathodic. Oxidation takes place at anode and reduction at cathode. It is also known as **voltaic cell.** It may be represented as shown in Fig. Zinc rod immersed in $ZnSO_4$ behaves as anode and copper rod immersed in $CuSO_4$ behaves as cathode.

- **Anode half cell reaction :**

$$Zn(s) \longrightarrow Zn^{2+}(aq) + 2e^- \text{ (oxidation)}$$

- **Cathode half cell reaction :**

$$Cu^{2+}(aq) + 2e^- \longrightarrow Cu(s) \text{ (reduction)}$$

- **Over all process :**

$$Zn(s) + Cu^{2+}(aq) \longrightarrow Cu(s) + Zn^{2+}(aq)$$

In galvanic cell like Daniel cell; electrons flow from anode (zinc rod) to the cathode (copper rod) through external circuit; zinc dissolves as Zn^{2+} ; Cu^{2+} ion in the cathode cell picks up two electron and become deposited at cathode.

Salt Bridge

It is U-shaped tube contains saturated solution of inert electrolyte like KCl, KNO_3, NH_4Cl and NH_4NO_3 etc. in agar-agar gel.

- Cation and anion of inert electrolyte have same mobility.
- Ions of inert electrolyte do not mix with electrolytic solution present in the half cell.
- Ions of inert electrolyte do not participate in electrochemical change.

Function of Salt Bridge

- It complete the electrical circuit.
- It maintained the two half cell electrically neutral by the flow of ions.

Representation of a cell (IUPAC conventions) : Let us explain the convention taking the example of Daniel cell.

(i) Anodic half cell is written on left and cathodic half cell on right hand side.

$$Zn(s)|ZnSO_4(aq)||CuSO_4(aq)|Cu(s)$$

(ii) Two half cells are separated by double vertical lines and it indicate salt bridge or any type of porous partition.

(iii) EMF (electromotive force) may be written on the right hand side of the cell.

(iv) Single vertical lines indicate the phase separation between electrode and electrolyte solution.

$$\underset{\text{Anode}}{\overset{}{Zn(s)\ \underset{(-)}{|}}}\ Zn^{+2}(aq)\,||\,Cu^{+2}(aq)\ \underset{\substack{(+)\\ \text{Catode}}}{|}\ Cu(s)$$

$$E^{\circ}_{cell} = E^{\circ}_{OP(LHS)} + E^{\circ}_{RP(RHS)}\ ;\quad E_{cell} = E_{OP(LHS)} + E_{RP(RHS)}\ ;\ E_{cell} = E_{RP(RHS)} - E_{RP(LHS)}$$

Nernst's Equation

$$E_{cell} = E^{\circ}_{cell} - \frac{2.303RT}{nF}\cdot \log Q$$

At 298 K temp. $E_{cell} = E^{\circ}_{cell} - \frac{0.0591}{n}\log Q$; where Q is reaction quotient.

Cell Thermodynamics

(i) $\Delta G = -nFE_{cell}$

(ii) $\Delta G^{o} = -nFE^{\circ}_{cell}$

(iii) Equilibrium constant (K), $\log K = \frac{n\times E^{\circ}_{cell}}{0.0591}$

(iv) Temp. coefficient of cell $= \left(\frac{\partial E}{\partial T}\right)_P$

(v) Enthalpy of reaction inside the cell, $\Delta H = -nFE + nFT\left(\frac{\partial E}{\partial T}\right)_P$

(vi) Entropy change inside the cell, $\Delta S = nF\left(\frac{\partial E}{\partial T}\right)_P$

Concentration Cell

Anode and cathode both are formed by same substance so $E^{\circ}_{cell} = 0$, such type of cell is known as concentration cell. Concentration gradient may arise either in electrode material or in electrolyte. Thus there are two types of concentration cells.

- **Electrode gas conc. cell :**

$$Pt, \underset{P_1\ atm}{H_2(g)}|H^+(C)|\underset{P_2\ atm}{H_2(g)}, Pt$$

$$E_{cell} = \frac{0.0591}{2}\log\frac{P_1}{P_2}$$

- **Electrolyte concentration cell :**

$$Zn(s)|ZnSO_4(C_1)||ZnSO_4(C_2)|Zn(s)$$

$$E_{cell} = \frac{0.0591}{2}\log\frac{C_2}{C_1}$$

Different Type of Electrodes

S.No.	Name of Electrode	Anode	Cathode
1.	Hydrogen electrode	$Pt(s)|H_2(g)|H^+(aq)$ $E_{OP} = E^{\circ}_{OP} - \frac{0.0591}{2}\log\frac{[H^+]^2}{P_{H_2}}$	$H^+(aq)|H_2(g)|Pt(s)$ $E_{RP} = E^{\circ}_{RP} - \frac{0.0591}{2}\log\frac{P_{H_2}}{[H^+]^2}$
2.	Metal-metal ion electrode	$M(s)|M^{+n}(aq)$ $E_{OP} = E^{\circ}_{OP} - \frac{0.0591}{n}\log[M^{+n}]$	$M^{+n}(aq)|M(s)$ $E_{RP} = E^{\circ}_{RP} - \frac{0.0591}{n}\log\frac{1}{[M^{+n}]}$

3.	Calomel electrode	$Hg\,(l), Hg_2Cl_2(s)\,\|\,Cl^-(aq)$ $E_{OP} = E^\circ_{OP} - \frac{0.0591}{2}\log\frac{1}{[Cl^-]^2}$	$Cl^-(aq)\,\|\,Hg_2Cl_2(s), Hg\,(l)$ $E_{RP} = E^\circ_{RP} - \frac{0.0591}{2}\log[Cl^-]^2$
4.	Redox electrode	$Pt\,(s)\|Fe^{+2}(aq), Fe^{+3}(aq)$ $E_{OP} = E^\circ_{OP} - \frac{0.0591}{1}\log\frac{[Fe^{+3}]}{[Fe^{+2}]}$	$Fe^{+3}(aq), Fe^{+2}(aq)\,\|\,Pt\,(s)$ $E_{RP} = E^\circ_{RP} - \frac{0.0591}{1}\log\frac{[Fe^{+2}]}{[Fe^{+2}]}$
5.	Metal insoluble salt anion electrode	$Ag\,(s)\|\,AgCl\,(s)\,\|\,Cl^-(aq)$ $E_{OP} = E^\circ_{OP} - \frac{0.0591}{1}\log\frac{1}{[Cl^-]}$	$Cl^-(aq)\,\|\,AgCl\,(s)\,\|\,Ag\,(s)$ $E_{RP} = E^\circ_{RP} - \frac{0.0591}{1}\log[Cl^-]$

Batteries

Electrochemical cells can be used as **batteries.** Batteries are of two kinds :

- **Primary battery :** Where the reaction occurs only once and can not be reused once it becomes dead over the course of time. For examples, dry cell (Leclanche cell), Mercury cell.
- **Secondary battery :** Which can be recharged by passing current through it in the opposite direction so that it can be used again. For example, lead-acid battery, Nickel-cadmium cell.

Fuel Cells

Fuel cells are the galvanic cells that convert the energy of combustion of fuels (*e.g.*, hydrogen, methane, methanol, etc.) directly into electrical energy.

Fuel cells are more efficient thermodynamically, and more of the energy of the reaction can be made available for useful work provided that the supply of reactants is maintained.

Methanol, ethanol, hydrazine, formaldehyde, carbon monoxide can be used as fuels in fuel cells apart from hydrogen.

Corrosion

When the metal exposed to some environment it gets converted to its oxides. The oxidative deterioration of metal is known as corrosion.

Example : Rusting of iron, tarnishing of silver, development of green coating on copper and bronze etc.

Oxidation $\qquad Fe\,(s) \longrightarrow Fe^{2+}\,(aq) + 2e^-$

Reduction $\qquad O_2(g) + 4H^+(aq) + 4e^- \longrightarrow 2H_2O(l)$

Atmospheric oxygen oxidised Fe^{2+}

$$2Fe^{2+}(aq) + 2H_2O(l) + \frac{1}{2}O_2(g) \longrightarrow Fe_2O_3(s) + 4H^+(aq)$$

Hydrated feric oxide $(Fe_2O_3 \cdot xH_2O(s))$ is known as rust.

- **Factors which enhance corrosion :**
  (i) Presence of impurities in the metal.
  (ii) Presence of moisture.
  (iii) Presence of electrolyte.
- **Prevention of corrosion :**
  (i) Barrier protection by oil/grease layer, paints or electroplating.
  (ii) Sacrificial protection by mixing the metal with more electropositive metal.

Conductance

- **Conductance (G):** It is defined as the reciprocal of the electrical resistance *i.e.*, $G = 1/R$. It measures the ease with which the current flows through a conductor.
  Unit : Siemens, S or Ω^{-1}.
- **Specific resistance or resistivity (ρ) :** The resistivity or specific resistance is defined as the resistance in ohm of a conductor having length equal to 1 cm and area of cross-section equal to 1 cm^2.

$$R \propto \frac{l}{A} \text{ or } R = \rho\frac{l}{A} \text{ where } \rho = \text{specific resistance, } l = \text{length of conductor}$$

A = area of cross-section of the conductor.

Unit : ohm cm.

- **Specific conductance or conductivity (k):** It is define as the reciprocal of specific resistance.

$$i.e., \quad k = \frac{1}{\rho} \quad \therefore \; R = \rho\frac{l}{A} \quad k = G \times \frac{l}{A}$$

$$k = G \times G^*$$

$$G^* = \frac{l}{A}, \quad \text{where } G^* = \text{cell constant}$$

When $l = 1$ cm and $A = 1\,cm^2$; $k = G$

Thus conductivity is the conductance of one centimeter cube of the solution of an electrolyte.
Unit : $\Omega^{-1}cm^{-1}, Scm^{-1}$.

- **Molar conductivity (Λ_M):** The conducting power of all the ions produced by dissolving 1 mole of an electrolyte in solution.

Mathematically, $\Lambda_M = k \times V$

$$\Lambda_M (Scm^2\,mol^{-1}) = k \times \frac{1000}{M} \qquad (M = \text{molarity})$$

Variation of conductivity and molar conductivity with concentration :

For weak and strong electrolytes

- Conductivity decreases with concentration.
- Molar conductivity increases with decrease in concentration.

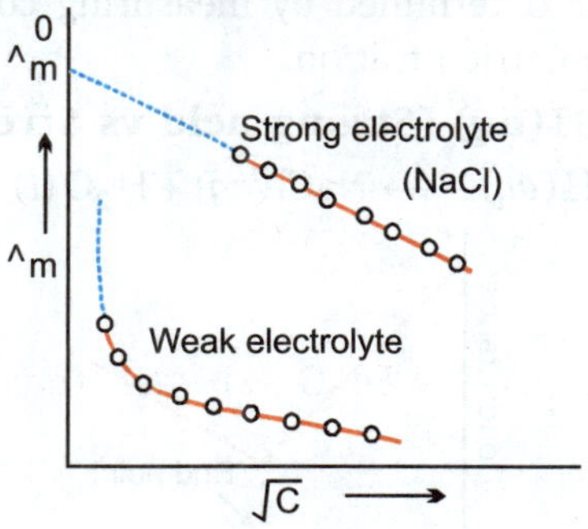

Kohlrausch's Law

According to this law, at infinite dilution, when the dissociation is complete, each ion makes a definite contribution towards molar conductivity of the electrolyte irrespective of the nature of the other ions present.

The molar conductivity of an electrolyte at infinite dilution is the sum of the ionic conductivities of the cations and the anions each multiplied by the number of ions present in one formula unit of the electrolyte

The equivalent conductivity of an electrolyte at infinite dilution is the sum of the equivalent conductivities of the cations and anions.

$$e.g., A_xB_y. \quad \Lambda_M^0(A_xB_y) = x\lambda_M^0(A^{y+}) + y\lambda_M^0(B^{x-})$$

$$\Lambda_{eq.}^0(A_xB_y) = \lambda_{eq.}^0(A^{y+}) + \lambda_{eq.}^0(B^{x-})$$

Application

(a) Determination of equivalent/molar conductivities of weak electrolytes at infinite dilution

$$\Lambda_M^0(CH_3COOH) = \Lambda_M^0(CH_3COONa) + \Lambda_M^0(HCl) - \Lambda_M^0(NaCl) - \Lambda_M^0(NaCl)$$

(b) Determination of degree of dissociation (α) and equilibrium constant (K) of an electrolyte at a given dilution

$$\alpha = \frac{\text{Molar conductivity at concentration } C}{\text{Molar conductivity at infinite dilution}} = \frac{\Lambda_M}{\Lambda_M^0}$$

	HA	$\rightleftharpoons$ H^+	+ A^-
Initial conc.	C	0	0
Conc. at equilibrium	$C - C\alpha$	$C\alpha$	$C\alpha$

$$K_a = \frac{C\alpha^2}{1-\alpha} = \frac{C\left(\frac{\Lambda_M}{\Lambda_M^0}\right)^2}{1 - \frac{\Lambda_M}{\Lambda_M^0}}$$

(c) Determination of the solubility of a sparingly soluble salt

Since the solution is saturated at infinite dilution $\Lambda_0 = \Lambda_M^0$ and molarity = solubility.

$$\therefore \quad \Lambda_M^0 = \frac{k \times 1000}{\text{molarity}}$$

$$\therefore \quad \text{Solubility } (S) = \frac{k \times 1000}{\Lambda_M^0}$$

For sparingly soluble salt A_xB_y : Solubility product $K_{sp} = x^x y^y \cdot S^{x+y}$

Conductometric Titration

Titration in which end point can be determined by measuring conductivity at each addition of the titre solution (for burette) is called conductometric titration.

Titration of HCl (*aq*) with NaOH (*aq*) (Strong acid vs Strong base)

$$HCl(aq) + NaOH(aq) \longrightarrow NaCl(aq) + H_2O(l)$$

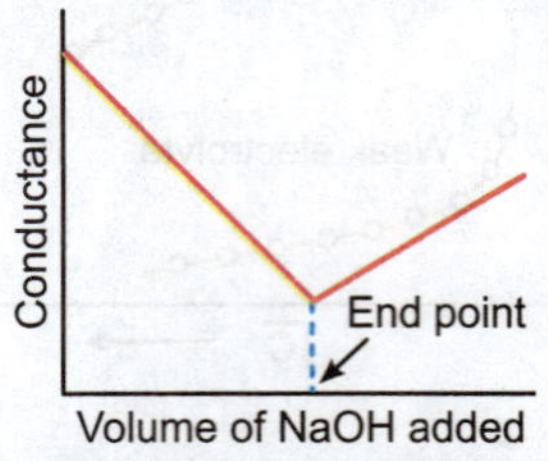

Titration of CH_3COOH (*aq*) with NaOH (*aq*) (Weak acid vs Strong base)

$$CH_3COOH(aq) + NaOH(aq) \longrightarrow CH_3COONa + H_2O(l)$$

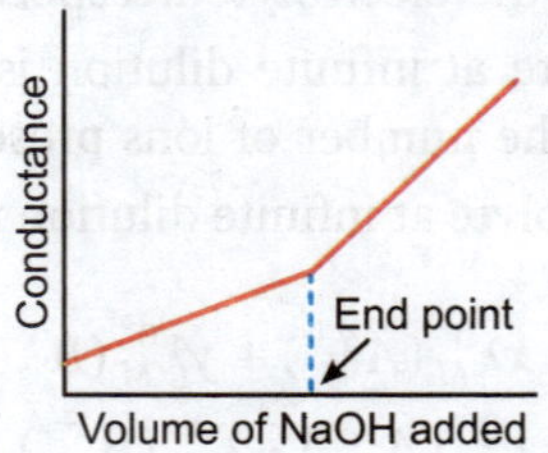

Level-1

1. A cell reaction would be spontaneous if the cell potential and $\Delta_r G$ are respectively :
   (a) positive and negative (b) negative, negative
   (c) zero, zero (d) positive, zero
2. Which of the following statement is correct?
   (a) Cathode is –ve terminal in both, galvanic and electrolytic cells
   (b) Anode is +ve terminal in both, galvanic and electrolytic cells
   (c) Cathode and anode are –ve terminal in electrolytic and galvanic cell.
   (d) Cathode and anode are +ve terminal in electrolytic and galvanic cell.
3. Electrolytes when dissolved in water dissociate into ions because :
   (a) they are unstable.
   (b) the water dissolves it.
   (c) the force of repulsion increases.
   (d) the force of electrostatic attraction is broken down by water.
4. The electric charge required for electrode deposition of one gram-equivalent of a substance is :
   (a) one ampere per second (b) 96500 coulombs per second
   (c) one ampere for one hour (d) charge on one mole of electrons
5. The amount of an ion liberated on an electrode during electrolysis does not depend upon :
   (a) conductance of the solution (b) current strength
   (c) time (d) electrochemical equivalent of the element
6. How many electrons are there in one coulomb electricity?
   (a) 6.023×10^{23} (b) 1.64×10^{-24}
   (c) 6.24×10^{18} (d) 6.24×10^{-24}
7. How many coulombs are provided by a current of 0.010 mA in the calculator battery that can operate for 1000 hours?
   (a) 1.0 (b) 10
   (c) 0.010 (d) 36
8. How many minutes are required to deliver 3.21×10^6 coulombs using a current of 500 A used in the commercial production of chlorine?
   (a) 8.3 (b) 5.3×10^4
   (c) 6420 (d) 107
9. Passage of a current for 548 seconds through a silver coulometer results in the deposition of 0.746 g of silver. What is the current (in A)?
   (a) 1.22 (b) 1.16
   (c) 1.07 (d) 1.00
10. Electrolysis can be used to determine atomic masses. A current of 0.550 A deposits 0.55 g of a certain metal in 100 minutes. Calculate the atomic mass of the metal if eq. mass = mol. mass/3
   (a) 100 (b) 45.0
   (c) 48.25 (d) 144.75
11. Beryllium occurs naturally in the form of beryl. The metal is produced from its ore by electrolysis after the ore has been converted to the oxide and then to the chloride. How many grams of Be(*s*) is deposited from a $BeCl_2$ solution by a current of 5.0 A that flows for 1.0 h? (Atomic weight : Be = 9)
   (a) 0.840 (b) 1.68
   (c) 1.42 (d) 1.08
12. How many minutes will it take to plate out 5.2 g of Cr from a $Cr_2(SO_4)_3$ solution using a current of 9.65 A? (Atomic mass : Cr = 52.0)

(a) 200 (b) 50
(c) 100 (d) 103

13. Calculate the current (in mA) required to deposit 0.195 g of platinum metal in 5.0 hours from a solution of $[PtCl_6^{2-}]$: (Atomic mass : Pt = 195)
(a) 310 (b) 31
(c) 21.44 (d) 5.36

14. How many Faradays are required to reduce 0.25 g of Nb (V) to the metal? (Atomic mass : Nb = 93)
(a) 2.7×10^{-3} (b) 1.3×10^{-2}
(c) 2.7×10^{-2} (d) 7.8×10^{-3}

15. One gm metal M^{3+} was discharged by the passage of 1.81×10^{23} electrons. What is the atomic mass of metal?
(a) 33.35 (b) 133.4
(c) 66.7 (d) None of these

16. Total charge required for the oxidation of two moles Mn_3O_4 into MnO_4^{2-} in presence of alkaline medium is :
(a) 5 F (b) 10 F
(c) 20 F (d) None of these

17. The electrolytic decomposition of dilute sulphuric acid with platinum electrode, cathodic reaction is :
(a) reduction of H^+ (b) oxidation of SO_4^{2-}
(c) reduction SO_3^{2-} (d) oxidation of H_2O

18. Which one of the following metals can not be obtained on electrolysis of aqueous solution of its salts ?
(a) Mg (b) Ag
(c) Cu (d) Cr

19. A highly concentrated solution of potassium sulphate in water is electrolysed using inert electrodes. The products at the cathode and anode are respectively.
(a) H_2, O_2 (b) O_2, H_2
(c) O_2, Na (d) $H_2, S_2O_8^{2-}$

20. The passage of current through a solution of certain electrolyte results in the evolution of $H_2(g)$ at cathode and $Cl_2(g)$ at anode. The electrolytic solution is :
(a) Water (b) *aq.* H_2SO_4
(c) *aq.* NaCl (d) *aq.* $CuCl_2$

21. When an aqueous solution of H_2SO_4 is electrolysed, the product at anode is :
(a) H^- (b) OH^-
(c) SO_4^{2-} (d) O_2

22. An aqueous solution of Na_2SO_4 is electrolysed using Pt electrodes.The products at the cathode and anode are respectively :
(a) H_2, SO_2 (b) O_2, NaOH
(c) H_2, O_2 (d) O_2, SO_2

23. The electrolysis of a solution resulted in the formation of $H_2(g)$ at the cathode and $O_2(g)$ at the anode. The solution is :
(a) AgCl(*aq*) (b) $H_2SO_4(aq)$
(c) highly concentrated NaCl(*aq*) solution (d) $CuCl_2(aq)$

24. If Pt is used as cathode in the electrolysis of aqueous NaCl solution, the ion produced at cathode is :
(a) H^+ (b) Na^+
(c) OH^- (d) Cl^-

25. A dilute aqueous solution of $CuSO_4$ is electrolysed using platinum electrodes.The products at the anode and cathode are :

(a) O_2, H_2 (b) H_2, O_2
(c) O_2, Cu (d) $S_2O_8^{2-}, H_2$

26. What products are formed during the electrolysis of concentrated aqueous solution of sodium chloride?

(I) $Cl_2(g)$ at anode (II) NaOH as electrolyte (III) $H_2(g)$ at cathode

(a) I only (b) I and II only
(c) I and III only (d) I, II and III

27. Which of the following aqueous solution produces metal after electrolysis ?

(a) $K_2Cr_2O_7$ (b) $KMnO_4$
(c) CH_3COONa (d) $CuCl_2$

28. How much time is required for complete decomposition of 4 moles of water using 4 ampere?

(a) 3.86×10^5 sec (b) 1.93×10^5 sec
(c) 96500 sec (d) 48250 sec

29. An aqueous solution containing 1 *M* each of Au^{3+}, Cu^{2+}, Ag^+, Li^+ is being electrolysed by using inert electrodes. The value of standard reduction potentials are :

$E^\circ_{Ag^+/Ag} = 0.80$ V, $E^\circ_{Cu^{2+}/Cu} = 0.34$ V, and $E^\circ_{Au^{3+}/Au} = 1.50$ V, $E^\circ_{Li^+/Li} = -3.03$ V,

With increasing voltage, the sequence of deposition of metals on the cathode will be :

(a) Li, Cu, Ag, Au (b) Cu, Ag, Au
(c) Au, Ag, Cu (d) Au, Ag, Cu, Li

30. If 0.50 L of a 0.60 *M* $SnSO_4$ solution is electrolysed for a period of 30.0 min using a current of 4.60 A. If inert electrodes are used, what is the final concentration of Sn^{2+} remaining in the solution? [at. mass of Sn = 119]

(a) 0.342 *M* (b) 0.544 *M*
(c) 0.389 *M* (d) 0.514 *M*

31. A 100.0 mL dilute solution of Ag^+ is electrolysed for 15.0 minutes with a current of 1.25 mA and the silver is removed completely. What was the initial $[Ag^+]$?

(a) 2.32×10^{-1} (b) 2.32×10^{-4}
(c) 2.32×10^{-3} (d) 1.16×10^{-4}

32. A 250.0 mL sample of a 0.20 *M* Cr^{3+} is electrolysed with a current of 96.5 A. If the remaining $[Cr^{3+}]$ is 0.1 *M*, the duration of process is :

(a) 25 sec (b) 225 sec
(c) 150 sec (d) 75 sec

33. The element indium is to be obtained by electrolysis of a molten halide of the element. Passage of a current of 3.20 A for a period of 40.0 min results in formation of 3.05 g of In. What is the oxidation state of indium in the halide melt? (Atomic mass of In = 114.8)

(a) 3 (b) 2
(c) 5 (d) 1

34. An electrolysis of a oxytungsten complex ion using 1.10 A for 40 min produces 0.838 g of tungsten. What is the charge on tungsten in the material? (Atomic mass of *W* = 184)

(a) 6 (b) 2
(c) 4 (d) 1

35. In the electrolysis of aqueous NaCl, what volume of $Cl_2(g)$ is produced in the time that it takes to liberate 5.0 liter of $H_2(g)$? Assume that both gases are measured at STP.

(a) 5.0 (b) 2.50
(c) 7.50 (d) 10.0

36. How many grams of Cr are deposited in the electrolysis of solution of $Cr(NO_3)_3$ in the same time that it takes to deposit 0.54 g of Ag in a silver coulometer arranged in series with the $Cr(NO_3)_3$ cell? (Atomic mass : Cr = 52.0; Ag = 108)

(a) 0.0866 (b) 0.0288
(c) 0.173 (d) 0.220

37. In the electrolysis of a $CuSO_4$ solution, how many grams of Cu are plated out on the cathode in the time that it takes to liberate 5.6 litre of $O_2(g)$, measured at 1 atm and 273 K, at the anode?

(a) 31.75 (b) 14.2
(c) 4.32 (d) None of these

38. Ammonium perchlorate, NH_4ClO_4, used in the solid fuel in the booster rockets on the space shuttle, is prepared from sodium perchlorate, $NaClO_4$, which is produced commercially by the electrolysis of a hot, stirred solution of sodium chloride. How many faradays are required to produce 1.0 kg of sodium perchlorate?

$$NaCl + 4H_2O \longrightarrow NaClO_4 + 4H_2$$

(a) 40.3 (b) 18.3
(c) 31.6 (d) 65.3

39. In the commercial preparation of aluminum, aluminum oxide (Al_2O_3) is electrolysed at 1000° C. How many coulombs of electricity are required to give 54 kg of aluminum? Assume following reaction takes place at cathode :

$$Al^{3+} + 3e^- \longrightarrow Al$$

(a) 17.3×10^8 (b) 3.21×10^7
(c) 1.82×10^4 (d) 57.6×10^7

40. When molten lithium chloride (LiCl) is electrolysed, lithium metal is formed at the cathode. If current efficiency is 75% then how many grams of lithium are liberated when 1930 C of charge pass through the cell ? (Atomic mass of Li = 7)

(a) 0.105 (b) 0.120
(c) 0.28 (d) 0.240

41. Sodium metal is produced commercially by the electrolysis of molten sodium chloride and chlorine is produced as a by product. How many litres of chlorine at 1.8 atm and 27° C will be produced if a current of 1.0×10^3 A is passed through NaCl (*l*) for 9.65 hr.?

(a) 2463 (b) 460
(c) 1800 (d) 1231.6

42. $H_2(g)$ and $O_2(g)$, can be produced by the electrolysis of water. What total volume (in L) of O_2 and H_2 are produced at 1 atm and 273 K when a current of 30 A is passed through a K_2SO_4 (*aq*) solution for 193 min.?

(a) 20.16 (b) 40.32
(c) 60.48 (d) 80.64

43. The cost of 2 Rs/kWh of operating an electric motor for 10 hours takes 10 amp at 110 V is :

(a) 79200 Rs (b) 22000 Rs
(c) 220 Rs (d) 22 Rs

44. A 1 *M* solution of H_2SO_4 is electrolysed. Select correct statement in respect of products obtained at anode and cathode respectively :

Given :
$$2SO_4^{2-} \longrightarrow S_2O_8^{2-} + 2e^-; E° = -2.01 \text{ V}$$
$$H_2O(l) \longrightarrow 2H^+(aq) + 1/2 O_2(g) + 2e^-; E° = -1.23 \text{ V}$$

(a) concentration of H_2SO_4 remain constant; H_2, O_2
(b) concentration of H_2SO_4 increases; O_2, H_2
(c) concentration of H_2SO_4 decreases; O_2, H_2
(d) concentration of H_2SO_4 remains constant; $S_2O_8^{2-}, H_2$

45. Cadmium amalgam is prepared by electrolysis of a solution of $CdCl_2$ using a mercury cathode. How long should a current of 4 A be passed in order to prepare 10% by mass Cd in Cd—Hg amalgam on cathode of 4.5 g Hg? (atomic mass of Cd = 112)

(a) 400 sec
(b) 215.40 sec
(c) 861.6 sec
(d) 430.8 sec

46. Use of electrolysis is in :

(a) electrorefining
(b) electroplating
(c) both (a) and (b)
(d) None of these

47. When a solution of $AgNO_3$ (1 *M*) is electrolysed using platinum anode and copper cathode, what are the products obtained at two electrodes?

Given : $E^\circ_{Cu^{2+}|Cu} = +0.34$ volt; $E^\circ_{O_2, H^+|H_2O} = +1.23$ volt; $E^\circ_{H^+|H_2} = +0.0$ volt;

$E^\circ_{Ag^+|Ag} = +0.8$ volt

(a) $Cu \longrightarrow Cu^{2+}$ at anode; $Ag^+ \longrightarrow Ag$ at cathode
(b) $H_2O \longrightarrow O_2$ at anode; $Cu^{2+} \longrightarrow Cu$ at cathode
(c) $H_2O \longrightarrow O_2$ at anode; $Ag^+ \longrightarrow Ag$ at cathode
(d) $NO_3^- \longrightarrow NO_2$ at anode; $Ag^+ \longrightarrow Ag$ at cathode

48. Which of the following statements is correct about Galvanic cell ?

(a) It converts chemical energy into electrical energy.
(b) It converts electrical energy into chemical energy.
(c) It converts metal from its free state to the combined state.
(d) It converts electrolyte into individual ions.

49. E° for $Cl_2(g) + 2e^- \longrightarrow 2Cl^-(aq)$ is 1.36 V; E° for $Cl^-(aq.) \longrightarrow 1/2Cl_2(g) + e^-$ is :

(a) 1.36 V
(b) −1.36 V
(c) −0.68 V
(d) 0.68 V

50. When two half-cells of electrode potential of E_1° and E_2° are combined to form a half cell of electrode potential E_3, then
(when n_1, n_2 and n_3 are no. of electrons exchanged in first, second and combined half-cells) :

(a) $E_3^\circ = E_2^\circ - E_1^\circ$
(b) $E_3^\circ = \dfrac{E_1^\circ n_1 + E_2^\circ n_2}{n_3}$
(c) $E_3^\circ = \dfrac{E_1^\circ n_1 - E_2^\circ n_2}{n_3^2}$
(d) $E_3^\circ = E_1^\circ + E_2^\circ$

51. The function of a salt bridge is to :

(a) maintain electrical neutrality of both half cells
(b) increase the cell potential at the positive electrode
(c) decrease the cell potential at the negative electrode
(d) eliminate the impurities present in the electrolyte

52. Saturated solution of KNO_3 with agar-agar is used to make 'salt bridge' because :

(a) size of K^+ is greater than that of NO_3^-
(b) velocity of NO_3^- is greater than that of K^+
(c) velocities of K^+ and NO_3^- are nearly the same
(d) both velocity and sizes of K^+ and NO_3^- ions are nearly same

53. A salt bridge may contain :

(a) a saturated solution of KCl and agar-agar
(b) a saturated solution of KNO_3 and agar-agar

(c) a saturated solution of NH_4NO_3 and agar-agar
(d) all of these

54. The nature of curve of E°_{cell} *vs.* log K_c is :
(a) straight line (b) parabola
(c) hyperbola (d) elliptical curve

55. Consider the following equations for a cell reaction

$$A + B \rightleftharpoons C + D; \quad E° = x \text{ volt}, K_{eq} = K_1$$
$$2A + 2B \rightleftharpoons 2C + 2D; \quad E° = y \text{ volt}, K_{eq} = K_2$$

then :
(a) $x = y, K_1 = K_2$ (b) $x = 2y, K_1 = 2K_2$
(c) $x = y, K_1^2 = K_2$ (d) $x^2 = y, K_1^2 = K_2$

56. Which graph correctly correlates E_{cell} as a function of concentrations for the cell

$$Zn(s) + 2Ag^+(aq) \longrightarrow Zn^{2+}(aq) + 2Ag(s), \quad E^\circ_{cell} = 1.56 \text{ V}$$

Y-axis : E_{cell}, X-axis : $\log_{10}\frac{[Zn^{2+}]}{[Ag^+]^2}$

(a) 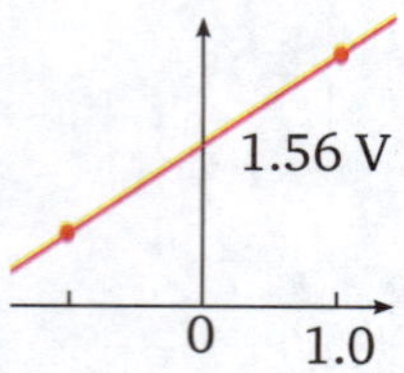

(b) 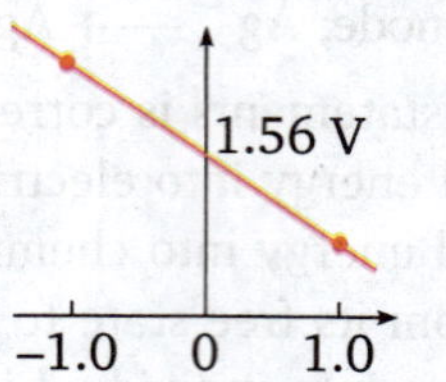

(c) 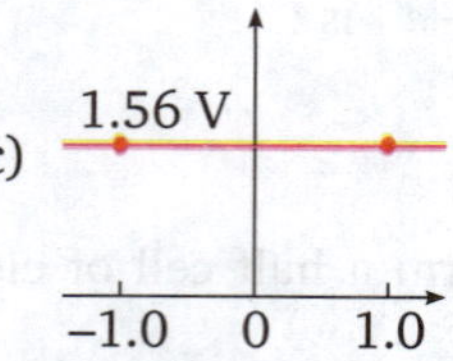

(d) 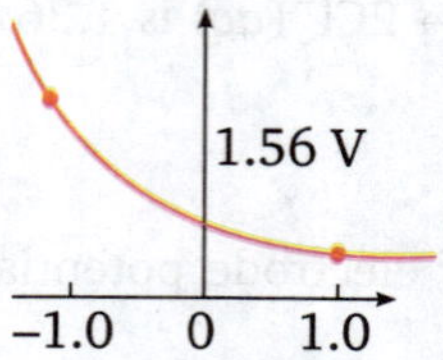

57. The Nernst equation $E = E° - RT/nF \ln Q$ indicates that the Q will be equal to equilibrium constant K_c when :
(a) $E = E°$ (b) $RT/nF = 1$
(c) $E =$ zero (d) $E° = 1$

58. The cell reaction $2Ag^+(aq) + H_2(g) \longrightarrow 2H^+(aq) + 2Ag(s)$, is best represented by :
(a) $Ag(s) | Ag^+(aq) || H^+(aq) | H_2(g) | Pt(s)$ (b) $Pt(s) | H_2(g) | H^+(aq) || Ag^+(aq) | Ag(s)$
(c) $Ag(s) | Ag^+(aq) || H_2(g) | H^+(aq) | Pt(s)$ (d) $Ag^+(aq) | Ag(s) || H_2(g) | H^+(aq)$

59. The cell reaction $Hg_2Cl_2(s) + Cu(s) \longrightarrow Cu^{2+}(aq) + 2Cl^-(aq) + 2Hg(l)$, is best represented by :
(a) $Cu(s) | Cu^{2+}(aq) || Hg_2Cl_2(s) | Hg(l)$
(b) $Cu(s) | Cu^{2+}(aq) || Hg(l) | HgCl_2(s)$
(c) $Cu(s) | Cu^{2+}(aq) || Cl^-(aq) | Hg_2Cl_2(s) | Hg(l) | Pt(s)$
(d) $Hg_2Cl_2(s) | Cl^-(aq) || Cu^{2+}(aq) | Cu(s)$

60. The cell reaction $Cr_2O_7^{2-}(aq) + 14H^+(aq) + 6Fe^{2+}(aq) \longrightarrow 6Fe^{3+}(aq) + 2Cr^{3+}(aq) + 7H_2O(l)$, is best represented by :
(a) $Pt(s) | Fe^{2+}(aq), Fe^{3+}(aq) || Cr_2O_7^{2-}(aq), Cr^{3+}(aq) | Pt(s)$
(b) $Pt(s) | Cr_2O_7^{2-}(aq), Cr^{+3}(aq) || Fe^{3+}(aq), Fe^{+2}(aq) | Pt(s)$

(c) $Fe^{2+}(aq)|Fe^{3+}(aq)||Cr_2O_7^{2-}(aq)|Cr^{3+}(aq)$

(d) $Cr_2O_7^{2-}(aq)|Cr^{3+}(aq)||Fe^{3+}(aq)|Fe^{2+}(aq)$

61. Select the correct cell reaction of the cell $Ag(s)|Ag^+(aq)||Cu^{2+}(aq)|Cu(s)$:

(a) $2Ag(s) + Cu(s) \longrightarrow Cu^{+2}(aq) + 2Ag^+(aq)$

(b) $Cu(s) + 2Ag^+(aq) \longrightarrow Cu^{2+}(aq) + 2Ag(s)$

(c) $2Ag(s) + Cu^{2+}(aq) \longrightarrow Cu(s) + 2Ag^+(aq)$

(d) $Cu^{2+}(aq) + 2Ag^+(aq) \longrightarrow 2Ag(s) + Cu(s)$

62. Select the correct cell reaction of the cell $Pt(s)|Cl_2(g)|Cl^-(aq)||Ag^+(aq)|Ag(s)$:

(a) $Cl_2(g) + Ag^+(aq) \longrightarrow Ag(s) + 2Cl^-(aq)$

(b) $Cl_2(g) + Ag(s) \longrightarrow 2Cl^-(aq) + Ag^+(aq)$

(c) $2Cl^-(aq) + 2Ag^+(aq) \longrightarrow 2Ag(s) + Cl_2(g)$

(d) $AgCl(s) \longrightarrow Ag^+(aq) + Cl^-(aq)$

63. Standard electrode potential of SHE at 298 K is :

(a) 0.05 V (b) 0.10 V

(c) 0.50 V (d) 0.00 V

64. The e.m.f. of the following galvanic cells :

(a) $Zn|Zn^{2+}(1M)||Cu^{2+}(1M)|Cu$ (b) $Zn|Zn^{2+}(0.1M)||Cu^{2+}(1M)|Cu$

(c) $Zn|Zn^{2+}(1M)||Cu^{2+}(0.1M)|Cu$ (d) $Zn|Zn^{2+}(0.1M)||Cu^{2+}(0.1M)|Cu$

are represented by E_1, E_2, E_3 and E_4 respectively. Which of the following statement is true?

(a) $E_1 > E_2 > E_3 > E_4$ (b) $E_3 > E_2 > E_1 > E_4$

(c) $E_3 > E_1 = E_4 > E_2$ (d) $E_2 > E_1 = E_4 > E_3$

65. Based on the cell notation for a spontaneous reaction, at the anode :

$$Ag(s)|AgCl(s)|Cl^-(aq)||Br^-(aq)|Br_2(l)|C(s)$$

(a) AgCl gets reduced (b) Ag gets oxidized

(c) Br^- gets oxidized (d) Br_2 gets reduced

66. Given the listed standard electrode potentials, what is $E°$ for the cell :

$$4BiO^+(aq) + 3N_2H_5^+(aq) \longrightarrow 4Bi(s) + 3N_2(g) + 4H_2O(l) + 7H^+(aq)$$

$$N_2(g) + 5H^+(aq) + 4e^- \longrightarrow N_2H_5^+(aq), E° = -0.23 \text{ V}$$

$$BiO^+(aq) + 2H^+(aq) + 3e^- \longrightarrow Bi(s) + H_2O(l), \quad E° = +0.32 \text{ V}$$

(a) +0.55 (b) +0.34

(c) +1.88 (d) +0.09

67. What is the standard electrode potential for the reduction of HClO?

$$HClO(aq) + H^+(aq) + 2e^- \longrightarrow Cl^-(aq) + H_2O(l)$$

Given : $Cr^{2+}(aq) \longrightarrow Cr^{3+}(aq) + e^-, \quad E° = 0.41 \text{ V}$

$$HClO(aq) + H^+(aq) + 2Cr^{2+}(aq) \longrightarrow 2Cr^{3+}(aq) + Cl^-(aq) + H_2O(l), E° = 1.80$$

(a) 1.39 (b) 1.54

(c) 1.22 (d) 0.90

68. The $E°$ for the following cell is +0.34 V. $In(s)|In(OH)_3(aq)||SbO_2^-(aq)|Sb(s)$.

Using $E° = -1.0$ V for the $In(OH)_3|$ In couple, calculate $E°$ for the $SbO_2^-|Sb$ half-reaction :

(a) −1.34 (b) +0.66

(c) +0.82 (d) −0.66

69. From the following half-cell reactions and their standard potentials, what is the smallest possible standard e.m.f. for spontaneous reactions?

$$PO_4^{3-}(aq) + 2H_2O(l) + 2e^- \longrightarrow HPO_3^{2-} + 3OH^-(aq); \quad E° = -1.05 \text{ V}$$

$$PbO_2(s) + H_2O(l) + 2e^- \longrightarrow PbO(s) + 2OH^-(aq); \quad E° = +0.28 \text{ V}$$

$$IO_3^-(aq) + 2H_2O(l) + 4e^- \longrightarrow IO^-(aq) + 4OH^-(aq); \quad E° = +0.56 \text{ V}$$

(a) +0.00 (b) +0.74
(c) +0.56 (d) +0.28

70. Determine which substance is the best reducing agent in Q. no. 69 :
(a) HPO_3^{2-} (b) PO_4^{3-}
(c) IO^- (d) IO_3^-

71. Which substance is the best oxidizing agent in Q. no. 69?
(a) IO_3^- (b) IO^-
(c) PbO (d) PO_4^{3-}

72. Consider the following half-cell reactions and associated standard half-cell potentials and determine the maximum voltage that can be obtained by combination resulting in spontaneous processes :

$$AuBr_4^-(aq) + 3e^- \longrightarrow Au(s) + 4Br^-(aq); \quad E° = -0.86 \text{ V}$$

$$Eu^{3+}(aq) + e^- \longrightarrow Eu^{2+}(aq); \quad E° = -0.43 \text{V}$$

$$Sn^{2+}(aq) + 2e^- \longrightarrow Sn(s); \quad E° = -0.14 \text{ V}$$

$$IO^-(aq) + H_2O(l) + 2e^- \longrightarrow I^-(aq) + 2OH^-(aq); \quad E° = +0.49 \text{ V}$$

(a) +0.72 (b) +1.54
(c) +1.00 (d) +1.35

73. The position of some metals in the electrochemical series in decreasing electropositive character is Mg > Al > Zn > Cu > Ag. What will happened if copper spoon is used to stirred a solution of aluminium nitrate ?
(a) The spoon gets coated with aluminium.
(b) An alloy of aluminium and copper is formed.
(c) No reaction occurs
(d) The solution starts turning blue

74. Zn can displace :
(a) Mg from its aqueous solution (b) Cu from its aqueous solution
(c) Na from its aqueous solution (d) Al from its aqueous solution

75. Based on the following informations arrange four metals *A*, *B*, *C* and *D* in order of decreasing ability to act as reducing agents :
(I) Only *A*, *B* and *C* react with 1 *M* HCl to give $H_2(g)$
(II) When *C* is added to solutions of the other metal ions, metallic *B* and *D* are formed
(III) Metal *C* does not reduce A^{n+}.
(a) $C > A > B > D$ (b) $C > A > D > B$
(c) $A > C > D > B$ (d) $A > C > B > D$

76. When an aqueous solution of $CuSO_4$ is stirred with a silver spoon then :
(a) Cu^+ will be formed (b) Ag^+ will be formed
(c) Cu^{2+} will be deposited (d) None of these

77. Based on the following informations arrange four metals *A*, *B*, *C* and *D* in order of increasing ability to act as reducing agents :
(I) Only *C* react with 1 *M* HCl to give $H_2(g)$

(II) When A is added to solution of the other metal ions, metallic D is formed but not B or C

(a) $D < A < C < B$
(b) $A < D < C < B$
(c) $B < D < A < C$
(d) $D < A < B < C$

78. In the reaction :

$$4Fe(s) + 3O_2(g) \longrightarrow 4Fe^{3+}(aq) + 6O^{2-}(aq)$$

which of the following statement is incorrect ?

(a) It is a redox reaction
(b) Fe is reducing agent
(c) O_2 is an oxidizing agent
(d) Fe is reduced to Fe^{3+}

79. Which of the following is displaced by Fe ?

(a) Ag
(b) Zn
(c) Na
(d) All of these

80. The standard potential at 25°C for the following half reactions is given :

$$Zn^{2+} + 2e^- \longrightarrow Zn; E° = -0.762 \text{ V}$$

$$Mg^{2+} + 2e^- \longrightarrow Mg; E° = -2.37 \text{ V}$$

When Zinc dust is added to the solution of $MgCl_2$.

(a) $ZnCl_2$ is formed
(b) Mg is precipitated
(c) Zn dissolved in the solution
(d) No reaction takes place

81. The element which can displace three other halogens from their compound is :

(a) F
(b) Cl
(c) Br
(d) I

82. Using the standard half-cell potential listed, calculate the equilibrium constant for the reaction :

$$Co(s) + 2H^+(aq) \longrightarrow Co^{2+}(aq) + H_2(g) \text{ at } 298 \text{ K}$$

$$Co^{2+}(aq) + 2e^- \longrightarrow Co(s) \quad E° = -0.277 \text{ V}$$

(a) 2.3×10^9
(b) 4.8×10^4
(c) 4.8×10^7
(d) 4.8×10^{11}

83. The $E°$ at 25° C for the following reaction is 0.22 V. Calculate the equilibrium constant at 25° C :

$$H_2(g) + 2AgCl(s) \longrightarrow 2Ag(s) + 2HCl(aq)$$

(a) 2.8×10^7
(b) 5.2×10^8
(c) 5.2×10^6
(d) 5.2×10^3

84. Electrode potential of the half cell $Pt(s) | Hg(l) | Hg_2Cl_2(s) | Cl^-(aq)$ can be increased by :

(a) Increasing $[Cl^-]$
(b) Decreasing $[Cl^-]$
(c) Increasing $Hg_2Cl_2(s)$
(d) Decreasing $Hg(l)$

85. The equilibrium constant for the following general reaction is 10^{30}. Calculate $E°$ for the cell at 298 K.

$$2X_2(s) + 3Y^{2+}(aq) \longrightarrow 2X_2^{3+}(aq) + 3Y(s)$$

(a) +0.105 V
(b) +0.2955 V
(c) 0.0985 V
(d) –0.2955 V

86. A solution containing H^+ and D^+ ions is in equilibrium with a mixture of H_2 and D_2 gases at 25° C. If the partial pressures of both gases are 1.0 atm, find the ratio of $[D^+]/[H^+]$:

(Given : $E°_{D^+/D_2} = -0.003$ V)

(a) 1.23
(b) 1.12
(c) 0.11
(d) 1.0

87. The $E°$ at 25°C for the following reaction is 0.55 V. Calculate the $\Delta G°$ in kJ/mol :

$$4BiO^+(aq) + 3N_2H_5^+ \longrightarrow 4Bi(s) + 3N_2(g) + 4H_2O(l) + 7H^+(aq)$$

(a) −637 (b) −424
(c) −106 (d) −318.5

88. Use the following standard electrode potentials, calculate $\Delta G°$ in kJ/mol for the indicated reaction :

$$5Ce^{4+}(aq) + Mn^{2+}(aq) + 4H_2O(l) \longrightarrow 5Ce^{3+}(aq) + MnO_4^-(aq) + 8H^+(aq)$$

$$MnO_4^-(aq) + 8H^+(aq) + 5e^- \longrightarrow Mn^{2+}(aq) + 4H_2O(l); \quad E° = +1.51\text{ V}$$

$$Ce^{4+}(aq) + e^- \longrightarrow Ce^{3+}(aq) \quad E° = +1.61\text{ V}$$

(a) −9.65 (b) −24.3
(c) −48.25 (d) −35.2

89. Consider an electrochemical cell in which the following reaction occurs and predict which changes will decrease the cell voltage :

$$Fe^{2+}(aq) + Ag^+(aq) \longrightarrow Ag(s) + Fe^{3+}(aq)$$

(I) decrease the $[Ag^+]$ (II) increase in $[Fe^{3+}]$ (III) increase the amount of Ag

(a) I (b) II and III
(c) II (d) I and II

90. Consider the following equation for an electrochemical cell reaction. Which of the following changes in condition will increase the cell voltage?

$$H_2(g) + PbCl_2(s) \longrightarrow Pb(s) + 2H^+(aq) + 2Cl^-(aq)$$

(I) addition of concentrated $HClO_4$ in the cell solution
(II) increase the pressure of $H_2(g)$ (III) increase the amount of Pb(s)
(a) III (b) I and II
(c) II and III (d) II

91. The standared electrode potential for the following reaction is +1.33 V. What is the potential at pH = 2.0?

$$Cr_2O_7^{2-}(aq, 1M) + 14H^+(aq) + 6e^- \longrightarrow 2Cr^{3+}(aq, 1M) + 7H_2O(l)$$

(a) +1.820 V (b) +1.990 V
(c) +1.60 V (d) +1.0542 V

92. The standard electrode potential for the following reaction is −0.57 V. What is potential at pH = 12.0?

$$TeO_3^{2-}(aq, 1M) + 3H_2O(l) + 4e^- \longrightarrow Te(s) + 6OH^-(aq)$$

(a) −0.17 V (b) +0.21 V
(c) −0.39 V (d) +1.95 V

93. $Co|Co^{2+}(C_2)||Co^{2+}(C_1)|Co$; for this cell, ΔG is negative if :

(a) $C_2 > C_1$ (b) $C_1 > C_2$
(c) $C_1 = C_2$ (d) unpredictable

94. What will be the emf for the given cell?

$$Pt|H_2(g, P_1)|H^+(aq)|H_2(g, P_2)|Pt$$

(a) $\frac{RT}{F}\ln\frac{P_1}{P_2}$ (b) $\frac{RT}{2F}\ln\frac{P_1}{P_2}$
(c) $\frac{RT}{F}\ln\frac{P_2}{P_1}$ (d) None of these

95. For the electrochemical cell $Pt(s)|\underset{1\ atm}{H_2(g)}|H^+(1M)||Cu^{2+}(1M)|Cu(s)$, which one of the following statements is true ?

(a) H^+ ions are formed at anode and Cu is deposited at cathode.

(b) H_2 liberated at cathode and Cu is deposited at anode.

(c) Oxidation occurs at cathode.

(d) Reduction occurs at anode.

96. In a concentration cell the same reagents are present in both the anode and the cathode compartments, but at different concentrations. Calculate the emf of a cell containing 0.040 M. Cr^{3+} in one compartment and 1.0 M Cr^{3+} in the other if Cr electrodes are used in both.

(a) 0.028 V (b) 0.249 V

(c) 0.083 V (d) 0.125 V

97. A 1.0 M solution of Cd^{2+} is added to excess iron and the system is allowed to reach equilibrium. What is the concentration of Cd^{2+} ?

$$Cd^{2+}(aq) + Fe(s) \longrightarrow Cd(s) + Fe^{2+}(aq); \quad E° = 0.037$$

(a) 0.195 (b) 0.097

(c) 0.053 (d) 0.145

98. The measured voltage for the reaction with the indicated concentrations is 1.50 V. Calculate $E°$.

$$Cr(s) + 3Ag^+(aq, 0.10\ M) \longrightarrow 3Ag(s) + Cr^{3+}(aq, 0.30\ M)$$

(a) 1.35 (b) 1.40

(c) 1.65 (d) 1.55

99. Calculate the standard voltage that can be obtained from an ethane oxygen fuel cell at 25° C.

$$C_2H_6(g) + 7/2\,O_2(g) \longrightarrow 2CO_2(g) + 3H_2O(l); \quad \Delta G° = -1467\ kJ/mol$$

(a) +0.91 (b) +0.54

(c) +0.72 (d) +1.08

100. $I_2(s)|I^-(0.1M)$ half cell is connected to a $H^+(aq)|H_2$ (1 bar)$|$Pt half cell and e.m.f. is found to be 0.7714 V. If $E°_{I_2|I^-} = 0.535$ V, find the pH of $H^+|H_2$ half-cell.

(a) 1 (b) 3

(c) 5 (d) 7

101. What is the SRP for $S^{2-}|CuS|Cu$ half-cell?

Given : $K_{sp}(CuS) = 10^{-35}$ and $E°_{Cu^{2+}|Cu} = 0.34$ V

(a) 1.034 V (b) 1.0 V

(c) –0.694 V (d) 0.694 V

102. Given the following standard electrode potentials, the K_{sp} for $PbBr_2$ is :

$$PbBr_2(s) + 2e^- \longrightarrow Pb(s) + 2Br^-(aq); \quad E° = -0.248\ V$$

$$Pb^{2+}(aq) + 2e^- \longrightarrow Pb(s); \quad E° = -0.126\ V$$

(a) 7.4×10^{-5} (b) 4.9×10^{-14}

(c) 5.2×10^{-6} (d) 2.3×10^{-13}

103. The standard free energy change for the following reaction is –210 kJ/mol. What is the standard cell potential?

$$2H_2O_2(aq) \longrightarrow 2H_2O(l) + O_2(g)$$

(a) +0.752 (b) +1.09

(c) +0.420 (d) +0.640

104. At equilibrium :
(a) $E^\circ_{cell} = 0, \Delta G^\circ = 0$
(b) $E_{cell} = 0, \Delta G = 0$
(c) both are correct
(d) none is correct

105. The $E°$ at 25° C for the following reaction at the indicated concentrations is 1.50 V. Calculate the ΔG in kJ/mol 25° C :

$$Cr(s) + 3Ag^+(aq, 0.1\,M) \longrightarrow 3Ag(s) + Cr^{3+}(aq, 0.1\,M)$$

(a) –140.94
(b) –295
(c) –212
(d) – 422.83

106. If $E^\circ_{Au^+/Au}$ is 1.69 V and $E^\circ_{Au^{3+}/Au}$ is 1.40 V, then $E^\circ_{Au^+/Au^{3+}}$ will be :
(a) 0.19 V
(b) 2.945 V
(c) 1.255 V
(d) None of these

107. Consider the following standard electrode potentials and calculate the equilibrium constant at 25° C for the indicated disproportionation reaction :

$$3Mn^{2+}(aq) \longrightarrow Mn(s) + 2Mn^{3+}(aq)$$

$$Mn^{3+}(aq) + e^- \longrightarrow Mn^{2+}(aq); \quad E^\circ = 1.51\,V$$

$$Mn^{2+}(aq) + 2e^- \longrightarrow Mn(s); \quad E^\circ = -1.185\,V$$

(a) 1.2×10^{-43}
(b) 2.4×10^{-73}
(c) 6.3×10^{-92}
(d) 1.5×10^{-62}

108. A galvanic cell is composed of two hydrogen electrodes, one of which is a standard one. In which of the following solutions should the other electrode be immersed to get maximum e.m.f. :
(a) 0.1 M HCl
(b) 0.1 M H_2SO_4
(c) 0.1 M NH_4OH
(d) 0.01 M HCOOH

109. $Ag|AgCl|Cl^-(C_2)||Cl^-(C_1)|AgCl|Ag$ for this cell ΔG is negative if :
(a) $C_1 = C_2$
(b) $C_1 > C_2$
(c) $C_2 > C_1$
(d) Both (a) and (c)

110. By how much is the oxidizing power of $Cr_2O_7^{2-}|Cr^{3+}$ couple decreased if the H^+ concentration is decreased from 1 M to 10^{-3} M at 25° C?
(a) 0.001 V
(b) 0.207 V
(c) 0.441 V
(d) 0.414 V

111. The temperature coefficient of a cell whose operation is based on the reaction

$$Pb(s) + HgCl_2(aq) \longrightarrow PbCl_2(aq) + Hg(l) \text{ is :}$$

$$\left(\frac{dE}{dT}\right)_p = 1.5 \times 10^{-4}\,VK^{-1} \text{ at } 298\,K$$

The change in entropy (in J/K mol) during the operation is :
(a) 8627
(b) 57.9
(c) 28.95
(d) 14.475

112. Thermodynamic efficiency of a cell is given by :
(a) $\dfrac{\Delta H}{\Delta G}$
(b) $\dfrac{nFE}{\Delta G}$
(c) $\dfrac{-nEF}{\Delta H}$
(d) nFE°

113. Calculate the value of equilibrium constant (K_f) for the reaction :

$$Zn^{2+}(aq) + 4OH^-(aq) \rightleftharpoons Zn(OH)_4^{2-}(aq)$$

Given : $Zn^{2+}(aq) + 2e^- \longrightarrow Zn(s);\ E° = -0.76\ V$

$$Zn(OH)_4^{2-}(aq) + 2e^- \longrightarrow Zn(s) + 4OH^-(aq);\ E° = -1.36\ V$$

$$2.303\frac{RT}{F} = 0.06$$

(a) 10^{10}
(b) 2×10^{10}
(c) 10^{20}
(d) none of these

114. Which of the following statement is false for fuel cells ?
(a) They are more efficient
(b) They are free from pollution
(c) They run till reactants are active
(d) Fuel burned with O_2

115. Which of the following reactions is used to make a fuel cell ?
(a) $Cd_{(s)} + 2Ni(OH)_{3(s)} \longrightarrow CdO_{(s)} + 2NiOH + H_2O_{(l)}$
(b) $Pb_{(s)} + PbO_{2(s)} + 2H_2SO_{4(aq)} \longrightarrow 2PbSO_{4(s)} + 2H_2O_{(l)}$
(c) $2H_{2(g)} + O_{2(g)} \longrightarrow 2H_2O_{(l)}$
(d) $2Fe_{(s)} + O_{2(g)} + 4H^+_{(aq)} \longrightarrow 2Fe^{2+}_{(aq)} + 2H_2O_{(l)}$

116. When a lead storage battery is charged it acts as :
(a) a fuel cell
(b) an electrolytic cell
(c) a galvanic cell
(d) a concentration cell

117. When lead storage battery is charged ?
(a) Lead dioxide dissolves
(b) Sulphuric acid is regenerated
(c) The lead electrode becomes coated with lead sulphate
(d) The amount of sulphuric acid decreases

118. The metal that forms a self-protecting film of oxide to prevent corrosion is :
(a) Na
(b) Al
(c) Cu
(d) Au

119. Rusting of iron is catalyzed by which of the following ?
(a) Fe
(b) Zn
(c) O_2
(d) H^+

120. The Zn acts as sacrificial or cathodic protection to prevent rusting of iron because :
(a) E°_{OP} of Zn < E°_{OP} of Fe
(b) E°_{OP} of Zn > E°_{OP} of Fe
(c) E°_{OP} of Zn = E°_{OP} of Fe
(d) Zn is cheaper than iron

121. In electrochemical corrosion of metals, the metal undergoing corrosion :
(a) acts as anode
(b) acts as cathode
(c) undergoes reduction
(d) None

122. When an acid cell is charged, then :
(a) voltage of cell increases
(b) resistance of cell increases
(c) electrolyte of cell dilutes
(d) None of these

123. Electrolytic conduction is due to the movement of :
(a) electrons
(b) ions
(c) atoms
(d) electrons as well as ions

124. Molten sodium chloride conducts electricity due to the presence of :
(a) free electron
(b) free ions
(c) free molecules
(d) atoms of sodium and chlorine

125. Pure water does not conduct electricity because it :

(a) is neutral
(b) is readily decomposed
(c) is almost totally unionized
(d) has a low boiling point

126. The relation among conductance (G), specific conductance (κ) and cell constant (l/A) is :

(a) $G = \kappa \frac{l}{A}$
(b) $G = \kappa \frac{A}{l}$
(c) $G\kappa = \frac{l}{A}$
(d) $G = \kappa A l$

127. If x is specific resistance (in $S^{-1}cm$) of the electrolyte solution and y is the molarity of the solution, then Λ_m (in $S\,cm^2 mol^{-1}$) is given by :

(a) $\frac{1000x}{y}$
(b) $1000\frac{y}{x}$
(c) $\frac{1000}{xy}$
(d) $\frac{xy}{1000}$

128. Equivalent conductivity can be expressed in terms of specific conductance (κ) and concentration (N) in gram equivalent dm^{-3} as :

(a) $\kappa \times N$
(b) $\frac{\kappa \times 1000}{N}$
(c) $\frac{\kappa \times N}{1000}$
(d) $\kappa \times N \times 1000$

129. Resistance of a decimolar solution between two electrodes 0.02 meter apart and 0.0004 m^2 in area was found to be 50 ohm. Specific conductance (κ) is :

(a) $0.1\,Sm^{-1}$
(b) $1\,S\,m^{-1}$
(c) $10\,S\,m^{-1}$
(d) $4 \times 10^{-4}\,S\,m^{-1}$

130. Resistance of 0.1 M KCl solution in a conductance cell is 300 ohm and conductivity is 0.013 Scm^{-1}. The value of cell constant is :

(a) $3.9\,cm^{-1}$
(b) $39\,m^{-1}$
(c) $3.9\,m^{-1}$
(d) None of these

131. Specific conductance of 0.1M nitric acid is $6.3 \times 10^{-2}\,ohm^{-1}\,cm^{-1}$. The molar conductivity of solution is:

(a) $630\,ohm^{-1}\,cm^2\,mol^{-1}$
(b) $315\,ohm^{-1}\,cm^2\,mol^{-1}$
(c) $100\,ohm^{-1}\,cm^2\,mol^{-1}$
(d) $6300\,ohm^{-1}\,cm^2\,mol^{-1}$

132. The molar conductivity of $N/10$ $BaCl_2$ solution at 20°C is 400 $S\,cm^2\,mol^{-1}$ and the resistance of the cell containing this solution at 20°C is 55 ohm. The cell constant is :

(a) $2.2\,cm^{-1}$
(b) $1.1\,cm^{-1}$
(c) $0.55\,cm^{-1}$
(d) $0.02\,cm^{-1}$

133. Dissociation constant of a weak acid (HA) in terms of Λ_m^∞ and Λ_m is :

(a) $K_a = \frac{C\Lambda_m^\infty}{(\Lambda_m - \Lambda^\infty)}$
(b) $K_a = \frac{C\Lambda_m^2}{\Lambda_m^\infty(\Lambda_m^\infty - \Lambda_m)}$
(c) $K_a = \frac{C(\Lambda_m^\infty)^2}{\Lambda_m^\infty(\Lambda_m^\infty - \Lambda_m)}$
(d) None of these

134. When a concentrated solution of an electrolyte is diluted
(a) its specific conductance increases
(b) its equivalent conductivity decreases
(c) its specific conductivity decreases and equivalent conductivity increases
(d) both specific and equivalent conductivity increase

135. The unit of equivalent conductivity is:
(a) ohm cm
(b) ohm^{-1} cm^2 $(g\text{ - equivalent})^{-1}$
(c) ohm cm^2 (g - equivalent)
(d) $S cm^{-2}$

136. Molar conductivity of a solution of an electrolyte AB_3 is 150 $Scm^2 mol^{-1}$. If it ionises as $AB_3 \longrightarrow A^{3+} + 3B^-$, its equivalent conductivity will be :
(a) 150 (in Scm^2eq^{-1})
(b) 75 (in Scm^2eq^{-1})
(c) 50 (in Scm^2eq^{-1})
(d) 80 (in Scm^2eq^{-1})

137. Equivalent conductivity of $Fe_2(SO_4)_3$ is related to molar conductivity by the expression :
(a) $\Lambda_{eq} = \Lambda_m$
(b) $\Lambda_{eq} = \Lambda_m/3$
(c) $\Lambda_{eq} = 3\Lambda_m$
(d) $\Lambda_{eq} = \Lambda_m/6$

138. The limiting equivalent conductivity of NaCl, KCl and KBr are 126.5, 150.0 and 151.5 S cm^2eq^{-1}, respectively. The limiting equivalent ionic conductivity for Br^- is 78 S cm^2eq^{-1}. The limiting equivalent ionic conductivity for Na^+ ions would be :
(a) 128
(b) 125
(c) 49
(d) 50

139. The specific conductance of a saturated solution of silver bromide is κ Scm^{-1}. The limiting ionic conductivity of Ag^+ and Br^- ions are x and y, respectively. The solubility of silver bromide in gL^{-1} is : (molar mass of AgBr = 188)
(a) $\dfrac{\kappa \times 1000}{x - y}$
(b) $\dfrac{\kappa}{x + y} \times 188$
(c) $\dfrac{\kappa \times 1000 \times 188}{x + y}$
(d) $\dfrac{x + y}{\kappa} \times \dfrac{1000}{188}$

140. The resistance of 0.1 N solution of formic acid is 200 ohm and cell constant is 2.0 cm^{-1}. The equivalent conductivity (in Scm^2eq^{-1}) of 0.1 N formic acid is :
(a) 100
(b) 10
(c) 1
(d) none of these

141. A conductance cell was filled with a 0.02 M KCl solution which has a specific conductance of 2.768×10^{-3} ohm^{-1} cm^{-1}. If its resistance is 82.4 ohm at 25°C, the cell constant is :
(a) 0.2182 cm^{-1}
(b) 0.2281 cm^{-1}
(c) 0.2821 cm^{-1}
(d) 0.2381 cm^{-1}

142. The conductance of a salt solution (AB) measured by two parallel electrodes of area 100 cm^2 separated by 10 cm was found to be $0.0001\,\Omega^{-1}$. If volume enclosed between two electrode contain 0.1 mole of salt, what is the molar conductivity (Scm^2mol^{-1}) of salt at same concentration :
(a) 10
(b) 0.1
(c) 1
(d) none of these

143. The equivalent ionic conductivity of Ba^{2+} and Cl^- at infinite dilution are 127 and 76 $ohm^{-1} cm^2 eq^{-1}$ respectively. The equivalent conductivity of $BaCl_2$ at infinity dilution (in $ohm^{-1} cm^2 eq^{-1}$) would be :

(a) 203 (b) 279

(c) 101.5 (d) 139.5

144. What will be the molar conductivity of Al^{3+} ions at infinite dilution if molar conductivity of $Al_2(SO_4)_3$ is 858 S $cm^2 mol^{-1}$ and ionic conductance of SO_4^{2-} is 160 S $cm^2 mol^{-1}$ at infinite dilution?

(a) 189 S $cm^2 mol^{-1}$ (b) 698 S $cm^2 mol^{-1}$

(c) 1018 S $cm^2 mol^{-1}$ (d) 429 S $cm^2 mol^{-1}$

145. Unit of ionic mobility is :

(a) $m V^{-1} s^{-1}$ (b) $m^2 V^{-2} s^{-1}$

(c) $m^2 V^{-1} s^{-1}$ (d) $m^{-2} V s^{-1}$

146. Λ^{∞}_{AgCl} can be obtained :

(a) by extraplotation of the graph Λ and $\sqrt{C}$ to zero concentration

(b) by known values of Λ^{∞} of $AgNO_3$, HCl and HNO_3

(c) both (a) and (b)

(d) None of these

147. Which of the following has the highest molar conductivity in solution?

(a) $[Pt(NH_3)_6]Cl_4$ (b) $[Pt(NH_3)_5Cl]Cl_3$

(c) $[Pt(NH_3)_5Cl_2]Cl_2$ (d) $[Pt(NH_3)_5Cl_3]Cl$

148. When a certain conductance cell was filled with KCl solution (conductivity $= 0.03 \Omega^{-1} cm^{-1}$) was it had a resistance of 100 Ω at 300 K temperature when filled with 0.01 *M* K_2SO_4, it has a resistance of 300 Ω. Calculate molar conductivity of K_2SO_4 solution in terms of $\Omega^{-1} m^2 mol^{-1}$.

(a) 10^{-5} (b) 10^5

(c) 10^{-1} (d) 10^{-3}

149. What is the equivalent conductivity (S $cm^2 eq^{-1}$) of 0.001 *M* $H_2SO_4(aq.)$ solution at 298 K? If measured conductivity of $H_2SO_4(aq.)$ solution was found to be $8 \times 10^{-4} ohm^{-1} cm^{-1}$.

(a) 200 (b) 300

(c) 400 (d) 600

150. Equivalent conductance of 0.1 *M* HA (weak acid) solution is 10 S cm^2 $equivalent^{-1}$ and that at infinite dilution is 200 S cm^2 $equivalent^{-1}$. Hence pH of HA solution is :

(a) 1.3 (b) 1.7

(c) 2.3 (d) 3.7

151. The conductivity of a saturated solution of $BaSO_4$ is 8×10^{-8} S cm^{-1} and its equivalent conductivity is 2.0 S cm^2 $equivalent^{-1}$. What is the value of K_{sp} of $BaSO_4$?

(a) 2×10^{-10} (b) 4×10^{-10}

(c) 8×10^{-10} (d) None of these

152. The conductivity of a strong electrolyte :

(a) increases on dilution (b) decreases on dilution

(c) does not change with dilution (d) depends upon density of electrolytes

153. The increase in equivalent conductivity of a weak electrolyte with dilution is due to :

(a) increase in degree of dissociation and decrease in ionic mobility

(b) decrease in degree of dissociation and decrease in ionic mobility

(c) increase in degree of dissociation and increase in ionic mobility
(d) decrease in degree of dissociation and increase in ionic mobility

154. Strong electrolytes are those which :
(a) conduct electricity
(b) dissolve readily in water
(c) dissociate into ions at high dilution
(d) completely dissociate into ions

155. The electric conduction of a salt solution in water depends on the :
(a) size of its molecules
(b) shape of its molecules
(c) size of solvent molecules
(d) extent of its ionization

156. A graph was plotted between molar conductivity of various electrolytes (NaCl, HCl and NH_4OH) and $\sqrt{C}$ (in $mol\,L^{-1}$). Correct set is :

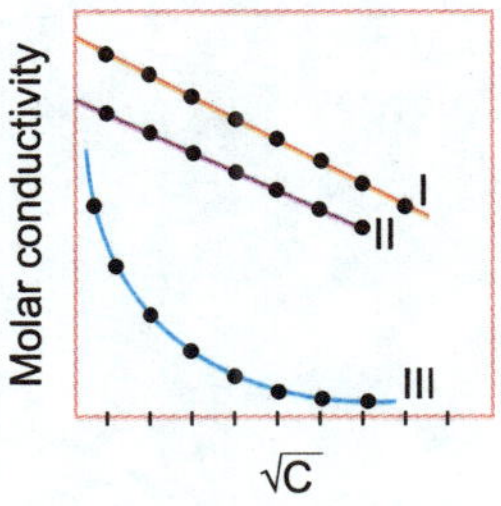

(a) I(NaCl), II(HCl), III(NH_4OH)
(b) I(HCl), II(NaCl), III(NH_4OH)
(c) I(NH_4OH), II(NaCl), III(HCl)
(d) I(NH_4OH), II(HCl), III(NaCl)

157. Which of the following is arranged in increasing order of ionic mobility ?
(a) $I^- < Br^- < Cl^- < F^-$
(b) $F^- < Cl^- < Br^- < I^-$
(c) $F^- < I^- < Cl^- < Br^-$
(d) $F^- < Cl^- < I^- < Br^-$

158. $HNO_3(aq)$ is titrated with NaOH(aq) conductometrically, graphical representation of the titration is :

(a)
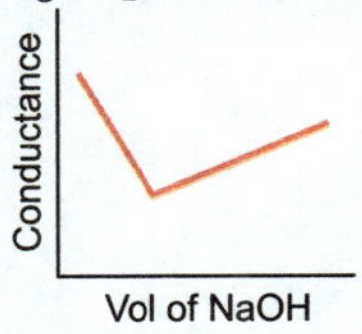

(b)
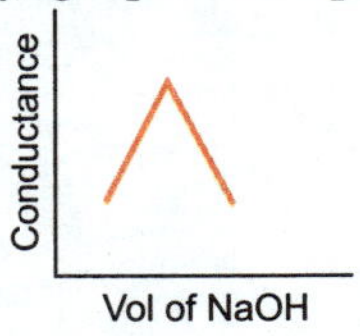

(c)
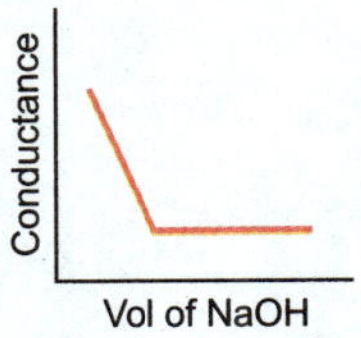

(d)
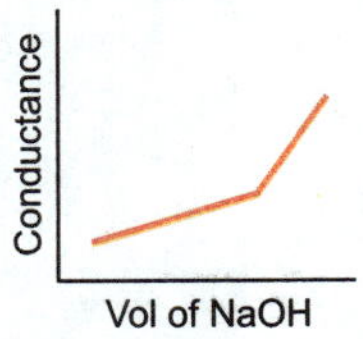

159. Which of the following plots will obtained for a conductometric titration of strong acid against a weak base?

(a)
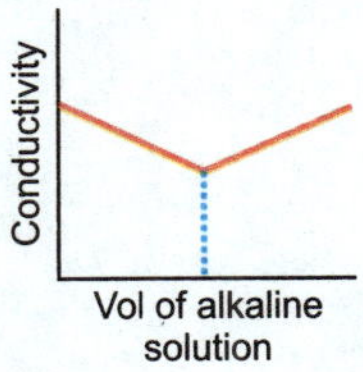

(b)
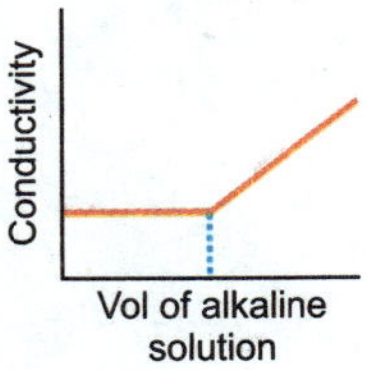

(c)
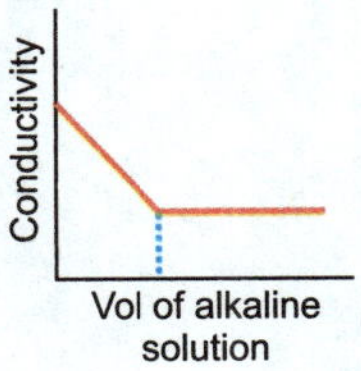

(d)
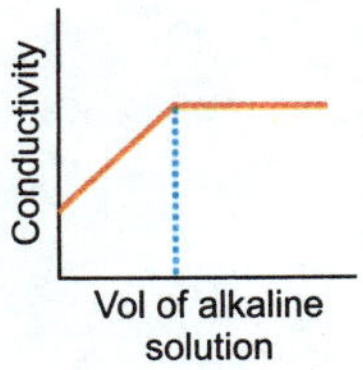

160. Conductometric titration curve of a equimolar mixture of a HCl and HCN with NaOH(aq) is :

(a)
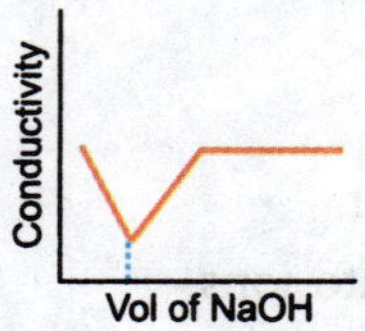

(b)
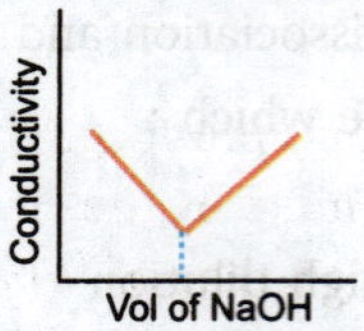

(c)
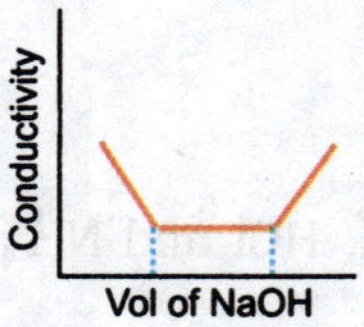

(d)
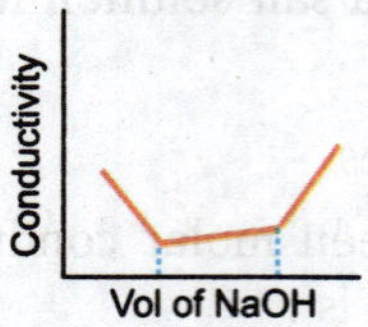

Level-2

1. In the Hall process, aluminium is produced by the electrolysis of molten Al_2O_3. How many second would it take to produce enough aluminum by the Hall process to make a case of 24 cans of aluminum soft-drink, if each can uses 5.0 g of Al, a current of 9650 amp is employed, and the current efficiency of the cell is 90.0% :
(a) 203.2 (b) 148.14
(c) 333 (d) 6.17

2. 108 g fairly concentrated solution of $AgNO_3$ is electrolysed by using 0.1 F charge. The mass of resulting solution is :
(a) 94 g (b) 11.6 g
(c) 96.4 g (d) None of these

3. The electrolysis of acetate solution produces ethane according to reaction :
$$2CH_3COO^- \longrightarrow C_2H_6(g) + 2CO_2(g) + 2e^-$$
The current efficiency of the process is 80%. What volume of gases would be produced at 27°C and 740 torr, if the current of 0.5 amp is passed through the solution for 96.5 min?
(a) 6.0 L (b) 0.60 L
(c) 1.365 L (d) 0.91 L

4. A layer of chromium metal 0.25 mm thick is to be plated on an auto bumper with a total area of $0.32\ m^2$ from a solution containing CrO_4^{2-}? What current flow is required for this electroplating if the bumper is to be plated in 60 s ? The density of chromium metal is $7.20\ g/cm^3$.
(a) 4.9×10^3 A (b) 1.78×10^3 A
(c) 5.3×10^4 A (d) 10.69×10^6 A

5. 100 mL of 0.05 M $CuSO_4(aq)$ solution was electrolysed using inert electrodes by passing current till the pH of the resulting solution was 2. The solution after electrolysis was neutralized and then treated with excess KI and formed I_2 titrated with 0.04 M $Na_2S_2O_3$. Calculate the required volume (in mL) of $Na_2S_2O_3$:
(a) 112.5 mL (b) 100 mL
(c) 125 mL (d) None of these

6. If the equilibrium constant for the reaction $H^+(aq) + OH^-(aq) \rightleftharpoons H_2O(l)$ is 10^{13} at certain temperature then what is the $E°$ for the reaction, $2H_2O(l) + 2e^- \rightleftharpoons H_2(g) + 2OH^-(aq)$
Given : $\frac{2.303\,RT}{F} = 0.066$
(a) 1.230 V (b) –0.858 V
(c) –0.80 V (d) –0.8274 V

7. A fuel cell develops an electrical potential from the combustion of butane at 1 bar and 298 K
$$C_4H_{10}(g) + 6.5O_2(g) \longrightarrow 4CO_2(g) + 5H_2O(l); \quad \Delta_rG° = -2746\ kJ/mol$$
What is $E°$ of a cell?
(a) 4.74 V (b) 0.547 V
(c) 4.37 V (d) 1.09 V

8. The cell $Pt|H_2(g, 0.1\,bar)|H^+\ (aq), pH = X||Cl^-(aq, 1\,M)|Hg_2Cl_2|Hg|Pt$,
has e.m.f. of 0.5755 V at 25°C. The SOP of calomel electrode is –0.28V, then pH of solution will be :
(a) 11 (b) 4.5
(c) 5.5 (d) None of these

9. For a cell reaction $2H_2(g) + O_2(g) \longrightarrow 2H_2O(l)$ $\Delta_r S^\circ_{298} = -0.32$ kJ/K. What is the value of $\Delta_f H^\circ_{298}(H_2O, l)$?

Given : $O_2(g) + 4H^+(aq) + 4e^- \longrightarrow 2H_2O(l)$; $E^\circ = 1.23$ V

(a) −285.07 kJ/mol (b) −570.14 kJ/mol
(c) 285.07 kJ/mol (d) None of these

10. What is the potential of an electrode which originally contained 0.1 M NO_3^- and 0.4 M H^+ and which has been treated by 80% of the cadmium necessary to reduce all the NO_3^- to NO(g) at 1 bar?

Given : $NO_3^- + 4H^+ + 3e^- \longrightarrow NO + 2H_2O$; $E^\circ = 0.96$ V; log 2 = 0.3

(a) 0.84 V (b) 1.08 V
(c) 1.23 V (d) 1.36 V

11. The standard reduction potential of normal calomel electrode and reduction potential of saturated calomel electrodes are 0.27 and 0.33 volt respectively. What is the concentration of Cl^- in saturated solution of KCl?

(a) 0.1 M (b) 0.01 M
(c) 0.001 (d) None

12. Determine the potential of the following cell :

$Pt|H_2(g, 0.1\ bar)|H^+(aq, 10^{-3}\ M)||MnO_4^-(aq, 0.1\ M),$
$Mn^{2+}(aq, 0.01\ M), H^+(aq, 0.01\ M)|Pt$

Given : $E^\circ_{MnO_4^-|Mn^{2+}} = 1.51$ V

(a) 1.54 V (b) 1.48 V
(c) 1.84 V (d) none of these

13. Copper reduces NO_3^- into NO and NO_2 depending upon concentration of HNO_3 in solution. Assuming $[Cu^{2+}] = 0.1\ M$, and $P_{NO} = P_{NO_2} = 10^{-3}$ bar, at which concentration of HNO_3, thermodynamic tendency for reduction of NO_3^- into NO and NO_2 by copper is same?

[Given : $E^\circ_{Cu^{2+}|Cu} = +0.34$ volt, $E^\circ_{NO_3^-|NO} = +0.96$ volt, $E^\circ_{NO_3^-|NO_2} = +0.79$ volt]

(a) $10^{1.23}$ M (b) $10^{0.56}$ M
(c) $10^{0.66}$ M (d) $10^{0.12}$ M

14. For the cell, $Pt|Cl_2(g, 0.4\ bar)|Cl^-(aq, 0.1\ M)||Cl^-(aq, 0.01\ M)|Cl_2(g, 0.2\ bar)|Pt$

The measured potential at 298 K is:

(a) 0.051 V (b) −0.051 V
(c) 0.102 V (d) 0.0255 V

15. The chlorate ion can disproportionate in basic solution according to reaction,

$$2ClO_3^- \rightleftharpoons ClO_2^- + ClO_4^-$$

What is the equilibrium concentration of perchlorate ions from a solution initially at 0.1 M in chlorate ions at 298 K?

Given : $E^\circ_{ClO_4^-|ClO_3^-} = 0.36$ V and $E^\circ_{ClO_3^-|ClO_2^-} = 0.33$ V at 298 K

(a) 0.019 M (b) 0.024 M
(c) 0.1 M (d) 0.19 M

16. A cell diagram shown below contains one litre of buffer solution of $HA(pK_a = 4)$ and NaA in both compartments. What is the cell e.m.f.?

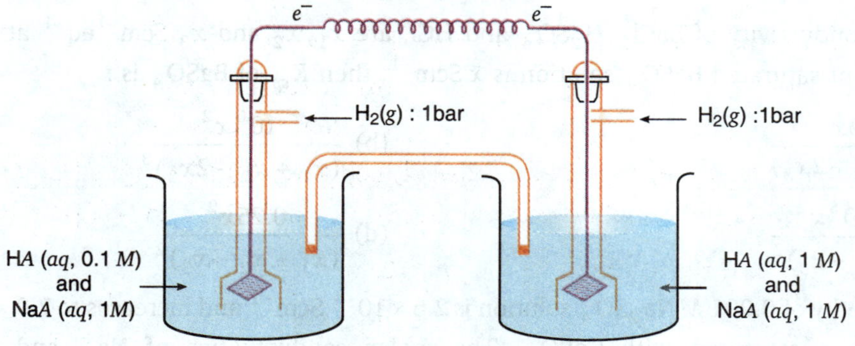

(a) 0.03 V (b) 0.06 V
(c) –0.06 V (d) None of these

17. Given the cell : $Cd(s)\,|\,Cd(OH)_2(s)\,|\,NaOH(aq, 0.01\,M)\,|\,H_2(g, 1\,bar)\,|\,Pt(s)$

with $E_{cell} = 0.0$ V. If $E^\circ_{Cd^{2+}|Cd} = -0.39$ V, then K_{sp} of $Cd(OH_2)$ is :

(a) 0.1 (b) 10^{-13}
(c) 10^{-15} (d) None of these

18. Calculate the e.m.f. (in V) of the cell :

$$Pt\,|\,\underset{0.1\text{ bar}}{H_2(g)}\,|\,\underset{1\,M}{BOH(aq)}\,||\,\underset{0.1\,M}{HA(aq)}\,|\,\underset{1\text{ bar}}{H_2(g)}\,|\,Pt;$$

Given : $K_a(HA) = 10^{-7}, K_b(BOH) = 10^{-6}$

(a) 0.39 V (b) 0.36 V
(c) 0.93 V (d) None of these

19. Calculate the potential of a half cell having reaction : $Ag_2S(s) + 2e^- \rightleftharpoons 2Ag(s) + S^{2-}(aq)$ in a solution buffered at pH = 3 and which is also saturated with $0.1\,M\,H_2S(aq)$:
[Given: $K_{sp}(Ag_2S) = 10^{-49}$, $K_{a_1} \cdot K_{a_2} = 10^{-21}$]

(a) 1.18 (b) 0.19
(c) –0.19 V (d) none of these

20. The conductivity of 0.1 N NaOH solution is 0.022 S cm^{-1}. When equal volume of 0.1 N HCl solution is added, the conductivity of resultant solution is decreased to 0.0055 Scm^{-1}. The equivalent conductivity in Scm^2equivalent^{-1} of NaCl solution is :

(a) 0.0055 (b) 0.11
(c) 110 (d) None of these

21. In above question after formation of NaCl, further 0.1 N HCl is added, the volume of which is double to that of the first portion added, the conductivity increases to 0.018 Scm^{-1}. The value of Λ_{eq}(HCl) is [assume no change in equivalent conductivity of NaCl (aq)] :

(a) 330 S cm^2eq^{-1} (b) 305 S cm^2eq^{-1}
(c) 415 S cm^2eq^{-1} (d) 360 S cm^2eq^{-1}

22. Given the following molar conductivities at infinite dilution (Λ_m^∞), HCl, 426 Ω^{-1} cm^2mol^{-1}; NaCl, 126 Ω^{-1} cm^2mol^{-1}; NaC (sodium crotonate), 83 $\Omega^{-1}cm^2mol^{-1}$. What is the dissociation constant of crotonic acid, if the conductivity of a 0.001 M crotonic acid solution is 3.83×10^{-5} $\Omega^{-1}cm^{-1}$?

(a) 10^{-5} (b) 1.11×10^{-5}

(c) 1.11×10^{-4} (d) 0.01

23. Equivalent conductivity of $BaCl_2, H_2SO_4$ and HCl, are x_1, x_2 and x_3 $Scm^{-1}eq^{-1}$ at infinite dilution. If conductivity of saturated $BaSO_4$ solution is x Scm^{-1}, then K_{sp} of $BaSO_4$ is :

(a) $\dfrac{500x}{(x_1+x_2-2x_3)}$ (b) $\dfrac{10^6x^2}{(x_1+x_2-2x_3)^3}$

(c) $\dfrac{2.5\times10^5x^2}{(x_1+x_2-x_3)^2}$ (d) $\dfrac{0.25x^2}{(x_1+x_2-x_3)^2}$

24. The conductivity of 0.001 M Na_2SO_4 solution is 2.6×10^{-4} Scm^{-1} and increases to 7.0×10^{-4} Scm^{-1}, when the solution is saturated with $CaSO_4$. The molar conductivities of Na^+ and Ca^{2+} are 50 and 120 Scm^2mol^{-1}, respectively. Neglect conductivity of used water.

What is the solubility product of $CaSO_4$?

(a) 4×10^{-6} (b) 1.57×10^{-3}

(c) 4×10^{-4} (d) 2.46×10^{-6}

25. The dissociation constant of a weak acid is 1.6×10^{-5} and the molar conductivity at infinite dilution is 380×10^{-4} $S\,m^2mol^{-1}$. If the cell constant is $0.01\,m^{-1}$ then conductance of 0.01 M acid solution is :

(a) 1.52×10^{-5} S (b) 1.52 S

(c) 1.52×10^{-3} S (d) 1.52×10^{-4} S

26. Three electrolytic cells X, Y, Z containing solution of 1M each NaCl, $AgNO_3$ and $CuSO_4$ respectively are connected in series combination. During electrolysis 21.6 gm of silver deposits at cathode in cell Y. Which is incorrect statement.

(a) 6.35 gm copper deposits at cathode in cell Z

(b) 2.24 litre Cl_2 is liberated (at 1 atm and 273 K) at anode in cell X

(c) 2.24 litre O_2 is liberated (at 1 atm and 273 K) at anode in cell Y

(d) 2.24 litre H_2 is liberated (at 1 atm and 273 K) at cathode in cell X

27. During electrolysis of $H_2SO_4(aq)$ with high charge density, $H_2S_2O_8$ formed as by product. In such electrolysis 22.4 L $H_2(g)$ and 8.4 L $O_2(g)$ liberated at 1 atm and 273 K at electrode. The moles of $H_2S_2O_8$ formed is :

(a) 0.25 (b) 0.50

(c) 0.75 (d) 1.00

28. $Zn(s)\,|\,Zn(CN)_4^{2-}(0.5M), CN^-(0.01)\,||\,Cu(NH_3)_4^{2+}(0.5M), NH_3(1M)\,|\,Cu(s)$

Given: K_f of $Zn(CN)_4^{-2}=10^{16}$, K_f of $Cu(NH_3)_4^{2+}=10^{12}$,

$E^\circ_{Zn|Zn^{+2}}=0.76\,V$; $E^\circ_{Cu^{+2}|Cu}=0.34\,V$, $\dfrac{2.303RT}{F}=0.06$

The emf of above cell is :

(a) 1.22 V (b) 1.10 V

(c) 0.98 V (d) None of these

29. The conductivity of a saturated solution of sparingly soluble salt, $Ba_3(PO_4)_2$ is 1.32×10^{-5} Scm^{-1}.

Calculate the value of K_{sp} of $Ba_3(PO_4)_2$ in M^5

[Given : $\Lambda^\circ_{eq}(BaCl_2)=170\,Scm^2\,eq^{-1}$; $\Lambda^\circ_{eq}(K_3PO_4)=150\,Scm^2\,eq^{-1}$; and $\Lambda^\circ_{eq}(KCl)=100\,Scm^2\,eq^{-1}$]

(a) 1.08×10^{-23} (b) 1.08×10^{-25}

(c) 10^{-25} (d) 2.7×10^{-23}

30. 20 mL of NaOH solution was titrated with 0.20 molL^{-1} H_2SO_4 solution in a conductivity cell. The data obtained were plotted to give the graph shown below :

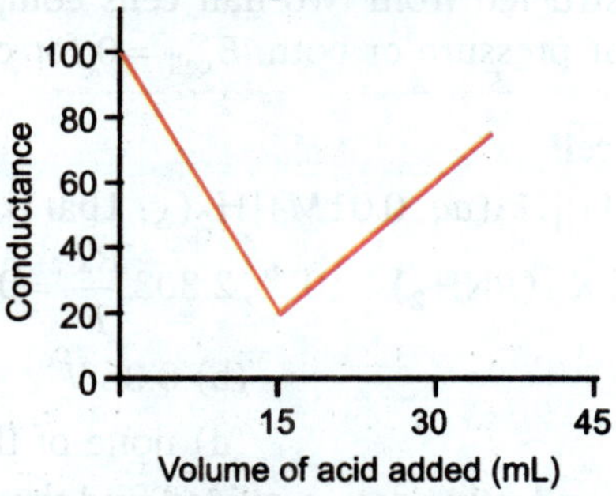

Calculate the concentration of the NaOH solution was used.

(a) 0.15 *M* (b) 0.30 *M*

(c) 0.45 *M* (d) 0.60 *M*

Level-3

Passage-1

Concentration cell is a cell constructed from two half cells composed of the same material but differing in ion concentrations or pressure or both. $E^\circ_{cell} = 0$ for concentration cells.

1. Calculate the EMF (in V) of the cell

$Pt|H_2(g, 1bar)|RNH_2(aq, 0.01M)||HA(aq, 0.01M)|H_2(g, 1bar)|Pt$

Given: at 298 K, $K_b(A^-) = 10^{-8}$, $K_b(RNH_2) = 10^{-6}$, $2.303\frac{RT}{F} = 0.06$, $\log 2 = 0.3$

(a) 0.126 V (b) 0.36 V

(c) 0.429 V (d) none of these

2. Carefully observe the given figure and using data provided find the EMF (in volt) of shown Galvanic cell:

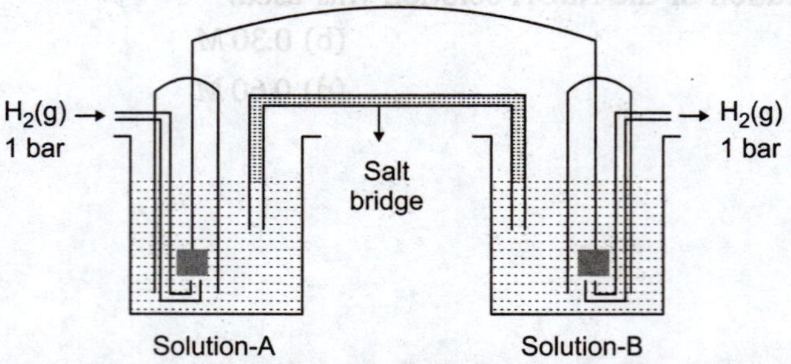

Solution A is $0.1M$ each in NH_4OH and NH_4Cl and solution B is $0.001M$ NH_4Cl.

[Given: $K_b(NH_4OH) = 10^{-5}$ and $\frac{2.303\,RT}{F} = 0.06$ volt]

(a) 0.12 V (b) 0.18 V

(c) 0.24 V (d) none of these

Passage-2

The cell potential (E_{cell}) of a reaction is related as $\Delta G = -nF\,E_{cell}$, where ΔG represents max. useful electrical work

n = no. of moles of electrons exchanged during the reaction

for reversible cell reaction $d(\Delta G) = (\Delta_r V)dp - (\Delta_r S).\,dT$

at constant pressure $d(\Delta G) = -(\Delta_r S)\cdot dT$

$\because$ At constant pressure $\Delta G = \Delta H - T\cdot\Delta S$ ...(1)

$\therefore$ $\Delta G = \Delta H + T\left(\frac{d(\Delta G)}{dT}\right)_p$ ...(2)

$\left(\frac{dE_{cell}}{dT}\right)_p$ is known as temperature coefficient of the e.m.f of the cell

1. The temperature coefficient of the e.m.f. of cell, $\left(\frac{dE}{dT}\right)_p$ is given by :

(a) $\frac{nF}{\Delta S}$ (b) $\frac{\Delta S}{nF}$

(c) $\frac{\Delta S}{nFT}$ (d) $-nFE$

2. At 300 K, ΔH for the reaction
$Zn(s) + AgCl(s) \longrightarrow ZnCl_2(aq) + 2Ag(s)$ is
-218 kJ/mol while the e.m.f. of the cell was 1.015 V. $\left(\frac{dE}{dT}\right)_p$ of the cell is :

(a) $-4.2 \times 10^{-4} VK^{-1}$ (b) $-3.81 \times 10^{-4} VK^{-1}$
(c) $0.11 VK^{-1}$ (d) $7.62 \times 10^{-4} VK^{-1}$

3. Calculate ΔS for the given cell reaction in Q. no. 2 :
(a) -73.53 J/K mol (b) 83.53 J/K mol
(c) 100 J/K mol (d) none of these

Passage-3

Molar conductivity (Λ_m) is defined as conducting power of the ions produced by 1 mole of an electrolyte in a solution. $\Lambda_m = \frac{\kappa}{C}$ where κ is conductivity (in S cm^{-1}) and C is molar concentration (in mol/cm^3)

The molar conductivity of 0.04 M solution of $MgCl_2$ is 200 S cm^2 mol^{-1} at 298 K. A cell with electrodes that are 2.0 cm^2 in surface area and 0.50 cm apart is filled with $MgCl_2$ solution.

1. Conductance of $MgCl_2$ solution is :
(a) 8×10^{-3} S (b) 32 S
(c) 0.032 S (d) None of these

2. How much current will flow when the potential difference between the two electrodes is 5.0 V?
(a) 156.25 A (b) 0.16 A
(c) 160 A (d) None of these

Passage-4

In a hydrogen oxygen fuel cell, electricity is produced. In this process $H_2(g)$ is oxidised at anode and $O_2(g)$ reduced at cathode.

Given : Cathode $O_2(g) + 2H_2O(l) + 4e^- \longrightarrow 4OH^-(aq)$

Anode $H_2(g) + 2OH^-(aq) \longrightarrow 2H_2O(l) + 2e^-$

4.48 litre H_2 at 1 atm and 273 K oxidised in 9650 sec.

1. The current produced is (in amp) :
(a) 1 A (b) 2 A
(c) 4 A (d) 8 A

2. The mass of water produced is :
(a) 7.2 g (b) 3.6 g
(c) 1.8 g (d) 0.9 g

3. If current produced in fuel cell, is used for the deposition of Cu^{+2} in 1 L, 2 M $CuSO_4(aq)$ solution for 241.25 sec using Pt electrode, the pH of solution after electrolysis is :
(a) 1 (b) 2
(c) 3 (d) 4

Passage-5

A saturated solution in $Agx(K_{sp} = 3 \times 10^{-12})$ and $Agy(K_{sp} = 10^{-12})$ has conductivity $0.4 \times 10^{-6}\ \Omega^{-1}\text{cm}^{-1}$.

Given : Limiting molar conductivity of $Ag^+ = 60\ \Omega^{-1}\text{cm}^2\text{mol}^{-1}$

Limiting molar conductivity of $x^- = 90\ \Omega^{-1}\text{cm}^2\text{mol}^{-1}$

1. The conductivity of y^- is (in $\Omega^{-1}\ \text{cm}^{-1}$):

(a) 1.45×10^{-7} (b) 1.45×10^{-5}

(c) 1.45×10^{-9} (d) None of these

2. The limiting molar conductivity of y^- is (in $\Omega^{-1}\ \text{cm}^2\ \text{mol}^{-1}$) :

(a) 290 (b) 2900

(c) 2.90 (d) None of these

One or More Answer is/are Correct

1. If the e.m.f of a galvanic cell is negative, it implies that :

(a) the cell reaction is spontaneous

(b) the cell reaction is non-spontaneous

(c) the cell reaction is exothermic

(d) the cell is working in reverse direction

2. Select correct statement(s) about electrolysis :

(a) Electric current is used to drive a non-spontaneous reaction

(b) ΔG is positive for chemical process during electrolysis

(c) Cations and anions are moved toward the anode and cathode respectively

(d) Over voltage is generally associated with evolution of O_2 gas

3. If the half-cell reaction $A + e^- \longrightarrow A^-$ has a large negative reduction potentials, it follows that :

(a) A is readily reduced (b) A is readily oxidised

(c) A^- is readily reduced (d) A^- is readily oxidised

4. Which of the following statement is correct?

If $E^\circ_{Cu^{2+}|Cu} = 0.34$ V and $E^\circ_{Sn^{2+}|Sn} = -0.136$ V, $E^\circ_{H^+|H_2} = -0.0$ V

(a) Cu^{2+} ions can be reduced by $H_2(g)$ (b) Cu can be oxidized by H^+

(c) Sn^{2+} ions can be reduced by H_2 (d) Sn can be oxidized by Cu^{2+}

5. The oxidation potential of hydrogen half-cell will be negative if :

(a) $p(H_2) = 1$ atm and $[H^+] = 1\,M$ (b) $p(H_2) = 1$ atm and $[H^+] = 2\,M$

(c) $p(H_2) = 0.2$ atm and $[H^+] = 1\,M$ (d) $p(H_2) = 0.2$ atm and $[H^+] = 0.2\,M$

6. Which of the following arrangement will produce oxygen at anode during electrolysis?

(a) Dilute H_2SO_4 with Pt electrodes

(b) Fused NaOH with inert electrodes

(c) Dilute H_2SO_4 with Cu electrodes

(d) Concentrated aq. NaCl with Pt electrodes

7. When an aqueous concentrated solution of lithium chloride is electrolysed using inert electrodes :

(a) Cl_2 is liberated at the anode

(b) Li is deposited at the cathode

(c) as the current flows, pH of the solution around the cathode remains constant
(d) as the current flows, pH of the solution around the cathode increases

8. Oxygen and hydrogen gas are produced at the anode and cathode during the electrolysis of fairly concentrated aqueous solution of :
(a) K_2SO_4
(b) $AgNO_3$
(c) H_2SO_4
(d) NaOH

9. During the purification of copper by electrolysis :
(a) the anode used is made of copper ore
(b) pure copper is deposited on the cathode
(c) the impurities such as Ag, Au present in solution as ions
(d) concentration of $CuSO_4$ solution remains constant during dissolution of Cu

10. When a lead storage battery is discharged :
(a) SO_2 is evolved
(b) lead sulphate is produced at both electrodes
(c) sulphuric acid is consumed
(d) water is formed

11. Which of the following is characteristic of the cathode in a voltaic cell?
(a) It may gain weight during reaction
(b) Electrons flow to it through the external circuit
(c) It is where oxidation occurs
(d) It receives electrons from ions in solution

12. In an electrochemical process, a salt bridge is used :
(a) to maintain electrical neutrality in each solution
(b) to complete the external circuit so that current can flow for long time
(c) to mix the solution of anodic and cathodic compartment
(d) to supply voltage

13. For a reaction in a galvanic cell the value of $-\Delta G°$ at certain temperature is not necessarily equal to :
(a) $nFE°$
(b) $RT \ln K$ at equilibrium
(c) $T \cdot \Delta S° - \Delta H°$
(d) zero

14. Standard electrode potential of two half-reactions are given below :

$$Fe^{2+} \rightleftharpoons Fe \quad ; \quad E° = -0.44 \text{ V}$$

$$Fe^{3+} \rightleftharpoons Fe^{2+} \quad ; \quad E° = +0.77 \text{ V}$$

If Fe^{2+}, Fe^{3+} and Fe are kept together :
(a) the concentration of Fe^{3+} increases
(b) the concentration of Fe^{3+} decreases
(c) the mass of Fe increases
(d) the concentration of Fe^{2+} decreases

15. Which of the following statements are correct regarding to galvanic cell?
(a) A reaction is spontaneous from left to right if $E_{cell} > 0$
(b) A reaction occurs from right to left if $E_{cell} < 0$
(c) If the system is at equilibrium no net reaction occurs
(d) E_{cell} is temperature-independent

16. Which of the following are concentration cells?
(a) $Pt|\underset{P_1}{H_2(g)}|HCl|\underset{P_2}{H_2(g)}|Pt$
(b) $\underset{a_1}{Cd, (Hg)}|\underset{(c)}{Cd^{2+}}|\underset{a_2}{(Hg), Cd}$
(c) $Zn(s)|\underset{c_1}{Zn^{2+}}||\underset{c_2}{Cu^{2+}}|Cu$
(d) $Ag|AgCl|\underset{c_1}{Cl^-(aq)}||\underset{c_2}{Br^-(aq)}|AgBr|Ag$

17. In electrolyte concentration cell :
(a) the electrode material and the solution in both half-cells are composed of the same substances
(b) only the concentrations of solutions of the same substances is different
(c) $E^\circ_{cell} = 0$
(d) the Nernst equation reduces to $E_{cell} = -\left(\frac{0.0591}{n}\right)\log Q$ at 25°C

18. The standard electrode potential of a metal-metal ion ($Ag|Ag^+$) and metal-sparingly soluble salt anion ($Ag|AgCl|Cl^-$) are related as :
(a) $E^\circ_{Ag^+|Ag} = E^\circ_{Cl^-|AgCl|Ag} + \frac{RT}{F}\ln K_{sp}$
(b) $E^\circ_{Cl^-|AgCl|Ag} = E^\circ_{Ag^+|Ag} + \frac{RT}{F}\ln K_{sp}$
(c) $E^\circ_{Cl^-|AgCl|Ag} = E^\circ_{Ag^+|Ag} - \frac{RT}{F}\ln\frac{[Cl^-]}{K_{sp}}$
(d) $E^\circ_{Cl^-|AgCl|Ag} = E^\circ_{Ag^+|Ag} - \frac{RT}{F}\ln\frac{K_{sp}}{[Cl^-]}$

19. Which of the following units is correctly matched?
(a) SI units of conductivity is $S\,m^{-1}$
(b) SI units of molar conductivity is $S\,cm^2 mol^{-1}$
(c) SI unit of conductance is S^{-1}
(d) All of these

20. Which of the following statements is/are correct?
(a) The conductance of one cm^3 (or 1 $unit^3$) of a solution is called specific conductance
(b) Specific conductance increases while molar conductivity decreases on progressive dilution
(c) The limiting equivalent conductivity of weak electrolyte cannot be determined exactly by extraplotation of the plot of Λ_{eq} against $\sqrt{c}$
(d) The conductance of metals is due to the movement of free electrons

21. Which is/are correct statement ?
(a) No corrosion takes place in vacuum
(b) Corrosion is protected by electroplating
(c) During rusting $Fe_2O_3 \cdot xH_2O$ formed
(d) In presence of electrolyte, corrosion takes place with greater rate

22. A dilute solution of KCl was placed between two Pt electrodes 10 cm apart across which a potential difference of 10 volt was applied. Which is/are correct statement (Given : molar conductivity of K^+ at infinite dilution is 96.5 $Scm^2\,mol^{-1}$.)
(a) Ionic mobility of K^+ is 10^{-3} $cm^2\,sec^{-1}\,volt^{-1}$.
(b) The speed of K^+ is 10^{-3} $cm\,sec^{-1}$.
(c) Distance travelled by K^+ in 5×10^3 sec is 5 cm.
(d) The potential gradient is 1.0 volt cm^{-1}.

23. Given : $Pt(s) \mid \underset{1\,bar}{H_2(g)} | 0.1M\,NH_4OH(aq) || 0.1M\,CH_3COOH(aq) \mid \underset{1\,bar}{H_2(g)} | Pt(s)$

$pK_b(NH_4OH) = 5$; $pK_a(CH_3COOH) = 5$; $\frac{2.303RT}{F} = 0.06$

Volume of $0.1M\,NH_4OH$ in anode half cell = 100 mL,
Volume of $0.1M\,CH_3COOH$ in cathode half cell = 100 mL
Which is/are correct statement ?
(a) The emf of given cell is 0.48 V.
(b) The emf of given cell is 0.36 V when 50 mL, 0.1 M NaOH added to cathode compartment
(c) The emf of given cell is 0.36 V when 50 mL 0.1 M HCl added to anode compartment
(d) The emf of given cell is 0.192 V when 100 mL 0.1 M NaOH added to anode compartment

24. Given : $Pt(s)|Cl_2(g)|Cl^-(C_1)||Cl^-(C_2)|Cl_2(g)|Pt(s)$
P_1 atm P_2 atm

Identify in which of following condition working of cell takes place :

(a) $C_1 > C_2$ and $P_1 = P_2$
(b) $P_2 > P_1$ and $C_1 = C_2$
(c) $C_1 < C_2$ and $P_1 = P_2$
(d) $P_2 < P_1$ and $C_1 = C_2$

25. 1000 mL 1*M* $CuSO_4(aq)$ is electrolysed by 9.65 A current for 100 sec using Pt-electrode. Which is/are correct statement ?

(a) Blue colour intensity decreases during electrolysis.
(b) Blue colour intensity remains constant if Cu-electrode used.
(c) pH of solution is 8 after electrolysis.
(d) 28 mL of CH_4 at 1 atm and 273 K required for its combustion by O_2, liberated during electrolysis.

Match the Column

Column–I and **Column–II** contains four entries each. Entries of Column-I are to be matched with some entries of Column-II. One or more than one entries of Column-I may have the matching with the same entries of Column-II

1.

Column–I	Column–II
(A) Dilute solution of HCl	(P) O_2 evolved at anode
(B) Dilute solution of NaCl	(Q) H_2 evolved at cathode
(C) Concentrated solution of NaCl	(R) Cl_2 evolved at anode
(D) Fairly concentrated solution of $AgNO_3$	(S) Ag deposition at cathode

2.

Column–I	Column–II
(A) If SOP of substance lies between −1.23 to −0.81 V	(P) Oxidation of substance is not possible
(B) If SOP of substance lies between −0.81 V to −0.40 V	(Q) Oxidation possible only in acidic medium
(C) If SOP is less than −1.23 V	(R) Oxidation possible in any medium
(D) If SOP is greater than −0.40 V	(S) Oxidation easily takes place

3.

Column–I	Column–II (SRP)
(A) $F_2 + 2e^- \rightleftharpoons 2F^-$	(P) 0.54
(B) $Cl_2 + 2e^- \rightleftharpoons 2Cl^-$	(Q) 1.09
(C) $Br_2 + 2e^- \rightleftharpoons Br^-$	(R) 1.36
(D) $I_2 + 2e^- \rightleftharpoons 2BI^-$	(S) 2.87

4.

Column–I	Column–II
(A) $Pt|Fe^{3+}, Fe^{2+}$	(P) Metal-metal ion half-cell
(B) $Pt|H_2|H^+$	(Q) Gas-gas ion half-cell
(C) $Pt|Hg|Hg_2^{2+}$	(R) Oxidation-reduction half-cell
(D) $Pb|PbSO_4|SO_4^{2-}$	(S) Metal-sparingly soluble salt half-cell

5.

Column–I (Property)	Column–II (Unit)
(A) Conductance	(P) Sm^{-1}
(B) Conductivity	(Q) $S^{-1}m$
(C) Molar conductivity	(R) Sm^2mol^{-1}
(D) Resistivity	(S) S

6.

Column–I (Ion)	Column–II (Molar Conductivity)
(A) H^+	(P) 350
(B) Na^+	(Q) 50
(C) Li^+	(R) 39
(D) Cs^+	(S) 77

7.

Column-I	Column-II
(A) Galvanic cell	(P) Used in space craft
(B) Electrolytic cell	(Q) No transformation of electrical energy into chemical energy
(C) Dead battery	(R) Cell reaction is spontaneous
(D) Fuel cell	(S) Cell reaction is non-spontaneous

Assertion-Reason Type Questions

Each question contains STATEMENT-1 (Assertion) and STATEMENT-2 (Reason).
Examine the statements carefully and mark the correct answer according to the instructions given below :
(A) If both the statements are TRUE and STATEMENT-2 is the correct explanation of STATEMENT-1
(B) If both the statements are TRUE but STATEMENT-2 is NOT the correct explanation of STATEMENT-1
(C) If STATEMENT-1 is TRUE and STATEMENT-2 is FALSE
(D) If STATEMENT-1 is FALSE and STATEMENT-2 is TRUE

1. **STATEMENT-1 :** E°_{cell} is negative for electrolytic cell.
STATEMENT-2 : ΔG° is +ve for electrolytic cell.

2. **STATEMENT-1 :** When 2 faraday of charge is passed through 0.1 M $H_2SO_4(aq)$, 11.2 litre O_2 evolved at STP.
STATEMENT-2 : Molecular mass of oxygen is 32.

3. **STATEMENT-1 :** Copper is dissolved at anode and deposited at cathode when Cu electrodes are used and electrolyte is 1 M $CuSO_4(aq)$ solution.
STATEMENT-2 : SOP of Cu is less than SOP of water and SRP of Cu is greater than SRP of water.

4. **STATEMENT-1 :** 1 coulomb charge deposits 1 g-equivalent of a substance.
STATEMENT-2 : 1 faraday is charge on 1 mole of electrons.

5. **STATEMENT-1 :** If SRP of substance is – 0.3 V, its reduction is possible at cathode.
STATEMENT-2 : Reduction potential of water lies between 0 to –0.8274 V at 25°C.

6. **STATEMENT-1 :** If SRP of substance is –0.5 V then reduction of substance is possible only in basic medium.

   STATEMENT-2 : SRP of water is –0.8274 V and at reduction potential is zero at $pH = 7$

7. **STATEMENT-1 :** The voltage of mercury cell remains constant for longer period of time.

   STATEMENT-2 : It is because net cell reaction does not involve ions.

8. **STATEMENT-1 :** Lead storage battery is a galvanic cell without salt bridge.

   STATEMENT-2 : A secondary cell is rechargeable cell.

9. **STATEMENT-1 :** The SRP of three metallic ions A, B, C are $-0.3, -0.5, 0.8$ volt respectively, so oxidizing power of ions is $C > A > B$.

   STATEMENT-2 : Higher the SRP, higher the oxidizing power.

10. **STATEMENT-1 :** If SOP of substance is less than –1.23 V and over voltage = 0 V, then its oxidation in its aqueous solution is not possible at 298 K.

    STATEMENT-2 : Standard reduction potential (SRP) of water is + 1.23 V.

11. **STATEMENT-1 :** We cannot add the electrode potential in order to get electrode potential of third electrode if no. of moles of electrons exchanged are not same.

    STATEMENT-2 : Electrode potential is an extensive property.

12. **STATEMENT-1 :** $E^\circ_{cell} = 0$ for a chloride ion concentration cell.

    STATEMENT-2 : For this concentration cell, $E_{cell} = \frac{RT}{F} \ln \frac{[Cl^-]_{LHS}}{[Cl^-]_{RHS}}$

13. **STATEMENT-1 :** If $\left(\frac{dE_{cell}}{dT}\right)_p > 0$ for a cell reaction then ΔS is positive.

    STATEMENT-2 : $\Delta S = nFT\left(\frac{dE}{dT}\right)_p$

14. **STATEMENT-1 :** Molar conductivity increases with decrease in concentration for weak electrolytes.

    STATEMENT-2 : No. of ions increases and no. of ions per unit volume decreases due to dilution.

15. **STATEMENT-1 :** Conductivity decreases with the decreases in concentration both the weak and strong electrolytes.

    STATEMENT-2 : No. of ions per unit volume linearly decreases in both electrolytes.

Level-4

Subjective Problems

1. How many faradays are required for reduction of 1 mol $C_6H_5NO_2$ into $C_6H_5NH_2$?
2. What is the equivalent mass of O_2 in the following reaction, $H_2O + \frac{1}{2}O_2 + 2e^- \longrightarrow 2OH^-$?
3. The amount of electricity which releases 2.0 g of gold from a gold salt is same as that which dissolves 0.967 g of copper anode during the electrolysis of copper sulphate solution. What is the oxidation number of gold in the gold ion ? (At. mass of Cu = 63.5; Au = 197)
4. When a molten salt was electrolysed for 5 min with 9.65 A current, 0.18 g of the metal was deposited. Calculate the Eq. mass of metal.
5. During the electrolysis of a concentrated brine solution, calculate the moles of chlorine gas produced by the passage of 4F electricity.
6. Calculate the cell potential (in V) if $\Delta G = -96.5$ kJ/mol and $n = 1$.
7. If K_c for the reaction
   $Cu^{2+}(aq) + Sn^{2+}(aq) \longrightarrow Sn^{4+}(aq) + Cu(s)$
   at 25°C is represented as 2.6×10^y then find the value of y.
   (Given : $E^\circ_{Cu^{2+}|Cu} = 0.34$ V; $E^\circ_{Sn^{4+}|Sn^{2+}} = 0.15$ V)
8. If ΔG° for the half cell $MnO_4^-|MnO_2$ in an acid solution is $-xF$ then find the value of x.
   (Given : $E^\circ_{MnO_4^-|Mn^{2+}} = 1.5$ V; $E^\circ_{MnO_2|Mn^{2+}} = 1.25$ V)
9. If the equilibrium constant for the reaction $Cd^{2+}(aq) + 4NH_3(aq) \rightleftharpoons Cd(NH_3)_4^{2+}(aq)$ is 10^x then find the value of x.
   (Given : $E^\circ_{Cd^{2+}|Cd} = -0.4$ V; $E^\circ_{Cd(NH_3)_4^{2+}|Cd} = -0.61$ V)
10. At what pH oxidation potential of water is -0.81 V ?
11. The resistance of a conductivity cell containing 0.001 M KCl solution at 298 K is 1500 Ω. What is the cell constant (in mm^{-1}) if the conductivity of 0.001 M KCl solution is 2×10^{-3} S mm^{-1}.
12. Molar conductivities at infinite dilution of KCl, HCl and CH_3COOK are 0.013, 0.038 and 0.009 Sm^2 mol^{-1} respectively at 291K. If conductivity of 0.001 M CH_3COOH is 2.72×10^{-3} Sm^{-1} then find % degree of dissociation of CH_3COOH.
13. A solution containing 1 M $XSO_4(aq)$ and 1 M $YSO_4(aq)$ is electrolysed. If conc. of X^{2+} is 10^{-z} M when deposition of Y^{2+} and X^{2+} starts simultaneously, calculate the value of Z.
    Given : $\frac{2.303RT}{F} = 0.06$
    $E^\circ_{X^{2+}|X} = -0.12$ V; $E^\circ_{Y^{2+}|Y} = -0.24$ V
14. Molar conductivity of aqueous solution of HA is 200 S cm^2 mol^{-1}, pH of this solution is 4.
    Calculate the value of pK_a(HA) at 25°C.
    Given : Λ_M^∞ (NaA) = 100 S cm^{2}mol^{-1}; Λ_M^∞ (HCl) = 425 S cm^{2}mol^{-1};
    Λ_M^∞ (NaCl) = 125 S cm^2 mol^{-1}
15. The standard reduction potential of a silver chloride electrode (metal-sparingly soluble salt electrode) is 0.209 V and for silver electrode is 0.80 V. If the moles of AgCl that can dissolve in 10 L of a 0.01 M NaCl solution is represented as 10^{-Z} then find the value of Z.

Numerical Problems

(If the numerical value has more than two decimal places, truncate/round-off the value to TWO decimal places.)

1. The electrolysis of 250 mL of a NaCl solution was carried out for a period of 20 minutes with a current of 2.0 A. The resulting solution was titrated with 0.62 M HNO_3 solution.
The volume of HNO_3 solution (in mL) required to achieve equivalence point is _____.

2. If 4 amp. current produced in fuel cell, use for the deposition of Ag^+ in 1 L $AgNO_3(aq)$ solution for 241.25 sec using Pt electrode.
Calculate the pOH of solution after electrolysis at 25°C.

3. A constant current was flown for 4 min through a solution of KI. At the end of experiment, liberated I_2 consumed 150 mL of 0.1M solution of $Na_2S_2O_3$ following the reaction:

$$I_2 + 2S_2O_3^{2-} \longrightarrow 2I^- + S_4O_6^{2-}$$

What was the average rate of current flow in ampere?

4. If electrolysis of $CuSO_4$ solution is carried out with 100 g impure Cu as anode of 90% purity and pure Cu of mass 100 g as cathode by passing 2 amp. current for 9650 sec then calculate mass of anode (in g) after electrolysis (At. mass Cu = 63)

5. Calculate E°_{cell} of following cell reaction (in volt).

$$MnO_4^-(aq.) + Mg(s) \longrightarrow Mg^{2+}(aq.) + MnO_2(s)$$

[Given : $E^\circ_{Mg^{2+}/Mg} = -2.35\,V$; $E^\circ_{MnO_4^-/MnO_3^-} = 0.83\,V$; $E^\circ_{MnO_3^-/MnO_2} = 4.27\,V$]

6. Calculate pK_{sp} of $PbSO_4$ at 25°C
$Pb(s) + SO_4^{2-}(aq.) \longrightarrow PbSO_4(s) + 2e^-$; $E° = +0.356\,V$
[Given : $E^\circ_{Pb^{2+}/Pb} = -0.124\,V$, $\dfrac{2.303\,RT}{F} = 0.06$]

7. Given : $Ag(s)|AgBr(s)|Br^-(1.0\,M)||Cl^-(0.5\,M)|AgCl(s)|Ag(s)$
If the emf of following cell is 0.12 V then the ratio of K_{sp}(AgCl) and K_{sp}(AgBr) is ________.
[Given : $2.303\dfrac{RT}{F} = 0.06$]

8. For the given cell, $Ag(s)|Ag^+$ (saturated solution of AgCl) $||Ag^+(0.1M)|Ag(s)$
Calculate the emf of cell, E (in volt)
[Given : $\dfrac{2.303\,RT}{F} = 0.06$, $K_{sp}(AgCl) = 10^{-10}$]

9. The overall formation constant for the reaction of 6 mole of CN^- with cobalt (II) is 1×10^{19}. If the formation constant for the reaction of 6 mole of CN^- with cobalt (III) is 10^x then find the value of x.
[Given : $Co(CN)_6^{3-} + e^- \longrightarrow Co(CN)_6^{4-}$; $E^\circ_{RP} = -0.85\,V$ and $Co^{3+} + e^- \longrightarrow Co^{2+}$; $E^\circ_{RP} = 1.85\,V$; $\dfrac{2.303\,RT}{F} = 0.06$]

10. The magnitude of E°_{OP} of following reaction is _____ volt.

$$Ag(s) + 2NH_3(aq) \longrightarrow Ag(NH_3)_2^+(aq) + e^-$$

[Given : $K_f(Ag(NH_3)_2^+) = 10^8$; $E^\circ_{Ag^+|Ag} = 0.80\,V$; $2.303\dfrac{RT}{F} = 0.06$]

11. The voltage of the cell given below is −0.61V.
$Pt|H_2(1bar)|NaHSO_3(0.4\,M), Na_2SO_3(3.8 \times 10^{-2}\,M)||Zn^{2+}(0.4\,M)|Zn$

[Given : $E^{\circ}_{Zn^{2+}|Zn} = -0.76$ V, $2.303\left(\frac{RT}{F}\right) = 0.06$]

If K_{a_2} of H_2SO_3 is 2×10^{-y} then find the value of y.

12. For the cell reaction : $Hg_2Br_2(s) + 2Ag(s) \longrightarrow 2Hg(l) + 2AgBr(s)$, the variation of emf with temperature is shown by the graph given below :

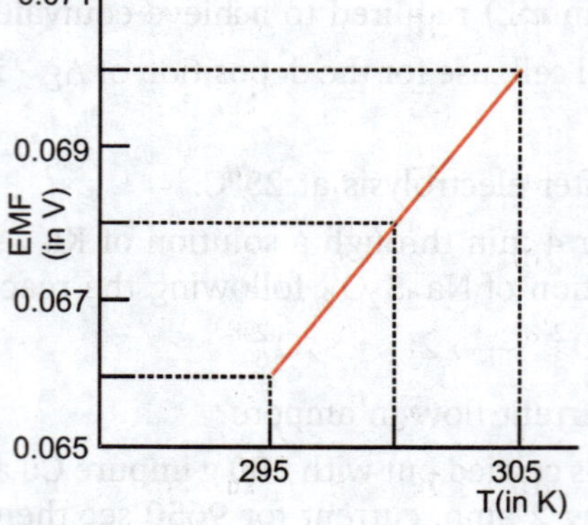

Calculate ΔH of cell reaction (in kJ) at 300 K.

13. Λ_{eq} of 0.10 N solution of CaI_2 is 100.0 $S\,cm^2\,eq^{-1}$ at 298 K. G* of the cell = 0.25 cm^{-1}.

How much current (in A) will flow when the potential difference between the electrode is 5 V?

14. If conductivity, κ (in Sm^{-1}) of saturated solution of AgCl is $z \times 10^{-6}$, then find the value of z.

[Given : limiting ionic conductance of Ag^+ and Cl^- ions is 60 $Scm^2\,mol^{-1}$ and 80 $Scm^2\,mol^{-1}$ respectively and K_{sp} of AgCl = 10^{-10}]

15. A sparingly soluble salt MX is dissolved in water to prepare 1 litre saturated solution. Now 10^{-6} mol NaX is added into this. Conductivity of this solution is 29×10^{-6} Sm^{-1}.

If K_{sp} of MX is $y \times 10^{-12}$ M^2 then find the value of y.

[Given : $\lambda^{\circ}_m(X^-) = 4 \times 10^{-3}\,Sm^2\,mol^{-1}$, $\lambda^{\circ}_m(Na^+) = 5 \times 10^{-3}\,Sm^2\,mol^{-1}$, $\lambda^{\circ}_m(M^+) = 6 \times 10^{-3}\,Sm^2\,mol^{-1}$]

Answers

Level 1

1. (a)	2. (c)	3. (d)	4. (d)	5. (a)	6. (c)	7. (d)	8. (d)	9. (a)	10. (c)
11. (a)	12. (b)	13. (c)	14. (b)	15. (d)	16. (c)	17. (a)	18. (a)	19. (d)	20. (c)
21. (d)	22. (c)	23. (b)	24. (c)	25. (c)	26. (d)	27. (d)	28. (b)	29. (c)	30. (d)
31. (d)	32. (d)	33. (a)	34. (a)	35. (a)	36. (a)	37. (a)	38. (d)	39. (d)	40. (a)
41. (a)	42. (c)	43. (d)	44. (b)	45. (b)	46. (c)	47. (c)	48. (a)	49. (b)	50. (b)
51. (a)	52. (c)	53. (d)	54. (a)	55. (c)	56. (b)	57. (c)	58. (b)	59. (c)	60. (a)
61. (c)	62. (c)	63. (d)	64. (d)	65. (b)	66. (a)	67. (a)	68. (d)	69. (d)	70. (a)
71. (a)	72. (d)	73. (c)	74. (b)	75. (d)	76. (d)	77. (d)	78. (d)	79. (a)	80. (d)
81. (a)	82. (a)	83. (a)	84. (a)	85. (b)	86. (b)	87. (a)	88. (c)	89. (d)	90. (d)
91. (c)	92. (c)	93. (b)	94. (b)	95. (a)	96. (a)	97. (c)	98. (d)	99. (d)	100. (b)
101. (c)	102. (a)	103. (b)	104. (b)	105. (d)	106. (d)	107. (c)	108. (c)	109. (c)	110. (d)
111. (c)	112. (c)	113. (c)	114. (d)	115. (c)	116. (b)	117. (b)	118. (b)	119. (d)	120. (b)
121. (a)	122. (a)	123. (b)	124. (b)	125. (c)	126. (b)	127. (c)	128. (b)	129. (b)	130. (a)
131. (a)	132. (b)	133. (b)	134. (c)	135. (b)	136. (c)	137. (d)	138. (d)	139. (c)	140. (a)
141. (b)	142. (b)	143. (a)	144. (a)	145. (c)	146. (b)	147. (a)	148. (c)	149. (c)	150. (c)
151. (b)	152. (b)	153. (c)	154. (d)	155. (d)	156. (b)	157. (b)	158. (a)	159. (c)	160. (d)

Level 2

1. (b)	2. (c)	3. (d)	4. (d)	5. (a)	6. (b)	7. (d)	8. (c)	9. (a)	10. (a)
11. (a)	12. (b)	13. (c)	14. (a)	15. (a)	16. (b)	17. (c)	18. (a)	19. (c)	20. (c)
21. (b)	22. (b)	23. (c)	24. (a)	25. (b)	26. (c)	27. (a)	28. (c)	29. (a)	30. (b)

Level 3

Passage–1:	1. (b)	2. (b)	
Passage–2:	1. (b)	2. (b)	3. (a)
Passage–3:	1. (c)	2. (b)	
Passage–4:	1. (c)	2. (b)	3. (b)
Passage–5:	1. (a)	2. (a)	

One or More Answers is/are correct

1.	(b,d)	2.	(a,b,d)	3.	(d)	4.	(a,d)	5.	(b,c)	6.	(a,b)	7.	(a,d)	8.	(a, c, d)
9.	(a,b,d)	10.	(b,c,d)	11.	(a,b)	12.	(a)	13.	(b,d)	14.	(b)	15.	(a,b,c)	16.	(a,b,d)
17.	(a,b,c,d)	18.	(b)	19.	(a)	20.	(a,c,d)	21.	(a,b,c,d)	22.	(a,b,c,d)	23.	(a,b,c)	24.	(a,b)
25.	(a,b,d)														

Match the Column

1. A → P, Q; B → P, Q; C → Q, R; D → P, S
2. A → Q; B → R; C → P; D → S, R
3. A → S; B → R; C → Q; D → P
4. A → R; B → Q; C → P; D → S
5. A → S; B → P; C → R; D → Q
6. A → P; B → Q; C → R; D → S
7. A → Q, R; B → S; C → Q; D → P, Q, R

Assertion-Reason Type Questions

1.	(A)	2.	(B)	3.	(C)	4.	(D)	5.	(A)	6.	(C)	7.	(A)	8.	(B)	9.	(A)	10.	(C)
11.	(C)	12.	(B)	13.	(C)	14.	(A)	15.	(C)										

Level-4

Subjective Problems

1.	6	2.	8	3.	3	4.	6	5.	2	6.	1	7.	6	8.	5	9.	7	10.	7
11.	3	12.	8	13.	4	14.	4	15.	7										

Numerical Problems

1.	40.00	2.	12.00	3.	6.03	4.	93.00	5.	4.33	6.	7.73	7.	50.00	8.	0.24
9.	64.00	10.	0.32	11.	4.00	12.	10.04	13.	0.20	14.	1.40	15.	6.00		

Hints & Solutions

Level 1

1. (a) For spontaneous

$$\Delta_r G = -nFE < 0$$

$\therefore \quad E > 0$

2. (c) Cathode and anode are (–) ve terminals in electrolytic and galvanic cell.

3. (d) Electrostatic attraction is broken down by polar solvent (water).

4. (d) $\underset{1\text{ mole}}{A^+} + e^- \longrightarrow \underset{1\text{ mole}}{A}$

$\therefore$ Charge of 1 mole electron is required for deposition of 1 equivalent of substance.

5. (a) According to stoichiometry liberated amount of ion depends upon $\left(\because w = \frac{EIt}{F}\right)$

(i) Electrolysis time (ii) Current strength

(iii) Electrochemical equivalent

6. (c) 1 F = 96500 C,

$$1\text{ C} = \frac{6.023 \times 10^{23}}{96500} = 6.24 \times 10^{18}$$

7. (d) Charge $= 0.01 \times 10^{-3} \times 1000 \times 60 \times 60 = 36$

8. (d) $3.21 \times 10^6 = 500 \times t$

$$\therefore \quad t = \frac{3.21 \times 10^6}{500 \times 60} = 107 \text{ min.}$$

9. (a) $W = ZIt$

$$0.746 = \frac{108}{1 \times F} \times 548 \times I \quad \Rightarrow \quad I = 1.22$$

10. (c) $\frac{0.55}{M} \times 3 = \frac{0.55 \times 100 \times 60}{96500}$

$\Rightarrow \quad M = 48.25 \text{ g/mol}$

11. (a) $w = \frac{9}{2} \times \frac{5 \times 1 \times 60 \times 60}{96{,}500} \approx 0.840 \text{ g}$

12. (b) $5.2 = \frac{52}{3} \times \frac{9.65}{96{,}500} \times t$ (sec.),

$t(\text{sec}) = 3000$

$t(\text{min}) = 50$

13. (c) $\frac{W}{E} = \frac{I \times t}{96500}$; $\frac{0.195}{195} \times 4 = \frac{5 \times 60 \times 60 \times I}{96500}$,

$I = 21.44$

14. (b) $0.25 = \frac{93}{5} \times \frac{Q}{F}$

$\therefore \quad Q = 1.3 \times 10^{-2}$

15. (d) When 1 mole of e^- is passed, mass of substance deposited

$$= \frac{1 \times 6.023 \times 10^{23}}{1.81 \times 10^{23}} \Rightarrow 3.34 \text{ g} = \text{eq. wt}$$

$\therefore$ atomic mass of metal $= 3.34 \times 3$

$= 10.03$ g

16. (c) 1 mole Mn_3O_4 lose $\left(6 - \frac{8}{3}\right) \times 3 = 10$ mole

e^-; so total charge required $= 2 \times 10 \Rightarrow 20$ F

17. (a) $2H^+ + 2e^- \longrightarrow H_2$

18. (a) $\because \quad E^\circ_{Mg^{2+}|Mg} < E^\circ_{Cr^{3+}|Cr} < E^\circ_{Cu^{2+}|Cu} < E^\circ_{Ag^+|Ag}$

19. (d) $\because \quad E^\circ_{K^+|K} < E^\circ_{H^+|H_2}$

$E^\circ_{H_2O|O_2} < E^\circ_{SO_4^{2-}|S_2O_8^{2-}}$

(High concentrate solution)

20. (c) $\because E^\circ_{Na^+|Na} < E^\circ_{H^+/H_2}$

$E^\circ_{Cu^{2+}|Cu} > E^\circ_{H^+/H_2}$

21. (d) $E^\circ_{SO_4^{2-}|S_2O_8^{2-}} < E^\circ_{H_2O|O_2}$

22. (c) $E^\circ_{Na^+|Na} < E^\circ_{H^+|H_2}$; $E^\circ_{SO_4^{2-}|S_2O_8^{2-}} < E^\circ_{H_2O|O_2}$

23. (b) In electrolysis of concentrated Cl^-, $Cl_2(g)$ is evolved in place of $O_2(g)$.

25. (c) $E^\circ_{Cu^{2+}|Cu} > E^\circ_{H_2O|H_2}$; $E^\circ_{SO_4^{2-}|S_2O_8^{2-}} < E^\circ_{H_2O|O_2}$

26. (d) Anode : $2Cl^-(aq) \longrightarrow Cl_2(g) + 2e^-$

Cathode: $2H_2O + 2e^- \longrightarrow H_2(g) + 2OH^-(aq)$

Net reaction

$$2Na^+(aq) + 2Cl^-(aq) + 2H_2O\ (l) \longrightarrow Cl_2(g) + H_2(g) + 2OH^-(aq) + 2Na^+(aq)$$

27. (d) $\because \quad E^\circ_{K^+|K} < E^\circ_{H_2O|H_2}$; $E^\circ_{Na^+|Na} < E^\circ_{H_2O|H_2}$;

$E^\circ_{Cu^{2+}|Cu} > E^\circ_{H_2O|H_2}$

28. (b) Let t sec be used; No. of faradays $= \frac{4 \times t}{96500}$

No. of moles of $H_2O = 4$, no. of equivalents of $H_2O = 4 \times 2$

$\therefore \quad \frac{4 \times t}{96500} = 4 \times 2$

$\Rightarrow \quad t = 1.93 \times 10^5$

29. (c) only Au^{3+}, Ag^+ and Cu^{2+} will deposit at cathode.

Li will not deposit at cathode because SRP of water is –0.8274 V so after Cu^{2+}; H_2 will evolve at cathode.

30. (d) $\frac{W}{M} \times 2 = \frac{4.6 \times 30 \times 60}{96500}$

moles of Sn^{2+} reduced $= 0.043$

Initial moles of $Sn^{2+} = 0.5 \times 0.6 \Rightarrow 0.30$;

remaining moles of $Sn^{2+} = 0.30 - 0.043$

$\therefore \quad [Sn^{2+}] = \frac{0.257}{0.5} = 0.514\ M$

31. (d) No. of moles of

$Ag^+ = \frac{15 \times 60 \times 1.25 \times 10^{-3}}{96,500} \times \frac{1}{1}$

$= 0.0116 \times 10^{-3}$

$\therefore \quad [Ag^+] = \frac{1.16 \times 10^{-5}}{\frac{100}{1000}} = 1.16 \times 10^{-4}$

32. (d) Initial moles of $Cr^{3+} = 0.25 \times 0.2 \Rightarrow 0.05$;

final moles of $Cr^{3+} = 0.25 \times 0.1 \Rightarrow 0.025$ moles of Cr^{3+} reduced, $0.05 - 0.025 \Rightarrow 0.025$;

$\therefore$ eq. of Cr^{3+} reduced, $0.025 \times 3 = \frac{t \times 96.5}{96500}$;

$t = 75$ sec

33. (a) $3.05 = \frac{114.8}{n} \times \frac{3.2 \times 40 \times 60}{96,500}$

$\therefore \quad n = 3$

34. (a) $\frac{W}{M} \times n = \frac{I \times t}{96500}$; $\frac{0.838}{184} \times n$

$= \frac{40 \times 60 \times 1.1}{96500}$

$\Rightarrow \quad n = 6$

35. (a) No. of equivalent of H_2 = No. of equivalent of Cl_2

$\frac{5 \times 1}{22.4} \times 2 = \frac{V_{Cl_2}}{22.4} \times 2$

$\therefore \quad V_{Cl_2(g)} = 5L$

36. (a) No. of equivalent of Ag $= \frac{0.54}{108} \times 1 =$ No. of equivalent of Cr $= \frac{W_{Cr}}{52} \times 3$

$\therefore \quad W_{Cr} = \frac{0.54}{108} \times \frac{52}{3} = 0.0866$

37. (a) No. of equivalent of Cu $= \frac{5.6 \times 1}{22.4} \times 4$

$= \frac{W_{Cu}}{63.5} \times 2$, $W_{Cu} = 31.75$

38. (d) no. of eq = no. of $F = \frac{1000}{122.5} \times 8 \Rightarrow 65.3$

39. (d) $54 \times 10^3 = \frac{27}{3} \times \frac{x}{96,500}$

$x = 57.9 \times 10^7$

40. (a) $\frac{W}{7} = \frac{1930 \times 0.75}{96500}$; $W = 0.105$ gm

41. (a) equivalents of Cl_2 produced

$= \frac{1000 \times 9.65 \times 3600}{96500} = 360$

moles of $Cl_2 = 180$

so $V = \frac{nRT}{P} \Rightarrow \frac{180 \times 0.0821 \times 300}{1.8}$

$= 2463$ L

42. (c) No. of equivalent of H_2 produced = Eq. of O_2

$= \frac{I \times t}{96500} = \frac{30 \times 193 \times 60}{96500}$

$= 3.6$

vol. of O_2 and H_2 produced are

$= 3.6 \times 5.6 + 3.6 \times 11.2$

$= 60.48$ litre

43. (d) 1 watt = 1 J/sec; 1 kWh $= \frac{V \times I \times t}{1000}$ (in hr)

Total kWh $= 10^{-3} \times 110 \times 10 \times 10 \Rightarrow 11$

Total cost $= 11 \times 2 \Rightarrow 22$ Rs.

44. (b) $E^\circ_{SO_4^{2-}|S_2O_8^{2-}} < E^\circ_{H_2O|O_2}$

45. (b) 10 g Cd should present with 90 g Hg.

mass of Cd required with 4.5 g of Hg at cathode

$= \frac{10}{90} \times 4.5 \Rightarrow 0.5$; $\quad (Cd^2 + 2e^- \rightarrow Cd)$

$\therefore \quad \frac{0.5}{112} \times 2 = \frac{4 \times t}{96500}$

$\Rightarrow \quad t = 215.40$ sec

46. (c) Use of electrolysis is in (i) Electrorefining (ii) Electroplating

47. (c) $E^\circ_{Ag^+|Ag} > E^\circ_{H^+|H_2}, E^\circ_{Cu^{2+}/Cu} < E^\circ_{O_2,H^+/H_2O}$

48. (a) Chemical energy into electrical energy.

49. (b) $\Delta G_1^\circ = -\Delta G_2^\circ$

$-2 \times FE^\circ_{Cl^-|Cl_2} = +2 \times FE^\circ_{Cl_2|Cl^-}$

$= -1.36$ V

50. (b) $\Delta G_3^\circ = \Delta G_1^\circ + \Delta G_2^\circ$

$-n_3FE_3^\circ = -n_1FE_1^\circ - n_2FE_2^\circ$

$\therefore \quad E_3^\circ = \frac{n_1E_1^\circ + n_2E_2^\circ}{n_3}$

51. (a) Maintain electrical neutrality of both half cells.

52. (c) Velocities of ions are nearly the same.

53. (d) A salt bridge contains strong electrolyte with agar-agar.

54. (a) $\log K_C = \frac{nFE^\circ}{2.303\,RT}$

$\therefore\ y = mx$

55. (c) $\Delta G_1^\circ = -1 \times F(x) = -RT \ln K_1$

$\Delta G_2^\circ = -1 \times F(y) = -RT \ln K_2$

$\Delta G_2^\circ = -2\Delta G_1^\circ = -2RT \ln K_1$

$\therefore\ K_2 = K_1^2,\ x = y$

56. (b) $E = E^\circ - \frac{0.0591}{2} \log \frac{[Zn^{2+}]}{[Ag^+]^2}$

$y = C - mx$

57. (c) When $Q = K_C$, $E = 0$

58. (b) $Pt(s)\,|H_2(g)\,|H^+(aq)\,||\,Ag^+(aq)\,|Ag(s)$,

Anode : $H_2|H^+$, Cathode : $Ag^+|Ag$

59. (c) $Cu(s)\,|Cu^{2+}(aq)\,||Cl^-(aq)\,|Hg_2Cl_2(s)\,|Hg(l)\,|Pt(s)$,

Anode : $Cu|Cu^{2+}$

Cathode : $Cl^-|Hg_2Cl_2|Hg$

60. (a) $Pt(s)\,|Fe^{2+}(aq), Fe^{3+}(aq)\,||Cr_2O_7^{2-}(aq), Cr^{3+}(aq)\,|Pt(s)$

Anode : $Fe^{2+}|Fe^{3+}$

Cathode : $Cr_2O_7^{2-}\,|Cr^{3+}$

61. (c) Anode : $2Ag(s) \rightleftharpoons 2Ag^+(aq) + 2e^-$

Cathode : $Cu^{2+}(aq) + 2e^- \rightleftharpoons Cu(s)$

Net reaction : $2Ag(s) + Cu^{2+}(aq) \rightleftharpoons 2Ag^+(aq) + Cu(s)$

62. (c) Anode : $Cl_2(g) \rightleftharpoons 2Cl^-(aq) + 2e^-$

Cathode : $2Ag^+(aq) + 2e^- \rightleftharpoons 2Ag(s)$

Net reaction : $Cl_2(g) + 2Ag^+(aq) \rightleftharpoons 2Cl^-(aq) + 2Ag(s)$

63. (d) $E^\circ_{H_2|H^+} = E^\circ_{H_2|H^+} = 0V$

64. (d) $E = E^\circ - \frac{0.0591}{2} \log \frac{[Zn^{2+}]}{[Cu^{2+}]}$

As $Q\downarrow$, $E\uparrow$

$Q_1 = 1,\ Q_2 = 0.1,\ Q_3 = 10,\ Q_4 = 1$

$E_2 > E_1 = E_4 > E_3$

65. (b) Ag becomes oxidized and Br^- becomes reduced.

66. (a) $N_2(g) + 5H^+(aq) + 4e^- \longrightarrow N_2H_5^+(aq)$, $E^\circ = -0.23$ V ...(1)

$BiO^+(aq) + 2H^+(aq) + 3e^- \longrightarrow Bi(S) + H_2O\,(l)$ $E^\circ = 0.32$ V ...(2)

[Reaction (1)] × −3 + [Reaction (2)] × 4, we get

$4BiO^+(aq) + 3N_2H_5^+(aq) \longrightarrow 4Bi\,(s) + 3N_2(g) + 4H_2O(l) + 7H^+(aq)$

$\Delta G^\circ = -3\,(-4FE^\circ) + 4(-3FE^\circ)$

$-12FE^\circ = 12F\,(-0.23) - 12F\,(0.32)$

$E^\circ = 0.23 + 0.32 = 0.55$

67. (a) $Cr^{2+}(aq) \longrightarrow Cr^{3+}(aq) + e^-$, $E^\circ = 0.41$ V ...(1)

$HClO(aq) + H^+(aq) + 2Cr^{2+}(aq) \longrightarrow 2Cr^{3+}(aq) + Cl^-(aq) + H_2O(l)$, $E^\circ = 1.8$ ...(2)

[Reaction (1)] × −2 + Reaction (2)] we get.

$HClO(aq) + H^+(aq) + 2e^- \longrightarrow Cl^-(aq) + H_2O(l)$

$\Delta G^\circ = -2FE^\circ - 2(-1 \times FE^\circ)$

$-2FE^\circ = -2F\,(1.8) + 2F\,(0.41)$

$E^\circ = +1.8 - 0.41 = 1.39$

68. (d) $0.34 = 1 + E^\circ_{SbO_2^-|Sb}$

$\Rightarrow\ E^\circ_{SbO_2^-|Sb} = 0.34 - 1 = -0.66$

69. (d) For spontaneous $E^\circ > 0$,

minimum possible spontaneous $E^\circ = 0.28$

70. (a) For best reducing agent, E°_{Red} is minimum.

71. (a) For best oxidizing agent, E°_{Red} is maximum.

72. (d) Maximum voltage $E^\circ = 0.49 + 0.86 = 1.35$

73. (c) $E^\circ_{Al^{3+}|Al} < E^\circ_{Cu^{2+}|Cu}$; $E^\circ_{Cu|Cu^{2+}} + E^\circ_{Al^{3+}|Al} = -ve$ (Non-spontaneous)

74. (b) $E^\circ_{Zn^{2+}|Cu} < E^\circ_{Cu^{2+}|Cu}$; $E^\circ_{Cu^{2+}|Cu} + E^\circ_{Zn|Zn^{2+}} = +ve$

75. (d) D does not react with HCl, so not easily oxidized at compared to A, B and C

C reduces B as well as D, but unable to reduce A

so A is oxidized easily as compared to other metals

76. (d) $E^\circ_{Cu^{2+}|Cu} < E^\circ_{Ag^+|Ag}$; $E^\circ_{Cu^{2+}|Cu} + E^\circ_{Ag^+|Ag^-} = -ve$ (Non-spontaneous)

77. (d) From (I) information, C is strong reducing agent.

From (II) information, A is strong reducing agent than D, B and C reducing agent than A.

$\therefore\ D < A < B < C$

78. (d) $4Fe(s) + 3O_2(g) \longrightarrow 4Fe^{3+} + 6O^{2-}(aq)$

Fe is reducing agent, O_2 is oxidizing agent.

79. (a) $E^\circ_{Na^+|Na} < E^\circ_{Zn^{2+}|Zn} < E^\circ_{Fe^{2+}|Fe} < E^\circ_{Ag^+|Ag}$

$E^\circ_{Fe|Fe^{2+}} + E^\circ_{Ag^+|Ag} = +ve$ (Spontaneous)

80. (d) $E^\circ_{Zn|Zn^{2+}} + E^\circ_{Mg^{2+}|Mg} = -1.608$ (Non-spontaneous)

81. (a) $E^\circ_{F_2|F^-} > E^\circ_{Cl_2|Cl^-} > E^\circ_{Br_2|Br^-} > E^\circ_{I_2|I^-}$

82. (a) $K = 10^{\frac{2(0.277)}{0.0591}} = 2.3 \times 10^9$

83. (a) $E^\circ = \frac{0.0591}{2} \log K_{eq}$; $\log K_{eq} = \frac{2 \times 0.22}{0.0591}$

$\Rightarrow\ 7.44$ or $K_{eq} \simeq 2.8 \times 10^7$

84. (a) $E = E° - \frac{0.0591}{2}\log\frac{1}{[Cl^-]^2}$; as $[Cl^-]\uparrow, E\uparrow$

85. (b) $K = 10^{\frac{6 \times FE°}{2.303RT}}$

$\log 10^{30} = \frac{6 \times E°}{0.0591} \Rightarrow E° = 0.2955$ V

86. (b) $E°_{cell} = \frac{0.0591}{2}\log K_{eq}$

$0.003 = \frac{0.0591}{2}\log\frac{[D^+]^2}{[H^+]^2}$

$= \frac{[D^+]}{[H^+]} = 1.123$

87. (a) $\Delta G° = -12 \times F \times E° = -636.9 \text{ kJ} = -637 \text{ kJ}$

88. (c) $E°_{cell} = E°_{RP(RHS)} - E°_{RP(LHS)}$

$= 1.61 - 1.51 \Rightarrow 0.10$ V

$\Delta G° = -nFE° \Rightarrow -5 \times 96500 \times 0.10$ J

$\Delta G° = -48.25$ kJ

89. (d) $E = E° - \frac{0.0591}{1}\log\frac{[Fe^{3+}]}{[Ag^+][Fe^{2+}]}$

as $Q\uparrow\{[Ag^+]\downarrow, [Fe^{3+}]\uparrow\}, E\downarrow$

90. (d) $E = E° - \frac{0.0591}{2}\log\frac{[H^+]^2[Cl^-]^2}{P_{H_2}}$

as $Q\downarrow\{(P_{H_2}\uparrow, [H^+]\downarrow\}, E\uparrow$

91. (c) $E_{Cr_2O_7^{2-}|Cr^{3+}} = E°_{Cr_2O_7^{2-}|Cr^{3+}} - \frac{0.0591}{6} \times \left(\log\frac{[Cr^{3+}]^2}{[Cr_2O_7^{2-}]} \times \frac{1}{[H^+]^{14}}\right)$

$E_{Cr_2O_7^{2-}|Cr^{3+}} = 1.33 - \frac{0.0591}{6}\log\frac{1}{(0.01)^{14}}$

$= 1.60$ V

92. (c) $E = -0.57 - \frac{0.0591}{4}\log\frac{[OH^-]^6}{[TeO_3^{2-}]}$

$= -0.57 - \frac{0.06}{4}\log(10^{-2})^6 = -0.39$

93. (b) $Co^{2+}(aq) \rightleftharpoons Co^{2+}(aq)$, $Q = \frac{C_2}{C_1}$

$C_1 \qquad C_2$

If $\frac{C_2}{C_1} < 1$, $\Delta G < 0$

94. (b) $\underset{P_1}{H_2(g)} \rightleftharpoons \underset{P_2}{H_2(g)}$

$E = \frac{RT}{F}\ln\left(\frac{P_1}{P_2}\right)$

95. (a) A: $H_2(g) \rightleftharpoons 2H^+(aq) + 2e^-$

C: $Cu^{2+}(aq) + 2e^- \rightleftharpoons Cu(s)$

96. (a) $E = \frac{0.0591}{3}\log\left(\frac{1}{0.04}\right) = 0.028$

97. (c) $Cd^{2+}(aq) + Fe(s) \rightleftharpoons Cd(s) + Fe^{2+}(aq)$

At eqm. $1 - x \quad - \quad - \quad x$

At equilibrium $E° = \frac{0.0591}{2}\log\left(\frac{Fe^{2+}}{Cd^{2+}}\right)$

$0.037 = \frac{0.0591}{2}\log\left(\frac{x}{1-x}\right)$

$x = [Fe^{2+}] \Rightarrow 0.947\ M$

$\therefore [Cd^{2+}] = 0.053\ M$

98. (d) $1.50 = E° - \frac{0.0591}{3}\log\left[\frac{0.3}{(0.1)^3}\right]$

$\therefore \quad E° = 1.55$

99. (d) $\Delta G° = -nFE°$

$E° = \frac{-1467 \times 10^3}{(-14 \times 96{,}500)} = +1.08$

100. (b) The cell reaction is

$H_2(g) + I_2(s) \rightleftharpoons 2H^+(aq) + 2I^-(aq)$

$0.7714 = 0.535 - \frac{0.0591}{2}\log\frac{[H^+]^2[I^-]^2}{P_{H_2}}$

pH = 3

101. (c) One design a cell in which the half cell reaction is $CuS(s) + 2e^- \rightleftharpoons Cu(s) + S^{2-}(aq)$ occurs

so $\quad Cu(s) \rightleftharpoons Cu^{2+} + 2e^- \quad \Delta G_1°$

$CuS + 2e^- \rightleftharpoons Cu(s) + S^{2-}(aq) \quad \Delta G_2°$

$CuS(s) \rightleftharpoons Cu^{2+}(aq) + S^{2-}(aq) \quad \Delta G_3°$

from $\Delta G_3° = \Delta G_1° + \Delta G_2°$

$E°_{S^{2-}|CuS|Cu} = E°_{Cu^{2+}Cu} + \frac{0.0591}{2}\log K_{sp}$

$\Rightarrow \quad 0.34 + \frac{0.0591}{2}\log(10^{-35})$

$= -0.694$ V

102. (a) $Pb \rightleftharpoons Pb^{2+} + 2e^-$, $\Delta G_1°$

$Pb^{2+} + 2Br^- \rightleftharpoons PbBr_2(s)$, $\Delta G_2°$

$Pb(s) + 2Br^-(aq) \rightleftharpoons PbBr_2(s) + 2e^-$

$\Delta G_3° = -2F(0.126) - RT\ln\frac{1}{K_{sp}}$

$-2F(0.248) = -2F(0.126) - RT\ln\frac{1}{K_{sp}}$

$\ln\left(\frac{1}{K_{sp}}\right) = \frac{2F}{RT}[0.248 - 0.126]$

$\log\left(\frac{1}{K_{sp}}\right) = \frac{2 \times (0.122)}{(0.0591)} = 4.128$

$K_{sp} = 10^{-4.128}$ or 7.4×10^{-5}

103. (b) $n = 2$ for given reaction so $\Delta G^\circ = -nFE^\circ$

$$\Rightarrow \quad -210 \times 1000 = -2 \times 96500 \times E^\circ$$

$$E^\circ = 1.09 \text{ V}$$

104. (b) At equilibrium, $\Delta G = 0$, $E_{cell} = 0$

105. (d) $\Delta G = \Delta G^\circ + RT \ln Q$

$$\Rightarrow \quad -nFE^\circ + 2.303\, RT \log \frac{[Cr^{3+}]}{[Ag^+]^3}$$

$$\Delta G = -3 \times 96500 \times 1.50 + 8.314 \times 298 \times 2.303 \log \frac{(0.1)}{(0.1)^3}$$

$$\Rightarrow -422838.3 \text{ J or } -422.83 \text{ kJ}$$

106. (d) $Au^+ + e^- \longrightarrow Au(s)$ $E^\circ = 1.69$ V ...(1); ΔG_1°

$Au^{3+} + 3e^- \longrightarrow Au(s)$ $E^\circ = 1.40$ V ...(2); ΔG_2°

From (2)-(1) $Au^{3+} + 2e^- \longrightarrow Au^+$;

$$\Delta G_3^\circ = \Delta G_2^\circ - \Delta G_1^\circ$$

$$-2 \times F \times E^\circ = -3 \times F \times 1.40 + 1 \times 1.69 \times F;$$

$$E^\circ = 1.255 \text{ V}$$

$$\therefore \quad E^\circ_{Au^+|Au^{3+}} = -1.255\text{V}$$

107. (c) $2Mn^{2+} \longrightarrow 2Mn^{3+} + 2e^-$, ΔG_1°

$Mn^{2+} + 2e^- \longrightarrow Mn$, ΔG_2°

$3Mn^{2+}(aq) \longrightarrow Mn(s) + 2Mn^{3+}(aq)$

$$-2 \times F \times E_3^\circ = -2 \times F \times [-1.51] - 2 \times F \times (-1.185)$$

$$E_3^\circ = -2.695$$

$$E_3^\circ = +\frac{0.0591}{2} \log K_{eq};\ K_{eq} \simeq 6.3 \times 10^{-92}$$

108. (c) For given concentration cell;

$$E_{cell} = \frac{RT}{F} \log \frac{[H^+]_{RHS}}{[H^+]_{LHS}}$$

for max. emf $[H^+]_{LHS}$ should be min. and that is for $NH_4OH(aq)$

109. (c) Given cell is electrolyte concentration cell

$$E_{cell} = \frac{RT}{F} \ln \frac{[Ag^+]_{RHS}}{[Ag^+]_{LHS}}$$

where $[Ag^+] = \frac{K_{sp}}{[Cl^-]}$. For spontaneous chemical change conc. of Ag^+ of cathodic compartment must be greater than anodic compartment.

110. (d) half cell reaction is

$$Cr_2O_7^{2-} + 14H^+ + 6e^- \longrightarrow 2Cr^{3+} + 7H_2O$$

$$E_{Cr_2O_7^{2-}|Cr^{3+}} - E^\circ_{Cr_2O_7^{2-}|Cr^{3+}}$$

$$= -\frac{0.0591}{n} \log \frac{[Cr^{3+}]^2}{[Cr_2O_7^{2-}][H^+]^{14}}$$

$$= \frac{0.0591}{6} \log [10^{-3}]^{14} = -0.414 \text{ V}$$

$\therefore$ Oxidising power reduced = 0.414 V

111. (c) $\Delta S = nF\left(\frac{dE}{dT}\right)_p \Rightarrow 2 \times 96500 \times 1.5 \times 10^{-4}$

$$\Rightarrow \quad 28.95 \text{ J/K-mol}$$

112. (c) Thermodynamic efficiency of a cell is given by the ratio of Gibbs function change to enthalpy change in the overall cell reaction.

113. (c) $E^\circ_{cell} = E^\circ_{RP(RHS)} - E^\circ_{RP(LHS)}$

$$= -0.76 - (-1.36) = 0.6$$

$$\Delta_r G^\circ = -RT \ln K_{eq};$$

$$\text{or} \quad \log K_{eq} = \frac{nFE^\circ}{RT \times 2.303}$$

$$\Rightarrow \quad \frac{2 \times 0.6}{0.06} \Rightarrow 20;\ K = 10^{20}$$

114. (d) Combustion takes place.

115. (c) Overall reaction; $2H_{2(g)} + O_{2(g)} \longrightarrow 2H_2O_{(l)}$

116. (b) $\because \quad \Delta G > 0$ (Electrolytic cell)

117. (b) During charging of a lead storage battery, the reaction at:

Anode : $PbSO_4 + 2e^- \longrightarrow Pb + SO_4^{2-}$

Cathode : $PbSO_4 + 2H_2O \longrightarrow PbO_2 + 4H^+ + SO_4^{2-} + 2e^-$

In both the reactions H_2SO_4 is regenerated.

118. (b) When aluminium surfaces are exposed to the atmosphere a thin invisible oxidise skin forms immediately, which protects the metal from further oxidation.

[$\because$ E°_{OP} of Al is much longer]

119. (d) Acidic medium.

120. (b) E°_{OP} of Zn > E°_{OP} of Fe

121. (a) Metal acts as anode

122. (a) Voltage of cell increases

123. (b) Movement of ions

124. (b) Free ions

125. (c) Water is weak electrolyte.

126. (b) $G = \kappa \times \left(\frac{A}{l}\right)$

127. (c) $\Lambda_m = \frac{\kappa \times 1000}{M} \Rightarrow \frac{1}{\rho} \times \frac{1}{M} \times 1000;$

$$\Lambda_m = \frac{1000}{xy}$$

128. (b) $\wedge_{eq} = \frac{\kappa \times 1000}{N}$

129. (b) $G^* = \frac{l}{A} = \frac{0.02}{0.0004} \Rightarrow 50\ m^{-1}$

$G = \frac{1}{R} = \frac{1}{50} \Rightarrow 0.02\ S$

$\kappa = G.G^* = 50 \times 0.02 = 1\ Sm^{-1}$

130. (a) Conductance $(G) = \frac{1}{R} \Rightarrow \frac{1}{300}$;

$\because \quad \kappa = G\left(\frac{l}{A}\right)$

Cell constant $\left(\frac{l}{A}\right) = \frac{0.013}{1} \times 300 = 3.9\ cm^{-1}$

131. (a) $\Lambda_m = \frac{\kappa \times 1000}{M} = \frac{6.3 \times 10^{-2} \times 1000}{0.1}$

$= 630\ ohm^{-1}\ cm^2\ mol^{-1}$

132. (b) $\Lambda_M = \frac{\kappa \times 1000}{M}$

$400 = \frac{\left(\frac{1}{55} \times G^*\right) \times 1000}{1/20}$

$G^* = 1.1\ cm^{-1}$

133. (b) $HA(aq) \rightleftharpoons H^+(aq) + A^-(aq)$

At eqm. $c(1-\alpha) \quad c\alpha \quad c\alpha$

$K_a = \frac{c\alpha^2}{1-\alpha}$; where $\alpha = \frac{\Lambda_m}{\Lambda_m^\infty}$

$\therefore \quad K_a = \frac{c\left(\frac{\Lambda_m}{\Lambda_m^\infty}\right)^2}{\left(1 - \frac{\Lambda_m}{\Lambda_m^\infty}\right)} = \frac{c\Lambda_m^2}{\Lambda_m^\infty(\Lambda_m^\infty - \Lambda_m)}$

134. (c) On dilution number of ions decrease in unit volume hence specific conductance decreases. But separation between ions also increase hence equivalent conductance increases.

136. (c) $\wedge_{eq} = \frac{150}{3} = 50\ S\ cm^2\ eq^{-1}$

137. (d) $\Lambda_m = n.\Lambda_{eq}$; when n = total cationic/Anionic charge

$\because \quad n = 6; \quad \therefore \quad \Lambda_{eq} = \Lambda_m/6$

138. (d) $\Lambda_m^\infty(NaBr) = \Lambda_m^\infty(NaCl) + \Lambda_m^\infty KBr - \Lambda_m^\infty(KCl)$

$\lambda_m^\infty(Na^+) + \lambda_m^\infty(Br^-) = 126.5 + 151.5 - 150$

$\lambda_m^\infty(Na^+) = 50$

139. (c) $\Lambda_m = \frac{\kappa}{M} \times 1000$; $(x+y) = \frac{\kappa}{M} \times 1000$

$\Rightarrow \quad M = \frac{1000 \times \kappa}{(x+y)}$

Solubility (in gL^{-1}) $= \frac{\kappa \times 1000 \times 188}{x+y}$

140. (a) $\kappa = G.G^* \Rightarrow \frac{1}{200} 2 \Rightarrow 0.01$;

$\Lambda_{eq} = \frac{\kappa}{N} \times 1000$

$\Rightarrow \frac{0.01}{0.1} \times 1000 \Rightarrow 100$

141. (b) $R = \rho\left(\frac{l}{A}\right)$

$82.4 = \left(\frac{1}{2.768}\right) \times 10^3 \left[\frac{l}{A}\right]$

$\frac{l}{A} = 0.2281$

142. (b) $G^* = \frac{l}{a} = \frac{10}{100} = 0.1$; $G = 0.0001S$

$V = 100 \times 10 = 1000\ cm^3 = 1$ litre

$\kappa = GG^* = 0.1 \times 0.0001 = 10^{-5}$

$\Lambda_m = \frac{\kappa \times 1000}{M}$

$= \frac{(0.1 \times 0.0001) \times 1000}{0.1} = 0.1\ Scm^2 mol^{-1}$

143. (a) Equivalent conductivity of $BaCl_2 = 127 + 76 = 203$

144. (a) $Al_2(SO_4)_3 \longrightarrow 2Al^{3+} + 3SO_4^{2-}$

$\Lambda_M^\infty = 2\lambda_{Al^{3+}}^\infty + 3\lambda_{SO_4^{2-}}^\infty$

$858 = 2\lambda_{Al^{3+}}^\infty + 3 \times 160$

$\therefore \quad \lambda_{Al^{3+}}^\infty = 189\ S\ cm^2\ mol^{-1}$

145. (c) Ionic mobility $(m^2 v^{-1} s^{-1})$

$= \frac{\text{Velocity}}{\text{Voltage / Distance}}$

146. (b) $\wedge_{AgCl}^\infty = \wedge_{AgNO_3}^\infty + \wedge_{HCl}^\infty - \wedge_{HNO_3}^\infty$

147. (a) $[Pt(NH_3)_6]Cl_4$ (1 mole) produces maximum number of ions.

148. (c) $K_{KCl} = G \times G^* \quad \Rightarrow \quad 0.03 = \frac{1}{100} \times G^*$

$G^* = 3$

$K_{K_2SO_4} = \frac{1}{300} \times 3 = 10^{-2}\ \Omega^{-1} cm^{-1} = 1\Omega^{-1} m^{-1}$

$\Lambda_M = \frac{1 \times 10^{-3}}{0.01} = 0.1\Omega^{-1}\ m^2 mol^{-1}$

149. (c) $\Lambda_{eq.} = \frac{1000 \times 8 \times 10^{-4}}{0.001 \times 2} = 400\ S\ cm^2 eq^{-1}$

150. (c) $\alpha = \frac{\Lambda_M}{\Lambda_M^\infty} = \frac{10}{200} = \frac{1}{20}$

$[H^+] = C\alpha = 0.1 \times \frac{1}{20} = 5 \times 10^{-3}$

$\therefore \quad pH = 3 - \log 5 = 2.3$

151. (b) $\Lambda_M = \frac{\kappa \times 1000}{M}$

$2 \times 2 = \frac{8 \times 10^{-8} \times 1000}{M}$

$M = 2 \times 10^{-5}$

$K_{sp}(BaSO_4) = M^2 = 4 \times 10^{-10}$

152. (b) On dilution, no. of ions decrease in per unit volume.

153. (c) On dilution degree of dissociation of a weak electrolyte increase, hence increase in ionic mobility.

154. (d) Completely dissociate into ions.

155. (d) Electric conduction of a salt solution in water depends on the extent of its ionization.

156. (b) Ionic molar conductivity of H^+ is very high and NH_4OH is a weak electrolyte.

157. (b) Hydrated size order $F^- > Cl^- > Br^- > I^-$

Ionic mobility $\alpha \dfrac{1}{\text{Hydrated size}}$

158. (a) Molar conductivity of H^+ and OH^- are very high as compare to other ions.

Initially conductance of solution sharply decreases due to consumption of free H^+. After complete neutralization further slightly increases due to presence of OH^-.

159. (c) Before neutralization conductance of solution sharply decreases due to consumption of free H^+. After neutralization further conductance almost remains same due to formation of buffer solution.

160. (d) Molar conductivity of H^+ and OH^- are very high as compared to other ions.

Initially conductance of solution sharply decreases due to consumption of free H^+ then increases due to formation of salt (NaCN) and after complete neutralization further sharply increases due to presence of OH^-

Level 2

1. (b) No. of equivalent of aluminum, $\dfrac{W}{E} = \dfrac{I \times \eta \times t}{96500}$

$$\frac{24 \times 5}{27} \times 3 = \frac{9650 \times 0.9 \times t}{96500}$$

$$t = 148.14 \text{ sec}$$

2. (c) At Anode: $2H_2O(l) \longrightarrow 4H^+(aq) + O_2(g) + 4e^-$

At Cathode : $Ag^+(aq) + e^- \longrightarrow Ag(s)$

Eq. of O_2 evolved = Eq. of Ag formed $\Rightarrow 0.1$

Total loss in mass $= \dfrac{0.1 \times 32}{4} + 0.1 \times 108$

$\Rightarrow \quad 0.8 + 10.8 \Rightarrow 11.6$

mass of final solution $= 108 - 11.6 \Rightarrow 96.4$ g

3. (d) Equivalents of CO_2 produced $= \dfrac{(I \times \eta) \times t}{96500}$

$$= \frac{0.5 \times 0.8 \times 96.5 \times 60}{96500} = 0.024$$

moles of CO_2 $(n = 1)$ produced $= 0.024$

moles of C_2H_6 $(n = 2)$ produced $= \dfrac{0.024}{2} = 0.012$

Total moles of gases produced $\Rightarrow 0.036$

$$V_{gases} = \frac{nRT}{P} = \frac{0.036 \times 0.0821 \times 300}{\left(\frac{740}{760}\right)} = 0.91 \text{ litre}$$

4. (d) Total volume of metal layer $= 0.25 \times 0.32 \times 10^{-3} \Rightarrow 0.08 \times 10^{-3}$ m^3

Total mass of chromium layer $= 0.08 \times 7.20 \times 10^6 \times 10^{-3} = 576$ g

$$\frac{576}{52} \times 6 = \frac{I \times 60}{96500}$$

$$I = 10.69 \times 10^4 \text{ A}$$

5. (a) Initial m moles of $Cu^{2+} = 5$;

m-eq. or m-moles of H^+ produced $= 100 \times 10^{-2} = 1$

$\Rightarrow$ m-moles of Cu^{2+} converted into $Cu = \frac{1}{2} = 0.5$

m-moles of Cu^{2+} remaining in solution $= 5 - 0.5 = 4.5$

$$2Cu^{2+} + 4I^- \longrightarrow Cu_2I_2 + I_2$$

and $I_2 + 2NaS_2O_3 \longrightarrow 2NaI + Na_2S_4O_6$

m-moles of Cu^{2+} remaining $= m$-moles of $Na_2S_2O_3$

$4.5 = 0.04 \times V \quad \Rightarrow V = 112.5$ mL

6. (b) $2H_2O(l) \rightleftharpoons 2H^+(aq) + 2OH^-(aq); \quad \Delta G_1^\circ$

$2H^+(aq) + 2e^- \rightleftharpoons H_2(g); \quad \Delta G_2^\circ$

$2H_2O(l) + 2e^- \rightleftharpoons H_2(g) + 2OH^-(aq); \Delta_r G_3^\circ$

$$\Delta_r G_3^\circ = \Delta G_1^\circ + \Delta G_2^\circ$$

$$-2 \times F \times E^\circ = -RT \ln(10^{-26}) - 2 \times F \times 0$$

$$E^\circ = \frac{0.066}{2} \log(10^{-26}) = -0.858 \text{ V}$$

7. (d) To find n we break the cell reaction into two half cell reactions

Anode : $13H_2O(l) \longrightarrow 6.5\,O_2(g) + 26H^+(aq.) + 26e^-$

Cathode : $4CO_2(g) + 26H^+(aq.) + 26e^- \longrightarrow C_4H_{10}(g) + 8H_2O(l)$

$$E^\circ = -\frac{\Delta G^\circ}{nF} = -\frac{(-2746) \times 1000}{26 \times 96500} = 1.09 \text{ V}$$

8. (c) For normal calomel electrode $E_{RP} = E_{RP}^\circ$;

$$\therefore E_{cell} = (0.28 - 0) - \frac{0.0591}{2} \log \frac{[H^+]^2}{P_{H_2}}$$

$$0.5755 = 0.28 + 0.0591 \text{ pH} + \frac{0.0591}{2} \log(0.1)$$

$$\text{pH} = 5.5$$

9. (a) We know

$$2H_2(g) \longrightarrow 4H^+ + 4e^-; \quad E^\circ = 0.0 \text{ V}$$

$$O_2(g) + 4H^+(aq) + 4e^- \longrightarrow 2H_2O(l) \; ; \; E^\circ = 1.23 \text{ V}$$

$$2H_2(g) + O_2(g) \longrightarrow 2H_2O(l) \; ; \; E_{cell}^\circ = 1.23 \text{ V}$$

$$\Delta G_{298}^\circ = -nFE^\circ = -4 \times 96500 \times 1.23 = -474.78 \text{ kJ}$$

$$\Delta_r H_{298}^\circ = \Delta_r G_{298}^\circ + T.\Delta_r S_{298}^\circ$$

$$= (-474.78) + 298 \times (-0.32) = -570.14 \text{ kJ/mol}$$

$$\therefore \quad \Delta_f H_{298}^\circ = -285.07 \text{ kJ/mol}$$

10. (a) After addition of Cd and its oxidation into Cd^{2+}.

$$NO_3^-(aq) + 4H^+(aq) + 3e^- \longrightarrow NO(g) + 2H_2O(l)$$

$$0.1 - x \qquad 0.4 - 4x;$$

where $x = 0.08$

$[NO_3^-]$ remaining $= 0.02\ M$; $[H^+]$ remaining $= 0.08\ M$

$$E_{NO_3^-|NO} = E^\circ_{NO_3^-|NO} - \frac{0.0591}{3}\log\frac{1}{[NO_3^-][H^+]^4}$$

$$= 0.96 - \frac{0.0591}{3}\log\frac{1}{(0.02)(0.08)^4} = 0.84\text{ V}$$

11. (a) E° for normal calomel electrode = E° for saturated calomel electrode

$$0.33 = 0.27 - \frac{0.06}{2}\log[Cl^-]^2;\quad [Cl^-] = 0.1\,M$$

12. (b) Anode : $H_2(g) \longrightarrow 2H^+(aq) + 2e^-$; $E^\circ = 0.0$ V

Cathode : $MnO_4^-(aq) + 8H^+(aq) + 5e^- \longrightarrow Mn^{2+}(aq) + 4H_2O(l)$; $E^\circ = 1.51$ V

$E_{cell} = E_{RP(RHS)} - E_{RP(LHS)}$

$$\Rightarrow \left(1.51 - \frac{0.06}{5}\log\frac{[Mn^{2+}]}{[MnO_4^-][H^+]^8}\right) - \left(0 - \frac{0.06}{2}\log\frac{P_{H_2}}{[H^+]^2}\right)$$

$$= 1.51 - \left(\frac{0.06}{5}\log\frac{0.01}{0.1\times(10^{-2})^8}\right) + \frac{0.06}{2}\log\left(\frac{0.1}{(10^{-3})^2}\right)$$

$$= 1.48\text{ V}$$

13. (c) $3Cu(s) + 8H^+ + 2NO_3^- \longrightarrow 3Cu^{2+} + 2NO + 4H_2O$

$Cu(s) + 4H^+ + 2NO_3^- \longrightarrow Cu^{2+} + 2NO_2 + 2H_2O$

Let concentration of HNO_3 is X so $[H^+] = X$

and $[NO_3^-] = X$

$$E_{NO_3^-|NO} - E_{Cu^{2+}|Cu} = E_{NO_3^-|NO_2} - E_{Cu^{2+}|Cu}$$

or

$$E_{NO_3^-|NO} = E_{NO_3^-|NO_2}$$

$$0.96 - \frac{0.0591}{3}\log\frac{10^{-3}}{x^5} = 0.79 - \frac{0.0591}{1}\log\left(\frac{10^{-3}}{x^3}\right)$$

$$\Rightarrow \quad 0.62 = \frac{0.0591}{6}\log\frac{10^{-9}}{x^{10}} = 0.45 - \frac{0.0591}{2}\log\frac{10^{-9}}{x^6}$$

$$\log x = 0.657 \simeq 0.66 \text{ or } x = 10^{+0.66}$$

14. (a) Anode : $2Cl^-(aq) \longrightarrow Cl_2(g) + 2e^-$

Cathode : $Cl_2(g) + 2e^- \longrightarrow 2Cl^-(aq)$

Cell reaction : $2Cl^-(aq) + Cl_2(g) \longrightarrow 2Cl^-(aq) + Cl_2(g)$

$$E_{cell} = -\frac{0.06}{2}\log\frac{[Cl^-]^2_{RHS}\cdot(P_{Cl_2})_{LHS}}{[Cl^-]^2_{LHS}\cdot(P_{Cl_2})_{RHS}} \qquad (\because E^\circ_{cell} = 0)$$

$$= -\frac{0.06}{2}\log\frac{(0.01)^2}{(0.1)^2}\times\left(\frac{0.4}{0.2}\right) \Rightarrow 0.051\text{ V}$$

15. (a)

$$H_2O(l) + ClO_3^-(aq) \longrightarrow ClO_4^-(aq) + 2H^+(aq) + 2e^-$$

$$2H^+(aq) + ClO_3^-(aq) + 2e^- \longrightarrow 2ClO_2^-(aq) + H_2O(l)$$

$$2ClO_3^-(aq) \longrightarrow ClO_2^-(aq) + ClO_4^-(aq)$$

$$E^\circ_{cell} = 0.33 - 0.36 = -0.03;\qquad E^\circ_{cell} = \frac{RT}{2F}\ln K$$

$$-0.03 = \frac{0.06}{2}\log K \quad\text{or}\quad K = 0.1$$

$$\underset{0.1-2x}{2ClO_3^-} \rightleftharpoons \underset{x}{ClO_4^-} + \underset{x}{ClO_2^-}$$

$$\frac{x^2}{(0.1-2x)^2}=\frac{1}{10};\ 3.16x=0.1-2x$$

$$5.16x=0.1$$

$$\Rightarrow \qquad x=0.1/5.16=0.0193\cong 1.9\times 10^{-2}$$

16. (b) $$E_{cell}=\frac{0.06}{1}\log\frac{[H^+]_{RHS}}{[H^+]_{LHS}}$$

or $$E_{cell}=0.06\,[(pH)_{LHS}-(pH)_{RHS}]$$

$$(pH)_{LHS}=pK_a+\log\frac{[A^-]}{[HA]}$$

$$=4+\log\left(\frac{1}{0.1}\right)=5;\quad (pH)_{RHS}=4$$

$$E_{cell}=0.06\,(5-4)=+0.06\text{ V}$$

17. (c) $$E_{cell}=E_{H^+|H_2}-E_{OH^-|Cd(OH)_2|Cd}$$

$$=E_{H^+|H_2}-E_{Cd^{2+}|Cd}$$

or $$E_{cell}=E^\circ-\frac{0.06}{2}\log\frac{[Cd^{2+}]}{[H^+]^2}$$

$\because$ $$E_{cell}=0,$$

$\therefore$ $$E^\circ_{cell}=0.03\log\times\frac{[Cd^{2+}][OH^-]^2}{K_w^2}$$

$$\log\frac{K_{sp}}{K_w^2}=\frac{0.39}{0.03}=13$$

$$K_{sp}=10^{13}\times(10^{-14})^2=10^{-15}$$

18. (a) $E_{cell}=\dfrac{0.06}{2}\log\dfrac{[H^+]^2_{RHS}[P_{H_2}]_{LHS}}{[H^+]^2_{LHS}[P_{H_2}]_{RHS}}$; $\because$ $\alpha<<1$

$\therefore$ $[OH^-]=\sqrt{K_b\times C}$; $[OH^-]=10^{-3}$

or $[H^+]_{LHS}=10^{-11}$

$[H^+]_{RHS}=\sqrt{K_a\times C}\Rightarrow 10^{-4}$

$\therefore$ $$E_{cell}=\frac{0.06}{2}\log\frac{(10^{-4})^2\times 0.1}{(10^{-11})^2}\times 1=0.39\text{ V}$$

19. (c) $$H_2S \rightleftharpoons 2H^+ + S^{2-}$$

$$K_{a_1}.K_{a_2}=\frac{[H^+]^2[S^{2-}]}{[H_2S]}=10^{-21}\times 0.1=(10^{-3})^2[S^{-2}]$$

$$[S^{-2}]=10^{-16}$$

$$E_{S^{-2}|Ag_2S|Ag}=E^\circ_{Ag^+|Ag}-\frac{0.06}{2}\log\frac{[S^{2-}]}{K_{sp}}$$

$$=0.80-\frac{0.06}{2}\log\frac{10^{-16}}{10^{-49}}=0.80-\frac{0.06}{2}\times 33\Rightarrow -0.19\text{ V}$$

20. (c) Normality of resultant solution $=\dfrac{0.1\times V}{V+V}=0.05\,N$

$$\Lambda_{eq}=1000\times\frac{0.0055}{0.05}=110\text{ S cm}^2\text{eq}^{-1}$$

21. (b) Resultant conc. of NaCl $= \frac{0.1 \times V}{4V} = 0.025\ N$

and $\quad HCl = \frac{0.1 \times 2V}{4V} = 0.05\ N$

$\therefore \quad \kappa_{total} = \kappa_{NaCl} + \kappa_{HCl}$

$$\kappa_{solution} = \frac{N(NaCl).\Lambda_{eq}(NaCl)}{1000} + \frac{N(HCl).\Lambda_{eq}(HCl)}{1000}$$

$$0.018 = \frac{0.025 \times 110}{1000} + \frac{0.05.\Lambda_{eq}(HCl)}{1000}$$

$$\Lambda_{eq}(HCl) = \frac{18 - 2.75}{0.05} = 305\ \text{S cm}^2\text{eq.}$$

22. (b) The molar conductivity of the dissociated form of crotonic acid is

$$\Lambda_m^{\infty}(HC) = \Lambda_m^{\infty}(HCl) + \Lambda_m^{\infty}(NaC) - \Lambda_m^{\infty}(NaCl)$$
$$= (426 + 83 - 126)\ \Omega^{-1}\ \text{cm}^2\text{mol}^{-1}$$
$$= 383\ \Omega^{-1}\ \text{cm}^2\text{mol}^{-1}$$

The molar conductivity of HC $= \frac{\kappa \times 1000}{m}$

$$\Lambda_m(HC) = \frac{3.83 \times 10^{-5}\Omega^{-1}\text{cm}^{-1}}{0.001} \times 1000$$
$$= 38.3\,\Omega^{-1}\ \text{cm}^2\text{mol}^{-1}$$

The degree of dissociation,

$$\alpha = \frac{\Lambda_m(HC)}{\Lambda_m^{\infty}(HC)} = \frac{(38.3\,\Omega^{-1}\ \text{cm}^2\text{mol}^{-1})}{(383\,\Omega^{-1}\ \text{cm}^2\text{mol}^{-1})} = 0.1$$

$$K_a = \frac{C\alpha^2}{1-\alpha} = \frac{(10^{-3})(0.1)^2}{1-0.1} = 1.11 \times 10^{-5}$$

23. (c) $\Lambda_m^{\infty}(BaSO_4) = 2\Lambda_{eq}^{\infty}(BaSO_4)$

$$\Lambda_{eq}^{\infty}(BaSO_4) = \Lambda_{eq}^{\infty}(Ba^{2+}) + \Lambda_{eq}^{\infty}(SO_4^{2-})$$
$$= \Lambda_{eq}^{\infty}(BaCl_2) + \Lambda_{eq}^{\infty}(H_2SO_4) - \Lambda_{eq}^{\infty}(HCl)$$
$$\Lambda_{eq}^{\infty}(BaSO_4) = x_1 + x_2 - x_3$$
$$\Lambda_m^{\infty} = 2(x_1 + x_2 - x_3)$$

for sparingly soluble salt

$$\Lambda_m^{\infty} = \frac{\kappa}{M} \times 1000$$

or $\quad M = \frac{x}{2(x_1 + x_2 - x_3)} \times 1000$

$$\Rightarrow \quad \frac{500x}{(x_1 + x_2 - x_3)}$$

$$K_{sp} = M^2 \quad \Rightarrow \quad \frac{2.5 \times 10^5 x^2}{(x_1 + x_2 - x_3)^2}$$

24. (a) Conductivity of $Na_2SO_4 = 2.6 \times 10^{-4}$

$$\Lambda_m(Na_2SO_4) = \frac{1000 \times 2.6 \times 10^{-4}}{0.001}$$
$$= 260\ \text{S cm}^2\text{mol}^{-1}$$

$$\lambda_m(SO_4^{2-}) = \Lambda_m(Na_2SO_4) - 2\lambda_m(Na^+)$$
$$= 260 - 2 \times 50$$
$$= 160\ \text{S cm}^2\text{mol}^{-1}$$

Conductivity of $CaSO_4$ solution

$$= 7 \times 10^{-4} - 2.6 \times 10^{-4}$$
$$= 4.4 \times 10^{-4}\ \text{S cm}^{-1}$$

$$\Lambda_m(CaSO_4) = \lambda_m(Ca^{2+}) + \lambda_m(SO_4^{2-})$$
$$= 120 + 160 = 280\ \text{S cm}^2\text{mol}^{-1}$$

$$\text{Solubility } C = \frac{1000 \times \kappa}{\Lambda_m} = \frac{1000 \times 4.4 \times 10^{-4}}{280}$$
$$= 1.57 \times 10^{-3}\ M$$

$$K_{sp} = [Ca^{2+}][SO_4^{2-}]_{total}$$
$$= (0.00157)(0.00157 + 0.001)$$
$$= 4.0 \times 10^{-6}\ M^2$$

25. (b) $K_a = \frac{c\alpha^2}{1-\alpha} \Rightarrow 1.6 \times 10^{-5} = \frac{0.01 \times \alpha^2}{1-\alpha}$

$$\alpha = \sqrt{\frac{1.6 \times 10^{-5}}{0.01}} \Rightarrow 0.04$$

$$\alpha = \frac{\Lambda_m}{\Lambda_m^{\infty}};\ \Lambda_m = 0.04 \times 380 \times 10^{-4}$$
$$= 15.2 \times 10^{-4}\ \text{S m}^2\text{mol}^{-1}$$

$$\kappa = \Lambda_m \times C = 15.2 \times 10^{-4} \times 10^3 \times 10^{-2}$$

$(\because 1\ \text{m}^3 = 1000\ \text{litre})$

$k = G.G^*$ and $G^* = 0.01\ \text{m}^{-1}$,

where $G^* = \frac{l}{A}$

$$\therefore G = \frac{1.52 \times 10^{-2}}{0.01} \Rightarrow 1.52\ \text{S}$$

26. (c) Cell X : At anode Cl_2 gas evolved;
At cathode H_2 gas evolved.
Cell Y: At anode O_2gas evolved;
At cathode Ag deposited.
Cell Z: At anode O_2gas evolved;
At cathode Cu deposited

$$\text{Eqn. of Ag deposited} = \left(\frac{W}{E}\right) = \frac{21.6}{108} = 0.20$$

= Eqn. of O_2 gas evolved; Vol. of O_2 gas evolved $= 0.2 \times 5.6 = 1.12$ L

Vol. of H_2 evolved = Vol. of Cl_2 evolved

$$= 0.2 \times 11.2 = 2.24\ \text{L}$$

27. (a) Eq. of H_2 = Eq. of O_2 + Eq. of $H_2S_2O_8$

$$2 \times \frac{22.4}{22.4} = \frac{4 \times 8.4}{22.4} + 2 \times n$$

$$2 = 1.5 + 2n \Rightarrow n = 0.25$$

29. (a) $\Lambda^\circ_{eq}[Ba_3(PO_4)_2] = \Lambda^\circ_{eq}(BaCl_2) + \Lambda_{eq}(K_3PO_4) - \Lambda^\circ_{eq}(KCl) = 170 + 150 - 100 = 220 \text{ Scm}^2 \text{eq}^{-1}$.

For $Ba_3(PO_4)_2$; $\Lambda_{eq} = \Lambda^\circ_{eq}$

$$\Lambda_{eq} = \frac{\kappa \times 1000}{N}; 220 = \frac{1.32 \times 10^{-5} \times 1000}{N}$$

$N = M \times n\text{-factor}$

$6 \times 10^{-5} = M \times 6$; $M = 10^{-5}$ mol/L

$K_{sp}[Ba_3(PO_4)_2] = 108 M^5 = 108(10^{-5})^5 = 1.08 \times 10^{-23}$

30. (b) m. eqn. of NaOH = m. eqn. of H_2SO_4

$1 \times$ millimol of NaOH $= 2 \times$ millimol of H_2SO_4

$20 \times M_{NaOH} = 2 \times 0.2 \times 15$

$M_{NaOH} = 0.3 \text{ mol L}^{-1}$

Level 3

Passage–1

1. (b)

A: $H_2 \longrightarrow 2H^+ + 2e^-$

C: $2H^+ + 2e^- \longrightarrow H_2$

$$E_{cell} = \frac{0.06}{2}\log\frac{[H^+]^2_{RHS}\cdot(P_{H_2})_{LHS}}{[H^+]^2_{LHS}\cdot(P_{H_2})_{RHS}} = \frac{0.06}{1}\log\frac{(H^+)_{RHS}}{(H^+)_{LHS}}$$

$\because$ α is negligible w.r.t. 1 for weak acid and weak base

In LHS | **In RHS**

$[OH^-] = \sqrt{10^{-6} \times 0.01}$ | $[H^+] = \sqrt{10^{-6} \times 0.01}$

$[OH^-] = 10^{-4}$ | $[H^+]_{RHS} = 10^{-4}$

$[H^+]_{LHS} = 10^{-10}$

$$E_{cell} = \frac{0.06}{1}\log\frac{(10^{-4})}{(10^{-10})}\times\frac{1}{1}$$

$$E_{cell} = \frac{0.06}{1}\log 10^6 = 0.06 \times 6 = 0.36 \text{ V}$$

2. (b) pH of solution (A): Solution A is a basic buffer

$$pOH = pK_b + \log\frac{[Salt]}{[Base]}$$

$\Rightarrow pOH = 5$; $pH = 9$

$\Rightarrow [H^+]_A = 10^{-9}\ M$

pH of solution (B):

$$pH = \frac{1}{2}[pK_w - pK_b - \log c]$$

$$\Rightarrow \frac{1}{2}[14 - 5 + 3] = 6$$

$\Rightarrow [H^+]_B = 10^{-6}$

Now, the cell is a concentration cell.

Cell reaction :

$$2H^+_B(aq) + 2e^- \longrightarrow H_{2B}(g)$$
$$H_{2A}(g) \longrightarrow 2H^+_A + 2e^-$$
$$2H^+_B + H_{2A}(g) \xrightarrow{n=2} 2H^+_A(aq) + H_{2B}(g)$$

Nernst's equation for cell

$$E_{cell} = 0 - \frac{0.06}{2}\log\frac{[H^+]^2_A}{[H^+]^2_B} = +\frac{0.06}{1}\log\frac{[H^+]_{RHS}}{[H^+]_{LHS}}$$

$$= +0.06\log\frac{10^{-6}}{10^{-9}} = 0.18 \text{ V}$$

Passage–2

1. (b) From equation 1 and 2, $\Delta S = \frac{-d(\Delta G)}{dT}$

$$\Delta S = \frac{d(nFE)}{dT} \Rightarrow nF\frac{dE}{dT} \text{ or } \left(\frac{dE}{dT}\right)_p = \frac{\Delta S}{nF}$$

2. (b) From equation 2;

$$\Delta H = nF\left(\frac{dE}{dT}\right)_p T - nFE_{cell}$$

$$-218 \times 1000 = 2 \times 96500 \times 300\left(\frac{dE}{dT}\right)_p - 2 \times 96500 \times 1.015$$

$$\left(\frac{dE}{dT}\right)_p = -3.81 \times 10^{-4} \text{VK}^{-1}$$

3. (a) $\Delta S = nF\left(\frac{dE}{dT}\right)_p = 2 \times 96500 \times -3.81 \times 10^{-4}$

$= -75.53$ J/mol-K

Passage–3

2. (b) $\kappa = \Lambda_m \cdot C \Rightarrow \frac{\Lambda_m \cdot M}{1000}$

$$\kappa = \frac{200 \times 0.04}{1000} \Rightarrow 8 \times 10^3 \text{ Scm}^{-1}$$

$$\kappa = G\left(\frac{l}{A}\right) \Rightarrow 8 \times 10^{-3} = G\left(\frac{0.50}{2}\right)$$

or $G = 0.032$ S; $V = IR \Rightarrow \frac{I}{G}$

$I = 5 \times 0.032 \Rightarrow 0.16$ A

Passage-4

1. (c) $2 \times \left(\frac{4.48 \times 1}{22.4}\right) = \frac{2}{2} \times \frac{I \times 9650}{96,500}$

I = 4 amp.

2. (b) Mass of $H_2O = \frac{4.48}{22.4} \times 18 = 3.6$ g

3. (b) $2H_2O(l) \longrightarrow 4H^+(aq) + O_2(g) + 4e^-$

$$2Cu^{2+} + 4e^- \longrightarrow 2Cu(s)$$
$$2Cu^{2+} + 2H_2O(l) \rightarrow 4H^+(aq) + O_2(g) + 2Cu(s)$$

$$\text{Moles}_{Cu} \times 2 = \frac{4 \times 241.25}{96,500}$$

$\therefore$ $\text{Mole}_{H^+} = 2 \times \text{mole of Cu} = 0.005 \times 2 = 0.01$

pH = 2

Passage-5

$$Agx(s) \rightleftharpoons \underset{(x+y)}{Ag^+(aq)} + \underset{x}{x^-(aq)}$$

$$Agy(s) \rightleftharpoons \underset{(x+y)}{Ag^+(aq)} + \underset{y}{y^-(aq)}$$

$$K_{sp}(Agx) = x(x+y) = 3 \times 10^{-12} \quad \ldots(1)$$

$$K_{sp}(Agy) = y(x+y) = 10^{-12} \quad \ldots(2)$$

$$\kappa_{(Agx+Agy)} = \kappa_{Ag^+} + \kappa_{x^-} + \kappa_{y^-} = 0.4 \times 10^{-6} \quad \ldots(3)$$

$$\lambda^{\infty}_{Ag^+} = \frac{1000 \times \kappa_{Ag^+}}{(x+y)} = 60 \quad \ldots(4)$$

$$\lambda^{\infty}_{x^-} = \frac{1000 \times \kappa_{x^-}}{x} = 90 \quad \ldots(5)$$

$$\therefore \; x + y = 2 \times 10^{-6} \quad \ldots(6)$$

$$\frac{x}{y} = 3 \quad \ldots(7)$$

$$x = 1.5 \times 10^{-6},\; y = 5 \times 10^{-7},\; \kappa_{Ag^+} = 1.2 \times 10^{-7}$$

$$\therefore \; \kappa_{y^-} = 1.45 \times 10^{-7}$$

$$\lambda^{\infty}_{y^-} = \frac{1000 \times 1.45 \times 10^{-7}}{5 \times 10^{-7}} = 290$$

$$\kappa_{x^-} = 1.35 \times 10^{-7}$$

Level 4

Subjective Problems

1. $6e^- +$ 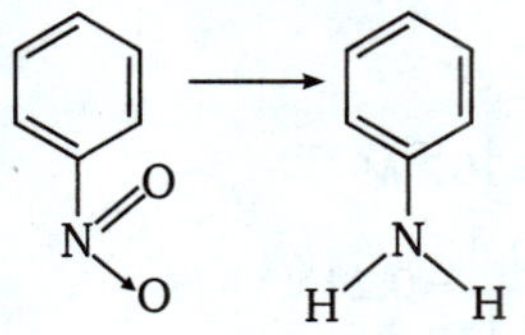

Required charge for reduction of 1 mole $C_6H_5NO_2 = 6F$

2. Equivalent mass $O_2 = \frac{32}{4} = 8$

3. $\frac{0.967}{63.5} \times 2 = \frac{2}{197} \times n_f$

$n_f = 3$

4. $\frac{0.18}{\text{Eq. mass}} = \frac{5 \times 60 \times 9.65}{96500}$, Eq. mass = 6

5. $2 \times \text{mole } Cl_2 = \frac{4F}{F}$

$\therefore$ mole $Cl_2 = 2$

6. $\Delta G = -96.5 \times 10^3 = -1 \times 96500 \times E$

$E = 1$

7. $K_C = 10^{\frac{2(0.34-0.15)}{0.0591}} = 2.6 \times 10^6$

8. $4H^+ + MnO_4^- + 3e^- \rightleftharpoons MnO_2 + 2H_2O$

$$\Delta G° = -3 \times F \times \left(\frac{1.5 \times 5 - 2 \times 1.25}{3}\right)$$

$$= -5F \;;\; x = 5$$

9.

$$Cd^{2+} + 2e^- \rightleftharpoons Cd \;,\; \Delta G_1^\circ$$

$$Cd + 4NH_3 \rightleftharpoons [Cd(NH_3)_4]^{2+} + 2e^- \;,\; \Delta G_2^\circ$$

$$Cd^{2+} + 4NH_3 \rightleftharpoons [Cd(NH_3)_4]^{2+} \;,\; \Delta G_3^\circ$$

$$\Delta G_1^\circ + \Delta G_2^\circ = \Delta G_3^\circ$$

$$-2 \times F \times (-0.40) - 2 \times F(+0.61) = RT \ln K_f$$

$$\therefore \; \ln K_f = \frac{2F}{RT}(0.61 - 0.4)$$

$$= -2 \times F(-0.4) - 2F(0.61)$$

$$\log K_f = \frac{2 \times 0.21}{0.0591} = 7 \quad \Rightarrow \quad K_f = 10^7$$

10. $2H_2O(l) \longrightarrow 4H^+(aq).+O_2(g) + 4e^-$

$$-0.81 = -1.23 - \frac{0.0591}{4} \log [H^+]^4$$

pH = 7

11. $1500 = \left(\frac{1}{2 \times 10^{-3}}\right) \times \left(\frac{l}{A}\right)$

$\therefore$ Cell constant $= 1500 \times 2 \times 10^{-3} = 3$

12. $\wedge^{\infty}_{CH_3COOH} = \wedge^{\infty}_{HCl} + \wedge^{\infty}_{CH_3COOK} - \wedge^{\infty}_{KCl}$

$$= 0.038 + 0.009 - 0.013$$

$$= 0.034$$

% ionisation of $CH_3COOH = \left(\frac{\wedge}{\wedge^{\infty}}\right)100$

$$= \frac{2.72 \times 10^{-3}}{0.034} \times 100 = 8$$

13. $-0.12 - \frac{0.0591}{2} \log\left(\frac{1}{x}\right) = -0.24$

$$\log \frac{1}{x} = \frac{0.12 \times 2}{0.06} = 4$$

$$x = 10^{-4}$$

14. $\Lambda^{\infty}_M(HA) = \Lambda^{\infty}_M(HCl) + \Lambda^{\infty}_M(NaA) - \Lambda^{\infty}_M(NaCl)$

$$= 425 + 100 - 125 = 400 \text{ S cm}^2 \text{mol}^{-1}$$

pH = 4, $[H^+] = 10^{-4} = \alpha C$

$$\alpha = \frac{\Lambda_m}{\Lambda^{\infty}_m} = \frac{200}{400} = 0.5;$$

$$K_a = \frac{(C\alpha) \cdot \alpha}{(1-\alpha)} = \frac{10^{-4}(0.5)}{(1-\alpha)} = 10^{-4};\; pK_a = 4$$

15. $E^{\circ}_{Cl^-/AgCl/Ag} = E^{\circ}_{Ag/Ag} + \frac{0.0591}{1}\log K_{sp}$

$$0.209 = 0.80 + \frac{0.0591}{1}\log K_{sp}$$

$K_{sp} = 10^{-10}$; Let solubility of AgCl in 0.01 M NaCl solution is x

$$10^{-10} = x(x + 0.01)$$

$$x = 10^{-8}$$

$\therefore$ Moles of AgCl dissolved in

$$10\,L = 10^{-8} \times 10 = 10^{-7}$$

■ Numerical Problems

1. Equivalent of OH^- produced $= \frac{2 \times 20 \times 60}{96500} = \frac{24}{965}$

Equivalent of NaOH produced

$$= \frac{24}{965} = \text{Equivalent of } HNO_3 \text{ needed}$$

$$\frac{24}{965} = 0.62 \times \frac{V}{1000}$$

$$V = 40\text{ mL}$$

2. At anode : $2H_2O \longrightarrow O_2 + 4H^+ + 4e^-$

Number of eqn. of H^+ produced $= \frac{4 \times 241.25}{96500}$

Moles of H^+ = eqn. of H^+ = 0.01

$$[H^+] = \frac{0.01}{1} = 0.01$$

$$pH = -\log(H^+) = -\log 0.01 = 2$$

$$pOH = 14 - 2 = 12$$

3. $2S_2O_3^{2-} \longrightarrow S_4O_6^{2-} + 2e^-$

m. eqn. of KI = m. eqn. of I_2 liberated

$= $ m. eqn. of $Na_2S_2O_3$

$= 150\text{ mL} \times 0.1 \times 1\,(n \text{ factor})$

$= 15$

No. of eqn. $= \frac{I \times t}{96500}$

$$15 \times 10^{-3} = \frac{I \times 4 \times 60}{96500}$$

$$I = \frac{96500\,C \times 15 \times 10^{-3}\,eqn}{4 \times 60\,s} = 6.03$$

4. No. of eqn. of Cu^{2+} reduced $= \frac{2 \times 9650}{96500} = 0.2$

At anode : $Cu \longrightarrow Cu^{2+} + 2e^-$

At cathode : $Cu^{2+} + 2e^- \longrightarrow Cu$

Weight of anode $= 100 - 0.1 \times 63 \times \frac{100}{90} = 93\text{ g}$

5. $MnO_4^- \xrightarrow{\Delta G_1^\circ} MnO_2$

$MnO_4^- \xrightarrow{\Delta G_2^\circ} MnO_3^- \xrightarrow{\Delta G_3^\circ} MnO_2$

$$\Delta G_1^\circ = \Delta G_2^\circ + \Delta G_3^\circ$$

$3E^\circ = 2 \times 0.83 + 1 \times 4.27$; $E^\circ_{MnO_4^-/MnO_2} = 1.976\text{ V}$

$$E^\circ_{cell} = E^\circ_{Mg/Mg^{2+}} + E^\circ_{MnO_4^-/MnO_2} = 2.35 + 1.976$$

$$= 4.326\text{ V} = 4.33\text{ V}$$

6. $E^\circ_{SO_4^{2-}/PbSO_4/Pb} = E^\circ_{Pb^{2+}/Pb} + \frac{0.06}{2}\log K_{sp}$

$$-0.356 = -0.124 - 0.03\,pK_{sp}$$

$$pK_{sp} = 7.73$$

7. $E_{cell} = E^\circ_{Ag/Ag^+} + \frac{0.06}{1} \times \log\frac{[Br^-]}{K_{sp}(AgBr)}$

$$+ E^\circ_{Ag^+/Ag} + \frac{0.06}{1} \times \log\frac{K_{sp}(AgCl)}{[Cl^-]}$$

$$0.12 = 0.06\log\frac{[Br^-]}{[Cl^-]} \times \frac{K_{sp}(AgCl)}{K_{sp}(AgBr)}$$

$$2 = \log\frac{1}{0.5} \times \frac{K_{sp}(AgCl)}{K_{sp}(AgBr)}$$

$$\frac{K_{sp}(AgCl)}{K_{sp}(AgBr)} = \frac{100}{2} = 50$$

8. Net cell reaction

$$\underset{(0.1\,M)}{Ag^+(aq)} \longrightarrow \underset{(S\,M)}{Ag^+(aq)}$$

where $K_{sp}(AgCl) = S^2$

$$S = 10^{-5}\,M$$

$$E = 0 - \frac{0.06}{1}\log\frac{10^{-5}}{0.1} = 0.24\text{ V}$$

9. Use $Co(CN)_6^{4-} \longrightarrow Co(CN)_6^{3-} + e$; $E^\circ_{OP} = +0.85\text{ V}$

$Co^{3+} + e \longrightarrow Co^{2+}$; $E^\circ_{RP} = 1.88\text{ V}$

$Co(CN)_6^{4-} + Co^{3+} \longrightarrow Co^{2+} + Co(CN)_6^{3-}$

and $E_{cell} = E^\circ_{cell} + \frac{0.059}{1}\log_{10}\frac{[Co^{3+}][Co(CN)_6^{4-}]}{[Co^{2+}][Co(CN)_6^{3-}]}$

or

$$E_{cell} = E^\circ_{cell} + \frac{0.059}{1}\log_{10}\frac{[Co^{3+}][Co(CN)_6^{4-}][CN^-]^6}{[Co^{2+}][Co(CN)_6^{3-}][CN^-]^6}$$

Also $6CN^- + Co^{2+} \rightleftharpoons Co(CN)_6^{4-}$

and $K_{f_1} = \frac{[Co(CN)_6^{4-}]}{[Co^{2+}][CN^-]^6}$

and $6CN^- + Co^{3+} \rightleftharpoons Co(CN)_6^{3-}$

and $K_{f_2} = \dfrac{[Co(CN)_6^{3-}]}{[Co^{3+}][CN^-]^6}$

$\therefore \quad E_{cell} = E^\circ_{cell} + \dfrac{0.059}{1}\log_{10}\dfrac{K_{f_1}}{K_{f_2}}$

$0 = 0.85 + 1.85 + \dfrac{0.06}{1}\log_{10}\dfrac{10^{19}}{K_{f_2}}$

(At equilibrium $E_{cell} = 0$)

$\therefore \quad \dfrac{K_{f_2}}{10^{19}} = 10^{45}; \quad K_{f_2} = 10^{64}$

$x = 64.00$

10. $Ag(s) \longrightarrow Ag^+(aq) + e^-, \qquad \Delta G_1^\circ = -1 \times FE^\circ$

$Ag^+(aq) + 2NH_3(aq) \longrightarrow [Ag(NH_3)_2^+](aq), \qquad \Delta G_2^\circ = -RT\ln K_f$

$Ag(s) + 2NH_3(aq) \longrightarrow [Ag(NH_3)_2^+](aq) + e^-, \qquad \Delta G^\circ = -1 \times FE^\circ_{OP}$

$\Delta G^\circ = \Delta G_1^\circ + \Delta G_2^\circ$

$E^\circ_{OP} = -0.8 + \dfrac{2.303\,RT}{F}\log 10^8$

$E^\circ_{OP} = -0.8 + 0.06 \times 8 = -0.32\text{ V}$

$|E^\circ_{OP}| = 0.32\text{ V}$

11. Anode : $H_2(g) \longrightarrow 2H^+(aq) + 2e^-$

Cathode : $Zn^{2+}(aq) + 2e^- \longrightarrow Zn(s)$

Cell reaction : $H_2(g) + Zn^{2+}(aq) \rightarrow 2H^+(aq) + Zn(s)$

$E_{cell} = E^\circ_{cell} - \dfrac{0.06}{2}\log\dfrac{[H^+]^2}{[Zn^{2+}]}$

$-0.61 = -0.76 - 0.03\log\dfrac{[H^+]^2}{(0.4)}$

$[H^+] = 2 \times 10^{-3} = x$

$HSO_3^-(aq) \rightleftharpoons H^+(aq) + SO_3^{2-}(aq)$

At eqm: $(0.4 - x) \qquad x \qquad (x + 3.8 \times 10^{-2})$

$K_{a_2} = \dfrac{2\times10^{-3}(2\times10^{-3} + 3.8\times10^{-2})}{(0.4 - 2\times10^{-3})}$

$K_{a_2} = 2\times10^{-4}, \quad y = 4.00$

12. $\left(\dfrac{\partial E}{\partial T}\right)_P = \dfrac{0.070 - 0.066}{10} = 4\times10^{-4}\text{ V/K}$

$\Delta_r H = -nF\left[E - T\left(\dfrac{\partial E}{\partial T}\right)_P\right]$

$\Delta_r H = -2\times96500\,[0.068 - 300\times4\times10^{-4}]$

$= 10036\text{ J} = 10.036\text{ kJ} \approx 10.04\text{ kJ}$

13. $\Lambda_{eq} = \dfrac{\kappa\times1000}{N}$

$\kappa = \dfrac{\Lambda_{eq}\times N}{1000} = \dfrac{100\times0.1}{1000} = 0.01\text{ S cm}^{-1}$

$G = \dfrac{\kappa}{\text{Cell constant }(G^*)} = \dfrac{0.01}{0.25} = 0.04\text{ ohm}^{-1}$ or S

$R = \dfrac{1}{G} = \dfrac{1}{0.04} = 25\text{ ohm}$

Current $(I) = \dfrac{V}{R} = \dfrac{5}{25} = 0.20\text{ A}$

14. $\Lambda_m^\circ(AgCl) = \dfrac{\kappa\times1000}{S}$

$140 = \dfrac{\kappa\times1000}{10^{-5}}$

$\kappa = 140\times10^{-8}\text{ S cm}^{-1} = 140\times10^{-8}\times100$

$= 1.40\times10^{-6}\text{ S m}^{-1}$

15. $MX(s) \rightleftharpoons M^+(aq) + X^-(aq)$

$\qquad x \qquad x + 10^{-6}$

$[Na] = 10^{-6}\,M$

$\kappa_{sol} = \kappa_{M^+} + \kappa_{X^-} + \kappa_{Na^+}$

$29\times10^{-6} = 10^3[6\times10^{-3}x + 4\times10^{-3}(x + 10^{-6}) + 5\times10^{-3}\times10^{-6}]$

$x = 2\times10^{-6}\,M$

$K_{sp} = [M^+][X^-] = 2\times10^{-6}\times3\times10^{-6}$

$= 6\times10^{-12}M^2, \quad y = 6$

CHAPTER 10

Dilute Solutions

Vapour Pressure : The pressure exerted by the vapours of liquid which are in equilibrium with it at a given temperature.

- Vapour pressure is not a surface phenomena.
- Vapour pressure increases with increase in temperature.

$$X(l) \rightleftharpoons X(g)$$

$$K_p = P_{x(g)}$$

$$\ln\frac{K_{P_2}}{K_{P_1}} = \ln\frac{P_2}{P_1} = \frac{\Delta H_{vap.}}{R}\left(\frac{1}{T_1} - \frac{1}{T_2}\right);$$

where $\Delta H_{vap.}$ is molar enthalpy of vaporisation (J/mol or cal/mol)

Raoult's Law

(i) Liquid solution containing non volatile solute :

Vapour pressure of liquid solution $\propto$ Mole fraction of solvent in liquid solution.

$$P_{Solution} \propto X_{Solvent}$$

$$P_{Solution} = P^{\circ}_{Solvent} \cdot X_{Solvent}$$

Where $P^{\circ}_{Solvent}$ = Vapour pressure of pure liquid solvent, $X_{Solvent}$ = Mole fraction of solvent.

For binary liquid solution ; $X_{Solute} + X_{Solvent} = 1$

Relative lowering of vapour pressure $= \frac{P^{\circ} - P}{P^{\circ}} = X_{Solute}$

(ii) Vapour Pressure of Solution containing volatile solute (B) and volatile solvent (A) :

The equilibrium vapour pressure of a volatile component is linearly proportional to the mole fraction of that component in liquid phase.

A solution contains volatile solute (B) and volatile solvent (A) :

$$P_A = P^{\circ}_A X_A \text{ and } P_B = P^{\circ}_B X_B$$

$$P = P^{\circ}_A X_A + P^{\circ}_B X_B$$

Where :

P_A and P_B are partial vapour pressure of A and B at given temperature.

P_A° = vapour pressure of pure A. ; P_B° = vapour pressure of pure B.

X_A = mole fraction of A in liquid phase. ; X_B = mole fraction of B in liquid phase.

y_A = mole fraction of A in vapour phase ; y_B = mole fraction of B in vapour phase.

Partial vapour pressure of 'A' = mole fraction of 'A' in vapour phase × total vapour pressure.

$$P_A = y_A P \quad \text{and} \quad P_B = y_B P$$

$$P_A^\circ X_A = y_A P \quad \text{and} \quad P_B^\circ X_B = y_B P$$

$$X_A + X_B = 1$$

$$\frac{y_A P}{P_A^\circ} + \frac{y_B P}{P_B^\circ} = 1; \qquad \frac{y_A}{P_A^\circ} + \frac{y_B}{P_B^\circ} = \frac{1}{P}$$

$$y_A = \frac{P_A^\circ X_A}{P_A^\circ X_A + P_B^\circ X_B}; \qquad y_B = \frac{P_B^\circ X_B}{P_A^\circ X_A + P_B^\circ X_B}$$

Ideal Solutions

The solutions which obey Raoult's Law are called ideal solutions. For ideality :

$F_{\text{Solute ...... Solvent}} \approx F_{\text{Solute...... Solute}}$ (F = Intermolecular attractive interactions)

$\approx F_{\text{Solvent ...... Solvent}}$

$\Delta H_{\text{mix.}} = 0$; $\Delta V_{\text{mix.}} = 0$

$\Delta S_{\text{mix.}} > 0$; $\Delta G_{\text{mix.}} < 0$

Examples : Benzene & Toluene, Hexane & Heptane, C_2H_5Br & C_2H_5I

Non-ideal solutions : The mixture which do not follow Raoult's law is known as non-ideal solutions.

Types of Real or Non-ideal Solutions

- Non-ideal solution with positive deviation.
- Non-ideal solution with negative deviation.

Solution shows positive deviation	Solution shows negative deviation
$P_{T,\text{exp}} > (P_A^\circ X_A + P_B^\circ X_B)$	$P_{T,\text{exp}} < (P_A^\circ X_A + P_B^\circ X_B)$
$A\text{----}B < \begin{matrix} A\text{----}A \\ B\text{----}B \end{matrix}$	$A\text{----}B > \begin{matrix} A\text{----}A \\ B\text{----}B \end{matrix}$
$\Delta H_{\text{mix.}} = +\text{ve}$	$\Delta H_{\text{mix.}} = -\text{ve}$
$\Delta V_{\text{mix.}} = +\text{ve}$	$\Delta V_{\text{mix.}} = -\text{ve}$
$\Delta S_{\text{mix.}} = +\text{ve}$	$\Delta S_{\text{mix.}} = +\text{ve}$
$\Delta G_{\text{mix.}} = -\text{ve}$	$\Delta G_{\text{mix.}} = -\text{ve}$
Ex. $H_2O + CH_3OH$	**Ex.** $H_2O + CH_3COOH$
$H_2O + C_2H_5OH$	$CHCl_3 + CH_3COCH_3$
$CHCl_3 + CCl_4$	$H_2O + HNO_3$

Azeotropic Mixture: During distillation, the mole fraction of more volatile component in vapour state is higher than that in liquid state. This makes distillation possible. However, there exist some solutions for particular compositions of which the mole fraction of components in liquid and vapour state is same. Thus, no distillation is possible, a mixture and it is termed as **azeotropic mixture.**

Azeotropes are binary mixtures having the same composition in liquid and vapour phase and boil at a constant temperature.

The solutions which show a large positive deviation from Raoult's law form minimum boiling azeotrope at a specific composition.

The solutions which show a large negative deviation from Raoult's law form maximum boiling azeotrope at a specific composition.

Completely Immiscible Liquids : When they are distilled, they distil in the ratio of their vapour pressure at that temperature.

$$P = P_A^\circ + P_B^\circ$$

For vapour of pure liquid we can apply : $P_A^\circ V = n_A RT$ and $P_B^\circ V = n_B RT$

So, $$\frac{P_A^\circ}{P_B^\circ} = \frac{n_A}{n_B}$$

When A and B are distilled mass ratio is given as $\dfrac{w_B}{w_A} = \dfrac{P_B^\circ \cdot M_B}{P_A^\circ \cdot M_A}$

Colligative Properties : Colligative Properties depend on the number of solute particles irrespective to their nature.

- Relative lowering of vapour pressure
- Elevation of boiling point
- Depression of freezing point
- Osmotic pressure

Normal Colligative Properties : When neither association nor dissociation of solute particles take place.

(i) Relative Lowering of Vapour Pressure : $\dfrac{P°-P}{P°} = X_{\text{Solute}}$

(ii) Elevation of boiling point : $\Delta T_b = K_b m$

(iii) Depression of freezing point : $\Delta T_f = K_f m$

(iv) Osmotic pressure : $\pi = CRT$

(i) Relative lowering of Vapour Pressure :

$$\frac{P°-P}{P°} = X_{\text{Solute}} = \frac{n}{n+N}$$

Where, n = moles of solute, N = moles of solvent

$$\frac{P°-P}{P} = \frac{n}{N}$$

$$\text{Molality } (m) = \left(\frac{P°-P}{P}\right) \times \frac{1000}{M(\text{in g/mol})}$$

Where, M = molar mass of solvent.

Ostwald-Walker Method : Experimental measure of relative lowering in V.P.

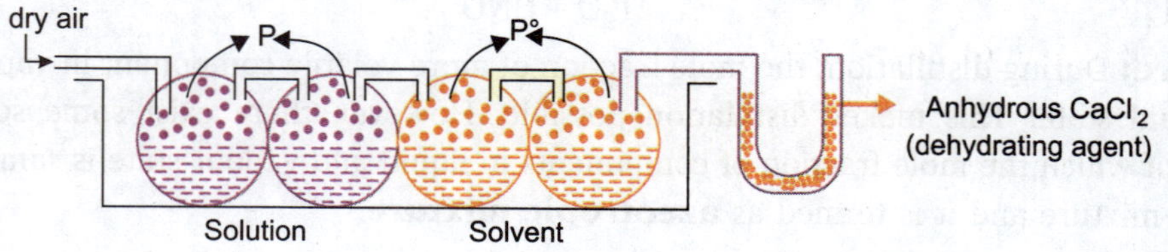

Loss in mass of solution containers $\alpha\, P$

Loss in mass of solvent containers $\alpha\, (P° - P)$

gain in mass of dehydrating agent $\alpha\, P°$

$$\frac{P° - P}{P°} = \frac{\text{Loss in mass of solvent}}{\text{Gain in mass of dehydrating agent}}$$

(ii) Elevation in Boiling Point :

$\Delta T_b = K_b m$

where $\Delta T_b = T_b - T_b^{\circ}$

K_b = Boiling point elevation constant or ebullioscopic constant

$$K_b = \frac{RT_b^{\circ 2}}{1000\, L_v}$$

L_v = Latent heat of vapourisation per gram

$$K_b = \frac{MRT_b^{\circ 2}}{1000\, \Delta H_{Vap}}$$

ΔH_{vap} = enthalpy of vapourisation per mole

M = molar mass of solvent (in g/mol)

T_b° = Boiling point of solvent

(iii) Depression in Freezing point :

$\Delta T_f = K_f m$

where $\Delta T_f = T_f - T_f^{\circ}$

K_f = Freezing point depression constant or cryoscopic constant

$$K_f = \frac{RT_f^{\circ 2}}{1000\, L_f}$$

L_f = Latent heat of fusion per gram

$$K_f = \frac{MRT_f^{\circ 2}}{1000\, \Delta H_{Fus}}$$

ΔH_{Fus} = enthalpy of fusion per mole

M = molar mass of solvent (in g/mol)

T_f° = freezing point of solvent

(iv) Osmotic Pressure (π) : The hydrostatic pressure built up on the solution which just stops osmosis. Alternatively, it may be defined as the pressure which must be applied to the concentrated solution in order to stop osmosis.

For dilute solutions $\pi = CRT = h\rho g$

where C is the total molar concentration of all the free species present in the solution, h is the height developed by the column of the concentrated solution and ρ is the density of the solution in the column.

On the basis of osmotic pressure, the solutions can be classified in three classes.

- **Isotonic solutions:** Two solutions having same osmotic pressures are called isotonic solutions. (This implies $C_1 = C_2$ at same temperature).
- **Hypertonic and hypotonic solution :** When two solutions are being compared, then the solution with higher osmotic pressure is termed as hypertonic. The solution with lower osmotic pressure is termed as **hypotonic.**
- **Abnormal Colligative Properties :** When solute particle associated or dissociated in solvent.

$$i = \text{Vant Hoff factor} = \frac{\text{Actual moles of solute}}{\text{Moles of solute without dissociation or association}}$$

$$= \frac{\text{Observed or experimental colligative properties}}{\text{Theoretical or calculated colligative properties}}$$

(i) Relative lowering of vapour pressure : $\dfrac{P° - P}{P°} = \dfrac{i.n_{solute}}{i.n_{solute} + n_{solvent}}$

(ii) Elevation of boiling point : $\Delta T_b = i.K_b m$

(iii) Depression of freezing point : $\Delta T_f = i.K_f m$

(iv) Osmotic pressure : $\pi = i.CRT$

For dissociation : $i = 1 + (n - 1)\alpha$

For association : $i = 1 + \left(\frac{1}{n} - 1\right)\alpha$

Degree of dissociation $\alpha = \frac{i-1}{n-1}$

Degree of association $\alpha = \frac{1-i}{1-1/n}$

Where n = Total number of particles of solute after dissociation or association.

Henry Law

This law deals with dissolution of gas in liquid. It states that mass of any gas dissolved in any solvent per unit volume is proportional to pressure of gas in equilibrium with liquid.

(i) $m = k \times P$

Where, m = mass of gas dissolved per unit volume.

k = proportionality constant.

P = pressure of gas.

(ii) $P = K_H \cdot X_{gas}$

Where, P = Pressure of gas.

K_H = Henry's constant

X_{gas} = Mole fraction of gas

Level-1

1. The vapour pressure of a given liquid will decrease if :
(a) surface area of liquid is decreased
(b) the volume of liquid in the container is decreased
(c) the volume of the vapour phase is increased
(d) the temperature is decreased

2. The normal boiling point of water is 373 K. Vapour pressure of water at temperature T is 19 mm Hg. If enthalpy of vaporisation is 40.67 kJ/mol, then temperature T would be
(Use : $\log 2 = 0.3, R : 8.3\ \text{JK}^{-1}\ \text{mol}^{-1}$):
(a) 250 K (b) 291.4 K
(c) 230 K (d) 290 K

3. A sample of liquid H_2O of mass 18.0 g is injected into an evacuated 7.6 L flask maintained at 27.0°C. If vapour pressure of H_2O at 27°C is 24.63 mm Hg, what weight percentage of the water will be vapourised when the system comes to equilibrium? Assume water vapours behaves as an ideal gas. The volume occupied by the liquid water is negligible compared to the volume of the container:
(a) 1% (b) 10%
(c) 18% (d) 20%

4. Raoult's law is obeyed by each constituent of a binary liquid solution when:
(a) the forces of attractions between like molecules are greater than those between unlike molecules
(b) the forces of attractions between like molecules are smaller than those between unlike molecules
(c) the forces of attractions between like molecules are identical with those between unlike molecules
(d) the volume occupied by unlike molecules are different

5. For a binary ideal liquid solution, the total vapour pressure of the solution is given as:
(a) $P_{\text{total}} = P_A^\circ + (P_A^\circ - P_B^\circ) X_B$ (b) $P_{\text{total}} = P_B^\circ + (P_A^\circ - P_B^\circ) X_A$
(c) $P_{\text{total}} = P_B^\circ + (P_B^\circ - P_A^\circ) X_A$ (d) $P_{\text{total}} = P_B^\circ + (P_B^\circ - P_A^\circ) X_B$

6. For an ideal binary liquid solution with $P_A^\circ > P_B^\circ$, which relation between X_A (mole fraction of A in liquid phase) and Y_A (mole fraction of A in vapour phase) is correct ?
(a) $Y_A < Y_B$ (b) $X_A > X_B$
(c) $\frac{Y_A}{Y_B} > \frac{X_A}{X_B}$ (d) $\frac{Y_A}{Y_B} < \frac{X_A}{X_B}$

7. An ideal solution is formed by mixing two volatile liquids A and B. X_A and X_B are the mole fractions of A and B respectively in the solution and Y_A and Y_B are the mole fractions of A and B respectively in the vapour phase. A plot of $\frac{1}{Y_A}$ along y-axis against $\frac{1}{X_A}$ along x-axis gives a straight line. What is the slope of the straight line ?
(a) $\frac{P_B^\circ}{P_A^\circ}$ (b) $\frac{P_A^\circ}{P_B^\circ}$
(c) $P_B^\circ - P_A^\circ$ (d) $P_A^\circ - P_B^\circ$

8. For a dilute solution, Raoult's law states that :
(a) the lowering of vapour pressure is equal to the mole fraction of solute
(b) the relative lowering of vapour pressure is equal to the mole fraction of solute
(c) the relative lowering of vapour pressure is proportional to the amount of solute in solution
(d) the vapour pressure of the solution is equal to the mole fraction of solvent

9. The solubility of a specific non-volatile salt is 4 g in 100 g of water at 25°C. If 2.0 g, 4.0 g and 6.0 g of the salt added of 100 g of water at 25°C, in system X, Y and Z. The vapour pressure would be in the order:

(a) $X < Y < Z$
(b) $X > Y > Z$
(c) $Z > X = Y$
(d) $X > Y = Z$

10. The boiling point of C_6H_6, CH_3OH, $C_6H_5NH_2$ and $C_6H_5NO_2$ are 80°C, 65°C, 184°C and 212°C respectively. Which will show highest vapour pressure at room temperature :
(a) C_6H_6
(b) CH_3OH
(c) $C_6H_5NH_2$
(d) $C_6H_5NO_2$

11. 6.0 g of urea (molecular mass = 60) was dissolved in 9.9 moles of water. If the vapour pressure of pure water is $P°$, the vapour pressure of solution is:
(a) $0.10\,P°$
(b) $1.10\,P°$
(c) $0.90\,P°$
(d) $0.99\,P°$

12. An ideal solution was found to have a vapour pressure of 80 torr when the mole fraction of a non-volatile solute was 0.2. What would be the vapour pressure of the pure solvent at the same temperature?
(a) 64 torr
(b) 80 torr
(c) 100 torr
(d) 400 torr

13. The vapour pressure of an aqueous solution of sucrose at 373 K is found to be 750 mm Hg. The molality of the solution at the same temperature will be :
(a) 0.26
(b) 0.73
(c) 0.74
(d) 0.039

14. Estimate the lowering of vapour pressure due to the solute (glucose) in a 1.0 M aqueous solution at 100°C :
(a) 10 torr
(b) 18 torr
(c) 13.45 torr
(d) 24 torr

15. Calculate the mass of non-volatile solute having molecular mass 40, which should be dissolved in 57 gm octane to reduce its vapour pressure to 80%:
(a) 47.2 g
(b) 5 g
(c) 106.2 g
(d) None of these

16. Equal mass of a solute are dissolved in equal mass of two solvents A and B and formed very dilute solution. The relative lowering of vapour pressure for the solution B has twice the relative lowering of vapour pressure for the solution A. If M_A and M_B are the molecular mass of solvents A and B respectively, then:
(a) $M_A = M_B$
(b) $M_B = 2 \times M_A$
(c) $M_A = 4M_B$
(d) $M_A = 2M_B$

17. An ideal solution has two components A and B. A is more volatile than B, *i.e.*, $P_A^\circ > P_B^\circ$ and also $P_A^\circ > P_{\text{total}}$. If X_A and Y_A are mole fractions of components A in liquid and vapour phases, then:
(a) $X_A = Y_A$
(b) $X_A > Y_A$
(c) $X_A < Y_A$
(d) Data insufficient

18. At 25°C, the vapour pressure of pure liquid A (mol. mass = 40) is 100 torr, while that of pure liquid B is 40 torr, (mol. mass = 80). The vapour pressure at 25°C of a solution containing 20 g of each A and B is:
(a) 80 torr
(b) 59.8 torr
(c) 68 torr
(d) 48 torr

19. Two liquids A and B form ideal solutions. At 300 K, the vapour pressure of solution containing 1 mole of A and 3 mole of B is 550 mm Hg. At the same temperature, if one more mole of B is added to this solution, the vapour pressure of the solution increases by 10 mm Hg. Determine the vapour pressure of A and B in their pure states (in mm Hg):
(a) 400, 600
(b) 500, 500
(c) 600, 400
(d) None of these

20. Two liquids A and B have vapour pressure in the ratio $P_A^\circ : P_B^\circ = 1:3$ at a certain temperature. Assume A and B form an ideal solution and the ratio of mole fractions of A to B in the vapour phase is 4 : 3, then the mole fraction of B in the solution at the same temperature is :

(a) $\frac{1}{5}$ (b) $\frac{2}{3}$

(c) $\frac{4}{5}$ (d) $\frac{1}{4}$

21. Two liquids A and B have P_A° and P_B° in the ratio of 1 : 3 and the ratio of number of moles of A and B in liquid phase are 1 : 3. Then mole fraction of 'A' in vapour phase in equilibrium with the solution is equal to:

(a) 0.1 (b) 0.2

(c) 0.5 (d) 1.0

22. Based on the given diagram, which of the following statements regarding the homogenous solutions of two volatile liquids are correct?

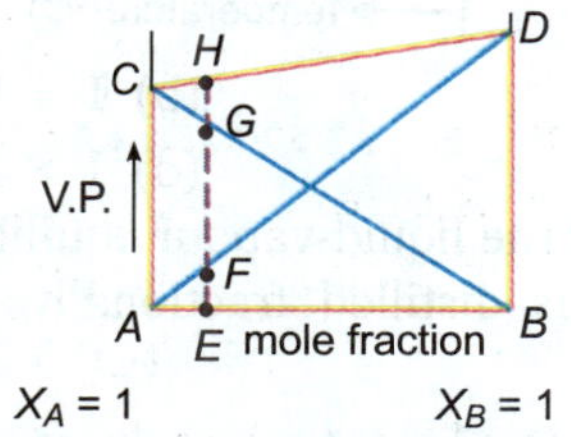

(1) Plots AD and BC show that Raoult's law is obeyed for the solution in which B is a solvent and A is the solute and as well as for that in which A is solvent and B is solute.

(2) Plot CD shows that Dalton's law of partial pressures is obeyed by the binary solution of components A and B.

(3) $EF + EG = EH$; and AC and BD correspond to the vapour pressures of the pure solvents A and B respectively.

(a) Only 1 (b) 2 and 3

(c) 1 and 3 (d) All

23. Which represents correct difference when non-volatile solute is present in an ideal solution?

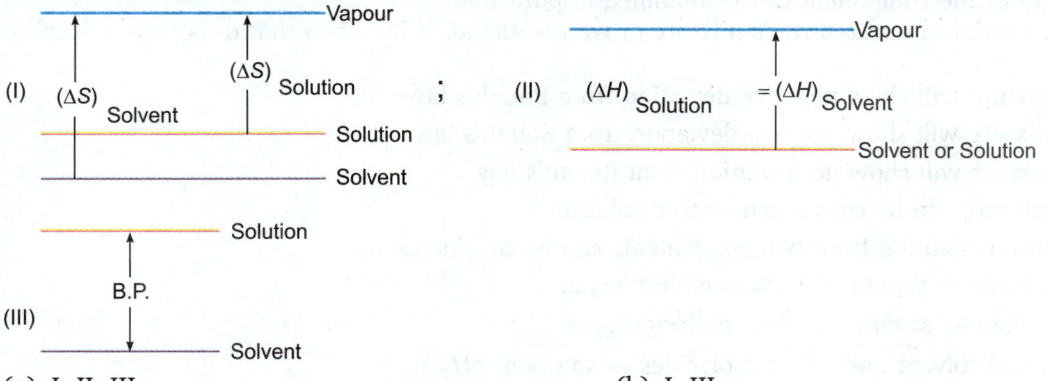

(a) I, II, III (b) I, III

(c) II, III (d) I, II

24. Select correct statement :

(a) Solution has more molecular randomness than a pure solvent. The entropy change between solution and solid is larger than the entropy change between pure solvent and solid

(b) Heat of fusion of solution and solvent are similar since similar forces of intermolecular forces are involved

(c) Solution containing sugar freezes at a lower temperature than pure water

(d) All are correct statements

25. Select correct statement?
(a) Heats of vaporisation for a pure solvent and for a solution are similar because similar intermolecular forces between solvent molecules must be overcome in both cases
(b) Entropy change between solution and vapour is smaller than the entropy change between pure solvent and vapour
(c) Boiling point of the solution is larger than that of the pure solvent
(d) All are correct statements

26. The vapour pressure curves of the same solute in the same solvent are shown below. The curves are parallel to each other and do not intersect. The concentrations of solutions are in order of:

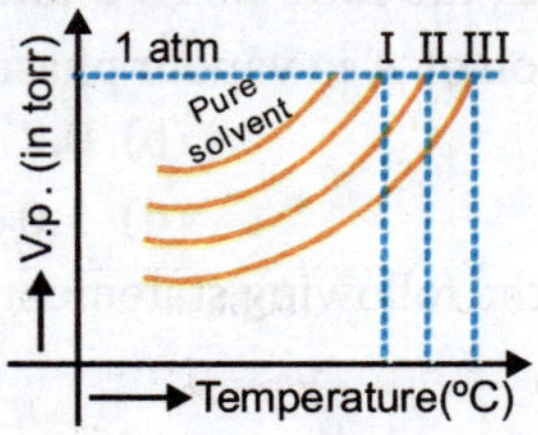

(a) I < II < III (b) I = II = III
(c) I > II > III (d) I > III > II

27. Boiling point composition diagram of the liquid-vapour equilibrium for *A* and *B* is shown in the figure. If a binary liquid mixture of *A* and *B* is distilled fractionally, which of the following would be correct observation?

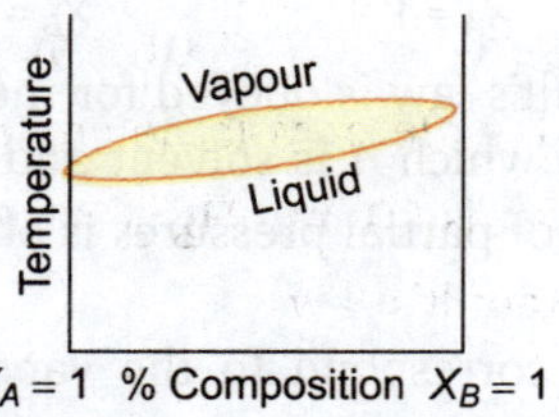

(a) Composition of the still (residue) will approach pure liquid *B* only
(b) Composition of the distillate will approach pure *A* only
(c) Composition of distillate and residue will approach pure *A* and *B* respectively
(d) Neither of the component can be obtained in pure state

28. The boiling point of an azeotropic mixture of water–ethanol is less than that of both water and ethanol. Then:
(a) the mixture will show negative deviation from Raoult's law
(b) the mixture will show positive deviation from Raoult's law
(c) the mixture will show no deviation from Raoult's law
(d) this mixture can be considered as true solution

29. Formation of a solution from two components can be considered as :
(i) Pure solvent → separated solvent molecules, ΔH_1
(ii) Pure solute → separated solute molecules, ΔH_2
(iii) separated solvent and solute molecules → solution, ΔH_3
Solution so formed will be ideal if :
(a) $\Delta H_{Soln} = \Delta H_1 + \Delta H_2 + \Delta H_3$ (b) $\Delta H_{Soln} = \Delta H_1 + \Delta H_2 - \Delta H_3$
(c) $\Delta H_{Soln} = \Delta H_1 - \Delta H_2 - \Delta H_3$ (d) $\Delta H_{Soln} = \Delta H_3 - \Delta H_1 - \Delta H_2$

30. Total vapour pressure of mixture of 1 mol X ($P_X^\circ = 150$ torr) and 2 mol Y ($P_Y^\circ = 300$ torr) is 240 torr. In this case:
(a) there is a negative deviation from Raoult's law
(b) there is a positive deviation from Raoult's law
(c) there is no deviation from Raoult's law
(d) can not be decided

31. In a mixture of A and B, components show positive deviation when:
(a) $A-B$ interaction is stronger than $A-A$ and $B-B$ interaction
(b) $A-B$ interaction is weaker than $A-A$ and $B-B$ interaction
(c) ΔV mix < 0, ΔS mix > 0
(d) ΔV mix $= 0$, ΔS mix > 0

32. A liquid mixture having composition corresponding to point Z in the figure shown is subjected to distillation at constant pressure. Which of the following statement is correct about the process?

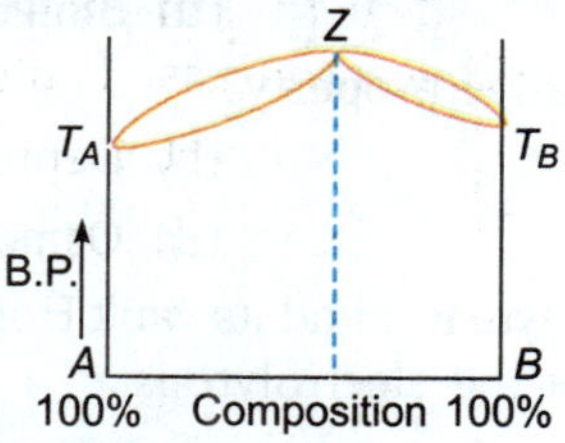

(a) The composition of distillate differs from the mixture
(b) The boiling point goes on changing
(c) The mixture has highest vapour pressure than for any other composition
(d) Composition of an azeotrope alters on changing the external pressure

33. Which will form maximum boiling azeotrope ?
(a) $C_6H_6 + C_6H_5CH_3$ solution
(b) $HNO_3 + H_2O$ solution
(c) $C_2H_5OH + H_2O$ solution
(d) *n*-hexane and *n*-heptane

34. Total vapour pressure of mixture of 1 mole of volatile component A ($P_A^\circ = 100$ mm Hg) and 3 mole of volatile component B ($P_B^\circ = 80$ mm Hg) is 90 mm Hg. For such case:
(a) there is positive deviation from Raoult's law
(b) boiling point has been lowered
(c) force of attraction between A and B is weaker than that between A and A or between B and B
(d) all the above statements are correct

35. The azeotropic mixture of water (B.P. = 100° C) and HCl (B.P. = 86°C) boils at about 120°C. During fractional distillation of this mixture, it is possible to obtain :
(a) pure HCl
(b) pure H_2O
(c) pure H_2O as well as pure HCl
(d) neither H_2O nor HCl

36. Azeotropic mixture of water and C_2H_5OH boils at 351 K. By distilling the mixture it is possible to obtain :
(a) Pure C_2H_5OH only
(b) Pure water only
(c) Neither C_2H_5OH nor water
(d) Both water and C_2H_5OH in pure state

37. An azeotropic mixture of two liquids has a boiling point higher than either of them when it:
(a) shows positive deviation from Raoult's law
(b) shows negative deviation from Raoult's law
(c) shows ideal behaviour
(d) is saturated

38. If two liquids A ($P_A^\circ = 100$ torr) and B ($P_B^\circ = 200$ torr) which are completely immiscible with each other (each one will behave independently of the other) are present in a closed vessel, the total vapour pressure of the system will be:
(a) less than 100 torr
(b) greater than 200 torr
(c) between 100 to 200 torr
(d) 300 torr

39. When a liquid that is immiscible with water was steam distilled at 95.2°C at a total pressure of 748 torr, the distillate contained 1.25 g of the liquid per gram of water. The vapour pressure of water is 648 torr at 95.2°C, what is the molar mass of liquid?

(a) 7.975 g/mol (b) 166 g/mol
(c) 145.8 g/mol (d) None of these

40. Water and chlorobenzene are immiscible liquids. Their mixture boils at 89°C under a reduced pressure of 7.7×10^4 Pa. The vapour pressure of pure water at 89°C is 7×10^4 Pa. Mass per cent of chlorobenzene in the distillate is :
(a) 50 (b) 60
(c) 78.3 (d) 38.46

41. Which of the following is not a colligative property?
(a) Vapour pressure (b) Depression in f.pt.
(c) Elevation in b.pt. (d) Osmotic pressure

42. The degree of dissociation of an electrolyte is α and its van't Hoff factor is i. The number of ions obtained by complete dissociation of 1 molecule of the electrolyte is :
(a) $\frac{i+\alpha-1}{\alpha}$ (b) $i-\alpha-1$
(c) $\frac{i-1}{\alpha}$ (d) $\frac{i+1+\alpha}{1-\alpha}$

43. One mole of a solute A is dissolved in a given volume of a solvent. The association of the solute take place as follows:

$$nA \rightleftharpoons A_n$$

If α is the degree of association of A, the van't Hoff factor i is expressed as:
(a) $i = 1-\alpha$ (b) $i = 1+\frac{\alpha}{n}$
(c) $i = \frac{1-\alpha+\frac{\alpha}{n}}{1}$ (d) $i = 1$

44. The van't Hoff factor i for an electrolyte which undergoes dissociation and association in solvent are respectively:
(a) greater than one and less than one (b) less than one and greater than one
(c) less than one and less than one (d) greater than one and greater than one

45. Which solution has the highest vapour pressure?
(a) 0.02 M NaCl at 50°C (b) 0.03 M sucrose at 15°C
(c) 0.005 M $CaCl_2$ at 50°C (d) 0.005 M $CaCl_2$ at 25°C

46. An aqueous solution is 1.00 molal in KI. Which change will cause the vapour pressure of the solution to increase ?
(a) addition of water (b) addition of NaCl
(c) addition of Na_2SO_4 (d) Addition of 1.0 molal KI

47. Four solutions of K_2SO_4 with the concentrations 0.1 m, 0.01 m, 0.001 m and 0.0001 m are available. The maximum value of colligative property corresponds to :
(a) 0.0001 m solution (b) 0.001 m solution
(c) 0.01 m solution (d) 0.1 m solution

48. Moles of Na_2SO_4 to be dissolved in 12 mole water to lower its vapour pressure by 10 mm Hg at a temperature at which vapour pressure of pure water is 50 mm is:
(a) 1.5 mole (b) 2 mole
(c) 1 mole (d) 3 mole

49. A very diluted saturated solution of a sparingly soluble salt X_3Y_4 has a vapour pressure of 20 mm Hg at temperature T, while pure water exerts a pressure of 20.0126 mm Hg at the same temperature. Calculate molality (m) at temperature T:

(a) 6.3×10^{-4} (b) 3.5×10^{-2}
(c) 5×10^{-3} (d) None of these

50. When 1 mole of a solute is dissolved in 1 kg of H_2O, boiling point of solution was found to be 100.5°C. K_b for H_2O is:
(a) 0.5 (b) 100
(c) 100.5 (d) 95.5

51. Chloroform, $CHCl_3$, boils at 61.7°C. If the K_b for chloroform is 3.63°C/molal, what is the boiling point of a solution of 15.0 kg of $CHCl_3$ and 0.616 kg of acenaphthalene, $C_{12}H_{10}$?
(a) 61.9 (b) 62.0
(c) 52.2 (d) 62.67

52. A compound has the empirical formula $C_{10}H_8Fe$. A solution of 0.26 g of the compound in 11.2 g of benzene (C_6H_6) boils at 80.26°C. The boiling point of benzene is 80.10°C; the K_b is 2.53°C/molal. What is the molecular formula of the compound?
(a) $C_{30}H_{24}Fe_3$ (b) $C_{10}H_8Fe$
(c) C_5H_4Fe (d) $C_{20}H_{16}Fe_2$

53. A solution of 0.640 g of azulene in 100.0 g of benzene boils at 80.23°C. The boiling point of benzene is 80.10°C; and K_b is 2.53°C/molal. What is the molecular mass of azulene?
(a) 108 (b) 99
(c) 125 (d) 134

54. One molal solution of a carboxylic acid in benzene shows the elevation of boiling point of 1.518 K. The degree of association for dimerization of the acid in benzene is (K_b for benzene $= 2.53$ K kg mol^{-1}) :
(a) 60% (b) 70%
(c) 75% (d) 80%

55. The boiling point elevation constant for toluene is 3.32 K kg mol^{-1}. The normal boiling point of toluene is 110.7°C. The enthalpy of vaporisation of toluene would be nearly:
(a) 17.0 kJ mol^{-1} (b) 34.0 kJ mol^{-1}
(c) 51.0 kJ mol^{-1} (d) 68.0 kJ mol^{-1}

56. Which one of the following aqueous solutions will exhibit highest boiling point ?
(a) 0.015 M urea (b) 0.01 M KNO_3
(c) 0.10 M Na_2SO_4 (d) 0.015 M glucose

57. Calculate the percentage degree of dissociation of an electrolyte XY_2 (Normal molar mass = 164) in water if the observed molar mass by measuring elevation in boiling point is 65.6:
(a) 75% (b) 25%
(c) 65% (d) None of these

58. If the elevation in boiling point of a solution of non-volatile, non-electrolytic and non-associating solute in solvent ($K_b = x$ K.kg.mol^{-1}) is y K, then the depression in freezing point of solution of same concentration would be
(K_f of the solvent $= z$ K.kg.mol^{-1})
(a) $\frac{2xz}{y}$ (b) $\frac{yz}{x}$
(c) $\frac{xz}{y}$ (d) $\frac{yz}{2x}$

59. When a solution containing non-volatile solute freezes, which equilibrium would exist?
(a) solid solvent $\rightleftharpoons$ liquid solvent (b) solid solute $\rightleftharpoons$ liquid solution
(c) solid solute $\rightleftharpoons$ liquid solvent (d) solid solvent $\rightleftharpoons$ liquid solution

60. Bromoform has a normal freezing point of 7.734°C and $K_f = 14.4$°C/m. A solution of 2.60 g of an unknown substance in 100 g of bromoform freezes at 5.43°C. What is the molecular mass of the unknown substance?
(a) 16.25 (b) 162.5
(c) 100 (d) none of these

61. C_6H_6 freezes at 5.5°C. At what temperature will a solution of 10.44 g of C_4H_{10} in 200 g of C_6H_6 freeze? K_f $(C_6H_6) = 5.12$°C/m
(a) 4.608°C (b) 0.892°C
(c) 5.5°C (d) none of these

62. How much ethyl alcohol must be added to 1.0 L of water so that solution will not freeze at –4°F? $(K_f = 1.86$°C/m)
(a) < 20 g (b) < 10.75 g
(c) < 494.5 g (d) > 494.5 g

63. The freezing point of a solution of 2.40 g of biphenyl $(C_{12}H_{10})$ in 75.0 g of benzene (C_6H_6) is 4.40°C. The normal freezing point of benzene is 5.50°C. What is the molal freezing point constant (°C/m) for benzene?
(a) 5.3 (b) 5.1
(c) 4.6 (d) 4.8

64. A solution containing 1.8 g of a compound (empirical formula CH_2O) in 40 g of water is observed to freeze at –0.465°C. The molecular formula of the compound is (K_f of water = 1.86 kg K mol^{-1}):
(a) $C_2H_4O_2$ (b) C_3H_6
(c) $C_4H_8O_4$ (d) $C_6H_{12}O_6$

65. Freezing point of the following equilibrium, liquid solvent $\rightleftharpoons$ solid solvent is :
(a) $\dfrac{\Delta H - \Delta G}{T\Delta S}$ (b) $\dfrac{\Delta H}{\Delta S}$
(c) $\dfrac{\Delta G}{\Delta S}$ (d) $\dfrac{\Delta S}{\Delta H}$

66. Freezing point of a solution is smaller than freezing point of a solvent. It is due to :
(a) ΔH of solution and solvent is almost identical since intermolecular forces between solvent molecules are involved
(b) ΔS of solution (between solution and solid) is larger than that of the ΔS of solvent (between solvent and solid)
(c) ΔS of the solution is smaller than that of the solvent
(d) ΔH of the solution is much higher than of solvent but ΔS of solution is smaller than that of the solvent

67. When 36.0 g of a solute having the empirical formula CH_2O is dissolved in 1.20 kg of water, the solution freezes at –0.93°C. What is the molecular formula of the solute? ($K_f = 1.86$°C kg mol^{-1})
(a) C_2H_4O (b) $C_2H_2O_2$
(c) $C_2H_4O_3$ (d) $C_2H_4O_2$

68. Calculate the molecular mass of a substance whose 7.0% by mass solution in water freezes at –0.93°C. The cryoscopic constant of water is 1.86°C kg mol^{-1}:
(a) 140 g mol^{-1} (b) 150.5 g mol^{-1}
(c) 160 g mol^{-1} (d) 155 g mol^{-1}

69. Camphor is often used in molecular mass determination because
(a) it is readily available (b) it has a very high cryoscopic constant
(c) it is volatile (d) it is solvent for organic substances

70. For 1 molal solution of each compound minimum freezing point will be assuming complete ionisation in each case:
(a) $[Fe(H_2O)_6]Cl_3$ (b) $[Fe(H_2O)_5Cl]Cl_2 \cdot H_2O$
(c) $[Fe(H_2O)_4Cl_2]Cl \cdot 2H_2O$ (d) $[Fe(H_2O)_3Cl_3] \cdot 3H_2O$

71. For 1 molal solution of each compound maximum freezing point will be assuming complete ionisation in each case:

(a) $[Fe(H_2O)_6]Cl_3$ (b) $[Fe(H_2O)_5Cl]Cl_2 \cdot H_2O$

(c) $[Fe(H_2O)_4Cl_2]Cl \cdot 2H_2O$ (d) $[Fe(H_2O)_3Cl_3] \cdot 3H_2O$

72. $PtCl_4 \cdot 6H_2O$ can exist as a hydrated complex. Its 1 molal aq. solution has depression in freezing point of 3.72. Assume 100% ionisation and $K_f(H_2O) = 1.86°C\ mol^{-1}$ kg, then complex is :

(a) $[Pt(H_2O)_6]Cl_4$ (b) $[Pt(H_2O)_4Cl_2]Cl_2 \cdot 2H_2O$

(c) $[Pt(H_2O)_3Cl_3]Cl \cdot 3H_2O$ (d) $[Pt(H_2O)_2Cl_4] \cdot 4H_2O$

73. A complex is represented as $CoCl_3 xNH_3$. Its 0.1 molal solution in water shows $\Delta T_f = 0.558$ K. K_f for H_2O is 1.86 K $molality^{-1}$. Assuming 100% ionisation of complex and co-ordination number of Co is six, calculate formula of complex:

(a) $[Co(NH_3)_6]Cl_3$ (b) $[Co(NH_3)_5Cl]Cl_2$

(c) $[Co(NH_3)_4Cl_2]Cl$ (d) none of these

74. The freezing point of equimolal aqueous solutions will be highest for:

(a) $C_6H_5NH_3Cl$ (b) $Ca(NO_3)_2$

(c) $La(NO_3)_2$ (d) $C_6H_{12}O_6$

75. The freezing point of a 4% aqueous solution of '*A*' is equal to the freezing point of 10% aqueous solution of '*B*'. If the molecular mass of '*A*' is 60, then the molecular mass of '*B*' will be:

(a) 160 (b) 90

(c) 45 (d) 180

76. Depression in freezing point of 0.01 molal aqueous HCOOH solution is 0.02046. 1 molal aqueous urea solution freezes at –1.86°C. Assuming molality equal to molarity, pH of HCOOH solution is:

(a) 2 (b) 3

(c) 4 (d) 5

77. When mercuric iodide is added to the aqueous solution of KI, then the:

(a) freezing point is raised (b) freezing point is lowered

(c) freezing point does not change (d) boiling point does not change

78. Dimer of acetic acid in liquid benzene is in equilibrium with acetic acid monomer at certain temperature and pressure. If 25% of the dimer molecules are separated out then

(a) freezing point of the solution reduces (b) average molar mass of solute increases

(c) boiling point of solution increases (d) molar mass of solute decreases

79. The temperature of a city was found to be –9.3°C. A car was used, whose radiator was filled with 5 L of water. What minimum quantity of antifreezing agent ethylene glycol were added to water of radiator in order to use the car for travelling? (K_f of water 1.86 K kg mol^{-1})

(a) 3200 g (b) 1670 g

(c) 1550 g (d) 2100 g

80. The cryoscopic constant of water is 1.86 K kg mol^{-1}. A 0.01 molal acetic acid solution produces a depression of 0.0194°C in the freezing point. The degree of dissociation of acetic acid is:

(a) zero (b) 0.043

(c) 0.43 (d) 1

81. If 1m solution of benzoic acid in benzene has a freezing point depression of 3.84°C. ($K_f = 5.12°C$ mol^{-1} kg) and boiling point elevation of 2.53°C ($K_b = 2.53°C\ mol^{-1}$ kg), then select the correct statement/s :

Statement I : there is dimer formation when undergoing freezing

Statement II : there is no change when undergoing boiling

Statement III : reverse of I and II

Statement IV : dimer formation in freezing and boiling state

(a) I, II (b) II, III
(c) III, I (d) only I

82. In a 0.5 molal solution KCl, KCl is 50% dissociated. The freezing point of solution will be (K_f = 1.86 K kg mol^{-1}):
(a) 274.674 K (b) 271.60 K
(c) 273 K (d) none of these

83. A 1.0 g sample of $Co(NH_2CH_2CH_2NH_2)_3Cl_3$ is dissolved in 25.0 g of water and the freezing point of the solution is –0.87°C. How many ions are produced per mole of compound? The K_f of water is 1.86°C/molal
(a) 2 (b) 3
(c) 4 (d) 5

84. An aqueous solution contains 3% and 1.8% by mass. urea and glucose respectively. What is the freezing point of solution? (K_f = 1.86°C/m)
(a) –1.172°C (b) –2.27°C
(c) –1.5°C (d) none of these

85. Phenol associates in benzene to a certain extent in dimerisation reaction. A solution containing 0.02 kg of phenol in 1.0 kg of benzene has its freezing point depressed 0.69 K. Hence, degree of association of phenol dimerized will be :
$[K_f(C_6H_6) = 5.12 \text{ K kg mol}^{-1}]$
(a) 0.63 (b) 0.73
(c) 0.83 (d) 0.93

86. Assuming complete ionisation, the solution having maximum freezing point will be:
(a) 1 *M* CaF_2 (b) 1.5 *M* $Al_2(SO_4)_3$
(c) 2 *M* NaCl (d) 1 *M* $AgNO_3$

87. In a 0.2 molal aqueous solution of a weak acid *HX* the degree of dissociation is 0.25. The freezing point of the solution will be nearest to: (K_f = 1.86 K kg mol^{-1})
(a) –0.26°C (b) 0.465°C
(c) –0.48°C (d) –0.465°C

88. An aqueous solution of 0.01 *M* KCl causes the same elevation in boiling point as an aqueous solution of urea. The concentration of urea solution is:
(a) 0.01 *M* (b) 0.005 *M*
(c) 0.02 *M* (d) 0.04 *M*

89. When some NaCl was dissolved in water, the freezing point depression was numerically equal to twice the molal f.p. depression constant. The relative lowering of vapour pressure of the solution is nearly :
(a) 0.036 (b) 0.018
(c) 0.0585 (d) 0.072

90. Which one of the following statements is false ?
(a) The correct order of osmotic pressure for 0.01 *M* aqueous solution of each compound is $BaCl_2$ > KCl > CH_3COOH > sucrose
(b) Isotonic solutions are those solutions which have the same osmotic pressure
(c) Raoult's law states that the vapour pressure of a component over a solution is proportional to its mole fraction in liquid state
(d) Two sucrose solutions of same molality prepared in different solvents will have the same freezing point depression

91. 0.1 molal aqueous solution of an electrolyte AB_3 is 90% ionised. The boiling point of the solution at 1 atm is : $(K_{b(H_2O)} = 0.52 \text{ K kg mol}^{-1})$
(a) 273.19 K (b) 374.92 K
(c) 376.4 K (d) 373.19 K

92. Which of the following aqueous solutions has osmotic pressure nearest to pure solvent?
(a) Na_2SO_4
(b) $BaCl_2$
(c) $Al_2(SO_4)_3$
(d) $C_{12}H_{22}O_{11}$

93. 0.1*M* NaCl and 0.05 *M* $BaCl_2$ solutions are separated by a semi-permeable membrane in a container. For this system, choose the correct answer :
(a) There is no movement of any solution across the membrane
(b) Water flows from $BaCl_2$ solution towards NaCl solution
(c) Water flows from NaCl solution towards $BaCl_2$ solution
(d) Osmotic pressure of 0.1*M* NaCl is lower than the osmotic pressure of $BaCl_2$ (assume complete dissociation)

94. Two aqueous solutions, *A* and *B*, are separated by a semi-permeable membrane. The osmotic pressure of solution *A* immediately begins to decrease. Which of the following statement is true ?
(a) The solvent molecules are moving from the solution of higher osmotic pressure to that of lower osmotic pressure.
(b) The initial osmotic pressure of solution *B* is greater than that of solution *A*.
(c) Solvent molecules are moving from solution *B* into solution *A*.
(d) Both (a) and (b) are true statements.

95. Which one of the following pairs of solution can we expect to be isotonic at the same temperature?
(a) 0.1 *M* urea and 0.1 *M* NaCl
(b) 0.1 *M* urea and 0.2 *M* $MgCl_2$
(c) 0.1 *M* NaCl and 0.1 *M* Na_2SO_4
(d) 0.1 *M* $Ca(NO_3)_2$ and 0.1 *M* Na_2SO_4

96. The empirical formula of a non-electrolyte is CH_2O. A solution containing 3 g L^{-1} of the compound exerts the same osmotic pressure as that of 0.05 *M* glucose solution. The molecular formula of the compound is :
(a) CH_2O
(b) $C_2H_4O_2$
(c) $C_4H_8O_4$
(d) $C_3H_6O_3$

97. A semipermeable membrane used in the measurement of osmotic pressure of a solution allows the passage of:
(a) solute molecules through it
(b) solvent molecules through it
(c) both solvent and solute molecules
(d) either solvent or solute

98. In the case of osmosis, solvent molecules move from :
(a) higher vapour pressure to lower vapour pressure
(b) higher concentration to lower concentration
(c) lower vapour pressure to higher vapour pressure
(d) higher osmotic pressure to lower osmotic pressure

99. The osmotic pressures of equimolar solutions of urea, $BaCl_2$ and $AlCl_3$ will be in the order :
(a) $AlCl_3 > BaCl_2 >$ urea
(b) $BaCl_2 > AlCl_3 >$ urea
(c) urea $> BaCl_2 > AlCl_3$
(d) $BaCl_2 >$ urea $> AlCl_3$

100. Which of the following solutions will have maximum osmotic pressure? Assume 90% dissociation of each salt:
(a) decimolar aluminium sulphate
(b) decimolar barium chloride solution
(c) decimolar sodium sulphate solution
(d) solution of equal volumes of decimolar barium chloride and decimolar sodium sulphate solutions

101. Consider 0.1 *M* solutions of two solutes *X* and *Y*. The solute *X* behaves as a univalent electrolyte while the solute *Y* dimerises in solution. Which of the following statements are correct regarding these solutions?
(1) The boiling point of the solution of *X* will be higher than that of *Y*
(2) The osmotic pressure of the solution of *Y* will be lower than that of *X*

(3) The freezing point of the solution of X will be lower than that of Y

(4) The relative lowering of vapour pressure of both the solutions will be the same

Select the correct answer from the option given below :

(a) 1, 2 and 3 (b) 2, 3 and 4
(c) 1, 2 and 4 (d) 1, 3 and 4

102. If M_{normal} is the normal molecular mass and α is the degree of ionization of $K_3[Fe(CN)_6]$, then the abnormal molecular mass of the complex in the solution will be:

(a) $M_{\text{normal}}(1+2\alpha)^{-1}$ (b) $M_{\text{normal}}(1+3\alpha)^{-1}$
(c) $M_{\text{normal}}(1+\alpha)^{-1}$ (d) equal to M_{normal}

103. Equal volumes of 0.1 *M* urea and 0.1 *M* glucose are mixed. The mixture will have:

(a) lower osmotic pressure (b) same osmotic pressure
(c) higher osmotic pressure (d) none of these

104. A 5% (w/V) solution of cane sugar (molecular mass $=342$) is isotonic with 1% (w/V) solution of a substance *X*. The molecular mass of *X* is:

(a) 34.2 (b) 171.2
(c) 68.4 (d) 136.8

105. Insulin is dissolved in a suitable solvent and the osmotic pressure (π) of solutions of various concentrations (g/cm^3) C is measured at 20°C. The slope of a plot of π against C is found to be 4.65×10^{-3}. The molecular mass of the insulin (g/mol) is:

(a) 3×10^5 (b) 9×10^5
(c) 4.5×10^5 (d) 5.16×10^6

106. An aqueous solution of sucrose ($C_{12}H_{22}O_{11}$) having a concentration of 34.2 gram/litre has an osmotic pressure of 2.38 atmospheres at 17°C. For an aqueous solution of glucose ($C_6H_{12}O_6$) to be isotonic with this solution, its concentration should be :

(a) 34.2 gram per litre (b) 17.1 gram per litre
(c) 18.0 gram per litre (d) 36.0 gram per litre

107. Which of the following experimental methods is adopted to determine osmotic pressure?

(a) Berkley-Hartely's method (b) Beckmann's method
(c) Landsberger's method (d) Differential method

108. Based upon the technique of reverse osmosis the approximate pressure required to desalinate sea water containing 2.5% (mass/volume) KNO_3 at 27°C will be

(a) 10.5 atm (b) 21 atm
(c) 12.2 atm (d) 6.09 atm

109. A 1% (mass/vol) KCl solution is ionised to the extent of 80%. The osmotic pressure at 27°C of the solution will be:

(a) 6.95 atm (b) 5.94 atm
(c) 2.71 atm (d) 3.30 atm

110. Osmotic pressure of blood is 7.40 atm, at 27°C. Number of moles of glucose to be used per litre for an intravenous injection that is to have same osmotic pressure of blood is:

(a) 0.3 (b) 0.2
(c) 0.1 (d) 0.4

111. The relationship between osmotic pressures (π_1, π_2 and π_3) at a definite temperature when 1 g glucose, 1 g urea and 1 g sucrose are dissolved in 1 litre of water is (assume $i=1$ for all):

(a) $\pi_1>\pi_2>\pi_3$ (b) $\pi_3>\pi_1>\pi_2$
(c) $\pi_2>\pi_1>\pi_3$ (d) $\pi_2>\pi_3>\pi_1$

112. van't Hoff proved that osmotic pressure (π) is a colligative property. For an ideal solution, osmotic pressure (π) is helpful to determine that molecular mass of solute using $M_B = \frac{W_B RT}{\pi \cdot V}$ Relation can be expressed by the curve (C = concentration):

(a) 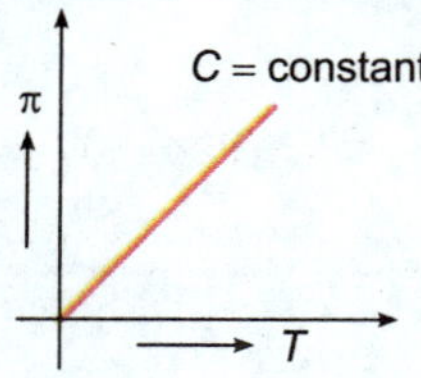

(b) 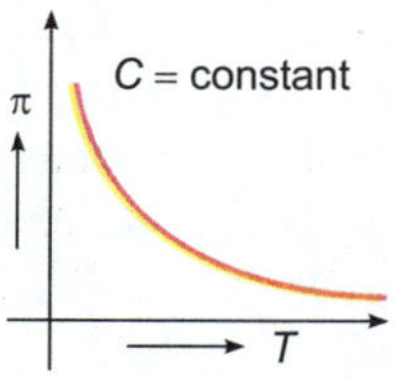

(c) 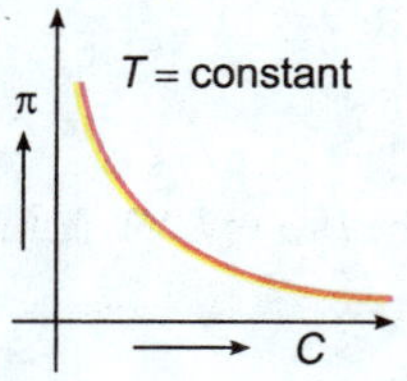

(d) 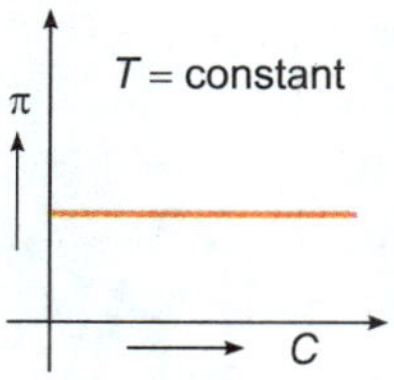

113. A solution containing 4.0 g of PVC in 2 litre of dioxane (industrial solvent) was found to have an osmotic pressure 3.0×10^{-4} atm at 27°C. The molar mass of the polymer (g/mol) will be :

(a) 1.6×10^4 (b) 1.6×10^5

(c) 1.6×10^3 (d) 1.6×10^2

114. The osmotic pressures of 0.010 M solutions of KI and of sucrose ($C_{12}H_{22}O_{11}$) are 0.432 atm and 0.24 atm respectively. The van't Hoff factor for KI is :

(a) 1.80 (b) 0.80

(c) 1.2 (d) 1.0

115. What is the correct sequence of osmotic pressure of 0.01 M *aq.* solution of :

(1) $Al_2(SO_4)_3$ (2) Na_3PO_4

(3) $BaCl_2$ (4) Glucose

(a) $\pi_4 > \pi_2 > \pi_3 > \pi_1$ (b) $\pi_3 > \pi_4 > \pi_2 > \pi_1$

(c) $\pi_3 > \pi_4 > \pi_1 > \pi_2$ (d) $\pi_1 > \pi_2 > \pi_3 > \pi_4$

116. 1.0 molar solution of the complex salt, $CrCl_3 \cdot 6H_2O$, displays an osmotic pressure of $3RT$. 0.5 L of the same solution on treatment with excess of $AgNO_3$ solution will yield (assume $\alpha = 1$) :

(a) 0.5 mole of AgCl (b) 1.0 mole of AgCl

(c) 1.5 mole of AgCl (d) 3.0 mole of AgCl

117. A 0.010 g sample of $Cr(NH_3)_4(SO_4)Cl$ is dissolved in 25.0 mL of water and the osmotic pressure of the solution is 59.1 torr at 25°C. How many moles of ions are produced per mole of compound?

(a) 1 (b) 4

(c) 2 (d) 3

118. Which of the following aqueous solutions should have the highest osmotic pressure?

(a) 0.011 M $AlCl_3$ at 50°C (b) 0.03 M NaCl at 25°C

(c) 0.012 M $(NH_4)_2SO_4$ at 25°C (d) 0.03 M NaCl at 50°C

119. $X_3Y_2 (i = 5)$ when reacted with $A_2B_3 (i = 5)$ in aqueous solution gives brown colour. These are separated by a semipermeable membrane AB as shown. Due to osmosis there is:

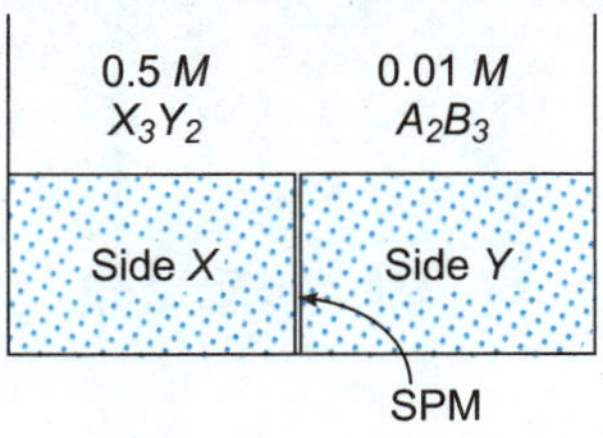

(a) brown colour formation in side X (b) brown colour formation in side Y
(c) formation in both of the sides X and Y (d) no brown colour formation

120. Which of the following curves represents the Henry's law?

(a)

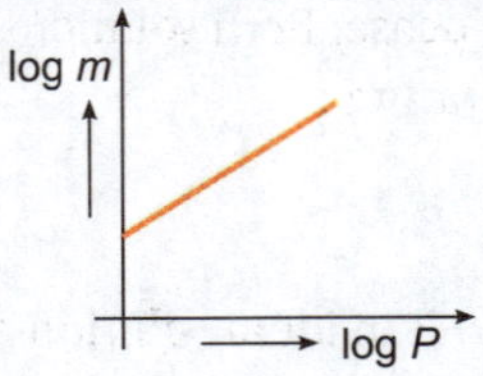

(b)

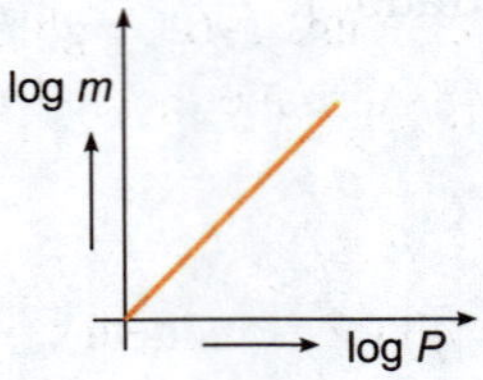

(c)

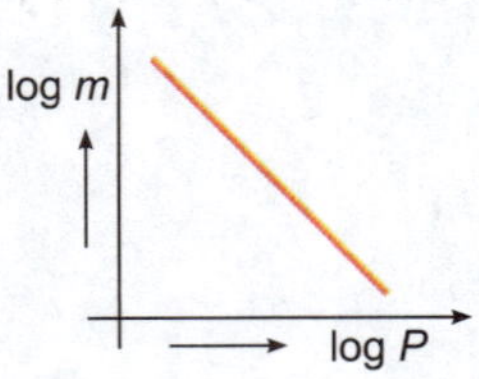

(d)

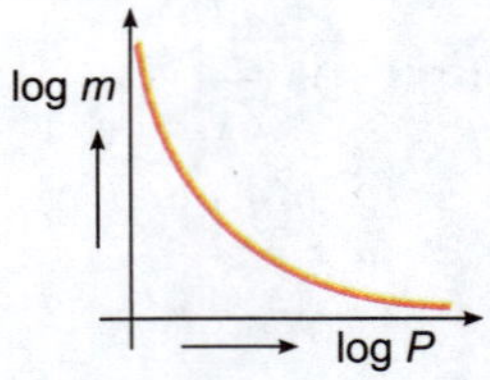

121. According to Henry's law, the solubility of a gas in a given volume of liquid increases with increase in:
(a) temperature (b) pressure
(c) Both (a) and (b) (d) None of these

122. At 300 K, 40 mL of $O_3(g)$ dissolves in 100 g of water at 1.0 atm. What mass of ozone dissolved in 400 g of water at a pressure of 4.0 atm at 300 K?
(a) 0.1 g (b) 1.24 g
(c) 0.48 g (d) 4.8 g

123. 1 kg of water under a nitrogen pressure of 1 atmosphere dissolves 0.02 gm of nitrogen at 293 K. Calculate Henry's law constant :
(a) 7.7×10^4 atm (b) 7.7×10^3 atm
(c) 2×10^{-5} atm (d) 2×10^{-2} atm

124. According to Henry's law, the partial pressure of gas (P_g) is directly proportional to mole fraction of gas in liquid solution, $P_{gas} = K_H . X_{gas}$, where K_H is Henry's constant. Which is incorrect ?
(a) K_H is characteristic constant for a given gas-solvent system
(b) Higher is the value of K_H, lower is solubility of gas for a given partial pressure of gas
(c) K_H has temperature dependence
(d) K_H decreases with increase of temperature

125. At 760 torr pressure and 20°C temperature, 1 L of water dissolves 0.04 gm of pure oxygen or 0.02 gm of pure nitrogen. Assuming that dry air is composed of 20% oxygen and 80% nitrogen (by volume), the masses (in g/L) of oxygen and nitrogen dissolved by 1 L of water at 20°C exposed to air at a total pressure of 760 torr are respectively :
(a) 0.008, 0.016 (b) 0.016, 0.008
(c) 0.16, 0.08 (d) 0.04, 0.02

Level-2

1. Two beaker A and B present in a closed vessel. Beaker A contains 152.4 g aqueous solution of urea, containing 12 g of urea. Beaker B contains 196.2 g glucose solution, containing 18 g of glucose. Both solutions allowed to attain the equilibrium. Determine mass % of glucose in its solution at equilibrium :
   (a) 6.71 (b) 14.49
   (c) 16.94 (d) 20
2. Two components A and B form an ideal solution. The mole fractions of A and B in ideal solution are X_A and X_B, while that of in vapour phase, these components have their mole fractions as Y_A and Y_B. Then, the slope and intercept of plot of $\frac{1}{Y_A}$ vs. $\frac{1}{X_A}$ will be :
   (a) $\frac{P_A^\circ}{P_B^\circ}, \frac{P_B^\circ - P_A^\circ}{P_B^\circ}$ (b) $\frac{P_B^\circ}{P_A^\circ}, \frac{P_A^\circ - P_B^\circ}{P_A^\circ}$
   (c) $\frac{P_B^\circ}{P_A^\circ}, \frac{P_B^\circ}{P_B^\circ - P_A^\circ}$ (d) $P_A^\circ - P_B^\circ, \frac{P_A^\circ}{P_B^\circ}$
3. At 48°C, the vapour pressure of pure CS_2 is 850 torr. A solution of 2.0 g of sulphur in 100 g of CS_2 has a vapour pressure 844.9 torr. Determine the atomicity of sulphur molecule:
   (a) 1 (b) 2
   (c) 4 (d) 8
4. An ideal solution contains two volatile liquids A ($P^\circ = 100$ torr) and B ($P^\circ = 200$ torr). If mixture contain 1 mole of A and 4 mole of B then total vapour pressure of the distillate is:
   (a) 150 (b) 180
   (c) 188.88 (d) 198.88
5. The vapour pressure of two pure liquids A and B, that form an ideal solution are 100 and 900 torr respectively at temperature T. This liquid solution of A and B is composed of 1 mole of A and 1 mole of B. What will be the pressure, when 1 mole of mixture has been vaporized?
   (a) 800 torr (b) 500 torr
   (c) 300 torr (d) None of these
6. Which of the following represents correctly the changes in thermodynamic properties during the formation of 1 mole of an ideal binary solution :

(a) 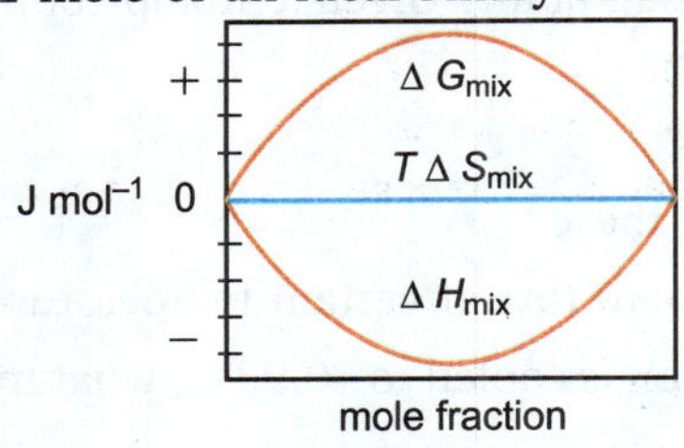

(b) 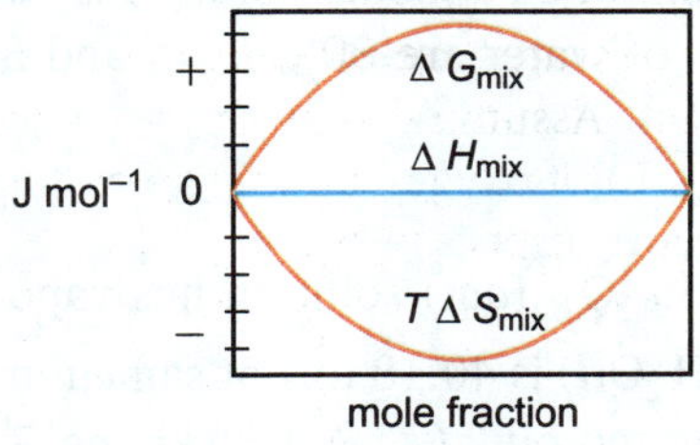

(c) 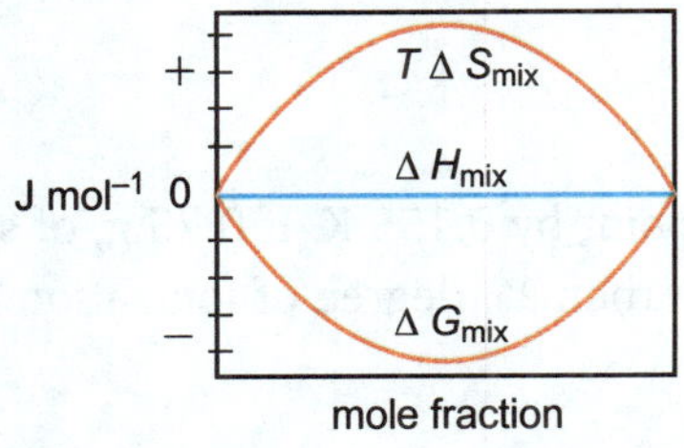

(d) 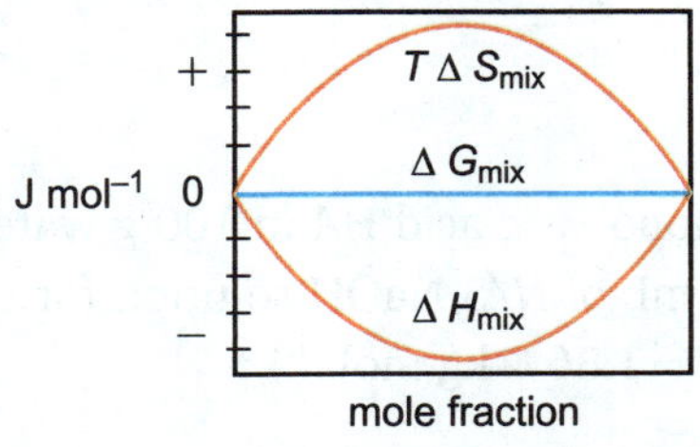

7. A saturated solution of XCl_3 has a vapour pressure 17.20 mm Hg at 20°C, while pure water vapour pressure is 17.25 mm Hg. Solubility product (K_{sp}) of XCl_3 at 20°C is:

(a) 9.8×10^{-2} (b) 10^{-5}

(c) 2.56×10^{-6} (d) 7×10^{-5}

8. A certain non-volatile electrolyte contains 40% carbon, 6.7% hydrogen and 53.3% oxygen. An aqueous solution containing 5% by mass of the solute boils at 100.15°C. Determine the molecular formula of the compound ($K_b = 0.51°C/m$):

(a) HCHO (b) CH_3OH

(c) C_2H_5OH (d) $C_6H_{12}O_6$

9. A 0.10 M solution of a monoprotic acid ($d = 1.01\ g/cm^3$) is 5% dissociated. What is the freezing point of the solution? The mol. mass of the acid is 300 and $K_f(H_2O) = 1.86°C/m$:

(a) −0.189°C (b) −0.194°C

(c) −0.199°C (d) None of these

10. An aqueous solution boils at 101°C. What is the freezing point of the same solution? (Given : $K_f = 1.86°C/m$ and $K_b = 0.51°C/m$)

(a) 3.647°C (b) −3.647°C

(c) −0.364°C (d) None of these

11. An industrial waste water is found to contain 8.2% Na_3PO_4 and 12% $MgSO_4$ by mass in solution. If % ionisation of Na_3PO_4 and $MgSO_4$ are 50 and 60 respectively then its normal boiling point is $[K_b(H_2O) = 0.50\ K\ kg\ mol^{-1}]$:

(a) 102.3°C (b) 103.35°C

(c) 101.785°C (d) None of these

12. Ratio of $\frac{\Delta T_b}{K_b}$ of 10 g AB_2 and 14 g A_2B per 100 g of solvent in their respective, solution (AB_2 and A_2B both are non-electrolytes) is 1 mol/kg in both cases. Hence, atomic wt. of A and B are respectively :

(a) 100, 40 (b) 60, 20

(c) 20, 60 (d) None of these

13. The freezing point of a solution containing 0.2 g of acetic acid in 20.0 g benzene is lowered by 0.45°C. Calculate the degree of dimerization of acetic acid in benzene. K_f for benzene is 5.12 K mol^{-1} kg :

(a) 0.527 (b) 0.80

(c) 0.945 (d) None of these

14. If boiling point of an aqueous solution is 100.1°C, what is its freezing point? Given, enthalpy of fusion and vaporisation of water are 80 cal g^{-1} and 540 cal g^{-1} respectively:

(a) 0.361°C (b) −0.361°C

(c) −3.61°C (d) None of these

15. 100 g of $C_6H_{12}O_6$ (*aq.*) solution has vapour pressure is equal to 40 torr at certain temperature. Vapour pressure of $H_2O(l)$ is 40.18 torr at same temperature. If this solution is cooled to −0.93°C, what mass of ice will be separated out? ($K_f = 1.86\ kg\ mol^{-1}$)

(a) 95.5 g (b) 4.5 g

(c) 45.5 g (d) 47.8 g

16. 1.0 g of a monobasic acid HA in 100 g water lowers the freezing point by 0.155 K. If 0.75 g, of same acid requires 25 mL of $N/5$ NaOH solution for complete neutralisation then %, degree of ionization of acid is (K_f of $H_2O = 1.86\ K\ kg\ mol^{-1}$) :

(a) 20% (b) 25%

(c) 40% (d) 50%

17. 0.1 *M* KI and 0.2 *M* $AgNO_3$ are mixed in 3 : 1 volume ratio. The depression of freezing point of the resulting solution will be $[K_f(H_2O) = 1.86 \text{ K kg mol}^{-1}]$:

(a) 3.72 K (b) 1.86 K
(c) 0.93 K (d) 0.279 K

18. If 0.1 *M* $H_2SO_4(aq.)$ solution shows freezing point –0.3906°C then what is the K_{a_2} for H_2SO_4? (Assume $m = M$ and $K_{f(H_2O)} = 1.86 \text{ K kg mol}^{-1}$)

(a) 0.122 (b) 0.0122
(c) 1.11×10^{-3} (d) None of these

19. A living cell contains a solution which is isotonic with 0.2 *M* glucose solution. What osmotic pressure develops when the cell is placed in 0.05 *M* $BaCl_2$ solution at 300 K?

(a) 1.23 atm (b) 3.69 atm
(c) 6.15 atm (d) None of these

20. What is the osmotic pressure of 0.2 *M* HX (*aq.*) solution at 300 K?
(Given : $K_{a(HX)} = 8 \times 10^{-5}$)

(a) 4.926 atm (b) 0.5024 atm
(c) 5.024 atm (d) None of these

21. A solution contain 8 g of a carbohydrate in 100 g of water has a density 1.025 g/mL and an osmotic pressure of 5 atm at 27°C. What is the molar mass of the carbohydrate?

(a) 387 (b) 374
(c) 3740 (d) None of these

22. Study the following figure and choose the correct options. Assuming complete dissociation of electrolyte:

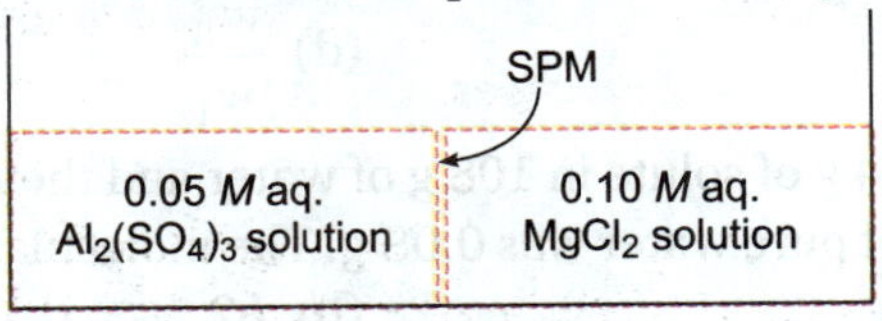

(a) There will be no net moment of any substance across the membrane
(b) $MgCl_2$ will flow towards the $Al_2(SO_4)_3$ solution
(c) $Al_2(SO_4)_3$ will flow towards the $MgCl_2$ solution
(d) The π (osmotic pressure) of 0.1 *M* $MgCl_2$ is higher than the π of 0.05 *M* $Al_2(SO_4)_3$

23. The total vapour pressure of a 4 mole % solution of NH_3 in water at 293 K is 50.0 torr. The vapour pressure of pure water is 17.0 torr at this temperature. Applying Henry's and Raoult's laws, calculate the total vapour pressure for a 5 mole % solution:

(a) 58.25 torr (b) 33 torr
(c) 42.1 torr (d) 52.25 torr

24.-25. The vapour pressure of two pure liquids *A* and *B* which form an ideal solution are 500 and 800 torr respectively at 300 K. A liquid solution of *A* and *B* for which the mole fraction of *A* is 0.60 is contained in a cylinder closed by a piston on which the pressure can be varied. The solution is slowly vaporized at 300 K by decreasing the applied pressure.

24. The composition of vapour when first bubble formed is:

(a) $y_A = 0.6; y_B = 0.4$ (b) $y_A = 0.48; y_B = 0.52$
(c) $y_A = 0.52; y_B = 0.48$ (d) $y_A = 0.5; y_B = 0.5$

25. What is the composition of last droplet of liquid remaining in equilibrium with vapour?

(a) $x_A = 0.6; x_B = 0.4$ (b) $x_A = 0.5; x_B = 0.5$
(c) $x_A = 0.7; x_B = 0.3$ (d) $x_A = 0.3; x_B = 0.7$

Level-3

Passage-1

Lowering in vapour pressure is determined by Ostwald and Walker dynamic method. It is based on the principle, that when air is allowed to pass through a solvent or solution, it takes up solvent vapour with it to get itself saturated at that temperature

I and II are weighted separately before and after passing dry air. Loss in mass of each set, gives the lowering of vapour pressure. The temperature of air, the solution and the solvent is kept constant.

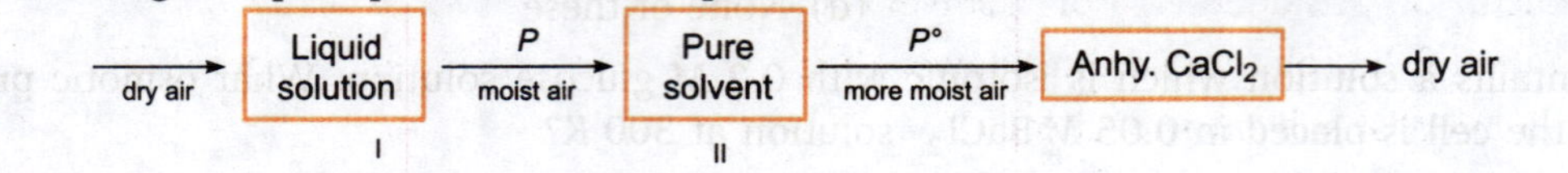

1. Loss in mass of solvent (w_{II}) will be proportional to:

(a) $P° - P$ (b) $P - P°$

(c) $\dfrac{P}{P°}$ (d) $P \times P°$

2. Gain in mass of anhydrous $CaCl_2$ is proportional to:

(a) P (b) $P°$

(c) $P - P°$ (d) $P° - P$

3. $\dfrac{P° - P}{P°}$ is equal to:

(a) $\dfrac{w_I}{w_I + w_{II}}$ (b) $\dfrac{w_{II}}{w_I + w_{II}}$

(c) $\dfrac{w_I}{w_I - w_{II}}$ (d) $\dfrac{w_{II}}{w_I}$

4. Dry air was passed through 9.24 g of solute in 108 g of water and then through pure water. The loss in mass of solution was 3.2 g and that of pure water was 0.08 g. The molecular mass (g/mol) of solute is nearly :

(a) 50 (b) 62

(c) 70 (d) 80

Passage-2

A dilute solution contains 'x' moles of solute A in 1 kg of solvent with molal elevation constant K_b. The solute dimerises in the solution according to the following equation. The degree of association is (α) :

$$2A \rightleftharpoons A_2$$

1. The van't Hoff factor will be:

(a) $i = 1 - 2\alpha$ (b) $i = 1 - \dfrac{\alpha}{2}$

(c) $i = 1 + \dfrac{\alpha}{2}$ (d) $i = 1 + \alpha$

2. The molecular mass observed will be:

(a) greater than actual molecular mass (b) lesser than actual molecular mass

(c) equal to the actual molecular mass (d) can be predicted by the data given

3. The degree of association is equal to:

(a) $\alpha = \dfrac{(K_b x - \Delta T_b)}{\Delta T_b 2}$ (b) $\alpha = \dfrac{2(K_b x - \Delta T_b)}{K_b x}$

(c) $\alpha = 2 + \dfrac{2\,\Delta T_b}{K_b x}$ (d) $\alpha = \dfrac{\Delta T_b}{2K_b x}$

One or More Answer is/are Correct

1. Which of the following statement(s) is/are correct, if intermolecular forces in liquids A, B and C are in the order of $A < B < C$?
(a) B evaporates more readily than A
(b) B evaporates more readily than C
(c) A evaporates more readily than C
(d) all evaporate at same rate at same temperature.

2. When non-volatile solute is added to a pure solvent, the:
(a) vapour pressure of the solution becomes lower than the vapour pressure of the pure solvent
(b) rate of evaporation of solvent is reduced
(c) rate of condensation of vapour is also reduced
(d) none of these

3. The total vapour pressure of a binary solution is given by

$$P = (100\, X_A + 260\, X_B) \text{ mm Hg}$$

where, X_A and X_B are the mole fractions of components A and B. This indicates that the:
(a) vapour pressure of solution is less than the pure B component
(b) vapour pressure of solution is more than that of pure A component
(c) vapour pressure of pure A is 100 mm Hg and that of pure B is 260 mm Hg
(d) the vapour pressure of pure A and B are 260 mm Hg and 100 mm Hg respectively

4. Which of the following is correct for an ideal solution?
(a) $\Delta H_{mix} = 0$ and $\Delta V_{mix} = 0$
(b) $\Delta V_{mix} = 0$ and $\Delta S_{mix} > 0$
(c) $\Delta H_{mix} > 0$ and $\Delta S_{mix} > 0$
(d) $\Delta G_{mix} < 0$ and $\Delta S_{mix} > 0$

5. For a non-ideal solution with a negative deviation:
(a) $\Delta H_{mix} = -$ ve
(b) $\Delta V_{mix} = -$ ve
(c) $\Delta S_{mix} = -$ ve
(d) $\Delta G_{mix} = -$ ve

6. A binary solution of liquids A and B will show positive deviation from Raoult's law if it fulfils the following condition:
(a) $P_A > X_A P_A^\circ$ and $P_B > X_B P_B^\circ$
(b) The intermolecular forces of $A - B < A - A, B - B$
(c) ΔH mixing is positive
(d) ΔV mixing is negative

7. Which of the following statements is/are correct about acetone and trichloromethane mixture?
(a) Mixtures of acetone and trichloromethane shows positive deviation from Raoult's law
(b) The forces of attraction acting between molecules of acetone and trichloromethane in a mixture are stronger than those acting between the molecules in pure acetone
(c) Pure acetone can be obtained by the careful fractional distillation of any mixture of acetone and trichloromethane
(d) When acetone and trichloromethane are mixed, the enthalpy change is negative

8. The azeotropic solutions of two miscible liquids:
(a) can be separated by simple distillation
(b) may show positive or negative deviation from Raoult's law
(c) are supersaturated solution
(d) behave like a single component and boil at a constant temperature

9. For exact determination of molecular mass through colligative properties measurement:
(a) solute must be volatile
(b) solution must be very dilute
(c) solution must be formed by addition of volatile solute
(d) solute must not be dissociated or associated

10. For the depression in freezing point experiment, the correct statement(s) is/are:
(a) vapour pressure of pure solvent is more than that of solution
(b) vapour pressure of pure solvent is less than that of solution
(c) only solute molecules solidify at the freezing point
(d) only solvent molecules solidify at the freezing point

11. The cryoscopic constant value depends upon:
(a) the molar mass of the solute in the solution
(b) the molar mass of the solvent in the solution
(c) the enthalpy of fusion of the solvent
(d) the freezing point of the solvent

12. Consider 0.1 *M* solutions of two solutes *X* and *Y*. The solute *X* behaves as univalent electrolyte, while the solute *Y* dimerises in solution. Select correct statement(s) regarding these solutions:
(a) The boiling point of solution of '*X*' will be higher than that of '*Y*'
(b) The osmotic pressure of solution of '*Y*' will be lower than that of '*X*'
(c) The freezing point of solution of '*X*' will be lower than that of '*Y*'
(d) The relative lowering of vapour pressure of both the solution will be the same

13. Consider following solutions:
(I) 1 *M* glucose (aq)
(II) 1 *M* sodium chloride (aq)
(III) 1 *M* acetic acid in benzene
(IV) 1 *M* ammonium phosphate (aq)
(a) all are isotonic solutions
(b) III is hypotonic of I, II, IV
(c) I, II, IV are hypertonic of III
(d) IV is hypertonic of I, II, III

14. Which of the following statements is (are) incorrect?
(a) 0.1 *M* KCl solution will have the same osmotic pressure as 0.1 *M* glucose solution
(b) 0.1 *M* KCl solution will have the same boiling point as 0.1 *M* urea solution
(c) 0.1 *m* glucose and 0.1 *m* urea are isotonic
(d) 0.1 *m* $MgCl_2$ solution will have less relative lowering of vapour pressure than 0.1 *m* NaCl

15. Consider following solutions:
0.1 *m* $C_6H_5NH_3^+Cl^-$; 0.1 *m* KCl; 0.1 m Glucose; 0.1 *m* $Na_2C_2O_4 \cdot 10H_2O$
(a) the solution with highest boiling point is 0.1 *m* $Na_2C_2O_4 \cdot 10H_2O$
(b) the solution with highest freezing point is 0.1 *m* glucose
(c) 0.1 *m* $C_6H_5NH_3Cl$ and 0.1 *m* NaCl will have the same osmotic pressure
(d) 0.1 *m* glucose solution will have the lowest osmotic pressure

Match the Column

Column-I and Column-II contains four entries each. Entries of column-I are to be matched with some entries of column-II. One or more than one entries of column-I may have the matching with the same entries of column-II.

1.

Column-I	Column-II
(A) *n*-hexane + *n*-heptane	(P) Can be separated by fractional distillation
(B) Acetone + chloroform	(Q) Maximum boiling azeotrope
(C) Chlorobenzene and bromobenzene	(R) Cannot be separated by fractional distillation completely
(D) Ethanol + water	(S) Minimum boiling azeotrope

2.

Column-I	Column-II
(A) $\pi_1 : 0.1 M$ glucose; $\pi_2 : 0.1 M$ urea	(P) π_1 and π_2 are isotonic
(B) $\pi_1 : 0.1 M$ NaCl; $\pi_2 : 0.1 M$ Na_2SO_4	(Q) No net migration of solvent across the membrane
(C) $\pi_1 : 0.1 M$ NaCl; $\pi_2 : 0.1 M$ KCl	(R) π_1 is hypertonic to π_2
(D) $\pi_1 : 0.1 M$ $CuSO_4$; $\pi_2 : 0.1 M$ sucrose	(S) π_1 is hypotonic to π_2

3.

Column-I	Column-II
(A) 0.1 M $Al_2(SO_4)_3$	(P) Solution with highest boiling point
(B) 0.1 M $AlPO_4$	(Q) van't Hoff factor is greater than 1
(C) 0.1 M urea	(R) Solution with lowest osmotic pressure
(D) 0.1 M $MgCl_2$	(S) Solution with lowest freezing point

4.

Column-I (Solute)	Column-II (Van't Hoff factor, *i*)
(A) $AlCl_3$ if $\alpha = 0.8$	(P) $i = 3.4$
(B) $BaCl_2$ if $\alpha = 0.9$	(Q) $i = 2.8$
(C) Na_3PO_4 if $\alpha = 0.9$	(R) $i = 3.8$
(D) $K_4[Fe(CN)_6]$ if $\alpha = 0.7$	(S) $i = 3.7$

5.

Column-I	Column-II
(A) Elevation of B.P.	(P) Colligative property
(B) Osmotic pressure	(Q) Ebullioscopic constant
(C) Relative lowering in V.P.	(R) Berkeley-Heartley method
(D) Depression of F.P.	(S) Ostwald and Walker method

Assertion-Reason Type Questions

Each question contains STATEMENT-1 (Assertion) and STATEMENT-2 (Reason).
Examine the statement carefully and mark the correct answer according to the instructions given below:
(A) If both the statements are TRUE and STATEMENT-2 is the correct explanation of STATEMENT-1
(B) If both the statements are TRUE but STATEMENT-2 is NOT the correct explanation of STATEMENT-1
(C) If STATEMENT-1 is TRUE and STATEMENT-2 is FALSE
(D) If STATEMENT-1 is FALSE and STATEMENT-2 is TRUE

1. STATEMENT-1 : An increase in surface area increases the rate of evaporation.
STATEMENT-2 : Stronger the intermolecular attraction forces, faster is the rate of evaporation at a given temperature.

2. STATEMENT-1 : An ideal solution obeys Raoult's law.
STATEMENT-2 : In an ideal solution, solute-solute as well as solvent-solvent, interactions are similar to solute-solvent interactions.

3. STATEMENT-1 : If a liquid solute more volatile than the solvent is added to the solvent, the vapour pressure of the solution is greater than vapour pressure of pure solvent.
STATEMENT-2 : Vapour pressure of solution is equal to vapour pressure of solvent.

4. STATEMENT-1 : ΔV_{mix} and ΔS_{mix} for an ideal solution is zero.
STATEMENT-2 : $A...B$ interaction in an ideal solution are same as between $A... A$ and $B...B$.

5. STATEMENT-1 : Elevation in boiling point will be high if the molal elevation constant of the liquid is high.
STATEMENT-2 : Elevation in boiling point is a colligative property.

6. STATEMENT-1 : The boiling point of 0.1 M urea solution is less than that of 0.1 M KCl solution.
STATEMENT-2 : Elevation of boiling point is directly proportional to the number of moles of non-volatile solute particles present in the solution.

7. STATEMENT-1 : The observed molar mass of acetic acid in benzene is more than the normal molar mass of acetic acid.
STATEMENT-2 : Molecules of acetic acid dimerise in benzene due to hydrogen bonding.

8. STATEMENT-1 : Addition of ethylene glycol to water lowers the freezing point of water, therefore, used as antifreeze substance.
STATEMENT-2 : Ethylene glycol is soluble in water.

9. STATEMENT-1 : Osmotic pressure is a colligative property.
STATEMENT-2 : Osmotic pressure is developed in a column due to osmosis.

10. STATEMENT-1 : Osmosis involves movement of solvent molecules from lower concentration to higher concentration of solute.
STATEMENT-2 : Solutions having the same osmotic pressure are called isotonic solutions.

11. STATEMENT-1 : Isotonic solutions must have the same molar concentration.
STATEMENT-2 : Solutions which have the same osmotic pressure are known as isotonic solution.

12. STATEMENT-1 : Isotonic solutions do not show phenomenon of osmosis.
STATEMENT-2 : Isotonic solutions have same molar concentration and same vant' Hoff factor at same temperature.

13. STATEMENT-1 : When dried fruits and vegetables are placed in water, they slowly get swollen.
STATEMENT-2 : It happens due to the phenomenon of osmosis.

14. STATEMENT-1 : Reverse osmosis is used to purify sea water.
STATEMENT-2 : Solvent molecules pass from concentrate solution to pure solvent through semipermeable membrane if high pressure ($> \pi$) is applied on solution side.

15. **STATEMENT-1 :** All solutes becomes more soluble in water at higher temperature.
STATEMENT-2 : Solubility of solute depends upon temperature.
16. **STATEMENT-1 :** Henry's law is always applicable for gases.
STATEMENT-2 : Raoult's law is a special case of Henry's law.
17. **STATEMENT-1 :** Increasing pressure on pure water decrease its freezing point.
STATEMENT-2 : Density of water is maximum at 273 K.
18. **STATEMENT-1 :** The molecular mass of acetic acid determined by depression in freezing point method in benzene and water was found to be different.
STATEMENT-2 : Water is polar and benzene is non-polar.
19. **STATEMENT-1 :** If red blood cells were removed from the body and placed in pure water, pressure inside the cell increases.
STATEMENT-2 : The concentration of the salt content in the cells increases.
20. **STATEMENT-1 :** Azeotrope is a binary mixture formed by ideal solutions.
STATEMENT-2 : Azeotrope boils with unchanged composition.

Level-4

Subjective Problems

1. The vapour pressure of two pure liquids A and B are 5 and 10 torr respectively. Calculate the total pressure of the solution (in torr) obtained by mixing 2 mole of A and 3 mole of B.
2. The vapour pressure of two pure liquids A and B are 50 and 40 torr respectively. If 8 moles of A is mixed with x moles of B, then vapour pressure of solution obtained is 48 torr. What is the value of x.
3. The vapour pressure of a liquid solution containing A and B is 99 torr. Calculate mole % of B in vapour phase.
   (Given : $P_A^\circ = 100$ torr ; $P_B^\circ = 80$ torr)
4. If 30 g of a solute of molecular mass 154 is dissolved in 250 g of benzene. What will be the elevation in boiling point of the resulting solution ?
   (Given : $K_{b(C_6H_6)} = 2.6\,\text{K kg mol}^{-1}$)
5. Calculate elevation in boiling point for 2 molal aqueous solution of glucose.
   (Given : $K_{b(H_2O)} = 0.5\,\text{K kg mol}^{-1}$)
6. Calculate depression of freezing point for 0.56 molal aq. solution of KCl.
   (Given : $K_{f(H_2O)} = 1.8\,\text{K kg mol}^{-1}$).
7. What is the maximum value of van't Hoff factor for $AlCl_3$?
8. A solution containing 500 g of a protein per litre is isotonic with a solution containing 3.42 g sucrose per litre. The molecular mass of protein in 5×10^x, hence x is.
9. An aqueous solution of urea has a freezing point of $-0.515°$ C. Predict the osmotic pressure (in atm) of the same solution at 37° C.
   (Given : $K_{f(H_2O)} = 1.86\,\text{K kg mol}^{-1}$)
10. 0.2 M aq. solution of KCl is isotonic with 0.2 M K_2SO_4 at same temperature.
    What is the van't Hoff factor of K_2SO_4 ?

Numerical Problems

(If the numerical value has more than two decimal places, truncate/round-off the value to TWO decimal places.)

1. When 50 g of non-volatile, non-electrolyte solute was dissolved in 390 g benzene, it causes decrease in vapour pressure by 5% then find the molar mass of solute (in g/mol).
2. Vapour pressure of a volatile substance A is 400 mm of Hg at 298 K. When some another volatile solvent B (V.P. = 500 mm at 298 K) is added to A, vapour pressure of the resulting solution was found to be 470 mm of Hg. To the resulting solution a third volatile solvent C (V.P. = 600 mm of Hg at 298 K) is added so that the vapour pressure of the resulting solution becomes equal to 496 mm of Hg.
   Determine mole fraction of A in the vapour phase in equilibrium with the final solution.
3. The vapour pressure of an aqueous solution is found to be 750 torr at a temperature T and the same solution shows an elevation in boiling point equal to 0.78 K. If T is the boiling point of pure water, then the atmospheric pressure (in torr) should be: [Given : K_b of water = 0.52 K-kg/mol]
4. A solution prepared by dissolving 38.2 g of $Na_2B_4O_7 \cdot xH_2O$ in 250 g water has boiling point 100.582°C. Find the value of x in the formula of salt.
   [K_b of the water is 0.52 K-kg mol^{-1}, Atomic mass of B = 11, Na = 23]

5. How many grams of sucrose (molecular weight = 342) should be dissolved in 100 g water in order to produce a solution with a 104.76°C difference between the normal freezing point and the normal boiling point temperature?
(Given : $K_f = 1.86$ K - kg mol^{-1}, $K_b = 0.52$ K - kg mol^{-1})

6. 18 g of glucose ($C_6H_{12}O_6$) is dissolved in 1 kg of water. If the solution is cooled to –0.2325° C, then how many grams of ice would separate?
[Given : $K_f(H_2O) = 1.86$ K - kg mol^{-1}]

7. The osmotic (in atm) pressure of 2.235% (w/V) KCl solution at 27°C is:

8. If 40% dimer formation of benzoic acid takes place in liquid benzene then find the observed molar mass of benzoic acid mixture (in g/mol).

9. For a sparingly solution salt AB_2, concentration of AB_2 in aqueous solution is written as $y \times 10^{-3}$ M. If saturated aqueous solution shows elevation in boiling point 0.0065 K then find 'y'.
[Given : Molal elevation constant (K_b) of H_2O = 0.5 K - kg/mol, Solubility product of AB_2 is 2.56×10^{-7}]

10. One litre solution is prepared by dissolving some solid lead-nitrate in water. The solution was found to boil at 100.15°C. To the resulting solution 0.2 mole NaCl was added, the resulting solution was found to freeze at –0.83° C. If the solubility product of $PbCl_2$ is $y \times 10^{-5}$ then find the value of y.
[Given : $K_b = 0.5$ K - kg mol^{-1} and $K_f = 1.86$ K - kg mol^{-1}. Assume molality to be equal to molarity]

Answers

Level 1

1. (d)	2. (b)	3. (a)	4. (c)	5. (b)	6. (c)	7. (a)	8. (b)	9. (d)	10. (b)
11. (d)	12. (c)	13. (c)	14. (c)	15. (b)	16. (b)	17. (c)	18. (a)	19. (a)	20. (a)
21. (a)	22. (d)	23. (a)	24. (d)	25. (d)	26. (a)	27. (c)	28. (b)	29. (a)	30. (a)
31. (b)	32. (d)	33. (b)	34. (d)	35. (d)	36. (c)	37. (b)	38. (d)	39. (c)	40. (d)
41. (a)	42. (a)	43. (c)	44. (a)	45. (c)	46. (a)	47. (d)	48. (c)	49. (c)	50. (a)
51. (d)	52. (d)	53. (c)	54. (d)	55. (b)	56. (c)	57. (a)	58. (b)	59. (d)	60. (b)
61. (b)	62. (d)	63. (a)	64. (d)	65. (b)	66. (b)	67. (d)	68. (b)	69. (b)	70. (a)
71. (d)	72. (c)	73. (b)	74. (d)	75. (a)	76. (b)	77. (a)	78. (b)	79. (c)	80. (b)
81. (a)	82. (b)	83. (c)	84. (a)	85. (b)	86. (d)	87. (d)	88. (c)	89. (a)	90. (d)
91. (d)	92. (d)	93. (b)	94. (c)	95. (d)	96. (b)	97. (b)	98. (a)	99. (a)	100. (a)
101. (a)	102. (b)	103. (b)	104. (c)	105. (d)	106. (c)	107. (a)	108. (c)	109. (b)	110. (a)
111. (c)	112. (a)	113. (b)	114. (a)	115. (d)	116. (b)	117. (c)	118. (d)	119. (d)	120. (a)
121. (b)	122. (b)	123. (a)	124. (d)	125. (a)					

Level-2

1. (b)	2. (b)	3. (d)	4. (c)	5. (c)	6. (c)	7. (d)	8. (d)	9. (c)	10. (b)
11. (c)	12. (b)	13. (c)	14. (b)	15. (d)	16. (b)	17. (d)	18. (b)	19. (a)	20. (c)
21. (b)	22. (d)	23. (a)	24. (b)	25. (c)					

Level-3

Passage–1:	1. (a)	2. (b)	3. (b)	4. (b)
Passage–2:	1. (b)	2. (a)	3. (b)	

One or More Answers is/are Correct

1.	(b,c)	2.	(a,b,c)	3.	(a,b,c)	4.	(a,b,d)	5.	(a,b,d)	6.	(a,b,c)	7.	(b,d)	8.	(b,d)
9.	(b,d)	10.	(a,d)	11.	(b,c,d)	12.	(a,b,c)	13.	(b,c,d)	14.	(a,b,d)	15.	(a,b,c,d)		

Match the Column

1. A→ P; B→ Q, R; C→ P; D→ R, S
2. A→ P, Q; B→ S; C→ P, Q; D→ R
3. A→ P, Q, S; B→ Q; C→ R; D→ Q
4. A→ P; B→ Q; C→ S; D→ R
5. A→ P, Q; B→ P, R; C→ P, S; D→ P

Assertion-Reason Type Questions

1.	(C)	2.	(A)	3.	(C)	4.	(D)	5.	(B)	6.	(A)	7.	(A)	8.	(B)	9.	(B)	10.	(B)
11.	(D)	12.	(A)	13.	(A)	14.	(B)	15.	(D)	16.	(D)	17.	(C)	18.	(A)	19.	(C)	20.	(D)

Level-4

Subjective Problems

1.	(8	2.	2	3.	4	4.	2	5.	1	6.	2	7.	4	8.	4	9.	7	10.	2

Numerical Problems

1.	190.00	2.	0.19	3.	770.25	4.	10.00	5.	68.40	6.	200.00	7.	14.78	8.	152.50
9.	01.00	10.	01.46												

Hints & Solutions

Level 1

1. (d) $x(l) \xrightleftharpoons{Eq^m} x(g)$, $K_p = Vp$ of liq, $\Delta_r H = +\text{ve}$

$T\downarrow, Vp\downarrow$

2. (b) Given $P_1 = 19$ mm Hg, $P_2 = 760$ mm Hg;

$\Delta H_{\text{vap.}} = 40670$ J/mol

Applying Clausius-Clapeyron's equation

$$\log \frac{P_2}{P_1} = \frac{\Delta H_{\text{vap}}}{2.303 \times R}\left(\frac{T_2 - T_1}{T_1 T_2}\right)$$

$$\text{or } \log \frac{760}{19} = \frac{40670}{2.303 \times 8.3}\left(\frac{373 - T_1}{T_1 \times 373}\right)$$

on solving, we get $T_1 = 291.4$ K

3. (a) Initial mass of $H_2O(l) = 18$ g;

$$n_{H_2O}(g) = \frac{\left(\frac{24.63}{760}\right) \times 7.6}{0.0821 \times 300} \Rightarrow 0.01$$

$$\% \text{ wt. of } H_2O \text{ vaporised} = \frac{0.01 \times 18}{18} \times 100$$

$\Rightarrow$ 1%

4. (c) Raoult's law is applicable for ideal solution.

5. (b) $P = P_A^\circ x_A + P_B^\circ x_B = P_B^\circ + x_A(P_A^\circ - P_B^\circ)$

$[\because x_B = 1 - x_A]$

6. (c) $$y_A = \frac{P_A^\circ \times x_A}{P_A^\circ x_A + P_B^\circ x_B} = \frac{x_A}{x_A + x_B\left(\frac{P_B^\circ}{P_A^\circ}\right)}$$

$[\because P_A^\circ > P_B^\circ]$

$$\therefore \frac{y_A}{x_A} = \frac{1}{x_A + x_B\left(\frac{P_B^\circ}{P_A^\circ}\right)} > 1 \Rightarrow y_A > x_A \quad \ldots(1)$$

$$\text{Similarly, } y_B = \frac{P_B^\circ \times x_B}{P_A^\circ x_A + P_B^\circ x_B} = \frac{x_B}{x_B + x_A\left(\frac{P_A^\circ}{P_B^\circ}\right)} \quad \ldots(2)$$

$\therefore \quad y_B < x_B$

$\therefore \quad \frac{y_A}{y_B} > \frac{x_A}{x_B}$

7. (a) $$y_A = \frac{P_A^\circ x_A}{P_A^\circ x_A + P_B^\circ x_B} = \frac{P_A^\circ x_A}{P_B^\circ + (P_A^\circ - P_B^\circ) x_A}$$

$$\frac{1}{y_A} = \frac{P_B^\circ x_B + P_A^\circ x_A}{P_A^\circ x_A} = 1 + \left(\frac{P_B^\circ}{P_A^\circ}\right)\left(\frac{1 - x_A}{x_A}\right)$$

$$= 1 - \frac{P_B^\circ}{P_A^\circ} + \frac{1}{x_A}\left(\frac{P_B^\circ}{P_A^\circ}\right)$$

$$\therefore \text{ Slope} = \frac{P_B^\circ}{P_A^\circ}$$

8. (b) $P_{\text{Solution}} = P^\circ_{\text{Solution}} X_{\text{Solvent}}$

$$\frac{P^\circ - P}{P^\circ} = X_{\text{Solute}}$$

9. (d) Solution X is unsaturated so v.p. will be more, solution Y and Z are saturated so v.p. of Y = v.p. of Z and 2 gm of solute would be present in form of solid in system Z.

10. (b) Highest vapour pressure for lowest boiling point.

11. (d) $$\frac{P^\circ - P}{P^\circ} = x_{\text{solute}} \text{ or } \frac{P^\circ - P}{P^\circ} = \frac{n}{n + N}$$

$$\Rightarrow \frac{0.1}{0.1 + 9.9} \Rightarrow 0.01; \; P = 0.99\, P^\circ$$

12. (c) $80 = P^\circ(0.8) \Rightarrow P^\circ = 100$ torr.

13. (c) Given $P_A = 750$ mm Hg

$\because$ 373 K is boiling point of water.

Thus, $P_A^\circ = 760$ mm Hg

$$m = \left(\frac{P^\circ - P}{P}\right) \times \frac{1000}{M_{\text{solvent}}}$$

$$\Rightarrow \frac{10}{750} \times \frac{1000}{18} \Rightarrow 0.74$$

14. (c) Normal boiling point of water is 100°C, hence $P^\circ(H_2O) = 760$ torr; $M(H_2O) = 18$ g/mol

$$m = \left(\frac{P^\circ - P}{P}\right) \times \frac{1000}{M_{H_2O}}$$

$\Rightarrow P = 746.5; \; \Delta P = 13.45$ torr

15. (b) $P_s = \left(\frac{80}{100}\right) P^\circ, w = ?,$

$M_{\text{solute}} = 40, \; w = 114$ g, $M_{\text{solvent}} = 114$

$$\because \frac{P^\circ - P_s}{P_s} = \frac{w \times M_{\text{solvent}}}{M_{\text{solute}} \times W}$$

$$\text{or } \frac{P^\circ - (80/100)P^\circ}{(80/100)P^\circ} = \frac{w \times 114}{40 \times 57};$$

$\therefore \quad w = 5$ g

16. (b) $\frac{P^\circ - P}{P^\circ} = \frac{n}{n+N} \approx \frac{n}{N}$ (for very dilute solution)

$$2 \times \frac{n}{N_A} = \frac{n}{N_B}$$

$$\Rightarrow \quad M_B = 2M_A$$

17. (c) We know that

$$Y_A = \frac{P_A^\circ X_A}{P_{\text{total}}} \quad \text{or} \quad \frac{Y_A}{X_A} = \frac{P_A^\circ}{P_{\text{total}}}$$

$$\because \; P_A^\circ > P_{\text{total}} \quad \text{so} \quad \frac{Y_A}{X_A} > 1 \quad \text{or} \quad Y_A > X_A$$

18. (a) Moles of $A = \frac{20}{40} \Rightarrow 0.5$;

moles of $B = \frac{20}{80} \Rightarrow 0.25$

$$x_A = \frac{0.5}{0.5 + 0.25} = 0.67; \quad x_B = 0.33$$

$$P_{\text{total}} = P_A^\circ X_A + P_B^\circ X_B$$

$$\Rightarrow 100 \times 0.67 + 40 \times 0.33 \Rightarrow 80 \text{ torr}$$

19. (a) Since, $P = X_A P_A^\circ + X_B P_B^\circ$, we have

$$\left(\frac{1}{1+3}\right) P_A^\circ + \left(\frac{3}{1+3}\right) P_B^\circ = 550 \text{ mm Hg};$$

$$\left(\frac{1}{1+4}\right) P_A^\circ + \left(\frac{4}{1+4}\right) P_B^\circ = 560 \text{ mm Hg}$$

That is, $0.25\, P_A^\circ + 0.75\, P_B^\circ = 550$ mm Hg;

$0.20\, P_A^\circ + 0.8\, P_B^\circ = 560$ mm Hg

Solving for P_A° and P_B°, we get;

$P_A^\circ = 400$ mm Hg and $P_B^\circ = 600$ mm Hg

20. (a) $y_A = \frac{P_A}{P} \Rightarrow \frac{P_A^\circ x_A}{P}$ and $y_B = \frac{P_B^\circ x_B}{P}$;

$$\frac{y_A}{y_B} = \frac{P_A^\circ}{P_B^\circ} \times \frac{x_A}{x_B} \quad \Rightarrow \quad \frac{4}{3} = \frac{1}{3} \times \frac{x_A}{(1 - x_A)}$$

$$x_A = \frac{4}{5} \quad \text{or} \quad x_B = \frac{1}{5}$$

21. (a) $x_A = \frac{1}{1+3} \Rightarrow \frac{1}{4}$; $x_B = \frac{3}{4}$;

$$\frac{y_A}{y_B} = \frac{P_A^\circ}{P_B^\circ} \times \frac{x_A}{x_B}$$

$$\frac{y_A}{(1 - y_A)} = \frac{1}{3} \times \frac{1}{3}$$

$$\Rightarrow \frac{1}{9} \quad \text{or} \quad y_A = \frac{1}{10}$$

22. (d)

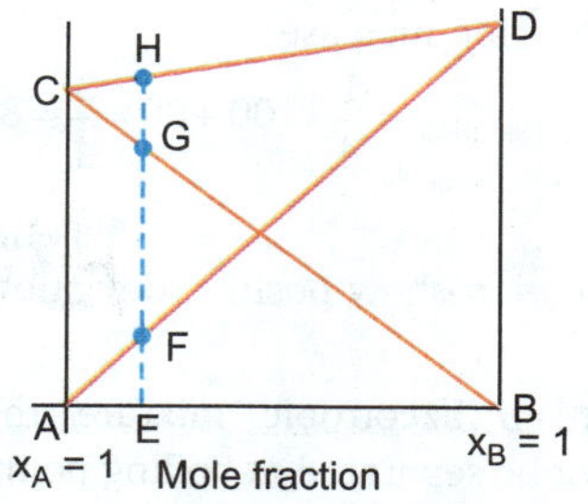

Plots AD show vapour pressure of B.

Plots BC show vapour pressure of A.

Plot CD show vapour pressure of liquid solution containing A and B volatile components.

$$P = P_A + P_B$$

$$\therefore \quad EH = EF + EG$$

23. (a) $\because \; (T_b)_{\text{Solvent}} < (T_b)_{\text{Solution}}$;

$\therefore \; \Delta S_{\text{Solv.}} > \Delta S_{\text{Solution}}$

$(\Delta H_{\text{Solution}}) = (\Delta H)_{\text{Solvent}}$

[Only solvent participate in vaporisation]

[Due to elevation in BP of solution]

24. (d) $(\Delta_f S)_{\text{Solution}} > (\Delta_f S)_{\text{Solvent}}$

[Since solution has more randomness than solvent]

$(\Delta_f H)_{\text{Solution}} \approx (\Delta_f H)_{\text{Solvent}}$

[Similar forces of intermolecular forces are involved in solution and solvent]

Due to presence of non-volatile solute (Sugar), freezing point of solution decrease.

25. (d) $(\Delta H_{\text{Vap}})_{\text{Solvent}} \approx (\Delta H_{\text{Vap}})_{\text{Solution}}$

[Because similar intermolecular forces are involved in both cases].

$\because \; (T_b)_{\text{Solvent}} > (T_b)_{\text{Solution}}$

$(\Delta S_{\text{Vap}})_{\text{Solution}} < (\Delta S_{\text{Vap}})_{\text{Solvent}}$

26. (a) Elevation in boiling point $\propto$ concentration of a solution. Thus, the order of concentration of solutions is I < II < III.

27. (c) $T_B^\circ > T_A^\circ$ [From diagram] $B \rightarrow$ Residue

$A \rightarrow$ Distillate

28. (b) The mixture will show positive deviation from Raoult's law hence boiling point of an azeotropic mixture of water-ethanol is less than that of both water and ethanol.

29. (a) $\Delta H_{\text{Solution}} = \Delta H_1 + \Delta H_2 + \Delta H_3$

30. (a) $P_{\text{Ideal solution}} = \frac{1}{3} \times 150 + \frac{2}{3} \times 300 = 250$

$\because \; P_{\text{Solution}} < P_{\text{Ideal solution}}$; (–) ve deviation.

31. (b) When $A - B$ interaction is weaker than $A - A$ and $B - B$ interaction, then mixture of A and B show positive deviation.

32. (d) On changing external pressure, composition of azeotrope alters.

33. (b) After mixing HNO_3 and H_2O, intermolecular force of attraction increase.

34. (d) $\because$ $P_{\text{Ideal solution}} = \frac{1}{4} \times 100 + 80 \times \frac{3}{4} = 85 \text{ mmHg}$

$< P_{\text{Solution}}$ (90 mm Hg)

$\therefore$ Solution show positive deviation from Raoult's law.

35. (d) From an azeotropic mixture the components cannot be separated at boiling point because it is a constant temperature boiling mixture.

36. (c) See solution of Q. 35.

37. (b) Show negative deviation from Raoult's law.

38. (d) $P = P_A^\circ + P_B^\circ = 300$ torr.

39. (c) For two immiscible liquid;

$$P_A^\circ = P_{\text{total}} - P_{H_2O}^\circ = 748 - 648 \Rightarrow 100$$

$$\frac{W_A}{W_B} = \frac{P_A^\circ M_A}{P_B^\circ M_B}; \quad M_A = \frac{1.25}{1} \times \frac{648 \times 18}{100}$$

$$\Rightarrow 145.8$$

40. (d) $\frac{W_A}{W_B} = \frac{P_A^\circ}{P_B^\circ} \times \frac{M_A}{M_B} \Rightarrow \frac{0.7}{7} \times \frac{112.5}{18}$

$$\Rightarrow 0.625$$

$$\% \frac{W_A}{W_A + W_B} \times 100 = \frac{0.625}{1.625} \times 100$$

$$\Rightarrow 38.46$$

41. (a) Lowering in vapour pressure is a colligative property.

42. (a) $i = 1 + (n-1)\alpha$ so $n = \frac{i + \alpha - 1}{\alpha}$

43. (c) $i = \frac{C(1-\alpha) + \frac{C\alpha}{n}}{C} \Rightarrow i = 1 - \alpha + \frac{\alpha}{n}$

44. (a) For dissociation $i = 1 + (n-1)\alpha > 1$

For dissociation $i = 1 + \left(\frac{1}{n} - 1\right)\alpha < 1$

45. (c) For high v.p. concentration of solute should be low and temperature should be high.

46. (a) On addition of strong electrolyte, vapour pressure decrease. On addition of water, vapour pressure increases due to increase in X_{Solvent}.

47. (d) Colligative property $\propto$ moles of non-volatile solute.

48. (c) $\frac{P^\circ - P}{P^\circ} = \frac{10}{50} = \frac{n}{n+12}$ $\quad \therefore n = 3$

(Number of ions of solute), moles of $K_2SO_4 = 1$

49. (c) $mi = \frac{P^\circ - P}{P} \times \frac{1000}{M_{\text{solvent}}}$;

$$7\,m = \frac{(20.0126 - 20) \times 1000}{20 \times 18};$$

$$m = 5 \times 10^{-3}$$

50. (a) $\Delta T_b = 100.5 - 100 = K_b \times 1 \quad \therefore K_b = 0.5$

51. (d) $\Delta T_b = K_b . m \Rightarrow 3.63 \times \frac{0.616/154}{15} \times 1000$;

$$T_b = 61.7 + 0.968$$

$$\Rightarrow 62.67°C$$

52. (d) $T_b = 80.26, \quad \Delta T_b = 0.16$;

$$0.16 = 2.53 \times \frac{0.26 / M}{11.20} \times 1000; \; M \approx 367$$

that is almost molar mass of $C_{20}H_{16}Fe_2$.

53. (c) $80.23 - 80.10 = 2.53 \times \left[\frac{0.64/x}{100} \times 10^3\right]$

$$x = 125$$

54. (d) ΔT_b (normal) $= K_b m = 2.53 \times 1 = 2.53$ K;

$$i = \frac{\Delta T_{b(\text{obs})}}{\Delta T_{b(\text{nor})}} = \frac{1.518}{2.53} = 0.6$$

$$i = 1 - \left(1 - \frac{1}{n}\right)\alpha; \quad \alpha = 0.8$$

55. (b) $\Delta H_{\text{vap.}} = \frac{M_{\text{solvent}} \times R \times T^{\circ 2}_{b\,(\text{solvent})}}{1000 \times K_b}$;

$$\therefore \Delta H_{\text{vap.}} = \frac{92 \times 8.314 \times (383.7)^2}{1000 \times 3.32}$$

$$= 33.91 \text{ kJ/mol} \approx 34 \text{ kJ/mol}$$

56. (c) B.P. $\propto$ moles of non-volatile solute.

57. (a)

	XY_2	$\rightleftharpoons$	X^{2+}	+	$2Y^-$
Initially	1		0		0
at equilibrium	$1-\alpha$		α		2α

Total no. of moles $= 1 - \alpha + \alpha + 2\alpha = 1 + 2\alpha$

$$i = \frac{\text{Normal molar mass}}{\text{Observed molar mass}}; \quad \frac{1+2\alpha}{1} = \frac{164}{65.6}$$

$$\therefore \quad \alpha = 0.75; \; \% \alpha = 75\%$$

58. (b) $\Delta T_b = K_b \times m$

$$\therefore m = \frac{y}{x} \quad \text{...(1)}$$

$$\therefore \Delta T_f = K_f m \quad \text{...(2)}$$

$$\Delta T_f = \left(\frac{y}{x}\right) z$$

59. (d) Liquid solution $\rightleftharpoons$ solid solvent.

60. (b) $\Delta T_f = 2.304; \; m = \frac{2.304}{14.4} \Rightarrow 0.16$;

$$m = \frac{n_{\text{solute}}}{W_{\text{solvent}}} \times 1000$$

$$0.16 = \frac{2.6}{M} \times \frac{1000}{100}; \; M \Rightarrow 162.5$$

61. (b) $n_{C_4H_{10}} = \frac{10.44}{58} \Rightarrow 0.18$;

$$m = \frac{0.18}{200} \times 1000 \Rightarrow 0.9$$

$$\Delta T_f = 5.12 \times 0.9 \Rightarrow 4.608;$$

$$T_f = 5.5 - 4.608 \Rightarrow 0.892°C$$

62. (d) $-4°F = -20°C$; $\therefore \Delta T_f = 20°C$

$\Delta T_f = K_f.m \Rightarrow m = \frac{20}{1.86} = 10.75$

mass of C_2H_5OH required

$\Rightarrow 10.75 \times 46 \Rightarrow 494.5$ g

63. (a) $4.4 - 5.5 = K_f\left[\frac{2.4/154}{75} \times 10^3\right]$ $\therefore K_f = 5.3$

64. (d) $0.465 = 1.86 \times \left[\frac{1.8/x}{40} \times 10^3\right]$

$x = 180$ g/mol $\therefore C_6H_{12}O_6$

65. (b) $T_f = \frac{\Delta H}{\Delta S}$

66. (b) $\because \left(\frac{\Delta H_f}{\Delta S_f}\right)_{Solution} < \left(\frac{\Delta H_f}{\Delta S_f}\right)_{Solvent}$

67. (d) $\Delta T_f = K_f.m \Rightarrow 0.93 = 1.86 \times \frac{36}{M} \times \frac{1}{1.2}$,

$M = 60$ and $30 \times n = 60$; $\therefore n = 2$

so compound is $C_2H_4O_2$.

68. (b) $0.93 = 1.86 \times \left[\frac{7/x}{93/1000}\right]$

$x = 150.5$ g mol^{-1}.

69. (b) It has a very high cryoscopic constant.

70. (a) Maximum ions minimum T_f.

71. (d) Maximum T_f for minimum ions.

72. (c) $3.72 = 1.86\,[im]$

$i = 2$ $[Pt(H_2O)_3Cl_3]Cl \cdot 3H_2O$

73. (b) $\Delta T_f = K_f \times \text{molality} \times i$;

$0.558 = 1.86 \times 0.1 \times i$ or $i = 3$

Thus complex is $[CoCl.xNH_3].Cl_2$. Since co-ordination number of CO is six, thus complex is $[CO(NH_3)_5Cl]Cl_2$.

74. (d) Depression in freezing point, $\Delta T_f = iK_f m$. The value of van't Hoff factor (i) is minimum for the glucose, which is a non-electrolyte. Hence, aqueous solution of glucose has highest freezing point.

75. (a) Molality of solution should be same $m_1 = m_2$;

$\frac{4}{60} \times \frac{1000}{96} = \frac{10}{M_B} \times \frac{1000}{90}$ or $M_B = 160$

76. (b) $\Delta T_f = K_f.m \Rightarrow 1.86 = K_f \times 1$

$\Rightarrow K_f(H_2O) = 1.86$

for HCOOH; $0.02046 = 1.86 \times 0.01(1+\alpha)$;

$\alpha = 0.1$; so $[H^+] = \alpha \Rightarrow 10^{-3}$; pH = 3

77. (a) The reaction when KI is added in HgI_2

$$HgI_2 + 2KI \longrightarrow K_2HgI_4 \rightleftharpoons 2K^+ + HgI_4^{2-}$$

Since, number of ions decreased, so freezing point increases.

78. (b) $\Delta T_f = mK_f i$

$\Delta T_b = mK_b i$

$M_{obs} = \frac{M}{i}$

$i\downarrow$, So M_{obs} increases

79. (c) To use car; $\Delta T_f = T_f^\circ - T_f = 9.3$ K

$\therefore \Delta T_f = K_f m \Rightarrow 9.3 = 1.86\,m \Rightarrow m = 5$

$\therefore$ number of moles of ethylene glycol added to 5 L of water of radiator $= 5 \times 5 = 25$

or $25 \times 62 \Rightarrow 1550$ g

80. (b) Given $K_1 = 1.86$ K kg mol^{-1}, $m = 0.01$, $\Delta T_f = 0.0194°C$

$\therefore \Delta T_f = i.K_f m$; $\frac{0.0194}{1.86 \times 0.01} = i$

or $i = 1.043$ or $\alpha = 0.043$

81. (a) $3.84 = i \times 5.12 \times 1$

$\therefore i = 0.75 < 1$

$2.53 = i \times 2.53 \times 1$

$i = 1$

82. (b) $\Delta T_f = (1+0.5) \times 1.86 \times 0.5 = 1.395$

$\therefore T_f = 273 - 1.395 = 271.60$

83. (c) $\Delta T_f = K_f.m.i$;

$0.87 = 1.86 \times i \times \left(\frac{1}{345.5} \times \frac{1000}{25}\right)$; $i = 4$

84. (a) $\Delta T_f = K_f\left(\frac{n_1 + n_2}{W_{solvent}} \times 1000\right)$;

$$\Delta T_f = 1.86\left(\frac{\frac{3}{60} + \frac{1.8}{180}}{95.2} \times 1000\right) \Rightarrow 1.172$$

$T_f^\circ - T_f = 1.172°C$; $T_f = -1.172°C$

85. (b) $\Delta T_f = K_f.m.i \Rightarrow i = 0.633$;

$i = 1 - \alpha + \frac{\alpha}{2} \Rightarrow \alpha = 0.73$

86. (d) Maximum freezing point for minimum ion.

87. (d) $i = 1 + \alpha \Rightarrow 1.25$

$\Delta T_f = K_f.m.i \Rightarrow 1.86 \times 0.2 \times 1.25$

or $\Delta T_f = 0.465$ or $T_f = -0.465°C$

88. (c) $0.01 \times K_b \times 2 = K_b \times x \times 1$

$x = 0.02\,M$

89. (a) $\frac{P^\circ - P}{P^\circ} = \frac{in}{N} = \frac{2 \times m \times M_{Solution}}{10^3}$

$= \frac{2 \times 1 \times 18}{1000} = \frac{9}{250} = 0.036$

$\Delta T_f = k_f(m), \frac{\Delta T_f}{K_f} = 2 = im$, $m = 1$, $i = 2$

90. (d) Depression in freezing point

$= K_f m$ (as $K_f\uparrow, \Delta T_f\uparrow$)

K_f depends upon the solvent.

91. (d) $T - 100 = 0.1(1 + 3 \times 0.9) \times 0.52$

$T = 373.19$ K

92. (d) $\pi \propto$ No. of ions for 1 mole solute.

93. (b) Water (solvent) flows from lower concentration to higher concentration.

94. (c) Solvent molecules are moving from solution B into solution A hence, osmotic pressure of solution A immediately begins to decrease.

95. (d) Solutions having same no. of ions are isotonic

Urea $(i=1)$, NaCl $(i=2)$, $MgCl_2 (i=3)$, $Na_2SO_4 (i=3)$, $Ca(NO_3)_2 (i=3)$.

96. (b) $\pi_{Glucose} = \pi_{Non\text{-}electrolyte}$

$$0.05 \times RT = \left(\frac{3}{x}\right) RT$$

$$\Rightarrow \quad x = 60, \; n = \frac{MFM}{EFM} = \frac{60}{30} = 2$$

$$\therefore \quad M.F. = (E.F.)_2, \; C_2H_4O_2$$

97. (b) A semipermeable membrane allow the passage of only solvent molecules.

98. (a) In osmosis, solvent molecules move from lower concentration (higher vapour pressure) to higher concentration (lower vapour pressure)

99. (a) $\pi \propto i$ [Equal concentration]

$AlCl_3 (i=4)$, Urea $(i=1)$, $BaCl_2 (i=3)$

100. (a) $\pi \propto i$ [Equal concentration]

Aluminium sulphate $Al_2(SO_4)_3$ $(i=5)$, Barium chloride $BaCl_2 (i=3)$,

Sodium sulphate $Na_2SO_4 (i=3)$.

101. (a) For $x (i=2)$ and $Y (i=1)$

$$\therefore \; (T_b)_x > (T_b)_y \; ; \; \pi_x > \pi_y \; ; \; (T_f)_x < (T_f)_y$$

102. (b) $K_3[Fe(CN)_6] \rightleftharpoons 3K^+ + [Fe(CN)_6]^{3-}$

At $t=0$ 1   0   0

After ionization $(1-\alpha)$ 3α α; $i = 1 + 3\alpha$

Since, $i = \dfrac{M_{normal}}{M_{abnormal}}$;

$$\therefore \quad \frac{1+3\alpha}{1} = \frac{M_{normal}}{M_{abnormal}}$$

103. (b) $\because$ [Urea] = [Glucose],

Isotonic solutions

104. (c) $\pi_1 = \pi_2$; $\dfrac{5}{342} \times \dfrac{1000}{100} = \dfrac{1}{X} \times \dfrac{1000}{100}$

$$\Rightarrow \quad X = 68.4$$

105. (d)

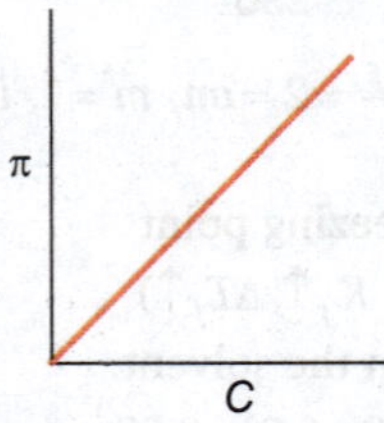

$$\pi = \frac{W}{M} \times \frac{RT}{V} \quad \text{or} \quad \frac{\pi}{C} = \frac{RT}{M}$$

$$\therefore \quad \text{slope} = \frac{RT}{M}$$

or $M = \dfrac{0.082 \times 293}{4.65 \times 10^{-3} \times 10^{-3}}$

$$\Rightarrow \quad 5.16 \times 10^6 \text{ g}$$

106. (c) $\pi_{Sucrose} = \pi_{Glucose}$

$$\frac{34.2}{342} \times RT = x \times RT$$

$\therefore$ Concentration of glucose $= 0.1 \times 180 = 18$ g/L

107. (a) Berkley-Hartely's method is adopted to determine osmotic pressure.

108. (c) $P = \left[\dfrac{2.5}{39+14+48} \times \dfrac{1000}{100}\right] \times 2 \times R \times 300$

$= 12.2$ atm.

109. (b) $\pi = i.C.RT$

$$\Rightarrow \quad 1.8 \times \left(\frac{1}{74.5} \times \frac{1}{0.1}\right) \times 24.63$$

$$\Rightarrow \quad 5.94$$

110. (a) $7.4 = n \times 0.0821 \times 300$

$$\therefore \quad n = 0.3$$

111. (c) $\pi = iCRT$; Where, $i = 1$ for each of the solution

$$\because \quad \pi \propto \frac{1}{\text{molecular mass of solute}};$$

$\therefore$ $\underset{\substack{\text{urea} \\ (60)}}{\pi_2} > \underset{\substack{\text{glucose} \\ (180)}}{\pi_1} > \underset{\substack{\text{sucrose} \\ (342)}}{\pi_3}$

112. (a) $\pi = CRT$

$\pi \propto T$ [C = const.]

$\pi \propto C$ [T = const.]

113. (b) $M_B = \dfrac{W_B \times R \times T}{\pi \times V} = \dfrac{4 \times 0.0821 \times 300}{3 \times 10^{-4} \times 2}$

$$\approx 1.6 \times 10^5$$

114. (a) $i = \dfrac{\text{Number of particles in KI solution}}{\text{Number of particles in sucrose solution}}$

$$\Rightarrow \quad \frac{\pi \text{ (KI solution)}}{\pi \text{ (sucrose solution)}} = \frac{0.432 \text{ atm}}{0.24 \text{ atm}} = 1.80$$

115. (d) $\pi = iCRT$; $i = 5$ for $Al_2(SO_4)_3$,

$i = 4$ for Na_3PO_4,

$i = 3$ for $BaCl_2$, $i = 1$ for glucose

So osmotic pressure of :

Glucose < $BaCl_2$ < Na_3PO_4 < $Al_2(SO_4)_3$

116. (b) $\pi = iCRT, i = 3 = 1 + (n-1)\alpha$;

For the salt $\alpha = 1$, hence $n = 3$

$$[Cr(H_2O)_5Cl]Cl_2 \cdot H_2O + 2AgNO_3 \longrightarrow [Cr(H_2O)_5Cl]^{2+} + 2AgCl(s)$$

0.5 L of 1m salt $= 0.5 \times 2 = 1.0$ mole of AgCl

117. (c) $\dfrac{59.1}{760} = \left[\dfrac{0.01}{(52 \times 4 \times 17 + 96 + 35.5)}\right] \times \dfrac{1000}{25} \times 0.0821 \times 298 \, (n)$

$$\therefore \quad n = 2$$

118. (d) $\pi \propto iCT$; iCT is maximum for 0.03 M NaCl at 50°C.

119. (d) Only solvent molecules can passed through SPM so only dilution is possible.

120. (a) Henry's law is $m = K \cdot P$; where, m = mass of gas absorbed by given volume of the solvent.

P = pressure of gas;

$\therefore \quad \log m = \log K + \log P$

121. (b) $W_{Gas} \propto P_{Gas}$ at constant T.

122. (b) $\because n_{O_3}$ dissolved in 100 g water at 1.0 atm

$$= \frac{40 \times 1 \times 10^{-3}}{0.0821 \times 300}$$

$\therefore n_{O_3}$ dissolved in 400 g water at 4 atm

$$= \frac{4 \times 4 \times 40 \times 10^{-3}}{24.63}$$

$$\therefore W_{O_3} = \frac{640 \times 10^{-3}}{24.63} \times 48 \approx 1.2 \text{ g}$$

123. (a) $P_{N_2} = (K_H)_{N_2} X_{Gas}$

$$1 = (K_H)_{N_2} \frac{0.02/28}{\frac{0.02}{28} + \frac{1000}{18}}$$

$(K_H)_{N_2} = 7.7 \times 10^4$ atm

124. (d) $P_{Gas} = \left(\frac{1}{K} \times n_{Solvent} \times MM_{Gas}\right) X_{Gas} = K_H X_{Gas}$

$$\left[\because K_H = \frac{n_{Solvent} \times MM_{Gas}}{K}\right]$$

As $T\uparrow$, $K_H\uparrow$.

125. (a) Mass of oxygen $= \frac{20}{100} \times 0.04 = 0.008$ g/L

Mass of nitrogen $= \frac{80}{100} \times 0.02 = 0.016$ g/L

Level 2

1. (b) Mole fraction of urea in its solution

$$= \frac{\frac{12}{60}}{\frac{12}{60} + \frac{140.4}{18}} \Rightarrow 0.025$$

Mole fraction of glucose

$$= \frac{\frac{18}{180}}{\frac{18}{180} + \frac{178.2}{18}} \Rightarrow 0.01$$

$\because$ Mole fraction of glucose is less so vapour pressure above the glucose solution will be higher than the pressure above urea solution, so some H_2O molecules will transfer from glucose to urea side in order to make the solutions of equal mole fraction to attain equilibrium, let x moles H_2O transferred

$$\therefore \frac{0.2}{0.2 + 7.8 + x} = \frac{0.1}{0.1 + 9.9 - x} \Rightarrow x = 4$$

now mass of glucose solution

$\Rightarrow 196.2 - 4 \times 18 \Rightarrow 124.2$

$$\text{wt. \% of glucose} = \frac{18}{124.2} \times 100 \Rightarrow 14.49$$

2. (b) $\because \quad P_A = P_A^\circ X_A,$
$P_B = P_B^\circ X_B$

(P_A° and P_B° = v.p. of pure A and B)

$$Y_A = \frac{P_A}{P_A + P_B}$$

$$\Rightarrow \frac{P_A^\circ X_A}{P_A^\circ X_A + P_B^\circ(1 - X_A)}$$

$$\Rightarrow Y_A = \frac{P_A^\circ X_A}{X_A(P_A^\circ - P_B^\circ) + P_B^\circ}$$

$$\Rightarrow \frac{1}{Y_A} = \left(\frac{P_A^\circ - P_B^\circ}{P_A^\circ}\right) + \frac{P_B^\circ}{P_A^\circ} \cdot \frac{1}{X_A}$$

So, slope is $\frac{P_B^\circ}{P_A^\circ}$ and intercept $= \frac{P_A^\circ - P_B^\circ}{P_A^\circ}$

3. (d) $\frac{P^\circ - P}{P} = \frac{n}{N} \Rightarrow \frac{W}{M \times N}$

$$\Rightarrow \frac{850 - 844.9}{844.9} = \frac{2 \times 76}{M \times 100}$$

$M \approx 252$; $n = \frac{252}{32} \approx 8$

$\therefore$ atomicity of sulphur is 8

4. (c) $P = P_A^\circ x_A + P_B^\circ x_B \Rightarrow 100 \times \frac{1}{5} + 200 \times \frac{4}{5}$

$\Rightarrow$ 180 torr

$$y_A(\text{composition of } A \text{ vapour phase}) = \frac{P_A^\circ x_A}{P}$$

$$\Rightarrow \frac{20}{180} \Rightarrow \frac{1}{9}, \quad \therefore Y_B = \frac{8}{9}$$

For condensation $y_A = x_A^1$; $y_B = x_B^1$

$P_{total} = P_A^\circ x_A^1 + P_B x_B^1$;

$P_{total} = 100 \times \frac{1}{9} + \frac{8}{9} \times 200 \Rightarrow 188.88$ torr

5. (c) Let n_B mole of B present in 1 mole of mixture that has been vaporized. Thus, $y_B = \frac{n_B}{1}$

Mole fraction of B in the remaining liquid phase will be $x_B = \frac{1 - n_B}{1}$

$$x_B = \frac{P - P_T^\circ}{P_B^\circ - P_T^\circ} \quad \ldots(1)$$

$$[\because P = P_T^\circ + (P_B^\circ - P_T^\circ)x_B]$$

and $\quad y_B = \frac{P_B}{P} \Rightarrow \frac{P_B^\circ x_B}{P} \quad \ldots(2)$

After substitution of values of x_B and y_B in (1) and (2)

we get $1 - n_B = \frac{P - P_T^\circ}{P_B^\circ - P_T^\circ} \quad \ldots(3)$

and $n_B = \frac{(1-n_B)P_B^\circ}{P}$ ...(4)

or $n_B = \frac{P_B^\circ}{P + P_B}$

so $1 - \frac{P_B^\circ}{P+P_B} = \frac{P - P_T^\circ}{P_B^\circ - P_T^\circ}$

$\Rightarrow P = \sqrt{P_B^\circ \cdot P_T^\circ} = \sqrt{100 \times 900}$

$\Rightarrow$ 300 torr

6. (c) For an ideal solution $\Delta H_{mix} = 0$ and ΔS_{mix} is always positive so ΔG_{mix} is negative.

7. (d) $mi = \frac{P^\circ - P}{P} \times \frac{1000}{M_{solvent}}$;

$4.m = \frac{(17.25 - 17.20) \times 1000}{17.2 \times 18}$; $m = 0.04$

for dilute solution molality ≈ molarity or S;

$K_{sp} \Rightarrow 27\,S^4 \simeq 7 \times 10^{-5}$

8. (d) Mole ratio of C : H : O is 1 : 2 : 1 so empirical formula is CH_2O

$m = \frac{\Delta T_b}{K_b} \Rightarrow \frac{0.15}{0.51} \Rightarrow 0.294;$

$0.294 = \frac{50}{M} \times \frac{1000}{950}$; $M \approx 180$

$(CH_2O)_n = 180$ or $30 \times n = 180$ or $n = 6$;

∴ molecular formula is $C_6H_{12}O_6$.

9. (c) Mass of 1 litre of solution = 1010 g;

mass of solvent = $1010 - 300 \times 0.1 \Rightarrow 980$ g

$m = \frac{0.1}{0.98} \Rightarrow 0.102;$

$\Delta T_f = K_f.m.i = (1+\alpha)K_f \cdot m$

$\Delta T_f = 1.05 \times 1.86 \times 0.102 = 0.199°C;$

$T_f = 0 - 0.199 = -0.199°C$

10. (b) $m = \frac{\Delta T_b}{K_b} \Rightarrow \frac{1}{0.51} = \frac{\Delta T_f}{K_f}$

$\Rightarrow \frac{1}{0.51} = \frac{\Delta T_f}{1.86}$

$\Delta T_f = 3.647$ or $T_f = -3.647°C$

11. (c) For Na_3PO_4, $i = 1 + 3\alpha = 1 + 3 \times 0.5 = 2.5$; for $MgSO_4$, $i = 1.6$

100 g solution contains 8.2 g Na_3PO_4 and 12 g $MgSO_4$

$\Delta T_b = K_b.m.i =$

$$K_b \cdot \left[\frac{\text{effective no. of moles of } (Na_3PO_4 + MgSO_4)}{\text{mass of solvent (in g)}} \times 1000\right]$$

$$\Delta T_b = 0.50 \left[\frac{\frac{8.2}{164} \times 2.5 + \frac{12}{120} \times 1.6}{79.8}\right] \times 1000$$

$= 1.785°C$

$T_b = 100 + 1.785 \Rightarrow 101.785°C$

12. (b) $\Delta T_b = K_b \left(\frac{w_{solute}}{W_{solvent}} \times \frac{1}{M_{solute}}\right) \times 1000$

$1 = \frac{1000 \times 10}{100 \times M_{AB_2}} \Rightarrow M_{AB_2} = 100$

similarly $M_{A_2B} = 140$

$100 = M_A + 2 \cdot M_B$

and $140 = 2.M_A + M_B;$

$M_A = 60$ and $M_B = 20$

13. (c) Given $w = 0.2$ g, $W = 20$ g, $\Delta T = 0.45$;

$\Delta T_f = \frac{1000 \times K_f \times w}{M \times W}$

or $0.45 = \frac{1000 \times 5.12 \times 0.2}{20 \times M}$

∴ M (observed) = 113.78

Now for $2CH_3COOH \rightleftharpoons (CH_3COOH)_2$

Before association	1	0
After association	$1-\alpha$	$\alpha/2$;

where α is the degree of association.

∴ $\frac{M_{normal}}{M_{observed}} = 1 - \alpha + \frac{\alpha}{2};$

or $\frac{60}{113.78} = 1 - \alpha + \frac{\alpha}{2}$; $\alpha = 0.945$

14. (b) Given $\Delta H_{fus.} = 80$ cal g^{-1},

$\Delta H_{vap.} = 540$ cal g^{-1}

We know, $\Delta T_b = K_b \times m$

and $\Delta T_f = K_f \times m;$

Also $K = \frac{RT^2}{1000 \times \Delta H}$

∴ $\frac{\Delta T_b}{\Delta T_f} = \frac{K_b}{K_f}$

$\Rightarrow \frac{RT_b^2}{1000 \times \Delta H_{vap.}} \times \frac{1000 \times \Delta H_{fus.}}{RT_f^2}$

$\Rightarrow \frac{\Delta T_b}{\Delta T_f} = \frac{T_b^2 \times \Delta H_{fus.}}{T_f^2 \times \Delta H_{vap.}}$

$\frac{0.1}{\Delta T_f} = \frac{373 \times 373 \times 80}{273 \times 273 \times 540}$

∵ $\Delta T_f = 0.361$ so,

$T_f = -0.361°C$

15. (d) Molality of solution $= \frac{P^\circ - P}{P} \times \frac{1000}{M}$

$\Rightarrow \frac{40.18 - 40}{40} \times \frac{1000}{18} \Rightarrow 0.25$

1000 g water present with 45 g glucose or 100 g solution has 4.31 g glucose and 95.69 g H_2O. Final molality is 0.5, 1000 g solvent contain 90 g glucose or 4.31 g glucose present with $\frac{1000}{90} \times 4.31$ = 47.88 g H_2O Mass of ice formed = 95.69 − 47.88 = 47.8 g

16. (b) $\Delta T_f = K_f \times \frac{w_{acid}}{M_{acid} \times W_{H_2O}} \times 1000;$

$\therefore \quad M_{acid}\text{ (observed)} = 120$

Normal molecular mass of acid can be calculated as milli-equivalents of acid = milli-equivalents of base (for HA and NaOH, $N = M$)

$$\frac{0.75}{M_{acid}} \times 1000 = 25 \times \frac{1}{5}$$

$$\Rightarrow \quad M_{acid\,(Normal)} = 150$$

$$\therefore \quad i = \frac{150}{120} \Rightarrow 1.25$$

$$i = 1 + \alpha; \quad \alpha = 0.25$$

$$\therefore \quad \%\,\alpha = 25$$

17. (d) $KI(aq) + AgNO_3(aq) \longrightarrow AgI(s)\downarrow + KNO_3(aq)$

	KI +	$AgNO_3$ $\longrightarrow$	KNO_3 +	$AgI\downarrow$
Initial :	0.3 mole	0.2 mole	—	—
Final :	0.1 mole	—	0.2	0.2

Total moles of solute ions $= (0.1 + 0.2) \times 2\ (\because i = 2)$

$$[\text{Solute}] = \frac{0.6}{4} M$$

$$\Delta T_f = K_f \cdot \frac{0.6}{4}$$

$$= 1.86 \times \frac{0.6}{4} = 0.279\text{ K}$$

18. (b) $\Delta T_f = K_f.m.i \Rightarrow 0.3906 = i \times 1.86 \times 0.1, \quad i = 2.1$

first step of H_2SO_4 is strong so

$$\underset{0}{H_2SO_4} \longrightarrow \underset{C}{H^+} + \underset{C(1-\alpha)}{HSO_4^-};$$

$$\underset{C(1-\alpha)}{HSO_4^-} \rightleftharpoons \underset{C\alpha}{H^+} + \underset{C\alpha}{SO_4^{2-}}$$

$$i = \frac{C(1-\alpha) + C\alpha + C\alpha + C}{C} \Rightarrow 2.1;$$

$$\alpha = 0.1$$

Now, $\underset{C(1-\alpha)}{HSO_4^-} \rightleftharpoons \underset{(C\alpha + C)}{H^+} + \underset{C\alpha}{SO_4^{2-}}$

$$K_{a_2} = \frac{C(1+\alpha) \times C\alpha}{C(1-\alpha)}$$

$$\Rightarrow \quad \frac{1.1 \times 0.1 \times 0.1}{0.9} = 0.0122$$

19. (a) $\pi = (iC_1 - i_2C_2)RT$

$\Rightarrow (1 \times 0.2 - 3 \times 0.05)\,0.0821 \times 300$

$\Rightarrow$ 1.23 atm

20. (c) $\alpha = \sqrt{\frac{8 \times 10^{-5}}{0.2}} \Rightarrow 0.02;$

$i = (1 + \alpha) \Rightarrow 1.02$

$\pi = i.C.R.T$

$\Rightarrow 1.02 \times 0.2 \times 0.0821 \times 300$

$\Rightarrow$ 5.024 atm

21. (b) 8 g of carbohydrate is present with 100 g of water in solution

or 108 g solution contains 8 g of carbohydrate;

$$V_{solution} = \frac{108}{1.025} \times 10^{-3}\text{ litre}$$

$$\pi = \frac{n}{V} \times R \times T$$

$$\Rightarrow \frac{8}{M} \times \frac{1.025}{108} \times 1000 \times 0.0821 \times 300 = 5;$$

$$M = 374\text{ g/mol}$$

22. (d) Only solvent molecule can pass through SPM.

Osmotic pressure $\pi \propto iC$ (at constant T).

23. (a) The given data are

$P_{water} = 17.0$ torr;

P_{total} (4 mole % solution)

$= P_{NH_3} + P_{water} = 50.0$ torr

$x_{NH_3} = 0.04$ and $x_{water} = 0.96$

Now according to Raoult's law;

$P_{water} = x_{water}\,P^\circ_{water}$

$= 0.96 \times 17.0\text{ torr} = 16.32$ torr

Now Henry's law constant for ammonia is

$$K_H(NH_3) = \frac{P_{NH_3}}{x_{NH_3}} = \frac{33.68\text{ torr}}{0.04} = 842\text{ torr}$$

Hence, for 5 mole % solution, we have

$P_{NH_3} = K_H(NH_3)\,x_{NH_3}$

$= (842\text{ torr})(0.05) = 42.1$ torr

$P_{water} = P^\circ_{water}\,x_{water}$

$= (17\text{ torr})(0.95) = 16.15$ torr

Thus, P_{total} (5 mole % solution)

$= P_{NH_3} + P_{water} = 42.1 + 16.15 = 58.25$ torr

24. (b) $P = P_A^\circ x_A + P_B^\circ x_B$

$\Rightarrow 500 \times 0.6 + 800 \times 0.4$

$\Rightarrow$ 620 torr

$$y_A = \frac{P_A}{P} \Rightarrow \frac{300}{620}$$

$\Rightarrow 0.48; \quad y_B = 0.52$

25. (c) $P_A^\circ = 500;\ P_B^\circ = 800$ torr

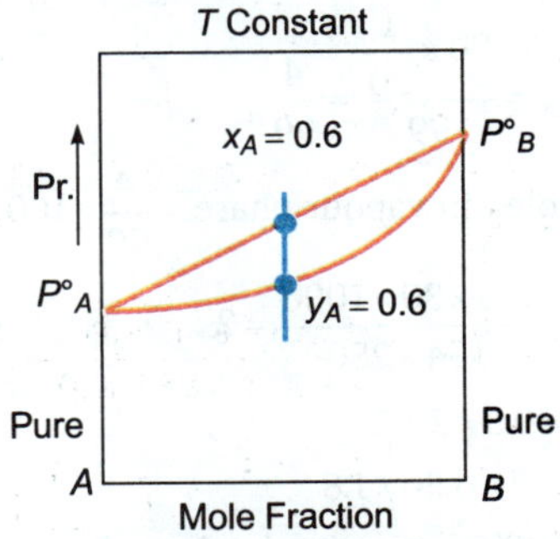

When most of the liquid has vaporized $x_A = 0.6$ (given) would be $y_A = 0.6$

$$y_A = \frac{P_A^\circ x_A}{P_A^\circ x_A + P_B^\circ(1 - x_A)}$$

$$\Rightarrow \quad 0.6 = \frac{500.x_A}{500x_A + 800(1-x_A)}$$

$$x_A = 0.70; \quad x_B = 0.30$$

Level 3

Passage-1

1. (a) Loss of mass of solvent $\propto$ lowering in vp $\propto (p°-P)$.
2. (b) Gain in mass of anhydrous $CaCl_2 \propto P°$.
3. (b) $\frac{P°-P}{P°} = \frac{w_{II}}{w_I + w_{II}}$ $\left\{\because p° \propto w_I + w_{II},\ P°-P \propto w_{II}\right.$
4. (b) $\frac{w_{II}}{w_I + w_{II}} = \frac{9.24/x}{\left(\frac{9.24}{x} + \frac{108}{8}\right)} = \frac{0.08}{3.28}$

$x = 62$

Passage-2

1. (b) $2A \rightleftharpoons A_2$; $n = 2$

$i = 1 + \left(\frac{1}{n} - 1\right)\alpha = 1 - \frac{\alpha}{2}$

2. (a) $i = \frac{M_T}{M_\alpha} < 1$
3. (b) $\Delta T_b = \left[1 + \left(\frac{1}{2} - 1\right)\alpha\right] K_b x$

$\therefore \quad \alpha = \frac{2(K_b x - \Delta T_b)}{K_b x}$

Level 4

Subjective Problems

1. $P = \frac{2}{5} \times 5 + \frac{3}{5} \times 10 = 8$ torr
2. $48 = \left(\frac{8}{8+x}\right)50 + \left(\frac{x}{x+8}\right)40$

$\therefore \quad x = 2$

3. $P = P_A^\circ + x_B(P_B^\circ - P_A^\circ)$

$99 = 100 + x_b(-20) \Rightarrow x_B = \frac{1}{20}$

$\therefore \quad y_B = \frac{80 \times \frac{1}{20}}{99} = \frac{4}{99}$

% mole_B in vapour phase $= \frac{4}{99} \times 100 = 4$

4. $\Delta T_b = \frac{2.6 \times 30 \times 1000}{154 \times 250} = 2$
5. $\Delta T_b = 0.5 \times 2 = 1$
6. $\Delta T_f = 2 \times 0.56 \times 1.8$
7. For 100% ionization $i = 4$
8. For isotonic solution

$\pi_{Protein} = \pi_{Sucrose}$

$\frac{500}{x} = \frac{3.42}{342} \Rightarrow x = 5 \times 10^4$

9. $\Delta T_f = 0.515 = 1.86 \times m$

$m = 0.276$

$\therefore \quad \pi = 0.276 \times 0.0821 \times 310 = 7$ atm

10. $\pi_{KCl} = \pi_{K_2SO_4}$

$0.2 \times 2 = 0.2 \times x$

$x = 2$

Numerical Problems

1. $\frac{P°-P_s}{P_s} = \frac{n_{solute}}{n_{solvent}}$; $\frac{100-95}{95} = \frac{\frac{50}{M}}{\frac{390}{78}}$

$M = 190.00$ g/mol

2. In mixture of A and B if mole fraction of $A = x$, then:

$400x + 500(1-x) = 470$; $x = 0.3$

In final mixture, if combined mole fraction of A and $B = y$, then

$470y + (1-y)600 = 496$; $y = 0.8$

Therefore, in the final mixture, mole fractions are: $x_A = 0.24$, $x_B = 0.56$, $x_C = 0.2$

Mole fraction of A in vapour phase:

$$y_A = \frac{x_A \cdot P_A^\circ}{P} = \frac{0.24 \times 400}{496} = 0.1935 \approx 0.19$$

3. $\Delta T_b = K_b \cdot m$, $0.78 = 0.52m$, $m = 1.5$

Molality $(m) = \left(\frac{P°-P}{P}\right) \times \frac{1000}{M_{Solvent}}$;

$1.5 = \left(\frac{P°-750}{750}\right) \times \frac{1000}{18}$

$P° = 770.25$ torr

4. Moles of solute $= \frac{38.2}{202 + 18x}$

Mass of $H_2O = \left(\frac{38.2 \times 18x}{202 + 18x} + 250\right)$ g

$$\Rightarrow 0.582 = 3 \times 0.52 \times \frac{38.2 \times 1000}{(202+18x)\left(250 + \frac{38.2 \times 18x}{202 + 18x}\right)}$$

$\Rightarrow x = 10$, formula $= Na_2B_4O_7 \cdot 10H_2O$

5. $\Delta T_b + \Delta T_f = (K_b + K_f)m$

$100 - T_b + T_f - 0 = (K_b + K_f)m$

$104.76 - 100 = (1.86 + 0.52)m$

$m = 2$; $2 = \frac{n_{Sucrose}}{100} \times 1000$

Mass of sucrose should be dissolved

$= 0.2 \times 342 = 68.40$ g

6. $\Delta T_f = K_f \cdot m$

$0.2325 = 1.86m$; molality $(m) = 0.125$

0.125 mole glucose present with 1000 g water

18 g glucose present with

$\frac{1000}{0.125 \times 180} \times 18 = 800$ g water

Weight of ice formed = Initial wt. of H_2O – Final wt. of water = 1000 – 800 = 200 g

7. $P = iCRT = 2 \times \frac{2.235}{74.5} \times \frac{1000}{100} \times 0.0821 \times 300$

$= 14.78 \text{ atm}$

8. $2C_6H_5COOH \rightleftharpoons (C_6H_5COOH)_2$

$\alpha = \frac{M_T - M_0}{M_0\left(\frac{1}{n} - 1\right)} \Rightarrow 0.40 = \frac{122 - M_0}{M_0\left(\frac{1}{2} - 1\right)}$;

$M_0 = 152.50$ g/mol

9. $AB_2(aq) \rightleftharpoons A^{2+}(aq) + 2B^-(aq)$

$S - x \qquad x \qquad 2x$

$K_{sp} = 2.56 \times 10^{-7} = 4x^3$; $x = 4 \times 10^{-3}$ M

$\Delta T_b = K_b(S + 2x)$

$6.5 \times 10^{-3} = 0.5(S + 2 \times 4 \times 10^{-3})$; $S = 5 \times 10^{-3}$

Concentration of $AB_2(aq) = S - x$

$= 5 \times 10^{-3} - 4 \times 10^{-3} = 1 \times 10^{-3}$ M

10. $\Delta T_b = iK_b m \Rightarrow 0.15 = 3 \times 0.5 \times m \Rightarrow m = 0.1$

Now,

$$\underset{\substack{0.1\\0}}{Pb(NO_3)_2} + \underset{\substack{0.2\\0}}{2NaCl} \longrightarrow \underset{\substack{0\\0.1}}{PbCl_2} + \underset{\substack{0\\0.2}}{2NaNO_3}$$

Now, the solution of two salts $NaNO_3$ and $PbCl_2$

$-\Delta T_f = 0.83 = K_f(2 \times 0.2 + 3S)$

Where S is molar solubility of $PbCl_2$.

$S = 1.54 \times 10^{-2}$

$\Rightarrow \quad K_{sp} = 4S^3 = 1.46 \times 10^{-5}$

CHAPTER 11

Solid State

Types of Solids

S. No.	Crystalline solids	Amorphous solids
1.	They have a definite crystal shape.	They do not have a definite geometrical shape
2.	Long-range order	Short-range order
3.	They have sharp melting points	They do not have sharp melting points.
4.	Enthalpy of fusion is definite	Enthalpy of fusion is not definite
5.	They are hard and rigid solids.	They are comparatively soft and not very rigid.
6.	Possess definite crystal system (cubic, tetragonal, hexagonal, octahedral, etc.)	Crystal system absent ; no regular repeating unit
7.	They are true solid	They are pseudo solid
8.	They split into two pieces when cut with a sharp-edged tool ; newly generated surfaces are plane and smooth.	They cut into two pieces with irregular surfaces when cut with a sharp-edged tool.
9.	They are anisotropic (refractive index and electrical conductivity are different in different directions) (*e.g.*, NaCl, CsBr, CaF_2 and ZnS	They are isotropic. (physical properties are same in all directions) (*e.g.*, Rubber, plastic and glass)

Types of Crystalline Solids

Type of solid	Intermolecular forces	Properties	Examples
Ionic	Ion-ion forces	Brittle, hard high melting	NaCl, KCl, $MgCl_2$
Molecular	Dispersion forces/Dipole-Dipole H-bond	Soft, low melting non-conducting	H_2O, Br_2, CO_2, CH_4

Covalent network	Covalent bonds	Hard: High melting	C-Diamond, SiO_2
Metallic	Metallic bonds	Variable hardness and melting point conducting	Na, Zn, Cu, Fe

Crystal System

	Crystal Systems	Bravais Lattice	Unit Cell Parameters	
			Intercepts	Crystal Angles
1.	Cubic	Primitive, face centered, body centered	$a = b = c$	$\alpha = \beta = \gamma = 90°$
2.	Orthorhombic	Primitive, face centered, body centered, end centered	$a \neq b \neq c$	$\alpha = \beta = \gamma = 90°$
3.	Rhombohedral	Primitive	$a = b = c$	$\alpha = \beta = \gamma \neq 90°$
4.	Monoclinic	Primitive, end centered	$a \neq b \neq c$	$\alpha = \gamma = 90°, \beta \neq 90°$
5.	Triclinic	Primitive	$a \neq b \neq c$	$\alpha \neq \beta \neq \gamma \neq 90°$
6.	Tetragonal	Primitive, body centered	$a = b \neq c$	$\alpha = \beta = \gamma = 90°$
7.	Hexagonal	Primitive	$a = b \neq c$	$\alpha = \beta = 90°, \gamma = 120°$

Some Important Formula

(i) $d = \dfrac{ZM}{N_0 a^3}$; (ii) Packing fraction $= \dfrac{Z \times V_{\text{Atom}}}{V_{\text{Unit cell}}}$; (iii) Void fraction = 1 – Packing fraction ;

(iv) C.No. = No. of first neighbours

where d = density ; M = Atomic mass of element ;

N_A = Avogadro's number ; a = Edge length of cube.

Cubic System

S.No.	Property	Simple Cubic Lattice	BCC Lattice	FCC Lattice
1.	Atomic radius (r), a = edge length of cube	$r = \dfrac{a}{2}$	$r = \dfrac{a\sqrt{3}}{4}$	$r = \dfrac{a\sqrt{2}}{4}$
2.	No of atoms per unit cell (Z)	$Z = 1$	$Z = 2$	$Z = 4$
3.	Co-ordination No. (C.No.)	6	8	12
4.	Packing fraction (P.F.)	0.52	0.68	0.74

Types of Packing

S.No.	Property	Hexagonal close packing (HCP)	Cubic close packing (CCP)
1.	Co-ordination No. (C.No.)	12	12
2.	No. of atoms per unit cell (z)	$z = 6$	$z = 4$
3.	Packing fraction (P.F.)	P.F. = 0.74	P.F. = 0.74
4.	Type of packing	*ABAB*......	*ABCABC*......

Types of Voids

S. No.	Name of Void	r_{Void}/r_{Sphere}	Co-ordination Number (C.No.)
1.	Triangular void	0.155	3
2.	Tetrahedral void	0.225	4
3.	Octahedral void	0.414	6
4.	Cubic void	0.732	8

Radius Ratio

S.No.	Limiting radius ratio $\left(\frac{r_{cation}}{r_{anion}}\right)$	C.No.	Shape
1.	< 0.155	2	Linear
2.	0.155 — 0.225	3	Planar triangle
3.	0.225 — 0.414	4	Tetrahedral
4.	0.414 — 0.732	6	Octahedral
5.	0.732 — 0.999	8	Like body centred cubic

Types of Ionic Structure

S.No.	Name of Structure	Location of Particle
1.	Rock salt (AB) NaCl	B^- : ccp lattice ; A^+ : Octahedral void
2.	Zinc blende (ZnS)	S^{2-} : ccp lattice ; Zn^{2+} : Alternate tetrahedral void
3.	CsCl	Cs^+ : Cube centre ; Cl^- : Corner of cube
4.	Fluorite structure (CaF_2)	Ca^{2+} : fcc lattice ; F^- : Tetrahedral void
5.	Antifluorite structure Na_2O	O^{2-} : fcc lattice ; Na^+ : Tetrahedral void

Bragg's Equation

$2d\sin\theta = n\lambda$

$n = 1$ First order reflection ; $n = 2$ Second order reflection ; θ = angle of reflection.

d = distance between planes ; λ = wavelength of x-ray

Defects in Crystals

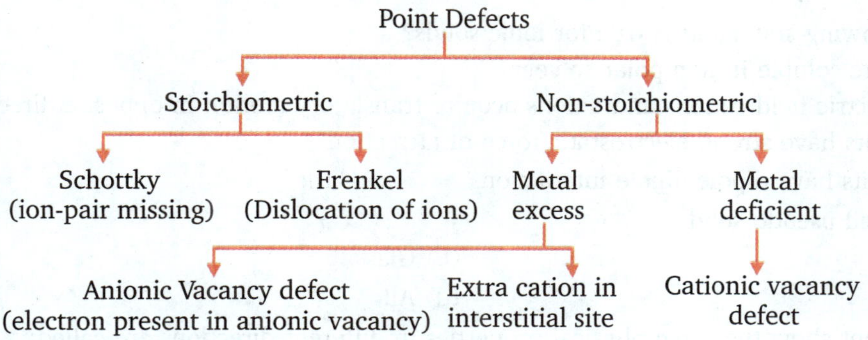

(1) Stoichiometric defects

- **Vacancy:** When constituent particle missing, then vacancy defects occurs.
- **Frenkel defect :** The defect in which an ion is displaced from its regular position to an interstitial position creating a vacancy. Such defect is known as Frenkel defects or dislocation defect.
- **Schottky defect :** Defect in which a pair of one cation and one anion of equal valence is missing from an ionic crystal leading to a pair of vacant sites. Such defect is known as Schottky defect.

(2) Non-Stoichiometric defects :

- **Metal excess defect :** The defect occurs due to anionic vacancies or due to the presence of extra cations at the interstitial site.
- **Metal deficiency defect :** The defect occurs due to absence of positive ions from lattice site or extra interstitial negative ions.

Electrical Properties

The solids can be **conductors, insulators** or **semiconductors.**

- Conductors conduct electricity through movement of electrons (metals) or ions (electrolytes).
- The atomic orbitals of metals form molecular orbitals which are so close in energy that they form bands. If the gap between the filled valence band the next higher unoccupied conduction band overlaps, the electrons can jump and are called conductors. If this gap is more, the electrons can not jump and the substance behaves as an insulator.
- In semiconductors, this gap is small. For example, silicon and germanium. Their conductivity can be increased by adding impurity (doping) which can be electron rich or electron deficient.
- When increase in conductivity is due to the electron-rich impurity, it is called n-type semiconductor.
- When the increase in conductivity is due to electron-deficient impurity, the conductors are called *p*-type semiconductors.

Magnetic Properties

- **Paramagnetic :** Substance weakly attracted in magnetic field and magnetized in same direction is known as paramagnetic substance.
- **Diamagnetic :** Substance weakly repelled in magnetic field is known as diamagnetic substance
- **Ferromagnetic :** Substance attracted very strongly by magnetic field is known as ferromagnetic substance.
- **Antiferromagnetic :** Substance in which oppositely located domains cancel out each other's magnetic moment is known as Antiferromagnetic substance.
- **Ferrimagnetic :** Substance in which magnetic moment of domains are arranged in parallel and antiparallel directions in unequal numbers is known as ferrimagnetic substance.

Level-1

1. Which of the following statement is true for ionic solids?
   (a) Ionic solids are soluble in non-polar solvent
   (b) Under the electric field cations and anions acquire translatory motion in opposite directions
   (c) Structural units have strong electrostatic force of attraction
   (d) Structural units have dipole-dipole interactions
2. Which one is called pseudo solid ?
   (a) CaF_2 (b) Glass
   (c) NaCl (d) All
3. Solids which do not show the same physical properties in different directions are called:
   (a) pseudo solids (b) isotropic solids
   (c) polymorphic solids (d) anisotropic solids
4. Graphite is an example of :
   (a) ionic solid (b) covalent solid
   (c) metallic crystal (d) none of these
5. Amorphous solids are:
   (a) isotropic and supercooled liquids (b) anisotropic and supercooled liquids
   (c) isoenthalpic and superheated liquids (d) isotropic and superheated solids
6. Crystals which are good conductor of electricity and heat are:
   (a) ionic crystals (b) covalent crystals
   (c) metallic crystals (d) molecular crystals
7. The bond length and bond angles in molecules in the solid state are calculated by:
   (a) X-ray diffraction technique (b) neutrons bombardment
   (c) protons bombardment (d) none of these
8. The unit cell with parameters $\alpha = \beta = \gamma = 90°$ and $a = b \neq c$ is :
   (a) cubic (b) triclinic
   (c) hexagonal (d) tetragonal
9. Triclinic crystal has the following unit cell parameters :
   (a) $a = b = c; \alpha = \beta = \gamma = 90°$ (b) $a = b \neq c; \alpha = \beta = \gamma = 90°$
   (c) $a \neq b \neq c; \alpha \neq \beta \neq \gamma \neq 90°$ (d) $a = b \neq c; \alpha = \beta = 90°, \gamma = 120°$
10. If all three interfacial angles defining the unit cell, are equal in magnitude, the crystal cannot be:
    (a) rhombohedral (b) cubic
    (c) hexagonal (d) tetragonal
11. In a hexagonal crystal :
    (a) $\alpha = \beta = \gamma \neq 90°; a = b = c$ (b) $\alpha = \beta = \gamma = 90°; a = b \neq c$
    (c) $\alpha = \beta = \gamma = 90°; a \neq b \neq c$ (d) $\alpha = \beta = 90°, \gamma = 120°; a = b \neq c$
12. Orthorhombic crystal has the following unit cell parameters :
    (a) $a = b = c; \alpha = \beta = \gamma = 90°$ (b) $a = b \neq c; \alpha = \beta = \gamma = 90°$
    (c) $a \neq b \neq c; \alpha = \beta = \gamma = 90°$ (d) $a = b \neq c; \alpha = \beta = 90°, \gamma = 120°$
13. Which of the crystal systems contains the maximum number of Bravais lattices?
    (a) Cubic (b) Hexagonal
    (c) Triclinic (d) Orthorhombic

14. The most unsymmetrical and symmetrical systems are, respectively :

(a) Tetragonal, Cubic (b) Triclinic , Cubic

(c) Rhombohedral, Hexagonal (d) Orthorhombic, Cubic

15. The crystal system of a compound with unit cell parameters, $a = 0.328$ nm, $b = 0.328$ nm, $c = 0.527$ nm and $\alpha = \beta = \gamma = 90°$ is :

(a) cubic (b) tetragonal

(c) monoclinic (d) rhombohedral

16. In the primitive cubic unit cell, the atoms are present at the:

(a) corners of the unit cell (b) centre of the unit cell

(c) centre of each face of the unit cell (d) one set of faces of the unit cell

17. In the body centered unit cell, the lattice points are present at the:

(a) corners of unit cell only

(b) corners and centre of unit cell

(c) corners and centre of each face of the unit cell

(d) corners and at one set of faces of unit cell

18. In the face centered unit cell, the lattice points are present at the:

(a) corners of unit cell only (b) corners and centre of the unit cell

(c) corners and face centres of the unit cell (d) face centres of the unit cell

19. The effective number of atoms per unit cell in a simple cube, face centred cube and body centred cube are respectively:

(a) 1, 4, 2 (b) 1, 2, 4

(c) 8, 14, 9 (d) 8, 4, 2

20. What would be the effective number of atoms per unit cell in end centered cubic unit cell, if this type of unit cell exist in nature ?

(a) 1 (b) 2

(c) 3 (d) 4

21. In the body centered cubic unit cell and simple cubic unit cell, the radius of atom in terms of edge length (a) of the unit cell is respectively:

(a) $\frac{a}{2}, \frac{a}{2\sqrt{2}}$ (b) $\frac{a}{\sqrt{2}}, \frac{a}{2}$

(c) $\frac{a}{2\sqrt{2}}, \frac{a}{2}$ (d) $\frac{\sqrt{3}\,a}{4}, \frac{a}{2}$

22. In the face centered cubic unit cell, the radius of atoms in terms of edge length (a) of unit cell is:

(a) $a/2$ (b) $a/\sqrt{2}$

(c) $a/2\sqrt{2}$ (d) $\sqrt{3}a/4$

23. The fraction of volume occupied by atoms in a primitive cubic unit cell is nearly:

(a) 0.48 (b) 0.52

(c) 0.55 (d) 0.68

24. The fraction of volume occupied by atoms in a body centered cubic unit cell is:

(a) 0.32 (b) 0.48

(c) 0.68 (d) 0.74

25. The fraction of volume occupied by atoms in a face centered cubic unit cell is:

(a) 0.32 (b) 0.48

(c) 0.68 (d) 0.74

26. Which of the following crystal lattice has the minimum empty space?
(a) Simple cubic (b) Body centred cubic
(c) Face centred cubic (d) Simple tetragonal

27. Which of the following has the smallest packing efficiency for atoms of a single type?
(a) Body centered cubic (b) Face centered cubic
(c) Simple cubic (d) None of these

28. Polonium crystallizes in a simple cubic structure. The edge of the unit cell is 0.236 nm. What is the radius of the polonium atom:
(a) 0.144 nm (b) 0.156 nm
(c) 0.118 nm (d) 0.102 nm

29. Lithium crystallizes as body centered cubic crystals. If the length of the side of unit cell is 350 pm, the atomic radius of lithium is:
(a) 303.1 pm (b) 606.2 pm
(c) 151.5 pm (d) 123.7 pm

30. Gold has a face centered cubic lattice with an edge length of the unit cube of 407 pm. The diameter of the gold atom is:
(a) 576.6 pm (b) 287.8 pm
(c) 352.5 pm (d) 704.9 pm

31. The neon atoms has a radius of 160 pm. What is the edge of the unit cell of a face centered structure of neon?
(a) 490 pm (b) 320 pm
(c) 453 pm (d) 481 pm

32. What are the number of atoms per unit cell and the number of nearest neighbours in a simple cubic structure?
(a) 1, 6 (b) 4, 12
(c) 2, 8 (d) 2, 6

33. What are the number of atoms per unit cell and the number of nearest neighbours in a face centered cubic structure?
(a) 4, 8 (b) 2, 8
(c) 2, 6 (d) 4, 12

34. What are the number of atoms per unit cell and the number of nearest neighbours in a body centered cubic structure?
(a) 4, 12 (b) 1, 6
(c) 2, 8 (d) 2, 5

35. Each edge of a cubic unit cell is 400 pm long. If atomic mass of the element is 120 and its density is 6.25 g/cm^3, the crystal lattice is : (use $N_A = 6 \times 10^{23}$)
(a) primitive (b) body centered
(c) face centered (d) end centered

36. Tungsten has an atomic radius of 0.136 nm. The density of tungsten is 19.4 g/cm^3. What is the crystal structure of tungsten? (Atomic mass $W = 184$)
(a) Simple cubic (b) Body centered cubic
(c) Face centered cubic (d) None of these

37. The density of argon (face centered cubic cell) is 1.83 g/cm^3 at 20°C. What is the length of an edge a unit cell? (Atomic mass : Ar = 40)
(a) 0.599 nm (b) 0.569 nm
(c) 0.525 nm (d) 0.551 nm

38. The density of nickel (face centered cubic cell) is 8.94 g/cm^3 at 20°C. What is the radius of the atom? (Atomic mass : Ni = 59)

(a) 0.124 nm (b) 0.136 nm
(c) 0.149 nm (d) 0.110 pm

39. The density of krypton (face centered cubic cell) is 3.19 g/cm^3. What is the radius of the atom? (Atomic mass : Kr = 84)
(a) 0.198 nm (b) 0.221 nm
(c) 0.206 nm (d) 0.225 nm

40. The face centered cubic cell of platinum has an edge length of 0.392 nm. Calculate the density of platinum (g/cm^3) : (Atomic mass : Pt = 195)
(a) 20.9 (b) 20.4
(c) 19.6 (d) 21.5

41. The body centered cubic cell of chromium has an edge length of 0.288 nm. Calculate the density of chromium (g/cm^3) : (Atomic mass : Cr = 52.0)
(a) 6.80 (b) 7.60
(c) 6.60 (d) 7.23

42. An element crystallizes in a face centered cubic lattice and the edge of the unit cell is 0.559 nm. The density is 3.19 g/cm^3. What is the atomic mass?
(a) 87.6 (b) 79.9
(c) 85.5 (d) 83.9

43. The element crystallizes in a body centered cubic lattice and the edge of the unit cell is 0.351 nm. The density is 0.533 g/cm^3. What is the atomic mass?
(a) 12.0 (b) 6.94
(c) 9.01 (d) 10.8

44. An element X (At. mass = 80 g/mol) has fcc structure. Calculate no. of unit cells in 8 gm of X:
(a) $0.4 \times N_A$ (b) $0.1 \times N_A$
(c) $4 \times N_A$ (d) none of these

45. Molybdenum (At. mass = 96 g mol^{-1}) crystallizes as bcc crystal. If density of crystal is 10.3 g/cm^3, then radius of Mo atom is (use $N_A = 6 \times 10^{23}$):
(a) 111 pm (b) 314 pm
(c) 135.96 pm (d) none of these

46. What is the coordination number of an atom for an element crystallizing with a cubic lattice? Calculate the corresponding coordination number for the simple, fcc and bcc lattices:
(a) 12 sc; 12 fcc; 8 bcc (b) 6 sc; 14 fcc; 8 bcc
(c) 8 sc; 12 fcc; 6 bcc (d) 6 sc; 12 fcc; 8 bcc

47. Which of the following layering pattern will have a void fraction of 0.260?
(a) ABCCBAABC (b) ABBAABBA
(c) ABCABCABC (d) ABCAABCA

48. The most malleable metals (Cu, Ag, Au) have close-packing of the type :
(a) Hexagonal close-packing
(b) Cubic close-packing
(c) Body-centred cubic packing
(d) Malleability is not related to type of packing

49. The co-ordination number of a metal crystallising in a hexagonal close-packed structure is:
(a) 12 (b) 4
(c) 8 (d) 6

50. If the ratio of coordination no. of A to that of B is $x:y$, then the ratio of no. of atoms of A to that no. of atoms of B in unit cell is:

(a) $x:y$ (b) $y:x$

(c) $x^2:y$ (d) $y:x^2$

51. The atomic radius of strontium (Sr) is 215 pm and it crystallizes with a cubic closest packing. Edge length of the cube is:

(a) 430 pm (b) 608.2 pm

(c) 496.53 pm (d) none of these

52. By X-ray diffraction it is found that nickel (at mass = 59 g mol^{-1}), crystallizes with ccp. The edge length of the unit cell is 3.5 Å. If density of Ni crystal is 9.0 g/cm^3, then value of Avogadro's number from the data is :

(a) 6.05×10^{23} (b) 6.11×10^{23}

(c) 6.02×10^{23} (d) 6.023×10^{23}

53. Stacking of square close packed layers give rise to :

(a) bcc structure (b) fcc structure

(c) Simple cubic structure (d) hcp structure

54. In a hexagonal close packed (hcp) structure of spheres, the fraction of the volume occupied by the sphere is A. In a cubic close packed structure the fraction is B. The relation for A and B is :

(a) $A=B$

(b) $A<B$

(c) $A>B$

(d) A is equal to the fraction in a simple cubic lattice.

55. The unit cell present in ABCABC, closest packing of atoms is:

(a) hexagonal (b) tetragonal

(c) face centered cubic (d) primitive cubic

56. The number of atoms present in a hexagonal close-packed unit cell is :

(a) 4 (b) 6

(c) 8 (d) 12

57. The unit cell present in ABAB, closest packing of atoms is:

(a) hexagonal (b) tetragonal

(c) face centered cubic (d) primitive cubic

58. The numbers of tetrahedral and octahedral holes in a hexagonal primitive unit cell are respectively:

(a) 8, 4 (b) 6, 12

(c) 2, 1 (d) 12, 6

59. Which one of the following schemes of ordering closed packed sheets of equal sized spheres do not generate closest packed lattice?

(a) ABCABC (b) ABACABAC

(c) ABBAABBA (d) ABCBCABCBC

60. In the closest packing of atoms, there are:

(a) one tetrahedral void and two octahedral voids per atom

(b) two tetrahedral voids and one octahedral void per atom

(c) two of each tetrahedral and octahedral voids per atom

(d) one of each tetrahedral and octahedral void per atom

61. Which of the following figures represents the cross-section of an octahedral site?

(a) 

(b) 

(c) 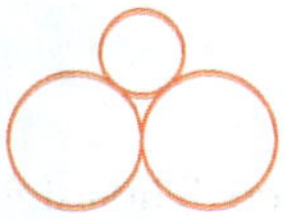

(d) 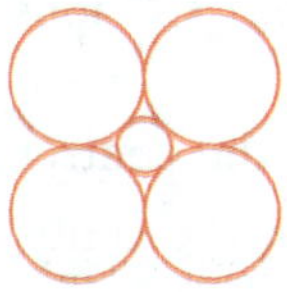

62. In which of the following pairs of structures, tetrahedral as well as octahedral holes are found?
(a) bcc and fcc
(b) hcp and simple cubic
(c) hcp and ccp
(d) bcc and hcp

63. The number of octahedral void in bcc structure is :
(a) 0
(b) 1
(c) 2
(d) 4

64. An ionic compound is expected to have octahedral structure if r_c/r_a $(r_c < r_a)$ lies in the range of:
(a) 0.414 to 0.732
(b) 0.732 to 0.82
(c) 0.225 to 0.414
(d) 0.155 to 0.225

65. An ionic compound is expected to have tetrahedral structure if r_c/r_a:
(a) lies in the range of 0.414 to 0.732
(b) lies in the range of 0.225 to 0.414
(c) lies in the range of 0.155 to 0.225
(d) is more than 0.732

66. An ionic compound is expected to have body centered type cubic unit cell if r_c/r_a:
(a) is greater than 0.732
(b) lies in the range of 0.414 to 0.732
(c) lies in the range of 0.255 to 0.414
(d) lies in the range of 0.155 to 0.225

67. In the closest packing of atoms *A* (radius : r_a), the radius of atom *B* that can be fitted into tetrahedral void is:
(a) 0.155 r_a
(b) 0.225 r_a
(c) 0.414 r_a
(d) 0.732 r_a

68. In the closest packing of atoms *A* (radius : r_a), the radius of atom *B* that can be fitted octahedral void is:
(a) 1.155 r_a
(b) 0.225 r_a
(c) 0.414 r_a
(d) 0.732 r_a

69. How many nearest neighbours are there in an atom or ion for an octahedral hole of a closed packed structure?
(a) 4
(b) 6
(c) 8
(d) 12

70. How many "nearest" and "next nearest" neighbours, respectively, does potassium have in bcc lattice ?
(a) 8, 8
(b) 8, 6
(c) 6, 8
(d) 6, 6

71. In the closest packing of atoms:
(a) the size of tetrahedral void is greater than that of octahedral void
(b) the size of tetrahedral void is smaller than that of octahedral void
(c) the size of tetrahedral void is equal to that of octahedral void
(d) the size of tetrahedral void may be greater or smaller or equal to that of octahedral void depending upon the size of atoms

72. In the ionic compound AB the ratio $r_{A^+} : r_{B^-}$ is 0.414. Indicate the correct statement among the following:
(a) Cations form close packing and anions exactly fit into the octahedral voids
(b) Anions form close packing and cations occupy precisely half of the tetrahedral voids
(c) Anions form close packing and cations occupy precisely all the octahedral voids
(d) Anions form close packing and cations fit into the octahedral voids loosely

73. In the unit cell of KCl (NaCl type), Cl^- ions constitute ccp and K^+ ions fall into the octahedral holes. These holes are:
(a) one at the centre and 6 at the centres of the faces
(b) one at the centre and 12 at the centres of the edges
(c) 8 at the centres of 8 small cubes forming the unit cell
(d) none of these

74. Which is incorrect statement ?
(a) In NaCl structure, tetrahedral voids are unoccupied
(b) In ZnS structure, octahedral voids are unoccupied
(c) In CaF_2 structure, all tetrahedral voids are occupied
(d) In Na_2O structure, all tetrahedral voids are unoccupied

75. If the radius of the anion in an ionic solid is 200 pm, what would be the radius of the cation that fits exactly into a cubic hole?
(a) 146.4 pm (b) 82.8 pm
(c) 45 pm (d) None of these

76. The CsCl type structure is exhibited by alkali halides only when the radius of the cation is large enough to keep touching its eight nearest neighbour anions. Below what minimum ratio of cation to anion radii (r^+/r^-) this contact is prevented?
(a) 0.225 (b) 0.414
(c) 0.632 (d) 0.732

77. MgO crystallizes in a cubic type crystal system. The ionic radii for Mg^{2+} and O^{2-} are 0.066 and 0.140 nm respectively. One can conclude that the Mg^{2+} ions occupy:
(a) a cubic hole in a simple cubic structure
(b) every tetrahedral hole in a close packed structure
(c) an octahedral hole in a cubic close packed structure
(d) every other tetrahedral hole in a close packed structure

78. The unit cell of diamond is made up of:
(a) 6 carbon atoms, 4 atoms constitute ccp and two atoms occupy half of octahedral voids
(b) 8 carbon atom, 4 atoms constitute ccp and 4 atoms occupy all the octahedral voids
(c) 8 carbon atoms, 4 atoms form fcc lattice and 4 atoms occupy half of the tetrahedral voids alternately
(d) 12 carbon atoms, 4 atoms form fcc lattice and 8 atoms occupy all the tetrahedral holes

79. In diamond, the coordination number of carbon is:
(a) four and its unit cell has eight carbon atoms
(b) four and its unit cell has six carbon atoms
(c) six and its unit cell has four carbon atoms
(d) four and its unit cell has four carbon atoms

80. Predict coordination number of the cation in crystals of the following compounds:
1. MgO : $r_c = 0.65$ Å; $r_a = 1.40$ Å; 2. MgS : $r_c = 0.65$ Å; $r_a = 1.84$ Å
(a) 6, 4 (b) 4, 6
(c) 3, 4 (d) 6, 8

81. In a cubic unit cell, seven of the eight corners are occupied by atoms A and centres of faces are occupied by atoms B. The general formula of the compound is :

(a) A_7B_6 (b) A_7B_{12}

(c) A_7B_{24} (d) $A_{24}B_7$

82. CaS exists in a cubic close packed arrangement of S^{2-} ions in which Ca^{2+} ions occupy 1/2 of the available tetrahedral holes. How many Ca^{2+} and S^{2-} ions are contained in the unit cell?

(a) 1, 1 (b) 2, 4

(c) 4, 4 (d) 4, 2

83. In the spinel structure, oxides ions are cubical-closest packed whereas 1/8th of tetrahedral voids are occupied by A^{2+} cation and 1/2 of octahedral voids are occupied by B^{3+} cations. The general formula of the compound having spinel structure is:

(a) $A_2B_2O_4$ (b) AB_2O_4

(c) $A_2B_4O_2$ (d) $A_4B_2O_2$

84. If the anions (A) form hexagonal closest packing and cations (C) occupy only 2/3 octahedral voids in it, then the general formula of the compound is:

(a) CA (b) CA_2

(c) C_2A_3 (d) C_3A_2

85. In a solid, oxide ions are arranged in ccp, cations A occupy $(1/8)^{th}$ of the tetrahedral voids and cations B occupy $(1/4)^{th}$ of the octahedral voids. The formula of the compound is:

(a) ABO_4 (b) AB_2O_3

(c) A_2BO_4 (d) AB_4O_4

86. In a face centered cubic arrangement of A and B atoms whose A atoms are at the corner of the unit cell and B atoms at the face centers. One of the B atoms missing from one of the face in unit cell. The simplest formula of compound is:

(a) AB_3 (b) A_8B_5

(c) A_2B_5 (d) $AB_{2/5}$

87. An alloy of copper, silver and gold is found to have copper constituting the ccp lattice. If silver atoms occupy the edge centres and gold is present at body centre, the alloy will have the formula:

(a) Cu_4Ag_2Au (b) Cu_4Ag_4Au

(c) Cu_4Ag_3Au (d) CuAgAu

88. Which of the following statements is correct in the rock-salt structure of an ionic compound?

(a) Co-ordination number of cation is four and anion is six

(b) Co-ordination number of cation is six and anion is four

(c) Co-ordination number of each cation and anion is four

(d) Co-ordination number of each cation and anion is six

89. Which of the following statements is correct in the body centered type cubic structure of an ionic compound?

(a) Co-ordination number of each cation and anion is two

(b) Co-ordination number of each cation and anion is four

(c) Co-ordination number of each cation and anion is six

(d) Co-ordination number of each cation and anion is eight

90. Which of the following statements is correct in the zinc blende type structure of an ionic compound?

(a) Co-ordination number of each cation and anion is two

(b) Co-ordination number of each cation and anion is four

(c) Co-ordination number of each cation and anion is six

(d) Co-ordination number of each cation and anion is eight

91. Which of the following expression is correct in case of a sodium chloride unit cell (edge length, a)?
(a) $r_c + r_a = a$
(b) $r_c + r_a = a/2$
(c) $r_c + r_a = 2a$
(d) $r_c + r_a = \sqrt{2}a$

92. In an ideal closest rock salt structure (edge length, a), which of the following expressions is correct?
(a) $r_a = \sqrt{2}a$
(b) $r_a = a/\sqrt{2}$
(c) $r_a = a/2\sqrt{2}$
(d) $r_a = a/4$

93. Which of the following expressions is correct in case of a CsCl unit cell (edge length, a)?
(a) $r_c + r_a = a$
(b) $r_c + r_a = a/\sqrt{2}$
(c) $r_c + r_a = \sqrt{3}a/2$
(d) $r_c + r_a = a/2$

94. In NaCl the centres of two nearest like-charged ions are present at a distance of:
(a) $\frac{1}{2}a\sqrt{2}$
(b) $\frac{1}{2}a$
(c) $\frac{\sqrt{3}}{2}a$
(d) $\frac{1}{\sqrt{2}}2a$

95. In sodium chloride crystal, the number of next nearest neighbours of each Na^+ ions is:
(a) 8 Cl^- ions
(b) 12 Na^+ ions
(c) 12 Cl^- ions
(d) 24 Cl^- ions

96. In an ionic compound A^+X^-, the radius of A^+ and X^- ions are 1.0 pm and 2.0 pm, respectively. The volume of the unit cell of the crystal AX will be :
(a) 27 pm^3
(b) 64 pm^3
(c) 125 pm^3
(d) 216 pm^3

97. The coordination number of cation and anion in fluorite CaF_2 and anti-fluorite Na_2O are respectively:
(a) 8 : 4 and 6 : 3
(b) 6 : 3 and 4 : 4
(c) 8 : 4 and 4 : 8
(d) 4 : 8 and 8 : 4

98. Select the incorrect statement in a CsCl crystal:
(a) Cs^+ forms a simple cubic lattice, Cl^- forms a simple cubic lattice
(b) Cl^- occupies body centre of Cs^+
(c) Cs^+ occupies body centre of Cl^-
(d) It is impossible for Cl^- to occupy body centre of Cs^+ because the body centre void of Cs^+ is smaller than Cl^- ion size

99. The radius of a divalent cation A^{2+} is 94 pm and of divalent anion B^{2-} is 146 pm. The compound AB has :
(a) Rock salt structure
(b) Zinc blende structure
(c) Antifluorite structure
(d) Caesium chloride like structure

100. A binary solid (AB) has rock salt structure. If the edge length is 400 pm, and radius of cation is 80 pm the radius of anion is:
(a) 100 pm
(b) 120 pm
(c) 250 pm
(d) 325 pm

101. For a solid with the structure shown in Fig., the coordination numbers of the points A and B, respectively, are

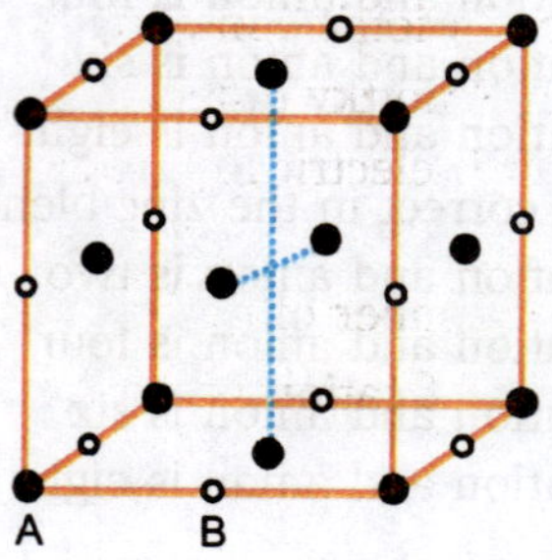

(a) 6, 8 (b) 8, 8
(c) 6, 6 (d) 4, 6

102. An ionic compound AB_2 has fluorite type structures. If the radius B^- is 200 pm, then the ideal radius of A^+ would be:
(a) 82.8 pm (b) 146.4 pm
(c) 40 pm (d) 45 pm

103. In which of the following structures, the anion has maximum coordination number ?
(a) NaCl (b) ZnS
(c) CaF_2 (d) Na_2O

104. CsCl has bcc structure with Cs^+ at the centre and Cl^- ion at each corner. If r_{Cs^+} is 1.69 Å and r_{Cl^-} is 1.81 Å, what is the edge length of the cube?
(a) 3.50 Å (b) 3.80 Å
(c) 4.04 Å (d) 4.50 Å

105. CsBr has bcc type structure with edge length 4.3 pm. The shortest inter ionic distance in between Cs^+ and Br^- is:
(a) 3.72 pm (b) 1.86 pm
(c) 7.44 pm (d) 4.3 pm

106. If the radius of Cl^- ion is 181 pm, and the radius of Na^+ ion is 101 pm then the edge length of unit cell is:
(a) 282 pm (b) 285.71 pm
(c) 512 pm (d) 564 pm

107. Ammonium chloride crystallizes in a body centered cubic lattice with edge length of unit cell equal to 387 pm. If the size of Cl^- ion is 181 pm, the size of NH_4^+ ion would be:
(a) 116 pm (b) 154 pm
(c) 174 pm (d) 206 pm

108. Salt *AB* has a zinc blende structure. The radius of A^{2+} and B^{2-} ions are 0.7 Å and 1.8 Å respectively. The edge length of *AB* unit cell is:
(a) 2.5 Å (b) 5.09 Å
(c) 5 Å (d) 5.77 Å

109. Transition metals, when they form interstitial compounds, the non-metals (H, B, C, N) are accomodated in:
(a) voids or holes in cubic-packed structure (b) tetrahedral voids
(c) octahedral voids (d) all of these

110. In diamond, each carbon atom is bonded to four other carbon atoms tetrahedrally. Alternate tetrahedral voids are occupied by carbon atoms. The number of carbon atoms per unit cell is :
(a) 4 (b) 6
(c) 8 (d) 12

111. Which of the following statements for crystals having Schottky defect is not correct?
(a) Schottky defect arises due to the absence of a cation and anion from the position which it is expected to occupy
(b) Schottky defect are more common in ionic compounds with high co-ordination numbers
(c) The density of the crystals having Schottky defect is larger than that of the perfect crystal
(d) The crystal having Schottky defect is electrically neutral as a whole

112. Which is correct statement ?
(a) When temperature increases then number of defects decreases.
(b) Schottky defect occurs when radius of cation is smaller

(c) Frenkel defect occurs when radius of cation is smaller

(d) None of these

113. Which of the following statements for crystals having Frenkel defect is not correct?

(a) Frenkel defects are observed where the difference in sizes of cations and anions is large

(b) The density of crystals having Frenkel defect is lesser than that of a pure perfect crystal

(c) In an ionic crystal having Frenkel defect may also contain Schottky defect

(d) Usually alkali halides do not have Frenkel defect

114. When anion leaves the normal lattice site and electron occupies interstitial sites in its crystal lattice, it is called:

(a) Schottky defect (b) Frenkel defect

(c) Metal excess defect (d) Stoichiometric defect

115. Which of the following defects does KBr show?

(a) Frenkel (b) Schottky

(c) Metal excess (d) Metal deficiency

116. Dopping of AgCl crystals with $CdCl_2$ results in:

(a) Schottky defect (b) Frenkel defect

(c) Substitutional cation vacancy (d) Formation of F-centres

117. NaCl shows Schottky defects and AgCl Frenkel defects. Their electrical conductivity is due to the:

(a) motion of ions and not the motion of electrons

(b) motion of electrons and not the motion of ions

(c) lower coordination number of NaCl

(d) higher coordination number of AgCl

118. Which one of the following crystals does not exhibit Frenkel defect?

(a) AgBr (b) AgCl

(c) CsCl (d) ZnS

119. Select the incorrect statement:

(a) Stoichiometry of crystal remains uneffected due to Schottky defect

(b) Frenkel defect is usually shown by ionic compounds having low coordination number

(c) F-centres generation is responsible factor for imparting the colour to the crystal

(d) Density of crystal always increases due to substitutional impurity defect

120. In diamond, carbon atom occupy fcc lattice points as well as alternate tetrahedral voids. If edge length of the unit cell is 356 pm, then diameter of carbon atom is:

(a) 77.07 pm (b) 154.14 pm

(c) 251.7 pm (d) 89 pm

121. When NaCl is dopped with 10^{-5} mole % of $SrCl_2$, what is the no. of cationic vacancies?

(a) $10^{-5} \times N_A$ (b) $10^{-7} \times N_A$

(c) $2 \times 10^{-7} N_A$ (d) None of these

122. The composition of a sample of Wustite is $Fe_{0.93}O$. What is the percentage of iron present as Fe^{3+} in total iron?

(a) 15.05% (b) 25%

(c) 35% (d) 45%

123. A certain sample of cuprous sulphide is found to have composition $Cu_{1.8}S$, because of incorporation of Cu^{2+} ions in the lattice. What is the mole % of Cu^{2+} in total copper content in this crystal?

(a) 99.8% (b) 11.11%

(c) 88.88% (d) None of these

124. Ferrimagnetism is in :

(a) ↑↑↑↑↑ (b) ↑↓↑↓

(c) ↑↑↑↓↓ (d) none of these

125. Fe_3O_4 is ferrimagnetic at room temperature but at 850 K it becomes :

(a) diamagnetic (b) ferromagnetic

(c) non-magnetic (d) paramagnetic

Level-2

1. When heated above 916°C, iron changes its bcc crystalline form to fcc without the change in the radius of atom. The ratio of density of the crystal before heating and after heating is:

(a) 1.069 (b) 0.918
(c) 0.725 (d) 1.231

2. $TlAl(SO_4)_2 \cdot xH_2O$ is bcc with 'a' = 1.22 nm. If the density of the solid is 2.32 g/cc, then the value of x is (Given : $N_A = 6 \times 10^{23}$; at. mass : Tl = 204, Al = 27, S = 32).

(a) 2 (b) 4
(c) 47 (d) 70

3. In an atomic bcc lattice what fraction of edge is not covered by atoms?

(a) 0.32 (b) 0.16
(c) 0.134 (d) 0.268

4. The packing efficiency of a simple cubic crystal with an interstitial atom exactly fitting at the body center is:

(a) 0.48 (b) 0.52
(c) 0.73 (d) 0.91

5. An atomic solid crystallizes in a body centre cubic lattice and the inner surface of the atoms at the adjacent corner are separated by 60.3 pm. If the atomic mass of A is 48, then density of the solid, is nearly:

(a) 2.7 g/cc (b) 5.07 g/cc
(c) 3.5 g/cc (d) 1.75 g/cc

6. Sodium (Na = 23) crystallizes in bcc arrangement with the interfacial separation between the atoms at the edge 53.6 pm. The density of sodium crystal is:

(a) 2.07 g/cc (b) 2.46 g/cc
(c) 1.19 g/cc (d) none of these

7. The density of solid argon (Ar = 40 g/mol) is 1.68 g/mL at 40 K. If the argon atom is assumed to be a sphere of radius 1.50×10^{-8} cm, what % of solid Ar is apparently empty space?
(use $N_A = 6 \times 10^{23}$)

(a) 35.64 (b) 64.36
(c) 74 (d) none of these

8. A body centered cubic lattice is made up of hollow spheres of B. Spheres of solid A are present in hollow spheres of B. Radius of A is half of radius of B. What is the ratio of total volume of spheres of B unoccupied by A in a unit cell and volume of unit cell?

(a) $\frac{7\sqrt{3}\pi}{64}$ (b) $\frac{7\sqrt{3}}{128}$
(c) $\frac{7.\pi}{24}$ (d) none of these

9. First three nearest neighbour distances for primitive cubic lattice are respectively (edge length of unit cell = a):

(a) $a, \sqrt{2}a, \sqrt{3}a$ (b) $\sqrt{3}a, \sqrt{2}a, a$
(c) $a, \sqrt{2}a, 2a$ (d) $a, \sqrt{3}a, 2a$

10. First three nearest neighbour distances for body centered cubic lattice are respectively:

(a) $\sqrt{2}a, a, \sqrt{3}a$ (b) $\frac{a}{\sqrt{2}}, a, \sqrt{3}a$
(c) $\frac{\sqrt{3}a}{2}, a, \sqrt{2}a$ (d) $\frac{\sqrt{3}a}{2}, a, \sqrt{3}a$

11. Given : The unit cell structure of compound is shown below.

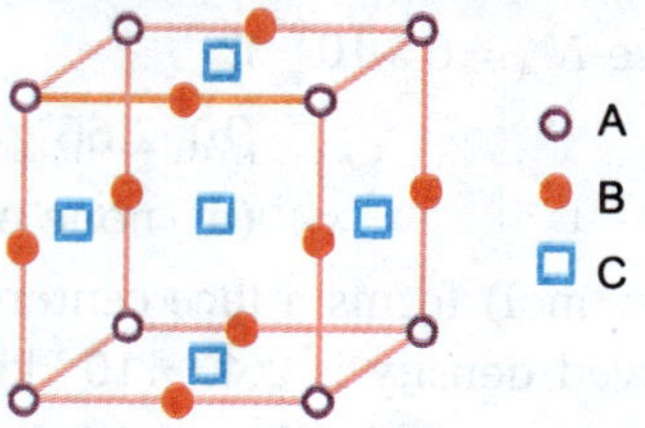

The formula of compound is :

(a) $A_8B_{12}C_5$ (b) AB_2C_3

(c) $A_2B_2C_5$ (d) ABC_5

12. The density of a pure substance '*A*' whose atoms are in cubic close pack arrangement is 1 g/cc. If all the tetrahedral voids are occupied by '*B*' atom, what is the density of resulting solid in g/cc. [Atomic mass $(A) = 30$ g/mol and atomic mass $(B) = 50$ g/mol]

(a) 3.33 (b) 4.33

(c) 2.33 (d) 5.33

13. In a planar tetra-atomic molecule, XY_3, X is at the centroid of the equilateral triangle formed by the atoms, Y. If the X—Y bond distance is 1 Å, what is the distance between the centres of any two Y atoms?

(a) $1/\sqrt{3}$ Å (b) $\sqrt{2}$ Å

(c) $\sqrt{3}$ Å (d) $1/\sqrt{2}$ Å

14. How many unit cells are present in 5.0 gm of crystal *AB* (formula mass of $AB = 40$) having rock salt type structure? (N_A = Avogadro's no.)

(a) N_A (b) $\frac{N_A}{10}$

(c) $4\,N_A$ (d) none of these

15. The density of CaF_2 (fluorite structure) is 3.18 g/cm^3. The length of the side of the unit cell is:

(a) 253 pm (b) 344 pm

(c) 546 pm (d) 273 pm

16. A crystal of lead (II) sulphide has NaCl structure. In this crystal the shortest distance between a Pb^{2+} ion and S^{2-} ion is 297 pm. What is the volume of unit cell in lead sulphide?

(a) 209.6×10^{-24} cm^3 (b) 207.8×10^{-23} cm^3

(c) 22.3×10^{-23} cm^3 (d) 209.8×10^{-23} cm^3

17. CdO has NaCl structures with density 8.27 g/cc. If the ionic radius of O^{2-} is 1.24 Å, determine ionic radius of Cd^{2+} :

(a) 1.5 Å (b) 1.1 Å

(c) 1.9 Å (d) 1.5 Å

18. KCl crystallizes in the same type of lattice as does NaCl. Given that $r_{Na^+}/r_{Cl^-} = 0.50$ and $r_{Na^+}/r_{K^+} = 0.70$, calculate the ratio of the side of the unit cell for KCl to that for NaCl:

(a) 1.143 (b) 1.224

(c) 1.414 (d) 0.875

19. Ferrous oxide has a cubic structure and edge length of the unit cell is 5.0 Å. Assuming the density of ferrous oxide to be 3.84 g/cm^3, the no. of Fe^{2+} and O^{2-} ions present in each unit cell be : (use $N_A = 6 \times 10^{23}$) :

(a) $4\,Fe^{2+}$ and $4\,O^{2-}$ (b) $2\,Fe^{2+}$ and $2\,O^{2-}$

(c) $1\,Fe^{2+}$ and $1\,O^{2-}$ (d) $3\,Fe^{2+}$ and $4\,O^{2-}$

20. If an element (at. mass = 50) crystallises in fcc lattice, with $a = 0.50$ nm. What is the density of unit cell if it contains 0.25% Schottky defects (use $N_A = 6 \times 10^{23}$)?
(a) 2.0 g/cc (b) 2.66 g/cc
(c) 3.06 g/cc (d) none of these

21. An element X (atomic mass = 24 gm/mol) forms a face centered cubic lattice. If the edge length of the lattice is 4×10^{-8} cm and the observed density is 2.40×10^3 kg/m^3, then the percentage occupancy of lattice points by element X is : (Use $N_A = 6 \times 10^{23}$) :
(a) 96 (b) 98
(c) 99.9 (d) none of these

22. In fcc lattice, A, B, C, D atoms are arranged at corner, face center, octahedral void and tetrahedral void respectively, then the body diagonal contains:
(a) $2A, C, 2D$ (b) $2A, 2B, 2C$
(c) $2A, 2B, D$ (d) $2A, 2D$

23. The distance between an octahedral and tetrahedral void in fcc lattice would be:
(a) $\sqrt{3}a$ (b) $\frac{\sqrt{3}a}{2}$
(c) $\frac{\sqrt{3}a}{3}$ (d) $\frac{\sqrt{3}a}{4}$

24. A_2B molecules (molar mass = 259.8 g/mol) crystallises in a hexagonal lattice as shown in figure. The lattice constants were $a = 5$ Å and $b = 8$ Å. If density of crystal is 5 g/cm^3 then how many molecules are contained in given unit cell? (use $N_A = 6 \times 10^{23}$)

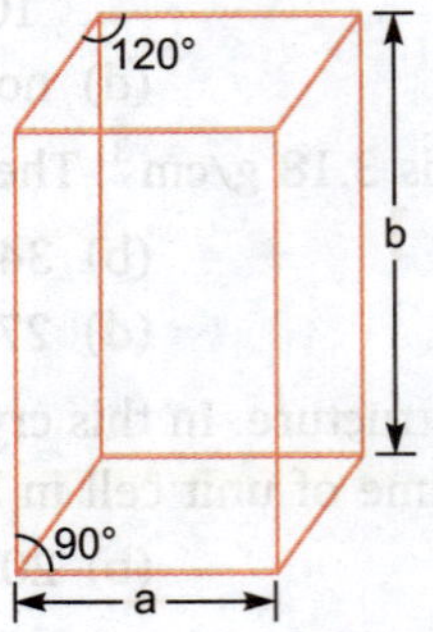

(a) 6 (b) 4
(c) 3 (d) 2

25. Graphite has h.c.p. arrangements of carbon atoms and the parallel planes are 3.35 Å apart. Determine density of graphite:
(a) 2.46 g/cc (b) 0.41 g/cc
(c) 1 g/cc (d) 1.41 g/cc

26. How many effective Na^+ and Cl^- ions are present respectively in a unit cell of NaCl solid (Rock salt structure) if all ions along line connecting opposite face centres are absent?
(a) 3, 3 (b) $\frac{7}{2}, 4$
(c) $\frac{7}{2}, \frac{7}{2}$ (d) $4, \frac{7}{2}$

27. A crystal is made of particles X and Y. X forms fcc packing and Y occupies all the octahedral voids. If all the particles along one body diagonal are removed then the formula of the crystal would be:
(a) X_4Y_3 (b) X_5Y_4
(c) X_4Y_5 (d) none of these

28. Select right expression for determining packing fraction (P. F.) of NaCl unit cell (assume ideal), if ions along an edge diagonal are absent:

(a) $\text{P.F.} = \dfrac{\frac{4}{3}\pi(r_+^3 + r_-^3)}{16\sqrt{2}\, r_-^3}$

(b) $\text{P.F.} = \dfrac{\frac{4}{3}\pi\left(\frac{5}{2}r_+^3 + 4r_-^3\right)}{16\sqrt{2}\, r_-^3}$

(c) $\text{P.F.} = \dfrac{\frac{4}{3}\pi\left(\frac{5}{2}r_+^3 + r_-^3\right)}{16\sqrt{2}\, r_-^3}$

(d) $\text{P.F.} = \dfrac{\frac{4}{3}\pi\left(\frac{7}{2}r_+^3 + r_-^3\right)}{16\sqrt{2}\, r_-^3}$

29. A crystal is made of particles X, Y and Z. X forms fcc packing. Y occupies all the octahedral voids of X and Z occupies all the tetrahedral voids of X. If all the particles along one body diagonal are removed then the formula of the crystal would be:

(a) XYZ_2

(b) X_2YZ_2

(c) $X_8Y_4Z_5$

(d) $X_5Y_4Z_8$

30. A crystal is made of particles A and B. A forms fcc packing and B occupies all the octahedral voids. If all the particles along the plane as shown in figure are removed, then, the formula of the crystal would be:

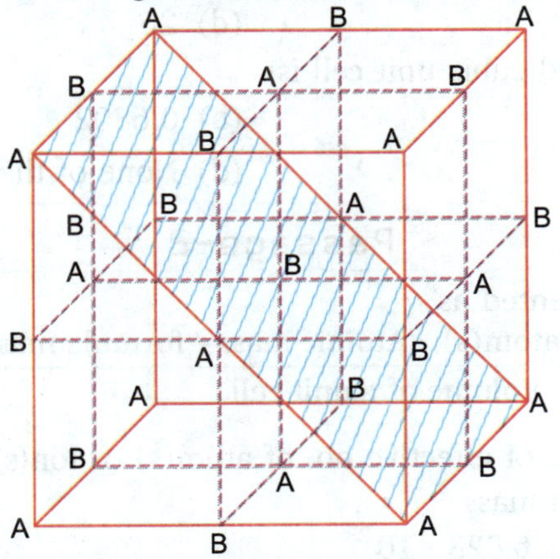

(a) AB

(b) A_5B_7

(c) A_7B_5

(d) none of these

31. In the rock salt AB, if C introduced in tetrahedral voids such that no distortion occurs, then formula of resultant compound is :

(a) ABC

(b) ABC_2

(c) A_4B_4C

(d) ABC_8

32. Given length of side of hexagonal unit cell is $\dfrac{100}{\sqrt{2}}$ pm. The volume of hexagonal unit cell is (in pm^3) :

(a) 8×10^6

(b) 1.5×10^6

(c) 64×10^6

(d) 36×10^6

Level-3

Passage-1

Packing fraction of a unit cell is defined as the fraction of the total volume of the unit cell occupied by the atom(s).

$$\text{P.F.} = \frac{\text{Volume of the atom(s) present in a unit cell}}{\text{Volume of unit cell}} = \frac{Z \times \frac{4}{3}\pi r^3}{a^3}$$

and % of empty space = 100 − P.F. × 100

where Z = effective number of atoms in a cube

r = radius of an atom

a = edge length of the cube

1. % of empty space in body centered cubic unit cell is nearly:
(a) 52.36 (b) 68
(c) 32 (d) 26

2. Packing fraction in face centered cubic unit cell is:
(a) 0.7406 (b) 0.6802
(c) 0.5236 (d) None of these

Passage-2

Density of a unit cell is represented as

$$\rho = \frac{\text{Effective no. of atom(s)} \times \text{Atomic mass / formula mass}}{\text{Volume of a unit cell}} = \frac{Z.M.}{N_A . a^3}$$

where, mass of unit cell = mass of effective no. of atom(s) or ion(s).

M = At. mass/formula mass

N_A = Avogadro's no. ⇒ 6.023×10^{23}

a = edge length of unit cell

1. Silver crystallizes in a fcc lattice and has a density of 10.6 g/cm^3. What is the length of an edge of the unit cell?
(a) 40.7 nm (b) 0.2035 nm
(c) 0.101 nm (d) 4.07 nm

2. An element crystallizes in a structure having fcc unit cell of an edge 200 pm. Calculate the density, if 100 g of this element contains 12×10^{23} atoms:
(a) 41.66 g/cm^3 (b) 4.166 g/cm^3
(c) 10.25 g/cm^3 (d) 1.025 g/cm^3

3. The density of KBr is 2.75 g/cm^{-3}. The length of the edge of the unit cell is 654 pm. To which type of cubic crystal, KBr belongs?
(a) simple cubic (b) bcc
(c) fcc (d) none of these

Passage-3

A spinel is an important class of oxides consisting of two types of metal ions with the oxide ions arranged in ccp layers. The normal spinel has one-eighth of the tetrahedral holes occupied by one type of metal ion and one-half of the octahedral holes occupied by another type of metal ion. Such a spinel is formed by Mg^{2+}, Al^{3+} and O^{2-}. The neutrality of the crystal is being maintained.

1. The formula of the spinel is:

(a) Mg_2AlO_4 (b) $MgAl_2O_4$

(c) $Mg_3Al_2O_6$ (d) none of these

2. Type of hole occupied by Al^{3+} ions is:

(a) tetrahedral (b) octahedral

(c) both (a) and (b) (d) none of these

3. Type of hole occupied by Mg^{2+} ions is:

(a) tetrahedral (b) octahedral

(c) both (a) and (b) (d) none of these

4. If oxide ion is replaced by $X^{-8/3}$, the number of anionic vacancy per unit cell is:

(a) 1 (b) 2

(c) 3 (d) 3/4

Passage-4

Ionic lattice has two major points defects, (1) Schottky (2) Frenkel defects. Schottky defects occurs due to the cation-anion pair's missing from the lattice sites. Frenkel defects occurs when an ion leaves its lattice site and fits into an interstitial space. The neutrality of the crystal is being maintained and we considered all losses from interstitial positions.

1. Which defect decreases density of the crystal?

(a) Frenkel defect (b) Schottky defect

(c) Both (a) and (b) (d) None of these

2. Structure shown here represents:

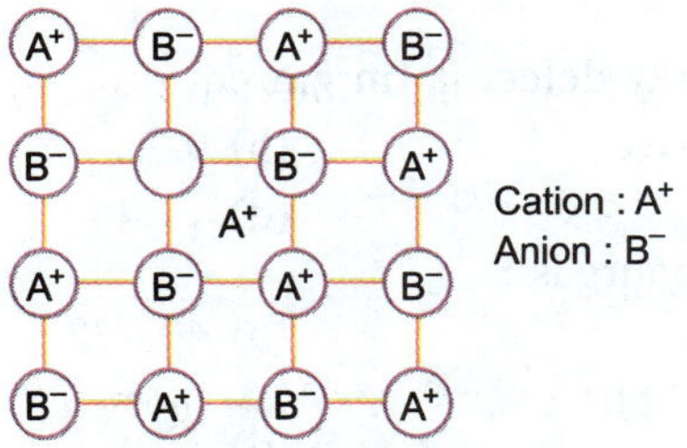

(a) Schottky defect (b) Frenkel defect

(c) Metal excess defect (d) None of these

3. Structure shown here represents:

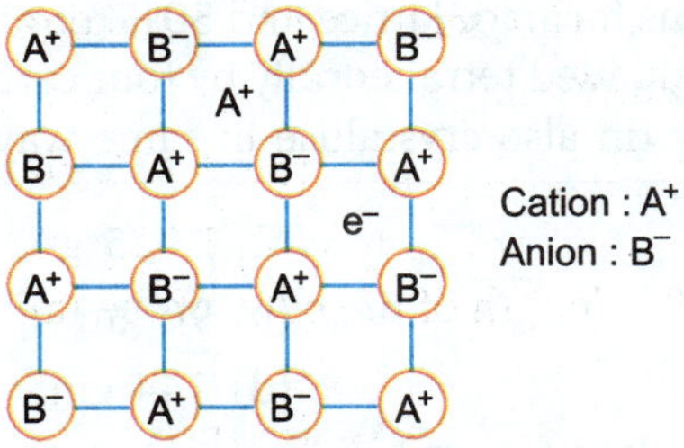

(a) Schottky defect (b) Frenkel defect

(c) Both defects (d) None of these

Passage-5

Doping means introduction of small amount of impurities like phosphorus, arsenic or boron into the pure crystal. In pure silicon, there are four valencies used in bonding with other four adjacent silicon atoms. When a silicon crystal is doped with a group-15 element (with five valence electrons) such as P, As, or Bi, the structure of the crystal lattice remains unchanged. Out of the five valence electrons of group-15 doped element, four electrons are used in normal covalent bonding with silicon while fifth electron is delocalised and thus conducts electricity.
Doping a silicon crystal with a group-13 element (with three valence electrons) such as B, Al, Ga or In products a semiconductor with three electrons in dopant. The place where fourth electron is missing is called an electron vacancy or hole. Such hole can move through the crystal like a positive charge giving rise conduction of electricity.

1. No. of valence electrons in silicon are:
(a) 3 (b) 4
(c) 5 (d) 6

2. Silicon that has been doped with group-15 elements is called:
(a) *p*-type semiconductor (b) *n*-type semiconductor
(c) electron vacancy or hole (d) none of these

3. Silicon that has been doped with group-13 elements is called:
(a) *p*-type semiconductor (b) *n*-type semiconductor
(c) electron vacancy or hole (d) none of these

Passage-6

Metallic Gold crystallise in fcc lattice and the length of cubic unit cell is 407 pm.
(Given : Atomic mass of Gold $= 197, N_A = 6 \times 10^{23}$)

1. The density if it have 0.2% Schottky defect is (in gm/cm^3) :
(a) 4.86 (b) 9.72
(c) 19.48 (d) 19.44

2. The distance between next neighbours is :
(a) 407 pm (b) $407\sqrt{2}$
(c) $\frac{407}{\sqrt{2}}$ (d) $\frac{407\sqrt{3}}{2}$

Passage-7

In diamond structure, carbon atoms form fcc lattice and 50% tetrahedral voids occupied by carbon atoms. Every carbon atoms is surrounded tetrahedrally by four carbon atoms with bond length 154 pm. Germanium, silicon and grey tin also crystallise in same way as diamond. ($N_A = 6 \times 10^{23}$)

1. The mass of diamond unit cell is :
(a) 96 amu (b) 96 gm
(c) 144 amu (d) 144 gm

2. The side length of diamond unit cell is (in pm) :
(a) 154 (b) 1422.63
(c) 711.32 (d) 355.66

3. The density of diamond unit cell is (in gm/cm^3) :
(a) 28.48 (b) 0.0556
(c) 0.445 (d) 3.56

One or More Answer is/are Correct

1. Select the correct statement(s):
(a) Co-ordination no. of an atom at a lattice point in simple cubic arrangement is 6
(b) Co-ordination no. of an atom at octahedral site is 8
(c) Co-ordination no. of an atom at a lattice point in hcp arrangement is 6
(d) Co-ordination no. of an atom at octahedral site is 6

2. Packing fraction of an identical solid sphere is 74% in:
(a) simple cubic structure (b) fcc structure
(c) hcp structure (d) bcc structure

3. In fcc structure octahedral voids are present at:
(a) edge centers (b) face centers
(c) body centers (d) corners

4. If the radius of Na^+ is 95 pm and that of Cl^- ions is 181 pm then:
(a) co-ordination no. of Na^+ is 6 (b) co-ordination no. of Cl^- is 8
(c) length of the unit cell is 552 pm (d) length of the unit cell is 380 pm

5. Select the correct statement(s):
(a) The co-ordination number of each type of ion in a CsCl crystal is twelve
(b) A metal that crystallizes in a bcc structure has a co-ordination number of twelve
(c) A unit cell of an ionic crystal shares some of its ions with other unit cells
(d) The length of the unit cell in NaCl in 552 pm (given that $r_{Na^+} = 85$ pm and $r_{Cl^-} = 181$ pm)

6. The hcp and ccp structure for a given element would be expected to have:
(a) the same co-ordination number (b) the same density
(c) the same packing fraction (d) all of the above

7. Select the correct statement(s) for the rock-salt structure (NaCl) :
(a) The tetrahedral voids are smaller than the octahedral sites
(b) The octahedral voids are occupied by cations and the tetrahedral sites are empty
(c) The radius ratio (r_c/r_a) is 0.225
(d) The radius ratio (r_c/r_a) is 0.732

8. Select the correct statement(s):
(a) The ionic crystal of AgBr may have Schottky defect
(b) The unit cell having crystal parameters, $a = b \neq c, \alpha = \beta = 90°, \gamma = 120°$ is hexagonal
(c) In ionic compounds having Frenkel defect the ratio r^+/r^- is high
(d) The co-ordination number of Na^+ ion in NaCl is 6

9. Which of the following represents the closest packed arrangement of uniform solid spheres:
(a) simple cubic unit cell (b) body centered cubic unit cell
(c) face centered cubic unit cell (d) hcp unit cell

10. Select the correct statement(s):
(a) For cubic system $a = b = c$ and $\alpha = \beta = \gamma = 90°$
(b) Hexagonal system $a = b \neq c$ and $\alpha = \beta = 90°$, $\gamma = 120°$
(c) For triclinic system $a \neq b \neq c$ and $\alpha \neq \beta \neq \gamma \neq 90°$
(d) The total no. of Bravais space lattice belonging to all the seven crystals are 14

11. Select the correct statement(s):

(a) Co-ordination no. of Cs^+ and Cl^- are 8, 8 in CsCl crystal

(b) If radius ratio $(r_c/r_a) < 0.225$ then shape of compound must be linear

(c) If radius ratio (r_c/r_a) lies between 0.414 to 0.732 then shape of ionic compound may be square planer (Ex. $PtCl_4^{2-}$)

(d) If radius ratio is less than 0.155 then shape of compound is linear

12. Select the correct statement(s):

(a) CsCl changes to NaCl structure on heating

(b) NaCl changes to CsCl structure on applying pressure

(c) Co-ordination number decreases on applying pressure

(d) Co-ordination number increases on heating

13. Select the correct statement(s):

(a) A NaCl type *AB* crystal lattice can be interpreted to be made up of two individual fcc type unit lattice of A^+ and B^- fused together in such a manner that the corner of one unit lattice becomes the edge centre of the other

(b) In a fcc unit cell, the body center is an octahedral void

(c) In an scc lattice, there can be no octahedral void

(d) In an scc lattice, the body center is the octahedral void

14. In an *AB* unit cell (Rock salt type) assuming A^+ forming fcc:

(a) The nearest neighbour of A^+ is $6B^-$ ion

(b) The nearest neighbour of B^- is $6A^+$ ion

(c) The second neighbour of A^+ is $12A^+$

(d) The packing fraction of *AB* crystal is $\frac{\sqrt{3}\pi}{8}$

15. Amorphous solids:

(a) do not have sharp melting points

(b) are isotropic

(c) have same physical properties in all directions

(d) are supercooled liquids

16. Which is/are correct statement about zinc blende structure ?

(a) The number of first neighbours of S^{2-} is 4

(b) The maximum distance between Zn^{2+} is $\frac{a\sqrt{3}}{2}$, where 'a' = edge length of unit cell

(c) If all tetrahedral voids are occupied by Zn^{2+} then C.N. of S^{2-} is 8.

(d) If all tetrahedral voids are occupied by Zn^{2+} then C.N. change from 4 : 4 to 8 : 8.

17. Compound X_2Y have antifluorite structure. Which is/are correct statement ?

(a) The minimum distance between X^+ is $\frac{a}{2}$, where 'a' = edge length of unit cell

(b) The co-ordination number ratio of x and y is 8 : 4

(c) If X^+ removed from alternate tetrahedral void then CN is 4 : 4.

(d) If X^+ removed from alternate tetrahedral void then CN is 4 : 8

18. Which is/are correct statement ?

(a) Packing fraction in *ABCABC* .... is 0.785

(b) Packing fraction in *AAA* .... is 0.52

(c) Packing fraction in *ABAB* .... is 0.74

(d) Void fraction in *ABCABC* .... is 0.26

19. Which is/are correct statement ?

(a) In simple cubic close packed arrangement no octahedral void is present at edge centre.

(b) In fcc unit cell octahedral void and tetrahedral void are vacant.
(c) Packing fraction : simple cubic unit cell < bcc unit cell < fcc unit cell
(d) Size of void: cubic void > octahedral void > tetrahedral void

20. Given : Radius of A^{2+} = 100 pm ; Radius of C^{+} = 240 pm ; Radius of B^{2-} = 300 pm ; Radius of D^{-} = 480 pm Which is/are correct statement ?
(a) Coordination number of A^{2+} in comp AB is 4
(b) Coordination number of A^{2+} in comp AB is 6.
(c) Coordination number of C^{+} in comp CD is 6
(d) Coordination number of C^{+} in comp CD is 8.

Match the Column

Column-I and **Column-II** contains four entries each. Entries of Column-I are to be matched with some entries of Column-II. One or more than one entries of Column-I may have the matching with the same entries of Column-II.

1.

Column-I	Column-II
(A) Tetragonal and Hexagonal	(P) are two crystal systems
(B) Cubic and Rhombohedral	(Q) $\alpha = \beta = \gamma$
(C) Monoclinic and Triclinic	(R) $a \neq b \neq c$
(D) Cubic and Orthorhombic	(S) $a = b = c$

2.

Column-I	Column-II
(A) If radius ratio $x = \left(\frac{r_c}{r_a}\right) < 0.155$	(P) Co-ordination no. is 8
(B) If $0.225 \leq x < 0.414$	(Q) Co-ordination no. is 4
(C) If $0.414 \leq x < 0.732$	(R) Co-ordination no. is 6
(D) If $0.732 \leq x < 1$	(S) Co-ordination no. is 2

3.

Column-I (Shape of Compound)	Column-II (Co-ordination No.)
(A) Linear	(P) 6
(B) Triangular planar	(Q) 4
(C) Square planar	(R) 2
(D) Octahedral	(S) 3

4.

Column-I	Column-II
(A) Rock salt structure	(P) general formula is AB
(B) Zinc blende structure	(Q) general formula is AB_3
(C) Fluorite structure	(R) general formula is A_2B
(D) Anti fluorite structure	(S) general formula is AB_2

5.

	Column-I		Column-II
(A)	Co-ordination no. of Ca^{2+} and F^- in fluorite structure	(P)	8, 4
(B)	C.No. of Zn^{2+} and S^{2-} in zinc blende structure	(Q)	8, 8
(C)	C.No. of Cs^+ and Cl^- in CsCl (bcc type) structure	(R)	4, 8
(D)	C.No. of Li^+ and O^{2-} in antifluorite structure	(S)	4, 4

6.

	Column-I [Bravais Lattice(s)]		Column-II (Crystal System)
(A)	Primitive, face centered, body centered, end centered	(P)	Cubic
(B)	Primitive, face centered, body centered	(Q)	Orthorhombic
(C)	Primitive, body centered	(R)	Hexagonal
(D)	Primitive only	(S)	Tetragonal

7.

	Column-I		Column-II
(A)	Magnetic moment in a paramagnetic substance	(P)	
(B)	Magnetic moment in a ferromagnetic	(Q)	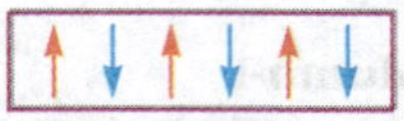
(C)	Magnetic moment in a antiferromagnetic	(R)	
(D)	Magnetic moment in a ferrimagnetic	(S)	

8.

	Column-I (Structure)		Column-II (Edge length of unit cell)
(A)	Rock salt (NaCl)	(P)	$a = (r_{\text{Cation}} + r_{\text{Anion}})$
(B)	CsCl	(Q)	$a = \frac{4}{\sqrt{3}}(r_{\text{Cation}} + r_{\text{Anion}})$
(C)	Zinc blende (ZnS)	(R)	$a = \frac{2}{\sqrt{3}}(r_{\text{Cation}} + r_{\text{Anion}})$
(D)	Anti fluorite (Na_2O)	(S)	$a = 2(r_{\text{Cation}} + r_{\text{Anion}})$

9.

	Column-I (Ideal structure)		Column-II (Packing fraction)
(A)	Rock salt	(P)	0.729
(B)	Zinc blende	(Q)	0.756
(C)	Fluorite	(R)	0.793
(D)	CsCl	(S)	0.748

10.

Column-I (Structure)	Column-II (Voids occupied)
(A) Rock salt	(P) 100% tetrahedral voids occupied by cation
(B) Zinc blende	(Q) 100% tetrahedral voids occupied by anion
(C) Fluorite	(R) 100% octahedral voids occupied by cation
(D) Anti fluorite (Na_2O)	(S) 50% tetrahedral voids occupied by cation

11.

Column-I (Void representation)	Column-II (Name of void)
(A) 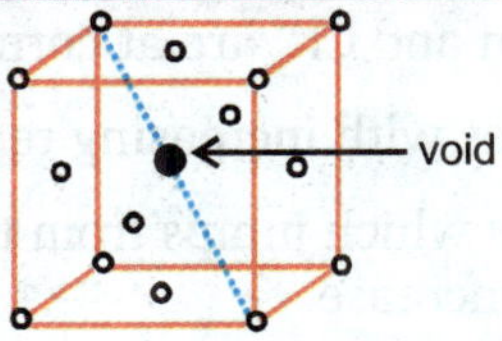	(P) Tetrahedral void
(B) 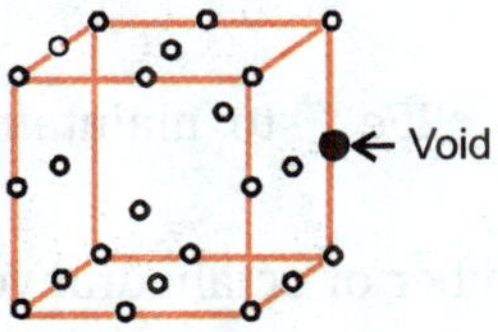	(Q) Octahedral void
(C) 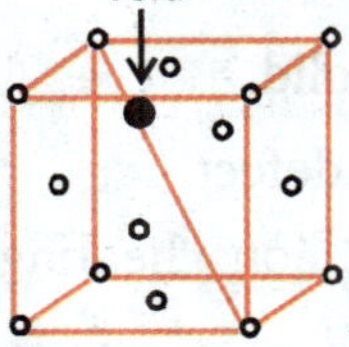	(R) Cubic void
(D) 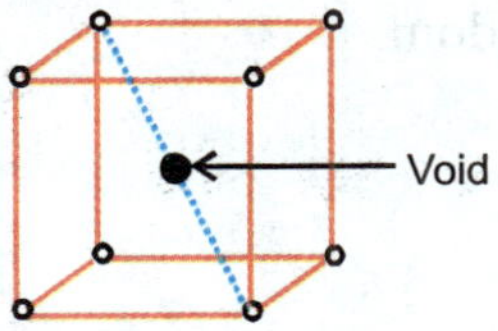	(S) Triangular void

Assertion-Reason Type Questions

Each question contains STATEMENT-1 (Assertion) and STATEMENT-2 (Reason).
Examine the statements carefully and mark the correct answer according to the instructions given below:
(A) If both the statements are TRUE and STATEMENT-2 is the correct explanation of STATEMENT-1
(B) If both the statements are TRUE but STATEMENT-2 is NOT the correct explanation of STATEMENT-1
(C) If STATEMENT-1 is TRUE and STATEMENT-2 is FALSE
(D) If STATEMENT-1 is FALSE and STATEMENT-2 is TRUE

1. **STATEMENT-1 :** In any ionic solid [*MX*] with Schottky defects, the number of positive and negative ions are the same.
   STATEMENT-2 : Equal number of cation and anion vacancies are present due to missing of few ions.
2. **STATEMENT 1 :** Amorphous solids are isotropic.
   STATEMENT 2 : Amorphous solids lack a regular three-dimensional arrangement of atoms.
3. **STATEMENT-1 :** Diamond is a covalent solid.
   STATEMENT-2 : The co-ordination number of each carbon atom in diamond is 4.

4. **STATEMENT-1 :** In NaCl structure, Na^+ ions occupy octahedral holes and Cl^- ions occupy ccp.

STATEMENT-2 : The distance of the nearest neighbours in NaCl structure is $a/2$ where a is the edge length of the cube.

5. **STATEMENT-1 :** For fluorite structure, the F^- ions occupy tetrahedral void and Ca^{2+} ions in ccp

STATEMENT-2 : The radius ratio of fluorite structure is 0.414.

6. **STATEMENT-1 :** In the body centered cubic type structure of CsCl the arrangement of Cs^+ is present in central octahedral void.

STATEMENT-2 : In CsCl, the Cs^+ ions are at body centered position and Cl^- are at corners.

7. **STATEMENT-1 :** Electrical conductivity of semi-conductors increases with increasing temperature.

STATEMENT-2 : With increase in temperature, number of electrons which jumps from the valence bond to the conduction band in semi-conductors increase.

8. **STATEMENT-1 :** FeO is non-stoichiometric with formula $Fe_{0.95}O$.

STATEMENT-2 : Some Fe^{2+} ions are replaced by Fe^{3+} as $3Fe^{2+} = 2Fe^{3+}$ to maintain electrical neutrality.

9. **STATEMENT-1 :** The number of tetrahedral voids is double the number of octahedral voids.

STATEMENT-2 : The size of the tetrahedral void is half of that of the octahedral void.

10. **STATEMENT-1 :** Due to Frenkel defect there is no effect on density of a solid.

STATEMENT-2 : Ions shift from lattice sites to interstitial sites in Frenkel defect.

11. **STATEMENT-1 :** Antiferromagnetic substances become paramagnetic on heating to high temperature.

STATEMENT-2 : Heating results in spins of electrons becoming random.

Level-4

Subjective Problems

1. In seven possible crystal system how many crystal system have $\alpha = \beta$?
2. In seven possible crystal system how many crystal system have more than one Bravais lattice ?
3. How many effective no. of atoms present in bcc unit cell ?
4. In solid X atoms goes to corner of the cube and two alternate face centre. Calculate effective number of atom of X in unit cell ?
5. Find the distance (in pm) between the body centered atom and one corner atom in an element ($a = 2.32$ pm)
6. The structure of MgO is similar to NaCl. What is the co-ordination number of Mg ?
7. How many number of formula units in a unit cell of CaF_2 ?
8. What is the co-ordination number of sodium in Na_2O ?
9. Density of Li crystal is 0.53 g/cm^3. The edge length of Li unit-call is 3.5 Å. Find out the number of Li atoms in a unit cell. ($N_A = 6.0 \times 10^{23} \text{ mol}^{-1}$, $M = 6.94 \text{ g mol}^{-1}$)
10. What is the co-ordination number of Cl^- in CsCl structure ?
11. In cubic system how many atoms arrangement exist in nature ?
12. The ionic radii of A^+ and B^- are 1.7 Å and 1.8 Å respectively. Find the co-ordination number of A^+.
13. If edge fraction unoccupied in ideal anti-fluorite structure is x. Calculate the value of Z. Where $Z = \dfrac{x}{0.097}$
14. Ionic solid Na^+A^- crystallise in rock salt type structure. 2.592 gm of ionic solid salt NaA dissolved in water to make 2 litre solution. The pH of this solution is 8. If distance between cation and anion is 300 pm, calculate density of ionic solid (in gm/cm^3).
    (Given : $pK_w = 13, pK_a(\text{HA}) = 5, N_A = 6 \times 10^{23}$)
15. Calculate the value of $\dfrac{Z}{10}$. Where
    Z = Co-ordination number of 2D-square close packing
    +
    Co-ordination number of 2D-hcp
    +
    Co-ordination number of 3D-square close packing
    +
    Co-ordination number of 3D, $ABCABC$ ...... Packing
    +
    Co-ordination number of 3D, $ABAB$ .... packing

Numerical Problems

If the numerical value has more than two decimal places, truncate/round-off the value to TWO decimal places.)

1. For a crystal, the unit cell axial lengths are found to be $a = 5.6$ Å, $b = 7.5$ Å and $c = 9.2$ Å.
   The three coordinate axes are mutually perpendicular.
   How many number of Bravais lattices present in this crystal system?
2. A metal x crystallizes in cubic system with unit cell edge length equal to 4.0 Å.
   If its density is 5.625 g/cm^3, then the atomic diameter of x (in Å)______.
   (Given : $N_A = 6 \times 10^{23}$, molar mass of $x = 54$ g/mol)

3. The normal sublimation temperature of a substance is –90° C. It exists as cubic close packed structure with unit cell edge length of 5.0 Å and density 3 g/cm^3 at –100° C and 1 atm.
Calculate the density of substance (in g/L) at 0°C and 1 atm.
(Given : $N_A = 6 \times 10^{23}$)

4. The distance between adjacent oppositely charged ions in rubidium chloride is 328.5 pm, in potassium chloride is 314 pm, and in potassium bromide is 329 pm. The distance between adjacent oppositely charged ions in rubidium bromide (in pm) is :

5. In A^+B^- type ionic compound (NaCl type) closest distance between oppositely charged ions is 200 pm. If B^- ions occupy each corner and face centre, A^+ ions occupy all octahedral voids.
Calculate (in pm) shortest distance between two anions.

6. In the fcc lattice of particle A, B particles occupy all tetrahedral void and C occupy all octahedral void. Calculate the closest distance (in Å) between particle B and C.
(Given : Radius of particle A is 141.4 Å ; $\sqrt{2} = 1.414$, $\sqrt{3} = 1.732$)

7. The crystalline structure of a solid is CCP for 'X' atoms. If 'Y' atoms occupy all the octahedral voids and 'Z' atoms occupy all the tetrahedral voids, without any distortion in the CCP arrangement of 'X' atoms, then the fraction of body diagonal not covered by the atoms is (Given : $\sqrt{2} = 1.414$)

8. Calculate packing fraction of NaCl solid if ions along an axis connecting opposite face centres are absent.

9. Consider a non-stoichiometric compound $Cu_{1.90}O$ due to metal deficient defect.
Calculate the value of molar ratio of Cu^+ and Cu^{2+} ions in the compound.

10. Calculate packing fraction in 2D-hexagonal arrangement of identical spheres.
(Given : $\pi = 3.14$, $\sqrt{2} = 1.41$, $\sqrt{3} = 1.73$)

Answers

Level 1

1. (c)	2. (b)	3. (d)	4. (b)	5. (a)	6. (c)	7. (a)	8. (d)	9. (c)	10. (c)
11. (d)	12. (c)	13. (d)	14. (b)	15. (b)	16. (a)	17. (b)	18. (c)	19. (a)	20. (b)
21. (d)	22. (c)	23. (b)	24. (c)	25. (d)	26. (c)	27. (c)	28. (c)	29. (c)	30. (b)
31. (c)	32. (a)	33. (d)	34. (c)	35. (b)	36. (b)	37. (c)	38. (a)	39. (a)	40. (d)
41. (d)	42. (d)	43. (b)	44. (d)	45. (c)	46. (d)	47. (c)	48. (b)	49. (a)	50. (b)
51. (b)	52. (b)	53. (c)	54. (a)	55. (c)	56. (b)	57. (a)	58. (d)	59. (c)	60. (b)
61. (d)	62. (c)	63. (a)	64. (a)	65. (b)	66. (a)	67. (b)	68. (c)	69. (b)	70. (b)
71. (b)	72. (c)	73. (b)	74. (d)	75. (a)	76. (d)	77. (c)	78. (c)	79. (a)	80. (a)
81. (c)	82. (c)	83. (b)	84. (c)	85. (a)	86. (c)	87. (c)	88. (d)	89. (d)	90. (b)
91. (b)	92. (c)	93. (c)	94. (a)	95. (b)	96. (d)	97. (c)	98. (d)	99. (a)	100. (b)
101. (c)	102. (d)	103. (d)	104. (c)	105. (a)	106. (d)	107. (b)	108. (d)	109. (d)	110. (c)
111. (c)	112. (c)	113. (b)	114. (c)	115. (b)	116. (c)	117. (a)	118. (c)	119. (d)	120. (b)
121. (b)	122. (a)	123. (b)	124. (c)	125. (d)					

Level 2

1. (b)	2. (c)	3. (c)	4. (c)	5. (d)	6. (c)	7. (b)	8. (d)	9. (a)	10. (c)
11. (b)	12. (b)	13. (c)	14. (d)	15. (c)	16. (a)	17. (b)	18. (a)	19. (a)	20. (b)
21. (a)	22. (a)	23. (d)	24. (d)	25. (b)	26. (a)	27. (b)	28. (b)	29. (d)	30. (a)
31. (b)	32. (b)								

Level 3

Passage-1:	1. (c)	2. (a)		
Passage-2:	1. (a)	2. (a)	3. (c)	
Passage-3:	1. (b)	2. (b)	3. (a)	4. (a)
Passage-4:	1. (b)	2. (b)	3. (d)	
Passage-5:	1. (b)	2. (b)	3. (a)	
Passage-6:	1. (d)	2. (a)		
Passage-7:	1. (a)	2. (d)	3. (d)	

One or More Answers is/are correct

1. (a,d)	**2.** (b,c)	**3.** (a,c)	**4.** (a,c)	**5.** (c)	**6.** (a,c)	**7.** (a,b)	**8.** (a,b,d)
9. (c,d)	**10.** (a,b,c,d)	**11.** (a,c,d)	**12.** (a,b)	**13.** (a,b,c)	**14.** (a,b,c)	**15.** (a,b,c,d)	**16.** (a,b,c)
17. (a,c)	**18.** (a,b,c,d)	**19.** (a,b,c,d)	**20.** (a,c)				

Match the Column

1. $A \rightarrow P$; $B \rightarrow P, Q, S$; $C \rightarrow P, R$; $D \rightarrow P, Q$
2. $A \rightarrow S$; $B \rightarrow Q$; $C \rightarrow Q, R$; $D \rightarrow P$
3. $A \rightarrow R$; $B \rightarrow S$; $C \rightarrow Q$; $D \rightarrow P$
4. $A \rightarrow P$; $B \rightarrow P$; $C \rightarrow S$; $D \rightarrow R$
5. $A \rightarrow P$; $B \rightarrow S$; $C \rightarrow Q$; $D \rightarrow R$
6. $A \rightarrow Q$; $B \rightarrow P$; $C \rightarrow S$; $D \rightarrow R$
7. $A \rightarrow P$; $B \rightarrow R$; $C \rightarrow Q$; $D \rightarrow S$
8. $A \rightarrow S$; $B \rightarrow R$; $C \rightarrow Q$; $D \rightarrow Q$
9. $A \rightarrow R$; $B \rightarrow S$; $C \rightarrow Q$; $D \rightarrow P$
10. $A \rightarrow R$; $B \rightarrow S$; $C \rightarrow Q$; $D \rightarrow P$
11. $A \rightarrow Q$; $B \rightarrow Q$; $C \rightarrow P$; $D \rightarrow R$

Assertion-Reason Type Questions

1. (A)	**2.** (A)	**3.** (B)	**4.** (B)	**5.** (C)	**6.** (D)	**7.** (A)	**8.** (A)	**9.** (C)	**10.** (A)
11. (A)									

Level 4

Subjective Problems

1. 5	**2.** 4	**3.** 2	**4.** 2	**5.** 2	**6.** 6	**7.** 4	**8.** 4	**9.** 2	**10.** 8
11. 3	**12.** 8	**13.** 3	**14.** 4	**15.** 4					

Numerical Problems

1. 4.00	**2.** 2.83	**3.** 2.51	**4.** 343.50	**5.** 282.80	**6.** 173.20	**7.** 0.24	**8.** 0.59
9. 18.00	**10.** 0.91						

Hints & Solutions

Level 1

13. (d) Orthorhombic contains

Simple, Body centre, Face centre & End centre bravais lattices.

14. (b) Cubic-symmetrical

$\alpha = \beta = \gamma = 90°, \quad a = b = c$

Triclinic-unsymmetrical

$\alpha \neq \beta \neq \gamma \neq 90°, \quad a \neq b \neq c$

15. (b) Tetragonal $\quad a = b \neq c, \quad \alpha = \beta = \gamma = 90°$

19. (a) Simple cubic cell "Z" $= 8 \times \frac{1}{8} = 1$

Face centred cubic unit cell

$$Z = 8 \times \frac{1}{8} + 6 \times \frac{1}{2} = 4$$

20. (b) Effective no. of atoms in End centred unit cell

$$= 8 \times \frac{1}{8} + 2 \times \frac{1}{2} = 2$$

atoms are present at corners as well as at end face's centre of unit cell.

21. (d) Simple cubic $\qquad$ Body centred cubic

$a = 2R \qquad \sqrt{3}a = 4R$

22. (c) Face centred cubic $\sqrt{2}a = 4R$

23. (b) Packing fraction (simple cubic) $= \dfrac{1 \times \frac{4}{3}\pi R^3}{a^3}$

$$= \frac{\pi}{6} = 0.52$$

"$a = 2R$"

24. (c) Body centred cubic $= \dfrac{2 \times \frac{4}{3}\pi R^3}{a^3} = \dfrac{\sqrt{3}\pi}{8}$

$\sqrt{3}a = 4R$

25. (d) Face centred cubic $= \dfrac{4 \times \frac{4}{3}\pi R^3}{a^3} = 0.74$

$\sqrt{2}a = 4R$

26, 27. (c, c)

Packing efficiency

In simple cubic $= 0.52$

Body centred cubic $= 0.68$

Face centred cubic $= 0.74$

28. (c) $a = 2R$

$$R = \frac{0.236}{2} = 0.118 \text{ nm}$$

29. (c) $\sqrt{3}a = 4R$

30. (b) $\sqrt{2}a = 4R$

31. (c) $\sqrt{2}a = 4R$

$$a = 2\sqrt{2}R$$

32. (a) $Z = 8 \times \frac{1}{8} = 1$

Nearest neighbour $= 6$

33. (d) $Z = 8 \times \frac{1}{8} + 6 \times \frac{1}{2} = 4$

Nearest neighbour $= 12$

34. (c) $Z = 8 \times \frac{1}{8} + 1 = 2$

Nearest neighbour $= 8$

35. (b) $D = \dfrac{ZM}{N_A a^3} \Rightarrow 6.25 = \dfrac{Z \times 12}{6.02 \times 10^{23} \times (4 \times 10^{-8})^3}$

$Z = 2$ body centred

36. (b) $D = \dfrac{ZM}{N_A a^3}$

Let W has *bcc* structures $\sqrt{3}a = 4r$

$\Rightarrow \quad Z = 2$

37. (c) $D = \dfrac{ZM}{N_A a^3} = \dfrac{4 \times 40}{6.02 \times 10^{23}\, a^3} = 1.83$

$a = 0.525$ nm

38. (a) $D = \dfrac{ZM}{N_A a^3} \qquad Z = 4$

$\sqrt{2}a = 4R$

39. (a) Refer Q.no. 38

40. (d) Refer Q.no. 38

41. (d) $D = \dfrac{ZM}{N_A a^3} \qquad Z = 2$

42. (d) Refer Q.no. 38

43. (b) Refer Q.no. 38

44. (d) Mole of 'x' $= \dfrac{8}{80} = 0.1$

No. of atoms $= 0.1\, N_A$

$\therefore$ No. of unit cell $= \dfrac{0.1\, N_A}{4} = 0.025\, N_A$

45. (c) $D = \frac{ZM_w}{N_A a^3}$

$Z = 2,\ M_w = 96,\ D = 10.3\ g/cm^3$

$\sqrt{3}a = 4R = \frac{\sqrt{3}\,a}{4}$

46. (d) Coordination no. of

simple cubic = 6

Face centred cubic = 12

Body centred cubic = 8

47. (c) F.C.C. unit cell *ABCABCABC* ....

packing efficiency = 74 %

50. (b) C.N. of $A = x(B)$

C.N. of $B = y(A)$

Formula A_yB_x

51. (b) $D = \frac{ZM}{N_A a^3}$

$Z = 4,\quad a = 2\sqrt{2}R$

52. (b) $D = \frac{ZM}{N_A a^3}$

$Z = 4,\quad a = 3.5 \times 10^{-8}\ \text{cm},\ Mw = 59$

54. (a) Packing efficiency of HCP & CCP is same.

56. (b) $Z = 12 \times \frac{1}{6} + 2 \times \frac{1}{2} + 3 = 6$

58. (d) No. of octahedral hole in a unit cell

= effective no. of atoms per unit cell

= Z

No. of tetrahedral holes = 2 × octahedral holes.

59. (c) If sheet is repeated than packing is poor.

60. (b) Refer Q. no. 58

62. (c) Tetrahedral & octahedral holes are present in hcp & ccp.

63. (a) Refer Q. no. 62

67. (b) $\frac{r_c}{r_a} = 0.225$ tetrahedral holes

68. (c) $\frac{r_c}{r_a} = 0.414$ octahedral holes

69. (b) an octahedral holes is formed by 6 sphere.

70. (b) No. of nearest neighbour = 8 (All body centre atom w.r.t. corner atom)

No. of next nearest neighbour = 6 (No. of corner atoms along edge w.r.t. to any corner).

71. (b) Refer Q. no. 67, 68

72. (c) Refer Q. no. 68

73. (b) Octahedral sites in fcc are present at each edge centre as well as at body centre.

74. (d) In Na_2O all tetrahedral void are occupied.

75. (a) $\frac{r_c}{r_a} = 0.732$

$r_c = 0.732 \times 200 = 146.4\ \text{pm}$

76. (d) for cubic void, $\frac{r_c}{r_a} = 0.732$

77. (c) $\frac{r_c}{r_a} = \frac{0.066}{0.140} = 0.47$

So, cation (Mg^{2+}) fitted in octahedral holes of O^{2-} are present at fcc lattice point.

78. (c) In diamond C-atoms are present in fcc lattice points as well as in alternate tetrahedral void.

79. (a) Refer Q. no. 78

80. (a) For MgO, $\frac{r_c}{r_a} = \frac{0.65}{1.4} = 0.47$

Mg^{2+} is in octahedral void so C.N. = 6

For MgS, $\frac{r_c}{r_a} = \frac{0.65}{1.84}$

Mg^{2+} is in tetrahedral void so C.N. = 4

81. (c) No. of $A = 7 \times \frac{1}{8} = \frac{7}{8}$

No. of $B = 6 \times \frac{1}{2} = 3$

$A_{7/8}B_3 = A_7B_{24}$

82. (c) No. of $S^{2-} = 8 \times \frac{1}{8} + 6 \times \frac{1}{2} = 4$

No. of $Ca^{2+} = \frac{1}{2} \times 8 = 4$

83. (b) No. of $A^{2+} = \frac{1}{8} \times 8 = 1$

No. of $B^{3+} = \frac{1}{2} \times 4 = 2$

No. of $O^{2-} = 8 \times \frac{1}{8} + 6 \times \frac{1}{2} = 4$

(AB_2O_4)

84. (c) No. of $C = 6 \times \frac{2}{3}$

No. of $A = 6$

(C_2A_3)

85. (a) No. of $O^{2-} = 8 \times \frac{1}{8} + 6 \times \frac{1}{2} = 4$

No. of $A = \frac{1}{8} \times 8 = 1$

No. of $B = \frac{1}{4} \times 4 = 1$

86. (c) No. of $A = 8 \times \frac{1}{8} = 1$

No. of $B = 5 \times \frac{1}{2} = \frac{5}{2}$

A_2B_5

87. (c) No. of Cu $= 8 \times \frac{1}{8} + 6 \times \frac{1}{2} = 4$

No. of Ag $= 12 \times \frac{1}{4} = 3$

No. of Au = 1

$Cu_4Ag_3Au_1$

88. (d) In NaCl structure

Cl^- ions are present at fcc lattice point

Na^+ ions are present in all octahedral void.

89. (d) In B.C.C. ionic structure

Cation is present in cubic void

Anions are present at each corner of cubic

90. (b) In ZnS, S^{2-} ions are present in F.C.C. Lattice point Zn^{2+} ions are present in alternate tetrahedral void.

91. (b) "edge"

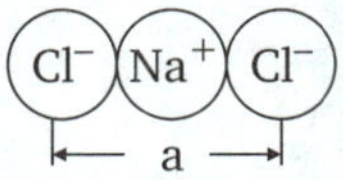

$r^+ + r^- = \frac{a}{2}$

92. (c) In ideal arrangement of anions

$$\sqrt{2}a = 4r_a$$

$$r_a = \frac{a}{2\sqrt{2}}$$

93. (c) $r_c + r_a = \frac{\sqrt{3}a}{2}$

Anion are at corners and cations are at body centre.

94. (a) Distance b/w two nearest $Cl^- = \frac{a}{\sqrt{2}}$

Distance b/w two nearest $Na^+ = \frac{a}{\sqrt{2}}$

95. (b) w.r.t. body centre Na^+ ion

all 12 Na^+ ions of edge centre.

96. (d) $\frac{r^+}{r^-} = \frac{1}{2} = 0.5$ F.C.C. lattice point

Cations are at octahedral void

$$(r^+ + r^-) = \frac{a}{2} = 3$$

$$a = 6$$

$$\text{volume} = (6)^3 = 216\ \text{pm}^3$$

97. (c) In CaF_2 cations are present at F.C.C. lattice points and anions are at all tetrahedral void, antiflourites structure is reverse arrangement of CaF_2.

98. (d) Refer Q. no. 89.

99. (a) $\frac{r_c}{r_a} = \frac{94}{146} = 0.64$

100. (b) $r_c + r_a = \frac{a}{2}$

102. (d) $\frac{r_A^+}{r_B^-} = 0.225$

$$r_A^+ = 45$$

103. (d) In Na_2O O^{2-} are at F.C.C. lattice point

Na^+ are at all tetrahedral void

C.N. of $O^{2-} = 8$

104. (c) $r_c + r_A = \frac{\sqrt{3}a}{2}$

$$a = \frac{2}{\sqrt{3}}[1.69 + 1.81]$$

105. (a) $r_c + r_a = \frac{\sqrt{3}a}{2}$

106. (d) $r_c + r_a = \frac{a}{2}$

107. (b) $r_c + r_a = \frac{\sqrt{3}a}{2}$

108. (d) $r_c + r_a = \frac{\sqrt{3}a}{4}$

110. (c) Refer Q. no. 78

111. (c) Due to missing of ions. Density decreases in Schottky defect.

112. (c) In Frenkel defect ions are shifted in its interstitial site but density remains unchanged.

113. (b) Refer Q.no. 112

115. (b) Alkali metal halide shows schottky defect.

116. (c) Two Ag^+ ions will replaced by one Cd^{2+},

so there is one vacancy for each Cd^{2+}

117. (a) In Schottky defect ions are missing from their position and in Frenkel defect ions are shifted in their interstitial site from their lattice point.

118. (c) Refer Q. no. 115.

119. (d) Density of crystal always increases due to substitutional impurity defect.

120. (b) $\frac{\sqrt{3}a}{4} = (2R)$

121. (b) No. of vacancy = No. of Sr^{2+}

No of Sr^{2+} in 1 mole NaCl $= 10^{-7} \times N_A$

122. (a) Fe^{3+} $\quad$ Fe^{2+}

$x\%$ $\quad$ $(100 - x)\%$

$$\frac{200}{93} = \frac{3x}{100} + \frac{2(100 - x)}{100}$$

$$\%Fe^{3+} = 15.05\%$$

123. (b) Cu^{2+} $\quad$ Cu^+

$x\%$ $\quad$ $(100 - x)\%$

$$\frac{20}{18} = \frac{2x}{100} + \frac{(100 - x)}{100}$$

$$x = 11.11$$

Level 2

1. (b) $\rho_1 = \frac{2 \times 56}{\left(\frac{4r}{\sqrt{3}}\right)^3}$; $\rho_2 = \frac{4 \times 56}{(2\sqrt{2}r)^3}$

$$\frac{\rho_1}{\rho_2} \Rightarrow 0.918$$

2. (c) $2.32 = \frac{2 \times M}{6 \times 10^{23}(1.22)^3 \times 10^{-21}}$

$$\Rightarrow M = 1264 \Rightarrow x \approx 47$$

3. (c) Edge not covered by atom $= a - 2r$

also in bcc, $4r = \sqrt{3}a$

Therefore, edge not covered

$$= a - \frac{\sqrt{3}}{2}a = a\left(\frac{2-\sqrt{3}}{2}\right)$$

$\Rightarrow$ Fraction of edge not covered by atoms

$$= \frac{2-\sqrt{3}}{2} = 0.134$$

4. (c) In a simple cubic crystal, $a = 2r$

Let the radius of the interstitial atom in simple cubic structure be x, then

$$2(r+x) = \sqrt{3}a = \sqrt{3} \times 2r$$

So, $x = \sqrt{3}r - r \Rightarrow 0.732\,r$

$$\text{Packing fraction} = \frac{\frac{4}{3}\pi r^3 + \frac{4}{3}\pi(0.732\,r)^3}{(2r)^3}$$

$$= \frac{\frac{4}{3}\pi r^3[1+(0.732)^3]}{8r^3}$$

$$= \frac{\pi}{6}[1+(0.732)^3] = 0.73$$

5. (d) Given $a - 2r = 60.3$ and for bcc, $4r = \sqrt{3}a$

$$\Rightarrow a - \frac{\sqrt{3}}{2}a = 60.3 \Rightarrow a = 450 \text{ pm}$$

$$\text{Density } (\rho) = \frac{2 \times 48}{6.023 \times 10^{23} \times (4.5)^3 \times 10^{-24}}$$

$$= 1.75 \text{ g/cc}$$

6. (c) $a - 2r = 53.6$ pm ...(i)

also $4r = \sqrt{3}a$

$$\Rightarrow a - \frac{\sqrt{3}}{2}a = 53.6$$

$$\Rightarrow a = \frac{53.6 \times 2}{2-\sqrt{3}} = 400 \text{ pm}$$

$$\text{Density } (\rho) = \frac{2 \times 23}{6.023 \times 10^{23} \times 4^3 \times 10^{-24}}$$

$$= 1.19 \text{ g/cc}$$

7. (b) Vol. of all atoms in 1.68 g of Ar

$$= \frac{1.68}{40} \times 6 \times 10^{23} \times \frac{4}{3} \times \frac{22}{7} \times (1.50^3 \times 10^{-8})$$

$$= 0.3564$$

Vol. of solid Ar $= 1 \text{ cm}^3$

% of empty space $= (1 - 0.3564) \times 100$

$$= 64.36$$

8. (d) Effective no. of atoms of B present in a unit cell $= 2$

Total volume of B unoccupied by A in a unit cell

$$= 2 \times \frac{4}{3}(R^3 - r^3) \times \pi$$

$$= \frac{7\pi R^3}{3} \qquad \left(\because r = \frac{R}{2}\right)$$

Volume of unit cell $= a^3$

$$\Rightarrow \left(\frac{4R}{\sqrt{3}}\right)^3 = \frac{64}{3\sqrt{3}}R^3 \qquad (\because \sqrt{3}a = 4R)$$

$$\text{Desired ratio} \Rightarrow \frac{\frac{7\pi R^3}{3}}{\frac{64}{3\sqrt{3}}R^3} = \frac{7\pi\sqrt{3}}{64}$$

9. (a)

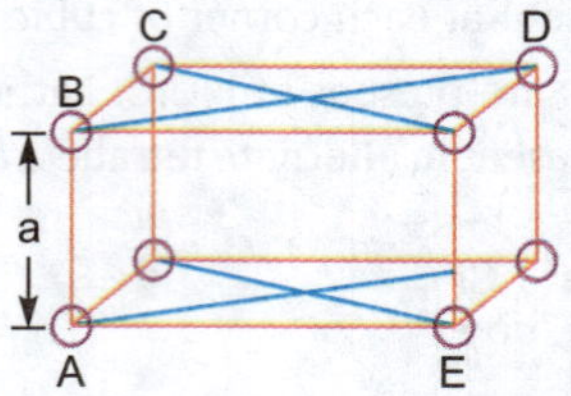

$AB = a$ (nearest)

$BD = \sqrt{2}a$ (next-nearest)

$CE = \sqrt{3}a$ (next-next-nearest)

10. (c)

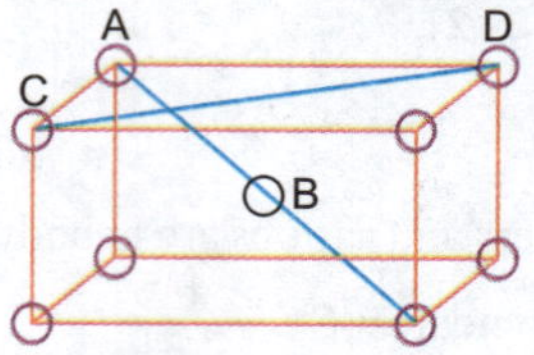

$AB = \frac{\sqrt{3}a}{2}$ (nearest)

$AC = a$ (next-nearest)

$CD = \sqrt{2}a$ (next-next-nearest)

11. (b) For atoms of $A = 8 \times \frac{1}{8} = 1$

For atoms of $B = 8 \times \frac{1}{4} = 2$

For atoms of $C = 4 \times \frac{1}{2} + 1 = 3$

12. (b) Let volume of fcc unit cell $= V$

$$\rho_A = \frac{4 \times M_A}{N_A V}$$

$$\rho_B = \frac{8 \times M_B}{N_A . V}$$

$$\frac{\rho_A}{\rho_B} = \frac{M_A}{2M_B} = \frac{30}{2 \times 50} = 0.3$$

$$\rho_B = \frac{\rho_A}{0.3} = 3.33 \text{ g/cc}$$

Total density $\rho_A + \rho_B = 4.33$ g/cc

13. (c)

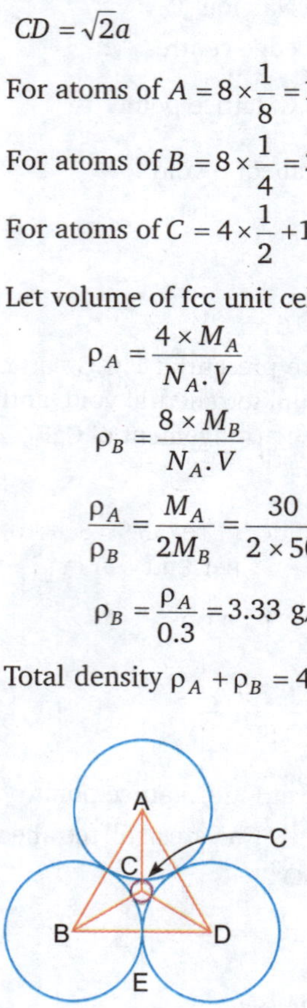

$$\frac{BE}{BC} = \cos 30° = \frac{\sqrt{3}}{2}$$

Distance between the centres of two B atoms

$$= 2 \times BE = \sqrt{3}\ \text{Å}$$

14. (d) 4 formula units of AB are present in a unit cell

$$\therefore \text{no. of unit cell} = \frac{4}{40} \times \frac{N_A}{4}$$

$$\Rightarrow 0.025\, N_A$$

15. (c) Use $\rho = \dfrac{Z.M}{N_A . a^3}$

$$Z = 4; \quad M = 40 + 2 \times 19$$

16. (a) $a = 2\,(r^+ + r^-)$

$$= 2 \times 297 \text{ pm}$$

Volume of unit cell $= a^3$

$$= 209.6 \times 10^{-24} \text{ cm}^3$$

17. (b) $8.27 = \dfrac{4 \times 128}{6.023 \times 10^{23}\, a^3}$

$$\Rightarrow a = 46.8 \times 10^{-9} \text{ cm} = 4.68\ \text{Å}$$

$$= 2\,[r_{O^{2-}} + r_{Cd^{2+}}]$$

$$\Rightarrow r_{Cd^{2+}} = 1.1\ \text{Å}$$

18. (a) $\dfrac{r_{Na^+} + r_{Cl^-}}{r_{Cl^-}} = 1.5$

and $\dfrac{r_{K^+}}{r_{Cl^-}} = \dfrac{5}{7}$ or $\dfrac{r_{K^+} + r_{Cl^-}}{r_{Cl^-}} = \dfrac{12}{7}$

$$\frac{a_1}{a_2} = \frac{2\,(r_{K^+} + r_{Cl^-})}{2\,(r_{Na^+ + Cl^-})}$$

$$= \frac{12}{7} \times \frac{1}{1.5} = 1.143$$

19. (a) Mass of unit cell $= d \times V$

$$= (5 \times 10^{-8})^3 \times 3.84 \times 6 \times 10^{23}$$

$$= 288 \text{ a.m.u.}$$

Mass of formula unit of FeO $= 56 + 16 = 72$ a.m.u.

$$\therefore \text{no. of formula unit } = \frac{288}{72} = 4$$

20. (b) Density of crystals decrease due to Schottky defects effective no. of atoms per unit cell in given lattice

$$= 4\left(1 - \frac{0.25}{100}\right) = 3.99$$

$$\rho = \frac{50 \times 3.99}{6 \times 10^{23}\,(0.50 \times 10^{-7})^3}$$

$$\Rightarrow = \frac{50 \times 3.99}{6 \times 125 \times 0.1}$$

$$\rho = 2.66 \text{ g/cm}^3$$

21. (a) Theoretical density $= \dfrac{Z.M}{N_A . a^3}$

$$= \frac{4 \times 24 \times 10^{-3}}{6 \times 10^{23} \times (4 \times 10^{-10})^3}$$

$$= 2.5 \times 10^3 \text{ kg/m}^3$$

$$\%\ \text{occupancy} = \frac{\text{Observed density}}{\text{Ideal density}} \times 100$$

$$= \frac{2.4 \times 10^3}{2.5 \times 10^3} \times 100 = 96\%$$

23. (d) Octahedral void present at the centre of cube and tetrahedral void is present at $(1/4)^{th}$ of the distance along each body diagonal.

$\therefore \dfrac{\sqrt{3}a}{2} = 2 \times$ distance between octahedral and tetrahedral void.

24. (d) Vol. of unit cell $= a^2 \sin 60° \times b$

$$= 173.2 \times 10^{-24} \text{ cm}^3$$

$$a = 5\ \text{Å}$$

Mass of unit cell

$$= 173.2 \times 10^{-24} \times 5 \times 6 \times 10^{23}$$

$$= 519.6 \text{ g}$$

No. of molecules present in given unit cell

$$= \frac{519.6}{259.8} = 2$$

25. (b) In hcp arrangement total height

$$h = 4 \times \sqrt{\frac{2}{3}}\, r$$

as per given

$$\frac{h}{2} = 2\sqrt{\frac{2}{3}}\, r = 3.35\ \text{Å} = 3.35 \times 10^{-8} \text{ cm.}$$

Volume of unit cell $= 24\sqrt{2}r^3$

$$= 24\sqrt{2}\left(\sqrt{\frac{3}{2}} \times \frac{1}{2} \times 3.35 \times 10^{-8}\right)^3$$

$$= 2.93 \times 10^{-22} \text{ cm}^3$$

Effective no. of atoms in hexagonal unit cell $= 6$

$$\therefore \text{ density} = \frac{6 \times 12}{6.023 \times 10^{23} \times 2.93 \times 10^{-22}}$$

$$= 0.41 \text{ g/cc}$$

26. (a) No. of $Na^+ = 4 - 1 = 3$

No. of $Cl^- = 4 - \left(2 \times \dfrac{1}{2}\right) = 3$

27. (b) Along one body diagonal $2X$ atoms from 2 corners, one Y particle (at the centre of cube) will be removed.

So, effective no. of X particles in a unit cell

$$= 4 - \left(2 \times \frac{1}{8}\right) = \frac{15}{4}$$

and effective no. of Y particles in a unit cell

$$= 4 - 1 = 3$$

$$X : Y \quad \frac{15}{4} : 3$$

or $\quad 5 : 4$

28. (b) Effective no. of $Na^+ = 4 - \left(1 + 2 \times \frac{1}{4}\right) = \frac{5}{2}$

For Cl^- $= 4\sqrt{2}a = 4r^-$

$$P.F. = \frac{\text{Vol. of effective no. of cations and anions}}{\text{Vol. of unit cell}}$$

29. (d) No. of X particles per unit cell = 4

No. of Y particles = 4

No. of Z particles = 8

Along one body diagonal $2X$ atoms from 2 corners, one Y particle and $2Z$ particles will be removed.

So, effective no. of X particles in a unit cell

$$= 4 - \frac{1}{8} \times 2 = \frac{15}{4}$$

Effective no. of Y particles in a unit cell

$$= 4 - 1 = 3$$

Effective no. of Z particles in a unit cell

$$= 8 - 2 = 6$$

$$X : Y : Z$$

$$\frac{15}{4} : 3 : 6$$

$$5 : 4 : 8$$

30. (a) Effective no. of A particles are removed

$$= 4 \times \frac{1}{8} + 2 \times \frac{1}{2} = \frac{3}{2}$$

Effective no. of A particles present in a unit cell

$$= 4 - \frac{3}{2} = \frac{5}{2}$$

Effective no. of B particles are removed

$$= 2 \times \frac{1}{4} + 1 = \frac{3}{2}$$

Effective no. of B particles present in a unit cell

$$= 4 - \frac{3}{2} = \frac{5}{2}$$

$$A : B$$

$$\frac{5}{2} : \frac{5}{2} \quad \text{or} \quad 1 : 1$$

Level 3

Passage-1

1. (c) $P.F. = \dfrac{2 \times \frac{4}{3}\pi r^3}{\left(\frac{4r}{\sqrt{3}}\right)^3} \simeq 0.68$

% of empty space $= 100 - 0.68 \times 100 = 32$

2. (a) $P.F. = \dfrac{4 \times \frac{4}{3}\pi r^3}{\left(\frac{4r}{\sqrt{2}}\right)^3} = 0.7406$

Passage-2

1. (a) $10.6 = \dfrac{4 \times 108}{a^3 \times 6.023 \times 10^{23}}$

$\therefore\ a = 40.7$ nm

2. (a) Mass of 12×10^{23} atoms $= 100$ gm

Mass of 6.022×10^{23} atom

$$= \frac{100}{12 \times 10^{23}} \times 6.023 \times 10^{23}$$

$$= 50.18$$

$$\therefore \quad \rho = \frac{4 \times 50.18 \times 10}{6.023 \times 10^{23} \times (200 \times 10^{-10})^3}$$

$$= 41.66 \text{ g/cm}^3$$

3. (c) By Hit and Trial : For fcc $Z = 4$ (4 pairs of KBr)

$$M = 39 + 80 = 119$$

Passage-3

3. (a) 1 to 3

c.c.p. generated f.c.c., octahedral voids = effective number of atoms in a unit cell, tetrahedral voids $= 2 \times$ effective no. of atoms in a unit cell.

4. (a) O^{2-} is replaced by $X^{8/3-}$ so formula of spinel is $MgAl_2X_3$ and deficiency by one anion.

Passage-6

1. (d) $D_{Theo} = \dfrac{ZM}{N_A \times a^3}$

$$= \frac{4 \times 197}{6 \times 10^{23} \times (407 \times 10^{-10})^3}$$

$$= \frac{4 \times 197}{6 \times 10^{23} \times 6.74 \times 10^{-23}}$$

$$= 19.485 \text{ g/cm}^3$$

2. (a) Distance between next neighbour $= a$

$$= 407 \text{ pm}$$

Passage-7

1. (a) Effective no. of C-atoms per unit cell = 8

mass $= 12 \times 8 = 96$ amu

2. (d) $\dfrac{\sqrt{3}a}{4} = B.L_{C-C}$

$$D = \frac{4 \times 154}{\sqrt{3}} = 355.66 \text{ pm}$$

3. (d) $D = \dfrac{ZM}{N_A a^3} = \dfrac{8 \times 12}{6 \times 10^{23} \times (355.66 \times 10^{-10})^3}$

$$= 3.56 \text{ g/cm}^3$$

One or More Answers is/are Correct

4. (a, c)

$$\frac{r_{Na^+}}{r_{Cl^-}} = \frac{95}{181} = 0.52$$

Cl$^-$ ion at F.C.C. lattice point

Na$^+$ ion are at octahedral void

C.No. of NaCl = 6 : 6

$r_c + r_a = \frac{a}{2}$

7. (a, b)

Refer Q. no. 4

11. (a, c, d)

CsCl ionic solid is B.C.C.

12. (a, b)

On increasing temperature C.No. ↓

C.No. change 8 : 8 to 6 : 6

On increasing pressure C.No.↑

14. (a, b, c)

Refer Q. no. 4

16. (a, b, c)

Refer Q. no. 90 (Level-1)

17. (a, c)

Refer Q. no. 97 (Level-1)

18. (a, b, c, d)

(A) F.C.C. (3D) $= \dfrac{4\times\frac{4}{3}\pi R^3}{(2\sqrt{2}R)^3} = 0.74$

(B) Simple cubic $= \dfrac{1\times\frac{4}{3}\pi R^3}{a^3}$

$a = 2R$

H.C.P. (3D) $= \dfrac{6\times\frac{4}{3}\pi R^3}{24\sqrt{2}R^3}$

Void fraction $= 1 - 0.74 = 0.26$

20. (a, c)

$\dfrac{r_{A^{2+}}}{r_{B^{2-}}} = \dfrac{100}{300} = 0.333$ $\qquad \dfrac{r_{C^+}}{r_{D^-}} = \dfrac{-240}{480} = 0.5$

B^{2-} are at F.C.C. lattice point and A^{2+} ions are alternate tetrahedral void $\qquad$ D^- at F.C.C. lattice point C^+ in all octahedral void

C.No. of AB = 4 : 4 $\qquad$ C.No. of CD = 6 : 6

Level 4

Subjective Problems

1. Cubic $\alpha = \beta = \gamma = 90$

Tetragonal $\alpha = \beta = \gamma = 90$

Orthorhombic $\alpha = \beta = \gamma = 90$

Rombohedral $\alpha = \beta = \gamma \neq 90$

Hexagonal $\alpha = \beta = 90°$

2. Cubic, tetragonal, orthorhombic, monoclinic

3. $Z = 8\times\frac{1}{8} + 1 = 2$

4. No. of x-atoms $= 8\times\frac{1}{8} + 2\times\frac{1}{2} = 2$

5. Distance b/w body centred and corner atom $= \frac{\sqrt{3}a}{2}$

6. Mg^{2+} is in octahedral void

So its C.No. = 6

7. No. Ca^{2+} per unit cell = 4

No. of F^- per unit cell = 8

No. of CaF_2 unit per unit cell = 4

8. Na^+ ions are present in tetrahedral void in Na_2O, So C.No. of Na^+ = 4

9. $D = \dfrac{ZM}{N_A a^3}$

$Z = 2$

10. Cl$^-$ ions are at corners and Cs$^+$ at body centre

So C.No. of Cl$^-$ = 8 Cs$^+$

11. Simple cubic

body centred cubic

face centred cubic

12. $\dfrac{r_{A^+}}{r_{B^-}} = \dfrac{1.7}{1.8} = 0.944$

B^- are at corner of cube A^+ at body centre

C.No. of $A^+ = 8$

13. Fraction of edge unoccupied $= \dfrac{a-2R}{a}$

$a = 2\sqrt{2}R \qquad X = \dfrac{2(\sqrt{2}-1)}{2\sqrt{2}}$

$X = \dfrac{0.414}{1.414} = 0.293$

$Z = \dfrac{X}{0.097} = \dfrac{0.293}{0.097} = 3$

14. NaA is salt of SBWA

$\text{pH} = \dfrac{pK_w}{2} + \dfrac{1}{2}(pK_a + \log c)$

$8 = 6.5\times\dfrac{1}{2}(5 + \log c)$

$1.5\times 2 = 5 + \log c$

$\log c = -2$

$c = 0.01 = \dfrac{2.592}{M\times 2}$

$M = \dfrac{2.592}{2\times 0.01} = 129.6$

$D = \dfrac{ZM}{N_A a^3} \qquad Z = 4$

$= 4\,g/cm^3 \qquad a = 2\times 300 = 600\,pm$

15. $Z = 4 + 6 + 6 + 12 + 12 = 40$

$\dfrac{Z}{10} = \dfrac{40}{10} = 4$

Numerical Problems

2. Hence, x exist in CCP (FCC)

$$d = \frac{Z \cdot M}{N_A \cdot V}$$

$$5.625 = \frac{Z \times 54}{(6 \times 10^{23}) \times (4 \times 10^{-8})^3} \quad ; \quad Z = 4$$

For FCC : $\sqrt{2}a = 4r$

$$r = \frac{\sqrt{2} \times 4.0}{4} = 1.414 \text{ Å}$$

$$d = 2r = 2.828 \approx 2.83 \text{ Å}$$

3. $d = \dfrac{Z \cdot M}{N_A \cdot V}$

$$3 = \frac{4 \times M}{(6 \times 10^{23}) \times (5 \times 10^{-8})^3}$$

$\Rightarrow \quad M = 56.25$ g/mol

At 0°C, the substance will exist as ideal gas and its density

$$= \frac{56.25 \text{ mol L}^{-1}}{22.4 \text{ L mol}^{-1}} = 2.51 \text{ g/L}$$

4. $r_{Rb^+} + r_{Cl^-} = 328.5 \text{ pm} = a$

$r_{K^+} + r_{Cl^-} = 314 \text{ pm} = b$

$r_{K^+} + r_{Br^-} = 329 \text{ pm} = c$

$\therefore \quad r_{Rb^+} + r_{Br^-} = a + c - b = 343.5 \text{ pm}$

5. $a = 2(r_+ + r_-) = 2 \times 200 = 400 \text{ pm}$

Shortest distance between two anions

$$= \frac{\sqrt{2}a}{2} = \frac{\sqrt{2} \times 400}{2} = 282.80 \text{ pm}$$

6. $4r = \sqrt{2}a,\ 4 \times 200 = \sqrt{2}a,\ a = 2\sqrt{2} \times 141.4 \text{ Å}$

Distance between octahedral void and tetrahedral void in FCC

$$= \frac{\sqrt{3}a}{4} = \frac{\sqrt{3}}{4} \times 2\sqrt{2} \times 141.4$$

$$= \frac{\sqrt{3} \times 141.4}{\sqrt{2}} \text{ Å} = \sqrt{3} \times 100 \text{ Å}$$

$$= 1.732 \times 100 \text{ Å} = 173.20 \text{ Å}$$

7. Let radius of X atom $= r$

Fraction of body diagonal covered by atoms

$$= \frac{2r + 2 \times 0.414r + 4 \times 0.225\, r}{\sqrt{3} \times \left(\dfrac{4r}{\sqrt{2}}\right)} = 0.76$$

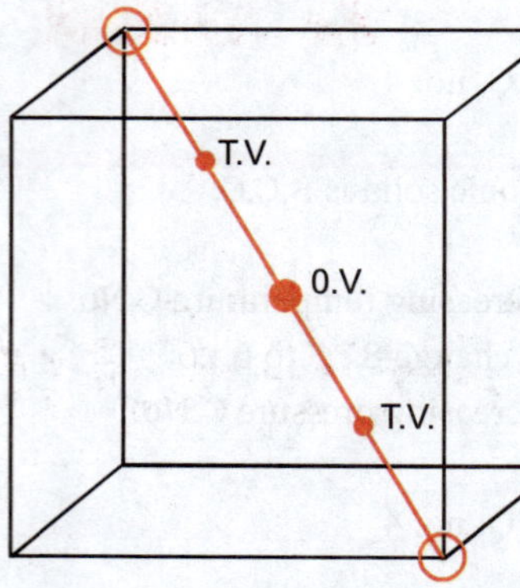

$\therefore$ Fraction, not covered by atoms of body diagonal $= 1 - 0.76 = 0.24$.

8. Number of $Na^+ = 3$,

Number of $Cl^- = 3$

where $a^3 = 2(r_c + r_a)^3$ and $\sqrt{2}a = 4r_a$

Packing fraction

$$= \frac{4\pi}{3}(3r_c^3 + 3r_a^3) \times \frac{1}{16\sqrt{2}r_a^3} = 0.593 \approx 0.59$$

9. Let x atoms are in $+2$ state

$$x \times 2 + (1.90 - x)1 - 2 = 0$$

$$x = 0.1$$

$$\frac{n_{Cu^{2+}}}{n_{Cu^+}} = \frac{0.1}{1.90 - 0.1} = \frac{1}{18} \Rightarrow \frac{n_{Cu^+}}{n_{Cu^{2+}}} = 18$$

10. $\text{P.F.} = \dfrac{\text{Area of effective no. of atoms}}{\text{Area of unit cell}}$

$$= \frac{3 \times \pi r^2}{6 \times \frac{1}{2} \times 2r \times 2r \sin 60°}$$

$$= \frac{\pi}{2 \times 1.73} = \frac{3.14}{2 \times 1.73} = 0.907 \approx 0.91$$

CHAPTER 12

Surface Chemistry

Terms Related to Adsorption and Absorption

- **Adsorption :** The accumulation of molecular species at the surface rather than bulk of solid or liquid is known as adsorption.
- **Adsorbate :** The molecular species or substance which accumulates at the surface is known as adsorbate.
- **Adsorbent :** The surface at which accumulation of adsorbate takes place is known as adsorbent.
  Ex : Water vapours (adsorbate) adsorb at the surface of silica gel (adsorbent).
- **Desorption :** The process of removing an adsorbed substance from adsorbent is known as desorption. When equilibrium is established then rate of adsorption and rate of desorption are equal.
- **Absorption :** When atoms, molecules or ions enter in bulk phase of solid or liquid is known as absorption.
  Ex. Water vapours absorbed by anhydrous $CaCl_2$.
- **Sorption :** When adsorption and absorption takes place simultaneously then phenomena is known as sorption.
  Ex. Dyeing of fabric

Characteristics of Adsorption

- Adsorption is surface phenomena.
- In adsorption enthalpy decreases.
- In adsorption entropy decreases.
- When decrease in free energy takes place then adsorption takes place.

Types of Adsorption

(1) Based on concentration :

- **Positive adsorption :** If the concentration of adsorbate is more at the surface as compared to its concentration in the bulk phase then it is called positive adsorption
- **Negative adsorption :** If concentration of adsorbate is less at the surface as compared to its concentration in the bulk phase then it is called negative adsorption

(2) Based on forces existing between adsorbate molecule and adsorbent :

- **Physical adsorption :** If the forces of attraction existing between adsorbate and adsorbent are van der Waal's forces, then adsorption is called physical adsorption.

- **Chemical adsorption :** If the forces of attraction existing between adsorbate particles and adsorbent are almost of the same strength as chemical bonds then adsorption is called chemical adsorption.

Comparison between physisorption and chemisorption

S.No.	Physisorption	Chemisorption
1.	Low heat of adsorption usually in range of 20 – 40 kJ/mol.	High heat of adsorption in the range of 80 – 240 kJ/mol.
2.	Forces of attraction are van der Waal's forces.	Forces of attraction are chemical bond forces.
3.	It is reversible in nature.	It is irreversible in nature.
4.	It usually takes place at low temp and it decreases with increasing temp.	It takes place at high temperature and it increase with increasing temperature.
5.	More easily liquefiable gases are adsorbed readily.	Gases which can react with the adsorbent show chemisorption.
6.	It forms multimolecular layers.	It forms monomolecular layer.
7.	It does not require any activation energy.	It requires activation energy.
8.	It is not specific.	It is highly specific.
9.	It increases with increase the surface area of adsorbent.	It also increases with increase the surface area of adsorbent.

Factors Affecting Adsorption

- **Nature of the gas :** Easily liquefiable gases adsorb to greater extent.
- **Effect of nature of adsorbent :** When adsorbent is activated then extent of adsorption increases.
- **Specific area of the solid :** Greater the specific area of the solid, greater would be its adsorbent power.
- **Effect of pressure of the gas :** The adsorption causes a net decrease in pressure as the gas gets adsorbed and thus the increase in pressure favours the process of adsorption.

Adsorption Isotherm

Graph between extent of the adsorption $\left(\frac{x}{m}\right)$ & pressure (P) at a given temperature is called adsorption isotherm. There are two types of adsorption isotherms

(i) Freundlich Adsorption isotherm :

Freundlich's Equation $\frac{x}{m} = k \times P^{1/n}$

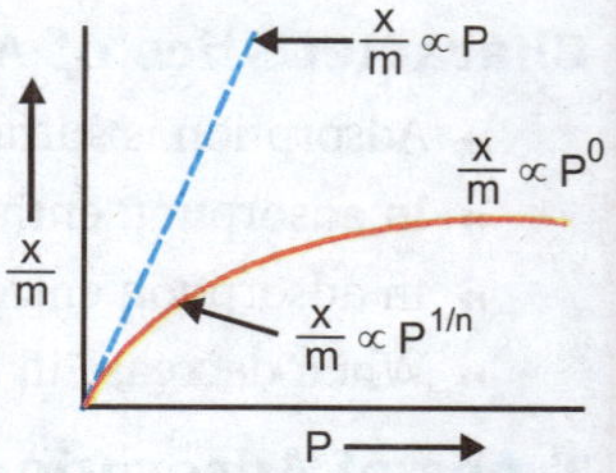

where x = mass of adsorbate adsorbed, m = mass of adsorbent, p = pressure

- **At low pressure :** $\frac{x}{m} = k \times P$
- **At intermediate pressure :** $\frac{x}{m} = k \times P^{1/n}$, where $n \geq 1$
- **At high pressure :** $\frac{x}{m}$ = constant

$$\log \frac{x}{m} = \log k + \frac{1}{n} \log p$$

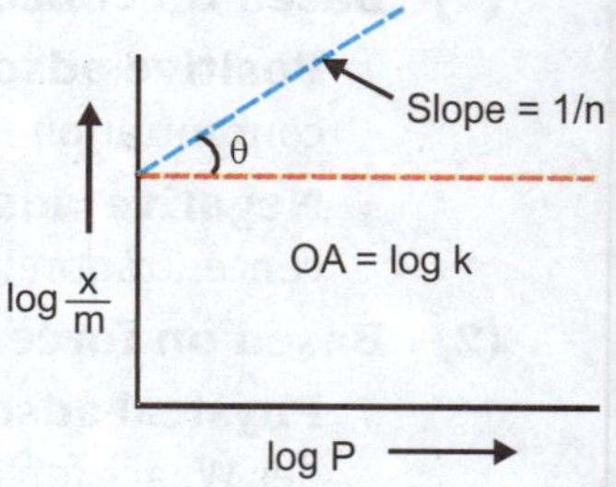

(ii) Langmuir Adsorption isotherm : According to Langmuir

- There is adsorption of gas molecules on the surface of the solid.
- There is desorption of the adsorbed molecules from the surface of the solid.
- There is formation of unimolecular layer and so it is chemisorption
- A dynamic equilibrium is attained when rate of adsorption = rate of desorption.

$$\frac{x}{m} = \frac{ap}{1+bp}$$

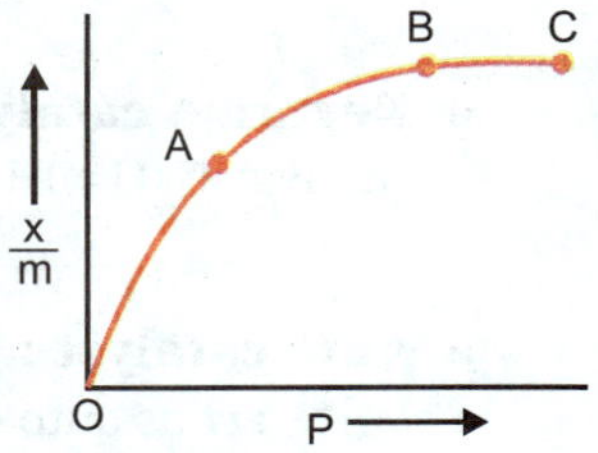

Where a & b Langmuir parameters.

Case-I At very high pressure

$$bp >>> 1,$$

$$\therefore \quad \frac{x}{m} = \frac{ap}{bp} = \frac{a}{b} = \text{constant}$$

Case-II At very low pressure

$$bp <<< 1$$

$$\therefore \quad \frac{x}{m} = ap$$

Adsorption Isobar

Graph between extent of the adsorption $\left(\frac{x}{m}\right)$ and temp. at a given pressure is called adsorption isobar.

A physical adsorption isobar shows a decrease in x/m as the temperature rises.

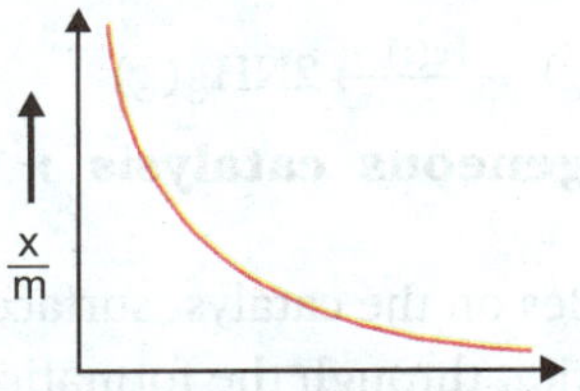

Physical adsorption isobar

The isobar of chemisorption shows an increase in the beginning and then decrease as the temperature rises.

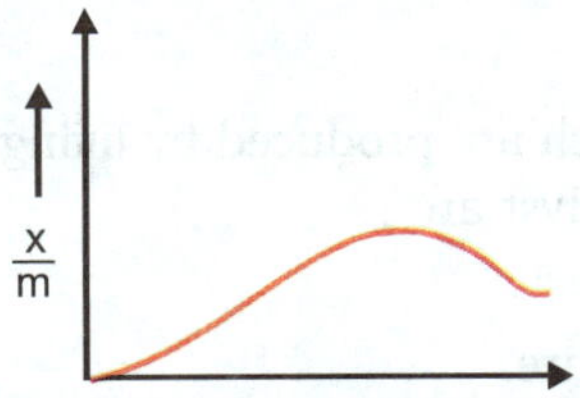

Chemical adsorption isobar

Application of Adsorption

- Activated charcoal is used in gas masks to remove poisonous gases such as CH_4, CO etc.
- Animal charcoal is used as decolorizer in manufacture of sugar.
- Silica gel is used for removing moisture and controlling humidity.
- Some industrial processes take place in the presence of catalysts which are based upon adsorption.
- Chromatographic purification of compounds which is based upon adsorption.
- The ion exchange resins used for removing hardness of water are also based upon adsorption.

Catalyst

1. Catalyst : A substance which influence the rate of reaction is known as catalyst provided it remains unchanged in amount a well as in composition.

- **Positive catalyst :** The catalyst which increase the rate of chemical reaction is known as positive catalyst. Fe act as positive catalyst in following reaction.

$$N_2 + 3H_2 \xrightarrow{Fe} 2NH_3$$

- **Negative catalyst :** The catalyst which decrease the rate of chemical reaction is known as negative catalyst. C_2H_5OH act as negative catalyst in following reaction.

$$CHCl_3 + O \xrightarrow{C_2H_5OH} COCl_2 + HCl$$

- **Auto catalyst :** When one of product act as catalyst then such type of catalyst is known as auto catalyst. Mn^{2+} act as auto catalyst in following reactions

$$2MnO_4^- + 5C_2O_4^{2-} + 16H^+ \longrightarrow 2Mn^{2+} + 10CO_2 + 8H_2O$$

2. Catalyst promoters : Substance which increase the activity of catalyst is known as catalyst promoters. Mo act as catalyst promoters in following reaction.

$$N_2 + 3H_2 \xrightarrow[Mo]{Fe} 2NH_3;$$

3. Homogeneous catalytic reaction : If catalyst is present in same phase as reactant then it is called homogenous catalytic reaction.

$$2SO_2(g) + O_2(g) \xrightarrow{NO(g)} 2SO_3(g)$$

4. Heterogeneous catalytic reaction : If catalyst is present in different phase as reactant then it is called heterogenous catalytic reaction.

$$N_2(g) + 3H_2(g) \xrightarrow{Fe(s)} 2NH_3(g)$$

5. Adsorption theory of heterogeneous catalysis : The mechanism of heterogenous catalysis involves the following steps :

- Adsorption of reactant molecules on the catalyst surface.
- Occurrence of a chemical reaction through the formation of an intermediate.
- Desorption of products from the catalyst surface.
- Diffusion of products away from the catalyst surface.

Enzyme Catalyst

Complex nitrogenous compounds which are produced by living plants and animals are known as enzymes. The characteristic features of enzyme catalyst are :

- They are highly efficient.
- They are highly specific in nature.
- They are highly active under optimum temperature.
- They are highly active under optimum pH.
- There is increase in their activity in the presence of activators and coenzymes.
- They are influenced by inhibitors and poisons.

$$C_{12}H_{22}O_{11} + H_2O \xrightarrow{Invertase} \underset{Glucose}{C_6H_{12}O_6} + \underset{Fructose}{C_6H_{12}O_6}$$

$$C_{12}H_{22}O_{11} + H_2O \xrightarrow{Maltase} \underset{Glucose}{2C_6H_{12}O_6}$$

$$\underset{Starch}{(C_6H_{10}O_5)_n} + nH_2O \xrightarrow{Diastase} \underset{Maltose}{C_{12}H_{22}O_{11}}$$

$$\underset{Glucose}{C_6H_{12}O_6} \xrightarrow{Zymase} 2C_2H_5OH + 2CO_2$$

Mechanism of Enzyme Catalysis

Step 1 : Binding of enzyme to substrate to form an activated complex $E + S \longrightarrow ES$

Step 2 : Decomposition of the activated complex to form products $ES \longrightarrow E + P$

Colloids

Depending on size of particles mixture are divided into three classes :

- **True solution :** In true solution size of particles is less than 1 nm.
- **Suspension :** In suspension size of particles is greater than 1000 nm.
- **Colloidal solution :** In colloidal solution size of particles lies from 1 nm to 1000 nm.

It consists two phase *i.e.,* dispersed phase and dispersion medium.

Dispersed Phase : It is the component present in small proportion and is just like a solute in a true solution. For example, in the colloidal state of sulphur in water, the former acts as a dispersed phase.

Dispersion Medium : It is normally the component present in excess and is just like a solvent in a solution.

The particles of the dispersed phase are scattered in the dispersion medium in a colloidal system.

Classification of Colloids

1. Based on Physical State of Dispersed Phase & Dispersion Medium :

Dispersed phase	Dispersion medium	Colloidal system	Examples
Gas	Liquid	Foam or froth	Soap sols, lemonade froth, whipped cream.
Gas	Solid	Solid foam/Sol	Pumice stone, styrene, foam, foam rubber.
Liquid	Gas	Aerosols of Liquids	Fog, clouds, fine insecticide sprays.
Liquid	Liquid	Emulsions	Milk
Liquid	Solid	Gels	Cheese, butter, boot polish, table jellies.
Solid	Gas	Aerosols of Solid	Smoke, dust
Solid	Liquid	Sols	Most paint starch dispersed in water, gold sol, muddy water, inks.
Solid	Solid	Solid sols	Ruby glass, some gem stones.

2. Based on Interaction or Affinity of Phases :

(a) Lyophilic Colloids : The colloidal system in which the particle of dispersed phase have great affinity for the dispersion medium, are called lyophilic

(b) Lyophobic colloids : The colloidal system in which the dispersed phase have no affinity for the dispersion medium are called lyophobic.

	Property	Lyophilic sols	Lyophobic sols
1.	Nature	Reversible	Irreversible
2.	Preparation	They are prepared very easily by shaking or warming the substance with dispersion medium.	They are difficult to prepare, Special methods are used.
3.	Stability	They are very stable and are not easily coagulated by electrolytes.	They are generally unstable and get easily coagulated on addition of electrolytes.

4.	Charge	Particles carry no or very little charge depending upon the pH of the medium.	Colloidal particles have characteristic charge (positive or negative)
5.	Viscosity	Viscosity is much higher than that of the medium.	Viscosity is nearly the same as that of the medium
6.	Surface Tension	Surface tension is usually less than that of the medium.	Surface tension is nearly the same as that of the medium.
7.	Migration in electric field	The particles may or may not migrate in an electric field.	The colloidal particles migrate either towards cathode or anode in an electric field.
8.	Solvation	Particles are heavily solvated.	Particles are not solvated.
9.	Visibility	The particles can not be seen under ultra microscope.	The particles though invisible, can be seen under ultra microscope.
10.	Tyndall effect	Less distinct.	More distinct.
11.	Action of electrolyte	Large amount of electrolyte is required to cause coagulation.	Small amount of electrolyte is sufficient to cause coagulation.
12.	Examples	Mostly organic substances *e.g.*, starch, gums, proteins, gelatin etc.	Generally inorganic substance *e.g.*, metal sols, sulphides and oxides sols.

3. Based on Type of Particles of the Dispersed Phase :

- **Multimolecular colloids :** The multimolecular colloidal particles consists of aggregate of atoms of small molecules with diameters less than 10^{-9} m or 1 nm.

  For example, a sol of gold contains particles of various sizes having several atoms.

- **Macromolecular colloids :** The macromolecular colloidal particles themselves are large molecules. They have very high molecular mass varying from thousands to millions. These substances are generally polymers.

  For example, starch, cellulose and proteins, polyethene, nylon etc.

- **Associated colloids :** Certain substances behave as strong electrolytes at low concentration but at higher concentrations these substances exhibit colloidal characteristics due to the formation of aggregated particles. These aggregated particles are called micelles, which are also known as associated colloids.

Kraft Temperature : The formation of micelles takes place only above a particular temperature called Kraft temperature.

Critical Micelle Concentration : Formation of micelle takes place above kraft temperature and above a particular concentration called the critical micelle concentration (CMC).

For soaps, the CMC is about $10^{-4} M$ to $10^{-3} M$.

4. Method of Preparation of Colloids :

- **Mechanical Dispersion :** In this method the substance is first finely powdered and a suspension is made by shaking the powdered substance with the dispersion medium. This suspension is then passed through a colloidal mill where suspension is converted into colloids. The colloidal solutions of rubber, ink, paints and varnishes are prepared by this method.
- **Electrical Dispersion (Bredig's arc method) :** In this method an electric arc is struck between two metallic rods and kept under the dispersion medium. The particles of the metal break away from the rods and disperse in the liquid. Colloidal solutions of Au, Pt, Ag, Cu and such other metals can be prepared by this method.
- **Peptization :** The process in which freshly formed precipitate converts into a sol by the addition of suitable electrolyte is known as peptization.

(a) Added electrolyte is known as peptizing agent.

(b) The peptization action is due to preferential adsorption of common ion.

Ex.

$$\underset{\text{ppt}}{AgI} + \underset{\text{Peptising agent}}{AgNO_3} \longrightarrow [AgI]\ Ag^+ : NO_3^-$$

$$\underset{\text{ppt}}{AgI} + \underset{\text{Peptising agent}}{KI} \longrightarrow [AgI]\ I^- : K^+$$

Chemical methods :

Double decomposition : $As_2O_3 + 3H_2S \longrightarrow \underset{\text{Yellow sol}}{As_2S_3} + 3H_2O$

Oxidation : $2H_2S + SO_2 \longrightarrow 2H_2O + \underset{\text{Sol}}{3S}$

Reduction : $2AuCl_3 + 3SnCl_2 \longrightarrow 3SnCl_4 + \underset{\text{Sol}}{2Au}$

Hydrolysis : $FeCl_3 + 3H_2O \longrightarrow \underset{\text{Red sol}}{Fe(OH)_3} + 3HCl$

5. Purification of Colloidal Solution :

- **Dialysis :** The process of separating the particles of colloid from those of crystalloid, by means of diffusion through a suitable membrane is called dialysis. Its principle is based upon the fact that colloidal particles can not pass through a parchment or cellophane membrane, while the ions of the electrolyte can pass through it.
- **Electrodialysis :** The ordinary process of dialysis is slow. To increase the process of purification, the dialysis is carried out by applying electric field. This process is called electrodialysis. The artificial kidney machine works on the principle of dialysis.

6. Properties of Colloidal Solution :

(i) Physical Properties :

- **Heterogeneity :** Colloidal solutions are heterogeneous in nature consisting of two phases viz, the dispersed phase and the dispersion medium.
- **Filterability :** Colloidal particles readily pass through ordinary filter papers. It is because the size of the pores of the filter paper is larger than that of the colloidal particles.
- **Non-settling nature :** Colloidal solutions are quite stable as the colloidal particles remain suspended in the dispersion medium indefinitely. Thus there is no effect of gravity on the colloidal particles.
- **Colour :** The colour of the colloidal solution is not always the same as the colour of the substances in the bulk. The colour of the colloidal solution depends upon the following factors :
  - (i) Size and shape of colloidal particles.
  - (ii) Wavelength of the source of light.
  - (iii) Method of preparation of the colloidal solution.
  - (iv) Nature of the colloidal solution.
  - (v) The way an observer receives the light.
- **Stability :** Colloidal solutions are quite stable. Only a few solutions of larger particles may settle but very slowly.

(ii) Mechanical Properties :

- **Brownian movement :** It involves motion of colloidal particles in zig-zag paths. It is due to the moving molecules of the dispersion medium which are constantly colliding with colloidal particles.

(i) It is not observed in ordinary suspension.

(ii) It offers a visible proof of the random kinetic motion of molecules in a liquid.

(iii) It provides a direct demonstration of ceaseless motion of molecule as postulated by kinetic theory.

(iv) It provide stability to colloidal sol by not allowing them to settle down.

- **Sedimentation :** Under the influence of gravity, colloids tends to settle down very slowly. This phenomena is known as sedimentation. It can be accelerated by ultracentrifugation.

(iii) Optical properties :

- **Tyndall Effect :** The scattering of light by colloidal particles is known as Tyndall effect.

The colloidal particles first absorb light and then a part of the absorbed light is scattered from the surface of the colloidal particles. Maximum scattered intensity is being in a plane at right angles to the plane of incident light. The path becomes visible when seen from that direction.

(iv) Electrical properties : Colloidal particles carry an electric charge which is equal and opposite to dispersion medium.

- **Electrophoresis :** The movement of colloidal particles towards a particular electrode in an electric field is known as electrophoresis. It provides information about the nature of the electrical charge on the colloidal particle.
- **Electro-osmosis :** When electrophoresis of dispersed particles in a colloidal system is prevented by suitable means, it is observed that the dispersion medium itself begins to move in an electric field. This phenomenon is called electro-osmosis.

Electrically charged sols

	Positively charged sols	Negatively charged sols
1.	Ferric hydroxide, aluminium hydroxide	Metals such as Pt, Cu, Au, Ag, Sb_2, S_3, CdS Metals sulphides, *e.g.*, arsenic sulphide
2.	Basic dyes such as methylene blue	Starch, clay, silicic acid, gum, gelatin, charcoal
3.	Haemoglobin	Acid dyes, such as eosin, congo Red

- **Electrical double layer :** The surface of a colloidal particle acquires a positive or a negative charge by selective adsorption of ions carrying +ve or –ve charges respectively. The charged layer attracts counter ions from the medium which forms a second layer. Thus, an electrical double layer is formed on the surface of the particles.

$[AgI]\ I^-$: K^+

$[AgI]\ Ag^+$: NO_3^-

$[Fe(OH)_3]\ Fe^{3+}$: $3Cl^-$

Selective adsorption of positive ions solid

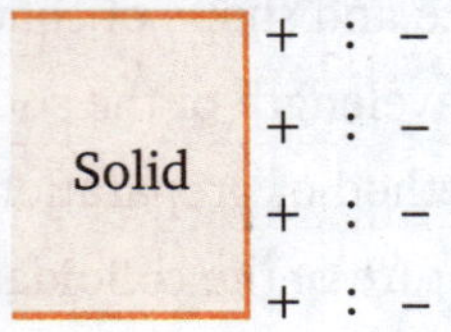

Electro kinetic potential : The potential difference between the fixed charge layer and diffused layer of opposite charge is called electro kinetic potential or zeta potential.

- **Coagulation :** The stability of the colloidal system is due to the existence of charged particles. If the charges are neutralised or destroyed, then the colloidal solution gets precipitated. This is called Coagulation or Flocculation. It can be done by

(i) adding electrolyte

(ii) mutual action of sols

(iii) persistent dialysis

(iv) By cooling

(v) By electrophoresis

Coagulation value : The minimum concentration of the electrolyte required in millimoles per litre of solution to cause coagulation or flocculation is called coagulation value.

Hardy Schulze Rule : According to this rule the coagulating power of the active ion increases with the valency of the active ion.

Coagulating power of cations : $Al^{3+} > Ba^{2+}$ or $Mg^{2+} > Na^+$ or K^+.

Coagulating power of anion : $[Fe(CN)_6]^{4-} > PO_4^{3-} > SO_4^{2-} > Cl^-$

- **Isoelectric Point of Colloid :** The hydrogen ion concentration at which the colloidal particles are neither positively charged nor negatively charged (*i.e.*, uncharged) is known as isoelectric point of the colloid.
- **Protective colloids :** In presence of lyophilic colloids like starch no coagulation of lyophobic colloids takes places and lyophilic colloids act as protective colloids. Greater protecting power of a lyophilic colloid lesser would be gold number.

Gold Number : It is defined as the minimum amount of the protective colloid in milligrams which must be added to 10 mL solution of standard gold sol, which prevents coagulation on adding one mL of 10% NaCl solution.

- **Emulsions :** An emulsion is a colloidal solution of a liquid. It may be defined as a heterogeneous system consisting of more than one immiscible liquids dispersed in one another in the form of droplets

Oil in water type emulsion (*O/W*) : In this emulsion, oil is the dispersed phase and water is the dispersion medium. It is denoted by *O/W* or *O* in *W*. For example, milk (liquid fat dispersed in water), vanishing cream, etc.

Water in oil type : In this emulsion, water is the dispersed phase and oil is the dispersion medium. It is denoted by *W/O* or *W* in *O*. For example, butter, cod liver oil, cold cream, etc.

Distinction between Two Types of Emulsions

Dye test : It involves the addition of oil soluble dye to the emulsion under experiment. If the emulsion acquires the colour of the dye readily, it is water-in-oil type emulsion and if the emulsion remains colourless, it is oil-in-water type emulsion.

Dilution test : As a general rule, an emulsion can be diluted with the dispersion medium while the addition of the dispersed phase forms a separate layer. Thus, if an emulsion can be diluted with oil, it is water-in-oil type.

Preparation of emulsion (Emulsification) : Emulsification is the process which involves the preparation of emulsion. Generally, an emulsion is prepared by subjecting a mixture of the immiscible liquid to a distinct layers upon standing. This substance which stabilizes the emulsion is called emulsifier or emulsifying agent. The commonly used emulsifying agents are soaps, detergents and lyophilic colloids, casein.

Function of emulsifier : The main function of emulsifier or emulsifying agents is to lower the interfacial tension between oil and water and thus helps the intermixing of two liquids.

Demulsification : The process which involves the breaking of an emulsion into two separate liquid layers is called demulsification.

- **Gels :** Colloidal system in which liquids are the dispersed phase and solid act as the dispersion medium are called gels. The common examples are : boot polishes, jelly, gum arabic, agar agar, processed cheese and silicic acid.

When the gels are allowed to stand for a long time, they give out small quantities of trapped liquids with accumulate on its surface. This action of gels is known as Synresis or Weeping.

» Uses of Colloids

(i) Medicines
(ii) Dyes
(iii) Rubber industry
(iv) Formation of delta
(v) Purification of water
(vi) Artificial rain
(vii) Smoke precipitation
(viii) Sewage disposal
(ix) Cleansing action of soap and detergent
(x) In Photography

Level-1

1. The size of particles in suspension, true solution and colloidal solution varies in the order :
   (a) suspension > colloidal > true solution
   (b) true solution > suspension > colloidal
   (c) suspension > colloidal = true solution
   (d) none of these
2. A colloidal system has what size of particles?
   (a) 10^{-4} m to 10^{-10} m (b) 10^{-5} m to 10^{-7} m
   (c) 10^{-9} m to 10^{-12} m (d) 10^{-6} m to 10^{-9} m
3. Which are not purely surface phenomena?
   (a) Adsorption, surface tension (b) Surface tension, viscosity
   (c) Adsorption, viscosity (d) Absorption, viscosity
4. Adsorbed acetic acid on activated carbon is :
   (a) adsorber (b) absorber
   (c) adsorbent (d) adsorbate
5. Adsorption due to strong chemical forces is called :
   (a) chemisorption (b) physisorption
   (c) both (a) and (b) (d) none of these
6. Adsorption of gases on solid surface is exothermic because :
   (a) free energy increases
   (b) entropy decreases
   (c) entropy increases
   (d) interaction developed between gas and solid particles
7. The nature of bonding forces in adsorption are :
   (a) purely physical such as van der Waals' forces
   (b) purely chemical
   (c) both chemical and physical are possible
   (d) none of these
8. Which one of the following is not applicable to chemisorption?
   (a) Heat of adsorption is negative (b) It takes place at high temperature
   (c) It is reversible (d) It forms mono-molecular layer
9. Which one of the following characteristics is correct for physical adsorption?
   (a) It is very specific
   (b) Adsorption on solids is irreversible
   (c) Adsorption decreases with decrease in temperature
   (d) Generally both enthalpy and entropy of adsorption are negative
10. Which of the following statement is not correct?
    (a) Physical adsorption is due to van der Waals' forces
    (b) Physical adsorption is irreversible
    (c) Chemical adsorption increases with increase in temperature upto certain limit than decreases
    (d) Enthalpy of adsorption ($|\Delta H|$) for a chemical adsorption is greater than that of physical adsorption
11. Which gas will be adsorbed on a solid to greater extent?
    (a) Having non-polar molecule
    (b) Having highest critical temperature

(c) Having lowest critical temperature
(d) Having lowest critical pressure

12. Which of the following factors affects the adsorption of a gas on solid?
(a) Critical temperature (T_c) (b) Temperature of gas
(c) Pressure of gas (d) All of these

13. Which gas is adsorbed to maximum amount by activated carbon?
(a) $H_2(g)$ (b) He(g)
(c) CO(g) (d) $CO_2(g)$

14. The volume of gases NH_3, CO_2 and H_2 adsorbed by one gram of charcoal at 300 K are in order of:
(a) $H_2 > CO_2 > NH_3$ (b) $NH_3 > H_2 > CO_2$
(c) $NH_3 > CO_2 > H_2$ (d) $CO_2 > NH_3 > H_2$

15. Which of the following is used to adsorb water?
(a) Silica gel (b) Calcium acetate
(c) Hair gel (d) Anhydrous $CaCl_2$

16. Absorption and adsorptions are respectively:
(a) surface phenomena, bulk phenomena
(b) bulk phenomena, surface phenomena
(c) both are bulk phenomena
(d) both are surface phenomena

17. Adsorption is multilayer in case of:
(a) physical adsorption (b) chemisorption
(c) both (a) and (b) (d) none of these

18. Reversible adsorption is:
(a) chemical adsorption (b) physical adsorption
(c) both (a) and (b) (d) none of these

19. The effect of pressure on adsorption is high if:
(a) temperature is low (b) temperature is high
(c) temperature is very high (d) larger charcoal piece is taken

20. Sorption is the phenomenon:
(a) reverse of adsorption
(b) reverse of absorption
(c) when adsorption and absorption takes place simultaneously
(d) none of these

21. Which of the following adsorption isotherms represents the adsorption of a gas by a solid involving multilayers of formation? (P_s = saturation pressure)

(a) 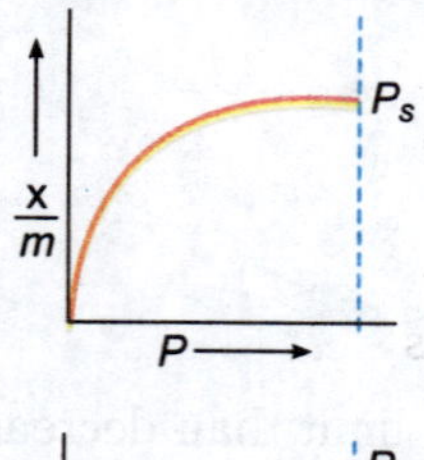

(b) 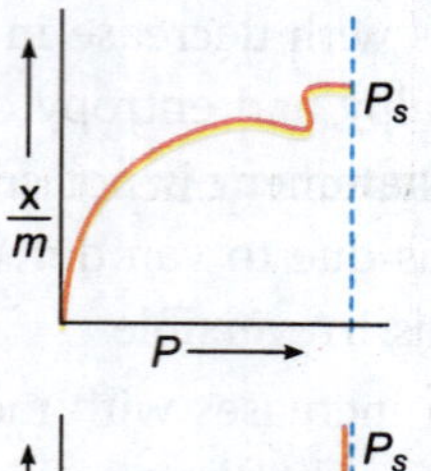

(c) 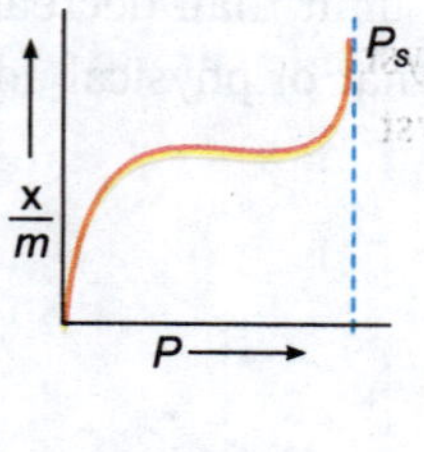

(d) 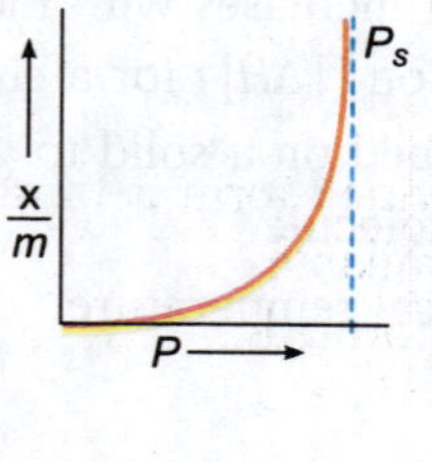

22. A plot of $\log\left(\frac{x}{m}\right)$ against $\log P$ for the adsorption of a gas on a solid gives a straight line with slope equal to:
(a) $\frac{1}{n}$ (b) n
(c) log K (d) K

23. The heat evolved in chemisorption lies in the range (in kJ/mol) of :
(a) 80 to 240 (b) 20 to 40
(c) 40 to 80 (d) 20 to 100

24. The heat evolved in physisorption lies in the range (in kJ/mol) of :
(a) 20–40 (b) 40–100
(c) 100–200 (d) 200–400

25. According to the adsorption theory of catalysis, the speed of the reaction increases because :
(a) in the process of adsorption, the activation energy of the molecules becomes large
(b) adsorption produces heat which increases the speed of the reaction
(c) adsorption lowers the activation energy of the reaction
(d) adsorption increases the activation energy of the reaction

26. 3.6 gram of oxygen is adsorbed on 1.2 g of metal powder. What volume of oxygen adsorbed per gram of the adsorbent at 1 atm and 273 K?
(a) $0.19\ L\ g^{-1}$ (b) $1\ L\ g^{-1}$
(c) $2.1\ L\ g^{-1}$ (d) None of these

27. A catalytic poison renders the catalyst ineffective because:
(a) It is preferentially adsorbed on the catalyst
(b) It adsorbs the molecules of the reactants
(c) It combines chemically with the catalyst
(d) It combines chemically with one of the reactants

28. The catalyst used in the hydrogenation of oils is:
(a) Fe (b) Ni
(c) Pt (d) V_2O_5

29. The function of zymase is to:
(a) change starch into sugar
(b) ferment glucose to alcohol and CO_2
(c) change malt sugar into glucose
(d) change starch into malt sugar and dextrin

30. The conversion of maltose to glucose is possible by the enzyme:
(a) zymase (b) lactase
(c) maltase (d) diastase

31. Shape selective catalysis is a reaction catalysed by:
(a) enzymes (b) Ziegler-Natta Catalyst
(c) zeolites (d) platinum

32. The process which is catalysed by one of the product is called:
(a) acid-base catalysis (b) autocatalysis
(c) negative catalysis (d) homogeneous catalysis

33. An inhibitor is essentially:
(a) a negative catalyst (b) a heterogeneous catalyst
(c) an auto catalyst (d) a homogeneous catalyst

34. A catalyst in the finely divided form is most effective because:
(a) less surface area is available
(b) more active sites are formed

(c) more energy gets stored in the catalyst
(d) none of these

35. Identify the correct statement regarding enzymes:
(a) Enzymes are specific biological catalysts that normally works at high temperature
(b) Enzymes are normally heterogeneous catalysts which decrease the reaction rate
(c) Enzymes are specific biological catalysts with low molar masses
(d) Enzymes are specific biological catalysts that are very specific in nature

36. A liquid leaves no residue when passed through the ultra-filter paper. The liquid is:
(a) a suspension (b) oil
(c) a colloidal sol (d) a true solution

37. Crystalloids differ from colloids mainly in respect of :
(a) electrical behaviour (b) particle nature
(c) particle size (d) solubility

38. Surface tension of lyophilic sols is:
(a) lower than water (b) more than water
(c) equal to water (d) none of these

39. Which one of the following is not used for preparing lyophilic sols?
(a) Starch (b) Gum
(c) Gelatin (d) Metal sulphide

40. Which one of the sols acts as protective colloid?
(a) As_2S_3 (b) Gelatin
(c) Au (d) $Fe(OH)_3$

41. Which one of the following is lyophilic colloid?
(a) Pt (b) Gum
(c) Fog (d) Blood

42. Small liquid droplets dispersed in another liquid is called:
(a) suspension (b) emulsion
(c) gel (d) true solution

43. Which of the following is an example of associated colloid?
(a) Protein in water (b) Soap in water
(c) Rubber in benzene (d) $FeCl_3$ in H_2O

44. Among the following, select the properties of lyophilic colloidal sols:
(a) viscosity same as that of the medium
(b) extensive hydration takes place
(c) Surface tension higher or same as that of the medium
(d) particle can be readly detected under ultramicroscope

45. Fog is an example of colloidal system of:
(a) liquid in a gas (b) gas in a liquid
(c) gas in a solid (d) solid in a liquid

46. Colloidal systems are not classified on the basis of :
(a) molecular size
(b) nature of the particles
(c) surface tension value
(d) interaction between disperse phase and dispersion medium

47. All colloids:
(a) are suspensions of one phase in another
(b) are two-phase systems
(c) contain only water-soluble particles
(d) are true solutions

48. Which of the following is a lyophobic colloid?
(a) Gelatin (b) Sulphur
(c) Starch (d) Gum arabic

49. Which of the following is not a gel?
(a) Cheese (b) Jellies
(c) Curd (d) Milk

50. Cleaning action of soap occurs because:
(a) non-polar tails of soap molecules dissolve in grease
(b) oil and grease dissolved into hydrophilic centres of soap micelles acid washed away
(c) hydrophilic heads dissolves in grease
(d) grease dissolve in soap solution

51. Arsenous sulphide sol is prepared by passing H_2S through arsenous oxide solution. The charge developed on the particles is due to adsorption of:
(a) H^+ (b) S^{2-}
(c) OH^- (d) O^{2-}

52. Bredig's arc method cannot be used for the preparation of colloidal sol of :
(a) copper (b) gold
(c) silver (d) sodium

53. As_2S_3 sol is:
(a) positive colloid (b) negative colloid
(c) neutral colloid (d) none of the above

54. Which of the following electrolyte will be most effective in coagulation of negative sol ?
(a) KNO_3 (b) $K_4[Fe(CN)_6]$
(c) Na_3PO_4 (d) $MgCl_2$

55. The minimum amount of an electrolyte required to cause coagulation of a sol is called :
(a) Coagulation value (b) Gold number
(c) Protective value (d) None of these

56. The electrical charge on a colloidal particle is indicated by:
(a) Brownian movement (b) electrophoresis
(c) ultramicroscope (d) molecular sieves

57. Colloidal particles in a sol can be coagulated by :
(a) heating (b) adding an electrolyte
(c) adding oppositely charged sol (d) any of the above methods

58. Peptization involves:
(a) precipitation of colloidal particles
(b) disintegration of colloidal aggregates
(c) purification of colloids
(d) impact of molecules of the dispersion medium on the colloidal particles

59. Hardy-Schulze law states that:
(a) solution must have higher gold number
(b) disperse phase and dispersion medium must be of the same sign
(c) micelles coagulate in presence of surfactants
(d) the ions carrying more opposite charge to that of sol particle are effective in coagulation

60. Given below are a few electrolytes, indicate which one among them will bring about the coagulation of a gold sol quickest and in the least of concentration?
(a) NaCl (b) $MgSO_4$
(c) $Al_2(SO_4)_3$ (d) $K_4[Fe(CN)_6]$

61. The ability of an ion to bring about coagulation of a given colloid depends upon:
(a) its size
(b) the magnitude of its charge only
(c) the sign of its charge alone
(d) both magnitude and sign of its charge

62. An emulsifying agent is a substance which
(a) stabilizes the emulsion
(b) de-stabilizes the emulsion
(c) coagulates the emulsion
(d) break the interfacial film between suspended particle and medium

63. Colloidal solution of gold is generally prepared by:
(a) colloidal mill
(b) double decomposition method
(c) Bredig's method
(d) peptization

64. The formation of colloid from suspension is called:
(a) peptisation (b) condensation
(c) sedimentation (d) fragmentation

65. The method usually employed for the destruction of a colloidal solution is:
(a) dialysis (b) addition of electrolytes
(c) diffusion through animal membrane (d) condensation

66. Which of the following has minimum flocculation value for positively charged sol?
(a) Cl^- (b) SO_4^{2-}
(c) PO_4^{3-} (d) $[Fe(CN)_6]^{4-}$

67. Which of the following will have the highest coagulating power for $Fe(OH)_3$ colloid?
(a) PO_4^{3-} (b) SO_4^{2-}
(c) Ca^{2+} (d) Al^{3+}

68. Artificial rain is caused by spraying:
(a) opposite charged collidal dust over a cloud
(b) same charged collidal dust over a cloud
(c) both
(d) none of these

69. Colloids can be purified by :
(a) condensation (b) peptization
(c) coagulation (d) dialysis

70. At what pH range the rate of enzyme catalyzed reaction is maximum ?
(a) 5 to 7 (b) 6 to 8
(c) 7 to 9 (d) 8 to 10

71. Protective sols are:
(a) lyophilic (b) lyophobic
(c) both (a) and (b) (d) none of these

72. Electro-osmosis is observed when:
(a) dispersion medium particles begins to move in an electric field
(b) dispersed phase begins to move in an electric field
(c) both (a) and (b)
(d) none of the above

73. On adding $AgNO_3$ solution into KI solution, a negatively charged colloidal sol is obtained when they are in:
(a) 50 mL of $0.1\,M$ $AgNO_3$ + 50 mL of $0.1\,M$ KI
(b) 50 mL of $0.1\,M$ $AgNO_3$ + 50 mL of $0.2\,M$ KI
(c) 50 mL of $0.2\,M$ $AgNO_3$ + 50 mL of $0.1\,M$ KI
(d) None of these

74. A sol is prepared by addition of excess of $AgNO_3$ solution in KI solution. The charge likely to develop on colloidal particles is:
(a) positive
(b) negative
(c) no charge
(d) both charges

75. The gold numbers of protective colloids A, B, C and D are 0.04, 0.004, 10 and 40 respectively. The protective powers of A, B, C and D are in the order:
(a) $A > B > C > D$
(b) $B > A > C > D$
(c) $D > C > A > B$
(d) $D > C > B > A$

76. In Brownian motion, the paths of the particles are:
(a) linear
(b) curved
(c) zig-zag
(d) uncertain

77. Which one of the following is not represented by sols?
(a) Adsorption
(b) Tyndall effect
(c) Flocculation
(d) Paramagnetism

78. The Tyndall effect associated with colloidal particles is due to:
(a) presence of electrical charges
(b) scattering of light
(c) absorption of light
(d) reflection of light

79. Blue colour of the sky is due to:
(a) absorption of light by dust particles
(b) reflection of light by dust particles
(c) scattering of light by dust particles
(d) presence of clouds

80. The apparatus used to coagulate carbon particles from smoke is called:
(a) cottrell smoker
(b) cottrell precipitator
(c) cottrell absorber
(d) none of these

Level-2

1. Select correct adsorption isobars for chemisorption and physisorption respectively :

(where $\frac{x}{m}$ = extent of adsorption, T = temperature)

(a) 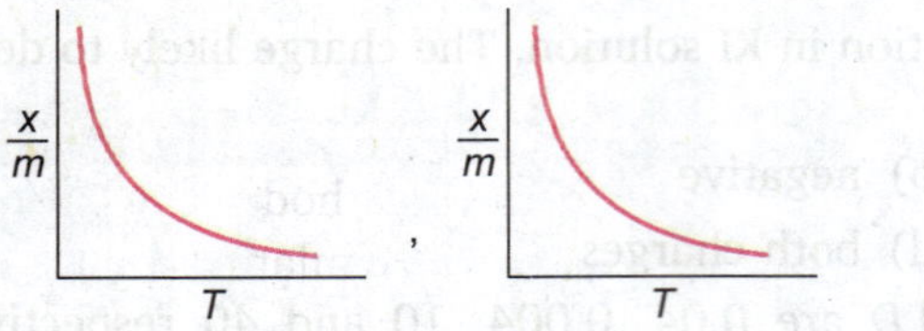

(b) 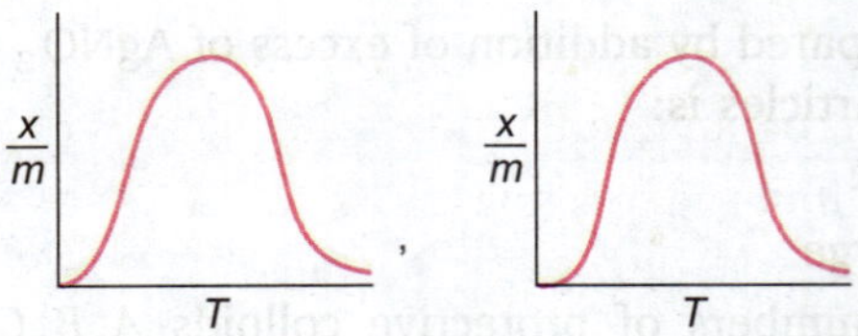

(c) 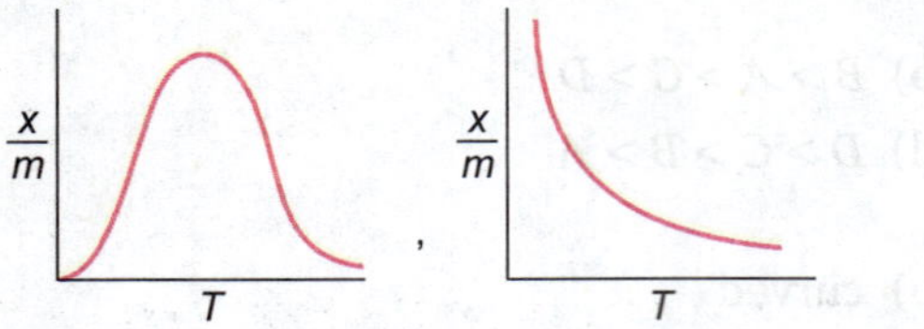

(d) 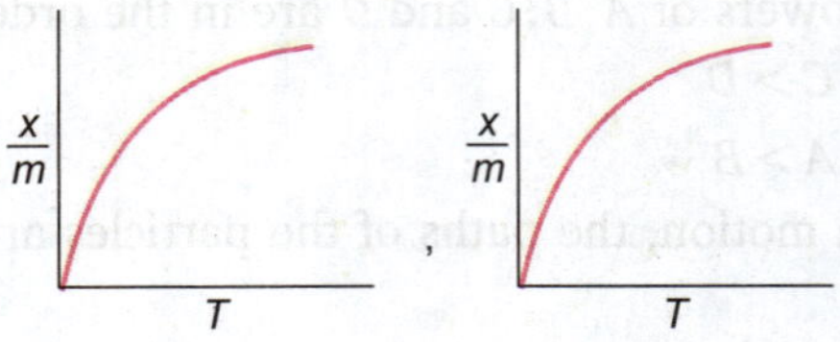

2. Which among the following statements is false?
(a) Increase of pressure increases the amount of adsorption
(b) Increase of temperature may decrease the amount of adsorption
(c) Adsorption may be monolayered or multilayered
(d) Particle size of the adsorbent will not affect the amount of adsorption

3. Select incorrect statement;
(a) Lyophilic sols are reversible
(b) Lyophilic sols are self stabilized
(c) Lyophobic sols are obtained from inorganic materials
(d) Lyophobic sols particles are hydrated

4. Which one of the following statements is not correct in respect of lyophilic sols?
(a) There is a considerable interaction between the dispersed phase and dispersion medium
(b) These are quite stable and are not easily coagulated
(c) They always carry significant characteristic charge
(d) The particles are hydrated

5. Alums purify muddy water by:
(a) dialysis (b) absorption
(c) coagulation (d) ultrafiltration

6. Lyophilic sols are more stable than lyophobic sols because:
(a) the colloidal particles have positive charge
(b) the colloidal particles have negative charge
(c) the colloidal particles are solvated
(d) there is strong electrostatic repulsion between the colloidal particles

7. Arrange the following electrolytes in the increasing order of coagulation power for the coagulation of As_2S_3 sol:

K_2SO_4	$CaCl_2$	Na_3PO_4	$AlCl_3$
(I)	(II)	(III)	(IV)

(a) I < II < III < IV (b) I = III < II < IV
(c) II < IV < I < II (d) II < III < IV < I

8. Equal volume each of two sols of AgI, one obtained by adding $AgNO_3$ to slight excess of KI and another obtained by adding KI to slight excess of $AgNO_3$, are mixed together. Then:
(a) the two sols will stabilize each other

(b) the sol particles will acquire more electric charge
(c) the sols will coagulate each other mutually
(d) a true solution will be obtained

9. Which of the following statements is not true about the oil-in-water type emulsion?
(a) Addition of small amount of water, no separate layer of water is formed
(b) Addition of a small amount of oil soluble dye renders the entire emulsion coloured
(c) Addition of oil results in the formation of two layers
(d) Addition of small amount of oil soluble dye renders the droplet coloured

10. Under the influence of an electric field, the particles in a sol migrate towards cathode. The coagulation of the same sol is studied using NaCl, Na_2SO_4 and Na_3PO_4 solutions. Their coagulating values will be in maximum for:
(a) NaCl (b) Na_2SO_4
(c) Na_3PO_4 (d) same for all

11. Select incorrect statement:
(a) Micelles are associated colloids
(b) The electrical charge on a colloid particle is indicated by electrophoresis
(c) Formation of micelles takes place above Kraft temperature
(d) Formation of micelles takes place below CMC

12. Bredig's arc method involves:
(a) dispersion of metal (b) condensation of metal
(c) dispersion as well as condensation (d) neither dispersion nor condensation

13. A freshly prepared $Fe(OH)_3$ precipitate is peptized by adding $FeCl_3$ solution. The charge on the colloidal particle is due to preferential adsorption of:
(a) Cl^- ions (b) Fe^{3+} ions
(c) OH^- ions (d) none of these

14. Select incorrect statement :
(a) Gold sol is multimolecular colloid
(b) Large number of particles of a substance aggregate together and formed multimolecular colloids
(c) Metal sulphides are lyophobic colloids
(d) Sulphur sol is multimolecular colloid and hydrophilic in nature

15. Associated colloids :
(a) raise both the surface tension and viscosity of water
(b) lower both the surface tension and viscosity of water
(c) lower the surface tension and raise the viscosity of water
(d) have greater concentration at the surface layer than the bulk of the solution

16. Below critical micelle concentration (CMC):
(a) the surfactant molecules of ions undergo association to form clusters
(b) the viscosity of solution increases abruptly
(c) substances like grease, fat, etc. dissolve colloidally
(d) salt behave as normal, strong electrolyte

17. During electro-osmosis of $Fe(OH)_3$ sol:
(a) sol particles move towards anode
(b) sol particles move towards cathode
(c) the dispersion medium moves towards anode
(d) the dispersion medium moves towards cathode

18. Select incorrect statement:
(a) Soap and detergent lower the interfacial surface tension between oil and water
(b) Basic principle of peptization is reverse of coagulation

(c) Soap and detergent used as emulsifiers
(d) Lyophilic sols need stabilizing agent

19. Smoke precipitator works on the principle of :
(a) centrifugation (b) neutralization of charge on colloids
(c) absorption (d) addition of electrolytes

20. What is the correct sequence of the decreasing coagulation value of the following electrolyte for the coagulation of ferric hydroxide sol?
(I) Na_3PO_4 (II) KCl (III) K_2SO_4 (IV) $K_4[Fe(CN)_6]$
(a) IV > I > III > II (b) II > III > I > IV
(c) I > II > III > IV (d) IV > III > II > I

21. Which of the following is true in respect of chemical adsorption?
(a) $\Delta H < 0$, $\Delta S > 0$, $\Delta G > 0$ (b) $\Delta H < 0$, $\Delta S < 0$, $\Delta G < 0$
(c) $\Delta H > 0$, $\Delta S > 0$, $\Delta G < 0$ (d) $\Delta H > 0$, $\Delta S < 0$, $\Delta G > 0$

22. For the graph below, select correct order of temperature?

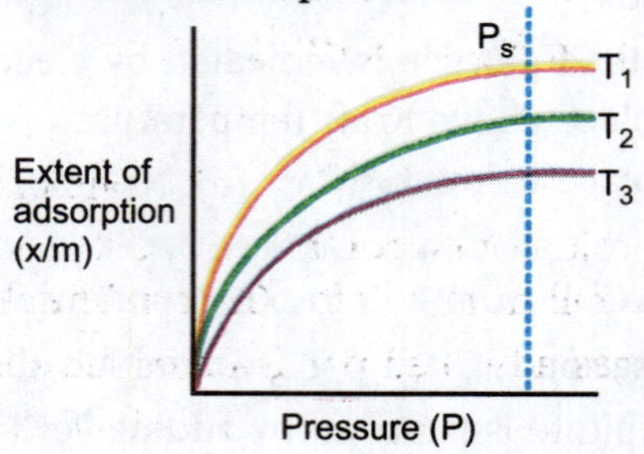

(a) $T_1 > T_2 > T_3$ (b) $T_2 > T_3 > T_1$
(c) $T_3 > T_2 > T_1$ (d) $T_1 = T_2 = T_3$

23. Although, nitrogen does not adsorb on a surface at room temperature, it adsorbs on the surface at 83 K. Which one of the following statements is correct ?
(a) At 83 K, there is formation of monolayer
(b) At 83 K, nitrogen is adsorbed as atoms
(c) At 83 K, nitrogen molecules are held by chemical bonds
(d) At 83 K, there is formation of multimolecular layers

24. For the coagulation of 50 mL of ferric hydroxide sol 10 mL of 0.5 M KCl is required. What is the coagulation value of KCl?
(a) 5 (b) 10
(c) 100 (d) None of these

25. 100 mL of 0.6 *M* acetic acid is shaken with 2 g activated carbon. The final concentration of the solution after adsorption is 0.5 *M*. What is the amount of acetic acid adsorbed per gram of carbon?
(a) 0.6 g (b) 0.3 g
(c) 1.2 g (d) None of these

26. A detergent ($C_{12}H_{25}SO_4^-Na^+$) solution becomes a colloidal sol at a concentration of 10^{-3} *M*. On an average 10^{13} colloidal particles are present in 1 mm^3. What is the average number of ions which are contained by one colloidal particle (micelle)? [Given: $N_A = 6 \times 10^{23}$]
(a) 6×10^7 (b) 10
(c) 60 (d) None of these

27. One gram of activated carbon has a surface area of 1000 m^2. Considering complete coverage as well as monomolecular adsorption, how much ammonia at 1 atm and 273 K would be adsorbed on the surface of

$\frac{44}{7}$ g carbon if radius of a ammonia molecules is 10^{-8} cm.

[Given: $N_A = 6 \times 10^{23}$]

(a) 7.46 L
(b) 0.33 L
(c) 44.8 L
(d) 23.5 L

28. At 1 atm and 273 K the volume of nitrogen gas required to cover a sample of silica gel, assuming monolayer adsorption, is found to be 1.30 $cm^3 g^{-1}$ of the gel. The area occupied by a nitrogen molecule is 0.16 nm^2. What is the surface area per gram of silica gel? [Given: $N_A = 6 \times 10^{23}$]

(a) 5.568 $m^2 g^{-1}$
(b) 3.48 $m^2 g^{-1}$
(c) 1.6 $m^2 g^{-1}$
(d) None of these

29. 10% sites of catalyst bed have adsorbed by H_2. On heating H_2 gas is evolved from sites and collected at 0.03 atm and 300 K in a small vessel of 2.46 cm^3.

No. of sites available is 5.4×10^{16} per cm^2 and surface area is 1000 cm^2. Find out the no. of surface sites occupied per molecule of H_2. (Given $N_A = 6 \times 10^{23}$)

(a) 1
(b) 2
(c) 3
(d) None of these

30. A sample of 16 g charcoal was brought into contact with CH_4 gas contained in a vessel of 1 litre at 27°C. The pressure of gas was found to fall from 760 to 608 torr. The density of charcoal sample is 1.6 g/cm^3. What is the volume of the CH_4 gas adsorbed per gram of the adsorbent at 608 torr and 27°C?

(a) 125 mL/g
(b) 16.25 mL/g
(c) 26 mL/g
(d) None of these

Level-3

Passage-1

A graph between x/m and the pressure P of the gas at a constant temperature is called adsorption isotherm. Where x is the no. of moles of the adsorbate and m is the mass of the adsorbent. Adsorption isotherms of different shapes have been experimentally observed. According to Frundlich adsorption isotherm.

$$x/m = kP^{1/n}$$

where k and n are constant parameters depending upon the nature of the solid and gas.

1. In the given isotherm select the incorrect statement :

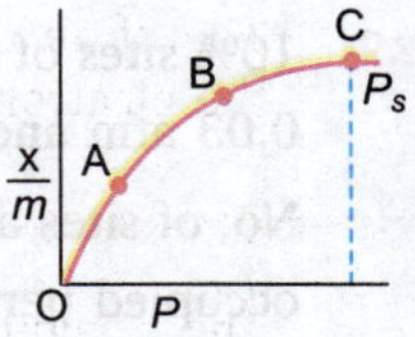

(a) $\frac{x}{m} \propto P^{1/n}$ along OA

(b) $\frac{x}{m} \propto P^{0}$ when point B is reached

(c) $\frac{x}{m}$ does not increase as rapidly with pressure along BC due to less surface area available for adsorption

(d) nature of isotherm is different for two gases for same adsorbent

2. Adsorption isotherm of $\log\left(\frac{x}{m}\right)$ and $\log P$ was found of the type :

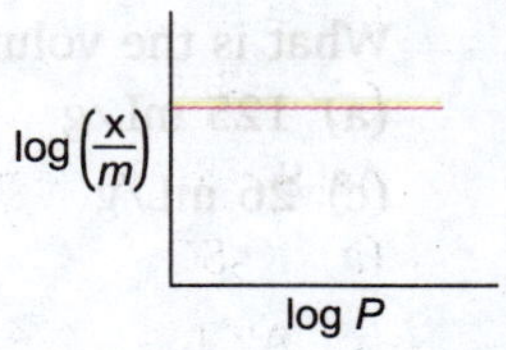

This is true when :

(a) $P = 0$ (b) $P = 1$

(c) $\frac{1}{n} = 0$ (d) $P < 1$

3. Graph between $\log\left(\frac{x}{m}\right)$ and $\log P$ is a straight line at angle 45° with intercept OA as shown.

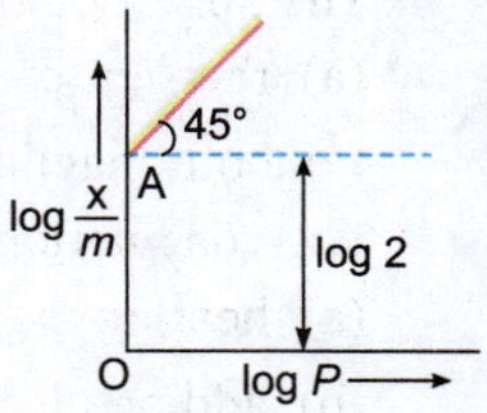

Hence, $\left(\frac{x}{m}\right)$ at a pressure of 2 atm is :

(a) 2 (b) 4

(c) 8 (d) 1

Passage-2

The protective power of the lyophilic colloids is expressed in terms of gold number, a term introduced by Zsigmondy. Gold number is the number of milli-gram of the protective colloid which prevent the coagulation of 10 mL of red gold sol, when 1 mL of a 10 per cent solution of sodium chloride is added to it. Thus, smaller the gold number of lyophilic colloid, the greater is the protective power.

1. On addition of one mL of solution of 10% NaCl to 10 mL of red gold sol in presence of 0.025 g of starch, the coagulation is just prevented. The gold number of starch is:

(a) 0.025 (b) 0.25

(c) 2.5 (d) 25

2. Which of the following statement(s) is/are correct?

(a) Higher the gold number, more protective power of colloid

(b) Lower the gold number, more the protective power

(c) Higher the coagulation value, more the coagulation power

(d) Lower the coagulation value, higher the coagulation power

3. Gold number gives an indication of:
(a) protective nature of colloids
(b) purity of gold in suspension
(c) the charge on a colloidal solution of gold
(d) gram of gold per litre of solution

Passage-3

Coagulation is the process by which the dispersed phase of a colloid is made to aggregate and thereby separate from the continuous phase. The minimum concentration of an electrolyte in milli-moles per litre of the electrolyte solution which is required to cause the coagulation of colloidal sol is called coagulation value.

Therefore higher is the coagulating power of effective ion, smaller will be the coagulation value of the electrolyte.

$$\text{Coagulation value} \propto \frac{1}{\text{coagulating power}}$$

The coagulation values of different electrolytes are different. This behaviour can be easily understood by Hardy-Schulze rule which states.

"The greater is the valency of the effective ion greater is its coagulating power."

1. Which one has the highest coagulating power?
(a) K^+ (b) Ca^{2+}
(c) Al^{3+} (d) Sn^{4+}

2. As_2S_3 sol is negatively charged. Capacity to precipitate it is highest in:
(a) K_2SO_4 (b) Na_3PO_4
(c) $AlCl_3$ (d) $CaCl_3$

3. The ability of an ion to bring coagulation of a given colloid depends upon:
(a) the sign of its charge (b) magnitude of its charge
(c) both magnitude and sign (d) none of these

4. The coagulation of colloidal particles of the sol can be caused by:
(a) heating
(b) adding electrolyte
(c) adding oppositely charged sol
(d) all of these

Passage-4

Emulsions are normally prepared by shaking the two components together vigorously although some kind of emulsifying agent usually has to added to stabilize the product. This emulsifying agent may be a soap or other surfactant (surface active) species or a lyophilic sol that forms a protective film around the dispersed phase.

Emulsions broadly classified into two types :
(i) Oil in water emulsions (O/W) : Oil acts as dispersed phase and water acts as dispersion medium.
(ii) Water in oil emulsions (W/O) : Water acts as dispersed phase and oil acts as dispersion medium. Dye test, dilution test may be employed for identification of emulsions.

1. Read two statements:
(1) Milk is an example of oil in water (O/W) type emulsion
(2) Cold cream is an example of water in oil (W/O) type emulsion

(a) Only statement 1 is correct
(b) Only statement 2 is correct
(c) Both are correct
(d) None of these

2. Select correct statement:
(a) Water in oil emulsions are less viscous than the aqueous emulsions
(b) Electrical conductance of aqueous emulsions is less than that of oil emulsions
(c) Deemulsification can be done by soap or detergent
(d) An emulsion can be diluted with H_2O then it is oil in water (O/W) type

One or More Answer is/are Correct

1. Select the correct statement(s) :
(a) Physical adsorption is specific in nature
(b) Chemical adsorption highly specific in nature
(c) Physical adsorption is due to free valence of atoms
(d) Chemical adsorption is due to stronger interaction or bond formation

2. Select the correct statement(s):
(a) Adsorption is a non-spontaneous process
(b) Surface energy decreases during the process of adsorption
(c) Adsorption takes place with decrease of entropy
(d) In general adsorption is exothermic process

3. Select the correct statement(s):
(a) Physisorption is favoured by low temperature
(b) Chemisorption is favoured by very high temperature because the process is endothermic
(c) Chemisorption increases with increase in temperature owing to high activation energy
(d) Oxygen adsorbed by charcoal can be desorbed by lowering pressure and increasing temperature

4. If adsorption of a gas on a solid is limited to monolayer formation, then which of the following statements are true?
(a) At low pressures, $\frac{x}{m}$ varies proportionately with p
(b) At moderate pressures, $\frac{x}{m}$ varies less than proportionately with p
(c) At high pressures, $\frac{x}{m}$ becomes independent of p
(d) At high pressures, $\frac{x}{m}$ varies more than proportionately with p

5. Which of the following are multimolecular colloids?
(a) Sulphur sol
(b) Starch
(c) Gold sol
(d) Soap solution

6. Which of the following is not lyophilic sols?
(a) Gelatin sol
(b) Silver sol
(c) Sulphur sol
(d) As_2S_3 sol

7. Which of the following is/are correct for sols?
(a) Its surface tension is lower than that of water
(b) Its viscosity is higher than that of water
(c) Its surface tension is higher than that of water
(d) Its viscosity is equal to that of water

8. Select the correct statement(s):
(a) Benzene is dispersed phase in benzosol
(b) Lyophobic sols are irreversible and not so stable
(c) Lyophobic sol can be produced by double decomposition
(d) When a solution of sulphur in alcohol is added in excess of water a sol of alcohol is formed

9. When negatively charged colloid like As_2S_3 sol is added to positively charged $Fe(OH)_3$ sol in stoichiometric amounts
(a) both the are precipitated simultaneously
(b) this process is called mutual coagulation
(c) they becomes positively charged colloid
(d) they become negatively charged colloid

10. Colloidal gold can be prepared by:
(a) Bredig's arc method
(b) reduction of $AuCl_3$
(c) hydrolysis
(d) peptization

11. The coagulation of sol particles may be brought about by:
(a) boiling
(b) persistent dialysis
(c) adding electrolyte
(d) adding oppositely charged sol

12. Select the correct statement(s) :
(a) A sol is prepared by addition of excess of $AgNO_3$ solution in KI solution. The charge likely to develop on colloidal particle is positive
(b) The effects of pressure on physical adsorption is high if temperature is low
(c) Ultracentrifugation process is used for preparation of colloids
(d) Gold number is the index for extent of gold plating done

13. Colloidal solution can be purified by:
(a) dialysis
(b) electrodialysis
(c) electrophoresis
(d) ultrafiltration

14. Coagulation of colloids can be achieved by:
(a) centrifugation
(b) adding electrolyte
(c) change in pH
(d) adding water

15. Which are the properties of colloidal solution ?
(a) Adsorption
(b) Tyndall effect
(c) Flocculation
(d) Depression of freezing point

16. In the aqueous solution of soaps above :
(a) the cations associate to form the aggregates
(b) the anions associate to form the clusters of colloidal dimension
(c) the polar ends of the ions forming the clusters are directed towards water
(d) the non-polar (hydrocarbon) ends are directed towards water

17. Amongst the following which is/are correct statement about the metal sulphide sols?
(a) The sol particles are positively charged due to preferential adsorption of metal ions
(b) The sol particles are negatively charged due to preferential adsorption of sulphide ions
(c) The cations of added electrolytes are effective in causing the coagulation of the sol
(d) The sol is stable due to both the electric charge and hydration of the particles

18. Emulsion can be destroyed by:
(a) the addition of an emulsifier
(b) electrophoresis with a high potential
(c) freezing
(d) all of these

19. Which of the following statement is/are correct for electrophoresis?
(a) Colloids are uncharged particles and do not migrate towards the electrodes when electric field is applied
(b) In electrophoresis, dispersed phase particle of sol migrates either to the anode or to the cathode depending on the positively or negatively charged sol
(c) Electrophoresis is useful for finding the charge on a sol
(d) All of these

20. Select the false statement(s):
(a) Brownian motion and Tyndall effect are shown by true solutions
(b) Sorption process is combinations of adsorption and adsorption process
(c) Law Hardy Schulze is related with coagulation of a sol
(d) Higher is the gold number greater will be the protective power of a colloid

21. Select correct statement(s)
(a) Lyophobic colloids are used to protect lyophilic colloids
(b) Dehydrating agent is used to coagulate lyophilic sols
(c) Rubber is obtained by coagulation of latex
(d) Sometimes, the rainfall occurs when two oppositely charged clouds meet

22. In which of the followings, Tyndall effect is/are not observed?
(a) Sugar solution (b) Emulsions
(c) Urea solution (d) Solution of proteins

23. Which of the following reactions are examples for heterogeneous catalysis?
(a) $2H_2O_2(l) \xrightarrow{MnO_2(s)} 2H_2O(l) + O_2(g)$
(b) $2SO_2(g) + O_2(g) \xrightarrow{NO(g)} 2SO_3(g)$
(c) $H_2(g) + C_2H_4(g) \xrightarrow{Ni(s)} C_2H_6(g)$
(d) $CH_3COOC_2H_5(aq) + H_2O(l) \xrightarrow{HCl(aq)} CH_3COOH(aq.) + C_2H_5OH(aq.)$

24. Select correct statement(s)
(a) The role of a catalyst in a reversible reaction is to allow the equilibrium to be achieved quickly
(b) Diffusion process is involved in mechanism of heterogeneous catalysis process
(c) Hydrolysis of cane sugar is catalysed by H^+
(d) Promotors enhance the activity of a catalyst

25. Select correct statement(s)
(a) Blood is a colloidal solution
(b) Alum is used in water purification
(c) River water is a colloidal solution of clay
(d) Colloidal medicines are more effective due to large surface area

Match the Column

Column-I and **Column-II** contains four entries each. Entries of Column-I are to be matched with some entries of Column-II. One or more than one entries of Column-I may have the matching with the same entries of Column-II.

1.

Column-I	Column-II
(A) Chemisorption	(P) Exothermic
(B) Physical adsorption	(Q) Endothermic
(C) Desorption	(R) Removal of adsorbed material
(D) Activation of adsorbent	(S) Highly specific in nature

2.

Column-I	Column-II
(A) Chemisorption	(P) Not specific and decreases with temperature
(B) Physisorption	(Q) Specific and increases with temperature
(C) Desorption of a solute on liquid surface	(R) Increases the surface tension of the liquid
(D) Adsorption of a solute on a liquid surface	(S) Decreases the surface tension of the liquid

3.

Column-I	Column-II
(A) Milk	(P) Aerosol
(B) Dust	(Q) Emulsion
(C) Cheese	(R) Gel
(D) Froth	(S) Foam

4.

Column-I	Column-II
(A) Liquid dispersed in gas	(P) Foam
(B) Gas dispersed in liquid	(Q) Emulsion
(C) Liquid dispersed in solid	(R) Aerosol
(D) Liquid dispersed in liquid	(S) Gel

5.

Column-I	Column-II
(A) As_2S_3 sol	(P) Lyophobic colloid
(B) Sulphur sol	(Q) Macromolecular colloid
(C) Starch	(R) Multimolecular colloid
(D) Soap	(S) Associated colloid

6.

Column-I	Column-II
(A) Coagulation	(P) Due to presence of charge
(B) Electrophoresis	(Q) Due to scattering of light
(C) Tyndall effect	(R) Due to neutralization of charge
(D) Brownian movement	(S) Due to impact of the molecules of the dispersion medium with colloidal particles

7.

Column-I	Column-II
(A) Peptization	(P) Preparation of sols
(B) Ultra centrifugation	(Q) Purification of sols
(C) Electrodialysis	(R) Preparation of metal sols
(D) Bredig's arc method	(S) Movement of ions across the membrane in presence of electric field

Assertion-Reason Type Questions

Each questions contains STATEMENT-1 (Assertion) and STATEMENT-2 (Reason).
Examine the statement carefully and mark the correct answer according to the instructions given below:
(A) If both the statement are TRUE and STATEMENT-2 is the correct explanation of STATEMENT-1
(B) If both the statements are TRUE and STATEMENT-2 is NOT the correct explanation of STATEMENT-1
(C) If STATEMENT-1 is TRUE and STATEMENT-2 is FALSE
(D) If STATEMENT-1 is FALSE and STATEMENT-2 is TRUE

1. **STATEMENT-1 :** For adsorption $\Delta G, \Delta H, \Delta S$ all have –ve values.
STATEMENT-2 : Adsorption is a exothermic process in which randomness decreases due to force of attraction between adsorbent and adsorbate.

2. **STATEMENT-1 :** The extent of adsorption of CO_2 is much more higher than of H_2.
STATEMENT-2 : $CO_2(g)$ has higher critical temperature and more van der Waals' force of attraction as compared to $H_2(g)$.

3. **STATEMENT-1 :** In absorption, the molecules of a substance are uniformly distributed throughout the body of other substance.
STATEMENT-2 : In some cases, both absorption and adsorption takes place simultaneously.

4. **STATEMENT-1 :** More heat evolved in physisorption than in chemisorption.
STATEMENT-2 : Molecules of adsorbate and adsorbent are held by van der Waals' forces in physisorption and by chemical bonds in chemisorption.

5. **STATEMENT-1 :** Colloidal solution is electrically neutral.
STATEMENT-2 : Due to similar nature of the charge carried by the dispersed phase particles, they repel each other and do not combine to form bigger particles.

6. **STATEMENT-1 :** Soap and detergent are macro-molecular colloids.
STATEMENT-2 : Soap and detergent are molecules of large size.

7. **STATEMENT-1 :** The conductivity of a solution having surfactant molecules decrease sharply at the CMC.

STATEMENT-2 : Micelles are formed by surfactant molecules above the critical micelle concentration (CMC).

8. **STATEMENT-1 :** The micelle formed by sodium stearate in water has —COO^- groups at the surface.
STATEMENT-2 : Surface tension of water is reduced by the addition of stearate.

9. **STATEMENT-1 :** Protein, starch are lyophilic colloids.
STATEMENT-2 : They have strong interaction with the dispersion medium.

10. **STATEMENT-1 :** Colloidal AgI is prepared by adding KI in slight excess to $AgNO_3$ solution, the sol particles migrate toward anode under electric field.
STATEMENT-2 : Colloidal particles adsorb ions and thus becomes electrically neutral.

11. **STATEMENT-1 :** When SnO_2 is reacted with NaOH, then its sol particles are attracted towards cathode.
STATEMENT-2 : When SnO_2 is reacted with NaOH, then it gives SnO_3^{2-} which is adsorbed by SnO_2, so it is negatively charged sol.

12. **STATEMENT-1 :** For coagulation of positively charged sols, $[Fe(CN)_6]^{4-}$ ion has higher coagulating power than that of $PO_4^{3-}, SO_4^{2-}, Cl^-$.
STATEMENT-2 : Because according to Hardy Schulze rule, higher is the valency of ions for the oppositely charged sol particles, better will be the precipitation.

13. **STATEMENT-1 :** Dispersed phase particles of colloidal solution cannot pass through ultra-filter paper.
STATEMENT-2 : The size of colloidal particles are larger than the size of true solution particles.

14. **STATEMENT-1 :** When a finely divided active carbon is stirred into a dilute solution of a dye, the intensity of colour in the solution is decreased.
STATEMENT-2 : The dye is adsorbed on the solid surface.

15. **STATEMENT-1 :** Silica gel is used for drying air.
STATEMENT-2 : Silica gel adsorbs moisture from air.

16. **STATEMENT-1 :** ZSM-5 is a type of zeolites used as a catalyst in petrochemical industries.
STATEMENT-2 : Zeolites are microporous aluminosilicates three dimensional network silicates in which some silicon atoms are replaced by aluminium atoms.

17. **STATEMENT-1 :** Lyophilic colloids are called as reversible sols.
STATEMENT-2 : Lyophilic sols are extensively hydrated.

18. **STATEMENT-1 :** A catalyst is more effective in finely divided form.
STATEMENT-2 : Finely divided form has more surface area.

19. **STATEMENT-1 :** Sky appears blue in colour.
STATEMENT-2 : Colloidal particles of dust along with water suspended in air scatter blue light.

20. **STATEMENT-1 :** The conversion of fresh precipitate to colloidal state is called peptization.
STATEMENT-2 : It is caused by addition of common ions.

21. **STATEMENT-1 :** Colloidal solutions are stable but colloidal particles do not settle down.
STATEMENT-2 : Brownian movement counters the force of gravity act on colloidal particles.

22. **STATEMENT-1 :** A colloid gets coagulated by addition of an electrolyte.
STATEMENT-2 : Coagulation depends on the valance and sign of the charge of the coagulant ion.

23. STATEMENT-1 : Suspensions are visible to naked eye.

STATEMENT-2 : Size of suspension particles are smaller than collidal particles.

24. STATEMENT-1 : At pH of isoelectric point, the sol particles of amino acids neither move towards anode nor towards cathode.

STATEMENT-2 : Because at the isoelectric point, the concentration of conjugate base and conjugate acid of the Zwitter ions becomes equal and so one's charge is counterbalanced by other.

25. STATEMENT-1 : Gold sol is multimolecular and hydrophobic in nature.

STATEMENT-2 : Gold sol is prepared by Bredig's arc method.

Level-4

Subjective Problems

1. How many colloidal systems exist in nature ?
2. How many colloidal systems exist in nature in which liquid is dispersed phase ?
3. How many colloidal systems exist in nature in which gas is dispersed phase ?
4. What is the minimum diameter (in nm) of colloidal particles ?
5. For soaps critical micelle concentration (CMC) is 10^{-x} (min.) to 10^{-y} (max.) mol/L. What is the value of x?

Numerical Problems

(If the numerical value has more than two decimal places, truncate/round-off the value to TWO decimal places.)

1. It was found that 0.20 mg of an adsorbate covered $0.08\ m^2$ of a solid adsorbent surface.
   If effective surface area of each adsorbated molecule is $y \times 10^{-20}\ m^2$ then find the value of y.
   [Given : $N_A = 6 \times 10^{23}$; $M_{\text{Adsorbate}} = 200$ g/mol]
2. A 0.10 M of an acid solution in benzene is dropped on a water surface, the benzene evaporates and the acid forms a monomolecular film of solid type. What volume of the above solution (in mL) would be required to cover a 3000 cm^2 surface area of water with monomolecular layer of acid?
   Area covered by single acid molecule is $0.2\ nm^2$.
3. 100 mL of 0.01M surfactant solution forms a monolayer covering 0.54 cm^2 on a adsorbent. If the polar head of surfactant is approximated as a cube, if its edge length is represented by $y \times 10^{-11}$ cm then find the value of y. (Given : $N_A = 6 \times 10^{23}$)
4. In an experiment, a graph between $\log\left(\frac{x}{m}\right)$ *vs* $\log P$ was found to be linear with a slope of 45°C. The intercept on the $\log\left(\frac{x}{m}\right)$ axis was found to be 0.60. Calculate the mass of gas (in g) adsorbed on 2.5 g of charcoal under a pressure of 1.75 atm of gas.
   (Given : $\log 2 = 0.30$)
5. In an experiment, addition of 40 mL of 0.05M AB_2 electrolyte to 30 mL of Sb_2S_3 sol just causes the complete coagulation. Calculate the coagulation value of AB_2 electrolyte.
6. The flocculation value of HCl for arsenic sulphide sol. is 30 m molL^{-1}. If H_2SO_4 is used for the flocculation of arsenic sulphide, the amount (in grams) of H_2SO_4 in 250 mL sol. required for the above purpose is______.
   (Molecular mass of $H_2SO_4 = 98$ g/mol)
7. In a coagulation experiment, 40 mL of As_2S_3 sol. is diluted with water and mixed with 0.2 M solution of an electrolyte X_2Y, 100 mL is the total volume of solution. It is found that 20 mL of X_2Y completely coagulate As_2S_3 sol. within 2 hrs. What is the coagulation value of X_2Y?
8. 7.45 mL of 20% $\left(\frac{w}{w}\right)$ KCl(aq) solution of density 1.1 g/mL is added for complete precipitation of 750 mL a metal sol. (neglect change in volume after addition of KCl solution).
   Calculate the coagulation value of KCl.
9. What is the coagulating power ratio of $BaCl_2$ with that of KCl?
   Given that their coagulation values are 0.70 and 50 respectively.
10. The average time for which an hydrogen atom remains adsorbed on a metal surface is 0.20 sec at 727°C and 0.80 sec at 227°C. Calculate the activation energy of desorption (in kcal/mol)
    (Given : $\log 2 = 0.30$)

Answers

Level 1

1. (a)	2. (d)	3. (d)	4. (d)	5. (a)	6. (d)	7. (c)	8. (c)	9. (d)	10. (b)
11. (b)	12. (d)	13. (d)	14. (c)	15. (a)	16. (b)	17. (a)	18. (b)	19. (a)	20. (c)
21. (a)	22. (a)	23. (a)	24. (a)	25. (c)	26. (c)	27. (a)	28. (b)	29. (b)	30. (c)
31. (c)	32. (b)	33. (a)	34. (b)	35. (d)	36. (d)	37. (c)	38. (a)	39. (d)	40. (b)
41. (b)	42. (b)	43. (b)	44. (b)	45. (a)	46. (c)	47. (b)	48. (b)	49. (d)	50. (a)
51. (b)	52. (d)	53. (b)	54. (d)	55. (a)	56. (b)	57. (d)	58. (b)	59. (d)	60. (c)
61. (d)	62. (a)	63. (c)	64. (a)	65. (b)	66. (d)	67. (a)	68. (a)	69. (d)	70. (a)
71. (a)	72. (a)	73. (b)	74. (a)	75. (b)	76. (c)	77. (d)	78. (b)	79. (c)	80. (b)

Level 2

1. (c)	2. (d)	3. (d)	4. (c)	5. (c)	6. (c)	7. (b)	8. (c)	9. (b)	10. (a)
11. (d)	12. (c)	13. (b)	14. (d)	15. (c)	16. (d)	17. (c)	18. (d)	19. (b)	20. (b)
21. (b)	22. (c)	23. (d)	24. (c)	25. (b)	26. (c)	27. (a)	28. (a)	29. (c)	30. (b)

Level 3

Passage–1	1. (b)	2. (c)	3. (b)	
Passage–2	1. (d)	2. (b)	3. (a)	
Passage–3	1. (d)	2. (c)	3. (c)	4. (d)
Passage–4	1. (c)	2. (d)		

One or More Answers is/are Correct

1. (b,d)	2. (b,c,d)	3. (a,c, d)	4. (a,b,c)	5. (a,c)	6. (b,c,d)	7. (a,b)	8. (b,c)
9. (a,b)	10. (a,b)	11. (a,b,c,d)	12. (a,b)	13. (a,b,d)	14. (a,b,c)	15. (a,b,c,d)	16. (b,c)
17. (b,c)	18. (b,c)	19. (b,c)	20. (a,d)	21. (b,c,d)	22. (a,c)	23. (a,c)	24. (a,b,c,d)
25. (a,b,c,d)							

Match the Column

1.	A→ P, S;	B→ P;	C→ Q, R;	D→ Q
2.	A→ Q;	B→ P;	C→ R;	D→ S
3.	A→ Q;	B→ P;	C→ R;	D→ S
4.	A→ R;	B→ P;	C→ S;	D→ Q
5.	A→ P, R;	B→ P, R;	C→ Q;	D→ S
6.	A→ R;	B→ P;	C→ Q;	D→ S
7.	A→ P;	B→ Q;	C→ Q, S;	D→ P, R

Assertion-Reason Type Questions

1. (A)	**2.** (A)	**3.** (B)	**4.** (D)	**5.** (B)	**6.** (D)	**7.** (A)	**8.** (B)	**9.** (A)	**10.** (C)
11. (D)	**12.** (A)	**13.** (A)	**14.** (A)	**15.** (A)	**16.** (B)	**17.** (B)	**18.** (A)	**19.** (A)	**20.** (B)
21. (A)	**22.** (B)	**23.** (C)	**24.** (A)	**25.** (B)					

Level 4

Subjective Problems

1. 8	**2.** 3	**3.** 2	**4.** 1	**5.** 4

Numerical Problems

1. 13.33	**2.** 2.50	**3.** 03.00	**4.** 17.50	**5.** 28.57	**6.** 0.37	**7.** 40.00	**8.** 29.33
9. 71.43	**10.** 2.76						

Hints & Solutions

Level 1

1. (a)

Solution	Particle size
True solution	Size < 1 nm
Colloidal solution	1 nm < size < 1000 nm
Suspension	Size > 1000 nm

2. (d) 10^{-6} m to 10^{-9} m
3. (d) Absorption is not surface phenomena.
4. (d) Adsorbate adsorb at the surface of adsorbent.
5. (a) Chemical adsorption.
6. (d) Attractive force developed at the surface between adsorbent and adsorbate.
7. (c) Physical adsorption and chemical adsorption.
8. (c) Chemical adsorption is irreversible in nature.
9. (d) $\Delta H < O$ $\quad \Delta S < O$
10. (b) Physical adsorption is reversible in nature.
11. (b) Critical temp. ↑ Attractive force ↑
    Extent of adsorption ↑.
12. (d) Adsorption↑ T_c ↑
    Adsorption↑ T ↓
    Adsorption↑ P ↑
13. (d) Van der Waal force ↑ adsorption↑
14. (c) Van der Waal force ↑ adsorption↑
15. (a) Silica gel adsorb water.
16. (b) Adsorption : surface phenomena
    Absorption : bulk phenomena
17. (a) Physical adsorption is multilayer adsorption.
18. (b) Physical adsorption is reversible adsorption.
19. (a) At low temperature, pressure have greater effect on adsorption.

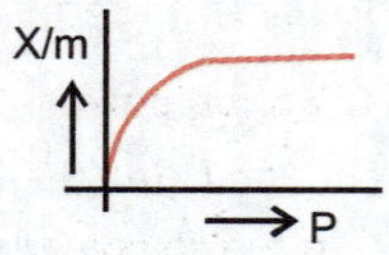

20. (c) In sorption, adsorption and absorption takes place simultaneously.
21. (a)

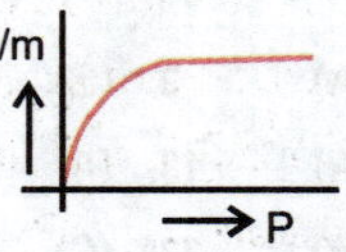

22. (a) $\frac{x}{m} = KP^{\frac{1}{n}}$

$$\log \frac{x}{m} = \log K + \frac{1}{n} \times \log P$$

$$\text{Slope} = \frac{1}{n}$$

23. (a) ΔH : 80 to 240 kJ mol^{-1}
24. (a) ΔH : 20 to 40 kJ mol^{-1}
25. (c) Adsorption lowers the activation energy of reaction because surface energy released during adsorption.
26. (c) Mass of O_2 per gram of adsorbent $= \frac{3.6}{1.2} = 3$

No. of moles of O_2 per gram of adsorbent

$$= \frac{3}{32}$$

Volume of O_2 per gram of adsorbent

$$= \frac{3}{32} \times \frac{0.0821 \times 273}{1}$$

$$= 2.10$$

27. (a) It is preferentially adsorb on the catalyst.
28. (b) Ni used as catalyst in hydrogenation of oil.
29. (b) $C_6H_{12}O_6 \xrightarrow{\text{Zymase}} CO_2 + C_2H_5OH$
30. (c) Maltose $\xrightarrow{\text{Maltase}}$ glucose
31. (c) Zeolites is shape selective catalyst.
32. (b) Autocatalyst
33. (a) Negative catalyst
34. (b) Surface area↑ No of sites ↑
35. (d) Enzyme are specific in nature.
36. (d) True solution.
37. (c) particle size
38. (a) Surface tension of lyophilic solution is lower than water.

39. (d) Metal sulphide are lyophobic solution.

40. (b) Gelatin act as protective colloids.

41. (b) Gum is lyophilic solution.

42. (b) Emulsion : D.P. liquid

D.M. liquid

43. (b) Soap formed associated colloids in water.

44. (b) D.P. have greater affinity towards D.M.

45. (a) fog : liquid in gas.

46. (c) Colloids divided on the basis of molecular size, affinity of D.P. towards D.M. and particle nature.

47. (b) Colloids have D.P. and D.M.

48. (b) Sulphur is lyophobic colloids.

49. (d) Milk is emulsion.

50. (a) Non-polar tails of soap molecules dissolve grease.

51. (b) $As_2S_3 + H_2S \longrightarrow As_2S_3 : S^{2-} : H^+$

Preferential adsorption of S^{2-} takes place. So negative charge developed.

52. (d) Bredig's arc method is used to prepare metal sol.

53. (b) Metal sulphides is negative colloids.

54. (d) Positive charge ↑ Coagulating power ↑

$MgCl_2$

55. (a) Millimoles of electrolyte required to coagulates one litre colloidal solution is called coagulation value.

56. (b) By electrophoresis, we can determine charge of colloidal solution.

57. (d) Coagulation takes place by heating, adding electrolyte and addition of opposite charge colloids.

58. (b) In peptisation, suspended state converts into colloidal state.

59. (d) Coagulating power α charge.

60. (c) Gold sol. have negative charge. So $Al_2(SO_4)_3$ is most effective for coagulation.

61. (d) Coagulating power depend upon sign of charge and magnitude of charge.

62. (a) Emulsifying agent stablise emulsion.

63. (c) Gold solution is prepared by Bredig's arc method.

64. (a) In peptisation, colloid formed from suspension.

65. (b) By addition of electrolyte.

66. (d) For positive charge colloids coagulating power↑ coagulation value↓. $[Fe(CN)_6]^{4-}$.

67. (a) $Fe(OH)_3$ is a positive sol. Hence, greater the charge on anion more is the coagulation power.

68. (a) Mutual coagulation

69. (d) Dialysis is a method of purification of colloids.

70. (a) 5 to 7

71. (a) Lyophilic solution are protective colloids.

72. (a) In electro osmosis D.M. move towards opposite charge electrode.

73. (b)
$$\begin{array}{cccc} AgNO_3 & + \; KI & \longrightarrow AgI(s) & + \; KNO_3 \\ 5\times10^{-3} & 10\times10^{-3} & & \\ 0 & 5\times10^{-3} & 5\times10^{-3} & \end{array}$$
$$AgI(s) + I^- \longrightarrow AgI : I^-$$

Negative charge colloids.

74. (a) $AgNO_3 + KI \longrightarrow AgI + KNO_3$

excess

$AgI + Ag^+ \longrightarrow AgI \cdot Ag^+$

75. (b) Smaller is the 'gold number' of protective colloid, greater is its protective power.

76. (c) *Zig-zag* path.

77. (d) paramagnetism.

78. (b) Tyndall effect takes place due to scattered of light.

79. (c) Light scattered by dust particles.

80. (b) Cottrell precipitator.

Level 2

1. (c)

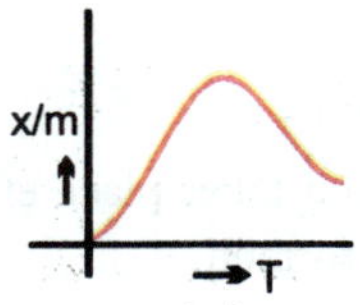

Chemical adsorption

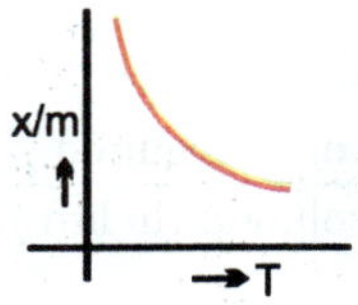

Physical adsorption

2. (d) Surface area↑ rate of adsorption↑

3. (d) In lyophobic sol D.P. have less affinity towards D.M.

4. (c) In lyophilic sol no charge present.

5. (c) Alum, purified to muddy water by coagulation.

6. (c) D.P. particles have great affinity towards D.M. in lyophilic colloids.

7. (b) Charge↑ coagulation power↑

8. (c) The sols obtained in the two cases will be oppositely charged so coagulate each other.

9. (b) Colour of dye appear in patches in *o/w* emulsion.

10. (a) The sol particles migrate towards cathode. So they are positively charged. Hence, anions would be

effective in coagulation. Greater is the valence of effective ion, smaller will be its coagulating value.

11. (d) Formation of micelles takes place above CMC.

12. (c) Bredig's arc method involve dispersion and condensation.

13. (b) Preferential adsorption of Fe^{3+} takes place. So positive charge develop.

14. (d) Sulphur solution is multimolecular by hydrophobic in nature.

15. (c) Associate colloids decrease surface tension and increase in viscosity.

16. (d) Below CMC, salt behave as normal electrolyte.

17. (c) In electro-osmosis, the sol particles are prevented from migration whereas the dispersion medium migrates in the direction opposite to those of particles. Here, the medium is negatively charged.

18. (d) No stabilising agent required for the preparation of lyophilic sol.

19. (b) Neutralisation of charge.

20. (b) Ferric hydroxide solution have positive charge. So order of coagulation value.

$KCl > K_2SO_4 > Na_3PO_4 > K_4[Fe(CN)_6]$

Charge↑ coagulation value↓

21. (b) $\Delta H < 0,\ \Delta S < 0,\ \Delta G < 0$

22. (c) Temp↑ Extent of adsorption↓

23. (d) At 83 K physical adsorption takes place effectively so multimolecular layers formed.

24. (c) Coagulation value

$$= \frac{\text{Number of millimoles of electrolyte required}}{\text{Volume of colloidal solution (in litre)}}$$

$$= \frac{10 \times 0.5}{50} \times 1000 = 100$$

25. (b) Mass of acetic acid adsorbed by 2 g charcoal

$$= 100 \times 10^{-3} \times (0.6 - 0.5) \times 60$$

$$\Rightarrow \quad 0.6;\ \frac{x}{m} = \frac{0.6}{2} \Rightarrow 0.3$$

26. (c) No. of sodium lauryl sulphate molecules $(CH_3(CH_2)_{11}SO_4^-Na^+)$ in 1 litre solution

$$= 10^{-3} \times 6 \times 10^{23}$$
$$= 6 \times 10^{20}$$

No. of sodium lauryl sulphate molecules per $mm^3 = 6 \times 10^{14}$

No. of colloidal particles per $mm^3 = 10^{13}$

No. of molecules per colloidal particle

$$= \frac{6 \times 10^{14}}{10^{13}}$$
$$= 60$$

27. (a) Total surface area of carbon $= \frac{44}{7} \times 10^7\ cm^2$

$r = 10^{-8}$ cm

Surface area of $NH_3 = \pi r^2$

$$= \frac{22}{7} \times 10^{-16}\ cm^2$$

No. of NH_3 molecules adsorbed $= \dfrac{\frac{44}{7} \times 10^7}{\frac{22}{7} \times 10^{-16}}$

$$= 2 \times 10^{23}$$

Vol. of NH_3 adsorbed at 1 atm 273 K

$$= \frac{2 \times 10^{23}}{6 \times 10^{23}} \times 22.4 = 7.46\ L$$

28. (a) No. of molecules per gram of N_2 in monolayer

$$= \frac{6 \times 10^{23}}{22{,}400} \times 1.30$$
$$= 3.48 \times 10^{19}$$

Cross-sectional area of a molecule

$$= 1.6 \times 10^{-19}\ m^2$$

∴ Area covered by molecules per gram

$$= 3.48 \times 10^{19} \times 1.6 \times 10^{-19} = 5.568\ m^2$$

∴ Surface area $= 5.568\ m^2 g^{-1}$

29. (c) Adsorbed moles of $H_2 = \dfrac{0.03 \times 2.46 \times 10^{-3}}{0.0821 \times 300}$

$$= 3 \times 10^{-6}$$

∴ No. of absorbed molecules of

$$H_2 = 3 \times 10^{-6} \times 6 \times 10^{23}$$
$$\Rightarrow \quad 18 \times 10^{17}$$

Total no. of surface sites available

$$= 5.4 \times 10^{16} \times 1000$$
$$\Rightarrow \quad 5.4 \times 10^{19}\ cm^2$$

No. of surface sites that is occupied by adsorption of

$$H_2 = \frac{10}{100} \times 5.4 \times 10^{19}$$
$$\Rightarrow \quad 5.4 \times 10^{18}$$

No. of surface sites occupied by one molecule

$$H_2 = \frac{5.4 \times 10^{18}}{18 \times 10^{17}} = 3$$

30. (b) Final volume of gas, at 608 torr pressure

$$V_2 = \frac{P_1 V_1}{P_2}$$

$$= \frac{760 \times 1}{608}$$

$\Rightarrow 1.25$ L or $V_2 = 1250$ mL

Volume occupied by gas = Volume of vessel − Volume occupied by charcoal

$$= 1000 - \frac{16}{1.6} = 990 \text{ mL}$$

Difference of volume is due to adsorption of gas by charcoal.

∴ Volume of gas adsorbed by charcoal

$= 1250 - 990$

$\Rightarrow 260$ mL

Volume of the gas adsorbed per gram of charcoal

$= \frac{260}{16} = 16.25$ mL/g at 608 torr and 27°C.

Level 3

Passage-1

1. (b) At point B $\frac{x}{m} \propto P^{1/n}$
2. (c) $\log \frac{x}{m} = \log K + \frac{1}{n} \times \log P$
3. (b) $\frac{x}{m} = K \cdot P^{1/n}$

   $\frac{x}{m} = 2 \times 2^1 = 4$

Passage-2

1. (d) Gold no. $= 0.025 \times 10^3 = 25$
2. (b) Gold no ↓ protecting power ↑
3. (a) Protective nature of colloids.

Passage-3

1. (d) Charge↑ coagulating power↑
2. (c) Al^{3+} charge ↑ coagulating power ↑
3. (c) Coagulation depend upon charge and magnitude.
4. (d) Coagulation takes place by heating, adding electrolyte and mutual coagulation.

Passage-4

1. (c) Milk : o/w emulsion

   Cold cream : w/o emulsion
2. (d) o/w emulsion diluted by addition of water.

One or More Answers is/are Correct

1. (b, d)

   Physical adsorption are not specific in nature due to van der Waal's forces and chemical adsorption are specific in nature and due to chemical bond.
2. (b,c,d)

   Adsorption is spontaneous process.

   Enthalpy of adsorption is –ve

   Change in entropy = –ve
3. (a, c, d)

   Low temperature is favourable for physical adsorption.

   Chemical adsorption required activation energy.
4. (a, b, c)

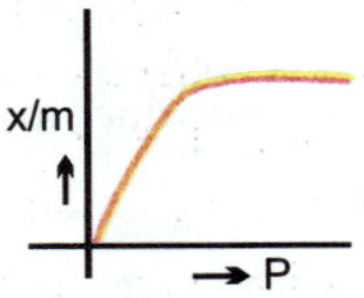

5. (a, c)

   Sulphur and gold are multimolecular colloids.
6. (b, c, d)

   Gelatin is lyophilic colloids.
7. (a, b)

   Surface tension is less than water and viscosity is greater than water in case of lyophilic sol.
8. (b, c)

   Lyophobic sol are ir-reversible and less stable and they produced by double decomposition.
9. (a, b)

   Mutual coagulation.
10. (a, b)

    Gold solution prepared by Bredig's arc method and by reduction of $AuCl_3$.
11. (a, b, c, d)

    Coagulation takes place by boiling,

    persistent dialysis

    add of electrolyte

    mutual coagulation.
12. (a, b)

    $AgNO_3 + KI \longrightarrow AgI + KNO_3$

    $AgI + AgNO_3 \longrightarrow AgIAg^+ : NO_3$

    At low temperature, extent of physical adsorption is greater due to increase in pressure.
13. (a, b, d)

    Colloidal solution can be purified by dialysis, electrolysis and ultrafiltration.

14. (a, b, c)

Coagulation occurs due to centrifugation, add of electrolyte and change in pH.

15. (a, b, c, d)

Sol shows Tyndall effect adsorption coagulation sol have colligative properties

16. (b, c)

Above CMC, associate colloids formed.

Non polar part directed towards water.

17. (b, c)

Charge develop due to preferential adsorption of ion addition of electrolyte cause coagulation.

18. (b, c)

Coagulation occurs due to electrophoresis and freezing.

19. (b, c)

In electrophoresis colloidal particle migrate towards opposite charge electrode.

20. (a, d)

Tyndall effect does not shown by true solution due to small size of particles.

Sorption involve adsorption and absorption coagulating power ∝ charge, smaller the gold number greater would be the protecting power.

21. (b, c, d)

Lyophilic colloids used as protecting colloids electrolyte used for coagulation. Rubber is obtained by coagulation of latex. By mutual coagulation rain fall occurs.

22. (a, c)

Due to size of particle

23. (a, c)

MnO_2, Ni

24. (a,b,c,d)

In presence of catalyst equilibrium obtained quickly.

Mechanism of hetrogenous catalyst is based on adsorption.

H^+ act as auto catalyst

Catalyst promotor increase the efficiency of catalyst.

25. (a,b, c,d)

Blood is colloidal solution, water purification by alum is based on coagulation. River water is colloidal solution of clay surface area↑ preferential adsorption↑.

Level 4

Subjective Problems

1. Eight gas colloidal system does not exists.

2. Aerosol, Emulsion, Gels.

3. Foam, Solid foam.

4. Size 1nm

5. CMC of soap

$10^{-4}\,M$ to $10^{-3}\,M$

$x = 4$

Numerical Problems

1. Total no. of adsorbate molecules

$$= \frac{0.20}{200} \times 10^{-3} \times 6 \times 10^{23}$$

$$= 6 \times 10^{17}$$

Effective surface area of each molecule

$$= \frac{0.08}{6 \times 10^{17}} = 13.33 \times 10^{-20}$$

$$\therefore y = 13.33$$

2. Let V mL of solution is required.

Milli moles of acid $= 0.1 \times V$

No. of molecules of acid

$$= 0.1 \times V \times 10^{-3} \times 6 \times 10^{23} = 6 \times 10^{17} \times V$$

Area covered by acid molecules

$$= 6 \times 10^{17} \times V \times 0.2 \times 10^{-14}\ \text{cm}^2$$

$$= 1.2 \times 10^{3} \times V\,\text{cm}^2$$

As per given : $1.2 \times 10^3 \times V = 3000$

$$V = 2.50\ \text{mL}$$

3. Milli moles of surfactant molecules

$$= 100 \times 0.01 = 1\,\text{m mol}$$

Mole of surfactant molecules $= 10^{-3}$ moles

No. of surfactant molecule

$$= 10^{-3} \times 6 \times 10^{23} = 6 \times 10^{20}\ \text{molecules}$$

Area occupied by a surfactant molecule

$$a^2 = \frac{0.54\ \text{cm}^2}{6 \times 10^{20}} = 9 \times 10^{-22}$$

$$= 9 \times 10^{-22}\ \text{cm}^2$$

$$a = 3 \times 10^{-11}\ \text{cm}$$

4. Freundlich equation, $\frac{x}{m} = KP^{1/n}$

$$\log\left(\frac{x}{m}\right) = \log K + \frac{1}{n}\log P\ ;$$

$$\text{slope} = \frac{1}{n} = \tan 45° = 1$$

So, $n = 1$

Intercept, $\log K = 0.60$; $\therefore K = 4$

$$\frac{x}{m} = KP^{1/n} = 4(1.75)^1 = 7$$

Mass of gas adsorbed $= 7 \times$ mass of charcoal

$$= 7 \times 2.5\ \text{g} = 17.50\ \text{g}$$

5. Coagulation value is no. of milli moles of electrolyte is required per litre of colloidal solution.

$\therefore$ Coagulation value of AB_2

$$= \frac{40 \times 0.05}{(40 + 30) \times 10^{-3}} = 28.57$$

6. Flocculation value of $H_2SO_4 = 15$,

$$15 = \text{m.mole of } H_2SO_4 \times \frac{1000}{250}$$

Moles of $H_2SO_4 = 3.75 \times 10^{-3}$

Mass of $H_2SO_4 = 3.75 \times 10^{-3} \times 98$

$$= 0.3675\ \text{g} \approx 0.37\ \text{g}$$

7. M. moles of X_2Y is required $= 20 \times 0.2 = 4$

Coagulation value of $X_2Y = \frac{4}{100} \times 1000$

$= 40$ m. moles per litre.

8. Mass of KCl $= 7.45 \times 1.1 \times \frac{20}{100}$

Milli moles of KCl $= \frac{74.5}{74.5} \times \frac{2.2}{100} \times 1000 = 22$

Milli moles of KCl required per litre colloidal solution

$$= \frac{22}{750} \times 1000 = 29.33.$$

9. $\frac{\text{Coagulation power of } BaCl_2}{\text{Coagulation power of KCl}}$

$$= \frac{\text{Coagulation value of KCl}}{\text{Coagulation value of } BaCl_2} = \frac{50}{0.70}$$

$$= 71.428 = 71.43$$

10. Average life for adsorption, $T = \frac{1}{k}$

So $k_1 t_1 = k_2 t_2$ for a given process

$$\ln\left(\frac{t_1}{t_2}\right) = \frac{E_a}{R}\left[\frac{1}{T_1} - \frac{1}{T_2}\right]; \ln\left(\frac{0.20}{0.80}\right) = \frac{E_a}{2}\left[\frac{1}{1000} - \frac{1}{500}\right]$$

or $$2.303 \log\left(\frac{0.20}{0.80}\right) = \frac{E_a}{2}\left[\frac{1}{1000} - \frac{1}{500}\right]$$

$E_a = 2763.6$ cal/mol ;

$E_a = 2.76$ kcal/mol

Note.....

Note.....

श्री
BalaJi

PHYSICAL CHEMISTRY CHAPTERWISE

15 YEARS (2011-2025)

PAPERS with Detailed Solutions

JEE

(MAIN & ADVANCED)

SHRI BALAJI PUBLICATIONS
(EDUCATIONAL PUBLISHERS & DISTRIBUTORS)
[AN ISO 9001-2008 CERTIFIED ORGANIZATION]
Muzaffarnagar (U.P.) - 251001

Published by:
SHRI BALAJI PUBLICATIONS
(EDUCATIONAL PUBLISHERS & DISTRIBUTORS)
[AN ISO 9001-2008 CERTIFIED ORGANIZATION]
6, Gulshan Vihar,
Jansath Road, Muzaffarnagar (U.P.)
Phone : 0131-2660440 (O)
website : www.shribalajibooks.com
email : sbjpub@gmail.com

Edition : 2025

Price : ₹ 350.00

CONTENTS

GROW GREEN

SAVE NATURE

CHAPTER

1

Some Basic Concept in Chemistry

A. Objective Questions

[Only one option is correct]

1. Given that the abundance of isotopes ^{54}Fe, ^{56}Fe and ^{57}Fe are 5%, 90% and 5%, respectively, the atomic mass of Fe is: **(IIT 2009)**

(a) 55.85 (b) 55.95 (c) 55.75 (d) 56.05

2. The ratio of masses of oxygen and nitrogen in a particular gaseous mixture is 1 : 4. The ratio of number of their molecule is : **[JEE (Mains) 2014]**

(a) 1 : 4 (b) 7 : 32 (c) 1 : 8 (d) 3 : 16

3. The molecular formula of a commercial resin used for exchanging ions in water softening is $C_8H_7SO_3Na$ (Mol. wt. 206). What would be the maximum uptake of Ca^{2+} ions by the resin when expressed in mole per gram resin ? **[JEE (Mains) 2015]**

(a) $\frac{2}{309}$ (b) $\frac{1}{412}$ (c) $\frac{1}{103}$ (d) $\frac{1}{206}$

4. 3 g of activated charcoal was added to 50 mL of acetic acid solution (0.06 *N*) in a flask. After an hour it was filtered and the strength of the filtrate was found to be 0.042 *N*. The amount of acetic acid adsorbed (per gram of charcoal) is : **[JEE (Mains) 2015]**

(a) 42 mg (b) 54 mg (c) 18 mg (d) 36 mg

5. 1 gram of a carbonate (M_2CO_3) on treatment with excess HCl produces 0.01186 mole of CO_2. The molar mass of M_2CO_3 in g mol^{-1} is: **[JEE (Mains) 2017]**

(a) 118.6 (b) 11.86 (c) 1186 (d) 84.3

6. The most abundant elements by mass in the body of a healthy human adult are: oxygen (61.4 %), carbon (22.9%), hydrogen (10.0%) and nitrogen (2.6%). The weight which a 75 kg person would gain if all ^1H atoms are replaced by ^2H atoms is: **[JEE (Mains) 2017]**

(a) 7.5 kg (b) 10 kg (c) 15 kg (d) 37.5 kg

7. 100 mL of a water sample contains 0.81 g of calcium bicarbonate and 0.73 g of magnesium bicarbonate. The hardness of this water sample expressed in terms of equivalents of $CaCO_3$ is :
(Molar mass of calcium bicarbonate is 162 g mol^{-1} and magnesium bicarbonate is 146 g mol^{-1}.)
[JEE (Mains) 2019]

(a) 10,000 ppm (b) 100 ppm (c) 1,000 ppm (d) 5,000 ppm

8. 0.27 g of a long chain fatty acid was dissolved in 100 cm^3 of hexane. 10 mL of this solution was added dropwise to the surface of water in a round watch glass. Hexane evaporates and a monolayer is formed. The distance from edge to centre of the watch glass is 10 cm. What is the height of the monolayer? [Density of fatty acid = 0.9 g cm^{-3}; $\pi = 3$] **[JEE (Mains) 2019]**

(a) 10^{-4}m (b) 10^{-2}m (c) 10^{-8}m (d) 10^{-6}m

9. The average molar mass of chlorine is 35.5 g mol^{-1}. The ratio of ^{35}Cl to ^{37}Cl in naturally occurring chlorine is close to: **[JEE Main (Sept.) 2020]**

(a) 4 : 1 (b) 3 : 1 (c) 2 : 1 (d) 1 : 1

10. The strengths of 5.6 volume hydrogen peroxide (of density 1 g/mL) in terms of mass percentage and molarity (M), respectively, are: **[JEE Main (Sept.) 2020]**

(Take molar mass of hydrogen peroxide as 34 g/mol)

(a) 0.85 and 0.25 (b) 0.85 and 0.5 (c) 1.7 and 0.5 (d) 1.7 and 0.25

11. A solution of two components containing n_1 moles of the 1st component and n_2 moles of the 2nd component is prepared. M_1 and M_2 are the molecular weights of component 1 and 2 respectively. If d is the density of the solution in g mL^{-1}, C_2 is the molarity and x_2 is the mole fraction of the 2nd component, then C_2 can be expressed as: **[JEE Main (Sept.) 2020]**

(a) $C_2 = \dfrac{1000x_2}{M_1 + x_2(M_2 - M_1)}$ (b) $C_2 = \dfrac{dx_2}{M_2 + x_2(M_2 - M_1)}$

(c) $C_2 = \dfrac{1000dx_2}{M_1 + x_2(M_2 - M_1)}$ (d) $C_2 = \dfrac{dx_1}{M_2 + x_2(M_2 - M_1)}$

12. Complete combustion of 1.80 g of an oxygen containing compound ($C_xH_yO_z$) gave 2.64 g of CO_2 and 1.08 g of H_2O. The percentage of oxygen in the organic compound is: **[JEE Main (Feb.) 2021]**

(a) 51.63 (b) 63.53 (c) 53.33 (d) 50.33

13. Haemoglobin contains 0.34% of iron by mass. The number of Fe atoms in 3.3 g of haemoglobin is:

(**Given :** Atomic mass of Fe is 56 u, N_A in 6.022×10^{23} mol^{-1}) **[JEE Main (July) 2022 (II)]**

(a) 1.21×10^5 (b) 12.0×10^{16} (c) 1.21×10^{20} (d) 3.4×10^{22}

14. Using the rules for significant figures, the correct answer for the expression $\dfrac{0.02858 \times 0.112}{0.5702}$ will be :

[JEE Main (June) 2022 (II)]

(a) 0.005613 (b) 0.00561 (c) 0.0056 (d) 0.006

15. Compound A contains 8.7% Hydrogen, 74% Carbon and 17.3% Nitrogen. The molecular formula of the compound is : (**Given :** Atomic masses of C, H and N are 12, 1 and 14 amu respectively.)

The molar mass of the compound A is 162 g mol^{-1}. **[JEE Main (June) 2022 (II)]**

(a) $C_4H_6N_2$ (b) C_2H_3N (c) C_5H_7N (d) $C_{10}H_{14}N_2$

16. $C(s) + O_2(g) \longrightarrow CO_2(g) + 400$ kJ

$C(s) + \frac{1}{2}O_2(g) \longrightarrow CO(g) + 100$ kJ

When coal of purity 60% is allowed to burn in presence of insufficient oxygen, 60% of carbon is converted into 'CO' and the remaining is converted into 'CO_2'. **[JEE Main (July) 2022 (II)]**

The heat generated when 0.6 kg of coal is burnt is ________.

(a) 1600 kJ (b) 3200 kJ (c) 4400 kJ (d) 6600 kJ

17. The number of molecules and moles in 2.8375 litres of O_2 at STP are respectively :

[JEE Main (April) 2023 (I)]

(a) 7.527×10^{22} and 0.250 mol (b) 1.505×10^{23} and 0.250 mol

(c) 7.527×10^{23} and 0.125 mol (d) 7.527×10^{22} and 0.125 mol

18. Which of the following have same number of significant figures?

(A) 0.00253 (B) 1.0003 (C) 15.0 (D) 163

Choose the correct answer from the options given below : **[JEE Main (April) 2023 (II)]**

(a) A, B and C only (b) C and D only (c) A, C and D only (d) B and C only

19. A metal chloride contains 55.0% of chlorine by weight. 100 mL vapours of the metal chloride at STP weight 0.57 g. The molecular formula of the metal chloride is :

(Given : Atomic mass of chlorine is 35.5 u) **[JEE Main (April) 2023 (I)]**

(a) MCl_2 (b) MCl_4 (c) MCl_3 (d) MCl

20. What is the mass ratio of ethylene glycol ($C_2H_6O_2$, molar mass = 62 g/mol) required for making 500 g of 0.25 molal aqueous solution and 250 mL of 0.25 molar aqueous solution?

[JEE Main (Jan.) 2023 (II)]

(a) 1 : 1 (b) 3 : 1 (c) 2 : 1 (d) 1 : 2

21. The quantity which changes with temperature is : **[JEE Main (Jan.) 2024]**

(a) Molarity (b) Mass percentage (c) Molality (d) Mole fraction

22. A sample of $CaCO_3$ and $MgCO_3$ weighed 2.21 g is ignited to constant weight of 1.152 g. The composition of mixture is : **[JEE Main (Jan.) 2024]**

(**Given :** Molar mass in g mol^{-1}; $CaCO_3$: 100, $MgCO_3$: 84)

(a) 1.187 g $CaCO_3$ + 1.023 g $MgCO_3$ (b) 1.023 g $CaCO_3$ + 1.023 g $MgCO_3$

(c) 1.187 g $CaCO_3$ + 1.187 g $MgCO_3$ (d) 1.023 g $CaCO_3$ + 1.187 g $MgCO_3$

23. Choose the correct statements. **[JEE Main (Jan.) 2025]**

(A) Weight of a substance is the amount of matter present in it.

(B) Mass is the force exerted by gravity on an object.

(C) Volume is the amount of space occupied by a substance.

(D) Temperatures below $0°C$ are possible in Celsius scale, but in Kelvin scale negative temperature is not possible.

(E) Precision refers to the closeness of various measurements for the same quantity.

(a) (B), (C) and (D) only (b) (A), (B) and (C) only

(c) (A), (D) and (E) only (d) (C), (D) and (E) only

24. 2.8×10^{-3} mol of CO_2 is left after removing 10^{21} molecules from its 'x' mg sample. The mass of CO_2 taken initially is :

Given : $N_A = 6.02 \times 10^{23} mol^{-1}$ **[JEE Main (Jan.) 2025]**

(a) 196.2 mg (b) 98.3 mg (c) 150.4 mg (d) 48.2 mg

25. 0.1M solution of KI reacts with excess of H_2SO_4 and KIO_3 solution. According to equation

$$5I^- + IO_3^- + 6H^+ \rightarrow 3I_2 + 3H_2O$$

Identify the correct statements.

(A) 200 mL of KI solution reacts with 0.004 mol of KIO_3.

(B) 200 mL of KI solution reacts with 0.006 mol of H_2SO_4.

(C) 0.5 L of KI solution produced 0.005 mol of I_2.

(D) Equivalent weight of KIO_3 is equal to $\left(\frac{\text{Molecular weight}}{5}\right)$.

Choose the correct answer from the options given below : **[JEE Main (Jan.) 2025]**

(a) (A) and (D) only (b) (B) and (C) only (c) (A) and (B) only (d) (C) and (D) only

B. Integer Type Problems

1. A student performs a titration with different burettes and finds titre values of 25.2 mL, 25.25 mL and 25.0 mL. The number of significant figures in the average titre value is : **(IIT 2010)**

2. Silver (atomic mass = 108 g mol^{-1}) has a density of 10.5 g cm^{-1}. The number of silver atoms on a surface of area $10^{-12} m^2$ can be expressed in scientific notation as $y \times 10^8$. The value of x is: **(IIT 2010)**

3. The volume (in mL) of 0.1M $AgNO_3$ required for complete precipitation of chloride ions present in 30 mL of 0.01M solution of $[Cr(H_2O)_5Cl]Cl_2$, as silver chloride is close to : **(IIT 2011)**

4. 29.2% (w/w) HCl stock solution has a density of 1.25 g mL^{-1}. The molecular weight of HCl is 36.5 g mol^{-1}. The volume (mL) of stock solution required to prepare a 200 mL solution of 0.4 M HCl is : **(IIT 2012)**

5. A compound H_2X with molar weight of 80 g is dissolved in a solvent having density of 0.4 g mL^{-1}. Assuming no change in volume upon dissolution, the molality of a 3.2 molar solution is: **[JEE (Advanced) 2014]**

6. The mole fraction of a solute in a solution is 0.1. At 298 K, molarity of this solution is the same as its molality. Density of this solution at 298 K is 2.0 g cm^{-3}. The ratio of the molecular weights of the solute and solvent, $\left(\frac{MW_{solute}}{MW_{solvent}}\right)$ is: **[JEE (Advanced) 2016]**

7. The ammonia prepared by treating ammonium sulphate with calcium hydroxide is completely used by $NiCl_2 \cdot 6H_2O$ to form a stable coordination compound. Assume that both the reactions are 100% complete. If 1584 g of ammonium sulphate and 952 g of $NiCl_2 \cdot 6H_2O$ are used in the preparation, the combined weight (in grams) of gypsum and the nickel-ammonia coordination compound thus produced is : **[JEE (Advanced) 2018]**

(Atomic weights in g mol^{-1} : H = 1, N = 14, O = 16, S = 32, Cl = 35.5, Ca = 40, Ni = 59)

8. The amount of water produced (in g) in the oxidation of 1 mole of rhombic sulphur by conc. HNO_3 to a compound with the highest oxidation state of sulphur is ____ .

(Given data : Molar mass of water = 18 g mol^{-1}) **[JEE (Advanced) 2019]**

9. $NaClO_3$ is used, even in spacecraft to produce O_2. The daily consumption of pure O_2 by a person is 492 L at 1 atm, 300 K. How much amount of $NaClO_3$, in grams, is required to produce O_2 for the daily consumption of a person at 1 atm, 300 K?

$$NaClO_3(s) + Fe(s) \longrightarrow O_2(g) + NaCl(s) + FeO(s)$$

$R = 0.082$ L atm mol^{-1} K^{-1}. **[JEE Main (Jan.) 2020]**

10. The mass of ammonia in grams produced when 2.8 kg of dinitrogen quantitatively reacts with 1 kg of dihydrogen is ________. **[JEE Main (Sept.) 2020]**

11. The molarity of HNO_3 in a sample which has density 1.4 g/mL and mass percentage of 63% is (Molecular Weight of HNO_3 = 63) **[JEE Main (Jan.) 2020]**

12. 5.00 mL of 0.10 M oxalic acid solution taken in a conical flask is titrated against NaOH from a burette using phenolphthalein indicator. The volume of NaOH required for the appearance of permanent faint pink colour is tabulated below for five experiments. What is the concentration, in molarity, of the NaOH solution? **[JEE (Advanced) 2020]**

Exp. No.	Vol. of NaOH (mL)
1	12.5
2	10.5
3	9.0
4	9.0
5	9.0

13. Aluminium reacts with sulphuric acid to form aluminium sulphate and hydrogen. What is the volume of hydrogen gas in litres (L) produced at 300 K and 1.0 atm pressure, when 5.4 g of aluminium and 50.0 mL of 5.0 M sulphuric acid are combined for the reaction ?

(Use molar mass of aluminium as 27.0 g mol^{-1}, $R = 0.082$ atm L $mol^{-1}K^{-1}$)

[JEE (Advanced) 2020]

14. 4.5 g of compound A (MW = 90) was used to make 250 mL of its aqueous solution. The molarity of the solution in M is $x \times 10^{-1}$. The value of x is ________. (Rounded off to the nearest integer) **[JEE Main (Feb.) 2021]**

15. The $NaNO_3$ weighed out to make 50 mL of an aqueous solution containing 70.0 mg Na^+ per mL is ______ g. (Rounded off to the nearest integer)
[**Given :** Atomic weight in g mol^{-1} – Na : 23; N : 14; O : 16] **[JEE Main (Feb.) 2021]**

16. A 6.50 molal solution of KOH (*aq.*) has a density of 1.89 g cm^{-3}. The molarity of the solution is ______ mol dm^{-3}. (Round off to the nearest integer).
[Atomic masses: K : 39.0 u; O : 16.0 u; H: 1.0 u] **[JEE Main (March) 2021]**

17. When 35 mL of 0.15 *M* lead nitrate solution is mixed with 20 mL of 0.12 *M* chromic sulphate solution, ______ $\times 10^{-5}$ moles of lead sulphate precipitate out. (Round off to the nearest integer). **[JEE Main (March) 2021]**

18. The mole fraction of a solute in a 100 molal aqueous solution.........$\times 10^{-2}$.
(Round off to the nearest integer). (**Given:** Atomic masses: H:1.0 u, O:16.0u) **[JEE Main (March) 2021]**

19. Chlorophyll extracted from the crushed green leaves was dissolved in water to make 2L solution of Mg of concentration 48 ppm. The number of atoms of Mg in this solution is $x \times 10^{20}$ atoms. The value of x is ______. (Nearest integer)
(**Given :** Atomic mass of Mg is 24 g mol^{-1}; $N_A = 6.02 \times 10^{23} mol^{-1}$) **[JEE Main (July) 2022 (I)]**

20. A protein 'A' contains 0.30% of glycine (molecular weight 75). The minimum molar mass of the protein 'A' is ______ $\times 10^3$ g mol^{-1} [Nearest integer] **[JEE Main (June) 2022 (II)]**

21. An organic compound gives 0.220 g of CO_2 and 0.126 g of H_2O on complete combustion. If the % of carbon is 24 then the % hydrogen is ______ $\times 10^{-1}$. (Nearest integer) **[JEE Main (April) 2023 (I)]**

22. A sample of a metal oxide has formula $M_{0.83}O_{1.00}$. The metal *M* can exist in two oxidation states +2 and +3. In the sample of $M_{0.83}O_{1.00}$, the percentage of metal ions existing in +2 oxidation state is ______% (Nearest integer) **[JEE Main (Jan.) 2023 (II)]**

23. 1 g of a carbonate (M_2CO_3) on treatment with excess HCl produces 0.01 mol of CO_2. The molar mass of M_2CO_3 is ______ g mol^{-1}. (Nearest integer) **[JEE Main (April) 2023 (II)]**

24. 0.5 g of an organic compound (*X*) with 60% carbon will produce ______ $\times 10^{-1}$ g of CO_2 on complete combustion. **[JEE Main (April) 2023 (I)]**

25. A solution of H_2SO_4 is 31.4% H_2SO_4 by mass and has a density of 1.25 g/mL. The molarity of the H_2SO_4 solution is ______ M (Nearest integer)
[**Given :** Molar mass of H_2SO_4 = 98 g mol^{-1}] **[JEE Main (Jan.) 2024]**

26. Number of moles of methane required to produce 22 g $CO_2(g)$ after combustion is $x \times 10^{-2}$ moles. The value of x is : **[JEE Main (Jan.) 2024]**

27. Some CO_2 gas was kept in a sealed container at a pressure of 1 atm and at 273 K. This entire amount of CO_2 gas was later passed through an aqueous solution of $Ca(OH)_2$. The excess unreacted $Ca(OH)_2$ was later neutralized with 0.1 *M* of 40 mL HCl. If the volume of the sealed container of CO_2 was x, then x is .... cm^3 (nearest integer).
[Given : The entire amount of $CO_2(g)$ reacted with exactly half the initial amount of $Ca(OH)_2$ present in the aqueous solution.] **[JEE Main (Jan.) 2025]**

28. When 81.0 g of aluminium is allowed to react with 128.0 g of oxygen gas, the mass of aluminium oxide produced in gram is .... (Nearest integer)
Given : Molar mass of Al is 27.0 g mol^{-1}, Molar mass of O is 16.0 g mol^{-1} **[JEE Main (Jan.) 2025]**

29. Consider the following reaction occurring in the blast furnace.

$$Fe_3O_{4(s)} + 4CO_{(g)} \longrightarrow 3Fe_{(l)} + 4CO_{2(g)}$$

'x' kg of iron is produced when 2.32×10^3 kg Fe_3O_4 and 2.8×10^2 kg CO are brought together in the furnace. The value of 'x' is .... (nearest integer)

(Given : Molar mass of $Fe_3O_4 = 232\ g\ mol^{-1}$, Molar mass of $CO = 28\ g\ mol^{-1}$, Molar mass of $Fe = 56\ g\ mol^{-1}$) **[JEE Main (Jan.) 2025]**

Answers

[A] Objective Questions (Only one option is correct)

1. (b)	**2.** (b)	**3.** (b)	**4.** (c)	**5.** (d)	**6.** (a)	**7.** (a)	**8.** (d)	**9.** (b)	**10.** (c)
11. (c)	**12.** (c)	**13.** (c)	**14.** (b)	**15.** (d)	**16.** (d)	**17.** (d)	**18.** (c)	**19.** (a)	**20.** (c)
21. (a)	**22.** (c)	**23.** (d)	**24.** (a)	**25.** (a)					

[B] Integer Type Problems

1. (3)	**2.** (7)	**3.** (6)	**4.** (8)	**5.** (8)
6. (9)	**7.** (2992)	**8.** (288)	**9.** (2130)	**10.** (3400)
11. (14)	**12.** (0.11)	**13.** (6.15)	**14.** (2)	**15.** (13)
16. (9)	**17.** (525)	**18.** (64)	**19.** (25)	**20.** (25)
21. (56)	**22.** (59)	**23.** (100)	**24.** (11)	**25.** (4)
26. (50)	**27.** (45)	**28.** (153)	**29.** (420)	

Hints and Solutions

A. Objective Questions

[Only one option is correct]

1. (b) Average atomic mass $= \dfrac{(54 \times 5) + (56 \times 90) + (57 \times 5)}{100} = 55.95$

2. (b) Let the mass of $O_2 = a$ g

$\therefore$ Mass of $N_2 = 4a$ g

Number of moles of $O_2 = \dfrac{a}{32}$

Number of moles of $N_2 = \dfrac{4a}{28} = \dfrac{a}{7}$

$\therefore$ Ratio of moles $= \dfrac{a}{32} : \dfrac{a}{7} = 7:32$

3. (b) $2C_8H_7SO_3^-Na^+ + Ca^{2+} \longrightarrow (C_8H_7SO_3^-)_2Ca^{2+} + 2Na^+$
2mole(2×206=412g) 1mole

1 mole of $Ca^{2+} \equiv 412$ g of resin

Maximum uptake of Ca^{2+} ions by the resin $= \frac{1 \text{ mol}}{412 \text{ g}} = \frac{1}{412}$ mol/g

4. (c) No. of milliequivalents of acetic acid initially taken $= (0.06\, N) \times (50 \text{ mL}) = 3$ meq.
No. of milliequivalents of acetic acid left in the filtrate $= (0.042\, N) \times (50 \text{ mL}) = 2.1$ meq.
No. of milliequivalents of acetic acid adsorbed by activated charcoal $= (3 - 2.1) = 0.9$ meq.
Amount of acetic acid adsorbed by 3 g of activated charcoal $= 0.9 \times 60 = 54$ mg
Amount of acetic acid adsorbed by 1 g of activated charcoal $= \frac{54}{3} = 18$ mg

5. (d) According to the question,
$M_2CO_3 + 2HCl \longrightarrow 2MCl + H_2O + CO_2$
In this equation, number of moles of M_2CO_3 is equal to that of CO_2.
i.e., $n_{M_2CO_3} = n_{CO_2}$

$$\frac{\text{wt. of } M_2CO_3}{\text{molar mass of } M_2CO_3} = n_{CO_2}$$

$$\frac{1\text{g}}{\text{molar mass of } M_2CO_3} = 0.01186 \text{ mol}$$

$$\text{Molar mass of } M_2CO_3 = \frac{1}{0.01186} \approx 84.3 \text{ g mol}^{-1}$$

6. (a) Mass of elements in the body of a healthy human adult are:
Oxygen (61.4%), carbon (22.9%), hydrogen (10%) and nitrogen (2.6%).
Weight of the person = 75 kg

Mass due to $^1H = 75 \times \frac{10}{100} = 7.5$ kg

On replacing ^1H by ^2H, 7.5 kg mass would replace with 15 kg.
∴ Net mass gained by person $= (15 - 7.5)$ kg $= 7.5$ kg

7. (a) ppm of $CaCO_3 = \frac{\left(\frac{0.73}{146} + \frac{0.81}{162}\right) \times 100}{100} \times 10^6 = 10^4$ ppm

8. (d) Surface area of watch glass $= \pi r^2 = 3 \times (10)^2$
(Given : $\pi = 3$) $= 300$ cm^2
0.27 g of fatty acid was dissolved in 100 cm^3 of hexane.
∴ Mass of fatty acid in 10 cm^3 of hexane $= \frac{10 \times 0.27}{100} = 0.027$ g

Given that, density of fatty acid $= 0.9$ g cm^{-3}

$$\text{Volume of fatty acid} = \frac{0.027 \text{ g}}{0.9 \text{ g cm}^{-3}} = 0.03 \text{ cm}^3$$

$$\text{Height of the monolayer} = \frac{\text{Volume of fatty acid}}{\text{Surface area of watch glass}}$$

$$= \frac{0.03 \text{ cm}^3}{300 \text{ cm}^2} = 10^{-4} \text{ cm or } 10^{-6} \text{ m}$$

9. (b) Let total moles = 1, Moles of $^{35}Cl = x$, Moles of $^{37}Cl = (1 - x)$

$$M_{avg} = \frac{\text{Total mass}}{\text{Total moles}}$$

$$35.5 = \frac{x(35) + (1-x)37}{1} \quad \Rightarrow x = 0.75$$

Molar ratio $= \frac{x}{1-x} = \frac{0.75}{0.25} = \frac{3}{1}$

10. (c) Volume strength = molarity × 11.2

Molarity $(M) = \frac{5.6}{11.2} = 0.5$

$0.5\,M\,H_2O_2$ means 0.5 moles of H_2O_2 is present in 1 L solution.

⇒ Weight of $H_2O_2 = 0.5 \times 34 = 17$ gm

⇒ Weight of solution = 1 × 1000 mL × (1 gm/mL) = 1000 gm

% by mass $= \frac{17}{1000} \times 100 = 1.7\%$

11. (c) Let total moles are 1.

So, moles of 2nd component $= n_2 = x_2$

Moles of 1st component $= n_1 = 1 - x_2$

Mass of 1st component = mole × molar mass $= n_1 \times M_1 = (1-x_2)M_1$

Mass of 2nd component $= n_2M_2 = x_2M_2$

Total mass of solution $= (1-x_2)M_1 + M_2x_2$

$$\Rightarrow \quad \text{Volume} = \frac{\text{Mass}}{\text{Density}} = \frac{(1-x_2)M_2 + M_2x_2}{d}$$

$$\Rightarrow \quad C_2 = \frac{n_2}{V(\text{mL})} \times 1000 = \frac{x_2 \cdot 1000 \cdot d}{(1-x_2)M_2 + M_2x_2} = \frac{d \cdot x_2 \cdot 1000}{M_1 + x_2(M_2 - M_1)}$$

12. (c) $C_xH_yO_z + \left(x + \frac{y}{4} - \frac{z}{2}\right)O_2 \longrightarrow xCO_2 + \frac{y}{2}H_2O$

⇒ Moles of C = Moles of $CO_2 = \frac{2.64}{44} = 0.06$ mol

Weight of carbon $= 0.06 \times 12 = 0.72$ g

⇒ Moles of H = 2 × moles of $H_2O = 2 \times \frac{1.08}{18} = 0.12$ g

Since, weight of hydrogen $= 0.12 \times 1 = 0.12$ g

⇒ Weight of oxygen = weight of compound – weight of H – weight of O

$= 1.80\text{ g} - 0.12\text{ g} - 0.72 = 0.96$ g

$$\%\text{ oxygen} = \left(\frac{0.96}{1.80}\right) \times 100 = 53.33\%$$

13. (c) Number of Fe atoms in 3.3 g of haemoglobin $= \frac{0.34}{100} \times \frac{3.3}{56} \times 6.022 \times 10^{23} = 1.206 \times 10^{20}$

14. (b) 0.02858 → 4 significant figures.

0.112 → 3 significant figures.

0.5702 → 4 significant figures.

The answer should have three significant figures.

15. (d)

C	74%	$\frac{74}{12} = 6.16$	$\frac{6.16}{1.23} = 5$
N	17.3%	$\frac{17.3}{14} = 1.23$	$\frac{1.23}{1.23} = 1$
H	8.7%	$\frac{8.7}{1} = 8.7$	$\frac{8.7}{1.23} = 7$

Empirical formula = C_5NH_7

Empirical weight = $(12\times 5)+(14\times 1)+(1\times 7)=81$ amu

Multiplying factor = $\dfrac{\text{Molecular weight}}{\text{Empirical weight}}=\dfrac{162}{81}=2$

Molecular formula = 2 × (Empirical formula) $C_{10}N_2H_{14}$

16. (d) $C(s)+O_2(g)\longrightarrow CO_2(g)+400\text{kJ}$

$C(s)+\frac{1}{2}O_2(g)\longrightarrow CO(g)+100\text{kJ}$

Weight of coal = 0.6 kg = 0.6×10^3 g

Weight of pure carbon = $0.6\times 10^3\times\dfrac{60}{100}$

$n_C=\dfrac{0.6\times 10^3\times 60}{100}\times\dfrac{1}{12}=\dfrac{6\times 60}{12}$

40% of C is converted into CO_2,

Energy released = $\dfrac{6\times 60}{12}\times\dfrac{40}{100}\times 400=4800$ kJ

60% of C is converted into CO,

Energy released = $\dfrac{6\times 60}{12}\times\dfrac{60}{100}\times 100=1800$ kJ

Total energy released = 4800 + 1800 = 6600 kJ

17. (d) Number of moles of $O_2=\dfrac{2.8375}{22.7}=0.125$

$\Rightarrow$ Number of molecules = $0.125\,N_A=7.525\times 10^{22}$

18. (c) (A) 0.00253, Significant figures = 3 (All non zero digits are significant.)
(B) 1.0003, Significant figures = 5 (Zeros between non-zero digit are significant.)
(C) 15.0, Significant number = 3.
(D) 163, Significant number = 3

19. (a) Molecular weight of metal chloride = $\dfrac{0.57}{100}\times 22700=129.39$

Weight of Cl = 129.39 × 0.55 = 71.1645

$\therefore$ Mole of Cl = $\dfrac{71.1645}{35.5}\cong 2$, Hence MCl_2

20. (c) Assuming, : Mass of solvent ≈ Mass of solution

Case I : $0.25=\dfrac{W_1}{62}\times\dfrac{1000}{500}$

Case II : $0.25=\dfrac{W_2}{62}\times\dfrac{1000}{250}\Rightarrow\dfrac{W_1}{W_2}=\dfrac{2}{1}$

21. (a) Molarity = $\dfrac{\text{Moles of solute}}{\text{Volume of solution}}$

Since volume depends on temperature, molarity will change upon change in temperature.

22. (a) $CaCO_3(s)\xrightarrow{\Delta}CaO(s)+CO_2(g)$

$MgCO_3(s)\xrightarrow{\Delta}MgO(s)+CO_2(g)$

Let the weight of $CaCO_3$ be x gm

$\therefore$ weight of $MgCO_3$ = $(2.21-x)$ gm

Moles of $CaCO_3$ decomposed = moles of CaO formed

$\dfrac{x}{100}$ = moles of CaO formed

$\therefore$ weight of CaO formed $= \dfrac{x}{100} \times 56$

Moles of $MgCO_3$ decomposed = moles of MgO formed

$$\frac{(2.21 - x)}{84} = \text{moles of MgO formed}$$

$\therefore$ weight of MgO formed $= \dfrac{2.21 - x}{84} \times 40$

$$\Rightarrow \quad \frac{2.21 - x}{84} \times 40 + \frac{x}{100} \times 56 = 1.152$$

$\therefore$ $x = 1.1886$ g = weight of $CaCO_3$ and weight of $MgCO_3$ = 1.0214 g

24. (a) $(\text{moles})_{\text{intial}} = \dfrac{x \times 10^{-3}}{44}$

$(\text{moles})_{\text{removal}} = \dfrac{10^{21}}{6.02 \times 10^{23}}$

$(\text{moles})_{\text{left}} = (\text{moles})_{\text{initial}} - (\text{moles})_{\text{removed}}$

$$2.8 \times 10^{-3} = \frac{x \times 10^{-3}}{44} - \frac{10^{21}}{6.02 \times 10^{23}}$$

$\Rightarrow \quad x = 196.2$ mg

B. Integer Type Problems

1. (3) Because, least precise data contains 3 significant figures.

Average value $= \dfrac{25.2 + 25.25 + 25.0}{3} = 25.15 \Rightarrow 25.2$

2. (7) Density = Mass of unit volume, *i.e.*, 1 cm^3 silver contains = 10.5 g

Number of atoms of Ag in 1 cm^3 or 10.5 g $= \dfrac{10.5}{108} \times N_A$ or number of Ag-atoms in 1 cm^2 $= \left(\dfrac{10.5}{108} \times N_A\right)^{2/3}$ or number of atoms of Ag in $10^{-12} m^2$ or $10^{-8} cm^2$

$$= \left(\frac{10.5}{108} \times N_A\right)^{2/3} \times 10^{-8} = \left(\frac{10.5 \times 6.022 \times 10^{24}}{108}\right)^{2/3} \times 10^{-8}$$

$= 1.5 \times 10^7 \quad \Rightarrow \quad x = 7$

3. (6) Number of ionisable Cl^- in $[Cr(H_2O)_5Cl]Cl_2$ is 2.

$\therefore$ Millimoles of $Cl^- = 30 \times 0.01 \times 2 = 0.6$

$\therefore$ Millimoles of Ag^+ required $= 0.6$

$\therefore$ $0.6 = 0.1V \Rightarrow V = 6$ mL

4. (8) 29.2% w/w of HCl

$d = 1.25$ g/mL, $M_{HCl} = 36.5 \text{ g mol}^{-1}$

Mass of HCl = 29.2 g

Volume of solution = 100 g/1.25 g mL^{-1} = 80 mL

$$\text{Molarity} = \frac{(\text{g/M.W.})}{\text{Volume of solution}} \times 1000 = \frac{(29.2/36.5)}{80} \times 1000 = 10\,M$$

Using $M_1V_1 = M_2V_2$

$$10 \times V_1 = 200 \times 0.4 \Rightarrow V_1 = \frac{200 \times 0.4}{10} = 8$$

5. (8) Here, $V_{\text{Solution}} \approx V_{\text{Solvent}}$

Since, in 1 L solution, 3.2 moles of solute are present,

So, $1\,\text{L solution} \approx 1\,\text{L solvent}\ (d = 0.4\,\text{g/mL}) \approx 0.4\,\text{kg}$

So, $\text{molality}\ (m) = \dfrac{\text{Moles of solute}}{\text{Mass of solvent (kg)}} = \dfrac{3.2}{0.4} = 8$

6. (9) $m = \dfrac{\chi_B \times 1000}{\chi_A \times M_A} \quad \Rightarrow \quad m = \dfrac{1000}{9M_A}$ ......(i)

Also using,

$m = \dfrac{1000M}{1000d - MM_B} \Rightarrow MM_B = 1000$ .....(ii)

$1 = \dfrac{1000}{1000 \times 2 - MM_B}$

Using (i) and (ii): $\dfrac{M_B}{M_A} = 9$ (Molality = molarity)

7. (2992)

$(NH_4)_2SO_4 + Ca(OH)_2 \longrightarrow 2NH_3 + CaSO_4{\cdot}2H_2O$

$\dfrac{1584}{132} = 12$ mole ; $2NH_3$: 24 mole ; $CaSO_4{\cdot}2H_2O$: 12 mole

$2NH_3$ (24 mole) $\xrightarrow{NiCl_2{\cdot}6H_2O \text{ (4 mole)}}$

$CaSO_4{\cdot}2H_2O \quad + \quad [Ni(NH_3)_6]Cl_2$

12 mole ; 4 mole

$12 \times 172 = 2064$; $4 \times (59 + 71 + 17 \times 6) = 928$

Total mass = 2992 g

8. (288) $32H_2O + S_8 \longrightarrow 8H_2SO_4 + 48H^{\oplus} + 48e^{\ominus}$

$[e^{\ominus} + H^{\oplus} + HNO_3 \longrightarrow NO_2 + H_2O] \times 48$

$S_8 + 48HNO_3 \longrightarrow 8H_2SO_4 + 48NO_2 + 16H_2O$

1 mol ; 16 mol

$\Rightarrow$ Mass of H_2O formed $= (16 \times 18)\ \text{g} = 288\ \text{g}$

9. (2130)

Moles of $O_2 = \dfrac{PV}{RT} = \dfrac{1 \times 492}{0.082 \times 300} = 20$

$\Rightarrow$ Since 1 mole of O_2 is produced by 1 mole of $NaClO_3$ consumed. So moles of $NaClO_3$ reacted = 20

$\Rightarrow$ Weight of $NaClO_3$ reacted $= 20 \times 106.5 = 2130$ gm

10. (3400)

$$\underset{\substack{\frac{2800}{28} \\ = 100\,\text{mol}}}{N_2} + \underset{\substack{\frac{1000}{2} \\ = 500\,\text{mol}}}{3H_2} \longrightarrow 2NH_3$$

$\Rightarrow$ 100 mol of N_2 will react with $3 \times 100 = 300$ mol H_2.

So, H_2 is excess reagent while N_2 is limiting reagent.

$\Rightarrow$ Moles of NH_3 formed $= 2 \times 100 = 200$ mol

$\Rightarrow$ Weight of NH_3 produced $= 200\ \text{mol} \times 17\ \text{gm/mol} = 3400$ gm

11. (14) 63% w/w HNO_3 means 63 gm of HNO_3 is present in 100 gm solution.

Moles of $HNO_3 = \dfrac{63}{63} = 1$

Volume of solution $= \dfrac{100}{1.4}$ mL

$\Rightarrow \quad \text{Molarity} = \dfrac{\text{Moles of HNO}_3}{\text{Volume of solution (L)}} = \dfrac{1 \times 1000}{\dfrac{100}{1.4}} = 14 \text{ mol/L}$

12. (0.11) No. of eq. of oxalic acid = No. of eq. of NaOH

or $\dfrac{5.00 \times 0.10}{1000} \times 2 = \dfrac{9.0 \times M}{1000} \times 1$

$\therefore$ Molarity of NaOH solution $= \dfrac{1}{9} = 0.11M$

13. (6.15) $2Al + 3H_2SO_4 \longrightarrow Al_2(SO_4)_3 + 3H_2$

Moles of Al taken $= \dfrac{5.4}{27} = 0.2$

Moles of H_2SO_4 taken $= \dfrac{50 \times 5.0}{1000} = 0.25$

As $\dfrac{0.2}{2} > \dfrac{0.25}{3}$, H_2SO_4 is limiting reagent

Now, moles of H_2 formed $= \dfrac{3}{3} \times 0.25 = 0.25$

$\therefore$ Volume of H_2 gas formed $= \dfrac{nRT}{P} = \dfrac{0.25 \times 0.082 \times 300}{1} = 6.15$ L

14. (2) Molarity $(M) = \dfrac{\text{mol}}{V(\text{Lt})} = \dfrac{\text{mol}}{V\text{ (mL)}} \times 1000$

Moles of solute $= \dfrac{4.5}{90}$

Molarity $(M) = \dfrac{4.5 \times 1000}{90 \times 250} = 0.2 = 2 \times 10^{-1} \dfrac{\text{mol}}{\text{L}}$

15. (13) Weight of Na^+ present in 50 mL of $Na^+ = 50 \times 70$

$= 3500 \text{ mg} = 3.5 \text{ gm}$

$\Rightarrow \quad$ Moles of $Na^+ = \dfrac{3.5}{23}$

Moles of $NaNO_3 = \dfrac{3.5}{23}$

Weight of $NaNO_3 = \dfrac{3.5}{23} \times 85 = 13$ gm

16. (9) 6.5 molal solution means 6.5 moles of solute (KOH) are present in 1 kg solvent (H_2O).

So, weight of KOH $= 6.5 \times 56 = 364$ g

Weight of solution $= 1000 + 364 = 1364$ g

Volume of solution $= \dfrac{1364}{1.89} \text{ cm}^3$

$\Rightarrow \quad$ Molarity $(M) = \dfrac{\text{Moles (solute)}}{\text{Volume (L)}}$

$= \dfrac{6.5}{\dfrac{1364}{1.89}} \times 1000 = \dfrac{6.5 \times 1.89 \times 1000}{1364} = 9 \text{ mol/L}$

17. (525)

Balanced reaction is

$$3Pb(NO_3)_2 + Cr_2(SO_4)_3 \longrightarrow 3PbSO_4 + 2Cr(NO_3)_3$$

Millimoles of $Pb(NO_3)_2 = 35 \times 0.15 = 5.25$

Millimoles of $Cr_2(SO_4)_3 = 20 \times 0.12 = 2.4$

Since, $Pb(NO_3)_2$ is limiting reagent.

Millimoles of $PbSO_4$ formed $= 5.25$

$\Rightarrow$ Moles of $PbSO_4 = 5.25 \times 10^{-3} = 525 \times 10^{-5}$

18. (64) 100 molal solution means 100 moles of solutes are present in 1kg solvent (water)

$$X = \frac{n_{solute}}{n_{solute} + n_{solvent}}$$

$$X = \frac{100}{100 + \frac{1000}{18}} = \frac{1800}{2800}$$

$$X = 0.6428 = 64.28 \times 10^{-2}$$

19. (25) Let the weight of Mg in the extract $= x$ g

$$\frac{x}{2000} \times 10^6 = 48 \quad \text{[Assuming 1000 mL} \approx \text{1000 kg for water]}$$

$$x = 96 \times 10^{-3}$$

$$n_{Mg} = \frac{\text{weight}}{\text{molar mass}} = \frac{96 \times 10^{-3}}{24} = 0.004$$

Number of Mg atom $= 0.004 \times 6.02 \times 10^{23} = 24.8 \times 10^{20} \approx 25 \times 10^{20}$

20. (25) There will be minimum one molecule of glycine.

0.30% glycine $\equiv 75$ g

1% of protein $\equiv \frac{75}{0.30}$ g

100% of protein $\equiv \frac{75}{0.30} \times 100 \text{ g} = 25000 \text{ g} = 25 \times 10^3$ g

21. (56) Moles of $CO_2 = \frac{0.22}{44} = \frac{1}{200}$

$\therefore$ Moles of carbon $=$ (Moles of CO_2) $\times 1 = \frac{1}{200}$

$\therefore$ Weight of C $= \frac{1}{200} \times 12 = 0.06$

% of C $= \frac{0.06}{W} \times 100 = 24$ (W = Weight of organic compound) $\Rightarrow W = 0.25$

Moles of $H_2O = \frac{0.126}{18} = 0.007$

$\therefore$ Moles of H atom $= 2 \times 0.007 = 0.014$

% of Hydrogen $= \frac{0.014 \times 1}{W} \times 100 = \frac{0.014 \times 1}{0.25} \times 100 = 5.6 = 56 \times 10^{-1}$

22. (59) Let the amount of M^{2+} in the sample be x then the amount of M^{3+} in the sample $= 0.83 - x$

$\therefore \quad 2x + 3(0.83 - x) = 2 \quad \Rightarrow \quad x = 0.49$

$$\% M^{2+} = \frac{0.49}{0.83} \times 100 = 59\%$$

23. (100)

$$\underset{1\text{g}}{M_2CO_3} + \underset{\text{Excess}}{2HCl} \longrightarrow \underset{0.02 \text{ mole}}{2MCl} + H_2O + \underset{0.01 \text{ mole}}{CO_2}$$

From principle of atomic conservation of carbon atom,

Mole of $M_2CO_3 \times 1 =$ Mole of $CO_2 \times 1$

$$\frac{1 \text{ g}}{\text{molar mass of } M_2CO_3} = 0.01 \times 1$$

$\therefore$ Molar mass of M_2CO_3 = 100 g/mole

24. (11) Percentage of Carbon $= \frac{12}{44} \times \frac{\text{mass of } CO_2 \text{ formed}}{\text{mass of compound taken}} \times 100$

$$60 = \frac{12}{44} \times \frac{\text{mass of } CO_2 \text{ formed}}{0.5} \times 100$$

Mass of CO_2 formed $= \frac{60 \times 44 \times 0.5}{12 \times 100}$ g $= 1.1$ g $= 11 \times 10^{-1}$ g

25. (4) $M = \frac{n_{solute}}{V} \times 1000 = \frac{\left(\frac{31.4}{98}\right)}{\left(\frac{100}{1.25}\right)} \times 1000 = 4.005 \approx 4$

26. (50) $CH_4(g) + 2O_2(g) \longrightarrow CO_2(g) + 2H_2O(l)$

$n_{CO_2} = \frac{22}{44} = 0.5$ moles

So moles of CH_4 required $= 0.5$ moles

i.e., 50×10^{-2} mole

$x = 50$

27. (45) Let moles of $CO_2 = n$

moles of $Ca(OH)_2$ total initially $= 2n$

excess $Ca(OH)_2 = n$

gm equivalent of $Ca(OH)_2$ = gm equivalent of HCl

$n \times 2 = 0.1 \times \frac{40}{1000} \times 1$

$n = 2 \times 10^{-3}$

Volume of $CO_2 = 2 \times 10^3 \times 22400 = 44.8 \text{ cm}^3$

28. (153) $4Al + 3O_2 \longrightarrow 2Al_2O_3$

$\frac{81}{27} = 3$ mole $\quad \frac{128}{32} = 4$ mole

Limiting reagent

$\therefore$ mole of Al_2O_3 formed $= \frac{1}{2} \times 3$ mole

$\therefore$ wt. of Al_2O_3 formed $= \frac{3}{2} \times 102 = 153$ gm

29. (420)

Moles of $Fe_3O_4 = \frac{2.32 \times 10^3 \times 10^3}{232} = 10000$ mol

Moles of CO $= \frac{2.8 \times 10^2 \times 10^3}{28} = 10000$ mol

$Fe_3O_4 + 4CO \longrightarrow 3Fe + 4CO_2$

10^4 mol $\quad 10^4$ mol

CO is L.R.

Mole of Fe $= \frac{3}{4} \times 10^4$

Mass of Fe $= \frac{3}{4} \times \frac{10^4 \times 56}{1000}$ kg $= 420$ kg

❑❑❑

CHAPTER

2

Atomic Structure

A. Objective Questions

[Only one option is correct]

1. The kinetic energy of an electron in the second Bohr orbit of a hydrogen atom : [a_0 is Bohr radius] **(IIT 2012)**

(a) $\frac{h^2}{4\pi^2 m a_0^2}$ (b) $\frac{h^2}{16\pi^2 m a_0^2}$ (c) $\frac{h^2}{32\pi^2 m a_0^2}$ (d) $\frac{h^2}{64\pi^2 m a_0^2}$

2. Energy of an electron is given by $E = -2.178 \times 10^{18}\,\text{J}\left(\frac{Z^2}{n^2}\right)$. Wavelength of light required to excite an electron in an hydrogen atom from level $n = 1$ to $n = 2$ will be : **(IIT 2013)**

($h = 6.62 \times 10^{-34}$ Js and $c = 3.0 \times 10^{-8}\ \text{ms}^{-1}$)

(a) 1.214×10^{-7} m (b) 2.816×10^{-7} m (c) 6.500×10^{-7} m (d) 8.500×10^{-7} m

3. The correct set of four quantum numbers for the valence electrons of rubidium atom ($Z = 37$) is : **[JEE (Mains) 2014]**

(a) $5, 0, 0, +\frac{1}{2}$ (b) $5, 1, 0, +\frac{1}{2}$ (c) $5, 1, 1, +\frac{1}{2}$ (d) $5, 0, 1, +\frac{1}{2}$

4. Which of the following is the energy of a possible excited state of hydrogen? **[JEE (Mains) 2015]**

(a) −3.4 eV (b) +6.8 eV (c) +13.6 eV (d) −6.8 eV

5. A stream of electrons from a heated filament was passed between two charged plates kept at a potential difference V esu. If e and m are charge and mass of an electron respectively, then the value of h/λ (where λ is wavelength associated with electron wave) is given by: **[JEE (Mains) 2016]**

(a) meV (b) $2meV$ (c) $\sqrt{meV}$ (d) $\sqrt{2meV}$

6. P is the probability of finding the 1s electron of hydrogen atom in a spherical shell of infinitesimal thickness dr, at a distance r from the nucleus. The volume of this shell is $4\pi r^2 dr$. The qualitative sketch of the dependence of P on r is: **[JEE (Advanced) 2016]**

(a) 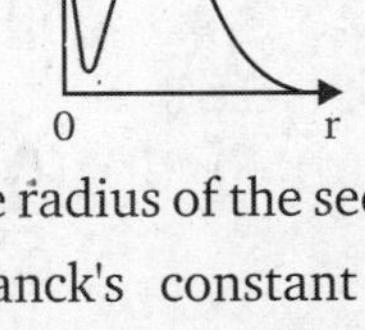

(b) 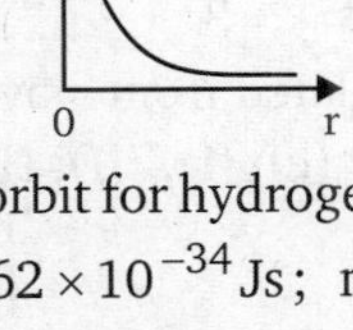

(c) 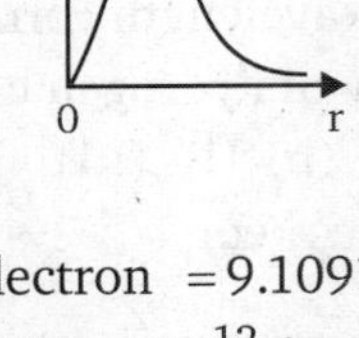

(d) 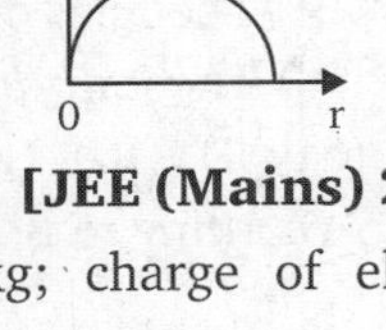

7. The radius of the second Bohr orbit for hydrogen atom is: **[JEE (Mains) 2017]**

(Planck's constant $(h) = 6.6262 \times 10^{-34}$ Js; mass of electron $= 9.1091 \times 10^{-31}$ kg; charge of electron $= 1.60210 \times 10^{-19}$ C; Permittivity of vacuum $(\varepsilon_0) = 8.854185 \times 10^{-12}\ \text{kg}^{-1}\text{m}^{-3}\text{A}^2$)

(a) 0.529 Å (b) 2.12 Å (c) 1.65 Å (d) 4.76 Å

8. The group having isoelectronic species is: **[JEE (Mains) 2017]**

(a) O^{2-}, F^-, Na, Mg^{2+} (b) O^-, F^-, Na^+, Mg^{2+}

(c) $O^{2-}, F^-, Na^+, Mg^{2+}$ (d) O^-, F^-, Na, Mg^{2+}

9. The quantum number of four electrons are given below:

I. $n=4, l=2, m_l=-2, m_s=-1/2$ II. $n=3, l=2, m_l=-2, m_s=+1/2$

III. $n=4, l=1, m_l=0, m_s=+1/2$ IV. $n=3, l=1, m_l=1, m_s=-1/2$

The correct order of their increasing energies will be : **[JEE (Mains) 2019]**

(a) IV < II < III < I (b) IV < III < II < I

(c) I < III < II < IV (d) I < II < III < IV

10. The size of the iso-electronic species Cl^-, Ar and Ca^{2+} is affected by : **[JEE (Mains) 2019]**

(a) principal quantum number of valence shell
(b) nuclear charge
(c) electron-electron interaction in the outer orbitals
(d) azimuthal quantum number of valence shell.

11. If p is the momentum of the fastest electron ejected from a metal surface after the irradiation of light having wavelength λ, then for 1.5 p momentum of the photoelectron, the wavelength of the light should be : (Assume kinetic energy of ejected photoelectron to be very high in comparison to work function) **[JEE (Mains) 2019]**

(a) $\frac{2}{3}\lambda$ (b) $\frac{4}{9}\lambda$ (c) $\frac{3}{4}\lambda$ (d) $\frac{1}{2}\lambda$

12. Amongst the following statements, that which was not proposed by Dalton was: **[JEE Main (Jan.) 2020]**

(a) matter consists of invisible atoms
(b) when gases combine or reproduced in a chemical reaction they do so in a simple ratio by volume provided all gases are at the same temperature and pressure
(c) chemical reactions involve reorganisation of atoms. These are neither created nor destroyed in a chemical reaction
(d) All the atoms of a given element have identical properties including identical mass. Atoms of different elements differ in mass

13. The radius of the second Bohr orbit, in terms of the Bohr radius, a_0, in Li^{2+} is: **[JEE Main (Jan.) 2020]**

(a) $\frac{4a_0}{3}$ (b) $\frac{2a_0}{9}$ (c) $\frac{2a_0}{3}$ (d) $\frac{4a_0}{9}$

14. The de-Broglie wavelength of an electron in the 4th Bohr orbit is: **[JEE Main (Jan.) 2020]**

(a) $6\pi a_0$ (b) $4\pi a_0$ (c) $2\pi a_0$ (d) $8\pi a_0$

15. For the Balmer series in the spectrum of H-atom, $\bar{\nu} = R_H\left\{\frac{1}{n_1^2} - \frac{1}{n_2^2}\right\}$ the correct statements among (I) to (IV) are : **[JEE Main (Jan.) 2020]**

(I) As wavelength decreases, the lines in the series converge
(II) The integer n_1 is equal to 2
(III) The lines of longest wavelength corresponds to $n_2 = 3$
(IV) The ionization energy of hydrogen can be calculated from wave number of these lines

(a) (II), (III), (IV) (b) (I), (III), (IV) (c) (I), (II), (III) (d) (I), (II), (IV)

16. According to Bohr's atomic theory:

(A) Kinetic energy of electron is $\propto \frac{Z^2}{n^2}$.

(B) The product of velocity (v) of electron and principal quantum number (n), 'vn' $\propto Z^2$.

(C) Frequency of revolution of electron in an orbit $\propto \dfrac{Z^3}{n^3}$.

(D) Coulombic force of attraction on the electron is $\propto \dfrac{Z^3}{n^4}$.

Choose the most appropriate answer from the options given below : **[JEE Main (Feb.) 2021]**

(a) (C) Only
(b) (A) Only
(c) (A), (C) and (D) only
(d) (A) and (D) only

17. The shortest wavelength of H-atom in the Lyman series is λ_1. The longest wavelength in the Balmer series of H^+ is: **[JEE Main (Sept.) 2020]**

(a) $\dfrac{5\lambda_1}{9}$
(b) $\dfrac{36\lambda_1}{5}$
(c) $\dfrac{9\lambda_1}{5}$
(d) $\dfrac{27\lambda_1}{5}$

18. The difference between the radii of 3rd and 4th orbits of Li^{2+} in ΔR_1. The difference between the radii of 3rd and 4th orbits of He^+ is ΔR_2. Ratio $\Delta R_1 : \Delta R_2$ is: **[JEE Main (Sept.) 2020]**

(a) 8 : 3
(b) 2 : 3
(c) 3 : 8
(d) 3 : 2

19. Hydrogen has three isotopes (A), (B) and (C). If the number of neutron(s) in (A), (B) and (C) respectively, are (X), (Y) and (Z), the sum of (X), (Y) and (Z) is: **[JEE Main (Jan.) 2020]**

(a) 4
(b) 3
(c) 2
(d) 1

20. The figure that is not a direct manifestation of the quantum nature of atom is: **[JEE Main (Sept.) 2020]**

(a) 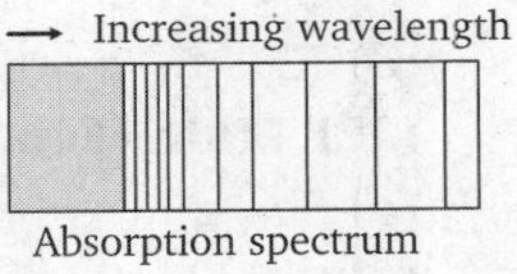

(b) 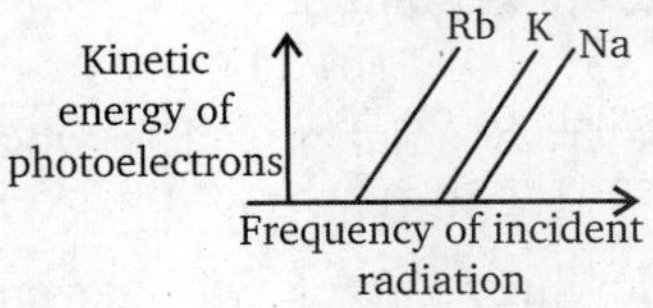

(c) 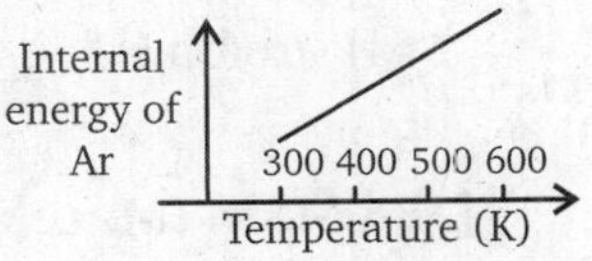

(d) 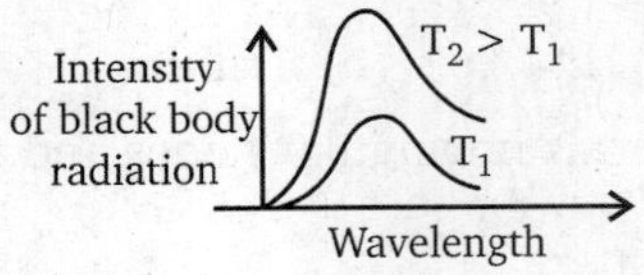

21. The correct statement about probability density (except at infinite distance from nucleus) is: **[JEE Main (Sept.) 2020]**

(a) It can be zero for $3p$ orbital
(b) It can be zero for $1s$ orbital
(c) It can be negative for $2p$ orbital
(d) It can never be zero for $2s$ orbital

22. In the sixth period, the orbitals that are filled are: **[JEE Main (Sept.) 2020]**

(a) $6s$, $5f$, $6d$, $6p$
(b) $6s$, $6p$, $6d$, $6f$
(c) $6s$, $4f$, $5d$, $6p$
(d) $6s$, $5d$, $5f$, $6p$

23. The number of subshells associated with $n = 4$ and $m = -2$ quantum numbers is: **[JEE Main (Sept.) 2020]**

(a) 2
(b) 16
(c) 8
(d) 4

24. Consider the hypothetical situation where the azimuthal quantum number, l, takes values 0, 1, 2,...$n + 1$, where n is the principal quantum number. Then, the element with atomic number: **[JEE Main (Sept.) 2020]**

(a) 9 is the first alkali metal
(b) 8 is the first noble gas
(c) 13 has a half-filled valence subshell
(d) 6 has a $2p$-valence subshell

25. Given below are two statements : **[JEE Main (March) 2021]**

Statement I : Bohr's theory accounts for the stability and line spectrum of Li^+ ion.

Statement II : Bohr's theory was unable to explain the splitting of spectral lines in the presence of a magnetic field.

In the light of the above statements, choose the most appropriate answer from the options given below :
(a) Both statement I and statement II are true.
(b) Statement I is false but statement II is true.
(c) Both statement I and statement II are false.
(d) Statement I is true but statement II is false.

26. The plots of radial distribution functions for various orbitals of hydrogen atom against 'r' are given below: **[JEE Main (Feb.) 2021]**

(1)
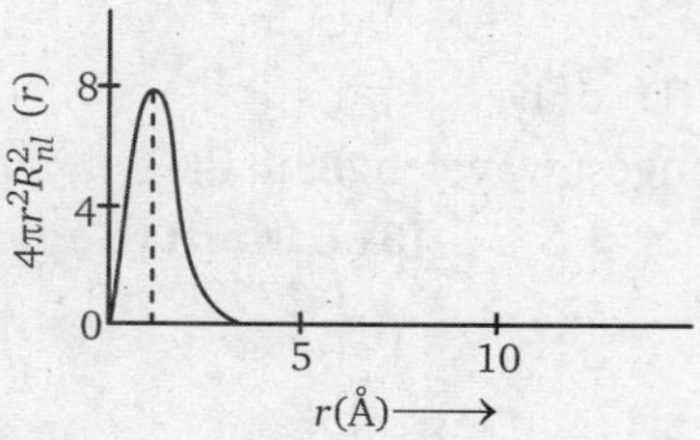

(2)
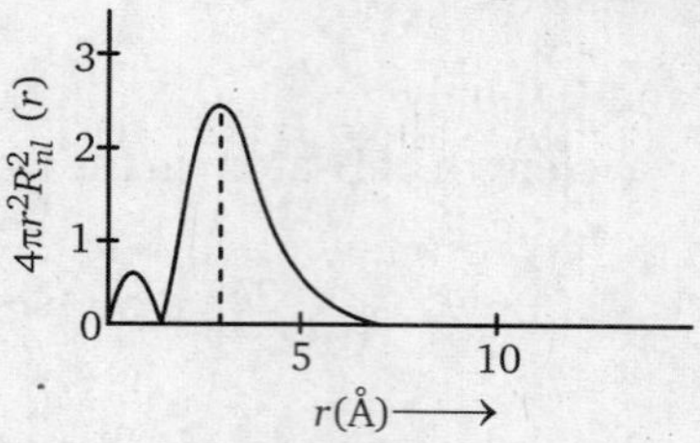

(3)
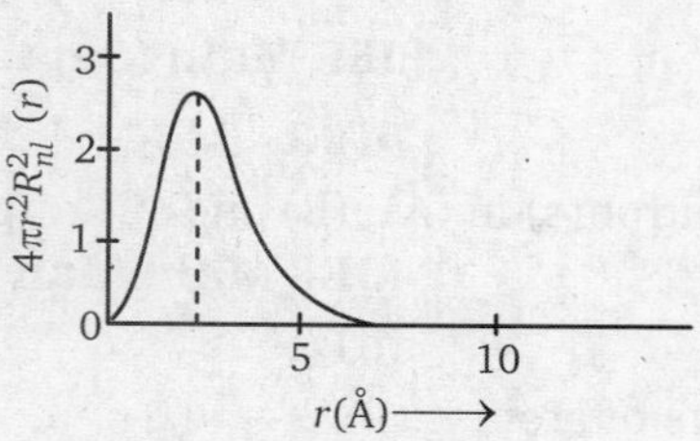

(4)
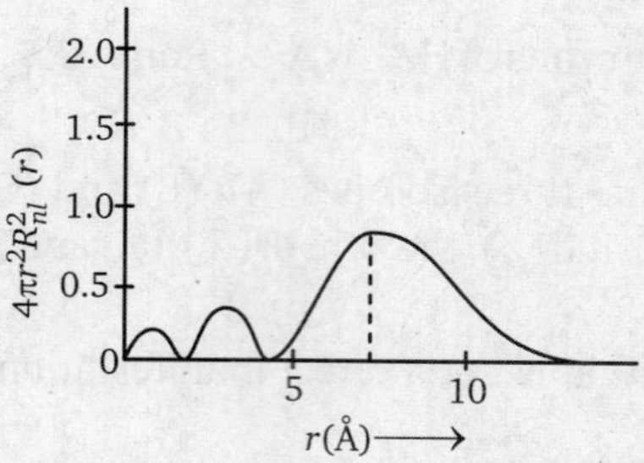

The correct plot for 3s-orbital is:
(a) (2) (b) (1) (c) (4) (d) (3)

27. The orbital having two radial as well as two angular nodes is: **[JEE Main (Feb.) 2021]**
(a) 3p (b) 4f (c) 4d (d) 5d

28. Which of the following forms of hydrogen emits low energy β^- particles? **[JEE Main (Feb.) 2021]**
(a) Deuterium ^2_1H (b) Tritium ^3_1H (c) Protium ^1_1H (d) Proton H^+

29. A certain orbital has no angular nodes and two radial nodes. The orbital is: **[JEE Main (March) 2021]**
(a) 2s (b) 3s (c) 3p (d) 2p

30. Identify the incorrect statement from the following : **[JEE Main (July) 2022 (I)]**
(a) A circular path around the nucleus in which an electron moves is proposed as Bohr's orbit
(b) An orbital is the one electron wave function (Ψ) in an atom
(c) The existence of Bohr's orbits is supported by hydrogen spectrum
(d) Atomic orbital is characterised by the quantum numbers n and l only

31. Hydrogen has three isotopes : protium (^1H), deuterium (^2H or D) and tritium (^3H or T). They have nearly same chemical properties but different physical properties. They differ in : **[JEE Main (June) 2022 (II)]**
(a) number of protons (b) atomic number
(c) electronic configuration (d) atomic mass

32. Given below are the quantum numbers for 4 electrons.
A. $n=3, l=2, m_l=1, m_s=+1/2$
B. $n=4, l=1, m_l=0, m_s=+1/2$
C. $n=4, l=2, m_l=-2, m_s=-1/2$
D. $n=3, l=1, m_l=-1, m_s=+1/2$

The correct order of increasing energy is : **[JEE Main (July) 2022 (II)]**
(a) D < B < A < C (b) D < A < B < C (c) B < D < A < C (d) B < D < C < A

33. The shortest wavelength of hydrogen atom in Lyman series is λ. The longest wavelength in Balmer series of He^+ is : **[JEE Main (Jan.) 2023 (I)]**

(a) $\frac{5}{9\lambda}$ (b) $\frac{9\lambda}{5}$ (c) $\frac{36\lambda}{5}$ (d) $\frac{5\lambda}{9}$

34. The energy of an electron in the first Bohr orbit of hydrogen atom is -2.18×10^{-18} J. Its energy in the third Bohr orbit is ________. **[JEE Main (April) 2023 (I)]**

(a) $\frac{1}{27}$ of this value (b) One third of this value

(c) Three times of this value (d) $\frac{1}{9}$th of this value

35. If the radius of the first orbit of hydrogen atom is a_0, then de-Broglie's wavelength of electron in 3rd orbit is: **[JEE Main (April) 2023 (II)]**

(a) $\frac{\pi a_0}{6}$ (b) $\frac{\pi a_0}{3}$ (c) $6\pi a_0$ (d) $3\pi a_0$

36. The wave function (ψ) of 2s is given by

$$\psi_{2s} = \frac{1}{2\sqrt{2}\,\pi}\left(\frac{1}{a_0}\right)^{1/2}\left(2 - \frac{r}{a_0}\right)e^{-r/2a_0}$$

At $r = r_0$, radial node is formed. Thus, r_0 in terms of a_0. **[JEE Main (Jan.) 2023 (II)]**

(a) $r_0 = a_0$ (b) $r_0 = 4a_0$ (c) $r_0 = \frac{a_0}{2}$ (d) $r_0 = 2a_0$

37. According to the wave-particle duality of matter by de-Broglie, which of the following graph plot presents most appropriate relationship between wavelength of electron (λ) and momentum of electron (p) ? **[JEE Main (Jan.) 2024]**

(a) 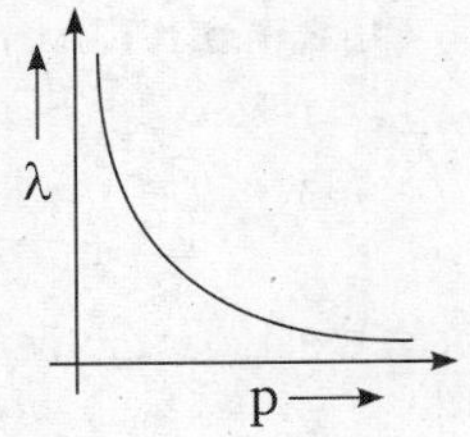

(b) 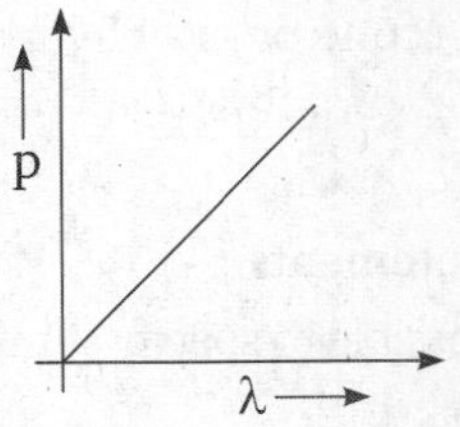

(c) 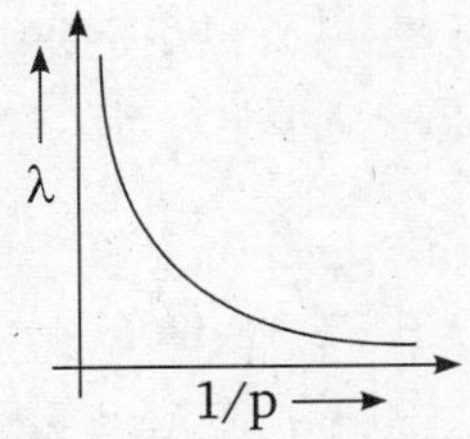

(d) 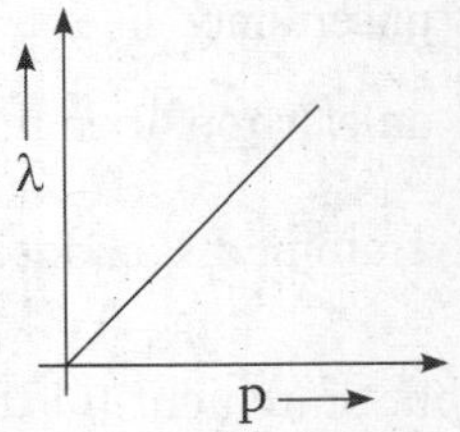

38. Given below are two statements :

Statement I : A spectral line will be observed for a $2p_x \rightarrow 2p_y$ transition.

Statement II : $2p_x$ and $2p_y$ are degenerate orbitals. **[JEE Main (Jan.) 2025]**

In the light of the above statements, choose the correct answer from the options given below :

(a) Both Statement I and Statement II are true.

(b) Both Statement I and Statement II are false.

(c) Statement I is true but Statement II is false.

(d) Statement I is false but Statement II is true.

39. Heat treatment of muscular pain involves radiation of wavelength of about 900 nm. Which spectral line of H atom is suitable for this? **[JEE Main (Jan.) 2025]**

Given : Rydberg constant, $R_H = 10^5\,cm^{-1}$, $h = 6.6 \times 10^{-34}$ Js, $c = 3 \times 10^8$ m/s

(a) Paschen series, $\infty \rightarrow 3$ (b) Lyman series, $\infty \rightarrow 1$
(c) Balmer series, $\infty \rightarrow 2$ (d) Paschen series, $5 \rightarrow 3$

40. Which of the following linear combination of atomic orbitals will lead to formation of molecular orbitals in homonuclear diatomic molecules (internuclear axis in z-direction)? **[JEE Main (Jan.) 2025]**

A. $2p_z$ and $2p_x$ B. $2s$ and $2p_x$ C. $3d_{xy}$ and $3d_{x^2-y^2}$ D. $2s$ and $2p_z$
E. $2p_z$ and $3d_{x^2-y^2}$

(a) E only (b) A and B only (c) D only (d) C and D only

41. For hydrogen atom, the orbital/s with lowest energy is/are :

(A) $4s$ (B) $3p_x$ (C) $3d_{x^2-y^2}$ (D) $3d_{z^2}$
(E) $4p_z$

Choose the correct answer from the options given below : **[JEE Main (Jan.) 2025]**

(a) (A) and (E) only (b) (B) only (c) (A) only (d) (B), (C) and (D) only

42. In a multielectron atom, which of the following orbitals described by three quantum numbers will have same energy in absence of electric and magnetic fields?

A. $n=1, l=0, m_l=0$ B. $n=2, l=0, m_l=0$ C. $n=2, l=1, m_l=1$ D. $n=3, l=2, m_l=1$
E. $n=3, l=2, m_l=0$

Choose the correct answer from the options given below : **[JEE Main (Jan.) 2025]**

(a) A and B only (b) B and C only (c) C and D only (d) D and E only

43. Which of the following is/are not correct with respect to energy of atomic orbitals of hydrogen atom?

(A) $1s < 2p < 3d < 4s$ (B) $1s < 2s = 2p < 3s = 3p$
(C) $1s < 2s < 2p < 3s < 3p$ (D) $1s < 2s < 4s < 3d$

Choose the correct answer from the options given below : **[JEE Main (Jan.) 2025]**

(a) (B) and (D) only (b) (A) and (C) only (c) (C) and (D) only (d) (A) and (B) only

44. If a_0 is denoted as the Bohr radius of hydrogen atom, then what is the de-Broglie wavelength (λ) of the electron present in the second orbit of hydrogen atom? (n : any integer). **[JEE Main (Jan.) 2025]**

(a) $\dfrac{2a_0}{n\pi}$ (b) $\dfrac{8\pi a_0}{n}$ (c) $\dfrac{4\pi a_0}{n}$ (d) $\dfrac{4n}{\pi a_0}$

45. Given below are two statements :

Statement-I : It is impossible to specify simultaneously with arbitrary precision, both the linear momentum and the position of a particle.

Statement-II : If the uncertainty in the measurement of position and uncertainty in measurement of momentum are equal for an electron, then the uncertainty in the measurement of velocity is $\geq \sqrt{\dfrac{h}{\pi}} \times \dfrac{1}{2m}$.

In the light of the above statements, choose the correct answer from the options given below :

[JEE Main (Jan.) 2025]

(a) Statement-I is true but Statement-II is false
(b) Both Statement-I and Statement-II are true
(c) Statement-I is false but Statement-II is true
(d) Both Statement-I and Statement-II are false

B. Objective Questions

[One or more than one option(s) is/are correct]

1. Which of the following statement(s) is (are) correct? **(IIT 1988)**

(a) The electronic configuration of Cr is $[Ar]3d^5 4s^1$. (Atomic number of Cr = 24)
(b) The magnetic quantum number may have a negative value
(c) In silver atom, 23 electrons have a spin of one type and 24 of the opposite type. (Atomic number of Ag = 47)

(d) The oxidation state of nitrogen in HN_3 is –3

2. Ground state electronic configuration of nitrogen atom can be represented by : **(IIT 1999)**

(a) [↑↓] [↑↓] [↑] [↑] [↑] (b) [↑↓] [↑↓] [↑] [↓] [↑]

(c) [↑↓] [↑↓] [↑] [↓] [↓] (d) [↑↓] [↑↓] [↓] [↓] [↓]

3. The ground state energy of hydrogen atom is –13.6 eV. Consider an electronic state Ψ of He^+ whose energy, azimuthal quantum number and magnetic quantum number are –3.4 eV, 2 and 0, respectively. Which of the following statement(s) is (are) true for the state ψ? **[JEE (Advanced) 2019]**

(a) It is a $4d$ state.

(b) The nuclear charge experienced by the electron in this state is less than $2e$, where e is the magnitude of the electronic charge.

(c) It has 3 radial nodes.

(d) It has 2 angular nodes.

C. Matching Type Problems

1. According to Bohr's theory

E_n = Total energy, K_n = Kinetic energy

V_n = Potential energy, r_n = Radius of nth orbit

Match the following : **(IIT 2006)**

	Column I		Column II
(a)	$V_n/K_n = ?$	(p)	0
(b)	If radius of nth orbit $\propto E_n^x$, $x = ?$	(q)	–1
(c)	Angular momentum in lowest orbital	(r)	–2
(d)	$\frac{1}{r^n} \propto Z^y$, $y = ?$	(s)	1

2. Match the entries in **Column I** with the **Column II.** **(IIT 2007)**

	Column I		Column II
(a)	Radius of nth orbital	(p)	Inversely proportional to Z
(b)	Energy of nth shell	(q)	Integral multiple of $h/2\pi$.
(c)	Angular momentum in lowest orbital	(r)	Proportional to n^2
(d)	Velocity of electron in nth	(s)	Inversely proportional to n.

3. Match the entries in **Column I** with the correctly related quantum number(s) in **Column II.** **(IIT 2008)**

	Column I		Column II
(a)	Orbital angular momentum of the electron in a hydrogen-like atomic orbital	(p)	Principal quantum number
(b)	A hydrogen-like one-electron wave function obeying Pauli's principle	(q)	Azimuthal quantum number

(c)	Shape, size and orientation of hydrogen- like atomic orbitals	(r)	Magnetic quantum number
(d)	Probability density of electron at the nucleus in hydrogen-like atom	(s)	Electron spin quantum number

4. Match **List I** with **List II.**

	List I (Spectral Series for Hydrogen)		List II (Spectral Region/ Higher Energy State)
(a)	Lyman	(p)	Infrared region
(b)	Balmer	(q)	UV region
(c)	Paschen	(r)	Infrared region
(d)	Pfund	(s)	Visible region

Choose the correct answer from the options given below : **[JEE Main (Jan.) 2024]**

(a) a—q, b—r, c—p, d—s
(b) a—p, b—r, c—q, d—s
(c) a—q, b—s, c—r, d—p
(d) a—p, b—q, c—r, d—s

D. Passage Based Problems

Passage-1

The hydrogen-like species Li^{2+} is in a spherically symmetric state S_1 with one radial node. Upon absorbing light the ion undergoes transition to a state S_2 has one radial node and its energy is equal to the ground state energy of the hydrogen atom. **(IIT 2010)**

1. The state S_1 is :

(a) 1*s* (b) 2*s* (c) 2*p* (d) 3*s*

2. Energy of the state S_1 in units of the hydrogen atom ground state energy is :

(a) 0.75 (b) 1.50 (c) 2.25 (d) 4.50

3. The orbital angular momentum quantum number of the state S_2 is :

(a) 0 (b) 1 (c) 2 (d) 3

Passage-2

Answer Q. 1-3 by appropriately matching the information given in the three columns of the following table.

The wave function, Ψ_{n,l,m_l} is a mathematical function whose value depends upon spherical polar coordinates (r,θ,ϕ) of the electron and characterized by the quantum numbers n, l and m_l. Here r is distance from nucleus, θ is colatitude and ϕ is azimuth. In the mathematical functions given in the table, Z is atomic number and a_0 is Bohr radius. **[JEE (Advanced) 2017]**

Column 1	Column 2	Column 3
(I) 1*s* orbital	(i) $\psi_{n,l,m_l} \propto \left(\frac{Z}{a_0}\right)^{\frac{3}{2}} e^{-(Zr/a_0)}$	(P) 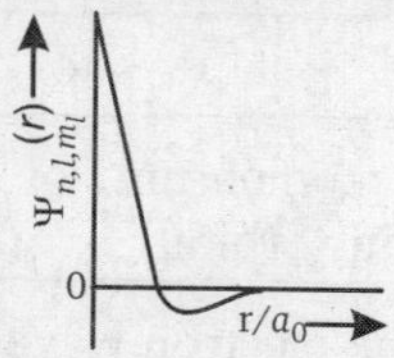

(II) 2*s* orbital	(ii) One radial node	(Q) Probability density at nucleus $\propto \frac{1}{a_0^3}$
(III) $2p_z$ orbital	(iii) $\psi_{n,l,m_l} \propto \left(\frac{Z}{a_0}\right)^{\frac{5}{2}} re^{-\left(\frac{Zr}{2a_0}\right)} \cos\theta$	(R) Probability density is maximum at nucleus
(iv) $3d_{z^2}$ orbital	(iv) *xy*- plane is a nodal plane	(S) Energy needed to excite electron from $n=2$ state to $n=4$ state is $\frac{27}{32}$ times the energy needed to excite electron from $n=2$ state to $n=6$ state.

1. For He^+ ion, the only incorrect combination is:
(a) (I) (i) (R) (b) (II) (ii) (Q) (c) (I) (i) (S) (d) (I) (iii) (R)

2. For the given orbital in Column 1, the only correct combination for any hydrogen-like species is:
(a) (I) (ii) (S) (b) (IV) (iv) (R) (c) (III) (iii) (P) (d) (II) (ii) (P)

3. For hydrogen atom, the only correct combination is:
(a) (II) (i) (Q) (b) (I) (iv) (R) (c) (I) (i) (P) (d) (I) (i) (S)

Passage-3

Consider the Bohr's model of a one-electron atom where the electron moves around the nucleus. In the following, List-I contains some quantities for the nth orbit of the atom and List-II contains options showing how they depend on n.
[JEE (Advanced) 2019]

	List-I		List-II
(I)	Radius of the nth orbit	(P)	$\propto n^{-2}$
(II)	Angular momentum of the electron in the nth orbit	(Q)	$\propto n^{-1}$
(III)	Kinetic energy of the electron in the nth orbit	(R)	$\propto n^{0}$
(IV)	Potential energy of the electron in the nth orbit	(S)	$\propto n^{1}$
		(T)	$\propto n^{2}$
		(U)	$\propto n^{1/2}$

1. Which of the following options has the correct combination considering List-I and List-II?
(a) (III), (S) (b) (IV), (Q) (c) (III), (P) (d) (IV), (U)

2. Which of the following options has the correct combination considering List-I and List-II?
(a) (II), (R) (b) (II), (Q) (c) (I), (P) (d) (I), (T)

E. Integer Type Problems

1. The work function (ϕ) of some metals is listed below. The number of metals which will show photoelectric effect when light of 300 nm wavelength falls on the metal is : **(IIT 2011)**

Metal	Li	Na	K	Mg	Cu	Ag	Fe	Pt	W
ϕ(eV)	2.4	2.3	2.2	3.7	4.8	4.3	4.7	6.3	4.75

2. The maximum number of electrons that can have principal quantum number, $n = 3$ and spin quantum number, $m_s = -\frac{1}{2}$, is : **(IIT 2011)**

3. The atomic masses of He and Ne are 4 and 20 amu, respectively. The value of the de-Broglie wavelength of He gas at –73°C is "M" times that of the de-Broglie wavelength of Ne at 727°C. M is :
[JEE (Advanced) 2013]

4. In an atom, the total number of electrons having quantum numbers $n = 4$, $|m_l| = 1$ and $m_s = -1/2$ is :
[JEE (Advanced) 2014]

5. Not considering the electronic spin, the degeneracy of the second excited state ($n = 3$) of H-atom is 9, while the degeneracy of the second excited state of H^- is : **[JEE (Advanced) 2015]**

6. The 1st, 2nd and the 3rd ionization enthalpies I_1, I_2 and I_3 of four atoms with atomic numbers $n, n+1$, $n+2$ and $n+3$, where $n < 10$, are tabulated below. What is the value of n?

Atomic number	Ionization Enthalpy (kJ/mol)		
	I_1	I_2	I_3
n	1681	3374	6050
$n+1$	2081	3952	6122
$n+2$	496	4562	6910
$n+3$	738	1451	7733

[JEE (Advanced) 2020]

7. The work function of sodium metal is 4.41×10^{-19} J. If photons of wavelength 300 nm are incident on the metal, the kinetics energy of the ejected electrons will be ($h = 6.63 \times 10^{-34}$ Js; $c = 3 \times 10^8$ m/s) ___ $\times 10^{-21}$ J. **[JEE Main (Sept.) 2020]**

8. A proton and a Li^{3+} nucleus are accelerated by the same potential. If λ_{Li} and λ_p denote the de-Broglie wavelengths of Li^{3+} and proton respectively, then the value of $\frac{\lambda_{Li}}{\lambda_p}$ is $x \times 10^{-1}$. The value of x is ________.
(Rounded off to the nearest integer)
(Mass of Li^{3+} = 8.3 mass of proton) **[JEE Main (Feb.) 2021]**

9. Electromagnetic radiation of wavelength 663 nm is just sufficient to ionise the atom of metal A. The ionization energy of metal A in kJ mol^{-1} is ________. (Rounded-off to the nearest integer)
[$h = 6.63 \times 10^{-34}$ Js, $c = 3.00 \times 10^8$ ms^{-1}, $N_A = 6.02 \times 10^{23}$ mol^{-1}] **[JEE Main (Feb.) 2021]**

10. When light of wavelength 248 nm falls on a metal of threshold energy 3.0 eV, the de-Broglie wavelength of emitted electrons is ______ Å. (Round off to the nearest integer).
[Use : $\sqrt{3} = 1.73$, $h = 6.63 \times 10^{-34}$ Js, $m_e = 9.1 \times 10^{-31}$ kg; $c = 3.0 \times 10^8$ ms^{-1}; 1eV = 1.6×10^{-19} J]
[JEE Main (March) 2021]

11. The number of orbitals with $n = 5, m_l = +2$ is ______. (Round off to the nearest integer).
[JEE Main (March) 2021]

12. In the ground state of atomic Fe($Z = 26$), the spin only-magnetic moment is ______ $\times 10^{-1}$ BM. (Round off to the nearest integer).
[Given: $\sqrt{3} = 1.73, \sqrt{2} = 1.41$] **[JEE Main (March) 2021]**

13. If the wavelength for an electron emitted from H-atom is 3.3×10^{-10}m, then energy absorbed by the electron in its ground state compared to minimum energy required for its escape from the atom, is ______ times. (Nearest integer)

[Given : $h = 6.626 \times 10^{-34}$ J s, Mass of electron $= 9.1 \times 10^{-31}$ kg] **[JEE Main (July) 2022 (II)]**

14. The orbital angular momentum of an electron in 3*s* orbital is $\frac{xh}{2\pi}$. The value of x is ______ .

[JEE Main (April) 2023 (II)]

15. Assume that the radius of the first Bohr orbit of hydrogen atom is 0.6 Å. The radius of the third Bohr orbit of He^+ is ______ picometer. (Nearest integer) **[JEE Main (Jan.) 2023 (II)]**

16. If wavelength of the first line of the Paschen series of hydrogen atom is 720 nm, then the wavelength of the second line of this series is ______ nm. (Nearest integer) **[JEE Main (Jan.) 2023 (I)]**

17. Number of spectral lines obtained in He^+ spectra, when an electron makes transition from fifth excited state to first excited state will be : **[JEE Main (Jan.) 2024]**

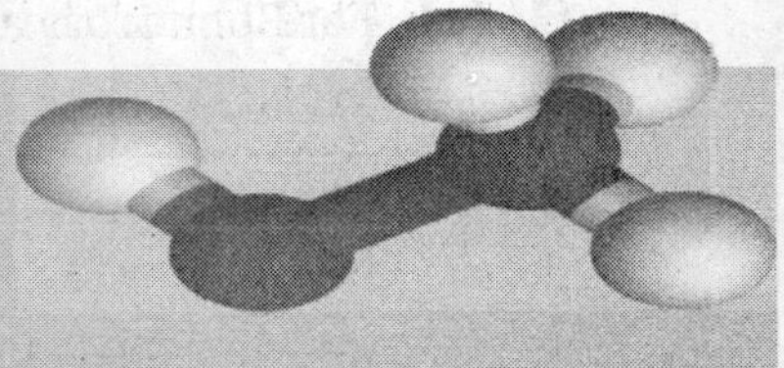

Answers

[A] Objective Questions (Only one option is correct)

1. (c)	**2.** (a)	**3.** (a)	**4.** (a)	**5.** (d)	**6.** (c)	**7.** (b)	**8.** (c)	**9.** (a)	**10.** (b)
11. (b)	**12.** (b)	**13.** (a)	**14.** (d)	**15.** (c)	**16.** (c)	**17.** (c)	**18.** (b)	**19.** (b)	**20.** (c)
21. (a)	**22.** (c)	**23.** (a)	**24.** (b)	**25.** (b)	**26.** (c)	**27.** (d)	**28.** (b)	**29.** (b)	**30.** (d)
31. (d)	**32.** (b)	**33.** (b)	**34.** (d)	**35.** (c)	**36.** (d)	**37.** (a)	**38.** (d)	**39.** (a)	**40.** (c)
41. (d)	**42.** (d)	**43.** (c)	**44.** (b)	**45.** (b)					

[B] Objective Questions (One or more than one option(s) is/are correct)

1. (a, b, c) **2.** (a, d) **3.** (a, d)

[C] Match the Columns

1. a → r; b → q; c → p; d → s
2. a → r, p; b → r; c → q; d → s
3. a → q; b → p, q, r, s; c → p, q, r; d → p, q, r
4. **(c)** a → q; b → s; c → r; d → p

[D] Passage Based Problems

Passage-1

1. (b) **2.** (c) **3.** (b)

Passage-2

1. (d) **2.** (d) **3.** (d)

Passage-3

1. (c) **2.** (d)

[E] Integer Type Problems

1. (4)	**2.** (9)	**3.** (5)	**4.** (6)	**5.** (3)
6. (9)	**7.** (222)	**8.** (2)	**9.** (180)	**10.** (9)
11. (3)	**12.** (49)	**13.** (2)	**14.** (0)	**15.** (270)
16. (492)	**17.** (10)			

Hints and Solutions

A. Objective Questions

[Only one option is correct]

1. (c) For Bohr orbit, angular momentum is

$$mvr_n = \frac{nh}{2\pi}; \quad v = \frac{nh}{2\pi m r_n} \qquad \text{...(i)}$$

Kinetic energy, $$\text{K.E.} = \frac{1}{2}mv^2 \qquad \text{...(ii)}$$

By putting the value of v from (i) to (ii),

$$\text{K.E.} = \frac{1}{2} m \times \frac{n^2h^2}{4\pi^2 m^2 r_n^2} = \frac{n^2h^2}{8\pi^2 m r_n^2}$$

For second Bohr orbit, $n = 2$

$$r_n = a_0 \times n^2 \qquad (a_0 = \text{Bohr radius})$$

$$r_n = 4a_0$$

$$\text{K.E.} = \frac{(2)^2 h^2}{8\pi^2 m (4a_0)^2}$$

Thus, $$\text{K.E.} = \frac{h^2}{32\pi^2 m a_0^2}$$

2. (a) $\frac{1}{\lambda} = R_H Z^2 \left(\frac{1}{n_1^2} - \frac{1}{n_2^2}\right)$

$Z = 1, n_1 = 1, n_2 = 2$

$\lambda = 1.214 \times 10^{-7}$ m

3. (a) $_{37}\text{Rb} \rightarrow 1s^2, 2s^2 2p^6, 3s^2 3p^6 3d^{10}, 4s^2 4p^6, 5s^1$

So last electron enters in 5s orbital

Hence $n = 5, l = 0, m = 0, s = \pm\frac{1}{2}$

4. (a) Energy of electron in the nth orbit of H-atom is,

$$E_n = \frac{-13.6}{n^2}\text{eV} = \frac{-13.6}{2^2} = -3.4\,\text{eV}$$

5. (d) $\lambda = \frac{h}{p} = \frac{h}{\sqrt{2m\text{K.E.}}} = \frac{h}{\sqrt{2meV}} \qquad (\because \text{K.E.} = eV)$

$\frac{h}{\lambda} = \sqrt{2meV}$

6. (c) $r = 0$ at nucleus

For 1*s* orbital probability of finding electron can not be zero for any value of *r*.

7. (b) Radius of nth orbit for H-atom is

$$r = \frac{n^2 a_0 \text{Å}}{Z}$$

$$r = \frac{(2)^2 \times 0.529}{1} \text{Å} \qquad [\because n = 2, \text{for second orbit}]$$

$$r = 2.12 \text{ Å}$$

8. (c) The species having same number of electrons are called isoelectronic species.

Number of e^-s in $O^{2-} = 8 + 2 = 10$

Number of e^-s in $F^- = 9 + 1 = 10$

Number of e^-s in $Na^+ = 11 - 1 = 10$

Number of e^-s in $Mg^{2+} = 12 - 2 = 10$

Therefore, the ions, given in option (c) are isoelectronic.

9. (a) Higher the value of n, higher will be the energy. If n is same, then higher the value of $(n + l)$, higher will be the energy. Thus the increasing order of energy is IV < II < III < I.

10. (b) As nuclear charge increases, the electrons are attracted more strongly and drawn inwards, so in isoelectronic ions the size of ion decreases as the nuclear charge increases.

11. (b) $h\nu - \phi = \text{KE}$

$$\left(\frac{hc}{\lambda}\right)_{\text{incident}} = \text{KE} + \phi$$

As KE is very high in comparison to work function (ϕ). So,

$$\left(\frac{hc}{\lambda}\right)_{\text{incident}} \approx \text{KE}$$

$$\text{KE} = \frac{p^2}{2m} = \frac{hc}{\lambda_{\text{incident}}} \quad \text{...(i)}, \qquad \Rightarrow \frac{(1.5\,p)^2}{2m} = \frac{hc}{\lambda'} \quad \text{...(ii)}$$

On dividing equation (ii) by (i), we get

$$(1.5)^2 = \lambda/\lambda' \Rightarrow \lambda' = \frac{4}{9}\lambda$$

12. (b) Statement (b) was according to Avogadro's law ($V \propto n$).

13. (a) $R_n = \dfrac{a_0 \times n^2}{z}$

For Li^{2+}, $z = 3$

In second Bohr orbit, $R_2 = \dfrac{a_0 \times z^2}{3} = \dfrac{4a_0}{3}$

14. (d) $2\pi r = n\lambda$

For fourth orbit ($n = 4$)

Radius (r) $= a_0 \times n^2 = 4^2 \times a_0 = 16\,a_0$

Since $2\pi r = n\lambda$

$2\pi \times 16a_0 = 4\lambda \Rightarrow \lambda = 8\pi a_0$

15. (c) $\bar{\nu} = \dfrac{1}{\lambda} = R_H\left\{\dfrac{1}{n_1^2} - \dfrac{1}{n_2^2}\right\}$

For Balmer series, $n_1 = 2, n_2 = 3, 4, 5 \ldots \infty$

For shortest wavelength, $n_1 = 2$ and $n_2 = \infty$

For longest wavelength, $n_1 = 2$ and $n_2 = 3$

16. (c) (A) $\text{KE} = -\text{TE} = 13.6\dfrac{Z^2}{n^2} \qquad \Rightarrow \text{KE} \propto \dfrac{Z^2}{n^2}$

(B) $v = V_0\dfrac{Z}{n} \qquad \Rightarrow v \times n \propto Z$

(C) $\nu = \dfrac{v}{2\pi r} \qquad \Rightarrow \nu \propto \dfrac{Z^2}{n^3}$

(D) $F = \frac{Kq_1q_2}{r^2}$ $\Rightarrow F \propto \frac{(e)(Ze)}{\left(\frac{n^2}{Z}\right)^2} \propto \frac{Z^3}{n^4}$

17. (c) Shortest wavelength of Lyman series in H-atom is wavelength when electron make a transition from $n_2 = \infty$ to $n_1 = 1$.

$$\frac{1}{\lambda_1} = R_H\left(\frac{1}{1^2} - \frac{1}{\infty^2}\right)$$

$$R_H = \frac{1}{\lambda_1}$$

Longest wavelength in Balmer series of He^+ forms when electron make transition from $n_2 = 3$ to $n_1 = 2$.

$$\lambda_2 = R_H\left(\frac{1}{n_1^2} - \frac{1}{n_2^2}\right) \times z^2 = \frac{1}{\lambda_1} \times \left(\frac{1}{2^2} - \frac{1}{3^2}\right) \times 2^2$$

$$\lambda_2 = \lambda_1 \times \frac{9}{5}$$

18. (b) Formula of radii $r_n = \frac{a_0 \times n^2}{z}$

For He^+, $z = 2$

Li^{2+}, $z = 3$

$$\frac{\Delta R_1}{\Delta R_2} = \frac{\left(\frac{4^2}{3} - \frac{3^2}{3}\right)a_0}{\left(\frac{4^2}{2} - \frac{3^2}{2}\right)a_0} = \frac{7/3}{7/2} = \frac{2}{3}$$

19. (b) Hydrogen have 3 isotopes:

Isotope	Neutrons
^1_1H	0
^2_1H	1
^3_1H	2

Total neutrons = 0 + 1 + 2 = 3

20. (c) Internal energy of a gas is function of temperature, this results is not manifestation of quantum nature of atom while other given options can be explained on basis of quantum theory of atom.

21. (a)

Orbital	Probability density curve	Orbital	Probability density curve
1*s*	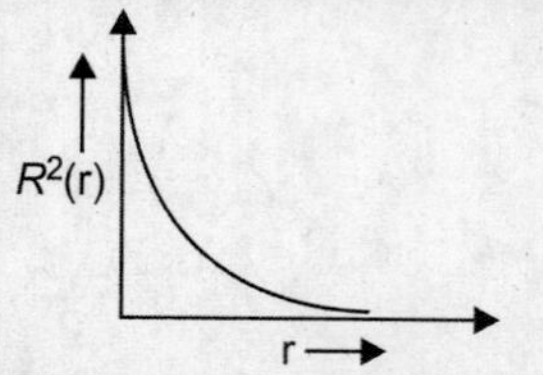	2*s*	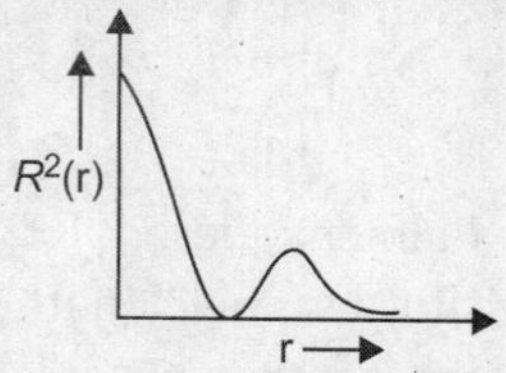
2*p*	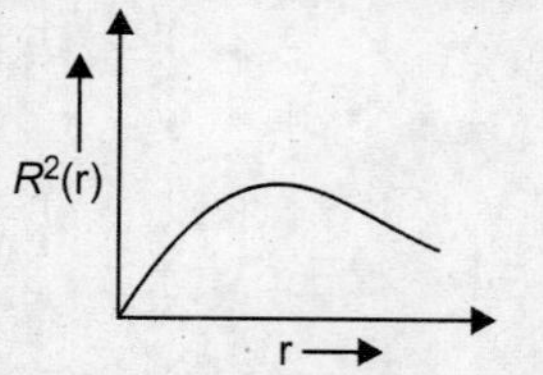	3*p*	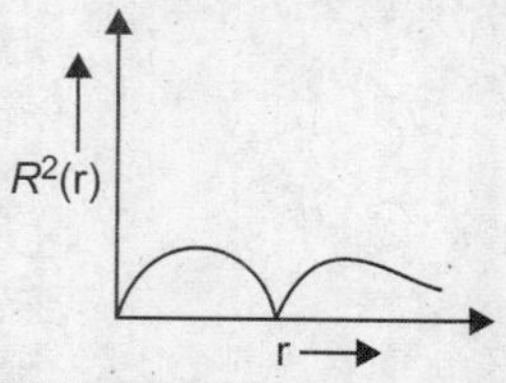

22. (c) From $(n+l)$ rule

Orbital	n	l	$(n+l)$
$6s$	6	0	6
$4f$	4	3	7
$5d$	5	2	7
$6p$	6	1	7

Order of filling electron in 6th group $6s$, $4f$, $5d$, $6p$

23. (a) For $n = 4$

Values of $l = 0, 1, 2, 3$

For $l = 0,\ m_l = 0$

$l = 1,\ m_l = -1, 0, +1$

$l = 2,\ m_l = -2, -1, 0, +1, +2$

$l = 3,\ m_l = -3, -2, -1, 0, +1, +2, +3$

So, possible subshells are two, which are:

$n = 4,\ l = 2,\ m_l = -2$

and $n = 4,\ l = 3,\ m_l = -2$

24. (b) $n = 1 \Rightarrow l = 0, 1, 2$

$n = 2 \Rightarrow l = 0, 1, 2, 3$

$n = 3 \Rightarrow l = 0, 1, 2, 3, 4$

n	l	$n+l$	subshell
1	0	1	1s
	1	2	1p
	2	3	1d
2	0	2	2s
	1	3	2p
	2	4	2d
	3	5	2f
3	0	3	3s
	1	4	3p
	2	5	3d
	3	6	3f
	4	7	3g

From $(n+l)$ rule, energy of subshell will be in order :

$$1s < 1p < 2s < 1d < 2p < 3s < 2d < 3p < 4s$$

Atomic number (Z)	Electronic configuration
9	$1s^2\ 1p^6\ 2s^1$
8	$1s^2\ 1p^6$
13	$1s^2\ 1p^6\ 2s^2\ 1d^3$
6	$1s^2\ 1p^4$

25. (b) Statement I is false. Since Bohr's theory accounts for stability and line spectrum of single electronic species, *e. g.*, H, He^{+}, Li^{2+} etc. Bohr's theory was unable to explain splitting of spectral lines in electric and magnetic field. So, statement I is true.

26. (c) For '3*s*' orbital:

Number of radial nodes $= n - l - 1 = 3 - 0 - 1 = 2$

So, graph will intersect at 2 points on horizontal axis.

27. (d) Number of angular nodes $= l = 2$ (*d*-subshell). Number of radial nodes $= n - l - 1 = 2$

$\Rightarrow \quad n = 5$

So, orbital will be 5*d*.

28. (b) β^- particles are emitted when $\frac{n}{P}$ ratio is high.

For tritium (^3_1H)

$n = 2; \quad P = 1 \quad \Rightarrow \frac{n}{P} = 2$

So, it will emit β^- particle to become stable.

29. (b) Number of angular nodes $= l = 0$

Since, $l = 0$ represents '*s*' orbital.

Number of radial nodes $= n - l - 1 = 2$

$$n - 0 - 1 = 2$$
$$n = 3$$

So, orbital will be 3*s*.

30. (d) Atomic orbital is characterised by *n*, *l*, *m*.

31. (d) Isotopes have same atomic number but different mass number. They have different number of neutrons which results in different mass number.

32. (b) Lower the value of $(n + l)$ for an orbital, lower is its energy. If two orbitals have same value of $(n + l)$ then lower value of *n* will have the lower energy

A	⇒	3*d*	⇒	$n + l = 3 + 2 = 5$
B	⇒	4*p*	⇒	$n + l = 4 + 1 = 5$
C	⇒	4*d*	⇒	$n + l = 4 + 2 = 6$
D	⇒	3*s*	⇒	$n + l = 3 + 0 = 3$

order is D < A < B < C

33. (b) For H :
$$\frac{1}{\lambda} = R_H \times 1^2\left(\frac{1}{1^2} - \frac{1}{\infty^2}\right) \quad \text{... (i)}$$
$$\frac{1}{\lambda_{He^+}} = R_H \times 2^2 \times \left(\frac{1}{4} - \frac{1}{9}\right) \quad \text{... (ii)}$$

From equations (i) and (ii) $\frac{\lambda_{He^+}}{\lambda} = \frac{9}{5}$

$$\lambda_{He^+} = \lambda \times \frac{9}{5} \quad \Rightarrow \quad \lambda_{He^+} = \frac{9\lambda}{5}$$

34. (d)
$$E_n = \frac{-2.18 \times 10^{-18} Z^2}{n^2}$$

i.e., $E_n \propto \frac{1}{n^2}$ ⇒ Energy in the 3rd Bohr orbit is $\frac{1}{9}$ th of energy in first Bohr orbit.

35. (c)
$$(r_3)_H = \frac{a_0 n^2}{Z} = a_0 \times 3^2 = 9a_0$$
$$2\pi r = n\lambda \quad \Rightarrow \quad 2\pi \times 9a_0 = 3\lambda \quad \Rightarrow \quad \lambda = 6\pi a_0$$

36. (d) At node $\psi_{2s} = 0$

$$\therefore \quad 2 - \frac{r_0}{a_0} = 0 \quad \Rightarrow \quad r_0 = 2a_0$$

37. (a)

$$\lambda = \frac{h}{p}\left[\lambda \propto \frac{1}{p}\right]$$

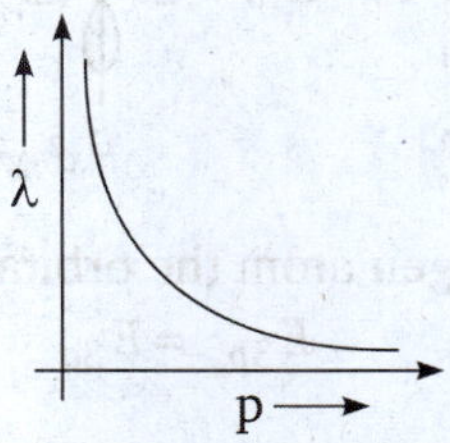

$\Rightarrow$ $\lambda p = h$ (constant)

So, the plot is a rectangular hyperbola.

38. (d) No spectral line will be observed for a $2p_x \rightarrow 2p_y$ transition because $2p_x$ and $2p_y$ orbitals are degenerate orbitals.

39. (a) $\lambda = 900$ nm H-atom ($Z = 1$)

$= 9 \times 10^{-5}$ cm

$R_H = 10^5 \text{cm}^{-1}$

Rydberg eq. $= \dfrac{1}{\lambda} = R_H Z^2 \times \left(\dfrac{1}{n_1^2} - \dfrac{1}{n_2^2}\right)$

$$\Rightarrow \quad \frac{1}{\lambda \times R_H} = \frac{1}{n_1^2} - \frac{1}{n_2^2}$$

$$\Rightarrow \quad \frac{1}{9 \times 10^{-5}\text{cm} \times 10^5 \text{cm}^{-1}} = \left(\frac{1}{n_1^2} - \frac{1}{n_2^2}\right)$$

$$\Rightarrow \quad \frac{1}{n_1^2} - \frac{1}{n_2^2} = \frac{1}{9}$$

It is possible when $n_1 = 3, n_2 = \infty$

Possible series : $\infty \rightarrow 3$

40. (c) **A :** 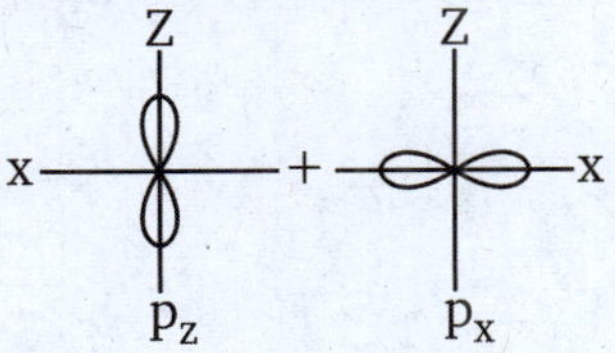

$\Rightarrow$ no molecular orbital

B : 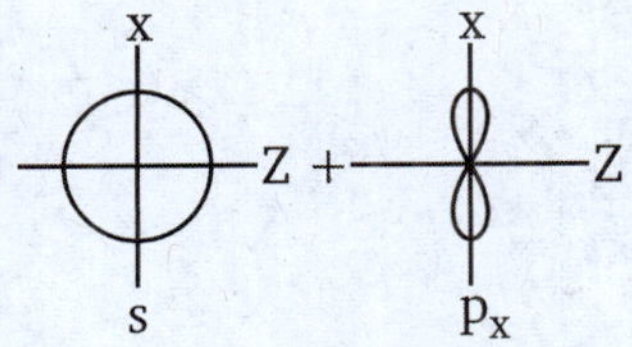

$\Rightarrow$ no molecular orbital

C : 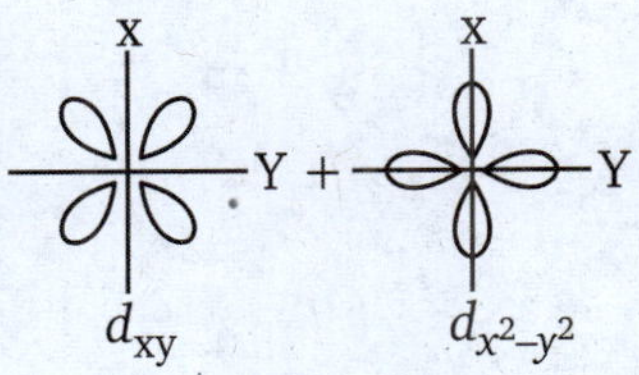

$\Rightarrow$ no molecular orbital

D : 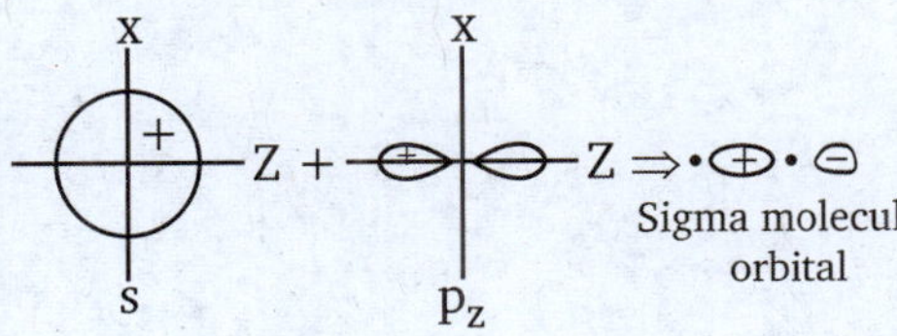

Sigma molecular orbital

E : 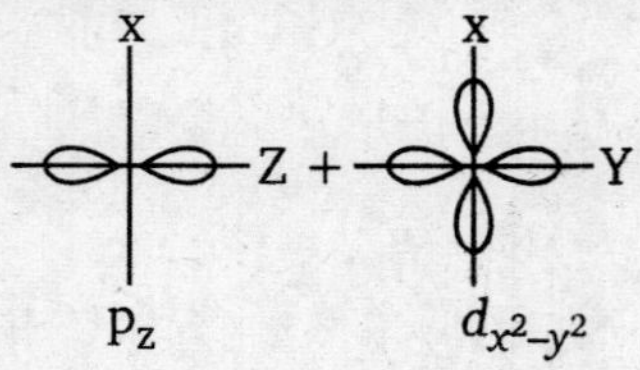

$\Rightarrow$ no molecular orbital

41. (d) In hydrogen atom the orbitals in a shell are degenerate means energy depends only on 'n'

$\therefore \quad E_{3p_x} = E_{3d_{x^2-y^2}} = E_{3d_{z^2}}$

42. (d)

	orbital
A : $n = 1, l = 0, m_l = 0$	1s
B : $n = 2, l = 0, m_l = 0$	2s
C : $n = 3, l = 1, m_l = 1$	3p
D : $n = 3, l = 2, m_l = 1$	3d
E : $n = 3, l = 2, m_l = 0$	3d

In absence of electric and magnetic fields, all orbitals of $3d$ are degenerate.

43. (c) For single electron species energy only depends on 'n' (principal quantum number)

So, energy of $2s = 2p$

and energy of $3d < 4s$

44. (b) $2\pi r_n = n\lambda$

$2\pi(4a_0) = n\lambda$

$= \lambda = \dfrac{8\pi a_0}{n}$

B. Objective Questions

[One or more than one option(s) is/are correct]

1. (a, b, c)

(a) $_{24}Cr$ = [Ar] $3d^5, 4s^1$

(b) $m = -l$ to $+l$ including zero.

(c) $_{47}Ag$ = $1s^2, 2s^2, 2p^6, 3s^2, 3p^6, 3d^{10}, 4s^2, 4p^6, 4d^{10}, 5s^1$

Thus 24 electron have $s = +\frac{1}{2}$ while rest 23 have $s = -\frac{1}{2}$.

(d) Oxidation number of N in NH_3 is $-\frac{1}{3}$.

2. (a, d)

(b) and (c) are against Hund's rule.

3. (a, d)

$$E_{He^+} = -13.6 \times \frac{(2)^2}{n^2} = -3.4 = \frac{-13.6}{4}$$

$n^2 = 16$ so $n = 4$

Quantum number are $n = 4, l = 2, m = 0$

So, subshell is d.

angular node $= l = 2$

Radial node $= [n - l - 1] = 4 - 2 - 1 = 1$

C. Matching Type Problems

1. a — r, b — q, c — p, d — s

(a) $$K_n = \frac{1}{2}.\frac{Z.e^2}{r_n}$$

$$V_n = -\frac{Z.e^2}{r_n}$$

$$\frac{V_n}{K_n} = \frac{-\frac{Z.e^2}{r_n}}{\frac{Ze^2}{2r_n}} = -2$$

(b) $$E_n = -\frac{2\pi^2.mZ^2.e^4}{n^2.h^2}$$

or $$E_n \propto \frac{1}{n^2}$$

but $$r_n = \frac{n^2 \times h^2}{4\pi^2.mZe^2} \Rightarrow r_n \propto n^2$$

$\therefore$ $$E_n \propto \frac{1}{r^n}$$

$\therefore$ $r_n \propto (E_n)^{-1} \Rightarrow x = -1$

(c) For lowest orbital $n = 1$ and $l = 0$

and Orbital angular momentum $= \frac{h}{2\pi}\sqrt{l(l+1)}$

$$\frac{h}{2\pi}\cdot\sqrt{0(0+1)} = 0$$

(d) $$r_n = \frac{n^2.h^2}{4\pi^2.mZe^2} \quad \text{or} \quad r_n \propto \frac{1}{Z}$$

or $$\frac{1}{r_n} \propto Z^y \quad \Rightarrow y = 1$$

3. a — q; b — p, q, r, s; c — p, q, r; d — p, q, r

(a) Orbital angular momentum $= \frac{h}{2\pi}\sqrt{l(l+1)}$

(b) An electron in an orbital is defined by using four quantum numbers n, l, m, s.

(c) Size of atomic orbital depends upon principal quantum number (n), shape upon azimuthal quantum number (l) and orientation upon magnetic quantum number (m).

(d) Probability density of electron at the nucleus in H-like atom depends upon square wave function which depends upon three quantum numbers n, l and m.

4. (c) a — q; b — s; c — r; d — p

Fact based.

D. Passage Based Problems

Passage-1

1. (b) For S_1 (spherically symmetrical) state, number of nodes = 1

$$1 = n - 1 \quad \therefore \; n = 2$$

i.e., it is $2s$

For S_2, radial node = 1

$$E_{S_2} = \frac{-13.6 \times z^2}{n^2} = E_H \text{ in ground state} = -13.6$$

$$E = \frac{-13.6 \times 9}{n^2} \quad \Rightarrow \quad n = 3$$

So, state S_1 is $2s$ and S_2 is $3p$

2. (c) $$\frac{E_{S_1}}{E_{H(\text{ground})}} = \frac{-13.6 \times 9}{4 \times (-13.6)} = 2.25$$

3. (b) Azimuthal quantum number for $S_2 = l = 1$

Passage-2

1. (d) 1s orbital can not have θ function (angular function). Therefore (d) is incorrect.

2. (d) 2s orbital — One radial node $(n - l - 1)$

(II) (ii)

(P)

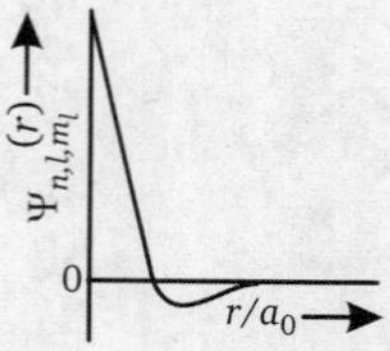

3. (d) For H-atom:

$$1s \text{ orbital} — \psi_{nlm} \propto \left(\frac{Z}{a_0}\right)^{3/2} e^{-\left(\frac{Zr}{a_0}\right)}, S$$

$$E_4 - E_2 = -\frac{13.6}{16} - \left(-\frac{13.6}{4}\right) = \frac{3 \times 13.6}{16}$$

$$E_6 - E_2 = -\frac{13.6}{36} - \left(-\frac{13.6}{4}\right) = \frac{8 \times 13.6}{36}$$

$E_4 - E_2$ is $\frac{27}{32}$ times of $E_6 - E_2$.

Passage-3

1. (c)

2. (d) $$r_n = 0.529\left(\frac{n^2}{Z}\right) \text{Å} \Rightarrow r_n \propto n^2$$

$$\text{Angular momentum } (l) = \left(\frac{nh}{2\pi}\right) \Rightarrow l \propto n^1$$

$$\text{K.E.} = \frac{1}{2}mv^2 = \frac{1}{2}m\left[2.18 \times 10^6 \frac{Z}{n}\right]^2$$

$$\Rightarrow \text{K.E.} \propto \frac{Z^2}{n^2} \Rightarrow \text{K.E.} \propto n^{-2}$$

$$P.E. = -2\,K.E. \Rightarrow P.E. \propto \frac{Z^2}{n^2}$$

$$\Rightarrow P.E. \propto n^{-2}$$

E. Integer Type Problems

1. (4) The energy associated with incident photon $= \dfrac{hc}{\lambda}$

$$\Rightarrow E = \frac{6.6 \times 10^{-34} \times 3 \times 10^8}{300 \times 10^{-9}}\,\text{J}$$

$$E \text{ in eV} = \frac{6.6 \times 10^{-34} \times 3 \times 10^8}{300 \times 10^{-9} \times 1.6 \times 10^{-19}} = 4.16\,\text{eV}$$

So, number of metals showing photo-electric effects will be (4), *i.e.*, Li, Na, K, Mg

2. (9) For principal quantum number ($n = 3$)

Number of orbitals $= n^2 = 9$

So, number of electrons with $m_s = -\dfrac{1}{2}$ will be 9.

3. (5)
$$\frac{\lambda_{He} = \dfrac{h}{m_{He}V_{He}}}{\lambda_{Ne} = \dfrac{h}{m_{Ne}V_{Ne}}} = \frac{m_{Ne}V_{Ne}}{m_{He}V_{He}}$$

$$M = \frac{20}{4} \times \frac{\sqrt{T/M}}{\sqrt{T/M}} = \frac{20}{4} \times \frac{\sqrt{1000/20}}{\sqrt{200/4}} = 5 \times \sqrt{\frac{50}{50}} = 5$$

4. (6) $n = 4,\ l = 0, 1, 2, 3$

$m_l = 1$ (only in p, d and f-orbitals)

2 electrons in each orbital have $m_s = -1/2$

Hence total no. of electrons = 6

5. (3) Single electron species don't follow the $(n + l)$ rule but multi electron species do.

Ground state of $H^- = 1s^2$

First excited state of $H^- = 1s^1, 2s^1$

Second excited state of $H^- = 1s^1, 2s^0, 2p^1$

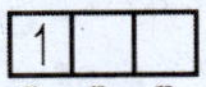

p_x p_y p_z

(3 degenerate orbitals)

6. (9)

Atomic number	Ionization Enthalpy (kJ/mol)		
	I_1	I_2	I_3
n	1681	3374	6050
$n+1$	2081	3952	6122
$n+2$	496	4562	6910
$n+3$	738	1451	7733

By observing the values of I_1, I_2 and I_3 for atomic number $(n + 2)$, it is observed that $I_2 >> I_1$.

This indicates that number of valence shell electrons is 1 and atomic number $(n + 2)$ should be an alkali metal.

Also for atomic number $(n + 3), I_3 >> I_2$.

This indicates that it will be an alkaline earth metal which suggests that atomic number $(n + 1)$ should be a noble gas and atomic number (n) should belong to Halogen family. Since $n < 10$; hence $\mathbf{n = 9}$ (F atom).

Note : $n = 1$ (H atom) cannot be the answer because it does not have I_2 and I_3 values.

7. (222)

Energy of incident photon $(E) = \dfrac{hc}{\lambda}$

$$= \frac{6.63 \times 10^{-34} \times 3 \times 10^{8}}{300 \times 10^{-19}} \text{ J} = 6.63 \times 10^{-19} \text{ J}$$

$$(K.E.)_{max} = E - W$$
$$= 6.63 \times 10^{-19} - 4.41 \times 10^{-19}$$
$$= 2.22 \times 10^{-19} \text{ J} = 222 \times 10^{-21} \text{ J}$$

8. (2) de-Broglie wavelength, $\lambda = \dfrac{h}{\sqrt{2mqV}}$

For Li^{2+}, $q = 3e$

$$m_{Li^{2+}} = 8.3 \times m_p$$

$$\frac{\lambda_{Li^{2+}}}{\lambda_p} = \sqrt{\frac{m_p \times q_p}{m_{Li^{2+}} \times q_{Li^{2+}}}} = \sqrt{\frac{m_p \times e}{8.3 m_p \times 3e}} = \frac{1}{5} = 0.2 = 2 \times 10^{-1}$$

9. (180)

$$E = \frac{hc}{\lambda} = \frac{6.63 \times 10^{-34} \times 3 \times 10^{8}}{663 \times 10^{-9}} \text{ J}$$

Energy in kJ/mol $= \dfrac{6.63 \times 10^{-34} \times 3 \times 10^{8} \times 6.02 \times 10^{23}}{663 \times 10^{-9} \times 1000} = 180 \text{ kJ}$

10. (9) Energy of incident photon $= \dfrac{hc}{\lambda} = \dfrac{6.63 \times 10^{-34} \times 3 \times 10^{8}}{248 \times 10^{-9}} \text{ J}$

$$= \frac{6.63 \times 10^{-34} \times 3 \times 10^{8}}{248 \times 10^{-9} \times 1.6 \times 10^{-19}} \text{ eV} = 5 \text{ eV}$$

Kinetic energy of photoelectrons $= 5 - 3 = 2 \text{ eV} = 2 \times 1.6 \times 10^{-19} \text{ J} = 3.2 \times 10^{-19} \text{ J}$

de-Broglie wavelength $(\lambda) = \dfrac{h}{\sqrt{2mk}} = \dfrac{6.63 \times 10^{-34}}{\sqrt{2 \times 9.1 \times 10^{-31} \times 3.2 \times 10^{-19}}}$

$$= 8.72 \times 10^{-10} \text{ m} = 8.72 \text{ Å}$$

11. (3) For $n = 5$, $l = 0, 1, 2, 3, 4$

$l = 0$, $m = 0$

$l = 1$, $m_l = -1, 0, +1$

$l = 2$, $m_l = -2, -1, 0, +1, +2$

$l = 3$, $m_l = -3, -2, -1, 0, +1, +2, +3$

$l = 4$, $m_l = -4, -3, -2, -1, 0, +1, +2, +3, +4$

⇒ So, number of orbitals with $n = 5$ and $m_l = 2$ are 3.

12. (49) Electronic configuration of Fe $= [Ar]4s^2 3d^6$

↿⇂	↿	↿	↿	↿

$(3d^6)$

Unpaired electrons $(n) = 4$

Magnetic moment $(\mu) = \sqrt{n(n+2)}$

$$= \sqrt{4(4+2)} = \sqrt{24} = 4.89\,\text{BM}$$

$$= 48.9 \times 10^{-1}\,\text{BM}$$

13. (2) $\lambda = \dfrac{h}{\sqrt{2m\,(\text{K.E.})}}$; $\text{K.E.} = \dfrac{h^2}{2m\lambda^2}$

$$\text{K.E.} = \frac{h^2}{2m\lambda^2} = \frac{43.9 \times 10^{-68}}{2 \times 9.1 \times 10^{-31} \times 10.89 \times 10^{-20}}$$

$$\text{K.E.} = 2.215 \times 10^{-18}$$

$$E_{abs} = E_{req} + \text{K.E.}$$

$$\frac{E_{abs}}{E_{req}} = 1 + \frac{\text{K.E.}}{E_{req}} = 1 + \frac{2.215 \times 10^{-18}}{13.6 \times 1.602 \times 10^{-19}} = 2.0166$$

14. (0) Orbital angular momentum $= \sqrt{1(1+1)}\,\dfrac{h}{2\pi}$

Value of 1 for s orbital = 0 $\Rightarrow$ orbital angular momentum = 0.

15. (270) $r \propto \dfrac{n^2}{Z}$ $\quad \therefore\ r_{He^+} = r_H \times \dfrac{n^2}{Z}$

$$r_{He^+} = 0.6 \times \frac{(3)^2}{2} = 2.7\ \text{Å} = 270\ \text{pm}$$

16. (492) For first line of Paschen series

$$\frac{1}{(\lambda_1)_p} = R_H Z^2\left(\frac{1}{9} - \frac{1}{16}\right)$$

For second line of Paschen series

$$\frac{1}{(\lambda_2)_p} = R_H Z^2\left(\frac{1}{9} - \frac{1}{25}\right)$$

$$\frac{(\lambda_2)_p}{(\lambda_1)_p} = \frac{\dfrac{7}{16 \times 9}}{\dfrac{16}{25 \times 9}} = \frac{25 \times 7}{16 \times 16}$$

$$(\lambda_2)_p = \frac{25 \times 7}{16 \times 16} \times 720$$

$$(\lambda_2)_p = 492\ \text{nm}$$

17. (10) 5th excited state $\Rightarrow$ $n_1 = 6$

1st excited state $\Rightarrow$ $n_2 = 2$

$$\Delta n = n_1 - n_2 = 6 - 2 = 4$$

Maximum number of spectral lines

$$= \frac{\Delta n(\Delta n + 1)}{2} = \frac{4(4+1)}{2} = 10$$

❑❑❑

CHAPTER

3

States of Matter (Gaseous and Liquid State)

A. Objective Questions

[Only one option is correct]

1. The term that corrects for the attractive forces present in a real gas in the van der Waals' equation is: **(IIT 2009)**

(a) nb (b) $\frac{an^2}{V^2}$ (c) $-\frac{an^2}{V^2}$ (d) $-nb$

2. The packing efficiency of the two-dimensional square unit cell shown below is : **(IIT 2010)**

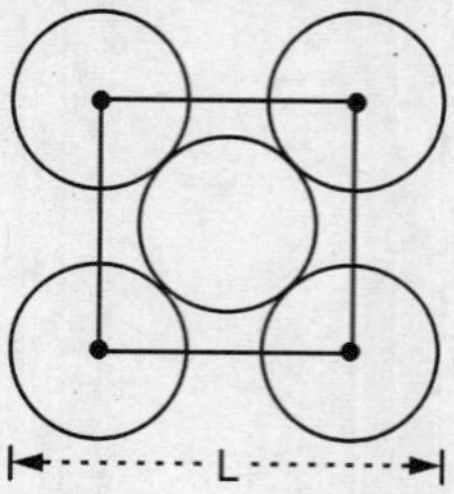

(a) 32.97% (b) 68.02% (c) 74.05% (d) 78.54%

3. For a gaseous state, if most probable speed is denoted by C^*, average speed by $\overline{C}$ and mean square speed by C, then for a large number of molecules the ratios of these speeds are : **[JEE (Mains) 2013]**

(a) $C^*:\overline{C}:C = 1.225 : 1.128 : 1$
(b) $C^*:\overline{C}:C = 1.128 : 1.225 : 1$
(c) $C^*:\overline{C}:C = 1 : 1.128 : 1.225$
(d) $C^*:\overline{C}:C = 1 : 1.225 : 1.128$

4. If Z is a compressibility factor, van der Waals' equation at low pressure can be written as : **[JEE (Mains) 2014]**

(a) $Z = 1 + \frac{RT}{Pb}$ (b) $Z = 1 - \frac{a}{VRT}$ (c) $Z = 1 - \frac{Pb}{RT}$ (d) $Z = 1 + \frac{Pb}{RT}$

5. Two closed bulbs of equal volume (V) containing an ideal gas initially at pressure p_i and temperature T_1 are connected through a narrow tube of negligible volume as shown in the figure below. The temperature of one of the bulbs is then raised to T_2. The final pressure p_f is : **[JEE (Mains) 2016]**

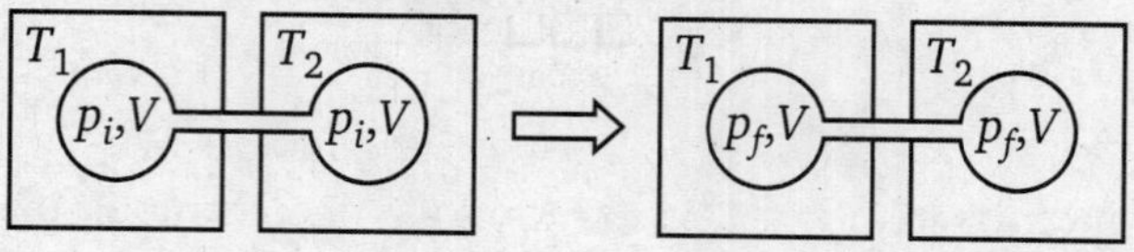

(a) $p_i\left(\frac{T_1T_2}{T_1+T_2}\right)$ (b) $2p_i\left(\frac{T_1}{T_1+T_2}\right)$ (c) $2p_i\left(\frac{T_2}{T_1+T_2}\right)$ (d) $2p_i\left(\frac{T_1T_2}{T_1+T_2}\right)$

6. The qualitative sketches I, II and III given below show the variation of surface tension with molar concentration of three different aqueous solutions KCl, CH_3OH and $CH_3(CH_2)_{11}OSO_3^-Na^+$ at room temperature. The correct assignment of the sketches is : **[JEE (Advanced) 2016]**

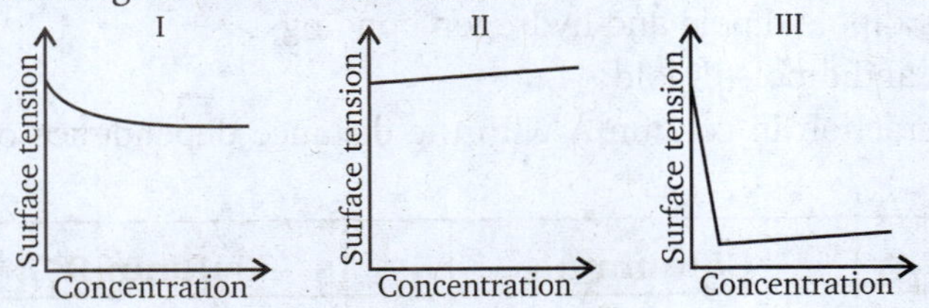

(a) I : KCl II : CH_3OH III : $CH_3(CH_2)_{11}OSO_3^-Na^+$

(b) I : $CH_3(CH_2)_{11}OSO_3^-Na^+$ II : CH_3OH III : KCl

(c) I : KCl II : $CH_3(CH_2)_{11}OSO_3^-Na^+$ III : CH_3OH

(d) I : CH_3OH II : KCl III : $CH_3(CH_2)_{11}OSO_3^-Na^+$

7. An open vessel at 27°C is heated until 2/5 of the air in it has been expelled. Assuming the volume of the vessel to be constant, find the temperature to which the vessel was heated: **[JEE Main 2019]**

(a) 750 K (b) 500 K (c) 750°C (d) 500°C

8. Points I, II and III in the following plot respectively correspond to: (u_{mp}: most probable velocity) **[JEE Main 2019]**

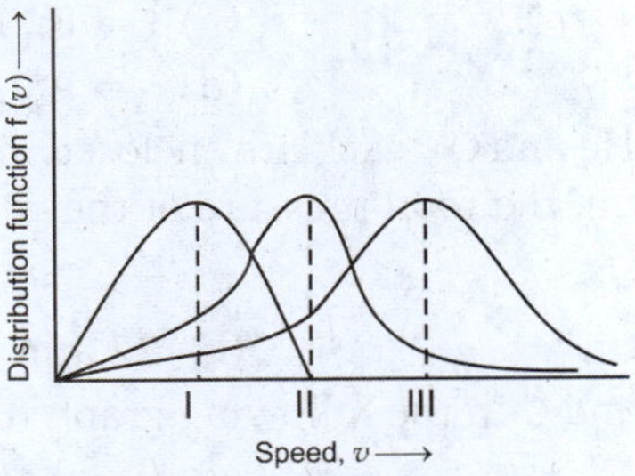

(a) u_{mp} of O_2 (400 K); u_{mp} of N_2 (300 K); u_{mp} of H_2 (300 K)

(b) u_{mp} of N_2 (300 K); u_{mp} of H_2 (300 K); u_{mp} of O_2 (400 K)

(c) u_{mp} of H_2 (300 K); u_{mp} of N_2 (300 K); u_{mp} of O_2 (400 K)

(d) u_{mp} of N_2 (300 K); u_{mp} of O_2 (400 K); u_{mp} of H_2 (300 K)

9. The volume of gas *A* is twice than that of gas *B*. The compressibility factor of gas *A* is thrice than that of gas *B* at same temperature. The pressures of the gases for equal number of moles are: **[JEE Main 2019]**

(a) $P_A = 2P_B$ (b) $2P_A = 3P_B$ (c) $3P_A = 2P_B$ (d) $P_A = 3P_B$

10. If the distribution of molecular speeds of a gas is as per the figure shown below, then the ratio of the most probable, the average and the root mean square speed, respectively, is : **[JEE (Advanced) 2020]**

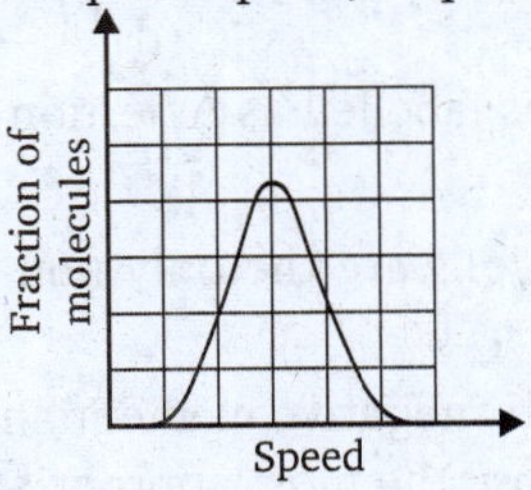

(a) 1 : 1 : 1 (b) 1 : 1 : 1.224 (c) 1 : 1.128 : 1.224 (d) 1 : 1.128 : 1

11. The relative strength of interionic/intermolecular forces in decreasing order is: **[JEE Main (Jan.) 2020]**

(a) ion-dipole > dipole-dipole > ion-ion (b) dipole-dipole > ion-dipole > ion-ion

(c) ion-ion > ion-dipole > dipole-dipole (d) ion-dipole > ion-ion > dipole-dipole

12. The predominant intermolecular forces present in ethyl acetate, a liquid are: **[JEE Main (Jan.) 2020]**
(a) hydrogen bonding and London dispersion
(b) Dipole-dipole and hydrogen bonding
(c) London dispersion, dipole-dipole and hydrogen bonding
(d) London dispersion and dipole-dipole

13. Match the type of interaction in column A with the distance dependence of their interaction energy in column B: **[JEE Main (Sept.) 2020]**

	Column-A		Column-B
(I)	ion-ion	(A)	$\frac{1}{r}$
(II)	dipole-dipole	(B)	$\frac{1}{r^2}$
(III)	London dispersion	(C)	$\frac{1}{r^3}$
		(D)	$\frac{1}{r^6}$

(a) I→ A; II→ C; III→ D
(b) I→ B; II→ D; III→ C
(c) I→ A; II→ B; III→ C
(d) I→ A; II→ B; III→ D

14. A mixture of one mole each of H_2, He and O_2 each are enclosed in a cylinder of volume V at temperature T. If the partial pressure of H_2 is 2 atm, the total pressure of the gases in the cylinder is:

[JEE Main (Sept.) 2020]

(a) 14 atm
(b) 6 atm
(c) 22 atm
(d) 38 atm

15. Identify the correct labels of A, B and C in the following graph from the options given below:

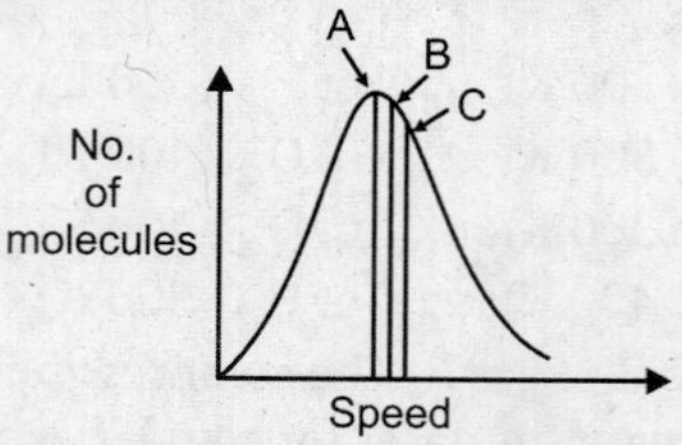

Root mean square speed (u_{rms}); most probable speed(u_{mp}); average speed ($u_{av.}$)

[JEE Main (Jan.) 2020]

(a) $A - u_{av}, B - u_{rms}, C - u_{mp}$
(b) $A - u_{rms}, B - u_{mp}, C - u_{av}$
(c) $A - u_{mp}, B - u_{rms}, C - u_{av}$
(d) $A - u_{mp}, B - u_{av}, C - u_{rms}$

16. Given below are two statements: one is labelled as Assertion A and the other is labelled as Reason R.

[JEE Main (Feb.) 2021]

Assertion A : Dipole-dipole interactions are the only non-covalent interactions, resulting in hydrogen bond formation.

Reason R : Fluorine is the most electronegative element and hydrogen bonds in HF are symmetrical. In the light of the above statements, choose the most appropriate answer from the options given below.
(a) A is false but R is true
(b) Both A and R are true and R is the correct explanation of A
(c) A is true R is false
(d) Both A and R are true but R is not the correct explanation of A.

17. An evacuated glass vessel weighs 40.0 g when empty, 135.0 g when filled with a liquid of density 0.95 g mL^{-1} and 40.5 g when filled with an ideal gas at 0.82 atm at 250 K. The molar mass of the gas in g mol^{-1} is : **[JEE Main (June) 2022 (I)]**

(a) 35 (b) 50 (c) 75 (d) 125

18. Arrange the following gases in increasing order of van der Waals constant '*a*' : **[JEE Main (April) 2023 (II)]**

A. Ar B. CH_4 C. H_2O D. C_6H_6

Choose the correct option from the following :

(a) B, C, D and A (b) C, D, B and A (c) A, B, C and D (d) D, C, B and A

19. Identify the mixture that shows positive deviations from Raoult's Law : **[JEE Main (Jan.) 2024]**

(a) $(CH_3)_2CO + C_6H_5NH_2$ (b) $CHCl_3 + C_6H_6$

(c) $CHCl_3 + (CH_3)_2CO$ (d) $(CH_3)_2CO + CS_2$

B. Objective Questions

[One or more than one option(s) is/are correct]

1. A gas described by van der Waals' equation : **(IIT 2008)**

(a) behaves similar to an ideal gas in the limit of large molar volumes

(b) behaves similar to an ideal gas in the limit of large pressure

(c) is characterized by van der Waals' coefficients that are dependent on the identity of the gas but are independent of the temperature

(d) has the pressure that is lower than the pressure exerted by the same gas behaving ideally

2. The correct statement(s) regarding defects in solids is(are): **(IIT 2009)**

(a) Frenkel defect is usually favoured by a very small difference in the sizes of cation and anion

(b) Frenkel defect is a dislocation defect

(c) Trapping of an electron in the lattice leads to the formation of F-center

(d) Schottky defects have no effect on the physical properties of solids

3. According to kinetic theory of gases : **(IIT 2011)**

(a) collisions are always elastic

(b) heavier molecules transfer more momentum to the wall of the container

(c) only a small number of molecules have very high velocity

(d) between collisions, the molecules move in straight lines with constant velocities

4. Which of the following statement(s) is (are) correct regarding the root mean square speed (u_{rms}) and average translational kinetic energy (E_{av}) of a molecule in a gas at equilibrium? **[JEE (Advanced) 2019]**

(a) u_{rms} is inversely proportional to the square root of its molecular mass.

(b) E_{av} is doubled when its temperature is increased four times.

(c) u_{rms} is doubled when its temperature is increased four times.

(d) E_{av} at a given temperature does not depend on its molecular mass.

C. Matching Type Problems

1. Match the gases under specified conditions listed in **Column I** with their properties/laws in **Column II.**

(IIT 2007)

	Column I		Column II
(a)	Hydrogen gas ($P = 200$ atm, $T = 273$ K)	(p)	compressibility factor $\neq 1$
(b)	Hydrogen gas ($P \sim 0, T = 273$ K)	(q)	attractive forces are dominant
(c)	CO_2 ($P = 1$ atm, $T = 273$ K)	(r)	$PV = nRT$
(d)	Real gas with very large molar volume	(s)	$P(V - nb) = nRT$

2. Match the crystal system/unit cells listed in **Column I** with their characteristic features in **Column II.**

(IIT 2007)

	Column I		Column II
(a)	Simple cubic and face-centred cubic	(p)	have these cell parameters $a = b = c$ and $\alpha = \beta = \gamma$
(b)	Cubic and rhombohedral	(q)	are law crystal systems
(c)	Cubic and tetragonal	(r)	have only two crystallographic angles of 90°
(d)	Hexagonal and monoclinic	(s)	belong to same crystal system

D. Passage Based Problems

Passage-1

In hexagonal system of crystals, a frequency encountered arrangement of atoms is described as a hexagonal prism. Here, the top and bottom of the cell are regular hexagons and three atoms are sandwiched in between them. A space-filling model of this structure, called hexagonal close-packed (hcp), is constituted of a sphere on a flat surface surrounded in the same plane by six identical spheres as closely as possible. Three spheres are then placed over the first layer so that they touch each other and represent the second layer. Each one of the three spheres touches three spheres of the bottom layer. Finally, the second layer is covered with a third layer that is identical to the bottom layer in relative position. Assume radius of every sphere to be r.

(IIT 2008)

1. The number of atoms in this hcp unit cell is :

(a) 4 (b) 6 (c) 12 (d) 17

2. The volume of this hcp unit cell is :

(a) $24\sqrt{2}\,r^3$ (b) $16\sqrt{2}\,r^3$ (c) $12\sqrt{2}\,r^3$ (d) $\frac{64}{3\sqrt{3}}r^3$

3. The empty space in this hcp unit cell is :

(a) 74% (b) 47.6 % (c) 32% (d) 26%

Passage-2

x and y are two volatile liquids with molar weights of 10 g mol^{-1} and 40 g mol^{-1} respectively. Two cotton plugs, one soaked in x and the other soaked in y, are simultaneously placed at the ends of a tube of length $L = 24$ cm, as shown in the figure. The tube is filled with an inert gas at 1 atm pressure and a temperature of 300 K. Vapours of x and y react to form a product which is first observed at a distance d cm from the plug soaked in x. Take x and y to have equal molecular diameters and assume ideal behaviour for the inert gas and the two vapours.

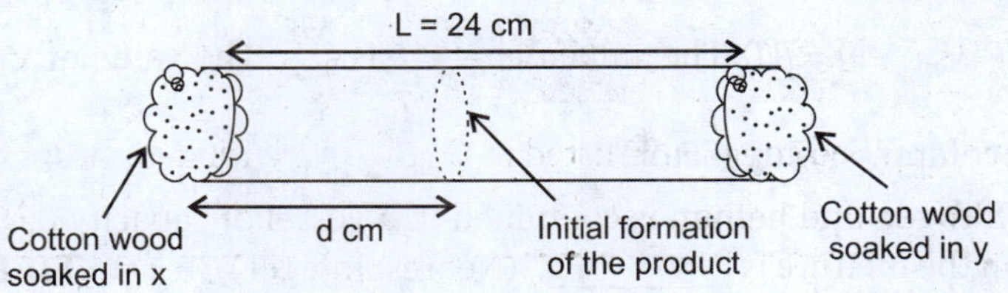

[JEE (Advanced) 2014]

1. The value of d in cm (shown in the figure) as estimated from Graham's law is :

(a) 8 (b) 12 (c) 16 (d) 20

2. The experimental value of d is found to be smaller than the estimate obtained using Graham's law. This is due to :

(a) larger mean free path of x as compared to that of y

(b) larger mean free path for y as compared to that of x

(c) increased collision frequency of y with the inert gas as compared to that of x with the inert gas

(d) increased collision frequency of x with the inert gas as compared to that of y with the inert gas

E. Assertion and Reason

Read the following questions and answer as per the direction given below :

(a) Statement 1 is true; Statement 2 is also true; Statement 2 is a correct explanation of Statement 1.

(b) Statement 1 is true; Statement 2 is true; Statement 2 is not the correct explanation of Statement 1.

(c) Statement 1 is true; Statement 2 is false.

(d) Statement 1 is false; Statement 2 is true.

1. Statement 1 : The pressure of a fixed amount of an ideal gas is proportional to its temperature.

Statement 2 : Frequency of collisions and their impact both increase in proportion to the square root of temperature. **(IIT 2000)**

2. Statement 1 : In any ionic solid [MX] with Schottky defects, the number of positive and negative ions are same.

Statement 2 : Equal number of cation and anion vacancies are present. **(IIT 2001)**

F. Integer Type Problems

1. At 400 K, the root mean square (rms) speed of a gas **X** (molecular weight = 40) is equal to the most probable speed of gas **Y** at 60 K. The molecular weight of the gas **Y** is: **(IIT 2009)**

2. The diffusion coefficient of an ideal gas is proportional to its mean free path and mean speed. The absolute temperature of an ideal gas is increased 4 times and its pressure is increased 2 times. As a result, the diffusion coefficient of this gas increases x times. The value of x is :

[JEE (Advanced) 2016]

3. A spherical balloon of radius 3 cm containing helium gas has a pressure of 48×10^{-3} bar. At the same temperature, the pressure of a spherical balloon of radius 12 cm containing the same amount of gas will be ________ $\times 10^{-6}$ bar. **[JEE Main (Sept.) 2020]**

4. The volume occupied by 4.75 g of acetylene gas at 50°C and 740 mm Hg pressure is _____ L. (Rounded off to the nearest integer)

[Given: $R = 0.0826$ L atm K^{-1} mol^{-1}] **[JEE Main (Feb.) 2021]**

5. A car tyre is filled with nitrogen gas at 35 psi at 27°C. It will burst if pressure exceeds 40 psi. The temperature in °C at which the car tyre will burst is _______. (Rounded-off to the nearest integer)

[JEE Main (Feb.) 2021]

6. A certain gas obeys $P(V_m - b) = RT$. The value of $\left(\frac{\partial Z}{\partial P}\right)_T$ is $\frac{xb}{RT}$. The value of x is ______.
(Integer answer) (Z: compressibility factor) **[JEE Main (Feb.) 2021]**

7. A 10 g mixture of hydrogen and helium is contained in a vessel of capacity 0.0125 m^3 at 6 bar and 27°C. The mass of helium in the mixture is ______ g. (Nearest integer) **[JEE Main (July) 2022 (II)]**

8. At constant temperature a gas is at a pressure of 940.3 mm Hg. The pressure at which its volume decreases by 40% is ______ mm Hg. (Nearest integer) **[JEE Main (April) 2023 (I)]**

9. A certain quantity of real gas occupies a volume of 0.15 dm^3 at 100 atm and 500 K when its compressibility factor is 1.07. Its volume at 300 atm and 300 K (When its compressibility factor is 1.4) is ______ $\times 10^{-4} dm^3$ (Nearest integer) **[JEE Main (April) 2023 (I)]**

10. The osmotic pressure of a dilute solution is 7×10^5 Pa at 273 K. Osmotic pressure of the same solution at 283 K is ______ $\times 10^4 Nm^{-2}$. **[JEE Main (Jan.) 2024]**

Answers

[A] Objective Questions (Only one option is correct)

1. (b) **2.** (d) **3.** (c) **4.** (b) **5.** (c) **6.** (d) **7.** (b) **8.** (d) **9.** (b) **10.** (b)
11. (c) **12.** (d) **13.** (a) **14.** (b) **15.** (d) **16.** (a) **17.** (d) **18.** (c) **19.** (d)

[B] Objective Questions (One or more than one option(s) is/are correct)

1. (a, c, d) **2.** (b, c) **3.** (a, b, c, d) **4.** (a, c, d)

[C] Match the Columns

1. a → p,s; b → r; c → p,q; d → r
2. a → p,s; b → p, q; c → q; d → q,r

[D] Passage Based Problems

Passage-1

1. (b) **2.** (a) **3.** (d)

Passage-2

1. (c) **2.** (d)

[E] Assertion and Reason

1. (b) **2.** (a)

[F] Integer Type Problems

1. (4) **2.** (4) **3.** (750) **4.** (5) **5.** (70)
6. (1) **7.** (8) **8.** (1567) **9.** (392) **10.** (72.56 or 73)

Hints and Solutions

A. Objective Questions

[Only one option is correct]

1. (b) The measure of force of attraction for n mole of real gas is $\left(\frac{n^2a}{V^2}\right)$.

$\because$ $$\left(P+\frac{n^2a}{V^2}\right)[V^2-nb]=nRT \qquad \text{(for } n \text{ mole gas)}$$

2. (d)

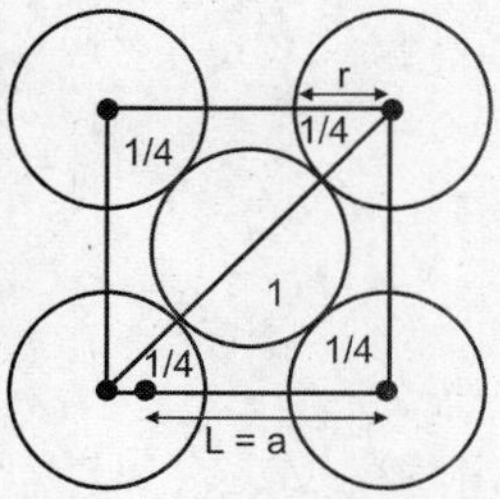

Number of particles in one unit cell $=1+4\times\frac{1}{4}=2$

If radius of particle is r, then total area covered by particles $=2\pi r^2$

Total area of unit cell $=L^2(\text{or } a^2)$

But, $a=2\sqrt{2}\,r$

and Packing efficiency $=\frac{\text{Area covered by particle}}{\text{Total area}}=\frac{2\pi r^2}{(2\sqrt{2}r)^2}=\frac{\pi}{4}$

$\Rightarrow$ $\frac{22}{7\times 4}\times 100=78.54\%$

3. (c) $C^*:\bar{C}:C=\sqrt{\frac{2RT}{M}}=\sqrt{\frac{8RT}{\pi M}}=\sqrt{\frac{3RT}{M}}$

$\sqrt{2}:\sqrt{\frac{8}{3.14}}=\sqrt{3}$

$1:1.128:1.225$

4. (b) Compressibility factor $(Z)=\frac{PV}{RT}$ (For one mole of real gas)

van der Waals' equation

$$\left(P+\frac{a}{V^2}\right)(V-b)=RT$$

At low pressure, $V-b\approx V$

$\therefore$ $$\left(P+\frac{a}{V^2}\right)V=RT$$

or $$PV+\frac{a}{V}=RT$$

$$\frac{PV}{RT} = 1 - \frac{a}{VRT}$$

So, $$Z = 1 - \frac{a}{VRT}$$

5. (c) Initially, number of moles of gas in each bulb is

$$n_1 = \frac{p_i V}{RT_1} \text{ and } n_2 = \frac{p_i V}{RT_1}$$

After the temperature of second bulb is raised to T_2 then the number of moles of gas in both the bulbs are

$$n_1' = \frac{p_f V}{RT_1} \text{ and } n_2' = \frac{p_f V}{RT_2}$$

Now, the total number of moles of gas in both the bulbs remains same in both the cases.

$$n_1 + n_2 = n_1' + n_2'$$

$$\frac{2p_i V}{RT_1} = \frac{p_f V}{RT_1} + \frac{p_f V}{RT_2}$$

$$\frac{2p_i V}{RT_1} = \frac{p_f V}{R}\left(\frac{T_2 + T_1}{T_1 T_2}\right)$$

$$p_f = \frac{2p_i T_2}{T_1 + T_2}$$

6. (d)

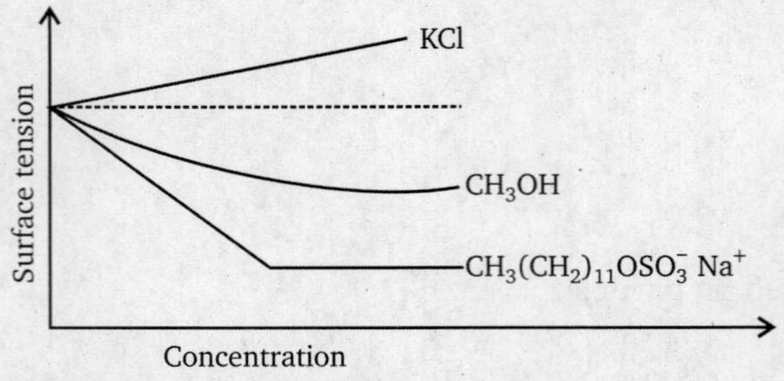

For KCl curve–Increase of surface tension for inorganic salts.

For CH_3OH curve–Decrease of surface tension progressively for alcohols.

For $CH_3(CH_2)_{11}OSO_3^-Na^+$ curve– Decrease of surface tension before CMC (Critical Micelle Concentration) and then almost unchanged.

7. (b) At constant P and V

$$nT = \text{constant}$$

$$n_1T_1 = n_2T_2$$

$$n \times 300 = \left(n - \frac{2n}{5}\right) \times T_2$$

$$T_2 = 300 \times \frac{5}{3} = 500\,\text{K}$$

8. (d) $$u_{mp} = \sqrt{\frac{2RT}{M}}$$

So, $$u_{mp} \propto \sqrt{\frac{T}{M}}$$

For O_2 at 400 K, $u_{mp} \propto \sqrt{\frac{400}{32}}$

For N_2 at 300 K, $u_{mp} \propto \sqrt{\frac{300}{28}}$

For H_2 at 300 K, $u_{mp} \propto \sqrt{\frac{300}{2}}$

So, order will be,

$$u_{mp}(N_2) < u_{mp}(O_2) < u_{mp}(H_2)$$

9. (b) Compressibility factor $(Z) = \frac{PV}{nRT}$

Given,
$$V_A = 2V_B$$
$$Z_A = 3Z_B$$
$$P = \frac{Z \times nRT}{V}$$
$$\frac{P_A}{P_B} = \frac{Z_A}{Z_B} \times \frac{V_B}{V_A} = 3 \times \frac{1}{2} = \frac{3}{2}$$
$$\Rightarrow \quad 2P_A = 3P_B$$

10. (b) Graph represents symmetrical distribution of speed and hence, the most probable and the average speed should be same. But the root mean square speed must be greater than the average speed.

11. (c) Order of strength is ion-ion > ion-dipole > dipole-dipole.

12. (d) $CH_3—\underset{\underset{O}{\|}}{C}—OC_2H_5$ $(\mu \neq 0)$ is polar molecule.

13. (a)

Type	Interaction energy
ion-ion	$E \propto \frac{1}{r}$
dipole-dipole	$E \propto \frac{1}{r^3}$
London dispersion	$E \propto \frac{1}{r^6}$

14. (b) Partial pressure (P_i) = Mole fraction $(X_i) \times P_{Total}$

$$P_{H_2} = 2 \text{ atm}$$
$$X_{H_2} = \left(\frac{n_{H_2}}{n_{H_2} + n_{He} + n_{O_2}}\right) = \frac{1}{3}$$
$$P_{Total} = \frac{P_{H_2}}{X_{H_2}} = \frac{2}{1/3} = 6 \text{ atm}$$

15. (d)
$$u_{avg} = \sqrt{\frac{8RT}{\pi M}}$$
$$u_{mp} = \sqrt{\frac{2RT}{M}}$$
$$u_{rms} = \sqrt{\frac{3RT}{M}}$$
$$u_{mp} : u_{avg} : u_{rms} = \sqrt{2} : \sqrt{\frac{8}{\pi}} : \sqrt{3}$$
$$= 1.4 : 1.59 : 1.732$$

$A \rightarrow u_{mp}$
$B \rightarrow u_{avg}$
$C \rightarrow u_{rms}$

16. (a) It is correct that non-covalent interactions like dipole-dipole interaction can lead to H-bond formation. But ion dipole can also lead to H-bond formation.

$K^+ F^{\ominus}$ --- H — F strongest H-bond

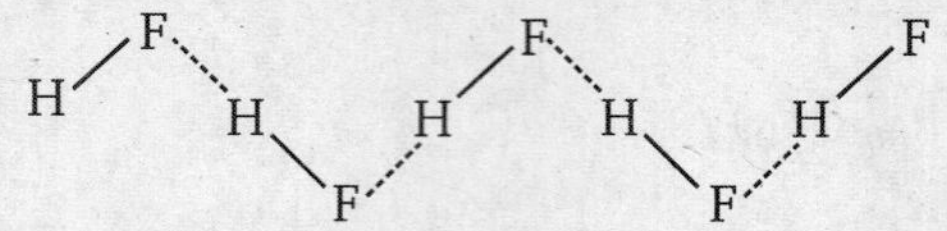

Symmetrical H-bonds are present.
Assertion is incorrect.
Reason is correct.

17. (d) Mass of liquid = 135 – 40 = 95 g

Volume of liquid $= \frac{\text{mass}}{\text{density}} = \frac{95}{0.95}$ mL = 100 mL = 0.1L

Mass of ideal gas = 40.5 – 40 g= 0.5 g

$$PV = nRT$$

$$0.82 \times 0.1 = \left(\frac{0.5}{M}\right) \times 0.082 \times 250 \quad \Rightarrow \quad M = 125$$

18. (c) van der Waal constant = 'a'

(A) Ar, a = 1.34

(B) CH_4, a = 2.25

(C) H_2O, a = 5.46

(D) C_6H_6, a = 18.57

'a' signify the force of attraction and it is directly proportional to surface area.

19. (d) $(CH_3)_2CO + CS_2$ exhibits positive deviations from Raoult's law.

B. Objective Questions

[One or more than one option(s) is/are correct]

1. (a, c, d)

$$\left(P + \frac{n^2a}{V^2}\right)(V - nb) = nRT$$

and a, c, d can be explained on the basis of this equation.

4. (a, c, d)

$$u_{rms} = \sqrt{\frac{3RT}{M}} \Rightarrow u_{rms} \propto \sqrt{\frac{1}{M}} \propto \sqrt{T}$$

Average kinetic energy $(E_{av}) = \frac{3}{2}kT \Rightarrow E_{av} \propto T$

C. Matching Type Problems

1. (a) → p, s (b) → r (c) → p, q (d) → r

According to van der Waals' equation

$$\left(P + \frac{n^2a}{V^2}\right)(V - nb) = nRT$$

(a) For H_2 gas, at P = 200 atm and T = 273 K *i.e.*, at high P and low T.

$$Z = \frac{PV}{nRT} > 1$$

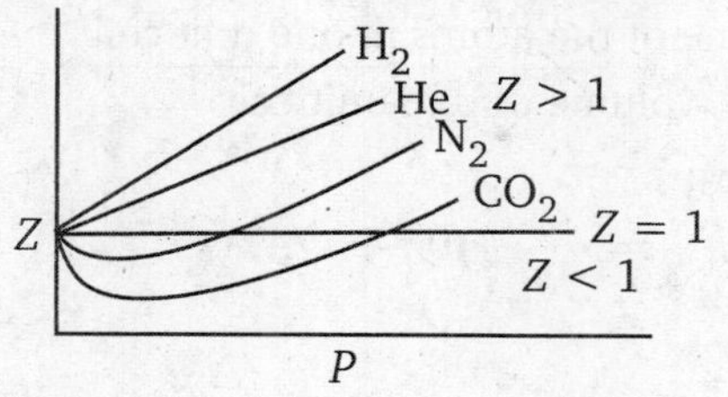

Also at high P, the factor $\frac{n^2a}{V^2}$ will be negligibly small in comparison to P and thus van der Waals' equation is changed to $P(V - nb) = nRT$

(b) At very low P and moderate temperature, H_2 behaves almost ideally thus under such condition, $PV = nRT$

(c) At moderate P and T, Z for CO_2 is less than one which reveals that attractive forces are dominant and $Z \neq 1$.

(d) When volume is very large (*i.e.*, P is very low and T is high), the intermolecular forces almost disappear and thus gas tend to behave ideally. $\frac{n^2a}{V^2}$ be small and hence can be neglected. Also nb will also be negligibly small in comparison to V, thus gas follow $PV = nRT$

2. (a) → p, s (b) → p, q (c) → q (d) → q, r

Intercepts and interfacial angles for various crystal systems are given below:

Cubic	$a = b = c$	$\alpha = \beta = \gamma = 90°$
Rhombohedral	$a = b = c$	$\alpha = \beta = \gamma \neq 90°$
Tetragonal	$a = b \neq c$	$\alpha = \beta = \gamma = 90°$
Hexagonal	$a = b \neq c$	$\alpha = \beta = 90°, \gamma = 120°$
Monoclinic	$a \neq b \neq c$	$\alpha = \beta = 90°, \gamma \neq 90°$

D. Passage Based Problems

Passage-1

1. (b) Total no. of atoms in 1 unit cell of h.c.p.

$$= (12 \times 1/6) + 3 + (2 \times 1/2) = 6$$

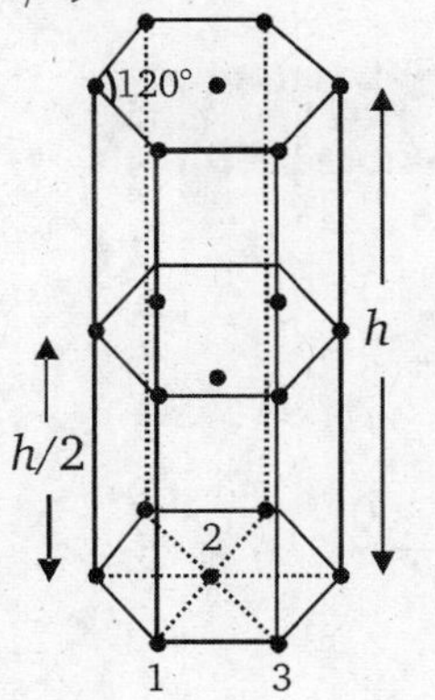

Base area of h.c.p. unit cell $= 6 \times \frac{\sqrt{3}}{4} \times 4r^2$

2. (a) Height of unit cell $= 4r\sqrt{\frac{2}{3}}$

Volume of unit cell = height × base area

$$= 4r\sqrt{\frac{2}{3}} \times 6 \times \frac{\sqrt{3}}{4} \times 4r^2 = 24\sqrt{2}r^3$$

3. (d) $\because$ Packing fraction $= \dfrac{\text{volume of the atoms in one unit cell}}{\text{volume of one unit cell}}$

$$= \frac{6 \times \frac{4}{3}\pi r^3}{24\sqrt{2}r^3} = \frac{\pi}{3\sqrt{2}} = 0.74 = 74\%$$

$\therefore$ Empty space = 26%

Pasasge-2

1. (c) According to Graham's law,

$$\frac{\gamma_x}{\gamma_y} = \sqrt{\frac{M_y}{M_x}} \Rightarrow \frac{d}{24-d} = \sqrt{\frac{40}{10}} = 2$$

$$d = 48 - 2d \Rightarrow 3d = 48$$

$$d = 16 \text{ cm}$$

2. (d) As the collision frequency increases, molecular speed decreases then the expected.

E. Assertion and Reason

1. (b) At constant volume

$$P \propto T \qquad \text{(if } n \text{ and } V \text{ are constant)}$$

and $\quad u_{rms} = \sqrt{\frac{3RT}{M}}$

or $\quad$ for a given gas

$$u_{rms} \propto \sqrt{T}$$

$\because$ Collision frequency $\propto u_{rms}$

$\therefore$ Collision frequency $\propto \sqrt{T}$

or $\quad$ Collision frequency $\propto \sqrt{P}$

Thus increase in the pressure leads to increased collision frequency.

2. (a) Schottky defect is a stoichiometric defect in which same number of cations and anions is missing from their respective site.

F. Integer Type Problems

1. (4) $\because \quad u_{rms} = u_{MP}$

$$\sqrt{\frac{3RT}{M_x}} = \sqrt{\frac{2RT}{M_y}}$$

$$\sqrt{\frac{3R \times 400}{40}} = \sqrt{\frac{2R \times 60}{M_y}}$$

$$M_y = 4$$

2. (4) Diffusion coefficient $\propto \lambda\, C_{mean}$

$\lambda \propto \frac{T}{P}$ and $C_{mean} \propto \sqrt{T}$

Diffusion coefficient $\propto \frac{T}{P}\sqrt{T}$

Diffusion coefficient $\propto \frac{T^{3/2}}{P}$

If T is increased four times and pressure is increased two times diffusion coefficient will become 4 times.

3. (750) For same amount of gas at constant temperature

$$P_1V_1 = P_2V_2 \text{ (Boyle's law)}$$

$$48 \times 10^{-3} \times \frac{4}{3}\pi(3)^3 = P_2 \times \frac{4}{3}\pi(12)^3$$

$$P_2 = 0.75 \times 10^{-3} \text{ bar} = 750 \times 10^{-6} \text{ bar}$$

4. (5) Acetylene is C_2H_2.

Molar mass = 26 g/mol

Temperature = (273 + 50)K = 323 K

Pressure $(P) = \frac{740}{760}$ atm

From ideal gas equation $V = \frac{nRT}{P} = \left(\frac{4.75}{26}\right) \times \frac{0.0826 \times 323}{(740/760)} = 5.006\text{ L}$

5. (70)

$$P \propto T$$

$$\frac{P_1}{T_1} = \frac{P_2}{T_2}$$

$$T_2 = P_2 \times \frac{T_1}{P_1} = 40 \times \frac{300}{35} = 342.85\text{K} = (342.85 - 273.15)°\text{C} = 69.7$$

6. (1) $P(V_m - b) = RT$

$$PV_m - Pb = RT$$

$$PV_m = RT + Pb$$

$$\frac{PV_m}{RT} = 1 + \frac{Pb}{RT}$$

$$Z = 1 + \frac{Pb}{RT}$$

$$\left(\frac{\partial Z}{\partial P}\right)_T = \frac{b}{RT}$$

So, $x = 1$

7. (8)

$$PV = n_{mix}RT$$

$$n_{mix} = \frac{6 \times 12.5}{0.083 \times 300} \approx 3$$

Let mole of He = x ; Mole of $H_2 = 3 - x$

$$4x + 2(3 - x) = 10; \quad x = 2 \text{ mol}$$

Mass of He = 8 g

8. (1567) $P_1V_1 = P_2V_2 \quad \Rightarrow \quad 940.3 \times 100 = P_2 \times 60$

P_2 = 1567 mm of Hg

9. (392) $Z = \frac{PV}{nRT}$; $n = \frac{PV}{ZRT}$

$Z_1 = 1.07$, $P_1 = 100$ atm, $V_1 = 0.15$ L, $T_1 = 500$ K

$Z_2 = 1.4$, $P_2 = 300$ atm, $T_2 = 300$ K, $V_2 = ?$

$$\frac{P_1V_1}{Z_1RT_1} = \frac{P_2V_2}{Z_2RT_2} = n \quad \Rightarrow \quad V_2 = \frac{1.4}{1.07} \times 0.03 = 392 \times 10^{-4}\text{dm}^3$$

10. (72.56 or 73)

$$\pi = CRT$$

$$\Rightarrow \quad \frac{\pi_1}{\pi_2} = \frac{T_1}{T_2}$$

$$\Rightarrow \quad \pi_2 = \frac{\pi_1 T_2}{T_1} = \frac{7 \times 10^5 \times 283}{273} = 72.56 \times 10^4 \text{ Nm}^{-2}$$

❑❑❑

CHAPTER

4

Thermodynamics

A. Objective Questions

[Only one option is correct]

1. The species which by definition has ZERO standard molar enthalpy of formation at 298 K is : **(IIT 2010)**

(a) $Br_{2(g)}$ (b) $Cl_{2(g)}$ (c) $H_2O_{(g)}$ (d) $CH_{4(g)}$

2. Using the data provided, calculate the multiple bond energy ($kJ\,mol^{-1}$) of a $C \equiv C$ bond in C_2H_2. That energy is : (take the bond energy of a C — H bond as $350\,kJ\,mol^{-1}$). **(IIT 2012)**

$$2C_{(s)} + H_{2(g)} \longrightarrow C_2H_{2(g)}; \quad \Delta H = 225\,kJ\,mol^{-1}$$

$$2C_{(s)} \longrightarrow 2C_{(g)}; \quad \Delta H = 1410\,kJ\,mol^{-1}$$

$$H_{2(g)} \longrightarrow 2H_{(g)}; \quad \Delta H = 330\,kJ\,mol^{-1}$$

(a) 1165 (b) 837 (c) 865 (d) 815

3. For one mole of a van der Waals' gas when $b = 0$ and $T = 300$ K, the PV vs $1/V$ plot is shown below. The value of the van der Waals constant a (atm litre^2 mol^{-2}) is : **(IIT 2012)**

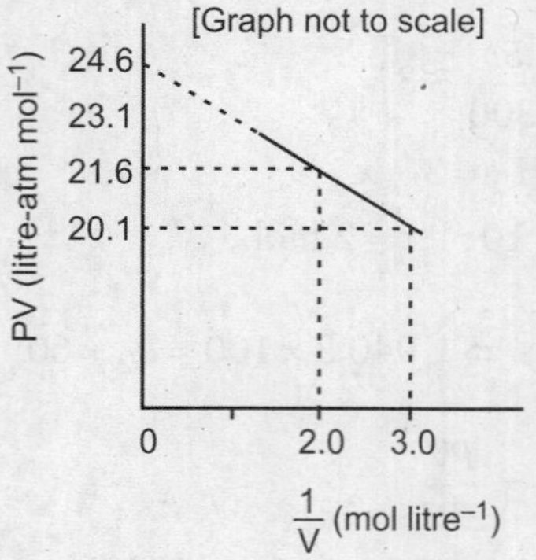

(a) 1.0 (b) 4.5 (c) 1.5 (d) 3.0

4. A piston filled with 0.04 mol of an ideal gas expands reversibly from 50.0 mL to 375 mL at a constant temperature of 37.0°C. As it does so, it absorbs 208 J of heat. The values of q and w for the process will be : ($R = 8.314$ J/mol K, ln 7.5 = 2.01) **[JEE (Mains) 2013]**

(a) $q = +208$ J, $w = -208$ J (b) $q = -208$ J, $w = -208$ J

(c) $q = -208$ J, $w = +208$ J (d) $q = +208$ J, $w = +208$ J

5. The standard enthalpies of formation of $CO_2(g)$, $H_2O(l)$ and glucose (s) at 25°C are –400 kJ/mol, –300 kJ/mol and –1300 kJ/mol, respectively. The standard enthalpy of combustion per gram of glucose at 25°C is: **[JEE (Advanced) 2013]**

(a) +2900 kJ (b) –2900 kJ (c) –16.11 kJ (d) +16.11 J

6. For complete combustion of ethanol,

$$C_2H_5OH(l) + 3O_2(g) \longrightarrow 2CO_2(g) + 3H_2O(l),$$

the amount of heat produced as measured in bomb calorimeter, is $1364.47\,kJ\,mol^{-1}$ at 25°C. Assuming ideality the enthalpy of combustion, $\Delta_c H$, for the reaction will be :
$(R = 8.314\,kJ\,mol^{-1})$ **[JEE (Mains) 2014]**

(a) $-1366.95\,kJ\,mol^{-1}$ (b) $-1361.95\,kJ\,mol^{-1}$
(c) $-1460.50\,kJ\,mol^{-1}$ (d) $-1350.50\,kJ\,mol^{-1}$

7. For the process, $H_2O(l) \longrightarrow H_2O(g)$ at $T = 100°C$ and 1 atm pressure, the correct choice is :
[JEE (Advanced) 2014]

(a) $\Delta S_{System} > 0$ and $\Delta S_{Surrounding} > 0$ (b) $\Delta S_{System} > 0$ and $\Delta S_{Surrounding} < 0$
(c) $\Delta S_{System} < 0$ and $\Delta S_{Surrounding} > 0$ (d) $\Delta S_{System} < 0$ and $\Delta S_{Surrounding} < 0$

8. One mole of a monoatomic real gas satisfies the equation $P(V - b) = RT$ where b is a constant. The relationship of interatomic potential $V(r)$ and interatomic distance r for the gas is given by:
[JEE (Advanced) 2015]

(a) 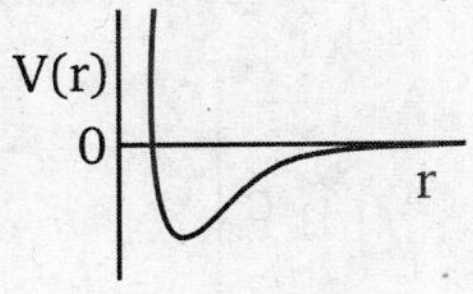

(b) 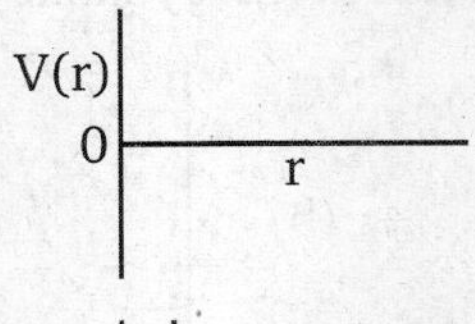

(c) 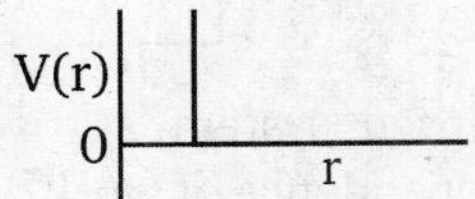

(d) 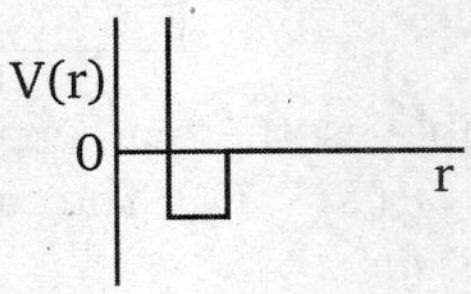

9. The heats of combustion of carbon and carbon monoxide are –393.5 and –283.5 $kJ\,mol^{-1}$ respectively. The heat of formation (in kJ) of carbon monoxide per mole is : **[JEE (Mains) 2016]**

(a) 110.5 (b) 676.5 (c) –676.5 (d) –110.5

10. One mole of an ideal gas at 300 K in thermal contact with surroundings expands isothermally from 1.0 L to 2.0L against a constant pressure of 3.0 atm. In this process, the change in entropy of surroundings (ΔS_{surr}) in $J\ K^{-1}$ is: (1 L atm = 101.3 J) **[JEE (Advanced) 2016]**

(a) 5.763 (b) 1.013 (c) –1.013 (d) –5.763

11. Given, $C_{(graphite)} + O_{2(g)} \longrightarrow CO_{2(g)}$; $\Delta_r H^\circ = -393.5\,kJ\,mol^{-1}$

$H_{2(g)} + \frac{1}{2}O_{2\,(g)} \longrightarrow H_2O_{(l)}$; $\Delta_r H^\circ = -285.9\,kJ\,mol^{-1}$

$CO_{2(g)} + 2H_2O_{(l)} \longrightarrow CH_{4(g)} + 2O_{2(g)}$; $\Delta_r H^\circ = +890.3\,kJ\,mol^{-1}$

Based on the above thermochemical equations, the value of $\Delta_r H°$ at 298 K for the reaction,
$C_{(graphite)} + 2H_{2(g)} \longrightarrow CH_{4(g)}$ will be: **[JEE (Mains) 2017]**

(a) $-74.8\ kJ\ mol^{-1}$ (b) $-144.0\ kJ\ mol^{-1}$ (c) $+74.8\ kJ\ mol^{-1}$ (d) $+144.0\ kJ\ mol^{-1}$

12. ΔU is equal to : **[JEE (Mains) 2017]**

(a) adiabatic work (b) isothermal work (c) isochoric work (d) isobaric work

13. The combustion of benzene (l) gives $CO_2(g)$ and $H_2O(l)$. Given that heat of combustion of benzene at constant volume is $-3263.9\ kJ\,mol^{-1}$ at 25° C; heat of combustion (in $kJ\,mol^{-1}$) of benzene at constant pressure will be : **[JEE (Mains) 2018]**
$(R = 8.314\ JK^{-1}\ mol^{-1})$

(a) 3260 (b) – 3267.6 (c) 4152.6 (d) –452.46

14. For silver, $C_p(\text{JK}^{-1}\,\text{mol}^{-1}) = 23 + 0.01T$. If the temperature (T) of 3 moles of silver is raised from 300 K to 1000 K at 1 atm pressure, the value of ΔH will be close to : **[JEE (Mains) 2019]**

(a) 13 kJ (b) 16 kJ (c) 62 kJ (d) 21 kJ

15. Which one of the following equations does not correctly represent the first law of thermodynamics for the given processes involving an ideal gas ?
(Assume non-expansion work is zero) **[JEE (Mains) 2019]**

(a) Adiabatic process : $\Delta U = -w$ (b) Isochoric process : $\Delta U = q$

(c) Isothermal process : $q = -w$ (d) Cyclic process : $q = -w$

16. 5 moles of an ideal gas at 100 K are allowed to undergo reversible compression till its temperature becomes 200 K. If $C_V = 28\ \text{JK}^{-1}\text{mol}^{-1}$, calculate ΔU and ΔPV for this process. $(R = 8.0\ \text{J K}^{-1}\text{mol}^{-1})$ **[JEE (Mains) 2019]**

(a) $\Delta U = 2.8\,\text{kJ}; \Delta(PV) = 0.8\ \text{kJ}$ (b) $\Delta U = 14\,\text{kJ}; \Delta(PV) = 4\ \text{kJ}$

(c) $\Delta U = 14\,\text{kJ}; \Delta(PV) = 18\ \text{kJ}$ (d) $\Delta U = 14\ \text{J}; \Delta(PV) = 0.8\ \text{J}$

17. For a diatomic ideal gas in a closed system, which of the following plots does not correctly describe the relation between various thermodynamic quantities? **[JEE (Mains) 2019]**

(a) 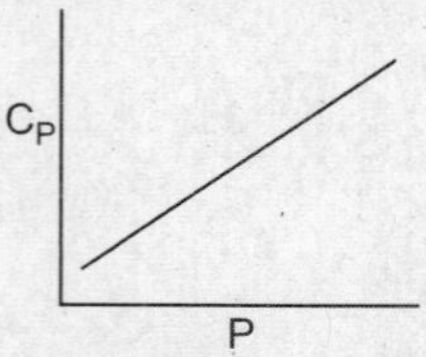

(b) 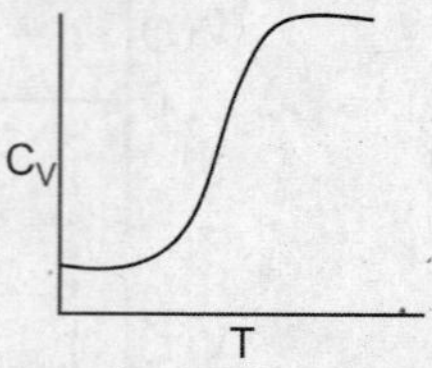

(c) 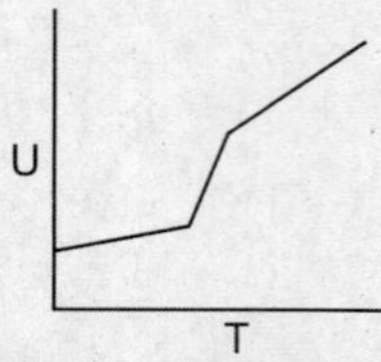

(d) 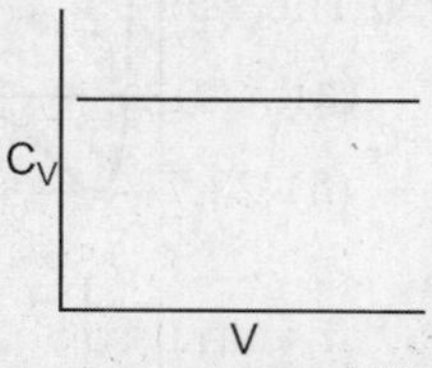

18. Consider the reversible isothermal expansion of an ideal gas in a closed system at two different temperatures T_1 and T_2 $(T_1 < T_2)$. The correct graphical depiction of the dependence of work done (W) on the final volume (V) is: **[JEE (Mains) 2019]**

(a) 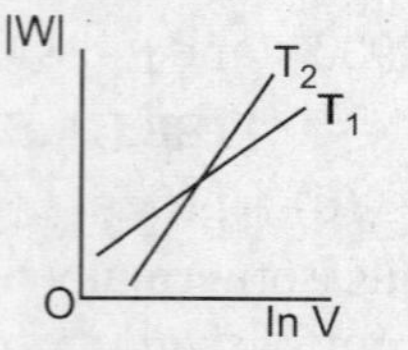

(b) 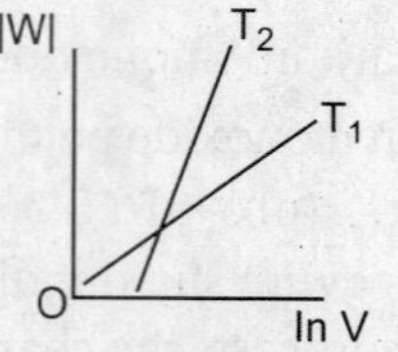

(c) 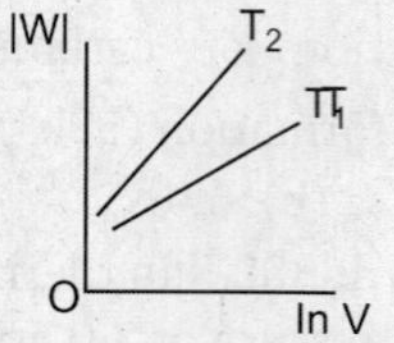

(d) 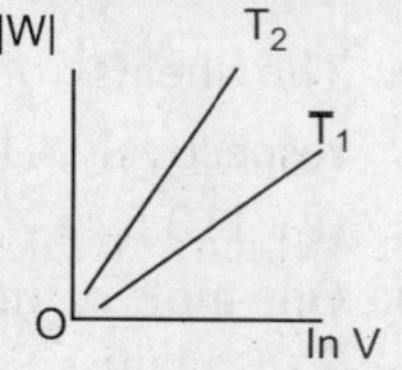

19. 5 mole of ideal gas at 100K $(C_{V,m} = 28\ \text{J/mol/K})$. It is heated upto 200 K. Calculate ΔU and $\Delta(PV)$ for the process. $(R = 8\ \text{J/mol-K})$ **[JEE (Mains) 2019]**

(a) $\Delta U = 28\,\text{kJ};\ \Delta(PV) = 8\text{kJ}$ (b) $\Delta U = 14\,\text{kJ};\ \Delta(PV) = 4\text{kJ}$

(c) $\Delta U = 14\,\text{kJ};\ \Delta(PV) = 8\text{kJ}$ (d) $\Delta U = 28\,\text{kJ};\ \Delta(PV) = 4\text{kJ}$

20. Two blocks of the same metal having same mass and at temperature T_1 and T_2, respectively, are brought in contact with each other and allowed to attain thermal equilibrium at constant pressure. The change in entropy, ΔS, for this process is: **[JEE (Mains) 2019]**

(a) $2C_p \ln\left|\dfrac{(T_1+T_2)^{1/2}}{T_1T_2}\right|$ (b) $2C_p \ln\left(\dfrac{T_1+T_2}{4T_1T_2}\right)$ (c) $2C_p \ln\left[\dfrac{T_1+T_2}{2T_1T_2}\right]$ (d) $C_p \ln\left[\dfrac{(T_1+T_2)^2}{4T_1T_2}\right]$

21. The entropy change associated with the conversion of 1kg of ice at 273K to water vapours at 383K is: (Specific heat of water liquid and water vapour are $4.2\,\text{kJK}^{-1}\,\text{kg}^{-1}$ and $2.0\,\text{kJK}^{-1}\,\text{kg}^{-1}$; heat of liquid fusion and vaporisation of water are $344\,\text{kJkg}^{-1}$ and $2491\,\text{kJkg}^{-1}$, respectively). $(\log 273 = 2.436,$ $\log 373 = 2.572,\ \log 383 = 2.583)$ **[JEE (Mains) 2019]**

(a) $7.90\,\text{kJkg}^{-1}\,\text{K}^{-1}$ (b) $2.64\,\text{kJkg}^{-1}\,\text{K}^{-1}$ (c) $8.49\,\text{kJkg}^{-1}\,\text{K}^{-1}$ (d) $9.26\,\text{kJkg}^{-1}\,\text{K}^{-1}$

22. Five moles of an ideal gas at 1 bar and 298 K is expanded into vacuum to double the volume. The work done is : **[JEE Main (Sept.) 2020]**

(a) $C_V(T_2 - T_1)$ (b) $-RT \ln\dfrac{V_2}{V_1}$ (c) zero (d) $-RT\dfrac{V_2}{V_1}$

23. For one mole of an ideal gas, which of these statements must be true? **[JEE Main (Sept.) 2020]**

(A) U and H each depends only on temperature.

(B) Compressibility factor Z is not equal to 1.

(C) $C_{P,m} - C_{V,m} = R$

(D) $dU = C_V dT$ for any process

(a) (A), (C) and (D) (b) (B), (C) and (D) (c) (A) and (C) (d) (C) and (D)

24. The true statement amongst the following is : **[JEE Main (Jan.) 2020]**

(a) S is a function of temperature but DS is not a function of temperature.

(b) Both DS and S are functions of temperature.

(c) S is not a function of temperature but DS is a function of temperature.

(d) Both S and DS are not functions of temperature.

25. The variation of equilibrium constant with temperature is given below:

Temperature	Equilibrium Constant
$T_1 = 25°C$	$K_1 = 10$
$T_2 = 100°C$	$K_2 = 100$

The value of $\Delta H°$, $\Delta G°$ at T_1 and $\Delta G°$ at T_2 (in kJ mol^{-1}) respectively, are close to:

[Use $R = 8.314\ JK^{-1}\ mol^{-1}$] **[JEE Main (Sept.) 2020]**

(a) 28.7, – 7.14 and –5.71 (b) 0.64, – 7.14 and –5.71

(c) 28.4, – 5.71 and –14.29 (d) 0.64, –5.71 and –14.29

26. If enthalpy of atomisation for $Br_2(A)$ is x kJ/mol and bond enthalpy for Br_2 is y kJ/mol, the relation between them: **[JEE Main (Jan.) 2020]**

(a) is $x = y$ (b) is $x > y$ (c) does not exist (d) is $x < y$

27. The process that is not endothermic in nature is: **[JEE Main (Sept.) 2020]**

(a) $H(g) + e^- \longrightarrow H^-(g)$ (b) $O^-(g) + e^- + O^{2-}(g)$

(c) $Na(g) \longrightarrow Na^+(g) + e^-$ (d) $Ar(g) + e^- \longrightarrow Ar^-(g)$

28. Which of the following relation is not correct? **[JEE Main (July) 2022 (I)]**

(a) $\Delta H = \Delta U - P\Delta V$ (b) $\Delta U = q + W$

(c) $\Delta S_{sys} = \Delta S_{surr} \geq 0$ (d) $\Delta G = \Delta H - T\Delta S$

29. At 25°C and 1 atm pressure, the enthalpy of combustion of benzene (l) and acetylene (g) are –3268 kJ mol^{-1} and –1300 kJ mol^{-1}, respectively. The change in enthalpy for the reaction $3C_2H_2(g) \longrightarrow C_6H_6(l)$, is : **[JEE Main (June) 2022 (II)]**

(a) +324 kJ mol^{-1} (b) +632 kJ mol^{-1} (c) –632 kJ mol^{-1} (d) –732 kJ mol^{-1}

30. What happens when methane undergoes combustion in systems A and B respectively ? **[JEE Main (April) 2023 (II)]**

Adiabatic system	Diathermic container
System A	System B

(a)

System A	System B
Temperature rises	Temperature remains same

(b)

System A	System B
Temperature falls	Temperature rises

(c)

System A	System B
Temperature falls	Temperature remains same

(d)

System A	System B
Temperature remains same	Temperature rises

31. Given, **[JEE Main (April) 2023 (I)]**

(A) $2CO(g) + O_2(g) \longrightarrow 2CO_2(g)$ $\quad \Delta H_1^\theta = -x \text{ kJ mol}^{-1}$

(B) $C(\text{graphite}) + O_2(g) \longrightarrow CO_2(g)$ $\quad \Delta H_2^\theta = -y \text{ kJ mol}^{-1}$

The ΔH^θ for the reaction

$C(\text{graphite}) + \frac{1}{2}O_2(g) \longrightarrow CO(g)$ is :

(a) $\frac{x-2y}{2}$ (b) $\frac{x+2y}{2}$ (c) $\frac{2x-y}{2}$ (d) $2y - x$

32. Choose the correct option for free expansion of an ideal gas under adiabatic condition from the following : **[JEE Main (Jan.) 2024]**

(a) $q=0,\ \Delta T \neq 0,\ w=0$ (b) $q=0,\ \Delta T < 0,\ w \neq 0$
(c) $q \neq 0,\ \Delta T = 0,\ w=0$ (d) $q=0,\ \Delta T = 0,\ w=0$

33. Which of the following is not correct? **[JEE Main (Jan.) 2024]**

(a) ΔG is negative for a spontaneous reaction (b) ΔG is positive for a spontaneous reaction
(c) ΔG is zero for a reversible reaction (d) ΔG is positive for a non-spontaneous reaction

34. A liquid when kept inside a thermally insulated closed vessel at 25°C was mechanically stirred from outside. What will be the correct option for the following thermodynamic parameters? **[JEE Main (Jan.) 2025]**

(a) $\Delta U > 0, q=0, w>0$ (b) $\Delta U = 0, q=0, w=0$ (c) $\Delta U < 0, q=0, w>0$ (d) $\Delta U = 0, q<0, w>0$

35. Ice at –5°C is heated to become vapour with temperature of 110°C at atmospheric pressure. The entropy change associated with this process can be obtained from : **[JEE Main (Jan.) 2025]**

(a) $\int_{268\text{ K}}^{383\text{ K}} C_p dT + \frac{\Delta H_{\text{melting}}}{273} + \frac{\Delta H_{\text{boiling}}}{373}$

(b) $\int_{268\text{K}}^{273\text{K}} \frac{C_{p,m}}{T} dT + \frac{\Delta H_{m,\text{fusion}}}{T_f} + \frac{\Delta H_{m,\text{ vaporisation}}}{T_b} + \int_{273\text{K}}^{373\text{K}} \frac{C_{p,m} dT}{T} + \int_{373\text{ K}}^{383\text{ K}} \frac{C_{p,m} dT}{T}$

(c) $\int_{268\text{K}}^{383\text{K}} C_p dT + \frac{q_{\text{rev}}}{T}$

(d) $\int_{268\text{K}}^{273\text{K}} C_{p,m} dT + \frac{\Delta H_{m,\text{fusion}}}{T_f} + \frac{\Delta H_{m,\text{ vaporisation}}}{T_b} + \int_{273\text{K}}^{373\text{K}} C_{p,m} dT + \int_{373\text{K}}^{383\text{K}} C_{p,m} dT$

36. The effect of temperature on spontaneity of reactions are represented as: **[JEE Main (Jan.) 2025]**

	ΔH	ΔS	**Temperature**	**Spontaneity**
(A)	+	–	any T	Non-spontaneous
(B)	+	+	low T	Spontaneous
(C)	–	–	low T	Non-spontaneous
(D)	–	+	any T	Spontaneous

(a) (B) and (D) only (b) (A) and (D) only (c) (B) and (C) only (d) (A) and (C) only

37. Let us consider an endothermic reaction which is non-spontaneous at the freezing point of water. However, the reaction is spontaneous at boiling point of water. Choose the correct option. **[JEE Main (Jan.) 2025]**

(a) Both ΔH and ΔS are (+ve). (b) ΔH is (–ve) but ΔS is (+ve).

(c) ΔH is (+ve) but ΔS is (–ve). (d) Both ΔH and ΔS are (–ve).

38. Ice and water are placed in a closed container at a pressure of 1 atm and temperature 273.15 K. If pressure of the system is increased 2 times, keeping temperature constant, then identify correct observation from following : **[JEE Main (Jan.) 2025]**

(a) Volume of system increases. (b) Liquid phase disappears completely.

(c) The amount of ice decreases. (d) The solid phase (ice) disappears completely.

39.

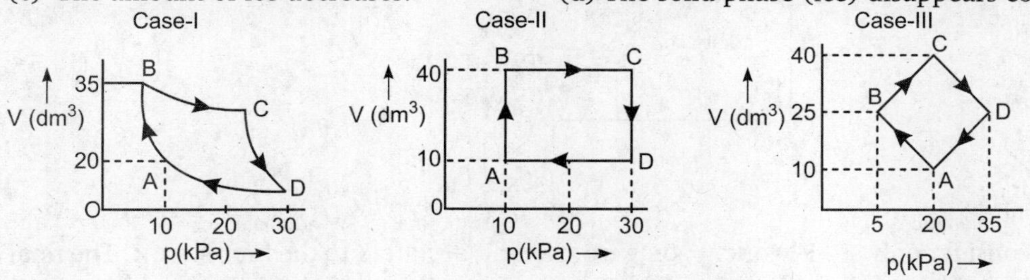

An ideal gas undergoes a cyclic transformation starting from the point A and coming back to the same point by tracing the path $A \to B \to C \to D \to A$ as shown in the three cases above.

Choose the correct option regarding ΔU. **[JEE Main (Jan.) 2025]**

(a) ΔU (Case-III) > ΔU (Case-II) > ΔU (Case-I)

(b) ΔU (Case-I) > ΔU (Case-II) > ΔU (Case-III)

(c) ΔU (Case-I) > ΔU (Case-III) > ΔU (Case-II)

(d) ΔU (Case-I) = ΔU (Case-II) = ΔU (Case-III)

40. 500 J of energy is transferred as heat to 0.5 mol of argon gas at 298 K and 1.00 atm. The final temperature and the change in internal energy respectively are :

Given : $R = 8.3\ JK^{-1}mol^{-1}$ **[JEE Main (Jan.) 2025]**

(a) 348 K and 300 J (b) 378 K and 300 J (c) 368 K and 500 J (d) 378K and 500 J

41. $S(g) + \frac{3}{2}O_2(g) \to SO_3(g) + 2x$ kcal

$SO_2(g) + \frac{1}{2}O_2(g) \to SO_3(g) + y$ kcal

The heat of formation of $SO_2(g)$ is given by : **[JEE Main (Jan.) 2025]**

(a) $\frac{2x}{y}$ kcal (b) $y - 2x$ kcal (c) $2x + y$ kcal (d) $x + y$ kcal

42. If C(diamond) $\to$ C(graphite) $+ X$ kJ mol^{-1}

C(diamond) + $O_2(g) \to CO_2(g) + Y$ kJ mol^{-1}

C(graphite) + $O_2(g) \to CO_2(g) + Z$ kJ mol^{-1}

At constant temperature, then : **[JEE Main (Jan.) 2025]**

(a) $X = Y + Z$ (b) $-X = Y + Z$ (c) $X = -Y + Z$ (d) $X = Y - Z$

B. Objective Questions

[One or more than one option(s) is/are correct]

1. Among the following, the state function(s) is (are): **(IIT 2009)**

(a) internal energy (b) irreversible expansion work

(c) reversible expansion work (d) molar enthalpy

2. Among the following, the intensive property is (properties are) : **(IIT 2010)**

(a) molar conductivity (b) electromotive force

(c) resistance (d) heat capacity

3. The reversible expansion of an ideal gas under adiabatic and isothermal conditions is shown in the figure. Which of the following statement(s) is(are) correct? **(IIT 2012)**

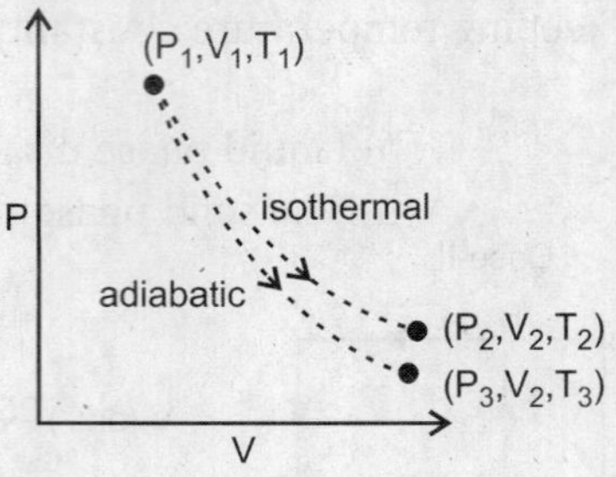

(a) $T_1 = T_2$ (b) $T_3 > T_1$

(c) $w_{\text{isothermal}} > w_{\text{adiabatic}}$ (d) $\Delta U_{\text{isothermal}} > \Delta U_{\text{adiabatic}}$

4. For an ideal gas, consider only P - V work in going from an initial state X to the final state Z. The final state Z can be reached by either of the two paths shown in the figure. Which of the following choice(s) is(are) correct? [Take ΔS as change in entropy and w as work done] **(IIT 2012)**

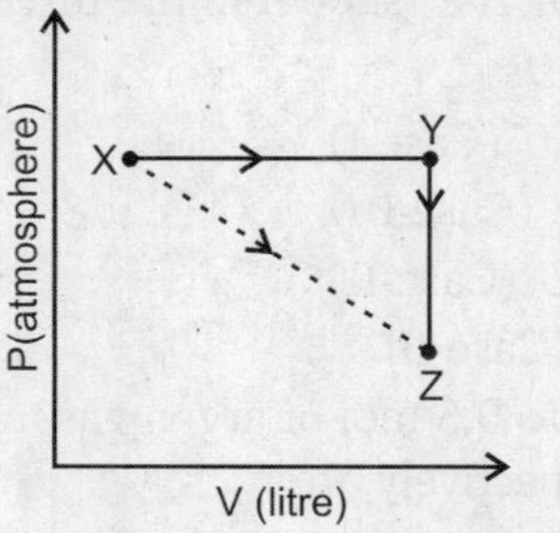

(a) $\Delta S_{X\to Z} = \Delta S_{X\to Y} + \Delta S_{Y\to Z}$ (b) $w_{X\to Z} = w_{X\to Y} + w_{Y\to Z}$

(c) $w_{X\to Y\to Z} = w_{X\to Y}$ (d) $\Delta S_{X\to Y\to Z} = \Delta S_{X\to Y}$

5. An ideal gas in a thermally insulated vessel at internal pressure $= P_1$, Volume $= V_1$ and absolute temperature $= T_1$ expands irreversibly against zero external pressure as shown in diagram. The final internal pressure, volume and absolute temperature of gas are P_2, V_2 and T_2 respectively. For this expansion

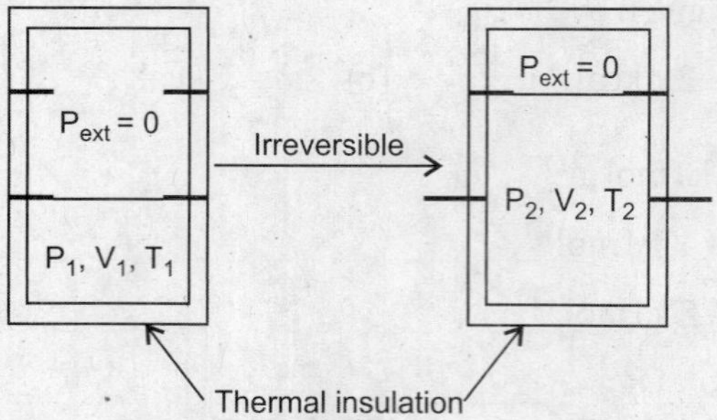

[JEE (Advanced) 2014]

(a) $q = 0$ (b) $T_2 = T_1$ (c) $P_2V_2 = P_1V_1$ (d) $P_2V_2^{\gamma} = P_1V_1^{\gamma}$

6. An ideal gas is expanded from (p_1, V_1, T_1) to (p_2, V_2, T_2) under different conditions. The correct statement(s) among the following is (are): **[JEE (Advanced) 2017]**

(a) If the expansion is carried out freely, it is simultaneously both isothermal as well as adiabatic.

(b) The work done by the gas is less when it is expanded reversibly from V_1 to V_2 under adiabatic conditions as compared to that when expanded reversibly from V_1 to V_2 under isothermal conditions.

(c) The work done on the gas is maximum when it is compressed irreversibly from (p_2, V_2) to (p_1, V_1) against constant pressure p_1.

(d) The change in internal energy of the gas is (i) zero, if it is expanded reversibly with $T_1 = T_2$, and (ii) positive, if it is expanded reversibly under adiabatic conditions with $T_1 \neq T_2$.

7. A reversible cyclic process for an ideal gas is shown below. Here, P, V and T are pressure, volume and temperature respectively. The thermodynamic parameters q, w, H and U are heat, work, enthalpy and internal energy, respectively.

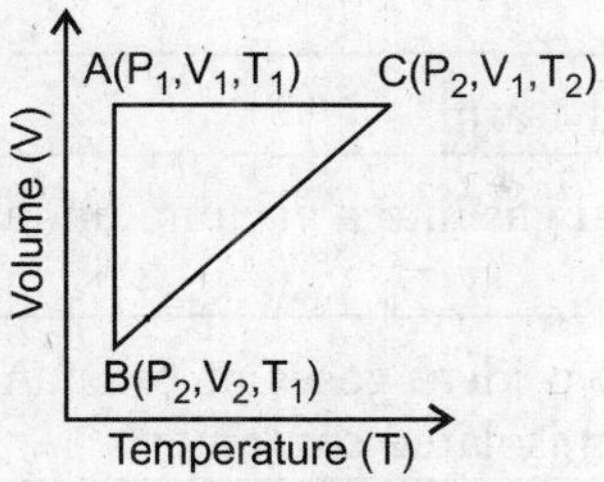

The correct option(s) is (are) : **[JEE (Advanced) 2018]**

(a) $q_{AC} = \Delta U_{BC}$ and $w_{AB} = P_2(V_2 - V_1)$
(b) $w_{BC} = P_2(V_2 - V_1)$ and $q_{BC} = \Delta H_{AC}$
(c) $\Delta H_{CA} < \Delta U_{CA}$ and $q_{AC} = \Delta U_{BC}$
(d) $q_{BC} = \Delta H_{AC}$ and $\Delta H_{CA} > \Delta U_{CA}$

8. Choose the reaction(s) from the following options, for which the standard enthalpy of reaction is equal to the standard enthalpy of formation. **[JEE (Advanced) 2019]**

(a) $\frac{1}{8}S_{8(s)} + O_{2(g)} \longrightarrow SO_{2(g)}$
(b) $2H_{2(g)} + O_{2(g)} \longrightarrow 2H_2O_{(l)}$
(c) $\frac{3}{2}O_{2(g)} \longrightarrow O_{3(g)}$
(d) $2C_{(g)} + 3H_{2(g)} \longrightarrow C_2H_{6(g)}$

9. In thermodynamics the P-V work done is given by $W = \int dVP_{ext}$.
For a system undergoing a particular process, the work done is,

$$W = -\int dV\left(\frac{RT}{V-b} - \frac{a}{V^2}\right).$$

This equation is applicable to a : **[JEE (Advanced) 2020]**

(a) System that satisfies the van der Waals' equation of state.
(b) Process that is reversible and isothermal.
(c) Process that is reversible and adiabatic.
(d) Process that is irreversible and at constant isothermal.

C. Matching Type Problems

1. Match the transformations in **column I** with appropriate options in **column II.** **(IIT 2011)**

	Column I		Column II
(a)	$CO_2(s) \rightarrow CO_2(g)$	(p)	phase transition
(b)	$CaCO_3(s) \rightarrow CaO(s) + CO_2(g)$	(q)	allotropic change
(c)	$2H\bullet \rightarrow H_2(g)$	(r)	ΔH is positive
(d)	$P_{(white,solid)} \rightarrow P_{(red,solid)}$	(s)	ΔS is positive
		(t)	ΔS is negative

2. Match the thermodynamic processes given under **Column-I** with the expression given under **Column-II**. **[JEE (Advanced) 2015]**

	Column-I		Column-II
(a)	Freezing of water at 273 K and 1 atm	(p)	$q = 0$
(b)	Expansion of 1 mole of an ideal gas into a vacuum under isolated conditions	(q)	$w = 0$
(c)	Mixing of equal volumes of two ideal gases at constant temperature and pressure in an isolated container	(r)	$\Delta S_{sys} < 0$
(d)	Reversible heating of $H_2(g)$ at 1 atm from 300 K to 600 K, followed by reversible cooling to 300 K at 1 atm	(s)	$\Delta U = 0$
		(t)	$\Delta G = 0$

3. Match List-I with List-II. **[JEE Main (Jan.) 2025]**

	List-I (Partial Derivatives)		List-II (Thermodynamic Quantity)
(A)	$\left(\frac{\partial G}{\partial T}\right)_P$	(I)	C_P
(B)	$\left(\frac{\partial H}{\partial T}\right)_P$	(II)	$-S$
(C)	$\left(\frac{\partial G}{\partial P}\right)_T$	(III)	C_V
(D)	$\left(\frac{\partial U}{\partial T}\right)_V$	(IV)	V

D. Assertion and Reason

Read the following questions and answer as per the direction given below :

(a) Statement 1 is true; Statement 2 is true; Statement 2 is a correct explanation of Statement 1.
(b) Statement 1 is true; Statement 2 is true; Statement 2 is not the correct explanation of Statement 1.
(c) Statement 1 is true; Statement 2 is false.
(d) Statement 1 is false; Statement 2 is true.

1. Statement 1 : The heat absorbed during the isothermal expansion of an ideal gas against vacuum is zero.

Statement 2: The volume occupied by the molecules of an ideal gas is zero at absolute zero temperature. **(IIT 2000)**

2. Statement 1 : For every chemical reaction at equilibrium, standard Gibbs energy of reaction is zero.

Statement 2 : At constant temperature and pressure, chemical reactions are spontaneous in the direction of decreasing Gibbs energy. **(IIT 2008)**

3. Statement 1 : There is a natural asymmetry between converting work to heat and converting heat to work.

Statement 2 : No process is possible in which the sole result in the absorption of heat from a reservoir and its complete conversion into work. **(IIT 2008)**

E. Integer Type Problems

1. In a constant volume calorimeter, 3.5 g of a gas with molecular weight 28 was burnt in excess oxygen at 298.0 K. The temperature of the calorimeter was found to increase from 298.0 K to 298.45 K due to the combustion process. Given that the heat capacity of the calorimeter is 2.5 kJ K^{-1}, the numerical value for the enthalpy of combustion of the gas in kJ mol^{-1} is: **(IIT 2009)**

2. One mole of an ideal gas is taken from a to b along two paths denoted by the solid and the dashed lines as shown in the graph below. If the work done along the solid line path is w_s, and that along the dotted line path is w_d, then the integer closest to the ratio w_d/w_s is : **(IIT 2010)**

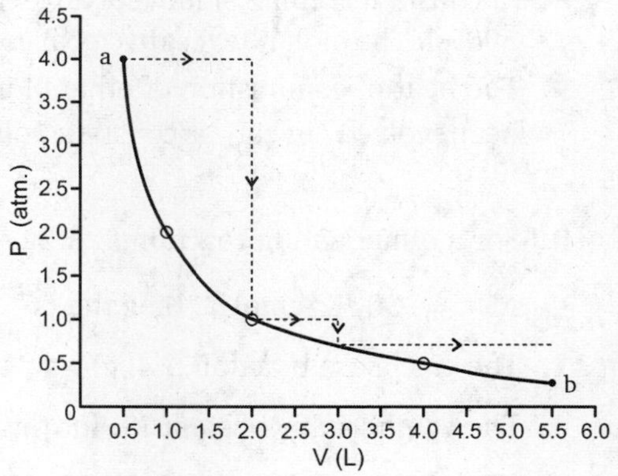

3. To an evacuated vessel with movable piston under external pressure of 1 atm, 0.1 mol of He and 1.0 mol of an unknown compound (vapour pressure 0.68 atm. at 0° C) are introduced. Considering the ideal gas behaviour, the total volume (in litre) of the gases at 0° C is close to : **(IIT 2011)**

4. A closed tank has two compartments A and B. Both filled with oxygen (assumed to be ideal gas). The partition separating the two compartments is fixed and is a perfect heat insulator (Figure 1). If the old partition is replaced by a new partition which can slide and conduct heat but does **NOT** allow the gas to leak across (Figure 2). The volume (in m^3) of the compartment A after the system attains equilibrium is: **[JEE (Advanced) 2018]**

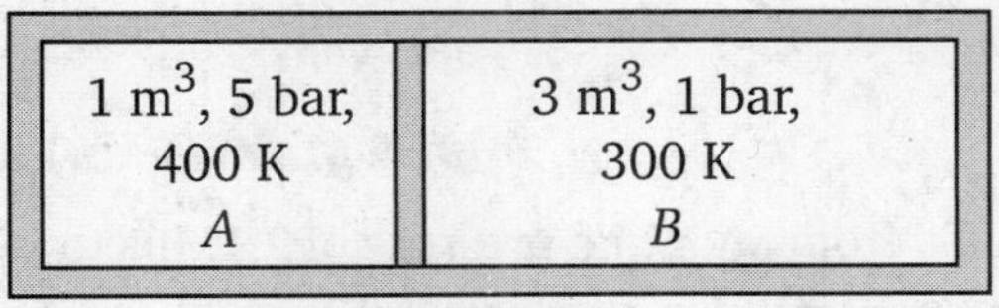

Figure 1

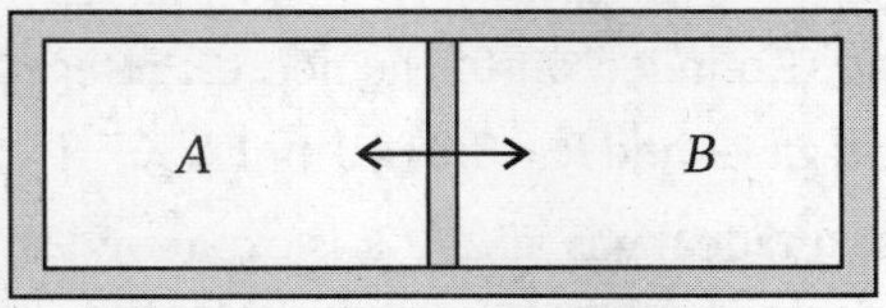

Figure 2

5. Tin is obtained from cassiterite by reduction with coke. Use the data given below to determine the minimum temperature (in K) at which the reduction of cassiterite by coke would take place.

At 298 K : $\Delta_f H°(SnO_2(s)) = -581.0$ kJ mol^{-1},

$\Delta_f H°(CO_2(g)) = -394.0$ kJ mol^{-1}

$S°(SnO_2(s)) = 56.0$ $JK^{-1}mol^{-1}$, $S°(Sn(s)) = 52.0$ J $K^{-1}mol^{-1}$,

$S°(C(s)) = 6.0$ $JK^{-1}mol^{-1}$, $S°(CO_2(g)) = 210.0$ J $K^{-1}mol^{-1}$.

Assume that the enthalpies and the entropies are temperature independent. **[JEE (Advanced) 2020]**

6. For the reaction $A(l) \longrightarrow 2B(g)$; $\Delta U = 2.1$ kcal, $\Delta S = 20$ cal K^{-1} at 300 K

Hence ΔG in kcal is: **[JEE Main (Jan.) 2020]**

7. The magnitude of work done by a gas that undergoes a reversible expansion along the path ABC shown in the figure is ———— **[JEE Main (Jan.) 2020]**

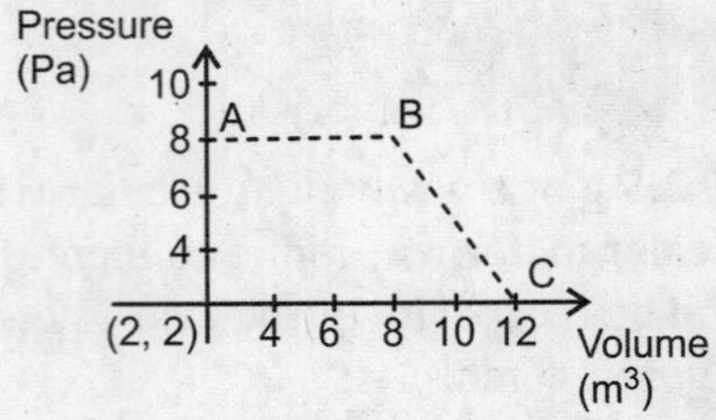

8. At constant volume, 4 mole of an ideal gas when heated from 300 K to 500 K changes its internal energy by 5000 J. The molar heat capacity at constant volume is____. **[JEE Main (Jan.) 2020]**

9. The heat of combustion of ethanol into carbon dioxides and water is –327 kcal at constant pressure. The heat evolved (in cal) at constant volume at 27°C (if all gases behave ideally) is ($R = 2$ cal mol^{-1} K^{-1}) :

[JEE Main (Sept.) 2020]

10. For a dimerization reaction, $2A(g) \longrightarrow A_2(g)$ at 298 K, $\Delta U^{\ominus} = -20$ kJ mol^{-1},

$\Delta S^{\ominus} = -30$ $JK^{-1}mol^{-1}$, then the $\Delta G^{\ominus}$ will be ______ J. **[JEE Main (Sept.) 2020]**

11. The gas phase reaction, $2A(g) \rightleftharpoons A_2(g)$ at 400 K has $\Delta G° = +25.2$ kJ mol^{-1}.

The equilibrium constant K_c for this reaction is _____ $\times 10^{-2}$. (Round off to the nearest integer)

[**Use :** $R = 8.3$ J $mol^{-1}K^{-1}$, $\ln 10 = 2.3$, $\log_{10} 2 = 0.30$, 1 atm = 1 bar]

[antilog (– 0.3) = 0.501] **[JEE Main (March) 2021]**

12. For the reaction $A(g) \rightarrow (B)(g)$, the value of he equilibrium constant at 300 K and 1 atm is equal to 100.0. The value of $\Delta_r G$ for the reaction at 300 K and 1 atm in J mol^{-1} is $-xR$, where x is _____ (Rounded off to the nearest integer) ($R = 8.31$ J $mol^{-1}K^{-1}$ and $\ln 10 = 2.3$) **[JEE Main (Feb.) 2021]**

13. The reaction of cyanamide, $NH_2CN(s)$ with oxygen was run in a bomb calorimeter and ΔU was found to be –742.24 kJ mol^{-1}. The magnitude of ΔH_{298} for the reaction

$$NH_2CN(s) + \frac{3}{2}O_2(g) \longrightarrow N_2(g) + O_2(g) + H_2O(l)$$

is ______ kJ. (Rounded off to the nearest integer)

[Assume ideal gases and $R = 8.314$ J mol^{-1} K^{-1}] **[JEE Main (Feb.) 2021]**

14. Five moles of an ideal gas at 293 K is expanded isothermally from an initial pressure of 2.1 MPa to 1.3 MPa against at constant external pressure 4.3 MPa. The heat transferred in this process is ______ kJ mol^{-1}. (Rounded-off to the nearest integer) [Use $R = 8.314$ J mol^{-1} K^{-1}] **[JEE Main (Feb.) 2021]**

15. At 25°C, 50 g of iron reacts with HCl to form $FeCl_2$. The evolved hydrogen gas expands against a constant pressure of 1 bar. The work done by the gas during this expansion is ______ J.

(Round off to the nearest integer)

[Given : $R = 8.314$ J mol^{-1} K^{-1}. Assume, hydrogen is an ideal gas]

[Atomic mass of Fe is 55.85 u] **[JEE Main (March) 2021]**

16. The standard enthalpies of formation of Al_2O_3 and CaO are –1675 kJ mol^{-1} and –635 kJ mol^{-1} respectively.

For the reaction, $3CaO + 2Al \longrightarrow 3Ca + Al_2O_3$ the standard reaction enthalpy $\Delta_r H° = ...$ kJ.

(Round off to the Nearest Integer). **[JEE Main (March) 2021]**

17. For the reaction $C_2H_6 \longrightarrow C_2H_4 + H_2$

the reaction enthalpy $\Delta_r H = ...$kJ mol^{-1}. (Round off to the nearest integer).

[**Given:** Bond enthalpies in kJ mol^{-1}: C — C: 347, C = C: 611; C — H: 414, H — H: 436]

[JEE Main (March) 2021]

18. 17.0 g of NH_3 completely vapourises at –33.42°C and 1 bar pressure and the enthalpy change in the process is 23.4 kJ mol^{-1}. The enthalpy change for the vaporisation of 85 g of NH_3 under the same conditions is ________ kJ. **[JEE Main (June) 2022 (I)]**

19. A fish swimming in water body when taken out from the water body is covered with a film of water of weight 36 g. When it is subjected to cooking at 100°C, then the internal energy for vaporisation in kJ mol^{-1} is ________ . [Nearest integer]

[Assume steam to be an ideal gas. Given $\Delta_{vap}H^{\ominus}$ for water at 373 K and 1 bar is 41.1 kJ mol^{-1}; $R = 8.31$ JK^{-1} mol^{-1}] **[JEE Main (June) 2022 (II)]**

20. When 600 mL of 0.2 M HNO_3 is mixed with 400 mL of 0.1 M NaOH solution in a flask, the rise in temperature of the flask is ______ $\times 10^{-2}$°C.

(Enthalpy of neutralisation = 57 kJ mol^{-1} and Specific heat of water = 4.2 JK^{-1} g^{-1})

(Neglect heat capacity of flask) **[JEE Main (July) 2022 (I)]**

21. When a 60 W electric heater is immersed in a gas for 100s in a constant volume container with adiabatic walls, the temperature of the gas rises by 5°C. The heat capacity of the given gas is ______ JK^{-1}. (Nearest integer) **[JEE Main (April) 2023 (I)]**

22. When 2 litre of ideal gas expands isothermally into vacuum to a total volume of 6 litre, the change in internal energy is ______ J. (Nearest integer) **[JEE Main (Jan.) 2023 (I)]**

23. 1 mole of ideal gas is allowed to expand reversibly and adiabatically from a temperature of 27°C. The work done is 3 kJ mol^{-1}. The final temperature of the gas is ______ K (Nearest integer).

Given $C_v = 20\ J\ mol^{-1}\ K^{-1}$ **[JEE Main (Jan.) 2023 (II)]**

24. Solid fuel used in rocket is a mixture of Fe_2O_3 and Al (in ratio 1 : 2). The heat evolved (kJ) per gram of the mixture is ________ (Nearest integer)

Given: $\Delta H_f^{\theta}(Al_2O_3) = -1700$ kJ mol^{-1}, $\Delta H_f^{\theta}(Fe_2O_3) = -840$ kJ mol^{-1}

Molar mass of Fe, Al and O are 56, 27 and 16 g mol^{-1} respectively. **[JEE Main (April) 2023 (I)]**

25. The value of entropy change, $S_\beta - S_\alpha$ (in J mol^{-1} K^{-1}), at 300 K is ______ .

[Use : ln 2 = 0.69

Given : $S_\beta - S_\alpha = 0$ at 0 K] **[JEE (Advanced) 2023]**

26. If 5 moles of an ideal gas expands from 10 L to a volume of 100 L at 300 K under isothermal and reversible condition then work, W is –x J. The value of x is ______ .

(Given: $R = 8.314\ JK^{-1}\ mol^{-1}$) **[JEE Main (Jan.) 2024 (I)]**

27. For a certain thermochemical reaction $M \longrightarrow N$ at $T = 400$ K, $\Delta H^{\ominus} = 77.2$ kJ mol^{-1}, $\Delta S = 122$ J K^{-1}, log equilibrium constant (log K) is – ______ $\times 10^{-1}$. **[JEE Main (Jan.) 2024]**

28. Consider the following reaction at 298 K, $\frac{3}{2}O_2(g) \rightleftharpoons O_3(g)$, $K_P = 2.47 \times 10^{-29}$.

$\Delta_r G^{\ominus}$ for the reaction is ________ kJ. (Given : $R = 8.314\ JK^{-1}\ mol^{-1}$) **[JEE Main (Jan.) 2024]**

29. The standard enthalpy and standard entropy of decomposition of N_2O_4 to NO_2 are 55.0 kJ mol^{-1} and 175.0 J/K/mol respectively. The standard free energy change for this reaction at 25°C in J mol^{-1} is .... (Nearest integer). **[JEE Main (Jan.) 2025]**

30. Standard entropies of X_2, Y_2 and XY_5 are 70, 50 and 110 $JK^{-1}mol^{-1}$ respectively. The temperature in kelvin at which the reaction.

$$\frac{1}{2}X_2 + \frac{5}{2}Y_2 \rightarrow XY_5; \Delta H^{-} = -35 \text{ kJ mol}^{-1}$$

Will be at equilibrium is .... (Nearest integer) **[JEE Main (Jan.) 2025]**

31. Consider the following cases of standard enthalpy of reaction (ΔH_r°, in kJ mol^{-1})

$$C_2H_6(g) + \frac{7}{2}O_2(g) \rightarrow 2CO_2(g) + 3H_2O(l), \quad \Delta H_1^\circ = -1550$$

$$C(\text{graphite}) + O_2(g) \rightarrow CO_2(g), \quad \Delta H_2^\circ = -393.5$$

$$H_2(g) + \frac{1}{2}O_2(g) \rightarrow H_2O(l), \quad \Delta H_3^\circ = -286$$

The magnitude of $\Delta H^\circ_{f\,C_2H_6(g)}$ is .... kJ mol^{-1} (Nearest integer). **[JEE Main (Jan.) 2025]**

32. The bond dissociation enthalpy of X_2 $\Delta H^\circ_{\text{bond}}$ calculated from the given data is .... kJ mol^{-1}. (Nearest integer)

$M^+X^-(s) \rightarrow M^+(g) + X^-(g) \Delta H^\circ_{\text{lattice}} = 800$ kJ mol^{-1}

$M(s) \rightarrow M(g)\ \Delta H^\circ_{\text{sub}} = 100$ kJ mol^{-1}

$M(g) \rightarrow M^+(g)^- + e^-(g)\ \Delta H_i^\circ = 500$ kJ mol^{-1}

$X(g) + e^-(g) \rightarrow X^-(g)\ \Delta H^\circ_{eg} = -300$ kJ mol^{-1}

$M(s) + \frac{1}{2}X_2(g) \rightarrow M^+X^-(s) \Delta H_f^\circ = -400$ kJ mol^{-1}

[Given : M^+X^- is a pure ionic compound and X forms a diatomic molecule X_2 is gaseous state.]

[JEE Main (Jan.) 2025]

33. The formation enthalpies, $\Delta H_f^\ominus$ for $H_{(g)}$ and $O_{(g)}$ are 220.0 and 250.0 kJ mol^{-1} respectively, at 298.15 K, and ΔH_f^- for $H_2O_{(g)}$ is –242.0 kJ mol^{-1} at the same temperature. The average bond enthalpy of the O—H bond in water at 298.15 K is .... kJ mol^{-1} (nearest integer). **[JEE Main (Jan.) 2025]**

34. Consider the following data :

Heat of formation of $CO_2(g) = -393.5$ kJ mol^{-1}

Heat of formation of $H_2O(l) = -286.0$ kJ mol^{-1}

Heat of combustion of benzene = –3267.0 kJ mol^{-1}

The heat of formation of benzene is .... kJ mol^{-1}.

(Nearest integer) **[JEE Main (Jan.) 2025]**

Answers

[A] Objective Questions (Only one option is correct)

1. (b) **2.** (d) **3.** (c) **4.** (a) **5.** (c) **6.** (a) **7.** (b) **8.** (a) **9.** (d) **10.** (c)
11. (a) **12.** (a) **13.** (b) **14.** (c) **15.** (a) **16.** (b) **17.** (a) **18.** (d) **19.** (b) **20.** (d)
21. (d) **22.** (c) **23.** (a) **24.** (b) **25.** (c) **26.** (b) **27.** (a) **28.** (a) **29.** (c) **30.** (a)
31. (a) **32.** (d) **33.** (b) **34.** (a) **35.** (b) **36.** (c) **37.** (a) **38.** (d) **39.** (d) **40.** (a)
41. (b) **42.** (d)

[B] Objective Questions (One or more than one option(s) is/are correct)

1. (a, d) **2.** (a, b) **3.** (a, d) **4.** (a, c) **5.** (a, b, c)
6. (a, b, c) **7.** (b, c) **8.** (a, c) **9.** (a, b, c)

[C] Matching Type Problems

1. (a)—p, r, s; (b)—r, s; (c)—t; (d)—p, q, t
2. (a)—r, t ; (b)—p, q, s ; (c)—p, q, s ; (d)—p, q, s, t
3. (A)-(II), (B)-(I), (C)-(IV), (D)-(III)

[D] Assertion and Reason

1. (b) **2.** (d) **3.** (a)

[E] Integer Type Problems

1. (9) **2.** (2) **3.** (7) **4.** (2.22) **5.** (935)
6. (–2.70) **7.** (48) **8.** (6.25) **9.** (–326400) **10.** (–13540)
11. (2) **12.** (1380) **13.** (741) **14.** (15) **15.** (2218)
16. (230) **17.** (128) **18.** (117) **19.** (38) **20.** (54)
21. (1200) **22.** (0) **23.** (150) **24.** (4) **25.** (0.31)
26. (28721) **27.** (37) **28.** (163) **29.** (2850) **30.** (700)
31. (95) **32.** (200) **33.** (466) **34.** (48)

Hints and Solutions

A. Objective Questions

[Only one option is correct]

1. (b) $Cl_{2(g)}$ is considered as standard state of Cl_2 and by assumption standard molar enthalpy of formation in standard state is taken as zero.

2. (d) Ethyne molecule breaks as

$$C_2H_{2(g)} \longrightarrow 2C_{(g)} + 2H_{(g)} \quad \text{...(i)}$$

Given equations are

$$2C_{(s)} + H_{2(g)} \longrightarrow C_2H_{2(g)}; \Delta H = 225\,\text{kJ mol}^{-1} \quad \text{...(ii)}$$

$$2C_{(s)} \longrightarrow 2C_{(g)}; \quad \Delta H = 1410\,\text{kJ mol}^{-1} \quad \text{...(iii)}$$

$$H_{2(g)} \longrightarrow 2H_{(g)}; \quad \Delta H = 330\,\text{kJ mol}^{-1} \quad \text{...(iv)}$$

Eqn. (i) can be obtained by adding equation (iii) and (iv) and then subtracting eq. (ii).

ΔH for (i) $= \Delta H_{(iii)} + \Delta H_{(iv)} - \Delta H_{(ii)}$

$= 1410 + 330 - 225 = 1515\,\text{kJ/mol}$

Equation (i) involves breaking of 2 C — H bonds and 1 C $\equiv$ C bonds, thus

ΔH of eq. (i) $= 2 \times$ bond energy of C — H + bond energy of C $\equiv$ C bond.

$1515 = 2 \times 350 +$ C $\equiv$ C bond energy

Bond energy of C $\equiv$ C $= 815\,\text{kJ/mol}$

3. (c) van der Waals' equation is

$$\left(P + \frac{a}{V^2}\right)(V - b) = RT$$

As given that $b = 0$

$$PV + \frac{a}{V} = RT \quad \text{or} \quad PV = RT - \frac{a}{V}$$

Comparing with $y = mx + c$

Intercept $(c) = RT$

Slope $(m) = -a$

$$\text{Slope} = \frac{y_2 - y_1}{x_2 - x_1} = \frac{20.1 - 21.6}{3 - 2} = -1.5$$

Thus, $\quad a = +1.5$

4. (a) $q = +208$ J (as it absorbs heat)

$$w_{\text{Rew}} = -2.303nRT \log_{10}\left(\frac{V_2}{V_1}\right)$$

$$= -2.303 \times (0.04) \times 8.314 \times (310) \log_{10}\left(\frac{375}{50}\right) = -208\text{ J}$$

5. (c) $C_6H_{12}O_6(s) + 6O_2(g) \longrightarrow 6CO_2(g) + 6H_2O(l)$

$$\Delta H = 6 \times (-400) + 6(-300) - (-1300)$$

$$= -2900\text{ kJ/mol} = -16.11\text{ kJ/g}$$

6. (a) $C_2H_5OH(l) + 3O_2(g) \longrightarrow 2CO_2(g) + 3H_2O(l)$

Bomb calorimeter gives ΔU of the reaction

but given that $\Delta H = \Delta U + \Delta n_g RT$

$$\Delta U = -1364.47 \text{ kJ mol}^{-1}$$

$$\Delta n_g = -1$$

$$= -1364.47 - \frac{1 \times 8.314 \times 298}{1000} = -1366.93 \text{ kJ mol}^{-1}$$

7. (b) $H_2O(l) \rightleftharpoons H_2O(v)$

Under these condition, system is in equilibrium, so

$$\Delta S_{\text{Total}} = 0$$

$$\Delta S_{\text{System}} + \Delta S_{\text{Surroundings}} = 0$$

$$\Delta S_{\text{System}} = -\Delta S_{\text{Surroundings}}$$

Hence $\Delta S_{\text{System}} > 0$; $\Delta S_{\text{Surroundings}} < 0$

8. (a) At large inter-ionic distances (because $a \to 0$) the P.E. would remain constant. However, when $r \to 0$; repulsion would suddenly increase.

9. (d) The required equation is

$$C_{(s)} + \frac{1}{2}O_{2(g)} \longrightarrow CO_{(g)}; \Delta H_f = ?$$

Given that :

$$C_{(s)} + O_{2(g)} \longrightarrow CO_{2(g)}; \Delta H_1 = -393.5 \text{ kJ mol}^{-1} \quad \text{...(i)}$$

$$CO_{(g)} + \frac{1}{2}O_{2(g)} \longrightarrow CO_{2(g)}; \Delta H_2 = -283.5 \text{ kJ mol}^{-1}$$

...(ii)

Subtracting eqn. (ii) from eqn. (i) will give the required equation.

$$\Rightarrow \quad \Delta H_f = \Delta H_1 - \Delta H_2 = -393.5 - (-283.5) = -110 \text{ kJ mol}^{-1}$$

10. (c) For isothermal expansion, $\Delta U = 0$

As pressure is constant therefore, process is irreversible.

$$\Rightarrow q_{\text{irrev}} = -w_{\text{irrev}} = -(-P\Delta V) = -[-3(2-1)]$$

$$= 3 \text{ L atm} = 3 \times 101.3 \text{ J}$$

$$\Delta S_{\text{surr}} = \frac{-q_{\text{irrev}}}{T} = -\frac{3 \times 101.3}{300 \text{ K}} = -1.013 \text{ J K}^{-1}$$

11. (a) $C_{(\text{graphite})} + O_{2(g)} \longrightarrow CO_{2(g)}; \quad \Delta_r H° = -393.5 \text{ kJ mol}^{-1}$ ...(i)

$H_{2(g)} + \frac{1}{2}O_{2(g)} \longrightarrow H_2O_{(l)}; \quad \Delta_r H° = -285.8 \text{ kJ mol}^{-1}$ ...(ii)

$CO_{2(g)} + 2H_2O_{(l)} \longrightarrow CH_{4(g)} + 2O_{2(g)}; \quad \Delta_r H° = +890.3 \text{ kJ mol}^{-1}$ ...(iii)

$C_{(\text{graphite})} + 2H_{2(g)} \longrightarrow CH_{4(g)}; \Delta_r H° = ?$

On applying the mathematical operation,

(i) + (ii) × 2 + (iii), we get

$$C_{(\text{graphite})} + O_{2(g)} \longrightarrow CO_{2(g)}; \quad \Delta_r H° = -393.5 \text{ kJ mol}^{-1}$$

$$2H_{2(g)} + O_{2(g)} \longrightarrow 2H_2O_{(l)}; \quad \Delta_r H° = -571.6 \text{ kJ mol}^{-1}$$

$$CO_{2(g)} + 2H_2O_{(l)} \longrightarrow CH_{4(g)} + 2O_{2(g)}; \quad \Delta_r H° = +890.3 \text{ kJ mol}^{-1}$$

$$C_{(\text{graphite})} + 2H_{2(g)} \longrightarrow CH_{4(g)};$$

$$\Delta_r H° = [-393.5 + (-571.6) + 890.3] \text{ kJ mol}^{-1} = -74.8 \text{ kJ mol}^{-1}$$

12. (a) According to 1st law of thermodynamics

$$\Delta U = q + w \quad \text{....(i)}$$

where, ΔU = change in internal energy

q = heat, w = work done

For adiabatic process, $q = 0$

$\therefore \quad \Delta U = w$

i.e., change in internal energy is equal to adiabatic work.

13. (b) $C_6H_6 + \frac{15}{2}O_2(g) \longrightarrow 6CO_2(g) + 3H_2O$

$$\Delta n_g = -\frac{3}{2}$$

$$\Delta H = \Delta U + \Delta n_{(g)}RT = -3263.9 - \frac{1.5 \times 8.314 \times 298}{1000}$$

$$= -3267.6 \text{ kJ/mol}$$

14. (c) $\Delta H = nC_p\Delta T = 3(23 + 0.01T)\Delta T$

ΔH for heating 3 moles of silver from 300 K (T_1) to 1000 K (T_2)

$$\Delta H = \int_{T_1}^{T_2} 3(23 + 0.01T)\Delta T = 3\left(23T + 0.01\frac{T^2}{2}\right)_{T_1}^{T_2}$$

$$= 3\left[\left(23 \times 1000 + 0.01 \times \frac{1000 \times 1000}{2}\right) - \left(23 \times 300 + 0.01 \times \frac{300 \times 300}{2}\right)\right]$$

$$= 3[(23000 + 5000) - (6900 + 450)]$$

$$= 3 \times 20650 = 61950 \text{ J} \approx 62 \text{ kJ}$$

15. (a) According to first law of thermodynamics

$$\Delta U = q + w$$

For adiabatic process, $q = 0$

$$\Delta U = w$$

For isochoric process $\Delta V = 0, w = 0$

thus, $\quad \Delta U = q$

For isothermal process, $\Delta U = 0$, thus $q = -w$

For cyclic process, $\Delta U = 0$; thus $q = -w$

16. (b) Number of moles $(n) = 5$

$T_1 = 100\text{ K}, T_2 = 200\text{ K}, C_v = 28 \text{ JK}^{-1}\text{mol}^{-1}$

$R = 8.0 \text{ JK}^{-1}\text{mol}^{-1}$

We know, $\Delta U = nC_v\Delta T = 5 \times 28 \times 100 = 14000$ J or 14 kJ

$\Delta(PV) = nR\Delta T = 5 \times 8 \times 100 = 4000$ J or 4 kJ

17. (a) Since C_P and C_V are functions of temperature only, they don't depend over P and V.

Graph (a) is incorrect.

Also internal energy of an ideal gas depends over temperature.

18. (d) $W = -nRT \ln \frac{V_f}{V_i}$

$|W| = nRT \ln \frac{V_f}{V_i}$

$|W| = nRT \ln V - nRT \ln V_i$

$|W| Vs \ln V$: straight line of form $y = mx + C$

Slope $(m) = nRT$

Intercept $(C) = -nRT \ln V_i$

$\because \quad T_2 > T_1$

Correct graphical representation option is (d).

19. (b) For ideal gas, $\Delta U = nC_{V,m}\Delta T$

$$= 5 \times 28 \times (200 - 100)\ \text{J} = 14000\ \text{J} = 14\ \text{kJ}$$

$$\Delta H = nC_{P,m}\Delta T$$

$$C_{P,m} = C_{V,m} + R = (28 + 8) = 36\ \text{J/mol/K}$$

$$\Delta H = 5 \times 36 \times 100\ \text{J} = 18000\ \text{J} = 18\ \text{kJ}$$

$$\Delta H = \Delta U + \Delta(PV)$$

$$\Delta(PV) = \Delta H - \Delta U = 18 - 14 = 4\ \text{kJ}$$

20. (d) Let final equilibrium temperature $= T$

Heat loss = heat gain.

$$C_p(T_2 - T) = C_p(T - T_1)$$

$$T = \frac{T_1 + T_2}{2}$$

$$\Delta S_{\text{Total}} = \Delta S_1 + \Delta S_2 = C_p \ln\frac{T}{T_1} + C_p \ln\frac{T}{T_2}$$

$$= C_p \ln\frac{T^2}{T_1T_2} = C_p \ln\left[\frac{(T_1 + T_2)^2}{4T_1T_2}\right]$$

21. (d)

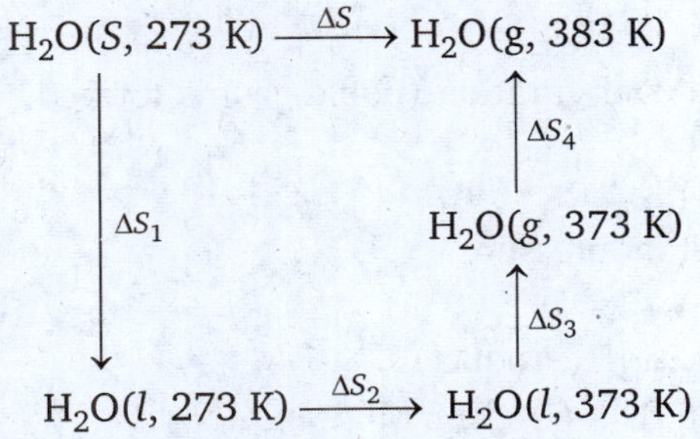

$$\Delta S = \Delta S_1 + \Delta S_2 + \Delta S_3 + \Delta S_4$$

$$\Delta S_1 = \frac{\Delta H_{\text{Fu.}}}{T_{\text{Fu.}}} = \frac{344 \times 1}{273}\ \text{kJ/K} = 1.26\ \text{kJ/kg-K}$$

$$\Delta S_2 = ms \ln\frac{T_f}{T_i} = 4.2 \times 1 \ln\left(\frac{373}{273}\right) = 1.31\ \text{kJ/kg-K}$$

$$\Delta S_3 = \frac{\Delta H_{\text{vap.}}}{T_b} = \frac{2491 \times 1}{373} = 6.68\ \text{kJ/Kg-K}$$

$$\Delta S_4 = ms \ln\frac{T_f}{T_i} = 2 \times 1 \ln\left(\frac{383}{373}\right) = 0.052\ \text{kJ/kg-K}$$

$$\Delta S = 1.26 + 1.31 + 6.68 + 0.052 = 9.3\ \text{kJ/kg-K}$$

22. (c) In vacuum, $P_{\text{ext}} = 0$

$$W = -P_{\text{ext}}(V_f - V_i)$$

$$W = 0$$

23. (a) For ideal gas

$$\Delta H = nC_{P,m}\Delta T$$

$$\Delta U = nC_{V,m}\Delta T$$

So, U and H are functions of temperature.

$Z = \dfrac{PV}{nRT} = 1$ (Always for ideal gas)

$C_{P,m} - C_{V,m} = R$ (Always for ideal gas)

24. (b) $dS = \int \frac{dq_{revs.}}{T}$

Both dS and S are functions of temperature.

25. (c) From van't Hoff equation

$$\ln \frac{K_2}{K_1} = \frac{\Delta H^\circ}{R}\left(\frac{1}{T_1} - \frac{1}{T_2}\right)$$

$$\cdot \Delta H^\circ = R \times \frac{T_1 T_2}{(T_2 - T_1)} \times \ln \frac{K_2}{K_1} = \frac{8.314 \times 298 \times 373}{1000 \times 75} \times 2.303$$

$$= 28.4 \text{ kJ/mol}$$

$$\Delta G^\circ = -RT \ln K = -2.303 RT \log K$$

at T_1, $\Delta G_1^\circ = -2.303 \times 8.314 \times 10^{-3} \times 298 \log(10) = -5.71$ kJ/mol

at T_2, $\Delta G_2^\circ = -2.303 \times 8.314 \times 10^{-3} \times 373 \log 100 = -14.276$ kJ/mol

26. (b)

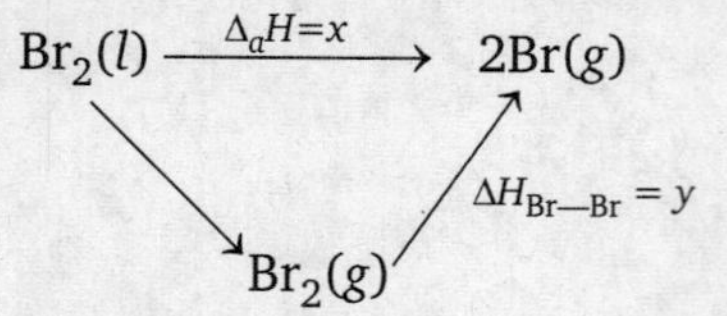

From Hess's law: $x = \Delta H_{vap.}(Br_2, l) + y$

$$x - y = \Delta H_{vap.}(Br_2, l) > 0$$

$$x > y$$

27. (a) $H(g) + e^- \longrightarrow H^-(g)$

During this process stable configuration (noble gas) is formed so, heat is released while other given processes are endothermic.

28. (a) $H = U + PV$ (By definition)

$\Delta H = \Delta U + \Delta(PV)$ at constant pressure

$\Delta H = \Delta U + P\Delta V$

29. (c) $\Delta_r H = \Sigma \Delta_c H(\text{Reactant}) - \Sigma \Delta_c H(\text{Product})$

$= 3 \times (-1300) - (-3268) = -632 \text{ kJ mol}^{-1}$

30. (a) Adiabatic boundary does not allow heat exchange thus heat generated in container can't escape out thereby increasing the temperature.

In case of diathermic container, heat flow can occur to maintain the constant temperature.

31. (a) $C(\text{graphite}) + \frac{1}{2}O_2(g) \longrightarrow CO(g)$; ... (i) $\quad \Delta H$

$C(\text{graphite}) + O_2(g) \longrightarrow CO_2(g)$; ... (ii) $\quad \Delta H_1 = -y$ kJ/mole

$CO_2(g) \longrightarrow CO(g) + \frac{1}{2}O_2(g)$; ... (iii) $\quad \Delta H_2 = \frac{x}{2}$ kJ/mole

eq. (i) = eq. (ii) + eq. (iii) $\quad \therefore \Delta H = \frac{x}{2} - y = \frac{x - 2y}{2}$

32. (d) During free expansion of an ideal gas under adiabatic condition $q = 0, \Delta T = 0, w = 0$.

33. (b) $(\Delta G)_{P,T} = (+)$ ve for non-spontaneous process

34. (a) Thermally insulated $\Rightarrow q = 0$

from I[st] law, $\Delta U = q + w$

$\Delta U = w$

$w > 0, \Delta U > 0$

35. (b) Ice → Ice ⇌ Water → Water ⇌ Water vapour → Water vapour

268 K (1) 273 K (2) 273 K (3) 273 K (4) 373 K (5) 383 K

$$\Delta S_{overall} = \Delta S_1 + \Delta S_2 + \Delta S_3 + \Delta S_4 + \Delta S_5$$

$$\Delta S_2 = \frac{\Delta H_{m,\text{fusion}}}{273} \qquad T_f = 273\,\text{K}$$

$$\Delta S_3 = \int_{273}^{373} \frac{C_{p,m}\,dT}{T}$$

$$\Delta S_4 = \frac{\Delta H_{m,\text{vaporisation}}}{373} \qquad T_b = 373\,\text{K}$$

$$\Delta S_5 = \int_{373\text{K}}^{383\text{ K}} \frac{C_{p,m}\,dT}{T}$$

36. (c) $\because \Delta G = \Delta H - T\Delta S$

For spontaneity of reaction : $\Delta G = -\text{ve}$

37. (a) Reaction is spontaneous at relatively high temperature and non-spontaneous at low temperature

$\Delta G = \Delta H - T\Delta S$

It is only possible when ΔH and ΔS both are positive.

38. (d)

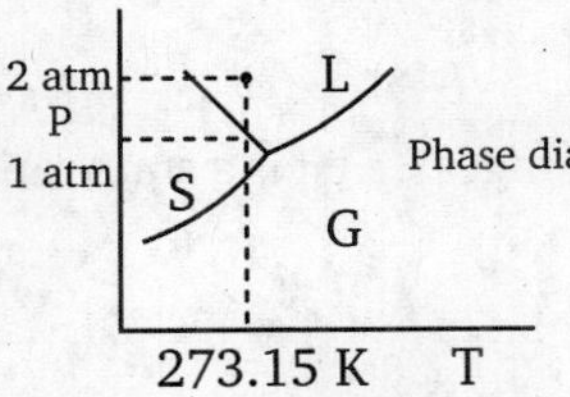

Phase diagram of H_2O

If pressure is made two times then mixture of ice and water will completely convert into water (liquid) form.

39. (d) As internal energy 'U' is a state function, its cyclic integral must be zero in a cyclic process.

$\therefore\ \Delta U$ case (I) $= \Delta U$ case (II) $= \Delta U$ case (III)

40. (a) $q_p = n \times c_p \times \Delta T$

$\Rightarrow 500 = 0.5 \times \frac{5}{2} \times 8.3\,(T_f - 298)$

$\Rightarrow T_f \simeq 346.2\text{ K}$

$\frac{\Delta H}{\Delta U} = \frac{C_p}{C_v} = \left(\frac{5}{3}\right)$

$\Rightarrow \Delta U = \frac{3}{5} \times 500 = 300\text{ J}$

41. (b) $\underset{(g)}{SO_2} + \frac{1}{2}\underset{(g)}{O_2} \longrightarrow \underset{(g)}{SO_3} \qquad \Delta H = -y$

$\Delta H_r = (\Delta H_f)_{SO_3} - (\Delta H_f)_{SO_2}$

$-y = -2x - (\Delta H_f)_{SO_2}$

$(\Delta H_f)_{SO_2} = y - 2x$

B. Objective Questions

[One or more than one option(s) is/are correct]

1. (a, d)

Work in path function whether it is reversible or irreversible.

2. (a, b)

(c) and (d) depends upon the size of system.

3. (a, d)

For isothermal expansion,

$\Delta T =$ constant, *i.e.*, $T_1 = T_2$

At constant temperature, internal energy (ΔU) of the system remains constant *i.e.*,

$\Delta U_{\text{isothermal}} = 0$

Reversible expansion work $= -2.303nRT \log \frac{V_2}{V_1}$

In adiabatic expansion, no heat is allowed to enter or leave the system, hence $q = 0$.

$\Delta E = q + w$. $\because q = 0$

$\Delta E = w$

In expansion, work is done by the system, hence w is negative, accordingly, $\Delta E = -\text{ve}$, *i.e.*, internal energy decreases and therefore the temperature of the system falls.

Thus, $T_3 < T_1$

$\Delta U_{\text{adiabatic}} = -\text{ve}$

Thus, $\Delta U_{\text{isothermal}} > \Delta U_{\text{adiabatic}}$

$w_{\text{adiabatic}} = \frac{R}{(\gamma - 1)}(T_2 - T_1)$

Thus, including sign, $w_{\text{isothermal}} < w_{\text{adiabatic}}$

4. (a, c)

As ΔS does not depend on path and only depends on initial and final stages *i.e.*, it is a state function thus

$$\Delta S_{X \to Z} = \Delta S_{X \to Y} + \Delta S_{Y \to Z}$$

and $\Delta S_{Y \to Z}$ is not zero thus

$$\Delta S_{X \to Y \to Z} \neq \Delta S_{X \to Y}$$

As we know that work is not a state function and depends on path,

Thus, $w_{X \to Z} = w_{X \to Y} + w_{Y \to Z}$

$w_{X \to Y} = PdV$ (*P* is constant)

$w_{Y \to Z} = 0$ (*V* is constant)

$$w_{X \to Z} = 2.303nRT \log \frac{V_2}{V_1}$$

$$w_{X \to Y \to Z} = w_{X \to Y} + w_{Y \to Z}$$

As $w_{Y \to Z} = 0$, hence $w_{X \to Y \to Z} = w_{X \to Y}$

5. (a, b, c)

Since vessel is isothermally insulated *i.e.*, the process is adiabatic hence $q = 0$.

Also $P_{\text{ext}} = 0$ hence $w = 0$

From 1st law of thermodynamics $\Delta E = q \times w$

$\therefore \quad \Delta E = 0$ (for ideal gas)

$\therefore \quad \Delta T = 0 \quad$ or $\quad T_2 = T_1$

Applying $PV = nRT$ where n_2R and T are constant then $P_1V_1 = P_2V_2$

PV^{γ} = constant is applicable to ideal gas in reversible adiabatic process. Hence $P_2V_2^{\gamma} = P_1V_1^{\gamma}$ is not applicable.

7. (b, c)

(a) $q_{AC} = U_{BC} = nc_V(T_2 - T_1)$

$w_{AB} = nRT_1 \ln \frac{V_2}{V_1}$

(b) $q_{BC} = \Delta H_{AC} = nc_P(T_2 - T_1)$ (Since BC is isobaric process)

$w_{BC} = -P_2(V_1 - V_2)$

(c) $nc_P(T_1 - T_2) < nc_V(T_1 - T_2)$

$\therefore \Delta H_{CA} < \Delta U_{CA}$

(d) It is opposites to option (c).

8. (a, c)

The standard enthalpy of formation of a compound is the change in the standard enthalpy when one mole of the compound is formed from the requisite amounts of elements in their stable state of aggregation.

9. (a, b, c)

For 1 mole van der Waals' gas

$$P = \frac{RT}{V-b} - \frac{a}{V^2}$$

If $P_{ext} = P$, means process is reversible. For van der Waals' gas, expression is correct for all reversible process.

C. Matching Type Problems

3. (b) (A) $dG = VdP - SdT$

Constant pressure

$dG = -SdT$

$$\left(\frac{\partial G}{\partial T}\right)_P = -S$$

(B) $dH = (dq)_P = nC_p dT$

$$\left(\frac{\partial H}{\partial T}\right)_P = C_P$$

(C) $dG = VdP - SdT$

At constant temperature

$dG = VdP$

$$\left(\frac{\partial G}{\partial P}\right)_T = V$$

(D) $dU = nC_V dT = (q)_V$

$$\left(\frac{\partial U}{\partial T}\right)_V = C_V$$

D. Assertion and Reason

1. (b) $W = -P_{ext.} \cdot \Delta V$

When gas expands against vacuum , then $P_{ext.} = 0$, so work done is zero.

Theoretically at $T = 0\,K$, $V = 0$ (Charles, law), however none of the gases known is ideal.

2. (d) ΔG° is not necessarily zero and these reactions are spontaneous for which $\Delta G = -$ ve.

3. (a) There is natural asymmetry between converting work into heat and vice-versa because of the lapse of energy in this process which is generalized as the statement of II law of thermodynamics.

E. Integer Type Problems

1. (9) $\because$ $C_P - C_V = R$

$\therefore$ $C_P = C_V + R$

$C_P = 2500 + 8.314$ $(\because C_V = 2.5\,kJK^{-1} = 2500\,JK^{-1})$

$= 2508.314\ JK^{-1}$

$q_p = C_P \times \Delta T$ $(\because \Delta T = 298.45 - 298 = 0.45)$

$$= 2508.314 \times 0.45 = 1128.74\ \text{J}$$

$$\Delta H = \frac{q_p}{n} = \frac{1128.74}{3.5/28} = 9029.92\ \text{Jmol}^{-1} = 9.0299\ \text{kJmol}^{-1}$$

2. (2)

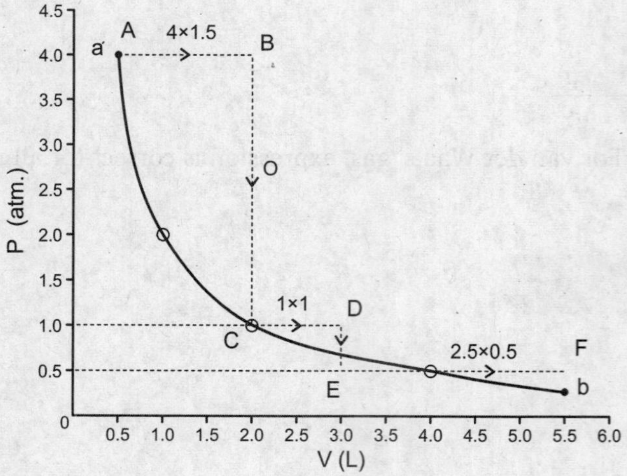

$$w_d = w_{AB} + w_{CD} + w_{EF}$$
$$= 4 \times 1.5 + 1 \times 1 + 0.5 \times 2.5 = 8.25\ \text{L} \cdot \text{atm}$$

$$w_s = 2.303\, nRT \log \frac{V_2}{V_1} = 2.303\, PV \log \frac{V_2}{V_1} \qquad (\because PV = nRT)$$

$\because$ $P = 4$ atm., $V = 0.5$ L, $V_1 = 0.5$ L, $V_2 = 5.5$ L

$$w_s = 2.303 \times 4 \times 0.5 \log \frac{5.5}{0.5} = 4.79\ \text{L} \cdot \text{atm.}$$

or $$\frac{w_d}{w_s} = \frac{8.25}{4.79} = 1.72 \approx 2$$

3. (7) For any ideal gas, $PV = nRT$

$$0.32 \times V = 0.1 \times 0.0821 \times 273$$
$$V = 7\ \text{litre}$$

(unknown compound X will not follow ideal gas equation)

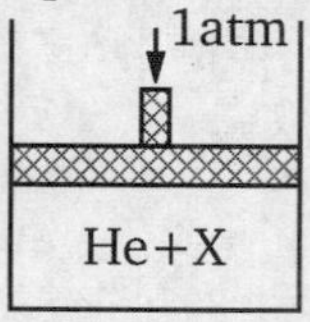

For He, $n = 0.1, P = 0.32$ atm., $V = ?, T = 273$

4. (2.22)

$t = 0$ $\quad A$ 1 m^3, 5 bar, 400 K $\quad B$ 3 m^3, 1 bar, 300 K

$t = t$ $\quad A$ $(1 + x)$ m^3, a bar, b K $\quad B$ $(3 - x)$ m^3, a bar, b K

Since no. of moles in side A will remain same

$$\frac{P_i V_i}{RT_i} = \frac{P_f V_f}{RT_f}$$

$$\frac{5 \times 1}{R \times 400} = \frac{(1 + x) \times a}{R \times b}$$

$$(1 + x)\, a = \frac{5}{400} b \qquad \ldots(1)$$

Similarly, $$\frac{3 \times 1}{R \times 300} = \frac{(3 - x) \times a}{R \times b}$$

$$(3-x)\,a=\frac{1}{100}b \quad ...(2)$$

(1) ÷ (2)

$$\frac{1+x}{3-x}=\frac{5}{4}$$

$$4+4x=15-5x$$

$$9x=11 \Rightarrow x=\frac{11}{9}$$

$\therefore$ Vol. (side A) $=1+x=1+\frac{11}{9}=2.22$

5. (935.00) $SnO_{2(s)}+C_{(s)}\longrightarrow Sn_{(s)}+CO_{2(g)}$

$$\Delta H^\circ_{rxn}=[-394]-[-581]=187\text{ kJ/mol}$$

$$\Delta S^\circ_{rxn}=[52+210]-[56+6]=200\text{ J/K-mol}$$

$$T=\frac{\Delta H^\circ}{\Delta S^\circ}=\frac{187\times 1000}{200}=935\text{ K}$$

6. (–2.70) $A(l)\longrightarrow 2B(g)$

$$\Delta H=\Delta U+(\Delta n_g)RT$$

$$=2.1+\frac{(2)\times 2}{1000}\times 300=3.3\text{ kcal/mol}$$

$$\Delta G=\Delta H-T\Delta S$$

$$=3.3-\frac{300\times 20}{1000}=-2.70\text{ kcal/mol}$$

7. (48)

Magnitude of work done = Area enclosed by curve

$$=\frac{1}{2}\times(10+6)\times 6=48\text{ J}$$

8. (6.25)

For an ideal gas, $\Delta U=nC_{V,m}\,\Delta T$

$$C_{V,m}=\frac{\Delta U}{n\times\Delta T}=\frac{5000}{4\times 200}=6.25\text{ J/mol-K}$$

9. (–326400)

$$C_2H_5OH(l)+3O_2(g)\longrightarrow 2CO_2(g)+3H_2O(l)$$

$$\Delta n_g=2-3=-1$$

$\Rightarrow$ Heat of reaction at constant pressure $(q_p)=\Delta H$

$\Rightarrow$ Heat of reaction at constant volume $(q_V)=\Delta U$

$$\Rightarrow \quad \Delta_c H=\Delta_c U+(\Delta n_g)RT$$

$$\Delta_c U=\Delta_c H+RT$$

$$=-327\times 1000+2\times 300=-326400\text{ cal}$$

10. (–13540) $2A(g)\longrightarrow A_2(g)$

$$\Delta n_g=1-2=-1$$

$$\Delta H^\circ=\Delta U^\circ+(\Delta n_g)RT=-20+(-1)\times\frac{8.314}{1000}\times 298=-22.48\text{ kJ/mol}$$

$$\Delta G^\circ=\Delta H^\circ-T\,\Delta S^\circ=-22.48-298\times\left(\frac{-30}{1000}\right)$$

$$=-13.54\text{ kJ/mol}=-13540\text{ J/mol}$$

11. (2) $2A_{(g)}\rightleftharpoons A_2(g)$

$$\Delta G^\circ_r=-RT\ln K_p$$

$$25200 = -2.3 \times 8.3 \times 400 \log K_p$$
$$K_p = 5.01 \times 10^{-4} \text{ bar}^{-1}$$
$$K_p = 5.01 \times 10^{-9} \text{ Pa}^{-1}$$
$$K_p = K_c(RT)^{\Delta n_g}$$
$$\Delta n_g = -1$$
$$K_c = K_p \times RT$$
$$K_c = 5.01 \times 10^{-9} \times 8.3 \times 400$$
$$K_c = 1.66 \times 10^{-5} \text{ m}^3/\text{mol} = 1.66 \times 10^{-2} \text{ L/mol}$$

12. (1380) $K_p = 100$

$$\Delta_r G° = -RT \ln K_p = -R(300)\ln(100) = -2R \times 300 \times \ln 10$$
$$= -2R \times 300 \times 2.3 = -1380\ R$$
$$x = 1380$$

13. (741)

$$NH_2CN(s) + \frac{3}{2}O_2(g) \longrightarrow N_2(g) + O_2(g) + H_2O(l)$$
$$\Delta n_g = \left(1 + 1 - \frac{3}{2}\right) = \frac{1}{2}$$
$$\Delta H = \Delta U + (\Delta n_g)RT = -742.24 + \frac{1}{2} \times \frac{8.314}{1000} \times 298 = -741 \text{ kJ/mol}$$

14. (15) $W = -P_{\text{ext}}(V_f - V_i) = -P_{\text{ext}}\left(\frac{nRT}{P_2} - \frac{nRT}{P_1}\right)$

$$= -4.3\left(\frac{5 \times 293R}{1.3} - \frac{5 \times 293\,R}{2.1}\right) = -5 \times 293 \times 8.314 \times 43 \times \left(\frac{1}{13} - \frac{1}{21}\right)$$
$$= -15347.7 \text{ J} = -15.34 \text{ kJ}$$

For isothermal process, $\Delta U = 0$

$$q = -W = 15.34 \text{ kJ}$$

15. (2218)

Reaction is $Fe(s) + 2HCl(aq) \longrightarrow FeCl_2(aq) + H_2(g)$; $\Delta n_g = 1$

$\Rightarrow$ moles of iron $= \frac{50}{55.85} = 0.895$ mol

So, moles of H_2 produced $= 0.895$ mol

Work done by 1 mol of $H_2(w) = P_{\text{ext}} \cdot \Delta V = (\Delta n_g)RT = 1 \times 8.314 \times 298$ J

Work done by 0.895 moles of $H_2 = 0.895 \times 8.314 \times 298 \text{ J} = 2218.06$ J

16. (230) For reaction : $3CaO_{(s)} + 2Al_{(s)} \longrightarrow 3Ca_{(s)} + Al_2O_{3(s)}$

$$\Delta_r H° = \Delta_f H° \text{ (products)} - \Delta_f H° \text{ (Reactants)}$$
$$= \Delta_f H°(Al_2O_3) - 3 \times \Delta_f H°(CaO) = -1675 - 3(-635) = 230 \text{ kJ/mol.}$$

17. (128)

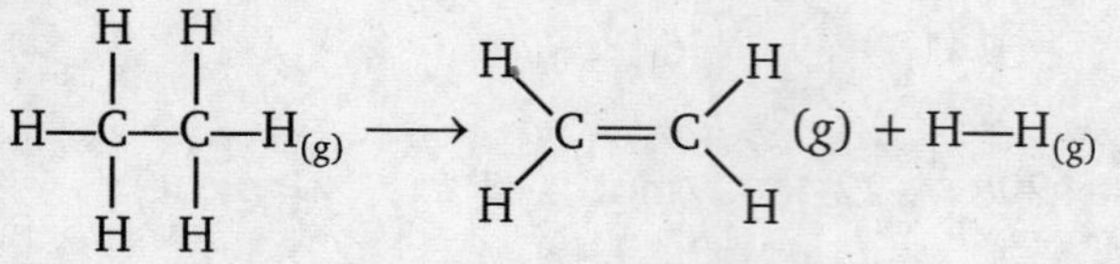

$$\Delta_r H = \Sigma \text{B.E (Reactant)} - \Sigma \text{B.E (Product)}$$
$$= (\Delta H_{C-C} + 6 \times \Delta H_{C-H}) - (\Delta H_{C=C}) + 4 \times \Delta H_{C-H} + \Delta H_{H-H})$$
$$= 347 + 2 \times 414 - 611 - 436 = 128 \text{ kJ/mol}$$

18. (117)

Molar mass of NH_3 = 17 g/mol

moles of NH_3 given $= \frac{17}{17} = 1$ mole.

For 1 mole enthalpy change = 23.4 kJ
5 moles enthalpy change = 23.4 × 5 = 117 kJ

19. (38)

$$H_2O(l) \longrightarrow H_2O(g)$$

$$\Delta n_g = \Sigma np - \Sigma nR = 1 - 0 = 1$$

$$\Delta H = \Delta U + \Delta n_g RT$$

$$\Delta U = \Delta H - \Delta n_g RT$$

$$= 41.1 - \frac{1 \times 8.31 \times 373}{1000} \text{ kJ/mol} = 38 \text{ kJ/mol}$$

20. (54)

HNO_3	NaOH
600 mL × 0.2 M	400 mL × 0.1 M
= 120 m mol	= 40 m mol

$$HNO_3 + NaOH \longrightarrow NaNO_3 + H_2O$$

	HNO_3	NaOH	$NaNO_3$
Start	120	40	
End	80	0	40 m mol

$$\Delta_r H = 40 \text{ m mol} \times (57 \times 10^3) \text{ J mol}^{-1}$$

$$= 40 \times 10^{-3} \text{mol} \times 57 \times 10^3 \text{ J mol}^{-1} = 2280 \text{ J}$$

$$mS\Delta T = 2280 \quad \Rightarrow \quad 1000 \text{ g} \times 4.2 \times \Delta T = 2280$$

$$\Delta T = \frac{2280}{4.2} \times 10^{-3} = \frac{22800}{42} \times 10^{-3} = 542.86 \times 10^{-3}$$

$$= 54.286 \times 10^{-2} \text{K} = 54.286 \times 10^{-2} \,^\circ\text{C}$$

21. (1200)

Power of heater = 60 W
Total energy emitted = 60 × 100 = 6000 J
Heat capacity × temperature difference = 6000

Heat capacity $= \frac{6000}{5} = 1200 \text{ J K}^{-1}$

22. (0) For an, ideal gas internal energy is a function of temperature and for isothermal process temperature is constant.

∴ The change in internal energy is zero.

23. (150)

$q = 0, \quad \Delta U = w$

$$1 \times 20 \times [T_2 - 300] = -3000$$

$$T_2 - 300 = -150 \quad \Rightarrow \quad T_2 = 150 \text{ K.}$$

24. (4)

$$Fe_2O_3 + 2Al \longrightarrow Al_2O_3 + 2Fe$$

Molar mass : 160 g 27 g

$$(\Delta H_f^\circ)_{\text{reaction}} = [(\Delta H_f^\circ)_{Al_2O_3} + 2(\Delta H_f^\circ)_{Fe}] - [(\Delta H_f^\circ)_{Fe_2O_3} + 2(\Delta H_f^\circ)_{Al}]$$

$$= [-1700 + 0] - [-840 + 0] = -860 \text{ kJ/mol}$$

Total mass of mixture = Fe_2O_3 + Al (1 : 2 molar ratio)
= (160) + (2 × 27) = 214 g/mol

Heat evolved per gram $= \frac{860}{214} = 4$ kJ/g

25. (0.31)

$$\alpha \rightarrow \beta \text{ and } S^0_{\beta(600)} - S^0_{\alpha(600)} = 6 - 5 = 1 \text{ (from graph)}$$

$$S^0_{\alpha(600)} - S^0_{\alpha(300)} + C_{P(\alpha)} \ln \frac{600}{300}$$

$$S^0_{\beta(600)} = S^0_{\beta(300)} + C_{P(\beta)} \ln \frac{600}{300}$$

$$S^0_{\beta(600)} - S^0_{\alpha(600)} = S^0_{\beta(300)} - S^0_{\alpha(300)} + C_{P(\beta)} - C_{P(\alpha)} \ln 2$$

$$6 - 5 = S^0_{\beta(300)} - S^0_{\alpha(300)} + [1 \times \ln 2]$$

$$1 = S^0_{\beta(300)} - S^0_{\alpha(300)} + 0.69$$

So, $S^0_{\beta(300)} - S^0_{\alpha(300)} = 0.31\ \text{J mol}^{-1}\ \text{K}^{-1}$

26. (28721)

It is isothermal reversible expansion, so work done negative

$$W = -2.303\ nRT \log\left(\frac{V_2}{V_1}\right)$$

$$= -2.303 \times 5 \times 8.314 \times 300 \log\left(\frac{100}{10}\right)$$

$$= -28720.713\ \text{J} = -28721\ \text{J}$$

27. (37) $\Delta G^\circ = \Delta H^\circ - T\Delta S^\circ$

$$= 77.2 \times 10^3 - 400 \times 122 = 28400\ \text{J}$$

$$\Delta G^\circ = -2.303\ RT \log K$$

$\Rightarrow$ $28400 = -2.303 \times 8.314 \times 400 \log K$

$\Rightarrow$ $\log K = -3.708 = -37.08 \times 10^{-1}$

28. (163)

$$\frac{3}{2} O_2(g) \rightleftharpoons O_3(g)\ ; \quad K_p = 2.47 \times 10^{-29}$$

$$\Delta_r G^\ominus = -RT \ln K_P$$

$$= -8.314 \times 10^{-3} \times 298 \times \ln(2.47 \times 10^{-29})$$

$$= -8.314 \times 10^{-3} \times 298 \times (-65.87) = 163.19\ \text{kJ}$$

29. (2850)

$\Delta H^\circ_{rxn} = 55$ kJ/mol, $\quad T = 298$ K

$\Delta S^\circ_{rxn} = 175$ J/mol

$\Delta G^\circ_{rxn} = \Delta H^\circ_{rxn} - T\Delta S^\circ_{rxn}$

$\Rightarrow \Delta G^\circ_{rxn} = 55000\ \text{J/mol} - 298 \times 175\ \text{J/mol}$

$\Rightarrow \Delta G^\circ_{rxn} = 55000 - 52150$

$\Rightarrow \Delta G^\circ_{rxn} = 2850\ \text{J/mol}$

30. (700)

$$\frac{1}{2} X_2 + \frac{5}{2} Y_2 \rightleftharpoons XY_5$$

$$\Delta S^\circ_{Rxn} = 110 - \left[\left(\frac{1}{2} \times 70\right) + \left(\frac{5}{2} \times 50\right)\right] = 110 - 160 = -50\ \text{JK}^{-1}\text{mol}^{-1}$$

$\Delta G^\circ = 0$ at eqb

$\Delta G^\circ = \Delta H^\circ - T\Delta S^\circ$

$0 = -35000 - T(-50)$

$T = 700$ kelvin

31. (95)

$$2C_{(graphite)} + 3H_2(g) \rightarrow C_2H_6(g), \quad \Delta H_f = ?$$

$$C_2H_6(g) + \frac{7}{2}O_2(g) \rightarrow 2CO_2(g) + 3H_2O(l), \quad \Delta H_1 = -1550$$

$$C_{(graphite)} + O_2(g) \rightarrow CO_2(g), \quad \Delta H_2 = -393.5$$

$$H_2(g) + \frac{1}{2}O_2(g) \rightarrow H_2O(l), \quad \Delta H_3 = -286$$

$$\Delta H_f = 2\Delta H_2 + 3\Delta H_3 - \Delta H_1 = 95 \text{ kJ/mol.}$$

32. (200)

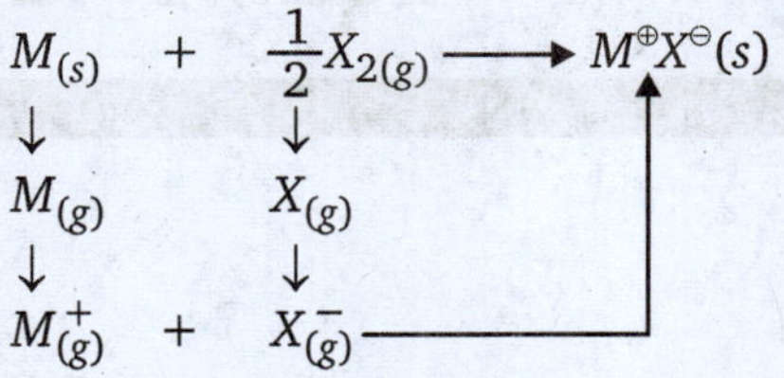

$$\therefore \Delta H_f(MX) = \Delta H_{sub}(M) + \text{I.E.}(M) + \frac{1}{2}[\text{B.E.}(X-X)] + EG(X) + \text{L.E.}(MX)$$

$$-400 = (100) + (500) + \frac{1}{2}(\text{B.E.}) + (-300) + (-800)$$

$$\therefore \quad \text{B.E.} = 200 \text{ kJ mol}^{-1}$$

33. (466)

$$\frac{1}{2}H_{2(g)} \rightarrow H(g)\,; \qquad \Delta_f H(H_{(g)}) = 220 \text{kJ/mol}$$

$$\frac{1}{2}O_{2(g)} \rightarrow O(g)\,; \qquad \Delta_f H(O_{(g)}) = 250 \text{kJ/mol}$$

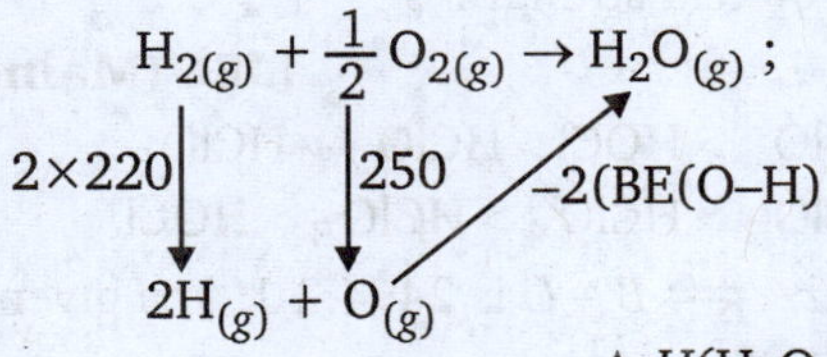

$$\Delta_f H(H_2O_{(g)}) = -242 \text{ kJ/mol}$$

$$\Delta H_f(H_2O_{(l)}) = -242 = 440 + 250 - 2\,[\text{B.E.}(O — H)]$$

$$\text{BE}(O — H) = 466 \text{ kJ/mol}$$

34. (48)

$$\Delta H_f[CO_2(g)] = -393.5 \text{kJ/mol}$$

$$\Delta H_f[H_2O(l)] = -286.0 \text{kJ/mol}$$

$$\Delta H_c[C_6H_6] = -3267.0 \text{kJ/mol}$$

$$\Delta H_f C_6H_6 = (?)$$

$$C_6H_6 + \frac{15}{2}O_2(g) \longrightarrow 6CO_2(g) + 3H_2O(l)$$

$$\Delta H_R = \Delta H_C = \sum \Delta H_f(P) - \sum \Delta H_f(R)$$

$$-3267 = 6 \times (-393.5) + 3(-286) - \Delta H_f(C_6H_6)$$

$$\Delta H_f(C_6H_6) = 48 \text{kJ/mol}$$

❑❑❑

CHAPTER

5

Equilibrium

A. Objective Questions

[Only one option is correct]

1. How many litres of water must be added to 1 litre of an aqueous solution of HCl with a pH of 1 to create an aqueous solution with pH of 2 ? **[JEE (Mains) 2013]**

(a) 0.1 L (b) 0.2 L (c) 2.0 L (d) 9.0 L

2. For the reaction $SO_{2(g)} + \frac{1}{2}O_{2(g)} \rightleftharpoons SO_{3(g)}$, if $K_P = K_C(RT)^x$ where the symbols have usual meaning then the value of x is : (assuming ideality) **[JEE (Mains) 2014]**

(a) -1 (b) $-\frac{1}{2}$ (c) $\frac{1}{2}$ (d) 1

3. Among the following oxoacids, the correct decreasing order of acid strength is : **[JEE (Mains) 2014]**

(a) $HOCl > HClO_2 > HClO_3 > HClO_4$
(b) $HClO_4 > HOCl > HClO_2 > HClO_3$
(c) $HClO_4 > HClO_3 > HClO_2 > HOCl$
(d) $HClO_2 > HClO_4 > HClO_3 > HOCl$

4. The standard Gibbs energy change at 300 K for the reaction $2A \rightleftharpoons B + C$ is 2494.2 J. At a given time, the composition of the reaction mixture is $[A] = \frac{1}{2}$, $[B] = 2$ and $[C] = \frac{1}{2}$. The reaction proceeds in the : $[R = 8.314\ J/K/mol, e = 2.718]$ **[JEE (Mains) 2015]**

(a) forward direction because $Q < K_c$
(b) reverse direction because $Q < K_c$
(c) forward direction because $Q > K_c$
(d) reverse direction because $Q > K_c$

5. The following reaction is performed at 298 K.

$$2NO_{(g)} + O_{2(g)} \rightleftharpoons 2NO_{2(g)}$$

The standard free energy of formation of $NO_{(g)}$ is 86.6 kJ/mol at 298 K. What is the standard free energy of formation of $NO_{2(g)}$ at 298 K? $(K_p = 1.6 \times 10^{12})$ **[JEE (Mains) 2015]**

(a) $8660 - \frac{\ln(1.6 \times 10^{12})}{R(298)}$
(b) $0.5[2 \times 86{,}600 - R(298)\ln(1.6 \times 10^{12})]$
(c) $R(298)\ln(1.6 \times 10^{12}) - 86600$
(d) $86600 + R(298)\ln(1.6 \times 10^{12})$

6. The % yield of ammonia as a function of time in the reaction

$$N_2(g) + 3H_2(g) \rightleftharpoons 2NH_3(g), \Delta H < 0$$

at (P, T_1) is given ahead :

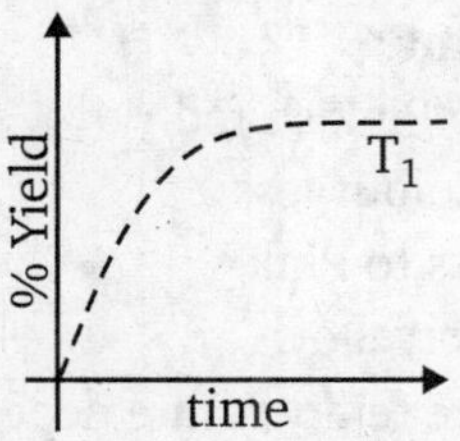

If this reaction is conducted at (P, T_2), with $T_2 > T_1$, the % yield of ammonia as a function of time is represented by : **[JEE (Advanced) 2015]**

(a)
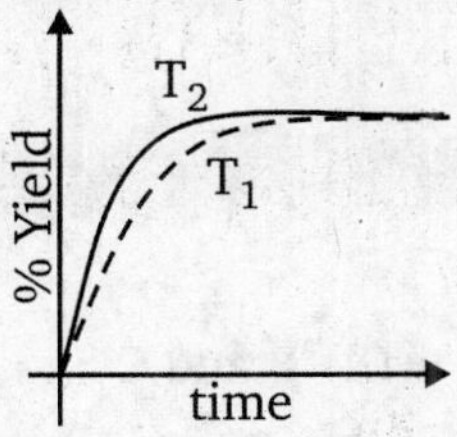

(b)
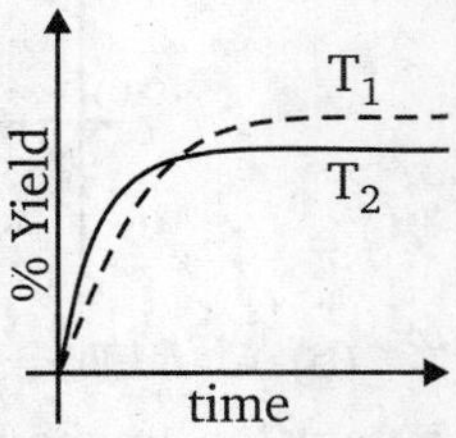

(c)
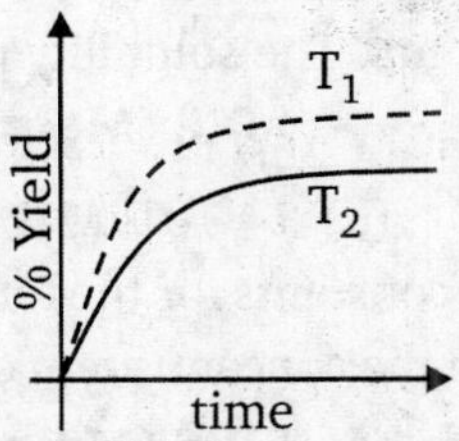

(d)
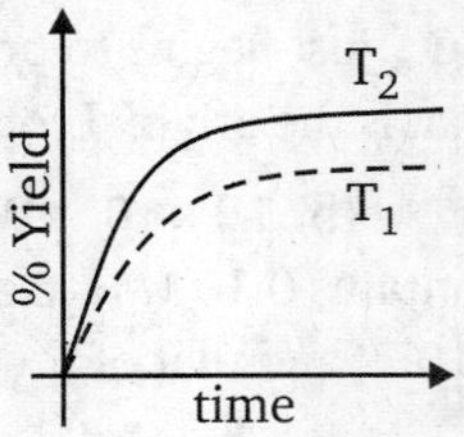

7. The equilibrium constant at 298 K for a reaction, $A + B \rightleftharpoons C + D$ is 100. If the initial concentration of all the four species were 1*M* each, then equilibrium concentration of *D* (in mol L^{-1}) will be : **[JEE (Mains) 2016]**

(a) 0.182 (b) 0.818 (c) 1.818 (d) 1.182

8. pK_a of a weak acid (*HA*) and pK_b of a weak base (*B*OH) are 3.2 and 3.4 respectively. The pH of their salt (*AB*) solution is: **[JEE (Mains) 2017]**

(a) 7.0 (b) 1.0 (c) 7.2 (d) 6.9

9. In the following reactions, ZnO is respectively acting as a/an: **[JEE (Mains) 2017]**

(A) $ZnO + Na_2O \longrightarrow Na_2ZnO_2$

(B) $ZnO + CO_2 \longrightarrow ZnCO_3$

(a) acid and acid (b) acid and base (c) base and acid (d) base and base

10. The standard state Gibbs free energies of formation of $C_{(graphite)}$ and $C_{(diamond)}$ at $T = 298$ K are

$$\Delta_f G^\circ [C_{(graphite)}] = 0 \text{ kJ mol}^{-1}$$

$$\Delta_f G^\circ [C_{(diamond)}] = 2.9 \text{ kJ mol}^{-1}$$

The standard state means that the pressure should be1 bar, and substance should be pure at a given temperature. The conversion of graphite $[C_{(graphite)}]$ to diamond $[C_{(diamond)}]$ reduces its volume by $2 \times 10^{-6} m^3 mol^{-1}$. If $C_{(graphite)}$ is converted to $C_{(diamond)}$ isothermally at $T = 298$ K, the pressure at which $C_{(graphite)}$ is in equilibrium with $C_{(diamond)}$ is:

[Useful information : $1 J = 1 kg\,m^2 s^{-2}$; $1 Pa = 1 kg\,m^{-1} s^{-2}$; $1\text{ bar} = 10^5$ Pa] **[JEE (Advanced) 2017]**

(a) 29001 bar (b) 58001 bar (c) 14500 bar (d) 14501 bar

11. Which of the following salts is the most basic in aqueous solution? **[JEE (Mains) 2018]**

(a) $FeCl_3$ (b) $Pb(CH_3COO)_2$ (c) $Al(CN)_3$ (d) CH_3COOK

12. An alkali is titrated against an acid with methyl orange as indicator, which of the following is a correct combination? **[JEE (Mains) 2018]**

	Base	Acid	End point
(a)	Weak	Strong	Yellow to Pinkish red
(b)	Strong	Strong	Pink to Colour less
(c)	Weak	Strong	Colourless to Pink
(d)	Strong	Strong	Pinkish to yellow

13. Which of the following lines correctly show the temperature dependence of equilibrium constant, K, for an exothermic reaction? **[JEE (Mains) 2018]**

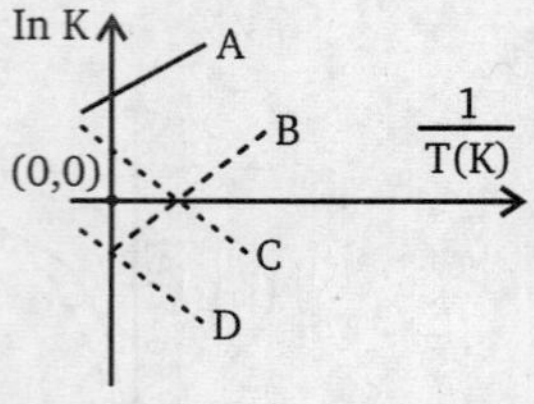

(a) C and D (b) A and D (c) A and B (d) B and C

14. An aqueous solution contains an unknown concentration of Ba^{2+}. When 50 mL of a 1 M solution of Na_2SO_4 is added, $BaSO_4$ just begins to precipitate. The final volume is 500 mL. The solubility product of $BaSO_4$ is 1×10^{-10}. What is the original concentration of Ba^{2+}? **[JEE (Mains) 2018]**

(a) $1.1 \times 10^{-9} M$ (b) $1.0 \times 10^{-10} M$ (c) $5 \times 10^{-9} M$ (d) $2 \times 10^{-9} M$

15. An aqueous solution contains 0.10 M H_2S and 0.20 M HCl. If the equilibrium constants for the formation of HS^- from H_2S is 1.0×10^{-7} and that of S^{2-} from HS^- ions is 1.2×10^{-13} then the concentration of S^{2-} ions in aqueous solution is: **[JEE (Mains) 2018]**

(a) 6×10^{-21} (b) 5×10^{-19} (c) 5×10^{-8} (d) 3×10^{-20}

16. If solubility product of $Zr_3(PO_4)_4$ is denoted by K_{sp} and its molar solubility is denoted by S, then which of the following relations between S and K_{sp} is correct? **[JEE (Mains) 2019]**

(a) $S = \left(\frac{K_{sp}}{144}\right)^{1/6}$ (b) $S = \left(\frac{K_{sp}}{216}\right)^{1/7}$ (c) $S = \left(\frac{K_{sp}}{929}\right)^{1/9}$ (d) $S = \left(\frac{K_{sp}}{6912}\right)^{1/7}$

17. For the following reactions, equilibrium constants are given :

$$S_{(s)} + O_{2(g)} \rightleftharpoons SO_{2(g)}; \quad K_1 = 10^{52}$$

$$2S_{(s)} + 3O_{2(g)} \rightleftharpoons 2SO_{3(g)}; K_2 = 10^{129}$$

The equilibrium constant for the reaction, $2SO_{2(g)} + O_{2(g)} \rightleftharpoons 2SO_{3(g)}$ is : **[JEE (Advanced) 2019]**

(a) 10^{154} (b) 10^{77} (c) 10^{181} (d) 10^{25}

18. For the equilibrium $A \rightleftharpoons B$, the variation of the rate of the forward (a) and reverse (b) reaction with time is given by: **[JEE Main (Sept.) 2020]**

(a)
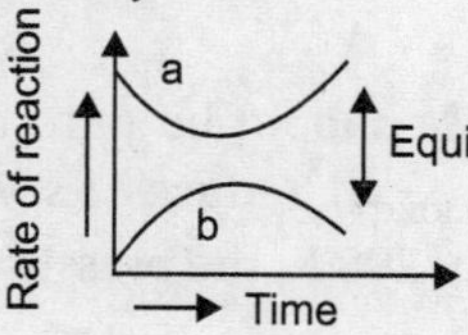

(b)
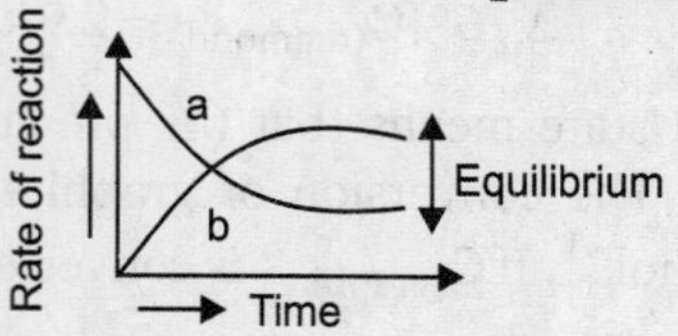

(c)
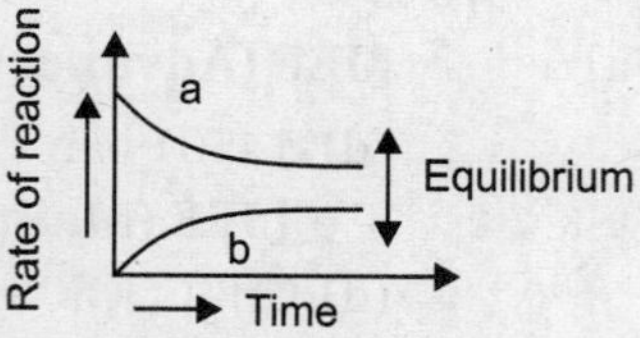

(d)
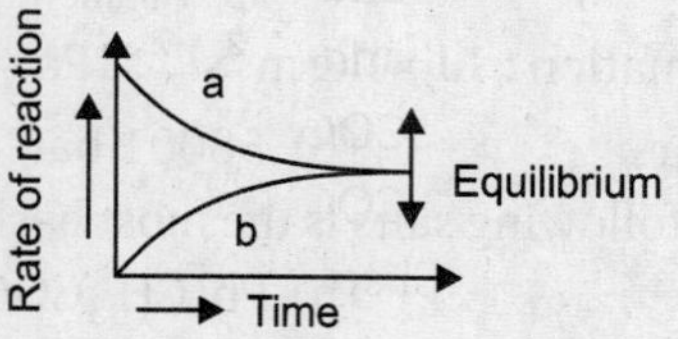

19. If the equilibrium constant for $A \rightleftharpoons B + C$ is $K_{(eq)}^{(1)}$ and that of $B + C \rightleftharpoons P$ is $K_{(eq)}^{(2)}$, the equilibrium constant for $A \rightleftharpoons P$ is: **[JEE Main (Sept.) 2020]**

(a) $K_{(eq)}^{(1)} K_{(eq)}^{(2)}$ (b) $K_{(eq)}^{(1)} / K_{(eq)}^{(1)}$ (c) $K_{(eq)}^{(1)} + K_{(eq)}^{(2)}$ (d) $K_{(eq)}^{(2)} - K_{(eq)}^{(1)}$

20. The value of K_C is 64 at 800 K for the reaction

$$N_2(g) + 3H_2(g) \rightleftharpoons 2NH_3(g)$$

The value of K_C for the following reaction is:

$$NH_3(g) \rightleftharpoons \frac{1}{2}N_2(g) + \frac{3}{2}H_2(g)$$ **[JEE Main (Sept.) 2020]**

(a) $\frac{1}{64}$ (b) 8 (c) $\frac{1}{4}$ (d) $\frac{1}{8}$

21. For the reaction

$$Fe_2N(s) + \frac{3}{2}H_2(g) \rightleftharpoons 2Fe(s) + NH_3(g)$$ **[JEE Main (Sept.) 2020]**

(a) $K_c = K_p(RT)$ (b) $K_c = K_p(RT)^{-1/2}$
(c) $K_c = K_p(RT)^{1/2}$ (d) $K_c = K_p(RT)^{3/2}$

22. If the figure shown below reactant A (represented by square) is in equilibrium with product B (represented by circle). The equilibrium constant is: **[JEE Main (Jan.) 2020]**

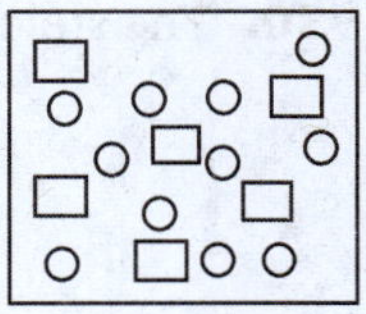

(a) 8 (b) 2 (c) 4 (d) 1

23. Consider the following reaction:

$$N_2O_4 \rightleftharpoons 2NO_2(g); \quad \Delta H° = +58\,kJ$$

For each of the following cases (A, B) the direction in which the equilibrium shifts is: **[JEE Main (Sept.) 2020]**

(A) Temperature is decreased
(B) Pressure is increased by adding N_2 at constant T
(a) (A) towards reactant, (B) towards product (b) (A) towards reactant, (B) no change
(c) (A) towards product, (B) no change (d) (A) towards reactant

24. For the following Assertion and Reason, the correct option is : **[JEE Main (Jan.) 2020]**
Assertion: The pH of water increase with increase in temperature.
Reason: The dissociation of water into H^+ and OH^- is an exothermic reaction.
(a) assertion is not true, but reason is true
(b) both assertion and reason are false
(c) but the reason is not the correct explanation for the assertion
(d) both assertion and reason are true, and the reason is the correct explanation for the assertion

25. Arrange the following solutions in the decreasing order of pOH: **[JEE Main (Sept.) 2020]**
(1) 0.01 M HCl (2) 0.01 M NaOH
(3) 0.01 M CH_3COONa (4) 0.01 M NaCl
(a) (1) > (3) > (4) > (2) (b) (1) > (4) > (3) > (2)
(c) (2) > (3) > (4) > (1) (d) (2) > (4) > (3) > (1)

26. An acidic buffer is obtained on mixing: **[JEE Main (Sept.) 2020]**
(a) 100 mL of 0.1M HCl and 200 mL of 0.1M NaCl
(b) 100 mL of 0.1M HCl and 200 mL of 0.1M CH_3COONa
(c) 100 mL of 0.1M CH_3COOH and 100 mL of 0.1M NaOH
(d) 100 mL of 0.1M CH_3COOH and 200 mL of 0.1M NaOH

27. While titration dilute HCl solution with aqueous NaOH, which of the following will not be required? **[JEE Main (Sept.) 2020]**
(a) Burette and porcelain tile (b) Clamp and phenolphthalein
(c) Pipette and distilled water (d) Bunsen burner and measuring cylinder

28. The solubility product of $Cr(OH)_3$ at 298 K is 6.0×10^{-31}. The concentration of hydroxide ions in a saturated solution of $Cr(OH)_3$ will be : **[JEE Main (Jan.) 2020]**

(a) $(18 \times 10^{-31})^{1/4}$ (b) $(4.86 \times 10^{-29})^{1/4}$ (c) $(18 \times 10^{-31})^{1/2}$ (d) $(2.22 \times 10^{-31})^{1/4}$

29. For the following assertion and reason, the correct option is: **[JEE Main (Sept.) 2020]**

Assertion(A): When Cu(II) and sulphide ions are mixed they react together extremely quickly to give a solid.

Reason(R): The equilibrium constant of

$$Cu^{2+}(aq) + S^{2-}(aq) \rightleftharpoons CuS(s)$$

is high because the solubility product is low.

(a) (A) is false and(R) is true.
(b) Both (A) and (R) are false.
(c) Both (A) and (R) are true but (R) is not the explanation for (A).
(d) Both (A) and (R) are true and (R) is the explanation for (A).

30. The stoichiometry and solubility product of a salt with the solubility curve given below is, respectively: **[JEE Main (Jan.) 2020]**

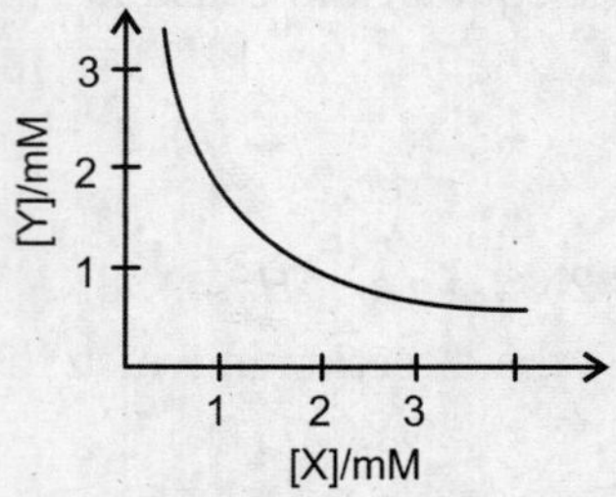

(a) $XY, 2 \times 10^{-5} M^3$ (b) $XY_3, 1 \times 10^{-9} M^3$ (c) $XY_2, 4 \times 10^{-9} M^3$ (d) $X_2Y, 2 \times 10^{-9} M^3$

31. The K_{sp} for the following dissociation is 1.6×10^{-5}

$$PbCl_2(s) \rightleftharpoons Pb^{2+}(aq) + 2Cl^-(aq)$$

Which of the following choices is correct for a mixture of 300 mL 0.134 M $Pb(NO_3)_2$ and 100 mL 0.4 M NaCl? **[JEE Main (Jan.) 2020]**

(a) Not enough data provided (b) $Q > K_{sp}$
(c) $Q < K_{sp}$ (d) $Q = K_{sp}$

32. Which of the following compound can not act as Lewis base? **[JEE Main (March) 2021]**

(a) NF_3 (b) PCl_5 (c) SF_4 (d) ClF_3

33. The solubility of $Ca(OH)_2$ in water is :

[Given : The solubility product of $Ca(OH)_2$ in water $= 5.5 \times 10^{-6}$] **[JEE Main (Feb.) 2021]**

(a) 1.77×10^{-6} (b) 1.11×10^{-6} (c) 1.11×10^{-2} (d) 1.77×10^{-2}

34. The solubility of AgCN in a buffer solution of pH = 3 is x. The value of x is:

[Assume: No cyano complex is formed; $K_{sp}(AgCN) = 2.2 \times 10^{-16}$ and $K_a(HCN) = 6.2 \times 10^{-10}$]

[JEE Main (Feb.) 2021]

(a) 0.625×10^{-6} (b) 1.9×10^{-5} (c) 2.2×10^{-16} (d) 1.6×10^{-6}

35. The equilibrium constant for the reversible reaction $2A(g) \rightleftharpoons 2B(g) + C(g)$ is K_1 and for the reaction $\frac{3}{2}A(g) \rightleftharpoons \frac{3}{2}B(g) + \frac{3}{4}C(g)$ is K_2

K_1 and K_2 are related as : **[JEE Main (June) 2022 (I)]**

(a) $K_1 = \sqrt{K_2}$ (b) $K_2 = \sqrt{K_1}$ (c) $K_2 = K_1^{3/4}$ (d) $K_1 = K_2^{3/4}$

36. 4.0 moles of argon and 5.0 moles of PCl_5 are introduced into an evacuated flask of 100 litre capacity at 610 K. The system is allowed to equilibrate. At equilibrium, the total pressure of mixture was found to be 6.0 atm. The K_p for the reaction is :
[Given : $R = 0.082\ \text{L atm K}^{-1}\ \text{mol}^{-1}$] **[JEE Main (June) 2022 (II)]**
(a) 2.25 (b) 6.24 (c) 12.13 (d) 15.24

37. Class XII students were asked to prepare one litre of buffer solution of pH 8.26 by their chemistry teacher. The amount of ammonium chloride to be dissolved by the student in 0.2 M ammonia solution to make one litre of the buffer is (Given : $pK_b(NH_3) = 4.74$; Molar mass of $NH_3 = 17\ \text{g mol}^{-1}$; Molar mass of $NH_4Cl = 53.5\ \text{g mol}^{-1}$) **[JEE Main (July) 2022 (II)]**
(a) 53.5 g (b) 72.3 g (c) 107.0 g (d) 126.0 g

38. The solubility of AgCl will be maximum in which of the following : **[JEE Main (June) 2022 (I)]**
(a) 0.01 *M* KCl (b) 0.01 *M* HCl (c) 0.01 *M* $AgNO_3$ (d) Deionised water

39. For a concentrated solution of a weak electrolyte (K_{eq} = equilibrium constant) A_2B_3 of concentration 'c', the degree of dissociation 'α' is : **[JEE Main (April) 2023 (I)]**
(a) $\left(\dfrac{K_{eq}}{108\,c^4}\right)^{\frac{1}{5}}$ (b) $\left(\dfrac{K_{eq}}{6\,c^5}\right)^{\frac{1}{5}}$ (c) $\left(\dfrac{K_{eq}}{5\,c^4}\right)^{\frac{1}{5}}$ (d) $\left(\dfrac{K_{eq}}{25\,c^2}\right)^{\frac{1}{5}}$

40. Which of the following statement(s) is/are correct? **[JEE Main (April) 2023 (I)]**
(A) The pH of 1×10^{-8} *M* HCl solution is 8.
(B) The conjugate base $H_2PO_4^-$ is HPO_4^{2-}
(C) K_w increase with increase in temperature
(D) When a solution of weak monoprotic acid is titrated against a strong base at half neutralisation point, $pH = \dfrac{1}{2}pK_a$
Choose the correct answer from the options given below.
(a) (B), (C), (D) (b) (A), (D) (c) (A), (B), (C) (d) (B), (C)

41. $A(g) \rightleftharpoons B(g) + \dfrac{C}{2}(g)$. The correct relationship between K_P, α and equilibrium pressure P is :
[JEE Main (Jan.) 2024]
(a) $K_P = \dfrac{\alpha^{1/2}P^{1/2}}{(2+\alpha)^{1/2}}$ (b) $K_P = \dfrac{\alpha^{3/2}P^{1/2}}{(2+\alpha)^{1/2}(1-\alpha)}$
(c) $K_P = \dfrac{\alpha^{1/2}P^{3/2}}{(2+\alpha)^{3/2}}$ (d) $K_P = \dfrac{\alpha^{1/2}P^{1/2}}{(2+\alpha)^{3/2}}$

42. A vessel at 1000 K contains CO_2 with a pressure of 0.5 atm. Some of CO_2 is converted into CO on addition of graphite. If total pressure at equilibrium is 0.8 atm, then K_P is : **[JEE Main (Jan.) 2025]**
(a) 0.18 atm (b) 1.8 atm
(c) 0.3 atm (d) 3 atm

43. For the reaction,

$$H_2(g) + I_2(g) \rightleftharpoons 2HI(g)$$

Attainment of equilibrium is predicted correctly by : **[JEE Main (Jan.) 2025]**

(a) 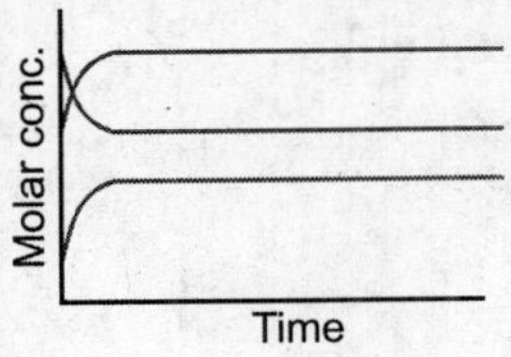

(b) 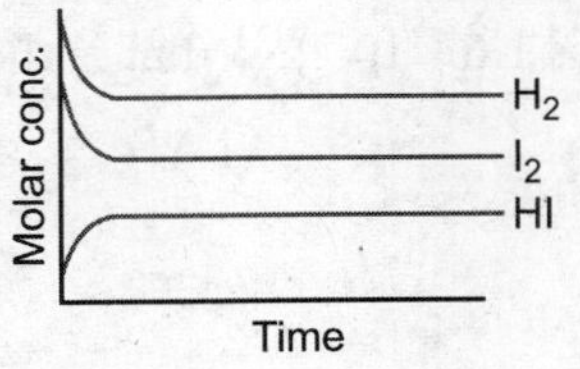

(c) 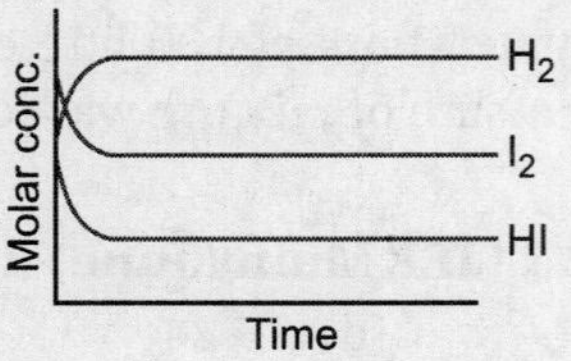

(d) 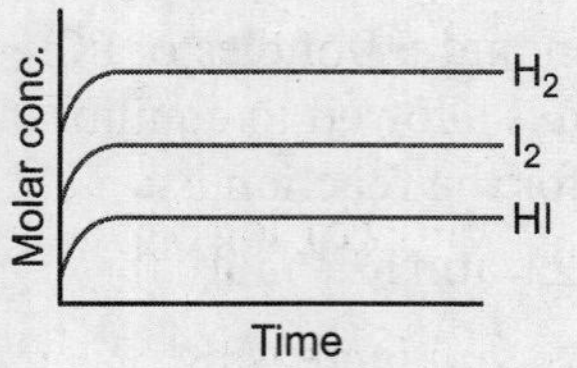

44. Consider the equilibrium

$$CO(g) + 3H_2(g) \rightleftharpoons CH_4(g) + H_2O(g)$$

If the pressure applied over the system increases by two fold at constant temperature then

(A) Concentration of reactants and products increases.

(B) Equilibrium will shift in forward direction.

(C) Equilibrium constant increases since concentration of products increases.

(D) Equilibrium constant remains unchanged as concentration of reactants and products remain same.

Choose the correct answer from the options given below: **[JEE Main (Jan.) 2025]**

(a) (A) and (B) only
(b) (A), (B) and (D) only
(c) (B) and (C) only
(d) (A), (B) and (C) only

45. The molar solubility(s) of zirconium phosphate with molecular formula $(Zr^{4+})_3(PO_4^{3-})_4$ is given by relation: **[JEE Main (Jan.) 2025]**

(a) $\left(\dfrac{K_{sp}}{6912}\right)^{1/7}$
(b) $\left(\dfrac{K_{sp}}{5348}\right)^{1/6}$
(c) $\left(\dfrac{K_{sp}}{8435}\right)^{1/7}$
(d) $\left(\dfrac{K_{sp}}{9612}\right)^{1/3}$

46. Which of the following happens when NH_4OH is added gradually to the solution containing $1M\ A^{2+}$ and $1M\ B^{3+}$ ions?

Given : $K_{sp}[A(OH)_2] = 9 \times 10^{-10}$ and $K_{sp}[B(OH)_3] = 27 \times 10^{-18}$ at 298 K **[JEE Main (Jan.) 2025]**

(a) $B(OH)_3$ will precipitate before $A(OH)_2$.
(b) $A(OH)_2$ and $B(OH)_3$ will precipitate together.
(c) $A(OH)_2$ will precipitate before $B(OH)_3$.
(d) Both $A(OH)_2$ and $B(OH)_3$ do not show precipitation with NH_4OH.

47. Consider the reaction

$$X_2Y(g) \rightleftharpoons X_2(g) + \frac{1}{2}Y_2(g)$$

The equation representing correct relationship between the degree of dissociation (x) of $X_2Y(g)$ with its equilibrium constant K_P is ....

Assume x to be very very small. **[JEE Main (Jan.) 2025]**

(a) $x = \sqrt[3]{\dfrac{2K_p}{P}}$
(b) $x = \sqrt[3]{\dfrac{2K_p^2}{P}}$
(c) $x = \sqrt[3]{\dfrac{K_p}{2P}}$
(d) $x = \sqrt[3]{\dfrac{K_p}{P}}$

48. pH of water is 7 at 25°C. If water is heated to 80°C, it's pH will :

(a) Decrease
(b) Remains the same
(c) H^+ concentration increases, OH^- concentration decreases
(d) Increase

49. K_{sp} for $Cr(OH)_3$ is 1.6×10^{-30}. What is the molar solubility of this salt in water?

(a) $\sqrt[4]{\dfrac{1.6 \times 10^{-30}}{27}}$
(b) $\dfrac{1.8 \times 10^{-30}}{27}$
(c) $\sqrt[5]{1.8 \times 10^{-30}}$
(d) $\sqrt[2]{1.6 \times 10^{-30}}$

50. A weak acid HA has degree of dissociation x. Which option gives the correct expression of $pH = pK_a$? **[JEE Main (Jan.) 2025]**

(a) $\log(1+2x)$ (b) $\log\left(\frac{1-x}{x}\right)$ (c) 0 (d) $\log\left(\frac{x}{1-x}\right)$

51. Arrange the following in increasing order of solubility product :

$Ca(OH)_2$, AgBr, PbS, HgS **[JEE Main (Jan.) 2025]**

(a) $PbS < HgS < Ca(OH)_2 < AgBr$ (b) $HgS < PbS < AgBr < Ca(OH)_2$

(c) $Ca(OH)_2 < AgBr < HgS < PbS$ (d) $HgS < AgBr < PbS < Ca(OH)_2$

52. At temperature T, compound $AB_{2(g)}$ dissociates as

$AB_{2(g)} \rightleftharpoons AB_{(g)} + \frac{1}{2}B_{2(g)}$ having degree of dissociation x (small compared to unity). The correct expression for x in terms of K_p and P is : **[JEE Main (Jan.) 2025]**

(a) $\sqrt[3]{\frac{2K_p}{P}}$ (b) $\sqrt[4]{\frac{2K_p}{P}}$

(c) $\sqrt[3]{\frac{2K_p^2}{P}}$ (d) $\sqrt{K_p}$

B. Objective Questions

[One or more than one option(s) is/are correct]

1. When two reactants, A and B are mixed to give products, C and D, the reaction quotient, (Q) at the initial stages of the reaction : **(IIT 2000)**

(a) is zero (b) decreases with time

(c) is independent of time (d) increases with time

2. Aqueous solutions of HNO_3, KOH, CH_3COOH and CH_3COONa of identical concentration are provided. The pair(s) of solutions which form a buffer upon mixing is(are) : **(IIT 2010)**

(a) HNO_3 and CH_3COOH (b) KOH and CH_3COONa

(c) HNO_3 and CH_3COONa (d) CH_3COOH and CH_3COONa

3. The equilibrium $2Cu^{I} \rightleftharpoons Cu^{\circ} + Cu^{II}$ in aqueous medium at 25° C shifts towards the left in the presence of : **(IIT 2011)**

(a) NO_3^- (b) Cl^- (c) SCN^- (d) CN^-

4. The option(s) with only amphoteric oxides is(are): **[JEE (Advanced) 2017]**

(a) Cr_2O_3, BeO, SnO, SnO_2 (b) ZnO, Al_2O_3, PbO, PbO_2

(c) NO, B_2O_3, PbO, SnO_2 (d) Cr_2O_3, CrO, SnO, PbO

5. For a reaction taking place in a container in equilibrium with its surroundings, the effect of temperature on its equilibrium constant K in terms of change in entropy is described by: **[JEE (Advanced) 2017]**

(a) with increase in temperature, the value of K for endothermic reaction increases because unfavourable change in entropy of the surroundings decreases

(b) with increase in temperature, the value of K for exothermic reaction decreases because favourable change in entropy of the surroundings decreases

(c) with increase in temperature, the value of K for exothermic reaction decreases because the entropy change of the system is positive

(d) with increase in temperature, the value of K for endothermic reaction increases because the entropy change of the system is negative.

C. Matching Type Problems

1. Match the K_a values

Compound	K_a
(a) Benzoic acid	3.3×10^{-5}
(b) 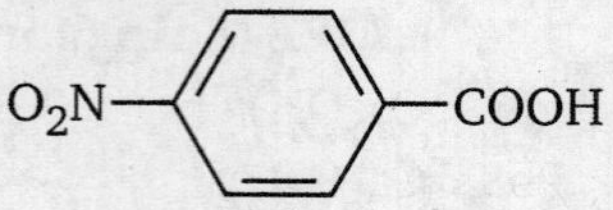	6.3×10^{-5}
(c) 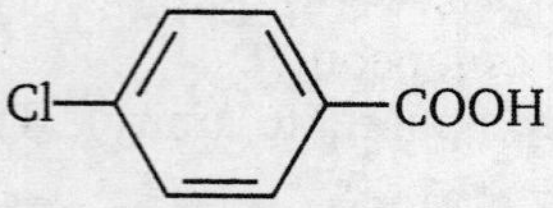	30.6×10^{-5}
(d) 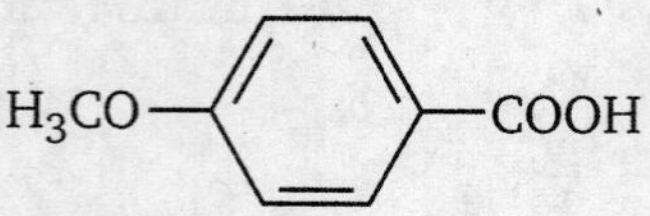	6.4×10^{-5}
(e) 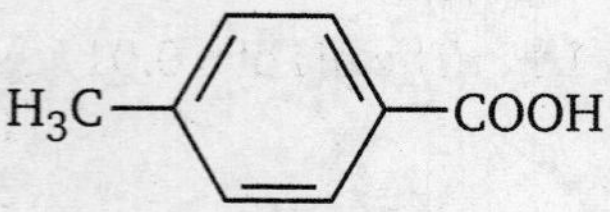	4.2×10^{-5}

(IIT 2003)

D. Passage Based Problems

Passage-1

When 100 mL of 1.0 *M* HCl was mixed with 100 mL of 1.0 *M* NaOH in an insulated beaker at constant pressure, a temperature increase of 5.7°C was measured for the beaker and its contents **(Expt. 1).** Because the enthalpy of neutralization of a strong acid with a strong base is a constant ($-57.0\ \text{kJ mol}^{-1}$), this experiment could be used to measure the calorimeter constant. In a second experiment **(Expt. 2)**, 100 mL of 2.0 *M* acetic acid ($K_a = 2.0 \times 10^{-5}$) was mixed with 100 mL of 1.0 *M* NaOH (under identical conditions to **Expt. 1)** where a temperature rise of 5.6°C was measured.

(Consider heat capacity of all solutions as $4.2\ \text{J g}^{-1}\text{K}^{-1}$ and density of all solutions as $1.0\ \text{g mL}^{-1}$)

[JEE (Advanced) 2015]

1. Enthalpy of dissociation (in kJ mol^{-1}) of acetic acid obtained from the **Expt. 2** is :

(a) 1.0 (b) 10.0 (c) 24.5 (d) 51.4

2. The pH of the solution after **Expt. 2** is :

(a) 2.8 (b) 4.7 (c) 5.0 (d) 7.0

Passage-2

Thermal decomposition of gaseous X_2 to gaseous X at 298 K takes place according to the following equation:

$$X_{2(g)} \rightleftharpoons 2X_{(g)}$$

The standard reaction Gibbs energy, $\Delta_r G^\circ$, of this reaction is positive. At the start of the reaction, there is one mole of X_2 and no. X. As the reaction proceeds, the number of moles of X formed is given by β. Thus, $\beta_{\text{equilibrium}}$ is the number of moles of X formed at equilibrium. The reaction is carried out at a constant total pressure of 2 bar. Consider the gases to behave ideally. (Given: $R = 0.083\ \text{L bar K}^{-1}\text{mol}^{-1}$)

[JEE (Advanced) 2016]

1. The equilibrium constant K_p for this reaction at 298 K, in terms of $\beta_{\text{equilibrium}}$, is:

(a) $\dfrac{8\beta^2_{\text{equilibrium}}}{2-\beta_{\text{equilibrium}}}$ (b) $\dfrac{8\beta^2_{\text{equilibrium}}}{4-\beta^2_{\text{equilibrium}}}$

(c) $\dfrac{4\beta^2_{\text{equilibrium}}}{2-\beta_{\text{equilibrium}}}$ (d) $\dfrac{4\beta^2_{\text{equilibrium}}}{4-\beta^2_{\text{equilibrium}}}$

2. The incorrect statement among the following, for this reaction, is:

(a) decrease in the total pressure will result in formation of more moles of gaseous X

(b) at the start of the reaction, dissociation of gaseous X_2 takes place spontaneouly

(c) $\beta_{\text{equilibrium}} = 0.7$

(d) $K_c < 1$

E. Integer Type Problems

1. The dissociation constant of a substituted benzoic acid at 25°C is 1.0×10^{-4}. The pH of a 0.01 M solution of its sodium salt is: **(IIT 2009)**

2. The total number of diprotic acids among the following is : **(IIT 2010)**

H_3PO_4 H_2SO_4 H_3PO_3 H_2CO_3 $H_2S_2O_7$

H_3BO_3 H_3PO_2 H_2CrO_4 H_2SO_3

3. In 1L saturated solution of AgCl $[K_{sp}(AgCl) = 1.6 \times 10^{-10}]$, 0.1 mol of CuCl $[K_{sp}(CuCl) = 1.0 \times 10^{-6}]$ is added. The resultant concentration of Ag^+ in the solution is 1.6×10^{-x}. The value of "x" is : **(IIT 2011)**

4. The conductance of a 0.0015 M aqueous solution of a weak monobasic acid was determined by using a conductivity cell consisting of platinized Pt electrodes, The distance between the electrodes is 120 cm with an area of cross section of 1 cm^2. The conductance of this solution was found to be 5×10^{-7} S. The pH of the solution is 4. The value of limiting molar conductivity (Λ°_m) of this weak monobasic acid in aqueous solution is $Z \times 10^2$ S $cm^2 mol^{-1}$. The value of Z is: **[JEE (Advanced) 2017]**

5. The solubility of a salt of weak acid (AB) at pH 3 is $Y \times 10^{-3}$ mol L^{-1}. The value of Y is :

(Given that the value of solubility product of $AB(K_{sp}) = 2 \times 10^{-10}$ and the value of ionization constant of $HB\,(K_a) = 1 \times 10^{-8}$) **[JEE (Advanced) 2018]**

6. For the following reaction, the equilibrium constant K_c at 298 K is 1.6×10^{17}.

$$Fe^{2+}_{(aq)} + S^{2-}_{(aq)} \rightleftharpoons FeS_{(s)}$$

When equal volumes of 0.06 M $Fe^{2+}_{(aq)}$ and 0.2M $S^{2-}_{(aq)}$ solutions are mixed, the equilibrium concentration of $Fe^{2+}_{(aq)}$ is found to be $Y \times 10^{-17}$ M. The value of Y is ____. **[JEE (Advanced) 2019]**

7. Consider the reaction $A \rightleftharpoons B$ at 1000 K. At time t', the temperature of the system was increased to 2000 K and the system was allowed to reach equilibrium. Throughout this experiment the partial pressure of A was maintained at 1 bar. Given below is the plot of the partial pressure of B with time. What is the ratio of the standard Gibbs energy of the reaction at 1000 K to that at 2000 K ? **[JEE (Advanced) 2020]**

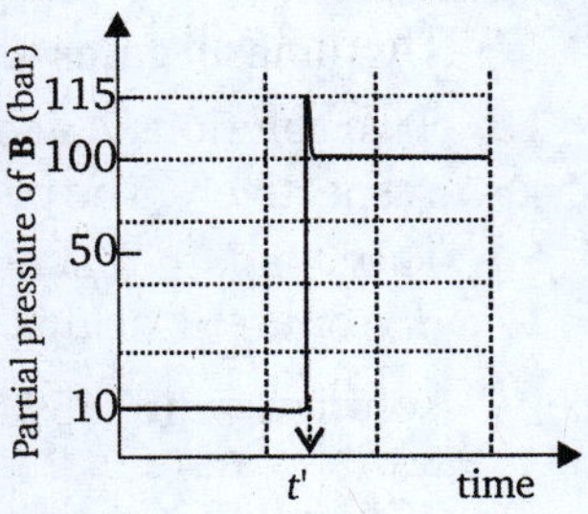

8. A solution of 0.1M weak base (B) is titrated with 0.1M of a strong acid (HA). The variation of pH of the solution with the volume of HA added is shown in the figure below. What is the pK_b of the base? The neutralization reaction is given by : $B + HA \longrightarrow BH^+ + A^-$.

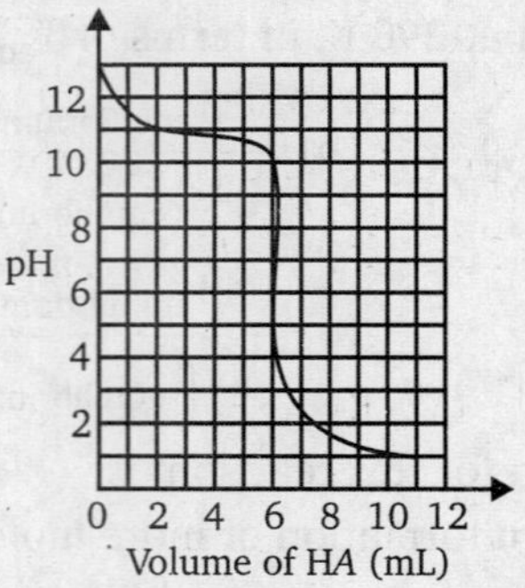

[JEE (Advanced) 2020]

9. An acidified solution of 0.05 M Zn^{2+} is saturated with 0.1 M H_2S. What is the minimum molar concentration (M) of H^+ required to prevent the precipitation of ZnS ?

Use $K_{sp}(ZnS) = 1.25 \times 10^{-22}$ and Overall dissociation constant of H_2S, $K_{NET} = K_1K_2 = 1 \times 10^{-21}$

[JEE (Advanced) 2020]

10. The gas phase reaction, $2A(g) \rightleftharpoons A_2(g)$ at 400 K has $\Delta G° = +25.2$ kJ mol^{-1}.

The equilibrium constant K_c for this reaction is ______ $\times 10^{-2}$. (Round off to the nearest integer)

[Use : $R = 8.3$ J mol^{-1}K^{-1}, $\ln 10 = 2.3$, $\log_{10} 2 = 0.30$, 1 atm = 1 bar]

[antilog (– 0.3) = 0.501] **[JEE Main (Feb.) 2021]**

11. For a reaction $X + Y = 2Z$, 1.0 mole of X, 1.5 mole of Y and 0.5 mole of Z were taken in a 1 L vessel and allowed to react. At equilibrium, the concentration of Z was 1.0 mol L^{-1}. The equilibrium constant of the reaction is $\frac{x}{15}$. The value of x is ________. **[JEE Main (Sept.) 2020]**

12. At 1990 K and 1 atm pressure, there are equal number of Cl_2 molecules and Cl atoms in the reaction mixture. The value K_P for the reaction $Cl_2(g) \rightleftharpoons 2Cl(g)$ under the above conditions is $x \times 10^{-1}$. The value of x is ______.

(Rounded off to the nearest integer) **[JEE Main (Feb.) 2021]**

13. For the reaction $A(g) \longrightarrow B(g)$, the value of the equilibrium constant at 300 K and 1 atm is equal to 100.0. The value of Δ_rG for the reaction at 300 K and 1 atm in J mol^{-1} is $-xR$, where x is ______ (Rounded off to the nearest integer) ($R = 8.31$ J mol^{-1}K^{-1} and $\ln 10 = 2.3$) **[JEE Main (Feb.) 2021]**

14. For the reaction $A(g) \rightleftharpoons B(g)$ at 495 K, $\Delta_rG° = -9.478$ kJ mol^{-1}

If we start the reaction in a closed container at 495 K with 22 millimoles of A, the amount of B is the equilibrium mixture is _______ millimoles. (Round off to the nearest integer).

[$R = 8.314$ J mol^{-1} K^{-1}; $\ln 10 = 2.303$] **[JEE Main (March) 2021]**

15. Consider the reaction $N_2O_4(g) \rightleftharpoons 2NO_2(g)$.

The temperature at which $K_c = 20.4$ and $K_p = 600.1$, is.......K. (Round off to the nearest integer).

[Assume all gases are ideal and $R = 0.0831$ L-bar K^{-1} mol^{-1}] **[JEE Main (March) 2021]**

16. Two solutions A and B, each of 100 L was made by dissolution of 4g of NaOH and 9.8 g of H_2SO_4 in water, respectively. The pH of the resultant solution obtained from mixing 40 L of solution A and 10 L of solution B is: **[JEE Main (Jan.) 2020]**

17. 3 g of acetic acid is added to 250 mL of 0.1 M HCl and the solution made up to 500 mL. To 20 mL of this solution $\frac{1}{2}$ mL of 5 M NaOH is added. The pH of the solution is_________. **[JEE Main (Jan.) 2020]**

[Given: pK_a of acetic acid = 4.75, molar mass of acetic acid = 60 g/mol, $\log 3 = 0.4771$]

Neglect any changes in volume.

18. A soft drink was bottled with a partial pressure of CO_2 of 3 bar over the liquid at room temperature. The partial pressure of CO_2 over the solution approaches a value of 30 bar when 44 g of CO_2 is dissolved in 1 kg of water at room temperature. The approximate pH of the soft drink is ___ $\times 10^{-1}$.

(First dissociation constant of $H_2CO_3 = 4.0 \times 10^{-7}$; log 2 = 0.3; density of the soft drink = 1 g mL^{-1})

[JEE Main (Sept.) 2020]

19. If the solubility product of AB_2 is 3.20×10^{-11} M^3, then the solubility of AB_2 in pure water is ________ $\times 10^{-4}$ mol L^{-1}. [Assuming that neither kind of ion reacts with water] **[JEE Main (Sept.) 2020]**

20. The solubility of $CdSO_4$ in water is 8.0×10^{-4} mol L^{-1}. Its solubility in 0.01 M H_2SO_4 solution is ______ $\times 10^{-6}$ mol L^{-1}. (Round off to the nearest integer) (Assume that solubility is much less than 0.01 M)

[JEE Main (March) 2021]

21. 10.0 mL of Na_2CO_3 solution is titrated against 0.2 *M* HCl solution. The following titre values were obtained in 5 readings.

4.8 mL, 4.9 mL, 5.0 mL, 5.0 mL and 5.0 mL

Based on these readings, and convention of titrimetric estimation of concentration of Na_2CO_3 solution is ________mM. (Round off to the nearest integer) **[JEE Main (March) 2021]**

22. The solubility product of PbI_2 is 8.0×10^{-9}. The solubility of lead iodide in 0.1 molar solution of lead nitrate is $x \times 10^{-6}$ mol / L. The value of x is ________. (Rounded off to the nearest integer)

[Given : $\sqrt{2} = 1.41$] **[JEE Main (Feb.) 2021]**

23. 0.4 g mixture of NaOH, Na_2CO_3 and some inert impurities was first titrated with $\frac{N}{10}$ HCl using phenolphthalein as an indicator, 17.5 mL of HCl was required at the end point. After his methyl orange was added and titrated. 1.5 mL of same HCl was required for the next end point. The weight percentage of Na_2CO_3 in the mixture is_______. (Rounded-off to the nearest integer) **[JEE Main (Feb.) 2021]**

24. Consider titration of NaOH solution *versus* 1.25 *M* oxalic acid solution. At the end point following burette readings were obtained. **[JEE Main (Feb.) 2021]**

(i) 4.5 mL (ii) 4.5 mL (iii) 4.4 mL (iv) 4.4 mL (v) 4.4 mL

If the volume of oxalic acid taken was 10.0 mL then the molarity of the NaOH solution is ______*M*. (Rounded-off to the nearest integer)

25. Two salts A_2X and MX have the same value of solubility product of 4.0×10^{-12}. The ratio of their molar solubilities, *i.e.*, $\frac{S(A_2X)}{S(MX)} =$ ______.

(Round off to the nearest integer). **[JEE Main (March) 2021]**

26. 0.01 moles of a weak acid HA ($K_a = 2.0 \times 10^{-6}$) is dissolved in 1.0L of 0.1*M* HCl solution. The degree of dissociation of HA is ...... $\times 10^{-5}$.

(Round off to the nearest integer).

[Neglect volume change on adding HA. Assume degree of dissociation << 1]

[JEE Main (March) 2021]

27. In order to prepare a buffer solution of pH 5.74, sodium acetate is added to acetic acid. If the concentration of acetic acid in the buffer is 1.0 *M*, the concentration of sodium acetate in the buffer is .... *M*. (Round off to the nearest integer).

[Given: pK_a (acetic acid) = 4.74] **[JEE Main (March) 2021]**

28. A box contains 0.90 g of liquid water in equilibrium with water vapour at 27°C. The equilibrium vapour pressure of water at 27°C 32.0 Torr. When the volume of the box is increased, some of the liquid water evaporates to maintain the equilibrium pressure. If all the liquid water evaporates, then the volume of the box must be ______ litre. [Nearest integer]

(Given : $R = 0.082$ L atm K^{-1} mol^{-1})

(Ignore the volume of the liquid water and assume water vapours behave as an ideal gas).

[JEE Main (June) 2022 (II)]

29. $2O_3(g) \rightleftharpoons 3O_2(g)$

At 300 K, ozone is fifty percent dissociated. The standard free energy change at this temperature and 1 atm pressure is (–) ____ J mol^{-1} (Nearest integer)

[Given : ln 1.35 = 0.3 and $R = 8.3$ J K^{-1} mol^{-1}] **[JEE Main (June) 2022 (I)]**

30. If the solubility product of PbS is 8×10^{-28}, then the solubility of PbS in pure water at 298 K is $x \times 10^{-16}$ mol L^{-1}. The value of x is ________ . (Nearest integer)

[Given : $\sqrt{2} = 1.41$] **[JEE Main (July) 2022 (I)]**

31. 50 mL of 0.1 M CH_3COOH is being titrated against 0.1 M NaOH. When 25 mL of NaOH has been added, the pH of the solution will be ________ $\times 10^{-2}$. (Nearest integer)

[Given : $pK_a(CH_3COOH) = 4.76$] **[JEE Main (June) 2022 (I)]**

log 2 = 0.30 log 3 = 0.48 log 5 = 0.69 log 7 = 0.84 log 11 = 1.04

32. A mixture of 1 mole of H_2O and 1 mole of CO is taken in a 10 litre container and heated to 725 K. At equilibrium 40% of water by mass reacts with carbon monoxide according to the equation :

$$CO(g) + H_2O(g) \rightleftharpoons CO_2(g) + H_2(g).$$

The equilibrium constant $K_C \times 10^2$ for the reaction is ________ . (Nearest integer)

[JEE Main (April) 2023 (I)]

33. 4.5 moles each of hydrogen and iodine is heated in a sealed ten litre vessel. At equilibrium, 3 moles of HI were found. The equilibrium constant for $H_2(g) + I_2(g) \rightleftharpoons 2HI(g)$ is : **[JEE Main (April) 2023 (II)]**

34. An analyst wants to convert. 1 L HCl of pH = 1 to a solution of HCl of pH 2. The volume of water needed to do this dilution is ________ mL. (Nearest integer) **[JEE Main (April) 2023 (I)]**

35. The following concentrations were observed at 500 K for the formation of NH_3 from N_2 and H_2. At equilibrium : $[N_2] = 2 \times 10^{-2}$M, $[H_2] = 3 \times 10^{-2}$M and $[NH_3] = 1.5 \times 10^{-2}$M. Equilibrium constant for the reaction is ________ . **[JEE Main (Jan.) 2024]**

36. K_a for CH_3COOH is 1.8×10^{-5} and K_b for NH_4OH is 1.8×10^{-5}. The pH of ammonium acetate solution will be : **[JEE Main (Jan.) 2024]**

37. If 1 mM solution of ethylamine produces pH = 9 then the ionization constant (K_b) of ethylamine is 10^{-x}. The value of x is .... (nearest integer). **[JEE Main (Jan.) 2025]**

(The degree of ionization of ethylamine can be neglected with respect to unity.)

38. The observed and normal masses of compound MX_2 are 65.6 and 164 respectively. The percent degree of ionisation of MX_2 is ....%. (Nearest integer) **[JEE Main (Jan.) 2025]**

39. Electrolysis of 600 mL aqueous solution of NaCl for 5 min changes the pH of the solution to 12. The current in ampere used for the given electrolysis is .... (Nearest integer).

[JEE Main (Jan.) 2025]

40. If A_2B is 30% ionised in an aqueous solution, then the value of van't Hoff factor (i) is .... $\times 10^{-1}$.

[JEE Main (Jan.) 2025]

Answers

[A] Objective Questions (Only one option is correct)

1. (d)	**2.** (b)	**3.** (c)	**4.** (d)	**5.** (b)	**6.** (b)	**7.** (c)	**8.** (d)	**9.** (b)	**10.** (c)
11. (d)	**12.** (a)	**13.** (c)	**14.** (a)	**15.** (d)	**16.** (d)	**17.** (d)	**18.** (d)	**19.** (a)	**20.** (d)
21. (c)	**22.** (b)	**23.** (b)	**24.** (b)	**25.** (b)	**26.** (b)	**27.** (d)	**28.** (a)	**29.** (d)	**30.** (c)
31. (b)	**32.** (b)	**33.** (c)	**34.** (b)	**35.** (c)	**36.** (a)	**37.** (c)	**38.** (d)	**39.** (a)	**40.** (d)
41. (b)	**42.** (b)	**43.** (b)	**44.** (a)	**45.** (a)	**46.** (a)	**47.** (b)	**48.** (a)	**49.** (a)	**50.** (d)
51. (b)	**52.** (c)								

[B] Objective Questions (One or more than one option(s) is/are correct)

1. (a, d)	**2.** (d)	**3.** (b, c, d)
4. (a, b)	**5.** (a, b)	

[C] Matching Type Problems

1. (a)–q, (b)–r, (c)–s, (d)–p, (e)–t

[D] Passage Based Problems

Passage-1

1. (a) **2.** (b)

Passage-2

1. (b) **2.** (c)

[E] Integer Type Problems

1. (8)	**2.** (6)	**3.** (7)	**4.** (6)	**5.** (4.47)
6. (8.93)	**7.** (0.25)	**8.** (3)	**9.** (0.2)	**10.** (2)
11. (16)	**12.** (5)	**13.** (1380)	**14.** (20)	**15.** (354)
16. (10.6)	**17.** (5.2271)	**18.** (37)	**19.** (2)	**20.** (64)
21. (50)	**22.** (141)	**23.** (4)	**24.** (6)	**25.** (50)
26. (2)	**27.** (10)	**28.** (29)	**29.** (747)	**30.** (282)
31. (476)	**32.** (44)	**33.** (1)	**34.** (9000)	**35.** (417)
36. (7)	**37.** (7)	**38.** (75)	**39.** (2)	**40.** (16)

Hints and Solutions

A. Objective Questions

[Only one option is correct]

1. (d) $0.1 \times 1 = (1+V) \times 0.01$

$$1 + V = \frac{0.1}{0.01} \Rightarrow 1 + V = 10$$

$$\Rightarrow \quad V = 10 - 1 = 9.0\text{ L}$$

2. (b) $SO_2(g) + \frac{1}{2}O_2(g) \rightleftharpoons SO_3(g)$

$$K_P = K_C(RT)$$

$x = \Delta n_g$ = no. of gaseous moles of product – no. of gaseous moles of reactant

$$= 1 - \left(1 + \frac{1}{2}\right) = 1 - \frac{3}{2} = -\frac{1}{2}$$

3. (c) In oxyacids of chlorine strength of acid increases with increase in O.N. of central Cl-atom.

4. (d) $2A \rightleftharpoons B + C$

Given : $T = 300\text{ K}, \Delta G° = 2494.2\text{ J}, R = 8.314\text{ JK}^{-1}\text{ mol}^{-1}$

$$\Delta G° = -2.303\, RT \log K_c$$

$$2494.2 = -2.303 \times 8.314 \times 300 \times \log K_c$$

$$\log K_c = \text{antilog}(-0.4342)$$

$$K_c = 0.3679$$

$$Q_c = \frac{[B][C]}{[A]^2} = \frac{2 \times \frac{1}{2}}{\left(\frac{1}{2}\right)^2} = 4$$

Here, $Q_c > K_c$ thus, the reaction will proceed in the backward direction.

5. (b) Given : $T = 298\text{ K}, \Delta G^\circ_{f(NO)} = 86.6\text{ kJ/mol}, \Delta G^\circ_{f(NO_2)} = ?, K_p = 1.6 \times 10^{12}$

$$2NO_{(g)} + O_{2(g)} \rightleftharpoons 2NO_{2(g)}$$

$$\Delta G^\circ_r = 2\Delta G^\circ_{f(NO_2)} - 2\Delta G^\circ_{f(NO)} - \Delta G^\circ_{f(O_2)}$$

$$\Delta G^\circ_r = 2\Delta G^\circ_{f(NO_2)} - 2 \times 86{,}600$$

$$\Delta G^\circ_r = -RT \ln K_p$$

$$2\Delta G^\circ_{f(NO_2)} - 2 \times 86{,}600 = -R(298)\ln(1.6 \times 10^{12})$$

$$\Delta G^\circ_{f(NO_2)} = \frac{2 \times 86{,}600 - R(298)\ln(1.6 \times 10^{12})}{2}$$

6. (b) $N_2(g) + 3H_2(g) \rightleftharpoons 2NH_3(g);\ \Delta H < 0$

Increasing the temperature lowers equilibrium yield of ammonia.

However, at higher temperature the initial rate of forward reaction would be greater than at lower temperature that is why the percentage yield of NH_3 too would be more initially.

7. (c) $$A + B \rightleftharpoons C + D$$

	A	B	C	D
Initial conc.	1	1	1	1
At equilibrium	$1-x$	$1-x$	$1+x$	$1+x$

Now, $K_c = \frac{[C][D]}{[A][B]}$

i.e., $100 = \frac{(1+x)^2}{(1-x)^2} \Rightarrow 10 = \frac{1+x}{1-x}$

$$10 - 10x = 1 + x \Rightarrow 9 = 11x \Rightarrow x = 0.818$$

So, concentration of D at equilibrium $= 1 + 0.818 = 1.818\ M$

8. (d) pH of a salt of a weak acid and a weak base is given by:

$$\text{pH} = 7 + \frac{1}{2}(pK_a - pK_b) = 7 + \frac{1}{2}(3.2 - 3.4) = 6.9$$

9. (b) ZnO is an amphoteric oxide. So, it can act either as an acid or as a base.

In, $\underset{\text{acid}}{ZnO} + \underset{\text{base}}{Na_2O} \longrightarrow \underset{\text{salt}}{Na_2ZnO_2}$

ZnO acts as an acidic oxide.

In, $\underset{\text{base}}{ZnO} + \underset{\text{acid}}{CO_2} \longrightarrow \underset{\text{salt}}{ZnCO_3}$

ZnO acts as a basic oxide.

10. (c) $\Delta G = PdV$

$$[\Delta_f G^\circ_{(\text{diamond})} - \Delta_f G^\circ_{(\text{graphite})}] = PdV$$

$$2.9 \times 10^3\ \text{J mol}^{-1} = P \times 2 \times 10^{-6}\ \text{m}^3\ \text{mol}^{-1}$$

$$P = 1.45 \times 10^9\ \text{Pa}$$

$$P = 1.45 \times 10^9 \times 10^{-5}\ \text{bar} = 1.45 \times 10^4\ \text{bar}$$

$$P = 14500\ \text{bar}$$

11. (d) CH_3COOK is a salt of a weak acid and a strong base.

$\therefore$ Most basic

12. (a) Methyl orange is used for titration of strong acid and weak base.

13. (c) $\Delta G = \Delta H - T\Delta S$

$$-RTnk = \Delta H - T\Delta S$$

$$nk = \frac{-\Delta H}{RT} + \frac{\Delta S}{R}$$

Slope is $\frac{-\Delta H}{R}$

Since ΔH is –ve

$\therefore$ Slope is positive.

14. (a) $BaSO_4(s) \rightleftharpoons Ba^{2+}(aq) + SO_4^{2-}(aq)$; $K_{sp} = 10^{-10}$

$$Na_2SO_4 \longrightarrow 2Na^+ + SO_4^{2-}$$

Conc. of SO_4^{2-} in final solution $= \frac{50 \times 1}{500} = 0.1M$

For final solution

$\Rightarrow [Ba^{2+}][SO_4^{2-}] = 10^{-10} \Rightarrow [Ba^{2+}] = 10^{-9}\ M$

$$M_i V_1 = M_f V_f$$

$$C \times 450 = 10^{-9} \times 500 \Rightarrow C = 1.1 \times 10^{-9}\ M$$

15. (d) $HCl \longrightarrow \underset{0.2M}{H^+} + \underset{0.2\,M}{Cl^-}$

$$H_2S \rightleftharpoons H^+ + HS^- \qquad K_1 = 10^{-7}$$

$HS^- \rightleftharpoons H^+ + S^{2-}$ $K_2 = 1.2 \times 10^{-13}$

$H_2S \rightleftharpoons 2H^+ + H^{2-}$ $K = K_1 \cdot K_2 = 1.2 \times 10^{-29}$

$$K = \frac{[H^+]^2[S^{2-}]}{[H_2S]} \qquad [H^+] = 0.2M, [H_2S] = 0.1$$

$$1.210^{-20} = \frac{(0.2)^2[S^{2-}]}{0.1} \Rightarrow [S^{2-}] = 3 \times 10^{-20} M$$

16. (d) $Zr_3(PO_4)_4 \rightleftharpoons \underset{3S}{3Zr^{4+}} + \underset{4S}{4PO_4^{3-}}$

$$K_{sp} = (3S)^3(4S)^4 = (27S^3)(256\,S^4) = 6912\,S^7$$

$$S = (K_{sp}/6912)^{1/7}$$

17. (d) For the reaction, $S_{(s)} + O_{2(g)} \rightleftharpoons SO_{2(g)}$

$$K_1 = \frac{[SO_2]}{[O_2]} = 10^{52}$$

Then for the reaction,

$$2SO_{2(g)} \rightleftharpoons 2S_{(s)} + 2O_{2(g)}$$

$$K_1' = \frac{[O_2]^2}{[SO_2]^2} = \left(\frac{1}{10^{52}}\right)^2 \qquad \text{...(i)}$$

For the reaction, $2\,S_{(s)} + 3O_{2(g)} \rightleftharpoons 2SO_{3(g)}$

$$K_2 = \frac{[SO_3]^2}{[O_2]^3} = 10^{129} \qquad \text{...(ii)}$$

For the reaction, $2SO_{2(g)} + O_{2(g)} \rightleftharpoons 2SO_{3(g)}$

$$K_3 = \frac{[SO_3]^2}{[SO_2]^2[O_2]}$$

By multiplying equation (i) and (ii), we get

$$K_3 = K_1' \times K_2 = \left(\frac{1}{10^{52}}\right)^2 \times 10^{129} = \frac{10^{129}}{10^{104}} = 10^{25}$$

18. (d) For reversible reaction, $A \rightleftharpoons B$

In forward reaction, $r_f \propto [A]$

as $[A]$ decreases with time, r_f decreases.

For backward reaction, $r_b \propto [B]$

As B forms during reaction, $[B]$ increases, so r_b also increases.

At equilibrium, both forward and reverse reaction rates become equal.

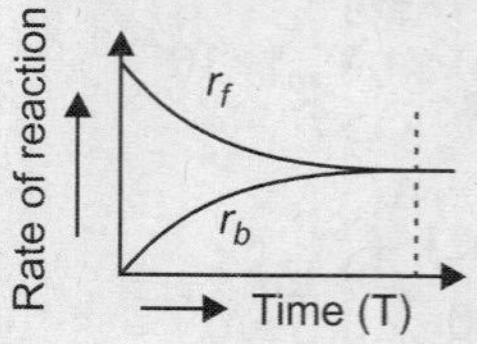

19. (a) $A \rightleftharpoons B + C,\ K_{(eq)}^{(1)}$

$$K_{(eq)}^{(1)} = \frac{[B][C]}{[A]} \qquad \text{...(i)}$$

$B + C \rightleftharpoons P,\ K_{(eq)}^{(2)}$

$$K_{(eq)}^{(2)} = \frac{[P]}{[B][C]} \qquad \text{...(ii)}$$

$$A \rightleftharpoons P, \; K_{eq} = \frac{[P]}{[A]}$$

Multiplying eqn. (i) and (ii)

we get, $K_{(eq)}^{(1)} \times K_{(eq)}^{(2)} = K_{eq.}$

20. (d) $N_2(g) + 3H_2(g) \rightleftharpoons 2NH_3(g); \quad K_c = 64$

$$2NH_3(g) \rightleftharpoons N_2(g) + 3H_2(g); \quad K'_c = \frac{1}{64}$$

$$NH_3(g) \rightleftharpoons \frac{1}{2}N_2(g) + \frac{3}{2}H_2(g); \quad K''_c = \left(\frac{1}{64}\right)^{1/2} = \frac{1}{8}$$

21. (c) $Fe_2N(s) + \frac{3}{2}H_2(g) \rightleftharpoons 2Fe(s) + NH_3(g)$

$$\Delta n_g = 1 - \frac{3}{2} = -\frac{1}{2}$$

$$K_p = K_c(RT)^{\Delta n_g} = K_c(RT)^{-1/2}$$

$$K_c = K_p(RT)^{1/2}$$

22. (b) $A \rightleftharpoons B$

Number of 'squares' $= 6\,(A)$

Number of 'circles' $= 11\,(B)$

$$K_{eq} = \frac{[B]}{[A]} = \frac{11}{6} \cong 2$$

23. (b) $N_2O_4(g) \rightleftharpoons 2NO_2(g), \; \Delta H° > 0$

(a) For endothermic ($\Delta H° > 0$) reaction

$T\uparrow$ cause forward shift

$T\downarrow$ cause backward shift

(b) Since, N_2 is inert gas here; it does not affect equilibrium position.

24. (b) $H_2O(l) \rightleftharpoons H^+(aq) + OH^-(aq); \; \Delta_r H > 0$

On increasing temperature reaction shifts in forward direction. So equilibrium constant K_w increases.

Neutral pH or pOH $= \frac{pK_w}{2}$

So neutral pH and pOH also decreases.

25. (b) pH + pOH = 14 at 25°C

$0.01M$ HCl	pH = 2
$0.01M$ NaOH	pH = 12
$0.01M$ CH_3COONa	7 < pH < 12
$0.01M$ NaCl	pH = 7

⇒ Order of pH: $0.01M$ NaOH > $0.01M$ CH_3COONa > $0.01M$ NaCl > $0.01M$ HCl

Order of pOH will be: $0.01M$ HCl > $0.01M$ NaCl > $0.01M$ CH_3COONa > $0.01M$ NaOH

⇒ 1 > 4 > 3 > 2

26. (b) Acidic buffer formed when solution contains ($CH_3COOH + CH_3COONa$)

So, option (a) cannot form buffer.

In option (b):

HCl +	CH_3COONa ⟶	CH_3COOH +	NaCl
100 × 0.1	200 × 0.1		
= 10 m mol	= 20 m mol	–	–
−10 m mol	−10 m mol	+10 m mol	10 m mol
–	10 m mol	10 m mol	–

27. (d) Bunsen burner is not required during titration.

28. (a) $Cr(OH)_3(s) \rightleftharpoons Cr^{3+}(aq) + 3OH^-(aq)$

$$\quad - \qquad\qquad S \qquad\qquad 3S$$

$$K_{sp} = [Cr^{3+}][OH^-]^3 = S(3S)^3 = 27S^4$$

$$27S^4 = 6 \times 10^{-31}$$

$$(3S)^4 = 18 \times 10^{-31}$$

$$[OH^-] = 3S = (18 \times 10^{-31})^{1/4}$$

29. (d) Equilibrium constant does not tells whether a reaction is slow or fast. Slow or fast process is a kinetic parameter but equilibrium constant is a thermodynamic parameter.

30. (c) From graph $[X] = 1 \times 10^{-3}\ M$

$$[Y] = 2 \times 10^{-3}\ M$$

$$XY_2(s) \rightleftharpoons X^{2+}_{(aq)} + 2Y^-_{(aq)}$$

$$K_{sp} = [X^{2+}][Y^-]^2 = (10^{-3})(2 \times 10^{-3})^2 = 4 \times 10^{-9}\ M^3$$

31. (b) $[Pb^{2+}] = \dfrac{300 \times 0.134}{(300 + 100)} = 0.1M$

$$[Cl^-] = \frac{100 \times 0.4}{(300 + 100)} = 0.1M$$

$$Q_{sp} = [Pb^{2+}][Cl^-]^2 = 0.1 \times (0.1)^2 = 10^{-3}$$

$$K_{sp} = 1.6 \times 10^{-5}$$

$$Q_{sp} > K_{sp}$$

32. (b) Since in PCl_5 no lone pair of electron present over central atom, it can not acts as Lewis base.

33. (c) Let solubility of $Ca(OH)_2$ in water is 'S' mol/L

$$Ca(OH)_2(s) \rightleftharpoons Ca^{2+}(aq) + 2OH^-(aq)$$

$$S \qquad\qquad S \qquad\qquad 2S$$

$$k_{sp} = [Ca^{2+}][OH^-]^2$$

$$5.5 \times 10^{-6} = (S)(2S)^2$$

$$4S^3 = 5.5 \times 10^{-6}$$

$$S = 1.11 \times 10^{-2}\ \text{mol/L}$$

34. (b) $AgCN(s) \rightleftharpoons Ag^+(aq) + CN^-(aq);\ K_{sp}$

$$x \qquad\qquad x \qquad\qquad x - y$$

$$CN^- + H^+ \rightleftharpoons HCN,\ \frac{1}{K_a}$$

$$x \qquad 10^{-3} \qquad -$$

$$x - y \qquad 10^{-3} \qquad y$$

Since $\dfrac{1}{K_a}$ is very high, 2nd reaction will go to completion, so, $x - y \cong 0 \Rightarrow \quad x \cong y$

Adding both equations, we get

$$AgCN(s) + H^+(aq) \rightleftharpoons Ag^+(aq) + HCN(aq),\ \frac{K_{sp}}{K_a}$$

$$- \qquad 10^{-3} \qquad\qquad x \qquad\qquad x$$

$$\frac{K_{sp}}{K_a} = \frac{x^2}{10^{-3}}$$

$$x = \sqrt{\frac{K_{sp}}{K_a} \times 10^{-3}} = \sqrt{\frac{2.2 \times 10^{-16} \times 10^{-3}}{6.2 \times 10^{-10}}} = 1.9 \times 10^{-5}\ M$$

35. (c) $2A(g) \rightleftharpoons 2B(g) + C(g)$ K_1 ... (i)

$\frac{3}{2}A(g) \rightleftharpoons \frac{3}{2}B(g) + \frac{3}{4}C(g)$ K_2 ... (ii)

eq.(ii) is $\frac{3}{4}$ times of eq.(i), hence, $K_2 = (K_1)^{3/4}$

36. (a) PCl_5 = 5 mole;   Ar = 4 mole

$$P_{Total} = \frac{9 \times 0.82 \times 610}{100} = 4.5 \text{ atm}$$

$$P_y = X_y P_T$$

(P_T → Total pressure; X_y → mole fraction of y)

$$X_{PCl_5} = \frac{5}{9} \;;\; X_{Ar} = \frac{4}{9}$$

$$P_{PCl_5} = \frac{5 \times 4.5}{9} = 2.5 \text{ atm} \;;\; P_{Ar} = \frac{4 \times 4.5}{9} = 2 \text{ atm}$$

$$\underset{2.5-P}{PCl_5} \rightleftharpoons \underset{P}{PCl_3} + \underset{P}{Cl_2}$$

$$P_{total} = 2.5 - P + P + P + P_{Ar} = 6 \text{ atm}$$

$P = 1.5$ atm

$$K_p = \frac{1.5 \times 1.5}{1} = 2.25$$

37. (c) pH + pOH = 14

pOH = 14 – 8.26

$$pOH = pK_b + \log\frac{[NH_4^+]}{[NH_3]}$$

$$\Rightarrow \quad 5.74 = 4.74 + \log\frac{[NH_4^+]}{0.2} \quad \Rightarrow \quad [NH_4^+] = 2\text{ M}$$

Molar mass of NH_4Cl = 53.5 g mol^{-1}

Amount of NH_4Cl = 2 × 53.5 = 107 g.

38. (d) In deionised water no common ion effect will take place so solubility of AgCl will be maximum in it.

$$KCl \longrightarrow K^+ + Cl^-$$

$$HCl \longrightarrow H^+ + Cl^-$$

$$AgNO_3 \longrightarrow Ag^+ + NO_3^-$$

39. (a)
$$\underset{C(1-\alpha)}{A_2B_3(aq.)} \rightleftharpoons \underset{2C\alpha}{2A^{3+}(aq.)} + \underset{3C\alpha}{3B^{2-}(aq.)}$$

$$K_{eq} = \frac{[A^{3+}]^2[B^{2-}]^3}{[A_2B_3]} = \frac{4c^2\alpha^2 \times 27c^3\alpha^3}{c(1-\alpha)}, \quad \alpha << 1$$

$$\Rightarrow \quad K_{eq} = \frac{108\,c^5\alpha^5}{c}, \quad \alpha = \left(\frac{K_{eq}}{108\,c^4}\right)^{\frac{1}{5}}$$

40. (d) (A) pH of 10^{-8} M HCl is in acidic range (6.98).

(B) Conjugate base of $H_2PO_4^-$ is HPO_4^{2-}

(C) K_w increases with increasing temperature, as the temperature increases, the dissociation of water increases.

(D) At half neutralisation point, half of the acid is present in the form of salt.

$$pH = pK_a + \log\frac{1}{1} = pK_a$$

41. (b) $A(g) \rightleftharpoons B(g) + \frac{C}{2}(g)$

$t = t_{eq}$ $\quad (1-\alpha) \quad \alpha \quad \frac{\alpha}{2}$

$$P_B = \frac{\alpha}{\left(1+\frac{\alpha}{2}\right)} \cdot P, \quad P_A = \frac{(1-\alpha)}{\left(1+\frac{\alpha}{2}\right)} \cdot P, \quad P_C = \frac{\frac{\alpha}{2}}{\left(1+\frac{\alpha}{2}\right)} \cdot P$$

$$K_P = \frac{P_B \cdot P_C^{1/2}}{P_A} = \frac{(\alpha)^{3/2}(P)^{1/2}}{(1-\alpha)(2+\alpha)^{1/2}}$$

42. (b) $CO_2(g) + C(s) \rightleftharpoons 2CO(g)$

$0.5 \quad\quad -$

$0.5-x \quad\quad 2x$

$$P_{total} = 0.5 + x = 0.8$$
$$x = 0.3$$
$$K_P = \frac{(0.6)^2}{0.2} = 1.8$$

43. (b) $H_2 + I_2 \rightleftharpoons 2HI$

Concentration of H_2 and I_2 decreases until equilibrium condition and concentration of HI increases till equilibrium condition and after equilibrium concentration of all the reactant and products remain constant.

45. (a) $Zr_3(PO_4)_4(s) \rightleftharpoons 3Zr^{4+}(aq) + 4PO_4^{3-}(aq)$

$— \quad\quad 3s \quad\quad 4s$

$$K_{sp} = (3s)^3(4s)^4 = 6912\, s^7$$
$$s = \left(\frac{K_{sp}}{6912}\right)^{1/7}$$

46. (a) Condition for precipitation $Q_{ip} > K_{sp}$

For $[A(OH)_2]$

$[A^{2+}][OH^-]^2 > 9 \times 10^{-10}$

$[A^{2+}] = 1M$

$\Rightarrow \quad [OH^-] > 3 \times 10^{-5} M$

For $[B(OH)_3]$

$[B^{3+}][OH^-]^3 > 27 \times 10^{-18}$

$[B^{3+}] = 1M$

$\Rightarrow \quad [OH^-] > 3 \times 10^{-6} M$

So, $B(OH)_3$ will precipitate before $A(OH)_2$.

47. (b) 1 mole

$X_2Y_{(g)} \rightleftharpoons X_{2(g)} + \frac{1}{2}Y_{2(g)}$

$1-x$ mole $\quad x$ mole $\quad \frac{x}{2}$ mole

$$\therefore \quad P_{x_2y} = \frac{1-x}{1+\frac{x}{2}} \times P$$

$$P_{x_2} = \frac{x}{1+\frac{x}{2}} \times P$$

$$P_{Y_2} = \frac{x/2}{1+\frac{x}{2}} \times P$$

$$\therefore \; K_p = \left(\frac{x}{1+\frac{x}{2}} P\right)\left(\frac{x}{2\left(1+\frac{x}{2}\right)} P\right)^{1/2} \Big/ \left(\frac{1-x}{1+\frac{x}{2}}\right) \times P$$

$$\therefore \; K_p = \left(\frac{x}{1-x}\right)\left(\frac{x}{2\left(1+\frac{x}{2}\right)}\right)^{1/2} \times P^{1/2}$$

$\because$ x to be very very small.

$$\therefore \; K_p = \frac{x^{3/2}}{(2)^{1/2}} \times P^{1/2}$$

$$\therefore \; x^{3/2} = \frac{K_p \times 2^{1/2}}{P^{1/2}}$$

$$\therefore \; x^3 = \frac{K_p^2 \times 2}{P}$$

$$x = \left(\frac{K_p^2 \times 2}{P}\right)^{1/3}$$

48. (a) With increase in temperature, K_w of water increases so, degree of dissociation of water increase $\therefore$ pH as well as pOH of water decreases.

49. (a) $Cr(OH)_{3(s)} \rightleftharpoons Cr^{+3}_{(aq)} + 3OH^-_{(aq)}$

At eq : s $3s$

$$K_{sp} = (s).(3s)^3 = 27s^4$$

$$27s^4 = 1.6 \times 10^{-30}$$

$$s = \left(\frac{1.6}{27} \times 10^{-30}\right)^{1/4}$$

50. (d) $HA \rightleftharpoons H^{\oplus} + A^{\ominus}$

$t = 0$ a

$t = t$ $a(1-x)$ ax ax

$$K_a = (ax)\frac{(x)}{1-x}; [H^+] = ax$$

$$-\log(K_a) = -\log(ax) - \log\left(\frac{x}{1-x}\right)$$

$$pK_a = pH - \log\left(\frac{x}{1-x}\right)$$

$$pH - pK_a = \log\left(\frac{x}{1-x}\right)$$

51. (b) Based on the K_{sp} values and salt analysis cation identification, we can say that order of K_{sp} value is :

$HgS < PbS < AgBr < Ca(OH)_2$

K_{sp} values

$HgS \rightarrow 4 \times 10^{-53}$

$PbS \rightarrow 8 \times 10^{-28}$

$AgBr \rightarrow 5 \times 10^{-13}$

$Ca(OH)_2 \rightarrow 5.5 \times 10^{-6}$

52. (c) $AB_{2(g)} \rightleftharpoons AB_{(g)} + \frac{1}{2}B_{2(g)}$

$$t_{eq.} \quad \frac{(1-x)}{1+\frac{x}{2}}P \quad \frac{xP}{1+\frac{x}{2}} \quad \frac{\left(\frac{x}{2}\right)P}{1+\frac{x}{2}}$$

$$\Rightarrow \quad x << 1 \Rightarrow 1+\frac{x}{2} \simeq 1 \text{ and } 1-x \simeq 1$$

$$\Rightarrow \quad K_p = \frac{(xP)\cdot\left(\frac{xP}{2}\right)^{1/2}}{P}$$

$$\Rightarrow \quad K_p^2 = x^2 \cdot \frac{xP}{2}$$

$$x = \sqrt[3]{\frac{2K_p^2}{P}}$$

B. Objective Questions

[One or more than one option(s) is/are correct]

1. (a, d)

$$Q = \frac{[C][D]}{[A][B]}$$

at $t = 0$

$$[C] = [D] = 0$$

$$\therefore \quad Q = 0$$

With time conc. of D and D increases, so Q increases with time.

2. (d) (c) may be a conditional buffer if $[HNO_3] < [CH_3COONa]$, but given that $[HNO_3] = [CH_3COONa]$, so (d), is the correct answer (a mixture of weak acid and its salt with strong base).

3. (b, c, d)

Cu^{2+} ions will react with CN^- and SCN^- forming $[Cu(CN)_4]^{3-}$ and $[Cu(SCN)_4]^{3-}$ leading the reaction in the backward direction.

$$Cu^{2+} + 2CN^- \longrightarrow Cu(CN)_2$$

$$2Cu(CN)_2 \longrightarrow 2CuCN + (CN)_2$$

$$CuCN + 3CN^- \longrightarrow [Cu(CN)_4]^{3-}$$

$$Cu^{2+} + 4SCN^- \longrightarrow [Cu(SCN)_4]^{3-}$$

Cu^{2+} also combines with $CuCl_2$ which reacts with Cu to produce CuCl pushing the reaction in the backward direction.

$$CuCl_2 + Cu \longrightarrow 2CuCl\downarrow$$

4. (a, b)

Amphoteric oxides are ZnO, Al_2O_3, PbO, PbO_2, BeO, SnO, SnO_2, Cr_2O_3.

NO is a neutral oxide.

CrO is a basic oxide.

B_2O_3 is an acidic oxide.

5. (a, b)

(a) With increase in temperature, the value of K for endothermic reaction increases because unfavourable change in entropy of the surroundings decreases.

(b) With increase in temperature, the value of K for exothermic reaction decreases because favourable change in entropy of the surroundings decreases.

C. Matching Type Problems

1. a —q, b — r, c — s, d — p, e — t

Order of acidic strength $\alpha . K_a$ and $-I$ producing groups increases the acidic strength while $+M$ or $-I$ producing groups decreases the acidic strength.

D. Passage Based Problems

Passage-1

1. (a) $HCl + NaOH \longrightarrow NaCl + H_2O$

$n = 100 \times 1 = 100$ m mole $= 0.1$ mole

Energy evolved due to neutralization of HCl and NaOH $= 0.1 \times 57 = 5.7$ kJ $= 5700$ joule

Energy used to increase temperature of solution $= 200 \times 4.2 \times 5.7 = 4788$ joule

Energy used to increase temperature of calorimeter $= 5700 - 4788 = 912$ joule

$$ms \cdot \Delta t = 912$$

$$ms \times 5.7 = 912$$

$$ms = 160 \text{ joule/°C}$$ [Calorimeter constant]

Energy evolved by neutralization of CH_3COOH and NaOH $= 200 \times 4.2 \times 5.6 + 160 \times 5.6$
$= 5600$ joule

So energy used in dissociation of 0.1 mole $CH_3COOH = 5700 - 5600 = 100$ joule

Enthalpy of dissociation $= 1$ kJ/mol

2. (b) $CH_3COOH = \dfrac{1 \times 100}{200} = \dfrac{1}{2}$

$CH_3CONa = \dfrac{1 \times 100}{200} = \dfrac{1}{2}$

$$pH = pK_a + \log \frac{[salt]}{[acid]}$$

$$pH = 5 - \log 2 + \log \frac{1/2}{1/2} \Rightarrow pH = 4.7$$

Passage-2

1. (b)

$$X_{2(g)} \rightleftharpoons 2X_{(g)}$$

At $t=0$: 1, 0

At eq.: $1 - \dfrac{\beta_{eq}}{2}$, β_{eq}

$$K_p = \frac{(P_X)^2}{(P_{X_2})}$$

$$P_X = \left(\frac{\beta_{eq}}{1 - \frac{\beta_{eq}}{2} + \beta_{eq}} \right) P_{total} = \left(\frac{\beta_{eq}}{1 + \frac{\beta_{eq}}{2}} \right) P_{total}$$

$$P_{X_2} = \left(\frac{1-\frac{\beta_{eq}}{2}}{1+\frac{\beta_{eq}}{2}}\right) P_{total}$$

$$K_p = \frac{\left[\left(\frac{\beta_{eq}}{1+\frac{\beta_{eq}}{2}}\right) P_{total}\right]^2}{\left(\frac{1-\frac{\beta_{eq}}{2}}{1+\frac{\beta_{eq}}{2}}\right) P_{total}} = \left(\frac{\beta_{eq}^2}{1-\frac{\beta_{eq}^2}{4}}\right) P_{total}$$

$$= \left(\frac{4\beta_{eq}^2}{4-\beta_{eq}^2}\right) P_{total} = \left(\frac{4\beta_{eq}^2}{4-\beta_{eq}^2}\right) \times 2 = \frac{8\beta_{eq}^2}{4-\beta_{eq}^2}$$

2. (c) (a) If the pressure on the system is decreased, the equilibrium will shift in the direction in which pressure increases *i.e.*, increase in no. of moles takes place *i.e.*, in forward direction.

(b) At the start of the reaction, $Q < K$ thus, the reaction will proceed in the forward direction *i.e.*, reaction is spontaneous.

(c) If $\beta_{eq} = 0.7$ then, $K_p = \frac{8 \times (0.7)^2}{4-(0.7)^2} > 1$

$\Delta G° = -RT \ln K_p$ so, $\Delta G° = -\text{ve}$ but given $\Delta G° = +\text{ve}$

So, K_p should be less than 1 hence, $\beta_{eq} \neq 0.7$.

(d) $K_p = K_c(RT)^{\Delta n}$

$K_c < K_p$ $\quad (\because RT > 1)$

If $K_p < 1$ then $K_c < 1$

E. Integer Type Problems

1. (8) Given $\quad K_a = 1.0 \times 10^{-4}$

$$pK_a = -\log(1 \times 10^{-4}) = 4, \text{ and } c = 0.01M$$

$\because$
$$\text{pH} = 7 + \frac{1}{2}(pK_a + \log c) \text{ at } 25°\text{C}$$

$$\text{pH} = 7 + \frac{1}{2}[4 + \log(0.01)] = 7 + 1 = 8$$

2. (6) $H_2SO_4, H_2CO_3, H_2S_2O_7, H_2CrO_4, H_3PO_3, H_2SO_3$

3. (7) Let the solubility of AgCl is x mol litre^{-1}

$$\underset{}{AgCl} \rightleftharpoons \underset{x}{Ag^+} + \underset{x}{Cl^-}$$

and that of CuCl is y mol litre^{-1}

$$CuCl \rightleftharpoons \underset{y}{Cu^+} + \underset{y}{Cl^-}$$

$\therefore$
$$K_{sp} \text{ of AgCl} = [Ag^+][Cl^-]$$

$$1.6 \times 10^{-10} = x(x+y) \quad \text{...(i)}$$

Similarly, K_{sp} of CuCl $= [Cu^+][Cl^-]$

$$1.6 \times 10^{-6} = y(x + y) \qquad ...(ii)$$

On solving (i) and (ii)

$$[Ag^+] = 1.6 \times 10^{-7}$$

$$\therefore \quad x = 7$$

4. (6) $\kappa = G \times \dfrac{l}{a}$

$$\kappa = 5 \times 10^{-7} \times \frac{120 \text{ cm}}{1 \text{ cm}^2} = 6 \times 10^{-5} \text{ S cm}^{-1}$$

$$\Lambda_m^c = \frac{\kappa \times 1000}{C} = \frac{6 \times 10^{-5} \times 1000}{0.0015}$$

pH = 4, $[H^+] = 10^{-4} = C\alpha$

$$\alpha = \frac{\Lambda_m^c}{\Lambda_m^\circ}$$

$$\frac{10^{-4}}{0.0015} = \frac{6 \times 10^{-5} \times 1000}{0.0015 \times \Lambda_m^\circ}$$

$$\Lambda_m^\circ = 6 \times 10^2$$

$$Z = 6$$

5. (4.47)

$$s = \sqrt{\frac{K_{sp}[H^+]}{K_a} + K_{sp}} = \sqrt{(2 \times 10^{-10})\left(\frac{10^{-3}}{10^{-8}} + 1\right)} = 4.47 \times 10^{-3}$$

$\therefore \; Y = 4.47$

6. (8.93)

$$Fe^{2+}_{(aq)} + S^{2-}_{(aq)} \rightleftharpoons FeS_{(s)}; \quad K_c = 1.6 \times 10^{17}$$

Before mixing : 0.06 M 0.2 M

or 0.03 M or 0.1 M

At eqm : x $0.1 - 0.03 = 0.07 M$

$$K_c = \frac{1}{[Fe^{2+}][S^{2-}]} \Rightarrow 1.6 \times 10^{17} = \frac{1}{(x) \times 0.07}$$

$$x = \frac{10^{-17}}{1.6 \times 0.07} = \frac{10^{-15}}{11.2} = \frac{100}{11.2} \times 10^{-17}$$

$$x = 8.928 \times 10^{-17} = Y \times 10^{-17} \quad \therefore Y = 8.93$$

7. (0.25) $K_{eq.} = \dfrac{[B]}{[A]}$

$$K_{1000} = \frac{10}{1} = 10 \text{ and } K_{2000} = \frac{100}{1} = 100$$

Now, $\dfrac{\Delta G^\circ_{1000}}{\Delta G^\circ_{2000}} = \dfrac{(-RT \ln K_{eq})_{1000}}{(-RT \ln K_{eq})_{2000}} = \dfrac{1000 \times \ln 10}{2000 \times \ln 100} = \mathbf{0.25}$

8. (3) At 3 mL pH of solution is 11.

$$B + HA \longrightarrow BH^+ + A^-$$

At 3 mL best buffer action is shown.

So, $pOH = pK_b = 3$ (as basic buffer is formed)

So, $pK_b = 3$

9. (0.20)

For ppt., $[Zn^{+2}][S^{-2}] = K_{sp}$

$$[S^{-2}] = \frac{1.25 \times 10^{-22}}{0.05} = 2.5 \times 10^{-21} M$$

$$H_2S \rightleftharpoons 2H^+ + S^{-2}$$

$$K_{Net} = 10^{-21} = \frac{[H^+]^2 \times 2.5 \times 10^{-21}}{0.1}$$

$$[H^+]^2 = \frac{1}{25}$$

$$[H^+] = \frac{1}{5} M = 0.2 M$$

10. (2)

$$2A_{(g)} \rightleftharpoons A_2(g)$$

$$\Delta_r G^\circ = -RT \ln K_p$$

$$25200 = -2.3 \times 8.3 \times 400 \log K_p$$

$$K_p = 5.01 \times 10^{-4} \text{ bar}^{-1}$$

$$K_p = 5.01 \times 10^{-9} \text{ Pa}^{-1}$$

$$K_p = K_c (RT)^{\Delta n_g}$$

$$\Delta n_g = -1$$

$$K_c = K_p \times RT$$

$$K_c = 5.01 \times 10^{-9} \times 8.3 \times 400$$

$$K_c = 1.66 \times 10^{-5} \text{ m}^3/\text{mol} = 1.66 \times 10^{-2} \text{ L/mol}$$

11. (16)

	x	+ y	$\rightleftharpoons$ $2z$
Initial	1	1.5	0.5
Change	$-x$	$-x$	$+2x$
equilibrium	$1-x$	$1.5-x$	$0.5+2x$

It is given, $[z]_{eq} = 1M$

$$0.5 + 2x = 1 \Rightarrow x = \frac{1}{4}$$

$$[x]_{eq} = 1 - x = 1 - \frac{1}{4} = 0.75 M$$

$$[y]_{eq} = 1.5 - x = 1.5 - \frac{1}{4} = 1.25 M$$

$$K_{eq} = \frac{[z]^2}{[x][y]} = \frac{1^2}{\left(\frac{3}{4}\right) \times \left(\frac{5}{4}\right)} = \frac{16}{15}$$

12. (5)

$$Cl_2(g) \rightleftharpoons 2Cl(g)$$

Let moles of Cl_2 and Cl are 'a' at equilibrium.

$$P_i = X_i \times P_{Total}$$

$$P_{Cl_2} = \left(\frac{a}{a+a}\right) \times 1 = \frac{1}{2}$$

$$P_{Cl} = \left(\frac{a}{a+a}\right) \times 1 = \frac{1}{2}$$

$$K_p = \frac{(P_{Cl})^2}{(P_{Cl_2})} = \frac{\left(\frac{1}{2}\right)\left(\frac{1}{2}\right)}{\left(\frac{1}{2}\right)} = 0.5 = 5 \times 10^{-1}$$

13. (1380)
$$K_p = 100$$
$$\Delta_r G° = -RT \ln K_p = -R(300)\ln(100)$$
$$= -2R \times 300 \times \ln 10 = -2R \times 300 \times 2.3 = -1380\ R$$
$$x = 1380$$

14. (20)
$$\Delta_r G° = -RT \ln K$$
$$\ln K = -\frac{\Delta_r G°}{RT} = \frac{9.478 \times 10^3}{8.314 \times 495} = 2.303 = \ln 10$$

So, $K = 10$

	A	$\rightleftharpoons$	B
$t = 0$	20		–
$t = t_{eq}$	$20 - x$		x

$$K = \frac{[B]}{[A]} = \frac{x}{22 - x} = 10$$
$$x = 20$$

Millimoles of B formed = 20

15. (354) $N_2O_4(g) \rightleftharpoons 2NO_2(g)$
$$\Delta n_g = 2 - 1 = 1$$
$$K_p = 600.1$$
$$K_c = 20.4$$
$$K_p = K_c(RT)^{\Delta n_g} = K_c RT$$
$$T = \frac{K_p}{K_c \cdot R} = \frac{600.1}{20.4 \times 0.83}$$
$$T = 354\ \text{K}$$

16. (10.6)
$$[\text{NaOH}] = \frac{(4/40)\ \text{mol}}{100\ \text{L}} = 10^{-3}\ M\ (\text{solution '}A\text{'})$$
$$[H_2SO_4] = \frac{(9.8/98)\ \text{mol}}{100\ \text{L}} = 10^{-3}\ M\ (\text{solution '}B\text{'})$$

Moles of NaOH in 40 L of solution 'A' $= 40 \times 10^{-3} = 4 \times 10^{-2}$ mol

NaOH	$\longrightarrow$	Na^+	+	OH^-
4×10^{-2} mol		–		–
–				4×10^{-2} mol

moles of H_2SO_4 in 10 L of solution 'B' $= 10^{-3} \times 10 = 10^{-2}$ mol

H_2SO_4	$\longrightarrow$	$2H^+$	+	SO_4^{2-}
10^{-2} mol		–		–
–		2×10^{-2} mol		–

H^+	+	OH^-	$\longrightarrow$	H_2O
2×10^{-2} mol		4×10^{-2} mol		–
–		2×10^{-2} mol		–

$$[OH^-] = \left(\frac{2 \times 10^{-2}}{10 + 40}\right) \text{mol/L} = 4 \times 10^{-4}\ M$$
$$\text{pOH} = 4 - \log 4 = 3.4$$
$$\text{pH} = 14 - 3.4 = 10.6$$

17. (5.2271)

Moles of acetic acid $= \frac{3}{60} = \frac{1}{20} = 0.05$

Moles of HCl $= 0.1 \times \frac{250}{1000} = 0.025$

Moles of acetic acid in 20 mL $= 0.05 \times \frac{20}{500} = 0.002$

Moles of HCl in 20 mL $= 0.025 \times \frac{20}{500} = 0.001$

Moles of NaOH $= 5 \times \frac{0.5}{1000} = 0.0025$

NaOH	+ HCl	$\longrightarrow$ NaCl	+ H_2O
0.0025	0.001	–	–
0.0015	–	0.001	0.001

CH_3COOH	+ NaOH	$\longrightarrow$ CH_3COONa	+ H_2O
0.002	0.0015	–	–
0.0005	–	0.0015	–

$$pH = pK_a + \log\frac{[CH_3COONa]}{[CH_3COOH]} = 4.75 + \log\left(\frac{0.0015}{0.0005}\right)$$

$$= 4.75 + 0.4771 = 5.2271$$

18. (37) Let moles of CO_2 in soft drink bottle $= a$

From Henry's law, $P = K_H \cdot x$

$$\frac{3}{30} = \frac{K_H \cdot a}{K_H \cdot 1}$$

$$a = 0.1 \text{ mol}$$

Volume of solution $= 1000\text{ g} \times (1\text{ gm/mL}) = 1000\text{ mL} = 1\text{L}$

$$[CO_2] = \frac{0.1}{1} = 0.1 \text{ mol / L}$$

H_2O	+ CO_2	$\longrightarrow$ H_2CO_3
	0.1 mol/L	0.1 mol/L

$$H_2CO_3(aq) \rightleftharpoons H^+(aq) + HCO_3^-(aq)$$

$$pH = \frac{1}{2}(pK_{a_1} - \log C) = \frac{1}{2}(6.4 - \log 0.1) = 3.7 = 37 \times 10^{-1}$$

19. (2) Let solubility is 'S' mol/L

$AB_2(s) \rightleftharpoons$	$A^{2+}(aq)$	$+ 2B^-(aq)$
–	S	$2S$

$$K_{sp} = [A^{2+}][B^-]^2 = (S)(2S)^2 = 4S^3$$

$$4S^3 = 3.2 \times 10^{-11}$$

$$S = 2 \times 10^{-4} \text{ mol/L}$$

20. (64) $K_{sp} = S^2 = (8 \times 10^{-4})^2 = 64 \times 10^{-8}\ M^2$

$H_2SO_4 \longrightarrow$	$2H^+$	$+ SO_4^{2-}$
0.01	0.02	0.01

$CdSO_4 \rightleftharpoons$	Cd^{2+}	$+ SO_4^{2-}$
x	x	$(0.01 + x)$

$$K_{sp} = 64 \times 10^{-8} = x(0.01 + x)$$

Solubility of $CdSO_4$ in H_2SO_4 decreases due to common ion effect.

So, $0.01 + x \cong 0.01$

$$0.01(x) = 64 \times 10^{-8}$$

$$x = 64 \times 10^{-6} \text{ mol/L}$$

21. (50) Most precise volume of HCl is 5 mL.

Equivalents (Na_2CO_3) = Equivalents (HCl)

$$M \times \frac{10}{1000} \times 2 = \frac{0.2 \times 5 \times 1}{1000}$$

$$M = 0.05 \text{ mol/L} = 0.05 \times 1000 = 50 \text{ mM}$$

22. (141)

Let solubility of PbI_2 is 'x' mol/L in $Pb(NO_3)_2$ solution.

$$\underset{0.1M}{Pb(NO_3)aq} \longrightarrow \underset{0.1M}{Pb^{2+}(aq)} + 2NO_3^-(aq)$$

$$\underset{x}{PbI_2(s)} \rightleftharpoons \underset{(0.1+x)}{Pb^{2+}(aq)} + \underset{2x}{2I^-(aq)}$$

$$K_{sp} = [Pb^{2+}][I^-]^2$$

$$K_{sp} = (0.1 + x)(2x)^2$$

$0.1 + x \cong 0.1$ (due to common ion)

$$0.1(2x)^2 = 8 \times 10^{-9}$$

$$x^2 = 2 \times 10^{-8}$$

$$x = \sqrt{2} \times 10^{-4}$$

$$x = 1.41 \times 10^{-4}$$

$$x = 141 \times 10^{-6}\ M$$

23. (4) Let moles of NaOH = x

Moles of $Na_2CO_3 = y$

At HPh end point,

$$\underset{n_f=1}{NaOH} + HCl \longrightarrow NaCl + H_2O$$

$$\underset{n_f=1}{Na_2CO_3} + HCl \longrightarrow NaHCO_3 + NaCl$$

Equivalents (NaOH + Na_2CO_3) = Equivalent (HCl)

$$x \times 1 + y \times 1 = \frac{1}{10} \times 17.5 = 1.75 \quad \text{...(i)}$$

At methyl orange end point

$$\underset{n_f=1}{NaOH} + HCl \longrightarrow NaCl + H_2O$$

$$\underset{n_f=2}{Na_2CO_3} + 2HCl \longrightarrow 2NaCl + H_2O + CO_2$$

Equivalents (NaOH + Na_2CO_3) = Equivalent (HCl)

$$x \times 1 + y \times 2 = \frac{1}{10} \times 19 = 1.9 \quad \text{...(ii)}$$

Solving eqn. (i) and (ii),

We get,

$$x = 1.60$$

$$y = 0.15$$

$$\% Na_2CO_3 = \frac{0.15 \times 10^{-3} \times 10^6}{0.4} \times 100 = 3.97\%$$

24. (6) Since most precise volume of NaOH is 4.4 mL Equivalents ($H_2C_2O_4$)

= Equivalents (NaOH)

$$M \times V \times n_f = M \times V \times n_f$$

$$1.25 \times 10 \times 2 = M \times 4.4 \times 1$$

$$M = 5.68 \text{ mol/L}$$

25. (50)

$$A_2X(s) \rightleftharpoons 2A^+(aq) + X^{2-}(aq)$$
$$S_1 \qquad 2S_1 \qquad S_1$$
$$K_{sp} = 4S_1^3 = 4\times10^{-12}$$
$$S_1 = 10^{-4}M$$
$$MX(s) \rightleftharpoons M^+(aq) + X^-(aq)$$
$$S_2 \qquad S_2 \qquad S_2$$
$$K_{sp} = S_2^2 = 4\times10^{-12}$$
$$S_2 = 2\times10^{-6}\,M$$
$$\frac{S_{A_2X}}{S_{MX}} = \frac{S_1}{S_2} = \frac{10^{-4}}{2\times10^{-6}} = 50$$

26. (2) $HCl(aq) \longrightarrow H^+(aq) + Cl^-(aq)$

0.1M	—	—
—	0.1M	0.1M

$HA(aq) \rightleftharpoons H^+(aq) + A^-(aq)$

0.01	0.1	—
$0.01 - x$	$0.1 + x$	x

$$K_a = \frac{[H^+][A^-]}{[HA]} = \frac{x(0.1+x)}{(0.01-x)} \qquad (\because\ 0.1 + x \cong 0.1 \text{ and } 0.01 - x \cong 0.01)$$
$$2\times10^{-6} = \frac{0.1\times x}{0.01}$$
$$x = 2\times10^{-7}$$
$$\alpha = \frac{x}{0.01} = \frac{2\times10^{-7}}{0.01} = 2\times10^{-5}$$

27. (10) From Henderson's equation,
$$pH = pK_a + \log\frac{[CH_3COONa]}{[CH_3COOH]}$$
$$5.74 = 4.74 + \log\frac{[CH_3COONa]}{1}$$
$$1 = \log_{10}[CH_3COONa]$$
$$[CH_3COONa] = 10\,M$$

28. (29) Moles of water $= \dfrac{0.90}{18}$

By ideal gas equation $PV = nRT$
$$V = \frac{nRT}{P} = \frac{0.90\times0.82\times300\times760}{18\times32} = 29.21 \approx 29$$

29. (747) $2O_3(g) \rightleftharpoons 3O_2(g)$

$t = 0$	2	0
$t = eq.$	$2 - 1 = 1$	3/2

Total moles at equilibrium = 1 + 3/2 = 5/2

Total pressure = 1 atm. (given)

Partial pressure = mole fraction × total pressure
$$P_{O_3} = \frac{1}{5/2}\times1 = \frac{2}{5}\ ; \qquad P_{O_2} = \frac{3/2}{5/2}\times1 = \frac{3}{5}$$

$$K_P = \frac{(P_{O_2})^3}{(P_{O_3})^2} = \frac{(3/5)^3}{(2/5)^2} = 1.35$$

Free energy change, $\Delta G = -RT \ln K_P = -8.3 \times 300 \times \ln 1.35 = -747 \text{ J mol}^{-1}$

30. (282) $K_{sp} = S^2$

$$S = \sqrt{K_{sp}} = \sqrt{8 \times 10^{-28}} = 2\sqrt{2} \times 10^{-14} = 2.82 \times 10^{-14} = 282 \times 10^{-16}$$

31. (476) Moles of CH_3COOH = 50 × 0.1 = 5 m mole

Moles of NaOH = 25 × 0.1 = 2.5 m mole

$$NaOH + CH_3COOH \longrightarrow CH_3COONa + H_2O$$

	NaOH	CH_3COOH	$CH_3COONa + H_2O$
	2.5 m mole	5 m mole	
	0	2.5 m mole	2.5 m mole

Total volume of this buffer solution = 75 mL

$$pH = pK_a + \log\left(\frac{2.5/75}{2.5/78}\right)$$

$\Rightarrow$ $pH = 4.76 = 476 \times 10^{-2}$

32. (44) $CO(g) + H_2O(g) \rightleftharpoons CO_2(g) + H_2(g)$

	$CO(g)$	$H_2O(g)$	$CO_2(g)$	$H_2(g)$
$t = 0$;	1 mol	1 mol	0	0
at equ. ;	$1-x$	$1-x$	x	x

at equilibrium 40% by mass water reacts with CO

$x = 0.4$, $1 - x = 0.6$

$$K_C = \frac{[CO_2][H_2]}{[CO][H_2O]} = \frac{0.4 \times 0.4}{0.6 \times 0.6} = 0.44$$

$$K_C \times 10^2 = 44$$

33. (1) $H_2(g) + I_2(g) \rightleftharpoons 2HI(g)$

	$H_2(g)$	$I_2(g)$	$2HI(g)$
$t = 0$;	4.5	4.5	–
t_{eq} ;	3	3	3

$$K_C = \frac{[HI]^2}{[H_2][I_2]} = \frac{(3)^2}{3 \times 3} = \frac{9}{9} = 1$$

34. (9000) $(M_1 \times V_1) = (M_2 \times V_2)$

$$10^{-1} \times 1 = 10^{-2} \times V_2$$

$$V_2 = 10 \text{ L}$$

Water added = 10 – 1 = 9 Litre = 9000 mL

35. (417)

$$K_C = \frac{[NH_3]^2}{[N_2][H_2]^3}$$

$$K_C = \frac{(1.5 \times 10^{-2})^2}{(2 \times 10^{-2}) \times (3 \times 10^{-2})^3}$$

$$K_C = 417$$

36. (7)

$$pH = \frac{pK_w + pK_a - pK_b}{2}$$

$$pK_a = pK_b$$

$\Rightarrow$ $$pH = \frac{pK_w}{2} = 7$$

37. (7) $C_2H_5NH_2(aq) + H_2O \rightleftharpoons C_2H_5NH_3^+ + \overset{\ominus}{O}H$

$C = 10^{-3}\,M \qquad — \qquad —$

$C(1-\alpha) \qquad C\alpha \qquad C\alpha$

$\Rightarrow \quad C = 10^{-3} \qquad = 10^{-5} \qquad = 10^{-5}$

$1 - \alpha \simeq 1$

Given, pH $= 9 \Rightarrow$ pOH $= 5 \Rightarrow [\overset{\ominus}{O}H] = 10^{-5}\,M$

Now, $K_b = \dfrac{[C_2H_5NH_3^+][\overset{\ominus}{O}H]}{[C_2H_5NH_2]}$

$\Rightarrow \quad K_b = \dfrac{10^{-5} \times 10^{-5}}{10^{-3}} = 10^{-7}$

38. (75)

$MX_2 \rightarrow M^{2+} + 2X^-$

$i = \dfrac{\text{normal molar mass}}{\text{observed molar mass}}$

$i = \dfrac{164}{65.6}$

$1 + (3-1)\alpha = \dfrac{164}{65.6}$

$2\alpha = \dfrac{98.4}{65.6}$

$\alpha = 0.75$

Percent dissociation = 75%

39. (2) Electrolysis of NaCl is

$$NaCl + H_2O(aq) \rightarrow NaOH(aq) + \frac{1}{2}Cl_2(g) + \frac{1}{2}H_2(g)$$

Since, during electrolysis pH changes to 12

So, $[OH^{\ominus}] = 10^{-2}$ and $[H^+] = 10^{-12}$

So, by Faraday law

Gram amount of substance deposited = Amount of electricity passed

$10^{-2} \times \dfrac{600}{1000} \times 96500 = I \times t$

$\dfrac{10^{-2} \times 600}{1000} \times 96500 = I \times 5 \times 60$

$I = \dfrac{10^{-2} \times 600 \times 96500}{1000 \times 5 \times 60}$

$I = 1.93$ ampere

So, $I = 2$ ampere (nearest integer)

40. (16)

$A_2B \rightarrow 2A^+ + B^{-2};\ y = 3$

$\alpha = 0.3$

$i = 1 + (y-1)\alpha$

$= 1 + (3-1)(0.3) = 1.6 = 16 \times 10^{-1}$

❑❑❑

CHAPTER

6

Redox Reactions

A. Objective Questions

[Only one option is correct]

1. In which of the following reactions H_2O_2 acts as a reducing agent? **[JEE (Mains) 2014]**

1. $H_2O_2 + 2H^+ + 2e^- \longrightarrow 2H_2O$
2. $H_2O_2 - 2e^- \longrightarrow O_2 + 2H^+$
3. $H_2O + 2e^- \longrightarrow 2OH^-$
4. $H_2O_2 + 2OH^- - 2e^- \longrightarrow O_2 + 2H_2O$

(a) 1, 2 (b) 3, 4 (c) 1, 3 (d) 2, 4

2. The molecular formula of a commercial resin used for exchanging ions in water softening is $C_8H_7SO_3Na$ (mol. wt. = 206). What would be the maximum uptake of Ca^{2+} ions by the resin when expressed in mole per gram resin? **[JEE Main 2015]**

(a) 1/103 (b) 1/206 (c) 2/309 (d) 1/412

3. The order of the oxidation state of the phosphorus atom in $H_3PO_2, H_3PO_4, H_3PO_3$ and $H_4P_2O_6$ is: **[JEE (Advanced) 2017]**

(a) $H_3PO_4 > H_3PO_2 > H_3PO_3 > H_4P_2O_6$
(b) $H_3PO_2 > H_3PO_3 > H_4P_2O_6 > H_3PO_4$
(c) $H_3PO_3 > H_3PO_2 > H_3PO_4 > H_4P_2O_6$
(d) $H_3PO_4 > H_4P_2O_6 > H_3PO_3 > H_3PO_2$

4. In order to oxidise a mixture of one mole of each of $FeC_2O_4, Fe_2(C_2O_4)_3, FeSO_4$ and $Fe_2(SO_4)_3$ in acidic medium, the number of moles of $KMnO_4$ required is : **[JEE (Mains) 2019]**

(a) 2 (b) 1.5 (c) 3 (d) 1

5. An example of a disproportionation reaction is: **[JEE (Mains) 2019]**

(a) $2NaBr + Cl_2 \longrightarrow 2NaCl + Br_2$
(b) $2KMnO_4 \longrightarrow 2KMnO_4 + MnO_2 + O_2$
(c) $2CuBr \longrightarrow CuBr_2 + Cu$
(d) $2MnO_4^- + 10I^- + 16H^+ \longrightarrow 2Mn^{2+} + 5I_2 + 8H_2O$

6. In the reaction of oxalate with permanganate in acidic medium, the number of electrons involved in producing one molecule of CO_2 is: **[JEE (Mains) 2019]**

(a) 10 (b) 5 (c) 2 (d) 1

7. The hardness of a water sample (in terms of equivalents of $CaCO_3$) containing 10^{-3} M $CaSO_4$ is : (molar mass of $CaSO_4 = 136\ g\ mol^{-1}$) **[JEE Main 2019]**

(a) 10 ppm (b) 90 ppm (c) 50 ppm (d) 100 ppm

8. 100 mL of a water sample contains 0.81 g of calcium bicarbonate and 0.73 g of magnesium bicarbonate. The hardness of this water sample expressed in terms of equivalents of $CaCO_3$ is: (molar mass of calcium bicarbonate is $162\ g\ mol^{-1}$ and magnesium bicarbonate is $146\ g\ mol^{-1}$) **[JEE Main 2019]**

(a) 100 ppm (b) 1,000 ppm (c) 5,000 ppm (d) 10,000 ppm

9. The redox reaction among the following is: **[JEE Main (Jan.) 2020]**
(a) reaction of $[Co(H_2O)_6]Cl_3$ with $AgNO_3$
(b) combination of dinitrogen with dioxygen at 2000 K
(c) reaction of H_2SO_4 with NaOH
(d) formation of ozone from atmospheric oxygen in the presence of sunlight

10. The compound that cannot act both as oxidising and reducing agent is: **[JEE Main (Jan.) 2020]**
(a) H_2O_2 (b) H_2SO_3 (c) HNO_2 (d) H_3PO_4

11. Oxidation number of potassium in K_2O, K_2O_2 and KO_2 respectively is: **[JEE Main (Jan.) 2020]**
(a) +1, +2 and +4 (b) +2, +1 and +½ (c) +1, +1 and +1 (d) +1, +4 and +2

12. The ammonia (NH_3) released on quantitative reaction of 0.6 g urea (NH_2CONH_2) with sodium hydroxide (NaOH) can be neutralized by: **[JEE Main (Jan.) 2020]**
(a) 100 mL of 0.1 *N* HCl (b) 100 mL of 0.2 *N* HCl
(c) 200 mL of 0.2 *N* HCl (d) 200 mL of 0.4 *N* HCl

13. The exact volumes of 1 *M* NaOH solution required to neutralise 50 mL of 1 *M* H_3PO_3 solution and 100 mL of 2 *M* H_3PO_2 solution, respectively, are : **[JEE Main (March) 2021]**
(a) 100 mL and 100 mL (b) 100 mL and 50 mL (c) 100 mL and 200 mL (d) 50 mL and 50 mL

14. Which of the following equation depicts the oxidizing nature of H_2O_2? **[JEE Main (Feb.) 2021]**
(a) $KIO_4 + H_2O_2 \longrightarrow KIO_3 + H_2O + O_2$ (b) $2I^- + H_2O_2 + 2H^+ \longrightarrow I_2 + 2H_2O$
(c) $I_2 + H_2O_2 + 2OH^- \longrightarrow 2I^- + 2H_2O + O_2$ (d) $Cl_2 + H_2O_2 \longrightarrow 2HCl + O_2$

15. The oxidation states of nitrogen in NO, N_2O and NO_3^- are in the order of : **[JEE Main (March) 2021]**
(a) $NO_3^- > NO_2 > NO > N_2O$ (b) $NO_2 > NO_3^- > NO > N_2O$
(c) $N_2O > NO_2 > NO > NO_3^-$ (d) $NO > NO_2 > N_2O > NO_3^-$

16. Which one of the following is an example of disproportionation reaction? **[JEE Main (June) 2022 (II)]**
(a) $3MnO_4^{2-} + 4H^+ \longrightarrow 2MnO_4^- + MnO_2 + 2H_2O$
(b) $MnO_4^{2-} + 4H^+ + 4e^- \longrightarrow MnO_2 + 2H_2O$
(c) $10I^- + 2MnO_4^- + 16H^+ \longrightarrow 2Mn^{2+} + 8H_2O + 5I_2$
(d) $8MnO_4^- + 3S_2O_3^{2-} + H_2O \longrightarrow 8MnO_2 + 6SO_4^{2-} + 2OH^-$

17. Which of the following options are correct for the reaction

$$2[Au(CN)_2]^-(aq) + Zn(s) \longrightarrow 2Au(s) + [Zn(CN)_4]^{2-}(aq)$$

[JEE Main (April) 2023 (I)]

A. Redox reaction B. Displacement reaction
C. Decomposition reaction D. Combination reaction

Choose the correct answer from the options given below :
(a) A and B only (b) A only (c) C and D only (d) A and D only

18. $2IO_3^- + xI^- + 12H^+ \longrightarrow 6I_2 + 6H_2O$
What is the value of x? **[JEE Main (April) 2023 (I)]**
(a) 12 (b) 2 (c) 6 (d) 10

19. The correct group of halide ions which can be oxidised by oxygen in acidic medium is : **[JEE Main (April) 2023 (II)]**
(a) Br^- only (b) Cl^-, Br^- and I^- only (c) Br^- and I^- only (d) I^- only

20. In acidic medium, $K_2Cr_2O_7$ shows oxidising action as represented in the half reaction

$$Cr_2O_7^{2-} + XH^+ + Ye^- \longrightarrow 2A + ZH_2O$$

X, Y, Z and A respectively are : **[JEE Main (Jan.) 2024]**
(a) 8, 6, 4 and Cr_2O_3 (b) 14, 7, 6 and Cr^{3+} (c) 8, 4, 6 and Cr_2O_3 (d) 14, 6, 7 and Cr^{3+}

21. Which of the following reactions are disproportionation reactions ?

(A) $Cu^+ \longrightarrow Cu^{2+} + Cu$ (B) $3MnO_4^{2-} + 4H^+ \longrightarrow 2MnO_4^- + MnO_2 + 2H_2O$

(C) $2KMnO_4 \longrightarrow K_2MnO_4 + MnO_2 + O_2$ (D) $2MnO_4^- + 3Mn^{2+} + 2H_2O \longrightarrow 5MnO_2 + 4H^+$

Choose the correct answer from the options given below : **[JEE Main (Jan.) 2024]**

(a) (A), (B) (b) (B), (C), (D) (c) (A), (B), (C) (d) (A), (D)

22. The species which does not undergo disproportionation reaction is : **[JEE Main (Jan.) 2025]**

(a) ClO_2^- (b) ClO_4^- (c) ClO^- (d) ClO_3^-

23. Given below are two statements :

Statement-I : In the oxalic acid *vs* $KMnO_4$ (in the presence of dil. H_2SO_4) titration the solution needs to be heated initially to 60°C, but no heating is required in ferrous ammonium sulphate (FAS) *vs* $KMnO_4$ titration (in the presence of dil. H_2SO_4).

Statement-II : In oxalic acid *vs* $KMnO_4$ titration, the initial formation of $MnSO_4$ takes place at high temperature, which then acts as catalyst for further reaction. In the case of FAS *vs* $KMnO_4$, heating oxidizes Fe^{2+} into Fe^{3-} by oxygen of air and error may be introduced in the experiment.

In the light of the above statements, choose the correct answer from the options given below :

[JEE Main (Jan.) 2025]

(a) Statment-I is false but Statement-II is true. (b) Both Statement-I and Statement-II are true.

(c) Statement-I is true but Statement-II is false. (d) Both Statement-I and Statement-II are false.

24. Which of the following oxidation reactions are carried out by both $K_2Cr_2O_7$ and $KMnO_4$ in acidic medium?

A. $I^- \rightarrow I_2$ B. $S^{2-} \rightarrow S$ C. $Fe^{2+} \rightarrow Fe^{3+}$ D. $I^- \rightarrow IO_3^-$

E. $S_2O_3^{2-} \rightarrow SO_4^{2-}$

Choose the correct answer from the options given below : **[JEE Main (Jan.) 2025]**

(a) B, C and D only (b) A, D and E only (c) A, B and C only (d) C, D and E only

B. Objective Questions

1. For the reaction,

$$I^- + ClO_3^- + H_2SO_4 \longrightarrow Cl^- + HSO_4^- + I_2$$

The correct statement(s) in the balanced equation is/are : **[JEE (Advanced) 2014]**

(a) stoichiometric coefficient of HSO_4^- is 6 (b) iodide is oxidized

(c) sulphur is reduced (d) H_2O is one of the product

C. Matching Type Problems

1. Match the reactions in **Column I** with the nature of the reactions/type of the products listed in **Column II.** **(IIT 2007)**

	Column I		Column II
(a)	$O_2^- \longrightarrow O_2 + O_2^{2-}$	(p)	redox reaction
(b)	$CrO_4^{2-} + 4H^+ \longrightarrow$	(q)	one of the products has trigonal planar structure
(c)	$MnO_4^- + NO_2^- + H^+ \longrightarrow$	(r)	dimeric bridged tetrahedral metal ion
(d)	$NO_3^- + H_2SO_4 + Fe^{2+} \longrightarrow$	(s)	disproportionation

2. Match the **List-I** with **List-II**. **[JEE Main (Jan.) 2025]**

	List-I (Redox Reaction)		List-II (Type of Redox Reaction)
(A)	$CH_{4(g)} + 2O_{2(g)} \xrightarrow{\Delta} CO_{2(g)} + 2H_2O_{(l)}$	(I)	Disproportionation reaction
(B)	$2NaH_{(s)} \xrightarrow{\Delta} 2Na_{(s)} + H_{2(g)}$	(II)	Combination reaction
(C)	$V_2O_{5(s)} + 5Ca_{(s)} \xrightarrow{\Delta} 2V_{(s)} + 5CaO_{(s)}$	(III)	Decomposition reaction
(D)	$2H_2O_{2(aq)} \xrightarrow{\Delta} 2H_2O_{(l)} + O_{2(g)}$	(IV)	Displacement reaction

D. Passage Based Problems

Passage-1

Bleaching powder and bleach solution are produced on a large scale and used in several house-hold products. The effectiveness of bleach solution is often measured by iodometry. **(IIT 2012)**

1. 25 mL of household bleach solution was mixed with 30 mL of 0.50 *M* KI and 10 mL of 4 *N* acetic acid. In the titration of the liberated iodine, 48 mL of 0.25 *N* $Na_2S_2O_3$ was used to reach the end point. The molarity of the household bleach solution is :

(a) 0.48 *M* (b) 0.96 *M* (c) 0.24 *M* (d) 0.024 *M*

2. Bleaching powder contains a salt of an oxoacid as one of its components. The anhydride of that oxoacid is :

(a) Cl_2O (b) Cl_2O_7 (c) ClO_2 (d) Cl_2O_6

E. Integer Type Problems

1. The oxidation number of Mn in the product of alkaline oxidative fusion of MnO_2 is: **(IIT 2009)**

2. Among the following, the number of elements showing only one non-zero oxidation state is :

O, Cl, F^-, N, P, Sn, Tl, NaTi **(IIT 2010)**

3. The difference in the oxidation numbers of the two types of sulphur atoms in $Na_2S_4O_6$ is: **(IIT 2011)**

4. Reaction of Br_2 with Na_2CO_3 in aqueous solution gives sodium bromide and sodium bromate with evolution of CO_2 gas. The number of sodium bromide molecules involved in the balanced chemical equation is : **(IIT 2011)**

5. Consider the following list of reagents :

Acidified $K_2Cr_2O_7$, alkaline $KMnO_4$, $CuSO_4$, H_2O_2, Cl_2, O_3, $FeCl_3$, HNO_3 and $Na_2S_2O_3$. The total number of reagents that can oxidise aqueous iodide to iodine is : **[JEE (Advanced) 2014]**

6. In neutral or faintly alkaline solution, 8 moles of permanganate anion quantitatively oxidize thiosulphate anions to produce *X* moles of a sulphur containing product. The magnitude of *X* is: **[JEE (Advanced) 2016]**

7. In the chemical reaction between stoichiometric quantities of $KMnO_4$ and KI in weakly basic solution, what is the number of moles of I_2 released for 4 moles of $KMnO_4$ consumed ? **[JEE (Advanced) 2020]**

8. The hardness of a water sample containing 10^{-3} *M* $MgSO_4$ expressed as $CaCO_3$ equivalents (in ppm) is_____. **[JEE Main (Jan.) 2020]**

(molar mass of $MgSO_4$ is 120.38 g/mol)

9. The oxidation states of iron atoms in compounds (*A*), (*B*) and (*C*), respectively, are *x*, *y* and *z*. The sum of *x*, *y* and *z* is .... **[JEE Main (Sept.) 2020]**

$Na_4[Fe(CN)_5(NOS)]$ $Na_4[FeO_4]$ $[Fe(CO_9)]$
(A) (B) (C)

10. The volume (in mL) of 0.1 N NaOH required to neutralise 10 mL of 0.1 N phosphinic acid is ______.
[JEE Main (Sept.) 2020]

11. Consider the following equations :

$$2Fe^{2+} + H_2O_2 \longrightarrow xA + yB \quad \text{(in basic medium)}$$

(in acidic medium)

$$2MnO_4^- + 6H^+ + 5H_2O_2 \longrightarrow x'C + y'D + z'E$$

(in acidic medium)

The sum of the stoichiometric coefficients x, y, x', y' and z' for products A, B, C, D and E, respectively, is ....
[JEE Main (Sept.) 2020]

12. The volume, in mL, of 0.02 M $K_2Cr_2O_7$ solution required to react with 0.288 g of ferrous oxalate in acidic medium is ______.
(Molar mass of Fe = 56 mol^{-1}) **[JEE Main (Sept.) 2020]**

13. In basic medium CrO_4^{2-} oxidises $S_2O_3^{2-}$ to form SO_4^{2-} and itself changes into $Cr(OH)_4^-$. The volume of 0.154 M CrO_4^{2-} required to react with 40 mL of 0.25 M $S_2O_3^{2-}$ is ______ mL.
(Rounded-off to the nearest integer) **[JEE Main (Feb.) 2021]**

14. 0.4 g mixture of NaOH, Na_2CO_3 and some inert impurities was first titrated with $N/10$ HCl using phenolphthalein as an indicator, 17.5 mL of HCl was required at the end point. After this methyl orange was added and titrated. 1.5 mL of same HCl was required for the next end point. The weight percentage of Na_2CO_3 in the mixture is______. (Rounded-off to the nearest integer)
[JEE Main (Feb.) 2021]

15. Consider titration of NaOH solution *versus* 1.25 M oxalic acid solution. At the end point following burette readings were obtained. **[JEE Main (Feb.) 2021]**
(i) 4.5 mL (ii) 4.5 mL (iii) 4.4 mL (iv) 4.4 mL (v) 4.4 mL
If the volume of oxalic acid taken was 10.0 mL then the molarity of the NaOH solution is ______ M. (Rounded-off to the nearest integer)

16. $2MnO_4^- + bC_2O_4^{2-} + cH^+ \longrightarrow xMn^{2+} + yCO_2 + zH_2O$
If the above equation is balanced with integer coefficients, the value of c is______.
(Round off to the nearest integer). **[JEE Main (March) 2021]**

17. 15 mL of aqueous solution of Fe^{2+} in acidic medium completely reacted with 20 mL of 0.03 M aqueous $Cr_2O_7^{2-}$. The molarity of the Fe^{2+} solution is.........$\times 10^{-2} M$ (Round off to the nearest integer).
[JEE Main (March) 2021]

18. On reaction with stronger oxidizing agent like KIO_4, hydrogen peroxide with the evolution of O_2. The oxidation number of I in KIO_4 changes to ______. **[JEE Main (July) 2022 (II)]**

19. See the following chemical reaction : **[JEE Main (April) 2023 (II)]**

$$Cr_2O_7^{2-} + XH^+ + 6Fe^{2+} \longrightarrow YCr^{3+} + 6Fe^{3+} + ZH_2O$$

The sum of X, Y and Z is ______ .

20. In ammonium phosphomolybdate, the oxidation state of Mo is + ______ .
[JEE Main (April) 2023 (I)]

Answers

[A] Objective Questions (Only one option is correct)

1. (d)	**2.** (d)	**3.** (d)	**4.** (a)	**5.** (c)	**6.** (d)	**7.** (d)	**8.** (b)	**9.** (b)	**10.** (d)
11. (c)	**12.** (b)	**13.** (c)	**14.** (b)	**15.** (a)	**16.** (a)	**17.** (a)	**18.** (d)	**19.** (d)	**20.** (d)
21. (a)	**22.** (b)	**23.** (b)	**24.** (c)						

[B] Objective Questions (One or more than one option(s) is/are correct)

1. (a, b, d)

[C] Matching Type Problems

1. (a) → p, s; (b) → r ; (c) → p, q; (d) → p **2.** A-II, B-III, C-IV, D-I

[D] Passage Based Problems

1. (c) **2.** (a)

[E] Integer Type Problems

1. (6)	**2.** (1)	**3.** (5)	**4.** (5)	**5.** (7)
6. (6)	**7.** (6)	**8.** (100)	**9.** (6)	**10.** (10)
11. (19)	**12.** (50)	**13.** (173)	**14.** (4)	**15.** (6)
16. (16)	**17.** (24)	**18.** (5)	**19.** (23)	**20.** (6)

Hints and Solutions

A. Objective Questions

[Only one option is correct]

1. (d) A reducing agent reduce others but oxidizes itself. And since oxidation is loss of electrons, so in option (2) and (4) H_2O_2 acts as reducing agent.

2. (d) Balanced equation is

$$2C_8H_7(SO_3)Na + Ca^{2+} \longrightarrow [C_8H_7(SO_3)]_2Ca + 2Na^+$$

Since, 1 mole of Ca^{2+} required 2 moles of resin $= 2 \times 206 = 412$ g resin.

So, moles of Ca^{2+} per gram of resin $= \frac{1}{412}$.

3. (d) Species — Oxidation state of P

Species	Oxidation state of P
H_3PO_4	+5
$H_4P_2O_6$	+4
H_3PO_3	+3
H_3PO_2	+1

4. (a) $FeC_2O_4 \rightleftharpoons Fe^{2+} + C_2O_4^{2-}$

For 1 mole of Fe^{2+}, 2/10 mole of $KMnO_4$ is needed and for 1 mole of $C_2O_4^{2-}$, 2/5 mole of $KMnO_4$ is needed. So for 1 mole FeC_2O_4,

moles of $KMnO_4$ required $= \frac{2}{10} + \frac{2}{5} = \frac{6}{10}$

$Fe_2(C_2O_4)_3 \rightleftharpoons 2Fe^{3+} + 3C_2O_4^{2-}$

Fe^{3+} is not affected by $KMnO_4$.

For $3C_2O_4^{2-}$,

moles of $KMnO_4$ needed $= 3 \times \frac{2}{5}$ moles

1 mole of $FeSO_4$ requires = 2/10 moles of $KMnO_4$ $Fe_2(SO_4)_3$ is not oxidised by $KMnO_4$.

So total moles of $KMnO_4 = \frac{6}{10} + \frac{6}{5} + \frac{2}{10} = \frac{6+12+2}{10} = 2$ moles

5. (c) In option (c), copper is present in following oxidation state

$$CuBr \longrightarrow +1$$
$$CuBr_2 \longrightarrow +2$$
$$Cu \longrightarrow 0$$

6. (d)
$$[MnO_4^- + 8H^+ + 5e^- \longrightarrow Mn^{2+} + 4H_2O] \quad \text{...(i)}$$
$$[C_2O_4^{2-} \longrightarrow 2CO_2 + 2e^-] \quad \text{...(ii)}$$

Multiplying reaction (i) through 2 and reaction (ii) with 5, we get

$$2MnO_4^- + 16H^+ + 5C_2O_4^{2-} \longrightarrow 10CO_2 + 2Mn^{2+} + 8H_2O$$

Since, overall 10 electrons are involved for 10 mol CO_2 formation so 1 mol electron will be involved per molecule of CO_2.

7. (d) 10^{-3} *M* means 10^{-3} mol $CaSO_4$ is present in one litre of solution.

10^{-3} mol $CaSO_4$ will contain 10^{-3} mol $CaCO_3$. (Since, valence factor for both is 2.)

Weight of $CaCO_3$ in mg $= 10^{-3} \times 100 \times 10^3 = 100$ mg

Since, 1 mg/L = 1ppm

100 mg/L of $CaCO_3$ = 100 ppm

8. (b) From law of equivalence,

$$\text{Eq. of } CaCO_3 = \text{Eq. of } Ca(HCO_3)_2 + \text{Eq. of } Mg(HCO_3)_2$$
$$\left(\frac{w}{100}\right) \times 2 = \left(\frac{0.81}{162}\right) \times 2 + \left(\frac{0.73}{146}\right) \times 2$$
$$w = 1\text{g}$$
$$\text{ppm} = \left(\frac{1}{1000}\right) \times 10^6 = 1000 \text{ ppm}$$

9. (b) $\overset{0}{N_2} + \overset{0}{O_2} \xrightarrow{2000K} 2\overset{+2\,-2}{NO}$

10. (d) H_3PO_4 can acts as oxidising agent only since it is present in its maximum oxidation state (+5). While in $H_2\overset{-1}{O_2}$, $H_2\overset{+4}{S}O_3$ and $H\overset{+3}{N}O_2$, elements are present in their intermediate oxidation state so they can acts as both oxidising agent as well as reducing agent.

11. (c) Alkali metal atoms always exist in +1 oxidation state in its compound.

12. (b) Moles of urea $= \frac{0.6}{60} = 0.01$ mole

So, moles of $NH_3 = 2 \times 0.01 = 0.02$ mole

$$NH_3 + HCl \longrightarrow NH_4Cl$$

0.02 mole of NH_3, require 0.02 mole of HCl for neutralisation.

13. (c)

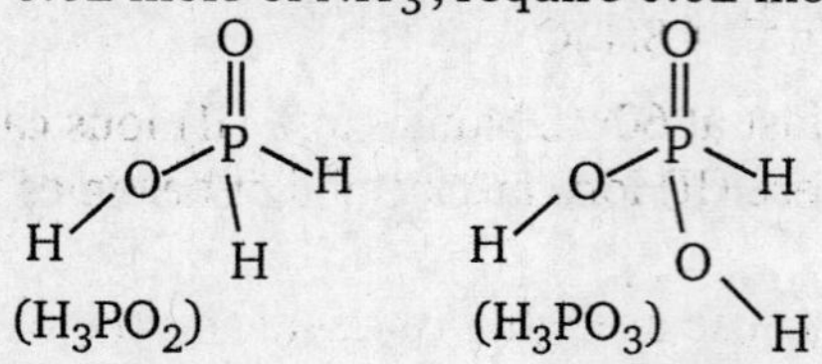

(H_3PO_2) $(n_f = 1)$ (H_3PO_3) $(n_f = 2)$

$$\text{Equivalents} = \text{Molarity} \times \text{Volume} \times n_f$$

Equivalents of H_3PO_3 = Equivalents of NaOH

$$1 \times 2 \times 50 = 1 \times 1 \times V$$
$$V = 100 \text{ mL}$$

Equivalents of H_3PO_2 = Equivalents of NaOH

$$2 \times 100 \times 1 = 1 \times V \times 1$$
$$V = 200 \text{ mL}$$

14. (b) $\overset{+7}{KIO_4} + \underset{(R.A)}{H_2O_2} \longrightarrow \overset{+5}{KIO_3} + H_2O + O_2$

$2\overset{-1}{I^-} + \underset{(O.A.)}{H_2O_2} + 2H^+ \longrightarrow \overset{0}{I_2} + 2H_2O$

$\overset{0}{I_2} + \underset{(R.A.)}{H_2O_2} + 2OH^- \longrightarrow 2\overset{-1}{I} + 2H_2O + O_2$

$\overset{0}{Cl_2} + \underset{(R.A.)}{H_2O_2} \longrightarrow 2H\overset{-1}{Cl} + O_2$

15. (a)

Compound	Oxidation state of 'nitrogen'
NO_3^-	+5
NO_2	+4
NO	+2
N_2O	+1

16. (a) In disproportion reaction one oxidation state is simultaneously oxidised and reduced.

Reduction

$$3\overset{+6}{Mn}O_4^{2-} + 4H^+ \longrightarrow 2\overset{+7}{Mn}O_4^- + \overset{+4}{Mn}O_2 + 2H_2O$$

Oxidation

17. (a) $2[\overset{+1}{Au}(CN)_2] + \overset{0}{Zn}(s) \longrightarrow 2\overset{0}{Au} + [\overset{+2}{Zn}(CN)_4]^{2-}$

Zn displaced Au^+

Reduction and Oxidation both are taking place.

18. (d) Number of atoms of iodine on reactant side = number of atoms of Iodine on product side

$$2 + x = 6 \times 2 \quad \Rightarrow \quad x = 10$$

$$2IO_3 + 10I^- + 12H^+ \longrightarrow 6I_2 + 6H_2O$$

19. (d) Only I^- among halides can be oxidised to Iodine by oxygen in acidic medium due to its reduction potential.

$$4I^-(aq) + 4H^+(aq) + O_2(g) \longrightarrow 2I_2(s) + 2H_2O(l)$$

22. (b) $ClO_4^- \rightarrow x + \{(-2) \times 4\} = -1 \Rightarrow x = +7$

Chlorine is in its maximum oxidation state, so disproportionation not possible in ClO_4^-.

23. (b) For the titration : Oxalic acid *vs* $KMnO_4$

$$2MnO_4^- + 5(COO)_2^{2-} + 16H^+ \longrightarrow 10CO_2 + 2Mn^{2+} + 8H_2O$$

This reaction is slow at room temperature, but becomes fast at 60° C. Manganese (II) ions catalyse the reaction; thus, the reaction is autocatalytic; once manganese (II) ions are formed, it becomes faster and faster.

The titration of FAS *vs* $KMnO_4$ do not require heating because at higher temperature the oxidation of Fe^{2+} to Fe^{3+} by atmospheric O_2 will be prominent.

24. (c) $I^- \xrightarrow{H^+} I_2$ $\qquad$ $I^- \xrightarrow{OH^-} IO_3^-$

$S^{2-} \xrightarrow{H^+} S$ $\qquad$ $S_2O_3^{2-} \xrightarrow{OH^-} SO_4^{2-}$

$Fe^{2+} \longrightarrow Fe^{3+}$ $\qquad$ $S_2O_3^{2-} \xrightarrow{H^+} S\downarrow + SO_4^{2-}$

B. Objective Questions

1. (a, b, d)

The balanced equation is,

$$ClO_3^- + 6I^- + 6H_2SO_4 \longrightarrow 3I_2 + Cl^- + 6HSO_4^- + 3H_2O$$

C. Matching Type Problems

1. (a) — p, s; (b) — r; (c) — p, q; (d) — p

(a) It is a redox reaction in which O_2^- is both oxidised and reduced and such redox reactions are known as disproportionation reactions.

(b) $2CrO_4^{2-} \xrightarrow{4H^+} Cr_2O_7^{2-}$

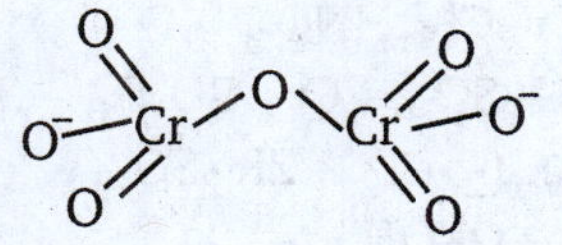

(c) $MnO_4^- + NO_2^- + H^+ \longrightarrow Mn^{2+} + NO_3^- + H_2O$

It is a redox reaction in which Mn^{7+} is reduced to Mn^{2+} and N^{3+} is oxidised to N^{5+}.

(d) $NO_3^- + H_2SO_4 + Fe^{2+} \longrightarrow Fe^{3+} + NO + SO_4^{2-} + H_2O$

In this redox reaction N^{5+} is reduced to N^{2+} while Fe^{2+} is oxidised to Fe^{3+}.

2. (a) (A) Combustion of hydrocarbon

(B) Decomposition into gaseous product

(C) Displacement of 'V' by 'Ca' atom.

(D) Disproportionation of $H_2O_2^{-1}$ into O^{-2} and O^0 oxidation states.

D. Passage Based Problems

1. (c) $2\,CH_3COOH + CaOCl_2 \longrightarrow (CH_3COO)_2Ca + H_2O + Cl_2$

$$2KI + Cl_2 \longrightarrow 2KCl + I_2$$

$$2Na_2S_2O_3 + I_2 \longrightarrow Na_2S_4O_6 + 2NaI$$

1 millimole of $CaOCl_2$ = 1 millimole of I_2

$= 1/2$ millimole of $Na_2S_2O_3$

$$\text{Millimoles of } CaOCl_2 = \frac{1}{2} \times 0.25 \times 48 = 6$$

$$\text{Molarity} \times \text{volume} = 6 \text{ millimoles}$$

$$\text{Molarity} \times 25 = 6 \text{ millimoles}$$

$$\text{Molarity} = \frac{6}{25} = 0.24\,M$$

2. (a) Bleaching powder is $Ca(OCl)Cl$.

It contains OCl^- ion *i.e.*, HOCl acid.

(H)OCl

(HO)Cl $\xrightarrow{-H_2O} Cl_2O$

Anhydride of hypochlorus acid

E. Integer Type Problems

1. (6) $2MnO_2 + 4KOH + O_2 \longrightarrow 2K_2MnO_4 + 2H_2O$

Let the Ox. No. of Mn in K_2MnO_4 is a.

$\therefore\ 2 \times 1 + a + 4 \times (-2) = 0 \Rightarrow a = \mathbf{6}$

2. (1) Na, show only one non-zero oxidation. Na shows only 1+ oxidation state.

Note : *Although most commonly reported answer may be two i.e., Na and F. But, in hypofluorous acid (HOF), F is in +1 oxidation state, however, acid is unstable but well reported.*

3. (5) 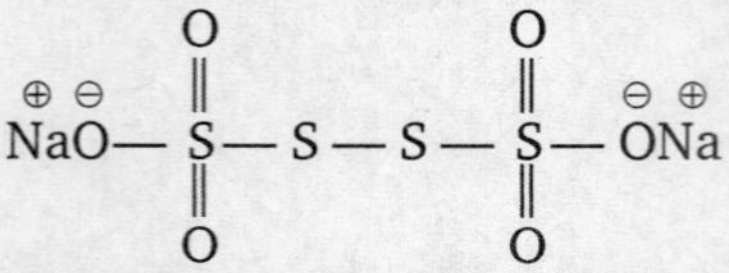

S will have oxidation number = +5, 0

Difference in oxidation number = 5

4. (5) $3Br_2 + 3Na_2CO_3 \longrightarrow 5NaBr + NaBrO_3 + 3CO_2$

So, number of NaBr molecules = 5

5. (7) $K_2Cr_2O_7, CuSO_4, H_2O_2, Cl_2, O_3, FeCl_3, HNO_3$

$$K_2Cr_2O_7 + 7H_2SO_4 + 6KI \longrightarrow 4K_2SO_4 + Cr_2(SO_4)_3 + 3I_2 + 7H_2O$$

$$2CuSO_4 + 4KI \longrightarrow Cu_2I_2 + I_2 + 2K_2SO_4$$

$$H_2O_2 + 2KI \longrightarrow I_2 + 2KOH$$

$$Cl_2 + 2KI \longrightarrow 2KCl + I_2$$

$$O_3 + 2KI + H_2O \longrightarrow 2KOH + I_2 + O_2$$

$$2FeCl_3 + 2KI \longrightarrow 2FeCl_2 + I_2 + 2KCl$$

$$8HNO_3 + 6KI \longrightarrow 6KNO_3 + 2NO + 4H_2O + 3I_2$$

$$2KMnO_4 + KI + H_2O \longrightarrow KIO_3 + 2MnO_2 + 2KOH$$

6. (6) $$\overset{}{MnO_4^-} + \overset{+2\ 2-}{S_2O_3} \xrightarrow[\text{Medium}]{\text{Neutral}} \overset{+4}{MnO_2} + \overset{+6\ 2-}{SO_4}$$

($S_2O_3^{2-} \to SO_4^{2-}$: *n*-factor=4; $MnO_4^- \to MnO_2$: *n*-factor=3)

$$\text{g eq. of } MnO_4^- = \text{g eq. of } SO_4^{2-}$$

$$3 \times 8 = n \times 4$$

$$n = 6$$

Mole of SO_4^{2-} formed = 6

7. (6) $KMnO_4 + KI \longrightarrow MnO_2 + I_2$

$$\text{Eq. of } KMnO_4 = \text{Eq. of } I_2$$
$$4 \times 3 = n \times 2$$
$$n = 6$$

8. (100)

$$\text{Equivalents of } MgSO_4 = \text{Equivalents of } CaCO_3$$
$$10^{-3} \times 2 = M \times 2$$
$$M = 10^{-3} \text{ mol/L}$$

10^{-3} mole $CaCO_3$ is present in 1 L solution.

$$\text{Weight of } CaCO_3 = 10^{-3} \times 100 = 0.1\text{g}$$
$$\text{ppm} = \frac{\text{wt. of } CaCO_3}{\text{wt. of solution}} \times 10^6 = \frac{0.1}{1000} \times 10^6 = 100$$

9. (6) $Na_4[Fe(CN)_5NOS]$

$$4(+1) + x + 5(-1) + (-1) = 0$$
$$x = +2$$

$Na_4[FeO_4]$

$$4(+1) + y + 4(-2) = 0$$
$$y = 4$$

$[Fe_2(CO)_9]$

$$2z + 9(0) = 0$$
$$z = 0$$
$$x + y + z = 2 + 4 + 0 = 6$$

10. (10) Phosphinic acid is H_3PO_2.

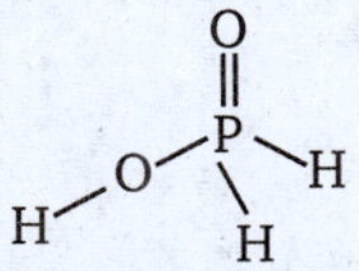

(Monoprotic acid)

$$\text{Equivalent}(H_3PO_2) = \text{Equivalent(NaOH)}$$
$$0.1 \times 10 = 0.1 \times V$$
$$V = 10 \text{ mL}$$

11. (19)

Balanced equations are $2Fe^{2+} + H_2O_2 \longrightarrow 2Fe^{3+} + 2OH^-$

So, $x = 2$ and $y = 2$

$$2MnO_4^- + 5H_2O_2 + 6H^+ \longrightarrow 2Mn^{2+} + 8H_2O + 5O_2$$
$$x' = 2,\ y' = 8,\ z' = 5$$

So, $x + y + x' + y' + z' = 2 + 2 + 2 + 8 + 5 = 19$

12. (50)

$$\overset{+6}{K_2Cr_2O_7} + \overset{+2\ +3}{FeC_2O_4} \longrightarrow Cr^{3+} + Fe^{3+} + CO_2$$
$$n_f = 6 \qquad n_f = 3$$

From law of chemical equivalence,

$$\text{Equivalents }(K_2Cr_2O_7) = \text{Equivalent }(Fe_2C_2O_4)$$
$$M \times V(\text{L}) \times n_f = \text{mole} \times n_f$$
$$0.02 \times V \times 6 = \frac{0.288}{144} \times 3$$
$$V = 0.05\text{L} = 50 \text{ mL}$$

13. (173)

$$\overset{+6}{CrO_4^{2-}} + \overset{+2}{S_2O_3^{2-}} \longrightarrow \overset{+3}{Cr(OH)_4^-} + \overset{+6}{SO_4^{2-}}$$

$$n_f = 3 \quad n_f = 8$$

Equivalents of CrO_4^{2-} = Equivalents of $S_2O_3^{2-}$

$$0.154 \times 3 \times V = 0.25 \times 40 \times 8$$

$$V = 173.16 \text{ mL}$$

14. (4) Let moles of NaOH = x

Moles of $Na_2CO_3 = y$

At HPh end point,

$$\underset{n_f=1}{NaOH} + HCl \longrightarrow NaCl + H_2O$$

$$\underset{n_f=1}{Na_2CO_3} + HCl \longrightarrow NaHCO_3 + NaCl$$

Equivalents $(NaOH + Na_2CO_3)$ = equivalent (HCl)

$$x \times 1 + y \times 1 = \frac{1}{10} \times 17.5 = 1.75 \quad \text{...(i)}$$

At methyl orange end point

$$\underset{n_f=1}{NaOH} + HCl \longrightarrow NaCl + H_2O$$

$$\underset{n_f=2}{Na_2CO_3} + 2HCl \longrightarrow 2NaCl + H_2O + CO_2$$

Equivalents $(NaOH + Na_2CO_3)$ = equivalent (HCl)

$$x \times 1 + y \times 2 = \frac{1}{10} \times 19 = 1.9 \quad \text{...(ii)}$$

Solving eqn. (i) and (ii),

We get $x = 1.60$

$y = 0.15$

$$\% \, Na_2CO_3 = \frac{0.15 \times 10^{-3} \times 10^6}{0.4} \times 100 = 3.97\%$$

15. (6) Since, most precise volume of NaOH is 4.4 mL $eq.(H_2C_2O_4) = eq.(NaOH)$

$$M \times V \times n_f = M \times V \times n_f$$

$$1.25 \times 10 \times 2 = M \times 4.4 \times 1$$

$$M = 5.68 \text{ mol/L}$$

16. (16)

(5) × 2

$$\overset{+7}{MnO_4^-} + \overset{+3}{C_2O_4^{2-}} \longrightarrow Mn^{2+} + \overset{+4}{2CO_2}$$

(2) × 5

$$2MnO_4^- + 5C_2O_4^{2-} \longrightarrow 2Mn^{2+} + 10CO_2$$

Balancing 'O' through H_2O :

$$2MnO_4^- + 5C_2O_4^{2-} \longrightarrow 2Mn^{2+} + 10CO_2 + 8H_2O$$

Balancing 'H' through H^+ :

$$2MnO_4^- + 5C_2O_4^{2-} + 16H^+ \longrightarrow 2Mn^{2+} + 10CO_2 + 8H_2O$$

17. (24) $\underset{n_f=1}{Fe^{2+}} + \underset{n_f=6}{Cr_2O_7^{2-}} \longrightarrow Fe^{3+} + Cr^{3+}$

$$\text{Equivalents }(Fe^{2+}) = \text{Equivalent }(Cr_2O_7^{2-})$$

$$M \times \frac{15}{1000} \times 1 = 0.03 \times \frac{20}{1000} \times 6$$

$$M = 0.24 = 24 \times 10^{-2}$$

18. (5) $IO_4^- + H_2O_2 \longrightarrow IO_3^- + O_2$

The oxidation number of I in KIO_4 is +7 and it changes to KIO_3, in which oxidation number of I is +5.

19. (23) $Cr_2O_7^{2-} + 14H^+ + 6Fe^{2+} \longrightarrow 6Fe^{3+} + 2Cr^{3+} + 7H_2O$

$x = 14, \quad y = 2, \quad z = 7$

Hence $(x + y + z) = 14 + 2 + 7 = 23$

20. (6) $(NH_4)_3PO_4 \cdot 12MoO_3$

Let X = oxidation state of Mo in MoO_3

$X + [(-2) \times 3] = 0$

$X = +6$

❑❑❑

CHAPTER

7

Solid State

A. Objective Questions

[Only one option is correct]

1. A compound M_pX_q has cubic close packing (*ccp*) arrangement of *X*. Its unit cell structure shown below. The empirical formula of the compound is : **(IIT 2012)**

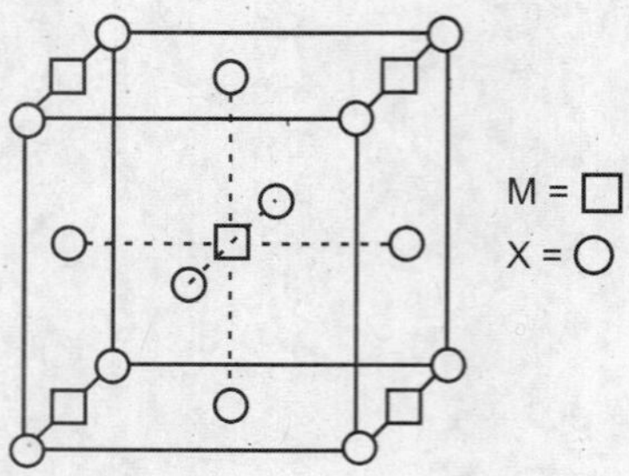

(a) MX (b) MX_2 (c) M_2X (d) M_5X_{14}

2. Which of the following exists as covalent crystals in the solid state? **[JEE (Mains) 2013]**

(a) Iodine (b) Silicon (c) Sulphur (d) Phosphorus

3. Experimentally it was found that a metal oxide has formula $M_{0.98}O$. Metal *M*, is present as M^{2+} and M^{3+} in its oxide. Fraction of the metal which exists as M^{3+} would be : **[JEE (Mains) 2013]**

(a) 7.01% (b) 4.08% (c) 6.05% (d) 5.08%

4. The arrangement of X^- ions around A^+ ion in solid *AX* is given in the figure (not drawn to scale). If the radius of X^- is 250 pm, the radius of A^+ is : **(JEE Advanced 2013)**

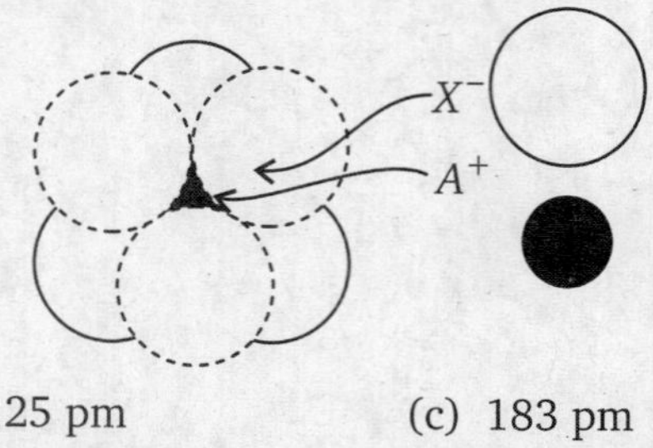

(a) 104 pm (b) 125 pm (c) 183 pm (d) 57 pm

5. CsCl crystallises in body centred cubic lattice. If '*a*' is its edge length then which of the following expressions is correct ? **[JEE (Mains) 2014]**

(a) $r_{Cs^+} + r_{Cl^-} = 3a$ (b) $r_{Cs^+} + r_{Cl^-} = \dfrac{3a}{2}$ (c) $r_{Cs^+} + r_{Cl^-} = \dfrac{\sqrt{3}}{2}a$ (d) $r_{Cs} + r_{Cl^-} = \sqrt{3}\,a$

6. Sodium metal crystallizes in a body centred cubic lattice with a unit cell edge of 4.29 Å. The radius of sodium atom is approximately : **[JEE (Mains) 2015]**

(a) 5.72 Å (b) 0.93 Å (c) 1.86 Å (d) 3.22 Å

7. A metal crystallises in a face centred cubic structure. If the edge length of its unit cell is 'a', the closest approach between two atoms in metallic crystal will be: **[JEE (Mains) 2017]**

(a) $\sqrt{2}a$ (b) $\frac{a}{\sqrt{2}}$ (c) $2a$ (d) $2\sqrt{2}a$

8. Which type of 'defect' has the presence of cations in the interstital sites? **[JEE (Mains) 2018]**

(a) Frenkel defect (b) Metal deficiency defect

(c) Schottky defect (d) Vacancy defect

9. Element 'B' forms *ccp* structure and 'A' occupies half of the octahedral voids, while oxygen atoms occupy all the tetrahedral voids. The structure of bimetallic oxide is : **[JEE (Mains) 2019]**

(a) AB_2O_4 (b) A_4B_2O (c) A_2B_2O (d) A_2BO_4

10. Consider the *bcc* unit cells of the solids 1 and 2 with the position of atoms as shown below. The radius of atom B is twice that of atom A. The unit cell edge length is 50% more in solid 2 than in 1. What is the approximate packing efficiency in solid 2 ? **[JEE (Mains) 2019]**

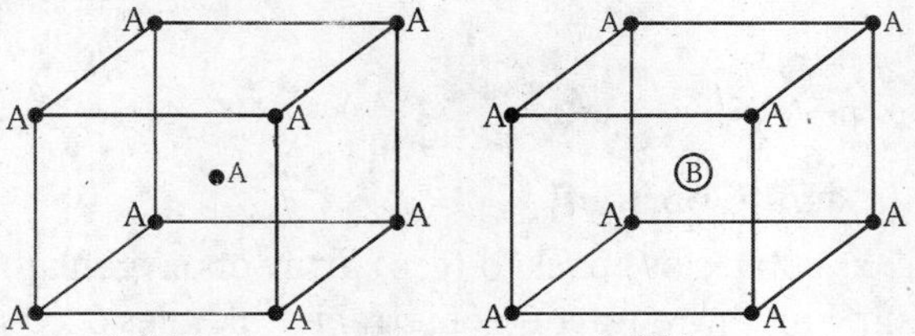

Solid 1 Solid 2

(a) 75% (b) 45% (c) 65% (d) 90%

11. 'X' melts at low temperature and is a bad conductor of electricity in both liquid and solid state. X is: **[JEE Main (Jan.) 2020]**

(a) silicon carbide (b) mercury (c) zinc sulphide (d) carbon tetrachloride

12. An element crystallises in a face-centred cubic (fcc) unit cell with cell edge a. The distance between the centres of two nearest octahedral voids in the crystal lattice is: **[JEE Main (Sept.) 2020]**

(a) $\frac{a}{2}$ (b) $\sqrt{2}a$ (c) a (d) $\frac{a}{\sqrt{2}}$

13. Which of the following compound is likely to show both Frenkel and Schottky defects in its crystalline form? **[JEE Main (Jan.) 2020]**

(a) CsCl (b) AgBr (c) ZnS (d) KBr

14. A hard substance melts at high temperature and is an insulator in both solid and in molten state. This solid is most likely to be a/an : **[JEE Main (March) 2021]**

(a) ionic solid (b) molecular solid (c) metallic solid (d) covalent solid

15. In a binary compound, atoms of element A form a hcp structure and those of element M occupy 2/3 of the tetrahedral voids of the hcp structure. The formula of the binary compound is: **[JEE Main (March) 2021]**

(a) M_2A_3 (b) M_4A_3 (c) M_4A (d) MA_3

16. The oxide that shows magnetic property is : **[JEE Main (March) 2021]**

(a) SiO_2 (b) Mn_3O_4 (c) Na_2O (d) MgO

17. An element X has a body centred cubic (bcc) structure with a cell edge of 200 pm. The density of the element is 5 g cm^{-3}. The number of atoms present in 300 g of the element X is ______ . **[JEE Main (June) 2022 (I)]**

[Given : Avogadro constant, $N_A = 6.0 \times 10^{23} \text{mol}^{-1}$.]

(a) $5N_A$ (b) $6N_A$ (c) $15N_A$ (d) $25N_A$

18. The incorrect statement about the imperfections in solids is : **[JEE Main (June) 2022 (I)]**

(a) Schottky defect decreases the density of the substance

(b) Interstitial defect increases the density of the substance

(c) Frenkel defect does not alter the density of the substance
(d) Vacancy defect increases the density of the substance

19. The correct relationship between unit cell edge length 'a' and radius of sphere 'r' for face centred and body centred cubic structures respectively are :

[JEE Main (April) 2023 (II)]

(a) $r = 2\sqrt{2}\, a$ and $\sqrt{3}\, r = 4a$
(b) $r = 2\sqrt{2}\, a$ and $4r = \sqrt{3}\, a$
(c) $2\sqrt{2}\, r = a$ and $4r = \sqrt{3}\, a$
(d) $2\sqrt{2}\, r = a$ and $\sqrt{3}\, r = 4a$

20. A cubic solid is made up of two elements X and Y. Atoms of X are present on every alternate corner and one at the center of cube. Y is at $\frac{1}{3}$ of the total faces. The empirical formula of the compound is :

[JEE Main (Jan.) 2023 (I)]

(a) $X_2Y_{1.5}$
(b) $X_{2.5}Y$
(c) $XY_{2.5}$
(d) $X_{1.5}Y_2$

B. Objective Questions

[One or more than one option(s) is/are correct]

1. If the unit cell of a mineral has cubic close packed (ccp) array of oxygen atoms with m fraction of octahedral holes occupied by aluminium ions and n fraction of tetrahedral holes occupied by magnesium ions, m and n, respectively, are :

[JEE (Advanced) 2015]

(a) $\frac{1}{2}, \frac{1}{8}$
(b) $1, \frac{1}{4}$
(c) $\frac{1}{2}, \frac{1}{2}$
(d) $\frac{1}{4}, \frac{1}{8}$

2. The correct statement(s) for cubic close packed (ccp) three dimensional structure is (are):

[JEE (Advanced) 2016]

(a) the number of the nearest neighbours of an atom present in the topmost layer is 12.
(b) the efficiency of atom packing is 74%.
(c) the number of octahedral and tetrahedral voids per atom are 1 and 2, respectively.
(d) the unit cell edge length is $2\sqrt{2}$ times the radius of the atom.

3. The cubic unit cell structure of a compound containing cation M and anion X is shown below. When compared to the anion, the cation has smaller ionic radius. Choose the correct statement(s).

[JEE (Advanced) 2020]

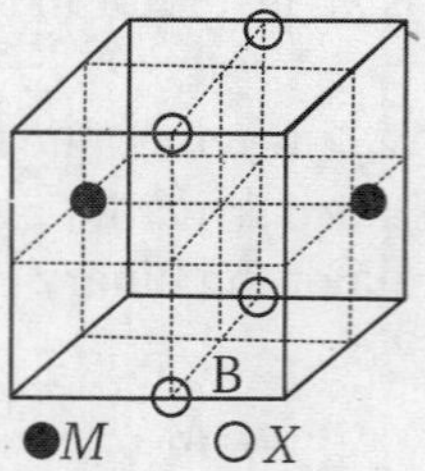

(a) The empirical formula of the compound is MX.
(b) The cation M and anion X have different coordination geometries.
(c) The ratio of M—X bond length to the cubic unit cell edge length is 0.866.
(d) The ratio of the ionic radii of cation M to anion X is 0.414.

C. Integer Answer Type Problems

1. The number of hexagonal faces that are present in a truncated octahedron is : **(IIT 2011)**

2. A crystalline solid of a pure substance has a facecentred cubic structure with a cell edge of 400 pm. If the density of the substance in the crystal is 8 g cm^{-3}, then the number of atoms present in 256 g of the crystal is $N \times 10^{24}$. The value of N is :

[JEE (Advanced) 2017]

3. Consider an ionic solid MX with NaCl structure. Construct a new structure (Z) whose unit cell is constructed from the unit cell of MX following the sequential instructions given below. Neglect the charge balance.

(i) Remove all the anions (X) except the central one

(ii) Replace all the face centered cations (M) by anions (X)

(iii) Remove all the corner cations (M)

(iv) Replace the central anion (X) with cation (M)

The value of $\left(\frac{\text{number of anions}}{\text{number of cations}}\right)$ in Z is .... **[JEE (Advanced) 2018]**

4. An element with molar mass 2.7×10^{-2} kg mol^{-1} forms a cubic unit cell with edge length 405 pm. If its density is 2.7×10^{3} kg m^{-3}, the radius of the element is approximately ______ $\times 10^{-12}$ m (to the nearest integer). **[JEE Main (Sept.) 2020]**

5. The coordination number of an atom in a bodycentered cubic structure is ______.
[Assume that the lattice is made up of atoms.] **[JEE Main (Feb.) 2021]**

6. The unit cell of copper corresponds to a face centered cube of edge length 3.596 Å with one copper atom at each lattice point. The calculated density of copper in kg/m^3 is ______. [Molar mass of Cu: 63.54 g; Avogadro number $= 6.022 \times 10^{23}$] **[JEE Main (Feb.) 2021]**

7. The number of octahedral voids per lattice site in a lattice is _____. **[JEE Main (Feb.) 2021]**
(Rounded off to the nearest integer)

8. A certain element crystallises in a bcc lattice of unit cell edge length 27 Å. If the same element under the same conditions crystallises in the fcc lattice, the edge length of the unit cell in Å will be ________.
(Round off to the nearest integer).
[Assume each lattice point has a single atom]
[Assume $\sqrt{3} = 1.73$, $\sqrt{2} = 1.41$] **[JEE Main (March) 2021]**

9. Ga (atomic mass 70 u) crystallizes in a hexagonal close packed structure. The total number of voids in 0.581 g of Ga is ______ $\times 10^{21}$. (Round off to the nearest integer). **[JEE Main (March) 2021]**

10. KBr is doped with 10^{-5} mole per cent of $SrBr_2$. The number of cationic vacancies in 1g of KBr crystal is .....10^{14}. (Round off to the nearest integer).
[Atomic mass : K: 39.1u, Br : 79.9u, $N_A = 6.023 \times 10^{23}$] **[JEE Main (March) 2021]**

11. Metal M crystallizes into a fcc lattice with the edge length of 4.0×10^{-8} cm. The atomic mass of the metal is ________ g/mol. (Nearest integer) **[JEE Main (July) 2022 (II)]**
(Use : $N_A = 6.02 \times 10^{23}$ mol^{-1}, density of metal M = 9.03 g cm^{-3})

12. The distance between Na^+ and Cl^- ions in solid NaCl of density 43.1 g cm^{-3} is _____ $\times 10^{-10}$ m.
(Nearest integer)
(Given : $N_A = 6.02 \times 10^{23}$ mol^{-1}) **[JEE Main (June) 2022 (I)]**

13. Sodium metal crystallizes in a body centred cubic lattice with unit cell edge length of 4 Å. The radius of sodium atom is ________ $\times 10^{-1}$ Å. (Nearest integer) **[JEE Main (April) 2023 (II)]**

14. Number of crystal system from the following where body centred unit cell can be found, is ...... Cubic, tetragonal, orthorhombic, hexagonal, rhombohedral, monoclinic, triclinic.

[JEE Main (April) 2023 (II)]

Answers

[A] Objective Questions (Only one option is correct)

1. (b) **2.** (b) **3.** (b) **4.** (a) **5.** (c) **6.** (c) **7.** (b) **8.** (a) **9.** (a) **10.** (d)
11. (d) **12.** (d) **13.** (b) **14.** (d) **15.** (b) **16.** (b) **17.** (d) **18.** (d) **19.** (c) **20.** (none)

[B] Objective Questions (One or more than one option(s) is/are correct)

1. (a) **2.** (b, c, d) **3.** (a, c)

[C] Integer Type Problems

1. (8) **2.** (2) **3.** (3) **4.** (143) **5.** (8)
6. (9076) **7.** (1) **8.** (33) **9.** (15) **10.** (5)
11. (87) **12.** (1) **13.** (17) **14.** (3)

Hints and Solutions

A. Objective Questions

[Only one option is correct]

1. (b) $8X$ atoms present at the corners.

Atoms contribute to 1 unit cell $= \frac{1}{8} \times 8 = 1$

$6X$ atoms present at the face centres.

Atoms contribute to 1 unit cell $= 6 \times \frac{1}{2} = 3$

Total X atoms $= 3 + 1 = 4$

$4M$ atoms present at edge centres.

Atoms present in 1 unit cell $= 4 \times \frac{1}{4} = 1$

$1M$ atom present at body centre and it contribute completely to 1 unit cell.

Thus, total M atoms in one unit cell $= 1 + 1 = 2$

Ratio is $M : X :: 2 : 4 :: 1 : 2$

Thus, empirical formula is MX_2.

2. (b) Silicon exists as covalent crystals in the solid state.

3. (b) $3x + 2(0.98 - x) = 2$

$$x + 1.96 = 2 \Rightarrow x = 0.04$$

$$\% \text{ of } M^{3+} = \frac{0.04}{0.98} \times 100 = 4.08\%$$

4. (a) $\frac{r_+}{r_-} = 0.414$ for octahedral void.

$$r_+ = 0.414 \times 250 = 103.5 \cong 104 \text{ pm}$$

5. (c)

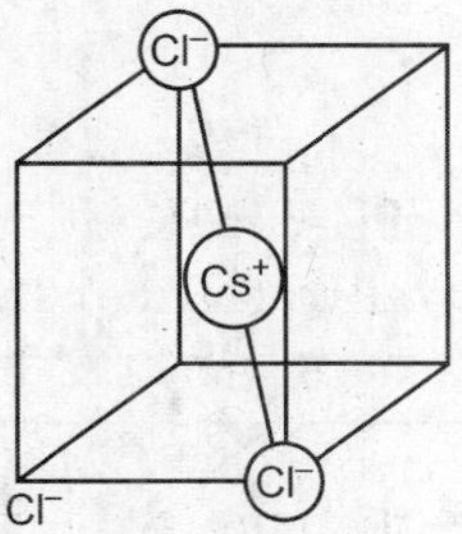

$$2r_{Cl^-} + 2r_{Cs^+} = \sqrt{3}\, a$$

$$r_{Cl^-} + r_{Cs^+} = \frac{\sqrt{3}\, a}{2}$$

6. (c) For *bcc*, $r = \frac{\sqrt{3}}{4} a$

$$r = \frac{\sqrt{3}}{4} \times 4.29 = 1.86 \text{ Å}$$

7. (b) For *fcc*,

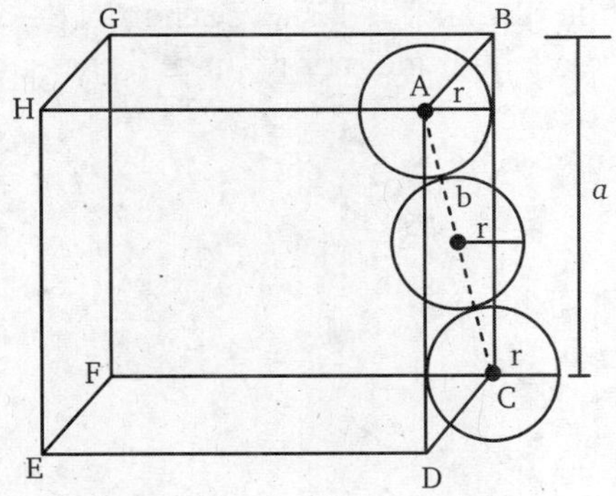

then $b = 4r = \sqrt{2}a$

$$a = \frac{4r}{\sqrt{2}} = 2\sqrt{2}r \Rightarrow r = \frac{a}{2\sqrt{2}}$$

Therefore, distance of closest approach $= 2r = 2 \times \frac{a}{2\sqrt{2}} = \frac{a}{\sqrt{2}}$

8. (a) Pressure of cation in interstitial sites is 'Frenkel' defect.

9. (a) Number of atoms per unit cell in *ccp* $= N = 4$

Number of octahedral voids $= N = 4$

A occupies half of the octahedral voids, thus *A* atoms per unit cell $= 4/2 = 2$

Tetrahedral voids $= 2 \times N = 8$

So number of oxygen atoms $= 8$

$$A:B:O :: 2 : 4 : 8$$

Thus, structure of bimetallic oxide is AB_2O_4.

10. (d) Packing efficiency $= \dfrac{\left(Z \times \frac{4}{3}\pi r_A^3\right) + \left(Z + \frac{4}{3}\pi r_B^3\right)}{a^3}$

For *bcc*, $2(r_A + r_B) = \sqrt{3}\, a$

Given that, $2(r_A + 2r_A) = \sqrt{3}\, a$

$\therefore \quad 2\sqrt{3}\, r_A = a$

$$\text{Packing efficiency} = \frac{1\times\frac{4}{3}\pi r_A^3 + 1\times\frac{4}{3}\pi(8r_A^3)}{8\times 3\sqrt{3}\, r_A^3} = \frac{9\times\frac{4}{3}\pi}{8\times 3\sqrt{3}} = \frac{\pi}{2\sqrt{3}}$$

$$\text{Packing efficiency} = \frac{\pi}{2\sqrt{3}}\times 100 \approx 90\%$$

11. (d) Carbon tetrachloride being molecular solid have weak van der Waals' forces. So, that it melts at low temperature and it is bad conductor of electricity in both liquid and solid state.

12. (d) Position of octahedral void in fcc unit cell is at all the edge centre and body centre. Assume a face *ABCD*

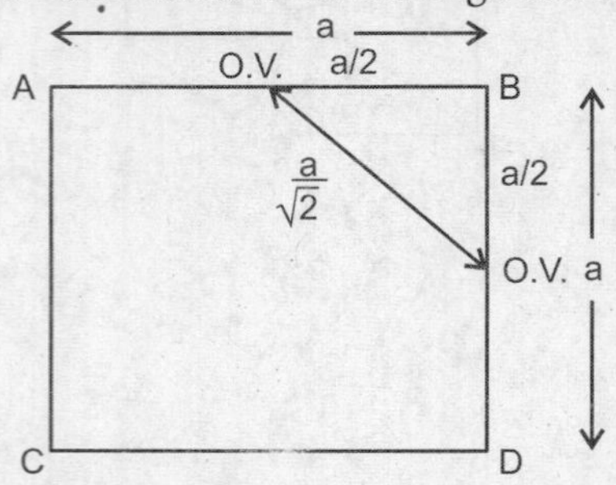

Minimum distance between two octahedral void (O.V.) $= \frac{a}{\sqrt{2}}$

13. (b) AgBr show both type of defect Schottky as well as Frenkel.
ZnS show Frenkel defect while KBr and CsCl show Schottky defect.

14. (d) Covalent or network solids have high melting point and they are insulator in their solid and molten form.

15. (b) In hcp unit cell effective atoms = 6
Number of tetrahedral voids $= 2\times 6 = 12$

M	A
$12\times\frac{2}{3}$	6
$=8$	$=6$

Simplest ratio = 4:3
So, formula is M_4A_3.

16. (b) Mn_3O_4 shows magnetic properties (paramagnetic).

17. (d) $$d = \frac{Z\times M}{a^3\times N_A}$$

$Z = 2$ for bcc

$$5\text{ g/cm}^3 = \frac{2\times M}{(200\times 10^{-10}\text{ cm})^3\times 6.0\times 10^{23}} \Rightarrow M = 12\text{ g}$$

12 g of element contain $= N_A$ atoms

300 g of element contains $= N_A\times\frac{300}{12} = 25\, N_A$

18. (d)

- Schottky defect is a type of vacancy defect in ionic compounds. It decrease the density of substance.
- Interstitial defect occurs when constituent particles occupy the interstitial site. This increase the density of substance.
- Frenkel defect occurs when constituent particle occupy the interstitial site. It create vacancy defect at original and interstitial defect at new location. It does not change density of crystal.
  Due to vacancy defect density of the substance will decrease.

19. (c) For FCC., $a\sqrt{2} = 4r \Rightarrow r = \frac{a\sqrt{2}}{4} \Rightarrow a = 2\sqrt{2}r$

For BCC., $4r = a\sqrt{3}$

20. (none)

$$X_{4\times\frac{1}{8}+1\times 1} \quad Y_{6\times\frac{1}{3}\times\frac{1}{2}}$$

$$\Rightarrow \quad X_{\frac{1}{2}+1} \quad Y_1 \quad \Rightarrow \quad X_{\frac{3}{2}}Y_1$$

$$\Rightarrow \quad X_{1.5} \quad Y_1 \quad \Rightarrow \quad X_3Y_2$$

B. Objective Questions

[One or more than one option(s) is/are correct]

1. (a) For ccp lattice :

Number of O-atoms = $Z = 4$

Number of Octahedral voids = 4

Number of tatrahedral voids = 8

Number of Al^{3+} ions = $4 \times m$

Number of Mg^{2+} ions = $8 \times n$

∴ The formula of mineral is $Al_{4m}Mg_{8n}O_4$.

$$4m(+3) + 8n(+2) + 4(-2) = 0$$

$$12m + 16n - 8 = 0$$

$$4(3m + 4n - 2) = 0$$

$$\therefore \quad m = \frac{1}{2} \text{ and } n = \frac{1}{8}$$

2. (b, c, d)

CCP has a packing efficiency of 74%.

The nearest neighbours in the top most layer is 9.

The number of tetrahedral and octahedral voids are 2 and 1 respectively per atom.

The edge length is $2\sqrt{2}$ times the radius of atom ($\sqrt{2}a = 4r$)

3. (a, c)

(a) $Z_M = 2 \times \frac{1}{2} = 1$

$Z_X = 4 \times \frac{1}{4} = 1$

∴ Empirical formula is MX

(b) Coordinate numbers of both M and X are 8.

(c) Bond length of M—X bond = $AB = \sqrt{3}\frac{a}{2} = 0.866a$

(d) $r_M : r_X = (\sqrt{3} - 1) : 1 = 0.732 : 1.000$

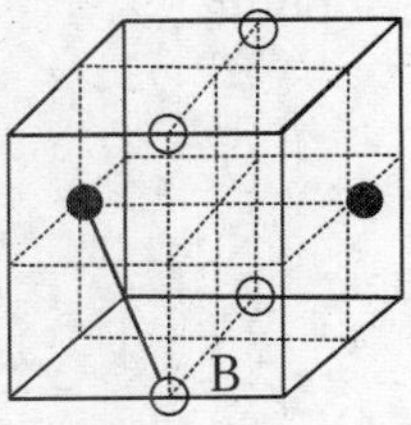

C. Integer Answer Type Problems

1. (8)

2. (2) $d = \dfrac{Z \times M}{N_A \times a^3}$

$$8 = \frac{4 \times M}{6.022 \times 10^{22} \times (400 \times 10^{-10})^3}$$

$M = 76.8 \text{ g mol}^{-1}$

76.8 g contain $= 6 \times 10^{23}$ atoms

$\therefore$ 256 g will contain $= 20 \times 10^{23}$ atoms $= 2 \times 10^{24}$ atoms

$\therefore \quad N = 2$

3. (3) X^- at octahedral void

M^+ at face centered cube

Now

	M^+	X^-
Step-1	4	1
Step-2	4 − 3	3 + 1
Step-3	4 − 3 − 1	3 + 1
Step-4	1	3

$Z = \dfrac{3}{1} = 3$

4. (143)

$$d = \frac{Z \times M}{N_A \times a^3}$$

$$2.7 \times 10^3 = \frac{Z \times 2.7 \times 10^{-2}}{6 \times 10^{23} \times (405 \times 10^{-12})^3}$$

$$Z = 3.98$$

So structure will be fcc.

For fcc unit cell, $\sqrt{2}a = 4R$

$$R = \frac{\sqrt{2} \times 405 \times 10^{-12}}{4}$$

$$R = \frac{1.414 \times 405 \times 10^{-12}}{4} = 143.16 \times 10^{-12} \text{ m}$$

5. (8) Coordination number of atom in bcc unit cell = 8

6. (9076)

$$d = \frac{Z \times M}{N_A \times a^3}$$

$$= \frac{4 \times 63.54 \times 10^{-3}}{6.02 \times 10^{23} \times (3.596 \times 10^{-10})^3} = 9076.2 \text{ kg/m}^3$$

7. (1) There are 1 octahedral voids per lattice point.

8. (33) $a = 27$ Å

For bcc lattice, $\sqrt{3}\, a = 4R$

$$R = \frac{\sqrt{3}\, a}{4}$$

For fcc lattice, $\sqrt{2}\, a = 4R$

$$a = \frac{4R}{\sqrt{2}} = \frac{4}{\sqrt{2}} \times \frac{\sqrt{3} \times 27}{4} = \frac{1.73}{1.41} \times 27 = 33$$

9. (15) If number of atoms $= N$

$$\text{Number of TV} = 2N$$
$$\text{Number of OV} = N$$
$$\text{Total (TV + OV)} = 3N$$
$$\text{Number of Ga atoms} = \frac{0.581}{70} \times 6.023 \times 10^{23}$$
$$\text{Total voids} = 3 \times \text{number of atoms}$$
$$= 3 \times \frac{0.581}{70} \times 6.023 \times 10^{23} = 15 \times 10^{21}$$

10. (5) 1 ion of Sr^{2+} corresponds to 1 cationic vacancy.

Moles of Sr^{2+} ions per mole (= 119 g) KBr $= \frac{10^{-5}}{100} = 10^{-7}$

$\because$ 119g KBr have 10^{-7} mole cationic vacancy.

$\therefore$ 1g KBr will have $= \frac{10^{-7}}{119} \times 1$ mole cationic vacancies.

$$= \frac{10^{-7}}{119} \times 6.023 \times 10^{23} = 5.06 \times 10^{14} \text{ cationic vacancies.}$$

11. (87) Edge length $(a) = 4 \times 10^{-8}$ cm

Density $(d) = 9.03$ g/mL

In FCC lattice; $Z = 4$

$$d = \frac{ZM}{N_A a^3} \quad ; \quad M = \frac{9.03 \times 6.02 \times 10^{23} \times 64 \times 10^{-24}}{4} = 86.97$$

12. (1) There are 4 formula units per unit cell.

$$d = \frac{Z \times M}{a^3 \times N_A} \Rightarrow 43.1 = \frac{4 \times 58.5}{a^3 \times 6.02 \times 10^{23}}$$

$\Rightarrow$ $a = 2.08 \times 10^{-8}$ cm $\Rightarrow$ $a = 2.08 \times 10^{-10}$ cm

Also $a = 2(r_{Na^+} + r_{Cl^-})$

$\Rightarrow$ $r_{Na^+} + r_{Cl^-} = 1.04 \times 10^{-10}$ m

$\therefore$ The answer is 1.

13. (17) $\sqrt{3}\,a = 4r$

$$\sqrt{3} \times 4 = 4r \Rightarrow r = 1.732\ \text{Å} = 17.32 \times 10^{-1}$$

14. (3) Cubic, tetragonal and orthorhombic have body centered unit cell.

❑❑❑

CHAPTER

8

Solutions

A. Objective Questions

[Only one option is correct]

1. Dissolving 120 g of urea (mol. wt. 60) in 1000 g of water gave a solution of density 1.15 g/mL. The molarity of the solution is : **(IIT 2011)**

(a) 1.78 *M* (b) 2.00 *M* (c) 2.05 *M* (d) 2.22 *M*

2. The freezing point (in °C) of a solution containing 0.1g of $K_3[Fe(CN)_6]$ (Mol. wt. 329) in 100 g of water ($K_f = 1.86\ K\ kg\ mol^{-1}$) is : **(IIT 2011)**

(a) -2.3×10^{-2} (b) -5.7×10^{-2} (c) -5.7×10^{-3} (d) -1.2×10^{-2}

3. For a dilute solution containing 2.5g of a non-volatile non-electrolyte solution in 100g of water, the elevation in boiling point at 1 atm pressure is 2° C. Assuming concentration of solute is much lower than the concentration of solvent, the vapour pressure (mm of Hg) of the solution is :
(take $K_b = 0.76\ K\ kg\ mol^{-1}$) **(IIT 2012)**

(a) 724 (b) 740 (c) 736 (d) 718

4. The molarity of a solution obtained by mixing 750 mL of 0.5 *M* HCl with 250 mL of 2 *M* HCl will be : **[JEE (Mains) 2013]**

(a) 0.875 *M* (b) 1.00 *M* (c) 1.75 *M* (d) 0.975 *M*

5. Consider separate solutions of 0.500 *M* $C_2H_5OH(aq)$, 0.100 *M* $Mg_3(PO_4)_2(aq)$, 0.250 *M* $KBr(aq)$ and 0.125 *M* $Na_3PO_4(aq)$ at 25° C. Which statement is **true** about these solutions, assuming all salts to be strong electrolytes ? **[JEE (Mains) 2014]**

(a) They all have the same osmotic pressure.
(b) 0.100 *M* $Mg_3(PO_4)_2(aq)$ has the highest osmotic pressure.
(c) 0.125 *M* $Na_3PO_4(aq)$ has the highest osmotic pressure.
(d) 0.500 *M* $C_2H_5OH(aq)$ has the highest osmotic pressure.

6. The vapour pressure of acetone at 20°C is 185 torr. When 1.2 g of a non-volatile substance was dissolved in 100 g of acetone at 20°C, its vapour pressure was 183 torr. The molar mass ($g\ mol^{-1}$) of the substance is: **[JEE (Mains) 2015]**

(a) 128 (b) 488 (c) 32 (d) 64

7. 18 g glucose ($C_6H_{12}O_6$) is added to 178.2 g water. The vapour pressure of water (in torr) for this aqueous solution is : **[JEE (Mains) 2016]**

(a) 7.6 (b) 76.0 (c) 752.4 (d) 759.0

8. The freezing point of benzene decreases by 0.45°C when 0.2 g of acetic acid is added to 20 g of benzene. If acetic acid associates to form a dimer in benzene, percentage association of acetic acid in benzene will be:
(K_f for benzene= $5.12\ K kg\ mol^{-1}$) **[JEE (Mains) 2017]**

(a) 74.6% (b) 94.6% (c) 64.6% (d) 80.4%

9. Pure water freezes at 273 K and 1 bar. The addition of 34.5 g of ethanol to 500 g of water changes the freezing point of the solution. Use the freezing point depression constant of water as 2 K kg mol^{-1}. The figures shown below represent plots of vapour pressure (V.P.) *versus* temperature (T). [Molecular weight of ethanol is 46 g mol^{-1}]

Among the following, the option representing change in the freezing point is: **[JEE (Advanced) 2017]**

(a) 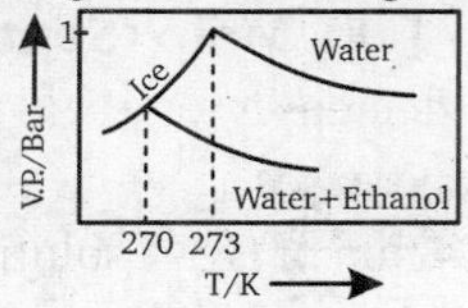

(b) 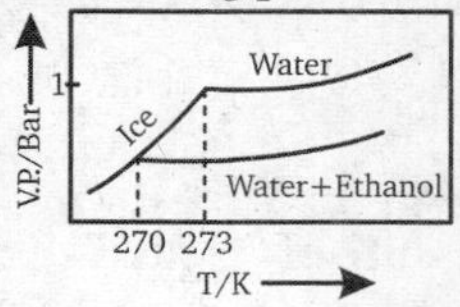

(c) 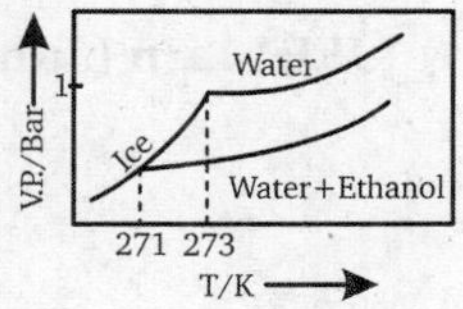

(d) 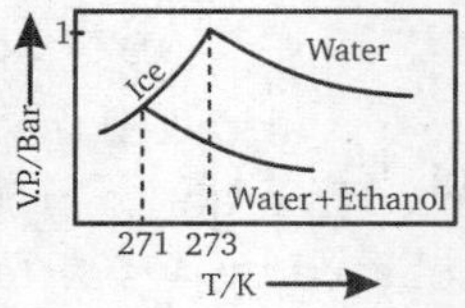

10. For 1 molal aqueous solution of the following compounds, which one will show the highest freezing point? **[JEE (Mains) 2018]**

(a) $[Co(H_2O)_4Cl_2]Cl \cdot 2H_2O$ (b) $[Co(H_2O)_3Cl_3] \cdot 3H_2O$

(c) $[Co(H_2O)_6]Cl_3$ (d) $[Co(H_2O)_5Cl]Cl_2 \cdot H_2O$

11. The vapour pressure of pure liquids A and B are 400 and 600 mmHg, respectively at 298 K. On mixing the two liquids, the sum of their initial volumes is equal to the volume of the final mixture. The mole fraction of liquid B is 0.5 in the mixture. The vapour pressure of the final solution, the mole fractions of components A and B in vapour phase, respectively are : **[JEE (Mains) 2019]**

(a) 450 mm Hg, 0.5, 0.5 (b) 450 mm Hg, 0.4, 0.6

(c) 500 mm Hg, 0.5, 0.5 (d) 500 mm Hg, 0.4, 0.6

12. For the solution of the gases w, x, y and z in water at 298 K, the Henry's law constants (K_H) are 0.5, 2,35 and 40 k bar, respectively. The correct plot for the given data is : **[JEE (Mains) 2019]**

(a) 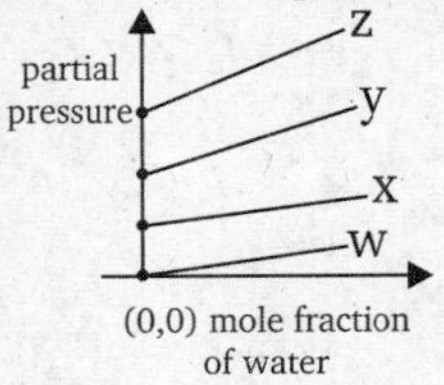

(b) 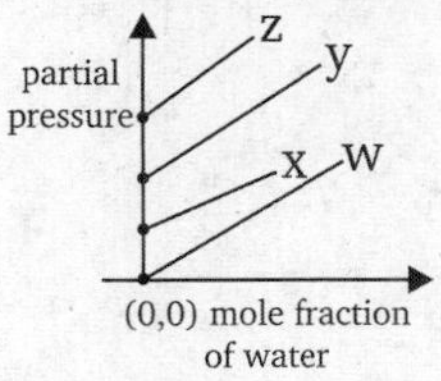

(c) 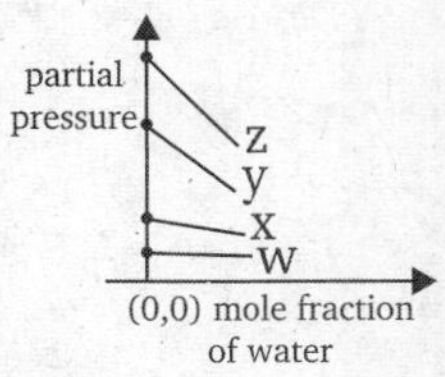

(d) 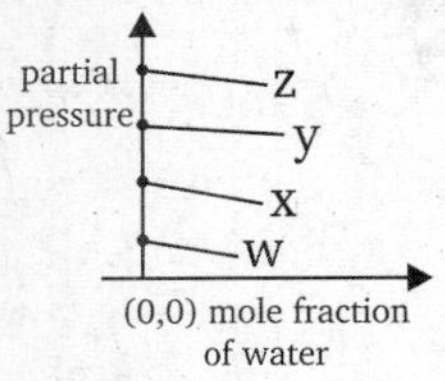

13. A graph of vapour pressure and temperature for three different liquids X, Y and Z is shown below: **[JEE Main (Jan.) 2020]**

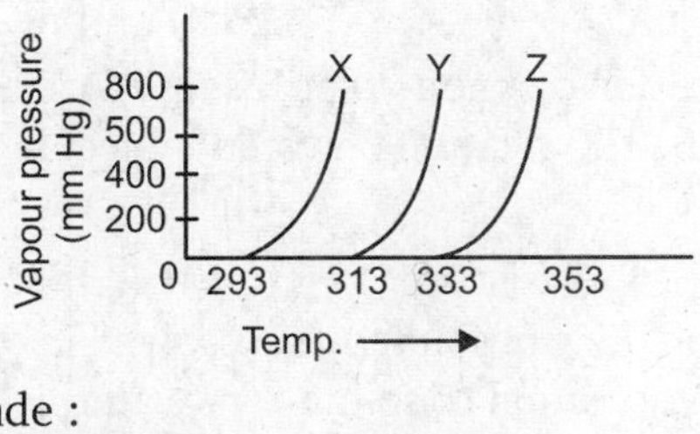

The following inferences are made :

(A) X has higher intermolecular interactions compared to Y.

(B) X has lower intermolecular interactions compared to Y.

(C) Z has lower intermolecular interactions compared to Y.

The correct inference(s) is/are:

(a) (C) (b) (A) and (C) (c) (B) (d) (A)

14. Two open beakers one containing a solvent and the other containing a mixture of that solvent with a non-volatile solute are together sealed in a container. Over time: **[JEE Main (Jan.) 2020]**

(a) the volume of the solution increases and the volume of the solvent decreases
(b) the volume of the solution and the solvent does not change
(c) the volume of the solution does not change and the volume of the solvent decreases
(d) the volume of the solution decreases and the volume of the solvent increases

15. An open beaker of water in equilibrium with water vapour is in a sealed container. When a few grams of glucose are added to the beaker of water, the rate at which water molecules :

[JEE Main (Sept.) 2020]

(a) leaves the vapour increases
(b) leaves the solution decreases
(c) leaves the solution increases
(d) leaves the vapour decreases

16. At 35°C, the vapour pressure of CS_2 is 512 mm of Hg and that of acetone is 344 mm of Hg. A solution of CS_2 in acetone has a total vapour pressure of 600 mm of Hg. The false statement among the following is:

[JEE Main (Jan.) 2020]

(a) CS_2 and acetone are less attracted to each other than to themselves.
(b) Heat must be absorbed in order to produce the solution at 35°C.
(c) Raoult's law is not obeyed by this system.
(d) a mixture of 100 mL CS_2 and 100 mL acetone has a volume of < 200 mL.

17. A set of solutions is prepared using 180 g of water as a solvent and 10 g of different non-volatile solutes A, B and C. The relative lowering of vapour pressure in the presence of these solutes are in the order [Given, molar mass of $A = 100\ \text{mol}^{-1}$; $B = 200\ \text{g mol}^{-1}$; $C = 10{,}000\ \text{g mol}^{-1}$] **[JEE Main (Sept.) 2020]**

(a) $B > C > A$
(b) $C > B > A$
(c) $A > B > C$
(d) $A > C > B$

18. The size of a raw mango shrinks to a much smaller size when kept in a concentrated salt solution. Which one of the following process can explain this? **[JEE Main (Sept.) 2020]**

(a) Dialysis
(b) Osmosis
(c) Diffusion
(d) Reverse osmosis

19. For a solution of the gases A, B C and D in water at 298 K, the values of Henry's law constant (K_H) are 30.40, 2.34, 1.56×10^{-5} and 0.513 k bar respectively. In the given graph, the lines marked as 'p' and 's' correspond respectively to : **[JEE Main (June) 2022 (I)]**

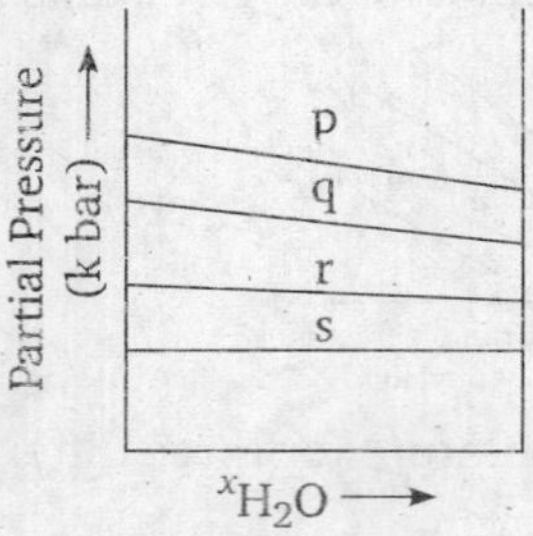

(a) A and C
(b) B and A
(c) D and A
(d) C and D

20. Boiling point of a 2% aqueous solution of a non-volatile solute A is equal to the boiling point of 8% aqueous solution of a non-volatile solute B. The relation between molecular weights of A and B is.

[JEE Main (July) 2022 (I)]

(a) $M_A = 4M_B$
(b) $M_B = 4M_A$
(c) $M_A = 8M_B$
(d) $M_B = 8M_A$

21. Solute A associates in water. When 0.7 g of solute A is dissolved in 42.0 g of water, it depresses the freezing point by 0.2°C. The percentage association of solute A in water, is

[Given : Molar mass of A = 93 g mol^{-1},

Molal depression constant of water is 1.86 K kg mol^{-1}] **[JEE Main (June) 2022 (II)]**

(a) 50%
(b) 60%
(c) 70%
(d) 80%

22. A solution is prepared by adding 2 g of "X" of 1 mole of water. Mass percent of "X" in the solution is :

[JEE Main (April) 2023 (II)]

(a) 20%
(b) 5%
(c) 2%
(d) 10%

23. What weight of glucose must be dissolved in 100 g of water to lower the vapour pressure by 0.20 mm Hg? (Assume dilute solution is being formed)
Given : Vapour pressure of pure water is 54.2 mm Hg at room temperature. Molar mass of glucose is 180 g mol^{-1}. **[JEE Main (April) 2023 (II)]**
(a) 4.69 g (b) 3.59 g (c) 2.59 g (d) 3.69 g

24. Evaluate the following statements for their correctness. **[JEE Main (Jan) 2023 (II)]**
(A) The elevation in boiling point temperature of water will be same for 0.1 M NaCl and 0.1 M urea
(B) Azeotropic mixtures boil without change in their composition
(C) Osmosis always takes place from hypertonic to hypotonic solution
(D) The density of 32% H_2SO_4 solution having molarity 4.09 M is approximately 1.26 g mL^{-1}
(E) A negatively charged sol is obtained when KI solution is added to silver nitrate solution
Choose the correct answer from the options given below :
(a) B, D and E only (b) A, B and D only (c) A and C only (d) B and D only

25. A solution of two miscible liquids showing negative deviation from Raoult's law will have : **[JEE Main (Jan.) 2024]**
(a) increased vapour pressure, increased boiling point
(b) increased vapour pressure, decreased boiling point
(c) decreased vapour pressure, decreased boiling point
(d) decreased vapour pressure, increased boiling point

26. The solution from the following with highest depression in freezing point/lowest freezing point is : **[JEE Main (Jan.) 2024]**
(a) 180 g of acetic acid dissolved in water (b) 180 g of acetic acid dissolved in benzene
(c) 180 g of benzoic acid dissolved in benzene (d) 180 g of glucose dissolved in water

27. We have three aqueous solutions of NaCl labelled as 'A', 'B' and 'C' with concentration 0.1 M, 0.01 M and 0.001 M respectively. The value of van't Hoff factor (i) for these solutions will be in the order. **[JEE Main (Jan.) 2024]**
(a) $i_A < i_B < i_C$ (b) $i_A < i_C < i_B$ (c) $i_A = i_B = i_C$ (d) $i_A > i_B > i_C$

28. Arrange the following solutions in order of their increasing boiling points.
(i) $10^{-4}M$ NaCl (ii) $10^{-4}M$ Urea (iii) $10^{-3}M$ NaCl (iv) $10^{-2}M$ NaCl **[JEE Main (Jan.) 2025]**
(a) (ii) < (i) < (iii) < (iv) (b) (ii) < (i) $\cong$ (iii) < (iv)
(c) (i) < (ii) < (iii) < (iv) (d) (iv) < (iii) < (i) < (ii)

29. Density of $3M$ NaCl solution is 1.25 g/mL. The molality of the solution is : **[JEE Main (Jan.) 2025]**
(a) 1.79 m (b) 2 m (c) 3 m (d) 2.79 m

30. When a non-volatile solute is added to the solvent, the vapour pressure of the solvent decreases by 10 mm of Hg. The mole fraction of the solute in the solution is 0.2. What would be the mole fraction of the solvent if decrease in vapour pressure is 20 mm of Hg ? **[JEE Main (Jan.) 2025]**
(a) 0.6 (b) 0.4 (c) 0.2 (d) 0.8

31. Consider a binary solution of two volatile liquid components 1 and 2 x_1 and y_1 are the mole fractions of component 1 in liquid and vapour phase, respectively. The slope and intercept of the linear plot of $\frac{1}{x_1}$ vs $\frac{1}{y_1}$ are given respectively as : **[JEE Main (Jan.) 2025]**
(a) $\frac{P_1^\circ}{P_2^\circ}, \frac{P_2^\circ - P_1^\circ}{P_2^\circ}$ (b) $\frac{P_2^\circ}{P_1^\circ}, \frac{P_1^\circ - P_2^\circ}{P_2^\circ}$
(c) $\frac{P_1^\circ}{P_2^\circ}, \frac{P_1^\circ - P_2^\circ}{P_2^\circ}$ (d) $\frac{P_2^\circ}{P_1^\circ}, \frac{P_2^\circ - P_1^\circ}{P_2^\circ}$

32. Consider the given plots of vapour pressure (VP) *vs* temperature (T/K) which amongst the following options is correct graphical representation showing ΔT_f, depression in the freezing point of solvent in a solution? **[JEE Main (Jan.) 2025]**

(a)
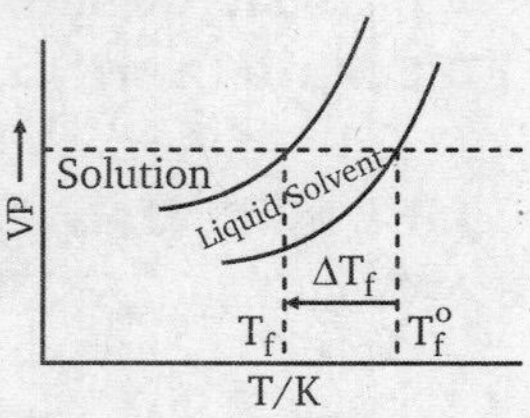

(b)
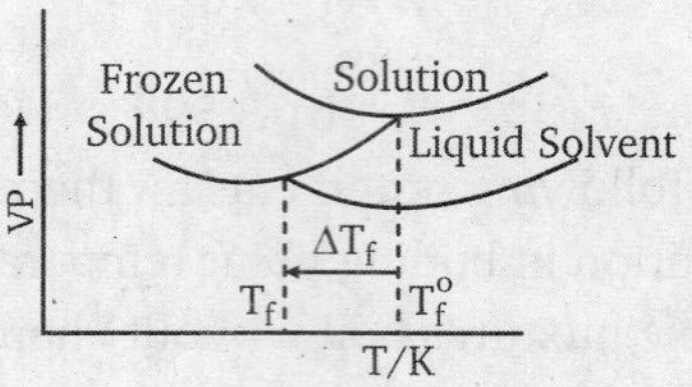

(c)
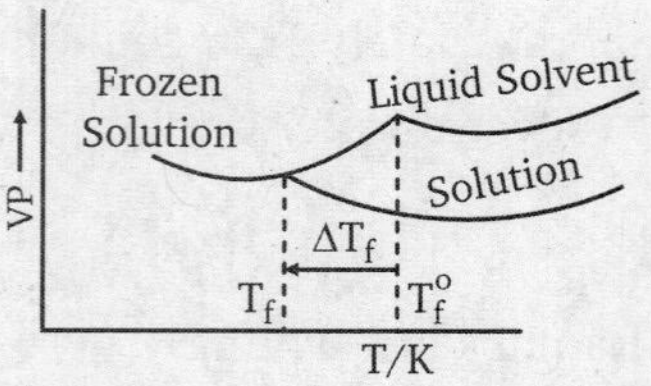

(d)
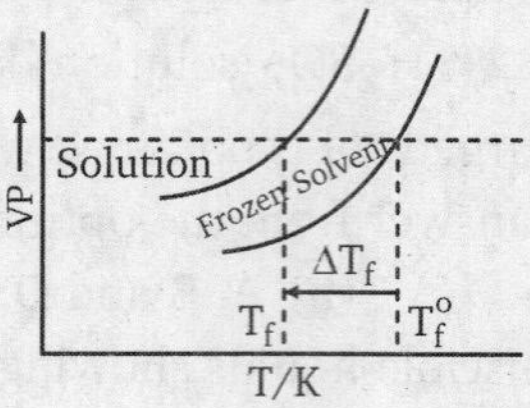

33. Which of the following mixing of 1*M* base and 1*M* acid leads to the largest increase in temperature?

(a) 30 mL HCl and 30 mL NaOH
(b) 30 mL CH_3COOH and 30 mL NaOH
(c) 50 mL HCl and 20 mL NaOH
(d) 45 mL CH_3COOH and 25 mL NaOH

34. What is the freezing point depression constant of a solvent, 50 g of which contain 1 g non-volatile solute (molar mass 256 g mol^{-1}) and the decrease in freezing point is 0.40 K? **[JEE Main (Jan.) 2025]**

(a) 5.12 K kg mol^{-1}
(b) 4.43 K kg mol^{-1}
(c) 1.86 K kg mol^{-1}
(d) 3.73 K kg mol^{-1}

35. Concentrated nitric acid is labelled as 75% by mass. The volume in mL of the solution which contains 30 g of nitric acid is ....

Given : Density of nitric acid solution is 1.25 g/mL **[JEE Main (Jan.) 2025]**

(a) 45
(b) 55
(c) 32
(d) 40

36. Assume a living cell with 0.9% (w/w) of glucose solution (aqueous). This cell is immersed in another solution having equal mole fraction of glucose and water.

(Consider the data upto first decimal place only)

The cell will : **[JEE Main (Jan.) 2025]**

(a) shrink since solution is 0.5% (w/w)
(b) shrink since solution is 0.45% (w/w) as a result of association of glucose molecules (due to hydrogen bonding)
(c) swell up since solution is 1% (w/w)
(d) show no change in volume since solution is 0.9% (w/w)

37. 1.24 g of AX_2 (molar mass 124 g mol^{-1}) is dissolved in 1 kg of water to form a solution with boiling point of 100.0156°C while 25.4 g of AY_2 (molar mass 250 g mol^{-1}) in 2 kg of water constitutes a solution with a boiling point of 100.0260°C.

$K_b(H_2O) = 0.52$ K kg mol^{-1}

Which of the following is correct? **[JEE Main (Jan.) 2025]**

(a) AX_2 and AY_2 (both) are completely unionised.
(b) AX_2 and AY_2 (both) are fully ionised.
(c) AX_2 is completely unionised while AY_2 is fully ionised.
(d) AX_2 is fully ionised while AY_2 is completely unionised.

38. Given below are two statements :

Statement-I : NaCl is added to the ice at 0°C, present in the ice cream box to prevent the melting of ice cream.

Statement-II : On addition of NaCl to ice at 0°C, there is a depression in freezing point.

In the light of the above statements, choose the correct answer from the options given below :

[JEE Main (Jan.) 2025]

(a) Statement-I is false but Statement-II is true
(b) Both Statement-I and Statement-II are true
(c) Both Statement-I and Statement-II are false
(d) Statement-I is true but Statement-II is false

B. Objective Questions

[One or more than one option(s) is/are correct]

1. Benzene and nepthalene form an ideal solution at room temperature. For this process, the true statement(s) is (are) : **[JEE (Advanced) 2013]**

(a) ΔG is positive
(b) ΔS_{system} is positive
(c) $\Delta S_{\text{surroundings}} = 0$
(d) $\Delta H = 0$

2. Mixture(s) showing positive deviation from Raoult's law at 35° C is (are): **[JEE (Advanced) 2016]**

(a) carbon tetrachloride + methanol
(b) carbon disulphide + acetone
(c) benzene + toluene
(d) phenol + aniline

3. For solution formed by mixing liquids L and M, the vapour pressure of L plotted against the mole fraction of M in solution is shown in the following figure. Here x_L and x_M represent mole fractions of L and M, respectively, in the solution. The correct statement(s) applicable to this system is (are):

[JEE (Advanced) 2017]

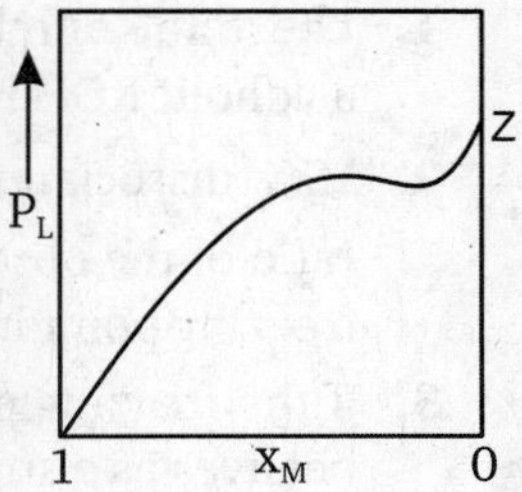

(a) Attractive intermolecular interactions between L-L in pure liquid L and M-M in pure liquid M are stronger than those between L-M when mixed in solution.
(b) The point Z represents vapour pressure of pure liquid M and Raoult's law is obeyed when $x_L \rightarrow 0$.
(c) The point Z represents vapour pressure of pure liquid M and Raoult's law is obeyed from $x_L = 0$ to $x_L = 1$.
(d) The point Z represents vapour pressure of pure liquid L and Raoult's law is obeyed when $x_L \rightarrow 1$.

C. Passage Based Problems

Passage

Properties such as boiling point, freezing point and vapour pressure of a pure solvent change when solute molecules are added to get homogeneous solution. There are called colligative properties. Applications of colligative properties are very useful in day-to-day life. One of its examples is the use of ethylene glycol and water mixture as anti-freezing liquid in the radiator of automobiles. A solution M is prepared by mixing ethanol and water. The mole fraction of ethanol in the mixture is 0.9.

Given : Freezing point depression constant of water $(K_f^{\text{water}}) = 1.86\ \text{K kg mol}^{-1}$

Freezing point depression constant of ethanol $(K_f^{\text{ethanol}}) = 2.0\ \text{K kg mol}^{-1}$

Boiling point elevation constant of water $(K_b^{\text{water}}) = 0.52\ \text{K kg mol}^{-1}$

Boiling point elevation constant of ethanol $(K_b^{\text{ethanol}}) = 1.2\ \text{K kg mol}^{-1}$

Standard freezing point of water = 273 K

Standard freezing point of ethanol = 155.7 K

Standard boiling point of water = 373 K

Standard boiling point of ethanol = 351.5 K

Vapour pressure of pure water = 32.8 mm Hg

Vapour pressure of pure ethanol = 40 mm Hg

Molecular weight of water = 18 g mol^{-1}

Molecular weight of ethanol = 46 g mol^{-1}

In answering the following questions, consider the solutions to be ideal dilute solutions and solutes to be non-volatile and non-dissociative. **(IIT 2008)**

1. The freezing point of the solution M is :

(a) 268.7 K (b) 268.5 K (c) 234.2 K (d) 150.9 K

2. The vapour pressure of the solution M is :

(a) 39.3 mm Hg (b) 36.0 mm Hg (c) 29.5 mm Hg (d) 28.8 mm Hg

3. Water is added to be solution M such that the mole fraction of water in the solution becomes 0.9. The boiling point of this solution is :

(a) 380.4 K (b) 376.2 K (c) 375.5 K (d) 354.7 K

D. Integer Type Problems

1. The total number of alkenes possible by dehydrobromination of 3-bromo-3-cyclopentylhexane using alcoholic KOH is : **(IIT 2011)**

2. MX_2 dissociates in M^{2+} and X^- ions in an aqueous solution, with a degree of dissociation (α) of 0.5. The ratio of the observed depression of freezing point of the aqueous solution to the value of the depression of freezing point in the absence of ionic dissociation is: **[JEE (Advanced) 2014]**

3. If the freezing point of a 0.01 molal aqueous solution of a cobalt (III) chloride-ammonia complex (which behaves as a strong electrolyte) is –0.0558° C, the number of chloride(s) in the co-ordination sphere of the complex is : [K_f of water = 1.86 K kg mol^{-1}] **[JEE (Advanced) 2015]**

4. Liquids A and B form ideal solution over the entire range of composition. At temperature T, equimolar binary solution of liquids A and B has vapour pressure 45 torr. At the same temperature, a new solution of A and B having mole fractions x_A and x_B, respectively, has vapour pressure of 22.5 torr. The value of x_A/x_B in the new solution is : (given that the vapour pressure of pure liquid A is 20 torr at temperature T) **[JEE (Advanced) 2018]**

5. The plot given below shows P-T curves (where P is the pressure and T is the temperature) for two solvents X and Y and isomolal solutions of NaCl in these solvents. NaCl completely dissociates in both the solvents.

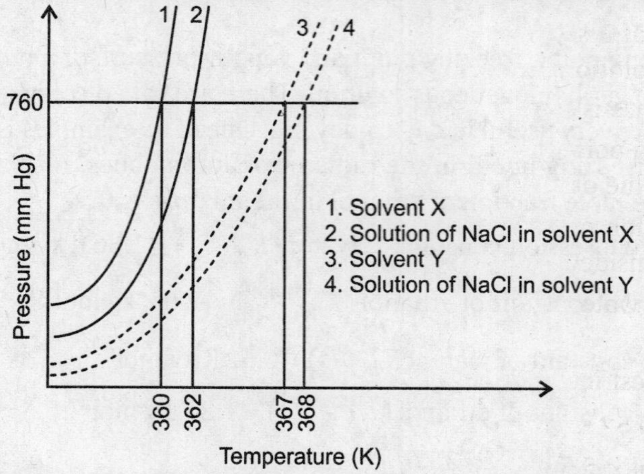

On addition of equal number of moles of a non-volatile solute S in equal amount (in kg) of these solvents, the elevation of boiling point of solvent X is three times that of solvent Y. Solute S is known to undergo dimerization in these solvents. If the degree of dimerization is 0.7 in solvent Y, the degree of dimerization in solvent X is : **[JEE (Advanced) 2018]**

6. On dissolving 0.5 g of a non-volatile non-ionic solute to 39 g of benzene, its vapour pressure decreases from 650 mm Hg to 640 mm Hg. The depression of freezing point of benzene (in K) upon addition of the solute is ___.
(Given data : Molar mass and the molal freezing point depression constant of benzene are 78 g mol^{-1} and 5.12 K kg mol^{-1}, respectively) **[JEE (Advanced) 2019]**

7. The mole fraction of urea in an aqueous urea solution containing 900 g of water is 0.05. If the density of the solution is 1.2 g cm^{-3}, the molarity of urea solution is ____.
(Given data : Molar masses of urea and water are 60 g mol^{-1} and 18 g mol^{-1}, respectively).
[JEE (Advanced) 2019]

8. Liquids *A* and *B* form ideal solution for all compositions of *A* and *B* at 25° C. Two such solutions with 0.25 and 0.50 mole fractions of *A* have the total vapour pressures of 0.3 and 0.4 bar, respectively. What is the vapour pressure of pure liquid *B* in bar ? **[JEE (Advanced) 2020]**

9. How much amount of NaCl should be added to 600 g of water ($\rho = 1.00$ g/mL) to decrease the freezing point of water to −0.2° C? _____ . **[JEE Main (Jan.) 2020]**

10. A cylinder containing an ideal gas(0.1 mol of 1.0 dm^3) is in thermal equilibrium with a large volume of 0.5 molal aqueous solution of ethylene glycol at its freezing point. If the stoppers S_1 and S_2 (as shown in the figure) are suddenly withdrawn, the volume of the gas in litres after equilibrium is achieved will be.
[JEE Main (Jan.) 2020]
(Given, K_f (water) = 2.0 K kg mol^{-1}, $R = 0.08$ dm^3 atm K^{-1} mol^{-1}).

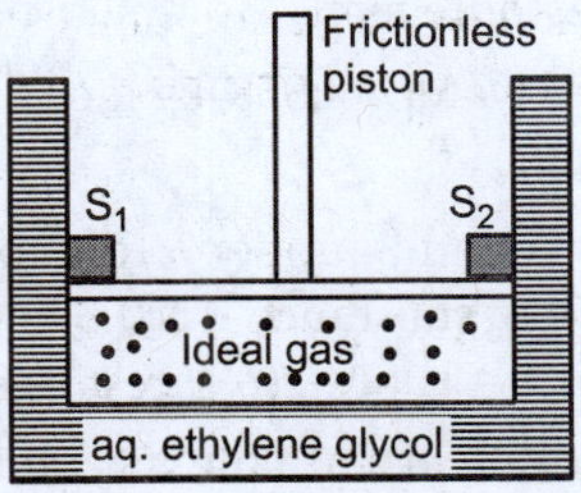

11. If 250 cm^3 of an aqueous solution containing 0.73 g of a protein *A* is isotonic with one litre of another aqueous solution containing 1.65 g of a protein *B*, at 298 K, the ratio of the molecular masses of *A* and *B* is _________ $\times 10^{-2}$ (to the nearest integer). **[JEE Main (Sept.) 2020]**

12. The osmotic pressure of a solution of NaCl is 0.10 atm and that of a glucose solution is 0.20 atm. The osmotic pressure of a solution formed by mixing 1 L of the sodium chloride solution with 2 L of the glucose solution is $x \times 10^{-3}$ atm. *x* is .... (nearest integer). **[JEE Main (Sept.) 2020]**

13. The elevation of boiling point of 0.10 m aqueous $CrCl_3xNH_3$ solution is two times that of 0.05 m aqueous $CaCl_2$ solution. The value of *x* is _________.
[Assume 100% ionisation of the complex and $CaCl_2$, coordination number of Cr as 6, and that all NH_3 molecules are present inside the coordination sphere] **[JEE Main (Sept.) 2020]**

14. Solute *A* dimerizes in water. The boiling point of a 2 molar solution of *A* is 100.52° C. The percentage association of *A* is ____. **[JEE Main (March) 2021]**
(Round off to the nearest integer)
[Use : K_b for water = 0.52 K kg mol^{-1}; Boiling point of water = 100° C]

15. When 9.45 g of $ClCH_2COOH$ is added to 500 mL of water, its freezing point drops by 0.5°C. The dissociation constant of $ClCH_2COOH$ is $x \times 10^{-3}$. The value of *x* is________.
(Rounded off to the nearest integer)
[$K_{f(H_2O)}$ = 1.86 $kgmol^{-1}$] **[JEE Main (Feb.) 2021]**

16. If a compound *AB* dissociates to the extent of 75% in an aqueous solution, the molality of the solution which shows a 2.5K rise in the boiling point of the solution is ______ molal.

(Rounded-off to the nearest integer)

$[K_b = 0.52 \text{ K kg mol}^{-1}]$ **[JEE Main (Feb.) 2021]**

17. 224 mL of $SO_2(g)$ at 298 K and 1 atm is passed through 100 mL of 0.1*M* NaOH solution. The non-volatile solute produced is dissolved in 36g of water. The lowering of vapour pressure of solution (assuming the solution is dilute) ($P_{(H_2O)} = 24$ mm of Hg) is $x \times 10^{-2}$ mm of Hg, the value of *x* is ______. (Integer answer)

[JEE Main (Feb.) 2021]

18. When 12.2 g of benzoic acid is dissolved in 100 g of water, the freezing point of solution was found to be $-0.93°$ C ($K_f(H_2O) = 1.86$K kg mol^{-1}). The number (*n*) of benzoic acid molecules associated (assuming 100% association) is ________. **[JEE Main (Feb.) 2021]**

19. AB_2 is 10% dissociated in water to A^{2+} and B^-. The boiling point of a 10.0 molal aqueous solution of AB_2 is ________ °C. (Round off to the nearest integer).

[Given : Molal elevation constant of water $K_b = 0.5$K kg mol^{-1}boiling point of pure water = 100° C]

[JEE Main (March) 2021]

20. A 1 molal $K_4Fe(CN)_6$ solution has a degree of dissociation of 0.4. Its boiling point is equal to that of another solution which contains 18.1 weight per cent of a non-electrolytic solute *A*. The molar mass of *A* is ......u. (Round off to the nearest integer).

[Density of water = 1.0 g cm^{-3}] **[JEE Main (March) 2021]**

21. 2 molal solution of a weak acid H*A* has a freezing point of 3.885°C. The degree of dissociation of this acid is ....$\times 10^{-3}$. (Round off to the nearest integer).

[**Given:** Molal depression constant of water = 1.85K kg mol^{-1}, Freezing point of pure water = 0° C]

[JEE Main (March) 2021]

22. A gaseous mixture of two substances A and B, under a total pressure of 0.8 atm is in equilibrium with an ideal liquid solution. The mole fraction of substance A is 0.5 in the vapour phase and 0.2 in the liquid phase. The vapour pressure of pure liquid A is ____ atm. (Nearest integer) **[JEE Main (July) 2022 (II)]**

23. When 800 mL of 0.5 M nitric acid is heated in a beaker, its volume is reduced to half and 11.5 g of nitric acid is evaporated. The molarity of the remaining nitric acid solution is $x \times 10^{-2}$M. (Nearest Integer)

(Molar mass of nitric acid is 63 g mol^{-1}) **[JEE Main (July) 2022 (I)]**

24. 150 g of acetic acid was contaminated with 10.2 g ascorbic acid ($C_6H_8O_6$) to lower down its freezing point by ($x \times 10^{-1}$)°C. The value of *x* is ________. (Nearest integer)

[Given : $K_f = 3.9$ K kg mol^{-1}; Molar mass of ascorbic acid = 176 g mol^{-1}]

[JEE Main (July) 2022 (I)]

25. The elevation in boiling point for 1 molal solution of non-volatile solute A is 3 K. The depression in freezing point for 2 molal solution of A in the same solvent is 6 K. The ratio of K_b and K_f *i.e.*, K_b/K_f is 1 : *X*. The value of *X* is : [Nearest integer] **[JEE Main (July) 2022 (II)]**

26. 5 g of NaOH was dissolved in deionized water to prepare a 450 mL stock solution. What volume (in mL) of this solution would be required to prepare 500 mL of 0.1 M solution?

[Given : Molar mass of Na, O and H is 23, 16 and 1 g mol^{-1} respectively]. **[JEE Main (Jan.) 2023 (I)]**

27. Consider the following pairs of solution which will be isotonic at the same temperature. The number of pairs of solutions is/are ________. **[JEE Main (April) 2023 (II)]**

A. 1 *M* aq. NaCl and 2 *M* aq. Urea

B. 1 *M* aq. $CaCl_2$ and 1.5 *M* aq. KCl

C. 1.5 *M* aq. $AlCl_3$ and 2 *M* aq. Na_2SO_4

D. 2.5 *M* aq. KCl and 1 *M* aq. $Al_2(SO_4)_3$

28. Mass of Urea (NH_2CONH_2) required to be dissolved in 1000 g of water to reduce the vapour pressure of water by 25% is .... g. (Nearest integer)

Given : Molar mass of N, C, O and H are 14, 12, 16 and 1 g mol^{-1} respectively.

[JEE Main (April) 2023 (I)]

29. Mass of ethylene glycol (antifreeze) to be added to 18.6 kg of water to protect the freezing point at –24°C is ________ kg (Molar mass in g mol^{-1} for ethylene glycol 62, K_f of water = 1.86 K kg mol^{-1})

[JEE Main (Jan.) 2024]

30. 20 mL of 2*M* NaOH solution is added to 400 mL of 0.5*M* NaOH solution. The final concentration of the solution is .... $\times 10^{-2}$ *M*. (Nearest integer). **[JEE Main (Jan.) 2025]**

31. The molarity of a 70% (mass/mass) aqueous solution of a monobasic acid (*X*) is ....*M* (Nearest integer) **[JEE Main (Jan.) 2025]**

[Given : Density of aqueous solution of (*X*) is 1.25 g mL^{-1}, Molar mass of the acid is 70 g mol^{-1}]

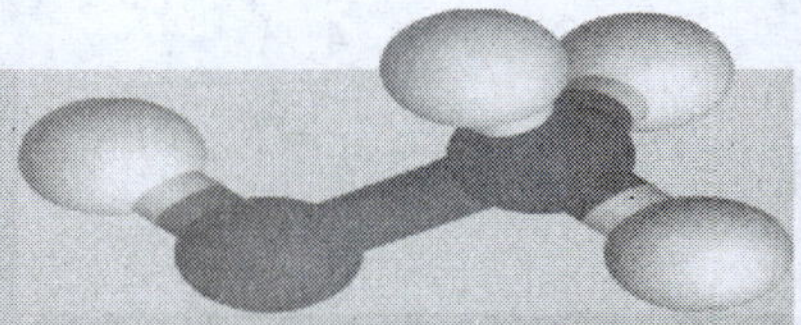

Answers

[A] Objective Questions (Only one option is correct)

1. (c)	**2.** (a)	**3.** (a)	**4.** (a)	**5.** (a)	**6.** (d)	**7.** (c)	**8.** (b)	**9.** (b)	**10.** (b)
11. (d)	**12.** (c)	**13.** (c)	**14.** (a)	**15.** (b)	**16.** (d)	**17.** (c)	**18.** (b)	**19.** (a)	**20.** (b)
21. (d)	**22.** (d)	**23.** (d)	**24.** (d)	**25.** (d)	**26.** (a)	**27.** (a)	**28.** (a)	**29.** (d)	**30.** (a)
31. (a)	**32.** (c)	**33.** (a)	**34.** (a)	**35.** (c)	**36.** (d)	**37.** (d)	**38.** (b)		

[B] Objective Questions (One or more than one option(s) is/are correct)

1. (b, c, d)	**2.** (a, b)	**3.** (a, d)

[C] Passage Based Problems

1. (d)	**2.** (b)	**3.** (b)

[D] Integer Type Problems

1. (5)	**2.** (2)	**3.** (1)	**4.** (19)	**5.** (0.05)
6. (1.03)	**7.** (2.98)	**8.** (0.20)	**9.** (1.755)	**10.** (2.176)
11. (1.77)	**12.** (167)	**13.** (5)	**14.** (100)	**15.** (35)
16. (3)	**17.** (18)	**18.** (2)	**19.** (106)	**20.** (85)
21. (50)	**22.** (2)	**23.** (54)	**24.** (15)	**25.** (1)
26. (180)	**27.** (4)	**28.** (1111)	**29.** (15)	**30.** (57)
31. (125)				

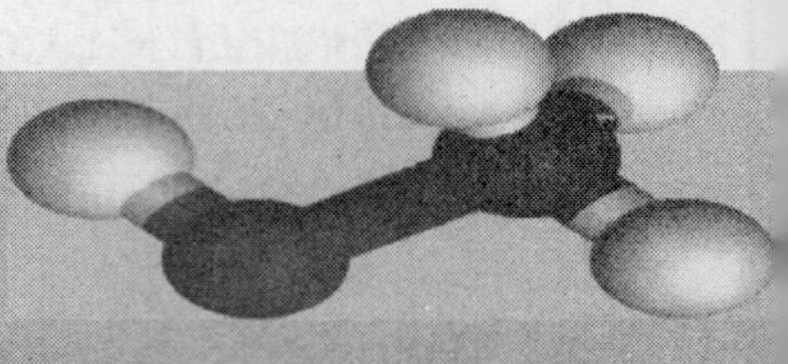

Hints and Solutions

A. Objective Questions

[Only one option is correct]

1. (c) Total mass of solution = 1000 + 120 = 1120 g

Total volume of solution in (L) $= \frac{1120}{1.15} \times 10^3$

$$\text{Molarity} = \frac{W}{M} \times \frac{1}{V\,(\text{in L})} = \frac{120}{60} \times \frac{1.15 \times 10^3}{1120} = 2.05\,M$$

2. (a) $K_3[Fe(CN)_6] \longrightarrow 3K^+ + [Fe(CN)_6]^{3-}$

$i = 4$

$$\Delta T_f = K_f \times i \times \frac{m}{M} \times \frac{1000}{W} = 1.86 \times 4 \times \frac{0.1}{329} \times \frac{1000}{100}$$

$$= 2.3 \times 10^{-2}$$

$T_f = -2.3 \times 10^{-2}$

3. (a) We know that, $\Delta T_b = K_b \times \text{molality}$

Also from Raoult's law,

$$\frac{P° - P_S}{P_S} = \frac{W \times M}{m \times w} = \frac{w \times 1000 \times m}{m \times w \times 1000}$$

$$\frac{P° - P_S}{P_S} = \text{molality} \times \frac{M}{1000}$$

$$\frac{P° - P_S}{P_S} = \frac{\Delta T_b}{K_b} \times \frac{M}{1000}$$

$P° = 1\,\text{atm} = 760\,\text{mm Hg}$; $\Delta T_b = 2$, $K_b = 0.76, M = 18$

$$\frac{760 - P_S}{P_S} = \frac{2}{0.76} \times \frac{18}{1000}$$

$$\frac{760 - P_S}{P_S} = 0.0474$$

$$P_S = 725.6\,\text{mm Hg}$$

4. (a) $(\text{Molarity})_{mix} = \frac{0.5 \times 750 + 250 \times 2}{1000} = 0.875\,M$

5. (a) $\pi = i\,CRT$

$\pi_{C_2H_5OH} = 1 \times 0.500 \times R \times T = 0.5\,RT$

$\pi_{Mg_3(PO_4)_2} = 5 \times 0.100 \times R \times T = 0.5\,RT$

$\pi_{KBr} = 2 \times 0.250 \times R \times T = 0.5\,RT$

$\pi_{Na_3PO_4} = 4 \times 0.125 \times RT = 0.5\,RT$

6. (d) $\frac{P° - P_s}{p°} = \frac{w_2 M_1}{w_1 M_2}$

Given: $P° = 185$ torr, $w_1 = 100\,g, w_2 = 1.2\,g, P_s = 183$ torr

$$M_1 = M_{CH_3COCH_3} = 58\,g\,mol^{-1}$$

$$\frac{185 - 183}{185} = \frac{1.2 \times 58}{100 \times M_2}$$

$$\Rightarrow \quad M_2 = \frac{1.2 \times 58 \times 185}{100 \times 2} = 64.38 \approx 64\,g\,mol^{-1}$$

7. (c) Number of moles of glucose $(n_{C_6H_{12}O_6}) = \frac{18}{180} = 0.1$ mol

Number of moles of water $(n_{H_2O}) = \frac{178.2}{18} = 9.9$ mol

Mole fraction of water in solution $(x_{H_2O}) = \frac{n_{H_2O}}{n_{H_2O} + n_{C_6H_{12}O_6}} = \frac{9.9}{10} = 0.99$

Vapour pressure of water in aqueous solution,

$$P_{H_2O} = P°_{H_2O}\, x_{H_2O} = 760 \text{ torr} \times 0.99 = 752.4 \text{ torr}$$

8. (b) $\Delta T_f = 0.45°C$

w_2 (acetic acid) $= 0.2$ g, w_1 (benzene) $= 20$ g, $K_f = 5.12$ K kg mol^{-1}

$$\Delta T_f = i \times K_f \times m$$

$$\therefore \quad i = \frac{\Delta T_f}{K_f \times m} = \frac{0.45 \times 20 \times 60}{5.12 \times 0.2 \times 1000}$$

$$i = 0.527$$

According to question,

$$2CH_3COOH \rightleftharpoons (CH_3COOH)_2$$

Initially:	1 mol	0
After time t:	$(1-\alpha)$ mol	$\frac{\alpha}{2}$

$$\Rightarrow \quad i = 1 - \alpha + \frac{\alpha}{2}$$

$$i = 1 - \frac{\alpha}{2} \quad .....(i)$$

On putting the value of i in equation (i), we get

$$0.527 = 1 - \frac{\alpha}{2} \Rightarrow -0.946 = -\alpha$$

$$\alpha = 0.946$$

$\therefore$ Percentage association of acetic acid in benzene = 94.6%

9. (b) $\Delta T_f = K_f \times m = 2 \times \frac{34.5 \times 2}{46} = 2 \times 1.5 = 3$

Freezing point of ethanol + water mixture = 273 − 3 = 270

10. (b) Depression in freezing pt.

$$\Delta T_f = iK_f m$$

Less the value of i,

Higher the value of freezing pt.

For (b) $i = 1$(min)

11. (d) According to Raoult's law :

$$P = P_A^\circ X_A + P_B^\circ X_B$$

If $X_B = 0.5$ then $X_A = 1 - 0.5 = 0.5$

$$P_A^\circ = 400\,mmHg, P_B^\circ = 600\,mmHg$$

$$P = 0.5(400) + 0.5(600)$$

$$P = 200 + 300 = 500\,mmHg$$

Mole fraction in vapour phase can be given as

$$Y_A = \frac{X_A P_A^\circ}{P} = \frac{0.5 \times 400}{500} = 0.4$$

$$Y_B = \frac{X_B P_B^\circ}{P} = \frac{0.5 \times 600}{500} = 0.6$$

12. (c) According to Henry's law,

$$p = K_H x \text{ or } p_{gas} = K_H \cdot x_{gas}$$

$$p_{gas} = K_H(1 - x_{H_2O}) = K_H - K_H \cdot x_{H_2O}$$

On comparing this with $y = mx + c$, we get the plot as given in option (c).

The slope will be –ve and the $p_{gas} \propto K_H$ *i.e.*, lower the value of K_H, lower is the partial pressure of the gas.

13. (c) Boiling point order : $Z > Y > X$

Higher is the b.p. of liquid, more will be intermolecular interaction between molecules.

So, order of intermolecular interaction is: $Z > Y > X$.

14. (a)

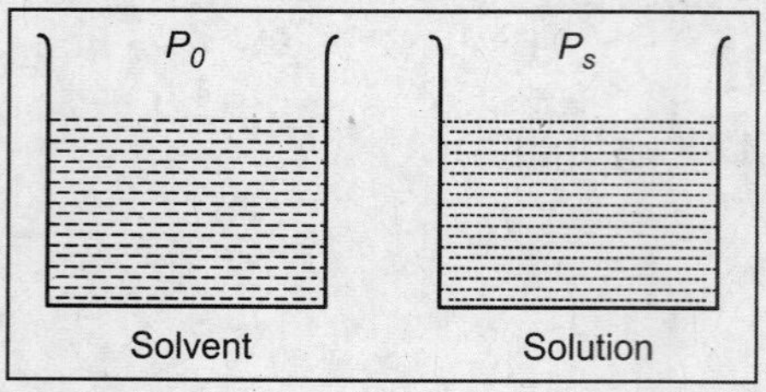

Since $P_0 > P_s$

There will be flow of solvent molecules towards solution to equalize pressure on both sides.

15. (b) On adding glucose in water, escaping tendency of water molecules from solution decreases since evaporation is a surface phenomenon.

16. (d) $P^\circ_{(CS_2)} = 512$ mm Hg

$P^\circ_{(Acetone)} = 344$ mm Hg

Total vapour pressure = 600 mm Hg

Since total vapour pressure is greater than either of vapour pressure of its pure component, solution will show positive deviation from Raoult's law.

So, $\Delta H_{mix} > 0$ (Heat must be absorbed).

$\Rightarrow$ $P_{Total} > P_{(CS_2)} + P_{(Acetone)}$ (Raoult's law is not obeyed)

$\Rightarrow$ $\Delta V_{mix} > 0$

Means $V_{Total} > V_{(CS_2)} + V_{(Acetone)}$

$V_{Total} > (100 + 100)$

$V_{Total} > 200$ mL

17. (c) Relative lowering of vapour pressure is a colligative property, it depends on number of solute particles present in solution.

$$n_A = \frac{10}{100}$$

$$n_B = \frac{10}{200}$$

$$n_C = \frac{10}{10{,}000}$$

Order of particles in solution : $A > B > C$

Same will be order of relative lowering of vapour pressure.

18. (b) Raw mango shrinks in concentrated salt solution due to osmosis.

19. (a) According to Henry's law

$$p = K_H \chi$$

for χ constant : $p \propto K_H$

p has maximum partial pressure.

So it should have maximum K_H and s have minimum partial pressure, so K_H should be minimum.

20. (b)

$$\Delta_A T_b = \frac{2}{M_A \times 100} \times 1000 \times K_b$$

$$\Delta_B T_b = \frac{8}{M_B \times 100} \times 1000 \times K_b$$

Given, $\Delta_A T_b = \Delta_B T_b$

$$\Rightarrow \quad \frac{2}{M_A} = \frac{8}{M_B} \quad \Rightarrow \quad M_B = 4M_A$$

21. (d)

$$\Delta T = i \times K_f \times m$$

$$0.2 = i \times 1.86 \times \frac{0.7}{93} \times \frac{1000}{42} \quad \Rightarrow \quad i = 0.60$$

$$\underset{1-\alpha}{2A} \rightleftharpoons \underset{\frac{\alpha}{2}}{A_2}$$

$$i = 1 - \alpha + \frac{\alpha}{2} \quad \Rightarrow \quad 1 - \frac{\alpha}{2} = 0.60 \quad \Rightarrow \quad \alpha = 0.80$$

22. (d) Solute (X) = 2 g; Solvent (H_2O) = 1 mole = 18 g

Total mass = 2 + 18 = 20 g

$$\% \text{ mass of } X = \frac{2}{20} \times 100 = 10\%$$

23. (d)

$$\frac{P^\circ - P_s}{P^\circ} = \frac{n}{N} \quad \text{(for dilute solution)}$$

$$\frac{0.2}{54.2} = \frac{n \times 18}{100} \quad \Rightarrow \quad n = \frac{1000}{271 \times 18}$$

$$w = \frac{100 \times 180}{271 \times 18} \quad \Rightarrow \quad w = 3.69 \text{ g}$$

24. (d)

$$\Delta T_b \propto i \times C$$

Osmosis always takes place from hypotonic to hypertonic solution.

When KI solution is added to $AgNO_3$ solution, positively charged solution results due to adsorption of Ag^+ ions from dispersion medium.

25. (d) Solution with negative deviation has

$$P_T < P_A^\circ X_A + P_B^\circ X_B$$

$$P_A < P_A^\circ X_A$$

$$P_B < P_B^\circ X_B$$

If vapour pressure decreases so boiling point increases.

26. (a) ΔT_f is maximum when $i \times m$ is maximum.

(1) $m_1 = \frac{180}{60} = 3, \quad i = 1 + \alpha$

Hence,

$$\Delta T_f = (1 + \alpha) \cdot k_f = 3 \times 1.86 = 5.58^\circ\text{C} \quad (\alpha << 1)$$

(2) $m_2 = \frac{180}{60} = 3, \quad i = 0.5, \quad \Delta T_f = \frac{3}{2} \times k'_f = 7.68^\circ\text{C}$

(3) $m_3 = \frac{180}{122} = 1.48, \quad i = 0.5, \quad \Delta T_f = \frac{1.48}{2} \times k'_f = 3.8^\circ\text{C}$

(4) $m_4 = \frac{180}{180} = 1, \quad i = 1, \quad \Delta T_f = 1 \cdot k'_f = 1.86^\circ\text{C}$

As per NCERT, $k'_f(H_2O) = 1.86 \text{ k} \cdot \text{kg mol}^{-1}$

$k'_f(\text{Benzene}) = 5.12 \text{ k} \cdot \text{kg mol}^{-1}$

27. (a)

Salt	Values of i (for different conc. of a salt)		
	0.1 M	0.01 M	0.001 M
NaCl	1.87	1.94	1.94

i approach 2 as the solution become very dilute.

28. (a) $\Delta T_b = iK_b \cdot m \propto i \cdot C$

where C = concentration

Options	$I \cdot C$
(i)	2×10^{-4}
(ii)	1×10^{-4}
(iii)	2×10^{-3}
(iv)	2×10^{-2}

B.P. order : (ii) < (i) < (iii) < (iv)

29. (d) $3M$ NaCl, $d_{sol} = 1.25 \text{g/mol}$

$$\text{Molality} = \frac{M \times 1000}{1000\,d - M \times M_w} = \frac{3000}{1250 - 175.5} = 2.79$$

30. (a) $\because P^o - P \propto X_{solute}$

and $10 \propto 0.2$

$\therefore \quad 20 \propto 0.4$

$\therefore \quad X_{solvent} = 1 - X_{solute} = 1 - 0.4 = 0.6$

31. (a) $\because$ For liquid solution of two liquids '1' and '2'

$$P_1 = P_T Y_1 = P_1^\circ x_1$$

$$\therefore \quad \frac{P_T}{x_1} = \frac{P_1^\circ}{y_1}$$

$$\therefore \quad \frac{P_2^\circ + x_1(P_1^\circ - P_2^\circ)}{x_1} = \frac{P_1^\circ}{y_1}$$

$$\therefore \quad \frac{P_2^\circ}{x_1} + (P_1^\circ - P_2^\circ) = \frac{P_1^\circ}{y_1}$$

$$\therefore \quad \frac{1}{x_1} = \left(\frac{P_1^\circ}{P_2^\circ}\right)\left(\frac{1}{y_1}\right) + \left(\frac{P_2^\circ - P_1^\circ}{P_2^\circ}\right)$$

$$\therefore \quad \text{Slope} = \left(\frac{P_1^\circ}{P_2^\circ}\right)$$

$$\therefore \quad \text{Intercept} = \left(\frac{P_2^\circ - P_1^\circ}{P_2^\circ}\right)$$

32. (c) On adding non-volatile solute in a solvent, the freezing point of solution decreases.

$$T_f < T_f^\circ$$

F.P. of solution < F.P. of pure solvent

Also V.P. of solution decreases on adding non-volatile solute in a solvent.

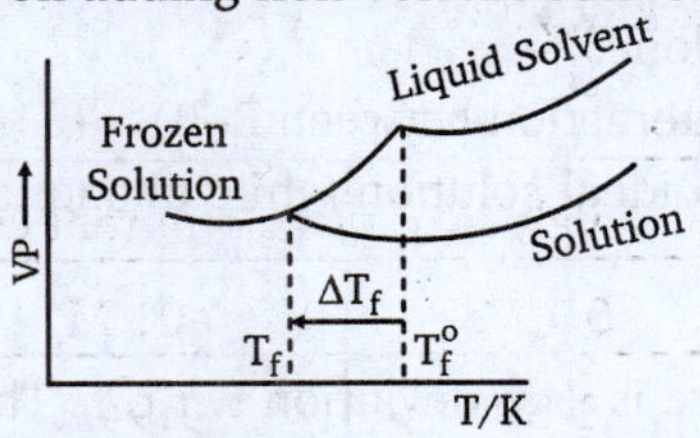

33. (a) Higher the number of millimoles of acid or base reacted higher will be temperature rise.

Option (d) n_{acid} or n_{base} reacted = 30 m mol

Option (b) n_{acid} or n_{base} reacted = 30 m mol

but less energy will be released by neutralisation reaction of weak acid hence option (b) can not be correct.

Option (c) ⇒ 20 m mol

Option (d) ⇒ 25 m mol

34. (a) $\Delta T_f = K_b . m$

$$0.4 = K_b \frac{1/256}{50 \times 10^{-3}}$$

$K_b = 5.12$ K kg/mol

35. (c) % w/w of $HNO_3 = 75\%$

means 100 gm of solution containing 75 g of HNO_3

and $\left(\frac{gm}{m_1}\right)_{solution} = 1.25 = \frac{100 \text{ gm}}{V}$

V_{mL} of 100 gm solution $= \frac{100}{1.25}$ mL

∵ 75 gm of HNO_3 present in $\frac{100}{1.25}$ mL solution

∴ 30 gm of HNO_3 present in $\frac{100}{1.25 \times 75} \times 30 = 32$ mL solution

36. (d) Living cell = 0.9 gm in 100 gm of solution % $w/w = 0.9$

Solution is have equal moles of glucose and water = 0.5

Weight of solution $= 0.5 \times 180 + 0.5 \times 18 = 99$ gm

% $w/w \simeq 90\%$

Concentrated solution

= Cell will shrink.

37. (d) For $AX_2 : \Delta T_b = K_b \times m \times i$

$$0.0156 = 0.52 \times \frac{0.01}{1} \times i_{AX_2}$$

$\Rightarrow i_{AX_2} = 3 \Rightarrow$ complete ionisation

For $AY_2 : \Delta T_b = K_b \times m \times i$

$0.026 = 0.52 \times 0.0508 \times i_{AY_2}$

$\Rightarrow i_{AY_2} \simeq 1$ ∴ complete unionisation

B. Objective Questions

[One or more than one option(s) is/are correct]

1. (b, c, d)

Condition for ideal solution $\Delta G < 0, \Delta S > 0$ & $\Delta H = 0$

2. (a, b)

Positive deviation is observed when the interaction between A and B in mixutre is weaker than the interaction between A and B alone.

(Interaction between A–A + interaction between B–B) > (2 × interaction between A–B).

Benzene and toluene forms an ideal solution while phenol and aniline form a solution showing –ve deviation from Raoult's law.

3. (a, d)

From graph it is clear that there is +ve deviation w.r.t. L. Therefore, option (a) is correct. When $x_L \rightarrow 1$, then z will have value equal to P_L° (vapour pressure of pure L). Therefore, option (d) is also correct.

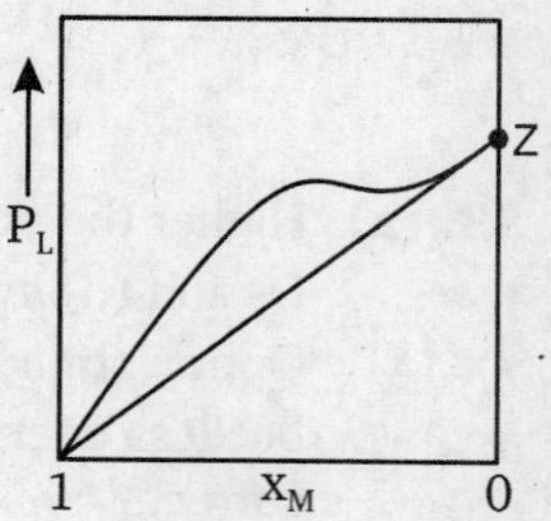

C. Passage Based Problems

1. (d) $\Delta T_f = K_f \times m = 2 \times \dfrac{0.1}{0.9 \times 46} \times 1000 = 4.83\text{K}$

Freezing point of solution (M),

$$T_f' = T_f^\circ - \Delta T_f = 155.7 - 4.83 = 150.9\text{ K}$$

2. (b) Total vapour pressure $P = P_A^\circ X_A$ (Solute is non-volatile)

$$P = 40 \times 0.9 = 36 \text{ mm of Hg}$$

($\because$ vapour pressure of pure ethanol = 40 mm of Hg)

3. (b) $\Delta T_b = K_b \times m = 0.52 \times \dfrac{0.1}{0.9 \times 18} \times 1000 = \dfrac{520}{9 \times 18} = 3.20$

The boiling point of the solution

$$T_b' = T_b^\circ + \Delta T_b$$

$$= 373 + 3.20 = 376.2 \text{ K}$$

($\because$ standard b.pt. of water = 373 K)

D. Integer Type Problems

1. (5) Total no. of alkenes will be = 5

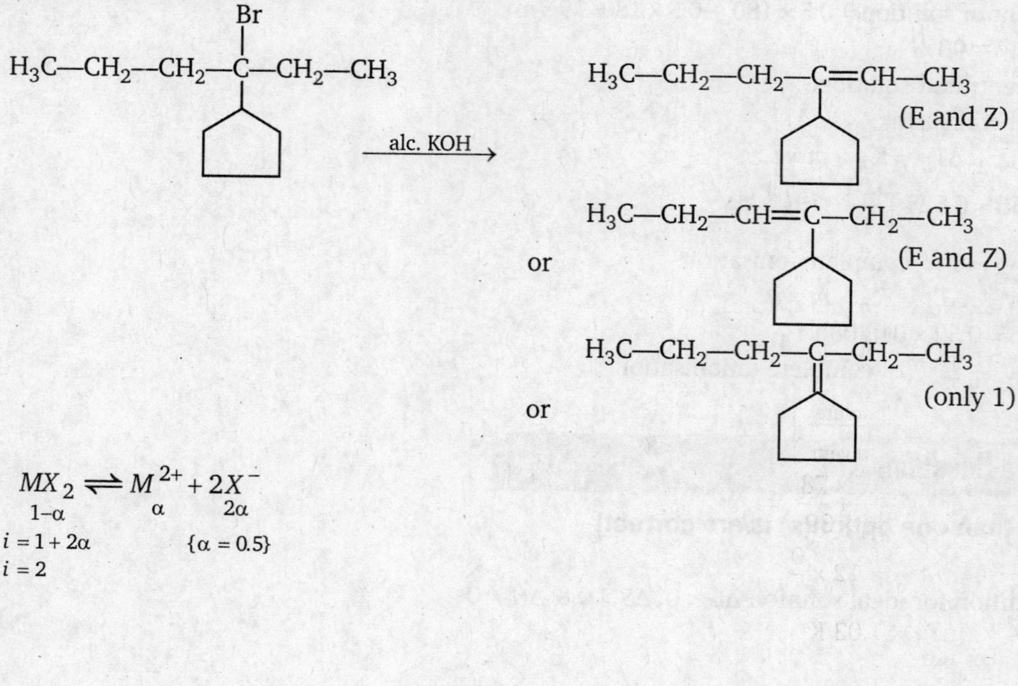

2. (2) $MX_2 \rightleftharpoons M^{2+} + 2X^-$

$\quad 1-\alpha \qquad \alpha \qquad 2\alpha$

$i = 1 + 2\alpha \qquad \{\alpha = 0.5\}$

$i = 2$

3. (1) $\Delta T_f = iK_f m$

$0.0558 = i \times 1.86 \times 0.01$

$i = 3$

$\therefore$ Molecular formula of the complex is $[Co(NH_3)_5Cl]Cl_2$. Thus only one Cl^- ion is in the Co-ordination sphere.

4. (19.00) $P_A^\circ \times \frac{1}{2} + P_B^\circ \times \frac{1}{2} = 45$

$\therefore \quad P_A^\circ + P_B^\circ = 90$

$20 + P_B^\circ = 90$

$\therefore \quad P_B^\circ = 70 \ \& \ P_A^\circ = 20$

Now, $\quad 20x_A + 70(1 - x_A) = 22.5$

$20x_A + 70 - 70x_A = 22.5$

$50x_A = 47.5$

$\therefore \quad x_A = \frac{47.5}{50}$

$\therefore \quad x_B = \frac{2.5}{50}$

$\therefore \quad \frac{x_A}{x_B} = \frac{47.5}{2.5} = 19$

5. (0.05)

From curve (1) & (2)

$\Delta T_{b(x)} = k_{b(x)} m_{NaCl}$

$2 = k_{b(x)} m_{NaCl}$ ...(1)

$\Delta T_{b(y)} = k_{b(y)} m_{NaCl}$

$1 = k_{b(y)} m_{NaCl}$ ...(2)

$\therefore \quad \frac{k_{b(x)}}{k_{b(y)}} = \frac{2}{1}$

Now if solute dimerises

$$\begin{array}{ccc} 2s & \longrightarrow & s \\ 1 & & \\ 1-\alpha & & \alpha/2 \end{array}$$

$\therefore \quad i = 1 - \frac{\alpha}{2}$

$\therefore \quad \Delta T_{b(x,\,s)} = \left(1 - \frac{\alpha}{2}\right) m_s \times k_{b(x)}$ ...(3)

$\Delta T_{b(y,\,s)} = \left(1 - \frac{0.7}{2}\right) m_s \times k_{b(y)}$ ...(4)

eq. (3) ÷ (4)

$\frac{3}{1} = \frac{(1 - \alpha/2)}{0.65} \times \frac{2}{1}$

$\therefore \quad \alpha = 0.05$

6. (1.03) $\frac{p^o - p_s}{p_s} = i\left[\frac{n_{solute}}{n_{solvent}}\right]$

$\frac{650 - 640}{640} = 1 \times \frac{0.5 \times 78}{M \times 39} \Rightarrow M_{solute} = 64 \text{ g}$

$\Delta T_f = K_f \times m = 5.12 \times \frac{0.5 \times 1000}{64 \times 39}$

$\Rightarrow \quad \Delta T_f = 1.03 \text{ K}$

7. (2.98)

Mole fraction of urea in aqueous solution = 0.05

Let number of moles of solution is 1 mole.

	Mole	Mass	Volume
Solute	0.05	3 g	
Solvent	0.95	17.1 g	
Solution	1	20.1 g	$\frac{20.1}{1.2}\ \text{cm}^3$

$$\text{Molarity} = \frac{n_{\text{solute}}}{V_{\text{solution (in mL)}}} \times 1000 = \frac{0.05 \times 1.2}{20.1} \times 1000 = \frac{60}{20.1} = 2.985$$

Answer after rounding off = 2.98

Answer after truncation = 2.98

8. (0.20)

$$P_T = P_A^\circ X_A + P_B^\circ X_B$$

$$0.3 = P_A^\circ \times 0.25 + P_B^\circ \times 0.75 \quad \text{...(i)}$$

$$0.4 = P_A^\circ \times 0.5 + P_B^\circ \times 0.5$$

$$0.8 = P_A^\circ + P_B^\circ \quad \text{...(ii)}$$

On solving equation (i) and (ii)

$$P_A^\circ = 0.6,\ P_B^\circ = 0.2$$

9. (1.755) Let 'w' gm NaCl is added

$$\Delta T_f = 0 - (-0.2) = 0.2°\text{C}$$

$$\Delta T_f = ik_f \times m$$

$$0.2 = 2 \times 2 \times \frac{w \times 1000}{58.5 \times 600}$$

$$w = \frac{0.2 \times 58.5 \times 600}{2 \times 2 \times 1000}\ \text{gm} = 1.755\ \text{gm}$$

10. (2.176)

$$\Delta T_f = K_f \times m$$

$$T_f^\circ - T_f = 2 \times 0.5 = 1$$

$$T_f = 273 - 1 = 272\,\text{K}$$

$$\text{Pressure due to gas } (P) = \frac{0.1 \times 0.08 \times 272}{1} = 2.176\ \text{atm.}$$

In initial and final state, $P_1V_1 = P_2V_2$

$$2.176 \times 1 = 1 \times V_2 \Rightarrow V_2 = 2.176\,\text{L}$$

11. (1.77)

Isotonic solutions have same osmotic pressure.

$$\pi_A = \pi_B$$

$$C_ART = C_BRT$$

$$\therefore \quad C_A = C_B$$

$$\frac{0.73}{M_A \times 0.25} = \frac{1.65}{M_B \times 1}$$

$$\frac{M_A}{M_B} = \frac{0.73}{0.25 \times 1.65} = 1.77$$

12. (167) $\pi = \dfrac{\pi_1V_1 + \pi_2V_2}{(V_1 + V_2)} = \dfrac{0.1 \times 1 + 0.2 \times 2}{(1+2)} = \dfrac{0.5}{3}\ \text{atm} = \dfrac{500}{3} \times 10^{-3}\ \text{atm} = 167 \times 10^{-3}\ \text{atm}$

13. (5) $\Delta T_b = iK_b \times m$

$$(\Delta T_b)_{Complex} = 2 \times (\Delta T_b)_{CaCl_2}$$

$$i \times K_b \times 0.10 = 2 \times 3 \times K_b \times 0.05 \; (\because i \text{ for } CaCl_2 = 3)$$

$$\Rightarrow \quad i_{Complex} = 3$$

So, formula of complex = $[Cr(NH_3)_5Cl]Cl_2$

14. (100) $\Delta T_b = 100.52 - 100 = 0.52°C$

$$\underset{\substack{1 \\ 1-\alpha}}{2A} \rightleftharpoons \underset{\substack{- \\ \frac{\alpha}{2}}}{A_2}$$

$$i = \frac{(1-\alpha) + \frac{\alpha}{2}}{1} = 1 - \frac{\alpha}{2}$$

$$\Delta T_b = iK_b \times m$$

$$i = \frac{\Delta T_b}{K_b \times m} = \frac{0.52}{0.52 \times 2} = 0.5$$

$$i = 1 - \frac{\alpha}{2} = 0.5$$

$$\alpha = 1 \Rightarrow \%\alpha = 100$$

15. (35) $\underset{\substack{C \\ C-C\alpha}}{ClCH_2COOH} \rightleftharpoons \underset{\substack{- \\ C\alpha}}{ClCH_2COO^-} + \underset{\substack{- \\ C\alpha}}{H^+}$

$$i = \frac{(C - C\alpha) + C\alpha + C\alpha}{C}$$

$$i = 1 + \alpha$$

$$\Delta T_f = i\,K_f \times m$$

$$0.5 = (1+\alpha) \times 1.86 \times \frac{9.45}{94.5 \times 500} \times 1000$$

$$(1+\alpha) = \frac{0.5 \times 94.5 \times 500}{1.86 \times 9.45 \times 1000} = 1.34$$

$$\alpha = 0.34$$

$$K_a = \frac{C\alpha^2}{1-\alpha} = \frac{0.2 \times (0.34)^2}{(1-0.34)} = 0.035 = 35 \times 10^{-3}$$

16. (3) $\Delta T_b = iK_b \times m$

$$i = 1 + \alpha = 1 + 0.75 = 1.75$$

$$\Delta T_b = 2.5\text{K}, \; K_b = 0.52\text{K kg mol}^{-1}$$

$$m = \frac{\Delta T_b}{i \times K_b} = \frac{2.5}{0.52 \times 1.75} = 2.75$$

17. (18) Moles of $SO_2 = \frac{PV}{RT} = \frac{1 \times 224 \times 10^{-3}}{0.0821 \times 298} = 9.2 \times 10^{-3}$ mol

$$\underset{9.2\text{ m mol}}{SO_2} + \underset{\substack{10\text{ mmol} \\ (\text{L.R})}}{2NaOH} \longrightarrow \underset{5\text{ m mol}}{Na_2SO_3} + H_2O$$

mmol of Na_2SO_3 formed = 5

$$Na_2SO_3 \longrightarrow 2Na^+ + SO_3^{2-}$$

$i = 3$

Relative lowering of V. P.

$$\Rightarrow \quad \frac{P° - P_s}{P°} = i \times X_{Solute} = i \times \left(\frac{n_B}{n_A + n_B}\right)$$

Since solution is dilute $n_A + n_B \cong n_A$

Lowering of vapour pressure $= P^\circ - P_s = P^\circ \times i \times \dfrac{n_B}{n_A} = 24 \times 3 \times \dfrac{5 \times 10^{-3}}{(36/18)} = 18 \times 10^{-2}$ mm Hg

18. (2) $\Delta T_f = i\, K_f \times m$

$$0 - (-0.93) = i \times 1.86 \times \frac{12.2}{122 \times 0.1}$$

$$i = \frac{0.93}{1.86} = 0.5$$

$$i = 1 + \left(\frac{1}{n} - 1\right)\alpha$$

$$0.5 = 1 + \left(\frac{1}{n} - 1\right) \times 1 \Rightarrow n = 2$$

19. (106)

$$\begin{array}{ccccc} AB_2 & \rightleftharpoons & A^{2+} & + & 2B^- \\ 1 & & - & & - \\ 1-\alpha & & \alpha & & 2\alpha \end{array}$$

$$i = \frac{(1-\alpha) + \alpha + 2\alpha}{1} = 1 + 2\alpha$$

$$i = 1 + 2(0.1) = 1.2$$

$$\Delta T_b = iK_b \times m$$

$$T_b - T_b^\circ = 1.2 \times 0.5 \times 10$$

$$T_b = 100 + 6 = 106^\circ\text{C} \qquad (\because T_b^\circ = 100^\circ\text{C})$$

20. (85)

$$\begin{array}{ccccc} K_4Fe(CN)_6 & \longrightarrow & 4K^+ & + & Fe(CN)_6^{4-} \\ 1 & & - & & - \\ 1-\alpha & & 4\alpha & & \alpha \end{array}$$

$$i = \frac{(1-\alpha) + 4\alpha + \alpha}{1} = 1 + 4\alpha = 1 + 4(0.4) = 2.6$$

Effective molality $= i \times m = 2.6 \times 1 = 2.6\,m$

So, molality of solute 'A' should be 2.6.

Weight of solute = 18.1 g

weight of solvent = (100 – 18.1) = 81.9 g

Let molar mass of 'A' = M

$$m = \frac{(18.1/M)}{81.9} \times 1000 = 2.6$$

$$M = \frac{18.1 \times 1000}{81.9 \times 2.6} = 85.31$$

21. (50)

$$\begin{array}{ccccc} HA_{(aq)} & \longrightarrow & H^+_{(aq)} & + & A^-_{(aq)} \\ 1 & & - & & - \\ 1-\alpha & & \alpha & & \alpha \end{array}$$

$$i = \frac{(1-\alpha) + \alpha + \alpha}{1} = 1 + \alpha$$

$$\Delta T_f = iK_f \times m$$

$$i = \frac{\Delta T_f}{K_f \times m} = \frac{3.885}{1.85 \times 2} = 1.05$$

$$i = 1 + \alpha = 1.05$$

$$\alpha = 0.05 = 50 \times 10^{-3}$$

22. (2) Mole fraction of substance A in liquid phase = 0.2

The Raoult's law equation is :

$$P_A = (P_A^\circ) \times X_A \Rightarrow 0.4 \text{ atm} = P_A^\circ \times 0.2$$

$\Rightarrow \quad P_A^\circ = \dfrac{0.4}{0.2} = 2$ atm

23. (54) Moles of nitric acid $= 800 \times 10^{-3} \times 0.5 = 0.4$

Moles of nitric acid evaporated $= \dfrac{11.5}{63} = 0.182$

Moles left $= 0.4 - 0.182 = 0.218$

Molarity $= \dfrac{0.218 \times 10^3}{400} = 0.545\text{M} = 54.5 \times 10^{-2}$ M

24. (15) Moles of ascorbic acid $= \dfrac{10.2 \text{ g}}{176 \text{ g mol}^{-1}} = 0.058$ mol

$\Delta T_f = (x \times 10^{-1})^\circ\text{C}$

$\Delta T_f = K_f$ molality $= 3.9 \times \dfrac{0.058}{150} \times 1000 = 1.5^\circ\text{C} = 15 \times 10^{-1}\,^\circ\text{C}$

25. (1) $\Delta T_b = iK_b m_1; \quad \Delta T_f = iK_f m_2$

$\dfrac{\Delta T_b}{\Delta T_f} = \dfrac{K_b \times 1}{K_f \times 2} \quad \Rightarrow \quad \dfrac{3}{6} = \dfrac{1}{2} = \dfrac{K_b}{K_f} \times \dfrac{1}{2}$

$\dfrac{K_b}{K_f} = \dfrac{1}{2} \quad \Rightarrow \quad x = 1$

26. (180) $\quad M = \dfrac{5}{40} \times \dfrac{1000}{450}; \quad M_1V_1 = M_2V_2$

$\left(\dfrac{5}{40} \times \dfrac{1000}{450}\right) \times V_1 = 0.1 \times 500 \quad \Rightarrow \quad V_1 = 180$

27. (4) $\quad \pi = icRT \quad \Rightarrow \quad \pi \propto i \times c$

$\therefore$ (A) $\pi_{NaCl} \propto 2 \times 1 \quad ; \quad \pi_{Urea} \propto 1 \times 2$

(B) $\pi_{CaCl_2} \propto 3 \times 1 \quad ; \quad \pi_{KCl} \propto 2 \times 1.5$

(C) $\pi_{AlCl_3} \propto 4 \times 1.5 \quad ; \quad \pi_{Na_2SO_4} \propto 3 \times 2$

(D) $\pi_{KCl} \propto 2 \times 2.5 \quad ; \quad \pi_{Al_2(SO_4)_3} \propto 5 \times 1$

$\therefore$ A, B, C and D are isotonic pairs.

28. (1111) $\quad \dfrac{P^\circ - P_s}{P_s} = \dfrac{n_{solute}}{n_{solvent}} = \dfrac{\frac{x}{60}}{\frac{1000}{18}} = \dfrac{P^\circ - 0.75P^\circ}{0.75P^\circ}$

$\Rightarrow \quad x = \dfrac{1000}{9} = 1111$ g

29. (15) $\quad \Delta T_f = iK_f \times \text{molality}$

$24 = (1) \times 1.86 \times \dfrac{W}{62 \times 18.6}$

$W = 14880 \text{ gm} = 14.880 \text{ kg}$

30. (57)

$M_F = \dfrac{M_1V_1 + M_2V_2}{V_1 + V_2} = \dfrac{2 \times 20 + 0.5 \times 400}{420} = 0.571M = 57.1 \times 10^{-2} M$

31. (125) Assuming 100 gm solution contain 70 gm solute.

Volume of 100 gm solution will be $\dfrac{100}{1.25}$ mL.

Molarity $= \dfrac{70/70}{100/1.25} \times 1000 = 12.5$ or 125×10^{-1}

CHAPTER

9

Electrochemistry

A. Objective Questions

[Only one option is correct]

1. $AgNO_3(aq.)$ was added to an aqueous KCl solution gradually and the conductivity of the solution was measured. The plot of conductance (Λ) *versus* the volume of $AgNO_3$ is : **(IIT 2011)**

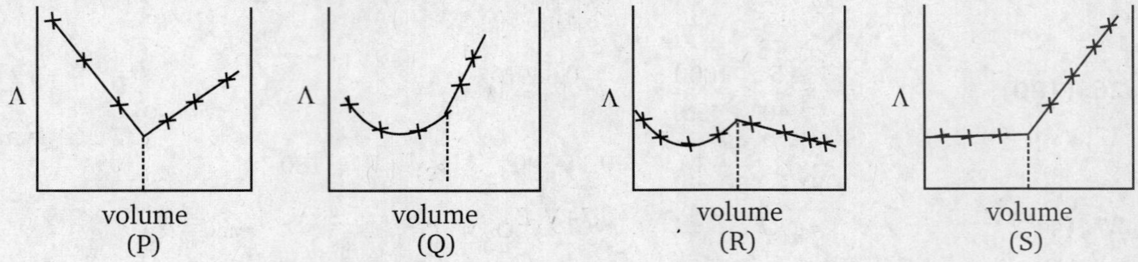

(a) (P) (b) (Q) (c) (R) (d) (S)

2. Consider the following cell reaction :

$$2Fe_{(s)} + O_{2(g)} + 4H^+_{(aq)} \longrightarrow 2Fe^{2+}_{(aq)} + 2H_2O_{(l)}, \qquad E° = 1.67\ V$$

At $[Fe^{2+}] = 10^{-3} M$, $P(O_2) = 0.1$ atm and pH = 3, the cell potential at 25° C is : **(IIT 2011)**

(a) 1.47 V (b) 1.77 V (c) 1.87 V (d) 1.57 V

3. Given : $E°_{Cr^{3+}/Cr} = -0.74$ V; $E°_{MnO_4^-/Mn^{2+}} = 1.51$ V

$E°_{Cr_2O_7^{2-}/Cr^{3+}} = 1.33$ V; $E°_{Cl/Cl^-} = 1.36$ V

Based on the data given above, strongest oxidising agent will be : **[JEE (Mains) 2013]**

(a) Cl^- (b) Cr^{3+} (c) Mn^{2+} (d) MnO_4^-

4. Resistance of 0.2 *M* solution of an electrolyte is 50 Ω. The specific conductance of the solution is 1.4 S m^{-1}. The resistance of 0.5 *M* solution of the same electrolyte is 280 Ω. The molar conductivity of 0.5 *M* solution of the electrolyte in S m^2 mol^{-1} is : **[JEE (Main) 2014]**

(a) 5×10^{-4} (b) 5×10^{-3} (c) 5×10^3 (d) 5×10^2

5. The equivalent conductance of NaCl at concentration C and at infinite dilution are λ_C and λ_∞, respectively. The correct relationship between λ_C and λ_∞ is given as : **[JEE (Mains) 2014]**
(Where the constant b is positive)

(a) $\lambda_C = \lambda_\infty + (b)C$ (b) $\lambda_C = \lambda_\infty - (b)C$ (c) $\lambda_C = \lambda_\infty - (b)\sqrt{C}$ (d) $\lambda_C = \lambda_\infty + (b)\sqrt{C}$

6. Given below are the half-cell reactions

$$Mn^{2+} + 2e^- \longrightarrow Mn; \qquad E° = -1.18\ V$$

$$(Mn^{3+} + e^- \longrightarrow Mn^{2+}); \qquad E° = +1.51\ V$$

The $E°$ for $3Mn^{2+} \longrightarrow Mn + 2Mn^{3+}$ will be : **[JEE (Mains) 2014]**

(a) –2.69 V; the reaction will not occur
(b) –2.69 V; the reaction will occur
(c) –0.33 V; the reaction will not occur
(d) –0.33 V; the reaction will occur

7. In a galvanic cell, the salt bridge **[JEE (Advanced) 2014]**

(a) does not participate chemically in the cell reaction.
(b) stops the diffusion of ions from one electrode to another.
(c) is necessary for the occurrence of the cell reaction.
(d) ensures mixing of the two electrolytic solution.

8. Two faradays of electricity are passed through a solution of $CuSO_4$. The mass of copper deposited at the cathode is (at. mass of Cu = 63.5 amu) : **[JEE (Mains) 2015]**

(a) 2 g
(b) 127 g
(c) 0 g
(d) 63.5 g

9. Galvanization is applying a coating of : **[JEE (Mains) 2016]**

(a) Pb
(b) Cr
(c) Cu
(d) Zn

10. For the following electrochemical cell at 298 K,

$$Pt_{(s)}|H_{2(g)},(1\text{ bar})|H^+_{(aq)},(1M)||M^{4+}_{(aq)},M^{2+}_{(aq)}|Pt_{(s)}$$

$$E_{cell} = 0.092\text{ V when } \frac{[M^{2+}_{(aq)}]}{[M^{4+}_{(aq)}]} = 10^x$$

Given : $E°_{M^{4+}/M^{2+}} = 0.151\text{ V};\ 2.303\frac{RT}{F} = 0.059\text{ V}$

The value of x is: **[JEE (Advanced) 2016]**

(a) –2
(b) –1
(c) 1
(d) 2

11. Given: $E°_{Cl_2/Cl^-} = 13.6\text{ V},\ E°_{Cr^{3+}/Cr} = -0.74\text{ V}, E°_{Cr_2O_7^{2-}/Cr^{3+}} = 1.33\text{ V},\ E°_{MnO_4^-/Mn^{2+}} = 1.51\text{ V},$

Among the following, the strongest reducing agent is : **[JEE (Mains) 2017]**

(a) Cr^{3+}
(b) Cl^-
(c) Cr
(d) Mn^{2+}

12. For the following cell, $Zn_{(s)}|\ ZnSO_{4(aq)}||CuSO_{4(aq)}|Cu_{(s)}$

When the concentration of Zn^{2+} is 10 times the concentration of Cu^{2+}, the expression of ΔG (in J mol^{-1}) is:
[F is Faraday constant; R is gas constant; T is temperature, $E°_{cell} = 1.1V$] **[JEE (Advanced) 2017]**

(a) $2.303RT - 2.2F$
(b) $-2.2F$
(c) $2.303RT + 1.1F$
(d) $1.1F$

13. How long (approximate) should water be electrolysed by passing through 100 amperes current so that the oxygen released can completely burn 27.66 g of diborane? (Atomic weight of B=10.8 u) **[JEE (Mains) 2018]**

(a) 3.2 hours
(b) 1.6 hours
(c) 6.4 hours
(d) 0.8 hours

14. Given that $E°_{O_2/H_2O} = +1.23\text{ V}; E°_{S_2O_8^{2-}/SO_4^{2-}} = 2.05\text{ V}; E°_{Br_2/Br^-} = +1.09\text{ V}; E°_{Au^{3+}/Au} = +1.4\text{ V}$

The strongest oxidizing agent is : **[JEE (Mains) 2019]**

(a) $S_2O_8^{2-}$
(b) O_2
(c) Br_2
(d) Au^{3+}

15. Calculate the standard cell potential (in V) of the cell in which following reaction takes place

$$Fe^{2+}_{(aq)} + Ag^+_{(aq)} \rightarrow Fe^{3+}_{(aq)} + Ag_{(s)}$$

Given that $E°_{Ag^+/Ag} = x\text{ V};\ E°_{Fe^{2+}/Fe} = y\text{ V};\ E°_{Fe^{3+}/Fe} = z\text{V}$ **[JEE (Mains) 2019]**

(a) $x - z$
(b) $x - y$
(c) $x + 2y - 3z$
(d) $x + y - z$

16. Molar conductivity (Λ_m) of aqueous solution of sodium stearate, which behaves as a strong electrolyte, is recorded at varying concentrations (C) of sodium stearate. Which one of the following plots provides the correct representation of micelle formation in the solution?
(critical micelle concentration (CMC) is marked with an arrow in the figures) **[JEE (Advanced) 2019]**

(a) 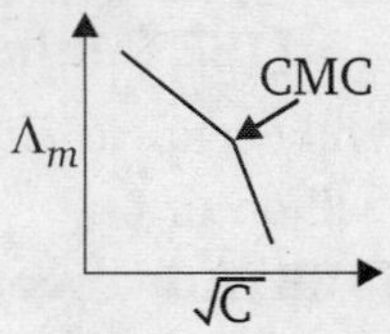

(b) 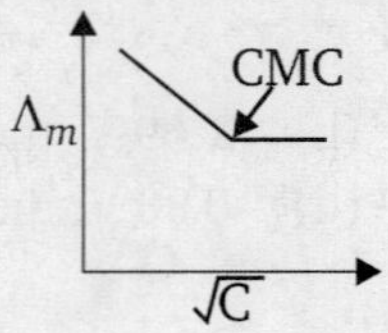

(c) 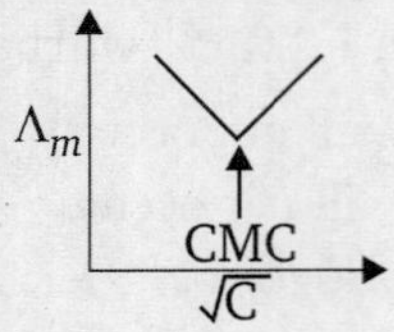

(d) 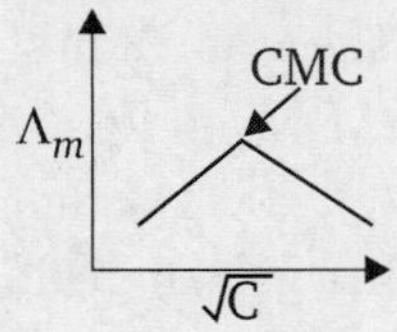

17. 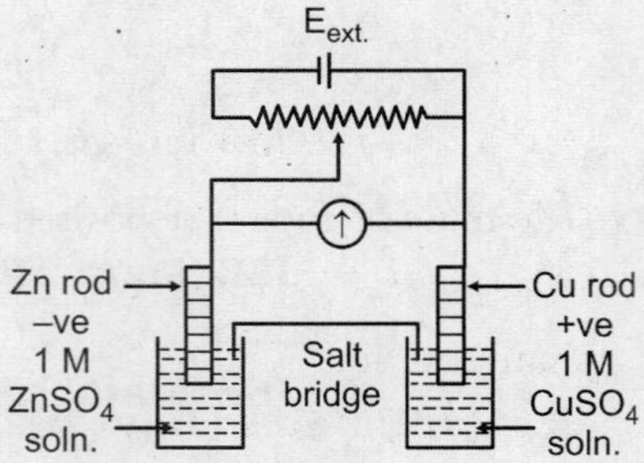

$E^\circ_{Cu^{2+}|Cu} = +0.34\ V$; $E^\circ_{Zn^{2+}|Zn} = -0.76\ V$

Identify the incorrect statement from the options below for the above cell:

[JEE Main (Sept.) 2020]

(a) If $E_{ext} < 1.1\ V$, Zn dissolves at anode and Cu deposits at cathode

(b) If $E_{ext} = 1.1\ V$, no flow of e^- or current occurs

(c) If $E_{ext} > 1.1\ V$, Zn dissolves at Zn electrode and Cu deposits at Cu electrode

(d) If $E_{ext} > 1.1\ V$, e^- flows from Cu to Zn

18. For the given cell;

$$Cu(s)\,|\,Cu^{2+}(C_1 M)\,||\,Cu^{2+}(C_2 M)\,|\,Cu(s)$$

Change in Gibbs energy (ΔG) is negative, if : **[JEE Main (Sept.) 2020]**

(a) $C_1 = C_2$ (b) $C_2 = \frac{C_1}{\sqrt{2}}$ (c) $C_1 = 2C_2$ (d) $C_2 = \sqrt{2}\, C_1$

19. 250 mL of a waste solution obtained from the workshop of a goldsmith contains $0.1M$ $AgNO_3$ and $0.1\ M$ AuCl. The solution was electrolyzed at 2V by passing a current of 1A for 15 minutes. The metal/metals electrodeposited will be :

($E^\circ_{Ag^+/Ag} = 0.8\ V$, $E^\circ_{Au^+/Au} = 1.69\ V$) **[JEE Main (Sept.) 2020]**

(a) silver and gold in equal mass proportion

(b) only silver

(c) only gold

(d) silver and gold in proportion to their atomic weights

20. Amongst the following, the form of water with the lowest ionic conductance at 298 K is :

[JEE Main (Jan.) 2020]

(a) sea water (b) distilled water

(c) saline water used for intravenous injection (d) water from a well

21. Let C_{NaCl} and C_{BaSO_4} be the conductances (in S) measured for saturated aqueous solution of NaCl and $BaSO_4$, respectively, at a temperature T.

Which of the following is false? **[JEE Main (Sept.) 2020]**

(a) $C_{NaCl}(T_2) > C_{NaCl}(T_1)$ for $T_2 > T_1$ (b) $C_{NaCl} >> C_{BaSO_4}$ at given T

(c) $C_{BaSO_4}(T_2) > C_{BaSO_4}(T_1)$ for $T_2 > T_1$

(d) Ionic mobilities of ions from both salts increase with T

22. The variation of molar conductivity with concentration of an electrolyte (x) in aqueous solution is shown in the given figure. **[JEE Main (Sept.) 2020]**

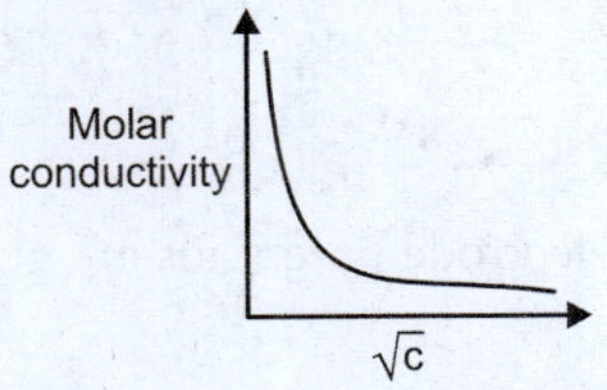

(a) KNO_3 (b) CH_3COOH (c) HCl (d) NaCl

23. The equation that is incorrect is: **[JEE Main (Jan.) 2020]**

(a) $(\Lambda^{\circ}_m)_{NaBr} - (\Lambda^{\circ}_m)_{NaI} = (\Lambda^{\circ}_m)_{KBr} - (\Lambda^{\circ}_m)_{NaBr}$ (b) $(\Lambda^{\circ}_m)_{NaBr} - (\Lambda^{\circ}_m)_{NaCl} = (\Lambda^{\circ}_m)_{KBr} - (\Lambda^{\circ}_m)_{KCl}$

(c) $(\Lambda^{\circ}_m)_{H_2O} = (\Lambda^{\circ}_m)_{HCl} + (\Lambda^{\circ}_m)_{NaOH} - (\Lambda^{\circ}_m)_{NaCl}$ (d) $(\Lambda^{\circ}_m)_{KCl} - (\Lambda^{\circ}_m)_{NaCl} = (\Lambda^{\circ}_m)_{KBr} - (\Lambda^{\circ}_m)_{NaBr}$

24. The product obtained from the electrolytic oxidation of acidified sulphate solution, is :

[JEE Mains (July) 2021 (I)]

(a) HSO_4^- (b) HO_3SOOSO_3H (c) HO_2SOSO_2H (d) HO_3SOSO_3H

25. The molar conductivity of a conductivity cell filled with 10 moles of 20 mL NaCl solution is Λ_{m_1} and that of 20 moles another identical cell having 80 mL NaCl solution is Λ_{m_2}. The conductivities exhibited by these two cells are same.

The relationship between Λ_{m_2} and Λ_{m_1} is : **[JEE Mains (July) 2022 (II)]**

(a) $\Lambda_{m_2} = 2\Lambda_{m_1}$ (b) $\Lambda_{m_2} = \Lambda_{m_1}/2$ (c) $\Lambda_{m_2} = \Lambda_{m_1}$ (d) $\Lambda_{m_2} = 4\Lambda_{m_1}$

26. In which of the following half cells, electrochemical reaction is pH dependent?

[JEE Mains (June) 2022 (I)]

(a) $Pt|Fe^{3+}, Fe^{2+}$ (b) $MnO_4^-|Mn^{2+}$ (c) $Ag|AgCl|Cl^-$ (d) $\frac{1}{2}F_2|F^-$

27. Plotting $1/\Lambda_m$ against $c\Lambda_m$ for aqueous solutions of a monobasic weak acid (HX) resulted in a straight line with y-axis intercept of P and slope of S. The ratio P/S is : **[Adv. 2023]**

[Λ_m = molar conductivity, Λ°_m = limiting molar conductivity, c = molar concentration,

K_a = dissociation constant of HX]

(a) $K_a\Lambda^{\circ}_m$ (b) $K_a\Lambda^{\circ}_m/2$ (c) $2K_a\Lambda^{\circ}_m$ (d) $1/(K_a\Lambda^{\circ}_m)$

28. The reaction, $\frac{1}{2}H_2(g) + AgCl(s) \rightleftharpoons H^+(aq) + Cl^-(aq) + Ag(s)$

occurs in which of the given galvanic cell. **[JEE Mains (April) 2023 (I)]**

(a) Pt | $H_2(g)$ | KCl(sol^n) | AgCl(s) | Ag (b) Pt | $H_2(g)$ | HCl(sol^n) | AgCl(s) | Ag

(c) Ag | AgCl(s) | KCl(sol^n) | AgCl(s) | Ag (d) Pt | $H_2(g)$ | HCl(sol^n) | $AgNO_3$(sol^n) | Ag

29. The standard electrode potential (M^{3+}/M^{2+}) for V, Cr, Mn and Co are –0.26V, – 0.41 V, +1.57 V and +1.97 V, respectively. The metal ions which can liberate H_2 from a dilute acid are :

[JEE Mains (June) 2023 (I)]

(a) V^{2+} and Mn^{2+} (b) Cr^{2+} and CO^{2+} (c) V^{2+} and Cr^{2+} (d) Mn^{2+} and CO^{2+}

30. Reduction potential of ions are given below :

ClO_4^-	IO_4^-	BrO_4^-
$E^o = 1.19$ V	$E^o = 1.65$ V	$E^o = 1.74$ V

The correct order of their oxidising power is : **[JEE Mains (Jan.) 2024]**

(a) $ClO_4^- > IO_4^- > BrO_4^-$ (b) $BrO_4^- > IO_4^- > ClO_4^-$ (c) $BrO_4^- > ClO_4^- > IO_4^-$ (d) $IO_4^- > BrO_4^- > ClO_4^-$

31. A solution of aluminium chloride is electrolysed for 30 minutes using a current of 2A. The amount of the aluminium deposited at the cathode is ....

[Given : molar mass of aluminium and chlorine are 27 g mol^{-1} and 35.5 g mol^{-1} respectively, Faraday constant = 96500 C mol^{-1}]. **[JEE Mains (Jan.) 2025]**

(a) 1.660 g (b) 1.007 g (c) 0.336 g (d) 0.441 g

32. $FeO_4^{2-} \xrightarrow{+2.0V} Fe^{3+} \xrightarrow{0.8\,V} Fe^{2+} \xrightarrow{-0.5\,V} Fe^0$

In the above diagram, the standard electrode potentials are given in volt (over the arrow). The value of $E^{\ominus}_{FeO_4^{2-}/Fe^{2+}}$ is : **[JEE Mains (Jan.) 2025]**

(a) 1.7 V (b) 1.2 V (c) 2.1 V (d) 1.4 V

33. Standard electrode potentials for a few half cells are mentioned below:

$E^{o}_{Cu^{2+}/Cu} = 0.34\ V, E^{o}_{Zn^{2+}/Zn} = -0.76\ V, E^{o}_{Ag^{+}/Ag} = 0.80\ V, E^{o}_{Mg^{2+}/Mg} = -2.37\ V$

Which one of the following cells gives the most negative value of ΔG^o? **[JEE Mains (Jan.) 2025]**

(a) $Zn|Zn^{2+}(1M)||Ag^{+}(1M)|Ag$ (b) $Zn|Zn^{2+}(1M)||Mg^{2+}(1M)|Mg$

(c) $Ag|Ag^{+}(1M)||Mg^{2+}(1M)|Mg$ (d) $Cu|Cu^{2+}(1M)||Ag^{+}(1M)|Ag$

34. For the given cell

$$Fe^{2+}(eq) + Ag^{+}(aq) \longrightarrow Fe^{3+}(aq) + Ag(s)$$

The standard cell potential of the above reaction is Given:

$$Ag^{+} + e^{-} \longrightarrow Ag \qquad E^o = x\ V$$

$$Fe^{2+} + 2e^{-} \longrightarrow Fe \qquad E^o = y\ V$$

$$Fe^{3+} + 3e^{-} \longrightarrow Fe \qquad E^o = z\ V$$

[JEE Main (Jan.) 2025]

(a) $x + y - z$ (b) $x + 2y - 3z$ (c) $y - 2x$ (d) $x + 2y$

35. Based on the data given below :

$E^{o}_{Cr_2O_7^{2-}/Cr^{3+}} = 1.33\ V, E^{o}_{Cl_2/Cl^{(-)}} = 1.36\ V, E^{o}_{MnO_4^{-}/Mn^{2+}} = 1.51\ V, E^{o}_{Cr^{3+}/Cr} = -0.74\ V$

The strongest reducing agent is : **[JEE Main (Jan.) 2025]**

(a) Mn^{2+} (b) Cr (c) MnO_4^{-} (d) Cl^{-}

36. The molar conductivity of a weak electrolyte when plotted against the square root of its concentration, which of the following is expected to be observed? **[JEE Main (Jan.) 2025]**

(a) A small decrease in molar conductivity is observed at infinite dilution.
(b) A small increase in molar conductivity is observed at infinite dilution.
(c) Molar conductivity increases sharply with increase in concentration.
(d) Molar conductivity decreases sharply with increase in concentration.

37. For a $Mg|Mg^{2+}(aq)||Ag^{+}(aq)|Ag$ the correct Nernst equation is : **[JEE Main (Jan.) 2025]**

(a) $E_{cell} = E^{o}_{cell} - \frac{RT}{2F}\ln\frac{[Ag^{+}]}{[Mg^{2+}]}$ (b) $E_{cell} = E^{o}_{cell} + \frac{RT}{2F}\ln\frac{[Ag^{+}]^2}{[Mg^{2+}]}$

(c) $E_{cell} = E^{o}_{cell} - \frac{RT}{2F}\ln\frac{[Mg^{2+}]}{[Ag^{+}]}$ (d) $E_{cell} = E^{o}_{cell} - \frac{RT}{2F}\ln\frac{[Ag^{+}]^2}{[Mg^{2+}]}$

38. Match List-I with List-II :

	List-I (Applications)		List-II (Batteries/Cell)
(A)	Transistors	(I)	Anode—Zn/Hg; Cathode—HgO + C
(B)	Hearing aids	(II)	Hydrogen fuel cell
(C)	Invertors	(III)	Anode—Zn; Cathode—Carbon
(D)	Apollo space ship	(IV)	Anode—Pb; Cathode—$Pb|PbO_2$

Choose the correct answer from the options given below : **[JEE Main (Jan.) 2025]**

(a) (A)-(III), (B)-(I), (C)-(IV), (D)-(II)
(b) (A)-(III), (B)-(II), (C)-(IV), (D)-(I)
(c) (A)-(IV), (B)-(III), (C)-(II), (D)-(I)
(d) (A)-(II), (B)-(III), (C)-(IV), (D)-(I)

39. O_2 gas will be evolved as a product of electrolysis of :
(A) an aqueous solution of $AgNO_3$ using silver electrodes
(B) an aqueous solution of $AgNO_3$ using platinum electrodes
(C) a dilute solution of H_2SO_4 using platinum electrodes
(D) a high concentration solution of H_2SO_4 using platinum electrodes
Choose the correct answer from the options given below : **[JEE Main (Jan.) 2025]**
(a) (B) and (C) only
(b) (A) and (D) only
(c) (B) and (D) only
(d) (A) and (C) only

B. Objective Questions

[One or more than one option(s) is/are correct]

1. For the reduction of NO_3^- ion in an aqueous solution, E^o is +0.96 V. Values of E^o for some metal ions are given below :

$$V^{2+}(aq) + 2e^- \longrightarrow V \quad E^o = -1.19\ V$$
$$Fe^{3+}(aq) + 3e^- \longrightarrow Fe \quad E^o = -0.04\ V$$
$$Au^{3+}(aq) + 3e^- \longrightarrow Au \quad E^o = +1.40\ V$$
$$Hg^{2+}(aq) + 2e^- \longrightarrow Hg \quad E^o = +0.86\ V$$

The pair(s) of metals that is (are) oxidized by NO_3^- in aqueous solution is (are): **(IIT 2009)**
(a) V and Hg
(b) Hg and Fe
(c) Fe and Au
(d) Fe and V

C. Passage Based Problems

Passage-1

Tollen's reagent is used for the detection of aldehydes. When a solution of $AgNO_3$ is added to glucose with NH_4OH, then gluconic acid is formed.

$$Ag^+ + e^- \longrightarrow Ag; \quad E^\circ_{red} = 0.8\ V$$
$$C_6H_{12}O_6 + H_2O \longrightarrow \underset{\text{gluconic acid}}{C_6H_{12}O_7} + 2H^+ + 2e^-; \quad E^\circ_{oxi} = -0.05\ V$$
$$[Ag(NH_3)_2]^+ + e^- \longrightarrow Ag(s) + 2NH_3; \quad E^\circ_{red} = 0.337\ V$$

[Use $2.303 \times \frac{RT}{F} = 0.0592$ and $\frac{F}{RT} = 38.92$ at 298 K] **(IIT 2006)**

1. For, $2Ag^+ + C_6H_{12}O_6 + H_2O \longrightarrow 2Ag(s) + C_6H_{12}O_7 + 2H^+$
Find $\log_{10} K$ of this reaction :
(a) 66.13
(b) 58.38
(c) 25.33
(d) 46.29

2. When ammonia is added to the solution, pH is raised to 11. Which half-cell reaction is affected by pH and by how much ?
(a) E_{oxi} glucose will increase by a factor of 0.65 from E°_{oxi}
(b) E_{oxi} glucose will decrease by a factor of 0.65 from E°_{oxi}
(c) E_{red} Ag^+ will increase by a factor of 0.65 from E°_{red}
(d) E_{red} Ag^+ will decrease by a factor of 0.65 from E°_{red}

3. Ammonia is always added in this reaction. Which of the following must be incorrect ?

(a) NH_3 combines with Ag^+ to form a complex

(b) $Ag(NH_3)_2^+$ is a weaker oxidising reagent than Ag^+

(c) In absence of NH_3 silver salt of gluconic acid is formed

(d) NH_3 has affected the standard reduction potential of glucose/gluconic acid electrode

Passage-2

Chemical reactions involve interaction of atoms and molecules. A large number of atoms/molecules (approximately 6.023×10^{23}) are present in a few grams of any chemical compound varying with their atomic/molecular masses. To handle such large numbers conveniently, the mole concept was introduced. This concept has implications in diverse areas such as analytical chemistry, biochemistry, electrochemistry and radiochemistry. The following example illustrates a typical case, involving chemical/electrochemical reaction, which requires a clear understanding of the mole concept. A 4.0 *M* aqueous solution of NaCl is prepared and 500 mL of this solution is electrolysed. This leads to the evolution of chlorine gas at one of the electrodes (atomic mass: Na = 23, Hg = 200 ; 1F = 96500C). **(IIT 2007)**

1. The total number of moles of chlorine gas evolved is :

(a) 0.5 (b) 1.0 (c) 2.0 (d) 3.0

2. If the cathode is a Hg electrode, the maximum weight (in gram) of amalgam formed from this solution is :

(a) 200 (b) 225 (c) 400 (d) 446

3. The total charge (coulombs) required for complete electrolysis is :

(a) 24125 (b) 48250 (c) 96500 (d) 193000

Passage-3

Redox reaction play a pivotal role in chemistry and biology. The values of standard redox potential ($E°$) of two half-cell reactions decide which way the reaction is expected to proceed. A simple example is a Daniel cell in which zinc goes into solution and copper gets deposited. Given below are a set of half-cell reactions (acidic medium) along with their $E°$ (*V* with respect to normal hydrogen electrode) values. Using this data obtain the correct explanations to Questions 1 to 3.

$$I_2 + 2e^- \longrightarrow 2I^-; \quad E° = 0.54$$

$$Cl_2 + 2e^- \longrightarrow 2Cl^-; \quad E° = 1.36$$

$$Mn^{3+} + e^- \longrightarrow Mn^{2+}; \quad E° = 1.50$$

$$Fe^{3+} + e^- \longrightarrow Fe^{2+}; \quad E° = 0.77$$

$$O_2 + 4H^+ + e^- \longrightarrow 2H_2O; \quad E° = 1.23$$

(IIT 2007)

1. Among the following, identify the correct statement

(a) chloride ion is oxidised by O_2 (b) Fe^{2+} is oxidised by iodine

(c) iodine ion is oxidised by chlorine (d) Mn^{2+} is oxidised by chlorine

2. While Fe^{3+} is stable, Mn^{3+} is not stable in acid solution because :

(a) O_2 oxidised Mn^{2+} to Mn^{3+}

(b) O_2 oxidised both Mn^{2+} to Mn^{3+} and Fe^{2+} to Fe^{3+}

(c) Fe^{3+} oxidised H_2O to O_2

(d) Mn^{3+} oxidised H_2O to O_2

3. Sodium fusion extract,obtained from aniline, on treatment with iron (II) sulphate and H_2SO_4 in presence of air gives a Prussian blue precipitate. The blue colour is due to the formation of :

(a) $Fe_4[Fe(CN)_6]_3$ (b) $Fe_3[Fe(CN)_6]_2$

(c) $Fe_4[Fe(CN)_6]_2$ (d) $Fe_3[Fe(CN)_6]_3$

Passage-4

The concentration of potassium ions inside a biological cell is at least twenty times higher than the outside. The resulting potential difference across the cell is important in several processes such as transmission of nerve impulses and maintaining the ion balance. A simple mode for such a concentration cell involving a metal M is :

$$M_{(s)}|M^+{(aq.)}\, 0.05\text{molar}\,||M^+{(aq.)}\, 1\text{molar}\,|M_{(s)}$$

For the above electrolytic cell the magnitude of the cell potential $|E_{cell}| = 70$ mV **(IIT 2010)**

1. For the above cell :

(a) $E_{cell} < 0;\ \Delta G > 0$ (b) $E_{cell} > 0;\ \Delta G < 0$

(c) $E_{cell} < 0;\ \Delta G° > 0$ (d) $E_{cell} > 0;\ \Delta G° > 0$

2. If the 0.05 molar solution of M^+ is replaced by 0.0025 molar M^+ solution, then the magnitude of the cell potential would be :

(a) 35 mV (b) 70 mV (c) 140 mV (d) 700 mV

Passage-5

The electrochemical cell shown below is a concentration cell.

M/M^{2+} (saturated solution of a sparingly soluble salt, MX_2)$||M^{2+}$ ($0.001\,\text{mol dm}^{-3}$)$|M$

The emf of the cell depends on the difference in concentrations of M^{2+} ions at the two electrodes. The emf of the cell at 298 K is 0.059 V. **(IIT 2012)**

1. The value of $\Delta G\,(\text{kJ mol}^{-1})$ for the given cell is : (take $1\text{F} = 96500\,\text{C mol}^{-1}$)

(a) −5.7 (b) 5.7 (c) 11.4 (d) −11.4

2. The solubility product (K_{sp}; $\text{mol}^3\,\text{dm}^{-9}$) of MX_2 at 298 K based on the information available for the given concentration cell is :

(take $2.303 \times R \times 298/\text{F} = 0.059$ V)

(a) 1×10^{-15} (b) 4×10^{-15} (c) 1×10^{-12} (d) 4×10^{-12}

D. Integer Type Problems

1. All the energy released from the reaction $X \rightarrow Y$. $\Delta_r G° = -193\,\text{kJ mol}^{-1}$ is used for oxidizing M^+ as $M^+ \rightarrow M^{3+} + 2e^-, E^0 = -0.25$ V. Under standard conditions, the number of moles of M^+ oxidized when one mole of X is converted to Y is :

[$\text{F} = 96500\,\text{C mol}^{-1}$] **[JEE (Advanced) 2015]**

2. The molar conductivity of a solution of a weak acid HX (0.01M) is 10 times smaller than the molar conductivity of a solution of a weak acid HY (0.10 M). If $\lambda^0_{X^-} \approx \lambda^0_{Y^-}$, the difference in their pK_a values, $pK_a(\text{HX}) - pK_a(\text{HY})$, is (consider degree of ionization of both acids to be $<<1$) :

[JEE (Advanced) 2015]

3. For the electrochemical cell, $\text{Mg}(s)\,|\,\text{Mg}^{2+}(aq, 1\,M)\,||\,\text{Cu}^{2+}(aq, 1\,M)\,|\,\text{Cu}(s)$ the standard emf of the cell is 2.70 V at 300 K. When the concentration of Mg^{2+} is changed to $x\,M$, the cell potential changes to 2.67 V at 300 K. The value of x is :

[Given, $\frac{F}{R} = 11500\,\text{KV}^{-1}$, where F is the Faraday constant and R is the gas constant, $\ln(10) = 2.30$]

[JEE (Advanced) 2018]

4. Consider a 70% efficient hydrogen-oxygen fuel cell working under standard conditions at 1 bar and 298 K. Its cell reaction is,

$$H_2(g) + \frac{1}{2}O_2(g) \longrightarrow H_2O(l).$$

The work derived from the cell on the consumption of 1.0×10^{-3} mol of $H_2(g)$ is used to compress 1.00 mol of a monoatomic ideal gas in a thermally insulated container. What is the change in the temperature (in K) of the ideal gas ?

The standard reduction potentials for the two half-cells are given below.

$$O_2(g) + 4H^+(aq.) + 4e^- \longrightarrow 2H_2O(l),\ E^\circ = 1.23\ V,$$

$$2H^+(aq.) + 2e^- \longrightarrow H_2(g), E^\circ = 0.00V.$$

Use $F = 96500$ C mol^{-1}, $R = 8.314$ J mol^{-1}K^{-1} **[JEE (Advanced) 2020]**

5. What would be the electrode potential for the given half cell reaction at pH = 5?

$$2H_2O \longrightarrow O_2 + 4H^+ + 4e^-,\ E^\circ_{Red} = 1.23\ V$$

($R = 8.314$ J mol^{-1} K^{-1}, $T = 198$ K, oxygen under std. atm pressure of 1 bar). **[JEE Main (Jan.) 2020]**

6. For an electrochemical cell

$Sn(s)\,|Sn^{2+}(aq.\ 1M)\,||Pb^{2+}(aq.\ 1M)\,|Pb(s)$ the ratio $\frac{[Sn^{2+}]}{[Pb^{2+}]}$ when this cell attains equilibrium is ______.

[JEE Main (Jan.) 2020]

(Given: $E^\circ_{Sn^{2+}/Sn} = -0.14$ V, $E^\circ_{Pb^{2+}/Pb} = -0.13$ V, $\frac{2.303\,RT}{F} = 0.06$)

7. 108 g of silver (molar mass 108 g mol^{-1}) is deposited at cathode from $AgNO_3(aq)$ solution by a certain quantity of electricity. The volume (in L) of oxygen gas produced at 273 K and 1 bar pressure from water by the same quantity of electricity is____. **[JEE Main (Jan.) 2020]**

8. For the disproportionation reaction

$2Cu^+(aq) \rightleftharpoons Cu(s) + Cu^{2+}$ at 298 K, ln K (where K is the equilibrium constant) is ______ $\times 10^{-1}$.

[JEE Main (Sept.) 2020]

Given : $\left(E^\circ_{Cu^{2+}/Cu^+} = 0.16\ V, E^\circ_{Cu^+/Cu} = 0.52\ V, \frac{RT}{F} = 0.025\right)$

9. An acidic solution of dichromate is electrolyzed for 8 minutes using 2A current. As per the following equation $Cr_2O_7^{2-} + 14H^+ + 6e^- \longrightarrow 2Cr^{3+} + 7H_2O$. The amount of Cr^{3+} obtained was 0.104 g. The efficiency of the process (in %) is :

(Take: $F = 96000$ C, At. mass of chromium = 52) ______. **[JEE Main (Sept.) 2020]**

10. Potassium chlorate is prepared by the electrolysis of KCl in basic solution

$$6HO^- + Cl^- \longrightarrow ClO_3^- + 3H_2O + 6e^-$$

If only 60% of the current is utilized in the reaction, the time (rounded to the nearest hour) required to produce 10 g of $KClO_3$ using a current of 2A is ______.

(Given : F = 96,500 C mol^{-1}; molar mass of $KClO_3$ = 122 g mol^{-1}) **[JEE Main (Sept.) 2020]**

11. The magnitude of the change in oxidising power of the MnO_4^- / Mn^{2+} couple is $x \times 10^{-4}$ V, if the H^+ concentration is decreased from $1M$ to 10^{-4} M at 25°C. (Assume concentration of MnO_4^- and Mn^{2+} to be same on change in H^+ concentration). The value of x is ______.

(Rounded off to the nearest integer)

[Given: $\frac{2.303\,RT}{F} = 0.059$] **[JEE Main (Feb.) 2021]**

12. Copper reduces NO_3^- into NO and NO_2 depending upon the concentration of HNO_3 in solution. (Assuming fixed $[Cu^{2+}]$ and $P_{NO} = P_{NO_2}$), the HNO_3 concentration at which the thermodynamic tendency for reduction of NO_3^- into NO and NO_2 by copper is same is 10^x M. The value of $2x$ is ______. (Rounded off to the nearest integer)

[Given, $E^\circ_{Cu^{2+}/Cu} = 0.34\,V, E^\circ_{NO_3^-/NO} = 0.96\,V, E^\circ_{NO_3^-/NO_2} = 0.79\,V$ and at 298 K, $\frac{2.303\,RT}{F} = 0.059$]

[JEE Main (Feb.) 2021]

13. A 5.0 mmol dm^{-3} aqueous solution of KCl has a conductance of 0.55 mS when measured in a cell constant 1.3 cm^{-1}. The molar conductivity of this solution is ______ mSm^2 mol^{-1}.

(Round off to the nearest integer) **[JEE Main (March) 2021]**

14. A KCl solution of conductivity of 0.14 Sm^{-1} shows a resistance of 4.19 Ω in a conductivity cell. If the same cell is filled with an HCl solution, the resistance drops to 1.03 Ω. The conductivity of the HCl solution is ______ $\times 10^{-2}$ S m^{-1}. (Round off to the nearest integer). **[JEE Main (March) 2021]**

15. For the reaction, $2Fe^{3+}(aq) + 2I^-(aq) \longrightarrow 2Fe^{2+}(aq) + I_2(s)$

the magnitude of the standard molar free energy change, $\Delta_r G^\circ_m = -$.......kJ (Round off to the nearest integer).

$$\left[\begin{array}{ll} E^\circ_{Fe^{2+}/Fe(s)} = -0.440\,V; & E^\circ_{Fe^{3+}/Fe(s)} = -0.036\,V \\ E^\circ_{I_2/2I^-} = 0.539\,V, & 1F = 96500\,C \end{array}\right]$$

[JEE Main (March) 2021]

16. The amount of charge in F (Faraday) required to obtain one mole of iron from Fe_3O_4 is ______. (Nearest integer) **[JEE Main (July) 2022 (I)]**

17. The solubility product of a sparingly soluble salt A_2X_3 is 1.1×10^{-23}. If specific conductance of the solution is 3×10^{-5} S m^{-1}, the limiting molar conductivity of the solution is $x \times 10^{-3}$ S m^2 mol^{-1}. The value of x is ______. **[JEE Main (June) 2022 (I)]**

18. The specific conductance of 0.0025 M acetic acid is 5×10^{-5} S cm^{-1} at a certain temperature. The dissociation constant of acetic acid is ______ $\times 10^{-7}$. (Nearest integer).

Consider limiting molar conductivity of CH_3COOH as 400 S cm^2 mol^{-1} **[JEE Main (April) 2023 (II)]**

19. At 298 K, the standard reduction potential for Cu^{2+}/Cu electrode is 0.34 V.

Given : $K_{sp}\,Cu(OH)_2 = 1 \times 10^{-20}$; Take $\frac{2.303\,RT}{F} = 0.059$ V

The reduction potential at pH = 14 for the above couple is (–) $x \times 10^{-2}$ V. The value of x is ______.

[JEE Main (April) 2023 (II)]

20. The standard reduction potential at 298 K for the following half cells are given below :

$NO_3^- + 4H^+ + 3e^- \longrightarrow NO(g) + 2H_2O$	$E^\circ = 0.97$ V
$V^{2+}(aq) + 2e^- \longrightarrow V$	$E^\circ = -1.19$ V
$Fe^{3+}(aq) + 3e^- \longrightarrow Fe$	$E^\circ = -0.04$ V
$Ag^+(aq) + e^- \longrightarrow Ag(s)$	$E^\circ = 0.80$ V
$Au^{3+}(aq) + 3e^- \longrightarrow Au(s)$	$E^\circ = 1.40$ V

The number of metal(s) which will be oxidized by NO_3^- is aqueous solution is ______.

[JEE Main (April) 2023 (II)]

21. One Faraday of electricity liberates $x \times 10^{-1}$ gram atom of copper sulphate, x is ______.

[JEE Main (Jan.) 2024]

22. The values of conductivity of some materials at 298.15 K in Sm^{-1} are $2.1 \times 10^3, 1.0 \times 10^{-16}, 1.2 \times 10, 3.91, 1.5 \times 10^{-2}, 1 \times 10^{-7}, 1.0 \times 10^3$. The number of conductors among the materials is ______.

[JEE Main (Jan.) 2024]

23. The hydrogen electrode is dipped in a solution of pH = 3 at 25°C. The potential of the electrode will be $-$ ________ $\times 10^{-2}$ V. $\left(\frac{2.303\,RT}{F} = 0.059\text{ V}\right)$ **[JEE Main (Jan.) 2024]**

24. A constant current was passed through a solution of $AuCl_4^-$ ion between gold electrodes. After a period of 10.0 minutes, the increase in mass of cathode was 1.314 g. The total charge passed through the solution is ________ $\times 10^{-2}$ F.

(Given atomic mass of Au = 197) **[JEE Main (Jan.) 2024]**

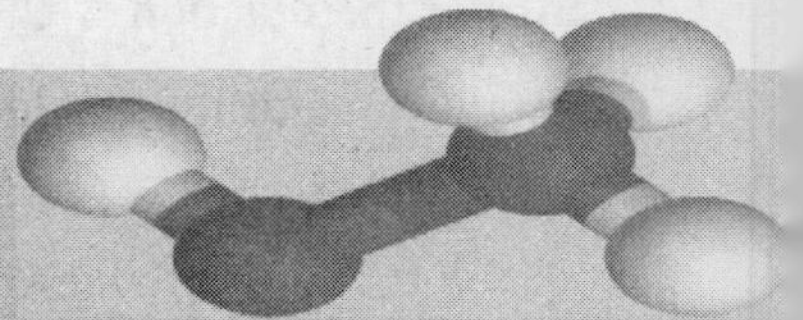

Answers

[A] Objective Questions (Only one option is correct)

1. (d)	**2.** (d)	**3.** (d)	**4.** (a)	**5.** (c)	**6.** (a)	**7.** (a)	**8.** (d)	**9.** (d)	**10.** (d)
11. (c)	**12.** (a)	**13.** (a)	**14.** (a)	**15.** (c)	**16.** (a)	**17.** (c)	**18.** (d)	**19.** (d)	**20.** (b)
21. (*)	**22.** (b)	**23.** (a)	**24.** (b)	**25.** (a)	**26.** (b)	**27.** (a)	**28.** (b)	**29.** (c)	**30.** (b)
31. (c)	**32.** (a)	**33.** (a)	**34.** (b)	**35.** (b)	**36.** (d)	**37.** (b)	**38.** (a)	**39.** (a)	

* *No option is incorrect.*

[B] Objective Questions (One or more than one option(s) is/are correct)

1. (a, b, d)

[C] Passage Based Problems

Passage-1

1. (c) **2.** (c) **3.** (d)

Passage-2

1. (b) **2.** (d) **3.** (d)

Passage-3

1. (c) **2.** (d) **3.** (a)

Passage-4

1. (b) **2.** (c)

Passage-5

1. (d) **2.** (b)

[D] Integer Type Problems

1. (4)	**2.** (3)	**3.** (10)	**4.** (13.32)	**5.** (–0.935 V)
6. (2.15)	**7.** (5.675)	**8.** (144)	**9.** (60)	**10.** (11)
11. (3776)	**12.** (4)	**13.** (14)	**14.** (57)	**15.** (45)
16. (3)	**17.** (3)	**18.** (66)	**19.** (25)	**20.** (3)
21. (5)	**22.** (4)	**23.** (18)	**24.** (2)	

Hints and Solutions

A. Objective Questions

[Only one option is correct]

1. (d) $AgNO_3 + KCl\ (aq.) \rightarrow AgCl\ (s) + KNO_3(aq.)$

Initially there is aq. KCl solution now as solution of $AgNO_3$ is added, AgCl(s) is formed. Hence conductivity of solution is almost compensated (or slightly increase) by the formation of KNO_3. After end point conductivity increases more rapidly because addition of excess $AgNO_3$ solution.

2. (d) $2Fe(s) + O_2(g) + 4H^+(aq) \longrightarrow 2Fe^{+2}(aq) + 2H_2O\ (l)$

$n = 4$ (no. of moles of electron involved)

From Nernst's equation,

$$E_{cell} = E^\circ_{cell} - \frac{0.0591}{n}\log Q = 1.67 - \frac{0.0591}{4}\log\frac{(10^{-3})^2}{0.1\times(10^{-3})^4} \qquad \{\because [H^+] = 10^{-pH}\}$$

$$= 1.67 - 0.106 = 1.57\ V$$

3. (d) MnO_4^- is strongest oxidizing agent.

4. (a) For 0.2 M solution, $R = 50\ \Omega$

$$\kappa = 1.4\ S\ m^{-1} = 1.4\times 10^{-2}\ S\ cm^{-1}$$

$$\Rightarrow \quad \rho = \frac{1}{\kappa} = \frac{1}{1.4\times 10^{-2}}\ \Omega\ cm$$

Now, $$R = \rho\frac{l}{a}$$

$$\Rightarrow \quad \frac{l}{a} = \frac{R}{\rho} = 50\times 1.4\times 10^{-2}$$

For 0.5 M solution, $R = 280\ \Omega$, $\kappa = ?$

$$\frac{l}{a} = 50\times 1.4\times 10^{-2}$$

$$\Rightarrow \quad R = \rho\frac{l}{a} \quad \Rightarrow \quad \frac{1}{\rho} = \frac{1}{R}\times\frac{l}{a}$$

$$\Rightarrow \quad \kappa = \frac{1}{280}\times 50\times 1.4\times 10^{-2}$$

$$= \frac{1}{280}\times 70\times 10^{-2} = 2.5\times 10^{-3}\ S\ cm^{-1}$$

Now, $$\Lambda_m = \frac{\sigma\times 1000}{M} = \frac{2.5\times 10^{-3}\times 1000}{0.5}$$

$$= 5\ S\ cm^2\ mol^{-1} = 5\times 10^{-4}\ S\ m^2\ mol^{-1}$$

5. (c) According to Debye Huckle onsager equation,

$$\lambda_C = \lambda_\infty - b\sqrt{C}$$

6. (a) Using $\Delta G^\circ = -nFE^\circ$

(1) $Mn^{2+} + 2e \longrightarrow Mn\ ;\ E^\circ = -1.18\ V;\qquad \Delta G^\circ_1 = -2F(-1.18) = 2.36\ F$

(2) $Mn^{3+} + e \longrightarrow Mn^{2+}\ ;\ E^\circ = +1.51V;\qquad \Delta G^\circ_2 = -F(1.51) = -1.51F$

(1) $-2\times$(2)

$$3Mn^{2+} \longrightarrow Mn + 2Mn^{3+};$$

$$\Delta G_3^\circ = \Delta G_1^\circ - 2\Delta G_2^\circ = [2.36 - 2(-1.51)]F$$
$$= (2.36 + 3.02)F = 5.38F$$

But $\Delta G_3^\circ = -2FE^\circ$

$$5.38F = -2FE^\circ \Rightarrow E^\circ = -2.69\text{ V}$$

As E° value is negative reaction is non-spontaneous.

8. (d) $Cu^{2+} + 2e^- \longrightarrow Cu$

2 F charge deposit 1 mole of Cu *i.e.*, 2 F of electricity deposit 63.5 g mass of Cu at the cathode.

9. (d) Galvanization is a method of rust prevention by applying zinc coating which acts as a sacrificial metal.

10. (d) For the given electrochemical cell, the reactions are

At anode: $H_{2(g)} \longrightarrow 2H^+_{(aq)} + 2e^-$

At cathode: $M^{4+}_{(aq)} + 2e^- \longrightarrow M^{2+}_{(aq)}$

$$H_{2(g)} + M^{4+}_{(aq)} \longrightarrow M^{2+}_{(aq)} + 2H^+_{(aq)}$$

$$E_{cell} = E^\circ_{cell} - \frac{0.059}{2}\log\frac{[M^{2+}][H^+]^2}{[M^{4+}]}$$

$$0.092 = (E^\circ_{M^{4+}/M^{2+}} - E^\circ_{H^+/H_2}) - \frac{0.059}{2}\log(10^x[H^+]^2)$$

$$0.092 = (0.151 - 0) - \frac{0.059}{2}\log(10^x \times 1^2)$$

$$0.092 = 0.151 - 0.0295\log 10^x$$

$$0.0295\log 10^x = 0.151 - 0.092$$

$$\log 10^x = \frac{0.059}{0.0295} = 2$$

$$10^x = \text{Antilog } 2 = 10^2$$

$$\therefore \quad x = 2$$

11. (c) More negative the E° value of the species, more stronger is the reducing agent.

Since, $E^\circ_{Cr^{3+}/Cr} = -0.74$ V (most negative among the given examples), Cr is the strongest reducing agent.

12. (a)

$$Zn - 2e \longrightarrow Zn^{2+}$$
$$Cu^{2+} + 2e \longrightarrow Cu$$
$$Zn + Cu^{2+} \longrightarrow Zn^{2+} + Cu$$

$$E_{cell} = E^\circ_{cell} - \frac{2.303RT}{nF}\log\left[\frac{Zn^{2+}}{Cu^{2+}}\right]$$

$$E_{cell} = 1.1 - \frac{2.303RT}{2F}\log 10 = 1.1 - \frac{2.303RT}{2F}$$

$$\Delta G = -nFE_{cell}$$
$$= -2F\left(1.1 - \frac{2.303RT}{2F}\right) = 2.303RT - 2.2F$$

13. (a) $B_2H_6 + 3O_2 \longrightarrow B_2O_3 + 3H_2O$

$$nB_2H_6 = \frac{27.66}{27.66} = 1$$

n_{O_2} required $= 3$

$$2H_2O \longrightarrow 2H_2 + O_2$$

n-factor for $O_2 = 4$

$\therefore$ Number of equivalent $= 3 \times 4 = 12F = 12 \times 96500C$

$$i \times t = 12 \times 96500$$

$$t = \frac{12 \times 96500}{100} s = \frac{12 \times 96500}{100 \times 3600} h = 3.2 \text{ hr}$$

14. (a) Greater is the reduction potential (+ve), more easily is the substance (element or ion) reduced or in other words, strongest oxidising agent it is. Thus the strongest oxidising agent among the given substances is $S_2O_8^{2-}$.

15. (c) Given reaction,

$$Fe^{2+}_{(aq)} + Ag^+_{(aq)} \longrightarrow Fe^{3+}_{(aq)} + Ag_{(s)}$$

$$Fe^{2+} + 2e^- \longrightarrow Fe; E^\circ_{Fe^{2+}/Fe} = y \text{ V}$$

$$\Delta G_1^\circ = -2Fy \quad \text{...(i)}$$

$$Fe^{3+} + 3e^- \longrightarrow Fe; E^\circ_{Fe^{3+}/Fe} = z \text{ V}$$

$$\Delta G_2^\circ = -3Fz \quad \text{...(ii)}$$

$$Ag^+ + e^- \longrightarrow Ag; E^\circ = +x \text{ V}$$

$$\Delta G_3^\circ = -Fx \quad \text{...(iii)}$$

On adding equations (i) and (iii) and subtracting (ii) from it, we get

$$Fe^{2+}_{(aq)} + Ag^+_{(aq)} \longrightarrow Ag_{(s)} + Fe^{3+}_{(aq)}$$

$$\therefore \quad \Delta G = \Delta G_1^\circ + \Delta G_3^\circ - \Delta G_2^\circ$$

$$-F E^\circ_{cell} = -2Fy - Fx - (-3Fz)$$

$$-E^\circ_{cell} = -2y - x + 3z \text{ or } E^\circ_{cell} = x + 2y - 3z$$

16. (a) As the concentration of sodium stearate increases beyond CMC, stearate ions get clubbed together and form micelles. This abruptly causes the concentration of the current carrier anions to decrease. This is reflected by the sharp change in Λ_m at CMC, followed by greater rate of decreases of Λ_m with $\sqrt{C}$.

17. (c) $E^\circ_{cell} = E^\circ_{RP}(\text{Cathode}) - E^\circ_{RP}(\text{Anode})$

$$= E^\circ_{Cu^{2+}/Cu} - E^\circ_{Zn^{2+}/Zn} = 0.34 - (-0.76) = 1.1 \text{V}$$

When $E_{Ext} < 1.1\text{V}$, cell will acts as galvanic cell.

Cell reaction in standard condition is:

$$Zn(s) + Cu^{2+}_{(aq)} \longrightarrow Zn^{2+}_{(aq)} + Cu(s)$$

❖ When $E_{ext} > 1.1\text{V}$, cell reaction will revert, cell will acts as electrolytic cell.

$$Zn^{2+}_{(aq)} + Cu(s) \longrightarrow Zn(s) + Cu^{2+}_{(aq)}$$

So, Zn^{2+} will deposit at Zn electrode, Cu will dissolve as Cu^{2+}.

18. (d) At anode: $Cu(s) \longrightarrow Cu^{2+}(C_1) + 2e^-$

At cathode: $Cu^{2+}(C_2) + 2e^- \longrightarrow Cu(s)$

$$Cu^{2+}(C_2) \longrightarrow Cu^{2+}(C_1)$$

$$\Delta G = -nFE_{cell}$$

if, $\Delta G < 0$ then $E_{cell} > 0$

For all given concentration, $E^\circ_{cell} = 0$

$$E_{cell} = E^\circ_{cell} - \frac{RT}{2F} \ln Q > 0$$

$$E_{cell} = 0 - \frac{RT}{2F}\ln\left(\frac{C_1}{C_2}\right) > 0$$

$$\ln\frac{C_1}{C_2} < 0$$

$$C_1 < C_2$$

So, only possibility is $C_2 = \sqrt{2}\,C_1$.

19. (d) As solution was electrolyzed by passing same charge for both the electrolyte, same equivalents of both metals will be deposited,

$$\text{equivalent (Ag)} = \text{equivalent (Au)}$$

$$\frac{W_{Ag}}{M_{Ag}} \times 1 = \frac{W_{Au}}{M_{Au}} \times 1$$

$$\frac{W_{Ag}}{W_{Au}} = \frac{M_{Ag}}{M_{Au}}$$

20. (b) Distilled water have lowest ionic conductance due to less number of ionic salts present in it.

21. (*) **(No option is incorrect)**

On increasing temperature conductance increases due to increased ionic mobility.

NaCl is completely soluble salt while $BaSO_4$ is partially soluble, so

$$C_{NaCl} >> C_{BaSO_4} \text{ (at any temp. } T)$$

22. (b) Graph shows behaviour of molar conductivity of weak electrolyte *vs* $\sqrt{c}$. For strong electrolyte molar conductivity varies linearly with $\sqrt{c}$. So correct answer will be CH_3COOH.

23. (a) $\Lambda^\circ_m(\text{NaBr}) - \Lambda^\circ_m(\text{NaI}) = \lambda^\circ_m(\text{Br}^-) - \lambda^\circ_m(\text{I}^-)$

$\Lambda^\circ_m(\text{KBr}) - \Lambda^\circ_m(\text{NaBr}) = \lambda^\circ_m(\text{K}^+) - \lambda^\circ_m(\text{Na}^+)$

So, $\Lambda^\circ_m(\text{NaBr}) - \Lambda^\circ_m(\text{NaI}) \neq \Lambda^\circ_m(\text{KBr}) - \Lambda^\circ_m(\text{NaBr})$

24. (b) At anode :

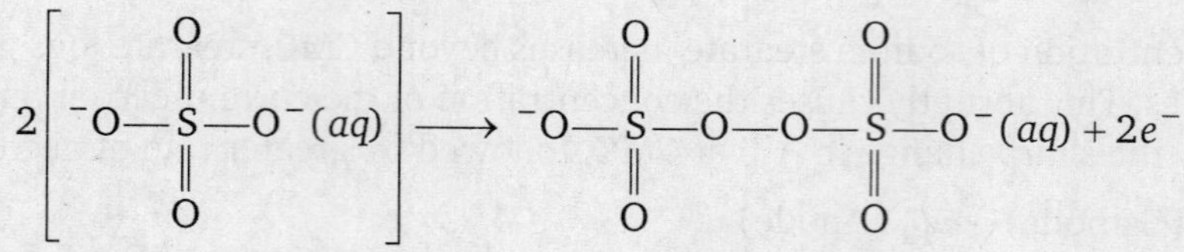

At cathode : $2H^+(aq) + 2e^- \longrightarrow H_2(g)$

Hence, the product will be $HO_3S—O—O—SO_3H$.

25. (a)
$$\Lambda_m = \frac{\kappa}{C} \quad \Rightarrow \quad \kappa = \Lambda_m \times C$$

from question, $\kappa_1 = \kappa_2$

$$\Lambda_{m_1} \times C_1 = \Lambda_{m_2} \times C_2; \quad \Lambda_{m_1} \times \frac{10 \times 10^3}{20} = \Lambda_{m_2} \times \frac{20 \times 10^3}{80}$$

$$2\Lambda_{m_1} = \Lambda_{m_2}$$

26. (b) Reduction of MnO_4^- is pH dependent.

In acidic medium

$$MnO_4^- + 5e^- \longrightarrow Mn^{2+}$$

In neutral medium

$$MnO_4^- + 3e^- \longrightarrow Mn^{4+}$$

In basic medium

$$MnO_4^- + e^- \longrightarrow Mn^{6+}$$

So, according to pH, the reaction and potential of cell changes.

27. (a) For weak acid, $\alpha = \frac{\Lambda_m}{\Lambda_m^\circ}$

$$K_a = \frac{c\alpha^2}{1-\alpha} \Rightarrow K_a(1-\alpha) = c\alpha^2$$

$$\Rightarrow K_a\left(1 - \frac{\Lambda_m}{\Lambda_m^\circ}\right) = c\left(\frac{\Lambda_m}{\Lambda_m^\circ}\right)^2$$

$$\Rightarrow K_a - \frac{\Lambda_m K_a}{\Lambda_m^\circ} = \frac{c\Lambda_m^2}{(\Lambda_m^\circ)^2} \Rightarrow K_a = \frac{c\Lambda_m^2}{(\Lambda_m^\circ)^2} + \frac{\Lambda_m K_a}{\Lambda_m^\circ}$$

Divide by 'Λ_m'

$$\Rightarrow \frac{K_a}{\Lambda_m} = \frac{c\Lambda_m}{(\Lambda_m^\circ)^2} + \frac{K_a}{\Lambda_m^\circ} \Rightarrow \frac{1}{\Lambda_m} = \frac{c\Lambda_m}{K_a(\Lambda_m^\circ)^2} + \frac{1}{\Lambda_0}$$

For the Plot $\frac{1}{\Lambda_m}$ vs $c\Lambda_m$, Slope $= \frac{1}{K_a(\Lambda_m^\circ)^2} = S$

y-intercept $= \frac{1}{\Lambda_m^\circ} = P$ $\therefore$ $\frac{P}{S} = \frac{\frac{1}{\Lambda_0}}{\frac{1}{K_a(\Lambda_m^\circ)^2}} = K_a\Lambda_m^\circ$

28. (b) Anode : $\frac{1}{2}H_2(g) \rightleftharpoons H^+(aq) + e^-$ Cathode

$AgCl(s) + e^- \longrightarrow Ag(s) + Cl^-(aq)$

29. (c) Metal cation with negative value of reduction potential (M^{3+}/M^{2+}) or with positive value of oxidation potential (M^{2+}/M^{3+}) will liberate H_2 from dilute acid.

30. (b) Higher the value of +ve SRP (Std. reduction potential) more is tendency to undergo reduction, so better is oxidising power of reactant.

Hence, ox. Power : $BrO_4^- > IO_4^- > ClO_4^-$

31. (c) gm equivalent of Al deposited $= \frac{It}{96500}$

$$\frac{w}{27} \times 3 = \frac{2 \times 30 \times 60}{96500}$$

$$w = 0.336 \text{ g}$$

32. (a) $FeO_4^{-2} \xrightarrow[n_1=3]{E_1^\circ=2V} Fe^{3+} \xrightarrow[n_2=1]{E_2^\circ=0.8V} Fe^{2+} \xrightarrow{E_3^\circ=-0.5V} Fe$

$FeO_4^{-2} \xrightarrow[n_4=4]{E_4^\circ=?} Fe^{2+}$

$$\Delta G_4^\circ = \Delta G_1^\circ + \Delta G_2^\circ$$

$$\Rightarrow -n_4FE_4^\circ = -n_1FE_1^\circ - n_2FE_2^\circ$$

$$\Rightarrow +4E_4^\circ = 3 \times 2 + (1 \times 0.8)$$

$$\Rightarrow E_4^\circ = \frac{6.8}{4} \text{ V}$$

$$\Rightarrow E_4^\circ = 1.7 \text{ V}$$

33. (a) $\because \Delta G^\circ = -nFE^\circ$

Option (a) $E^\circ = 0.8 + 0.76 = 1.56$ V

$\therefore \Delta G^\circ = -2 \times F \times 1.56 = -3.12$ V

Option (b) $E^\circ = -2.37 + 0.76 = -1.61$ V

$\therefore \Delta G^\circ = -2 \times F \times (-1.61) = +3.22$ V

Option (c) $E^\circ = -2.37 - 0.8 = -3.17$ V

$\therefore \Delta G^\circ = -2 \times F \times (-3.17) = +6.34$

Option (d) $E^\circ = 0.8 - 0.34 = 0.46$ V

$\Delta G^\circ = -2 \times F \times 0.46 = 0.92$ V

34. (b) $Fe^{2+}(aq) + Ag^+(aq) \rightarrow Fe^{3+}(aq) + Ag(s)$

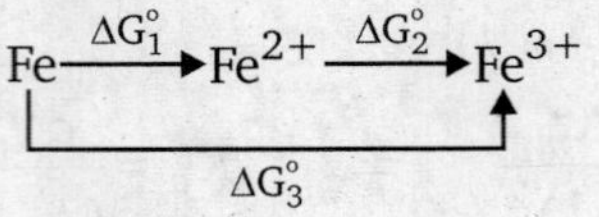

$$\Delta G_3^\circ = \Delta G_1^\circ + \Delta G_2^\circ$$

$$-3F(-z) = -2F(-y) + \Delta G_2^\circ$$

$$\Delta G_2^\circ = 3Fz - 2Fy$$

Also $\Delta G_2^\circ = -nFE^\circ_{Fe^{2+}/Fe^{3+}}$

$$3Fz - 2Fy = -1F\,(E^\circ_{Fe^{2+}/Fe^{3+}})$$

$$E^\circ_{Fe^{2+}/Fe^{3+}} = 2y - 3z$$

E°_{cell} for reaction will be

$$E^\circ_{Ag^+/Ag} + E^\circ_{Fe^{2+}/Fe^{3+}} = x + 2y - 3z$$

35. (b) For strongest reducing agent

Reduction potential should be lowest

Hence, Cr is strongest reducing agent.

36. (d)

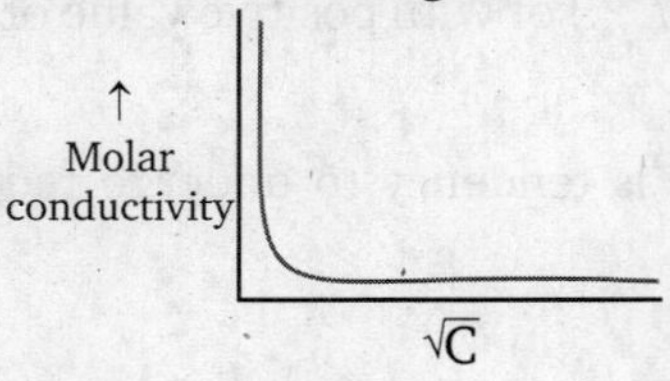

37. (b) According to Nernst equation :

$$E = E^\circ - \frac{RT}{nF} \ln Q$$

Cell reaction : $Mg_{(s)} + 2Ag^+_{(aq)} \rightleftharpoons 2Ag_{(s)} + Mg^{2+}_{(aq)}$

$\Rightarrow$ $$Q = \frac{[Mg^{2+}]}{[Ag^+]^2}$$

$\Rightarrow$ $$E = E^\circ_{cell} - \frac{RT}{2F} \ln\left[\frac{[Mg^{2+}]}{[Ag^+]^2}\right]$$

B. Objective Questions

[One or more than one option(s) is/are correct]

1. (a, b, d)

The species having less reduction potential with respect to NO_3^- ($E^\circ = 0.96$V) will be oxidised by NO_3^-.

C. Passage Based Problems

Passage-1

1. (c). For reaction

$$E° = E°_{O.P.(glucose)} + E°_{R.P. Ag^+/Ag}$$

$$E° = -0.05 + 0.8 = 0.75 \text{ V}$$

Now $\quad nFE^{o} = 2.303\ RT \log_{10} K$

$$\log_{10} K = \frac{nFE^{o}}{2.303RT} = \frac{nE^{o}}{2.303\ RT / F} = \frac{2 \times 0.75}{0.0592} = 25.33$$

2. (c). On increasing concentration of NH_3, the concentration of H^+ ion decreases, thus using Nernst equation, $E_{red} = E^{o}_{red} - \frac{0.0592}{n} \log_{10}[H^+]^2$

$$0 = -\frac{0.0592}{2} \times 2 \log_{10} 10^{-11} \qquad (\because pH = 11, \therefore [H^+] = 10^{-11}\ M)$$

$$= -0.0592 \times 11 = 0.6512 \text{ V} = 0.65 \text{ V}$$

E_{red} increased by a factor of 0.65 from E^{o}_{red}.

3. (d) Addition of NH_3 affects the value of E_{oxi} for glucose/gluconic acid and not $E°$. Thus, this statement is incorrect.

$$\underset{\text{Strong oxidant}}{E°_{RPAg^+}} > \underset{\text{Weak oxidant}}{E°_{RPAg(NH_3)_2^+}}$$

Passage-2

1. (b) $\quad NaCl \longrightarrow Na^+ + Cl^-$

At anode: $\quad 2Cl^- \longrightarrow Cl_2$

Therefore 1 mole of Cl_2 evolves.

2. (d) Na — Hg (amalgam) formed = 2 moles at cathode.

2 moles of amalgam = 23 × 2 + 2 × 200 = 446

3. (d) 2 moles of electrons (2 Faraday) are required.

$\because$ 1 F = 96500 coulombs

$\therefore$ 2 F = 193000 coulombs

Passage-3

1. (c) Reduction potential of I_2 is less than Cl_2.

2. (d) Reaction of Mn^{3+} with H_2O is spontaneous.

3. (a)

$$Na + C + N \longrightarrow NaCN$$

$$Fe^{2+} + 6CN^- \longrightarrow [Fe(CN_6)]^{4-}$$

$$4Fe^{3+} + 3[Fe(CN)_6]^{4-} \longrightarrow \underset{\text{Prussian blue}}{Fe_4[Fe(CN)_6]_3 \downarrow}$$

Passage-4

1. (b) It is an electrolyte concentration cell in which the electrode with concentrated electrolyte solution acts as cathode and second one as anode. For a concentration cell, applying Nernst's equation,

$$E_{cell} = \frac{0.059}{n} \log_{10} \frac{[M^{n+}_{aq.}]_{cathode}}{[M^{n+}_{aq.}]_{anode}}$$

$$\therefore \quad E_{cell} = \frac{0.059}{0.1}\log_{10}\frac{[1.0]}{5\times10^{-2}}$$

$$= 0.076 \text{ or } 76\,\text{mV} \qquad (i.e., \text{positive value})$$

Also $\Delta G = -nFE$, because E is positive so $\Delta G < 0$.

2. (c) In this case

$$E_{cell} = \frac{0.059}{1}\log_{10}\frac{[1.0]}{2.5\times10^{-3}} = 0.152\,\text{V or} \approx 140\,\text{mV}$$

Passage-5

1. (d) $\Delta G = -nEF$

As $M - 2e^- \longrightarrow M^{2+}$

Thus, $n = 2$

$\Delta G = -2\times0.059\times96500 = -11{,}387\,\text{J mol}^{-1} = -11.4\,\text{kJ mol}^{-1}$

2. (b) For concentration cell,

$$E_{cell} = \frac{0.0591}{n}\log\frac{C_{2(\text{RHS})}}{C_{1(\text{LHS})}}$$

$$E_{cell} = 0.059\,\text{V}, C_{2(\text{RHS})} = 0.001$$

$$0.059 = \frac{0.0591}{2}\log\frac{0.001}{C_1}$$

$$\Rightarrow \quad \frac{2\times0.059}{0.0591} = \log\frac{0.001}{C_1} \Rightarrow \text{antilog } 2 = \frac{0.001}{C_1}$$

$$\Rightarrow \quad C_1 = \frac{0.001}{100} = 10^{-5}$$

C_1 = concentration or solubility of $M^{2+} = 10^{-5}$

$$MX_2 \rightleftharpoons \underset{S}{M^{2+}} + \underset{2S}{2X^-}$$

$$K_{sp} = S(2S)^2 = 4S^3$$

$$K_{SP} = 4\times(10^{-5})^3 = 4\times10^{-15}\,\text{mol}^3\,\text{dm}^{-9}$$

D. Integer Type Problems

1. (4) $X \longrightarrow Y;\ \Delta_r G° = -193\,\text{kJ/mol}$

$M^+ \longrightarrow M^{3+} + 2e^-;\ E° = -0.25\,\text{V}$

$\Delta G°$ for this reaction is

$\Delta G° = -nFE° = -2\times96500\times(-0.25) = 48250\,\text{J/mol} = 48.25\,\text{kJ/mol}.$

So, the number of moles of M^+ oxidized when one mole of X is converted to $Y = \frac{193}{48.25} = 4$ moles

2. (3) $HX \rightleftharpoons H^+ + X^-; \quad K_a = \frac{[H^+][X^-]}{[HX]}$

$HY \rightleftharpoons H^+ + Y^-; \quad K_a = \frac{[H^+][Y^-]}{[HY]}$

Λ_m for HX $= \Lambda_{m_1}$; Λ_m for HY $= \Lambda_{m_2}$

$$\Lambda_{m_1} = \frac{1}{10}\Lambda_{m_2}$$

$$K_a = C\alpha^2$$

$$K_{a_1} = C_1 \times \left(\frac{\Lambda_{m_1}}{\Lambda^0_{m_1}}\right)^2$$

$$K_{a_2} = C_2 \times \left(\frac{\Lambda_{m_2}}{\Lambda^0_{m_2}}\right)^2$$

$$\frac{K_{a_1}}{K_{a_2}} = \frac{C_1}{C_2} \times \left(\frac{\Lambda_{m_1}}{\Lambda_{m_2}}\right)^2 = \frac{0.01}{0.1} \times \left(\frac{1}{10}\right)^2 = 0.001$$

$$pK_{a_1} - pK_{a_2} = 3$$

3. (10)

$$E_{cell} = E^\circ_{cell} - \frac{RT}{nf} \ln \frac{[Mg^{2+}]}{[Cu^{2+}]}$$

$$2.67 = 2.70 - \frac{300}{2 \times 11500} \ln x$$

$$\therefore \quad \frac{3}{230} \ln x = 0.03$$

$$\therefore \quad \ln x = \frac{0.03 \times 230}{3} = 2.3$$

$$\therefore \quad x = 10$$

4. (13.32)

$$E^\circ_{cell} = 1.23 - 0.00 = 1.23 \text{ V}$$

$$\Delta G^\circ_{cell} = -nF\, E^\circ_{cell} = -2 \times 96500 \times 1.23 \text{ J}$$

$\therefore$ Work derived from this fuel cell $= \frac{7}{100} \times (-\Delta G^\circ_{cell}) \times 10^{-3} = x\text{J}$

Since insulated vessel, hence $q = 0$

From equation, for monoatomic gas,

$$W = \Delta U \Rightarrow x = nC_{V,m}\Delta T \left\{C_{V,m} = \frac{3R}{2}\right\}$$

or, $$\frac{70}{100} \times (2 \times 96500 \times 1.23) \times 10^{-3} = 1 \times \frac{3}{2} \times 8.314 \times \Delta T$$

$$\therefore \quad \Delta T = 13.32$$

5. (–0.935 V)

$$2H_2O \longrightarrow O_2 + 4H^+ + 4e^-$$

$$E_{cell} = E^\circ_{cell} - \frac{RT}{nF} \ln Q$$

$$E_{cell} = -1.23 - \frac{0.059}{4} \log(10^{-5})^4$$

$$= -1.23 + 0.059 \times 5 = -0.935 \text{ V}$$

6. (2.15)

Anode reaction is: $Sn(s) \longrightarrow Sn^{2+}(aq) + 2e^-$

Cathode reaction is : $Pb^{2+}(aq) + 2e^- \longrightarrow Pb(s)$

Overall reaction is : $Sn(s) + Pb^{2+}(aq) \longrightarrow Sn^{2+}(aq) + Pb(s)$

$$E^\circ_{cell} = E^\circ_{Pb^{2+}/Pb} - E^\circ_{Sn^{2+}/Sn}$$

$$= -0.13 - (-0.14) = 0.01 \text{V}$$

At equilibrium, $$\log K = \frac{nE^\circ}{0.06} = \frac{2 \times 0.01}{0.06}$$

$$\log K = \frac{1}{3}$$

$$K = \frac{[Sn^{2+}]}{[Pb^{2+}]} = 10^{1/3} = 2.15$$

7. (5.675)

$$Ag^+(aq) + e^- \longrightarrow Ag(s)$$

$$2H_2O(l) \longrightarrow O_2(g) + 4H^+(aq) + 4e^-$$

From Faraday's second law, same equivalents of substance deposited/dissolved by passing same quantity of electricity.

$$\text{Equivalents of Ag} = \text{Equivalents of } O_2$$

$$\frac{108}{108} \times 1 = n \times 4$$

$$n(O_2) = \frac{1}{4}$$

$$V(O_2) = \frac{1}{4} \times 22.7 = 5.675\,\text{L}$$

8. (144)

$$2Cu^+ \rightleftharpoons Cu(s) + Cu^{2+}(aq)$$

At cathode: $Cu^+(aq) + e^- \rightleftharpoons Cu(s)$

An anode: $Cu^+(aq) \rightleftharpoons Cu^{2+}(aq) + e^-$

$$2Cu^+(aq) \rightleftharpoons Cu^{2+}(aq) + Cu(s)$$

$$E^\circ_{cell} = E^\circ_{Cu^+/Cu} - E^\circ_{Cu^{2+}/Cu^+}$$

$$= 0.52 - 0.16 = 0.36\text{ V}$$

At equilibrium, $E_{cell} = 0$

and $Q = K$

From Nernst's equation,

$$E_{cell} = E^\circ_{cell} - \frac{RT}{nF}\ln Q$$

$$0 = 0.36 - \frac{0.025}{1}\ln K$$

$$\ln K = 14.4 = 144 \times 10^{-1}$$

9. (60)

$$Cr_2O_7^{2-} + 14H^+ + 6e^- \longrightarrow 2Cr^{3+} + 7H_2O$$

Since, 6 mol e^- (= 6F) charge produces 2 mol (= 2 × 52 gm) Cr^{3+}.

$\Rightarrow$ 0.104 gm Cr^{3+} is produced by $= \frac{6}{104} \times 0.104\,\text{F}$

$$= \frac{6}{104} \times 0.104 \times 96000\,\text{C} = 576\,\text{C}$$

$$\%\,\eta = \frac{Q_{Actual}}{Q_{Theoretical}} \times 100 = \frac{576}{2 \times 8 \times 60} \times 100 = 60\%$$

10. (11)

$$Cl^- + 6OH^- \longrightarrow ClO_3^- + 3H_2O + 6e^-$$

1 mol (= 122 gm) $KClO_3$ is produced by 6*F* electricity.

10 gm $KClO_3$ will be produced by $= \frac{6}{122} \times 10\,F = \frac{6}{122} \times 10 \times 96500\,\text{C}$

Charge passed through the battery $= \frac{6}{122} \times 10 \times 96500 \times \frac{100}{60}$

$$\Rightarrow \quad Q = i \times t = 2 \times t$$
$$= \frac{6}{122} \times 10 \times 96500 \times \frac{100}{60}$$
$$t = \frac{6 \times 10 \times 96500 \times 100}{122 \times 60 \times 3600 \times 2} \text{ hr} = 10.98 \text{ hr} \cong 11 \text{hr}$$

11. (3776)

Balance half cell reaction is

$$MnO_4^- + 8H^+ + 5e^- \longrightarrow Mn^{2+} + 4H_2O$$
$$E_{cell} = E° - \frac{0.059}{5} \log \frac{[Mn^{2+}]}{[H^+]^8[MnO_4^-]}$$

In first case:

$$[H^+] = 1M$$
$$E_1 = E° - \frac{0.059}{5} \log \frac{[Mn^{2+}]}{[MnO_4^-]}$$

In second case:

$$[H^+] = 1 \times 10^{-4} M$$
$$E_2 = E° - \frac{0.059}{5} \log \frac{[Mn^{2+}]}{[MnO_4^-](10^{-4})^8}$$
$$E_2 = E° - \frac{0.059}{5} \log \frac{[Mn^{2+}]}{[MnO_4^-]} - \frac{0.059}{5} \log \frac{1}{10^{-32}}$$
$$E_2 = E_1 - \frac{0.059}{5} \times 32 = E_1 - 0.3776$$
$$E_1 - E_2 = 3776 \times 10^{-4} V$$

12. (4) When Cu reduces HNO_3 to NO

$$(Cu \longrightarrow Cu^{2+} + 2e^-) \times 3$$
$$(NO_3^- + 4H^+ + 3e^- \longrightarrow NO + 2H_2O) \times 2$$

$$3Cu(s) + 8H^+(aq) + 2NO_3^-(aq) \longrightarrow 3Cu^{2+}(aq) + 2NO(g) + 4H_2O(l)$$
$$E_1 = 1.3 - \frac{0.059}{6} \log \frac{[Cu^{2+}]^3 (P_{NO})^2}{[NO_3^-]^2 \times [H^+]^8} \quad \ldots(i)$$

When Cu reduces HNO_3 to NO_2

$$Cu \longrightarrow Cu^{2+} + 2e^-$$
$$(NO_3^- + 2H^+ + e^- \longrightarrow NO_2 + H_2O) \times 2$$

$$Cu(s) + 4H^+(aq) + 2NO_3^-(aq) \longrightarrow 2NO_2(g) + Cu^{2+}(aq) + 2H_2O(l)$$
$$E_2 = 1.13 - \frac{0.059}{2} \log \frac{[Cu^{2+}](P_{NO_2})^2}{(NO_3^-)^2 [H^+]^4}$$

Since, $\quad E_1 = E_2$

$$\Rightarrow \quad 1.3 - \frac{0.059}{6} \log Q_1 = 1.13 - \frac{0.059}{2} \log Q_2$$
$$\Rightarrow \quad 0.17 = \frac{0.059}{6} \{\log Q_1 - \log Q_2\}$$

$$\Rightarrow \quad 0.17 = \frac{0.059}{6}\left\{\log \frac{[Cu^{2+}]^3 \times (P_{NO})^2 \times [NO_3^-]^6 \times [H^+]^{12}}{[NO_3^-]^2 \times [H^+]^8 \times [Cu^{2+}]^3 \times (P_{NO_2})^6}\right\}$$

$$\Rightarrow \quad 0.17 = \frac{0.059}{6}\left\{\log \frac{[NO_3^-]^4 [H^+]^4}{(P_{NO_2})^4}\right\}$$

$$\Rightarrow \quad 0.17 = \frac{0.059}{6} \times 8 \log(HNO_3)$$

$$\Rightarrow \quad \log[HNO_3] = 2.16$$

$$[HNO_3] = 10^{2.16} = 10^x$$

$$x = 2.16$$

$$2x = 4.32$$

13. (14)

$$\text{Conductance } (G) = 0.55 \text{ mS}$$

$$\text{Cell constant } (G^*) = 1.3 \text{ cm}^{-1}$$

$$\text{Molarity } (M) = 5 \times 10^{-3} \text{ mol/L}$$

$$\text{Conductivity} = GG^* = 0.55 \text{ mS} \times 1.3 \times 100 \text{ m}^{-1}$$

$$= 55 \times 1.3 \text{ mSm}^{-1}$$

$$\text{Molar conductivity } (\Lambda_m) = \frac{1}{1000} \times \frac{K}{M}$$

$$= \frac{1}{1000} \times \frac{55 \times 1.3}{5 \times 10^{-3}} \text{ mSm}^2/\text{mol} = 14.3 \text{ mSm}^2/\text{mol}$$

14. (57) $\kappa = GG^* = \dfrac{G^*}{R}$

$$\kappa \cdot R = G^*$$

G^* is constant for given conductivity cell.

$$\kappa_1 R_1 = \kappa_2 R_2$$

$$\kappa_2 = \frac{0.14 \times 4.19}{1.03}$$

$$= 0.5695 \text{ Sm}^{-1} = 56.95 \times 10^{-2} \text{ Sm}^{-1}$$

15. (45)

$$Fe^{3+} + 3e^- \longrightarrow Fe \quad \ldots(i)$$

$$Fe^{2+} + 2e^- \longrightarrow Fe \quad \ldots(ii)$$

$$Fe^{3+} + e^- \longrightarrow Fe^{2+} \quad \ldots(iii)$$

Eqn. (iii) = Eqn. (i) – Eqn. (ii)

$$\Delta G_3^\circ = \Delta G_1^\circ - \Delta G_2^\circ$$

$$(-2)\, FE^\circ = (-3)\, F\, (-0.036) - (-2)\, F\, (-0.44)$$

$$E^\circ = 3(-0.036) + 2\,(10.44) = 0.772 \text{ volt}$$

$$2Fe^{3+} + 2I^- \longrightarrow 2Fe^{2+} + I_2$$

$$E^\circ_{cell} = E^\circ_{Fe^{3+}/Fe^{2+}} - E^\circ_{I_2/I^-}$$

$$= 0.772 - 0.539 = 0.233 \text{ V}$$

$$\Delta G^\circ = -nFE^\circ_{cell}$$

$$= \frac{(-2) \times 96500 \times 0.233}{1000} = -44.96 \text{ kJ}$$

16. (3) Fe_3O_4 is a mixture of FeO and Fe_2O_3. The oxidation state of Fe in FeO and Fe_2O_3 are +2 and +3 respectively.

$\overset{(+2)}{FeO} \xrightarrow{2e^-} Fe$

$\overset{(+3)}{Fe_2O_3} \xrightarrow{6e^-} 2Fe$

$Fe_3O_4 \xrightarrow{8F} 3Fe$

3 mole Fe requires = 8F

1 mole Fe requires $\frac{8}{3}$ F = 2.6 F

17. (3) $A_2X_3(s) \rightleftharpoons 2A^{+3}(aq) + 3X^{-2}(aq)$

Solubility = sM $\qquad 2s \qquad 3s$

$(2s)^2(3s)^3 = 1.1\times10^{-23}$; $\quad 108\,s^5 = 1.1\times10^{-23}$

$$s \simeq 10^{-5} M = 10^{-5}\,\frac{mol}{L} = 0.01\frac{mol}{m^3} \qquad \therefore\ [1L = 0.001\ m^3]$$

Now, $\Lambda_m = \Lambda_m^{\infty} = \frac{k}{m} = \frac{k}{s}$

$\Rightarrow \quad \Lambda_m^{\infty} = \frac{3\times10^{-5}}{0.01} = 3\times10^{-3} S/m^2/mol$

18. (66) $\Lambda_m = \frac{k}{C}\times 1000$

Given, $k = 5\times10^{-5}$ S cm^{-1}; $C = 0.0025$ M

$$\Lambda_m = \frac{5\times10^{-5}\times10^3}{0.0025} = \frac{5\times10^{-2}}{2.5\times10^{-3}} = 20\ S\ cm^2\ mol^{-1}$$

$$\alpha = \frac{20}{400} = \frac{1}{20} \qquad \left(\alpha = \frac{\Lambda_m}{\Lambda_m^{o}}\right)$$

$$K_a = \frac{C\alpha^2}{1-\alpha} = \frac{0.0025\times\frac{1}{20}\times\frac{1}{20}}{\frac{19}{20}}$$

$$= \frac{0.0025}{19\times20} = 6.6\times10^{-6} = 66\times10^{-7}$$

19. (25) $Cu(OH)_2(s) \rightleftharpoons Cu^{2+}(aq) + 2OH^-(aq)$

$K_{sp} = [Cu^{2+}][OH^-]^2$

pH = 14; pOH = 0; $[OH^-] = 1M$

$\therefore \quad [Cu^{2+}] = \frac{K_{sp}}{[1]^2} = 10^{-20}\ M$

$Cu^{2+}(aq) + 2e^- \longrightarrow Cu(s)$

$$E = E^o - \frac{0.059}{2}\log_{10}\frac{1}{[Cu^{2+}]}$$

$$= 0.34 - \frac{0.059}{2}\log_{10}\frac{1}{10^{-20}}$$

$$= -0.25 = -25\times10^{-2}$$

20. (3) Metal having lower standard reduction potential than 0.97 V will be oxidised by NO_3^-.

21. (5) $Cu^{2+} + 2e^- \longrightarrow Cu$

2 Faraday $\longrightarrow$ 1 mol Cu

1 Faraday $\longrightarrow$ 0.5 mol Cu deposit

$0.5 \text{ mol} = 0.5 \text{ g atom} = 5 \times 10^{-1}$

$x = 5$

22. (4) Conductivity ($S\ m^{-1}$)

$\left.\begin{array}{l} 2.1 \times 10^3 \\ 1.2 \times 10 \\ 3.91 \\ 1 \times 10^3 \end{array}\right\}$ conductors at 298.15 K

1×10^{-16} } Insulator at 298.15 K

$\left.\begin{array}{l} 1.5 \times 10^{-2} \\ 1 \times 10^{-7} \end{array}\right\}$ Semiconductor at 298.15 K

Therefore number of conductors is 4.

23. (18) $2H^+(aq) + 2e^- \longrightarrow H_2(g)$

$$E_{cell} = E^{\circ}_{cell} - \frac{0.059}{2} \log \frac{P_{H_2}}{[H^+]^2}$$

$$= 0 - 0.059 \times 3 = -0.177 \text{ volts} = -17.7 \times 10^{-2} \text{ V}$$

24. (2) $\frac{W}{E} = \frac{\text{charge}}{1F}$

$$\frac{1.314}{\frac{197}{3}} = \frac{Q}{1F}$$

$$Q = 2 \times 10^{-2} F$$

❑❑❑

CHAPTER

10

Chemical Kinetics

A. Objective Questions

[Only one option is correct]

1. Plots showing the variation of the rate constant (K) with temperature (T) are given below. The plot that follows Arrhenius equation is : **(IIT 2010)**

(a) 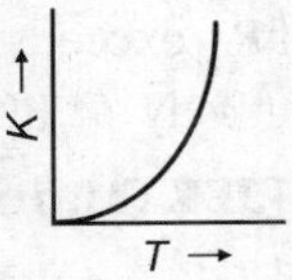

(b) 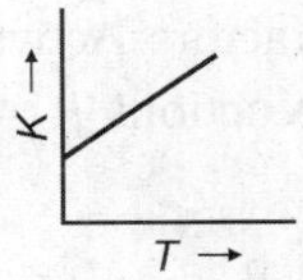

(c) 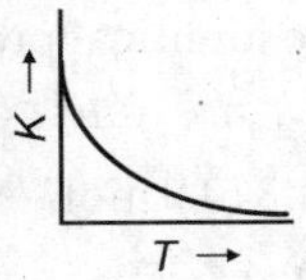

(d) 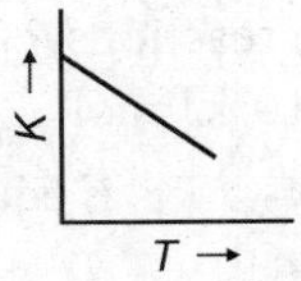

2. The rate of a reaction doubles when its temperature changes from 300 K to 310 K. Activation of such a reaction will be : ($R = 8.314\ JK^{-1}$ and $\log 2 = 0.301$) **[JEE (Mains) 2013]**

(a) 53.6 kJ mol^{-1} (b) 48.6 kJ mol^{-1} (c) 58.5 kJ mol^{-1} (d) 60.5 kJ mol^{-1}

3. In the reaction, $P + Q \longrightarrow R + S$

The time taken for 75% reaction of P is twice the time taken for 50% reaction of P. The concentration of Q varies with reaction time as shown in the figure. The overall order of the reaction is :

[JEE (Advanced) 2013]

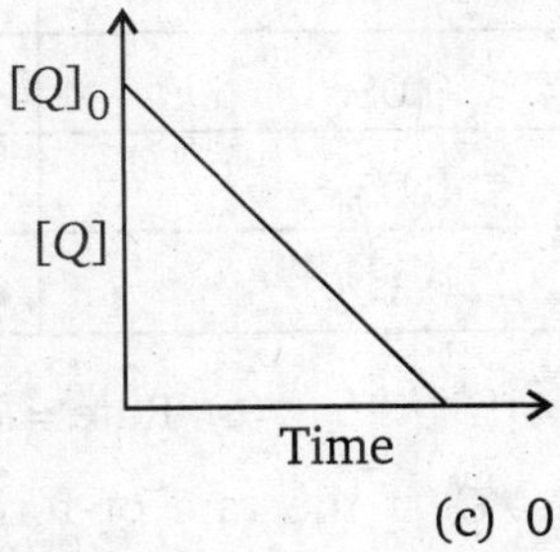

(a) 2 (b) 3 (c) 0 (d) 1

4. For the non-stoichiometric reaction $2A + B \longrightarrow C + D$, the following kinetic data were obtained in three separate experiments, all at 298 K.

Initial Concentration (A)	Initial Concentration (B)	Initial rate of formation of C ($mol\,L^{-1}s^{-1}$)
0.1 M	0.1 M	1.2×10^{-3}
0.1 M	0.2 M	1.2×10^{-3}
0.2 M	0.1 M	2.4×10^{-3}

The rate law for the formation of C is : **[JEE (Mains) 2014]**

(a) $\frac{dC}{dt} = k[A][B]$ (b) $\frac{dC}{dt} = k[A]^2[B]$

(c) $\frac{dC}{dt} = k[A][B]^2$ (d) $\frac{dC}{dt} = k[A]$

5. For the elementary reaction $\mathbf{M} \to \mathbf{N}$, the rate of disappearance of **M** increases by a factor of 8 upon doubling the concentration of **M**. The order of the reaction with respect of **M** is : **[JEE (Advanced) 2014]**

(a) 4 (b) 3 (c) 2 (d) 1

6. Higher order (>3) reactions are rare due to : **[JEE (Mains) 2015]**

(a) shifting of equilibrium towards reactants due to elastic collisions

(b) loss of active species on collision

(c) low probability of simultaneous collision of all the reacting species

(d) increase in entropy and activation energy as more molecules are involved.

7. Decomposition of H_2O_2 follows a first order reaction. In fifty minutes the concentration of H_2O_2 decreases from 0.5 to 0.125 M in one such decomposition. When the concentration of H_2O_2 reaches 0.05 M, the rate of formation of O_2 will be : **[JEE (Mains) 2016]**

(a) 6.93×10^{-2} mol min^{-1} (b) 6.93×10^{-4} mol L^{-1} min^{-1}

(c) 2.66 L min^{-1} at STP (d) 1.34×10^{-2} mol min^{-1}

8. Two reactions R_1 and R_2 have identical preexponential factors. Activation energy of R_1 exceeds that of R_2 by 10 kJ mol^{-1}. If k_1 and k_2 are rate constants for reactions R_1 and R_2 respectively at 300 K, then $\ln(k_2 / k_1)$ is equal to : ($R = 8.314$ J mol^{-1}K^{-1}) **[JEE (Mains) 2017]**

(a) 6 (b) 4 (c) 8 (d) 12

9. A 518° C, the rate of decomposition of a sample of gaseous acetadehyde, initially at a pressure of 363 torr, was 1.00 torr s^{-1} when 5% had reacted and 0.6 torr s^{-1} when 33% had reached. The order of the reaction is: **[JEE (Mains) 2018]**

(a) 1 (b) 0 (c) 2 (d) 3

10. For the reaction $2A + B \longrightarrow C$, the values of initial rate at different reactant concentrations are given in the table below. The rate law for the reaction is : **[JEE (Mains) 2019]**

[A] (mol L^{-1})	**[B] (mol L^{-1})**	**Initial Rate (mol L^{-1} s^{-1})**
0.05	0.05	0.045
0.10	0.05	0.090
0.20	0.10	0.72

(a) Rate $= k[A]^2[B]^2$ (b) Rate $= k[A][B]$ (c) Rate $= k[A][B]^2$ (d) Rate $= k[A]^2[B]$

11. For a reaction scheme, $A \xrightarrow{k_1} B \xrightarrow{k_2} C$, if the rate of formation of B is set to be zero then the concentration of B is given by : **[JEE (Mains) 2019]**

(a) $(k_1 + k_2)[A]$ (b) $\left(\frac{k_1}{k_2}\right)[A]$ (c) $k_1 k_2 [A]$ (d) $(k_1 - k_2)[A]$

12. For the reaction $2A + 3B + \frac{3}{2}C \longrightarrow 3P$, which statement is correct? **[JEE Main (Sept.) 2020]**

(a) $\frac{dn_A}{dt} = \frac{2}{3}\frac{dn_B}{dt} = \frac{3}{4}\frac{dn_C}{dt}$ (b) $\frac{dn_A}{dt} = \frac{dn_B}{dt} = \frac{dn_C}{dt}$

(c) $\frac{dn_A}{dt} = \frac{3}{2}\frac{dn_B}{dt} = \frac{3}{4}\frac{dn_C}{dt}$ (d) $\frac{dn_A}{dt} = \frac{2}{3}\frac{dn_B}{dt} = \frac{4}{3}\frac{dn_C}{dt}$

13. Consider the following reactions

$$A \rightarrow P_1;\ B \rightarrow P_2;\ C \rightarrow P_3;\ D \rightarrow P_4$$

The order of the above reactions are a, b, c and d, respectively. The following graph is obtained when log[rate] *vs.* log [conc.] are plotted:

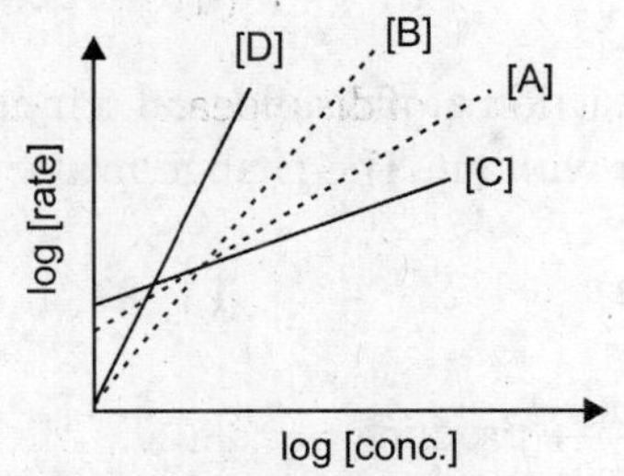

Among the following, the correct sequence for the order of the reactions is: **[JEE Main (Sept.) 2020]**

(a) D > A > B > C
(b) A > B > C > D
(c) C > A > B > D
(d) D > B > A > C

14. A flask contains a mixture of compounds A and B. Both compounds decompose by first –order kinetics. The half lives for A and B are 300 s and 180 s, respectively. If the concentrations of A and B are equal initially, the line required for the concentration of A to be four times that of B (in s) is : (Use $\ln 2 = 0.693$)

[JEE Main (Sept.) 2020]

(a) 120
(b) 300
(c) 900
(d) 180

15. The results given in the below table were obtained during kinetic studies of the following reaction :

[JEE Main (Sept.) 2020]

$$2A + B \longrightarrow C + D$$

Experiment	[A]/mol L^{-1}	[B]/mol L^{-1}	Initial rate/mol L^{-1} min^{-1}
I	0.1	0.1	6.00×10^{-3}
II	0.1	0.2	2.40×10^{-2}
III	0.2	0.1	1.20×10^{-2}
IV	X	0.2	7.20×10^{-2}
V	0.3	Y	2.88×10^{-1}

X and Y in the given table are respectively:

(a) 0.3, 0.4
(b) 0.4, 0.3
(c) 0.4, 0.4
(d) 0.3, 0.3

16. It is true that: **[JEE Main (Sept.) 2020]**

(a) A first order reaction is always a single step reaction.
(b) A zero order reaction is a single step reaction.
(c) A second order reaction is always a multistep reaction.
(d) A zero order reaction is a multistep reaction.

17. Consider the following plots of rate constant *versus* $\frac{1}{T}$ for four different reactions. Which of the following order is correct for the activation energies of these reactions?

[JEE Main (Jan.) 2020]

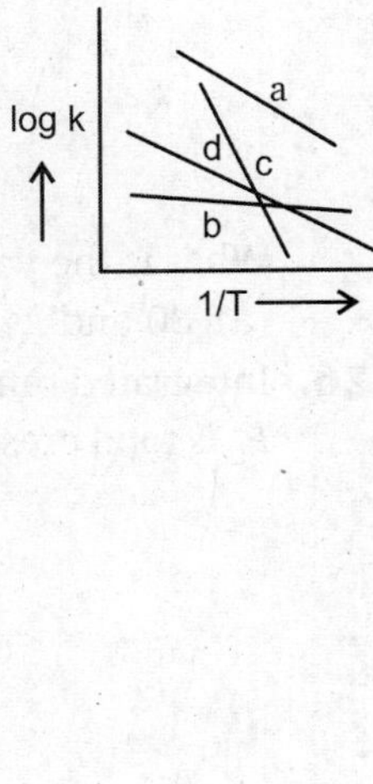

(a) $E_c > E_a > E_d > E_b$
(b) $E_b > E_d > E_c > E_a$
(c) $E_a > E_c > E_d > E_b$
(d) $E_b > E_a > E_d > E_c$

18. The rate constant (k) of a reaction is measured at different temperatures (T), and the data are plotted in the given figure. The activation energy of the reaction in kJ mol^{-1} is: (R is gas constant) **[JEE Main (Sept.) 2020]**

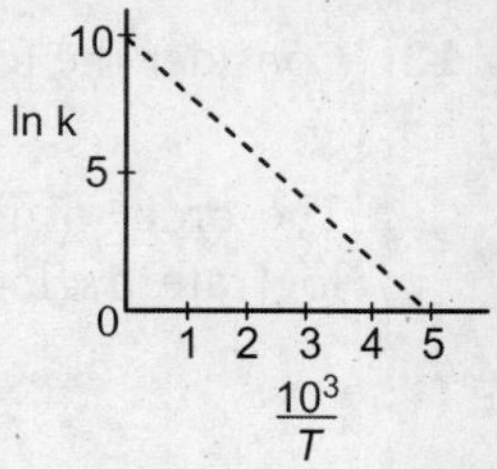

(a) $2R$ (b) $2/R$
(c) R (d) $1/R$

19. The rate of a certain biochemical reaction at physiological temperature (T) occurs 10^6 times faster with enzyme than without. The change in the activation energy upon adding enzyme is: **[JEE Main (Jan.) 2020]**

(a) $-6(2.303)RT$ (b) $+6RT$ (c) $-6RT$ (d) $+6(2.303)RT$

20. For following reactions

$$A \xrightarrow{700\text{K}} \text{product}$$
$$A \xrightarrow[\text{Catalyst}]{500\text{K}} \text{product}$$

it was found that the E_a is decreased by 30 kJ/mol in the presence of catalyst. If the rate remains unchanged, the activation energy for catalysed reaction is (Assume pre exponential factor is same): **[JEE Main (Jan.) 2020]**

(a) 135 kJ/mol (b) 105 kJ/mol (c) 75 kJ/mol (d) 198 kJ/mol

21. For the reaction $2H_2(g) + 2NO(g) \longrightarrow N_2(g) + 2H_2O$, the observed rate expression is, rate $= K_f[NO]^2[H_2]$. The rate expression for the reverse reaction is: **[JEE Main (Jan.) 2020]**

(a) $K_b[N_2][H_2O]$ (b) $K_b[N_2][H_2O]^2 / [NO]$
(c) $K_b[N_2][H_2O]^2 / [H_2]$ (d) $K_b[N_2][H_2O]^2$

22. For a reaction of order n, the unit of the rate constant is : **[JEE Main (July) 2021 (I)]**

(a) $mol^{1-n}\, L^{1-n}\, s$ (b) $mol^{1-n}\, L^{2n}\, s^{-1}$ (c) $mol^{1-n}\, L^{n-1}\, s^{-1}$ (d) $mol^{1-n}\, L^{1-n}\, s^{-1}$

23. At 30°C, the half life for the decomposition of AB_2 is 200 s and is independent of the initial concentration of AB_2. The time required for 80% of the AB_2 to decompose is :
(Given : log 2 = 0.30; log 3 = 0.48) **[JEE Main (June) 2022 (I); (July) 2022 (II)]**

(a) 200 s (b) 323 s (c) 467 s (d) 532 s

24. A student has studied the decomposition of a gas AB_3 at 25°C. He obtained the following data. **[JEE Main (Jan.) 2023 (II)]**

p(mm Hg)	50	100	200	400
Relative $t_{1/2}$(s)	4	2	1	0.5

The order of the reaction is :

(a) 0.5 (b) 2 (c) 1 (d) 0 (zero)

25. For a chemical reaction $A + B \longrightarrow$ Product, the order is 1 with respect to A and B. **[JEE Main (April) 2023 (II)]**

Rate mol L^{-1} s^{-1}	**[A] mol L^{-1}**	**[B] mol L^{-1}**
0.10	20	0.5
0.40	x	0.5
0.80	40	y

What is the value of x and y ?

(a) 80 and 2 (b) 40 and 4 (c) 160 and 4 (d) 80 and 4

26. Integrated rate law equation for a first order gas phase reaction is given by (where P_i is initial pressure and P_t is total pressure at time t) : **[JEE Main (Jan.) 2024]**

(a) $k = \frac{2.303}{t} \times \log \frac{P_i}{(2P_i - P_t)}$ (b) $k = \frac{2.303}{t} \times \log \frac{2P_i}{(2P_i - P_t)}$

(c) $k = \frac{2.303}{t} \times \log \frac{(2P_i - P_t)}{P_i}$ (d) $k = \frac{2.303}{t} \times \frac{P_i}{(2P_i - P_t)}$

27. Consider the given figure and choose the correct option : **[JEE Main (Jan.) 2025]**

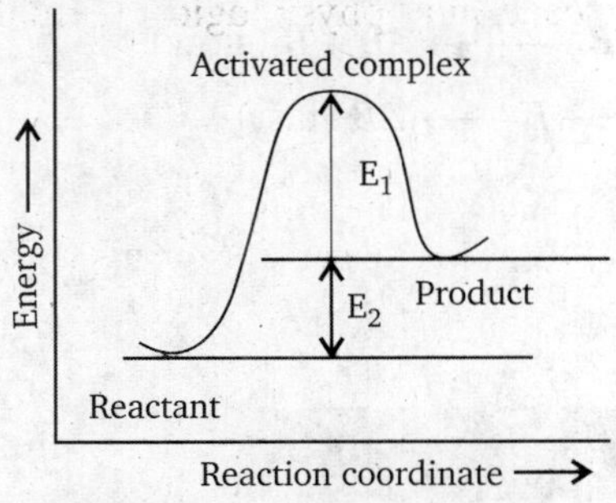

(a) Activation energy of backward reaction is E_1 and product is more stable than reactant.
(b) Activation energy of forward reaction is $E_1 + E_2$ and product is more stable than reactant.
(c) Activation energy of forward reaction is $E_1 + E_2$ and product is lesss stable than reactant.
(d) Activation energy of both forward and backward reaction is $E_1 + E_2$ and reactant is more stable than product.

28. For a reaction, $N_2O_{5(g)} \rightarrow 2NO_{2(g)} + \frac{1}{2}O_{2(g)}$ in a constant volume container, no products were present initially. The final pressure of the system when 50% of reaction gets completed is : **[JEE Main (Jan.) 2025]**

(a) 7/2 times of initial pressure (b) 5 times of initial pressure
(c) 5/2 times of initial pressure (d) 7/4 times of initial pressure

29.

$[A]_0/\text{mol L}^{-1}$	$t_{1/2}/\text{min}$
0.100	200
0.025	100

For a given reaction $R \rightarrow P, t_{1/2}$ is related to $[A]_0$ as given in table :
Given : $\log 2 = 0.30$
Which of the following is true ?

A. The order of the reaction is $\frac{1}{2}$.
B. If $[A]_0$ is $1M$, then $t_{1/2}$ is $200\sqrt{10}$ min.
C. The order of the reaction changes to 1 if the concentration of reactant changes from 0.100 M to 0.500 M.
D. $t_{1/2}$ is 800 min for $[A]_0 = 1.6M$.

Choose the correct answer from the options given below : **[JEE Main (Jan.) 2025]**

(a) A and C only (b) A and B only
(c) A, B and D only (d) C and D only

30. Consider the elementary reaction.

$$A(g) + B(g) \longrightarrow C(g) + D(g)$$

If the volume of reaction mixture is suddenly reduced to $\frac{1}{3}$ of its initial volume, the reaction rate will become 'x' times of the original reaction rate. The value of x is: **[JEE Main (Jan.) 2025]**

(a) $\frac{1}{9}$ (b) 9 (c) $\frac{1}{3}$ (d) 3

31. The reaction $A_2 + B_2 \longrightarrow 2AB$ follows the mechanism

$$A_2 \underset{k_{-1}}{\overset{k_1}{\rightleftharpoons}} A + A \text{ (fast)}$$

$$A + B_2 \xrightarrow{k_2} AB + B \text{ (slow)}$$

$$A + B \longrightarrow AB \text{ (fast)}$$

The overall order of the reaction is : **[JEE Main (Jan.) 2025]**

(a) 1.5 (b) 3 (c) 2.5 (d) 2

B. Objective Questions

[One or more than one option(s) is/are correct]

1. For the first order reaction

$$2N_2O_5(g) \longrightarrow 4NO_2(g) + O_2(g)$$ **(IIT 2011)**

(a) the concentration of the reactant decreases exponentially with time
(b) the half-life of the reaction decreases with increasing temperature
(c) the half-life of the reaction depends on the initial concentration of the reactant
(d) the reaction proceeds to 99.6% completion in eight half-life duration

2. According to the Arrhenius equation, **[JEE (Advanced) 2016]**

(a) a high activation energy usually implies a fast reaction
(b) rate constant increases with increase in temperature. This is due to a greater number of collisions whose energy exceeds the activation energy
(c) higher the magnitude of activation energy, stronger is the temperature dependence of the rate constant
(d) the pre-exponential factor is a measure of the rate at which collisions occur, irrespective of their energy.

3. In a biomolecular reaction, the steric factor P was experimentally determined to be 4.5. The correct option (*s*) among the following is (are): **[JEE (Advanced) 2017]**

(a) experimentally determined value of frequency factor is higher than that predicted by Arrhenius equation.
(b) the value of frequency factor predicted by Arrhenius equation is higher than that determined experimentally.
(c) the activation energy of the reaction is unaffected by the value of the steric factor.
(d) since $P = 4.5$, the reaction will not proceed unless an effective catalyst is used.

C. Assertion and Reason

Read the following questions and answer as per the direction given below :

(a) statement 1 is true; Statement 2 is true; Statement 2 is a correct explanation of Statement 1.
(b) Statement 1 is true; Statement 2 is true; Statement 2 is not the correct explanation of Statement 1.
(c) Statement 1 is true; Statement 2 is false.
(d) Statement 1 is false; Statement 2 is true.

1. Statement 1 : Corrosion is an electrochemical phenomenon in which pure metal acts as an anode and impure metal as a cathode.

Statement 2 : The rate of corrosion is more in alkaline medium than in acidic medium.

[JEE Main (Jan.) 2025]

2. Statement-I :

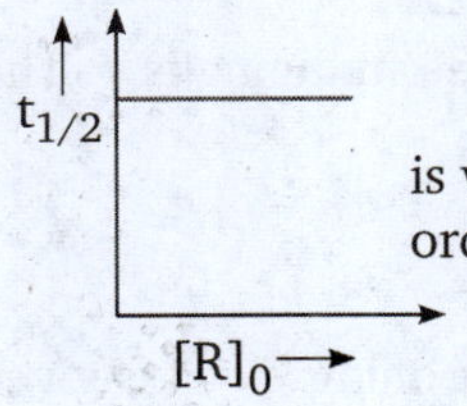

is valid for first order reaction

Statement-II :

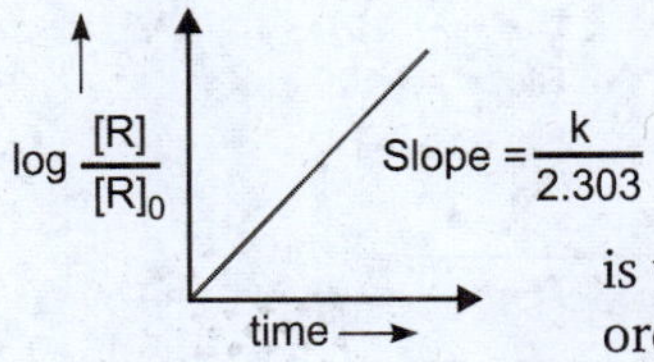

is valid for first order reaction

[JEE Main (Jan.) 2025]

D. Integer Type Problems

1. The concentration of R in the reaction, $R \rightarrow P$, was measured as a function of time and the following data is obtained :

$[R]$ (molar)	1.0	0.75	0.40	0.10
t (min.)	0.0	0.05	0.12	0.18

The order of the reaction is : **(IIT 2010)**

2. An organic compound undergoes first-order decomposition. The time taken for decomposition to 1/8 and 1/10 of its initial concentration are $t_{1/8}$ and $t_{1/10}$ respectively. What is the value of $\frac{[t_{1/8}]}{[t_{1/10}]} \times 10$?

(take $\log_{10} 2 = 0.3$) **(IIT 2012)**

3. The decomposition reaction,

$$2N_2O_{5(g)} \xrightarrow{\Delta} 2N_2O_{4(g)} + O_{2(g)}$$

is started in a closed cylinder under isothermal isochoric condition at an initial pressure of 1 atm. After $Y \times 10^3$ s, the pressure inside the cylinder is found to be 1.45 atm. If the rate constant of the reaction is $5 \times 10^{-4} s^{-1}$, assuming ideal gas behaviour, the value of Y is_____. **[JEE (Advanced) 2019]**

4. A sample of milk splits after 60 min. at 300 K and after 40 min. at 400 K when the population of lactobacillus acidophilus in it doubles. The activation energy (in kJ/mol) for this process is closest to _____.

(Given, $R = 8.3\ J\,mol^{-1}\,K^{-1}, \ln\left(\frac{2}{3}\right) = 0.4, e^{-3} = 4.0$) **[JEE Main (Jan.) 2020]**

5. If 75% of a first order reaction was completed in 90 minutes, 60% of the same reaction would be completed in approximately (in minutes) ________.

(Take: $\log 2 = 0.30$; $\log 2.5 = 0.40$) **[JEE Main (Sept.) 2020]**

6. The number of molecules with energy greater than the threshold energy for a reaction increases five fold by a rise of temperature from 27°C to 42°C. Its energy of activation in J/mol is .... (Take $\ln 5 = 1.6094$; $R = 8.314\ J\,mol^{-1}\,K^{-1}$) **[JEE Main (Sept.) 2020]**

7. A reaction has a half life of 1 min. The time required for 99.9% completion of the reaction is ______ min. (Round off to the nearest integer)

[Use : ln 2 = 0.69, ln 10 = 2.3] **[JEE Main (March) 2021]**

8. Sucrose hydrolyses in acid solution into glucose and fructose following first order rate law with a half-life of 3.33 h at 25°C. After 9 h, the fraction of sucrose remaining is f. The value of $\log_{10}\left(\frac{1}{f}\right)$ is ______ $\times 10^{-2}$.

(Rounded off to the nearest integer)

[Assume : $\ln 10 = 2.303$, $\ln 2 = 0.693$] **[JEE Main (Feb.) 2021]**

9. For the reaction, $aA + bB \longrightarrow cC + dD$, the plot of $\log k$ vs $\frac{1}{T}$ is given below :

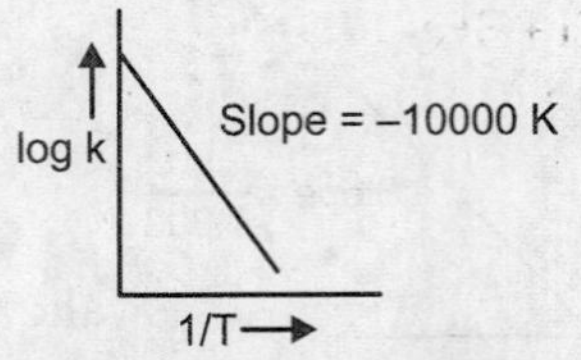

The temperature at which the rate constant of the reaction is $10^{-4}\ s^{-1}$ is ______ K.

(Rounded-off to the nearest integer)

[Given : The rate constant of the reaction is $10^{-5}\ s^{-1}$ at 500 K]. **[JEE Main (Feb.) 2021]**

10. The rate constant of a reaction increases by five times on increase in temperature from 27°C to 52°C. The value of activation energy in kJ mol^{-1} is______ (Rounded-off to the nearest integer). [$R = 8.314\ JK^{-1}\ mol^{-1}$] **[JEE Main (Feb.) 2021]**

11. If the activation energy of a reaction is 80.9 kJ mol^{-1}, the fraction of molecules at 700 K, having enough energy to react to form products is e^{-x}. The value of x is ______. **[JEE Main (Feb.) 2021]**

(Rounded off to the nearest integer) [Use $R = 8.31\ JK^{-1}\ mol^{-1}$]

12. The decomposition of formic acid on gold surface follows first order kinetics. If the rate constant at 300 K is $1.0 \times 10^{-3}\ s^{-1}$ and the activation energy $E_a = 11.488\ kJ\ mol^{-1}$, the rate constant at 200 K is ______ $\times 10^{-5}\ s^{-1}$. (Round off to the nearest integer).

(Given : $R = 8.314\ J\ mol^{-1}\ K^{-1}$) **[JEE Main (March) 2021]**

13. A and B decompose *via* first order kinetics with half-lives 54.0 min and 18.0 min respectively. Starting from an equimolar non reactive mixture of A and B, the time taken for the concentration of A to become 16 times that of B is ______ min. (Round off to the nearest integer). **[JEE Main (March) 2021]**

14. The reaction $2A + B_3 \longrightarrow 2AB$ is an elementary reaction.

For a certain quantity of reactants, if the volume of the reaction vessel is reduced by a factor of 3, the rate of the reaction increases by a factor of ....(Round off to the nearest integer).

[JEE Main (March) 2021]

15. $2NO(g) + Cl_2(g) \rightleftharpoons 2NOCl(s)$

This reaction was studied at –10° C and the following data was obtained.

run	$[NO]_0$	$[Cl_2]_0$	r_0
1	0.10	0.10	0.18
2	0.10	0.20	0.35
3	0.20	0.20	1.40

$[NO]_0$ and $[Cl_2]_0$ are the initial concentrations and r_0 is the initial reaction rate.

The overall order of the reaction is....... **[JEE Main (March) 2021]**

(Round off to the nearest integer).

16. For a reaction $A \longrightarrow 2B + C$ the half lives are 100 s and 50 s when the concentration of reactant A is 0.5 and 1.0 mol L^{-1} respectively. The order of the reaction is ______.

(Nearest integer) **[JEE Main (July) 2022 (I)]**

17. It has been found that for a chemical reaction with rise in temperature by 9 K the rate constant gets doubled. Assuming a reaction to be occurring at 300 K, the value of activation energy is found to be ________ kJ mol^{-1}. [Nearest integer] **[JEE Main (June) 2022 (II)]**
(Given : ln 10 = 2.3, $R = 8.3\ JK^{-1}\ mol^{-1}$, log 2 = 0.30)

18. $A(g) \longrightarrow 2B(g) + C(g)$ is a first order reaction. The initial pressure of the system was found to be 800 mm Hg which increased to 1600 mm Hg after 10 min. The total pressure of the system after 30 min will be ______ mm Hg. (Nearest integer) **[JEE Main (April) 2023 (II)]**

19. $KClO_3 + 6FeSO_4 + 3H_2SO_4 \longrightarrow KCl + 3Fe_2(SO_4)_3 + 3H_2O$
The above reaction was studied at 300 K by monitoring the concentration of $FeSO_4$ in which initial concentration was 10 M and after half an hour became 8.8 M. The rate of production of $Fe_2(SO_4)_3$ is ________ $\times 10^{-6}\ mol\ L^{-1}\ s^{-1}$. (Nearest integer) **[JEE Main (April) 2023 (I)]**

20. $A \longrightarrow B$
The rate constants of the above reaction at 200 K and 300 K are 0.03 min^{-1} and 0.05 min^{-1} respectively. The activation energy for the reaction is ________ J. (Nearest integer)
(Given : ln 10 = 2.3, $R = 8.3\ J\ K^{-1}\ mol^{-1}$; log 5 = 0.70, log 3 = 0.48, log 2 = 0.30)
[JEE Main (Jan.) 2023 (I)]

21. $r = k[A]$ for a reaction, 50% of A is decomposed in 120 minutes. The time taken for 90% decomposition of A is ________ minutes. **[JEE Main (Jan.) 2024]**

22. For a reaction taking place in three steps at same temperature, overall rate constant $K = \frac{K_1K_2}{K_3}$. If E_{a_1}, E_{a_2} and E_{a_3} are 40, 50 and 60 kJ/mol respectively, the overall E_a is ______ kJ/mol.
[JEE Main (Jan.) 2024]

23. NO_2 required for a reaction is produced by decomposition of N_2O_5 in CCl_4 as by equation

$$2N_2O_5(g) \longrightarrow 4NO_2(g) + O_2(g)$$

The initial concentration of N_2O_5 is 3 mol L^{-1} and it is 2.75 mol L^{-1} after 30 minutes.
The rate of formation of NO_2 is $x \times 10^{-3}$ mol L^{-1} min^{-1}, value of x is ________ .
[JEE Main (Jan.) 2024]

24. $A \rightarrow B$
The molecule A changes into its isomeric form B by following a first order kinetics at a temperature of 1000 K. If the energy barrier with respect to reactant energy for such isomeric transformation is 191.48 kJ mol^{-1} and the frequency factor is 10^{20}, the time required for 50% molecules of A to become B is .... picoseconds (nearest integer). [$R = 8.314\ J\ K^{-1} mol^{-1}$] **[JEE Main (Jan.) 2025]**

25. For the thermal decomposition of $N_2O_5(g)$ at constant volume, the following table can be formed, for the reaction mentioned below :

$$2N_2O_5(g) \longrightarrow 2N_2O_4(g) + O_2(g)$$

S.No.	Time/s	Total pressure/(atm)
1.	0	0.6
2.	100	‘x’

x =............$\times 10^{-3}$ atm [nearest integer]

Given : Rate constant for the reaction is $4.606 \times 10^{-2} s^{-1}$. **[JEE Main (Jan.) 2025]**

26. 37.8 g N_2O_5 was taken in a 1 L reaction vessel and allowed to undero the following reaction at 500 K
$2N_2O_{5(g)} \rightarrow 2N_2O_{4(g)} + O_{2(g)}$
The total pressure at equilibrium was found to be 18.65 bar.

Then, K_p =..........$\times 10^{-2}$ [nearest integer]

Assume N_2O_5 to behave ideally under these conditions.

Given : $R = 0.082$ bar L $mol^{-1}K^{-1}$ **[JEE Main (Jan.) 2025]**

27. Consider a complex reaction taking place in three steps with rate constants k_1, k_2 and k_3 respectively. The overall rate constant k is given by the expression $k = \sqrt{\frac{k_1 k_3}{k_2}}$. If the activation energies of the three steps are 60, 30 and 10 kJ mol^{-1} respectively, then the overall energy of activation in kJ mol^{-1} is .... (Nearest integer). **[JEE Main (Jan.) 2025]**

Answers

[A] Objective Questions (Only one option is correct)

1. (a) **2.** (a) **3.** (d) **4.** (d) **5.** (b) **6.** (c) **7.** (b) **8.** (b) **9.** (c) **10.** (c)
11. (b) **12.** (d) **13.** (d) **14.** (c) **15.** (a) **16.** (d) **17.** (a) **18.** (a) **19.** (a) **20.** (c)
21. (c) **22.** (c) **23.** (c) **24.** (b) **25.** (a) **26.** (a) **27.** (c) **28.** (d) **29.** (c) **30.** (b)
31. (a)

[B] Objective Questions (One or more than one option(s) is/are correct)

1. (a, b, d) **2.** (b, c, d) **3.** (a, c)

[C] Assertion and Reason

1. (c) **2.** (c)

[D] Integer Answer Type Problems

1. (0) **2.** (9) **3.** (2.30) **4.** (3.984) **5.** (60)
6. (84297) **7.** (10) **8.** (81) **9.** (526) **10.** (52)
11. (14) **12.** (10) **13.** (108) **14.** (27) **15.** (3)
16. (2) **17.** (59) **18.** (2200) **19.** (333) **20.** (2520)
21. (399) **22.** (30) **23.** (17) **24.** (69) **25.** (900)
26. (962) **27.** (20)

Hints and Solutions

A. Objective Questions

[Only one option is correct]

1. (a) According to Arrhenius equation

$$K = A \cdot e^{-E_a/RT}$$

Thus, K increases exponentially, when T is raised.

2. (a) $$0.3010 = \frac{E_c}{2.303R}\left[\frac{T_2 - T_1}{T_1T_2}\right]$$

$$\Rightarrow \quad 0.3010 = \frac{E_a}{2.303 \times 8.314 \times 10^{-3}}\left[\frac{310-300}{310 \times 300}\right]$$

$$E_a = 53.6 \text{ kJ mol}^{-1}$$

3. (d) $P + Q \longrightarrow R + S$

For 1st order reaction $t_{75\%} = 2 \times t_{1/2}$

$\therefore$ order w.r.t. P is 1.

For zero order reaction,

Integrated rate law is $(a_0 - x) = -kt + a_0$

$\therefore$ Q follows zero order.

Hence overall order is 1.

4. (d) $2A + B \longrightarrow C + D$

Rate of Reaction $= \dfrac{-1}{2}\dfrac{d[A]}{dt} = -\dfrac{d[B]}{dt} = \dfrac{d[C]}{dt} = \dfrac{d[D]}{dt}$

Let rate of Reaction $= k[A]^a[B]^b$

Or, $$\frac{d[C]}{dt} = k[A]^a[B]^b$$

Using the data given

$$1.2 \times 10^{-3} = k[0.1]^a[0.1]^b \quad \text{...(i)}$$

$$1.2 \times 10^{-3} = k[0.1]^a[0.2]^b \quad \text{...(ii)}$$

$$2.4 \times 10^{-3} = k[0.2]^a[0.1]^b \quad \text{...(iii)}$$

From equation (i) and (ii)

$$\frac{1.2 \times 10^{-3}}{1.2 \times 10^{-3}} = \frac{k[0.1]^a[0.1]^b}{k[0.1]^a[0.2]^b} \Rightarrow b = 0$$

From equation (i) and (iii)

$$\frac{1.2 \times 10^{-3}}{2.4 \times 10^{-3}} = \frac{k[0.1]^a[0.1]^b}{k[0.2]^a[0.1]^b} \Rightarrow a = 1$$

Thus, $$\frac{d[C]}{dt} = k[A]^1[B]^0$$

5. (b) $\frac{r_1}{r_2} = \frac{1}{8} = \frac{[M]^n}{[2M]^n} \Rightarrow n = 3$

6. (c) The reactions of higher order are very rare because of the less chances of the molecules to come together simultaneously and collide.

7. (b) Decomposition of H_2O_2 is represented as

$$H_2O_2 \longrightarrow H_2O + \frac{1}{2}O_2$$

Concentration of H_2O_2 decreases from 0.5 *M* to 0.125 *M* in 50 minutes *i.e.*, reduced to $\frac{1}{4}$th. So it can be represented as

$$0.5 \xrightarrow{t_{1/2}} 0.25 \xrightarrow{t_{1/2}} 0.125$$

$$\underbrace{\qquad\qquad\qquad}_{50\text{ min}}$$

$\Rightarrow \quad 2t_{1/2} = 50\text{ min} \Rightarrow t_{1/2} = 25\text{ min}$

Now, for a first order reaction, $k = \frac{0.693}{t_{1/2}} = \frac{0.693}{25}\text{ min}^{-1}$

Rate of H_2O_2 decomposition $= k[H_2O_2] = \frac{0.693}{25} \times (0.05) = 1.386 \times 10^{-3}\text{ mol L}^{-1}\text{ min}^{-1}$

Rate of formation of $O_2 = \frac{1}{2} \times$ rate of H_2O_2 decomposition

$$= \frac{1}{2} \times 1.386 \times 10^{-3} = 6.93 \times 10^{-4}\text{ mol L}^{-1}\text{ min}^{-1}$$

8. (b) According to the Arrhenius equation,

$$k = Ae^{-E_a/RT}$$

For reaction R_1; $\quad k_1 = Ae^{-E_{a_1}/RT}$

$$\ln k_1 = \ln A - \frac{E_{a_1}}{RT}$$

For reaction R_2, $\quad k_2 = Ae^{-E_{a_2}/RT}$

$$\ln k_2 = \ln A - \frac{E_{a_2}}{RT}$$

Now, $\quad \ln\left(\frac{k_2}{k_1}\right) = \frac{E_{a_1}}{RT} - \frac{E_{a_2}}{RT} = \frac{E_{a_1} - E_{a_2}}{RT} = \frac{\Delta E_a}{RT}$

$$= \frac{10 \times 10^3\text{ J mol}^{-1}}{8.314 \times 300} = 4$$

9. (c)

$$r = K[A]^2$$

$$1 = k(363 \times 0.95)^n \quad \text{.....(i)}$$

$$0.5 = k(363 \times 0.67)^n \quad \text{.....(ii)}$$

From (i) and (ii)

$$n = 2$$

10. (c) From the given experiments,

$$\text{Rate}_1 = k(0.05)^a(0.05)^b = 0.045 \quad \text{...(i)}$$

$$\text{Rate}_2 = k(0.10)^a(0.05)^b = 0.090 \quad \text{...(ii)}$$

$$\text{Rate}_3 = k(0.20)^a(0.10)^b = 0.72 \quad \text{...(iii)}$$

From eqs. (i) and (ii),

$$\frac{0.090}{0.045} = \frac{(0.10)^a}{(0.05)^a} \Rightarrow 2^1 = 2^a \Rightarrow a = 1$$

From eqs. (ii) and (iii),

$$\frac{0.72}{0.090} = \frac{(0.20)^a(0.10)^b}{(0.10)^a(0.05)^b}$$

$$8 = 2^a 2^b \qquad (\because 2^a = 2)$$

$$\frac{8}{2} = 2^b \Rightarrow 2^b = 4 \Rightarrow b = 2$$

Thus rate law is, Rate $= k[A][B]^2$

11. (b) $A \xrightarrow{k_1} B \xrightarrow{k_2} C$

According to given reaction scheme, the rate of formation of B is set to zero.

$$\frac{d[B]}{dt} = k_1[A] - k_2[B] = 0 \Rightarrow k_1[A] = k_2[B]$$

$$\therefore \qquad [B] = \frac{k_1}{k_2}[A]$$

12. (d) $2A + 3B + \frac{3}{2}C \longrightarrow 3P$

$$-\frac{1}{2}\frac{d[A]}{dt} = -\frac{1}{3}\frac{d[B]}{dt} = -\frac{1}{3/2}\frac{d[C]}{dt} = \frac{1}{3}\frac{d[P]}{dt}$$

So, $$\frac{1}{2}\frac{d[A]}{dt} = \frac{1}{3}\frac{d[B]}{dt} = \frac{2}{3}\frac{d[C]}{dt}$$

Since, $$C = n/V$$

$$\frac{dn_A}{dt} = \frac{2}{3}\frac{dn_B}{dt} = \frac{4}{3}\frac{dn_C}{dt}$$

13. (d) Let order $= n$

rate $(r) = K(\text{conc.})^n$

$\log(\text{rate}) = n\log(\text{conc.}) + \log K$

$\log(\text{rate})$ *vs* $\log(\text{conc.})$ will be straight line.

Slope $(m) = n$

Higher the slope, more will be order.

Slope: $D > B > A > C$

Order sequence : $D > B > A > C$

14. (c) $[A]_t = 4 \times [B]_t$

$$[A]_0 e^{-K_A \times t} = 4[B]_0 e^{-K_B \times t}$$

$$e^{(K_B - K_A)\times t} = 4 \qquad (\because [A]_0 = [B]_0)$$

$$(K_B - K_A) \times t = \ln 4$$

$$\left(\frac{\ln 2}{180} - \frac{\ln 2}{300}\right) \times t = 2\ln 2$$

$$t = \frac{2 \times 180 \times 300}{120} = 900\ \text{s}$$

15. (a) Differential rate law:

$$r = -\frac{1}{2}\frac{d[A]}{dt} = -\frac{d[B]}{dt} = K[A]^m[B]^n$$

$$6 \times 10^{-3} = K(0.1)^m(0.1)^n \qquad \ldots(i)$$

$$2.4 \times 10^{-2} = K(0.1)^m(0.2)^n \qquad \ldots(ii)$$

$$1.20 \times 10^{-2} = K(0.2)^m(0.1)^n \qquad \ldots(iii)$$

$$7.2 \times 10^{-2} = K(X)^m(0.2)^n \qquad \ldots(iv)$$

$$2.88 \times 10^{-1} = K(0.3)^m(Y)^n \qquad \ldots(v)$$

Dividing eqn. (iii) by eqn. (i),

$$m = 1$$

Dividing eqn. (ii) by eqn. (iii)

$$n = 2$$

So, rate law will be $r = K[A][B]^2$

Divide eqn. (iv) by eqn. (iii)

$$\left(\frac{X}{0.2}\right) \times \left(\frac{0.2}{0.1}\right)^2 = \frac{7.2 \times 10^{-2}}{1.2 \times 10^{-2}}$$

$$X = 0.3\,M$$

Divide eqn. (v) by eqn. (iv)

$$\left(\frac{Y}{0.2}\right)^2 = \frac{2.88 \times 10^{-1}}{7.2 \times 10^{-2}} \Rightarrow Y = 0.4\,M$$

16. (d) Zero order reaction is always complex (multi step).

❖ First and second order reaction can be elementary (single step) or complex (multi step).

17. (a) From Arrhenius equation $K = Ae^{-E_a/RT}$

$$\ln K = \ln A - \left(\frac{E_a}{R}\right)\frac{1}{T}$$

$$\log K = \log A - \left(\frac{E_a}{2.303\,R}\right)\frac{1}{T}$$

$\log K$ vs $\frac{1}{T}$ graph will be a straight line of slope $\frac{-E_a}{2.303\,R}$.

Higher the slope higher will be activation energy.

Order of slope : $c > a > d > b$

So order of E_a will be : $E_c > E_a > E_d > E_b$

18. (a) From Arrhenius equation

$$K = Ae^{-E_a/RT}$$

$$\ln K = \ln A - \frac{E_a}{RT}$$

$\ln K$ vs $1/T$ graph will be straight line of slope $\left(\frac{-E_a}{R}\right)$

$$\text{Slope} = \frac{-E_a}{R} = \frac{y_2 - y_1}{x_2 - x_1}$$

$$\frac{-E_a}{R} = \frac{-10}{5} \Rightarrow E_a = 2R$$

19. (a) $\ln\frac{K_c}{K_a} = \frac{E_a - E_c}{RT}$

K_c is rate constant in presence of catalyst K_a is rate constant in absence of catalyst.

$$2.303 \log_{10}\left(\frac{K_c}{K_a}\right) = \frac{E_a - E_c}{RT}$$

$$E_a - E_c = 2.303\,RT \log_{10}(10^6) = 6(2.303)\,RT$$

$$E_c - E_a = -6(2.303)\,RT$$

20. (c) $A \xrightarrow{700\text{K}}$ product

$A \xrightarrow[\text{Catalyst}]{500\text{K}}$ product

$$(K_1)_{\text{Uncatalyze}} = (K_2)_{\text{Catalyze}}$$

$$Ae^{-E_{a_1}/RT_1} = Ae^{-E_{a_2}/RT_2}$$

$$\frac{E_{a_1}}{T_1} = \frac{E_{a_2}}{T_2}$$

$$\frac{E_a}{700} = \frac{E_a - 30}{500}$$

$$500E_a = 700E_a - 21000$$

$$E_a = 105 \text{ kJ/mol}$$

So, activation energy for uncatalyse reaction $= E_a - 30 = 105 - 30 = 75$ kJ/mol

21. (c) $$K_{eq} = \frac{K_f}{K_b} = \frac{[N_2][H_2O]^2}{[H_2]^2[NO]^2}$$

$$K_f[H_2][NO]^2 = K_b\frac{[N_2][H_2O]^2}{[H_2]}$$

❖ Since at equilibrium, rate of forward reactions = rate of backward reaction

$$r_b = \frac{K_b[N_2][H_2O]^2}{[H_2]}$$

22. (c) Rate $= k[A]^n$

comparing units

$$\frac{(\text{mol/L})}{\text{sec}} = k\left(\frac{\text{mol}}{\text{L}}\right)^n \Rightarrow k = \text{mol}^{(1-n)}\,\text{L}^{(n-1)}\,\text{s}^{-1}$$

23. (c) For 1st order reaction

$$k = \frac{2.303}{t}\log\frac{a_0}{0.2\,a_0}$$

also $t_{1/2} = \dfrac{0.693}{k}$

$$k = \frac{0.693}{200} \Rightarrow \frac{0.693}{200} = \frac{2.303}{t}\log\frac{1}{0.2}$$

$$t = \frac{2.303}{0.693} \times 200\log\frac{1}{0.2} = 466.675 \approx 467 \text{ sec}$$

24. (b) $$t_{1/2} \propto (P_0)^{1-n}$$

$$\frac{(t_{1/2})_1}{(t_{1/2})_2} = \frac{(P_0)_1^{1-n}}{(P_0)_2^{1-n}} \Rightarrow \left(\frac{4}{2}\right) = \left(\frac{50}{100}\right)^{1-n}$$

$$\Rightarrow 2 = \left(\frac{1}{2}\right)^{1-n} \Rightarrow 2 = (2)^{n-1}$$

$$\Rightarrow n - 1 = 1 \Rightarrow n = 2$$

25. (a) $$r = K[A]^1[B]^1$$

$$0.1 = K(20)^1(0.5)^1 \quad \text{... (i)}$$

$$0.40 = K(x)^1(0.5)^1 \quad \text{... (ii)}$$

$$0.80 = K(40)^1(y)^1 \quad \text{... (iii)}$$

From (i) and (ii), $x = 80$

From (i) and (iii), $y = 2$

26. (a)

A	$\longrightarrow$	B	+	C
P_i		0		0
$P_i - x$		x		x

$$P_t = P_i + x$$

$$P_i - x = P_i - P_t + P_i = 2P_i - P_t$$

$$K = \frac{2.303}{t}\log\frac{P_i}{2P_i - P_t}$$

27. (c) Activation energy of forward reaction $= E_1 + E_2$

Energy of product > Energy of reactant Stability : Reactant > Product

28. (d)

$$N_2O_{5(g)} \longrightarrow 2NO_{2(g)} + \frac{1}{2}O_{2(g)}$$

$t = 0$	P_0	—	—
$t = t$	$P_0 - x$	$2x$	$\frac{x}{2}$

$$x = \frac{P_0}{2}$$

$$P_{total} = P_0 - \frac{P_0}{2} + P_0 + \frac{P_0}{4} = \frac{7}{4}P_0$$

29. (c) $t_{1/2} \propto \frac{1}{A_0^{n-1}}$

$$\frac{(t_{1/2})_1}{(t_{1/2})_2} = \frac{(A_0)_2^{n-1}}{(A_0)_1^{n-1}}$$

$$\frac{200}{100} = \left(\frac{0.025}{0.100}\right)^{n-1}$$

$$2 = \left(\frac{1}{4}\right)^{n-1} \Rightarrow n - 1 = -\frac{1}{2} \Rightarrow n = \frac{1}{2} \text{ (order)}$$

$$\Rightarrow \quad t_{1/2} \propto \sqrt{A_0}$$

$$\frac{200}{t_{1/2}} = \frac{(0.1)^{1/2}}{(1)^{1/2}} \quad \text{when } A_0 = 1M$$

$$t_{1/2} = 200\sqrt{10} \text{ min}$$

*Ist order kinetics have $t_{1/2}$ independent of their concentration. So upon changing the concentration $t_{1/2}$ should not change for first order reaction.

$$\frac{200}{t_{1/2}} = \frac{(0.1)^{1/2}}{(1.6)^{1/2}} \quad \text{when } A_0 = 1.6M$$

$$t_{1/2} = 800 \text{ min}$$

30. (b) $R_1 = K[A]^1[B]^1$

$$R_1 = K\left[\frac{n_A}{V}\right]^1\left[\frac{n_B}{V}\right]^1$$

$$R_2 = K\left[\frac{3n_A}{V}\right]^1\left[\frac{3n_B}{V}\right]^1$$

$$R_2 = 9R_1$$

31. (a)

$$\text{Rate} = k_2[A][B_2] \quad \text{...(1)}$$

$$\left(\frac{k_1}{k_{-1}}\right) = \left(\frac{[A]^2}{[A_2]}\right)$$

$$\Rightarrow \quad [A] = \sqrt{\frac{k_1}{k_{-1}}} \cdot \sqrt{[A_2]}$$

Substituting in (1); we get

$$\text{Rate} = k_2\sqrt{\frac{k_1}{k_{-1}}}\cdot[A_2]^{1/2}\cdot[B_2]$$

$$\therefore \quad \text{Order} = \left(\frac{3}{2}\right) = 1.5$$

B. Objective Questions

[One or more than one option(s) is/are correct]

1. (a, b, d)

For first order reaction

$$[A] = [A]_0 e^{-kt}$$

Hence concentration of $[NO_2]$ decreases exponentially.

Also, $t_{1/2} = \frac{0.693}{K}$. Which is independent of concentration and $t_{1/2}$ decreases with the increase of temperature.

$$t_{99.6} = \frac{2.303}{K}\log\left(\frac{100}{0.4}\right)$$

$$t_{99.6} = \frac{2.303}{K}(2.4) = 8\times\frac{0.693}{K} = 8t_{1/2}$$

2. (b, c, d)

$$K = Ae^{-\frac{Ea}{RT}}$$

High activation energy means lower value of K and slow reaction.

On increasing temperature energy of particles increases hence greater number of collisions occur whose energy exceeds the activation energy.

Rate of increase of K with temperature is higher when E_a has a large value.

Pre-exponential foctor is a measure of frequency of collisions.

3. (a, c)

C. Assertion and Reason

1. (c) **Statement 1 :** Corrosion is an example of electrochemical phenomenon in which pure metal act as anode and impure metal (rusted metal) act as cathode.

Statement 2 : Corrosion is more favourable in acid medium than alkaline so rate of corrosion is high in acid medium than alkaline.

2. (c) For first order reaction $t_{1/2} = \frac{\ln 2}{k}$

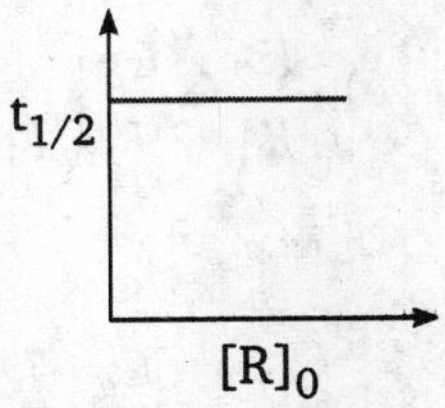

For first order reaction

$$\log\frac{[R]_0}{[R]} = \frac{1}{2.303}kt$$

$$\log\frac{[R]_0}{[R]} = \left(\frac{k}{2.303}\right)\times t$$

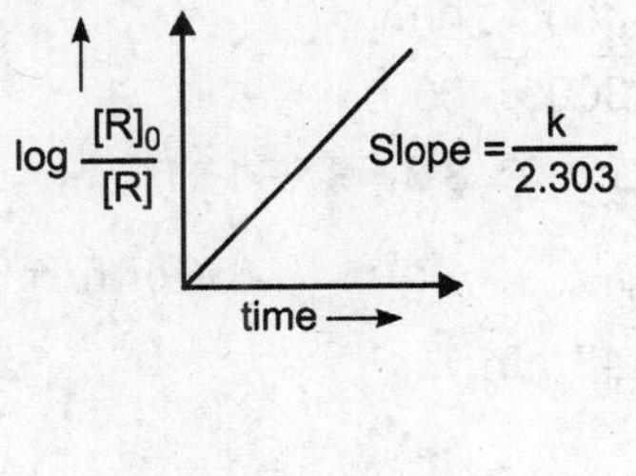

D. Integer Type Problems

1. (0) For zero order reaction, the rate law is written as

$$K_I = \frac{x}{t} = \frac{0.25}{0.05} = 5$$

$$K_{II} = \frac{x}{t} = \frac{0.60}{0.12} = 5$$

2. (9) $k = \frac{2.303}{t}\log\frac{a_0}{a}$. $\quad a = a_0/8$ at $t_{1/8}$

$$t_{1/8} = \frac{2.303}{k}\log\frac{a_0}{a_0/8} = \frac{2.303}{k}\log 8 \quad \text{...(i)}$$

When $t = t_{1/10}$, $a = a_0/10$

$$t_{1/10} = \frac{2.303}{k}\log\frac{a_0}{a_0/10} = \frac{2.303}{k}\log 10 \quad \text{...(ii)}$$

From eq. (i) and (ii),

$$\frac{[t_{1/8}]}{[t_{1/10}]} \times 10 = \frac{2.303}{k}\log 8 \times \frac{k}{2.303 \times \log 10} \times 10$$

$$= \frac{\log 8}{\log 10} \times 10 = \frac{\log 2^3}{\log 10} \times 10 = \frac{3\log 2}{\log 10} \times 10$$

$$= \frac{3 \times 0.3 \times 10}{1} = 9$$

3. (2.30)

From unit of k reaction is first order.

$$2N_2O_5 \xrightarrow{\Delta} 2N_2O_4 + O_2$$

	$2N_2O_5$	$2N_2O_4$	O_2
$t = 0$	1 atm	0	0
$t = t$	$(1-P)$	P	$\frac{P}{2}$
$t = \infty$	0	1 atm	0.5 atm

$P_0 = 1$ atm, $P_t = 1.45$ atm, $P_\infty = 1.5$ atm

$$t = \frac{1}{2K}\ln\left(\frac{P_\infty - P_0}{P_\infty - P_t}\right)$$

$$t = \frac{2.303}{2 \times 5 \times 10^{-4}}\log\left(\frac{1.5 - 1}{1.5 - 1.45}\right)$$

$$t = 2.303 \times 10^3 = Y \times 10^3$$

so, $\quad Y = 2.303$

4. (3.984) $\quad \ln\frac{K_2}{K_1} = \frac{E_a}{R}\left(\frac{1}{T_1} - \frac{1}{T_2}\right)$

Since $K \propto \frac{1}{t}$

$$\ln\left(\frac{t_1}{t_2}\right) = \frac{E_a}{R}\left(\frac{T_2 - T_1}{T_1T_2}\right)$$

$$\ln\left(\frac{60}{40}\right) = \frac{E_a}{R}\left(\frac{400 - 300}{300 \times 400}\right)$$

$$E_a = 0.4 \times 1200 \times 8.3 = 3.984 \text{ kJ/mol}$$

5. (60) $\quad t_{75\%} = 2 \times t_{50\%}$ (for first order)

So, $\quad t_{50\%} = \frac{90}{2} = 45$ min

$$t = \frac{1}{K}\ln\frac{a_0}{a_t}$$

$$t_{60\%} = \frac{45}{\ln 2} \times \ln\left(\frac{100}{40}\right) \qquad \left(\because K = \frac{\ln 2}{t_{1/2}}\right)$$

$$= 45 \times \left(\frac{\ln 2.5}{\ln 2}\right) = 45 \times \left(\frac{\log 2.5}{\log 2}\right) = 45 \times \frac{0.4}{0.3} = 60 \text{ min}$$

6. (84297) $$\ln\left(\frac{K_2}{K_1}\right) = \frac{E_a}{R}\left(\frac{1}{T_1} - \frac{1}{T_2}\right)$$

$$\ln 5 = \frac{E_a}{R}\left(\frac{1}{300} - \frac{1}{315}\right) \qquad \left(\because \frac{K_2}{K_1} = 5\right)$$

$$\ln 5 = \frac{E_a}{R} \times \left(\frac{15}{300 \times 315}\right)$$

$$E_a = \frac{1.6094 \times 8.314 \times 300 \times 315}{15} = 84297 \text{ J/mol}$$

7. (10) From 1st order kinetics :

$$t = \frac{1}{K}\ln\frac{a_0}{a_t}$$

$$t_{99.9\%} = \frac{t_{50\%}}{\ln 2}\ln\left(\frac{100}{0.1}\right) = \frac{1}{\ln 2} \times \ln 1000$$

$$= \frac{3\ln 10}{\ln 2} = \frac{3 \times 2.3}{0.69} = 10$$

8. (81) From 1st order kinetics, $k \times t = \ln\frac{a_0}{a_t}$

$$\left(\frac{\ln 2}{t_{1/2}}\right) \times t = 2.303 \log\frac{a_0}{a_t} \qquad \left(\because K = \frac{\ln 2}{t_{1/2}}\right)$$

$$a_0 = 1, \quad a_t = f$$

$$\frac{\ln 2 \times 9}{3.33 \times 2.303} = \log\left(\frac{1}{f}\right) = 0.81246 = 81.246 \times 10^{-2}$$

9. (526) $$\log K = \log A - \frac{E_a}{2.303\,RT}$$

$$\text{Slope} = \frac{-E_a}{2.303\,R} = -10000$$

$$E_a = 10000 \times 2.303\,R$$

$$\log\frac{K_2}{K_1} = \frac{E_a}{2.303\,R}\left(\frac{1}{T_1} - \frac{1}{T_2}\right)$$

$$\log\left(\frac{10^{-4}}{10^{-5}}\right) = 10000\left(\frac{1}{500} - \frac{1}{T_2}\right)$$

$$T_2 = 526.31\text{K}$$

10. (52) $$\ln\frac{K_2}{K_1} = \frac{E_a}{R}\left(\frac{1}{T_1} - \frac{1}{T_2}\right)$$

$$\ln 5 = \frac{E_a}{R}\left(\frac{1}{300} - \frac{1}{325}\right) \qquad \left(\because \frac{K_2}{K_1} = 5\right)$$

$$E_a = \frac{0.7 \times 2.303 \times 8.314 \times 300 \times 325}{25} = 52271 \text{J} = 52.27 \text{kJ}$$

11. (14)

Fraction of molecules $= e^{-E_a/RT}$

$$x = \frac{E_a}{RT} = \frac{80.9 \times 10^3}{8.31 \times 700} = 13.9$$

12. (10)

$$\ln\frac{K_{300}}{K_{200}} = \frac{E_a}{R}\left(\frac{1}{T_1} - \frac{1}{T_2}\right)$$

$$= \frac{11.488 \times 10^3}{8.314} \times \left(\frac{1}{200} - \frac{1}{300}\right)$$

$$= \frac{11.488 \times 10^3}{8.314} \times \frac{100}{200 \times 300} = 2.303 = \ln 10$$

$$\frac{K_{300}}{K_{200}} = 10$$

$$K_{200} = \frac{1}{10} \times K_{300} = \frac{1}{10} \times 10^{-3} = 10^{-4} = 10 \times 10^{-5}\ \text{s}^{-1}$$

13. (108)

$$[A_t] = 16[B]_t$$

$$[A_0]e^{-K_A \times t} = 16[B_0]e^{-K_B \times t}$$

$$e^{(K_B - K_A) \times t} = 16 \qquad (\because [A_0] = [B_0])$$

$$(K_B - K_A) \times t = \ln 16$$

$$\left(\frac{\ln 2}{18} - \frac{\ln 2}{54}\right) \times t = 4 \ln 2 \qquad \left(\because K = \frac{\ln 2}{t_{1/2}}\right)$$

$$\frac{2}{54} \times t = 4 \Rightarrow t = \frac{54 \times 4}{2} \text{ min} = 108 \text{ min}$$

14. (27) For elementary reaction order and coefficients are same $2A + B_2 \longrightarrow 2AB$

So, rate $(r) = K[A]^2[B_2]$

Since concentration $= \frac{\text{mol}}{\text{volume}}$

On decreasing volume by a factor of 3, concentration of A & B_2 increased by 3 times so new rate become $3 \times 3^2 = 27$ times.

15. (3) Let order with respect to NO is x and with respect to Cl_2 is y.

$$r = K\ [NO]^x\ [Cl_2]^y$$

$$0.18 = K(0.1)^x(0.1)^y \qquad \text{...(i)}$$

$$0.35 = K(0.1)^x(0.2)^y \qquad \text{...(ii)}$$

$$1.40 = K(0.20)^x(0.2)^y \qquad \text{...(iii)}$$

Dividing eqn. (ii) by eqn. (i)

$$2 = 2^y \Rightarrow y = 1$$

Dividing eqn. (iii) by eqn. (ii)

$$4 = 2^x \Rightarrow x = 2$$

Overall order $= x + y = 2 + 1 = 3$

16. (2) For zero order, $t_{1/2}$ decrease with decrease in concentration of reactions.

For first order, $t_{1/2}$ is independent of concentration of reaction.

For second order, $t_{1/2}$ increases with decrease in concentration of reactant.

$$\frac{(t_{1/2})_1}{(t_{1/2})_2} = \left(\frac{[A_0]_2}{[A_0]_1}\right)^{1-n} \Rightarrow \frac{100}{50} = \left(\frac{0.5}{1.0}\right)^{1-n}$$

$\Rightarrow \quad 2 = \left(\frac{1}{2}\right)^{1-n} \quad \Rightarrow \quad 2 = 2^{n-1} \quad \Rightarrow \quad n - 1 = 1 \quad \Rightarrow \quad n = 2$

17. (59) $\log_{10} \frac{K_2}{K_1} = \frac{E_a}{2.303\,R}\left[\frac{1}{T_1} - \frac{1}{T_2}\right]$

$T_1 = 300\text{ K}, T_2 = 309\text{ K}$

$$\log_{10} \frac{K_2}{K_1} = \frac{E_a}{2.303\,R}\left(\frac{1}{300} - \frac{1}{309}\right)$$

$$0.3 = \frac{E_a}{2.303 \times 8.3}\left(\frac{9}{300 \times 309}\right)$$

$$E_a = \frac{0.3 \times 2.303 \times 8.3 \times 300 \times 309}{9} = 59065.04\text{ J/mol}$$

$$E_a = 59.06\text{ kJ/mol}$$

18. (2200) $t_{1/2} = 10$ minutes

$$(P_A)_{30\,\text{min}} = (P_A)_0\left(\frac{1}{2}\right)^{30/10} = 10\text{ mm Hg}$$

	$A(g) \longrightarrow$	$2B(g)$ +	$C(g)$
at $t = 0$,	800 mm	0	0
at $t = 30$,	100 mm	1400 mm	700 mm

Total pressure after 30 minutes = 2200 mm Hg

19. (333)

$$KClO_3 + 6FeSO_4 + 3H_2SO_4 \longrightarrow KCl + 3Fe_2(SO_4)_3 + 3H_2O$$

$$\text{Rate} = \frac{-\Delta[KClO_3]}{\Delta t} = \frac{-1}{6}\frac{\Delta[FeSO_4]}{\Delta t} = \frac{+1}{3}\frac{\Delta[Fe_2(SO_4)_3]}{\Delta t}$$

$$\Rightarrow \quad \frac{\Delta[Fe_2(SO_4)_3]}{\Delta t} = \frac{-1}{2}\frac{\Delta[FeSO_4]}{\Delta t} = \frac{1}{2}\frac{(10 - 8.8)}{30 \times 60}$$

$$= 0.333 \times 10^{-3} = 333 \times 10^{-6}\text{ mol litre}^{-1}\text{ sec}^{-1}$$

20. (2520) $\log \frac{K_{300}}{K_{200}} = \frac{E_a}{2.3 \times 8.314}\left(\frac{1}{T_1} - \frac{1}{T_2}\right)$

$$\log \frac{0.05}{0.03} \frac{E_a}{2.305 \times 8.314} \times \left[\frac{1}{200} - \frac{1}{300}\right]$$

$$E_a = 2519.88\text{ J} \quad \Rightarrow \quad E_a = 2520\text{ J}$$

21. (399) $r = k[A]$

So, order of reaction = 1

$t_{1/2} = 120$ min

For 90% completion of reaction

$$\Rightarrow \quad k = \frac{2.303}{t}\log\left(\frac{a}{a-x}\right)$$

$$\Rightarrow \quad \frac{0.693}{t_{1/2}} = \frac{2.303}{t}\log\frac{100}{10}$$

$$\therefore \quad t = 399\text{ min}$$

22. (30) $K = \frac{K_1 \cdot K_2}{K_3} = \frac{A_1 \cdot A_2}{A_3} \cdot e^{\frac{(E_{a_1} + E_{a_2} - E_{a_3})}{RT}}$

$$A \cdot e^{-E_a/RT} = \frac{A_1 A_2}{A_3} \cdot e^{\frac{(E_{a_1} + E_{a_2} - E_{a_3})}{RT}}$$

$$E_a = E_{a_1} + E_{a_2} - E_{a_3} = 40 + 50 - 60 = 30\text{ kJ/mole}$$

23. (17) Rate of reaction (ROR)

$$= -\frac{1}{2}\frac{\Delta[N_2O_5]}{\Delta t} = \frac{1}{4}\frac{[NO_2]}{\Delta t} = \frac{\Delta[O_2]}{\Delta t}$$

$$\text{ROR} = -\frac{1}{2}\frac{\Delta[N_2O_5]}{\Delta t} = -\frac{1}{2}\frac{(2.75-3)}{30}\text{ mol L}^{-1}\text{ min}^{-1}$$

$$\text{ROR} = -\frac{1}{2}\frac{(-0.25)}{30}\text{ mol L}^{-1}\text{ min}^{-1}$$

$$\text{ROR} = \frac{1}{240}\text{ mol L}^{-1}\text{ min}^{-1}$$

$$\text{Rate of formation of } NO_2 = \frac{\Delta[NO_2]}{\Delta t} = 4 \times \text{ROR}$$

$$= \frac{4}{240} = 16.66 \times 10^{-3}\text{ mol L}^{-1}\text{ min}^{-1}$$

$$\simeq 17 \times 10^{-3}$$

24. (69)

$$t_{1/2} = \frac{0.693}{K}$$

$$K = Ae^{-E_a/RT}$$

$$= 10^{20} \times e^{-\frac{191.48 \times 10^3}{8.314 \times 1000}} = 10^{20} \times e^{-23.031} = 10^{20} \times -e^{\ln 10 \times 10}$$

$$= \frac{10^{20}}{10^{10}} = 10^{10}\text{ sec.}$$

$$t_{1/2} = \frac{0.693}{10^{10}} = 6.93 \times 10^{-11} = 69.3 \times 10^{-12}\text{ sec.}$$

25. (900)

$$K_{N_2O_5} = 2 \times 4.606 \times 10^{-2}\text{ s}^{-1}$$

	$2N_2O_5(g) \longrightarrow$	$2N_2O_4(g)$	$+ O_2(g)$
P_i	0.6	0	0
P_f	$0.6 - P$	P	$\frac{P}{2}$

$$2 \times 4.606 \times 10^{-2} = \frac{2.303}{100}\log\frac{0.6}{0.6-P}$$

$$9.212 \times 10^{-2} = \frac{2.303}{100}\log\frac{0.6}{0.6-P}$$

$$\frac{9.212 \times 10^{-2} \times 10^2}{2.303} = \log\frac{0.6}{0.6-P}$$

$$4 = \log_{10}\frac{0.6}{0.6-P}\ (\log_{10} x = y \text{ then } 10^y = x)$$

$$10^4 = \frac{0.6}{0.6-P}$$

$$\Rightarrow \quad 0.6 \times 10^4 - 10^4 P = 0.6$$

$$\Rightarrow \quad 10^4 P = 0.6(10^4 - 1)$$

$$P = (6000 - 0.6) \times 10^{-4}$$

$$= 5999.4 \times 10^{-4} = 0.59994$$

$$P_{\text{Total}} = 0.6 + \frac{P}{2} = 0.6 + 0.29997$$

$$= 0.89997 = 899.97 \times 10^{-3}$$

26. (962)

$$\text{Initial pressure of } N_2O_5 = \frac{\frac{37.8}{108} \times 0.082 \times 500}{1} = 14.35 \text{ bar}$$

$$2N_2O_5 \rightleftharpoons 2N_2O_4 + O_2$$

	$2N_2O_5$	$2N_2O_4$	O_2
$t = 0$	14.35		
$t = eq$	$14.35 - 2P$	$2P$	P

P_{Total} at eqb $= 14.35 + P = 18.65$

$$P = 4.3$$

$$P_{N_2O_5} = 5.75 \text{ bar}$$

$$P_{N_2O_4} = 8.6 \text{ bar}$$

$$P_{O_2} = 4.3 \text{ bar}$$

$$K_p = \frac{(8.6)^2 \times (4.3)}{(5.75)^2} = 9.619 = x \times 10^{-2}$$

$$x = 961.9 \approx 962$$

27. (20)

$$k = \sqrt{\frac{k_1 k_3}{k_2}}$$

$$A.e^{-E_a/RT} = \sqrt{\frac{A_1 e^{-E_{a_1}/RT} \times A_3 e^{-E_{a_3}/RT}}{A_2 e^{-E_{a_2}/RT}}}$$

By comparing exponential term

$$\frac{E_a}{RT} = \frac{1}{2} \times \left(\frac{E_{a_1}}{RT} + \frac{E_{a_3}}{RT} - \frac{E_{a_2}}{RT} \right)$$

$$E_a = (E_{a_1} + E_{a_3} - E_{a_2}) / 2$$

$$E_a = (60 + 10 - 30) / 2 = 20 \text{ kJ mol}^{-1}$$

❑❑❑

CHAPTER

11

Surface Chemistry

A. Objective Questions

[Only one option is correct]

1. Among the electrolytes $Na_2SO_4, CaCl_2, Al_2(SO_4)_3$ and NH_4Cl, the most effective coagulating agent for Sb_2S_3 sol is : **(IIT 2009)**

(a) Na_2SO_4 (b) $CaCl_2$ (c) $Al_2(SO_4)_3$ (d) NH_4Cl

2. The coagulating power of electrolytes having ions Na^+, Al^{3+} and Ba^{2+} for arsenic sulphide sol increases in the order : **[JEE (Mains) 2013]**

(a) $Al^{3+} < Ba^{2+} < Na^+$ (b) $Na^+ < Ba^{2+} < Al^{3+}$

(c) $Ba^{2+} < Na^+ < Al^{3+}$ (d) $Al^{3+} < Na^+ < Ba^{2+}$

3. Methylene blue, from its aqueous solution, is adsorbed on activated charcoal at 25°C. For this process, the correct statement is: **(JEE Advanced 2013)**

(a) The adsorption requires activation at 25°C.

(b) The adsorption is accompanied by a decrease in enthalpy.

(c) The adsorption increases with increase of temperature.

(d) The adsorption is irreversible.

4. Match the catalysts to the correct processes. **[JEE (Mains) 2015]**

	Catalyst		Process
(A)	$TiCl_4$	(i)	Wacker process
(B)	$PdCl_2$	(ii)	Ziegler-Natta polymerization
(C)	$CuCl_2$	(iii)	Contact process
(D)	V_2O_5	(iv)	Deacon's process

(a) (A)-(ii), (B)-(iii), (C)-(iv), (D)-(i) (b) (A)-(iii), (B)-(i), (C)-(ii), (D)-(iv)

(c) (A)-(iii), (B)-(ii), (C)-(iv), (D)-(i) (d) (A)-(ii), (B)-(i), (C)-(iv), (D)-(iii)

5. For a linear plot of $\log(x/m)$ *versus* $\log P$ in a Freundlich adsorption isotherm, which of the following statements is correct ? (k and n are constants) **[JEE (Mains) 2016]**

(a) Both k and $1/n$ appear in the slope term. (b) $1/n$ appears as the intercept.

(c) Only $1/n$ appears as the slope. (d) $\log(1/n)$ appears as the intercept.

6. The Tyndall effect is observed only when following conditions are satisfied: **[JEE (Mains) 2017]**

(A) the diameter of the dispersed particle is much smaller than the wavelength of the light used.

(B) the diameter of the dispersed particles is not much smaller than the wavelength of the light used.

(C) the refractive indices of the dispersed phase and dispersion medium are almost similar in magnitude.

(D) the refractive indices of the dispersed phase and the dispersion medium differ greatly in magnitude.

(a) (A) and (C) (b) (B) and (C) (c) (A) and (D) (d) (B) and (D)

7. Adsorption of a gas follows Freundlich adsorption isotherm. x is the mass of the gas adsorbed on mass m of the adsorbent.

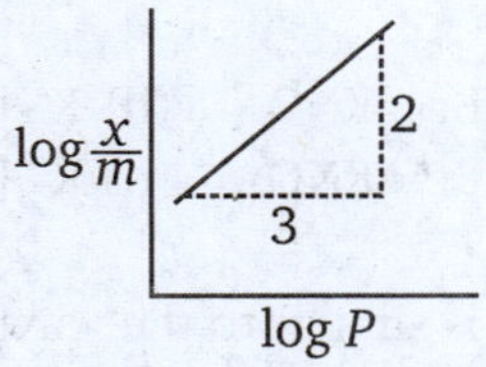

The plot of $\log \frac{x}{m}$ *versus* $\log P$ is shown in the given graph. $\frac{x}{m}$ is proportional to:

[JEE (Mains) 2019]

(a) P^2 (b) P^3 (c) $P^{2/3}$ (d) $P^{3/2}$

8. Amongst the following statements regarding adsorption, those that are valid are:

[JEE Main (Sept.) 2020]

(A) ΔH becomes less negative as adsorption proceeds.
(B) On a given adsorbent, ammonia is adsorbed more than nitrogen gas.
(C) On adsorption, the residual force acting along the surface of the adsorbent increases.
(D) With increase in temperature, the equilibrium concentration of adsorate increases.

(a) (C) and (D) (b) (A) and (B) (c) (D) and (A) (d) (B) and (C)

9. Adsorption of a gas follows Freundlich adsorption isotherm. If x is the mass of the gas adsorbed on mass m of the adsorbent, the correct plot of $\frac{x}{m}$ *versus* P is: **[JEE Main (Sept.) 2020]**

(a) 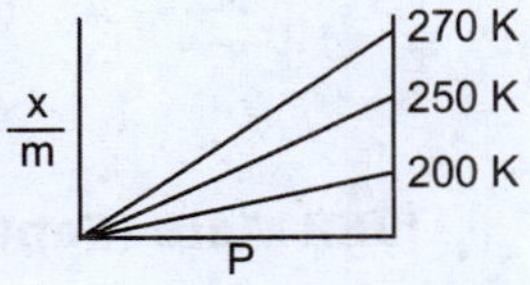

(b) 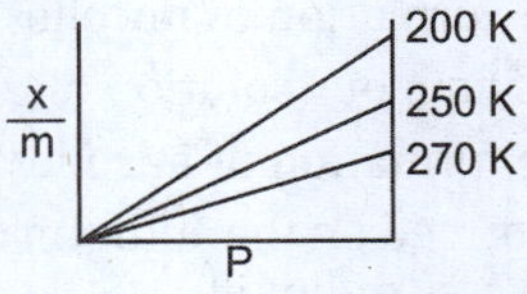

(c) 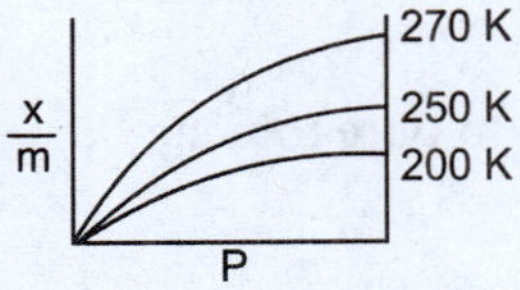

(d) 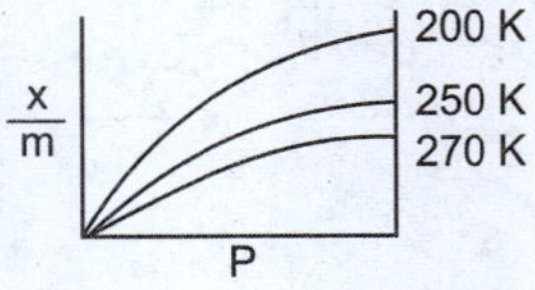

10. A mixture of gases O_2, H_2 and CO are taken in a closed vessel containing charcoal. The graph that represents the correct behaviour of pressure with time is: **[JEE Main (Jan.) 2020]**

(a) 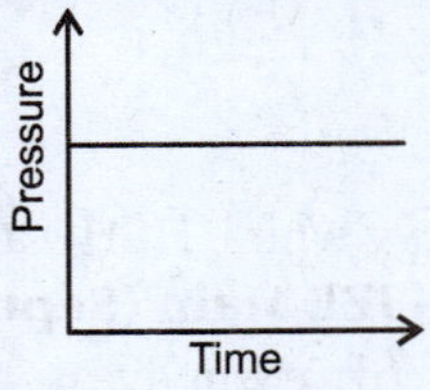

(b) 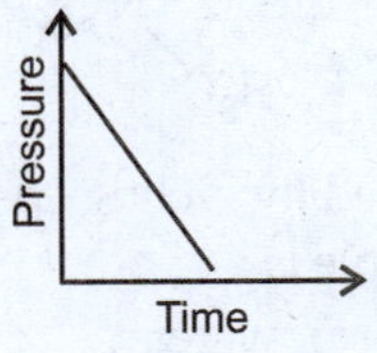

(c) 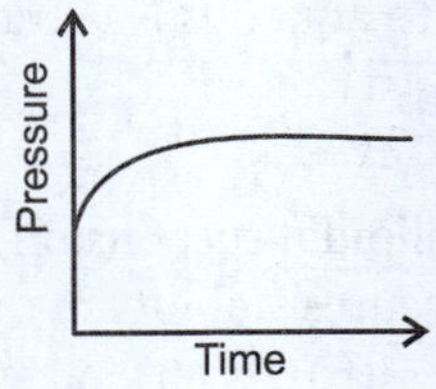

(d) 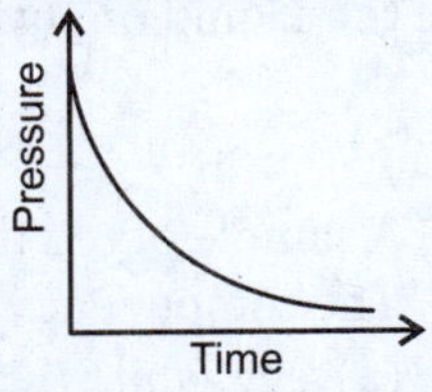

11. Match the following : **[JEE Main (Sept.) 2020]**

	Column-I		Column-II
(i)	Foam	(A)	smoke
(ii)	Gel	(B)	cell fluid
(iii)	Aerosol	(C)	jellies
(iv)	Emulsion	(D)	rubber
		(E)	froth
		(F)	milk

(a) (i)–(D), (ii)–(B), (iii)–(A), (iv)–(E) (b) (i)–(D), (ii)–(B), (iii)–(E), (iv)–(F)
(c) (i)–(B), (ii)–(C), (iii)–(E), (iv)–(D) (d) (i)–(E), (ii)–(C), (iii)–(A), (iv)–(F)

12. As per Hardy-Schulze formulation, the flocculation values of the following for ferric hydroxide sol are in the order: **[JEE Main (Jan.) 2020]**

(a) $AlCl_3 > K_3[Fe(CN)_6] > K_2CrO_4 > KBr = KNO_3$ (b) $K_3[Fe(CN)_6] < K_2CrO_4 < KBr = KNO_3 = AlCl_3$
(c) $K_3[Fe(CN)_6] < K_2CrO_4 < AlCl_3 < KBr < KNO_3$ (d) $K_3[Fe(CN)_6] > AlCl_3 > K_2CrO_4 > KBr > KNO_3$

13. Tyndall effect is observed when: **[JEE Main (Sept.) 2020]**

(a) The diameter of dispersed particles is similar to the wavelength of light used.
(b) The diameter of dispersed particles is much larger than the wavelength of light used.
(c) The refractive index of dispersed phase is greater than that of the dispersion medium.
(d) The diameter of dispersed particles is much smaller than the wavelength of light used.

14. Identify the correct molecular picture showing what happens at the critical micellar concentration (CMC) of an aqueous solution of a surfactant.
(o Polar head; ~ non-polar tail; • water) **[JEE Main (Sept.) 2020]**

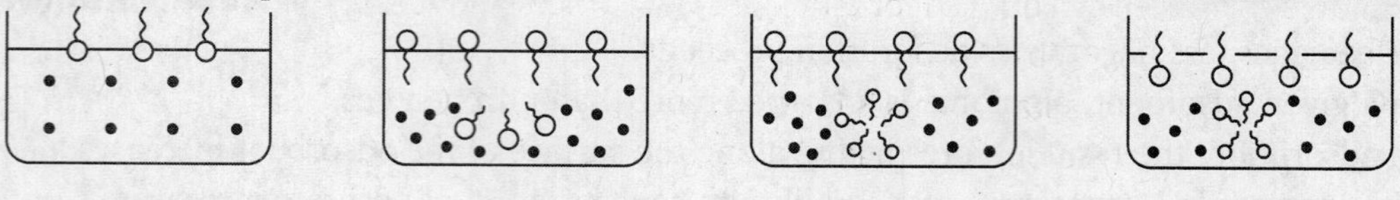

(A) (B) (C) (D)

(a) (B) (b) (A) (c) (C) (d) (D)

15. Kraft temperature is the temperature: **[JEE Main (Sept.) 2020]**

(a) below which the aqueous solution of detergents starts freezing
(b) below which the formation of micelles takes place
(c) above which the aqueous solution of detergents starts boiling
(d) above which the formation of micelles takes place.

16. An ionic micelles is formed on the addition of: **[JEE Main (Sept.) 2020]**

(a) Excess water to liquid

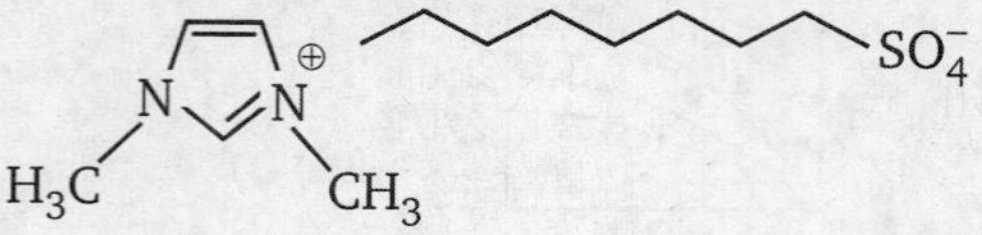

(b) Sodium stearate to pure toluene

(c) Liquid diethyl ether to aqueous NaCl solution

(d) Excess water to liquid

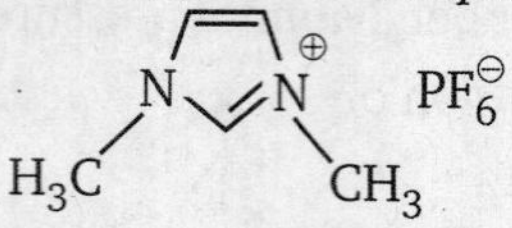

17. A sample of red ink (a colloidal suspension) is prepared by mixing eosin dye, egg white, HCHO and water. The component which ensures stability of the ink sample is : **[JEE Main (Sept.) 2020]**

(a) egg white (b) HCHO (c) water (d) eosin dye

18. In Freundlich adsorption isotherm, slope of *AB* line is : **[JEE Main (Feb.) 2021]**

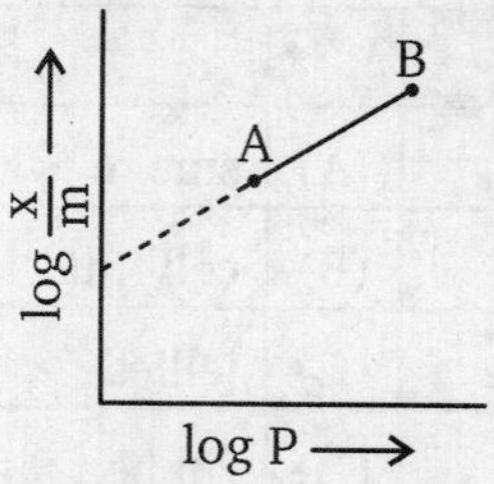

(a) $\log n$ with $(n > 1)$ (b) n with $(n, 0.1 \text{ to } 0.5)$

(c) $\log \frac{1}{n}$ with $(n < 1)$ (d) $\frac{1}{n}$ with $\left(\frac{1}{n} = 0 \text{ to } 1\right)$

19. In Freundlich adsorption isotherm at moderate pressure, the extent of adsorption $\left(\frac{x}{m}\right)$ is directly proportional to P^x. The value of x is: **[JEE Main (Feb.) 2021]**
(a) zero (b) $1/n$ (c) 1 (d) ∞

20. A colloidal system consisting of a gas dispersed in a solid is called a/an: **[JEE Main (March) 2021]**
(a) solid sol (b) gel (c) aerosol (d) foam

21. The charges on the colloidal CdS sol and TiO_2 sol are, respectively : **[JEE Main (March) 2021]**
(a) positive and positive (b) positive and negative
(c) negative and negative (d) negative and positive

22. The nature of charge on resulting colloidal particles when $FeCl_3$ is added to excess of hot water is : **[JEE Main (Feb.) 2021]**
(a) Positive (b) Sometimes positive and sometimes negative
(c) Neutral (d) Negative

23. Most suitable salt which can be used for efficient clotting of blood will be: **[JEE Main (Feb.) 2021]**
(a) $NaHCO_3$ (b) $FeSO_4$ (c) $Mg(HCO_3)_2$ (d) $FeCl_3$

24. Which one of the following statements is false for hydrophilic sols? **[JEE Main (Feb.) 2021]**
(a) Their viscosity is of the order of that of H_2O. (b) The sols cannot be easily coagulated.
(c) They do not require electrolytes for stability. (d) These sols are reversible in nature.

25. The incorrect statements below regarding colloidal solutions is : **[JEE Main (March) 2021]**
(a) A colloidal solution shows colligative properties.
(b) An ordinary filter paper can stop the flow of colloidal particles.
(c) The flocculating power of Al^{3+} is more than that of Na^+.
(d) A colloidal solution shows Brownian motion of colloidal particles.

26. 100 mL of 5% (w/v) solution of NaCl in water was prepared in 250 mL beaker. Albumin from the egg was poured into NaCl solution and stirred well. This resulted in a/an : **[JEE Main (July) 2022 (I)]**
(a) Lyophilic sol (b) Lyophobic sol (c) Emulsion (d) Precipitate

27. For micelle formation, which of the following statements are correct? **[JEE Main (July) 2022 (II)]**
(A) Micelle formation is an exothermic process
(B) Micelle formation is an endothermic process
(C) The entropy change is positive
(D) The entropy change is negative
(a) A and D only (b) A and C only (c) B and C only (d) B and D only

28. A 42.12% (w/v) solution of NaCl causes precipitation of a certain sol in 10 hours. The coagulating value of NaCl for the sol is :
[Given : Molar mass : Na = 23.0 g mol^{-1}; Cl = 35.5 g mol^{-1}] **[JEE Main (June) 2022 (II)]**
(a) 36 mmol L^{-1} (b) 36 mol L^{-1} (c) 1440 mol L^{-1} (d) 1440 mmol L^{-1}

29. Four gases A, B, C and D have critical temperatures 5.3, 33.2, 126.0 and 154.3 K respectively. For their adsorption on fixed amount of charcoal, the correct order is : **[JEE Main (April) 2023 (I)]**
(a) C > B > D > A (b) C > D > B > A (c) D > C > A > B (d) D > C > B > A

30. Which of the following represents the Freundlich adsorption isotherms? **[JEE Main (April) 2023 (I)]**

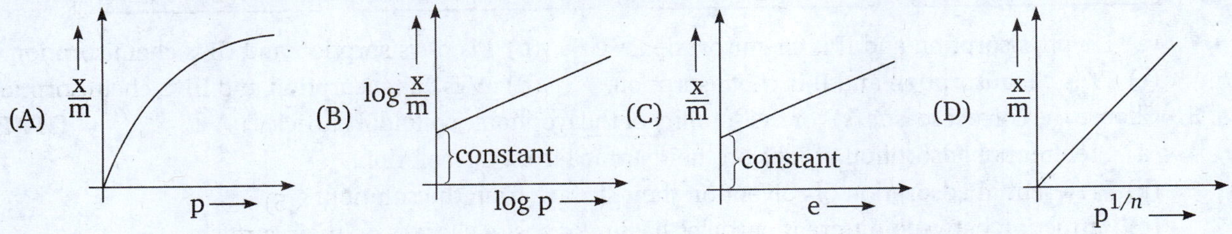

Choose the correct answer from the options given below :
(a) B, C, D only (b) A, B, D only (c) A, B only (d) A, C, D only

31. What happens when a lyophilic sol is added to a lyophobic sol? **[JEE Main (April) 2023 (I)]**
(a) Lyophilic sol is dispersed in lyophobic sol.
(b) Film of lyophobic sol is formed over lyophilic sol.
(c) Lyophobic sol is coagulated.
(d) Film of lyophilic sol is formed over lyophobic sol.

32. Ferric chloride is applied to stop bleeding because: **[JEE Main (April) 2023 (II)]**
(a) Cl^- ions cause coagulation of blood.
(b) Blood absorbs $FeCl_3$ and forms a complex.
(c) Fe^{3+} ions coagulate blood which is a negatively charged sol.
(d) $FeCl_3$ reacts with the constituents of blood which is a positively charged sol.

33. 'Adsorption' principle is used for which of the following purification method? **[JEE Main (Jan.) 2024]**
(a) Extraction (b) Chromatography (c) Distillation (d) Sublimation

34. Which of the following statements is not correct about rusting of iron? **[JEE Main (Jan.) 2024]**
(a) Coating of iron surface by tin prevents rusting, even if the tin coating is peeling off.
(b) When pH lies above 9 or 10, rusting of iron does not take place.
(c) Dissolved acidic oxides SO_2, NO_2 in water act as catalyst in the process of rusting.
(d) Rusting of iron is envisaged as setting up of electrochemical cell on the surface of iron object.

B. Objective Questions

[One or more than one option(s) is/are correct]

1. The correct statement(s) pertaining to the adsorption of a gas on a solid surface is (are) : **(IIT 2011)**
(a) Adsorption is always exothermic
(b) Physisorption may transform into chemisorption at high temperature
(c) Physisorption increases with increasing temperature but chemisorption decreases with increasing temperature
(d) Chemisorption is more exothermic than physisorption, however it is very slow due to higher energy of activation

2. The given graphs/data I, II, III and IV represent general trends observed for different physisorption and chemisorption processes under mild conditions of temperature and pressure. Which of the following choice(s) about I, II, III and IV is(are) correct ? **(IIT 2012)**

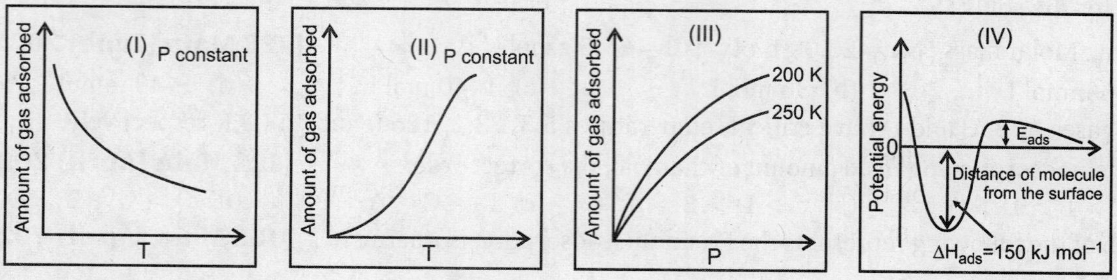

(a) I is physisorption and II is chemisorption. (b) I is physisorption and III is chemisorption.
(c) IV is chemisorption and II is chemisorption. (d) IV is chemisorption and III is chemisorption.

3. Choose the correct reason(s) for the stability of the lyophobic colloidal particles : **(IIT 2012)**
(a) Preferential adsorption of ions on their surface from the solution.
(b) Preferential adsorption of solvent on their surface from the solution.
(c) Attraction between different particles having opposite charges on their surface.

(d) Potential difference between the fixed layer and the diffused layer of opposite charges around the colloidal particles.

4. When O_2 is adsorbed on a metallic surface, electron transfer occurs from the metal to O_2. The TRUE statement(s) regarding this adsorption is(are) : **[JEE (Advanced) 2015]**

(a) O_2 is physisorbed (b) heat is released

(c) occupancy of π^*_{2p} of O_2 is increased (d) bond length of O_2 is increased

5. The correct statement(s) about surface properties is (are): **[JEE (Advanced) 2017]**

(a) cloud is an emulsion type of colloid in which liquid is dispersed phase and gas is dispersion medium.

(b) the critical temperatures of ethane and nitrogen are 563 K and 126 K, respectively. The adsorption of ethane will be more than that of nitrogen on same amount of activated charcoal at a given temperature.

(c) adsorption is accompanied by decrease in enthalpy and decrease in entropy of the system.

(d) brownian motion of colloidal particles does not depend on the size of the particles but depends on viscosity of the solution.

C. Assertion and Reason

Read the following questions and answer as per the direction given below :

(a) statement 1 is true; Statement 2 is true; Statement 2 is a correct explanation of Statement 1.

(b) Statement 1 is true; Statement 2 is true; Statement 2 is not the correct explanation of Statement 1.

(c) Statement 1 is true; Statement 2 is false.

(d) Statement 1 is false; Statement 2 is true.

1. Statement 1 : Micelles are formed by surfactant molecules above the critical micellar concentration (CMC).

Statement 2 : The conductivity of a solution having surfactant molecules decreases sharply at the CMC. **(IIT 2007)**

D. Integer Type Problems

1. The flocculation value of HCl for arsenic sulphide sol is 30 m mol L^{-1}. If H_2SO_4 is used for the flocculation of arsenic sulphide, the amount, in grams of H_2SO_4 in 250 mL required for the above purpose is ________

(Molecular mass of $H_2SO_4 = 98$ g/mol) **[JEE Main (Jan.) 2020]**

2. The mass of gas adsorbed, x, per unit mass of adsorbate, m, was measured at various pressures, P. A graph between $\log \frac{x}{m}$ and $\log P$ gives a straight line with slope equal to 2 and the intercept equal to 0.4771. The value of $\frac{x}{m}$ at a pressure of 4 atm is:

(Given $\log 3 = 0.4771$) **[JEE Main (Sept.) 2020]**

3. For Freundlich adsorption isotherm, a plot of $\log(x/m)$ (y-axis) and $\log P$ (x-axis) gives a straight line. The intercept and slope for the line is 0.4771 and 2, respectively. The mass of gas adsorbed per gram of adsorbent if the initial pressure is 0.04 atm, is________ $\times 10^{-4}$ g. ($\log 3 = 0.4771$)

[JEE Main (Sept.) 2020]

4. 3.12 g of oxygen is adsorbed on 1.2 g of platinum metal. The volume of oxygen adsorbed per gram of the adsorbent at 1 atm and 300 K in L is______.

[$R = 0.0821$ L atm K^{-1} mol^{-1}] **[JEE Main (Feb.) 2021]**

5. 2.0 g of H_2 gas is adsorbed on 2.5 g of platinum powder at 300 K and 1 bar pressure. The volume of the gas adsorbed per gram of the adsorbent is ______ mL.

(Given : R = 0.083 L bar K^{-1} mol^{-1}) **[JEE Main (June) 2022 (I)]**

6. When 200 mL of 0.2 M acetic acid is shaken with 0.6 g of wood charcoal, the final concentration of acetic acid after adsorption is 0.1 M. The mass of acetic acid adsorbed per gram of carbon is _____ g.

[JEE Main (June) 2022 (II)]

7. The graph of $\log \frac{x}{m}$ vs $\log p$ for an adsorption process is a straight line inclined at an angle of 45° with intercept equal to 0.6020. The mass of gas adsorbed per unit mass of adsorbent at the pressure of 0.4 atm is _________ $\times 10^{-1}$ (Nearest integer)

(Given : $\log 2 = 0.3010$) **[JEE Main (Jan.) 2023 (II)]**

8. The number of statement/s which are the characteristics of physisorption is _________ .

[JEE Main (Jan.) 2023 (II)]

(A) It is highly specific in nature.

(B) Enthalpy of adsorption is high.

(C) It decreases with increase in temperature.

(D) It results into unimolecular layer.

(E) No activation energy is needed.

9. 20 mL of 0.5 M NaCl is required to coagulate 200 mL of As_2S_3 solution in 2 hours. The coagulating value of NaCl is _________ . **[JEE Main (April) 2023 (I)]**

Answers

[A] Objective Questions (One option is correct)

1. (c)	**2.** (b)	**3.** (b)	**4.** (d)	**5.** (c)	**6.** (d)	**7.** (c)	**8.** (b)	**9.** (d)	**10.** (d)
11. (d)	**12.** (b)	**13.** (a)	**14.** (d)	**15.** (d)	**16.** (a)	**17.** (a)	**18.** (d)	**19.** (b)	**20.** (a)
21. (d)	**22.** (a)	**23.** (d)	**24.** (a)	**25.** (b)	**26.** (a)	**27.** (c)	**28.** (none)	**29.** (d)	**30.** (b)
31. (d)	**32.** (c)	**33.** (b)	**34.** (a)						

[B] Objective Questions (One or more than one option(s) is/are correct)

1. (a, b, d)	**2.** (a, c)	**3.** (a, d)	**4.** (a, b, c, d)	**5.** (b, c)

[C] Assertion and Reason

1. (b)

[D] Integer Type Problems

1. (0.3675)	**2.** (48)	**3.** (48)	**4.** (2)	**5.** (9960)
6. (2)	**7.** (16)	**8.** (2)	**9.** (50)	

Hints and Solutions

A. Objective Questions

[Only one option is correct]

1. (c) The most effective coagulating agent for Sb_2S_3 sol is $Al_2(SO_4)_3$ due to higher charge density on cation. (Hardy-Schulze ruel)

$$Al^{3+} > Ca^{2+} + Na^{+} > NH_4^{+}$$

2. (b) Factual.

3. (b) This is a process of physical adsorption it results in release of energy.

5. (c) For the Freundlich adsorption isotherm, equation is $\log\left(\frac{x}{m}\right) = \log k + \frac{1}{n}\log P$.

Now, comparing this equation with $y = mx + c$, slope $(m) = \frac{1}{n}$; intercept $(c) = \log k$

7. (c) For a plot between $\log x/m$ and $\log P$ slope is given by $1/n$.

Thus from the plot $\frac{1}{n} = \frac{y_2 - y_1}{x_2 - x_1} = 2/3$ and $x/m \propto P^{1/n} \propto P^{2/3}$

8. (b) As phenomenon of adsorption proceeds, residual forces on adsorbent surface decreases due to less number of active sites present. Also adsorption being an exothermic process, equilibrium concentration of adsorbate decreases due to shift of reaction in backward direction.

Also, NH_3 being polar adsorbed more than N_2.

9. (d) From Freundlich's adsorption isotherm,

$$\frac{x}{m} = kP^{1/n}$$

$\frac{x}{m}$ increases on increasing pressure (at constant temperature). Since adsorption is an exothermic process, it decreases while increasing temperature. So, extent of adsorption order will be maximum at 200 K and minimum at 270K.

10. (d) Gaseous adsorption will cause decrease of pressure exponentially with time.

11. (d) Correct combination is:

Foam	Froth
Gel	Jellies
Aerosol	Smoke
Emulsion	milk

12. (b) Ferric hydroxide sol is positive charged sol.

According to Hardy-Schulze rule, more will be charge on anion better will be its coagulating power and lower will be its coagulating value.

Order of coagulating power

$$Fe(CN)_6^{3-} > CrO_4^{2-} > Br^{-} = Cl^{-} = NO_3^{-}$$

Order of flocculation value

$$K_3[Fe(CN)_6] < K_2CrO_4 < KBr = KNO_3 = AlCl_3$$

13. (a) For Tyndall effect to show diameter of dispersed phase particles should be similar to the wave length of light used.

14. (d) At critical micelle concentration (CMC), molecules are dragged into bulk and take globular shape and form micelle. At surface molecules polar head should be dissolve in water and non-polar tail should protrude out of water.

15. (d) Kraft temperature (T_K) is temperature above which formations of micelles takes place.

16. (a)

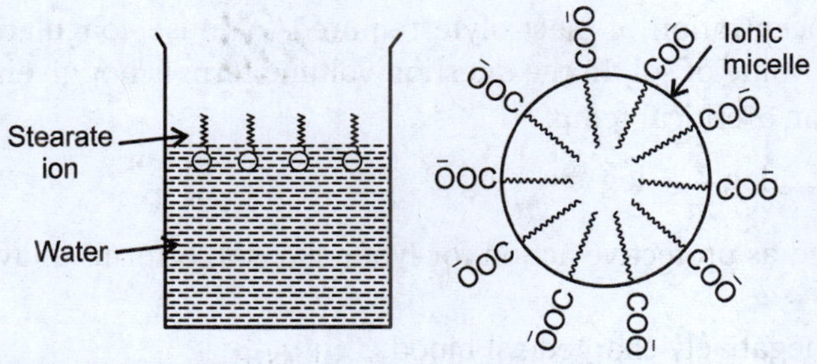

Ionic micelle is formed by hydrophilic part and hydrophobic part.

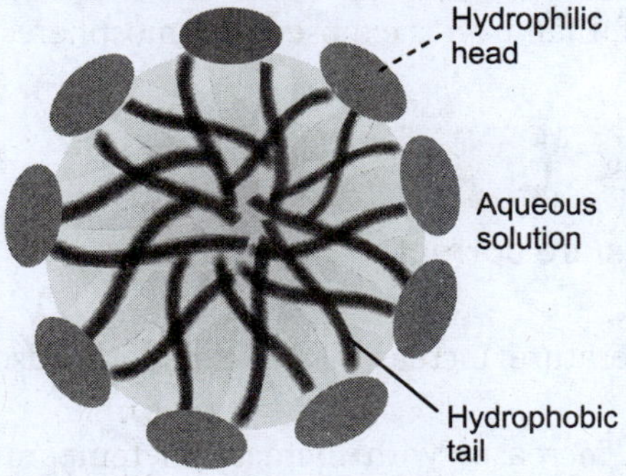

So only (a) option with such characteristics is hydrophilic part.

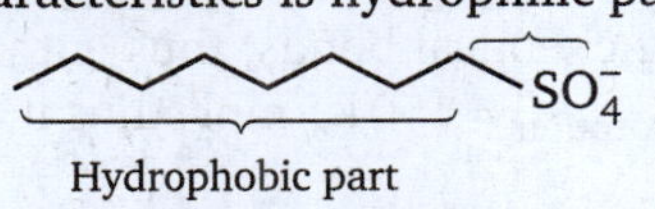

17. (a) Eosin dye imparts red colour to ink. While it is stabilizes through egg white which helps in prevent clotting by adhering to molecules and moving them past each other, giving ink a smoother flow.

18. (d) From Freundlich's Adsorption isotherm :

$$\frac{x}{m} = KP^{1/n} \qquad \left(0 \le \frac{1}{n} \le 1\right)$$

$$\log\frac{x}{m} = \frac{1}{n}\log P + \log K$$

$$y = mx + C$$

$$\text{Slope } (m) = \frac{1}{n}$$

19. (b) From Freundlich's adsorption isotherm

$$\frac{x}{m} = KP^{1/n}$$

So, $$x = \frac{1}{n}$$

20. (a) Gas dispersed in solid is called solid sol.

21. (d) CdS is negative sol.
TiO_2 is positive sol.

22. (a) Hydrated ferric hydroxide sol, a positively charged. Sol is formed due to adsorption of Fe^{3+} ions.

23. (d) Blood is a negatively charged sol. So according to Hardy Schulz's rule cation that have maximum valency will be most effective coagulating agent.
Most suitable salt for clotting will be $FeCl_3$.

24. (a) Viscosity of hydrophilic sols are greater than water since they are extensively hydrated.

25. (b) An ordinary filter paper cannot stop the flow of colloidal particles. It can be stopped through ultrafilter.

26. (a) It is the standard method for the preparation of lyophilic sol.

27. (c) For micelle formation, $\Delta S > 0$ (hydrophobic effect). This is possible because, the decrease in entropy due to clustering is offset by increase in entropy due to desolvation of the surfactant, Also $\Delta H > 0$.

28. (none)

The minimum concentration of electrolyte required to cause coagulation in milli-moles per litre is called coagulating value of sol. In the question volume term is not given hence data is insufficient.

29. (d) Extent of adsorption α critical temp.

30. (b) $\frac{x}{m} = kp^{1/n} \quad \Rightarrow \quad \log\frac{x}{m} = \log k + \frac{1}{n}\log p$

31. (d) Lyophilic sol is used as protective action for lyophobic sol. It forms a layer/film around the lyophobic sol.

32. (c) Fe^{3+} coagulation negatively charged sol blood.

33. (b) Principle used in chromatography is adsorption.

34. (a) As tin coating is peeled off, then iron is exposed to atmosphere.

B. Objective Questions

[One or more than one option(s) is/are correct]

2. (a, c)

In physisorption as temperature increases, the extent of adsorption decreases thus, I and III are physisorptions.

Chemical adsorption first increases with increase in temperature upto a certain extent and then decreases regularly thus II is chemisorption. In chemisorption, attractive forces between adsorbent and adsorbate molecules are strong chemical bonds, thus enthalpy of adsorption is high and of the order 80-240 kJ/mol. As in IV $\Delta H_{\text{adsorption}} = 150\,\text{kJ/mol}$, thus it also represents chemical adsorption.

4. (a, b, c, d)

*Adsorption of O_2 on metal surface is exothermic.

*During electron transfer from metal to O_2 electron occupies π^*_{2p} orbital of O_2.

*Due to electron transfer to O_2 the bond order of O_2 decreases hence bond length increases.

5. (b, c)

In adsorption process both ΔH & ΔS is –ve. Higher the critical temperature of a gas higher the extent of adsorption.

Cloud is not an emulsion.

Brownian motion depends on the size of the particles.

C. Assertion and Reason

1. (b) Both the statements are true but both explain different facts.

D. Integer Type Problems

1. (0.3675)

Since arsenic sulphide is negative charged sol, H^+ ion will coagulate it.

$\Rightarrow$ 30 m-mole of H^+ ions are required for coagulation of 1Lt As_2S_3 sol.

$\Rightarrow$ 30 mmol H^+ will be given by 15 mmol H_2SO_4 in 1Lt.

$\Rightarrow$ In 250 mL H_2SO_4 required $= \frac{15}{4} = 3.75$ milli-mole

$\Rightarrow$ Weight of $H_2SO_4 = \frac{3.75}{1000} \times 98 = 0.3675 \text{ gm}$

2. (48) From Freundlich adsorption isotherm,

$$\frac{x}{m} = KP^{1/n}$$

$$\log \frac{x}{m} = \log K + \frac{1}{n} \log P$$

$$\text{Slope} = \frac{1}{n} = 2$$

$$\text{Intercept} = \log K = 0.4771 = \log 3 \Rightarrow K = 3$$

So, equation is: $\frac{x}{m} = 3P^2$

When, $P = 4$ atm

$$\frac{x}{m} = 3 \times 4^2 = 48$$

3. (48) From Freundlich's adsorption isotherm, $\frac{x}{m} = KP^{1/n}$

$$\log \frac{x}{m} = \frac{1}{n} \log P + \log K$$

$$\text{Slope} = \frac{1}{n} = 2$$

$$\text{Intercept} = \log K = 0.4771 \Rightarrow K = 3$$

So, $\frac{x}{m} = 3P^2 \Rightarrow \frac{x}{1} = 3(0.04)^2$

$$x = 3 \times 16 \times 10^{-4} \text{ g} = 48 \times 10^{-4} \text{ g}$$

4. (2) Moles of $O_2 = \frac{3.12}{32}$; $V_{(O_2)} = \frac{nRT}{P}$

$$= \frac{3.12}{32} \times \frac{0.0821 \times 300}{1} = 2.4 \text{ L}$$

Volume of O_2 adsorbed per gram of adsorbent $= \frac{2.4}{1.2} = 2$

5. (9960) Volume of $H_2 = \frac{nRT}{p} = \frac{2}{2} \times \frac{0.083 \times 300}{1} = 24.9 \text{ L} = 24900 \text{ mL}$

Vol. of H_2 adsorbed per g of Pt $= \frac{249000}{2.5} \text{ mL} = 9960 \text{ mL}$

6. (2) Weight of wood charcoal = 0.6 g

Mass of acetic acid adsorbed $= \frac{M_1V_1 - M_2V_2}{1000} \times 60 = \frac{0.2 \times 200 - 0.1 \times 200}{1000} \times 60 = 1.2 \text{ g}$

Mass of acetic acid adsorbed per gram of carbon $= \frac{1.2}{0.6} = 2$

7. (16)

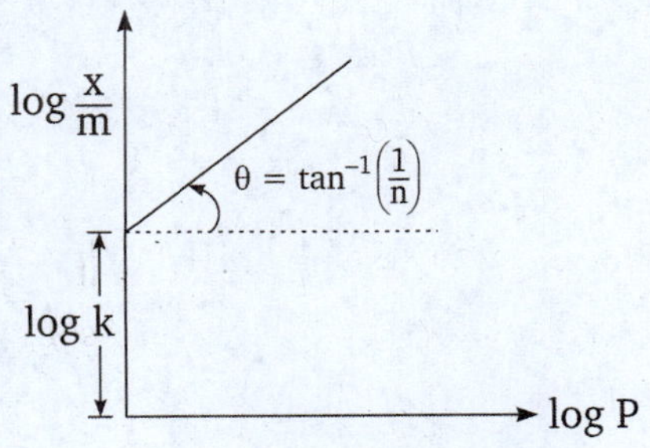

$$\log\frac{x}{m} = \log k + \frac{1}{n}\log P \quad \Rightarrow \quad \frac{1}{n} = \tan 45^\circ = 1$$

$$\log k = 0.6020 = \log 4 \quad \Rightarrow \quad k = 4$$

$$\therefore \quad \frac{x}{m} = k \cdot P^{1/n} \quad \Rightarrow \quad \frac{x}{m} = 4\ (0.4) = 1.6$$

$$\frac{x}{m} = 1.6 = 16 \times 10^{-1}$$

8. (2) Physisorption is not specific in nature and decreases with increase in temperature. Physisorption requires less or no activation energy.

9. (50) Coagulating value is required millimole of electrolyte needed to coagulate 1 L sol in 2 hours.

$$\text{Coagulating value} = \frac{20 \times 0.5}{200} \times 1000 = 50$$

❑❑❑

NOTE

NOTE